Complete Solutions M
PRECALCULUS

Phillip W. Bean
Mercer University

Jack C. Sharp
Floyd College

Thomas J. Sharp
West Georgia College

Produced by
Brett Shoelson

PWS
Publishing Company

PWS
Publishing Company

20 Park Plaza
Boston, Massachusetts 02116

PWS Publishing Company is a division of Wadsworth, Inc.

ISBN: 0-534-93296-7

Printed in the United States of America.
Printer: Edwards Brothers

1 2 3 4 5 6 7 8 9 -- 98 97 96 95 94 93

CONTENTS

CONTENTS

EXERCISES 1.1

[1] $A \cup B = \left\{-3, -1, 0, \frac{1}{2}, 2, 7\right\}$; $A \cap B = \{-3, 7\}$

[2] $S \cup T = T; S \cap T = S$

[3] $\varnothing, \{-1\}, \{0\}, \{1\}, \{-1, 0\}, \{0, 1\}, \{-1, 1\}, \{-1, 0, 1\}$

[4] $\{M, I, S, P\}$ 　　　　　　　　　　　　[5] $A' = \{1, 2, 3, \cdots, 49\}$

[6]   a) $A \cup B = \{1, 2, 5, 7, 9\}$

　　　b) $A \cap C = \{2, 7\}$

　　　c) $A \cap B' = A \cap \{1, 2, 3, 4, 6, 8, 10\} = \{1, 2\}$

　　　d) $A \cup (B \cap C) = A \cup \{7\} = \{1, 2, 7, 9\}$

　　　e) $(A \cup B) \cap (A \cup C) = \{1, 2, 5, 7, 9\} \cap \{1, 2, 3, 7, 9\} = \{1, 2, 7, 9\}$

　　　f) $A' \cap B = \{3, 4, 5, 6, 8, 10\} \cap B = \{5\}$

　　　g) $A \cup C' = A \cup \{1, 4, 5, 6, 8, 9, 10\} = \{1, 2, 4, 5, 6, 7, 8, 9, 10\}$

　　　h) $(B \cup C)' = $ complement of the set $\{2, 3, 5, 7, 9\} = \{1, 4, 6, 8, 10\}$

　　　i) $A \cap (B \cup C)' = A \cap \{1, 4, 6, 8, 10\} = \{1\}$

　　　j) $(A \cap B') \cup (A \cap C') = \{1, 2\} \cup \{1, 9\} = \{1, 2, 9\}$

[7]   T 　　　　　　[8]   T 　　　　　　[9]   F 　　　　　　[10]   T

[11]   F 　　　　　[12]   F 　　　　　[13]   F 　　　　　[14]   F

[15]   T 　　　　　[16]   F

[17]   a) $2^2 = 4$ 　　　b) $2^1 = 2$ 　　　c) $2^0 = 1$ 　　　d) $2^1 = 2$

　　　e) $2^{10} = 1024$ 　　f) 2^{41} 　　　g) $2^4 = 16$ 　　　h) 2^{26}

[18]

$(A \cap B)'$

[19]

$(A \cap B) \cup C$

[20]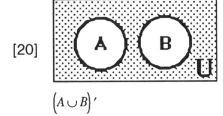

$(A \cup B)'$

[21]

$A \cap (B \cup C)$

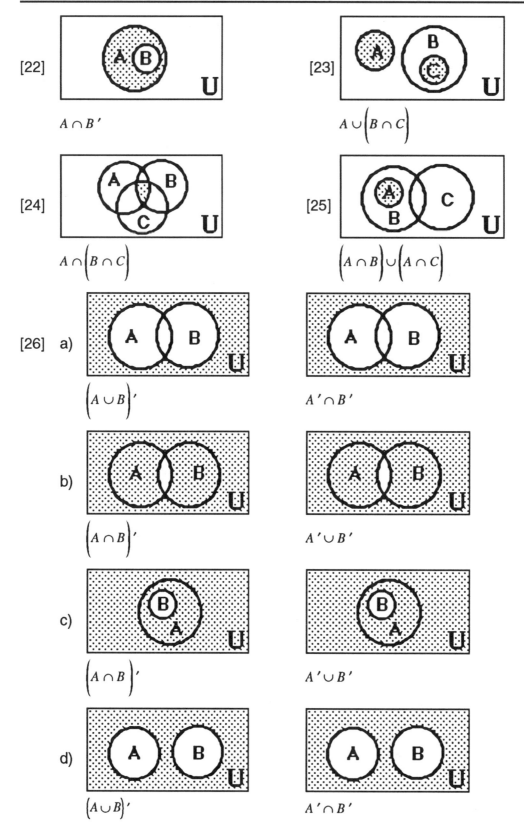

[22] $A \cap B\,'$

[23] $A \cup \left(B \cap C \right)$

[24] $A \cap \left(B \cap C \right)$

[25] $\left(A \cap B \right) \cup \left(A \cap C \right)$

[26]   a) $\left(A \cup B \right)'$ $A\,' \cap B\,'$

b) $\left(A \cap B \right)'$ $A\,' \cup B\,'$

c) $\left(A \cap B \right)'$ $A\,' \cup B\,'$

d) $\left(A \cup B \right)'$ $A\,' \cap B\,'$

EXERCISES 1.2

[1]   T   [2]   F   [3]   F   [4]   T

[5]   T   [6]   F   [7]   T   [8]   T

[9]   F   [10]   T   [11] $2.09 = \dfrac{209}{100}$ [12] $-10.71 = -\dfrac{1071}{100}$

[13] $x = 0.818181..$ and $100x = 81.818181... \Rightarrow 100x - x = 81.818181... - 0.818181... \Rightarrow$

$99x = 81 \Rightarrow x = \dfrac{81}{99} = \dfrac{9}{11}$

[14] $x = 0.252525...$ and $100x = 25.252525... \Rightarrow 100x - x = 25.252525... - 0.252525... \Rightarrow$

$99x = 25 \Rightarrow x = \dfrac{25}{99}$

[15] $x = 0.252525...$ and $100x = 25.252525... \Rightarrow 100x - x = 25.252525... - 0.252525... \Rightarrow$

$x = 7.444...$ and $10x = 74.444... \Rightarrow 10x - x = 74.444... - 7.444... \Rightarrow 9x = 67 \Rightarrow x = \dfrac{69}{7}$

[16] $x = -3.272727..$ and $100x = -327.272727.. \Rightarrow 100x - x = -327.272727... + 3.272727... \Rightarrow$

$x = 0.252525...$ and $100x = 25.252525... \Rightarrow 100x - x = 25.252525... - 0.252525... \Rightarrow$

$99x = -324 \Rightarrow x = -\dfrac{324}{99} = -\dfrac{36}{11}$

[17] $10x = 8.514514514..$ and $10{,}000x = 8514.514514514.. \Rightarrow$

$10{,}000x - 10x = 8514.514514514... - 8.514514514.. \Rightarrow 9990x = 8506 \Rightarrow x = \dfrac{8506}{9990}$

[18] $100x = 502.373737..$ and $10{,}000x = 50237.373737.. \Rightarrow$

$10{,}000x - 100x = 50237.373737... - 502.373737.. \Rightarrow 9900x = 49735 \Rightarrow x = \dfrac{49735}{9900} = \dfrac{9947}{1980}$

[19]   Distributive property   [20] Inverse property of addition

[21]   Identity property of addition   [22]   Commutative property of addition

[23]   Inverse property of multiplication   [24]   Commutative property of multiplication

[25]   Associative property of addition   [26]   Commutative property of multiplication

[27]   Inverse property of addition   [28]   Identity property of multiplication

[29]   Distributive property   [30]   Associative property of addition

[31]   Associative property of addition   [32]   Inverse property of addition

[33]   Identity property of addition   [34]   Associative property of multiplication

[35]   Inverse property of multiplication   [36]   Identity property of multiplication

[37] $|-2| = 2$ [38] $\left| \dfrac{1}{5} - 0.3 \right| = |-0.1| = 0.1$

[39] $r \geq 1 \Rightarrow r - 1 \geq 0.$ Thus $|r - 1| = r - 1$ [40] $3 - \pi < 0 \Rightarrow |3 - \pi| = -(3 - \pi) = \pi - 3$

[41] $y < 0 \Rightarrow -y > 0.$ Thus $|-y| = -y$

[42] $k < 5 \Rightarrow k - 5 < 0.$ Thus $|k - 5| = -(k - 5) = 5 - k$

[43] $[-2, 1) = \left\{ x \mid -2 \leq x < 1 \right\}$

[44] $(3, \infty) = \left\{ x \mid x > 3 \right\}$

[45] $[-4, \infty) = \left\{ x \mid x \geq -4 \right\}$

[46] $\left(-\frac{5}{3}, \frac{1}{2} \right) = \left\{ x \mid -\frac{5}{3} < x < \frac{1}{2} \right\}$

[47] $\left(-\infty, -\sqrt{2} \right] = \left\{ x \mid x \leq -\sqrt{2} \right\}$

[48] $(-\pi, \pi) = \left\{ x \mid -\pi < x < \pi \right\}$

[49] $(-\infty, 0) \cup \left(1, \frac{13}{8} \right) = \left\{ x \mid x < 0 \text{ or } 1 < x < \frac{13}{8} \right\}$

[50] $(-\infty, -3) \cup \left[-\frac{1}{2}, \infty \right) = \left\{ x \mid x < -3 \text{ or } x \geq -\frac{1}{2} \right\}$

[51] $d(-3, 4) = \left| 4 - (-3) \right| = \left| 7 \right| = 7$ [52] $d(2, -5) = \left| -5 - 2 \right| = \left| -7 \right| = 7$

[53] $d\left(\frac{4}{9}, \frac{7}{3} \right) = \left| \frac{7}{3} - \frac{4}{9} \right| = \left| \frac{17}{9} \right| = \frac{17}{9}$

[54] $d\left(\sqrt{6}, \sqrt{3} \right) = \left| \sqrt{3} - \sqrt{6} \right| = -\left(\sqrt{3} - \sqrt{6} \right) = \sqrt{6} - \sqrt{3}$

[55] $d\left(-\pi, \frac{\pi}{4} \right) = \left| \frac{\pi}{4} - (-\pi) \right| = \left| \frac{5\pi}{4} \right| = \frac{5\pi}{4}$ [56] $d(2.5, -1.8) = \left| -1.8 - 2.5 \right| = \left| -4.3 \right| = 4.3$

[57]   Case 1: Assume $a < 0$. Then $|a| = -a$. Since $-a > 0$ and $(-a)^2 = a^2$, $\sqrt{a^2} = -a$. Thus $\sqrt{a^2} = |a|$.

        Case 2: Assume $a = 0$. Then $\sqrt{a^2} = |a| = 0$.

        Case 3: Assume $a > 0$. Then $|a| = a$ and $\sqrt{a^2} = a$. Thus $\sqrt{a^2} = |a|$.

[58]   Case 1: Assume $a < 0$. Then $|a| = -a$. Also, since $-a > 0$, $|-a| = -a$. Thus $|-a| = |a|$.

        Case 2: Assume $a = 0$. Then $|-a| = |a| = 0$.

        Case 3: Assume $a > 0$. Then $|a| = a$. Also, since $-a < 0$, $|-a| = -(-a) = a$. Thus $|-a| = |a|$.

[59]   Case 1: Assume a and b agree in sign. Then $ab > 0 \Rightarrow |ab| = ab$. If a and b are both positive, $|a||b| = ab$. If a and b are both negative, $|a||b| = (-a)(-b) = ab$. Thus $|ab| = |a||b|$.

        Case 2: Assume a and b have opposite signs. Then $ab < 0 \Rightarrow |ab| = -ab$. If $a > 0$ and $b < 0$, $|a||b| = (a)(-b) = -ab$. If $a < 0$ and $b > 0$, $|a||b| = (-a)(b) = -ab$. Thus $|ab| = |a||b|$.

        Case 3: If at least one of a and b is equal to 0 then $|ab| = 0 = |a||b|$.

[60] a) Let $a = 2$ and $b = 0$. Then $|a + b| = |2 + 0| = 2$ and $|a| + |b| = 2 + 0 = 2$.

Thus $|a + b| = |a| + |b|$.

b) Let $a = 2$ and $b = -1$. Then $|a + b| = |2 + (-1)| = |1| = 1$ and $|a| + |b| = |2| + |-1| = 3$.

Thus $|a + b| < |a| + |b|$.

c) No, since $|a + b| \leq |a| + |b|$, for all real numbers a and b, by the Triangle Inequality.

EXERCISES 1.3

[1]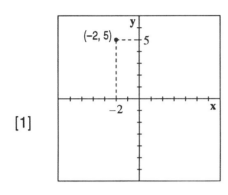

[2]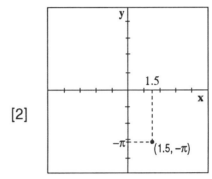

[3]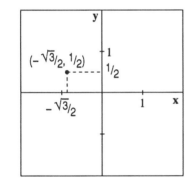

[4]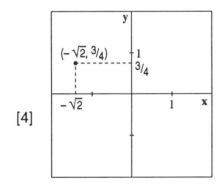

[5]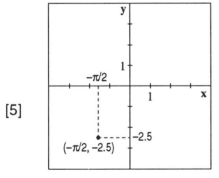

[6]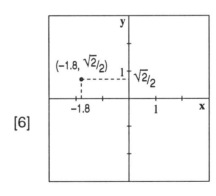

[7] $\sqrt{(-1-1)^2 + (2+2)^2} = \sqrt{20} = 2\sqrt{5}$

[8] $\sqrt{(-1-0)^2 + (5-3)^2} = \sqrt{5}$

[9] $\sqrt{\left(\dfrac{3}{2} - \dfrac{2}{1}\right)^2 + (0+2)^2} = \sqrt{5}$

[10] $\sqrt{(1-3)^2 + \left(\sqrt{2} + \sqrt{2}\right)^2} = \sqrt{12} = 2\sqrt{3}$

[11] $\sqrt{\left(0-\frac{\sqrt{3}}{2}\right)^2+\left(\frac{5}{4}-\frac{1}{2}\right)^2}=\sqrt{\frac{21}{16}}=\frac{\sqrt{21}}{4}$ [12] $\sqrt{(-1+1)^2+(3\pi+\pi)^2}=\sqrt{16\pi^2}=4\pi$

[13] $\sqrt{(a+a)^2+(0-2a)^2}=\sqrt{8a^2}=2\sqrt{2}\,a$ [14] $\sqrt{\left(a-\frac{a}{4}\right)^2+(2a-a)^2}=\sqrt{\frac{25a^2}{16}}=\frac{5a}{4}$

[15] $M=\left(\frac{-2+2}{2},\frac{4-6}{2}\right)=(0,-1)$ [16] $M=\left(\frac{3-2}{2},\frac{-1+7}{2}\right)=\left(\frac{1}{2},3\right)$

[17] $M=\left(\frac{1.5+3.7}{2},\frac{-3+1}{2}\right)=(2.6,-1)$ [18] $M=\left(\frac{\frac{7}{8}-\frac{1}{8}}{2},\frac{-\frac{3}{4}+0}{2}\right)=\left(\frac{3}{8},-\frac{3}{8}\right)$

[19] $M=\left(\frac{\frac{\sqrt{3}}{2}+\frac{\sqrt{3}}{2}}{2},\frac{\frac{1}{2}-\frac{1}{2}}{2}\right)=\left(\frac{\sqrt{3}}{2},0\right)$ [20] $M=\left(\frac{\frac{\pi}{2}+\frac{\pi}{3}}{2},\frac{-1+0}{2}\right)=\left(\frac{5\pi}{12},-\frac{1}{2}\right)$

[21] $M=\left(\frac{-a+3a}{2},\frac{a-a}{2}\right)=(a,0)$ [22] $M=\left(\frac{\frac{a}{8}+\frac{a}{4}}{2},\frac{\frac{a}{2}+a}{2}\right)=\left(\frac{3a}{16},\frac{3a}{4}\right)$

[23] The midpoint of $\overline{AB}$ is $M=\left(\frac{-1+5}{2},\frac{3+17}{2}\right)=(2,10)$.

$C=$ midpoint of $\overline{BM}=\left(\frac{2+5}{2},\frac{10+17}{2}\right)=\left(\frac{7}{2},\frac{27}{2}\right)$.

[24] The midpoint of $\overline{AB}$ is $M=\left(\frac{2-\frac{1}{2}}{2},\frac{-3+7}{2}\right)=\left(\frac{3}{4},2\right)$.

$C=$ midpoint of $\overline{AM}=\left(\frac{2+\frac{3}{4}}{2},\frac{-3+2}{2}\right)=\left(\frac{11}{8},-\frac{1}{2}\right)$.

[25] Perimeter of triangle $RST=d(R,S)+d(S,T)+d(R,T)$.   Using the distance formula, $d(R,S)=5\sqrt{2}$, $d(S,T)=\sqrt{17}$ and $d(R,T)=\sqrt{17}$. Thus, perimeter $=5\sqrt{2}+2\sqrt{17}$.

[26] $\sqrt{(x-1)^2+[(x+3)-2]^2}=2\Rightarrow(x-1)^2+(x+1)^2=4\Rightarrow$

$2x^2+2=4\Rightarrow x^2-1=0\Rightarrow x=\pm1$.

[27] Let $P=(x,0)$ be a point on the x-axis 5 units away from the point $Q=(0,4)$. Then $d(P,Q)=5\Rightarrow$

$\sqrt{(0-x)^2+(4-0)^2}=5\Rightarrow x^2+16=25\Rightarrow x^2-9=0\Rightarrow x=\pm3$

[28] $d(P,Q)=d(P,R)\Rightarrow\sqrt{(a+1)^2+(a-2)^2}=\sqrt{(a-10)^2+(a+1)^2}\Rightarrow$

$(a+1)^2+(a-2)^2=(a-10)^2+(a+1)^2\Rightarrow a^2-4a+4=a^2-20a+100\Rightarrow 16a=96\Rightarrow a=6$

[29] $\sqrt{[(y-1)-4]^2+(y+5)^2} = 2\sqrt{[(y-1)+3]^2+(y-2)^2} \Rightarrow$

$(y-5)^2+(y+5)^2 = 4[(y+2)^2+(y-2)^2] \Rightarrow 2y^2+50 = 4(2y^2+8) \Rightarrow y^2-3=0 \Rightarrow y = \pm\sqrt{3}$

[30] The center C of the circle is the midpoint of $\overline{AB}$. Thus $C = \left(\dfrac{2+0}{2}, \dfrac{4-4}{2}\right) = (1, 0)$. The radius r of

the circle is given by $r = d(C, A) = \sqrt{(1-2)^2+(0-4)^2} = \sqrt{17}$.

[31] Inside the circle since $d(C, P) = \sqrt{5} < 4 = r$. [32] On the circle since $d(C, P) = 4 = r$.

[33] On the circle since $d(C, P) = 4 = r$.

[34] Outside the circle since $d(C, P) = \dfrac{3\sqrt{17}}{2} > 4 = r$.

[35] The triangle is equilateral since $d(A, B) = d(B, C) = d(A, C) = 8$. Using $\overline{AB}$ as the base, the line

segment from the midpoint M of $\overline{AB}$ to C is an altitude. Since $M = (5, 0)$,

the altitude is $h = d(M, C) = 4\sqrt{3}$. Thus, the area of the triangle is

$A = \dfrac{1}{2}bh = \dfrac{1}{2}(8)\left(4\sqrt{3}\right) = 16\sqrt{3}$ square units.

[36] $d(P, Q) = d(Q, R) = \sqrt{17}$ and $d(P, R) = 3\sqrt{2}$. Thus, triangle PQR is isosceles. Using $\overline{PR}$ as the

base, the line segment from the midpoint M of $\overline{PR}$ to Q is an altitude. Since $M = \left(\dfrac{7}{2}, -\dfrac{7}{2}\right)$, it follows

that $h = d(M, Q) = \dfrac{5\sqrt{2}}{2}$. Thus the area of the triangle is $A = \dfrac{1}{2}bh = \dfrac{1}{2}(3\sqrt{2})\left(\dfrac{5\sqrt{2}}{2}\right) = \dfrac{15}{2}$.

[37] $d(A, B) = d(B, C) = d(A, C) = 2\sqrt{2}$. Equilateral.

[38] $d(A, B) = d(B, C) = \sqrt{17}$ and $d(A, C) = 3\sqrt{2}$. Isosceles.

[39] $d(A, B) = 3\sqrt{29}, d(B, C) = 17$ and $d(A, C) = \sqrt{610}$. Neither.

[40] $d(A, B) = d(B, C) = 17$ and $d(A, C) = \sqrt{578}$. Isosceles.

[41] $d(A, B) = 3\sqrt{2}, d(B, C) = \sqrt{17}$ and $d(A, C) = \sqrt{17}$. Isosceles.

[42] $d(A, B) = \sqrt{41}, d(B, C) = \sqrt{113}$ and $d(A, C) = \sqrt{130}$. Neither.

[43] Since $d(A, P) = d(B, P) = 2\sqrt{29}$, P lies on the perpendicular bisector of $\overline{AB}$.

[44] $d(P, Q) = 5\sqrt{2}, d(Q, R) = 5\sqrt{2}$ and $d(P, R) = 10\sqrt{2}$. Since $d(P, Q) + d(Q, R) = d(P, R)$,

the points P, Q, and R are collinear.

[45] Let $A = (3, 2), B = (4, 6)$ and $C = (0, -8)$. Then $d(A, B) = \sqrt{17}, d(B, C) = 2\sqrt{53}$

and $d(A, C) = \sqrt{109}$. The points are not collinear.

[46] Let $A = (-2, 1), B = (-4, 0)$ and $C = (2, 3)$. Then $d(A, B) = \sqrt{5}, d(B, C) = 3\sqrt{5}$

and $d(A, C) = 2\sqrt{5}$. Since $d(A, B) + d(A, C) = d(B, C)$, the points are collinear.

[47] Let $A = (0, -1)$, $B = (-2, 4)$ and $C = \left(1, -\frac{7}{2}\right)$. Then $d(A, B) = \sqrt{29}$, $d(B, C) = \frac{3\sqrt{29}}{2}$

and $d(A, C) = \frac{\sqrt{29}}{2}$. Since $d(A, B) + d(A, C) = d(B, C)$, the points are collinear.

[48] Let $A = (3, 2)$, $B = (6, 0)$ and $C = \left(-\frac{3}{2}, 5\right)$. Then $d(A, B) = \sqrt{13}$, $d(B, C) = \frac{5\sqrt{13}}{2}$

and $d(A, C) = \frac{3\sqrt{13}}{2}$. Since $d(A, B) + d(A, C) = d(B, C)$, the points are collinear.

[49] $d(T, U) = d(U, V) = d(V, W) = d(T, W) = 13$ and $d(T, V) = d(U, W) = \sqrt{338}$.

[50] $d(A, B) = d(B, C) = d(C, D) = d(A, D) = 3\sqrt{2}$ and $d(A, C) = d(B, D) = 6$.

[51] $d(A, B) = 2\sqrt{5}$, $d(B, C) = 10\sqrt{2}$ and $d(A, C) = 6\sqrt{5}$. $\left[d(A, B)\right]^2 + \left[d(A, C)\right]^2 = 200 = \left[d(B, C)\right]^2 \Rightarrow$

triangle ABC is a right triangle. Using $\overline{AB}$ as base and $\overline{AC}$ as altitude,

$A = \frac{1}{2}bh = \frac{1}{2}\left(2\sqrt{5}\right)\left(6\sqrt{5}\right) = 30$ square units.

[52] Area of inside square + area of 4 triangles = area of outside square.

$(a - b)^2 + 4\left[\frac{1}{2}ab\right] = c^2 \Rightarrow a^2 - 2ab + b^2 + 2ab = c^2 \Rightarrow a^2 + b^2 = c^2$.

EXERCISES 1.4

[1]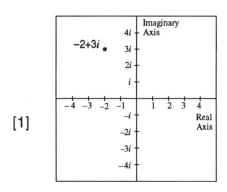

[2]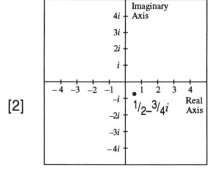

[3]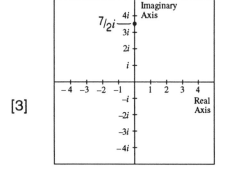

[4]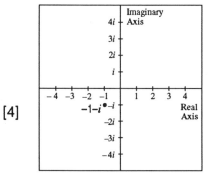

[5]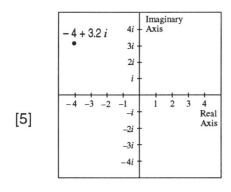

[6]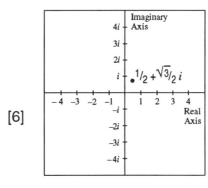

[7]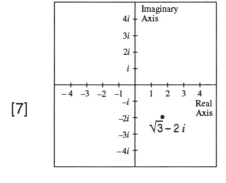

[8]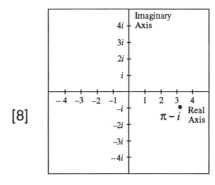

[9] $\sqrt{-81} = \sqrt{81}\, i = 9i$

[10] $\sqrt{-18} = \sqrt{18}\, i = 3\sqrt{2}\, i$

[11] $-\sqrt{-32} = -\sqrt{32}\, i = -4\sqrt{2}\, i$

[12] $-\sqrt{-14} = -\sqrt{14}\, i$

[13] $\sqrt{-12} \cdot \sqrt{-12} = \left(\sqrt{12}\, i\right)\left(\sqrt{12}\, i\right) = 12i^2 = -12$

[14] $\left(\sqrt{-9}\right)\left(-\sqrt{9}\right) = \left(\sqrt{9}\, i\right)(-3) = -9i$

[15] $-\sqrt{-3} \cdot \sqrt{-12} = -\sqrt{3}\, i\, \sqrt{12}\, i = -\sqrt{36}\, i^2 = (-6)(-1) = 6$

[16] $\sqrt{-6} \cdot \sqrt{-8} = \sqrt{6}\, i\sqrt{8}\, i = \sqrt{48}\, i^2 = -4\sqrt{3}$

[17] $\dfrac{\sqrt{-27}}{\sqrt{-3}} = \dfrac{\sqrt{27}\, i}{\sqrt{3}\, i} = \sqrt{9} = 3$

[18] $\dfrac{\sqrt{-56}}{\sqrt{-7}} = \dfrac{\sqrt{56}\, i}{\sqrt{7}\, i} = \sqrt{8} = 2\sqrt{2}$

[19] $(2 + 3i) + (5 - 7i) = 7 - 4i$

[20] $(-1 + i) - (4 - i) = -5 + 2i$

[21] $\left(\dfrac{1}{4} - 2i\right) - \left(-8 + \dfrac{3}{2}\, i\right) = \dfrac{33}{4} - \dfrac{7}{2}\, i$

[22] $\left(-\dfrac{7}{2} - 3i\right) + \dfrac{3}{2}\, i = -\dfrac{7}{2} - \dfrac{3}{2}\, i$

[23] $(-7 - i)(3 + 5i) = -21 - 3i - 35i - 5i^2 = -16 - 38i$

[24] $(2 - 4i)(-3 + 2i) = -6 + 12i + 4i - 8i^2 = 2 + 16i$

[25] $(1 - 2i)^2 = 1 - 4i + 4i^2 = -3 - 4i$

[26] $(-5 + 8i)^2 = 25 - 80i + 64i^2 = -39 - 80i$

[27] $(-5 + i)^3 = (-5 + i)\left(25 - 10i + i^2\right) = (-5 + i)(24 - 10i) = -110 + 74i$

[28] $\left(\sqrt{3} - 2i\right)^3 = \left(\sqrt{3} - 2i\right)\left(3 - 4\sqrt{3}\, i + 4i^2\right) = \left(\sqrt{3} - 2i\right)\left(-1 - 4\sqrt{3}\, i\right) = -9\sqrt{3} - 10i$

[29] $(3 - i)(3 + i)(-i) = \left(9 - i^2\right)(-i) = 10(-i) = -10i$

[30] $\left(2+\sqrt{2}\,i\right)\left(2-\sqrt{2}\,i\right)\left(\sqrt{2}\,i\right)=\left(4-2i^2\right)\left(\sqrt{2}\,i\right)=\left(6\right)\left(\sqrt{2}\,i\right)=6\sqrt{2}\,i$

[31] $\dfrac{1}{1-4i}\cdot\dfrac{1+4i}{1+4i}=\dfrac{1+4i}{17}=\dfrac{1}{17}+\dfrac{4}{17}\,i$

[32] $\dfrac{1}{-3+i}\cdot\dfrac{-3-i}{-3-i}=\dfrac{-3-i}{10}=-\dfrac{3}{10}-\dfrac{1}{10}\,i$

[33] $\dfrac{7+6i}{-i}\cdot\dfrac{i}{i}=\dfrac{-6+7i}{1}=-6+7i$

[34] $\dfrac{2i}{-1+4i}\cdot\dfrac{-1-4i}{-1-4i}=\dfrac{8-2i}{17}=\dfrac{8}{17}-\dfrac{2}{17}\,i$

[35] $\dfrac{1-\pi i}{1+\pi i}\cdot\dfrac{1-\pi i}{1-\pi i}=\dfrac{1-\pi^2-2\pi i}{1+\pi^2}=\dfrac{1-\pi^2}{1+\pi^2}-\dfrac{2\pi}{1+\pi^2}\,i$

[36] $\dfrac{1}{i}-\dfrac{i}{2}=\dfrac{1}{i}\cdot\dfrac{-i}{-i}-\dfrac{1}{2}\,i=-i-\dfrac{1}{2}\,i=-\dfrac{3}{2}\,i$

[37] $-\left(-2+4i\right)=2-4i$

[38] $-\left(1-i\right)=-1+i$

[39] $-\left(-\dfrac{1}{2}\,i\right)=\dfrac{1}{2}\,i$

[40] $-\left(\dfrac{1}{2}+\dfrac{\sqrt{3}}{2}\,i\right)=-\dfrac{1}{2}-\dfrac{\sqrt{3}}{2}\,i$

[41] $\dfrac{1}{1-3i}=\dfrac{1}{1-3i}\cdot\dfrac{1+3i}{1+3i}=\dfrac{1+3i}{10}=\dfrac{1}{10}+\dfrac{3}{10}\,i$

[42] $\dfrac{1}{-6i}=\dfrac{1}{-6i}\cdot\dfrac{6i}{6i}=\dfrac{6i}{36}=\dfrac{1}{6}\,i$

[43] $\dfrac{1}{\frac{\sqrt{2}}{2}+2i}=\dfrac{1}{\frac{\sqrt{2}}{2}+2i}\cdot\dfrac{\frac{\sqrt{2}}{2}-2i}{\frac{\sqrt{2}}{2}-2i}=\dfrac{\frac{\sqrt{2}}{2}-2i}{\frac{1}{2}+4}=\dfrac{2}{9}\left(\dfrac{\sqrt{2}}{2}-2i\right)=\dfrac{\sqrt{2}}{9}-\dfrac{4}{9}\,i$

[44] $\dfrac{1}{-5+\frac{\sqrt{3}}{2}\,i}=\dfrac{1}{-5+\frac{\sqrt{3}}{2}\,i}\cdot\dfrac{-5-\frac{\sqrt{3}}{2}\,i}{-5-\frac{\sqrt{3}}{2}\,i}=\dfrac{-5-\frac{\sqrt{3}}{2}\,i}{25+\frac{3}{4}}=\dfrac{4}{103}\left(-5-\dfrac{\sqrt{3}}{2}\,i\right)=-\dfrac{20}{103}-\dfrac{2\sqrt{3}}{103}\,i$

[45] $i^{12}=\left(i^2\right)^6=\left(-1\right)^6=1$

[46] $i^{37}=i^{36}\cdot i=\left(i^2\right)^{18}\cdot i=\left(-1\right)^{18}\cdot i=i$

[47] $i^{239}=i^{238}\cdot i=\left(i^2\right)^{119}\cdot i=\left(-1\right)^{119}\cdot i=-i$

[48] $i^{1388}=\left(i^2\right)^{694}=\left(-1\right)^{694}=1$

[49] $\left(-i\right)^{42}=\left(-1\right)^{42}\,i^{42}=\left(1\right)\left(i^2\right)^{21}=\left(-1\right)^{21}=-1$

[50] $\left(-2i\right)^9=\left(-2\right)^9\,i^9=\left(-512\right)\cdot i^8\cdot i=-512\left(i^2\right)^4\cdot i=-512\left(-1\right)^4\cdot i=-512i$

[51] $-i^{223}=-i^{222}\cdot i=-\left(i^2\right)^{111}\cdot i=-\left(-1\right)^{111}\cdot i=i$

[52] $-i^{80}=-\left(i^2\right)^{40}=-\left(-1\right)^{40}=-1$

[53] $i^{-28}=\dfrac{1}{i^{28}}=\dfrac{1}{\left(i^2\right)^{14}}=\dfrac{1}{\left(-1\right)^{14}}=1$

[54] $i^{-119}=\dfrac{1}{i^{119}}=\dfrac{1}{i^{118}\cdot i}=\dfrac{1}{\left(i^2\right)^{59}\cdot i}=\dfrac{1}{\left(-1\right)^{59}\cdot i}=\dfrac{1}{-i}=\dfrac{1}{-i}\cdot\dfrac{i}{i}=i$

[55] $\dfrac{i^{31}}{i^{19}}=i^{12}=\left(i^2\right)^6=\left(-1\right)^6=1$

[56] $i^{26}\cdot i^{-37}=i^{-11}=\dfrac{1}{i^{11}}=\dfrac{1}{i^{10}\cdot i}=\dfrac{1}{\left(i^2\right)^5\cdot i}=\dfrac{1}{\left(-1\right)^5\cdot i}=\dfrac{1}{-i}=\dfrac{1}{-i}\cdot\dfrac{i}{i}=i$

[57] $\left| -3+7i \right| = \sqrt{(-3)^2+(7)^2} = \sqrt{58}$

[58] $\left| 12-8i \right| = \sqrt{(12)^2+(-8)^2} = \sqrt{208} = 4\sqrt{13}$

[59] $\left| \dfrac{1}{2}-\dfrac{3}{4}i \right| = \sqrt{\left(\dfrac{1}{2}\right)^2+\left(-\dfrac{3}{4}\right)^2} = \dfrac{\sqrt{13}}{4}$

[60] $\left| -\dfrac{\sqrt{3}}{2}+\dfrac{1}{2}i \right| = \sqrt{\left(-\dfrac{\sqrt{3}}{2}\right)^2+\left(\dfrac{1}{2}\right)^2} = 1$

[61] $\left| -16\,i \right| = \sqrt{(-16)^2} = \sqrt{256} = 16$

[62] $\left| -9 \right| = 9$

[63] $\left| (-3+4i)^2 \right| = \left| -7-24i \right| = \sqrt{(-7)^2+(-24)^2} = 25$

[64] $\left| (1-i)^3 \right| = \left| (1-i)(1-2i+i^2) \right| = \left| (1-i)(-2i) \right| = \left| -2-2i \right| = \sqrt{(-2)^2+(-2)^2} = 2\sqrt{2}$

[65] $\left| i^{51} \right| = \left| i^{50}\cdot i \right| = \left| (i^2)^{25}\cdot i \right| = \sqrt{(-1)^{25}\cdot i} = \sqrt{-i} = \sqrt{(-1)^2} = 1$

[66] $\left| i^{104} \right| = \left| (i^2)^{52} \right| = \left| (-1)^{52} \right| = \left| 1 \right| = 1$

[67] $\left| -\dfrac{1}{3i} \right| = \left| -\dfrac{1}{3i}\cdot\dfrac{-3i}{-3i} \right| = \left| \dfrac{3i}{9} \right| = \left| \dfrac{1}{3}i \right| = \sqrt{\left(\dfrac{1}{3}\right)^2} = \dfrac{1}{3}$

[68] $\left| \dfrac{-2+3i}{-4i} \right| = \left| \dfrac{-2+3i}{-4i}\cdot\dfrac{4i}{4i} \right| = \left| \dfrac{-12-8i}{16} \right| = \left| -\dfrac{3}{4}-\dfrac{1}{2}i \right| = \sqrt{\left(-\dfrac{3}{4}\right)^2+\left(-\dfrac{1}{2}\right)^2} = \dfrac{\sqrt{13}}{4}$

[69] $\left| \overline{12-6i} \right| = \left| 12+6i \right| = \sqrt{(12)^2+(6)^2} = 6\sqrt{5}$

[70] $\left| \overline{3+4i} \right| = \left| 3-4i \right| = \sqrt{(3)^2+(-4)^2} = 5$

[71] Since $-2+5i = x-yi$, it follows that $x=-2$ and $y=-5$.

[72] $(1+i)^2 = 1+2i+i^2 = 2i$. Thus

    a) $(1+i)^6 = \left[(1+i)^2\right]^3 = (2i)^3 = 8i^3 = 8i^2\cdot i = -8i$

    b) $(1+i)^9 = (1+i)^8(1+i) = \left[(1+i)^2\right]^4(1+i) = (2i)^4(1+i) = 16i^4(1+i) = 16(i^2)^2(1+i) =$
      $16(-1)^2(1+i) = 16+16i$

    c) $(1+i)^{10} = \left[(1+i)^2\right]^5 = (2i)^5 = 32i^5 = 32i^4\cdot i = 32(i^2)^2\cdot i = 32(-1)^2 i = 32i$

[73]   a) $\left(-\dfrac{1}{2}+\dfrac{\sqrt{3}}{2}i\right)^3 = \left(-\dfrac{1}{2}+\dfrac{\sqrt{3}}{2}i\right)\left(\dfrac{1}{4}-\dfrac{\sqrt{3}}{2}i-\dfrac{3}{4}\right) = \left(-\dfrac{1}{2}+\dfrac{\sqrt{3}}{2}i\right)\left(-\dfrac{1}{2}-\dfrac{\sqrt{3}}{2}\right) = \dfrac{1}{4}+\dfrac{3}{4} = 1$

    b) $i^8 = (i^2)^4 = (-1)^4 = 1$

[74] Let $z_1 = a+bi$, $z_2 = c+di$ and $z_3 = e+fi$

    a) $z_1+z_2 = (a+c)+(b+d)i = (c+a)+(d+b)i = z_2+z_1$

b) $z_1(z_2 \cdot z_3) = (a+bi)\big[(ce-df)+(cf+de)\,i\big] = \big[a(ce-df)+bi^2(cf+de)\big] +$

$\qquad [a(cf+de)\,i + bi(ce-df)] = (ace-adf-bcf-bde) +$

$\qquad (acf+ade+bce-bdf)\,i$

$(z_1 \cdot z_2)z_3 = \big[(ac-bd)+(ad+bc)\,i\big](e+fi) = \big[(ac-bd)\,e +(ad+bc)f\,i^2\big] +$

$\qquad [(ac-bd)fi + (ad+bc)\,i] = (ace-bde-adf-bcf) +$

$\qquad (acf-bdf+ade+bce)\,i = (ace-adf-bcf-bde)+(acf+ade+bce-bdf)\,i.$

Thus $z_1(z_2 \cdot z_3) = (z_1 \cdot z_2)z_3$.

c) $z_1(z_2+z_3) = (a+bi)[(c+e)+(d+f)\,i] = (a+bi)(c+e)+(a+bi)(di+fi) =$

$\qquad ac+ae+bci+bei+adi+afi-bd-bf =$

$\qquad (ac+ae-bd-bf)+(bc+be+ad+af)\,i$

$z_1 \cdot z_2 + z_1 \cdot z_3 = \big[(ac-bd)+(ad+bc)\,i\big]+\big[(ae-bf)+(af+be)\,i\big] =$

$\qquad (ac+ae-bd-bf)+(bc+be+ad+af)\,i.$

Thus $z_1(z_2+z_3) = z_1 z_2 + z_1 z_3$.

[75] $\bar{z}=a-bi \Rightarrow |\bar{z}| = \sqrt{a^2+(-b)^2} = \sqrt{a^2+b^2} = |z|$

[76] a) $z_1+z_2 = (a+c)+(b+d)\,i \Rightarrow \overline{z_1+z_2} = (a+c)-(b+d)\,i$

$\qquad \bar{z_1}+\bar{z_2} = (a-bi)+(c-di) = (a+c)-(b+d)\,i.$

$\qquad$ Thus $\overline{z_1+z_2} = \bar{z_1}+\bar{z_2}$.

b) $z_1 \cdot z_2 = (ac-bd)+(ad+bc)\,i \Rightarrow \overline{z_1 \cdot z_2} = (ac-bd)-(ad+bc)\,i$

$\qquad \bar{z_1} \cdot \bar{z_2} = (a-bi)(c-di) = (ac-bd)-(ad+bc)\,i.$

$\qquad$ Thus $\overline{z_1 \cdot z_2} = \bar{z_1} \cdot \bar{z_2}$.

c) $\overline{z_1{}^n} = \underbrace{\overline{z_1 \cdot z_1 \cdot \ldots \cdot z_1}}_{n\ \text{factors}} = \underbrace{\bar{z_1} \cdot \bar{z_1} \cdot \ldots \cdot \bar{z_1}}_{n\ \text{factors}}$ (by part (b)) $= \bar{z_1}{}^n$

d) If $z=a$, where $a \in \mathbf{R}$, then $\bar{z} = \overline{a+0i} = a-0i = a = z$.

EXERCISES 1.5

[1] $3x+15 = 0 \Rightarrow x = -5$ $\qquad\qquad\qquad$ $\bullet\{-5\}$

[2] $\dfrac{x-2}{3} - \dfrac{x}{6} = \dfrac{7}{5} + 2x \Rightarrow 30\left(\dfrac{x-2}{3}\right) - 30\left(\dfrac{x}{6}\right) = 30\left(\dfrac{7}{5}\right) + 30\,(2x) \Rightarrow$

$\qquad 10x - 20 - 5x = 42 + 60x \Rightarrow x = -\dfrac{62}{55}$ $\qquad$ $\bullet\left\{-\dfrac{62}{55}\right\}$

[3] $2(3x-1)-4 = 5+7(x-1) \Rightarrow 6x-2-4 = 5+7x-7 \Rightarrow x = -4$ $\qquad$ $\bullet\{-4\}$

[4] $\dfrac{3x-1}{5} - \dfrac{2x+1}{10} = \dfrac{1-x}{4} - 3 \Rightarrow 40\left(\dfrac{3x-1}{5}\right) - 40\left(\dfrac{2x+1}{10}\right) = 40\left(\dfrac{1-x}{4}\right) - 40\,(3) \Rightarrow$

$24x - 8 - 8x - 4 = 10 - 10x - 120 \Rightarrow x = -\dfrac{98}{26} = -\dfrac{49}{13}$ $\bullet\left\{-\dfrac{49}{13}\right\}$

[5] $\dfrac{3x}{4} - 6 = 2 + \dfrac{x}{3} \Rightarrow 12\left(\dfrac{3x}{4}\right) - 12\,(6) = 12\,(2) + 12\left(\dfrac{x}{3}\right) \Rightarrow \quad 9x - 72 = 24 + 4x \Rightarrow x = \dfrac{96}{5}$ $\bullet\left\{\dfrac{96}{5}\right\}$

[6] $2\,(x-3) + 6x = 2\,(x+2) \Rightarrow 2x - 6 + 6x = 2x + 4 \Rightarrow x = \dfrac{10}{6} = \dfrac{5}{3}$ $\bullet\left\{\dfrac{5}{3}\right\}$

[7] $0.01x - 0.1\,(x-10) = 0.02x + 1 \Rightarrow 100\,(0.01x) - 100\,(0.1)\,(x-10) = 100\,(0.02x) + 100\,(1) \Rightarrow$

$x - 10x + 100 = 2x + 100 \Rightarrow x = 0$ $\bullet\{0\}$

[8] $0.03x + 0.5\,(2x-3) = 10 \Rightarrow 100\,(0.03x) + 100\,(0.5)\,(2x-3) = 100\,(10) \Rightarrow$

$3x + 100\,x - 150 = 1000 \Rightarrow x = \dfrac{1150}{103}$ $\bullet\left\{\dfrac{1150}{103}\right\}$

[9] $P = 2L + 2W \Rightarrow 2L = P - 2W \Rightarrow L = \dfrac{P - 2W}{2}$

[10] $A = P\,(1 + rt) \Rightarrow A = P + Prt \Rightarrow r = \dfrac{A - P}{Pt}$

[11] $A = LW \Rightarrow W = \dfrac{A}{L}$ [12] $A = \dfrac{1}{2}\,bh \Rightarrow 2A = bh \Rightarrow h = \dfrac{2A}{b}$

[13] $C = 2\pi r \Rightarrow r = \dfrac{C}{2\pi}$

[14] $F = \dfrac{9}{5}\,C + 32 \Rightarrow F - 32 = \dfrac{9}{5}\,C \Rightarrow 5F - 160 = 9C \Rightarrow C = \dfrac{5F - 160}{9} = \dfrac{5}{9}\,(F - 32)$

[15] $C = \dfrac{5}{9}\,(F - 32) \Rightarrow 9C = 5F - 160 \Rightarrow 5F = 9C + 160 \Rightarrow F = \dfrac{9}{5}\,C + 32$

[16] $S = 2\pi rh + 2\pi r^2 \Rightarrow 2\pi rh = S - 2\pi r^2 \Rightarrow h = \dfrac{S - 2\pi r^2}{2\pi r}$

[17] $V = \pi r^2 h \Rightarrow h = \dfrac{V}{\pi r^2}$ [18] $I = Prt \Rightarrow t = \dfrac{I}{Pr}$

[19] $Ax + By + C = 0 \Rightarrow By = -Ax - C \Rightarrow y = \dfrac{-Ax - C}{B} \Rightarrow y = -\dfrac{A}{B}x - \dfrac{C}{B}$

[20] $S = B + Bxy \Rightarrow S = B\,(1 + xy) \Rightarrow B = \dfrac{S}{1 + xy}$

[21] Let x be the number of student tickets sold. Then $3x$ is the number of adult tickets sold.

$2.75x + 5\,(3x) = 798.75 \Rightarrow 100\,(2.75x) + 100\,(15x) = 100\,(798.75) \Rightarrow \quad 1775\,x = 79875 \Rightarrow x = 45$

So, 45 student tickets were sold and $3\,(45) = 135$ adult tickets were sold.

[22] Let w be the width (in meters) of the lot. Then $l = 2w - 5$ is the length.

$2l + 2w = 440 \Rightarrow 2(2w - 5) + 2w = 440 \Rightarrow 6w = 450 \Rightarrow w = 75\text{m} \Rightarrow l = 2(75) - 5 \Rightarrow l = 145\text{ m}.$

So, the width of the lot is 75 m and the length is 145 m. Hence, the area of the lot is

$A = l\,w = (145)(75) = 10,875\ m^2$

[23] Let x be her standard hourly salary (in $). Then

$x(40) + \dfrac{3x}{2}(7) = 454.50 \Rightarrow 2(40x) + 2\left(\dfrac{3}{2}x\right)(7) = 2(454.50) \Rightarrow 80x + 21x = 909 \Rightarrow$

$101\,x = 909 \Rightarrow x = 9.$ So, her standard hourly salary is $9.

[24] Let x be the original price (in $) of the boots. Then $x - 40\ \% \ x = 48 \Rightarrow$

$100x - 100\,(40\ \%\ x) = 100\,(48) \Rightarrow 100\,x - 40x = 4800 \Rightarrow 60x = 4800 \Rightarrow x = 80$
So, the original price of the boots was $80.

[25] Let x be the dollar amount of sales in the indicated week. Then $250 + 10\ \%\ x = 302 \Rightarrow$

$100\,(250) + 100\,(10\ \%\ x) = 100\,(302) \Rightarrow 25000 + 10x = 30200 \Rightarrow 10x = 5200 \Rightarrow x = \$520$.

So, the dollar amount of sales that week was $520.

[26] Let x be the original price (in $) of the calculator. Then $x - 20\ \%\ x = 30 \Rightarrow$

$100x - 100\,(20\ \%\ x) = 100\,(30) \Rightarrow 100x - 20x = 3000 \Rightarrow 80x = 3000 \Rightarrow x = 37.50$

So, the original price of the calculator was $37.50.

[27] Let x be the required number of days of renting. Then $12x = 168 \Rightarrow x = 14.$

So, after 14 days of renting, the rental fee will be the same as the sale price of the new mower.

[28] Let x be the required number of years. Then

$25,000 + 1500\,x = 31,750 \Rightarrow 1500\,x = 6750 \Rightarrow x = 4.5.$

So, in four and one-half years, the population will be 31,750.

[29] $6x^2 + 7x - 3 = 0 \Rightarrow (3x - 1)(2x + 3) = 0 \Rightarrow 3x - 1 = 0 \text{ or } 2x + 3 = 0 \Rightarrow$

$x = \dfrac{1}{3} \text{ or } x = -\dfrac{3}{2}$ •$\left\{-\dfrac{3}{2}, \dfrac{1}{3}\right\}$

[30] $x^2 = -3(15 + 6x) \Rightarrow x^2 = -45 - 18x \Rightarrow x^2 + 18x + 45 = 0 \Rightarrow (x + 3)(x + 15) = 0 \Rightarrow$

$x = -3 \text{ or } x = -15$ •$\left\{-15,\ -3\right\}$

[31] $x^2 - 42 = -19x \Rightarrow x^2 + 19x - 42 = 0 \Rightarrow (x + 21)(x - 2) = 0 \Rightarrow x = -21 \text{ or } x = 2$ •$\left\{-21,\ 2\right\}$

[32] $2(x + 3) + 3x = 2x^2 - 6 \Rightarrow 2x + 6 + 3x = 2x^2 - 6 \Rightarrow 2x^2 - 5x - 12 = 0 \Rightarrow (2x + 3)(x - 4) = 0 \Rightarrow$

$x = -\dfrac{3}{2} \text{ or } x = 4$ •$\left\{-\dfrac{3}{2}, 4\right\}$

[33] $x(6x - 5) = 4 \Rightarrow 6x^2 - 5x - 4 = 0 \Rightarrow (3x - 4)(2x + 1) = 0 \Rightarrow x = \dfrac{4}{3} \text{ or } x = -\dfrac{1}{2}$ •$\left\{-\dfrac{1}{2}, \dfrac{4}{3}\right\}$

[34] $x(3x + 1) = 2 \Rightarrow 3x^2 + x - 2 = 0 \Rightarrow (3x - 2)(x + 1) = 0 \Rightarrow x = \dfrac{2}{3} \text{ or } x = -1$ •$\left\{-1, \dfrac{2}{3}\right\}$

[35] $2x^2 - 4x - 3 = 0 \Rightarrow 2x^2 - 4x = 3 \Rightarrow x^2 - 2x = \dfrac{3}{2} \Rightarrow x^2 - 2x + 1 = \dfrac{3}{2} + 1 \Rightarrow$

$(x-1)^2 = \dfrac{5}{2} \Rightarrow x - 1 = \pm\sqrt{\dfrac{5}{2}} = \pm\dfrac{\sqrt{10}}{2} \Rightarrow x = 1 \pm \dfrac{\sqrt{10}}{2} = \dfrac{2 \pm \sqrt{10}}{2}$ $\bullet\left\{\dfrac{2-\sqrt{10}}{2}, \dfrac{2+\sqrt{10}}{2}\right\}$

[36] $3x + 2 = 2x^2 \Rightarrow 2x^2 - 3x = 2 \Rightarrow x^2 - \dfrac{3}{2}x = 1 \Rightarrow x^2 - \dfrac{3}{2}x + \dfrac{9}{16} = 1 + \dfrac{9}{16} \Rightarrow$

$\left(x - \dfrac{3}{4}\right)^2 = \dfrac{25}{16} \Rightarrow x - \dfrac{3}{4} = \pm\sqrt{\dfrac{25}{16}} = \pm\dfrac{5}{4} \Rightarrow x = \dfrac{3}{4} \pm \dfrac{5}{4} = \dfrac{3 \pm 5}{4} \Rightarrow$

$x = \dfrac{3+5}{4} = 2 \text{ or } x = \dfrac{3-5}{4} = -\dfrac{1}{2}$ $\bullet\left\{-\dfrac{1}{2}, 2\right\}$

[37] $3x(x-4) = -13 \Rightarrow 3x^2 - 12x = -13 \Rightarrow x^2 - 4x = -\dfrac{13}{3} \Rightarrow x^2 - 4x + 4 = -\dfrac{13}{3} + 4 \Rightarrow$

$(x-2)^2 = -\dfrac{1}{3} \Rightarrow x - 2 = \pm\sqrt{-\dfrac{1}{3}} = \pm\dfrac{i\sqrt{3}}{3} \Rightarrow x = 2 \pm \dfrac{\sqrt{3}}{3}i$ $\bullet\left\{2 - \dfrac{\sqrt{3}}{3}i, 2 + \dfrac{\sqrt{3}}{3}i\right\}$

[38] $x^2 - 2x = 1 \Rightarrow x^2 - 2x + 1 = 2 \Rightarrow (x-1)^2 = 2 \Rightarrow x - 1 = \pm\sqrt{2} \Rightarrow x = 1 \pm \sqrt{2}$ $\bullet\left\{1-\sqrt{2}, 1+\sqrt{2}\right\}$

[39] $4x^2 = 3(4x - 3) \Rightarrow 4x^2 = 12x - 9 \Rightarrow 4x^2 - 12x = -9 \Rightarrow x^2 - 3x = -\dfrac{9}{4} \Rightarrow x^2 - 3x + \dfrac{9}{4} = -\dfrac{9}{4} + \dfrac{9}{4} \Rightarrow$

$\left(x - \dfrac{3}{2}\right)^2 = 0 \Rightarrow x = \dfrac{3}{2}$ $\bullet\left\{\dfrac{3}{2}\right\}$

[40] $x(3x + 6) = 9 \Rightarrow 3x^2 + 6x = 9 \Rightarrow x^2 + 2x = 3 \Rightarrow x^2 + 2x + 1 = 4 \Rightarrow (x+1)^2 = 4 \Rightarrow$

$x + 1 = \pm 2 \Rightarrow x = -1 \pm 2 \Rightarrow x = -1 + 2 = 1 \text{ or } x = -1 - 2 = -3$ $\bullet\left\{-3, 1\right\}$

[41] $2x^2 = 4x - 1 \Rightarrow 2x^2 - 4x + 1 = 0 \Rightarrow x = \dfrac{4 \pm \sqrt{8}}{4} = \dfrac{2 \pm \sqrt{2}}{2}$ $\bullet\left\{\dfrac{2-\sqrt{2}}{2}, \dfrac{2+\sqrt{2}}{2}\right\}$

[42] $x^2 = 4x - 13 \Rightarrow x^2 - 4x + 13 = 0 \Rightarrow x = \dfrac{4 \pm \sqrt{-36}}{2} = 2 \pm 3i$ $\bullet\left\{2-3i, 2+3i\right\}$

[43] $x^2 = 2(x-1) \Rightarrow x^2 - 2x + 2 = 0 \Rightarrow x = \dfrac{2 \pm \sqrt{-4}}{2} = 1 \pm i$ $\bullet\left\{1-i, 1+i\right\}$

[44] $4x^2 = 4x - 1 \Rightarrow 4x^2 - 4x + 1 = 0 \Rightarrow x = \dfrac{4 \pm \sqrt{0}}{8} = \dfrac{1}{2}$ $\bullet\left\{\dfrac{1}{2}\right\}$

[45] $9x^2 = -8(3x + 2) \Rightarrow 9x^2 + 24x + 16 = 0 \Rightarrow x = \dfrac{-24 \pm \sqrt{0}}{18} = -\dfrac{4}{3}$ $\bullet\left\{-\dfrac{4}{3}\right\}$

[46] $x^2 = 4(x+2) \Rightarrow x^2 - 4x - 8 = 0 \Rightarrow x = \dfrac{4 \pm \sqrt{48}}{2} = 2 \pm 2\sqrt{3}$ $\bullet\left\{2 - 2\sqrt{3}, 2 + 2\sqrt{3}\right\}$

[47] $x^2 - 2x - 4 = 0 \Rightarrow D = b^2 - 4ac = (-2)^2 - 4(1)(-4) = 20 > 0$.

Since $D > 0$, there are two distinct real roots.

[48] $3x^2 + 3x + 1 = 0 \Rightarrow D = b^2 - 4ac = (3)^2 - 4(3)(1) = -3 < 0.$

Since $D < 0$, there are two complex conjugate roots.

[49] $9x^2 + 6x + 1 = 0 \Rightarrow D = b^2 - 4ac = (6)^2 - 4(9)(1) = 0.$

Since $D = 0$, there is one real (double) root.

[50] $4x^2 + 3x - 10 = 0 \Rightarrow D = b^2 - 4ac = (3)^2 - 4(4)(-10) = 169 > 0.$

Since $D > 0$, there are two distinct real roots.

[51] $5x^2 - 2x + 1 = 0 \Rightarrow D = b^2 - 4ac = (-2)^2 - 4(5)(1) = -16 < 0.$

Since $D < 0$, there are two complex conjugate roots.

[52] $x^2 + 5x + 4 = 0 \Rightarrow D = b^2 - 4ac = (5)^2 - 4(1)(4) = 9 > 0.$

Since $D > 0$, there are two distinct real roots.

[53] $3x^2 + x = 4(3x+1) \Rightarrow 3x^2 + x = 12x + 4 \Rightarrow 3x^2 - 11x - 4 = 0 \Rightarrow (3x+1)(x-4) = 0 \Rightarrow$

$x = -\dfrac{1}{3}$ or $x = 4$ $\qquad\qquad \bullet \left\{ -\dfrac{1}{3},\ 4 \right\}$

[54] $16x(2-x) = 8 \Rightarrow 32x - 16x^2 = 8 \Rightarrow 16x^2 - 32x + 8 = 0 \Rightarrow 2x^2 - 4x + 1 = 0 \Rightarrow$

$x = \dfrac{4 \pm \sqrt{8}}{4} = 2 \pm \sqrt{2}$ $\qquad\qquad \bullet \left\{ 2 - \sqrt{2},\ 2 + \sqrt{2} \right\}$

[55] $7x = 2x^2 - 1 \Rightarrow 2x^2 - 7x - 1 = 0 \Rightarrow x = \dfrac{7 \pm \sqrt{57}}{4}$ $\qquad \bullet \left\{ \dfrac{7 - \sqrt{57}}{4},\ \dfrac{7 + \sqrt{57}}{4} \right\}$

[56] $3x^2 = 4x + 5 \Rightarrow 3x^2 - 4x - 5 = 0 \Rightarrow x = \dfrac{4 \pm \sqrt{76}}{6} = \dfrac{2 \pm \sqrt{19}}{3}$ $\qquad \bullet \left\{ \dfrac{2 - \sqrt{19}}{3},\ \dfrac{2 + \sqrt{19}}{3} \right\}$

[57] $5x^2 = 4x \Rightarrow 5x^2 - 4x = 0 \Rightarrow x(5x - 4) = 0 \Rightarrow x = 0$ or $5x = 4 \Rightarrow x = 0$ or $x = \dfrac{4}{5}$ $\qquad \bullet \left\{ 0,\ \dfrac{4}{5} \right\}$

[58] $x(5x-2) = 4(2-5x) \Rightarrow 5x^2 + 18x - 8 = 0 \Rightarrow (5x-2)(x+4) = 0 \Rightarrow x = \dfrac{2}{5}$ or $x = -4$ $\quad \bullet \left\{ -4,\ \dfrac{2}{5} \right\}$

[59] $9x^2 = 5(6x - 5) \Rightarrow 9x^2 - 30x + 25 = 0 \Rightarrow (3x - 5)^2 = 0 \Rightarrow x = \dfrac{5}{3}$ $\qquad \bullet \left\{ \dfrac{5}{3} \right\}$

[60] $x(6-x) = 18 \Rightarrow x^2 - 6x + 18 = 0 \Rightarrow x = \dfrac{6 \pm \sqrt{-36}}{2} = 3 \pm 3i$ $\qquad \bullet \{ 3 - 3i,\ 3 + 3i \}$

[61] $A = \pi r^2 \Rightarrow r^2 = \dfrac{A}{\pi} \Rightarrow r = \pm\sqrt{\dfrac{A}{\pi}} = \pm\dfrac{\sqrt{A}}{\sqrt{\pi}} = \pm\dfrac{\sqrt{A\pi}}{\pi}.$ Since $r \geq 0$, $r = \dfrac{\sqrt{A\pi}}{\pi}.$

[62] $x^2 + y^2 = r^2 \Rightarrow y^2 = r^2 - x^2 \Rightarrow y = \pm\sqrt{r^2 - x^2}$

[63] $D = b^2 - 4ac \Rightarrow b^2 = D + 4ac \Rightarrow b = \pm\sqrt{D + 4ac}$

[64] $V = \pi r^2 h = \Rightarrow r^2 = \dfrac{V}{\pi h} \Rightarrow r = \pm\sqrt{\dfrac{V}{\pi h}} = \dfrac{\pm\sqrt{V}}{\sqrt{\pi h}} = \dfrac{\pm\sqrt{V\pi h}}{\pi h}.$ Since $r \geq 0$, $r = \dfrac{\sqrt{V\pi h}}{\pi h}.$

[65] $A = 4h^2 - 3h - 5 \Rightarrow 4h^2 - 3h - (A+5) = 0 \Rightarrow h = \dfrac{-(-3) \pm \sqrt{(-3)^2 - 4(4)\left[-(A+5)\right]}}{2(4)} =$

$\dfrac{3 \pm \sqrt{9 + 16(A+5)}}{8} = \dfrac{3 \pm \sqrt{16A + 89}}{8}.$

[66] $S = 2\pi rh + 2\pi r^2 \Rightarrow 2\pi r^2 + 2\pi rh - S = 0 \Rightarrow r = \dfrac{-2\pi h \pm \sqrt{(2\pi h)^2 - 4(2\pi)(-S)}}{2(2\pi)} =$

$\dfrac{-2\pi h \pm \sqrt{4\pi^2 h^2 + 8\pi S}}{4\pi} = \dfrac{-\pi h \pm \sqrt{\pi^2 h^2 + 2\pi S}}{2\pi}.$

[67] We have $r_1 + r_2 = \left(\dfrac{-b + \sqrt{b^2 - 4ac}}{2a}\right) + \left(\dfrac{-b - \sqrt{b^2 - 4ac}}{2a}\right) = \dfrac{-b + \sqrt{b^2 - 4ac} - b - \sqrt{b^2 - 4ac}}{2a} =$

$\dfrac{-2b}{2a} = -\dfrac{b}{a}$ and $r_1 \cdot r_2 = \left(\dfrac{-b + \sqrt{b^2 - 4ac}}{2a}\right)\left(\dfrac{-b - \sqrt{b^2 - 4ac}}{2a}\right) = \dfrac{(-b)^2 - \left(\sqrt{b^2 - 4ac}\right)^2}{4a^2} =$

$\dfrac{b^2 - (b^2 - 4ac)}{4a^2} = \dfrac{b^2 - b^2 + 4ac}{4a^2} = \dfrac{4ac}{4a^2} = \dfrac{c}{a}.$

[68] Using the definition of absolute value, we must consider two cases.

If $a \geq 0$, then $|a| = a$ and we have $\pm 2|a| = \pm 2(a) = \pm 2a$. If $a < 0$, then $|a| = -a$ and we have $\pm 2|a| = \pm 2(-a) = \pm 2a$. So, in either case, $\pm 2|a| = \pm 2a$. (Note that in the derivation of the quadratic formula, $a \neq 0$.)

[69] If x is the length (in meters) of the longer leg, then $x - 21$ is the length of the shorter leg. By the Pythagorean Theorem, $x^2 + (x - 21)^2 = (39)^2 \Rightarrow x^2 + (x^2 - 42x + 441) = 1521 \Rightarrow$

$2x^2 - 42x - 1080 = 0 \Rightarrow x^2 - 21x - 540 = 0 \Rightarrow (x - 36)(x + 15) = 0 \Rightarrow \quad x = 36 \text{ or } x = -15$.

Since length cannot be negative, $x = 36$. So, the length of the longer leg is 36 m and the length of the shorter leg is $36 - 21 = 15$ m.

[70] Let x be the integer. Then $2x^2 = x + 15 \Rightarrow 2x^2 - x - 15 = 0 \Rightarrow (2x + 5)(x - 3) = 0 \Rightarrow$

$x = -\dfrac{5}{2}$ or $x = 3$. Since x is an integer, we know that $x \neq -\dfrac{5}{2}$. So, the integer is 3.

[71] If l is the length (in feet) of the lot, then $w = 2l + 6$ is the width. Since the area is 360 square feet and $A = lw$, we can write $lw = 360 \Rightarrow l(2l + 6) = 360 \Rightarrow 2l^2 + 6l - 360 = 0 \Rightarrow$

$l^2 + 3l - 180 = 0 \Rightarrow (l - 12)(l + 15) = 0 \Rightarrow l = 12 \text{ or } l = -15$. Since length cannot be negative, $l = 12$. So the length of the lot is 12 feet and the width is $2(12) + 6 = 30$ feet. Hence, the perimeter of the lot is $P = 2l + 2w = 2(12) + 2(30) = 84$ feet.

[72] Let x be the length (in inches) of each side of the original square piece of cardboard. Since $V = l\,w\,h$, we can write $V = (x-4)(x-4)(2)$. Since $V = 338$ cubic inches, it follows that

$(x-4)(x-4)(2) = 338 \Rightarrow (x^2 - 8x + 16)(2) = 338 \Rightarrow x^2 - 8x + 16 = 169 \Rightarrow$

$x^2 - 8x - 153 = 0 \Rightarrow (x-17)(x+9) = 0 \Rightarrow x = 17$ or $x = -9$. Since length cannot be negative, $x = 17$. So, the original piece of cardboard was a square 17 in. on each side. Hence, the area of the original piece of cardboard is $A = (17)(17) = 289\,\text{in}^2$.

[73] If x is one number, then $2x - 9$ is the other number. Since their product is 35, $x(2x-9) = 35 \Rightarrow$

$2x^2 - 9x - 35 = 0 \Rightarrow (2x+5)(x-7) = 0 \Rightarrow x = -\dfrac{5}{2}$ or $x = 7$. Since the numbers were given to be

positive, 7 is one of the numbers and $2(7) - 9 = 5$ is the other.

[74] If x is the first of two consecutive positive even integers, then $x + 2$ is the second. Since the sum of

their squares is 340, $x^2 + (x+2)^2 = 340 \Rightarrow x^2 + (x^2 + 4x + 4) = 340 \Rightarrow 2x^2 + 4x - 336 = 0 \Rightarrow$

$x^2 + 2x - 168 = 0 \Rightarrow (x-12)(x+14) = 0 \Rightarrow x = 12$ or $x = -14$. Since x is positive, 12 is the first integer, and $12 + 2 = 14$ is the second.

[75] Since the ball strikes the ground when its height is zero, solve for t:

$-16t^2 + 32t + 240 = 0 \Rightarrow t^2 - 2t - 15 = 0 \Rightarrow (t-5)(t+3) = 0 \Rightarrow t = 5$ or $t = -3$.

Since time cannot be negative, $t = 5$. So, the ball strikes the ground 5 seconds after it is thrown.

[76] Let x be the length (in cm) of the piece of cardboard which is removed from both the length and width of the rectangular sheet. Since the original area was $(46)(32) = 1472\,\text{cm}^2$, we can write

$(46 - x)(32 - x) = 1472 - 432 \Rightarrow 1472 - 78x + x^2 = 1040 \Rightarrow x^2 - 78x + 432 = 0 \Rightarrow$

$(x-6)(x-72) = 0 \Rightarrow x = 6$ or $x = 72$. Since the original sheet of cardboard was only 46 cm by 32 cm, it is impossible to remove a 72 cm piece from both the length and the width. So, when 6 cm is subtracted from both the length and the width, the new dimensions are $46 - 6 = 40$ cm by $32 - 6 = 26$ cm and the new area is $(40)(26) = 1{,}040\,\text{cm}^2$ which is $432\,\text{cm}^2$ less than the original area.

[77] Let l and w be the length and width (in meters) of the rectangle. Since $P = 2l + 2w$ is the perimeter and $A = l\,w$ is the area, we can write $40 = 2l + 2w$ and $91 = l\,w$. Since $40 = 2l + 2w$, $l + w = 20$, or $l = 20 - w$. Substituting $20 - w$ for l in the equation $91 = l\,w$, we obtain $91 = (20 - w)w$, or

$w^2 - 20w + 91 = 0 \Rightarrow (w-7)(w-13) = 0 \Rightarrow w = 7$ or $w = 13$. If $w = 7$, then $l = 20 - 7 = 13$. On the other hand, if $w = 13$, then $l = 20 - 13 = 7$. So, the rectangle is 13 m by 7 m.

[78] If b is the length (in inches) of the base of the triangle, then $h = b - 14$ is the length of the height.

Since $A = \dfrac{1}{2}b\,h$, we can write $120 = \dfrac{1}{2}b\,h \Rightarrow \dfrac{1}{2}(b)(b-14) = 120 \Rightarrow b^2 - 14b = 240 \Rightarrow$

$b^2 - 14b - 240 = 0 \Rightarrow (b-24)(b+10) = 0 \Rightarrow b = 24$ or $b = -10$. Since length cannot be negative, discard $b = -10$. So, 24 in. is the length of the base and $24 - 14 = 10$ in. is the length of the height.

[79] Let x be the positive integer. Then $3x^2 = 6x + 45 \Rightarrow 3x^2 - 6x - 45 = 0 \Rightarrow x^2 - 2x - 15 = 0 \Rightarrow$

$(x-5)(x+3) = 0 \Rightarrow x = 5$ or $x = -3$. Since x is positive, discard $x = -3$. So, the positive integer is 5.

[80] Let x be the length (in yds) of the horizontal sides of each of the two pens and let y be the length (in yds) of each of the vertical sides, as shown in the given figure. Then $4x + 3y = 600$ and $2xy = 15,000$. Solving the equation $4x + 3y = 600$ for y in terms of x, yields $3y = 600 - 4x$, or

$y = 200 - \dfrac{4}{3}x$. Since $2xy = 15,000$, $xy = 7,500$. Substituting $200 - \dfrac{4}{3}x$ for y in the last equation,

we have $x\left(200 - \dfrac{4}{3}x\right) = 7500 \Rightarrow 200x - \dfrac{4}{3}x^2 = 7500 \Rightarrow 600x - 4x^2 = 22,500 \Rightarrow$

$4x^2 - 600x + 22,500 = 0 \Rightarrow x^2 - 150x + 5625 = 0 \Rightarrow (x - 75)^2 = 0 \Rightarrow x = 75$. So, the dimensions of each pen are: 75 yd on the 4 horizontal sides and $200 - \dfrac{4}{3}(75) = 100$ yd on the three vertical sides.

EXERCISES 1.6

[1] $-4x - 16 \geq 0 \Rightarrow -4x \geq 16 \Rightarrow x \leq -4$

$\left(-\infty, -4\right]$

[2] $-4(x-2) + 3(x-4) \leq 6(x-7) + 3 \Rightarrow -4x + 8 + 3x - 12 \leq 6x - 42 + 3 \Rightarrow -7x \leq -35 \Rightarrow x \geq 5$

$\left[5, \infty\right)$

[3] $\dfrac{2x-1}{2} - \dfrac{1-x}{6} > \dfrac{x}{3} + 2 \Rightarrow 6\left(\dfrac{2x-1}{2}\right) - 6\left(\dfrac{1-x}{6}\right) > 6\left(\dfrac{x}{3}\right) + 6(2) \Rightarrow 6x - 3 - 1 + x > 2x + 12 \Rightarrow$

$5x > 16 \Rightarrow x > \dfrac{16}{5}$

$\left(\dfrac{16}{5}, \infty\right)$

[4] $-2x + 1 \leq 4 + x \Rightarrow -3x \leq 3 \Rightarrow x \geq -1$

$\left[-1, \infty\right)$

[5] $-5 \leq 1 - 2x < 9 \Rightarrow -6 \leq -2x < 8 \Rightarrow 3 \geq x > -4 \Rightarrow -4 < x \leq 3$

$\left(-4, 3\right]$

[6] $0 < 2 - 3x \leq 4 \Rightarrow -2 < -3x \leq 2 \Rightarrow \dfrac{2}{3} > x \geq -\dfrac{2}{3} \Rightarrow -\dfrac{2}{3} \leq x < \dfrac{2}{3}$

$\left[-\dfrac{2}{3}, \dfrac{2}{3}\right)$

[7] $4(x-6)-5x>11 \Rightarrow 4x-24-5x>11 \Rightarrow -x>35 \Rightarrow x<-35$

$\bullet \left(-\infty, -35\right)$

[8] $2(4x-1)-6(2x-5)<0 \Rightarrow 8x-2-12x+30<0 \Rightarrow -4x<-28 \Rightarrow x>7$

$\bullet \left(7, \infty\right)$

[9] $\dfrac{x-2}{3}+\dfrac{x+1}{8}>\dfrac{5}{6} \Rightarrow 24\left(\dfrac{x-2}{3}\right)+24\left(\dfrac{x+1}{8}\right)>24\left(\dfrac{5}{6}\right) \Rightarrow 8x-16+3x+3>20 \Rightarrow$

$11x>33 \Rightarrow x>3$

$\bullet \left(3, \infty\right)$

[10] $\dfrac{2x-3}{10}-\dfrac{3x-1}{4} \leq x-1 \Rightarrow 20\left(\dfrac{2x-3}{10}\right)-20\left(\dfrac{3x-1}{4}\right) \leq 20(x-1) \Rightarrow$

$4x-6-15x+5 \leq 20x-20 \Rightarrow -31x \leq -19 \Rightarrow x \geq \dfrac{19}{31}$

$\bullet \left[\dfrac{19}{31}, \infty\right)$

[11] $-1 \leq 3x+4 < 10 \Rightarrow -5 \leq 3x < 6 \Rightarrow -\dfrac{5}{3} \leq x < 2$

$\bullet \left[-\dfrac{5}{3}, 2\right)$

[12] $-6 < -4x-5 \leq -2 \Rightarrow -1 < -4x \leq 3 \Rightarrow \dfrac{1}{4} > x \geq -\dfrac{3}{4} \Rightarrow -\dfrac{3}{4} \leq x < \dfrac{1}{4}$

$\bullet \left[-\dfrac{3}{4}, \dfrac{1}{4}\right)$

[13] $\dfrac{4x-3}{-2} \leq 12 \Rightarrow -2\left(\dfrac{4x-3}{-2}\right) \geq -2(12) \Rightarrow 4x-3 \geq -24 \Rightarrow 4x \geq -21 \Rightarrow x \geq -\dfrac{21}{4}$

$\bullet \left[-\dfrac{21}{4}, \infty\right)$

[14] $-30x-2 \geq 5(2-3x) \Rightarrow -30x-2 \geq 10-15x \Rightarrow -15x \geq 12 \Rightarrow x \leq -\dfrac{12}{15} \Rightarrow x \leq -\dfrac{4}{5}$

$\bullet \left(-\infty, -\dfrac{4}{5}\right]$

[15] $6-3x \geq -14+2x \Rightarrow -5x \geq -20 \Rightarrow x \leq 4$

$\bullet \left(-\infty, 4\right]$

[16] $0 \leq 1 - 5x < 7 \Rightarrow -1 \leq -5x < 6 \Rightarrow \frac{1}{5} \geq x > -\frac{6}{5} \Rightarrow -\frac{6}{5} < x \leq \frac{1}{5}$

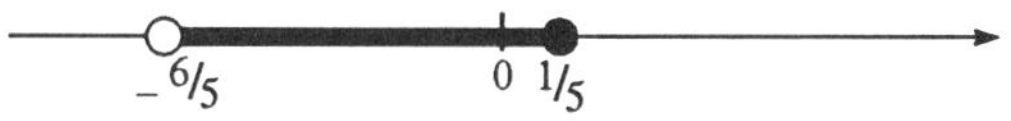

(number line graph: open circle at $-\frac{6}{5}$, closed circle at $\frac{1}{5}$)

$\bullet \left(-\frac{6}{5}, \frac{1}{5}\right]$

[17] $-7 \leq 2x - 3 \leq -4 \Rightarrow -4 \leq 2x \leq -1 \Rightarrow -2 \leq x \leq -\frac{1}{2}$

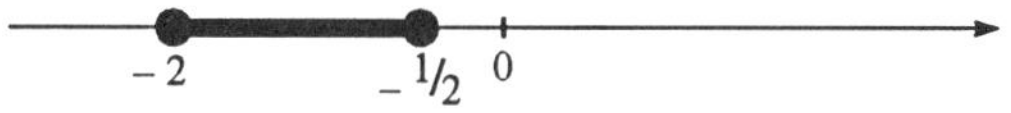

(number line graph: closed circles at -2 and $-\frac{1}{2}$)

$\bullet \left[-2, -\frac{1}{2}\right]$

[18] $\frac{2x}{3} + \frac{x-3}{2} < \frac{2x}{5} + \frac{3x+6}{10} \Rightarrow 30\left(\frac{2x}{3}\right) + 30\left(\frac{x-3}{2}\right) < 30\left(\frac{2x}{5}\right) + 30\left(\frac{3x+6}{10}\right) \Rightarrow$

$20x + 15x - 45 < 12x + 9x + 18 \Rightarrow 14x < 63 \Rightarrow x < \frac{63}{14} \Rightarrow x < \frac{9}{2}$

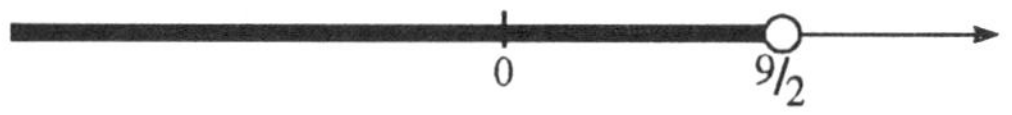

(number line graph: open circle at $\frac{9}{2}$)

$\bullet \left(-\infty, \frac{9}{2}\right)$

[19] $\frac{3}{8} + \frac{x-4}{2} \geq \frac{x+3}{4} \Rightarrow 8\left(\frac{3}{8}\right) + 8\left(\frac{x-4}{2}\right) \geq 8\left(\frac{x+3}{4}\right) \Rightarrow 3 + 4x - 16 \geq 2x + 6 \Rightarrow 2x \geq 19 \Rightarrow x \geq \frac{19}{2}$

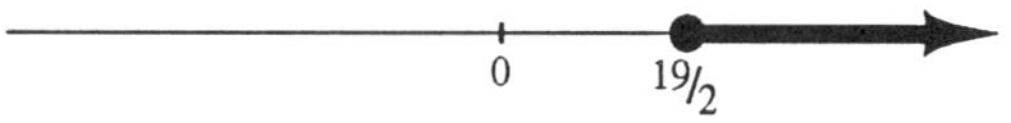

(number line graph: closed circle at $\frac{19}{2}$)

$\bullet \left[\frac{19}{2}, \infty\right)$

[20] $-3 < \frac{2x}{3} + 3 < 9 \Rightarrow -6 < \frac{2x}{3} < 6 \Rightarrow \frac{3}{2}(-6) < x < \frac{3}{2}(6) \Rightarrow -9 < x < 9$

(number line graph: open circles at -9 and 9)

$\bullet (-9, 9)$

[21] $x^2 + x - 12 \leq 0$

$x^2 + x - 12 = 0 \Rightarrow (x+4)(x-3) = 0 \Rightarrow x = -4$ or $x = 3$ (two cut points). (See table below).

(number line graph: closed circles at -4 and 3)

$\bullet [-4, 3]$

[22] $x^2 - 6x - 5 > 0$

$x^2 - 6x - 5 = 0 \Rightarrow x = \frac{6 \pm \sqrt{56}}{2} = 3 \pm \sqrt{14}$. So $x = 3 + \sqrt{14} \approx 6.74$ or $x = 3 - \sqrt{14} \approx -0.74$ (two cut

points). (See table below).

(number line graph: open circles at $3 - \sqrt{14} \approx -0.74$ and $3 + \sqrt{14} \approx 6.74$)

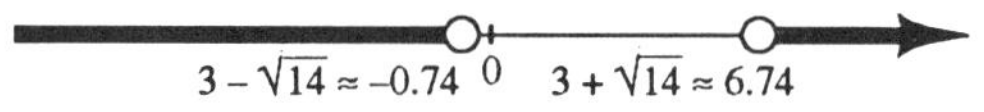

$\bullet \left(-\infty, 3 - \sqrt{14},\right) \cup \left(3 + \sqrt{14}, \infty\right)$

INT.	T.P.	VALUE	SIGN
$(-\infty, -4)$	-5	8	+
$(-4, 3)$	0	-12	−
$(3, \infty)$	4	8	+

Figure 21

INT.	T.P.	VALUE	SIGN
$(-\infty, 3 - \sqrt{14})$	-1	2	+
$(3 - \sqrt{14}, 3 + \sqrt{14})$	0	-5	−
$(3 + \sqrt{14}, \infty)$	7	2	+

Figure 22

[23] $4x^2 - 4x \geq 3 \Rightarrow 4x^2 - 4x - 3 \geq 0$

$4x^2 - 4x - 3 = 0 \Rightarrow (2x - 3)(2x + 1) = 0 \Rightarrow x = \dfrac{3}{2}$ or $x = -\dfrac{1}{2}$ (two cut points). (See table below).

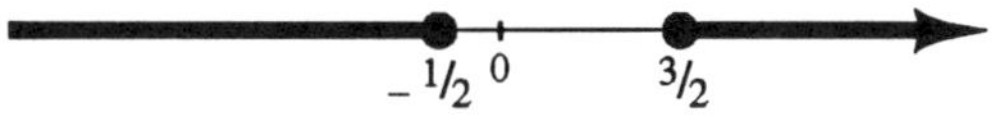

$\bullet \left(-\infty, -\dfrac{1}{2}\right] \cup \left[\dfrac{3}{2}, \infty\right)$

[24] $x^2 < 3 - 2x \Rightarrow x^2 + 2x - 3 < 0$

$x^2 + 2x - 3 = 0 \Rightarrow (x + 3)(x - 1) = 0 \Rightarrow x = -3$ or $x = 1$ (two cut points). (See table below).

$\bullet \left(-3, \, 1\right)$

INT.	T.P.	VALUE	SIGN
$\left(-\infty, -\dfrac{1}{2}\right)$	-1	5	$+$
$\left(-\dfrac{1}{2}, \dfrac{3}{2}\right)$	0	-3	$-$
$\left(\dfrac{3}{2}, \infty\right)$	2	5	$+$

Figure 23

INT.	T.P.	VALUE	SIGN
$(-\infty, -3)$	-4	5	$+$
$(-3, 1)$	0	-3	$-$
$(1, \infty)$	2	5	$+$

Figure 24

[25] $x^2 \leq 4x + 2 \Rightarrow x^2 - 4x - 2 \leq 0$

$x^2 - 4x - 2 = 0 \Rightarrow x = \dfrac{4 \pm \sqrt{24}}{2} = 2 \pm \sqrt{6}$. So, $x = 2 + \sqrt{6} \approx 4.45$ or $x = 2 - \sqrt{6} \approx -0.45$ (two cut

points). (See table below).

$\bullet \left[2 - \sqrt{6}, \, 2 + \sqrt{6}\right]$

[26] $-2x^2 > 20x + 52 \Rightarrow -2x^2 - 20x - 52 > 0 \Rightarrow x^2 + 10x + 26 < 0$

$x^2 + 10x + 26 = 0 \Rightarrow x = \dfrac{-10 \pm \sqrt{-4}}{2} \notin \mathbf{R}$ (no cut points). T.P.: $x = 0$.

$\bullet \varnothing$

INT.	T.P.	VALUE	SIGN
$\left(-\infty, \dfrac{-1 - \sqrt{2}}{3}\right)$	-1	2	$+$
$\left(\dfrac{-1 - \sqrt{2}}{3}, \dfrac{-1 + \sqrt{2}}{3}\right)$	0	-1	$-$
$\left(\dfrac{-1 + \sqrt{2}}{3}, \infty\right)$	1	14	$+$

INT.	T.P.	VALUE	SIGN
$(-\infty, 2 - \sqrt{6})$	-1	3	$+$
$(2 - \sqrt{6}, 2 + \sqrt{6})$	0	-2	$-$
$(2 + \sqrt{6}, \infty)$	5	3	$+$

Figure 25   **Figure 27**

[27] $3x^2 + 2x < \dfrac{1}{3} \Rightarrow 9x^2 + 6x - 1 < 0$

$9x^2 + 6x - 1 = 0 \Rightarrow x = \dfrac{-6 \pm \sqrt{72}}{2(9)} = \dfrac{-1 \pm \sqrt{2}}{3}$. So, $x = \dfrac{-1 + \sqrt{2}}{3} \approx 0.14$, or $x = \dfrac{-1 - \sqrt{2}}{3} \approx -0.80$

(two cut points). (See table above).

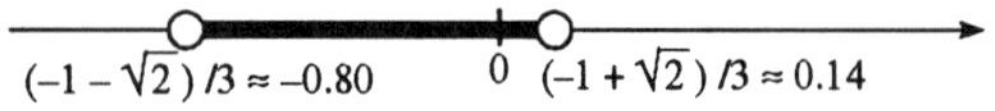

$\bullet \left(\dfrac{-1 - \sqrt{2}}{3}, \dfrac{-1 + \sqrt{2}}{3}\right)$

[28] $9 - x^2 < 0$

$9 - x^2 = 0 \Rightarrow 9 = x^2 \Rightarrow x = \pm 3$ (two cut points). (See table below).

● $\left(-\infty, -3\right) \cup \left(3, \infty\right)$

[29] $-x^2 - 4 > 0$

$-x^2 - 4 = 0 \Rightarrow x^2 + 4 = 0 \Rightarrow x^2 = -4 \Rightarrow x \notin \mathbf{R}$ (no cut points). T.P.: $x = 0$.

● $\varnothing$

INT.	T.P.	VALUE	SIGN
$(-\infty, -3)$	-4	-7	$-$
$(-3, 3)$	0	9	$+$
$(3, \infty)$	4	-7	$-$

INT.	T.P.	VALUE	SIGN
$(-\infty, -5)$	-6	7	$+$
$(-5, 1)$	0	-5	$-$
$(1, \infty)$	2	7	$+$

Figure 28   **Figure 30**

[30] $x^2 + 4x \le 5 \Rightarrow x^2 + 4x - 5 \le 0$

$x^2 + 4x - 5 = 0 \Rightarrow (x + 5)(x - 1) = 0 \Rightarrow x = -5$ or $x = 1$ (two cut points). (See table above).

● $[-5, 1]$

[31] $12 < x(x + 1) \Rightarrow 12 < x^2 + x \Rightarrow x^2 + x - 12 > 0$

$x^2 + x - 12 = 0 \Rightarrow (x + 4)(x - 3) = 0 \Rightarrow x = -4$ or $x = 3$ (two cut points). (See table below).

● $\left(-\infty, -4\right) \cup \left(3, \infty\right)$

INT.	T.P.	VALUE	SIGN
$(-\infty, -4)$	-5	8	$+$
$(-4, 3)$	0	-12	$-$
$(3, \infty)$	4	8	$+$

INT.	T.P.	VALUE	SIGN
$\left(-\infty, -\dfrac{7}{3}\right)$	-3	8	$+$
$\left(-\dfrac{7}{3}, 1\right)$	0	-7	$-$
$(1, \infty)$	2	13	$+$

Figure 31   **Figure 32**

[32] $x(3x + 4) < 7 \Rightarrow 3x^2 + 4x < 7 \Rightarrow 3x^2 + 4x - 7 < 0$

$3x^2 + 4x - 7 = 0 \Rightarrow (x - 1)(3x + 7) = 0 \Rightarrow x = 1$ or $x = -\dfrac{7}{3}$ (two cut points). (See table above).

● $\left(-\dfrac{7}{3}, 1\right)$

[33] $2x^2 \le -3x - 4 \Rightarrow 2x^2 + 3x + 4 \le 0$

$D = 9 - 4(2)(4) = -23 < 0$ (no cut points). T.P.: $x = 0$.

● $\varnothing$

[34] $1 - 2x^2 \ge -5 + 4x \Rightarrow -2x^2 - 4x + 6 \ge 0$

$-2x^2 - 4x + 6 = 0 \Rightarrow x^2 + 2x - 3 = 0 \Rightarrow (x + 3)(x - 1) = 0 \Rightarrow x = -3$ or $x = 1$ (two cut points).

(See table below).

● $[-3, 1]$

[35] $-\left(x^2 - 6x\right) < -16 \Rightarrow x^2 - 6x > 16 \Rightarrow x^2 - 6x - 16 > 0$

$x^2 - 6x - 16 = 0 \Rightarrow (x - 8)(x + 2) = 0 \Rightarrow x = 8$ or $x = -2$ (two cut points). (See table below).

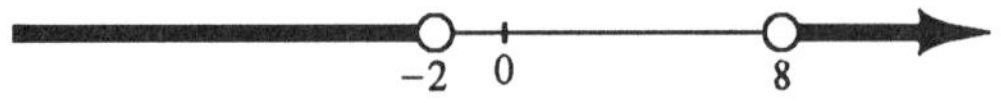

● $(-\infty, -2) \cup (8, \infty)$

INT.	T.P.	VALUE	SIGN
$(-\infty, -3)$	-4	-10	$-$
$(-3, 1)$	0	6	$+$
$(1, \infty)$	2	-10	$-$

Figure 34

INT.	T.P.	VALUE	SIGN
$(-\infty, -2)$	-3	11	$+$
$(-2, 8)$	0	-16	$-$
$(8, \infty)$	9	11	$+$

Figure 35

[36] $-x^2 \geq 1 - 3x \Rightarrow -x^2 + 3x - 1 \geq 0$

$-x^2 + 3x - 1 = 0 \Rightarrow x^2 - 3x + 1 = 0 \Rightarrow x = \dfrac{3 \pm \sqrt{5}}{2}$. So, $x = \dfrac{3 + \sqrt{5}}{2} \approx 2.62$, or $x = \dfrac{3 - \sqrt{5}}{2} \approx 0.38$ (two cut points). (See table below).

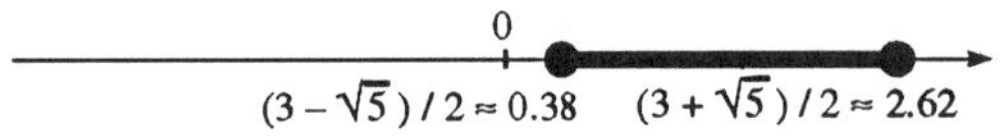

● $\left[\dfrac{3 - \sqrt{5}}{2}, \dfrac{3 + \sqrt{5}}{2}\right]$

INT.	T.P.	VALUE	SIGN
$\left(-\infty, \dfrac{3 - \sqrt{5}}{2}\right)$	0	-1	$-$
$\left(\dfrac{3 - \sqrt{5}}{2}, \dfrac{3 + \sqrt{5}}{2}\right)$	1	1	$+$
$\left(\dfrac{3 + \sqrt{5}}{2}, \infty\right)$	3	-1	$-$

Figure 36

INT.	T.P.	VALUE	SIGN
$\left(-\infty, -\dfrac{5}{3}\right)$	-2	6	$+$
$\left(-\dfrac{5}{3}, 4\right)$	0	-20	$-$
$(4, \infty)$	5	20	$+$

Figure 37

[37] $3x^2 \geq 7x + 20 \Rightarrow 3x^2 - 7x - 20 \geq 0$

$3x^2 - 7x - 20 = 0 \Rightarrow (x - 4)(3x + 5) = 0 \Rightarrow x = 4$ or $x = -\dfrac{5}{3}$ (two cut points). (See table above).

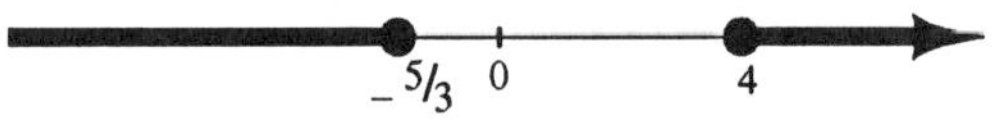

● $\left(-\infty, -\dfrac{5}{3}\right] \cup \left[4, \infty\right)$

[38] $x^2 \leq -7x - 6 \Rightarrow x^2 + 7x + 6 \leq 0$

$x^2 + 7x + 6 = 0 \Rightarrow (x + 6)(x + 1) = 0 \Rightarrow x = -6$ or $x = -1$ (two cut points). (See table below).

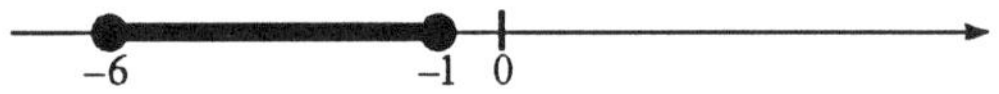

● $[-6, -1]$

INT.	T.P.	VALUE	SIGN
$(-\infty, -6)$	-7	6	$+$
$(-6, -1)$	-2	-4	$-$
$(-1, \infty)$	0	6	$+$

Figure 38

[39] $2x^2 > -2x - 1 \Rightarrow 2x^2 + 2x + 1 > 0$

$2x^2 + 2x + 1 = 0 \Rightarrow D = (2)^2 - 4(2)(1) = 4 - 8 = -4 < 0$ (no cut points). T.P.: $x = 0$.   ● R

[40] $x^2 < -x - 5 \Rightarrow x^2 + x + 5 < 0$

$x^2 + x + 5 = 0 \Rightarrow D = (1)^2 - 4(1)(5) = 1 - 20 = -19 < 0$ (no cut points). T.P.: $x = 0$.   ● ∅

[41]   Since $-20 \le C \le 30$ and $C = \frac{5}{9}(F - 32)$, we have $-20 \le \frac{5}{9}(F - 32) \le 30 \Rightarrow$

$\frac{9}{5}(-20) \le \frac{9}{5}\left[\frac{5}{9}(F - 32)\right] \le \frac{9}{5}(30) \Rightarrow -36 \le F - 32 \le 54 \Rightarrow -4 \le F \le 86.$ Hence, the temperature F,

in degrees Fahrenheit, has the average annual range: $-4 \le F \le 86$.

[42]   Since $8 \le A \le 150$ and $A = \frac{1}{2}bh$, it follows that $8 \le \frac{1}{2}bh \le 150$. Since all of the triangles have

$b = 12, 8 \le \frac{1}{2}(12)h \le 150 \Rightarrow 8 \le 6h \le 150 \Rightarrow \frac{4}{3} \le h \le 25.$ Hence, the range for the heights

(in cm) of the triangles is $\frac{4}{3} \le h \le 25$.

[43]   Since $90 \le IQ \le 150$ and $IQ = \frac{100M}{C}$, $90 \le \frac{100M}{C} \le 150$. Since everyone in the group is 10 years

old, $C = 10$. Hence, $90 \le \frac{100M}{10} \le 150 \Rightarrow 90 \le 10M \le 150 \Rightarrow 9 \le M \le 15$

Therefore, the mental age range for this group satisfies: $9 \le M \le 15$.

[44]   Let x be the number of bird houses made and sold per week. For the company to make a profit,

$R > C$. Since $R = 5x$ and $C = 30 + 3.5x$, we have: $5x > 30 + 3.5x \Rightarrow 1.5x > 30 \Rightarrow x > 20$. Hence,

to make a profit, the company must make and sell at least 21 bird houses per week.

[45]   Let x be the number of pots made and sold per week. For the potter to make a profit, $R > C$. Since

$R = 3x$ and $C = 25 + 2.5x$, we have: $3x > 25 + 2.5x \Rightarrow 0.5x > 25 \Rightarrow x > 50$.   Hence, to make a

profit, he must make and sell at least 51 pots per week.

[46]   a)   The projectile is above the ground if and only if $h > 0 \Rightarrow -16t^2 + 96t > 0 \Rightarrow t^2 - 6t < 0$.

To find the cut points, we solve the equation $t^2 - 6t = 0 \Rightarrow t(t - 6) = 0 \Rightarrow t = 0$ or $t - 6 = 0 \Rightarrow$

$t = 0$ or $t = 6$. Therefore, the projectile is above the ground during the time interval $(0, 6)$. (See

table below).

b)   The projectile is at least 128 feet high if and only if $h \ge 128 \Rightarrow -16t^2 + 96t \ge 128 \Rightarrow$

$-16t^2 + 96t - 128 \ge 0 \Rightarrow t^2 - 6t + 8 \le 0$. To find the cut points, we solve the equation

$t^2 - 6t + 8 = 0 \Rightarrow (t - 4)(t - 2) = 0 \Rightarrow t - 4 = 0$ or $t - 2 = 0 \Rightarrow t = 4$ or $t = 2$. Hence, the projectile

is at least 128 feet high during the time interval $[2, 4]$. (See table below).

INT.	T.P.	VALUE	SIGN
$(-\infty, 0)$	−1	7	+
$(0, 6)$	1	−5	−
$(6, \infty)$	7	7	+

Figure 46 a

INT.	T.P.	VALUE	SIGN
$(-\infty, 2)$	0	8	+
$(2, 4)$	3	−1	−
$(4, \infty)$	5	3	+

Figure 46 b

[47]　a)　The rocket is above the ground if and only if $h > 0 \Rightarrow -16t^2 + 88t > 0 \Rightarrow 2t^2 - 11t < 0$. To find the cut points, we solve the equation $2t^2 - 11t = 0 \Rightarrow t(2t - 11) = 0 \Rightarrow t = 0$ or $2t - 11 = 0 \Rightarrow$

$t = 0$ or $2t = 11 \Rightarrow t = 0$ or $t = \dfrac{11}{2} = 5.5$. Hence, the rocket is above the ground during the time

interval $\left(0, \dfrac{11}{2}\right)$. (See table below).

b)　The rocket is at least 120 feet high if and only if $h \geq 120 \Rightarrow -16t^2 + 88t \geq 120 \Rightarrow$

$-16t^2 + 88t - 120 \geq 0 \Rightarrow 2t^2 - 11t + 15 \leq 0$.　To find the cut points, we solve the equation

$2t^2 - 11t + 15 = 0 \Rightarrow (t - 3)(2t - 5) = 0 \Rightarrow t - 3 = 0$ or $2t - 5 = 0 \Rightarrow t = 3$ or $t = \dfrac{5}{2}$.

Therefore, the rocket is at least 120 feet high during the time interval $\left[\dfrac{5}{2}, 3\right]$. (See table below).

INT.	T.P.	VALUE	SIGN
$(-\infty, 0)$	−1	13	+
$\left(0, \dfrac{11}{2}\right)$	1	−9	−
$\left(\dfrac{11}{2}, \infty\right)$	6	6	+

Figure 47 a

INT.	T.P.	VALUE	SIGN
$\left(-\infty, \dfrac{5}{2}\right)$	0	15	+
$\left(\dfrac{5}{2}, 3\right)$	2.75	−0.125	−
$(3, \infty)$	4	3	+

Figure 47 b

[48]　a)　The frog is above the ground if and only if $h > 0 \Rightarrow -16t^2 + 8t > 0 \Rightarrow 2t^2 - t < 0$. To find the

cut points: $2t^2 - t = 0 \Rightarrow t(2t - 1) = 0 \Rightarrow t = 0$ or $2t - 1 = 0 \Rightarrow t = 0$ or $t = \dfrac{1}{2}$. Hence, the frog is

above the ground during the time interval $\left(0, \dfrac{1}{2}\right)$. (See table below).

b)　The frog is at most one foot high if and only if $h \leq 1 \Rightarrow -16t^2 + 8t \leq 1 \Rightarrow$

$-16t^2 + 8t - 1 \leq 0 \Rightarrow 16t^2 - 8t + 1 \geq 0$. To find the cut points: $16t^2 - 8t + 1 = 0 \Rightarrow (4t - 1)^2 = 0 \Rightarrow$

$4t - 1 = 0 \Rightarrow 4t = 1 \Rightarrow t = \dfrac{1}{4}$. Hence, the frog is at most one foot high for all values of t. Since time

cannot be negative, the frog is at most one foot high during the time interval $\left[0, \infty\right)$. (See table below).

INT.	T.P.	VALUE	SIGN
$(-\infty, 0)$	-1	3	$+$
$\left(0, \dfrac{1}{2}\right)$	$\dfrac{1}{4}$	-0.125	$-$
$\left(\dfrac{1}{2}, \infty\right)$	1	1	$+$

INT.	T.P.	VALUE	SIGN
$\left(-\infty, \dfrac{1}{4}\right)$	0	1	$+$
$\left(\dfrac{1}{4}, \infty\right)$	1	9	$+$

Figure 48 a **Figure 48 b**

EXERCISES 1.7

[1] $\sqrt{x-3} = 5 \Rightarrow \left(\sqrt{x-3}\right)^2 = (5)^2 \Rightarrow x-3 = 25 \Rightarrow x = 28$

$\underline{\text{Check}}$: If $x = 28$, then $\sqrt{x-3} = \sqrt{28-3} = \sqrt{25} = 5$. Hence, $S = \{28\}$.

[2] $\sqrt{2x-4} = 4 \Rightarrow \left(\sqrt{2x-4}\right)^2 = (4)^2 \Rightarrow 2x-4 = 16 \Rightarrow x = 10$

$\underline{\text{Check}}$: If $x = 10$, then $\sqrt{2x-4} = \sqrt{2(10)-4} = \sqrt{16} = 4$. Hence, $S = \{10\}$.

[3] $\sqrt[3]{5x-2} = -2 \Rightarrow \left(\sqrt[3]{5x-2}\right)^3 = (-2)^3 \Rightarrow 5x-2 = -8 \Rightarrow x = -\dfrac{6}{5}$

$\underline{\text{Check}}$: If $x = -\dfrac{6}{5}$, then $\sqrt[3]{5x-6} = \sqrt[3]{5\left(-\dfrac{6}{5}\right)-2} = \sqrt[3]{-8} = -2$. Hence, $S = \left\{-\dfrac{6}{5}\right\}$.

[4] $\sqrt[3]{3x-4} = -4 \Rightarrow \left(\sqrt[3]{3x-4}\right)^3 = (-4)^3 \Rightarrow 3x-4 = -64 \Rightarrow x = -20$

$\underline{\text{Check}}$: If $x = -20$, then $\sqrt[3]{3x-4} = \sqrt[3]{3(-20)-4} = \sqrt[3]{-64} = -4$. Hence, $S = \{-20\}$.

[5] $\sqrt{x-1} = 3 + \sqrt{x+1} \Rightarrow \left(\sqrt{x-1}\right)^2 = \left(3+\sqrt{x+1}\right)^2 \Rightarrow x-1 = 9 + 6\sqrt{x+1} + (x+1) \Rightarrow$

$x-1 = 10 + x + 6\sqrt{x+1} \Rightarrow -11 = 6\sqrt{x+1} \Rightarrow 121 = 36(x+1) \Rightarrow 121 = 36x + 36 \Rightarrow$

$36x = 85 \Rightarrow x = \dfrac{85}{36}$

$\underline{\text{Check}}$: If $x = \dfrac{85}{36}$, $\sqrt{x-1} = \sqrt{\dfrac{85}{36}-1} = \sqrt{\dfrac{49}{36}} = \dfrac{7}{6}$. On the other hand, $3 + \sqrt{x+1} =$

$3 + \sqrt{\dfrac{85}{36}+1} = 3 + \sqrt{\dfrac{121}{36}} = 3 + \dfrac{11}{6} = \dfrac{29}{6}$. Since $\dfrac{7}{6} \neq \dfrac{29}{6}$, $S = \varnothing$.

[6] $\sqrt{9x^2-2} = 3x-1 \Rightarrow \left(\sqrt{9x^2-2}\right)^2 = (3x-1)^2 \Rightarrow 9x^2-2 = 9x^2 - 6x + 1 \Rightarrow 6x = 3 \Rightarrow x = \dfrac{1}{2}$

$\underline{\text{Check}}$: If $x = \dfrac{1}{2}$, $\sqrt{9x^2-2} = \sqrt{9\left(\dfrac{1}{4}\right)-2} = \sqrt{\dfrac{1}{4}} = \dfrac{1}{2}$. On the other hand,

$3x - 1 = 3\left(\dfrac{1}{2}\right) - 1 = \dfrac{1}{2}$. Hence, $S = \left\{\dfrac{1}{2}\right\}$.

[7] $\sqrt{x-4} = 4 - x \Rightarrow \left(\sqrt{x-4}\right)^2 = (4-x)^2 \Rightarrow x - 4 = x^2 - 8x + 16 \Rightarrow x^2 - 9x + 20 = 0 \Rightarrow$

$(x-5)(x-4) = 0 \Rightarrow x = 5 \text{ or } x = 4.$

Check: If $x = 5$, then $\sqrt{x-4} = \sqrt{5-4} = \sqrt{1} = 1$. On the other hand, $4 - x = 4 - 5 = -1$

Since $1 \neq -1$, 5 is not a solution. If $x = 4$ then $\sqrt{x-4} = \sqrt{4-4} = \sqrt{0} = 0$ and $4 - x = 4 - 4 = 0$.

So, $S = \{4\}$.

[8] $\sqrt{x^2 + 5} = x - 2 \Rightarrow \left(\sqrt{x^2+5}\right)^2 = (x-2)^2 \Rightarrow x^2 + 5 = x^2 - 4x + 4 \Rightarrow 4x = -1 \Rightarrow x = -\dfrac{1}{4}$

Check: If $x = -\dfrac{1}{4}$, then $\sqrt{x^2+5} = \sqrt{\dfrac{1}{16} + 5} = \sqrt{\dfrac{81}{16}} = \dfrac{9}{4}$. On the other hand, $x - 2 = -\dfrac{1}{4} - 2 =$

$-\dfrac{9}{4}$. Since $\dfrac{9}{4} \neq -\dfrac{9}{4}$, $S = \emptyset$.

[9] $\sqrt{x^2 + 3x + 4} = x + 1 \Rightarrow \left(\sqrt{x^2+3x+4}\right)^2 = (x+1)^2 \Rightarrow x^2 + 3x + 4 = x^2 + 2x + 1 \Rightarrow x = -3$

Check: If $x = -3$, then $\sqrt{x^2+3x+4} = \sqrt{9 - 9 + 4} = \sqrt{4} = 2$. On the other hand,

$x + 1 = -3 + 1 = -2$. Since $2 \neq -2$, $S = \emptyset$.

[10] $\sqrt[3]{2x+1} = -1 \Rightarrow \left(\sqrt[3]{2x+1}\right)^3 = (-1)^3 \Rightarrow 2x + 1 = -1 \Rightarrow 2x = -2 \Rightarrow x = -1$

Check: If $x = -1$, then $\sqrt[3]{2x+1} = \sqrt[3]{-2+1} = \sqrt[3]{-1} = -1$. Hence, $S = \{-1\}$.

[11] $\sqrt{x^2 - 5x + 10} = 2 \Rightarrow \left(\sqrt{x^2-5x+10}\right)^2 = (2)^2 \Rightarrow x^2 - 5x + 10 = 4 \Rightarrow x^2 - 5x + 6 = 0 \Rightarrow$

$(x-3)(x-2) = 0 \Rightarrow x = 3 \text{ or } x = 2$

Check: If $x = 3$, then $\sqrt{x^2-5x+10} = \sqrt{9 - 15 + 10} = \sqrt{4} = 2$. Hence, 3 is a solution. If $x = 2$,

$\sqrt{x^2-5x+10} = \sqrt{4 - 10 + 10} = \sqrt{4} = 2$. Hence, 2 is also a solution. It follows that the solution

set is $S = \{2, 3\}$.

[12] $\sqrt{x+25} = \sqrt{x} + 1 \Rightarrow (\sqrt{x+25})^2 = (\sqrt{x}+1)^2 \Rightarrow x + 25 = x + 2\sqrt{x} + 1 \Rightarrow 24 = 2\sqrt{x} \Rightarrow \sqrt{x} = 12 \Rightarrow$

$\left(\sqrt{x}\right)^2 = (12)^2 \Rightarrow x = 144$

Check: If $x = 144$, then $\sqrt{x+25} = \sqrt{144+25} = \sqrt{169} = 13$. On the other hand,

$\sqrt{x} + 1 = \sqrt{144} + 1 = 12 + 1 = 13$. Hence, $S = \{144\}$.

[13] $\sqrt{x^2+9} = x + 3 \Rightarrow \left(\sqrt{x^2+9}\right)^2 = (x+3)^2 \Rightarrow x^2 + 9 = x^2 + 6x + 9 \Rightarrow 6x = 0 \Rightarrow x = 0$

Check: If $x = 0$, then $\sqrt{x^2+9} = \sqrt{0+9} = 3$ and $x + 3 = 0 + 3 = 3$. Hence, $S = \{0\}$.

[14] $\sqrt[3]{x^2 - 91} = -3 \Rightarrow \left(\sqrt[3]{x^2 - 91}\right)^3 = (-3)^3 \Rightarrow x^2 - 91 = -27 \Rightarrow x^2 = 64 \Rightarrow x = \pm 8$

       <u>Check</u>: If $x = 8$, then $\sqrt[3]{x^2 - 91} = \sqrt[3]{64 - 91} = \sqrt[3]{-27} = -3$. Hence, 8 is a solution.

       If $x = -8$, then $\sqrt[3]{x^2 - 91} = \sqrt[3]{64 - 91} = \sqrt[3]{-27} = -3$. Therefore, $S = \{-8, 8\}$

[15] $\sqrt{x} + 1 = \sqrt{2x + 2} \Rightarrow \left(\sqrt{x} + 1\right)^2 = \left(\sqrt{2x + 2}\right)^2 \Rightarrow x + 2\sqrt{x} + 1 = 2x + 2 \Rightarrow 2\sqrt{x} = x + 1 \Rightarrow$

       $\left(2\sqrt{x}\right)^2 = (x + 1)^2 \Rightarrow 4x = x^2 + 2x + 1 \Rightarrow x^2 - 2x + 1 = 0 \Rightarrow (x - 1)^2 = 0 \Rightarrow x = 1$

       <u>Check</u>: If $x = 1$, $\sqrt{x} + 1 = \sqrt{1} + 1 = 2$ and $\sqrt{2x + 2} = \sqrt{2 + 2} = \sqrt{4} = 2$. Hence, $S = \{1\}$.

[16] $1 - \sqrt{x - 1} = \sqrt{x + 1} \Rightarrow \left(1 - \sqrt{x - 1}\right)^2 = \left(\sqrt{x + 1}\right)^2 \Rightarrow 1 - 2\sqrt{x - 1} + (x - 1) = x + 1 \Rightarrow$

       $-2\sqrt{x - 1} = 1 \Rightarrow \left(-2\sqrt{x - 1}\right)^2 = (1)^2 \Rightarrow 4(x - 1) = 1 \Rightarrow 4x - 4 = 1 \Rightarrow x = \dfrac{5}{4}$

       <u>Check</u>: If $x = \dfrac{5}{4}$, then $1 - \sqrt{x - 1} = 1 - \sqrt{\dfrac{5}{4} - 1} = 1 - \sqrt{\dfrac{1}{4}} = 1 - \dfrac{1}{2} = \dfrac{1}{2}$. On the other hand,

       $\sqrt{x + 1} = \sqrt{\dfrac{5}{4} + 1} = \sqrt{\dfrac{9}{4}} = \dfrac{3}{2}$. Since $\dfrac{1}{2} \neq \dfrac{3}{2}$, $S = \varnothing$.

[17] $\sqrt{x + 5} = x - 1 \Rightarrow \left(\sqrt{x + 5}\right)^2 = (x - 1)^2 \Rightarrow x + 5 = x^2 - 2x + 1 \Rightarrow x^2 - 3x - 4 = 0 \Rightarrow$

       $(x - 4)(x + 1) = 0 \Rightarrow x = 4$ or $x = -1$

       <u>Check</u>: If $x = 4$, then $\sqrt{x + 5} = \sqrt{4 + 5} = \sqrt{9} = 3$ and $x - 1 = 4 - 1 = 3$. Hence, 4 is a solution.

       If $x = -1$, then $\sqrt{x + 5} = \sqrt{-1 + 5} = \sqrt{4} = 2$ and $x - 1 = -1 - 1 = -2$. Since $2 \neq -2$, -1 is not

       a solution. Hence, $S = \{4\}$.

[18] $\sqrt{x + 1} - \sqrt{x} = 3 \Rightarrow \sqrt{x - 1} = 3 + \sqrt{x} \Rightarrow \left(\sqrt{x + 1}\right)^2 = (3 + \sqrt{x})^2 \Rightarrow x + 1 = 9 + 6\sqrt{x} + x \Rightarrow$

       $6\sqrt{x} = -8 \Rightarrow \left(6\sqrt{x}\right)^2 = (-8)^2 \Rightarrow 36x = 64 \Rightarrow x = \dfrac{64}{36} = \dfrac{16}{9}$

       <u>Check</u>: If $x = \dfrac{16}{9}$, then $\sqrt{x + 1} - \sqrt{x} = \sqrt{\dfrac{16}{9} + 1} - \sqrt{\dfrac{16}{9}} = \sqrt{\dfrac{25}{9}} - \dfrac{4}{3} = \dfrac{5}{3} - \dfrac{4}{3} = \dfrac{1}{3}$.

       Since $\dfrac{1}{3} \neq 3$, $S = \varnothing$.

[19] $2x^{2/3} + x^{1/3} - 3 = 0$. If $u = x^{1/3}$ then $u^2 = x^{2/3}$ and the original equation becomes the following quadratic

       equation: $2u^2 + u - 3 = 0 \Rightarrow (2u + 3)(u - 1) = 0 \Rightarrow 2u + 3 = 0$ or $u - 1 = 0 \Rightarrow 2u = -3$ or $u = 1 \Rightarrow$

       $u = -\dfrac{3}{2}$ or $u = 1 \Rightarrow x^{1/3} = -\dfrac{3}{2}$ or $x^{1/3} = 1 \Rightarrow x = \left(-\dfrac{3}{2}\right)^3$ or $x = (1)^3 \Rightarrow x = -\dfrac{27}{8}$ or $x = 1$.

       So, $S = \left\{-\dfrac{27}{8}, 1\right\}$.

[20] $5\left(x^2-9\right)^2+44\left(x^2-9\right)+32=0$. If $u=x^2-9$, then $u^2=\left(x^2-9\right)^2$ and the given equation becomes

the following quadratic equation: $5u^2+44u+32=0\Rightarrow\left(u+8\right)\left(5u+4\right)=0\Rightarrow u+8=0$ or

$5u+4=0\Rightarrow u=-8$ or $5u=-4\Rightarrow u=-8$ or $u=-\dfrac{4}{5}\Rightarrow x^2-9=-8$ or $x^2-9=-\dfrac{4}{5}\Rightarrow x^2=1$ or

$x^2=-\dfrac{4}{5}+9\Rightarrow x=\pm1$ or $x^2=\dfrac{41}{5}\Rightarrow x=\pm1$ or $x=\pm\sqrt{\dfrac{41}{5}}=\pm\dfrac{\sqrt{205}}{5}$.

$\therefore S=\left\{-1,1-\dfrac{\sqrt{205}}{5},\dfrac{\sqrt{205}}{5}\right\}$.

[21] $x^4+5x^2-36=0$. If $u=x^2$, then $u^2=x^4$ and the original equation becomes $u^2+5u-36=0\Rightarrow$

$\left(u+9\right)\left(u-4\right)=0\Rightarrow u+9=0$ or $u-4=0\Rightarrow u=-9$ or $u=4\Rightarrow x^2=-9$ or $x^2=4\Rightarrow$

$x=\pm3i$ or $x=\pm2$. So, $S=\left\{-3i,3i,-2,2\right\}$.

[22] $\dfrac{x-1}{x+1}-2\left(\dfrac{x+1}{x-1}\right)+1=0$. If $u=\dfrac{x+1}{x-1}$, then $\dfrac{1}{u}=\dfrac{x-1}{x+1}$ and the original equation becomes

$\dfrac{1}{u}-2u+1=0\Rightarrow u\left(\dfrac{1}{u}\right)-u\left(2u\right)+u=0\Rightarrow 1-2u^2+u=0\Rightarrow-2u^2+u+1=0\Rightarrow2u^2-u-1=0\Rightarrow$

$\left(2u+1\right)\left(u-1\right)=0\Rightarrow 2u+1=0$ or $u-1=0\Rightarrow 2u=-1$ or $u=1\Rightarrow u=-\dfrac{1}{2}$ or $u=1\Rightarrow$

$\dfrac{x+1}{x-1}=-\dfrac{1}{2}$ or $\dfrac{x+1}{x-1}=1\Rightarrow 2x+2=-x+1$ or $x+1=x-1\Rightarrow 3x=3$ or

$1=-1$ (which is impossible). So, $S=\left\{1\right\}$.

[23] $x^6+9x^3+8=0$. If $u=x^3$ then $u^2=x^6$ and $u^2+9u+8=0\Rightarrow\left(u+8\right)\left(u+1\right)=0\Rightarrow u+8=0$ or

$u+1=0\Rightarrow u=-8$ or $u=-1\Rightarrow x^3=-8$ or $x^3=-1\Rightarrow x^3+8=0$ or $x^3+1=0\Rightarrow$

$(x+2)\left(x^2-2x+4\right)=0$ or $(x+1)\left(x^2-x+1\right)=0\Rightarrow x=-2$ or $x=\dfrac{2\pm\sqrt{4-4\left(1\right)\left(4\right)}}{2\left(1\right)}$, or

$x=-1$ or $x=\dfrac{1\pm\sqrt{1-4\left(1\right)\left(1\right)}}{2\left(1\right)}\Rightarrow x=-2$ or $x=\dfrac{2\pm\sqrt{-12}}{2}$, or $x=-1$ or $x=\dfrac{1\pm\sqrt{-3}}{2}\Rightarrow$

$x=-2$ or $x=\dfrac{2\pm2i\sqrt{3}}{2}$, or $x=-1$ or $x=\dfrac{1\pm i\sqrt{3}}{2}\Rightarrow x=-2$ or $x=1\pm\sqrt{3}\,i$, or $x=-1$ or

$x=\dfrac{1\pm\sqrt{3}\,i}{2}$. So, $S=\left\{-2,1-\sqrt{3}\,i,1+\sqrt{3}\,i,-1,\dfrac{1}{2}-\dfrac{\sqrt{3}}{2}i,\dfrac{1}{2}+\dfrac{\sqrt{3}}{2}i\right\}$.

[24] $x-\sqrt{x}-2=0$. If $u=\sqrt{x}$, then $u^2=x$ and the original equation becomes $u^2-u-2=0\Rightarrow$

$\left(u-2\right)\left(u+1\right)=0\Rightarrow u-2=0$ or $u+1=0\Rightarrow u=2$ or $u=-1\Rightarrow\sqrt{x}=2$ or $\sqrt{x}=-1\Rightarrow x=4$ or

$x=1$ (which will not check). So, $S=\left\{4\right\}$.

[25] $2\left(\dfrac{1-x}{x+2}\right)-3\left(\dfrac{x+2}{1-x}\right)-1=0$. If $u=\dfrac{x+2}{1-x}$, then $\dfrac{1}{u}=\dfrac{1-x}{x+2}$ and the given equation becomes

$2\left(\dfrac{1}{u}\right)-3u-1=0\Rightarrow u\left(\dfrac{2}{u}\right)-u(3u)-u=0\Rightarrow 2-3u^2-u=0\Rightarrow -3u^2-u+2=0\Rightarrow$

$3u^2+u-2=0\Rightarrow(3u-2)(u+1)=0\Rightarrow 3u-2=0$ or $u+1=0\Rightarrow 3u=2$ or $u=-1\Rightarrow u=\dfrac{2}{3}$ or

$u=-1\Rightarrow\dfrac{x+2}{1-x}=\dfrac{2}{3}$ or $\dfrac{x+2}{1-x}=-1\Rightarrow 2-2x=3x+6$ or $x+2=x-1\Rightarrow -4=5x\Rightarrow x=-\dfrac{4}{5}$ or

$2=-1$ (impossible). So, $S=\left\{-\dfrac{4}{5}\right\}$.

[26] $15(x+1)+\dfrac{4}{x+1}-23=0$. If $u=x+1$, then the given equation becomes $15u+\dfrac{4}{u}-23=0\Rightarrow$

$u(15u)+u\left(\dfrac{4}{u}\right)-23u=0\Rightarrow 15u^2+4-23u=0\Rightarrow 15u^2-23u+4=0\Rightarrow(5u-1)(3u-4)=0\Rightarrow$

$5u-1=0$ or $3u-4=0\Rightarrow 5u=1$ or $3u=4\Rightarrow u=\dfrac{1}{5}$ or $u=\dfrac{4}{3}\Rightarrow x+1=\dfrac{1}{5}$ or $x+1=\dfrac{4}{3}\Rightarrow$

$x=\dfrac{1}{5}-1$ or $x=\dfrac{4}{3}-1\Rightarrow x=-\dfrac{4}{5}$ or $x=\dfrac{1}{3}$. So, $S=\left\{-\dfrac{4}{5},\dfrac{1}{3}\right\}$.

[27] $(x^2+2)^2-4(x^2+2)-5=0$. If $u=x^2+2$, then $u^2=(x^2+2)^2$ and the given equation becomes

$u^2-4u-5=0\Rightarrow(u-5)(u+1)=0\Rightarrow u-5=0$ or $u+1=0\Rightarrow u=5$ or $u=-1\Rightarrow x^2+2=5$ or

$x^2+2=-1\Rightarrow x^2=3$ or $x^2=-3\Rightarrow x=\pm\sqrt{3}$ or $x=\pm\sqrt{3}\,i$. So, $S=\{-\sqrt{3},\sqrt{3},-\sqrt{3}\,i,\sqrt{3}\,i\}$.

[28] $x^{2/5}-x^{1/5}-2=0$. If $u=x^{1/5}$, then $u^2=x^{2/5}$ and the given equation becomes $u^2-u-2=0\Rightarrow$

$(u-2)(u+1)=0\Rightarrow u-2=0$ or $u+1=0\Rightarrow u=2$ or $u=-1\Rightarrow x^{1/5}=2$ or $x^{1/5}=-1\Rightarrow$

$\left(x^{1/5}\right)^5=(2)^5$ or $\left(x^{1/5}\right)^5=(-1)^5\Rightarrow x=32$ or $x=-1$. So, $S=\{-1,32\}$.

[29] $x^{2/3}-x^{1/3}-12=0$. If $u=x^{1/3}$, then $u^2=x^{2/3}$ and the given equation becomes $u^2-u-12=0\Rightarrow$

$(u-4)(u+3)=0\Rightarrow u-4=0$ or $u+3=0\Rightarrow u=4$ or $u=-3\Rightarrow x^{1/3}=4$ or $x^{1/3}=-3\Rightarrow$

$\left(x^{1/3}\right)^3=(4)^3$ or $\left(x^{1/3}\right)^3=(-3)^3\Rightarrow x=64$ or $x=-27$. So, $S=\{-27,64\}$

[30] $2x^4+17x^2-9=0$. If $u=x^2$, then $u^2=x^4$ and the given equation becomes $2u^2+17u-9=0\Rightarrow$

$(2u-1)(u+9)=0\Rightarrow 2u-1=0$ or $u+9=0\Rightarrow 2u=1$ or $u=-9\Rightarrow u=\dfrac{1}{2}$ or $u=-9\Rightarrow x^2=\dfrac{1}{2}$ or

$x^2=-9\Rightarrow x=\pm\sqrt{\dfrac{1}{2}}$ or $x=\pm 3\,i\Rightarrow x=\pm\dfrac{1}{\sqrt{2}}$ or $x=\pm 3\,i\Rightarrow x=\pm\dfrac{\sqrt{2}}{2}$ or $x=\pm 3\,i$.

So, $S=\left\{-\dfrac{\sqrt{2}}{2},\dfrac{\sqrt{2}}{2},-3\,i,3\,i\right\}$.

[31] $\dfrac{x}{x-2} + \dfrac{2}{2-x} = -\dfrac{2}{3}$. Notice that 2 is the only excluded value for the given equation. Assuming

that $x \neq 2$, we multiply both sides of the given equation by $3(x-2)$ (the LCD) getting

$$3(x-2)\left(\dfrac{x}{x-2}\right) + 3(x-2)\left(\dfrac{2}{2-x}\right) = 3(x-2)\left(-\dfrac{2}{3}\right) \Rightarrow 3x - 6 = -2x + 4 \Rightarrow 5x = 10 \Rightarrow x = 2.$$

Since 2 is an excluded value, the solution set is $\varnothing$.

[32] $\dfrac{3}{2x-1} + 4 = \dfrac{6x}{2x-1}$. Notice that $\dfrac{1}{2}$ is the only excluded value. Assuming that $x \neq \dfrac{1}{2}$,

we multiply both sides of the given equation by the LCD $2x - 1$, getting

$$(2x-1)\left(\dfrac{3}{2x-1}\right) + (2x-1)4 = (2x-1)\left(\dfrac{6x}{2x-1}\right) \Rightarrow 3 + 8x - 4 = 6x \Rightarrow 2x = 1 \Rightarrow x = \dfrac{1}{2}.$$

Since $\dfrac{1}{2}$ is an excluded value, $S = \varnothing$.

[33] $2 - \dfrac{4}{x} = \dfrac{9}{x+5}$. Notice that 0 and –5 are the excluded values for this equation. Assuming that

$x \neq 0$ and $x \neq -5$, we multiply both sides of the given equation by the LCD $x(x+5)$:

$$x(x+5)(2) - x(x+5)\left(\dfrac{4}{x}\right) = x(x+5)\left(\dfrac{9}{x+5}\right) \Rightarrow 2x(x+5) - 4(x+5) = 9x \Rightarrow$$

$$2x^2 + 10x - 4x - 20 - 9x = 0 \Rightarrow 2x^2 - 3x - 20 = 0 \Rightarrow (x-4)(2x+5) = 0 \Rightarrow x - 4 = 0 \text{ or}$$

$2x + 5 = 0 \Rightarrow x = 4 \text{ or } x = -\dfrac{5}{2}$. Since neither 4 nor $-\dfrac{5}{2}$ is an excluded value,

the solution set is $\left\{-\dfrac{5}{2}, 4\right\}$.

[34] $\dfrac{2x}{x-1} - 3 = \dfrac{7-3x}{x-1}$. Notice that 1 is the only excluded value. Assuming that $x \neq 1$, we multiply

both sides by the LCD $x - 1$: $(x-1)\left(\dfrac{2x}{x-1}\right) - (x-1)3 = (x-1)\left(\dfrac{7-3x}{x-1}\right) \Rightarrow 2x - 3x + 3 =$

$7 - 3x \Rightarrow 2x = 4 \Rightarrow x = 2$. Since 2 is not an excluded value, the solution set is $\{2\}$.

[35] $\dfrac{2}{3x} - \dfrac{5}{x+1} = \dfrac{1}{2x}$. Notice that 0 and –1 are the only excluded values. Assuming that $x \neq 0$ and

$x \neq -1$, we multiply both sides by the LCD $6x(x+1)$: $6x(x+1)\left(\dfrac{2}{3x}\right) - 6x(x+1)\left(\dfrac{5}{x+1}\right) =$

$6x(x+1)\left(\dfrac{1}{2x}\right) \Rightarrow 4x + 4 - 30x = 3x + 3 \Rightarrow -26x + 4 = 3x + 3 \Rightarrow -29x = -1 \Rightarrow x = \dfrac{1}{29}.$

Since $\dfrac{1}{29}$ is not an exluded value, the solution set is $\left\{\dfrac{1}{29}\right\}$.

[36] $\dfrac{3x}{x-2} - 4 = \dfrac{14-4x}{x-2}$. Notice that 2 is the only excluded value. Assuming that $x \neq 2$, we multiply

both sides by the LCD $x - 2$: $(x-2)\left(\dfrac{3x}{x-2}\right) - (x-2)4 = (x-2)\left(\dfrac{14-4x}{x-2}\right) \Rightarrow 3x - 4x + 8 =$

$14 - 4x \Rightarrow -x + 8 = 14 - 4x \Rightarrow 3x = 6 \Rightarrow x = 2$. Since 2 is an excluded value, the solution set

is $\varnothing$.

[37] $4 + \dfrac{17}{x} = \dfrac{15}{x^2}$. Notice that 0 is the only excluded value. Assuming that $x \neq 0$, we multiply

both sides by the LCD x^2: $x^2(4) + x^2\left(\dfrac{17}{x}\right) = x^2\left(\dfrac{15}{x^2}\right) \Rightarrow 4x^2 + 17x = 15 \Rightarrow 4x^2 + 17x - 15 = 0 \Rightarrow$

$(4x-3)(x+5) = 0 \Rightarrow 4x - 3 = 0$ or $x + 5 = 0 \Rightarrow x = \dfrac{3}{4}$ or $x = -5$. Since neither $\dfrac{3}{4}$ nor -5 is an

excluded value, the solution set is $\left\{-5, \dfrac{3}{4}\right\}$.

[38] $15x - \dfrac{1}{x} = -2$. Notice that 0 is the only excluded value. Assuming that $x \neq 0$, we multiply both

sides by the LCD x: $x(15x) - x\left(\dfrac{1}{x}\right) = x(-2) \Rightarrow 15x^2 - 1 = -2x \Rightarrow 15x^2 + 2x - 1 = 0 \Rightarrow$

$(5x-1)(3x+1) = 0 \Rightarrow 5x - 1 = 0$ or $3x + 1 = 0 \Rightarrow x = \dfrac{1}{5}$ or $x = -\dfrac{1}{3}$. Since neither $\dfrac{1}{5}$ nor $-\dfrac{1}{3}$

[39] $\dfrac{2x}{3x-1} - \dfrac{1}{x+2} = \dfrac{3-2x}{3x^2+5x-2}$. Since $3x^2 + 5x - 2 = (3x-1)(x+2)$, the only excluded values

are $\dfrac{1}{3}$ and -2. Assuming that $x \neq \dfrac{1}{3}$ and $x \neq -2$, we multiply both sides by the LCD

$(3x-1)(x+2)$: $(3x-1)(x+2)\left(\dfrac{2x}{3x-1}\right) - (3x-1)(x+2)\left(\dfrac{1}{x+2}\right) =$

$(3x-1)(x+2)\left(\dfrac{3-2x}{3x^2+5x-2}\right) \Rightarrow 2x(x+2) - (3x-1) = 3 - 2x \Rightarrow$

$2x^2 + 4x - 3x + 1 - 3 + 2x = 0 \Rightarrow 2x^2 + 3x - 2 = 0 \Rightarrow (2x-1)(x+2) = 0 \Rightarrow 2x - 1 = 0$ or $x + 2 =$

$0 \Rightarrow x = \dfrac{1}{2}$ or $x = -2$. Since -2 is an excluded value while $\dfrac{1}{2}$ is not, the solution set is $\left\{\dfrac{1}{2}\right\}$.

[40] $\dfrac{2x}{x-2}+\dfrac{5}{x+4}=\dfrac{6x-6}{x^2+2x-8}$. Since $x^2+2x-8=(x-2)(x+4)$, the only excluded values are 2

and -4. Assuming that $x\neq 2$ and $x\neq -4$, we multiply both sides by the LCD $(x-2)(x+4)$:

$$(x-2)(x+4)\left(\dfrac{2x}{x-2}\right)+(x-2)(x+4)\left(\dfrac{5}{x+4}\right)=(x-2)(x+4)\left(\dfrac{6x-6}{x^2+2x-8}\right)\Rightarrow$$

$$2x(x+4)+5(x-2)=6x-6\Rightarrow 2x^2+8x+5x-10=6x-6\Rightarrow 2x^2+7x-4=0\Rightarrow$$

$$(2x-1)(x+4)=0\Rightarrow 2x-1=0 \text{ or } x+4=0\Rightarrow x=\dfrac{1}{2} \text{ or } x=-4. \text{ Since } -4 \text{ is an excluded value}$$

while $\dfrac{1}{2}$ is not, the solution set is $\left\{\dfrac{1}{2}\right\}$.

[41] $|x-4|=6\Rightarrow x-4=\pm 6\Rightarrow x-4=-6 \text{ or } x-4=6\Rightarrow x=-2 \text{ or } x=10$

$\bullet\{-2,\ 10\}$

[42] $|2x+5|=13\Rightarrow 2x+5=\pm 13\Rightarrow 2x+5=-13 \text{ or } 2x+5=13\Rightarrow 2x=-18 \text{ or } 2x=8\Rightarrow$
$x=-9 \text{ or } x=4$

$\bullet\{-9,\ 4\}$

[43] $|2-5x|\geq 3\Rightarrow 2-5x\leq -3 \text{ or } 2-5x\geq 3\Rightarrow -5x\leq -5 \text{ or } -5x\geq 1\Rightarrow x\geq 1 \text{ or } x\leq -\dfrac{1}{5}$

$\bullet\left(-\infty,\ -\dfrac{1}{5}\right]\cup\left[1,\ \infty\right)$

[44] $|5x+2|\leq 3\Rightarrow -3\leq 5x+2\leq 3\Rightarrow -5\leq 5x\leq 1\Rightarrow -1\leq x\leq \dfrac{1}{5}$

$\bullet\left[-1,\ \dfrac{1}{5}\right]$

[45] $|6-7x|<1\Rightarrow -1<6-7x<1\Rightarrow -7<-7x<-5\Rightarrow 1>x>\dfrac{5}{7}\Rightarrow \dfrac{5}{7}<x<1$

$\bullet\left(\dfrac{5}{7},\ 1\right)$

[46] $|5-2x|>7\Rightarrow 5-2x<-7 \text{ or } 5-2x>7\Rightarrow -2x<-12 \text{ or } -2x>2\Rightarrow x>6 \text{ or } x<-1$

$\bullet\left(-\infty,-1\right)\cup\left(6,\infty\right)$

[47] $|5x+2|=|3x-4|\Rightarrow 5x+2=\pm(3x-4)\Rightarrow 5x+2=3x-4 \text{ or } 5x+2=-(3x-4)\Rightarrow$
$2x=-6 \text{ or } 5x+2=-3x+4\Rightarrow x=-3 \text{ or } x=\dfrac{2}{8}=\dfrac{1}{4}$

$\bullet\left\{-3,\ \dfrac{1}{4}\right\}$

[48] $|3x+15|=|1-6x|\Rightarrow 3x+15=\pm(1-6x)\Rightarrow 3x+15=1-6x \text{ or } 3x+15=-(1-6x)\Rightarrow$
$9x=-14 \text{ or } 3x+15=-1+6x\Rightarrow x=-\dfrac{14}{9} \text{ or } -3x=-16\Rightarrow x=-\dfrac{14}{9} \text{ or } x=\dfrac{16}{3}$

$\bullet\left\{-\dfrac{14}{9},\ \dfrac{16}{3}\right\}$

[49] $|7 - 6x| \geq 5 \Rightarrow 7 - 6x \leq -5$ or $7 - 6x \geq 5 \Rightarrow -6x \leq -12$ or $-6x \geq -2 \Rightarrow$

$x \geq 2$ or $x \leq \dfrac{-2}{-6} = \dfrac{1}{3} \Rightarrow x \geq 2$ or $x \leq \dfrac{1}{3}$

● $\left(-\infty, \dfrac{1}{3}\right] \cup \left[2, \infty\right)$

[50] $|2 - x| \geq 10 \Rightarrow 2 - x \leq -10$ or $2 - x \geq 10 \Rightarrow -x \leq -12$ or $-x \geq 8 \Rightarrow x \geq 12$ or $x \leq -8$

● $\left(-\infty, -8\right] \cup \left[12, \infty\right)$

[51] $|2 - 3x| - |-7| < 1 \Rightarrow |2 - 3x| - 7 < 1 \Rightarrow |2 - 3x| < 8 \Rightarrow -8 < 2 - 3x < 8 \Rightarrow$

$-10 < -3x < 6 \Rightarrow \dfrac{10}{3} > x > -2 \Rightarrow -2 < x < \dfrac{10}{3}$

● $\left(-2, \dfrac{10}{3}\right)$

[52] $|3x| - |-4| > 5 \Rightarrow |3x| - 4 > 5 \Rightarrow |3x| > 9 \Rightarrow 3x < -9$ or $3x > 9 \Rightarrow x < -3$ or $x > 3$

● $\left(-\infty, -3\right) \cup \left(3, \infty\right)$

[53] $|2x| - |-3| \geq -|-1| \Rightarrow |2x| - 3 \geq -1 \Rightarrow |2x| \geq 2 \Rightarrow 2x \leq -2$ or $2x \geq 2 \Rightarrow x \leq -1$ or $x \geq 1$

● $\left(-\infty, -1\right] \cup \left[1, \infty\right)$

[54] $|3x| - |-5| < -|-2| \Rightarrow |3x| - 5 < -2 \Rightarrow |3x| < 3 \Rightarrow -3 < 3x < 3 \Rightarrow -1 < x < 1$

● $\left(-1, 1\right)$

[55] $\left|\dfrac{x}{2} - 3\right| < 5 \Rightarrow -5 < \dfrac{x}{2} - 3 < 5 \Rightarrow -2 < \dfrac{x}{2} < 8 \Rightarrow -4 < x < 16$

● $\left(-4, 16\right)$

[56] $|1 - 10x| - |-6| > -|-3| \Rightarrow |1 - 10x| - 6 > -3 \Rightarrow |1 - 10x| > 3 \Rightarrow 1 - 10x < -3$ or

$1 - 10x > 3 \Rightarrow -10x < -4$ or $-10x > 2 \Rightarrow x > \dfrac{4}{10} = \dfrac{2}{5}$ or $x < -\dfrac{2}{10} = -\dfrac{1}{5} \Rightarrow x > \dfrac{2}{5}$ or $x < -\dfrac{1}{5}$

● $\left(-\infty, -\dfrac{1}{5}\right) \cup \left(\dfrac{2}{5}, \infty\right)$

[57] $\left|3 - \dfrac{x}{4}\right| \geq 10 \Rightarrow 3 - \dfrac{x}{4} \leq -10$ or $3 - \dfrac{x}{4} \geq 10 \Rightarrow \dfrac{-x}{4} \leq -13$ or $\dfrac{-x}{4} \geq 7 \Rightarrow x \geq 52$ or $x \leq -28$.

● $\left(-\infty, -28\right] \cup \left[52, \infty\right)$

[58] $\left|1 - \dfrac{x}{2}\right| - |-10| \leq -|-5| \Rightarrow \left|1 - \dfrac{x}{2}\right| - 10 \leq -5 \Rightarrow \left|1 - \dfrac{x}{2}\right| \leq 5 \Rightarrow -5 \leq 1 - \dfrac{x}{2} \leq 5 \Rightarrow$

$-6 \leq -\dfrac{x}{2} \leq 4 \Rightarrow 12 \geq x \geq -8 \Rightarrow -8 \leq x \leq 12$

● $\left[-8, 12\right]$

[59] $\dfrac{2x+3}{6-x} \geq 1 \Rightarrow \dfrac{2x+3}{6-x} - 1 \geq 0 \Rightarrow \dfrac{2x+3-6+x}{6-x} \geq 0 \Rightarrow \dfrac{3x-3}{6-x} \geq 0$. The cut points are 1 and 6

where 6 is an excluded value. From the chart below we see that $\dfrac{3x-3}{6-x}$ is positive on the open

interval $(1,\, 6)$. The cut point 1 satisfies the equation $\dfrac{3x-3}{6-x} = 0$ and hence satisfies the

inequality $\dfrac{3x-3}{6-x} \geq 0$. The cut point 6 is an excluded value. So $S = \big[1, 6\big)$.

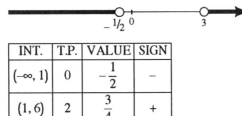

[60] $\dfrac{x+4}{3-x} < 1 \Rightarrow \dfrac{x+4}{3-x} - 1 < 0 \Rightarrow \dfrac{x+4-3+x}{3-x} < 0 \Rightarrow \dfrac{2x+1}{3-x} < 0$. The cut points are $-\dfrac{1}{2}$ and 3, where 3 is

an excluded value. From the table (shown below), $\dfrac{2x+1}{3-x}$ is negative on the open intervals $\left(-\infty,\, -\dfrac{1}{2}\right)$

and $(3, \infty)$. The cut point $-\dfrac{1}{2}$ does not satisfy the inequality $\dfrac{2x+1}{3-x} < 0$ since it causes the quotient on

the left side to be 0. The cut point 3 is an excluded value. Hence, the solution set is

$\left(-\infty,\, -\dfrac{1}{2}\right) \cup (3, \infty)$.

INT.	T.P.	VALUE	SIGN
$(-\infty, 1)$	0	$-\dfrac{1}{2}$	$-$
$(1, 6)$	2	$\dfrac{3}{4}$	$+$
$(6, \infty)$	7	-18	$-$

INT.	T.P.	VALUE	SIGN
$\left(-\infty, -\dfrac{1}{2}\right)$	-1	$-\dfrac{1}{4}$	$-$
$\left(-\dfrac{1}{2}, 3\right)$	0	$\dfrac{1}{3}$	$+$
$(3, \infty)$	4	-9	$-$

Figure 59   **Figure 60**

[61] $\dfrac{x^2+x-2}{2-x} \geq 0 \Rightarrow \dfrac{(x+2)(x-1)}{2-x} \geq 0$. Since $x^2+x-2 = (x+2)(x-1)$ we see that the cut points

are -2, 1, and 2 where 2 is an excluded value. From the table below we see that $\dfrac{x^2+x-2}{2-x}$ is

positive on the open intervals $\left(-\infty,\, -2\right)$ and $(1, 2)$. Since the cut points -2 and 1 satisfy the equation

$\dfrac{x^2+x-2}{2-x} = 0$ they also satisfy the inequality $\dfrac{x^2+x-2}{2-x} \geq 0$. The cut point 2 is an excluded

value. Hence, the solution set is $\left(-\infty,\, -2\right] \cup \big[1, 2\big)$.

[62] $\dfrac{x^2+x-12}{5-x}\le 0 \Rightarrow \dfrac{(x+4)(x-3)}{5-x}\le 0$. We see that the cut points are –4, 3, and 5 where 5 is an

excluded value. From the table below we see that $\dfrac{(x+4)(x-3)}{5-x}$ is negative on the open intervals

$(-4,\,3\,)$ and $\left(5,\,\infty\right)$. Since the cut points –4 and 3 satisfy the equation $\dfrac{(x+4)(x-3)}{5-x}=0,$ they also

satisfy the inequality $\dfrac{(x+4)(x-3)}{5-x}\le 0$. The cut point 5 is an excluded value. Hence, the solution

set is $\left[-4,\,3\,\right]\cup\left(5,\,\infty\right)$.

INT.	T.P.	VALUE	SIGN
$(-\infty,-2)$	-3	$\dfrac{4}{5}$	$+$
$(-2,1)$	0	-1	$-$
$(1,2)$	$\dfrac{3}{2}$	$\dfrac{7}{2}$	$+$
$(2,\infty)$	3	-10	$-$

Figure 61

INT.	T.P.	VALUE	SIGN
$(-\infty,-4)$	-5	$\dfrac{4}{5}$	$+$
$(-4,3)$	0	$-\dfrac{12}{5}$	$-$
$(3,5)$	4	8	$+$
$(5,\infty)$	6	-30	$-$

Figure 62

[63] $\dfrac{x^2-2x-3}{x^2+2x-8}\le 0 \Rightarrow \dfrac{(x-3)(x+1)}{(x+4)(x-2)}\le 0$. We see that the cut points are 3, –1, –4, and 2, where –4 and

2 are excluded values. From the table below we see that $\dfrac{(x-3)(x+1)}{(x+4)(x-2)}$ is negative on the open

intervals $(-4,-1)$ and $(2,3)$. Since the cut points 3 and –1 satisfy the equation $\dfrac{(x-3)(x+1)}{(x+4)(x-2)}=0,$

they also satisfy the inequality $\dfrac{(x-3)(x+1)}{(x+4)(x-2)}\le 0$. The cut points –4 and 2 are excluded values.

Hence, the solution set is $(-4,-1]\cup(2,3]$.

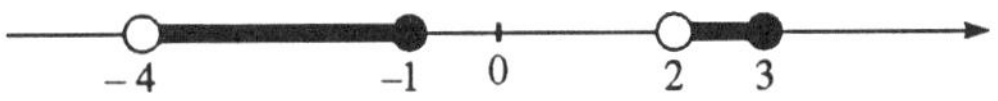

[64] $\dfrac{x^2-x-12}{x^2+x-2}\ge 0 \Rightarrow \dfrac{(x-4)(x+3)}{(x+2)(x-1)}\ge 0.$ We see that the cut points are 4, –3, –2, and 1, where –2 and

1 are excluded values. From the table below we see that $\dfrac{(x-4)(x+3)}{(x+2)(x-1)}$ is positive over the open

intervals $\left(-\infty,\,-3\right)$, $(-2,\,1)$, and $\left(4,\,\infty\right)$. Since the cut points 4 and –3 satisfy the equation

$\dfrac{(x-4)(x+3)}{(x+2)(x-1)}=0$ they also satisfy the inequality $\dfrac{(x-4)(x+3)}{(x+2)(x-1)}\ge 0$. The cut points –2 and 1 are

excluded values. Hence, the solution set is $\left(-\infty,\,-3\right]\cup\left(-2,\,1\right)\cup\left[4,\,\infty\right)$.

37

INT.	T.P.	VALUE	SIGN
$(-\infty, -4)$	-5	$\dfrac{32}{7}$	$+$
$(-4, -1)$	-2	$-\dfrac{5}{8}$	$-$
$(-1, 2)$	0	$\dfrac{3}{8}$	$+$
$(2, 3)$	$\dfrac{5}{2}$	$-\dfrac{7}{13}$	$-$
$(3, \infty)$	4	$\dfrac{5}{16}$	$+$

INT.	T.P.	VALUE	SIGN
$(-\infty, -3)$	-4	$\dfrac{4}{5}$	$+$
$(-3, -2)$	$-\dfrac{5}{2}$	$-\dfrac{13}{7}$	$-$
$(-2, 1)$	0	6	$+$
$(1, 4)$	2	$-\dfrac{5}{2}$	$-$
$(4, \infty)$	5	$\dfrac{2}{7}$	$+$

Figure 63   **Figure 64**

[65] $x^4 - 5x^2 > -4 \Rightarrow x^4 - 5x^2 + 4 > 0$. To find the cut points, we consider the equation $x^4 - 5x^2 + 4 = 0$. If we let $u = x^2$, then $u^2 = x^4$ and the last equation becomes the quadratic equation

$u^2 - 5u + 4 = 0 \Rightarrow (u - 4)(u - 1) = 0 \Rightarrow u - 4 = 0$ or $u - 1 = 0 \Rightarrow u = 4$ or $u = 1 \Rightarrow x^2 = 4$ or $x^2 = 1 \Rightarrow$

$x = \pm 2$ or $x = \pm 1$. Hence, there are four cut points: $-2, -1, 1,$ and 2. From the table below we see that

the solution set is $(-\infty, -2) \cup (-1, 1) \cup (2, \infty)$.

[66] $x^4 - 41x^2 < -400 \Rightarrow x^4 - 41x^2 + 400 < 0$. To find the cut points, we consider the

equation $x^4 - 41x^2 + 400 = 0$. If $u = x^2$, then $u^2 = x^4$ and this equation becomes the quadratic

equation $u^2 - 41u + 400 = 0 \Rightarrow (u - 25)(u - 16) = 0 \Rightarrow u - 25 = 0$ or $u - 16 = 0 \Rightarrow u = 25$ or $u = 16 \Rightarrow$

$x^2 = 25$ or $x^2 = 16 \Rightarrow x = \pm 5$ or $x = \pm 4$. Hence, there are four cut points: $-5, -4, 4,$ and 5. From the

table below, we see that the solution set is $(-5, -4) \cup (4, 5)$.

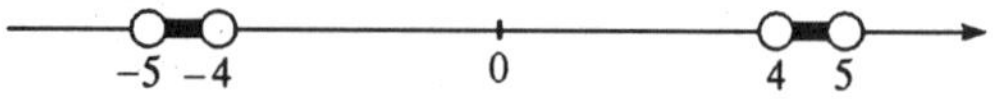

INT.	T.P.	VALUE	SIGN
$(-\infty, -2)$	-3	40	$+$
$(-2, -1)$	$-\dfrac{3}{2}$	$\dfrac{35}{16}$	$-$
$(-1, 1)$	0	4	$+$
$(1, 2)$	$\dfrac{3}{2}$	$\dfrac{35}{16}$	$-$
$(2, \infty)$	3	40	$+$

INT.	T.P.	VALUE	SIGN
$(-\infty, -5)$	-6	220	$+$
$(-5, -4)$	$-\dfrac{9}{2}$	$-\dfrac{323}{16}$	$-$
$(-4, 4)$	0	400	$+$
$(4, 5)$	$\dfrac{9}{2}$	$-\dfrac{323}{16}$	$-$
$(5, \infty)$	6	220	$+$

Figure 65   **Figure 66**

[67] $\dfrac{x^2-4}{1-x}\le 0 \Rightarrow \dfrac{(x-2)(x+2)}{1-x}\le 0$. We see that the cut points are 2, –2, and 1, where 1 is an excluded

value. From the table below we see that $\dfrac{(x-2)(x+2)}{1-x}$ is negative on the open intervals $(-2,\ 1)$

and $\left(2,\ \infty\right)$. Since the cut points –2 and 2 satisfy the equation $\dfrac{x^2-4}{1-x}=0$ they also satisfy the

inequality $\dfrac{x^2-4}{1-x}\le 0$. The cut point 1 is an excluded value. Hence, the solution set is

$\left[-2,1\right)\cup\left[2,\infty\right)$.

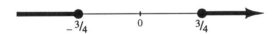

[68] $16x^4\ge 9-7x^2 \Rightarrow 16x^4+7x^2-9\ge 0$ To find the cut points, we consider the equation

$16x^4+7x^2-9=0$. If $u=x^2$, then $u^2=x^4$ and this equation becomes the quadratic equation

$16u^2+7u-9=0 \Rightarrow (16u-9)(u+1)=0 \Rightarrow 16u-9=0$ or $u+1=0 \Rightarrow 16u=9$ or $u=-1\Rightarrow$

$u=\dfrac{9}{16}$ or $x^2=-1$ (which has no real solutions) $\Rightarrow x^2=\dfrac{9}{16} \Rightarrow x=\pm\dfrac{3}{4}$. So, there are only two cut

points: $-\dfrac{3}{4},\dfrac{3}{4}$. From the table below we see that $16x^4+7x^2-9$ is positive on the open intervals

$\left(-\infty,\ -\dfrac{3}{4}\right)$ and $\left(\dfrac{3}{4},\ \infty\right)$. Since the cut points $-\dfrac{3}{4}$ and $\dfrac{3}{4}$ satisfy the equation $16x^4+7x^2-9=0$, they

also satisfy the inequality $16x^4+7x^2-9\ge 0$. Hence, the solution set is $\left(-\infty,\ -\dfrac{3}{4}\right]\cup\left[\dfrac{3}{4},\ \infty\right)$.

INT.	T.P.	VALUE	SIGN
$(-\infty,-2)$	-3	$\dfrac{5}{4}$	$+$
$(-2,1)$	0	-4	$-$
$(1,2)$	$\dfrac{3}{2}$	$\dfrac{7}{2}$	$+$
$(2,\infty)$	3	$-\dfrac{5}{2}$	$-$

INT.	T.P.	VALUE	SIGN
$\left(-\infty,-\dfrac{3}{4}\right)$	-1	14	$+$
$\left(-\dfrac{3}{4},\dfrac{3}{4}\right)$	0	-9	$-$
$\left(\dfrac{3}{4},\infty\right)$	1	14	$+$

Figure 67   **Figure 68**

REVIEW EXERCISES

[1] a) $H\cap K=K$ b) $J\cap K=\varnothing$ c) $H\cup J=H$

d) $J\cup K=H$ e) $H\cap(J\cup K)=H\cap H=H$

f) $(H\cap J)\cup(H\cap K)=J\cup K=H$

[2] $V=\{a,\ e,\ i,\ o,\ u\}$ [3] $\varnothing,\{-2\},\{0\},\{2\},\{-2,0\},\{0,2\},\{-2,2\},\{-2,0,2\}$

[4] a) $A'=\{1,\ 2,\ 3,\ ...\}$ b) $B'=\{0,\ 1,\ 2,\ ...,\ 10\}$ c) $U'=\varnothing$

d) $\emptyset' = U$

[5]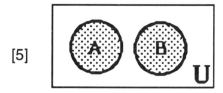

$A \cup B$

[6]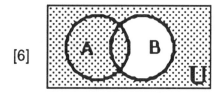

$A \cup B'$

[7]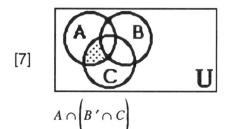

$A \cap \left(B' \cap C \right)$

[8]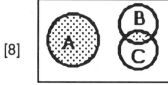

$A \cup \left(B \cap C \right)$

[9]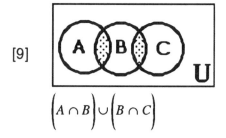

$\left(A \cap B \right) \cup \left(B \cap C \right)$

[10]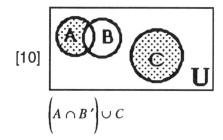

$\left(A \cap B' \right) \cup C$

[11] $x = 2.137137137\ldots$ and $1000x = 2137.137137137\ldots \Rightarrow 1000x - x =$

$2137.137137137\ldots - 2.137137137\ldots \Rightarrow 999x = 2135 \Rightarrow x = \dfrac{2135}{999}$

[12] $10x = 0.484848\ldots$ and $1000x = 48.484848\ldots \Rightarrow 1000x - 10x =$

$48.484848\ldots - 0.484848\ldots \Rightarrow 990x = 48 \Rightarrow x = \dfrac{48}{990} = \dfrac{24}{445}$

[13] Associative property of addition

[14] Distributive property

[15] Identity property of addition

[16] Inverse property of addition

[17] Identity property of multiplication

[18] Inverse property of multiplication

[19] $\pi > 3.14 \Rightarrow \pi - 3.14 > 0 \Rightarrow |\pi - 3.14| = \pi - 3.14$

[20] $y > 3 \Rightarrow 3 - y < 0 \Rightarrow |3 - y| = -(3 - y) = y - 3$

[21] a) $(-2, 4] = \left\{ x \mid -2 < x \leq 4 \right\}$

   b) $[-1, \infty) = \left\{ x \mid x \geq -1 \right\}$

c) $(-3, -1) \cup [1, \infty) = \left\{ x \mid -3 < x < -1 \text{ or } x \geq 1 \right\}$

d) $(-\infty, -\pi] \cup \left(0, \dfrac{5}{4}\right] = \left\{ x \mid x \leq -\pi \text{ or } 0 < x \leq \dfrac{5}{4} \right\}$

[22] a) $d(a, b) = |-1 + 8| = |7| = 7$ b) $d(a, b) = \left| \dfrac{7}{8} + \dfrac{3}{4} \right| = \left| \dfrac{13}{8} \right| = \dfrac{13}{8}$

[23]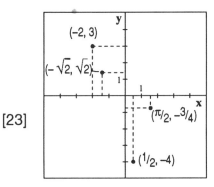

[24] $d(a, b) = \sqrt{(2 + 3)^2 + (-1 - 4)^2} = \sqrt{50} = 5\sqrt{2}; \; M = \left(\dfrac{-3 + 2}{2}, \dfrac{4 - 1}{2} \right) = \left(-\dfrac{1}{2}, \dfrac{3}{2} \right)$

[25] $d(A, B) = \sqrt{\left(\dfrac{1}{2} - \dfrac{5}{4} \right)^2 + (-1 - 3)^2} = \sqrt{\dfrac{265}{16}} = \dfrac{\sqrt{265}}{4}; \; M = \left(\dfrac{\frac{5}{4} + \frac{1}{2}}{2}, \dfrac{3 - 1}{2} \right) = \left(\dfrac{7}{8}, 1 \right)$

[26] Since $d(P_1, Q) = d(P_2, Q) = \sqrt{17}$, Q lies on the perpendicular bisector of $\overline{P_1 P_2}$

[27] The midpoint M of $\overline{JK}$ is $(1, -5)$. Thus $P = $ midpoint of $\overline{JM} = (-1, -2)$

[28] $d(A, B) = \sqrt{10}$, $d(B, C) = 10$ and $d(A, C) = 3\sqrt{10}$. Since $\left[d(A,B) \right]^2 + \left[d(A, C) \right]^2 =$

$\left(\sqrt{10} \right)^2 + \left(3\sqrt{10} \right)^2 = 100 = [d(B, C)]^2$, triangle ABC is a right triangle.

[29] Since $xy > 0$ if and only if x and y agree in sign, we shade quadrants I and III of the plane. (See graph below).

[30] (See graph below).

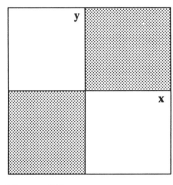

Figure 29

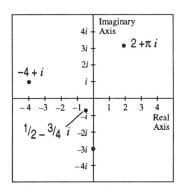

Figure 30

[31] $(2+3i)-(5-i)=-3+4i$

[32] $(-2+i)+(7-4i)=5-3i$

[33] $(1-2i)(1+3i)=1+i-6i^2=7+i$

[34] $i(5-3i)=5i-3i^2=3+5i$

[35] $\dfrac{2+3i}{1-i}\cdot\dfrac{1+i}{1+i}=\dfrac{2+5i-3}{2}=-\dfrac{1}{2}+\dfrac{5}{2}i$

[36] $\dfrac{1}{-9i}\cdot\dfrac{9i}{9i}=\dfrac{9i}{81}=\dfrac{1}{9}i$

[37] $|2-3i|=\sqrt{(2)^2+(-3)^2}=\sqrt{13}$

[38] $\left|\dfrac{3}{2}-\dfrac{1}{2}i\right|=\sqrt{\left(\dfrac{3}{2}\right)^2+\left(-\dfrac{1}{2}\right)^2}=\dfrac{\sqrt{10}}{2}$

[39] $i^{23}=i^{22}\cdot i=\left(i^2\right)^{11}\cdot i=(-1)^{11}\cdot i=-i$

[40] $i^{253}=i^{252}\cdot i=\left(i^2\right)^{126}\cdot i=(-1)^{126}\cdot i=i$

[41] $i^{17}=i^{16}\cdot i=\left(i^2\right)^8\cdot i=(-1)^8\cdot i=i\Rightarrow\left|i^{17}\right|=|i|=\sqrt{(0)^2+(1)^2}=\sqrt{1}=1$

[42] $\dfrac{1}{i^{39}}=\dfrac{1}{i^{38}\cdot i}=\dfrac{1}{\left(i^2\right)^{19}\cdot i}=\dfrac{1}{(-1)^{19}\cdot i}=\dfrac{1}{-i}\cdot\dfrac{i}{i}=i$

[43] $-(-2+6i)=2-6i$

[44] $z_1\cdot z_2=(ac-bd)+(ad+bc)i=(ca-db)+(ad+bc)i=z_2\cdot z_1$

[45] $\dfrac{1}{-2+3i}\cdot\dfrac{-2-3i}{-2-3i}=\dfrac{-2-3i}{13}=-\dfrac{2}{13}-\dfrac{3}{13}i$

[46] $\left(\dfrac{1}{2}+\dfrac{\sqrt{3}}{2}i\right)^3=\left(\dfrac{1}{2}+\dfrac{\sqrt{3}}{2}i\right)\left(\dfrac{1}{4}+\dfrac{\sqrt{3}}{2}i-\dfrac{3}{4}\right)=\left(\dfrac{1}{2}+\dfrac{\sqrt{3}}{2}i\right)\left(-\dfrac{1}{2}+\dfrac{\sqrt{3}}{2}i\right)=-\dfrac{1}{4}-\dfrac{3}{4}=-1$

[47] $3(x-1)-2(1-x)=2x-3(x+2)\Rightarrow 3x-3-2+2x=2x-3x-6\Rightarrow 6x=-1\Rightarrow x=-\dfrac{1}{6}$

$\bullet\left\{-\dfrac{1}{6}\right\}$

[48] $\dfrac{2x+1}{5}-\dfrac{x}{2}=\dfrac{1}{10}-x\Rightarrow 10\left(\dfrac{2x+1}{5}\right)-10\left(\dfrac{x}{2}\right)=10\left(\dfrac{1}{10}\right)-10x\Rightarrow 4x+2-5x=1-10x\Rightarrow$

$4x+2-5x=1-10x\Rightarrow 9x=-1\Rightarrow x=-\dfrac{1}{9}$
$\bullet\left\{-\dfrac{1}{9}\right\}$

[49] $\dfrac{3}{x+2}-\dfrac{5}{x-1}=\dfrac{2}{x+2}$. The excluded values are -2 and 1. Assuming that $x\neq-2$ and $x\neq1$, we

multiply both sides of the equation by the LCD $(x+2)(x-1)$:

$(x+2)(x-1)\left(\dfrac{3}{x+2}\right)-(x+2)(x-1)\left(\dfrac{5}{x-1}\right)=(x+2)(x-1)\left(\dfrac{2}{x+2}\right)\Rightarrow$

$3(x-1)-5(x+2)=2(x-1)\Rightarrow 3x-3-5x-10=2x-2\Rightarrow-4x=11\Rightarrow x=-\dfrac{11}{4}$.

Since $-\dfrac{11}{4}$ is not an excluded value, the solution set is $\left\{-\dfrac{11}{4}\right\}$.
$\bullet\left\{-\dfrac{11}{4}\right\}$

[50] $\dfrac{x}{x-1}-\dfrac{1}{1-x}=\dfrac{2}{x-1}$. The only excluded value is 1. Assuming that $x\neq 1$, we multiply both sides

by the LCD $x-1$: $(x-1)\left(\dfrac{x}{x-1}\right)-(x-1)\left(\dfrac{1}{1-x}\right)=(x-1)\left(\dfrac{2}{x-1}\right)\Rightarrow x+1=2\Rightarrow x=1.$ Since

1 is an excluded value, the solution set is $\varnothing$.   ● $\varnothing$

[51] $\dfrac{x}{x-3}+\dfrac{3}{3-x}=-1$. The only excluded value is 3. Assuming that $x\neq 3$ we multiply both sides of

the equation by the LCD $x-3$: $(x-3)\left(\dfrac{x}{x-3}\right)+(x-3)\left(\dfrac{3}{3-x}\right)=(x-3)(-1)\Rightarrow$

$x-3=-x+3\Rightarrow x=3.$ Since 3 is an excluded value, the solution set is $\varnothing$.   ● $\varnothing$

[52] $0.03\,x-0.4\,(2x-20)=30$. Multiplying both sides of the equation by the LCD 100, we get

$100\,(0.03x)-100\,(0.4)\,(2x-20)=100\,(30)\Rightarrow 3x-40\,(2x-20)=3000\Rightarrow$

$3x-80x+800=3000\Rightarrow-77x=2200\Rightarrow x=-\dfrac{2200}{77}=-\dfrac{200}{7}.$ ● $\left\{-\dfrac{200}{7}\right\}$

[53] $Ax+By+C=0\Rightarrow By=-Ax-C\Rightarrow B=\dfrac{-Ax-C}{y}\left(\text{for } y\neq 0\right)\Rightarrow B=-\dfrac{Ax+C}{y}\left(\text{for } y\neq 0\right)$

[54] $S=4\,\pi\,r^{2}\Rightarrow r^{2}=\dfrac{S}{4\,\pi}\Rightarrow r=\pm\sqrt{\dfrac{S}{4\,\pi}}=\pm\dfrac{\sqrt{S\,\pi}}{2\,\pi}.$ Since r cannot be negative, $r=\dfrac{\sqrt{S\,\pi}}{2\,\pi}.$

[55] $A=\dfrac{1}{2}\left(b_{1}+b_{2}\right)h\Rightarrow 2A=2\left(\dfrac{1}{2}\right)\left(b_{1}+b_{2}\right)h\Rightarrow 2A=b_{1}\,h+b_{2}\,h\Rightarrow b_{1}\,h=2A-b_{2}\,h\Rightarrow$

$b_{1}=\dfrac{2A-b_{2}\,h}{h}\left(\text{for } h\neq 0\right).$

[56] $D=b^{2}-4\,a\,c\Rightarrow 4\,a\,c=b^{2}-D\Rightarrow a=\dfrac{b^{2}-D}{4c}\left(\text{for } c\neq 0\right).$

[57] $y=2x^{2}-4x+1\Rightarrow 2x^{2}-4x+\left(1-y\right)=0.$ Using the quadratic formula with $a=2,\ b=-4$

and $c=1-y$, we have $x=\dfrac{-(-4)\pm\sqrt{(-4)^{2}-4\,(2)\,(1-y)}}{2\,(2)}=\dfrac{4\pm\sqrt{16-8+8y}}{4}=$

$\dfrac{4\pm\sqrt{8+8y}}{4}=\dfrac{4\pm\sqrt{4\,(2+2y)}}{4}=\dfrac{2\pm\sqrt{2+2y}}{2}.$

[58] $\dfrac{1}{x}-\dfrac{2}{y}=\dfrac{3}{w}.$ Assuming that $x\neq 0,\ y\neq 0,$ and $w\neq 0$, we multiply both sides of the equation by the

LCD $w\,x\,y$: $(w\,x\,y)\left(\dfrac{1}{x}\right)-(w\,x\,y)\left(\dfrac{2}{y}\right)=(w\,x\,y)\left(\dfrac{3}{w}\right)\Rightarrow w\,y-2\,w\,x=3\,x\,y\Rightarrow$

$w\,(y-2x)=3x\,y\Rightarrow w=\dfrac{3x\,y}{y-2x}\,(\text{for } y\neq 2x).$

[59] $-2(x-1)-3(x-4) \leq 4(x-1)+2 \Rightarrow -2x+2-3x+12 \leq 4x-4+2 \Rightarrow -9x \leq -16 \Rightarrow x \geq \dfrac{16}{9}$

● $\left[\dfrac{16}{9}, \infty\right)$

[60] $-4 \leq 2-3x < 5 \Rightarrow -6 \leq -3x < 3 \Rightarrow 2 \geq x > -1 \Rightarrow -1 < x \leq 2$

● $(-1, 2]$

[61] $-1 \leq 5-2x < 3 \Rightarrow -6 \leq -2x < -2 \Rightarrow 3 \geq x > 1 \Rightarrow 1 < x \leq 3$

● $(1, 3]$

[62] $-4 \leq -\dfrac{x}{2} < -1 \Rightarrow 8 \geq x > 2 \Rightarrow 2 < x \leq 8$

● $(2, 8]$

[63] $x^2 + 3x + 5 = 0$. Since $b^2 - 4ac = (3)^2 - 4(1)(5) = -11 < 0$, there are two complex conjugate roots.

[64] $4x^2 + 6x - 1 = 0$. Since $b^2 - 4ac = (6)^2 - 4(4)(-1) = 52 > 0$, there are two distinct real roots.

[65] $2x^2 - x - 3 = 0$. Since $b^2 - 4ac = (-1)^2 - 4(2)(-3) = 25 > 0$, there are two distinct real roots.

[66] $25x^2 - 30x + 9 = 0$. Since $b^2 - 4ac = (-30)^2 - 4(25)(9) = 0$, there is one real (double) root.

[67] $6x^2 - x - 12 = 0 \Rightarrow (2x-3)(3x+4) = 0 \Rightarrow x = \dfrac{3}{2}$ or $x = -\dfrac{4}{3}$.

● $\left\{-\dfrac{4}{3}, \dfrac{3}{2}\right\}$

[68] $x^2 - 4x - 4 = 0$. By the quadratic formula, $x = \dfrac{4 \pm \sqrt{32}}{2} = 2 \pm 2\sqrt{2}$

● $\left\{2 - 2\sqrt{2}, 2 + 2\sqrt{2}\right\}$

[69] $x(x-2) = -1 \Rightarrow x^2 - 2x + 1 = 0 \Rightarrow (x-1)^2 = 0 \Rightarrow x = 1$

● $\{1\}$

[70] $\dfrac{5x^2}{x+1} = 3$. The only excluded value is -1. Assuming that $x \neq -1$, we multiply both sides of the

equation by the LCD $x+1$: $(x+1)\left(\dfrac{5x^2}{x+1}\right) = (x+1)(3) \Rightarrow 5x^2 = 3x+3 \Rightarrow 5x^2 - 3x - 3 = 0$.

Using the quadratic formula we get $x = \dfrac{3 \pm \sqrt{69}}{10}$. Since neither value of x is an excluded value,

the solution set is $\left\{\dfrac{3 - \sqrt{69}}{10}, \dfrac{3 + \sqrt{69}}{10}\right\}$

● $\left\{\dfrac{3 - \sqrt{69}}{10}, \dfrac{3 + \sqrt{69}}{10}\right\}$

[71] $\dfrac{x^2}{2} - \dfrac{x}{3} + 1 = 0 \Rightarrow 6\left(\dfrac{x^2}{2}\right) - 6\left(\dfrac{x}{3}\right) + 6(1) = 6(0) \Rightarrow 3x^2 - 2x + 6 = 0$. By the quadratic formula,

$x = \dfrac{2 \pm \sqrt{-68}}{6} = \dfrac{2 \pm 2\sqrt{17}\, i}{6} = \dfrac{1 \pm \sqrt{17}\, i}{3}$

● $\left\{\dfrac{1}{3} - \dfrac{\sqrt{17}}{3}\, i, \dfrac{1}{3} + \dfrac{\sqrt{17}}{3}\, i\right\}$

[72] $3x(x+1)=20 \Rightarrow 3x^2 + 3x - 20 = 0$. Using the quadratic formula, we have

$$x = \frac{-3 \pm \sqrt{249}}{6}$$

$\bullet \left\{ \dfrac{-3 - \sqrt{249}}{6}, \dfrac{-3 + \sqrt{249}}{6} \right\}$

[73] $1 - \dfrac{1}{x} = \dfrac{2}{3x^2}$. The only excluded value is 0. Assuming that $x \neq 0$, we multiply both sides of the

equation by the LCD $3x^2$: $3x^2(1) - 3x^2\left(\dfrac{1}{x}\right) = 3x^2\left(\dfrac{2}{3x^2}\right) \Rightarrow 3x^2 - 3x = 2 \Rightarrow 3x^2 - 3x - 2 = 0$. Using

the quadratic formula, we have $x = \dfrac{3 \pm \sqrt{33}}{6}$. Since neither value of x is an excluded value, the

solution set is $\left\{ \dfrac{3 - \sqrt{33}}{6}, \dfrac{3 + \sqrt{33}}{6} \right\}$.

$\bullet \left\{ \dfrac{3 - \sqrt{33}}{6}, \dfrac{3 + \sqrt{33}}{6} \right\}$

[74] $x^2 = -1(x+3) \Rightarrow x^2 + x + 3 = 0$. By the quadratic formula, $x = \dfrac{-1 \pm \sqrt{-11}}{2} = \dfrac{-1 \pm \sqrt{11}\,i}{2}$.

$\bullet \left\{ -\dfrac{1}{2} - \dfrac{\sqrt{11}}{2}\,i, -\dfrac{1}{2} + \dfrac{\sqrt{11}}{2}\,i \right\}$

[75] $2x^2 - 2x + 1 = 0$. By the quadratic formula, $x = \dfrac{2 \pm \sqrt{-4}}{4} = \dfrac{2 \pm 2i}{4} = \dfrac{1 \pm i}{2}$ $\quad \bullet \left\{ \dfrac{1}{2} - \dfrac{1}{2}i, \dfrac{1}{2} + \dfrac{1}{2}i \right\}$

[76] $2x^2 - 3x - 5 = 0 \Rightarrow (2x-5)(x+1) = 0 \Rightarrow x = \dfrac{5}{2} \text{ or } x = -1$ $\quad \bullet \left\{ -1, \dfrac{5}{2} \right\}$

[77] $x(4x+2) = 5 \Rightarrow 4x^2 + 2x - 5 = 0$. By the quadratic formula, we have $x = \dfrac{-2 \pm \sqrt{84}}{8} =$

$\dfrac{-2 \pm \sqrt{(4)(21)}}{8} = \dfrac{-1 \pm \sqrt{21}}{4}$. $\quad \bullet \left\{ \dfrac{-1 - \sqrt{21}}{4}, \dfrac{-1 + \sqrt{21}}{4} \right\}$

[78] $x^2 = 3(3x - 1) \Rightarrow x^2 - 9x + 3 = 0$. By the quadratic formula, $x = \dfrac{9 \pm \sqrt{69}}{2}$ $\bullet \left\{ \dfrac{9 - \sqrt{69}}{2}, \dfrac{9 + \sqrt{69}}{2} \right\}$

[79] $x^2 = 4x + 1 \Rightarrow x^2 - 4x - 1 = 0$. By the quadratic formula, $x = \dfrac{4 \pm \sqrt{20}}{2} = 2 \pm \sqrt{5}$

$\bullet \left\{ 2 - \sqrt{5}, 2 + \sqrt{5} \right\}$

[80] $2(x^2 - 2) = -7x \Rightarrow 2x^2 + 7x - 4 = 0 \Rightarrow (2x-1)(x+4) = 0 \Rightarrow x = \dfrac{1}{2} \text{ or } x = -4$ $\quad \bullet \left\{ -4, \dfrac{1}{2} \right\}$

[81] $x^2 < x + 6 \Rightarrow x^2 - x - 6 < 0$

$x^2 - x - 6 = 0 \Rightarrow (x-3)(x+2) = 0 \Rightarrow x = 3 \text{ or } x = -2$ (two cut points). (See table below).

$\bullet (-2, 3)$

[82] $2x^2 \geq 6 - 4x \Rightarrow 2x^2 + 4x - 6 \geq 0 \Rightarrow x^2 + 2x - 3 \geq 0$

$x^2 + 2x - 3 = 0 \Rightarrow (x+3)(x-1) = 0 \Rightarrow x+3 = 0$ or $x-1 = 0 \Rightarrow x = -3$ or $x = 1$. (Two cut points). (See table below).

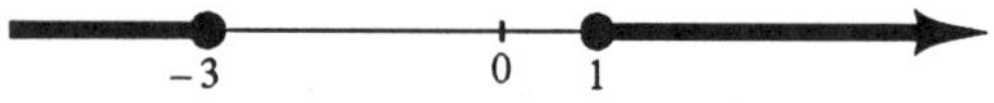

$\bullet \left(-\infty, -3\right] \cup \left[1, \infty\right)$

INT.	T.P.	VALUE	SIGN
$(-\infty, -2)$	-3	6	$+$
$(-2, 3)$	0	-6	$-$
$(3, \infty)$	4	6	$+$

Figure 81

INT.	T.P.	VALUE	SIGN
$(-\infty, -3)$	-4	5	$+$
$(-3, 1)$	0	-3	$-$
$(1, \infty)$	2	5	$+$

Figure 82

[83] $6x^2 > 2(5x+2) \Rightarrow 6x^2 - 10x - 4 > 0 \Rightarrow 3x^2 - 5x - 2 > 0$

$3x^2 - 5x - 2 = 0 \Rightarrow (3x+1)(x-2) = 0 \Rightarrow x = -\dfrac{1}{3}$ or $x = 2$.   (Two cut points). (See table below).

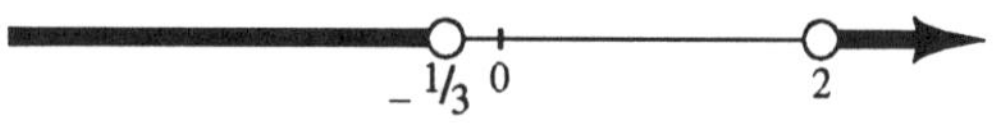

$\bullet \left(-\infty, -\dfrac{1}{3}\right) \cup \left(2, \infty\right)$

[84] $-2x^2 < -1 - x \Rightarrow -2x^2 + x + 1 < 0 \Rightarrow 2x^2 - x - 1 > 0$

$2x^2 - x - 1 = 0 \Rightarrow (x-1)(2x+1) = 0 \Rightarrow x = 1$ or $x = -\dfrac{1}{2}$. (Two cut points). (See table below).

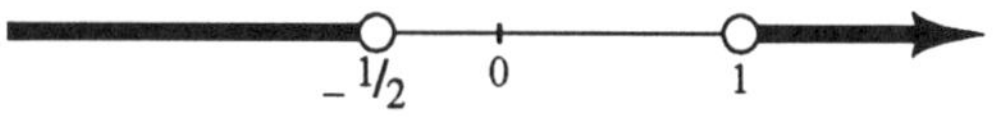

$\bullet \left(-\infty, -\dfrac{1}{2}\right) \cup \left(1, \infty\right)$

INT.	T.P.	VALUE	SIGN
$\left(-\infty, -\dfrac{1}{3}\right)$	-1	6	$+$
$\left(-\dfrac{1}{3}, 2\right)$	0	-2	$-$
$(2, \infty)$	3	10	$+$

Figure 83

INT.	T.P.	VALUE	SIGN
$\left(-\infty, -\dfrac{1}{2}\right)$	-1	2	$+$
$\left(-\dfrac{1}{2}, 1\right)$	0	-1	$-$
$(1, \infty)$	2	5	$+$

Figure 84

[85] $x(x+4) < 1 \Rightarrow x^2 + 4x - 1 < 0$

$x^2 + 4x - 1 = 0 \Rightarrow x = \dfrac{-4 \pm \sqrt{20}}{2} = -2 \pm \sqrt{5} \Rightarrow x = -2 - \sqrt{5} \approx -4.24$ or $x = -2 + \sqrt{5} \approx 0.24$ (Two cut points). (See table below).

$\bullet \left(-2 - \sqrt{5}, -2 + \sqrt{5}\right)$

[86] $x^2 < -2(2x+1) \Rightarrow x^2 + 4x + 2 < 0$

$x^2 + 4x + 2 = 0 \Rightarrow x = \dfrac{-4 \pm \sqrt{8}}{2} = -2 \pm \sqrt{2}$. So, $x = -2 + \sqrt{2} \approx -0.59$ and $x = -2 - \sqrt{2} \approx -3.41$ are the two cut points. (See table below).

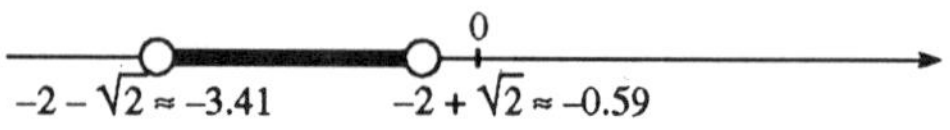

$\bullet \left(-2 - \sqrt{2}, -2 + \sqrt{2}\right)$

INT.	T.P.	VALUE	SIGN
$\left(-\infty, -2-\sqrt{5}\right)$	-5	4	$+$
$\left(-2-\sqrt{5}, -2+\sqrt{5}\right)$	0	-1	$-$
$\left(-2+\sqrt{5}, \infty\right)$	1	4	$+$

Figure 85

INT.	T.P.	VALUE	SIGN
$\left(-\infty, -2-\sqrt{2}\right)$	-4	2	$+$
$\left(-2-\sqrt{2}, -2+\sqrt{2}\right)$	-1	-1	$-$
$\left(-2+\sqrt{2}, \infty\right)$	0	2	$+$

Figure 86

[87] $x^2 \geq 4\left(2x-5\right) \Rightarrow x^2 - 8x + 20 \geq 0$

$x^2 - 8x + 20 = 0$. Since $D = \left(-8\right)^2 - 4\left(1\right)\left(20\right) = -16 < 0$, there are no cut points. (T.P.: $x=0$)

● R

[88] $x\left(x+4\right) > -1 \Rightarrow x^2 + 4x + 1 > 0$

$x^2 + 4x + 1 = 0 \Rightarrow x = \dfrac{-4 \pm \sqrt{12}}{2} = -2 \pm \sqrt{3}$. So $x = -2 - \sqrt{3} \approx -3.73$ and $x = -2 + \sqrt{3} \approx -0.27$ are

the two cut points. (See table below).

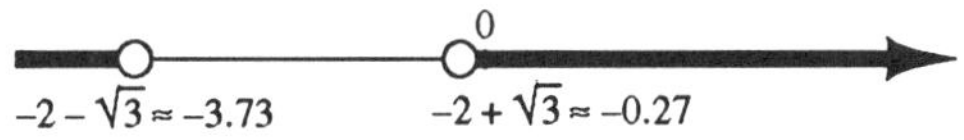

● $\left(-\infty, -2-\sqrt{3}\right) \cup \left(-2+\sqrt{3}, \infty\right)$

INT.	T.P.	VALUE	SIGN
$\left(-\infty, -2-\sqrt{3}\right)$	-4	1	$+$
$\left(-2-\sqrt{3}, -2+\sqrt{3}\right)$	-1	-2	$-$
$\left(-2+\sqrt{3}, \infty\right)$	0	1	$+$

Figure 88

[89] $\left|1-2x\right| = 3 \Rightarrow 1-2x = \pm 3 \Rightarrow 1-2x = -3$ or $1-2x = 3 \Rightarrow -2x = -4$ or $-2x = 2 \Rightarrow$
$x = 2$ or $x = -1$

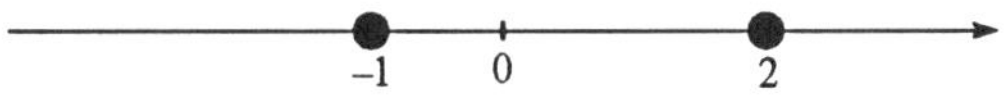

● $\left\{-1, 2\right\}$

[90] $\left|2x-7\right| = \left|1-x\right| \Rightarrow 2x-7 = \pm\left(1-x\right) \Rightarrow 2x-7 = 1-x$ or $2x-7 = -1+x \Rightarrow 3x = 8$ or $x = 6 \Rightarrow$
$x = \dfrac{8}{3}$ or $x = 6$

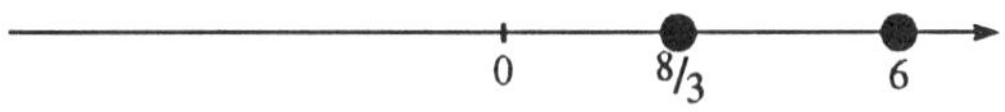

● $\left\{\dfrac{8}{3}, 6\right\}$

[91] $\left|3-2x\right| \geq 1 \Rightarrow 3-2x \leq -1$ or $3-2x \geq 1 \Rightarrow -2x \leq -4$ or $-2x \geq -2 \Rightarrow x \geq 2$ or $x \leq 1$

● $\left(-\infty, 1\right] \cup \left[2, \infty\right)$

[92] $\left|1-x\right| > 15 \Rightarrow 1-x < -15$ or $1-x > 15 \Rightarrow -x < -16$ or $-x > 14 \Rightarrow x > 16$ or $x < -14$

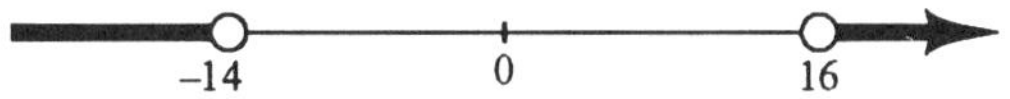

● $\left(-\infty, -14\right) \cup \left(16, \infty\right)$

[93] $\left|3-5x\right| \leq 3 \Rightarrow -3 \leq 3-5x \leq 3 \Rightarrow -6 \leq -5x \leq 0 \Rightarrow \dfrac{6}{5} \geq x \geq 0 \Rightarrow 0 \leq x \leq \dfrac{6}{5}$

● $\left[0, \dfrac{6}{5}\right]$

47

[94] $\left|\,2x+10\,\right|-\left|\,-12\,\right|<-\left|\,-3\,\right|\Rightarrow\left|\,2x+10\,\right|-12<-3\Rightarrow\left|\,2x+10\,\right|<9\Rightarrow-9<2x+10<9\Rightarrow$

$-19<2x<-1\Rightarrow-\dfrac{19}{2}<x<-\dfrac{1}{2}$

$\bullet\left(-\dfrac{19}{2},\,-\dfrac{1}{2}\right)$

[95] $\left|\dfrac{1-x}{2}\right|>5\Rightarrow\dfrac{1-x}{2}<-5\ \text{or}\ \dfrac{1-x}{2}>5\Rightarrow1-x<-10\ \text{or}\ 1-x>10\Rightarrow$

$-x<-11\ \text{or}\ -x>9\Rightarrow x>11\ \text{or}\ x<-9$

$\bullet\left(-\infty,-9\right)\cup\left(11,\infty\right)$

[96] $\left|\dfrac{x}{2}-1\right|-\left|\,-2\,\right|>\left|\,-3\,\right|\Rightarrow\left|\dfrac{x}{2}-1\right|-2>3\Rightarrow\left|\dfrac{x}{2}-1\right|>5\Rightarrow\dfrac{x}{2}-1<-5\ \text{or}\ \dfrac{x}{2}-1>5\Rightarrow$

$\dfrac{x}{2}<-4\ \text{or}\ \dfrac{x}{2}>6\Rightarrow x<-8\ \text{or}\ x>12$

$\bullet\left(-\infty,-8\right)\cup\left(12,\infty\right)$

[97] $\sqrt{x-4}=\sqrt{x}-4\Rightarrow\left(\sqrt{x-4}\right)^{2}=\left(\sqrt{x}-4\right)^{2}\Rightarrow x-4=x-8\sqrt{x}+16\Rightarrow-20=-8\sqrt{x}\Rightarrow$

$400=64x\Rightarrow x=\dfrac{400}{64}\Rightarrow x=\dfrac{25}{4}.$ <u>Check</u>: If $x=\dfrac{25}{4}$, then $\sqrt{x-4}=\sqrt{\dfrac{25}{4}-4}=\sqrt{\dfrac{9}{4}}=\dfrac{3}{2}.$ On

the other hand, $\sqrt{x}-4=\sqrt{\dfrac{25}{4}}-4=\dfrac{5}{2}-4=-\dfrac{3}{2}.$ Since $\dfrac{3}{2}\neq-\dfrac{3}{2},\ S=\varnothing.$ $\bullet\varnothing$

[98] $\sqrt{2x}-4=\sqrt{2x+4}\Rightarrow\left(\sqrt{2x}-4\right)^{2}=\left(\sqrt{2x+4}\right)^{2}\Rightarrow2x-8\sqrt{2x}+16=2x+4\Rightarrow-8\sqrt{2x}=-12\Rightarrow$

$128x=144\Rightarrow x=\dfrac{144}{128}\Rightarrow x=\dfrac{9}{8}.$ <u>Check</u>: If $x=\dfrac{9}{8},\ \sqrt{2x}-4=\sqrt{2\left(\dfrac{9}{8}\right)}-4=\dfrac{3}{2}-4=-\dfrac{5}{2},$

while $\sqrt{2x+4}=\sqrt{2\left(\dfrac{9}{8}\right)+4}=\sqrt{\dfrac{9}{4}+4}=\sqrt{\dfrac{25}{4}}=\dfrac{5}{2}.$ Since $-\dfrac{5}{2}\neq\dfrac{5}{2},\ S=\varnothing.$ $\bullet\varnothing$

[99] $\sqrt{x-5}=5-\sqrt{x}\Rightarrow\left(\sqrt{x-5}\right)^{2}=\left(5-\sqrt{x}\right)^{2}\Rightarrow x-5=25-10\sqrt{x}+x\Rightarrow-30=-10\sqrt{x}\Rightarrow$

$3=\sqrt{x}\Rightarrow x=9.$ <u>Check</u>: If $x=9,\ \sqrt{x-5}=\sqrt{9-5}=\sqrt{4}=2$ and $5-\sqrt{x}=5-\sqrt{9}=5-3=2.$

Hence, $S=\{9\}.$ $\bullet\{9\}$

[100] $\sqrt{x^{2}-4}=x-2\Rightarrow\left(\sqrt{x^{2}-4}\right)^{2}=\left(x-2\right)^{2}\Rightarrow x^{2}-4=x^{2}-4x+4\Rightarrow4x=8\Rightarrow x=2$

<u>Check</u>: If $x=2,\ \sqrt{x^{2}-4}=\sqrt{4-4}=0$ and $x-2=2-2=0.$ Hence, $S=\{2\}.$ $\bullet\{2\}$

[101] $\sqrt{x^{2}+25}=x+5\Rightarrow\left(\sqrt{x^{2}+25}\right)^{2}=\left(x+5\right)^{2}\Rightarrow x^{2}+25=x^{2}+10x+25\Rightarrow0=10x\Rightarrow x=0$

<u>Check</u>: If $x=0,\ \sqrt{x^{2}+25}=\sqrt{0+25}=5$ and $x+5=0+5=5.$ Hence, $S=\{0\}.$ $\bullet\{0\}$

[102] $\sqrt{x+4} = \sqrt{x} + 2 \Rightarrow \left(\sqrt{x+4}\right)^2 = \left(\sqrt{x}+2\right)^2 \Rightarrow x+4 = x+4\sqrt{x}+4 \Rightarrow 0 = 4\sqrt{x} \Rightarrow \sqrt{x} = 0 \Rightarrow x = 0$

Check: If $x = 0$, $\sqrt{x+4} = \sqrt{0+4} = 2$ and $\sqrt{x}+2 = \sqrt{0}+2 = 2$. So, $S = \{0\}$.   ● $\{0\}$

[103] $x^4 + 5x^2 + 4 = 0$. If $u = x^2$, then $u^2 = x^4$ and the given equation becomes $u^2 + 5u + 4 = 0 \Rightarrow$

$(u+4)(u+1) = 0 \Rightarrow u+4 = 0$ or $u+1 = 0 \Rightarrow u = -4$ or $u = -1 \Rightarrow x^2 = -4$ or $x^2 = -1 \Rightarrow$

$x = \pm 2i$ or $x = \pm i$.   ● $\{-2i, -i, i, 2i\}$

[104] $x^{2/3} - x^{1/3} - 6 = 0$. If $u = x^{1/3}$, then $u^2 = x^{2/3}$ and the given equation becomes $u^2 - u - 6 = 0 \Rightarrow$

$(u-3)(u+2) = 0 \Rightarrow u-3 = 0$ or $u+2 = 0 \Rightarrow u = 3$ or $u = -2 \Rightarrow x^{1/3} = 3$ or $x^{1/3} = -2 \Rightarrow$

$x = 27$ or $x = -8$.   ● $\{-8, 27\}$

[105] $x - 6\sqrt{x} + 8 = 0$. If $u = \sqrt{x}$, then $u^2 = x$ and the given equation becomes $u^2 - 6u + 8 = 0 \Rightarrow$

$(u-4)(u-2) = 0 \Rightarrow u-4 = 0$ or $u-2 = 0 \Rightarrow u = 4$ or $u = 2 \Rightarrow \sqrt{x} = 4$ or $\sqrt{x} = 2 \Rightarrow x = 16$ or $x = 4$

● $\{4, 16\}$

[106] $2x^4 - x^2 - 3 = 0$. If $u = x^2$, then $u^2 = x^4$, and the given equation becomes $2u^2 - u - 3 = 0 \Rightarrow$

$(2u-3)(u+1) = 0 \Rightarrow 2u-3 = 0$ or $u+1 = 0 \Rightarrow 2u = 3$ or $u = -1 \Rightarrow u = \dfrac{3}{2}$ or $x^2 = -1 \Rightarrow$

$x^2 = \dfrac{3}{2}$ or $x = \pm i \Rightarrow x = \pm\sqrt{\dfrac{3}{2}}$ or $x = \pm i \Rightarrow x = \dfrac{\pm\sqrt{6}}{2}$ or $x = \pm i$.   ● $\left\{-i, i, \dfrac{-\sqrt{6}}{2}, \dfrac{\sqrt{6}}{2}\right\}$

[107] $(x+1)^2 - 8(x+1) + 15 = 0$. If $u = x+1$ then $u^2 = (x+1)^2$ and the given equation becomes

$u^2 - 8u + 15 = 0 \Rightarrow (u-5)(u-3) = 0 \Rightarrow u-5 = 0$ or $u-3 = 0 \Rightarrow u = 5$ or $u = 3 \Rightarrow$

$x+1 = 5$ or $x+1 = 3 \Rightarrow x = 4$ or $x = 2$.   ● $\{2, 4\}$

[108] $\left(\dfrac{2}{x-1}\right)^2 + 2\left(\dfrac{2}{x-1}\right) - 3 = 0$. If $u = \dfrac{2}{x-1}$, then $u^2 = \left(\dfrac{2}{x-1}\right)^2$ and the given equation becomes

$u^2 + 2u - 3 = 0 \Rightarrow (u+3)(u-1) = 0 \Rightarrow u+3 = 0$ or $u-1 = 0 \Rightarrow u = -3$ or $u = 1 \Rightarrow$

$\dfrac{2}{x-1} = -3$ or $\dfrac{2}{x-1} = 1 \Rightarrow -3x-3 = 2$ or $x-1 = 2 \Rightarrow -3x = -1$ or $x = 3 \Rightarrow x = \dfrac{1}{3}$ or $x = 3$.   ● $\left\{\dfrac{1}{3}, 3\right\}$

[109] $\dfrac{x-1}{2-x} \le 0$. The cut points are 1 and 2, where 2 is an excluded value. From the table below, we see

that $\dfrac{x-1}{2-x}$ is negative on the intervals $\left(-\infty, 1\right)$ and $\left(2, \infty\right)$. Since 1 satisfies the equation $\dfrac{x-1}{2-x} = 0$,

it also satisfies the inequality $\dfrac{x-1}{2-x} \le 0$. Since 2 is an excluded value, $S = \left(-\infty, 1\right] \cup \left(2, \infty\right)$.

● $\left(-\infty, 1\right] \cup \left(2, \infty\right)$

[110] $\dfrac{x^2+4x+3}{3-x} \geq 0 \Rightarrow \dfrac{(x+3)(x+1)}{3-x} \geq 0$. The cut points are –3, –1, and 3, where 3 is an excluded

value. Since –3 and –1 satisfy the equation $\dfrac{x^2+4x+3}{3-x} = 0$, they also satisfy the inequality

$\dfrac{x^2+4x+3}{3-x} \geq 0$, but since 3 is an excluded value, $S = \left(-\infty,\ -3\right] \cup \left[-1,\ 3\right)$. (See table below).

● $\left(-\infty,\ -3\right] \cup \left[-1,\ 3\right)$

INT.	T.P.	VALUE	SIGN
$(-\infty, 1)$	0	$-\dfrac{1}{2}$	–
$(1, 2)$	$\dfrac{3}{2}$	1	+
$(2, \infty)$	3	–2	–

Figure 109

INT.	T.P.	VALUE	SIGN
$(-\infty, -3)$	–4	$\dfrac{3}{7}$	+
$(-3, -1)$	–2	$-\dfrac{1}{5}$	–
$(-1, 3)$	0	1	+
$(3, \infty)$	4	–35	–

Figure 110

[111] $\dfrac{x^2+x-6}{x^2-x-6} > 0 \Rightarrow \dfrac{(x+3)(x-2)}{(x-3)(x+2)} > 0$. The cut points are –3, 2, 3, and –2, where 3 and –2 are

excluded values. Thus $S = \left(-\infty,\ -3\right) \cup \left(-2,\ 2\right) \cup \left(3,\ \infty\right)$. (See table below).

● $\left(-\infty,\ -3\right) \cup \left(-2,\ 2\right) \cup \left(3,\ \infty\right)$

[112] $\dfrac{x+3}{5-x} < 1 \Rightarrow \dfrac{x+3}{5-x} - 1 < 0 \Rightarrow \dfrac{x+3-5+x}{5-x} < 0 \Rightarrow \dfrac{2x-2}{5-x} < 0$. So, the cut points are 1 and 5, where

5 is an excluded value. Hence, $S = \left(-\infty,\ 1\right) \cup \left(5,\ \infty\right)$. (See table below).

● $\left(-\infty,\ 1\right) \cup \left(5,\ \infty\right)$

INT.	T.P.	VALUE	SIGN
$(-\infty, -3)$	–4	$\dfrac{3}{7}$	+
$(-3, -2)$	$-\dfrac{5}{2}$	$-\dfrac{9}{11}$	–
$(-2, 2)$	0	1	+
$(2, 3)$	$\dfrac{5}{2}$	$-\dfrac{11}{9}$	–
$(3, \infty)$	4	$\dfrac{7}{3}$	+

Figure 111

INT.	T.P.	VALUE	SIGN
$(-\infty, 1)$	0	$-\dfrac{2}{5}$	–
$(1, 5)$	2	$\dfrac{2}{3}$	+
$(5, \infty)$	6	–10	–

Figure 112

[113] $x^4 < -13x^2 - 36 \Rightarrow x^4 + 13x^2 + 36 < 0$.

$x^4 + 13x^2 + 36 = 0$. If $u = x^2$, then $u^2 = x^4$ and the equation becomes $u^2 + 13u + 36 = 0 \Rightarrow$

$(u+9)(u+4) = 0 \Rightarrow u + 9 = 0$ or $u + 4 = 0 \Rightarrow u = -9$ or $u = -4 \Rightarrow x^2 = -9$ (no cut points) or

$x^2 = -4$ (no cut points). (T.P.: $x = 0$).   ● ∅

[114] $x^4 - 34x^2 \leq -225 \Rightarrow x^4 - 34x^2 + 225 \leq 0$. If $u = x^2$, then $u^2 = x^4$ and we can write

$u^2 - 34u + 225 = 0 \Rightarrow (u - 25)(u - 9) = 0 \Rightarrow u = 25$ or $u = 9 \Rightarrow x^2 = 25$ or $x^2 = 9 \Rightarrow x = \pm 5$ or

$x = \pm 3$. Since all four cut points satisfy the equation $x^4 - 34x^2 + 225 = 0$, they satisfy the

inequality $x^4 - 34x^2 + 225 \leq 0$. (See table below). $\bullet\, S = [-5, -3] \cup [3, 5]$

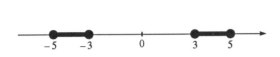

INT.	T.P.	VALUE	SIGN
$(-\infty, -5)$	-6	297	$+$
$(-5, -3)$	-4	-63	$-$
$(-3, 3)$	0	225	$+$
$(3, 5)$	4	-63	$-$
$(5, \infty)$	6	297	$+$

Figure 114

[115] If x is the number of adult tickets sold, then $\frac{4}{3}x$ is the number of student tickets sold. Since the

total income was \$4400, $4x + 2.50\left(\frac{4}{3}x\right) = 4400 \Rightarrow 30(4x) + 30(2.50)\left(\frac{4}{3}x\right) = 30(4400) \Rightarrow$

$120x + 100x = 132000 \Rightarrow 220x = 132000 \Rightarrow x = 600$.   Hence, 600 adult tickets and $\frac{4}{3}(600) = 800$

student tickets were sold.

[116] If w is the width (in meters) of the lot, then $l = 2w + 10$ is its length.   Since the perimeter is

380 m and $P = 2l + 2w$, $2(2w + 10) + 2w = 380 \Rightarrow 4w + 20 + 2w = 380 \Rightarrow 6w = 360 \Rightarrow$

$w = 60$ m. So, $l = 2(60) + 10 = 130$. Hence, $A = lw = (60)(130) = 7800\ \text{m}^2$.

[117] If his standard hourly salary is x (\$), then $2x$ is his double-time salary. Hence,

$40(x) + 12(2x) = 608 \Rightarrow 64x = 608 \Rightarrow x = \$9.50$.   Hence, his standard hourly salary is \$9.50.

[118] If x is one of the numbers, then $\frac{1}{x}$ is its reciprocal and we can write $x + \frac{1}{x} = \frac{58}{21} \Rightarrow$

$21x^2 + 21 = 58x \Rightarrow 21x^2 - 58x + 21 = 0 \Rightarrow (3x - 7)(7x - 3) = 0 \Rightarrow 3x - 7 = 0$ or $7x - 3 = 0 \Rightarrow$

$x = \frac{7}{3}$ or $x = \frac{3}{7}$. So, the numbers are $\frac{3}{7}$ and $\frac{7}{3}$.

[119] If l is the length (in feet) of the lot, then $w = 2l + 10$ is its width. Since the area is 1500 square

feet and $A = lw$, we can write $lw = 1500 \Rightarrow l(2l + 10) = 1500 \Rightarrow 2l^2 + 10l - 1500 = 0 \Rightarrow$

$l^2 + 5l - 750 = 0 \Rightarrow (l - 25)(l + 30) = 0 \Rightarrow l - 25 = 0$ or $l + 30 = 0 \Rightarrow l = 25$ or $l = -30$.

We discard the latter, since length cannot be negative. So, the length of the lot is 25 ft and its width

is $2(25) + 10 = 60$ ft.

[120] If x is the integer, then we can write $2x^2 = x + 45 \Rightarrow 2x^2 - x - 45 = 0 \Rightarrow (x - 5)(2x + 9) = 0 \Rightarrow$

$x - 5 = 0$ or $2x + 9 = 0 \Rightarrow x = 5$ or $x = -\dfrac{9}{5}$ (which is not an integer).   The integer is 5.

EXERCISES 2.1

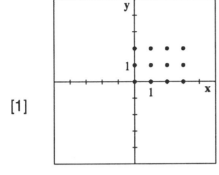

[1]

$dom(A) = \{0, 1, 2, 3\},\ ran(A) = \{0, 1, 2\}$

[2]

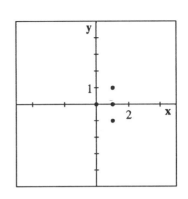

$dom(B) = \{0, 1\},\ ran(B) = \{-1, 0, 1\}$

[3]

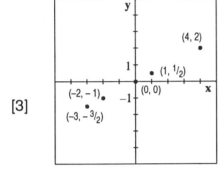

$dom(C) = \{-3, -2, 0, 1, 4\},$

$ran(C) = \left\{-\dfrac{3}{2}, -1, 0, \dfrac{1}{2}, 2\right\}$

[4]

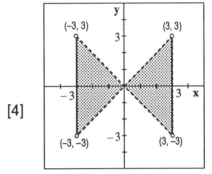

$dom(D) = \left\{x \mid -3 \le x \le 3\right\},$

$ran(D) = \left\{y \mid -3 < y < 3\right\}$

[5]   Only C.

[6] $(x - h)^2 + (y - k)^2 = r \Rightarrow x^2 - 2hx + h^2 + y^2 - 2ky + k^2 = r^2 \Rightarrow x^2 + y^2 - 2hx - 2ky + h^2 +$
       $k^2 - r^2 = 0.$ Letting $D = -2h,\ E = -2k$ and $F = h^2 + k^2 - r^2$, we get $x^2 + y^2 + Dx + Ey + F = 0.$

[7] $(x - 3)^2 + (y + 4)^2 = 100$ [8] $\left(x + \sqrt{3}\right)^2 + (y - 1)^2 = 6$

[9] $r^2 = d^2(P, C) = (2 + 1)^2 + (-1 - 5)^2 = 9 + 36 = 45.$ Thus, $(x - 2)^2 + (y + 1)^2 = 45.$

[10] $r^2 = d(P, C) = \left(\dfrac{3}{4} - \dfrac{5}{4}\right)^2 + \left(-\dfrac{1}{10} - 2\right)^2 = \dfrac{1}{4} + \dfrac{441}{100} = \dfrac{466}{100} = \dfrac{233}{50}.$ Thus, $\left(x - \dfrac{3}{4}\right)^2 + \left(y + \dfrac{1}{10}\right)^2 = \dfrac{233}{50}$

[11] $(x - 1)^2 + (y + 2)^2 = 1$ [12] $(x - 1)^2 + (y + 2)^2 = 4$

[13] $r^2 = \dfrac{1}{4}d^2(P, Q) = \dfrac{1}{4}[(5 + 1)^2 + (3 - 1)^2] = \dfrac{1}{4}(6^2 + 2^2) = \dfrac{1}{4}(40) = 10.$ The center is the midpoint of
       the line segment $\overline{PQ}$. Thus, the equation of the circle is $(x - 2)^2 + (y - 2)^2 = 10.$

[14] $r^2 = \frac{1}{4} d^2(P, Q) = \frac{1}{4}[(-1+3)^2 + (1-6)^2] = \frac{1}{4}[2^2 + (-5)^2] = \frac{1}{4}(29)$. The center is the midpoint of

line segment $\overline{PQ}$. The center has coordinates $\left(\frac{-3-1}{2}, \frac{6+1}{2}\right) = \left(-2, \frac{7}{2}\right)$. Thus, the equation of

the circle is $(x+2)^2 + \left(y - \frac{7}{2}\right)^2 = \frac{29}{4}$.

[15]   Center, (0, 0); radius 6   [16] Center (0, 0); radius 8

[17] $x^2 + y^2 - 2x - 2y + 1 = 0 \Rightarrow (x - 2x) + (y - 2y) = -1 \Rightarrow x^2 - 2x + 1 + y - 2y + 1 = 1 \Rightarrow$

   $(x-1)^2 + (y-1)^2 = 1$. Center (1, 1); radius, 1.

[18] $x^2 + y^2 + 6y = 25 \Rightarrow x^2 + y^2 + 6y + 9 = 25 + 9 = 34 \Rightarrow x^2 + (y-3)^2 = 34$. Center, (0, 3); radius, $\sqrt{34}$.

[19] $9x^2 + 9y^2 - 6x - 225 = 0 \Rightarrow x^2 + y^2 - \frac{2}{3}x - 25 = 0 \Rightarrow x^2 - \frac{2}{3}x + \frac{1}{9} + y^2 = 25 + \frac{1}{9} \Rightarrow$

   $\left(x - \frac{1}{3}\right)^2 + y^2 = \frac{226}{9}$. Center, $\left(\frac{1}{3}, 0\right)$; radius, $\frac{\sqrt{226}}{3}$.

[20] $16x^2 + 16y^2 + 16x - 8y + 3 = 0 \Rightarrow x^2 + y^2 + x - \frac{1}{2}y + \frac{3}{16} = 0 \Rightarrow (x^2 + x) + \left(y^2 - \frac{1}{2}y\right) = -\frac{3}{16} \Rightarrow$

   $x^2 + x + \frac{1}{4} + y^2 - \frac{1}{2}y + \frac{1}{16} = -\frac{3}{16} + \frac{1}{4} + \frac{1}{16} = \frac{1}{8}$. Center, $\left(-\frac{1}{2}, \frac{1}{4}\right)$; radius, $\frac{\sqrt{2}}{4}$.

[21] $4x^2 + 4y^2 - 4x - 8y - 11 = 0 \Rightarrow x^2 + y^2 - x - 2y - \frac{11}{4} = 0 \Rightarrow (x^2 - x) + (y^2 - 2y) = \frac{11}{4} \Rightarrow$

   $x^2 - x + \frac{1}{4} + y^2 - 2y + 1 = \frac{11}{4} + \frac{1}{4} + 1 = 4 \Rightarrow \left(x - \frac{1}{2}\right)^2 + (y-1)^2 = 4$. Center, $\left(\frac{1}{2}, 1\right)$; radius, 2

[22] $\frac{1}{2}x^2 + \frac{1}{2}y^2 - 6x + 4y + 8 = 0 \Rightarrow x^2 + y^2 - 12x + 8y + 16 = 0 \Rightarrow (x^2 - 12x) + (y^2 + 8y) = -16 \Rightarrow$

   $x^2 - 12x + 36 + y^2 + 8y + 16 = -16 + 36 + 16 = 36 \Rightarrow (x-6)^2 + (y+4)^2 = 36$.

   Center, $(6, -4)$; radius 6.

[23] $dom(R) = (-\infty, 0]$; $ran(R) = \mathbf{R}$; not a function.

[24] $dom(R) = \mathbf{R}$; $ran(R) = \mathbf{R}$; not a function.

[25] $dom(R) = [-3, 3]$; $ran(R) = [-3, 3]$; not a function.

[26] $dom(R) = [-3, 1]$; $ran(R) = [-3, 1]$; not a function.

[27] $dom(R) = (-\infty, 0) \cup (0, \infty)$; $ran(R) = \{-1, 1\}$; a function.

[28] $dom(R) = \mathbf{R}$; $ran(R) = [-2, \infty)$; a function.

[29] $f(0) = 3(0)^2 - (0) - 1 = -1; f(-1) = 3(-1)^2 - (-1) - 1 = 3; f(2) = 3(2)^2 - 2 - 1 = 9$

[30] $f(6) = \sqrt{6-2} - 3(6) = \sqrt{4} - 18 = -16; f(4) = \sqrt{4-2} - 3(4) = \sqrt{2} - 12;$
   $f(2) = \sqrt{2-2} - 3(2) = -6$

[31] $f(0) = \frac{0}{0^2 - 0 - 6} = 0; f(2) = \frac{2}{2^2 - 2 - 6} = -\frac{1}{2}; f(3)$ is undefined.

[32] $f(-1) = |-1 - 2| + 3 = |-3| + 3 = 3 + 3 = 6; \; f(0) = |0 - 2| + 3 = |-2| + 3 = 2 + 3 = 5;$

$f(1) = |1 - 2| + 3 = |-1| + 3 = 1 + 3 = 4$

[33] a) $f(a) = \sqrt{6a - 5}$　　　　　　　　　b) $f(a + h) = \sqrt{6(a + h) - 5}$

c) $\dfrac{f(a + h) - f(a)}{h} = \dfrac{\sqrt{6(a + h) - 5} - \sqrt{6a - 5}}{h}$

d) $f\left(\dfrac{1}{a}\right) = \sqrt{6\left(\dfrac{1}{a}\right) + 5} = \sqrt{\dfrac{6 + 5a}{a}}$　　　e) $\dfrac{1}{f(a)} = \dfrac{1}{\sqrt{6a - 5}}$

f) $f(a^2) = \sqrt{6a^2 - 5}$

[34] a) $f(a) = \dfrac{2a + 3}{a^2 + a - 1}$　　　　　　　　b) $f(a + h) = \dfrac{2(a + h) + 3}{(a + h)^2 + (a + h) - 1}$

c) $\dfrac{f(a + h) - f(a)}{h} = \dfrac{\dfrac{2(a + h) + 3}{(a + h)^2 + (a + h) - 1} - \dfrac{2a + 3}{a^2 + a - 1}}{h}$

d) $f\left(\dfrac{1}{a}\right) = \dfrac{2\left(\dfrac{1}{a}\right) + 3}{\left(\dfrac{1}{a}\right)^2 + \dfrac{1}{a} - 1} = \dfrac{\dfrac{2 + 3a}{a}}{\dfrac{1 + a - a^2}{a^2}} = \dfrac{2a + 3a^2}{1 + a - a^2}$

e) $\dfrac{1}{f(a)} = \dfrac{1}{\dfrac{2a + 3}{a^2 + a - 1}} = \dfrac{a^2 + a - 1}{2a + 3}$　　　　f) $f(a^2) = \dfrac{2a^2 + 3}{(a^2)^2 + a^2 - 1} = \dfrac{2a^2 + 3}{a^4 + a^2 - 1}$

[35] a) $f(a) = \dfrac{a + 1}{a^3 + 4a}$　　　　　　　　b) $f(a + h) = \dfrac{a + h + 1}{(a + h)^3 + 4(a + h)}$

c) $\dfrac{f(a + h) - f(a)}{h} = \dfrac{\dfrac{a + h + 1}{(a + h)^3 + 4(a + h)} - \dfrac{a + 1}{a^3 + 4a}}{h}$

d) $f\left(\dfrac{1}{a}\right) = \dfrac{\dfrac{1}{a} + 1}{\left(\dfrac{1}{a}\right)^3 + \dfrac{4}{a}} = \dfrac{\dfrac{1 + a}{a}}{\dfrac{1 + 4a^2}{a^3}} = \dfrac{a^2 + a^3}{1 + 4a^2}$　　e) $\dfrac{1}{f(a)} = \dfrac{1}{\dfrac{a + 1}{a^3 + 4a}} = \dfrac{a^3 + 4a}{a + 1}$

f) $f(a^2) = \dfrac{a^2 + 1}{(a^2)^3 + 4a^2} = \dfrac{a^2 + 1}{a^6 + 4a^2}$

[36] a) $f(a) = \dfrac{\sqrt{a}}{2a^2 - 10a + 5}$　　　　　　b) $f(a + h) = \dfrac{\sqrt{a + h}}{2(a + h)^2 - 10(a + h) + 5}$

c) $\dfrac{f(a + h) - f(a)}{h} = \dfrac{\dfrac{\sqrt{a + h}}{2(a + h)^2 - 10(a + h) + 5} - \dfrac{\sqrt{a}}{2a^2 - 10a + 5}}{h}$

d) $f\left(\dfrac{1}{a}\right)=\dfrac{\sqrt{\left(\frac{1}{a}\right)}}{2\left(\frac{1}{a}\right)^2-10\left(\frac{1}{a}\right)+5}=\dfrac{\sqrt{\frac{1}{a}}}{\frac{2-10a+5a^2}{a^2}}=\dfrac{a^{3/2}}{2-10a+5a^2}$

e) $\dfrac{1}{f(a)}=\dfrac{1}{\frac{\sqrt{a}}{2a^2-10a+5}}=\dfrac{2a^2-10a+5}{\sqrt{a}}$

f) $f\left(a^2\right)=\dfrac{\sqrt{a^2}}{2\left(a^2\right)^2-10a^2+5}=\dfrac{|a|}{2a^4-10a^2+5}$

[37] $\dfrac{f(x+h)-f(x)}{h}=\dfrac{[2(x+h)^2-(x+h)+1]-(2x^2-x+1)}{h}=$

$\dfrac{2x^2+4xh+2h^2-x-h+1-2x^2+x-1}{h}=\dfrac{4xh+2h^2-h}{h}=4x+2h-1$

[38] $\dfrac{f(x+h)-f(x)}{h}=\dfrac{5(x+h)^2+2(x+h)-1-(x^2-2x-1)}{h}=$

$\dfrac{5x^2+10xh+5h^2+2x+2h-1-5x^2+2x+1}{h}=\dfrac{10xh+5h^2+2h}{h}=10x+5h+2$

[39] $\dfrac{f(x+h)-f(x)}{h}=\dfrac{\sqrt{x+h}-\sqrt{x}}{h}=\dfrac{\sqrt{x+h}-\sqrt{x}}{h}\cdot\dfrac{\sqrt{x+h}+\sqrt{x}}{\sqrt{x+h}+\sqrt{x}}=\dfrac{x+h-x}{h\left(\sqrt{x+h}+\sqrt{x}\right)}=\dfrac{1}{\sqrt{x+h}+\sqrt{x}}$

[40] $\dfrac{f(x+h)-f(x)}{h}=\dfrac{\frac{1}{\sqrt{2(x+h)}}-\frac{1}{\sqrt{2x}}}{h}=\dfrac{\sqrt{2x}-\sqrt{2(x+h)}}{h\left(\sqrt{2(x+h)}\sqrt{2x}\right)}=$

$\dfrac{\sqrt{2x}-\sqrt{2(x+h)}}{h\left(\sqrt{2(x+h)}\sqrt{2x}\right)}\cdot\dfrac{\sqrt{2x}+\sqrt{2(x+h)}}{\sqrt{2x}+\sqrt{2(x+h)}}=\dfrac{2x-2x-2h}{h\left(\sqrt{2(x+h)}\sqrt{2x}\right)\left(\sqrt{2x}+\sqrt{2(x+h)}\right)}=$

$\dfrac{-2}{\left(\sqrt{2(x+h)}\sqrt{2x}\right)\left(\sqrt{2x}+\sqrt{2(x+h)}\right)}$

[41] $\dfrac{f(x+h)-f(x)}{h}=\dfrac{\frac{10}{x+h}-\frac{10}{x}}{h}=\dfrac{\frac{10x-10(x+h)}{x(x+h)}}{h}=\dfrac{-10h}{hx(x+h)}=\dfrac{-10}{x(x+h)}$

[42] $\dfrac{f(x+h)-f(x)}{h}=\dfrac{\frac{5}{2(x+h)-3}-\frac{5}{2x-3}}{h}=\dfrac{\frac{10x-15-10(x+h)+15}{[2(x+h)-3](2x-3)}}{h}=$

$\dfrac{-10h}{h[2(x+h)-3](2x-3)}=\dfrac{-10}{[2(x+h)-3](2x-3)}$

[43] $dom(f)=\mathbf{R}$　　　　　　　　[44] $dom(f)=\mathbf{R}$

[45] $dom(f)=\left[-\dfrac{3}{2},\infty\right)$　　　[46] $dom(f)=[0,5)\cup(5,\infty)$

[47] $dom(f) = \left(-\infty, -\frac{1}{2}\right) \cup \left(-\frac{1}{2}, \infty\right)$ [48] $dom(f) = (-\infty, 0) \cup (0, \infty)$

[49] $dom(f) = \mathbf{R}$ [50] $dom(f) = \left(-\infty, -2\right) \cup \left(-2, 3\right) \cup \left(3, \infty\right)$

[51] Let x be the width of the rectangular garden. Then the length is $y = \dfrac{100 - 2x}{2}$.

Thus, the area $A = x\left(\dfrac{100 - 2x}{2}\right)$.

[52] Let y be the width of the enclosure. Then $x\,y = 1000$ or $y = \dfrac{1000}{x}$.

Hence the cost $C = \$2.00x + 2 \cdot \$5.00\left[\dfrac{1000}{x}\right]$.

[53] If we place the semicircle in a Cartesian coordinate system with the center at the origin, the

equation of the semicircle is $y = \sqrt{a^2 - x^2}$. Hence, the rectangle has height $y = \sqrt{a^2 - x^2}$. and

length $2x$. Thus, the area $A = 2x\sqrt{a^2 - x^2}$.

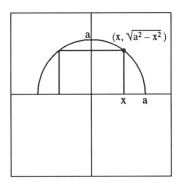

[54] $d(A, D) = \sqrt{x^2 + 4}$ and $d(D, B) = 6 - x$. If T_1 is the time required for the man to row from point

A to point D and T_2 is the time required for him to run from point D to point B, then

$$T = T_1 + T_2 = \frac{\sqrt{x^2 + 4}}{4} + \frac{6 - x}{6}.$$

[55] A, E [56] A, E [57] A, B, E [58] A, E

[59] $R = \{(x, y) \mid x$ and y are people and x and y are members of the same family$\}$.

[60] $R = \{(x, y) \mid x$ and y are integers and y is a multiple of $x\}$.

[61] $R = \{(x, y) \mid x$ and y are females and x and y are sorority sisters$\}$.

[62] Yes. Since R contains only one ordered pair, the reflexive and symmetric properties are satisfied.

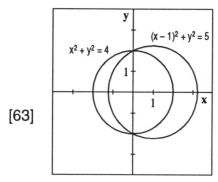

[63]

Points of intersection: $(0, -2), (0, 2)$

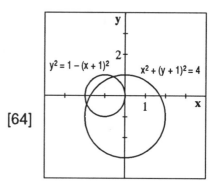

[64]

Points of intersection:
$(-1.91, -0.41), (-0.59, 0.91)$

EXERCISES 2.2

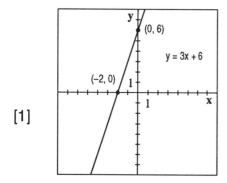

[1]

Slope 3; x-intercept $(-2, 0)$;
y-intercept $(0, 6)$

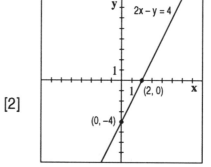

[2]

Slope 2; x-intercept $(2, 0)$;
y-intercept $(0, -4)$

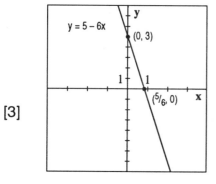

[3]

Slope -6; x-intercept $\left(\frac{5}{6}, 0\right)$;

y-intercept $(0, 5)$

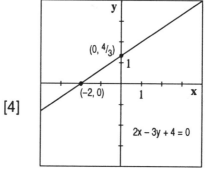

[4]

Slope $\frac{2}{3}$; x-intercept $(-2, 0)$;

y-intercept $\left(0, \frac{4}{3}\right)$

[5] $3y - 4 = 2x + 3 \Rightarrow 3y = 2x + 7 \Rightarrow$

$y = \dfrac{2}{3}x + \dfrac{7}{3}$

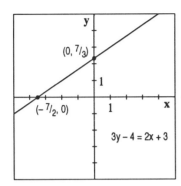

Slope $\dfrac{2}{3}$; x-intercept $\left(-\dfrac{7}{2}, 0\right)$;

y-intercept $\left(0, \dfrac{7}{3}\right)$

[6] $5x - 4 = 2y - 3 \Rightarrow 2y = 5x - 1 \Rightarrow$

$y = \dfrac{5}{2}x - \dfrac{1}{2}$

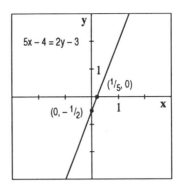

Slope $\dfrac{5}{2}$; x-intercept $\left(\dfrac{1}{5}, 0\right)$;

y-intercept $\left(0, -\dfrac{1}{2}\right)$

[7] $x = 8$

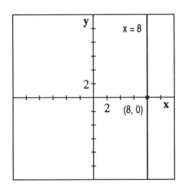

Slope undefined; x-intercept $(8, 0)$;

no y-intercept

[8] $-y = 10$

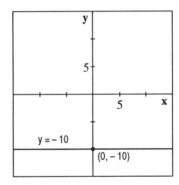

Slope 0; no x-intercept;

y-intercept $\left(0, -10\right)$

[9]   Slope $= \dfrac{-3 - 2}{4 + 1} = \dfrac{-5}{5} = -1.\ y - 2 = -1(x + 1) \Rightarrow y = -x + 1, \text{or} f(x) = -x + 1$

[10] $y + 1 = -2x \Rightarrow y = -2x - 1 \text{ or } f(x) = -2x - 1$

[11]   Since the graph of the function we seek is perpendicular to the line $3x - 2y + 1 = 0$, which has

slope $\dfrac{3}{2}$, the slope of the graph of the function is $-\dfrac{2}{3}$. Since the graph passes through $P = (-3, -1)$,

the equation is $y + 1 = -\dfrac{2}{3}(x + 3) \Rightarrow y = -\dfrac{2}{3}x - 3, \text{or} f(x) = -\dfrac{2}{3}x - 3.$

[12]   Since the graph of the function we seek is parallel to the x–axis, the slope of the graph of the

function is 0. Moreover, since the point $P = (2, -5)$ is on the graph, the equation is

$y = -5, \text{or} f(x) = -5.$

[13] The graph passes through the points $(5, 3)$ and $(-2, 7)$. Thus, the slope is $\dfrac{7-3}{-2-5} = -\dfrac{4}{7}$. Hence,

$$y - 3 = -\frac{4}{7}(x-5) \Rightarrow y = -\frac{4}{7}x + \frac{41}{7}, \text{ or } f(x) = -\frac{4}{7}x + \frac{41}{7}.$$

[14] The slope is $\dfrac{0+4}{-1-3} = -1$ since $(3, 4)$ and $(-1, 0)$ lie on the line. Thus the equation of the linear

function is $y = -1(x+1) \Rightarrow y = -x - 1$, or $f(x) = -x - 1$.

[15] The slope of the line $x - y + 4 = 0$ is 1. Since the graph contains the point $(1, -\sqrt{2})$,

$$y - 1 = 1(x + \sqrt{2}) \Rightarrow y = x + \sqrt{2} + 1, \text{ or } f(x) = x + \sqrt{2} + 1.$$

[16] The line determined by P_1 and P_2 has slope $-\dfrac{3}{7}$. Hence, the perpendicular bisector has slope $\dfrac{7}{3}$.

Moreover, the midpoint of the line segment $\overline{P_1P_2}$ is $\left(\dfrac{1}{2}, \dfrac{1}{2}\right)$. Thus, an equation of the function we

seek is $y - \dfrac{1}{2} = \dfrac{7}{3}\left(x - \dfrac{1}{2}\right) \Rightarrow y = \dfrac{7}{3}x - \dfrac{2}{3}$, or $f(x) = \dfrac{7}{3}x - \dfrac{2}{3}$.

[17]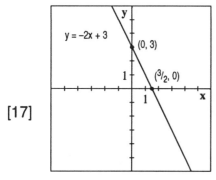

$y = -2x + 3$; x–intercept $\left(\dfrac{3}{2}, 0\right)$

[18]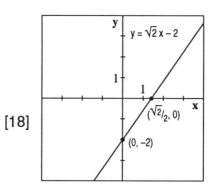

$y = \sqrt{2}x - 2$; x–intercept $\left(\dfrac{\sqrt{2}}{2}, 0\right)$

[19]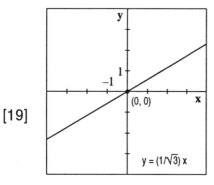

$y = \dfrac{1}{\sqrt{3}}x$; x–intercept $(0, 0)$

[20]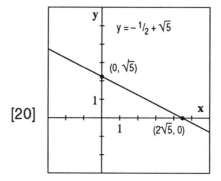

$y = -\dfrac{1}{2}x + \sqrt{5}$; x–intercept $(2\sqrt{5}, 0)$

[21]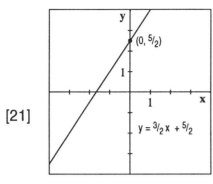

$y = \dfrac{3}{2}x + \dfrac{5}{2}$; slope $\dfrac{3}{2}$; y–intercept $\left(0, \dfrac{5}{2}\right)$

[22]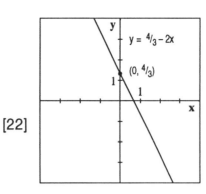

$y = \dfrac{4}{3} - 2x$; slope -2; y–intercept $\left(0, \dfrac{4}{3}\right)$

[23]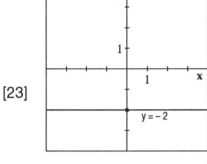

$y = -2$; slope 0; y–intercept $\left(0, -2\right)$

[24]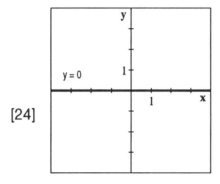

$y = 0$; slope 0; y–intercept $\left(0, 0\right)$

[25]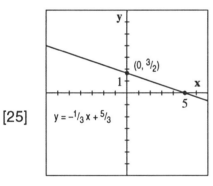

$y = -\dfrac{1}{3}x + \dfrac{5}{3}$; slope $-\dfrac{1}{3}$; y–intercept $\left(0, \dfrac{5}{3}\right)$

[26]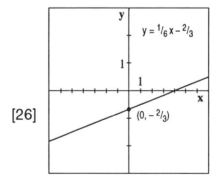

$y = \dfrac{1}{6}x - \dfrac{2}{3}$; slope $\dfrac{1}{6}$; y–intercept $\left(0, -\dfrac{2}{3}\right)$

[27] Since the point $P = \left(-2, \, -3\right)$ lies on the line $Cx + y + 9 = 0$, the coordinates of P must satisfy the equation. Hence, $-2C - 3 + 9 = 0$, or $C = 3$.

[28] The point $(-2, 0)$ lies on the line $2x + Cy + 4 = 0$. Hence, $2(-2) + C(0) + 4 = 0$, or $0 = 0$. Thus, C can be chosen arbitrarily.

[29] The slope of the line passing through $P_1 - (x_1, y_1)$ and $P_2 = (x_2, y_2)$ is $m = \dfrac{y_2 - y_1}{x_2 - x_1}$. $[x_1 \neq x_2$ since the line is nonvertical $]$. Using the point-slope formula with $P = P_1$, we have $y - y_1 = \dfrac{y_2 - y_1}{x_2 - x_1}(x - x_1)$, or $(y - y_1)(x_2 - x_1) = (y_2 - y_1)(x - x_1)$.

61

[30] a) $y - 3 = -\frac{1}{4}(x+1)$ or $y = -\frac{1}{4}x + \frac{11}{4}$. Thus, $f(x) = -\frac{1}{4}x + \frac{11}{4}$.

   b) The line has slope $\frac{1}{3\sqrt{2}}$ or $\frac{\sqrt{2}}{6}$. The equation is $y = \frac{\sqrt{2}}{6}(x - \sqrt{2})$ or $y = \frac{\sqrt{2}\,x}{6} - \frac{1}{3}$. Hence,

   $f(x) = \frac{\sqrt{2}\,x}{6} - \frac{1}{3}$.

   c) The line has slope $\frac{2}{3}$ and passes through the point $P = (1, -6)$.   Hence, an equation for the

   line is $y + 6 = \frac{2}{3}(x - 1)$ or $y = \frac{2}{3}x - \frac{20}{3}$. Thus, $f(x) = \frac{2}{3}x - \frac{20}{3}$.

[31] a) The linear function contains the ordered pairs $(250,\ 400)$ and $(550, 800\ )$. Using exercise 29

   above, we have $(R - 400)(300) = 400\ (C - 250) \Rightarrow R - 400 = \frac{4}{3}(C - 250) \Rightarrow$

   $R = \frac{4}{3}C + \frac{200}{3}$, or $R = \frac{1}{3}(4C + 200)$.

   b) $R = \frac{1}{3}[4\,(300) + 200] = \frac{1400}{3} \approx \$466.67$.

   c) $1000 = \frac{1}{3}(4C + 200) \Rightarrow 3000 = 4C + 200 \Rightarrow 2800 = 4C$ or $C = \$700$.

[32] a) If we express temperature measured in degrees Celsius (°C) as a linear function of
   temperature measued in degrees Fahrenheit (°F), the function contains the ordered pairs

   (32,0) and (212,100). Thus, the slope of the resulting line is $\frac{100}{180} = \frac{5}{9}$. Hence,

   $C - 0 = \frac{5}{9}(F - 32)$, or $C = \frac{5}{9}(F - 32)$.

   b) When $F = 100°$, $C = \frac{5}{9}(100 - 32) = \frac{5}{9}(68) = \frac{340}{9} \approx 37.8°$.

   c) If $C = -40°$, $-40 = \frac{5}{9}(F - 32) \Rightarrow -360 = 5F - 160 \Rightarrow 5F = -200$, or $F = -40°$.

[33] a) The function contains the ordered pairs $(0,\ 100{,}000\ )$ and $(20,\ 0\ )$. If we assume that the
   value V of the rental property is a linear function of the time t, we know that the slope of

   the graph of V is $-\frac{100000}{20} = -5000$. Thus $V(t) = -5000(t - 20)$, or $V(t) = -5000t + 100{,}000$.

   b) $42{,}500 = -5000\,t + 100000 \Rightarrow -5000\,t = -57500$, or $t = 11.5$ years.

[34] The desired linear function contains the ordered pairs $(50,\ 60)$ and $(80, 100\ )$. Thus, an equation

   for this function is $(y - 60)(80 - 50) = (100 - 60)(x - 50) \Rightarrow 30\,(y - 60) = 40\,(x - 50) \Rightarrow$

   $y = \frac{4}{3}(x - 50) + 60 \Rightarrow y = \frac{4}{3}(x - 5)$.

[35] $15 = 5k$ or $k = 3$. Thus, the force function is $f(x) = 3x$. Hence, $f(2) = 6$ lb.

[36] a) $S = 500 + 200n$ b) $dom(S) = \{n \mid n \text{ is a non-negative integer}\}$

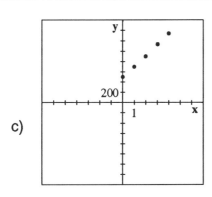

c)

[37]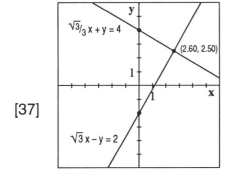

[38]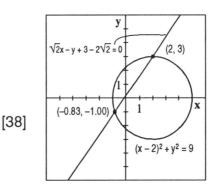

EXERECISES 2.3

[1]   a)   The equation $f(x) = 2(x+1)^2 - 3$ has the form of equation (4) on page 84 with $a = 2 > 0$ and $h = 1$. Thus, the parabola opens upward. The minimum value of f is $f(-h) = f(-1) = -3$.

      b)   The vertex is $(-1, -3)$. The axis of symmetry is $x = -1$. To determine the x-intercepts, we solve the equation $f(x) = 0$ as follows: $2(x+1)^3 - 3 = 0 \Rightarrow 2x^2 + 4x - 1 = 0 \Rightarrow x = \dfrac{-2 \pm \sqrt{6}}{2} \Rightarrow$ the x–intercepts $\left(\dfrac{-2+\sqrt{6}}{2}, 0\right)$ and $\left(\dfrac{-2-\sqrt{6}}{2}, 0\right)$. The y–intercept is $(0, f(0)) = (0, -1)$.

      c) $ran(f) = [-3, \infty)$ (See graph below).

[2]   a)   The equation $f(x) = -3(x-1)^2 + 2$ has the form of equation (4) on page 84 with $a = -3 < 0$ and $h = -1$. Thus, the parabola opens downward. The maximum value of f is $f(-h) = f(1) = 2$.

      b)   The vertex is $(1, 2) \Rightarrow$ the axis of symmetry is $x = 1$. To determine the x–intercepts, we solve the equation $f(x) = 0$, as follows: $-3(x-1)^2 + 2 = 0 \Rightarrow -3x^2 + 6x - 1 = 0 \Rightarrow$

$x = \dfrac{3 \pm \sqrt{6}}{3} \Rightarrow x$–intercepts are $\left(\dfrac{3+\sqrt{6}}{3}, 0\right)$ and $\left(\dfrac{3-\sqrt{6}}{3}, 0\right)$. The y–intercept is $(0, f(0)) = (0, -1)$.

      c) $ran(f) = (-\infty, 2]$.   (See graph below).

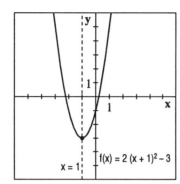

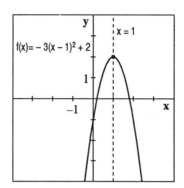

Figure 1   **Figure 2**

[3] **a)** In the equation $f(x) = 4x^2 - x + 10$, $a = 4$, $b = -1$, and $c = 10$. Since $a = 4 > 0$, the parabola

opens upward. The minimum value of f is $f\left(-\dfrac{b}{2a}\right) = f\left(\dfrac{1}{8}\right) = 4\left(\dfrac{1}{8}\right)^2 - \dfrac{1}{8} + 10 =$

$\dfrac{1}{16} - \dfrac{1}{8} + 10 = \dfrac{159}{16}$.

b) The vertex is $\left(\dfrac{1}{8}, f\left(\dfrac{1}{8}\right)\right) = \left(\dfrac{1}{8}, \dfrac{159}{16}\right) \Rightarrow$ the axis of symmetry is $x = \dfrac{1}{8}$.   Since the minimum

value of f is $\dfrac{159}{16} > 0$, there are no x–intercepts. The y–intercept is $(0, f(0)) = (0, 10)$.

c) $ran(f) = \left[\dfrac{159}{16}, \infty\right)$.   (See graph below).

[4] **a)** In the equation $f(x) = 3x^2 - x - 10$, $a = 3$, $b = -1$ and $c = -10$. Since $a = 3 > 0$, the parabola

opens upward. The minimum value of f is $f\left(-\dfrac{b}{2a}\right) = f\left(\dfrac{1}{6}\right) = 3\left(\dfrac{1}{6}\right)^2 - \dfrac{1}{6} - 10 =$

$\dfrac{1}{12} - \dfrac{1}{6} - 10 = -\dfrac{121}{12}$.

b) The vertex is $\left(\dfrac{1}{6}, f\left(\dfrac{1}{6}\right)\right) = \left(\dfrac{1}{6}, -\dfrac{121}{12}\right) \Rightarrow$ the axis of symmetry is $x = \dfrac{1}{6}$.   To determine the

x–intercepts we solve the equation $f(x) = 0$, as follows: $3x^2 - x - 10 = 0 \Rightarrow x = \dfrac{1 \pm 11}{6} \Rightarrow$

$x = -\dfrac{5}{6}$ or $x = 2 \Rightarrow x$–intercepts $\left(-\dfrac{5}{6}, 0\right)$ and $(2, 0)$. The y–intercept is $(0, f(0)) = (0, -10)$.

c) $ran(f) = \left[-\dfrac{121}{12}, \infty\right)$.   (See graph below).

[5] **a)** $f(x) - 5 = -(x + 4)^2 + 6 \Rightarrow f(x) = -(x + 4)^2 + 11$ which has the form of equation (4) on page

84 with $a = -1$ and $h = 4$. Since $a = -1 < 0$, the parabola opens downward. The maximum

value of f is $f(-h) = f(-4) = 11$.

b) The vertex is $(-4, 11) \Rightarrow$ the axis of symmetry is $x = -4$. We solve the equation $f(x) = 0$ to determine the x–intercepts: $f(x) = 0 \Rightarrow -(x+4)^2 + 11 = 0 \Rightarrow x^2 + 8x + 5 = 0 \Rightarrow$ $x = -4 \pm \sqrt{11} \Rightarrow x$–intercepts are $(-4 - \sqrt{11}, 0)$ and $(-4 + \sqrt{11}, 0)$. The y–intercept is the point $(0, f(0)) = (0, -5)$.

c) $ran(f) = (-\infty, 11]$ (See graph below).

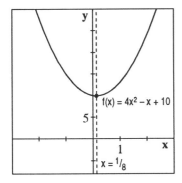

Figure 3

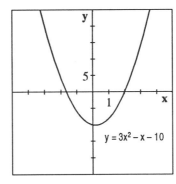

Figure 4

[6] a) The equation $f(x) = -(x-3)^2 + 5\pi$ has the form of equation (4) on page 84 with $a = -1$ and $h = -3$. Since $a = -1 < 0$, the parabola opens downward. The maximum value of f is $f(-h) = f(3) = 5\pi$.

b) The vertex is $(3, 5\pi) \approx (3, 15.71) \Rightarrow$ the axis of symmetry is $x = 3$. To determine the x–intercepts, we solve the equation $f(x) = 0$, as follows: $f(x) = 0 \Rightarrow -(x-3)^2 + 5\pi = 0 \Rightarrow$ $x^2 - 6x + 9 - 5\pi = 0 \Rightarrow x = 3 \pm \sqrt{5\pi}$. x–intercepts are $(3 + \sqrt{5\pi}, 0)$ and $(3 - \sqrt{5\pi}, 0)$.   The y–intercept is $(0, f(0)) = (0, -9 + 5\pi)$.

c) $ran(f) = (-\infty, 5\pi]$ (See graph below).

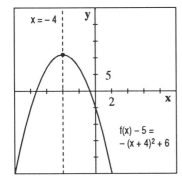

Figure 5

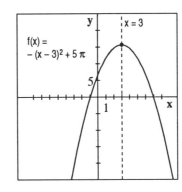

Figure 6

[7] a) In the equation $f(x) = \frac{1}{3}x^2 + \frac{1}{3}x - \frac{1}{3}$, $a = \frac{1}{3}$, $b = \frac{1}{3}$ and $c = -\frac{1}{3}$. Since $a = \frac{1}{3} > 0$, the parabola

opens upward. The minimum value of f is $f\left(-\frac{b}{2a}\right) = f\left(-\frac{1}{2}\right) = \frac{1}{3}\left(-\frac{1}{2}\right)^2 + \frac{1}{3}\left(-\frac{1}{2}\right) - \frac{1}{3} =$

$\frac{1}{12} - \frac{1}{6} - \frac{1}{3} = -\frac{5}{12}$.

b) The vertex is $\left(-\frac{1}{2}, -\frac{5}{12}\right) \Rightarrow$ the axis of symmetry is $x = -\frac{1}{2}$. To determine the x–intercepts,

we solve the equation $f(x) = 0$, as follows: $f(x) = 0 \Rightarrow \frac{1}{3}x^2 + \frac{1}{3}x - \frac{1}{3} = 0 \Rightarrow$

$x^2 + x - 1 = 0 \Rightarrow x = \frac{-1 \pm \sqrt{5}}{2} \Rightarrow$ the x–intercepts are $\left(\frac{-1+\sqrt{5}}{2}, 0\right)$ and $\left(\frac{-1-\sqrt{5}}{2}, 0\right)$. The

y–intercept is the point $(0, f(0)) = \left(0, -\frac{1}{3}\right)$.

c) $ran(f) = \left[-\frac{5}{12}, 0\right)$ (See graph below).

[8] a) In the equation $f(x) = \sqrt{2}x^2 + 5$, $a = \sqrt{2}$, $b = 0$ and $c = 5$. Since $a = \sqrt{2} > 0$, the parabola opens

upward. The minimum value of f is $f\left(-\frac{b}{2a}\right) = f(0) = 5$.

b) The vertex is $(0, 5) \Rightarrow$ the axis of symmetry is $x = 0$ (the y–axis). Since the minimum value

of f is $5 > 0$, there are no x–intercepts. The y–intercept is $(0, f(0)) = (0, 5)$ (the vertex).

c) $ran(f) = [5, \infty)$. (See graph below).

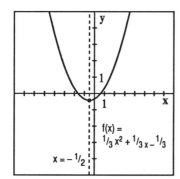

Figure 7   **Figure 8**

[9] a) In the equation $f(x) = -9x^2 + 12x - 4$, $a = -9$, $b = 12$ and $c = -4$. Since $a = -9 < 0$, the

parabola opens downward. The maximum value of f is $f\left(-\frac{b}{2a}\right) = f\left(\frac{2}{3}\right) =$

$-9\left(\frac{2}{3}\right)^2 + 12\left(\frac{2}{3}\right) - 4 = -4 + 8 - 4 = 0$.

b) The vertex is $\left(\frac{2}{3}, 0\right) \Rightarrow$ the axis of symmetry is $x = \frac{2}{3}$ and the only x–intercept is $\left(\frac{2}{3}, 0\right)$.

The y–intercept is $(0, f(0)) = (0, -4)$.

c) $ran(f) = (-\infty, 0]$. (See graph below).

[10] a) $f(x) = 2x^2 + \sqrt{2}\,x$, $a = 2$, $b = \sqrt{2}$ and $c = 0$. Since $a = 2 > 0$, the parabola opens upward.

The minimum value of f is $f\left(-\dfrac{b}{2a}\right) = f\left(\dfrac{-\sqrt{2}}{4}\right) = 2\left(-\dfrac{\sqrt{2}}{4}\right)^2 + \sqrt{2}\left(-\dfrac{\sqrt{2}}{4}\right) = \dfrac{1}{4} - \dfrac{1}{2} = -\dfrac{1}{4}$.

b) The vertex is $\left(-\dfrac{\sqrt{2}}{4}, -\dfrac{1}{4}\right) \Rightarrow$ the axis of symmetry is $x = -\dfrac{\sqrt{2}}{4}$. To determine the

x–intercepts, we solve the equation $f(x) = 0$, as follows: $f(x) = 0 \Rightarrow 2x^2 + \sqrt{2}\,x = 0 \Rightarrow$

$x = 0$ or $x = -\dfrac{\sqrt{2}}{2} \Rightarrow$ x–intercepts are $\left(-\dfrac{\sqrt{2}}{2}, 0\right)$ and $(0, 0)$. The y–intercept is $\left(0, f(0)\right) = (0, 0)$.

c) $ran(f) = \left[-\dfrac{1}{4}, \infty\right)$.

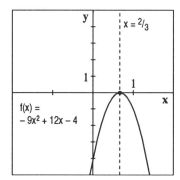

Figure 9   **Figure 10**

[11] a) $3[f(x) + 1] = (x - 8)^2 \Rightarrow f(x) = \dfrac{1}{3}(x - 8)^2 - 1$. The latter equation has the form of equation

(4) on page 84 with $a = \dfrac{1}{3}$ and $h = -8$. Since $a = \dfrac{1}{3} > 0$, the parabola opens upward. The

minimum value of f is $f(-h) = f(8) = -1$.

b) The vertex is $(8, -1) \Rightarrow$ the axis of symmetry is $x = 8$. To determine the x–intercepts, we

solve the equation $f(x) = 0$, as follows: $f(x) = 0 \Rightarrow \dfrac{1}{3}(x - 8)^2 - 1 = 0 \Rightarrow x^2 - 16x + 61 = 0 \Rightarrow$

$x = \dfrac{16 \pm \sqrt{12}}{2} = 8 \pm \sqrt{3} \Rightarrow$ x–intercepts $(8 - \sqrt{3}, 0)$ and $(8 + \sqrt{3}, 0)$. The y–intercept is

$\left(0, f(0)\right) = \left(0, \dfrac{61}{3}\right)$.

c) $ran(f) = [-1, \infty)$. (See graph below).

[12] a) $\dfrac{1}{2}[f(x) - 4] = (x + 3)^2 \Rightarrow f(x) = 2(x + 3)^2 + 4$. The latter equation has the form of equation

(4) on page 84 with $a = 2$ and $h = 3$. Since $a = 2 > 0$, the parabola opens upward. The

minimum value of f is $f(-h) = f(-3) = 4$.

b) The vertex is $(-3, 4) \Rightarrow$ the axis of symmetry is $x = -3$. Since the minimum of f is $4 > 0$,

there are no x–intercepts. The y–intercept is $\left(0, f(0)\right) = (0, 22)$.

c) $ran(f) = [4, \infty)$. (See graph below).

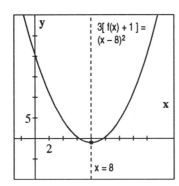

Figure 11

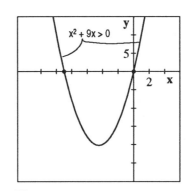

Figure 12

[13]   We graph the equation $y = x^2 - 4x$ shown below. The x–intercepts of this parabola are $(0, 0)$ and $(4, 0)$. Thus, the solution set for the inequality $x^2 - 4x < 0$ is the open interval $(0, 4)$.

● $S = (0, 4)$

[14]   The graph of $y = x^2 + 9x$ is shown below. The x–intercepts of this parabola are $(-9, 0)$ and $(0, 0)$. Thus, the solution set is $(-\infty, -9) \cup (0, \infty)$.

● $S = (-\infty, -9) \cup (0, \infty)$

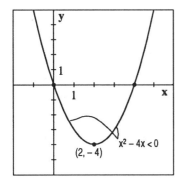

Figure 13

Figure 14

[15]   The associated equation for the inequality $x^2 + 2x \geq 3$ and the equivalent inequality $x^2 + 2x - 3 \geq 0$ is $y = x^2 + 2x - 3$. The graph of this function is the parabola shown below. The x–intercepts are $(-3, 0)$ and $(1, 0)$. Thus,   the solution set for the inequality $x^2 + 2x \geq 3$ is $(-\infty, -3] \cup [1, \infty)$.

● $S = (-\infty, -3] \cup [1, \infty)$

[16]   The associated equation for the inequality $x^2 + 16 \geq 0$ is $y = x^2 + 16$ which is the parabola shown below. Since $y \geq 0$ for all real numbers x, the solution set is **R**.

● $S = \mathbf{R}$

[17]   The associated equation for the inequality $x^2 + 10x + 25 > 0$ is $y = x^2 + 10x + 25$, whose graph is shown below. The solution set is $(-\infty, -5) \cup (-5, \infty)$.

● $S = (-\infty, -5) \cup (-5, \infty)$

[18]   The inequality $14x > x^2 + 49$ is equivalent to $x^2 - 14x + 49 < 0$. The associated equation $y = x^2 - 14x + 49$ is graphed below.   Since $y \geq 0$ for all $x \in \mathbf{R}$, the solution set is the empty set.

● $S = \varnothing$.

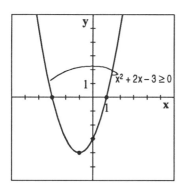

Figure 15

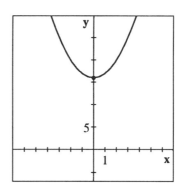

Figure 16

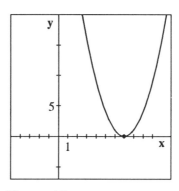

Figure 17   **Figure 18**

[19]   The inequality $-10 \geq 3x - x^2$ is equivalent to the inequality $10 + 3x - x^2 \leq 0$. The graph of the

associated equation $y = 10 + 3x - x^2$ is shown below. The x–intercepts are $(-2, 0)$ and $(5, 0)$.

Thus, the solution set for the inequality $-10 > 3x - x^2$ is $(-\infty, -2] \cup [5, \infty)$.   ● $S = (-\infty, -2] \cup [5, \infty)$

[20]   The inequality $x^2 + 2x \leq 24$ is equivalent to $x^2 + 2x - 24 \leq 0$. The associated equation

$y = x^2 + 2x - 24$ is graphed below. The x–intercepts are $(-6, 0)$ and $(4, 0)$. Hence, the solution

set is $[-6, 4]$.   ● $S = [-6, 4]$

[21]   The x–coordinates of the points of intersection of the graph of $f(x) = x^2 - 2x + 5$ and

$g(x) = 11 + 2x - x^2$ satisfy the equation $x^2 - 2x + 5 = 11 + 2x - x^2$. Solving for x we have

$x^2 - 2x + 5 = 11 + 2x - x^2 \Rightarrow 2x^2 - 4x - 6 = 0 \Rightarrow x^2 - 2x - 3 = 0 \Rightarrow (x - 3)(x + 1) = 0 \Rightarrow x = 3$ or

$x = -1$. Thus, the points of intersection are $(3, f(3)) = (3, 8)$ and $(-1, f(-1)) = (-1, 8)$.   ● $(3, 8), (-1, 8)$

[22]   In the equation $y = f(x) = x^2 - 4x + 5$, $a = 1$ and $b = -4$. Hence, the vertex for this parabola is the

point $V = \left(-\dfrac{b}{2a}, f\left(-\dfrac{b}{2a}\right)\right) = (2, f(2)) = (2, 1)$.   Thus, the distance from the origin $O = (0, 0)$ to V is

$d(O, V) = \sqrt{(2 - 0)^2 + (1 - 0)^2} = \sqrt{5}$.   ● $\sqrt{5}$

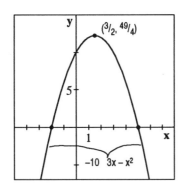

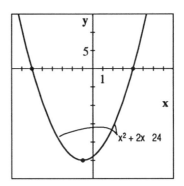

Figure 19   **Figure 20**

[23]   Since the function is quadratic, it has defining equation $f(x) = ax^2 + bx + c$ where $a \neq 0$.   Since the

graph passes through the origin, $f(0) = a(0)^2 + b(0) + c = 0 \Rightarrow c = 0$, and $f(x) = ax^2 + bx$. Since the

vertex is $(4, 4)$, $-\dfrac{b}{2a} = 4$, or $8a = -b$. In addition, $f\left(-\dfrac{b}{2a}\right) = 4 \Rightarrow a\left(-\dfrac{b}{2a}\right)^2 + b\left(-\dfrac{b}{2a}\right) = 4 \Rightarrow$

$\dfrac{b^2}{4a} - \dfrac{b^2}{2a} = 4 \Rightarrow -\dfrac{b^2}{4a} = 4 \Rightarrow b^2 = -16a$. Since $8a = -b$, $b^2 = 2b \Rightarrow b^2 - 2b = 0 \Rightarrow b(b-2) = 0$.

Thus $b = 0$ or $b = 2$. If $b = 0$, then $a = 0$. This is a contradiction, since $a \neq 0$. Therefore

$b = 2$ and $a = -\dfrac{1}{4}$. Hence, $f(x) = -\dfrac{1}{4}x^2 + 2x$.

[24]   Since there is only one x–intercept, this intercept must be the vertex. Since the line of symmetry is

the line $x = -2$, the vertex is $(-2, 0)$.   We know that the graph of the parabola is symmetric with

respect to its axis of symmetry $x = -2$. Thus, since the point $(0, 4)$ is on   the graph, so is the

point $(-4, 4)$. An equation for f has the form $f(x) = ax^2 + bx + c$ where $a \neq 0$. The three points

$(-2, 0), (0, 4)$ and $(-4, 4)$ must each satisfy this equation. For the point $(-2, 0)$, we have

$a(-2)^2 + b(-2) + c = 0$, or $4a - 2b + c = 0$ $\boxed{1}$. For the point $(0, 4)$, we have

$a(0)^2 + b(0) + c = 4$ $\boxed{2}$. From equation $\boxed{2}$, we get $c = 4$. Thus, equation $\boxed{1}$ gives us

$4a - 2b + 4 = 0$, or $2a - b + 2 = 0$ $\boxed{3}$. Using the point $(-4, 4)$, we get $a(-4)^2 + b(-4) + 4 = 4 \Rightarrow$

$16a - 4b + 4 = 4 \Rightarrow 16a - 4b = 0$, or $4a - b = 0$ $\boxed{4}$. From equation $\boxed{3}$ it follows that $b = 2a + 2$.

Substituting $2a + 2$ for b in equation $\boxed{4}$ yields $4a - (2a + 2) = 0 \Rightarrow 2a - 2 = 0 \Rightarrow 2a = 2 \Rightarrow a = 1$.

Hence, $b = 2(1) + 2 = 4$. Therefore, $f(x) = x^2 + 4x + 4$.

[25] Since the vertex of the parabola is $V = (-1, 4)$, the line $x = -1$ is the axis of symmetry. Since the graph of the parabola is symmetric with respect to its axis of symmetry, and since one x–intercept is the point $(2, 0)$, the second x–intercept is the point $(-4, 0)$. We know that $f(x) = ax^2 + bx + c$, where $a \neq 0$. Since $f(2) = 0$, $a(2)^2 + b(2) + c = 0 \Rightarrow 4a + 2b + c = 0 \Rightarrow c = -4a - 2b$ $\boxed{1}$. Since $f(-1) = 4$, $a(-1)^2 + b(-1) + c = 4 \Rightarrow a - b + c = 4$. Thus, $c = 4 - a + b$ $\boxed{2}$. Therefore, by equation $\boxed{1}$ and equation $\boxed{2}$, $-4a - 2b = 4 - a + b \Rightarrow 3a + 3b = -4$ $\boxed{3}$. Since $f(-4) = 0$, $a(-4)^2 + b(-4) + c = 0$. Thus, $16a - 4b + c = 0$ $\boxed{4}$. Substituting $-4a - 2b$ for c in equation $\boxed{4}$ and simplifying, we get $16a - 4b - 4a - 2b = 0 \Rightarrow 12a - 6b = 0 \Rightarrow 2a - b = 0$. Hence, $b = 2a$ $\boxed{5}$. Substituting $2a$ for b in $\boxed{3}$ yields $3a + 3(2a) = -4 \Rightarrow 9a = -4 \Rightarrow a = -\frac{4}{9}$. Substituting $-\frac{4}{9}$ for a in equation $\boxed{5}$ yields $b = -\frac{8}{9}$.

Using equation $\boxed{2}$, $c = 4 + \frac{4}{9} - \frac{8}{9} = \frac{32}{9}$. Thus, $f(x) = -\frac{4}{9}(x^2 + 2x - 8)$.

[26] Let x be the length of the rectangle. Then the width is $4 - x$. Hence, the area A of the rectangle is $A(x) = x(4 - x) = 4x - x^2$. Since the area is given by the quadratic equation $A(x) = 4x - x^2$ where $a = -1 < 0$, the maximum area is found by determining A at the x–coordinate of the vertex. Thus, the maximum area is $A\left(-\frac{b}{2a}\right) = A(2) = 4(2) - (2)^2 = 8 - 4 = 4$.

● 4

[27] Profit = Revenue − Expenses. Suppose the rent per unit is $r = 350 + 25n$, where n is the number of $25 increases in the rent. Thus $100 - n$ units are rented. Hence, the revenue generated is $R = (350 + 25n)(100 - n) = 35000 + 2150n - 25n^2$. Now the expenses incurred in renting $100 - n$ apartments are $E = 15n + 50(100 - n) = 5000 - 35n$. Hence, the profit is $P = R - E = 35000 + 2150n - 25n^2 - (5000 - 35n) = 30000 + 2185n - 25n^2$. Since the graph of the profit function is a parabola which opens downward, the maximum profit occurs at the x coordinate of the vertex which is $-\frac{b}{2a} = \frac{2185}{50} = 43.7$. Thus, we choose $n = 44$. Hence, the rent to be charged to maximize profit is $350 + 25(44) = 350 + 1100 = \$1450$. ● $1450

[28] Suppose $x + y = S$. Then $y = S - x$, and $P = xy = x(S - x) = Sx - x^2$. Hence, the maximum value of P occurs at $x = \frac{S}{2}$. Since $y = S - x$, $y = S - \frac{S}{2} = \frac{S}{2}$.

[29] a) The graph of the height function $h(t) = -16t^2 + 30t + 8$ is a parabola which opens downward. Thus, the maximum height obtained by the shot is the y–coordinate of the vertex of the

parabola, which is $h\left(\dfrac{15}{16}\right) = -16\left(\dfrac{15}{16}\right)^2 + 30\left(\dfrac{15}{16}\right) + 8 = -\dfrac{225}{16} + \dfrac{450}{16} + \dfrac{128}{16} = \dfrac{353}{16} \approx 22.06$ ft.

● 22.06 feet

b) To find how long the shot is airborne, we find the time t when $h(t)$, the height of the shot,

is 0. Thus, we must determine the solution of the equation $h(t) = 0 \Rightarrow -16t^2 + 30t + 8 = 0 \Rightarrow$

$8t^2 - 15t - 4 = 0 \Rightarrow t = \dfrac{15 \pm \sqrt{353}}{16}$. Since $\dfrac{15 - \sqrt{353}}{16} < 0$, we choose $t = \dfrac{15 + \sqrt{353}}{16} \approx 2.11$.

● 2.11 sec

[30] a) Let us introduce a coordinate system with the origin at the left end of the bridge as shown in

the figure below. The shortest distance from the bridge to the cable is $h\left(-\dfrac{b}{2a}\right) = h\left(\dfrac{8}{5}\right) =$

$\dfrac{1}{4}\left(\dfrac{8}{5}\right)^2 - \dfrac{4}{5}\left(\dfrac{8}{5}\right) + 10 = \dfrac{234}{25} \approx 9.36$ meters.
● 9.36 meters

b) The length of the bridge is $2\left(\dfrac{8}{5}\right) = \dfrac{16}{5} = 3.2$ meters
● 3.2 meters

[31] a) Since $x = 12,000 - 40p$ and $C(x) = 200,000 + 50x$, $C(p) = 200,000 + 50(12000 - 40p) =$
200,000 $+ 600,000 - 2000p = 800,000 - 2000p$.

b) $R(p) = (12,000 - 40p)p = 12,000p - 40p^2$

d) \$72.00 and \$278.00.

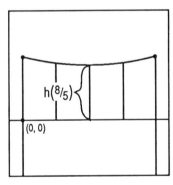

Figure 30

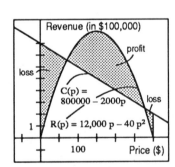

Figure 31 (c)

[32] We introduce a coordinate system with the origin located at the vertex of the parabola as shown in the figure below. The equation of the central parabola has the form $y = ax^2 + bx + c$ where $a \neq 0$. However since the vertex $(0, 0)$ lies on this parabola, $0 = a(0)^2 + b(0) + c$. Hence, $c = 0$, and thus $y = ax^2 + bx$. Moreover, since the point $(-200, 200)$ lies on the parabola $200 = a(-200)^2 + b(-200) \Rightarrow$ $200 = 40000a - 200b \Rightarrow 1 = 200a - b$. Also, since $(200, 200)$ lies on the parabola

$200 = a(200)^2 + b(200) \Rightarrow 1 = 200a + b$. It follows that $a = \dfrac{1}{200}$, $b = 0$ and $y = \dfrac{x^2}{200} \Rightarrow$ length of

strut is $\dfrac{(150)^2}{200} = 112.5$.
● 112.5 feet

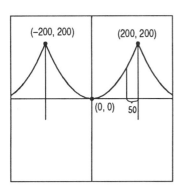

Figure 32

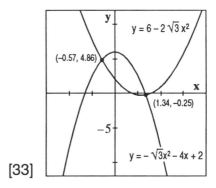

[33]

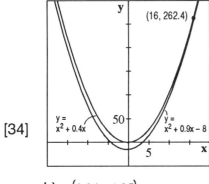

[34]

b) $(0.94, -1.35)$

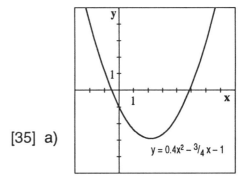

[35] a)

c) $S = [-0.90, \ 2.77]$

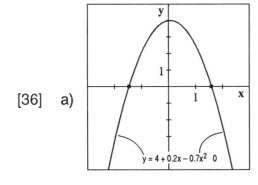

[36] a)

b) $(0.14, 4.01)$ c) $S = (-\infty, -2.25] \cup [2.54, \infty)$

EXERCISES 2.4

[1] $f(-x) = 4(-x) = -4x = -f(x)$. Thus f is odd and the graph is symmetric with respect to the origin.

[2] $f(-x) = 3(-x) - 4 = -3x - 4$. Thus $f(-x) \neq f(x)$ and $f(-x) \neq -f(x) \Rightarrow f$ is neither even nor odd.

[3] $f(-x) = (-x)^2 - 5(-x) + 2 = x^2 + 5x + 2$. Thus, $f(-x) \neq -f(x)$ and $f(-x) \neq f(x)$. Therefore f is neither even nor odd.

[4] $f(-x) = 3(-x)^5 - 2(-x)^4 = -3x^5 - 2x^4$. Thus $f(-x) \neq f(x)$ and $f(-x) \neq -f(x) \Rightarrow f$ is neither even nor odd.

[5] $f(-x) = \dfrac{1}{(-x)^6} = \dfrac{1}{x^6} = f(x)$. Thus, f is   even and its graph is symmetric with respect to the y–axis.

[6] $f(-x) = \dfrac{(-x)^2}{(-x)^2 + 4} = \dfrac{x^2}{x^2 + 4} = f(x)$. Hence, f is even and its graph is symmetric with respect to the

   y–axis.

[7] $f(-x) = \sqrt{-x + 2}$. Therefore, $f(-x) \neq f(x)$ and $f(-x) \neq -f(x)$. Thus f is neither even nor odd.

[8] $f(-x) = -x + \dfrac{1}{(-x)^2} = -x + \dfrac{1}{x^2}$. Hence, $f(-x) \neq f(x)$ and $f(-x) \neq -f(x)$ and f is neither even nor odd.

[9] $f(-x) = |-x - 3|$. Therefore, $f(-x) \neq f(x)$ and $f(-x) \neq -f(x)$. Hence, f is neither even nor odd.

[10] $f(-x) = (-x)^5 - 2(-x)^{1/5} = -\left(x^5 - 2x^{1/5}\right) = -f(x)$. Hence, f is odd and its graph is symmetric with respect to the origin.

[11]   a function; domain, **R**; range, $(-\infty, -1]$ [12]   a function; domain, **R**; range, $[1, \infty)$

[13]   not a function; domain, **R**; range, $(-2, \infty)$

[14]   not a function; domain, **R**; range, $(-\infty, -1] \cup [1, \infty)$

[15]   a function; domain, $(-\infty, 0] (0, \infty)$; range, **R**; an even function

[16]   not a function; domain, $[-3, 3]$; range, $[0, 6]$

[17]   To obtain the graph of $f(x) = x^2 - 4$, we translate the graph of $g(x) = x^2$ downward four units. (See graph below).

[18]   To obtain the graph of $f(x) = (x - 3)^2$, we translate the graph of $g(x) = x^2$ three units to the right. (See graph below).

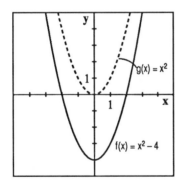

Figure 17   **Figure 18**

[19]   To obtain the graph of $f(x) = (x + 4)^2$, we translate the graph of $g(x) = x^2$ four units to the left. (See graph below).

[20]   To obtain the graph of $f(x) = -(x + 2)^2$, first we translate the graph of $g(x) = x^2$ two units to the left and then reflect the resulting graph through the x–axis. (See graph below).

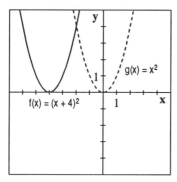

Figure 19

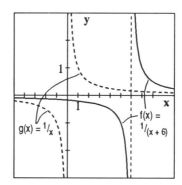

Figure 20

[21]　To obtain the graph of $f(x) = \dfrac{1}{x} + 5$ we translate the graph of $g(x) = \dfrac{1}{x}$　five units upward. (See graph below).

[22]　To obtain the graph of $f(x) = \dfrac{1}{x-6}$ we translate the graph of $g(x) = \dfrac{1}{x}$　six units to the right. (See graph below).

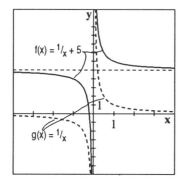

Figure 21

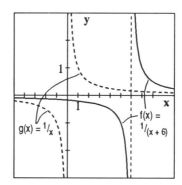

Figure 22

[23]　To obtain the graph of $f(x) = -\dfrac{1}{x} - 2$ we reflect the graph of $g(x) = \dfrac{1}{x}$ through the x–axis and translate the resulting graph two units downward. (See graph below).

[24]　$f(x) = \dfrac{1-4x}{x} = \dfrac{1}{x} - 4$. Hence, to obtain the graph of f, we translate the graph of $g(x) = \dfrac{1}{x}$ four units downward. (See graph below).

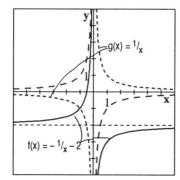

Figure 23

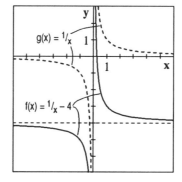

Figure 24-

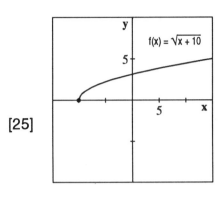

[25]

$dom\,(f) = [-10, \infty);\ ran\,(f) = [0, \infty)$

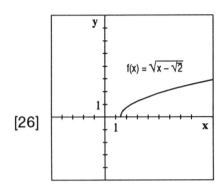

[26]

$dom\,(f) = [-\sqrt{2}, \infty);\ ran\,(f) = [0, \infty)$

[27]   Since $f(-x) = (-x)^3 + (-x) = -(x^3 + x) = -f(x), f$ is odd and its graph is symmetric with respect to the origin. (See graph and table below).

x	$f(x)$
0	0
1	2
2	10
3	30

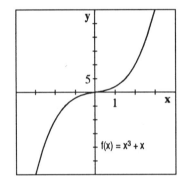

$dom\,(f) = \mathbf{R};\ ran(f) = \mathbf{R}$

[28] $f(-x) = (-x)^4 - (-x)^2 - 8 = x^4 - x^2 - 8 = f(x)$. Thus f is even and its graph is symmetric with respect to the y-axis.

x	$f(x)$
0	-8
1	-8
2	4
3	64
$\dfrac{1}{2}$	$-\dfrac{131}{16}$
$\dfrac{\sqrt{2}}{2}$	$-\dfrac{33}{4}$

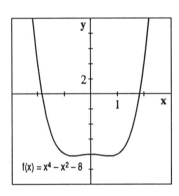

$dom\,(f) = \mathbf{R};$

$ran\,(f) = \left[-\dfrac{33}{4}, \infty\right)$

[29] $dom\,(f) = \mathbf{R};\ ran\,(f) = (-\infty, 4]$ (See graph below).

[30]   If $x \geq 0, f(x) = x + |\,x\,| = x + x = 2x$. If $x < 0, f(x) = x + |\,x\,| = x + (-x) = 0$.

Hence, $f(x) = \begin{cases} 2x, \text{if } x \geq 0 \\ 0, \text{if } x < 0 \end{cases}$. $dom\,(f) = \mathbf{R};\ ran\,(f) = [0, \infty)$. (See graph below).

[31] $dom\,(f) = \mathbf{R};\ ran\,(f) = [-3, \infty)$ (See graph below).

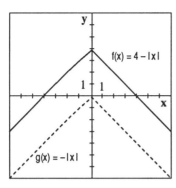

Figure 29

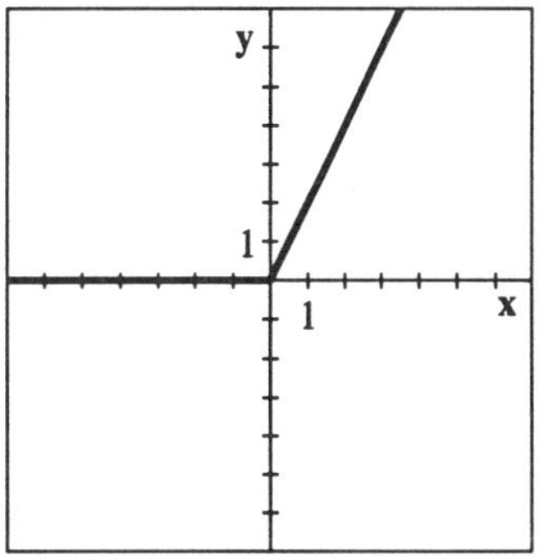

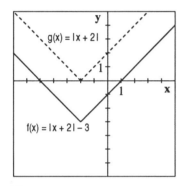

Figure 30

[32] If $x > \sqrt{3}, f(x) = \dfrac{|x - \sqrt{3}|}{x - \sqrt{3}} = \dfrac{x - \sqrt{3}}{x - \sqrt{3}} = 1.$ If $x < \sqrt{3}, f(x) = \dfrac{|x - \sqrt{3}|}{x - \sqrt{3}} = \dfrac{-(x - \sqrt{3})}{x - \sqrt{3}} = -1.$

$dom(f) = (-\infty, \sqrt{3}) \cup (\sqrt{3}, \infty); ran(f) = \{-1, 1\}.$ (See graph below).

[33] $dom(f) = \mathbf{R}; ran(f) = \mathbf{R}.$ (See graph below).

[34] $dom(f) = \mathbf{R}; ran(f) = \mathbf{R}.$ (See graph below).

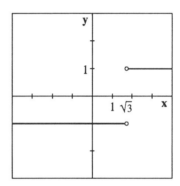

Figure 31

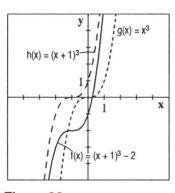

Figure 32

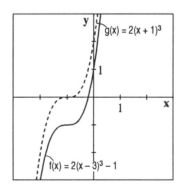

Figure 33

Figure 34

[35] $f(-x) = \dfrac{1}{(-x)^6} - 2 = \dfrac{1}{x^6} - 2 = f(x)$. Hence f is   even and its graph is symmetric with respect to the

y–axis.

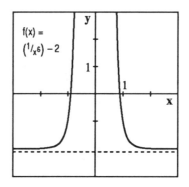

x	$\dfrac{1}{4}$	$\dfrac{1}{3}$	$\dfrac{1}{2}$	1	2
$f(x)$	4096	729	32	−1	$-\dfrac{127}{64}$

$dom(f) = (-\infty, 0) \cup (0, \infty);\ ran(f) = (-2, \infty)$

[36] $f(-x) = \dfrac{(-x)^2}{(-x)^2 + 4} = \dfrac{x^2}{x^2 + 4} = f(x)$. Thus f is even and its graph is symmetric about the y–axis.

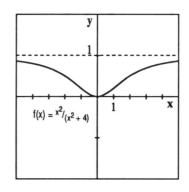

x	0	1	2	3	4	5	6
$f(x)$	0	$\dfrac{1}{5}$	$\dfrac{1}{2}$	$\dfrac{9}{13}$	$\dfrac{4}{5}$	$\dfrac{25}{29}$	$\dfrac{9}{10}$

$dom(f) = \mathbf{R};\ ran(f) = [0, 1)$

[37] $dom(f) = \mathbf{R};\ ran(f) = \{-2,\ 1\}$ (See graph below).

[38] $f(x) = \begin{cases} x+1, & \text{if } x \geq -1 \\ x, & \text{if } -2 < x < -1 \\ 1, & \text{if } x \leq -2 \end{cases}$. $dom(f) = \mathbf{R};\ ran(f) = (-2, -1) \cup [0, \infty)$.   (See graph below).

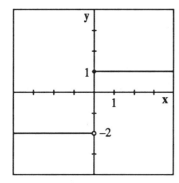

Figure 37

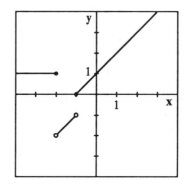

Figure 38

[39] $f(x) = \begin{cases} x^2, & \text{if } x \text{ is an integer} \\ 0, & \text{otherwise} \end{cases}$. $dom(f) = \mathbf{R};\ ran(f) = \{\, x^2 \mid x \text{ is an integer}\,\}$.   (See graph below).

[40] If $x \neq 3, f(x) = \dfrac{x^2 - 9}{x - 3} = \dfrac{(x + 3)(x - 3)}{x - 3} = x + 3.$ If $x = 3, f(3) = 2.$

$dom(f) = \mathbf{R}; ran(f) = (-\infty, 6) \cup (6, \infty).$ (See graph below).

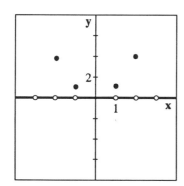

Figure 39

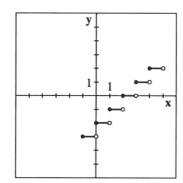

Figure 40

[41] $dom(f) = \mathbf{R}; ran(f) = \{ x \mid x \text{ is an integer}\}.$ (See graph below).

[42] Neither even nor odd as we see from the graph of Exercise 41 (below).

[43] To obtain the graph of $f(x) = [[x - 2]]$, we translate the graph of $g(x) = [[x]]$ two units to the right. (See graph below).

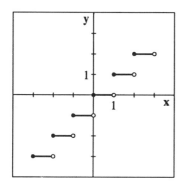

Figure 41

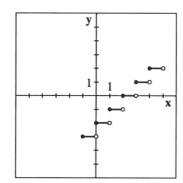

Figure 43

[44] To obtain the graph of $f(x) = [[x]] - 4$, we translate the graph of $g(x) = [[x]]$ four units downward. (See graph below).

[45] To obtain the graph of $f(x) = -2[[x]]$, we first graph the function $h(x) = 2[[x]]$ by observing that for each x, $h(x)$ is $2 \cdot g(x)$ where $g(x) = [[x]]$. We then reflect the graph of h through the x-axis to obtain the graph of f (below).

[46] n is even; n is odd.

[47] Let $f(x) = 0$. Then $f(x) = f(-x) = -f(x) = 0$. Thus, f is   even and odd. Assume that g is not the zero function and g is both even and odd. Since g is   not the zero function, there is a number $x \in dom(g)$ such that $g(x) \neq 0$. Since g is even, $g(-x) = g(x)$. Moreover, since g is odd, $g(-x) = -g(x)$. Hence $g(x) = -g(x)$. This is a contradiction since $g(x) \neq 0$. Thus, g is not both even and odd.

Figure 44   **Figure 45**

[48]   Yes, the zero function f for if x is a real number $f(x) = 0$ and $(x, 0)(x, -0)$ is on the graph
of f. Any function g different   from the zero function cannot have its graph symmetric with respect
to the x–axis as we shall now show. Suppose $g(x_0) \neq 0$. For the graph of g to be symmetric with
respect to the x–axis, the graph must contain the points $(x_0, g(x_0))$ and $(x_0, -g(x_0))$. This is
impossible since g is a function.

[49]   Let (x, y) be a point on the graph. Since the graph is symmetric with respect to the x–axis, the
point $(x, -y)$ is on the graph. In addition, since the graph is symmetric with respect to the y–axis
and since the point $(x, -y)$ is on the graph, so is the point $(-x, -y)$. Hence, if the point (x, y) is
on the graph so is the point $(-x, -y)$. Therefore, the graph is symmetric with respect to the origin.

[50]   In Exercise 49 we proved that if a graph is symmetric with respect to the x–axis and the y–axis, it
is symmetric with respect to the origin. We now prove that if a graph is symmetric with respect to
the x–axis and the origin, it is symmetric with respect to the y–axis. Suppose the point (x, y) is on
the graph. Symmetry with respect to the x–axis $\Rightarrow (x, -y)$ is on the graph. Moreover, symmetry
with respect to the origin $\Rightarrow (-x, y)$ is on the graph. Hence (x, y) on the graph $\Rightarrow (-x, y)$ on the
graph $\Rightarrow$ symmetry with respect to the y–axis.   The proof of the last case is similar.

[51]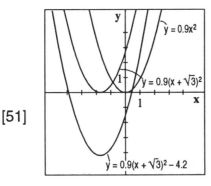

[52]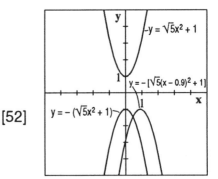

EXERCISES 2.5

[1]   a) $(f + g)(x) = f(x) + g(x) = (2 + x) + (2 - x) = 4.\ dom(f + g) = \mathbf{R}.$

   b) $(f - g)(x) = f(x) - g(x) = (2 + x) - (2 - x) = 2x.\ dom(f - g) = \mathbf{R}.$

   c) $(fg)(x) = f(x)\,g(x) = (2 + x)(2 - x) = 4 - x^2.\ dom(fg) = \mathbf{R}.$

d) $\left(\dfrac{f}{g}\right)(x) = \dfrac{f(x)}{g(x)} = \dfrac{(2+x)}{(2-x)}$. $dom\left(\dfrac{f}{g}\right) = (-\infty, 2) \cup (2, \infty)$.

[2] a) $(f+g)(x) = f(x) + g(x) = 4 - x + 3x + 2 = 2x + 6$. $dom(f+g) = \mathbf{R}$.

b) $(f-g)(x) = f(x) - g(x) = 4 - x - (3x + 2) = -4x + 2$. $dom(f-g) = \mathbf{R}$.

c) $(fg)(x) = f(x)g(x) = (4-x)(3x+2) = -3x^2 + 10x + 8$. $dom(fg) = \mathbf{R}$.

d) $\left(\dfrac{f}{g}\right)(x) = \dfrac{f(x)}{g(x)} = \dfrac{4-x}{3x+2}$. $dom\left(\dfrac{f}{g}\right) = \left(-\infty, -\dfrac{2}{3}\right) \cup \left(-\dfrac{2}{3}, \infty\right)$.

[3] a) $(f+g)(x) = f(x) + g(x) = (x^2 - x + 15) + (6 - x^2) = -x + 21$. $dom(f+g) = \mathbf{R}$.

b) $(f-g)(x) = f(x) - g(x) = (x^2 - x + 15) - (6 - x^2) = 2x^2 - x + 9$. $dom(f-g) = \mathbf{R}$.

c) $(fg)(x) = f(x)g(x) = (x^2 - x + 15)(6 - x^2) = -x^4 + x^3 - 9x^2 - 6x + 90$. $dom(fg) = \mathbf{R}$.

d) $\left(\dfrac{f}{g}\right)(x) = \dfrac{f(x)}{g(x)} = \dfrac{x^2 - x + 15}{6 - x^2}$. $dom\left(\dfrac{f}{g}\right) = \left(-\infty, -\sqrt{6}\right) \cup \left(-\sqrt{6}, \sqrt{6}\right) \cup \left(\sqrt{6}, \infty\right)$.

[4] a) $(f+g)(x) = f(x) + g(x) = 1 - x^2 - x^3 + x^3 - 8 = -7 - x^2$. $dom(f+g) = \mathbf{R}$.

b) $(f-g)(x) = f(x) - g(x) = 1 - x^2 - x^3 - (x^3 - 8) = 9 - x^2 - 2x^3$. $dom(f-g) = \mathbf{R}$.

c) $(fg)(x) = f(x)g(x) = (1 - x^2 - x^3)(x^3 - 8) = -8 + 8x^2 + 9x^3 - x^5 - x^6$. $dom(fg) = \mathbf{R}$.

d) $\left(\dfrac{f}{g}\right)(x) = \dfrac{f(x)}{g(x)} = \dfrac{1 - x^2 - x^3}{x^3 - 8}$. $dom\left(\dfrac{f}{g}\right) = (-\infty, 2) \cup (2, \infty)$.

[5] a) $(f+g)(x) = f(x) + g(x) = \dfrac{x+4}{5} + \dfrac{2}{x} = \dfrac{x^2 + 4x + 10}{5x}$. $dom(f+g) = (-\infty, 0) \cup (0, \infty)$.

b) $(f-g)(x) = f(x) - g(x) = \dfrac{x+4}{5} - \dfrac{2}{x} = \dfrac{x^2 + 4x - 10}{5x}$. $dom(f-g) = (-\infty, 0) \cup (0, \infty)$.

c) $(fg)(x) = f(x)g(x) = \left(\dfrac{x+4}{5}\right)\left(\dfrac{2}{x}\right) = \dfrac{2x+8}{5x}$. $dom(fg) = (-\infty, 0) \cup (0, \infty)$.

d) $\left(\dfrac{f}{g}\right)(x) = \dfrac{f(x)}{g(x)} = \dfrac{\dfrac{x+4}{5}}{\dfrac{2}{x}} = \dfrac{x^2 + 4x}{10}$. $dom\left(\dfrac{f}{g}\right) = (-\infty, 0) \cup (0, \infty)$.

[6] a) $(f+g)(x) = f(x) + g(x) = 4x^6 - 5x^3 + |x|$. $dom(f+g) = \mathbf{R}$.

b) $(f-g)(x) = f(x) - g(x) = 4x^6 - 5x^3 - |x|$. $dom(f-g) = \mathbf{R}$.

c) $(fg)(x) = f(x)g(x) = (4x^6 - 5x^3)|x|$. $dom(fg) = \mathbf{R}$.

d) $\left(\dfrac{f}{g}\right)(x) = \dfrac{f(x)}{g(x)} = \dfrac{4x^6 - 5x^3}{|x|}$. $\dfrac{f}{g}(x) = \begin{cases} 4x^5 - 5x^2, \text{if } x > 0 \\ -(4x^5 - 5x^2), \text{if } x < 0 \end{cases}$ $dom\left(\dfrac{f}{g}\right) = (-\infty, 0) \cup (0, \infty)$.

[7] a) $(f+g)(x) = f(x) + g(x) = x^5 + x^{1/5}$. $dom(f+g) = \mathbf{R}$.

b) $(f-g)(x) = f(x) - g(x) = x^5 - x^{1/5}$. $dom(f-g) = \mathbf{R}$.

c) $(fg)(x) = f(x)g(x) = (x^5)(x^{1/5}) = x^{26/5}$. $dom(fg) = \mathbf{R}$.

d) $\left(\dfrac{f}{g}\right)(x) = \dfrac{f(x)}{g(x)} = \dfrac{x^5}{x^{1/5}} = x^{24/5}$. $dom\left(\dfrac{f}{g}\right) = (-\infty, 0) \cup (0, \infty)$.

[8] a) $(f+g)(x) = f(x) + g(x) = \dfrac{1}{x-1} + \dfrac{x^2-1}{x} = \dfrac{x^3 - x^2 + 1}{x^2 - x}$.

$dom(f+g)(x) = (-\infty, 0] \cup (0, 1) \cup (1, \infty)$.

b) $(f-g)(x) = f(x) - g(x) = \dfrac{1}{x-1} - \dfrac{x^2-1}{x} = \dfrac{-x^3 + x^2 + 2x - 1}{x^2 - x}$.

$dom(f-g) = (-\infty, 0) \cup (0, 1) \cup (1, \infty)$.

c) $(fg)(x) = f(x)\,g(x) = \left(\dfrac{1}{x-1}\right)\left(\dfrac{x^2-1}{x}\right) = \dfrac{x+1}{x}$. $dom(fg) = (-\infty, 0) \cup (0, 1) \cup (1, \infty)$.

d) $\left(\dfrac{f}{g}\right)(x) = \dfrac{f(x)}{g(x)} = \dfrac{\frac{1}{x-1}}{\frac{x^2-1}{x}} = \dfrac{x}{x^3 - x^2 - x + 1}$. $dom\left(\dfrac{f}{g}\right) = (-\infty, -1) \cup (-1, 0) \cup (0, 1) \cup (1, \infty)$.

[9] a) $(f+g)(x) = f(x) + g(x) = \sqrt{x+3} + |x-2|$. $dom(f+g) = [-3, \infty)$.

b) $(f-g)(x) = f(x) - g(x) = \sqrt{x+3} - |x-2|$. $dom(f+g) = [-3, \infty)$.

c) $(fg)(x) = f(x)\,g(x) = \left(\sqrt{x+3}\right)|x-2|$. $dom(fg) = [-3, \infty)$.

d) $\left(\dfrac{f}{g}\right)(x) = \dfrac{f(x)}{g(x)} = \dfrac{\sqrt{x+3}}{|x-2|}$. $dom\left(\dfrac{f}{g}\right) = [-3, 2) \cup (2, \infty)$.

[10] a) $(f+g)(x) = f(x) + g(x) = |x-5| + x^2$. $dom(f+g) = \mathbf{R}$.

b) $(f-g)(x) = f(x) - g(x) = |x-5| - x^2$. $dom(f+g) = \mathbf{R}$.

c) $(fg)(x) = f(x)\,g(x) = |x-5|\,x^2$. $dom(fg) = \mathbf{R}$.

d) $\left(\dfrac{f}{g}\right)(x) = \dfrac{f(x)}{g(x)} = \dfrac{|x-5|}{x^2}$. $dom\left(\dfrac{f}{g}\right) = (-\infty, 0) \cup (0, \infty)$.

[11] a) $(f+g)(x) = f(x) + g(x) = \dfrac{x+3}{x-2} + \dfrac{1}{x+2} = \dfrac{(x+3)(x+2) + (x-2)}{(x-2)(x+2)} = \dfrac{x^2 + 5x + 6 + x - 2}{x^2 - 4} =$

$\dfrac{x^2 + 6x + 4}{x^2 - 4}$. $dom(f+g) = (-\infty, -2) \cup (-2, 2) \cup (2, \infty)$.

b) $(f-g)(x) = f(x) - g(x) = \dfrac{x+3}{x-2} - \dfrac{1}{x+2} = \dfrac{(x+3)(x+2) - (x-2)}{(x-2)(x+2)} = \dfrac{x^2 + 5x + 6 - x + 2}{x^2 - 4} =$

$\dfrac{x^2 + 4x + 8}{x^2 - 4}$. $dom(f-g) = (-\infty, -2) \cup (-2, 2) \cup (2, \infty)$.

c) $(fg)(x) = f(x)\,g(x) = \left(\dfrac{x+3}{x-2}\right)\left(\dfrac{1}{x+2}\right) = \dfrac{x+3}{x^2 - 4}$. $dom(fg) = (-\infty, -2) \cup (-2, 2) \cup (2, \infty)$.

d) $\left(\dfrac{f}{g}\right)(x) = \dfrac{f(x)}{g(x)} = \dfrac{\frac{x+3}{x-2}}{\frac{1}{x+2}} = \dfrac{(x+3)(x+2)}{x-2} = \dfrac{x^2+5x+6}{x-2}.$ $dom\left(\dfrac{f}{g}\right) = (-\infty, -2) \cup (-2, 2) \cup (2, \infty).$

[12] a) $(f+g)(x) = f(x) + g(x) = \dfrac{1}{\sqrt{x^2-5}} + \sqrt{x^2-5} = \dfrac{x^2-4}{\sqrt{x^2-5}}.$ $dom(f+g) = \left(-\infty, -\sqrt{5}\right) \cup \left(\sqrt{5}, \infty\right).$

b) $(f-g)(x) = f(x) - g(x) = \dfrac{1}{\sqrt{x^2-5}} - \sqrt{x^2-5} = \dfrac{6-x^2}{\sqrt{x^2-5}}.$ $dom(f-g) = \left(-\infty, -\sqrt{5}\right) \cup \left(\sqrt{5}, \infty\right).$

c) $(fg)(x) = f(x)\,g(x) = \left(\dfrac{1}{\sqrt{x^2-5}}\right)\left(\sqrt{x^2-5}\right) = 1.\; dom(fg)(x) = \left(-\infty, -\sqrt{5}\right) \cup \left(\sqrt{5}, \infty\right).$

d) $\left(\dfrac{f}{g}\right)(x) = \dfrac{f(x)}{g(x)} = \dfrac{\frac{1}{\sqrt{x^2-5}}}{\sqrt{x^2-5}} = \dfrac{1}{x^2-5}.\; dom\left(\dfrac{f}{g}\right)(x) = \left(-\infty, -\sqrt{5}\right) \cup \left(\sqrt{5}, \infty\right).$

[13] a) $(f+g)(1) = f(1) + g(1) = 4 - 3 = 1$ 　　b) $(f-g)(1) = f(1) - g(1) = 4 - (-3) = 7$

c) $(fg)(1) = f(1)\,g(1) = 4(-3) = -12$ 　　d) $\left(\dfrac{f}{g}\right)(1) = \dfrac{f(1)}{g(1)} = \dfrac{4}{-3} = -\dfrac{4}{3}$

[14] a) $(f+g)(-2) = f(-2) + g(-2) = 3 + (-5) = -2$

b) $\left(\dfrac{f-h}{g}\right)(-2) = \dfrac{(f-h)(-2)}{g(-2)} = \dfrac{f(-2) - h(-2)}{g(-2)} = \dfrac{3 - (-2)}{-5} = -1$

c) $\left(\dfrac{fg}{h-g}\right)(-2) = \dfrac{(fg)(-2)}{(h-g)(-2)} = \dfrac{f(-2)\,g(-2)}{h(-2) - g(-2)} = \dfrac{3(-5)}{-2 - (-5)} = \dfrac{-15}{3} = -5$

d) $\left(\dfrac{f-h}{g-h}\right)(-2) = \dfrac{(f-h)(-2)}{(g-h)(-2)} = \dfrac{f(-2) - h(-2)}{g(-2) - h(-2)} = \dfrac{3 - (-2)}{-5 - (-2)} = -\dfrac{5}{3}$

[15] a) $(f \circ g)(x) = f[g(x)] = 4 - g(x) = 4 - (5 + 2x) = -1 - 2x.$
$(g \circ f)(x) = g[f(x)] = 5 + 2f(x) = 5 + 2(4 - x) = 5 + 8 - 2x = 13 - 2x.$

b) $dom(f \circ g) = dom(g \circ f) = \mathbf{R}.$

[16] a) $(f \circ g)(x) = f[g(x)] = \sqrt{5} + g(x) = \sqrt{5} + 10 - \pi x$
$(g \circ f)(x) = g[f(x)] = 10 - \pi f(x) = 10 - \pi\left(\sqrt{5} + x\right) = 10 - \sqrt{5}\,\pi - \pi x.$

b) $dom(f \circ g) = dom(g \circ f) = \mathbf{R}.$

[17] a) $(f \circ g)(x) = f[g(x)] = 4g(x) - 3 = 4(1 - x^2) - 3 = 4 - 4x^2 - 3 = 1 - 4x^2$
$(g \circ f)(x) = g[f(x)] = 1 - [f(x)]^2 = 1 - (4x - 3)^2 = 1 - (16x^2 - 24x + 9) =$
$1 - 16x^2 + 24x - 9 = -8 + 24x - 16x^2$

b) $dom(f \circ g) = dom(g \circ f) = \mathbf{R}.$

[18] a) $(f \circ g)(x) = f[g(x)] = \dfrac{g(x)}{[g(x)]^2 - 1} = \dfrac{\dfrac{x-5}{2}}{\left(\dfrac{x-5}{2}\right)^2 - 1} = \dfrac{\dfrac{x-5}{2}}{\dfrac{x^2 - 10x + 25}{4} - 1} = \dfrac{2(x-5)}{x^2 - 10x + 21}$

$(g \circ f)(x) = g[f(x)] = \dfrac{f(x) - 5}{2} = \dfrac{\dfrac{x}{x^2 - 1} - 5}{2} = \dfrac{x - 5(x^2 - 1)}{2(x^2 - 1)} = \dfrac{-5x^2 + x + 5}{2x^2 - 2}$

b) $dom(f \circ g) = (-\infty, 3) \cup (3, 7) \cup (7, \infty).\ dom(g \circ f) = (-\infty, -1) \cup (-1, 1) \cup (1, \infty).$

[19] a) $(f \circ g)(x) = f[g(x)] = \dfrac{g(x)}{g(x) + 1} = \dfrac{\dfrac{1}{x}}{\dfrac{1}{x} + 1} = \dfrac{\dfrac{1}{x}}{\dfrac{1 + x}{x}} = \dfrac{1}{1 + x}$

$(g \circ f)(x) = g[f(x)] = \dfrac{1}{f(x)} = \dfrac{1}{\dfrac{x}{x + 1}} = \dfrac{x + 1}{x}$

b) $dom(f \circ g) = (-\infty, -1) \cup (-1, 0) \cup (0, \infty).\ dom(g \circ f) = (-\infty, -1) \cup (-1, 0) \cup (0, \infty).$

[20] a) $(f \circ g)(x) = f[g(x)] = \dfrac{|\,g(x) - 1\,|}{2} = \dfrac{|\,3x^2 - 4x + 5 - 1\,|}{2} = \dfrac{|\,3x^2 - 4x + 4\,|}{2} = \dfrac{3x^2 - 4x + 4}{2}$

$(g \circ f)(x) = g[f(x)] = 3[f(x)]^2 - 4f(x) + 5 = 3\left(\dfrac{|\,x - 1\,|}{2}\right)^2 - 4\left(\dfrac{|\,x - 1\,|}{2}\right) + 5 =$

$\dfrac{3}{4}(x^2 - 2x + 1) - 4\left(\dfrac{|\,x - 1\,|}{2}\right) + 5$

b) $dom(f \circ g) = dom(g \circ f) = \mathbf{R}.$

[21] a) $(f \circ g)(x) = f[g(x)] = 6 - |\,g(x)\,| = 6 - \left|\,\dfrac{x^2 - 1}{x^2}\,\right|$

$(g \circ f)(x) = g[f(x)] = \dfrac{[f(x)]^2 - 1}{[f(x)]^2} = \dfrac{(6 - |\,x\,|)^2 - 1}{(6 - |\,x\,|)^2} = \dfrac{36 - 12|\,x\,| + x^2 - 1}{36 - 12|\,x\,| + x^2} = \dfrac{35 - 12|\,x\,| + x^2}{36 - 12|\,x\,| + x^2}$

b) $dom(f \circ g) = (-\infty, 0) \cup (0, \infty).\ dom(g \circ f) = (-\infty, -6) \cup (-6, 6) \cup (6, \infty).$

[22] a) $(f \circ g)(x) = f[g(x)] = \sqrt{6[g(x)] - 2} = \sqrt{6\left(\dfrac{1}{x + 1}\right) - 2} = \sqrt{\dfrac{6 - 2(x + 1)}{x + 1}} =$

$\sqrt{\dfrac{6 - 2x - 2}{x + 1}} = \sqrt{\dfrac{4 - 2x}{x + 1}}$

$(g \circ f)(x) = g[f(x)] = \dfrac{1}{f(x) + 1} = \dfrac{1}{\sqrt{6x - 2} + 1}$

b) $dom(f \circ g) = (-1, 2].\ dom(g \circ f) = \left(\dfrac{1}{3}, \infty\right).$

[23] a) $(f \circ g)(x) = f[g(x)] = \sqrt{g(x)} = \sqrt{x^4 + 3x^2}$

$(g \circ f)(x) = g[f(x)] = (\sqrt{x})^4 + 3(\sqrt{x})^2 = x^2 + 3x$

b) $dom(f \circ g) = \mathbf{R}.\ dom(g \circ f) = [0, \infty).$

[24] a) $(f \circ g)(x) = f[g(x)] = \dfrac{g(x)+1}{[g(x)]^2} = \dfrac{\frac{1}{x+1}+1}{\left(\frac{1}{x+1}\right)^2} = \dfrac{\frac{1+x+1}{x+1}}{\frac{1}{(x+1)^2}} = (x+2)(x+1) = x^2+3x+2.$

$(g \circ f)(x) = g[f(x)] = \dfrac{1}{f(x)+1} = \dfrac{1}{\frac{x+1}{x^2}+1} = \dfrac{1}{\frac{x+1+x^2}{x^2}} = \dfrac{x^2}{x^2+x+1}$

b) $dom(f \circ g) = (-\infty,-1) \cup (-1,\infty).\ dom(g \circ f) = (-\infty,0) \cup (0,\infty).$

[25] a) $(f \circ g \circ h)(x) = f\{g[h(x)]\} = f\left(2-\dfrac{x}{3}\right) = \left(2-\dfrac{x}{3}\right)^2 - 1 = 4 - \dfrac{4}{3}x + \dfrac{x^2}{9} - 1 = 3 - \dfrac{4}{3}x + \dfrac{x^2}{9}$

b) $(g \circ f \circ h)(x) = g\{f[h(x)]\} = g\left(\dfrac{x^2}{9}-1\right) = 2 - \dfrac{x^2}{9} + 1 = 3 - \dfrac{x^2}{9}$

c) $(g \circ h \circ f)(x) = g\{h[f(x)]\} = g\left(\dfrac{x^2-1}{3}\right) = 2 - \dfrac{x^2-1}{3} = \dfrac{6-x^2+1}{3} = \dfrac{7-x^2}{3}$

d) $(h \circ f \circ g)(x) = h\{f[g(x)]\} = h[(2-x)^2-1] = h(4-4x+x^2-1) = h(3-4x+x^2) = \dfrac{3-4x+x^2}{3}$

e) $(f \circ h \circ h)(x) = f\{h[h(x)]\} = f\left(\dfrac{x}{9}\right) = \dfrac{x^2}{81} - 1$

f) $(g \circ g \circ f)(x) = g\{g[f(x)]\} = g(3-x^2) = 2 - 3 + x^2 = x^2 - 1$

[26] a) $(f \circ g \circ h)(x) = f\{g[h(x)]\} = f\left(\dfrac{\frac{2}{x}-1}{3}\right) = f\left(\dfrac{2-x}{3x}\right) = 1 + \left(\dfrac{2-x}{3x}\right)^2 = 1 + \dfrac{4-4x+x^2}{9x^2} =$

$\dfrac{9x^2+4-4x+x^2}{9x^2} = \dfrac{10x^2-4x+4}{9x^2}$

b) $(g \circ f \circ h)(x) = g\{f[h(x)]\} = g\left[1 + \left(\dfrac{2}{x}\right)^2\right] = g\left(\dfrac{x^2+4}{x^2}\right) = \dfrac{\frac{x^2+4}{x^2}-1}{3} = \dfrac{x^2+4-x^2}{3x^2} = \dfrac{4}{3x^2}$

c) $(g \circ h \circ f)(x) = g\{h[f(x)]\} = g\left(\dfrac{2}{1+x^2}\right) = \dfrac{\frac{2}{1+x^2}-1}{3} = \dfrac{2-1-x^2}{3+3x^2} = \dfrac{1-x^2}{3+3x^2}$

d) $(h \circ f \circ g)(x) = h\{f[g(x)]\} = h\left[1 + \left(\dfrac{x-1}{3}\right)^2\right] = h\left(1 + \dfrac{x^2-2x+1}{9}\right) = h\left(\dfrac{9+x^2-2x+1}{9}\right) =$

$h\left(\dfrac{10-2x+x^2}{9}\right) = \dfrac{2}{\frac{10-2x+x^2}{9}} = \dfrac{18}{10-2x+x^2}$

[27] a) $(h \circ f \circ g)(2) = h\{f[g(2)]\} = h[f(-1)] = h(5) = -4$

b) $(g \circ h \circ f)(\sqrt{7}) = g\{h[f(\sqrt{7})]\} = g[h(-2)] = g(6) = \pi$

[28] a) $(f \circ g)(2) = f[g(2)] = f(4) = 5$ \qquad b) $(g \circ h)(-1) = g[h(-1)] = g(5) = 0$

c) $h[g(4)] = h(1) = 5$

d) $(f \circ h \circ g)(2) = f\{h[g(2)]\} = f[h(4)] = f(2) = 1$

[29] a) $(f \circ g)(4) = f[g(4)] = f(0) = 4$ b) $(g \circ f)(4) = g[f(4)] = g(25) = 4$

c) $(f \circ g)(3) = f[g(3)] = f(-5) = 3$ d) $(g \circ f)(3) = g[f(3)] = g(18) = 3$

e) $(f \circ g)(16) = f[g(16)] = f(2) = 16$ f) $(g \circ f)(-6) = g[f(-6)] = g(-4) = -6$

[30] Let $f(x) = \sqrt{x}$ and $g(x) = 2x^2 - x + 1$. Then $(f \circ g)(x) = f[g(x)] = \sqrt{g(x)} = \sqrt{2x^2 - x + 1} = h(x)$.

[31] Let $f(x) = \sqrt{x}$ and $g(x) = x + 3$. Then $(f \circ g)(x) = f[g(x)] = \sqrt{g(x)} = \sqrt{x+3} = h(x)$. ($f$ and g are not unique).

[32] Let $f(x) = |x|$ and $g(x) = 10 - x^2$. Then $(f \circ g)(x) f[g(x)] = |g(x)| = |10 - x^2| = h(x)$.

[33] Let $f(x) = \dfrac{5}{x^3}$ and $g(x) = x - 3$. Then $(f \circ g)(x) = f[g(x)] = \dfrac{5}{[g(x)]^3} = \dfrac{5}{(x-3)^3} = h(x)$.

[34] Let $f(x) = \dfrac{|x|}{2} + \sqrt{\dfrac{1}{2}x}$ and $g(x) = 2x$. Then $(f \circ g)(x) = f[g(x)] =$

$\dfrac{|2x|}{2} + \sqrt{\dfrac{1}{2}(2x)} = \dfrac{2|x|}{2} + \sqrt{x} = |x| + \sqrt{x} = h(x)$.

[35] Let $f(x) = \sqrt[5]{x}$ and $g(x) = (x - 2)^4$. Then $(f \circ g)(x) = f[g(x)] = \sqrt[5]{g(x)} = \sqrt[5]{(x-2)^4} = h(x)$.

[36] Let $f(x) = \dfrac{5|x| + 6}{|x|}$ and $g(x) = 1 - x^2$. Then $(f \circ g)(x) = f[g(x)] = \dfrac{5|g(x)| + 6}{|g(x)|} =$

$\dfrac{5|1 - x^2| + 6}{|1 - x^2|} = h(x)$.

[37] Let $f(x) = x^{-5}$ and $g(x) = \dfrac{10x}{8 - x}$. Then $(f \circ g)(x) = f[g(x)] = [g(x)]^{-5} = \left(\dfrac{10x}{8-x}\right)^{-5} = h(x)$.

[38] The function under consideration is defined by $(f + g)(x) = |x - 1| + x + 1$. If $x \geq 1$, $|x - 1| = x - 1$, and in this case, $(f + g)(x) = 2(x + 1)$. However, if $x < 1$, $|x - 1| = -(x - 1)$, and

$(f + g)(x) = 0$. Thus $(f + g)(x) = \begin{cases} 2(x+1), \text{ if } x \geq 1 \\ 0, \text{ if } x < 1 \end{cases}$

[39] If $x < 1$, $|x - 1| = -(x - 1)$. If $x \geq 1$, $|x - 1| = x - 1$. Thus, if $x < 1$, $(fg)(x) = |x - 1|(x - 1) =$

$-(x - 1)(x - 1) = -(x^2 - 2x + 1) = -x^2 + 2x - 1$. If $x \geq 1$, $(fg)(x) = |x - 1|(x - 1) =$

$(x - 1)(x - 1) = x^2 - 2x + 1$. Thus $f(x) = \begin{cases} -x^2 + 2x - 1, \text{ if } x < 1 \\ x^2 - 2x + 1, \text{ if } x \geq 1 \end{cases}$.

[40] $V(t) = \pi r^2(t) h(t)$. Thus $h(t) = \dfrac{V(t)}{\pi r^2(t)} = \dfrac{t^3 - t^2 + t + 5}{\pi(10 + 3t)^2}$

[41] $V(t) = \dfrac{4}{3}\pi r^3(t)$. $S(t) = 4\pi r^2(t)$

[42] a) $(N \circ T)(t) = (5t + 20)^3 + 10,000$.

b)   When $t = 0, (N \circ T)(0) = 20^3 + 10000 = 18000$. When $t = 1, (N \circ T)(1) =$
$25^3 + 10000 = 25625$. When $t = 2, (N \circ T)(2) = 30^3 + 10000 = 37000$.

[43]   Suppose f and g are even functions. Then $(fg)(-x) = f(-x) g(-x) = f(x) g(x) = (fg)(x)$. $\therefore fg$ is even.

[44]   Suppose f and g are odd functions. Then $(fg)(-x) = f(-x) g(-x) = [-f(x)][-g(x)] =$
$f(x) g(x) = (fg)(x)$. $\therefore fg$ is an even function.

[45]   Suppose f is odd and g is even. Then $(fg)(-x) = f(-x) g(-x) = -f(x) g(x) = $ $-(fg)(x)$. $\therefore fg$ is odd.

[46]   If $g(x) = \frac{1}{2}[f(x) + f(-x)]$ and $h(x) = \frac{1}{2}[f(x) - f(-x)]$, then $g(x) + h(x) = f(x)$. Also,

$g(-x) = \frac{1}{2}[f(-x) + f(x)] = g(x) \Rightarrow g$ is even. Now $h(-x) = \frac{1}{2}[f(-x) - f(x)] =$

$-\frac{1}{2}[f(x) - f(-x)] = -h(x) \Rightarrow h$ is odd.

EXERCISES 2.6

[1]

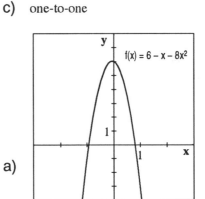

b)   decreasing on $(-\infty, \infty)$.

c)   one-to-one

[2]

b)   decreasing on $\left(-\infty, -3\right)$ and $\left(-3, \infty\right)$.

c)   one-to-one

[3]   a)

b)   increasing on $\left[-2, \infty\right)$,

       decreasing on $\left(-\infty, -2\right]$.

c)   not one-to-one

[4]   a)

b)   increasing on $\left(-\infty, -\frac{1}{16}\right]$,

       decreasing on $\left[-\frac{1}{16}, \infty\right)$,

c)   not one-to-one

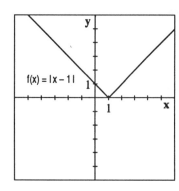

[5] a)

b)   increasing on $\left[1, \infty\right)$,

decreasing on $\left(-\infty, 1\right]$.

c)   not one-to-one

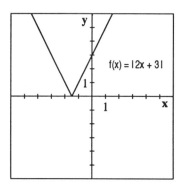

[6] a)

b)   increasing on $\left[-\frac{3}{2}, \infty\right)$,

decreasing on $\left(-\infty, -\frac{3}{2}\right]$.

c)   not one-to-one

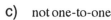

[7] a)

b)   increasing on $\left(-\infty, \infty\right)$.

c)   one-to-one

[8] a)

b)   decreasing on $\left[1, \infty\right)$.

d)   one-to-one.

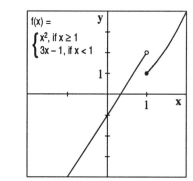

[9] a)

b)   increasing on $\left(-\infty, 1\right)$ and $\left[1, \infty\right)$.

c)   not one-to-one.

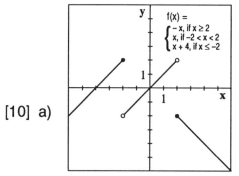

[10] a)

b)   increasing on $\left(-\infty, -2\right]$ and $\left(-2, 2\right)$,

decreasing on $\left[2, \infty\right)$.

c)   not one-to-one.

[11] $(f \circ g)(x) = f[g(x)] = 8 - 3[g(x)] = 8 - 3\left(\dfrac{8-x}{3}\right) = 8 - 8 + x = x$

$(g \circ f)(x) = g[f(x)] = \dfrac{8 - f(x)}{3} = \dfrac{8 - (8 - 3x)}{3} = \dfrac{3x}{3} = x.$

$dom\,(f) = ran(g) = \mathbf{R};\ ran(f) = dom\,(g) = \mathbf{R}.$ (See graph below).

[12] $(f \circ g)(x) = f[g(x)] = \pi + 5\,g(x) = \pi + 5\,\dfrac{(x-\pi)}{5} = x.$

$(g \circ f)(x) = g[f(x)] = \dfrac{f(x) - \pi}{5} = \dfrac{\pi + 5x - \pi}{5} = x.$

$dom\,(f) = ran(g) = \mathbf{R};\ ran(f) = dom\,(g) = \mathbf{R}.$ (See graph below).

[13] $(f \circ g)(x) = f[g(x)] = f(x^3 - 4) = \sqrt[3]{x^3 - 4 + 4} = \sqrt[3]{x^3} = x.$

$(g \circ f)(x) = g[f(x)] = [f(x)]^3 - 4 = \left(\sqrt[3]{x+4}\right)^3 - 4 = x + 4 - 4 = x.$

$dom\,(f) = ran(g) = \mathbf{R};\ ran(f) = dom\,(g) = \mathbf{R}.$ (See graph below).

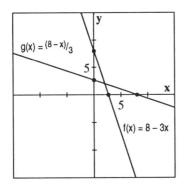

Figure 11

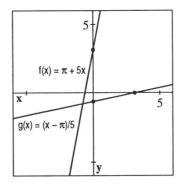

Figure 12

[14] $(f \circ g)(x) = f[g(x)] = [g(x)]^2 - 3 = \left(\sqrt{x+3}\right)^2 - 3 = x + 3 - 3 = x.$

$(g \circ f)(x) = g[f(x)] = \sqrt{f(x) + 3} = \sqrt{(x^2 - 3) + 3} = \sqrt{x^2} = x \text{ for } x \ge 0.$

$dom\,(f) = ran\,(g) = [0, \infty);\ ran\,(f) = dom\,(g) = [-3, \infty).$ (See graph below).

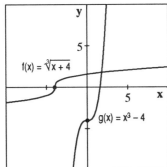

Figure 13

Figure 14

[15] $(f \circ g)(x) = f[g(x)] = \dfrac{1}{g(x)} = \dfrac{1}{\frac{1}{x}} = x$; $(g \circ f)(x) = g[f(x)] = \dfrac{1}{f(x)} = \dfrac{1}{\frac{1}{x}} = x.$

$dom(f) = ran(g) = (-\infty, 0) \cup (0, \infty);\ ran(f) = dom(g) = (-\infty, 0) \cup (0, \infty).$ (See graph below).

[16] $(f \circ g)(x) = f[g(x)] = -\dfrac{3}{g(x)} = -\dfrac{3}{-\frac{3}{x}} = x.$ $(g \circ f)(x) = g[f(x)] = -\dfrac{3}{f(x)} = -\dfrac{3}{-\frac{3}{x}} = x.$

$dom(f) = ran(g) = (-\infty, 0) \cup (0, \infty);\ ran(f) = dom(g) = (-\infty, 0) \cup (0, \infty).$ (See graph below).

[17] Suppose $f(x_1) = f(x_2)$. Then $9x_1 - 10 = 9x_2 - 10 \Rightarrow 9x_1 = 9x_2 \Rightarrow x_1 = x_2.$ Thus f is one-to-one. Let

$y = 9x - 10$. We interchange x and y getting $x = 9y - 10 \Rightarrow x + 10 = 9y \Rightarrow y = \dfrac{x + 10}{9}$. Therefore

$f^{-1}(x) = \dfrac{x + 10}{9}$

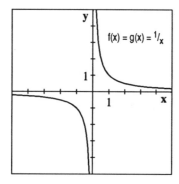

Figure 15

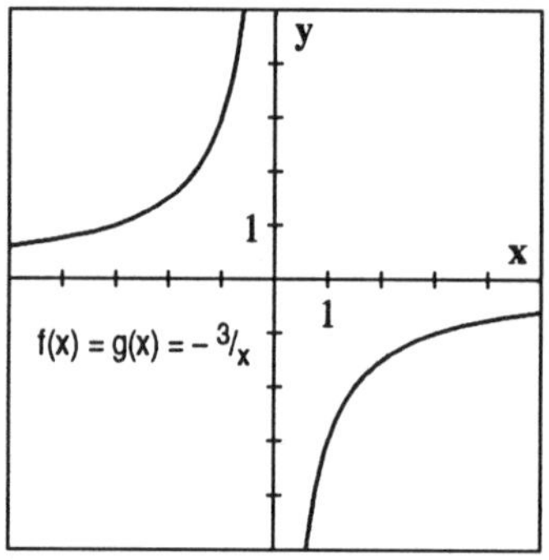

Figure 16

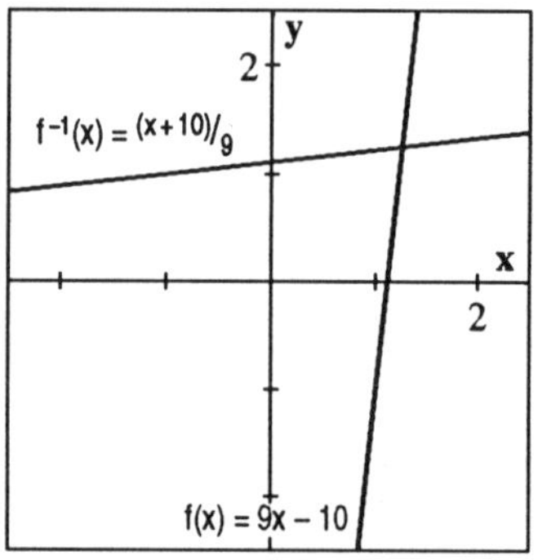

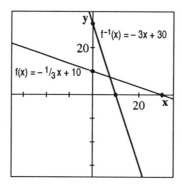

Figure 17

Figure 18

[18] Suppose $f(x_1) = f(x_2)$. Then $-\dfrac{1}{3}x_1 + 10 = -\dfrac{1}{3}x_2 + 10 \Rightarrow -\dfrac{1}{3}x_1 = -\dfrac{1}{3}x_2 \Rightarrow x_1 = x_2.$ Thus, f is one

to-one. Let $y = -\dfrac{1}{3}x + 10$. Interchanging x and y we get $x = -\dfrac{1}{3}y + 10 \Rightarrow -3(x - 10) = y.$

Therefore, $f^{-1}(x) = -3x + 30.$ (See graph above).

[19] Suppose $f(x_1) = f(x_2)$. Then $x_1^{3/5} = x_2^{3/5} \Rightarrow x_1^3 = x_2^3 \Rightarrow x_1 = x_2.$ Thus, f is one–to–one. Let $y = x^{3/5}$.

Interchanging x and y we get $x = y^{3/5} \Rightarrow x^5 = y^3 \Rightarrow y = x^{5/3}.$ Hence, $f^{-1}(x) = x^{5/3}.$ (See graph below).

[20] Suppose $f(x_1) = f(x_2)$. Then $-\sqrt{3x_1 - 2} = -\sqrt{3x_2 - 2} \Rightarrow 3x_1 - 2 = 3x_2 - 2 \Rightarrow x_1 = x_2$. Hence, f is one-to-one. Let $y = -\sqrt{3x - 2}$. Interchanging x and y we get $x = -\sqrt{3y - 2} \Rightarrow$

$x^2 = 3y - 2 \Rightarrow y = \dfrac{x^2 + 2}{3}$. Since $ran(f) = (-\infty, 0]$, $f^{-1}(x) = \dfrac{x^2 + 2}{3}$, $x \le 0$. (See graph below).

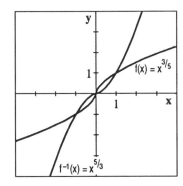

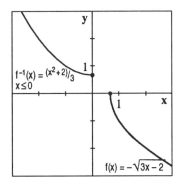

Figure 19　　　　　　　**Figure 20**

[21] Suppose $f(x_1) = f(x_2)$. Then $2 + \sqrt{x_1} = 2 + \sqrt{x_2} \Rightarrow \sqrt{x_1} = \sqrt{x_2} \Rightarrow x_1 = x_2$.　　Hence, f is one-to-one. Let $y = 2 + \sqrt{x}$, where $y \ge 2$. Interchanging x and y, we get $x = 2 + \sqrt{y} \Rightarrow x - 2 = \sqrt{y} \Rightarrow y = (x - 2)^2$.

Since $ran(f) = [2, \infty)$, $f^{-1}(x) = (x - 2)^2$, $x \ge 2$. (See graph below).

[22] Suppose $f(x_1) = f(x_2)$. Then $-(x_1)^2 - 2 = -(x_2)^2 - 2 \Rightarrow -(x_1)^2 = -(x_2)^2 \Rightarrow x_1^2 = x_2^2$. Since $x_1 \ge 0$ and

$x_2 \ge 0$, $x_1 = x_2$. Therefore, f is one-to-one. Let $y = -x^2 - 2$, for $x \ge 0$. Interchanging x and y,

we get $x = -y^2 - 2 \Rightarrow x + 2 = -y^2 \Rightarrow y^2 = -(x + 2)$. We choose $y = \sqrt{-(x + 2)}$ since $dom(f) = [0, \infty)$

and thus $f^{-1}(x) = \sqrt{-(x + 2)}$.　(See graph below).

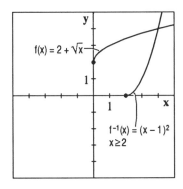

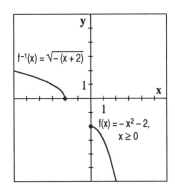

Figure 21　　　　　　　**Figure 22**

[23] Suppose $f(x_1) = f(x_2)$. Then $(x_1 + 2)^2 - 3 = (x_2 + 2)^2 - 3 \Rightarrow (x_1 + 2)^2 = (x_2 + 2)^2$. Since $x_1 \le -2$ and

$x_2 \le -2$, $x_1 + 2 \le 0$ and $x_2 + 2 \le 0$. If $(x_1 + 2)^2 = (x_2 + 2)^2$, then $x_1 + 2 = x_2 + 2 \Rightarrow x_1 = x_2$. Thus, f is

one-to-one. Let $y = (x + 2)^2 - 3$. Interchanging x and y we have $x = (y + 2)^2 - 3$. Solving　　for

y we get $x + 3 = (y + 2)^2 \Rightarrow y + 2 = \pm\sqrt{x + 3}$. Since $dom(f) = (-\infty, -2]$, we choose

$y = -2 - \sqrt{x + 3}$. Hence, $f^{-1}(x) = -2 - \sqrt{x + 3}$. (See graph below).

[24] Let $f(x_1) = f(x_2)$. Then $\dfrac{2x_1 - 7}{x_1 + 1} = \dfrac{2x_2 - 7}{x_2 + 1} \Rightarrow (2x_1 - 7)(x_2 + 1) = (2x_2 - 7)(x_1 + 1) \Rightarrow$

$2x_1 x_2 - 7x_2 + 2x_1 - 7 = 2x_1 x_2 - 7x_1 + 2x_2 - 7 \Rightarrow 2x_1 - 7x_2 = 2x_2 - 7x_1 \Rightarrow 9x_1 = 9x_2 \Rightarrow x_1 = x_2.$ Hence,

f is one-to-one. Let $y = \dfrac{2x - 7}{x + 1}$. Interchanging x and y and then solving the resulting equation for

y, we get $x = \dfrac{2y - 7}{y + 1} \Rightarrow x(y + 1) = 2y - 7 \Rightarrow xy + x = 2y - 7 \Rightarrow (2 - x)y = 7 + x \Rightarrow y = \dfrac{7 + x}{2 - x}.$

Thus $f^{-1}(x) = \dfrac{7 + x}{2 - x}$. (See graph below).

[25] a) Let $y = \dfrac{2 - x}{5}$. Interchanging x and y and solving the resulting equation for y, we get

$x = \dfrac{2 - y}{5} \Rightarrow 5x = 2 - y \Rightarrow y = 2 - 5x.$ Hence, $f^{-1}(x) = 2 - 5x.$

b) $f^{-1}(1) = -3$ and $\dfrac{1}{f(1)} = \dfrac{1}{\frac{1}{5}} = 5.$ This shows that $f^{-1}(x) \neq \dfrac{1}{f(x)}.$

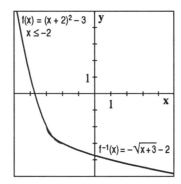

$f(x) = (x + 2)^2 - 3$
$x \leq -2$

$f^{-1}(x) = -\sqrt{x + 3} - 2$

$f(x) = {(2x - 7)}/{(x + 1)}$

$f^{-1}(x) = {(7 + x)}/{(2 - x)}$

Figure 23   **Figure 24**

[26] a) $\left(f \circ f^{-1}\right)(1) = 1$ **b)** $f^{-1}[f(2)] = 2$

c) $f(-1) = (-1)^3 + (-1) - 1 = -3.$ Thus, $f^{-1}(-3) = -1$

d) $f(0) = -1.$ Hence $f^{-1}(-1) = 0$

[27] Let $y = ax + b \, (a \neq 0)$. Interchanging x and y in this equation and solving the new equation for

y, we have $x = ay + b \Rightarrow x - b = ay \Rightarrow y = \dfrac{x - b}{a}.$ Hence, $f^{-1}(x) = \dfrac{x - b}{a}.$

[28] Let $y = x^2 - 2x + 4$ where $x \leq 1$. Interchanging x and y in this equation and solving the new

equation for y, we get $x = y^2 - 2y + 4 \Rightarrow y^2 - 2y - x + 4 = 0.$ From the quadratic formula, we

know $y = \dfrac{2 \pm \sqrt{4 - 4(4 - x)}}{2} = \dfrac{2 \pm 2\sqrt{x - 3}}{2} = 1 \pm \sqrt{x - 3}$ for $x \geq 3$. Since $dom\,(f) = \left(-\infty,\, 1\right]$, we

choose $y = 1 - \sqrt{x - 3}$. Therefore, $f^{-1}(x) = 1 - \sqrt{x - 3}.$

[29] By the definition of an even function f, if $(x_0,\, y_1) \in f$ so is $(-x_0,\, y_1)$. If $x_0 \neq 0$, then f contains two

ordered pairs with the same first components. Thus a one-to-one function that contains at least two

ordered pairs cannot be even.

[30]   Suppose $(f \circ g)(x) = f[g(x)] = x$ for each $x \in ran(f)$ and $(f \circ h)(x) = f[h(x)] = x$ for each

$x \in ran(f)$. Then $f[g(x)] = f[h(x)]$. Since f is one-to-one and since $ran(f) = dom(g) = dom(h)$,

$g(x) = h(x)$ for all $x \in dom(g) = dom(h)$. Therefore, $g = h$ and thus the inverse of f is unique.

[31]   Let R be a relation, and let $S = R^{-1}$. Then $(x, y) \in S^{-1}$ if and only if $(y, x) \in S$. Now $(y, x) \in S$ if

and only if $(x, y) \in R$. Thus, $(x, y) \in S^{-1}$ if and only if $(x, y) = R$. Hence, $S^{-1} = (R^{-1})^{-1} = R$.

[32]   a)

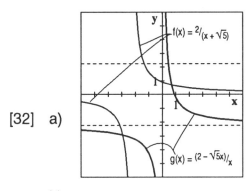

[33]   a)

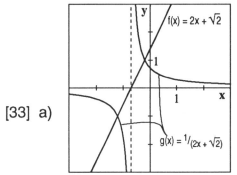

b)   yes.

b)   no.

[34]   a)

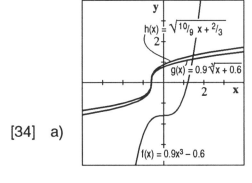

b)   h

REVIEW EXERCISES

[1]   not a function; domain, $[-4, 4]$; range, $[-2, 2]$.

[2]   a function; domain, **R**; range, $[-2, \infty)$.   [3]   a function; domain, **R**; range, $(0, \infty)$.

[4]   not a function; domain, $(-\infty, -1] \cup [1, \infty)$; range, **R**.

[5]   not a function; domain, $[-1, 1]$; range, $[-1, 1]$.

[6]   a function; domain, **R**; range, $[0, 3]$.   [7]   a function; domain, **R**; range, $[-2, \infty)$.

[8]   a function; domain, $[-4, 4)$; range, $\{-2, -1, 0, 1\}$.

[9]   center, $(0, 0)$; radius, 7.   [10]   center, $(0, 0)$; radius, $\sqrt{15}$.

[11] $\dfrac{(x-5)^2}{12} + \dfrac{(y+2)^2}{12} = 1 \Rightarrow (x-5)^2 + (y+2)^2 = 12$ ● center, $(5, -2)$; radius, $\sqrt{12}$

[12] $\dfrac{\left(x+\sqrt{7}\right)^2}{5}+\dfrac{\left(y-2\sqrt{3}\right)^2}{5}=1 \Rightarrow \left(x+\sqrt{7}\right)^2+\left(y-2\sqrt{3}\right)^2=5$ ● center, $\left(-\sqrt{7},2\sqrt{3}\right)$; radius, $\sqrt{5}$

[13] $x^2+y^2-10x-10y+35=0 \Rightarrow x^2-10x+25+y^2-10y+25=-35+25+25$ (completing the

square in x and y) $\Rightarrow (x-5)^2+(y-5)^2=15$ ● center, $(5,5)$; radius, $\sqrt{15}$

[14] $x^2+y^2-2\sqrt{3}\,x+2\sqrt{2}\,y+4=0 \Rightarrow x^2-2\sqrt{3}\,x+3+y^2+2\sqrt{2}\,y+2=-4+3+2$ (completing

the square in x and y) $\Rightarrow \left(x-\sqrt{3}\right)^2+\left(y+\sqrt{2}\right)^2=1$ ● center, $\left(\sqrt{3},-\sqrt{2}\right)$; radius, 1

[15] $x^2+y^2-1.4x-0.2y-3.5=0 \Rightarrow x^2-1.4x+0.49+y^2-0.2y+0.01=3.5+0.49+0.01$

(completing the square in x and y) $\Rightarrow (x-0.7)^2+(y-0.1)^2=4.$ ● center, $(0.7,0.1)$; radius, 2

[16] $x^2+y^2-2\pi x+4\pi y+5\pi^2-1=0 \Rightarrow x-2\pi x+\pi^2+y^2+4\pi y+4\pi^2=1-5\pi^2+\pi^2+4\pi^2$

(completing the square in x and y) $\Rightarrow (x-\pi)^2+(y+2\pi)^2=1$ ● center, $(\pi,-2\pi)$; radius, 1

[17] $dom(f) = \mathbf{R}$ (See graph below).   [18] $dom(f) = \mathbf{R}$ (See graph below).

[19] $dom(f) = \mathbf{R}$; symmetric with respect to the y–axis   (See graph below).

[20] $dom(f) = \mathbf{R}$ (See graph below).

[21] $dom(f) = (-\infty,-3)\cup(-3,\infty)$ (See graph below).

[22] $dom(f) = (-\infty,-4)\cup(-4,-3)\cup(-3,\infty)$ (See graph below).

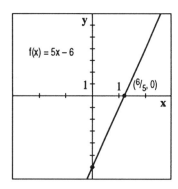

Figure 17

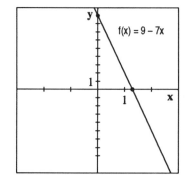

Figure 18

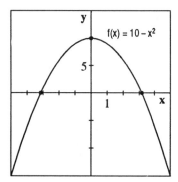

Figure 19

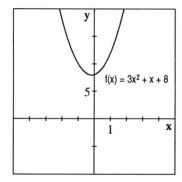

Figure 20

[23] $dom(f) = \mathbf{R}$; symmetric with respect to the y–axis   (See graph below).

[24] $dom(f) = \mathbf{R}$ (See graph below).

[25] $dom(f) = [-5, \infty)$ (See graph below).

[26] $dom(f) = [3, \infty)$ (See graph above).

[27] $dom(f) = \mathbf{R}$ (See graph below).

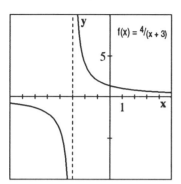

Figure 21

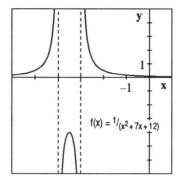

Figure 22

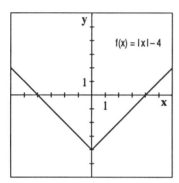

Figure 23

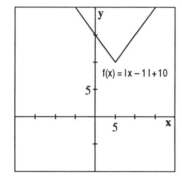

Figure 24

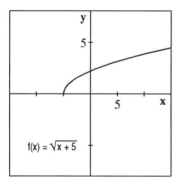

Figure 25

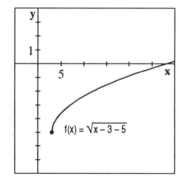

Figure 26

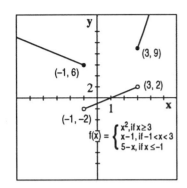

Figure 27

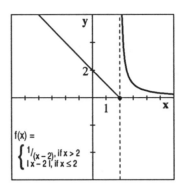

Figure 28

[28] $dom(f) = \mathbf{R}$ (See graph above).

[29] $dom(f) = \mathbf{R}$ (See graph below).

[30] $dom(f) = \mathbf{R}$ (See graph below).

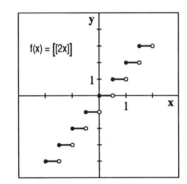

Figure 29

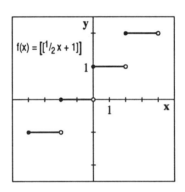

Figure 30

[31] a) $f(2) = 11 - 3(2) - (2)^2 = 11 - 6 - 4 = 1$ b) $f(-1) = 11 - 3(-1) - (-1)^2 = 11 + 3 - 1 = 13$

c) $f(a+2) = 11 - 3(a+2) - (a+2)^2 = 11 - 3a - 6 - a^2 - 4a - 4 = 1 - 7a - a^2$

d) $\dfrac{f(a+h) - f(a)}{h} = \dfrac{11 - 3(a+h) - (a+h)^2 - (11 - 3a - a^2)}{h} =$

$\dfrac{11 - 3a - 3h - a^2 - 2ah - h^2 - 11 + 3a + a^2}{h} = \dfrac{-3h - 2ah - h^2}{h} = -3 - 2a - h$

e) $f\left(\dfrac{1}{a}\right) = 11 - 3\left(\dfrac{1}{a}\right) - \left(\dfrac{1}{a}\right)^2 = 11 - \dfrac{3}{a} - \dfrac{1}{a^2} = \dfrac{11a^2 - 3a - 1}{a^2}$

[32] a) $f(2) = \sqrt{6(2) + 1} = \sqrt{13}$ b) $f(-1)$ is undefined

c) $f(a+2) = \sqrt{6(a+2) + 1} = \sqrt{6a + 13}$ d) $\dfrac{f(a+h) - f(a)}{h} = \dfrac{\sqrt{6(a+h) + 1} - \sqrt{6a + 1}}{h}$

e) $f\left(\dfrac{1}{a}\right) = \sqrt{6\left(\dfrac{1}{a}\right) + 1} = \sqrt{\dfrac{6+a}{a}}$

[33] a) $f(2) = \dfrac{1}{2(2)} = \dfrac{1}{4}$ b) $f(-1) = \dfrac{1}{2(-1)} = -\dfrac{1}{2}$

c) $f(a+2) = \dfrac{1}{2(a+2)} = \dfrac{1}{2a+4}$

d) $\dfrac{f(a+h)-f(a)}{h} = \dfrac{\dfrac{1}{2(a+h)} - \dfrac{1}{2a}}{h} = \dfrac{2a-2a-2h}{h[4a(a+h)]} = -\dfrac{1}{2a^2+2ah}$

e) $f\left(\dfrac{1}{a}\right) = \dfrac{1}{2\left(\dfrac{1}{a}\right)} = \dfrac{a}{2}$

[34] a) $f(2) = 2|2+3| = 2(5) = 10$ b) $f(-1) = 2|-1+3| = 2(2) = 4$

   c) $f(a+2) = 2|(a+2)+3| = 2|a+5|$

   d) $\dfrac{f(a+h)-f(a)}{h} = \dfrac{2|(a+h)+3| - 2|a+3|}{h} = \dfrac{2|a+h+3| - 2|a+3|}{h}$

[35] $3x - y = 4 \Rightarrow y = 3x - 4$ (See graph below). ● slope, 3; x–intercept, $\left(\dfrac{4}{3}, 0\right)$; y–intercept, $(0, -4)$

[36] $4x + 2y - 15 = 0 \Rightarrow y = -2x + \dfrac{15}{2}$ (See graph below).

● slope, –2; x–intercept, $\left(\dfrac{15}{4}, 0\right)$; y–intercept, $\left(0, \dfrac{15}{2}\right)$

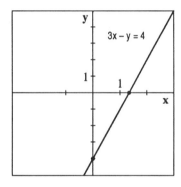

Figure 35

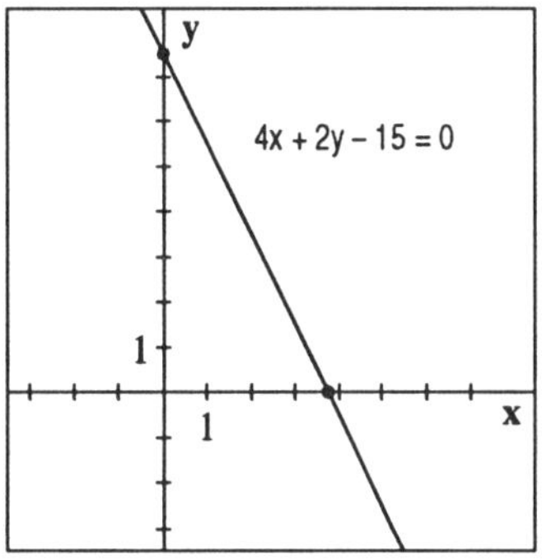

Figure 36

[37] $x = -3$ (See graph below).   ● slope, undefined; x–intercept, $(-3, 0)$; no y–intercept

[38] $4x = 3y - 5 \Rightarrow y = \dfrac{4}{3}x + \dfrac{5}{3}$ (See graph below). ● slope, $\dfrac{4}{3}$; x–intercept, $\left(-\dfrac{5}{4}, 0\right)$; y–intercept, $\left(0, \dfrac{5}{3}\right)$

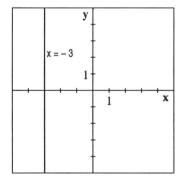
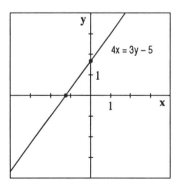

Figure 37

Figure 38

[39] $\frac{2}{5}y = -x + 1 \Rightarrow y = -\frac{5}{2}x + \frac{5}{2}$ (See graph below). ● slope, $-\frac{5}{2}$; x–intercept, $(1, 0)$; y–intercept, $\left(0, \frac{5}{2}\right)$

[40] $4y = 9 \Rightarrow y = \frac{9}{4}$ (See graph below). ● slope, 0; no x–intercept; y–intercept, $\left(0, \frac{9}{4}\right)$

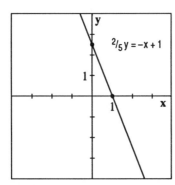

Figure 39

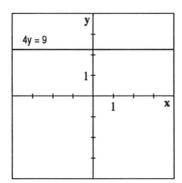

Figure 40

[41] The slope is $\frac{2-5}{3+1} = -\frac{3}{4}$. Thus, an equation for the line is $y - 5 = -\frac{3}{4}(x + 1) \Rightarrow$

$4y - 20 = -3x - 3 \Rightarrow 3x + 4y - 17 = 0$ ● $3x + 4y - 17 = 0$

[42] Since the x–coordinate of both P and Q is –2, this line is vertical. The equation of the line is

$x - 2 = 0$.   ● $x - 2 = 0$

[43] The slope of the line $3x - 2y + 4 = 0$ is $\frac{3}{2}$. Hence, the slope of the line we seek is $-\frac{2}{3}$. An

equation of the line is $y - 5 = -\frac{2}{3}x \Rightarrow 2x + 3y - 15 = 0$.   ● $2x + 3y - 15 = 0$.

[44] Since the tangent line to the circle $x^2 + y^2 = 16$ at the point $(0, -4)$ is horizontal, an equation for

this line is $y = -4 \Rightarrow y - 4 = 0$.   ● $y - 4 = 0$

[45] minimum value, –9. (See graph below).   [46] maximum value, $\frac{25}{12}$. (See graph below).

[47] maximum value, 4. (See graph below).   [48] maximum value, 4. (See graph below).

[49] $2x^2 - 1 = 4f(x) + 3 \Rightarrow 4f(x) = 2x^2 - 4 \Rightarrow f(x) = \frac{1}{2}x^2 - 1$. minimum value, –1. (See graph below).

[50] minimum value, π. (See graph below).

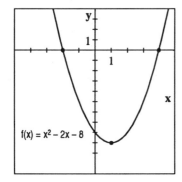

Figure 45

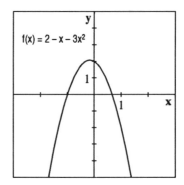

Figure 46

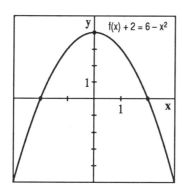

Figure 47

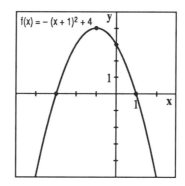

Figure 48

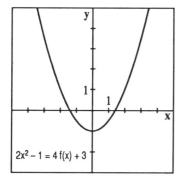

Figure 49

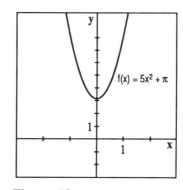

Figure 50

[51]   a)

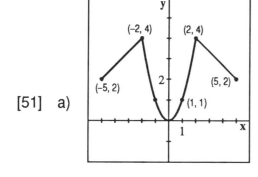

b)

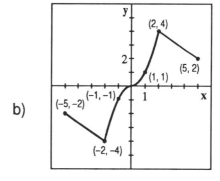

[52]   An equation for a quadratic function is $f(x) = y = ax^2 + bx + c$, where $a \neq 0$. Since the coordinates of each of the given points satisfy this equation, we have $9 = a - b + c$, $3 = c$, and $3 = 4a + 2b + c$. Since $c = 3$, it follows that $6 = a - b$ and $4a + 2b = 0$. Since $6 = a + b$, $a = b + 6$. Hence, $4(b + 6) + 2b = 0 \Rightarrow 6b = -24 \Rightarrow b = -4$. Since $a = b + 6$, $a = 2$. Therefore the equation is $f(x) = 2x^2 - 4x + 3$.   ● $f(x) = 2x^2 - 4x + 3$.

[53]   increasing on $(-\infty, -1]$; decreasing on $[-1, \infty)$.

[54]   decreasing on $(-\infty, 0)$ and on $(0, \infty)$.

[55]   increasing on $(-\infty, -1]$; decreasing on $[1, \infty)$.

[56]   increasing on $(-\infty, -1)$ and on $(-1, 0]$; decreasing on $[0, 1)$ and on $(1, \infty)$.

[57] $\left(\dfrac{f+g}{h-g}\right)(0) = \dfrac{(f+g)(0)}{(h-g)(0)} = \dfrac{f(0)+g(0)}{h(0)-g(0)} = \dfrac{-1+2}{4-2} = \dfrac{1}{2}$ ● $\dfrac{1}{2}$

[58] $\left[(f-g)+h\right](0) = (f-g)(0)+h(0) = f(0)-g(0)+h(0) = -1-2+4 = 1$ ● 1

[59] $\left(\dfrac{f}{g}+h\right)(0) = \left(\dfrac{f}{g}\right)(0)+h(0) = \dfrac{-1}{2}+4 = \dfrac{7}{2}.$ ● $\dfrac{7}{2}$

[60] $(4f+3g)(0) = (4f)(0)+(3g)(0) = 4f(0)+3g(0) = 4(-1)+3(2) = -4+6 = 2.$ ● 2

[61] $(f \circ g)(x) = f[g(x)] = \dfrac{g(x)-4}{3} = \dfrac{x^2-x-4}{3}$; $dom(f \circ g) = \mathbf{R}.$

[62] $(f \circ g)(x) = f[g(x)] = \dfrac{1}{g(x)} = \dfrac{1}{\dfrac{1}{x}} = x$; $dom(f \circ g) = (-\infty, 0) \cup (0, \infty).$

[63] $(f \circ g)(x) = f[g(x)] = \dfrac{1}{\sqrt{g(x)}-4} = \dfrac{1}{\sqrt{x^2-6x}-4}$;

$dom(f \circ g) = (-\infty, -2) \cup (-2, 0] \cup [6, 8) \cup (8, \infty).$

[64] $(f \circ g)(x) = f[g(x)] = \sqrt{g(x)+1} = \sqrt{(x^2-2x-9)+1} = \sqrt{x^2-2x-8}$;

$dom(f \circ g) = (-\infty, -2) \cup [4, \infty).$

[65] To determine if f is one-to-one, assume that $f(x_1) = f(x_2)$. Thus, $x_1^3+2 = x_2^3+2 \Rightarrow x_1^3 = x_2^3 \Rightarrow$

$x_1 = x_2$. Therefore, f is one-to-one.   ● one-to-one

[66] To determine if f is one-to-one, assume that $f(x_1) = f(x_2)$, where $x_1 \geq 2$ and $x_2 \geq 2$. Then

$(x_1-2)^2+3 = (x_2-2)^2+3 \Rightarrow (x_1-2)^2 = (x_2-2)^2 \Rightarrow x_1-2 = x_2-2$ since $x_1 \geq 2$ and $x_2 \geq 2$. Thus,

$x_1 = x_2$, and f is one-to-one.   ● one-to-one

[67] To determine if f is one-to-one, assume that $f(x_1) = f(x_2)$, where $x_1 \leq -3$ and $x_2 \leq -3$. Hence,

$|x_1+3|+1 = |x_2+3|+1 \Rightarrow |x_1+3| = |x_2+3|$. Since $x_1 \leq -3$ and $x_2 \leq -3$, $x_1+3 \leq 0$ and $x_2+3 \leq 0$.

Thus, it follows from the equation $|x_1+3| = |x_2+3|$ that $x_1+3 = x_2+3 \Rightarrow x_1 = x_2$. Therefore,

f is one-to-one.   ● one-to-one

[68] To determine if f is one-to-one, we assume that $f(x_1) = f(x_2)$. Hence, $\dfrac{2}{3x_1-1} = \dfrac{2}{3x_2-1} \Rightarrow$

$3x_1-1 = 3x_2-1 \Rightarrow x_1 = x_2$. Thus, f is one-to-one.   ● one-to-one

[69] Assume $f(x_1) = f(x_2)$. Then $\sqrt[5]{x_1} = \sqrt[5]{x_2} \Rightarrow x_1 = x_2$. Therefore, f is one-to-one. Let $y = \sqrt[5]{x}$.

Interchanging x and y in the latter equation and then solving the new equation for y, we get

$x = \sqrt[5]{y}$. Thus $f^{-1}(x) = x^5$. (See graph below).

[70] Assume $f(x_1) = f(x_2)$. It follows that $\dfrac{x_1+4}{9} = \dfrac{x_2+4}{9} \Rightarrow x_1+4 = x_2+4 \Rightarrow x_1 = x_2$. Hence, f is one-

to-one. Let $y = \dfrac{x+4}{9}$. Interchanging x and y in the latter equation and then solving the new

equation for y, we have $x = \dfrac{y+4}{9} \Rightarrow 9x = y+4 \Rightarrow y = 9x-4$. Thus $f^{-1}(x) = 9x-4$. (See graph

below).

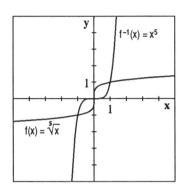

Figure 69

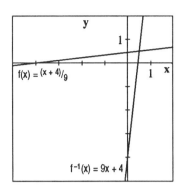

Figure 70

[71] Assume $f(x_1) = f(x_2)$, where $-3 \leq x_1 \leq 0$ and $-3 \leq x_2 \leq 0$.   It follows that $\sqrt{9 - x_1^2} = \sqrt{9 - x_2^2} \Rightarrow$

$9 - x_1^2 = 9 - x_2^2 \Rightarrow x_1^2 = x_2^2$. Since $-3 \leq x_1 \leq 0$ and $-3 \leq x_2 \leq 0$, it follows that $x_1 = x_2$. Let

$y = \sqrt{9 - x^2}$ where $-3 \leq x \leq 0$. Interchanging x and y in the latter equation and then solving the

new equation for y yields $x = \sqrt{9 - y^2} \Rightarrow x^2 = 9 - y^2 \Rightarrow y^2 = 9 - x^2 \Rightarrow y = \pm\sqrt{9 - x^2}$. Since

$dom(f) = ran(f^{-1}) = [-3, 0]$, we choose $y = -\sqrt{9 - x^2}$. Hence, $f^{-1}(x) = -\sqrt{9 - x^2}$ for $x \in [0, 3]$.

$(dom(f^{-1}) = ran(f) = [0, 3])$.   (See graph below).

[72] Assume $f(x_1) = f(x_2)$. Then $\sqrt[3]{x_1 + 1} = \sqrt[3]{x_2 + 1} \Rightarrow x_1 + 1 = x_2 + 1 \Rightarrow x_1 = x_2$. Hence, f is one-to-one.

Let $y = \sqrt[3]{x + 1}$. Interchanging x and y in the latter equation and solving the new equation for y,

we find that $x = \sqrt[3]{y + 1} \Rightarrow x^3 = y + 1 \Rightarrow y = x^3 - 1$. Hence $f^{-1}(x) = x^3 - 1$. (See graph below).

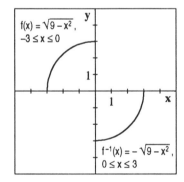

Figure 71

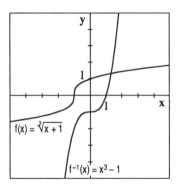

Figure 72

[73] Letting x represent the shorter side of the rectangle, the area A can be expressed as $A(x) =$

$x\left(\dfrac{20 - 2x}{2}\right) = x(10 - x) = 10x - x^2$. The graph of A is a parabola which opens downward. Thus,

the maximum value of A is the y–coordinate of the vertex, which is $A(5) = 25$.

[74] We introduce a coordinate system as shown in the figure below and find an equation of the given parabola. This equation has the form $y = ax^2 + bx + c$, where $a \neq 0$. Since the point $(0, 15)$ lies on this parabola, its coordinates satisfy the equation $y = ax^2 + bx + c \Rightarrow 15 = c$. Also, the coordinates of the points $(-15, 0)$ satisfy the latter equation. Hence, $0 = a(-15)^2 + b(-15) + 15$ and $0 = a(15)^2 + b(15) + 15$. Therefore, $0 = 225a - 15b + 15$ and $0 = 225a + 15b + 15$. Thus, $225a - 15b + 15 = 225a + 15b + 15 \Rightarrow -b = b \Rightarrow b = 0$. It follows that $a = -\frac{1}{15}$ and the equation of the parabola is $y = -\frac{1}{15}x^2 + 15$. Since the point $(x, 5)$ is on the parabola, $5 = -\frac{1}{15}x^2 + 15 \Rightarrow x^2 = 150 \Rightarrow x = 5\sqrt{6}$. Hence, the width of the arch is $10\sqrt{6}$.   ● $10\sqrt{6}$

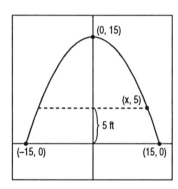

(0, 15)

(x, 5)

5 ft

(−15, 0)   (15, 0)

Figure 74

[75] The maximum height of the ball is the y–coordinate of the vertex of the parabola $y = 64t - 16t^2$. Hence, the height is $y = \left(-\frac{b}{2a}\right) = y(2) = 64$ feet.

[76] The dimensions can be expressed as x and $\frac{32 - 2x}{2} = 16 - x$. Hence, the area A is $A(x) = x(16 - x) = 16x - x^2$. Thus, the maximum possible area is the y–coordinate of the vertex of the parabola $A(x) = = 16x - x^2$, and the maximum possible area is $A\left(-\frac{b}{2a}\right) = A(8) = 64$. To find the dimensions that produce the maximum area, we solve the equation $64 = 16x - x^2$, or $x^2 - 16x + 64 = 0 \Rightarrow (x - 8)^2 = 0 \Rightarrow x = 8$. Hence, the dimensions are 8 and $16 - 8 = 8$.

    ● 8 ft, 8 ft; maximum area is 64 sq. ft.

[77] The slope of the line containing the points $(0, -1)$ and $(1, 3)$ is $\frac{3 - (-1)}{1} = 4$. The slope of the line containing the points $(0, -1)$ and $(4, -2)$ is $\frac{-2 - (-1)}{4} = -\frac{1}{4}$. Hence, the two lines are perpendicular and therefore, the triangle is a right triangle.

[78] a) $C(x) = \$250 + \$0.30\, x$ b) $C(x) = \$50\,n + \$0.30\, x$

[79] **a)** The linear demand function f contains the ordered pairs $(300,\ 50,000)$ and $(400,\ 30,000)$.

Thus, the slope of the graph of f is $\dfrac{50,000 - 30,000}{300 - 400} = -200$. Hence, an equation of the line

is $f(p) - 50,000 = -200(p - 300)$ or $f(p) = 110,000 - 200p$.

b) $20,000 = 110,000 - 200p \Rightarrow p = \$450$

c) $f(600) = 110,000 - 200(600) = 110,000 - 120,000 = -10,000$.

No. There is no demand for the VCRs if each costs \$600.

[80] Let x represent the income. Then $f(x) = \begin{cases} 0, \text{if } 0 \le x \le 999 \\ 10 + 0.05(x - 1000), \text{if } 1000 \le x \le 4,999 \\ 210 + 0.06(x - 5000), \text{if } 5000 \le x \le 9,999 \\ 510 + 0.03(x - 10,000), \text{if } x \ge 10,000 \end{cases}$

(See graph below).

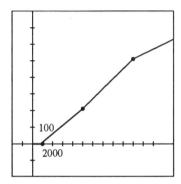

100

2000

Figure 80

EXERCISES 3.2

[1]
$$\begin{array}{r} 3m \quad - \ 2 \\ 2m+3 \overline{\smash{)}6m^2 + 5m - 1} \\ \underline{6m^2 + 9m} \\ -4m - 1 \\ \underline{-4m - 6} \\ 5 \end{array}$$

$Q(m) = 3m - 2; \ R(m) = 5$

[2]
$$\begin{array}{r} x \ + \ 7 \\ x-3 \overline{\smash{)}x^2 + 4x - 2} \\ \underline{x^2 - 3x} \\ 7x - 2 \\ \underline{7x - 21} \\ 19 \end{array}$$

$Q(x) = x + 7; \ R(m) = 19$

[3]
$$\begin{array}{r} y^3 + 2y^2 + 4y \ + \ 8 \\ y-2 \overline{\smash{)}y^4 + 0y^3 + 0y^2 + 0y + 1} \\ \underline{y^4 - 2y^3} \\ 2y^3 + 0y^2 + 0y + 1 \\ \underline{2y^3 - 4y^2} \\ 4y^2 + 0y + 1 \\ \underline{4y^2 - 8y} \\ 8y + 1 \\ \underline{8y - 16} \\ 17 \end{array}$$

$Q(y) = y^3 + 2y^2 + 4y + 8; \ R(y) = 17$

[4]
$$\begin{array}{r} -3t \ + \ 1 \\ 4t^2 + 0t + 1 \overline{\smash{)}-12t^3 + 4t^2 + 0t - 1} \\ \underline{-12t^3 + 0t^2 - 3t} \\ 4t^2 + 3t - 1 \\ \underline{4t^2 + 0t + 1} \\ 3t - 2 \end{array}$$

$Q(t) = -3t + 1; \ R(t) = 3t - 2$

[5]
$$\begin{array}{r} a^3 - 3a^2 + a \ - \ 3 \\ a^2 + 0a - 1 \overline{\smash{)}a^5 - 3a^4 + 0a^3 + 0a^2 + 3a - 1} \\ \underline{a^5 + 0a^4 - a^3} \\ -3a^4 + a^3 + 0a^2 + 3a - 1 \\ \underline{-3a^4 - 0a^3 + 3a^2} \\ a^3 - 3a^2 + 3a - 1 \\ \underline{a^3 + 0a^2 - a} \\ -3a^2 + 4a - 1 \\ \underline{-3a^2 + 0a + 3} \\ 4a - 4 \end{array}$$

$Q(a) = a^3 - 3a^2 + a - 3; \ R(a) = 4a - 4$

[6]
$$\begin{array}{r} 2y^2 + 3y \ - \ \dfrac{5}{2} \\ y + \dfrac{1}{2} \overline{\smash{)}2y^3 + 4y^2 \ - \ y + 2} \\ \underline{2y^3 + y^2} \\ 3y^2 - y + 2 \\ \underline{3y^2 + \dfrac{3}{2}y} \\ -\dfrac{5}{2}y + 2 \\ \underline{-\dfrac{5}{2}y - \dfrac{5}{2}} \\ \dfrac{9}{2} \end{array}$$

$Q(y) = 2y^2 + 3y - \dfrac{5}{2}; \ R(y) = \dfrac{9}{2}$

[7]
$$2x^2+x-2\overline{\smash{\big)}-x^4-6x^3+2x^2+x-1}$$
quotient on top: $-\dfrac{1}{2}x^2-\dfrac{11}{4}x+\dfrac{15}{8}$

$$-x^4-\dfrac{1}{2}x^3+x^2$$
$$-\dfrac{11}{2}x^3+x^2+x-1$$
$$-\dfrac{11}{2}x^3-\dfrac{11}{4}x^2+\dfrac{11}{2}x$$
$$\dfrac{15}{4}x^2-\dfrac{9}{2}x-1$$
$$\dfrac{15}{4}x^2+\dfrac{15}{8}x-\dfrac{15}{4}$$
$$-\dfrac{51}{8}x+\dfrac{11}{4}$$

$Q(x)=-\dfrac{1}{2}x^2-\dfrac{11}{4}x+\dfrac{15}{8}; \; R(x)=-\dfrac{51}{8}x+\dfrac{11}{4}$

[8]
$$t^2+\sqrt{2}\,t-1\overline{\smash{\big)}t^4+0t^3+0t^2+0t-1}$$
quotient on top: $t^2-\sqrt{2}\,t+3$

$$t^4+\sqrt{2}\,t^3-t^2$$
$$-\sqrt{2}\,t^3+t^2+0t-1$$
$$-\sqrt{2}\,t^3-2t^2+\sqrt{2}\,t$$
$$3t^2-\sqrt{2}\,t-1$$
$$3t^2+3\sqrt{2}\,t-3$$
$$-4\sqrt{2}\,t+2$$

$Q(t)=t^2-\sqrt{2}\,t+3; \; R(t)=-4\sqrt{2}\,t+2$

[9]
$$x-i\overline{\smash{\big)}x^5+0x^4+x^3+0x^2+0x}$$
quotient on top: x^4+ix^3

$$x^5-ix^4$$
$$ix^4+x^3+0x^2+0x$$
$$ix^4+x^3$$
$$0$$

$Q(x)=x^4+ix^3; \; R(x)=0$

[10]
$$r^2+0r-k\overline{\smash{\big)}r^5+0r^4+0r^3+0r^2+0r-k}$$
quotient on top: r^3+kr

$$r^5+0r^4-kr^3$$
$$kr^3+0r^2+0r-k$$
$$kr^3+0r^2-k^2r$$
$$k^2r-k$$

$Q(r)=r^3+kr; \; R(r)=k^2r-k$

[11]
$$\begin{array}{r|rrrr} -2 & 3 & 0 & -8 & 1 \\ & & -6 & 12 & -8 \\ \hline & 3 & -6 & 4 & -7 \end{array}$$

$Q(x)=3x^2-6x+4; \; R(x)=-7$

[12]
$$\begin{array}{r|rrrr} 2 & 5 & -6 & -28 & -2 \\ & & 10 & 8 & -40 \\ \hline & 5 & 4 & -20 & -42 \end{array}$$

$Q(m)=5m^2+4m-20; \; R(m)=-42$

[13]
$$\begin{array}{r|rrrrr} 3 & 1 & -3 & -5 & 2 & -16 \\ & & 3 & 0 & -15 & -39 \\ \hline & 1 & 0 & -5 & -13 & -55 \end{array}$$

$Q(y)=y^3-5y-13; \; R(y)=-55$

[14]
$$\begin{array}{r|rrrrrr} -1 & 1 & 0 & 4 & -6 & 0 & -9 \\ & & -1 & 1 & -5 & 11 & -11 \\ \hline & 1 & -1 & 5 & -11 & 11 & -20 \end{array}$$

$Q(p)=p^4-p^3+5p^2-11p+11; \; R(p)=-20$

[15]
$$\begin{array}{r|rrrr} 5 & 1 & -8 & 17 & -10 \\ & & 5 & -15 & 10 \\ \hline & 1 & -3 & 2 & 0 \end{array}$$

$Q(x)=x^2-3x+2; \; R(x)=0$

[16]
$$\begin{array}{r|rrrrr} -2 & 1 & 2 & -1 & -2 & 2 & 4 \\ & & -2 & 0 & 2 & 0 & -4 \\ \hline & 1 & 0 & -1 & 0 & 2 & 0 \end{array}$$

$Q(t)=t^4-t^2+2; \; R(p)=0$

[17]
$$\begin{array}{r|rrrrrrrr} 1 & 1 & 0 & 0 & 0 & -1 & 0 & 1 & 0 \\ & & 1 & 1 & 1 & 1 & 0 & 0 & 1 \\ \hline & 1 & 1 & 1 & 1 & 0 & 0 & 1 & 1 \end{array}$$

$Q(x)=x^6+x^5+x^4+x^3+1; \; R(x)=1$

[18]
$$\begin{array}{r|rrrr} -6 & 2 & 9 & -17 & 6 \\ & & -12 & 18 & -6 \\ \hline & 2 & -3 & 1 & 0 \end{array}$$

$Q(x)=2x^2-3x+1; \; R(x)=0$

[19]
$$\begin{array}{r|rrrrrrr} 1 & 1 & 0 & 0 & 0 & 0 & 0 & -1 \\ & & 1 & 1 & 1 & 1 & 1 & 1 \\ \hline & 1 & 1 & 1 & 1 & 1 & 1 & 0 \end{array}$$

$Q(x)=x^5+x^4+x^3+x^2+x+1; \; R(x)=0$

[20]
$$\begin{array}{r|rrrrr} -1 & 1 & 0 & 0 & 0 & 1 \\ & & -1 & 1 & -1 & 1 \\ \hline & 1 & -1 & 1 & -1 & 2 \end{array}$$

$Q(x)=x^3-x^2+x-1; \; R(x)=2$

[21]
$$\begin{array}{r|rrrrrr} -1 & -1 & 0 & 0 & 0 & 0 & 1 \\ & & -1 & -1 & -1 & -1 & -1 \\ \hline & -1 & -1 & -1 & -1 & -1 & 0 \end{array}$$

$Q(x)=-x^4-x^3-x^2-x-1; \; R(x)=0$

[22]
$$\begin{array}{r|rrrr} 2 & 3 & 0 & 0 & -1 & 0 & 1 \\ & & 6 & 12 & 24 & 46 & 92 \\ \hline & 3 & 6 & 12 & 23 & 46 & 93 \end{array}$$

$Q(y)=3y^4+6y^3+12y^2+23y+46; \; R(y)=93$

[23]

$$\begin{array}{r|rrrrr}
a & 1 & a^3 & a^2 & a^3 & a^4 \\
  &   & a   & a^4+a^2 & a^5+2a^3 & a^6+3a^4 \\ \hline
  & 1 & a^3+a & a^4+2a^2 & a^5+3a^3 & a^6+4a^4
\end{array}$$

$Q(a) = x^3 + (a^3+a)x^2 + (a^4+2a^2)x + a^5 + 3a^3;$
$R(a) = a^6 + 4a^4$

[24]

$$\begin{array}{r|rrrrr}
-a & 2 & 0 & -a^2 & 0 & 1 \\
   &   & -2a & 2a^2 & -a^3 & a^4 \\ \hline
   & 2 & -2a & a^2 & -a^3 & a^4+1
\end{array}$$

$Q(y) = 2y^3 - 2ay^2 + a^2y - a^3$
$R(a) = a^4 + 1$

[25]

$$\begin{array}{r|rrrr}
\frac{1}{2} & 8 & -4 & -2 & 1 \\
 &   & 4 & 0 & -1 \\ \hline
 & 8 & 0 & -2 & 0
\end{array}$$

$Q(x) = 8x^2 - 2;\ R(x) = 0$

[26]

$$\begin{array}{r|rrrr}
-\frac{1}{2} & 2 & -1 & 3 & -1 \\
 &   & -1 & 1 & -2 \\ \hline
 & 2 & -2 & 4 & -3
\end{array}$$

$Q(y) = 2y^2 - 2y + 4;\ R(y) = -3$

[27]

$$\begin{array}{r|rrrr}
-\frac{1}{2} & 1 & 1 & \frac{1}{2} & \frac{1}{8} \\
 &   & -\frac{1}{2} & -\frac{1}{4} & -\frac{1}{8} \\ \hline
 & 1 & \frac{1}{2} & \frac{1}{4} & 0
\end{array}$$

$Q(x) = x^2 + \frac{1}{2}x + \frac{1}{4};\ R(x) = 0$

[28]

$$\begin{array}{r|rrrr}
\frac{1}{3} & 3 & -\frac{2}{9} & \frac{1}{27} & 1 \\
 &   & \frac{1}{9} & -\frac{1}{27} & 0 \\ \hline
 & \frac{1}{3} & -\frac{1}{9} & 0 & 1
\end{array}$$

$Q(x) = \frac{1}{3}x^2 - \frac{1}{9}x;\ R(x) = 1$

[29]

$$\begin{array}{r|rrrr}
-2i & 1 & -1 & 4 & -4 \\
 &   & -2i & -4+2i & 4 \\ \hline
 & 1 & -1-2i & 2i & 0
\end{array}$$

$Q(x) = x^2 + (-1-2i)x + 2i;\ R(x) = 0$

[30]

$$\begin{array}{r|rrrrr}
-i & 1 & 2i & 0 & -i & 5 \\
 &   & -i & 1 & -i & -2 \\ \hline
 & 1 & i & 1 & -2i & 3
\end{array}$$

$Q(y) = y^3 + iy^2 + y - 2i;\ R(y) = 3$

[31]

$$\begin{array}{r|rrrrr}
1-i & 1 & 0 & -i & 0 & 1 \\
 &   & 1-i & -2i & -3-3i & -6 \\ \hline
 & 1 & 1-i & -3i & -3-3i & -5
\end{array}$$

$Q(x) = x^3 + (1-i)x^2 - 3ix - 3 - 3i;\ R(x) = -5$

[32]

$$\begin{array}{r|rrrr}
-1+i & 1 & 2 & 0 & -1 \\
 &   & -1+i & -2 & 2-2i \\ \hline
 & 1 & 1+i & -2 & 1-2i
\end{array}$$

$Q(x) = x^2 + (1+i)x - 2;\ R(x) = 1 - 2i$

[33]

$$\begin{array}{r|rrrrrr}
i & 1 & 0 & -1 & 0 & 0 & 0 \\
 &   & i & -1 & -2i & 2 & 2i \\ \hline
 & 1 & i & -2 & -2i & 2 & 2i
\end{array}$$

$Q(x) = x^4 + ix^3 - 2x^2 - 2ix + 2;\ R(x) = 2i$

[34]

$$\begin{array}{r|rrrrr}
-i & 1 & 0 & 3 & 0 & -1 \\
 &   & -i & -1 & -2i & -2 \\ \hline
 & 1 & -i & 2 & -2i & -3
\end{array}$$

$Q(x) = x^3 - ix^2 + 2x - 2i;\ R(x) = -3$

[35]

$$\begin{array}{r|rrrrr}
\frac{1}{2}i & 2 & -7i & -7 & 2i & 0 \\
 &   & i & 3 & -2i & 0 \\ \hline
 & 2 & -6i & -4 & 0 & 0
\end{array}$$

$Q(y) = 2y^3 - 6iy^2 - 4y;\ R(y) = 0$

[36]

$$\begin{array}{r|rrrrr}
-\frac{1}{4}i & 4 & 5i & -1 & 8i & 0 \\
 &   & -i & 1 & 0 & 2 \\ \hline
 & 4 & 4i & 0 & 8i & 2
\end{array}$$

$Q(x) = 4x^3 + 4ix^2 + 8i;\ R(x) = 2$

[37]

$$\begin{array}{r|rrrrrr}
-2i & 1 & 0 & 1-i & 0 & 0 & -2+i \\
 &   & -2i & -4 & -2+6i & 12+4i & 8-24i \\ \hline
 & 1 & -2i & -3-i & -2+6i & 12+4i & 6-23i
\end{array}$$

$Q(m) = m^4 - 2im^3 + (-3-i)m^2 + (-2+6i)m + 12 + 4i;\ R(m) = 6 - 23i$

[38]

$$\begin{array}{r|rrrr} i & 1 & 1-i & 2+i & 2 \\ & & i & i & -2+2i \\ \hline & 1 & 1 & 2+2i & 2i \end{array}$$

$Q(x) = x^2 + x + 2 + 2i \, ; \, R(x) = 2i$

[39]

$$\begin{array}{r|rrrrr} 3 & 1 & -2 & 1 & -10 & -6 \\ & & 3 & 3 & 12 & 6 \\ \hline & 1 & 1 & 4 & 2 & 0 \end{array}$$

Since $R(y) = 0$, $y - 3$ is a factor of $y^4 - 2y^3 + y^2 - 10y - 6$.

[40]

$$\begin{array}{r|rrrr} -1 & 2 & -4 & -3 & 3 \\ & & -2 & 6 & -3 \\ \hline & 2 & -6 & 3 & 0 \end{array}$$

Since $R(x) = 0$, $x + 1$ is a factor of $2x^3 - 4x^2 - 3x + 3$.

[41]

$$\begin{array}{r|rrrr} -\frac{1}{2} & 1 & 1 & 1 & \frac{3}{8} \\ & & -\frac{1}{2} & -\frac{1}{4} & -\frac{3}{8} \\ \hline & 1 & \frac{1}{2} & \frac{3}{4} & 0 \end{array}$$

Since $R(x) = 0$, $x + \frac{1}{2}$ is a factor of $x^3 + x^2 + x + \frac{3}{8}$.

[42]

$$\begin{array}{r|rrrr} \frac{1}{2} & 2 & 3 & -8 & 3 \\ & & 1 & 2 & -3 \\ \hline & 2 & 4 & -6 & 0 \end{array}$$

Since $R(y) = 0$, $y - \frac{1}{2}$ is a factor of $2y^3 + 3y^2 - 8y + 3$.

[43]

$$\begin{array}{r|rrrrr} \frac{1}{2}i & 2 & -7i & -7 & 2i & 0 \\ & & i & 3 & -2i & 0 \\ \hline & 2 & -6i & -4 & 0 & 0 \end{array}$$

Since $R(x) = 0$, $y - \frac{1}{2}i$ is a factor of $2x^4 - 7ix^3 - 7x^2 + 2ix$.

[44]

$$\begin{array}{r|rrrr} -i & 1 & 3i & -3 & -i \\ & & -i & 2 & i \\ \hline & 1 & 2i & -1 & 0 \end{array}$$

Since $R(x) = 0$, $x + i$ is a factor of $x^3 + 3ix^2 - 3x - i$.

[45]

$$\begin{array}{r|rrrr} 1+i & 2 & 3 & -2i & 2-8i \\ & & 2+2i & 3+7i & -2+8i \\ \hline & 2 & 5+2i & 3+5i & 0 \end{array}$$

Since $R(m) = 0$, $m - 1 - i$ is a factor of $2m^3 + 3m^2 - 2im + 2 - 8i$.

[46]

$$\begin{array}{r|rrrr} 1-i & 1 & i & -3+i & 2-2i \\ & & 1-i & 1-i & -2+2i \\ \hline & 1 & 1 & -2 & 0 \end{array}$$

Since $R(x) = 0$, $x - 1 + i$ is a factor of $x^3 + ix^2 + (-3 + i)x + 2 - 2i$.

[47]

$$\begin{array}{r|rrrrrr} \sqrt{2} & \sqrt{2} & 0 & -\frac{\sqrt{2}}{2} & -1 & 0 & -4 \\ & & 2 & 2\sqrt{2} & 3 & 2\sqrt{2} & 4 \\ \hline & \sqrt{2} & 2 & \frac{3\sqrt{2}}{2} & 2 & 2\sqrt{2} & 0 \end{array}$$

Since $R(x) = 0$, $x - \sqrt{2}$ is a factor of $\sqrt{2}\,x^5 - \frac{\sqrt{2}}{2}x^3 - x^2 - 4$.

[48]

$$\begin{array}{r|rrrrr} -\sqrt{2} & -1 & 0 & 1 & 0 & 2 \\ & & \sqrt{2} & -2 & \sqrt{2} & -2 \\ \hline & -1 & \sqrt{2} & -1 & \sqrt{2} & 0 \end{array}$$

Since $R(x) = 0$, $x + \sqrt{2}$ is a factor of $-x^4 + x^2 + 2$.

[49] By synthetic division, $R(x) = -2k^2 + k + 6$. $-2k^2 + k + 6 = 0 \Rightarrow k = -\frac{3}{2}$ or $k = 2$.

[50] By synthetic division, $R(x) = -k + 3$. $-k + 3 = 0 \Rightarrow k = 3$.

[51] By synthetic division, $R(x) = k^2 + 3k + 2$. $k^2 + 3k + 2 = 0 \Rightarrow k = -2$ or $k = -1$.

[52] By synthetic division, $R(x) = 13k + 26$. $13k + 26 = 0 \Rightarrow k = -2$.

[53] By synthetic division, $R(x) = -3c^2 + 9$. $-3c^2 + 9 = 0 \Rightarrow c = \pm\sqrt{3}$.

[54] By synthetic division, $R(x) = ar^2 + br + c$.

[55]
$$\begin{array}{r|rrr} a & 1 & 2a & -3b^2 \\ & & a & 3a^2 \\ \hline & 1 & 3a & 3a^2-3b^2 \end{array}$$

$R(x) = 0 \Rightarrow 3a^2 - 3b^2 = 0 \Rightarrow a^2 = b^2$.

[56] By synthetic division, $R(x) = a^4 + 2a^2 + 1 = \left(a^2 + 1\right)^2 \neq 0$ for all $a \in \mathbf{R}$.

[57] By synthetic division, $R(x) = k + 3$. $k + 3 = -5 \Rightarrow k = -8$.

[58] By synthetic division, $R(x) = 2k - 2$. $2k - 2 = 0 \Rightarrow k = 1$.

EXERCISES 3.3

[1] $P(-5) = (-5)^3 + 3(-5)^2 - 9(-5) + 5 = 0$.   [2] $P(20) = -(20)^2 + 3(20) - 10 = -350$.

[3] $P(-1) = 2(-1)^4 - 3(-1)^3 + 3(-1) + 4 = 6$.   [4] $P\left(\dfrac{1}{2}\right) = 8\left(\dfrac{1}{2}\right)^3 - 16\left(\dfrac{1}{2}\right) + 7 = 0$

[5] $P(i) = i^5 - i \cdot i + 1 = 2 + i$.

[6] $P(-2i) = 3(-2i)^3 - 7(-2i)^2 + 27(-2i) - 63 = -35 - 30i$.

[7] $P\left(\sqrt{2}\right) = 1 + 2\left(\sqrt{2}\right) + \left(\sqrt{2}\right)^2 - \left(\sqrt{2}\right)^4 = -1 + 2\sqrt{2}$.

[8] $P(1 - 2i) = (1 - 2i)^4 - 2(1 - 2i)^3 + 6(1 - 2i)^2 - 2(1 - 2i) + 5 = 0$.

[9] $P(-1 + i) = (-1 + i)^4 - (-1 + i)^2 + 1 = -3 + 2i$.   [10] $P\left(\dfrac{1}{2}i\right) = \left(\dfrac{1}{2}i\right)^4 - 1 = -\dfrac{15}{16}$.

[11] $P(5) = 0 \Rightarrow x - 5$ is a factor of $P(x) = x^3 - 8x^2 + 17x - 10$.

[12] $P(-6) = 0 \Rightarrow y + 6$ is a factor of $P(y) = 2y^3 + 9y^2 - 17y + 6$.

[13] $P\left(-\dfrac{1}{3}\right) = 0 \Rightarrow x + \dfrac{1}{3}$ is a factor of $P(x) = 6x^4 - 19x^3 - 25x^2 + 18x + 8$.

[14] $P\left(\dfrac{1}{2}\right) = \dfrac{1}{2} \neq 0 \Rightarrow t - \dfrac{1}{2}$ is not a factor of $P(t) = 2t^4 - 3t^2 + 1$.

[15] $P\left(-\sqrt{2}\right) = 6 \neq 0 \Rightarrow x + \sqrt{2}$ is not a factor of $P(x) = x^4 + 3x^2 - 4$

[16] $P\left(\sqrt{2}\right) = 0 \Rightarrow x - \sqrt{2}$ is a factor of $P(x) = \sqrt{2}\,x^5 - \dfrac{\sqrt{2}}{2}x^3 - x^2 - 4$.

[17] $P(-3i) = 0 \Rightarrow x + 3i$ is a factor of $P(x) = x^4 + 3ix^3 - 2x^2 - 6ix$.

[18] $P(i) = 2i \neq 0 \Rightarrow x - i$ is not a factor of $P(x) = x^3 + (1 - i)x^2 + (2 + i)x + 2$.

[19] $P(1 - i) = -1 - 6i \neq 0 \Rightarrow m - 1 + i$ is not a factor of $P(m) = m^3 + 2m^2 + 1$.

[20] $P(-1 + 2i) = 0 \Rightarrow x + 1 - 2i$ is a factor of $P(x) = x^3 + 2x^2 + 5x$.

[21] $P\left(\dfrac{2}{3}\right) = -4 \neq 0 \Rightarrow 3x - 2 = 3\left(x - \dfrac{2}{3}\right)$ is not a factor of $P(x) = 3x^4 + x^3 - 5x^2 - x - 2$.

[22]
$$\begin{array}{r|rrrrrr} -\dfrac{1}{2} & 4 & 0 & -17 & 0 & 0 & -2 \\ & & -2 & 1 & 8 & -4 & 2 \\ \hline & 4 & -2 & -16 & 8 & -4 & 0 \end{array}$$

$4y^5 - 17y^3 - 2 = \left(y + \dfrac{1}{2}\right)\left(4y^4 - 2y^3 - 16y^2 + 8y - 4\right) = \left(y + \dfrac{1}{2}\right) \cdot 2 \cdot \left(2y^4 - y^3 - 8y^2 + 4y - 2\right) =$

$(2y + 1)\left(2y^4 - y^3 - 8y^2 + 4y - 2\right)$

[23] $(x-2)(x+1)(x-3)=x^3-4x^2+x+6.$

[24] $(x-5)(x+\sqrt{2})(x-\sqrt{2})=x^3-5x^2-2x+10.$

[25] $(x+5)(x)(x-2)=x^3+3x^2-10x.$ [26] $(x+1)(x-1)(x-\tfrac{1}{2})\cdot 2 = 2x^3-x^2-2x+1.$

[27] $(x+i)(x-i)(x+2)(x-1)=x^4+x^3-x^2+x-2.$

[28] $[x-(1-2i)][x-(1+2i)](x+3)^2=x^4+4x^3+2x^2+12x+45.$

[29] $(x+\tfrac{1}{2})(x-\tfrac{3}{4})(x)(x-2)\cdot 8 = 8x^4-18x^3+x^2+6x.$

[30] $(x+\tfrac{2}{3})(x-\tfrac{1}{8})(x+i)(x-i)\cdot 24 = 24x^4+13x^3+22x^2+13x-2.$

[31] $(x+2i)(x-2i)(x-4)(x+1)^2=x^5-2x^4-3x^3-12x^2-28x-16.$

[32] $(x+\tfrac{1}{3}i)(x-\tfrac{1}{3}i)(x+2)(x-2)(x)\cdot 9 = 9x^5-35x^3-4x.$

[33]
$$\begin{array}{r|rrrr} 2 & 1 & -6 & 12 & -8 \\ & & 2 & -8 & 8 \\ \hline & 1 & -4 & 4 & 0 \end{array}$$

$x^3-6x^2+12x-8=(x-2)(x^2-4x-4)=(x-2)(x-2)(x-2).$ Thus, 2 is a triple root.

[34]
$$\begin{array}{r|rrrr} -1 & 1 & 7 & 11 & 5 \\ & & -1 & -6 & -5 \\ \hline & 1 & 6 & 5 & 0 \end{array}$$

$t^3+7t^2+11t+5=(t+1)(t^2+6t+5)=(t+1)(t+1)(t+5).$ Thus, -5 and -1 are the remaining roots.

[35]
$$\begin{array}{r|rrrr} -\tfrac{3}{2} & 2 & 5 & -1 & -6 \\ & & -3 & -3 & 6 \\ \hline & 2 & 2 & -4 & 0 \end{array}$$

$2y^3+5y^2-y-6=(y+\tfrac{3}{2})(2y^2+2y-4)=2(y+\tfrac{3}{2})(y^2+y-2)=2(y+\tfrac{3}{2})(y+2)(y-1).$

Thus, -2 and 1 are the remaining roots.

[36]
$$\begin{array}{r|rrrr} \tfrac{2}{5} & 5 & -2 & 5 & -2 \\ & & 2 & 0 & 2 \\ \hline & 5 & 0 & 5 & 0 \end{array}$$

$5x^3-2x^2+5x-2=(x-\tfrac{2}{5})(5x^2-5)=5(x-\tfrac{2}{5})(x^2-1)=5(x-\tfrac{2}{5})(x+1)(x-1).$

Thus, -1 and 1 are the remaining roots.

[37]
$$\begin{array}{r|rrrrrr} -i & 1 & 0 & 1 & -1 & 0 & -1 \\ & & -i & -1 & 0 & i & 1 \\ \hline & 1 & -i & 0 & -1 & i & 0 \end{array} \rightarrow \begin{array}{r|rrrrr} i & 1 & -i & 0 & -1 & i \\ & & i & 0 & 0 & -i \\ \hline & 1 & 0 & 0 & -1 & 0 \end{array}$$

$m^5+m^3-m^2-1=(m+i)(m-i)(m^3-1)=(m+1)(m-i)(m-1)(m^2+m+1).$ Using the

quadratic formula, the roots of $m^2+m+1=0$ are $-\tfrac{1}{2}\pm\tfrac{\sqrt{3}}{2}i.$ Thus, the remaining roots are

1 and $-\tfrac{1}{2}\pm\tfrac{\sqrt{3}}{2}i.$

[38]
$$\begin{array}{r|rrrrrr} -3i & 1 & 0 & 9 & 1 & 0 & 9 \\ & & -3i & -9 & 0 & -3i & -9 \\ \hline & 1 & -3i & 0 & 1 & -3i & 0 \end{array} \rightarrow \begin{array}{r|rrrrr} 3i & 1 & -3i & 0 & 1 & -3i \\ & & 3i & 0 & 0 & 3i \\ \hline & 1 & 0 & 0 & 1 & 0 \end{array}$$

$x^5 + 9x^3 + x^2 + 9 = (x + 3i)(x - 3i)(x^3 + 1) = (x + 3i)(x - 3i)(x + 1)(x^2 - x + 1)$.   Using the

quadratic formula, the roots of $x^2 - x + 1 = 0$ are $\frac{1}{2} \pm \frac{\sqrt{3}}{2} i$. Thus, the remaining roots are

-1 and $\frac{1}{2} \pm \frac{\sqrt{3}}{2} i$.

[39]
$$\begin{array}{r|rrrrr} -\sqrt{3} & 4 & 0 & -13 & 0 & 3 \\ & & -4\sqrt{3} & 12 & \sqrt{3} & -3 \\ \hline & 4 & -4\sqrt{3} & -1 & \sqrt{3} & 0 \end{array} \rightarrow \begin{array}{r|rrrr} \sqrt{3} & 4 & -4\sqrt{3} & -1 & \sqrt{3} \\ & & 4\sqrt{3} & 0 & -\sqrt{3} \\ \hline & 4 & 0 & -1 & 0 \end{array}$$

$4x^4 - 13x^2 + 3 = (x + \sqrt{3})(x - \sqrt{3})(4x^2 - 1) = (x + \sqrt{3})(x - \sqrt{3})(2x - 1)(2x + 1)$. Thus, the

remaining roots are $\pm \frac{1}{2}$.

[40] $6x^5 + 5x^4 - 29x^3 - 25x^2 - 5x = x(6x^4 + 5x^3 - 29x^2 - 25x - 5)$.

$$\begin{array}{r|rrrrr} -\frac{1}{2} & 6 & 5 & -29 & -25 & -5 \\ & & -3 & -1 & 15 & 5 \\ \hline & 6 & 2 & -30 & -10 & 0 \end{array} \rightarrow \begin{array}{r|rrrr} -\frac{1}{3} & 6 & 2 & -30 & -10 \\ & & -2 & 0 & 10 \\ \hline & 6 & 0 & -30 & 0 \end{array}$$

$x(6x^4 + 5x^3 - 29x^2 - 25x - 5) = x\left(x + \frac{1}{2}\right)\left(x - \frac{1}{3}\right)(6x^2 - 30) = 6x\left(x + \frac{1}{2}\right)\left(x - \frac{1}{3}\right)(x^2 - 5)$.

Thus, the remaining roots are 0 and $\pm\sqrt{5}$.

[41]
$$\begin{array}{r|rrrrr} 1-2i & 1 & -2 & 6 & -2 & 5 \\ & & 1-2i & -5 & 1-2i & -5 \\ \hline & 1 & -1-2i & 1 & -1-2i & 0 \end{array} \rightarrow \begin{array}{r|rrrr} 1+2i & 1 & -1-2i & 1 & -1-2i \\ & & 1+2i & 0 & 1+2i \\ \hline & 1 & 0 & 1 & 0 \end{array}$$

$x^4 - 2x^3 + 6x^2 - 2x + 5 = (x - 1 + 2i)(x - 1 - 2i)(x^2 + 1)$. Thus, the remaining roots are $\pm i$.

[42] $2y^4 - 7y^3 + 11y^2 + 3y + 7 = 0$.

$$\begin{array}{r|rrrrr} 2-\sqrt{3}\,i & 2 & -7 & 11 & 3 & 7 \\ & & 4-2\sqrt{3}\,i & -12-\sqrt{3}\,i & -5-\sqrt{3}\,i & -7 \\ \hline & 2 & -3-2\sqrt{3}\,i & -1-\sqrt{3}\,i & -2-\sqrt{3}\,i & 0 \end{array}$$

$$\begin{array}{r|rrrr} 2+\sqrt{3}\,i & 2 & -3-2\sqrt{3}\,i & -1-\sqrt{3}\,i & -2-\sqrt{3}\,i \\ & & 4+2\sqrt{3}\,i & 2+\sqrt{3}\,i & 2+\sqrt{3}\,i \\ \hline & 2 & 1 & 1 & 0 \end{array}$$

$2y^4 - 7y^3 + 11y^2 + 3y + 7 = (x - 2 + \sqrt{3}\,i)(x - 2 - \sqrt{3}\,i)(2x^2 + x + 1) = 0$. Using the quadratic

formula, the roots of $2x^2 + x + 1 = 0$ are $-\frac{1}{4} \pm \frac{\sqrt{7}}{4} i$. Thus, the remaining roots are $-\frac{1}{4} \pm \frac{\sqrt{7}}{4} i$.

[43]
$$\begin{array}{r|rrrrr} c & 1 & 0 & 3 & 0 & 4 \\ & & c & c^2 & c^3 + 3c & c^4 + 3c^2 \\ \hline & 1 & c & c^2 + 3 & c^3 + 3c & c^4 + 3c^2 + 4 \end{array}$$

Since $R(x) = c^4 + 3c^2 + 4 \ne 0$ for all $c \in \mathbf{R}$, $x^4 + 3x^2 + 4$ has no factor of the form $x - c$.

[44]
$$\begin{array}{r|rrrrr} -1 & k^2 & 0 & 3k & 0 & 2 \\ & & -k^2 & k^2 & -k^2 - 3k & k^2 + 3k \\ \hline & k^2 & -k^2 & k^2 + 3k & -k^2 - 3k & k^2 + 3k + 2 \end{array}$$

$R(x) = 0 \Rightarrow k^2 + 3k + 2 = 0 \Rightarrow (k + 2)(k + 1) = 0 \Rightarrow k = -2$ or $k = -1$.

[45]
$$\begin{array}{r|rrrr} 2 & k & 3 & 0 & -4k^2 \\ & & 2k & 4k+6 & 8k+12 \\ \hline & k & 2k+3 & 4k+6 & -4k^2+8k+12 \end{array}$$

$R(x)=0 \Rightarrow -4k^2+8k+12=0 \Rightarrow -4(k+1)(k-3)=0 \Rightarrow k=-1 \text{ or } k=3.$

[46]
$$\begin{array}{r|rrrr} 1 & 1 & 1 & -a & b \\ & & 1 & 2 & -a+2 \\ \hline & 1 & 2 & -a+2 & -a+b+2 \end{array} \qquad \begin{array}{r|rrrr} 1 & 1 & -1 & a & b \\ & & 1 & 0 & a \\ \hline & 1 & 0 & a & a+b \end{array}$$

$R(x)=a+b=0 \Rightarrow b=-a.$ Substituting $-a$ for b in $R(x)=-a+b+2=0 \Rightarrow -2a+2=0 \Rightarrow$ $2a=2 \Rightarrow a=1.$ Thus $a=1$ and $b=-1.$

[47] Let $P(x)=x^n+a^n.$ Since n is odd, $P(-a)=(-a)^n+a^n=-a^n+a^n=0.$ Hence, by the factor theorem, $x+a$ is a factor of $x^n+a^n.$

[48] Let c be a zero of the polynomial $P(x)=x^3-x+1.$ Then $P(c)=c^3-c+1=0 \Rightarrow c^3=c-1.$ Thus, $c^4=c^3 \cdot c=(c-1)c=c^2-c,$ and $c^6=c^3 \cdot c^3=(c-1)(c-1)=c^2-2c+1.$ Let $Q(x)=$ $-x^6+2x^4-x^2+1.$ Then $Q(c)=-c^6+2c^4-c^2+1=-(c^2-2c+1)+2(c^2-c)-c^2+1=$ $-c^2+2c-1+2c^2-2c-c^2+1=0 \Rightarrow c$ is a zero of $Q(x).$

[49] Since $P(x)$ has degree 3, we can write $P(x)=(x+1)(x-2)(x-a),$ for some number $a.$ Since $P(1)=6,$ it follows that $(1+1)(1-2)(1-a)=6.$ Thus, $2(-1)(1-a)=6,$ which means that $2a-2=6$ or that $a=4.$ Hence, $P(x)=(x+1)(x-2)(x-4)=x^3-5x^2+2x+8.$

[50] Let $P(x)=\left(x+\dfrac{1}{2}\right)(x+2i)(x-2i)(x-a),$ for some number $a.$ Then $P(0)=4 \Rightarrow$

$\left(0+\dfrac{1}{2}\right)(0+2i)(0-2i)(0-a)=4 \Rightarrow -2a=4 \Rightarrow a=-2.$ Thus,

$P(x)=\left(x+\dfrac{1}{2}\right)(x+2i)(x-2i)(x+2)=x^4+\dfrac{5}{2}x^3+5x^2+10x+4.$

[51] $P(-1)=3(-1)^{101}+9(-1)^{78}+2(-1)^{15}-4=-3+9-2-4=0.$ Since $P(-1)=0,$ $x+1$ is a factor of $P(x)=3x^{101}+9x^{78}+2x^{15}-4$ by the factor theorem.

[52] $P(-1)=(-1)^{84}-12(-1)^{43}+7(-1)-5=1+12-7-5=1.$ By the remainder theorem, 1 is the remainder when $P(x)=x^{84}-12x^{43}+7x-5$ is divided by $x+1.$

EXERCISES 3.4

[1] degree $=4$; roots are 1, 5 and $-\dfrac{3}{4}$ (multiplicity 2).

[2] degree $=8$; roots are $-3, -2$ (multiplicity 5) and $\dfrac{1}{2}$ (multiplicity 2)

[3] degree $=5$; roots are $\sqrt{3}$ (multiplicity 3) and $\pm 2i$

[4] degree $=4$; roots are $\pm\sqrt{5}$ and $-1\pm i$

[5] degree $=8$; $x^3-8=(x-2)(x^2+2x+4)$ and roots are 2 (multiplicity 4), $-1\pm\sqrt{3}\,i$ and 1 (multiplicity 2)

[6] degree $=8$; $x^4-16=(x+2)(x-2)(x^2+4)$ and roots are $\pm 2i$ (each with multiplicity 3) and ± 2

[7]
$$\begin{array}{r|rrrr} 2 & 1 & -3 & 0 & 4 \\ & & 2 & -2 & -4 \\ \hline & 1 & -1 & -2 & 0 \end{array} \rightarrow \begin{array}{r|rrr} 2 & 1 & -1 & -2 \\ & & 2 & 2 \\ \hline & 1 & 1 & 0 \end{array} \rightarrow \begin{array}{r|rr} 2 & 1 & 1 \\ & & 2 \\ \hline & 1 & 3 \end{array}$$

Thus, 2 is a double root of $x^3-3x^2+4=0$

[8]

$$\begin{array}{r|rrrrrr} 1 & 1 & 1 & -5 & -1 & 8 & -4 \\ & & 1 & 2 & -3 & -4 & 4 \\ \hline & 1 & 2 & -3 & -4 & 4 & 0 \end{array} \rightarrow \begin{array}{r|rrrrr} 1 & 1 & 2 & -3 & -4 & 4 \\ & & 1 & 3 & 0 & -4 \\ \hline & 1 & 3 & 0 & -4 & 0 \end{array} \rightarrow \begin{array}{r|rrrr} 1 & 1 & 3 & 0 & -4 \\ & & 1 & 4 & 4 \\ \hline & 1 & 4 & 4 & 0 \end{array} \rightarrow \begin{array}{r|rrr} 1 & 1 & 4 & 4 \\ & & 1 & 5 \\ \hline & 1 & 5 & 9 \end{array}$$

Thus, 1 is a triple root of $x^5 + x^4 - 5x^3 - x^2 + 8x - 4 = 0$

[9]

$$\begin{array}{r|rrrrrrr} i & 1 & 0 & 4 & 0 & 5 & 0 & 2 \\ & & i & -1 & 3i & -3 & 2i & -2 \\ \hline & 1 & i & 3 & 3i & 2 & 2i & 0 \end{array} \rightarrow \begin{array}{r|rrrrrr} i & 1 & i & 3 & 3i & 2 & 2i \\ & & i & -2 & i & -4 & -2i \\ \hline & 1 & 2i & 1 & 4i & -2 & 0 \end{array} \rightarrow \begin{array}{r|rrrrr} i & 1 & 2i & 1 & 4i & -2 \\ & & i & -3 & -2i & -2 \\ \hline & 1 & 3i & -2 & 2i & -4 \end{array}$$

Thus, i is a double root of $x^6 + 4x^4 + 5x^2 + 2 = 0$.

[10]

$$\begin{array}{r|rrrrrr} -\frac{1}{2} & 32 & 16 & -48 & -56 & -22 & -3 \\ & & -16 & 0 & 24 & 16 & 3 \\ \hline & 32 & 0 & -48 & -32 & -6 & 0 \end{array} \rightarrow \begin{array}{r|rrrrr} -\frac{1}{2} & 32 & 0 & -48 & -32 & -6 \\ & & -16 & 8 & 20 & 6 \\ \hline & 32 & -16 & -40 & -12 & 0 \end{array} \rightarrow$$

$$\begin{array}{r|rrrr} -\frac{1}{2} & 32 & -16 & -40 & -12 \\ & & -16 & 16 & 12 \\ \hline & 32 & -32 & -24 & 0 \end{array} \rightarrow \begin{array}{r|rrr} -\frac{1}{2} & 32 & -32 & -24 \\ & & -16 & 24 \\ \hline & 32 & -48 & 0 \end{array} \rightarrow \begin{array}{r|rr} -\frac{1}{2} & 32 & -48 \\ & & -16 \\ \hline & 32 & -64 \end{array}$$

Thus, $-\frac{1}{2}$ is a root of multiplicity 4 *of* $32x^5 + 16x^4 - 48x^3 - 56x^2 - 22x - 3 = 0$.

[11]

$$\begin{array}{r|rrrrrr} \frac{1}{3} & 81 & -108 & -27 & 96 & -53 & 12 & -1 \\ & & 27 & -27 & -18 & 26 & -9 & 1 \\ \hline & 81 & -81 & -54 & 78 & -27 & 3 & 0 \end{array} \rightarrow \begin{array}{r|rrrrrr} \frac{1}{3} & 81 & -81 & -54 & 78 & -27 & 3 \\ & & 27 & -18 & -24 & 18 & -3 \\ \hline & 81 & -54 & -72 & 54 & -9 & 0 \end{array} \rightarrow$$

$$\begin{array}{r|rrrrr} \frac{1}{3} & 81 & -54 & -72 & 54 & -9 \\ & & 27 & -9 & -27 & 9 \\ \hline & 81 & -27 & -81 & 27 & 0 \end{array} \rightarrow \begin{array}{r|rrrr} \frac{1}{3} & 81 & -27 & -81 & 27 \\ & & 27 & 0 & -27 \\ \hline & 81 & 0 & -81 & 0 \end{array} \rightarrow \begin{array}{r|rrr} \frac{1}{3} & 81 & 0 & -81 \\ & & 27 & 9 \\ \hline & 81 & 27 & -72 \end{array}$$

Thus, $\frac{1}{3}$ is a root of multiplicity 4 of $81x^6 - 108x^5 - 27x^4 + 96x^3 - 53x^2 + 12x - 1 = 0$.

[12]

$$\begin{array}{r|rrrrrrr} -2i & 1 & 0 & 12 & 0 & 48 & 0 & 64 \\ & & -2i & -4 & -16i & -32 & -32i & -64 \\ \hline & 1 & -2i & 8 & -16i & 16 & -32i & 0 \end{array} \rightarrow \begin{array}{r|rrrrrr} -2i & 1 & -2i & 8 & -16i & 16 & -32i \\ & & -2i & -8 & 0 & -32 & 32i \\ \hline & 1 & -4i & 0 & -16i & -16 & 0 \end{array} \rightarrow$$

$$\begin{array}{r|rrrrr} -2i & 1 & -4i & 0 & -16i & -16 \\ & & -2i & -12 & 24i & 16 \\ \hline & 1 & -6i & -12 & 8i & 0 \end{array} \rightarrow \begin{array}{r|rrrr} -2i & 1 & -6i & -12 & 8i \\ & & -2i & -16 & 56i \\ \hline & 1 & -8i & -28 & 64i \end{array}$$

Thus, $-2i$ is a triple root of $x^6 + 12x^4 + 48x^2 + 64 = 0$.

[13] $P(x) = (x+1)^2(x-3)(x-5) = x^4 - 6x^3 + 22x + 15$

[14] $P(x) = \left(x + \frac{1}{2}\right)\left(x - \frac{1}{3}\right)^3(x-6) = x^5 - \frac{13}{2}x^4 + \frac{17}{6}x^3 + \frac{61}{54}x^2 - \frac{43}{54}x + \frac{1}{9}$

[15] $P(x) = x(x+3)^2\left(x - \frac{1}{2}\right)^2 = x^5 + 5x^4 + \frac{13}{4}x^3 - \frac{15}{2}x^2 + \frac{9}{4}x$

[16] $P(x) = \left(x - \frac{2}{3}\right)\left(x + \frac{5}{6}\right)(x+2)^3 = x^5 + \frac{37}{6}x^4 + \frac{112}{9}x^3 + \frac{20}{3}x^2 - \frac{16}{3}x - \frac{40}{9}$

[17] Since c must be a factor of 5 and d a factor of 2, the values of $\frac{c}{d}$ are $\pm 1, \pm 5, \pm \frac{1}{2}$ and $\pm \frac{5}{2}$.

[18] Since c must be a factor of 3 and d a factor of 1, the values of $\frac{c}{d}$ are ± 1, and ± 3.

[19] Since c must be a factor of 8 and d a factor of 16, the values of $\frac{c}{d}$ are $\pm 1, \pm 2, \pm 4, \pm 8, \pm \frac{1}{2}, \pm \frac{1}{4}$, $\pm \frac{1}{8}$ and $\pm \frac{1}{16}$.

[20] Since c and d must each be factors of 1, the only values of $\frac{c}{d}$ are ± 1.

[21] Multiplying the given equation by 4 gives us $2m^4 - 3m^2 + 8m - 36 = 0$. Since c must be a factor of 36 and d a factor of 2, the values of $\frac{c}{d}$ are $\pm 1, \pm 2, \pm 3, \pm 4, \pm 6, \pm 9, \pm 18, \pm 36, \pm \frac{1}{2}, \pm \frac{3}{2}, \pm \frac{9}{2}$.

[22] $8\left(\frac{7}{8}x^4 - \frac{3}{4}x^2 + 1 = 0\right) \Rightarrow 7x^4 - 6x^2 + 8 = 0$. Since c must be a factor of 8 and d a factor of 7, the values of $\frac{c}{d}$ are $\pm 1, \pm 2, \pm 4, \pm 8, \pm \frac{1}{7}, \pm \frac{2}{7}, \pm \frac{4}{7}$, and $\pm \frac{8}{7}$.

[23] $-2r^9 + r^6 + 8r^4 - r = 0 \Rightarrow -r\left(2r^8 - r^5 - 8r^3 + 1\right) = 0$. Since c must be a factor of 1 and d a factor of 2, the values of $\frac{c}{d}$ are ± 1 and $\pm \frac{1}{2}$. Of course, 0 is also a rational root.

[24] $\frac{1}{2}\left(12y^3 - 2y^2 + 8y - 16 = 0\right) \Rightarrow 6y^3 - y^2 + 4y - 8 = 0$. Since c must be a factor of 8 and d a factor of 6, the values of $\frac{c}{d}$ are $\pm 1, \pm 2, \pm 4, \pm 8, \pm \frac{1}{2}, \pm \frac{1}{3}, \pm \frac{1}{6}, \pm \frac{2}{3}, \pm \frac{4}{3}$ and $\pm \frac{8}{3}$.

In Exercises 25 – 30, the values of c, d and $\frac{c}{d}$ are given. In each case, synthetic division and the remainder theorem will verify that none of possible rational roots $\frac{c}{d}$ is actually a zero of $P(x)$.

[25] $P(x) = 2x^3 - 9x^2 - 2x + 5.$ $c = 5, d = 2, \frac{c}{d} = \pm 1, \pm 5, \pm \frac{1}{2}, \pm \frac{5}{2}.$

[26] $P(x) = -x^4 + 5x^2 - 8.$ $c = 8, d = 1, \frac{c}{d} = \pm 1, \pm 2, \pm 4, \pm 8.$

[27] $P(y) = y^3 + y^2 - 8.$ $c = 8, d = 1, \frac{c}{d} = \pm 1, \pm 2, \pm 4, \pm 8.$

[28] $P(m) = m^6 + 2m^4 - 11m^2 - 4.$ $c = 4, d = 1, \frac{c}{d} = \pm 1, \pm 2, \pm 4.$

[29] $P(x) = 3x^4 - x^3 + x - 6.$ $c = 6, d = 3, \frac{c}{d} = \pm 1, \pm 2, \pm 3, \pm 6, \pm \frac{1}{3}, \pm \frac{2}{3}.$

[30] $P(t) = 4t^3 + 9t^2 - 14t + 2.$ $c = 2, d = 4, \frac{c}{d} = \pm 1, \pm 2, \pm \frac{1}{4}, \pm \frac{1}{2}.$

In Exercises 31 – 38, PRR denotes possible rational roots.

[31] $x^4 - 4x^3 - 2x^2 + 21x - 18 = 0.$ PRR $= \pm 1, \pm 2, \pm 3, \pm 6, \pm 9, \pm 18.$ ● $2, 3$

[32] $8x^3 - 4x^2 - 2x + 1 = 0.$ PRR $= \pm 1, \pm \frac{1}{2}, \pm \frac{1}{4}, \pm \frac{1}{8}.$ ● $-\frac{1}{2}, \frac{1}{2}$ (double root)

[33] $2y^4 - 13y^3 + 30y^2 - 28y + 8 = 0.$ PRR $= \pm 1, \pm 2, \pm 4, \pm 8, \pm \frac{1}{2}.$ ● $\frac{1}{2}, 2$ (triple root)

[34] $12t^3 - 16t^2 - 5t + 3 = 0.$ PRR $= \pm 1, \pm 3, \pm \frac{1}{2}, \pm \frac{1}{3}, \pm \frac{1}{4}, \pm \frac{1}{6}, \pm \frac{1}{12}, \pm \frac{3}{2}, \pm \frac{3}{4}.$ ● $-\frac{1}{2}, \frac{1}{3}, \frac{3}{2}$

[35] $x^5 + x^4 - 5x^3 - x^2 + 8x - 4 = 0.$ PRR $= \pm 1, \pm 2, \pm 4.$ ● -2 (double root), 1 (triple root)

[36] $8x^3 - 4x^2 - 2x + 1 = 0.$ PRR $= \pm 1, \pm \frac{1}{2}, \pm \frac{1}{4}, \pm \frac{1}{8}.$ ● $-\frac{1}{2}, \frac{1}{2}$ (double root)

[37] $6\left(2m^3 - \frac{16}{3}m^2 + \frac{1}{2}m + \frac{5}{6} = 0\right) \Rightarrow 12m^3 - 32m^2 + 3m + 5 = 0.$ PRR $= \pm 1, \pm 5, \pm \frac{1}{2}, \pm \frac{1}{3}, \pm \frac{1}{4},$ $\pm \frac{1}{6}, \pm \frac{1}{12}, \pm \frac{5}{2}, \pm \frac{5}{3}, \pm \frac{5}{4}, \pm \frac{5}{6}, \pm \frac{5}{12}.$ ● $-\frac{1}{3}, \frac{1}{2}, \frac{5}{2}$

[38] $8\left(4y^5 - \frac{1}{2}y^4 - \frac{1}{8}y^3 - 4y^2 + \frac{1}{2}y + \frac{1}{8} = 0\right) \Rightarrow 32y^5 - 4y^4 - y^3 - 32y^2 + 4y + 1 = 0.$ PRR $= \pm 1,$ $\pm \frac{1}{2}, \pm \frac{1}{4}, \pm \frac{1}{8}, \pm \frac{1}{16}, \pm \frac{1}{32}.$ ● $-\frac{1}{8}, \frac{1}{4}, 1$

In Exercises 39 – 50, PRR denotes possible rational roots, RR denotes rational roots, and RE denotes reduced equation.

[39] $\frac{1}{2}\left(2x^3-10x^2+12x-4=0\right)\Rightarrow x^3-5x^2+6x-2=0.$ PRR $=\pm 1,\pm 2.$ RR $=1.$ Solving the RE,

$x^2-4x+2=0,$ using the quadratic formula, we obtain the roots $2\pm\sqrt{2}.$ ●$1,2\pm\sqrt{2}$

[40] $x^4+3x^3+3x^2+3x+2=0.$ PRR $=\pm 1,\pm 2.$ RR $=-2$ and $1.$ The roots of the RE, $x^2+1=0$ are $\pm i$.

●$-2,1,\pm i$

[41] $3t^4-4t^3-8t^2+9t-2=0.$ PRR $=\pm 1,\pm 2,\pm\frac{1}{3},\pm\frac{2}{3}.$ RR $=\frac{1}{3}$ and $2.$ Using the quadratic formula,

the roots of the RE, $3x^2+3x-3=0,$ are $\pm\dfrac{-1\pm\sqrt{5}}{2}.$ ●$\frac{1}{3},2,\dfrac{-1\pm\sqrt{5}}{2}$

[42] $18y^3-21y^2-10y+8=0.$ PRR $=\pm 1,\pm 2,\pm 4,\pm 8,\pm\frac{1}{2},\pm\frac{1}{3},\pm\frac{1}{6},\pm\frac{1}{9},\pm\frac{1}{18},\pm\frac{2}{3},\pm\frac{2}{9},\pm\frac{4}{3},\pm\frac{4}{9},$

$\pm\frac{8}{3},\pm\frac{8}{9}.$ RR $=\frac{1}{2},-\frac{2}{3},\frac{4}{3}.$ ●$-\frac{2}{3},\frac{1}{2},\frac{4}{3}$

[43] $3x^3+8x^2+19x+10=0.$ PRR $=\pm 1,\pm 2,\pm 5,\pm 10,\pm\frac{1}{3},\pm\frac{2}{3},\pm\frac{5}{3},\pm\frac{10}{3}.$ RR $=-\frac{2}{3}.$ Using the

quadratic formula, the roots of the RE, $x^2+2x+5=0,$ are $-1\pm 2i$. ●$-\frac{2}{3},-1\pm 2i$

[44] $2x^5-3x^4-2x+3=0.$ PRR $=\pm 1,\pm 3,\pm\frac{1}{2},\pm\frac{3}{2}.$ RR $=-1,1,\frac{3}{2}.$ The roots of the RE, $2x^2+2=0,$

are $\pm i$. ●$-1,1,\frac{3}{2},\pm i$.

[45] $6\left(x^3-\frac{11}{6}x^2-\frac{2}{3}x+\frac{2}{3}=0\right)\Rightarrow 6x^3-11x^2-4x+4=0.$ PRR $=\pm 1,\pm 2,\pm 4,\pm\frac{1}{2},\pm\frac{1}{3},\pm\frac{1}{6},\pm\frac{2}{3},$

$\pm\frac{4}{3}.$ RR $=2,\frac{1}{2},-\frac{2}{3}.$ ●$-\frac{2}{3},\frac{1}{2},2$

[46] $2\left(4x^4+3x^3-\frac{15}{2}x^2-6x-1=0\right)\Rightarrow 8x^4+6x^3-15x^2-12x-2=0.$ PRR $=\pm 1,\pm 2,\pm\frac{1}{2},\pm\frac{1}{4},$

$\pm\frac{1}{8}.$ RR $=-\frac{1}{2},-\frac{1}{4}.$ Using the quadratic formula, the roots of the RE, $8x^2-16=0,$ *are* $\pm\sqrt{2}.$

●$-\frac{1}{2},-\frac{1}{4},\pm\sqrt{2}$

[47] $9x^3-7x+2=0.$ RR $=-1,\frac{1}{3},\frac{2}{3}.$ Thus, $9x^3-7x+2=9(x+1)\left(x-\frac{1}{3}\right)\left(x-\frac{2}{3}\right).$

[48] $x^4-7x^3+18x^2-20x+8=0.$ RR $=1,2$ (triple root). Thus, $x^4-7x^3+18x^2-20x+8=(x-1)\cdot$
$(x-2)^3.$

[49] $3x^4+4x^3+2x^2+8x-8=0.$ *RR* $=-2,\frac{2}{3}.$ Using the quadratic formula, the roots of the RE,

$3x^2+6=0$ are $\pm\sqrt{2}\,i.$ Thus, $3x^4+4x^3+2x^2+8x-8=3(x+2)\left(x-\frac{2}{3}\right)\left(x+\sqrt{2}\,i\right)\left(x-\sqrt{2}\,i\right).$

[50] $x^5+10x^2-x-10=0.$ RR $=-2,\pm 1.$ Using the quadratic formula, the roots of the RE,

$x^2-2x+5=0$ are $1\pm 2i.$ Thus, $x^5+10x^2-x-10=(x+2)(x+1)(x-1)\left(x-(1-2i)\right)$

$\left(x-(1+2i)\right).$

[51] a) Since $\left(-\sqrt{2}\right)^2-2=2-2=0,$ and $\left(\sqrt{2}\right)^2-2=2-2=0,-\sqrt{2}$ and $\sqrt{2}$ are solutions of the equation

$x^2-2=0.$ By Theorem 3.2, the equation $x^2-2=0$ has no other solutions.

b)   If $\frac{c}{d}$ is a rational root of the equation $x^2 - 2 = 0$, then c must be a factor of 2 and d a factor of 1.

Thus ± 1 and ± 2 are the only possible rational roots. By part (a), we know that ± 1 and ± 2 are not roots of $x^2 - 2 = 0$. Since $x^2 - 2 = 0$ has no rational roots, both $-\sqrt{2}$ and $\sqrt{2}$ must be irrational numbers.

[52]   As in Exercise[51], we can verify that $-\sqrt{5}$ and $\sqrt{5}$ are the only solutions of the equation $x^2 - 5 = 0$. By the rational root theorem, the only possible rational roots of $x^2 - 5 = 0$ are ± 1 and ± 5.

Since ± 1 and ± 5 are not roots of $x^2 - 5 = 0$, it follows that $-\sqrt{5}$ and $\sqrt{5}$ are both irrational numbers.

[53]   By the rational root theorem, the only possible rational roots of $x^3 - k^2 x^2 + 3k\ x - 1 = 0$ are ± 1. The synthetic division by $x + 1$ yields the remainder $k^2 - 3k - 2$ and the synthetic division by $x - 1$ yields the remainder $-k^2 + 3k$. Solving the equation $k^2 - 3k - 2 = 0$ and $-k^2 + 3k = 0$ by factoring, we find that the possible values of k are $-2, -1, 0$ and 3.

[54]   By the rational root theorem, the only possible rational roots of $x^3 + p\ x - p = 0$ are ± 1 and $\pm p$. Synthetic division by $x - 1$ gives us a remainder of 1. Thus 1 is not a root. Synthetic division by $x + 1$ gives us a remainder of $-2p - 1$. If $-2p - 1 = 0$, it follows that $p = \frac{1}{2}$. Since p is a prime this is impossible. Thus -1 is not a root. Synthetic division by $x - p$ yields the remainder $p^3 + p^2 - p$.

Now, $p^3 + p^2 - p = 0 \Rightarrow p\left(p^2 + p - 1\right) = 0 \Rightarrow p^2 + p - 1 = 0 \Rightarrow p^2 + p = 1$. Since p is a prime, $p \geq 2$, and $p^2 + p \geq 6$. Thus $p^2 + p \neq 1$ and p is not a root. Synthetic division by $x + p$ gives us the remainder $-p^3 - p^2 - p$. Now $-p^3 - p^2 - p = 0 \Rightarrow p^3 + p^2 + p = 0 \Rightarrow p\left(p^2 + p + 1\right) = 0 \Rightarrow p^2 + p + 1 = 0 \Rightarrow p^2 + p = -1$, which is impossible. Thus $-p$ is not a root and the original equation has no rational roots.

[55]   The x-intercepts, if any are the real solutions of $6x^3 + 13x^2 + x - 2 = 0$. The only possible rational roots are $\pm 1, \pm 2, \pm \frac{1}{2}, \pm \frac{1}{3}, \pm \frac{1}{6}$ and $\pm \frac{2}{3}$. Using synthetic division and the remainder theorem, we find that -2 is a root. Solving the reduced equation $6x^2 + x - 1 = 0$ by factoring, we obtain the roots $-\frac{1}{2}$ and $\frac{1}{3}$. Thus, the x-intercepts of the graph are $\left(-2, 0\right)\left(-\frac{1}{2}, 0\right)$ and $\left(\frac{1}{3}, 0\right)$.

[56]   The x-intercepts, if any, are the real solution of $x^4 + 2x^3 - 6x^2 - 16x - 8 = 0$. The only possible rational roots are $\pm 1, \pm 2, \pm 4$ and ± 8. Using synthetic division and the remainder theorem, we find that -2 is a double root. Solving the reduced equation $x^2 - 2x - 2 = 0$ using the quadratic formula, we obtain the roots $1 \pm \sqrt{3}$. Thus the x-intercepts of the graph are $\left(-2, 0\right), \left(1 - \sqrt{3}, 0\right)$ and $\left(1 + \sqrt{3}, 0\right)$.

[57] $f(x) = g(x) \Rightarrow 5x^3 - 6x^2 - x + 1 = x^3 + 6x^2 - 10x + 3 \Rightarrow 4x^3 - 12x^2 + 9x - 2 = 0$. The only possible rational roots are $\pm 1, \pm 2, \pm \frac{1}{2}$ and $\frac{1}{4}$. Using synthetic division and the remainder theorem, we find that 2 is a root. Solving the reduced equation $4x^2 - 4x + 1 = 0$ by factoring we find that $\frac{1}{2}$ is a double root. Thus the graphs intersect at the points $\left(\frac{1}{2}, -\frac{3}{8}\right)$ and $\left(2, 15\right)$.

[58] $f(x) = g(x) \Rightarrow 5x^3 + 3x^2 - 2x - 10 = 2x^4 + 4x^3 - 7x^2 + 2 \Rightarrow 2x^4 - x^3 - 10x^2 + 2x + 12 = 0.$

The only possible rational roots are $\pm 1, \pm 2, \pm 3, \pm 4, \pm 6, \pm 12, \pm\frac{1}{2}$ and $\pm\frac{3}{2}$. Using synthetic division and the remainder theorem, we find that 2 and $-\frac{3}{2}$ are roots. From the reduced equation $2x^2 - 4 = 0$, or $x^2 - 2 = 0$, we obtain the roots $\pm\sqrt{2}$. Thus the graphs intersect at the points $\left(-\frac{3}{2}, -\frac{137}{8}\right)$, $\left(-\sqrt{2}, -4 - 8\sqrt{2}\right), \left(\sqrt{2}, -4 + 8\sqrt{2}\right)$ and $(2, 38)$.

[59] $f(x) = g(x) \Rightarrow x^5 + 4x^3 - 3x - 5 = 3x^3 + 2x^2 + 9x + 3 \Rightarrow x^5 + x^3 - 2x^2 - 12x - 8 = 0.$ The only possible rational roots are $\pm 1, \pm 2, \pm 4$ and ± 8. Synthetic division and the remainder theorem will confirm that 2 is a root and that -1 is a double root. From the reduced equation $x^2 + 4 = 0$, we obtain the (nonreal) roots $\pm 2i$. Thus, the graphs of f and g intersect at $(-1, -7)$ and $(2, 53)$.

[60] $f(x) = g(x) \Rightarrow x^5 + 4x^4 - 9x^3 - 21x + 13 = 5x^5 - 12x^4 + 16x^3 - 25x^2 + 4 \Rightarrow 4x^5 - 16x^4 + 25x^3 - 25x^2 + 21x - 9 = 0.$ The only possible rational roots are $\pm 1, \pm 3, \pm 9, \pm\frac{1}{2}, \pm\frac{1}{4}, \pm\frac{3}{2}, \pm\frac{3}{4}, \pm\frac{9}{2}$ and $\pm\frac{9}{4}$. Synthetic division and the remainder theorem will confirm that 1 is a root and that $\frac{3}{2}$ is a double root. From the reduced equation $4x^2 + 4 = 0$ *or* $x^2 + 1 = 0$, we obtain the (nonreal) roots $\pm i$. Thus the graphs of f and g intersect at $\left(1, -12\right)$ and $\left(\frac{3}{2}, -\frac{673}{32}\right)$.

EXERCISES 3.5

In Exercises 1 – 8, $\text{VSP}(x)$ and $\text{VSP}(-x)$ denote the variations in sign in $P(x)$ and $P(-x)$, respectively. PR, NR and NC denote positive real roots, negative real roots and nonreal complex roots, respectively.

[1] $\text{VSP}(x) = 1, \text{VSP}(-x) = 2 \Rightarrow$

PR	NR	NC
1	2	0
1	0	2

[2] $\text{VSP}(x) = 1, \text{VSP}(-x) = 2 \Rightarrow$

PR	NR	NC
1	2	0
1	0	2

[3] $\text{VSP}(x) = 5, \text{VSP}(-x) = 0 \Rightarrow$

PR	NR	NC
5	0	0
3	0	2
1	0	4

[4] $\text{VSP}(x) = 3, \text{VSP}(-x) = 1 \Rightarrow$

PR	NR	NC
3	1	0
1	1	2

[5] $\text{VSP}(x) = 2, \text{VSP}(-x) = 1 \Rightarrow$

PR	NR	NC
2	1	2
0	1	4

[6] $\text{VSP}(x) = 0, \text{VSP}(-x) = 1 \Rightarrow$

PR	NR	NC
0	1	6

[7] $\text{VSP}(x) = 2, \text{VSP}(-x) = 1 \Rightarrow$

PR	NR	NC
2	1	0
0	1	2

[8] $\text{VSP}(x) = 2, \text{VSP}(-x) = 2 \Rightarrow$

PR	NR	NC
2	2	4
2	0	6
0	2	6
0	0	8

In Exercises 9 – 16, LB and UB denote lower bound and upper bound, respectively.

[9] $LB = -1, UB = 9$

[10] $LB = -1, UB = 2$

[11] $LB = -2, UB = 2$

[12] $LB = -2, UB = 5$

[13] $LB = -1, UB = 3$

[14] $LB = -1, UB = 6$

[15] $LB = -3, UB = 2$

[16] $LB = -2, UB = 1$

[17] Since $P(x) = 2x^5 + 6x^3 + 3x - 1$ has one variation in sign $P(x) = 0$ has only one positive real root. Since $P(-x) = -2x^5 - 6x^3 - 3x - 1$ has no variation in sign, there are no negative real roots. Since $P(x)$ has only one real zero, the graph of $y = P(x)$ crosses the x-axis exactly once.

[18] Since both $P(x) = 3x^4 + 8x^2 + 1$ and $P(-x) = 3x^4 + 8x^2 + 1$ have no variation in sign, $P(x) = 0$ has no real roots. Since $P(x)$ has no real zeros, the graph of $y = P(x)$ never crosses the x-axis.

[19] Since i is a zero of $P(x)$, its conjugate $-i$ must also be a zero. Thus, $P(x) = (x+2)(x-2) \cdot (x+i)(x-i) = x^4 - 3x^2 - 4$.

[20] The conjugate of $3+i$, which is $3-i$, must also be a zero of $P(x)$. Thus, $P(x) = x(x - (3+i)) \cdot (x - (3-i)) = x^3 - 6x^2 + 10x$.

[21] The conjugate of $-1-i$, namely $-1+i$, must also be a zero of $P(x)$. Thus, $P(x) = \left(x + \frac{3}{2}\right)(x-4) \cdot (x-(-1+i))(x-(-1-i)) = x^4 - \frac{1}{2}x^3 - 9x^2 - 17x - 12$.

[22] The conjugates of i and $-\sqrt{5}\,i$, which are $-i$ and $\sqrt{5}\,i$, respectively, must also be zeros of $P(x)$.

Thus, $P(x) = \left(x + \frac{\sqrt{3}}{2}\right)\left(x - \frac{\sqrt{3}}{2}\right)(x+i)(x-i)(x+\sqrt{5}\,i)(x-\sqrt{5}i) = x^6 + \frac{21}{4}x^4 + \frac{1}{2}x^2 - \frac{15}{4}$.

[23] The conjugates of $-\frac{1}{2}i$ and $1+4i$, which are $\frac{1}{2}i$ and $1-4i$, respectively, must also be zeros of $P(x)$. Thus, $P(x) = x\left(x + \frac{1}{2}i\right)\left(x - \frac{1}{2}i\right)(x-(1+4i))(x-(1-4i)) = x^5 - 2x^4 + \frac{69}{4}x^3 - \frac{1}{2}x^2 + \frac{17}{4}x$.

[24] The conjugate of $3-5i$, namely $3+5i$, must also be a zero of $P(x)$. Thus, $P(x) = \left(x + \frac{\sqrt{2}}{2}\right) \cdot \left(x - \frac{\sqrt{2}}{2}\right)(x-(3+5i))(x-(3-5i))\left(x - \frac{5}{2}\right) = x^5 - \frac{17}{2}x^4 + \frac{97}{2}x^3 - \frac{323}{4}x^2 - \frac{49}{2}x + \frac{85}{2}$.

[25] The conjugate of $\frac{1}{2} + \frac{\sqrt{3}}{2}i$, which is $\frac{1}{2} - \frac{\sqrt{3}}{2}i$, must also be a zero of $P(x)$. Thus, $P(x) = \left(x - \left(\frac{1}{2} + \frac{\sqrt{3}}{2}i\right)\right)\left(x - \left(\frac{1}{2} - \frac{\sqrt{3}}{2}i\right)\right)\left(x - \frac{\sqrt{3}}{2}\right) = x^3 - \left(1 + \frac{\sqrt{3}}{2}\right)x^2 + \left(1 + \frac{\sqrt{3}}{2}\right)x - \frac{\sqrt{3}}{2}$.

[26] The conjugate of $\frac{2}{3} - \frac{\sqrt{10}}{3}i$ and $-1+3i$, which are $\frac{2}{3} + \frac{\sqrt{10}}{3}i$ and $-1-3i$, respectively, must also be zeros of $P(x)$. Thus $P(x) = \left(x - \left(\frac{2}{3} + \frac{\sqrt{10}}{3}i\right)\right)\left(x - \left(\frac{2}{3} - \frac{\sqrt{10}}{3}i\right)\right)(x-(-1+3i)) \cdot (x-(-1-3i)) = x^4 + \frac{2}{3}x^3 + \frac{80}{9}x^2 - \frac{92}{9}x + \frac{140}{9}$.

[27] The only rational root of $2x^3 - 5x^2 + 1 = 0$ is $\frac{1}{2}$. Using the quadratic formula, we solve the reduced equation $2x^2 - 4x - 2 = 0$ to obtain the roots $1 \pm \sqrt{2}$.　　●$\frac{1}{2}$, $1 - 2\sqrt{2} \approx -1.8$, $1 + 2\sqrt{2} \approx 3.8$

[28] The only rational root of $2x^3 - 5x^2 + 6x - 3 = 0$ is 1. Using the quadratic formula, we solve the reduced equation $2x^2 - 3x + 3 = 0$ to obtain the roots $\frac{3}{4} \pm \frac{\sqrt{13}}{4}i$.　　●$1$

[29]   The only rational root of $x^4 - x^3 - 2 = 0$ is -1. Since each of $P(x) = x^4 - x^3 - 2$ and $P(-x) = x^4 + x^3 - 2$ has 1 variation in sign, $P(x) = 0$ has one positive real root and one negative real root. (The remaining 2 roots must be nonreal complex numbers.) Since $P(1) = -2$ and $P(2) = 6$ have opposite signs, the positive real root lies in the interval $(1, 2)$. Since $P(1.5) \approx -0.3$ and $P(1.6) \approx 0.5$ have opposite signs the root lies between 1.5 and 1.6. Since $P(1.55) \approx 0.05$ and $P(1.5)$ have opposite signs, the root is closer to 1.5.   ●$-1, 1.5$

[30]   It can be verified that $-3x^4 - x^3 + 2x^2 + 4 = 0$ has no rational roots. Since both $P(x) = -3x^4 - x^3 + 2x^2 + 4$ and $P(-x) = -3x^4 + x^3 + 2x^2 + 4$ have 1 variation in sign, there is 1 negative real root and 1 positive real root. Since $P(-2) = -28$ and $P(-1) = 4$ have opposite signs, the negative real root lies in the interval $(-2, -1)$. Since $P(1) = 2$ and $P(2) = -44$ have opposite signs, the positive root lies in the interval $(1, 2)$. Since $P(-1.4) \approx -0.9$ and $P(-1.3) \approx -1.0$ have opposite signs, the negative root is in the interval $(-1.4, -1.3)$. Since $P(-1.35) \approx 0.1$ and $P(-1.4)$ have opposite signs, the root is closer to -1.4. Similarly, $P(1.1) \approx 0.7$ and $P(1.2) \approx -1.1$ have opposite signs and the positive root lies in the interval $(1.1, 1.2)$. Since $P(1.15) \approx -0.1$ and $P(1.1)$ have opposite signs, the root is closer to 1.1.   ●$-1.4, 1.1$

[31]   It can be shown the $P(x) = 3x^5 + 2x + 9$ has no rational roots. Since $P(x)$ has no variation in sign, $P(x) = 0$ has no positive real roots. Since $P(-x) = -3x^5 - 2x + 9$ has 1 variation in sign, there is exactly 1 negative real root. From the fact that $P(-2) = -91$ and $P(-1) = 4$ have opposite signs, we know the root lies between -2 and -1. Since $P(-1.2) \approx -0.9$ and $P(-1.1) \approx 2.0$ have opposite signs, the root lies in the interval $(-1.2, -1.1)$. Since $P(-1.15) \approx 0.7$ and $P(-1.2)$ have opposite signs, the root is closer to -1.2. (Since there is only 1 real root, the remaining 4 roots must be nonreal complex numbers).   ●-1.2

[32]   Since $P(x) = 2x^5 + x^3 - 8$ and $P(-x) = -2x^5 - x^3 - 8$ have 1 and 0 variations in sign, respectively, $P(x) = 0$ has 1 positive real root and no negative real roots. (The remaining 4 roots must be nonreal complex numbers.) Since $P(1) = -5$ and $P(2) = 64$ have opposite signs, the root lies between 1 and 2. From the fact that $P(1.2) \approx -1.3$ and $P(1.3) \approx 1.6$ have opposite signs, we know that the root lies between 1.2 and 1.3. Since $P(1.25) \approx 0.1$ and $P(1.2)$ have opposite signs, the root is closer to 1.2.   ●1.2

[33]   Since $P(x) = x^4 - 7$ has only 1 variation in sign, the equation $x^4 - 7 = 0$ has only 1 positive real root, namely $\sqrt[4]{7}$. The pairs of opposite signs: $P(1) = -6$ and $P(2) = 9$; $P(1.6) \approx -0.4$ and $P(1.7) \approx 1.4$; and $P(1.62) \approx -0.1$ and $P(1.63) \approx 0.1$ place the root in the interval $(1.62, 1.63)$. Since $P(1.625) \approx -0.03$ and $P(1.63)$ have opposite signs, the root is closer to 1.63. Thus, $\sqrt[4]{7} \approx 1.63$.

[34]   Since $P(x) = x^5 - 12$ has only 1 variation in sign, the equation $x^5 - 12 = 0$ has only 1 positive real root, namely $\sqrt[5]{12}$. As in Exercise 33, the pairs of opposite signs: $P(1) = -11$ and $P(2) = 20$; $P(1.6) \approx -1.5$ and $P(1.7) \approx 2.2$; $P(1.64) \approx -0.1$ and $P(1.65) \approx 0.2$ place the root in the interval $(1.64, 1.65)$. Since $P(1.645) \approx 0.05$ and $P(1.64)$ have opposite signs, the root is closer to 1.64. Thus, $\sqrt[5]{12} \approx 1.64$.

[35] Let $P(x) = 3x^4 - x^3 + 2ix^2 + x - 3$. Since $P(i) = 0$ and $P(-i) = -4i \neq 0$, i is a root of $P(x) = 0$ and $-i$ is not a root of $P(x) = 0$. This is not a contradiction to Theorem 3.4 since the hypothesis of this Theorem requires that $P(x)$ have real coefficients.

[36] Let $P(x)$ be a polynomial of degree n, where n is odd, and assume that $P(x)$ has real coefficients. By Theorem 3.2 (p.135) we know that $P(x) = 0$ has exactly n roots, counting multiplicities. Let k denote the number of real roots of $P(x) = 0$ and let m denote the number of nonreal complex roots of $P(x) = 0$. Since $k + m = n$, and since m must be even by Theorem 3.4, it follows that k must be an odd positive integer. Thus $P(x)$ must have at least one real zero.

[37] Let $P(x) = x^3 - x^2 + 5x + 3$, and let $Q(x) = (x + 1)P(x) = x^4 + 4x^2 + 8x + 3$. Since $Q(x)$ has no variation in sign, the equation $Q(x) = 0$ has no positive real roots. Since $Q(x) = (x + 1) \cdot P(x)$, we see that any root of $P(x) = 0$ would also be a root of $Q(x) = 0$. Thus, the equation $P(x) = 0$ has no positive real roots.

[38] Let $P(x) = x^4 + x^3 + 2x^2 - 1$, and let $Q(x) = (x - 1) \cdot P(x) = x^5 + x^3 - 2x^2 - x + 1$. Since $Q(-x) = -x^5 - x^3 - 2x^2 + x + 1$ has only 1 variation in sign, the equation $Q(x) = 0$ has exactly 1 negative real root. Since $Q(x) = (x - 1) \cdot P(x)$, the negative root must also be a root of $P(x) = 0$. If $P(x) = 0$ had any additonal negative roots, they would also be negative roots of $Q(x) = 0$ and, hence, a contradiction to the fact that $Q(x) = 0$ has exactly 1 negative real root. Thus $P(x) = 0$ has exactly 1 negative real root.

[39] $C(x) = R(x) \Rightarrow x^3 + 12x - 16 = 14x - 10 \Rightarrow x^3 - 2x - 6 = 0$. Let $P(x) = x^3 - 2x - 6$. Since $P(x)$ has only 1 variation in sign, $P(x) = 0$ has only 1 positive real root. From the pairs of opposite signs: $P(2) = -2$ and $P(3) = 15$; and $P(2.1) \approx -0.9$ and $P(2.2) \approx 0.3$, we know that the root lies in the interval $(2.1, 2.2)$. Since $P(2.15) \approx -0.4$ and $P(2.2)$ have opposite signs, the root is closer to 2.2. Thus, the company must manufacture and sell $(2.2)(100) = 220$ compact disks just to break even.

[40] Let $z = a + bi$ be a root of $P(x) = 0$, where $P(x) = a_n x^n + a_{n-1}x^{n-1} + \ldots + a_1 x + a_0$ is a polynomial of degree n with real coefficients. Since $P(z) = 0$, it follows that $a_n z^n + a_{n-1}z^{n-1} + \ldots + a_1 z + a_0 = 0$. Taking the conjugate of both sides of this equation and applying the results of Exercise 76 of Section 1.4, we have: $\overline{a_n z^n + a_{n-1}z^{n-1} + \ldots + a_1 z + a_0} = \overline{0} \Rightarrow \overline{a_n z^n} + \overline{a_{n-1}z^{n-1}} + \ldots + \overline{a_1 z} + \overline{a_0} = \overline{0} \Rightarrow \overline{a_n}\, \overline{z}^n + \overline{a_{n-1}}\,\overline{z}^{n-1} + \ldots + \overline{a_1}\,\overline{z} + \overline{a_0} = \overline{0} \Rightarrow a_n \overline{z}^n + a_{n-1}\overline{z}^{n-1} + \ldots + a_1 \overline{z} + a_0 = 0 \Rightarrow P(\overline{z}) = 0$. Thus $\overline{z}$, the conjugate of z, is also a root of $P(x) = 0$.

[41] Since $s(t) = 2t^4 + 3t^3 + 6t^2 - t - 15$ has only 1 variation in sign, the equation $s(t) = 0$ has only 1 positive real root. From the pairs of opposite signs: $s(1) = -5$ and $s(2) = 63$; and $s(1.1) \approx -1.9$ and $s(1.2) \approx 1.8$, we know that the root lies in the interval $(1.1, 1.2)$. Since $s(1.15) \approx -0.2$ and $s(1.2)$ have opposite signs, the root is closer to 1.2. Thus, the displacement is equal to zero when t is approximately 1.2 seconds.

[42]　Substituting 2 for r and 0.25 for s in the equation $x^3 - 2rx^2 - 4r^3s = 0$, we get $x^3 - 4x^2 - 8 = 0$.

Since $P(x) = x^3 - 4x^2 - 8$ has 1 variation in sign, the equation $P(x) = 0$ has only 1 positive real

root. From the pairs of opposite signs: $P(4) = -8$ and $P(5) = 17$; and $P(4.4) \approx -0.3$ and $P(4.5) \approx$

2.1, we see that the root lies in the interval $(4.4, 4.5)$. Since $P(4.45) \approx 0.9$ and $P(4.4)$ have opposite

signs, the root is closer to 4.4. Thus, the styrofoam buoy will sink in water to a depth of

approximately 4.4 centimeters.

[43]　a)　$P(x) = x^4 - 3x^3 + 6x^2 - 12$. The prime 3 is a factor of $a_0 = -12$, $a_1 = 0$, $a_2 = 6$ and $a_3 = -3$.

Also, 3 is not a factor of $a_4 = 1$ and $3^2 = 9$ is not a factor of $a_0 = -12$.

　　　b)　$P(x) = 5x^3 + 12x^2 - 6x + 2$. The prime 2 is a factor of $a_0 = 2$, $a_1 = -6$, and $a_2 = 12$ but is

not a factor of $a_3 = 5$. Also $2^2 = 4$ is not a factor of $a_0 = 2$.

[44]　Since $P(x) = c_1x^4 + c_2x^2 + c_3x - c_4$ has only 1 variation in sign, the equation $P(x) = 0$ has only

one positive real root. Since $P(-x) = c_1x^4 + c_2x^2 - c_3x - c_4$ also has only one variation in sign,

$P(x) = 0$ has exactly 1 negative real root. Since $P(x) = 0$ has only 2 real roots, the remaining two

roots must be nonreal complex numbers.

[45]　Substituting x^3 for y in the equation $x^2 + y^2 = 4$, we get $x^2 + x^6 = 4$ or $x^2 + x^6 - 4 = 0$. Since $P(x) =$

$x^2 + x^6 - 4$ has only 1 variation sign, the equation $P(x) = 0$ has exactly one positive real root. From

the pairs of opposite signs: $P(1) = -2$ and $P(2) = 64$; and $P(1.1) \approx -0.6$ and $P(1.2) \approx 1.7$,

we know that the root lies between 1.1 and 1.2. Since $P(1.15) \approx 0.4$ and $P(1.1)$ have opposite

signs, the root is closer to 1.1. Thus, the x – coordinate of the point of intersection of the two graphs

is approximately 1.1.

[46]　The x – coordinate of A satisfies the equation $x^3 = x^2 + 2x + 3$, or $x^3 - x^2 - 2x - 3 = 0$. Since

$P(x) = x^3 - x^2 - 2x - 3$ has only 1 variation in sign, the equation $P(x) = 0$ has exactly one

positive real root. From the pairs of opposite signs: $P(2) = -3$ and $P(3) = 9$; and $P(2.3) \approx -0.7$

and $P(2.4) \approx 0.3$, we know that the root lies between 2.3 and 2.4. Since $P(2.35) \approx -0.2$ and $P(2.4)$

have opposite signs, the root is closer to 2.4. Thus, the x – coordinate of the point of intersection

of the two graphs is approximately 2.4.

[47]　$1 + \left(\frac{1}{2}\right)\left(\frac{1}{2}\right) - \left(\frac{1}{8}\right)\left(\frac{1}{2}\right)^2 + \left(\frac{1}{16}\right)\left(\frac{1}{2}\right)^3 \approx 1.227$; $\sqrt{1 + \frac{1}{2}} = \sqrt{1.5} \approx 1.225$

[48]　$1 - \frac{1}{2} + \left(\frac{1}{2}\right)^2 - \left(\frac{1}{2}\right)^3 = 0.625$; $\dfrac{1}{1 + \frac{1}{2}} = \frac{2}{3} \approx 0.667$

[49]　$1 + \left(\frac{3}{2}\right)\left(\frac{1}{2}\right) + \left(\frac{3}{8}\right)\left(\frac{1}{2}\right)^2 - \left(\frac{1}{16}\right)\left(\frac{1}{2}\right)^3 \approx 1.836$; $\left(1 + \frac{1}{2}\right)^{\frac{3}{2}} \approx 1.837$

[50]　a)　$y = x^3 - 5x - 1$. The smallest negative real zero is approximately -2.1.

　　　b)　$y = x^4 - 4x^3 - 2x^2 + 12x - 4$. The largest positive real zero is approximately 3.7.

EXERCISES 3.6

[1]	yes	[2]	no (sharp corner)
[3]	no (not a function)	[4]	yes
[5]	yes	[6]	no (not a function)
[7]	no (break in graph)	[8]	no (break in graph)

[9] $f(x) = -x^3$ (See graph below).

[10] $f(x) = \frac{1}{2}x^4$ (See graph below).

[11] $f(x) = 5x^2$ (See graph below).

[12] $f(x) = 12x^3$ (See graph below).

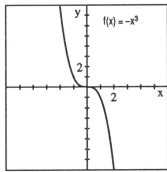

Figure 9

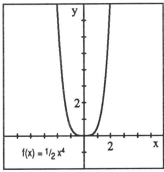

Figure 10

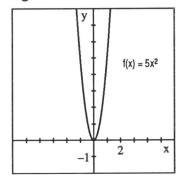

Figure 11

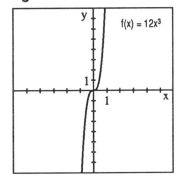

Figure 12

[13] $f(x) = -\frac{2}{3}x^5$ (See graph below).

[14] $f(x) = x^8$ (See graph below).

[15] $f(x) = 3x^6$ (See graph below).

[16] $f(x) = -\frac{1}{4}x^5$ (See graph below).

[17] $f(x) = 1 - x^4$ (See graph below).

[18] $f(x) = x^3 + 2$ (See graph below).

[19] $f(x) = \frac{1}{2}x^3 - 3$ (See graph below).

[20] $f(x) = (x-4)^3$ (See graph below).

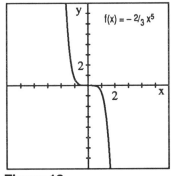

Figure 13

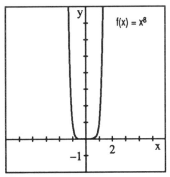

Figure 14

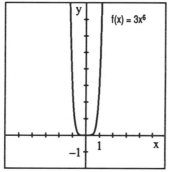

Figure 15

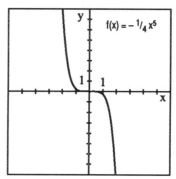

Figure 16

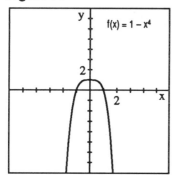

Figure 17

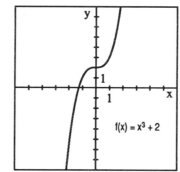

Figure 18

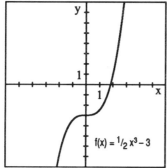

Figure 19

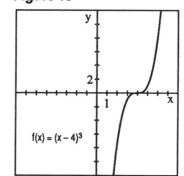

Figure 20

[21] $f(x) = (x + 1)^2 + 2$ (See graph below).

[22] $f(x) = 1 + \left(x - \frac{3}{2}\right)^2$ (See graph below).

[23] $f(x) = 3 - \left(x + \frac{5}{2}\right)^5$ (See graph below).

[24] $f(x) = 2 - (2 - x)^3$ (See graph below).

[25] $f(x) = \frac{4}{5}(x - 3)^3$ (See graph below).

[26] $f(x) = \frac{2}{3}(x - 4)^5$ (See graph below).

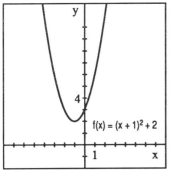

Figure 21

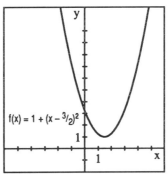

Figure 22

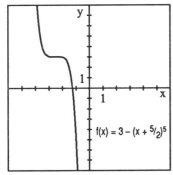

Figure 23

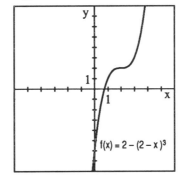

Figure 24

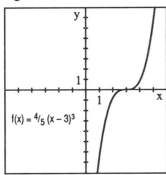

Figure 25

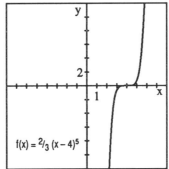

Figure 26

[27] (e)

[28] (a)

[29] (c)

[30] (d)

[31] (f)

[32] (b)

[33] $y = (x + 3)(x^2 - 1)$ (See graph below).

[34] $y = (x^2 - 9)(x + 1)$ (See graph below).

[35] $y = -2(x^2 - 1)(2x + 5)$ (See graph below).

[36] $y = \frac{1}{2}(2x - 3)(x + 2)$ (See graph below).

[37] $y = x(x^2 - 4)(x^2 + 2)$ (See graph below).

[38] $y = -x(x + 4)(x^2 - 4)$ (See graph below).

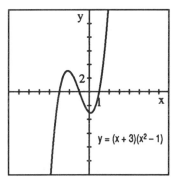

Figure 33

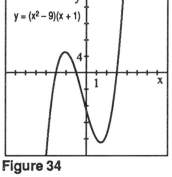

Figure 34

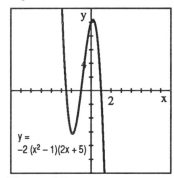

Figure 35

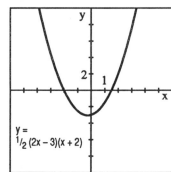

Figure 36

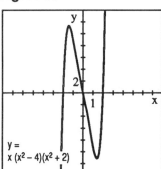

Figure 37

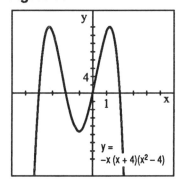

Figure 38

[39] $y = \frac{1}{4}\left(x^2 - 9\right)\left(x + 5\right)\left(x - 1\right)$ (See graph below).

[40] $y = \frac{1}{3}\left(x^2 + 2\right)\left(x^2 - 1\right)\left(2x - 5\right)$ (See graph below).

[41] $y = x^3 + 4x^2 - 4x - 16$ (See graph below). [42] $y = 2x^3 - x^2 - 10x + 5$ (See graph below).

[43] $y = x^4 - 9x^2 + 8$ (See graph below). [44] $y = x^5 + x^3 - 6x$ (See graph below).

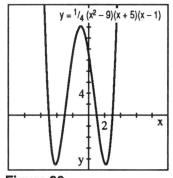

Figure 39

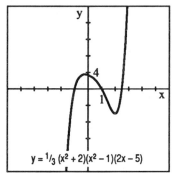

Figure 40

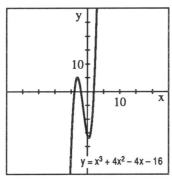

Figure 41

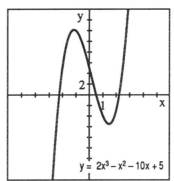

Figure 42

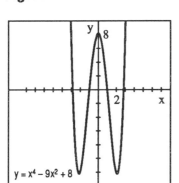

Figure 43

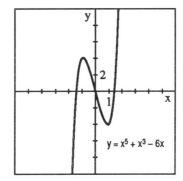

Figure 44

[45] $y = -2x^3 - 8x^2 + 3x + 12$ (See graph below).

[46] $y = -x^4 - 3x^2 + 4$ (See graph below).

[47] $y = x^4 - 2x^3 - x^2 + 6x - 8$ (See graph below).

[48] $y = x^4 - x^3 - x - 6$ (See graph below).

[49] $y = -x^4 + 4x^3 - 3x^2 - x + 3$ (See graph below).

[50] $y = -3x^4 + 4x^3 + x^2 + 4x + 4$ (See graph below).

[51] $y = \frac{1}{2}x^5 - \frac{3}{2}x^3 - 2x$ (See graph below).   [52] $y = \frac{3}{2}x^5 - 6x^2 - 12x$ (See graph below).

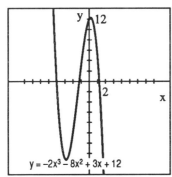

Figure 45

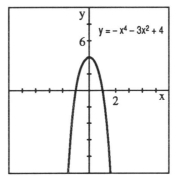

Figure 46

125

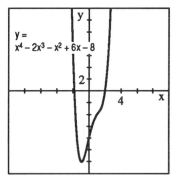

Figure 47

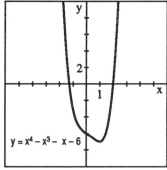

Figure 48

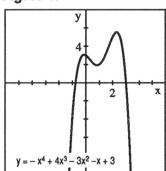

Figure 49

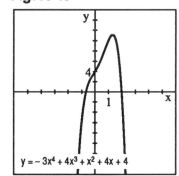

Figure 50

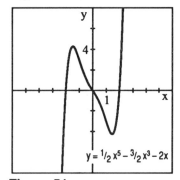

Figure 51

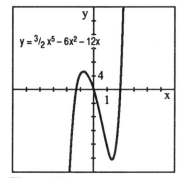

Figure 52

[53] $y = -x^4 + 3x^2 + 5$; 2 real zeros (See graph below).

[54] $y = 1.2x^3 - 3.5x + 4$; 1 real zero (See graph below).

[55] $y = x^5 - 12x^2 + 16$; 3 real zeros (See graph below).

[56] $y = -x^6 - x^3 + 8$; 2 real zeros (See graph below).

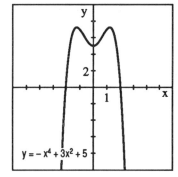

Figure 53

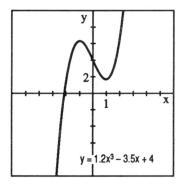

Figure 54

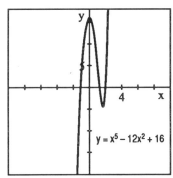

Figure 55

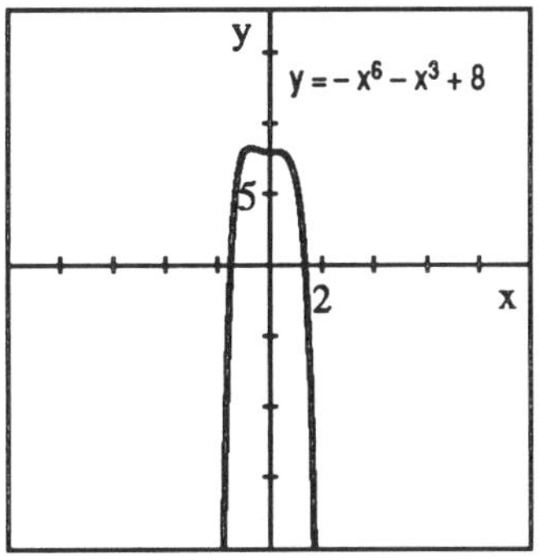

Figure 56

[57] $y = 2x^3 - 3x + 1$ and $y = -x^3 + x^2 - 5$; point of intersection is $(-1.4, -0.3)$ (See graph below).

[58] $y = 3x^4 - 8x^2 - 4x - 2$ and $y = -2x^2 + 3x + 6$; points of intersection are $(-1.3, -1.5)$ and

$(2.0, 4.2)$ (See graph below).

[59] $y = \frac{1}{3}x^3 - x^2 - 3x + 4$; Increases on $(-\infty, -1) \cup (3, \infty)$; decreases on $(-1, 3)$ (See graph below).

[60] $y = -2x^3 + 3x^2 + 12x - 7$; Increases on $(-1, 2)$; decreases on $(-\infty, -1) \cup (2, \infty)$

   (See graph below).

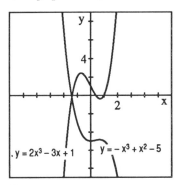

Figure 57

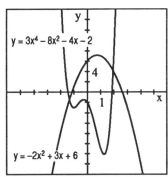

Figure 58

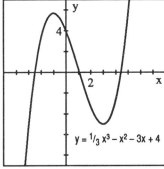

Figure 59

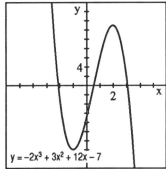

Figure 60

EXERCISES 3.7

[1] $f(x) = -\frac{1}{x^3}$ (See graph below).   [2] $f(x) = \frac{1}{x^2 - 4}$ (See graph below).

[3] $f(x) = \frac{4}{x + 5}$ (See graph below).   [4] $f(x) = \frac{3}{x - 2}$ (See graph below).

[5] $f(x) = \frac{x - 3}{2x + 4}$ (See graph below).   [6] $f(x) = \frac{x}{2 - x}$ (See graph below).

[7] $f(x) = -\dfrac{2}{(x-2)^2}$ (See graph below).

[8] $f(x) = -\dfrac{3}{(x-4)^2}$ (See graph below).

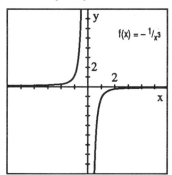

Figure 1

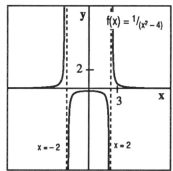

Figure 2

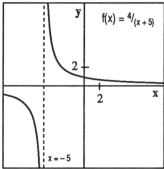

Figure 3

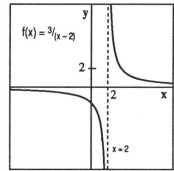

Figure 4

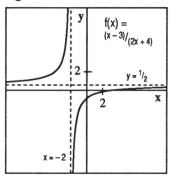

Figure 5

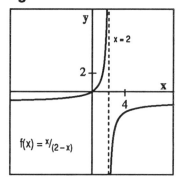

Figure 6

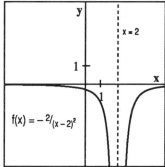

Figure 7

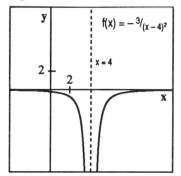

Figure 8

[9] $f(x) = \dfrac{1}{(x+3)(x-2)}$ (See graph below).

[10] $f(x) = \dfrac{2}{x^2-9}$ (See graph below).

128

[11] $f(x) = \dfrac{8x}{(x+1)^2}$ (See graph below).

[12] $f(x) = \dfrac{3x}{(x-2)^2}$ (See graph below).

[13] $f(x) = \dfrac{x^2-1}{x(x+4)}$ (See graph below).

[14] $f(x) = \dfrac{x+1}{(x+4)(x-1)}$ (See graph below).

[15] $f(x) = \dfrac{8x^2}{(2x+1)^2}$ (See graph below).

[16] $f(x) = \dfrac{16x^2}{(3-2x)^2}$ (See graph below).

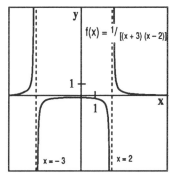

Figure 9

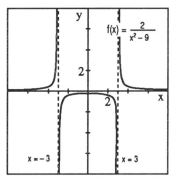

Figure 10

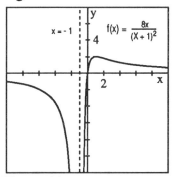

Figure 11

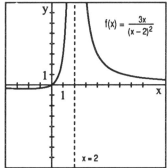

Figure 12

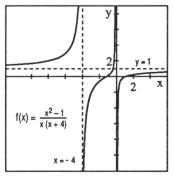

Figure 13

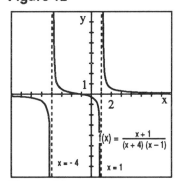

Figure 14

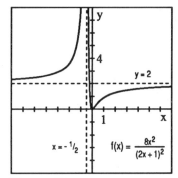

Figure 15

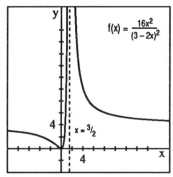

Figure 16

[17] $f(x) = \dfrac{x^2}{x^2 - x - 2}$ (See graph below).

[18] $f(x) = \dfrac{x - 5}{x^2 - 8x + 12}$ (See graph below).

[19] $f(x) = \dfrac{x^4 - 1}{\left(x^2 + 1\right)^2}$ (See graph below).

[20] $f(x) = \dfrac{\left(x^2 - 1\right)^2}{x^2\left(x^2 + 4\right)}$ (See graph below).

[21] $f(x) = \dfrac{x^2 - 1}{x + 1} = \dfrac{(x + 1)(x - 1)}{x + 1} = x - 1, \text{ if } x \ne -1.$ (See graph below).

[22] $f(x) = \dfrac{x^2 - 9}{x - 3} = \dfrac{(x + 3)(x - 3)}{x - 3} = x + 3, \text{ if } x \ne 3.$ (See graph below).

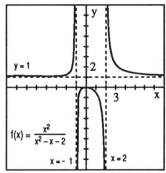

Figure 17

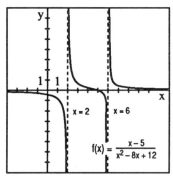

Figure 18

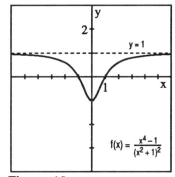

Figure 19

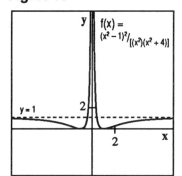

Figure 20

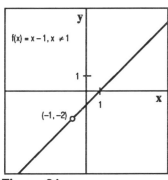

Figure 21

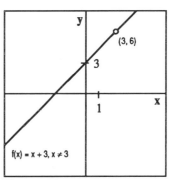

Figure 22

[23] $f(x) = \dfrac{x^2 - x - 6}{x - 3} = \dfrac{(x-3)(x+2)}{x-3} = x + 2, \text{ if } x \neq 3.$ (See graph below).

[24] $f(x) = \dfrac{3x^2 + 2x - 1}{3x - 1} = \dfrac{(3x-1)(x+1)}{3x-1} = x + 1, \text{ if } x \neq \dfrac{1}{3}.$ (See graph below).

[25] $f(x) = \dfrac{x^3 - 8}{x - 2} = \dfrac{(x-2)(x^2 + 2x + 4)}{x - 2} = x^2 + 2x + 4, \text{ if } x \neq 2.$ (See graph below).

[26] $f(x) = \dfrac{x^2 - 16}{2x - 8} = \dfrac{(x+4)(x-4)}{2(x-4)} = \dfrac{x+4}{2} = \dfrac{1}{2}x + 2, \text{ if } x \neq 4.$ (See graph below).

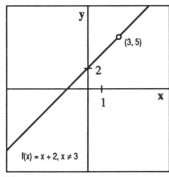

Figure 23

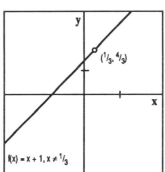

Figure 24

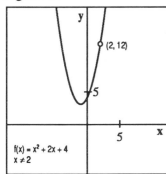

Figure 25

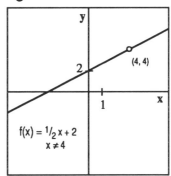

Figure 26

[27] $f(x) = \dfrac{x^2 - 4}{x + 1} = x - 1 - \dfrac{3}{x + 1} \Rightarrow y = x - 1 \text{ is oblique asymptote.}$ (See graph below).

[28] $f(x) = \dfrac{4x^2 - 1}{2x} = 2x - \dfrac{1}{2x} \Rightarrow y = 2x \text{ is oblique asymptote.}$ (See graph below).

[29] $f(x) = \dfrac{x^2 + 2x - 8}{x - 1} = x + 3 - \dfrac{5}{x - 1} \Rightarrow y = x + 3 \text{ is oblique asymptote.}$ (See graph below).

[30] $f(x) = \dfrac{x^2 - x + 1}{x + 1} = x - 2 + \dfrac{3}{x + 1} \Rightarrow y = x - 2$ is oblique asymptote.   (See graph below).

[31] $f(x) = \dfrac{x^3 - 3x^2 - x}{x^2 - 1} = x - 3 - \dfrac{3}{x^2 - 1} \Rightarrow y = x - 3$ is oblique asymptote.   (See graph below).

[32] $f(x) = \dfrac{x^3 - 1}{1 - x^2} = -x + \dfrac{x - 1}{1 - x^2} = -x - \dfrac{x - 1}{x^2 - 1} = -x - \dfrac{x - 1}{(x+1)(x-1)} = -x - \dfrac{1}{x + 1} \Rightarrow y = -x$

is oblique asymptote.   (See graph below).

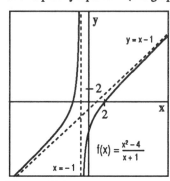

Figure 27

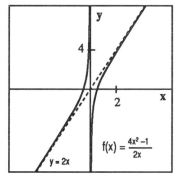

Figure 28

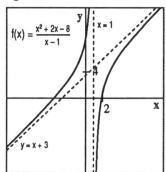

Figure 29

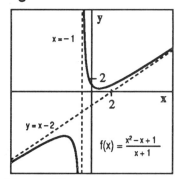

Figure 30

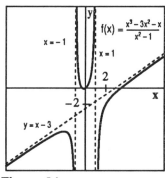

Figure 31

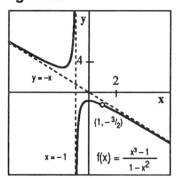

Figure 32

[33] $\overline{C}(x) = \dfrac{C(x)}{x} = \dfrac{x^2 + 2x + 15}{x} = x + 2 + \dfrac{15}{x} \Rightarrow y = x + 2$ is oblique asymptote.   (See graph below).

[34] $\overline{C}(x) = \dfrac{C(x)}{x} = \dfrac{2x^2 - x + 2}{x} = 2x - 1 + \dfrac{2}{x} \Rightarrow y = 2x - 1$ is oblique asymptote.   (See graph below).

[35] $f(x) = \dfrac{5x + 20}{x} = 5 + \dfrac{20}{x}$. The horizontal asymptote $y = 5$ tells the researcher that all but 5mg of drug is eventually absorbed by the body.   (See graph below).

[36] $f(x) = \dfrac{20(x+1)}{x+5} = \dfrac{20x+20}{x+5} = 20 - \dfrac{80}{x+5} \Rightarrow y = 20$ is a horizontal asymptote. To the electronics

firm, this means that the maximum number of components a worker can assemble is 20, regardless of the amount of training.   (See graph below).

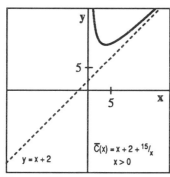

Figure 33

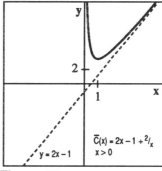

Figure 34

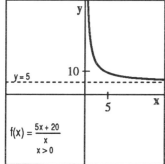

Figure 35

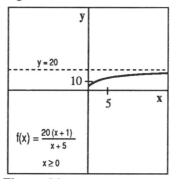

Figure 36

[37]   No. If the graph had both a horizontal asymptote and an oblique asymptote, it would fail the vertical line test and hence, would not be the graph of a function.

[38]   Dividing numerator and denominator by x^m yields

$$f(x) = \dfrac{a_n\left(\dfrac{x^n}{x^m}\right) + a_{n-1}\left(\dfrac{x^{n-1}}{x^m}\right) + \ldots + a_1\left(\dfrac{1}{x^{m-1}}\right) + a_0\left(\dfrac{1}{x^m}\right)}{b_m + b_{m-1}\left(\dfrac{1}{x}\right) + \ldots + b_1\left(\dfrac{1}{x^{m-1}}\right) + b_0\left(\dfrac{1}{x^m}\right)}. \text{ Let NUM denote the numerator}$$

and DEN denote the denominator of $f(x)$.

i)   If $n < m$, NUM $= a_n\left(\dfrac{1}{x^{m-n}}\right) + a_{n-1}\left(\dfrac{1}{x^{m-n+1}}\right) + \ldots + a_1\left(\dfrac{1}{x^{m-1}}\right) + a_0\left(\dfrac{1}{x^m}\right)$. Thus, as $|x| \to \infty$,

NUM $\to 0$ and DEN $\to b_m$. Hence, $y = \dfrac{0}{b_m}$ or $y = 0$ (the x - axis) is a horizontal asymptote.

ii)   If $n = m$, NUM $= a_n + a_{n-1}\left(\dfrac{1}{x}\right) + \ldots + a_1\left(\dfrac{1}{x^{m-1}}\right) + a_0\left(\dfrac{1}{x^m}\right)$. Thus, as $|x| \to \infty$, NUM $\to$

a_n and DEN $\to b_m$. Hence, $y = \dfrac{a_n}{b_m}$ is a horizontal asymptote.

iii)   If $n > m$, NUM $= a_n x^{n-m} + a_{n-1}x^{n-m-1} + \ldots + a_1\left(\dfrac{1}{x^{m-1}}\right) + a_0\left(\dfrac{1}{x^m}\right)$ Thus, as $|x| \to \infty$,

NUM $\to \pm\infty$ and DEN $\to b_m$. Hence, the graph of f has no horizontal asymptote.

[39] $f(x) = \dfrac{x^2 - x - 6}{x^2 - 2x}$ (See graph below).

[40] $f(x) = \dfrac{-x^3 + x^2 + 4}{x^2} = -x + 1 + \dfrac{4}{x^2} \Rightarrow y = 1 - x$ is an oblique asymptote.   (See graph below).

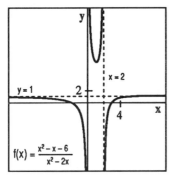

Figure 39

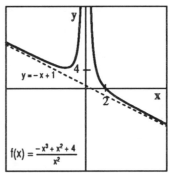

Figure 40

CHAPTER 3 REVIEW EXERCISES

[1]
$$\begin{array}{r} x - 2 \\ 2x+3\overline{\smash{\big)}\,2x^2 - x + 5} \\ \underline{2x^2 + 3x} \\ -4x + 5 \\ \underline{-4x - 6} \\ 11 \end{array}$$

$Q(x) = x - 2;\ R(x) = 11$

[2]
$$\begin{array}{r} y^2 - 3y + 9 \\ y+3\overline{\smash{\big)}\,y^3 + 0y^2 + 0y - 1} \\ \underline{y^3 + 3y^2} \\ -3y^2 + 0y - 1 \\ \underline{-3y^2 - 9y} \\ 9y - 1 \\ \underline{9y + 27} \\ -28 \end{array}$$

$Q(y) = y^2 - 3y + 9;\ R(y) = -28$

[3]
$$\begin{array}{r} x + 2 \\ x^2-x+1\overline{\smash{\big)}\,x^3 + x^2 + x + 1} \\ \underline{x^3 - x^2 + x} \\ 2x^2 + 0x + 1 \\ \underline{2x^2 - 2x + 2} \\ 2x - 1 \end{array}$$

$Q(x) = x + 2;\ R(x) = 2x - 1$

[4]
$$\begin{array}{r} 3x + 5 \\ 2x^2+0x+1\overline{\smash{\big)}\,6x^3 + 10x^2 + x - 8} \\ \underline{6x^3 + 0x^2 + 3x} \\ 10x^2 - 2x - 8 \\ \underline{10x^2 + 0x + 5} \\ -2x - 13 \end{array}$$

$Q(x) = 3x + 5;\ R(x) = -2x - 13$

[5]
$$\begin{array}{r|rrrrr} -3 & 3 & 2 & 0 & -4 & -1 \\ & & -9 & 21 & -63 & 201 \\ \hline & 3 & -7 & 21 & -67 & 200 \end{array}$$

$Q(x) = 3x^3 - 7x^2 + 21x - 67;\ R(x) = 200$

[6]
$$\begin{array}{r|rrrrr} -1 & -1 & 2 & -3 & 5 \\ & & 1 & -3 & 6 \\ \hline & -1 & 3 & -6 & 11 \end{array}$$

$Q(x) = -x^2 + 3x - 6;\ R(x) = 11$

[7]
$$\begin{array}{r|rrrr} \frac{1}{3} & 6 & -1 & 2 & 2 \\ & & 2 & \frac{1}{3} & \frac{7}{9} \\ \hline & 6 & 1 & \frac{7}{3} & \frac{25}{9} \end{array}$$

$Q(x) = 6x^2 + x + \dfrac{7}{3};\ R(x) = \dfrac{25}{9}$

[8]
$$\begin{array}{r|rrrrr} \frac{1}{4} & 4 & 0 & \frac{3}{4} & 0 & 1 \\ & & 1 & \frac{1}{4} & \frac{1}{4} & \frac{1}{16} \\ \hline & 4 & 1 & 1 & \frac{1}{4} & \frac{17}{16} \end{array}$$

$Q(x) = 4x^3 + x^2 + x + \dfrac{1}{4};\ R(x) = \dfrac{17}{16}$

[9]
$$\begin{array}{r|rrrr} i & 1 & 0 & 2 & -4 \\ & & i & -1 & i \\ \hline & 1 & i & 1 & -4+i \end{array}$$

$Q(x) = x^2 + ix + 1;\ R(x) = -4 + i$

[10]
$$\begin{array}{r|rrrr} -2i & 1 & -1 & 4 & -4 \\ & & -2i & -4+2i & 4 \\ \hline & 1 & -1-2i & 2i & 0 \end{array}$$

$Q(x) = x^2 - (1 + 2i)x + 2i;\ R(x) = 0$

[11]
$$\begin{array}{r|rrrrr} 1-i & 1 & 0 & i & 0 & 1 \\ & & 1-i & -2i & -1-i & -2 \\ \hline & 1 & 1-i & -i & -1-i & -1 \end{array}$$

$Q\,(x)=x^3+(1-i)\,x^2-i\,x-(1+i);$
$R\,(x)=-1$

[12]
$$\begin{array}{r|rrrr} 2+i & 1 & -4 & 5 & 0 \\ & & 2+i & -5 & 0 \\ \hline & 1 & -2+i & 0 & 0 \end{array}$$

$Q\,(x)=x^2-(2-i)\,x;\,R\,(x)=0$

[13]
$$\begin{array}{r|rrrr} 4 & 6 & -23 & -5 & 4 \\ & & 24 & 4 & -4 \\ \hline & 6 & 1 & -1 & 0 \end{array}$$

$R\,(x)=0\Rightarrow x-4$ is a factor of
$6\,x^3-23\,x^2-5\,x+4.$

[14]
$$\begin{array}{r|rrrr} -\frac{1}{2} & 2 & 1 & 8 & 4 \\ & & -1 & 0 & -4 \\ \hline & 2 & 0 & 8 & 0 \end{array}$$

$R\,(x)=0\Rightarrow x+\frac{1}{2}$ is a factor of
$2\,x^3+x^2+8\,x+4.$

[15]
$$\begin{array}{r|rrrrr} 3\,i & 1 & -3 & 11 & -27 & 18 \\ & & 3\,i & -9-9\,i & 27+6\,i & -18 \\ \hline & 1 & -3+3\,i & 2-9\,i & 6\,i & 0 \end{array}$$

$R\,(x)=0\Rightarrow x-3\,i$ is a factor of $x^4-3\,x^3+11\,x^2-27\,x+18.$

[16]
$$\begin{array}{r|rrrr} 1-2\,i & 1 & -5 & 11 & -15 \\ & & 1-2\,i & -8+6\,i & 15 \\ \hline & 1 & -4-2\,i & 3+6\,i & 0 \end{array}$$

$R\,(x)=0\Rightarrow x-1+2\,i$ is a factor of $x^3-5\,x^2+11\,x-15.$

[17] Synthetic division by $x-2$ yields $R\,(x)=2k+12.\,R\,(x)=0\Rightarrow 2k+12=0\Rightarrow k=-6.$

[18] Synthetic division by $x-1$ yields $R\,(x)=6-b\,.\,R\,(x)=0\Rightarrow 6-b=0\Rightarrow b=6.$

[19] $P\,(3)=2\,(3)^3-5\,(3)^2+3\,(3)-4=14\Rightarrow R\,(x)=14.$

[20] $P\,\left(-\sqrt{3}\right)=\left(-\sqrt{3}\right)^3+3\left(-\sqrt{3}\right)^2-3\left(-\sqrt{3}\right)+1=10\Rightarrow R\,(x)=10.$

[21] $P\,(-2\,i)=3\,(-2\,i)^4-2\,(-2\,i)^2+1=57\Rightarrow R\,(x)=57.$

[22] $P\,(2\,i)=(2\,i)^4+6\,(2\,i)^2+8=0\Rightarrow R\,(x)=0.$

[23] Synthetic division by $x+3$ yields $R\,(x)=0.$ Thus $x+3$ is a factor of $5\,x^3+17\,x^2+3\,x-9.$

[24] Synthetic division by $x-\sqrt{2}$ yields $R\,(x)=4\neq0.$ Thus $x-\sqrt{2}$ is not a factor of $x^3+x^2-2\,x+2.$

[25] Synthetic division by $x+\frac{1}{4}$ yields $R\,(x)=-\frac{19}{4}\neq0.$ Thus $x+\frac{1}{4}$ is not a factor of $8x^3-6x^2+5x-3.$

[26] Synthetic division by $x-i$ yields $R\,(x)=0.$ Thus $x-i$ is a factor of $2x^3-x^2+2x-1.$

[27] $P\,(x)=(x+1)(x-1)(x-3)=x^3-3x^2-x+3.$

[28] $P\,(x)=2\left[x\,(x+4)\left(x-\frac{1}{2}\right)\right]=2\,x^3+7\,x^2-4\,x.$

[29] $P\,(x)=\left[x-\left(1+\sqrt{2}\right)\right]\left[x-\left(1-\sqrt{2}\right)\right](x+2)(x-2)=x^4-2\,x^3-5\,x^2+8\,x+4.$

[30] $P\,(x)=\left[x-(3+i)\right]\left[x-(3-i)\right]x^2=x^4-6\,x^3+10\,x^2.$

[31] Dividing $6\,x^3+7\,x^2-1$ by $x+1$ yields the reduced equation $6\,x^2+x-1=0.$ The solutions of the reduced equation (by factoring) are $-\frac{1}{2}$ and $\frac{1}{3}.$ $\qquad\qquad\bullet-\frac{1}{2},\frac{1}{3}$

[32] Dividing $2\,x^3-4\,x^2-3\,x-9$ by $x-3$ yields the reduced equation $2\,x^2+2\,x+3=0.$ Using the quadratic formula, the solutions of the reduced equation are $-\frac{1}{2}\pm\frac{\sqrt{5}}{2}\,i\,.$ $\qquad\qquad\bullet-\frac{1}{2}\pm\frac{\sqrt{5}}{2}\,i$

[33] The conjugate of $-3i$, namely $3i$, must also be a solution. After division by $x + 3i$ and $x - 3i$, the reduced equation $x - 2 = 0$ yields the solution 2. ● $3i, 2$

[34] The conjugate of $3 + \sqrt{2}\,i$, which is $3 - \sqrt{2}\,i$ must also be a solution. After division by $x - \left(3 + \sqrt{2}\,i\right)$ and $x - \left(3 - \sqrt{2}\,i\right)$, the reduced equation $2x - 1 = 0$ yields the solution $\frac{1}{2}$. ● $3 - \sqrt{2}\,i, \frac{1}{2}$

[35] Let $P\left(x\right) = x^n - 1$, where n is an even positive integer. Then, $P\left(-1\right) = \left(-1\right)^n - 1 = 1 - 1 = 0 \Rightarrow$ $x + 1$ is a factor of $P\left(x\right)$ by the factor theorem.

[36] Let $P\left(x\right) = x^n - c^n$, where n is a positive integer. Since $P\left(c\right) = c^n - c^n = 0$, $x - c$ is a factor of $P\left(x\right)$ by the factor theorem.

[37] Dividing $3x^3 + bx^2 - 2b + 5$ by $x + 1$ yields a remainder of $-b + 2$. $R\left(x\right) = 0 \Rightarrow$ $-b + 2 = 0 \Rightarrow b = 2$.

[38] $P\left(-1\right) = 2\left(-1\right)^{51} - 13\left(-1\right)^{34} + 2\left(-1\right)^{17} - 8 = -25 \Rightarrow R\left(x\right) = -25$.

[39] degree 4; roots are $-\frac{4}{3}, -1$, and $\frac{1}{2}$ (double root)

[40] degree 8; roots are $-2, 0$ (triple root) and 3 (multiplicity 4)

[41] degree 3; roots are 0 and $\pm\sqrt{3}\,i$

[42] degree 4; roots are $2 \pm i$ and -1 (double root)

[43] $P\left(x\right) = \left(x + 2\right)\left(x - 1\right)^2\left(x - 4\right) = x^4 - 4x^3 - 3x^2 + 14x - 8$.

[44] $P\left(x\right) = \left(x + 5\right)\left(x - 4\right)\left(x - \sqrt{2}\right)\left(x + \sqrt{2}\right) = x^4 + x^3 - 22x^2 - 2x + 40$.

[45] $P\left(x\right) = \left(x + \frac{1}{3}\right)\left(x - \frac{1}{2}\right)^2 = x^3 - \frac{2}{3}x^2 - \frac{7}{12}x - \frac{1}{12}$.

[46] $P\left(x\right) = x\left(x + \frac{2}{5}\right)\left(x - 2\right)^3 = x^5 - \frac{28}{5}x^4 + \frac{48}{5}x^3 - \frac{16}{5}x^2 - \frac{16}{5}x$.

[47] $c = d = 3 \Rightarrow \frac{c}{d} = \pm 1, \pm 3$ or $\pm\frac{1}{3}$.

[48] $c = 18$ and $d = 8 \Rightarrow \frac{c}{d} = \pm 1, \pm 2, \pm 3, \pm 6, \pm 9, \pm 18, \pm\frac{1}{2}, \pm\frac{1}{4}, \pm\frac{1}{8}, \pm\frac{3}{2}, \pm\frac{3}{4}, \pm\frac{3}{8}, \pm\frac{9}{2}, \pm\frac{9}{4}$ and $\pm\frac{9}{8}$.

[49] $c = 12$ and $d = 18 \Rightarrow \frac{c}{d} = \pm 1, \pm 2, \pm 3, \pm 4, \pm 6, \pm 12, \pm\frac{1}{2}, \pm\frac{1}{3}, \pm\frac{1}{6}, \pm\frac{1}{9}, \pm\frac{1}{18}, \pm\frac{2}{3}, \pm\frac{2}{9}, \pm\frac{3}{2}, \pm\frac{4}{3}$ and $\pm\frac{4}{9}$.

[50] $c = 20$ and $d = 3 \Rightarrow \frac{c}{d} = \pm 1, \pm 2, \pm 4, \pm 5, \pm 10, \pm 20, \pm\frac{1}{3}, \pm\frac{2}{3}, \pm\frac{4}{3}, \pm\frac{5}{3}, \pm\frac{10}{3}$, and $\pm\frac{20}{3}$.

In Exercises 51 – 56, PRR denotes possible rational roots and RR denotes rational roots.

[51] PRR $= \pm 1, \pm 2, \pm 4, \pm\frac{1}{2}, \pm\frac{1}{3}, \pm\frac{1}{6}, \pm\frac{2}{3}$ and $\pm\frac{4}{3}$; RR $= \pm 1, -\frac{4}{3}$, and $\frac{1}{2}$.

[52] PRR $= \pm 1, \pm 2, \pm\frac{1}{3}, \pm\frac{1}{9}, \pm\frac{2}{3}$ and $\pm\frac{2}{9}$; RR $= -\frac{1}{3}$, and $\frac{2}{3}$.

[53] PRR $= \pm 1, \pm\frac{1}{2}$; no rational roots.

[54] PRR $= \pm 1, \pm 5, \pm\frac{1}{2}, \pm\frac{1}{4}, \pm\frac{5}{2}$ and $\pm\frac{5}{4}$; RR $= \pm\frac{1}{2}$.

[55] $6\left(\dfrac{2}{3}x^3+\dfrac{3}{2}x^2-x-\dfrac{1}{2}=0\right)\Rightarrow 4x^3+9x^2-6x-3=0.\ \text{PRR}=\pm 1,\pm 3,\pm\dfrac{1}{2},\pm\dfrac{1}{4},\pm\dfrac{3}{2},\pm\dfrac{3}{4};$

no rational roots.

[56] $20\left(\dfrac{1}{4}x^3-\dfrac{6}{5}x^2+\dfrac{3}{10}x+\dfrac{2}{5}=0\right)\Rightarrow 5x^3-24x^2+6x+8=0.\ \text{PRR}=\pm 1,\pm 2,\pm 4,\pm 8,\pm\dfrac{1}{5},\pm\dfrac{2}{5},\pm\dfrac{4}{5}$

and $\pm\dfrac{8}{5};\ \text{RR}=\dfrac{4}{5}.$

[57] By the rational root theorem, the only possible rational roots of the equation $x^3-2=0$ are ± 1 and ± 2. Synthetic division and the remainder theorem will verify that ± 1 and ± 2 are not roots of this equation. Since $\left(\sqrt[3]{2}\right)^3-2=2-2=0,\ \sqrt[3]{2}$ is a root of the equation $x^3-2=0$. Thus $\sqrt[3]{2}$ is an irrational number.

[58] $f(x)=g(x)\Rightarrow 2x^3-x^2-x+3=3x^3-x^2+2x-1\Rightarrow x^3+3x-4=0.$ The only rational root is 1. Solving the reduced equation $x^2+x+4=0$, using the quadratic formula, we obtain the nonreal complex roots $-\dfrac{1}{2}\pm\dfrac{\sqrt{15}}{2}i$. Thus, the only point of intersection of the two graphs is $(1,3)$.

[59] $f(x)=0\Rightarrow 3x^4+5x^2-8=0.$ The only rational roots are ± 1. The roots of the reduced equation $3x^2+8=0$ are the nonreal complex numbers $\pm\dfrac{2\sqrt{6}}{3}i$. Thus, the x–intercepts of the graph are -1 and 1.

[60] By the Pythagorean Theorem $y=\sqrt{(x+2)^2-x^2}=\sqrt{4x+4}=2\sqrt{x+1}.\ A=2\sqrt{3}=\dfrac{1}{2}(x)\cdot$

$\left(2\sqrt{x+1}\right)\Rightarrow 2\sqrt{3}=x\sqrt{x+1}\Rightarrow\left(2\sqrt{3}\right)^2=\left(x\sqrt{x+1}\right)^2\Rightarrow 12=x^2(x+1)\Rightarrow x^3+x^2-12=0.$

The only real root of this equation is 2. (The remaining 2 roots are nonreal complex numbers.) Thus, $x=2,\ x+2=4$ and $y=2\sqrt{3}.$ (See figure below).

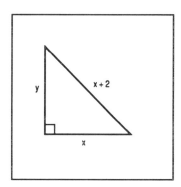

Figure 60

In Exercises 61–64, $\text{VSP}(x)$ and $\text{VSP}(-x)$ denote the variations in sign in $P(x)$ and $P(-x)$, respectively. PR, NR and NC denote positive real roots, negative real roots and nonreal complex roots, respectively.

[61] $\text{VSP}(x)=3$ and $\text{VSP}(-x)=1.$

PR	NR	NC
3	1	0
1	1	2

[62] $\text{VSP}(x)=3$ and $\text{VSP}(-x)=0.$

PR	NR	NC
3	0	0
1	0	2

[63] $\text{VSP}(x)=1$ and $\text{VSP}(-x)=1.$

PR	NR	NC
1	1	4

[64] $\text{VSP}(x)=2$ and $\text{VSP}(-x)=3.$

PR	NR	NC
2	3	0
2	1	2
0	3	2
0	1	4

In Exercises 65–68, LB and UB denote lower bound and upper bound, respectively.

[65] $\text{LB}=-2;\ \text{UB}=5$ [66] $\text{LB}=-2;\ \text{UB}=1$

[67] $LB = -1; UB = 2$ [68] $LB = -2; UB = 2$

In Exercises 69 – 72, RR denotes rational roots and RE denotes reduced equation.

[69] RR : none. Since $P(x) = 5x^3 + 8x - 6$ has 1 variation in sign and $P(-x) = -5x^3 - 8x - 6$ has no variation in sign, the equation $P(x) = 0$ has only one real root, which is positive. Since $P(0) = -6$ and $P(1) = 7$ have opposite signs, the root lies between 0 and 1. Since $P(0.6) \approx -0.1$ and $P(0.7) \approx$ 1.3 have opposite signs the root lies between 0.6 and 0.7. Since $P(0.65) \approx 0.6$ and $P(0.6)$ have opposite signs, the root is closer to 0.6.   ● 0.6

[70] $RR = -\frac{1}{2}$ and 1. The solutions of the RE, $2x^2 + 8 = 0$ are the nonreal complex numbers $\pm 2i$.

● $-\frac{1}{2}, 1$

[71] RR : 2 only. Since each of $P(x) = -2x^4 + 2x^3 + 3x^2 + 4$ and $P(-x) = -2x^4 - 2x^3 + 3x^2 + 4$ has 1 variation in sign, the equation has 1 positive real root (namely 2) and 1 negative real root. From the pairs of opposite signs: $P(-2) = -32$ and $P(-1) = 3$; and $P(-1.3) \approx -1.0$ and $P(-1.2) \approx 0.7$, we know the root lies between -1.3 and -1.2. Since $P(-1.25) \approx -0.1$ and $P(-1.2)$ have opposite signs, the root is closer to -1.2.   ● $-1.2, 2$

[72] RR : none. Using the procedure outlined in the solution of Exercise 71, we find that the only real root is approximately 1.7.   ● 1.7

[73] Let $P(x) = x^3 - 5$. Since $P(1) = -4$ and $P(2) = 3$ have opposite signs, the root lies between 1 and 2. From the pairs of opposite signs: $P(1.7) \approx -0.1$ and $P(1.8) \approx 0.8$; and $P(1.70) \approx -0.1$ and $P(1.71) \approx 0.0002$, the root lies between 1.70 and 1.71. Since $P(1.705) \approx -0.04$ and $P(1.71)$ have opposite signs, the root is closer to 1.71. Thus $\sqrt[3]{5} \approx 1.71$.

[74] $V = 40 \Rightarrow L \cdot W \cdot H = 40 \Rightarrow (14 - 2x)(6 - 2x)x = 40 \Rightarrow 84x - 40x^2 + 4x^3 = 40 \Rightarrow x^3 - 10x^2 + 21x - 10 = 0$. The only rational root is 2. The roots of the reduced equation $x^2 - 8x + 5 = 0$, which are $\frac{8 \pm \sqrt{44}}{2}$ are not feasible. Thus, each cut – out side has length 2 inches.   (See graph below).

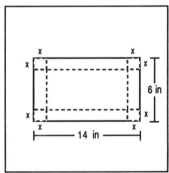

Figure 74   **Figure 77**

[75] Dividing $t^3 - 2t^2 - 5t + 6$ by $t - 1$ yields the reduced equation $x^2 - x - 6 = 0$. By factoring, the roots of the reduced equation are -2 and 3.   ● 3 seconds

[76] Let $P(x) = 1 - x^2 + x^4 - x^6 + x^8$. Then $P\left(\frac{1}{2}\right) = 1 - \left(\frac{1}{2}\right)^2 + \left(\frac{1}{2}\right)^4 - \left(\frac{1}{2}\right)^6 + \left(\frac{1}{2}\right)^8 \approx 0.801$, and

$$\frac{1}{\left(\frac{1}{2}\right)^2 + 1} = 0.800.$$

[77] $f(x) = -\frac{1}{2}x^3$ (See graph above).   [78] $f(x) = (x^2 - 1)(x + 3)^2$ (See graph below).

[79] $f(x) = 4x(x^2 - 1)$ (See graph below).   [80] $f(x) = -x^4 + 5x^2 - 4$ (See graph below).

[81] $f(x) = -\frac{5}{2}(x + 1)^2(x - 1)$ (See graph below).

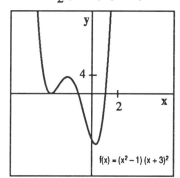

Figure 78

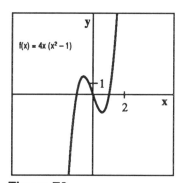

Figure 79

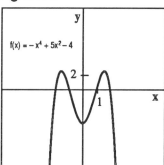

Figure 80

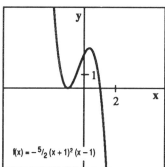

Figure 81

[82] $f(x) = x^4 - 6x^3 + 11x^2 - 6x$ (See graph below).

[83] $f(x) = 2x^3 - 5x^2 - 4x + 3$ (See graph below).

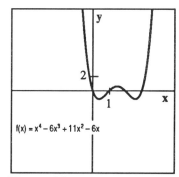

Figure 82

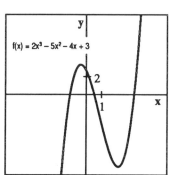

Figure 83

[84] $f(x) = x^3 + 2x^2 - x - 2$ (See graph below).   [85] $f(x) = \frac{x}{x - 2}$ (See graph below).

[86] $f(x) = \dfrac{3}{(x+1)(x-2)}$ (See graph below).

[87] $f(x) = \dfrac{2x}{1-x^2}$ (See graph below).

[88] $f(x) = \dfrac{x-2}{x+2}$ (See graph below).

[89] $f(x) = \dfrac{3x-1}{x^2-4}$ (See graph below).

[90] $f(x) = \dfrac{1}{x^2-9}$ (See graph below).

[91] $f(x) = \dfrac{3x^2-3x-6}{x^2+8x+16}$ (See graph below).

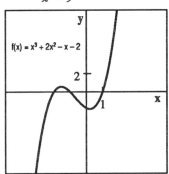

Figure 84

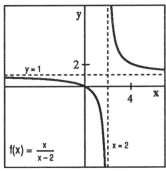

Figure 85

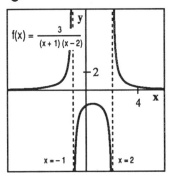

Figure 86

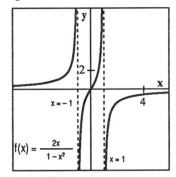

Figure 87

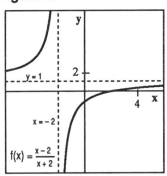

Figure 88

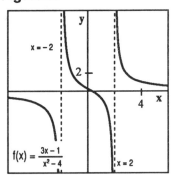

Figure 89

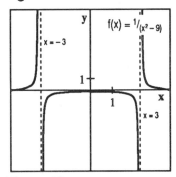

Figure 90

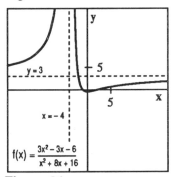

Figure 91

[92] $f(x) = \dfrac{x^2 - x - 2}{x^2 + 1}$ (See graph below).

[93] $f(x) = \dfrac{x^2 - 4}{x} = x - \dfrac{4}{x} \Rightarrow y = x$ is oblique asymptote.   (See graph below).

[94] $f(x) = \dfrac{x^2 + 2}{x} = x + \dfrac{2}{x} \Rightarrow y = x$ is oblique asymptote.   (See graph below).

[95] $f(x) = \dfrac{x^2 - 8}{x + 1} = x - 1 - \dfrac{7}{x + 1} \Rightarrow y = x - 1$ is oblique asymptote.   (See graph below).

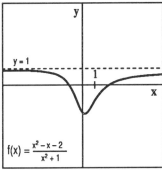

Figure 92

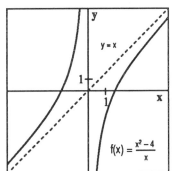

Figure 93

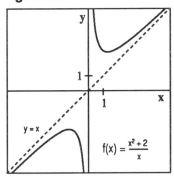

Figure 94

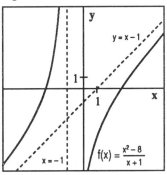

Figure 95

[96] $f(x) = \dfrac{x^2 - x - 2}{x - 1} = x - \dfrac{2}{x - 1} \Rightarrow y = x$ is oblique asymptote.   (See graph below).

[97] $\overline{C}(x) = \dfrac{C(x)}{x} = \dfrac{\frac{1}{2}x^2 + 2x + 5}{x} = \dfrac{1}{2}x + 2 + \dfrac{5}{x} \Rightarrow y = \dfrac{1}{2}x + 2$ is oblique asymptote. (See graph below).

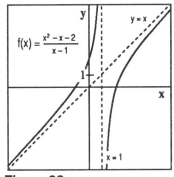

Figure 96

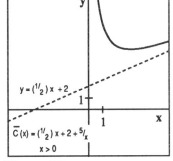

Figure 97

[98] $f(t) = \dfrac{4t + 12}{t} = 4 + \dfrac{12}{t} \Rightarrow y = 4$ is a horizontal asymptote. To the psychologist, the horizontal

asymptote $y = 4$ means that the average number of symbols recalled by a subject is 4, regardless of

the number of days the subject participates in the experiment.   (See graph below).

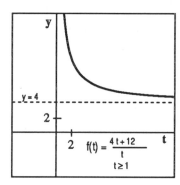

Figure 98

Exercises 4.1

[1] $dom(f) = \mathbf{R}$; $ran(f) = (0, \infty)$; horizontal asymptote, $y = 0$. (See graph below).

[2] $dom(f) = \mathbf{R}$; $ran(f) = (0, \infty)$; horizontal asymptote, $y = 0$. (See graph below).

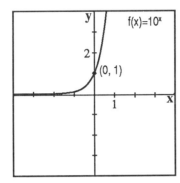

Figure 1

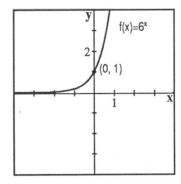

Figure 2

[3] $dom(f) = \mathbf{R}$; $ran(f) = (-\infty, 0)$; horizontal asymptote, $y = 0$. (See graph below).

[4] $dom(f) = \mathbf{R}$; $ran(f) = (-\infty, 0)$; horizontal asymptote, $y = 0$. (See graph below).

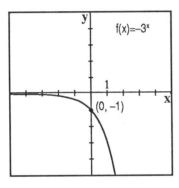

Figure 3

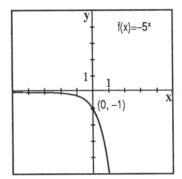

Figure 4

[5] $dom(f) = \mathbf{R}$; $ran(f) = (-\infty, 0)$; horizontal asymptote, $y = 0$. (See graph below).

[6] $dom(f) = \mathbf{R}$; $ran(f) = (-\infty, 0)$; horizontal asymptote, $y = 0$. (See graph below).

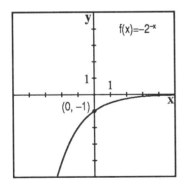

Figure 5

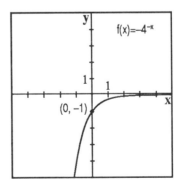

Figure 6

[7] $dom(f) = \mathbf{R}$; $ran(f) = (0, \infty)$; horizontal asymptote, $y = 0$. (See graph below).

[8] $dom(f) = \mathbf{R}$; $ran(f) = (0, \infty)$; horizontal asymptote, $y = 0$. (See graph below).

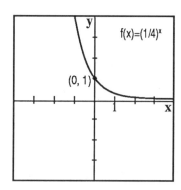

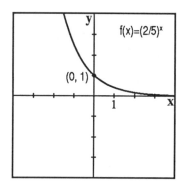

Figure 7   **Figure 8**

[9] $dom(f) = \mathbf{R}$; $ran(f) = [1, \infty)$.   (See graph below).

[10] $dom(f) = \mathbf{R}$; $ran(f) = (0, 1]$; horizontal asymptote, $y = 0$.   (See graph below).

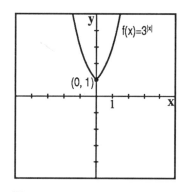

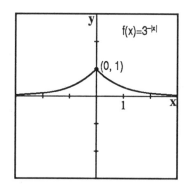

Figure 9   **Figure 10**

[11] $dom(f) = \mathbf{R}$; $ran(f) = (0, \infty)$; horizontal asymptote, $y = 0$.   (See graph below).

[12] $dom(f) = \mathbf{R}$; $ran(f) = (0, \infty)$; horizontal asymptote, $y = 0$.   (See graph below).

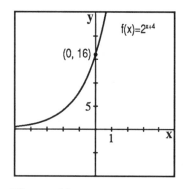

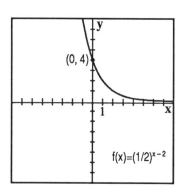

Figure 11   **Figure 12**

[13] $dom(f) = \mathbf{R}$; $ran(f) = (2, \infty)$; horizontal asymptote, $y = 2$.   (See graph below).
[14] $dom(f) = \mathbf{R}$; $ran(f) = (-2, \infty)$; horizontal asymptote, $y = -2$.   (See graph below).
[15] $dom(f) = \mathbf{R}$; $ran(f) = (0, \infty)$; horizontal asymptote, $y = 0$.   (See graph below).
[16] $dom(f) = \mathbf{R}$; $ran(f) = (0, e^2]$; horizontal asymptote, $y = 0$.   (See graph below).
[17] $dom(f) = \mathbf{R}$; $ran(f) = (3, \infty)$; horizontal asymptote, $y = 3$.   (See graph below).
[18] $dom(f) = \mathbf{R}$; $ran(f) = (-3, e^2 - 3]$; horizontal asymptote, $y = -3$.   (See graph below).

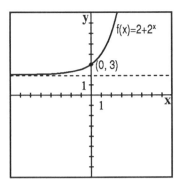

Figure 13

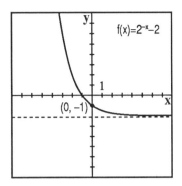

Figure 14

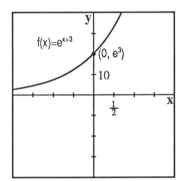

Figure 15

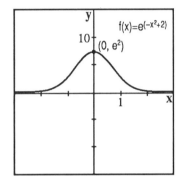

Figure 16

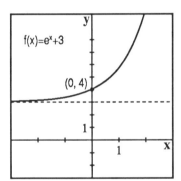

Figure 17

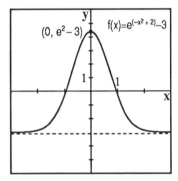

Figure 18

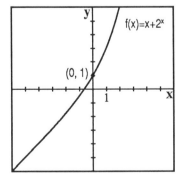

Figure 19

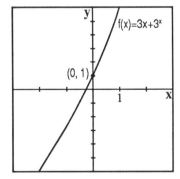

Figure 20

[19] $dom(f) = \mathbf{R}$; $ran(f) = \mathbf{R}$.   (See graph above).

[20] $dom(f) = \mathbf{R}$; $ran(f) = \mathbf{R}$.   (See graph above).

[21] $dom(f) = (-\infty, 0) \cup (0, \infty)$; $ran(f) = (0, 1) \cup (1, \infty)$; horizontal asymptote, $y = 1$.   (See graph below).

[22] $dom(f) = \mathbf{R}$; $ran(f) = (0, \infty)$; horizontal asymptote, $y = 0$.   (See graph below).

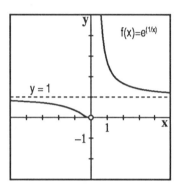

Figure 21

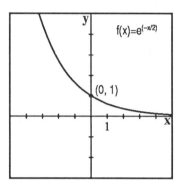

Figure 22

[23] $dom(f) = \mathbf{R}$; $ran(f) = (0, \infty)$; horizontal asymptote, $y = 0$.   (See graph below).

[24] $dom(f) = \mathbf{R}$; $ran(f) = (0, \infty)$; horizontal asymptote, $y = 0$.   (See graph below).

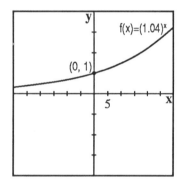

Figure 23

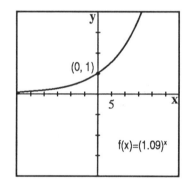

Figure 24

[25] $dom(f) = \mathbf{R}$; $ran(f) = (0, 16]$; horizontal asymptote, $y = 0$.   (See graph below).

[26] $dom(f) = \mathbf{R}$; $ran(f) = (0, 3]$; horizontal asymptote, $y = 0$.   (See graph below).

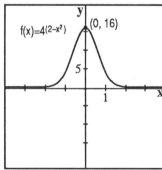

Figure 25

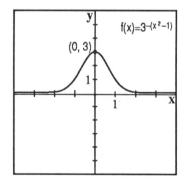

Figure 26

[27] For each $x \in$ **R**, $3^x > x^3$. Moreover, $3^x > 0$ for each x, while $x^3 > 0$ if and only if $x > 0$.

(See graph below).

[28] (See graph below).

[29] (See graph below).

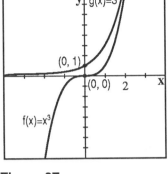

Figure 27

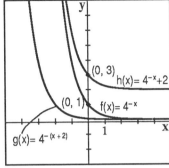

Figure 28

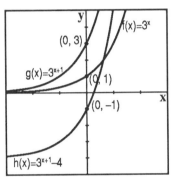

Figure 29

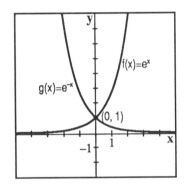

Figure 30, part a

[30] a) (See graph above).

b) $h(-x) = \dfrac{e^{-x} + e^{-(-x)}}{2} = \dfrac{e^{-x} + e^{x}}{2} = h(x) \Rightarrow h$ is even. (See graph below).

c) $k(-x) = \dfrac{e^{-x} - e^{-(-x)}}{2} = \dfrac{e^{-x} - e^{x}}{2} = -\dfrac{e^{x} - e^{-x}}{2} = -k(x) \Rightarrow k$ is odd. (See graph below).

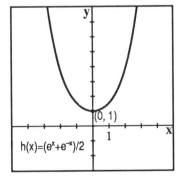

Figure 30, part b

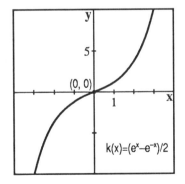

Figure 30, part c

[31]

x	3	3.1	3.14	3.141	3.1415	3.14159	π
3^x	27	30.13533	31.48914	31.52375	31.54107	31.54419	31.54428

As x approaches π, 3^x approaches 3^π.

[32]

x	1	1.4	1.41	1.414	1.4142	1.41421	$\sqrt{2}$
3^x	3	4.65554	4.70697	4.72770	4.72873	4.72879	4.72880

As x approaches $\sqrt{2}$, 3^x approaches $3^{\sqrt{2}}$.

[33] $dom\,(f) = ran\,(f^{-1}) = \mathbf{R}$; $ran\,(f) = dom\,(f^{-1}) = (0, \infty)$.   (See graph below).

[34] $dom\,(f) = ran\,(f) = \mathbf{R}$; $ran\,(f) = dom\,(f^{-1}) = (0, \infty)$.   (See graph below).

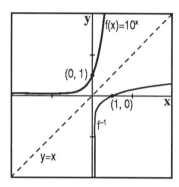

Figure 33

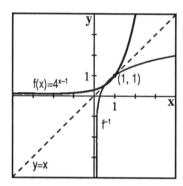

Figure 34

[35] $dom\,(f) = ran\,(f^{-1}) = \mathbf{R}$; $ran\,(f) = dom\,(f^{-1}) = (0, \infty)$.   (See graph below).

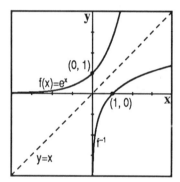

Figure 35

[36]   a) $(f \circ g)(x) = f\,[g\,(x)] = e^{g\,(x)} = e^{x-4}$; $(g \circ f)(x) = g\,[f\,(x)] = f\,(x) - 4 = e^x - 4$.

     b) $dom\,(f \circ g) = dom\,(g \circ f) = \mathbf{R}$.　　　c)   (See graphs below).

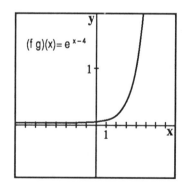

Figure 36 c, part 1

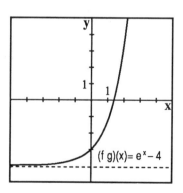

Figure 36 c, part 2

[37] $\dfrac{f\,(x+h) - f\,(x)}{h} = \dfrac{e^{x+h} - e^x}{h} = \dfrac{e^x e^h - e^x}{h} = \dfrac{e^x (e^h - 1)}{h}$.

[38] a) $F(1) - f(1) = e^1 - \left(1 + 1 + \frac{1}{2} + \frac{1}{6}\right) \approx 0.052$

b) $F(0.5) - f(0.5) = e^{0.5} - \left(1 + 0.5 + \frac{(0.5)^2}{2} + \frac{(0.5)^3}{6}\right) \approx 0.003$

c) $F(0.1) - f(0.1) = e^{0.1} - \left(1 + 0.1 + \frac{(0.1)^2}{2} + \frac{(0.1)^3}{6}\right) \approx 4.25 \cdot 10^{-6}$.

d) $F(0.01) - f(0.01) = e^{0.01} - \left(1 + 0.01 + \frac{(0.01)^2}{2} + \frac{(0.01)^3}{6}\right) \approx 4.17 \cdot 10^{-10}$.

e) $F(0.001) - f(0.001) = e^{0.001} - \left(1 + 0.001 + \frac{(0.001)^2}{2} + \frac{(0.001)^3}{6}\right) \approx 0.00$.

f) $F(0.0001) - f(0.0001) \approx 0.00$.

[39] We use the formula $A = P\left(1 + \frac{r}{n}\right)^{nt}$, where $P = 2{,}000$, $r = 0.07$, $n = 2$ and $t = 5$.

Thus, $A = 2{,}000\left(1 + \frac{0.07}{2}\right)^{10} = 2{,}000(1.035)^{10} \approx 2821.20$.   ● $2821.20

[40] Using the formula $A = P\left(1 + \frac{r}{n}\right)^{nt}$, where $P = 10{,}000$, $r = 0.0925$, $n = 12$ and $t = 2$,

we have $A = 10{,}000\left(1 + \frac{0.0925}{12}\right)^{24} \approx 10{,}000(1.008)^{24} \approx 12{,}107.45$.   ● $12,107.45

[41] Using the formula $A = P e^{rt}$, where $P = 6{,}000$, $r = 0.045$, and $t = 3$,

we get $A = 6000\, e^{(0.045)(3)} = 6000\, e^{0.135} \approx 6867.22$.   ● $6867.22

[42] We use the formula $A = \dfrac{P\left[(1+r)^n - 1\right]}{r}$, where $A = 300{,}000$, $r = \frac{0.08}{2} = 0.04$ and $n = 40$.

Thus, $300{,}000 = \dfrac{P\left[(1.04)^{40} - 1\right]}{0.04} \Rightarrow P = \dfrac{0.04\,(300{,}000)}{(1.04)^{40} - 1} \approx 3157.05$.   ● $3157.05

[43] At the end of one year, the amount in Janice's account will be $A = 5000\left(1 + \frac{0.10}{365}\right)^{365} \approx \$5525.78$.

At the end of one year, the value of Mary's account will be $A = 5000\, e^{0.1} \approx \$5525.85$. Hence, we see

that Mary earns approximately 7 cents more than Janice.

[44] Using the formula $A = \dfrac{P\left[(1+r)^n - 1\right]}{r}$, where $A = 20{,}000$, $r = \frac{0.09}{12} = 0.0075$ and $n = 60$.

We get $20{,}000 = \dfrac{P\left[(1.0075)^{60} - 1\right]}{0.0075} \Rightarrow P = \dfrac{0.0075\,(20{,}000)}{(1.0075)^{60} - 1} \approx 265.17$.   ● $265.17

[45] The amount in the account after one year is $A = \dfrac{200\left[(1.0075)^{12} - 1\right]}{0.0075} \approx 2501.52$. The amount

deposited into the account is \$2,400. Hence, the interest earned in the first year is approximately

\$2501.52 − \$2400.00 = \$101.52. The value of the account after two years is

$A = \dfrac{200\left[(1.0075)^{24} - 1\right]}{0.0075} \approx 5237.69$. The increase in the value of the account during the second year

was approximately 5237.69 − 2501.52 = 2736.17. Hence, the interest earned on the account during

the second year is approximately \$2736.17 − \$2400.00 = \$336.17. For the third year, we have

$A = \dfrac{200\left[(1.0075)^{36} - 1\right]}{0.0075} \approx 8230.54$. The increase in the account during the third year is

approximately 8230.54 − 5237.69 = 2992.85. Hence, the interest earned is approximately

\$2992.85 − \$2400.00 = \$592.85.   ● \$101.52; \$336.17; \$592.85

[46] To find the amount of money in Mary's account after 3 years, we use the formula $A = \dfrac{P\left[(1-r)^n - 1\right]}{r}$,

where $P = 150$, $r = \dfrac{0.09}{12} = 0.0075$ and $n = 36$. $A = \dfrac{150\left[(1.0075)^{36} - 1\right]}{.0075} \approx \$6172.91$. Therefore,

Mary does not have enough money in this account to buy a car which costs \$7000.00.

[47] a) $f(x) = 0 \Rightarrow xe^x + 5x^2 e^x = 0 \Rightarrow e^x(x + 5x^2) = 0 \Rightarrow e^x[x(1 + 5x)] = 0 \Rightarrow x = 0$ or $x = -\dfrac{1}{5}$.

     ● $-\dfrac{1}{5}, 0$

b) $g(x) = 0 \Rightarrow e^{2x-3} - e^{-x^2} = 0 \Rightarrow e^{2x-3} = e^{-x^2} \Rightarrow 2x - 3 = -x^2 \Rightarrow x^2 + 2x - 3 = 0 \Rightarrow x = -3$ or $x = 1$.

     ● −3, 1

c) $h(x) = 0 \Rightarrow 4^x - 2^x - 56 = 0 \Rightarrow \left(2^2\right)^x - 2^x - 56 = 0 \Rightarrow \left(2^x\right)^2 - 2^x - 56 = 0$. Letting $u = 2^x$, we

     have $u^2 - u - 56 = 0 \Rightarrow u = -7$ or $u = 8$. Thus $2^x = -7$ or $2^x = 8$. Since $2^x > 0$ for all $x \in \mathbf{R}$,

     $2^x = -7$ is impossible. Hence, $2^x = 8$ or $x = 3$.   ● 3

[48] (See graph below).   **[49]** (See graph below).

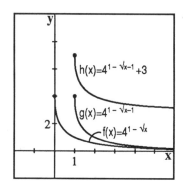

Figure 48

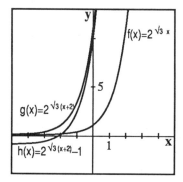

Figure 49

[50] (See graph below).   ● (−0.57, 4.81)

[51]   (See graph below).   ● $(-2.41, 0.19), (3.61, 12.22)$

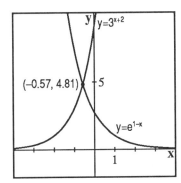

Figure 50

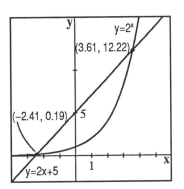

Figure 51

Exercises 4.2

[1] $\log_6 216 = 3.$

[2] $\log_8 64 = 2.$

[3] $\log_{1/3} 9 = -2.$

[4] $\log_{1/4} 1024 = -5.$

[5] $\log_5 \sqrt{5} = \frac{1}{2}.$

[6] $\log \sqrt[4]{10} = \frac{1}{4}.$

[7] $\ln 10 = 4t$

[8] $\ln 14 = 2 - 6t$

[9] $4^{-1} = \frac{1}{4}$

[10] $3^{-4} = \frac{1}{81}$

[11] $e^{1/4} = \sqrt[4]{e}$

[12] $e^{1/5} = \sqrt[5]{e}$

[13] $x = e^a$

[14] $2x + 3 = b^5$

[15] $\log_2 128 = \log_2 2^7 = 7.$

[16] $\log_2 512 = \log_2 2^9 = 9.$

[17] $\log_6 1296 = \log_6 6^4 = 4.$

[18] $\log_6 \frac{1}{216} = \log_6 6^{-3} = -3.$

[19] $\log_2 \frac{\sqrt[3]{32}}{8} = \log_2 \frac{2^{5/3}}{2^3} = \log_2 2^{-4/3} = -\frac{4}{3}.$

[20] $\log_4 \frac{\sqrt[3]{4}}{64} = \log_4 \frac{4^{1/3}}{4^3} = \log_4 4^{-8/3} = -\frac{8}{3}.$

[21] $5^{\log_5 25} = 25$

[22] $6^{\log_6 9^2} = 9^2 = 81.$

[23]   a) $dom(f) = (2, \infty)$ b)   (See graph below).

　　　 c)   vertical asymptote: x = 2

[24]   a) $dom(f) = (-3, \infty)$ b)   (See graph below).

　　　 c)   vertical asymptote: x = -3

[25]   a) $dom(f) = (-\infty, 0) \cup (0, \infty)$ b)   (See graph below).

　　　 c)   vertical asymptote: x = 0

[26]   a) $dom(f) = (-\infty, 1) \cup (1, \infty)$ b)   (See graph below).

　　　 c)   vertical asymptote: x = 1

153

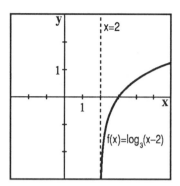

Figure 23 b

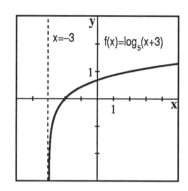

Figure 24 b

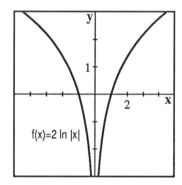

Figure 25 b

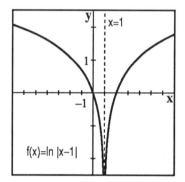

Figure 26 b

[27] a) $dom(f) = (2, \infty)$ b)   (See graph below).

c)   vertical asymptote: $x = 2$

[28] a) $dom(f) = (-5, \infty)$ b)   (See graph below).

c)   vertical asymptote: $x = -5$

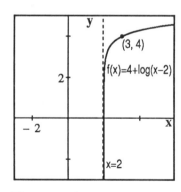

Figure 27 b

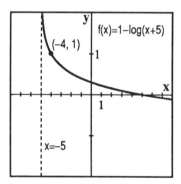

Figure 28 b

[29] a) $dom(f) = (-\infty, 0)$ b)   (See graph below).

c)   vertical asymptote: $x = 0$

[30] a) $dom(f) = (-\infty, 0)$ b)   (See graph below).

c)   vertical asymptote: $x = 0$

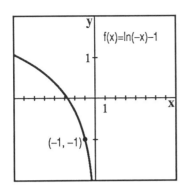

Figure 29 b

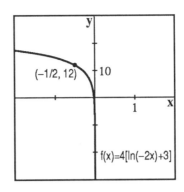

Figure 30 b

[31] a) $dom(f) = [1, \infty)$

b) (See graph below).

c) no vertical asymptote

[32] a) $dom(f) = (0, 1]$

b) (See graph below).

c) vertical asymptote: $x = 0$

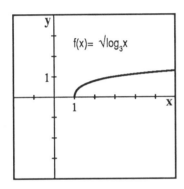

Figure 31 b

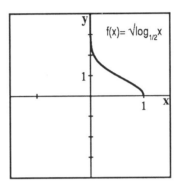

Figure 32 b

[33] a) $dom(f) = [1, \infty)$

b) (See graph below).

c) no vertical asymptote

[34] a) $dom(f) = [1, \infty)$

b) (See graph below).

c) no vertical asymptote

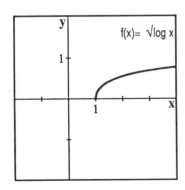

Figure 33 b

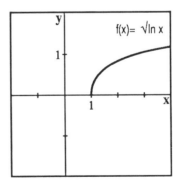

Figure 34 b

[35] $\log_x 10,000 = 4 \Rightarrow x^4 = 10,000 \Rightarrow x = 10$ ● 10

[36] $\log_x \frac{1}{256} = -8 \Rightarrow x^{-8} = \frac{1}{256} \Rightarrow x = 2$ ● 2

[37] $\log_3 x = -2 \Rightarrow x = 3^{-2} = \frac{1}{9}$ ● $\frac{1}{9}$

[38] $\log_9 729 = x \Rightarrow 9^x = 729 \Rightarrow x = 3$ ● 3

[39] $\log_5 x^2 = 2 \Rightarrow x^2 = 5^2 = 25 \Rightarrow x = \pm 5$ ● ± 5

[40] $\log_2 (x^2 + 2x) = 8 \Rightarrow x^2 + 2x = 2^8 = 256 \Rightarrow x^2 + 2x - 256 = 0.$ Using the quadratic formula,

we find that $x \approx -17.03$ or $x \approx 15.03.$ ● $-17.03, 15.03$

[41] $\log (x^2 - 5x + 14) = 1 \Rightarrow x^2 - 5x + 14 = 10 \Rightarrow x^2 - 5x + 4 = 0 \Rightarrow x = 1$ or $x = 4.$ ● 1, 4

[42] $\log_5 (2x^2 - 3x + 26) = 2 \Rightarrow 2x^2 - 3x + 26 = 5^2 = 25 \Rightarrow 2x^2 - 3x + 1 = 0 \Rightarrow x = \frac{1}{2}$ or $x = 1.$ ● $\frac{1}{2}, 1$

[43] $\ln e^{x^2} = 6 \Rightarrow x^2 = 6 \Rightarrow x = \pm\sqrt{6}.$ ● $\pm\sqrt{6}$

[44] $e^{\ln|x|} = 2 \Rightarrow |x| = 2 \Rightarrow x = \pm 2.$ ● ± 2

[45] a) $x \in dom(f)$ if and only if $3x + 1 > 0 \Rightarrow x > -\frac{1}{3}.$ ● $\left(-\frac{1}{3}, \infty\right)$

b) $x \in dom(f)$ if and only if $8 - 3x > 0 \Rightarrow x < \frac{8}{3}.$ ● $\left(-\infty, \frac{8}{3}\right)$

c) $(0, \infty)$ d) $(-\infty, 0) \cup (0, \infty)$

[46] a) $x \in dom(f)$ if and only if $6 + x - x^2 > 0 \Rightarrow (3 - x)(2 + x) > 0 \Rightarrow -2 < x < 3.$ ● $(-2, 3)$

b) $x \in dom(f)$ if and only if $\frac{x+2}{2x-3}$ is defined and is not equal to zero $\Rightarrow x \neq -2$ and $x \neq \frac{3}{2}.$

● $(-\infty, -2) \cup \left(-2, \frac{3}{2}\right) \cup \left(\frac{3}{2}, \infty\right)$

c) $x \in dom(f)$ if and only if $x > 0$ and $\ln x \neq -2,$ or $x \neq e^{-2}.$ ● $(0, e^{-2}) \cup (e^{-2}, \infty)$

d) $(0, \infty)$

In Exercises 47 - 54, the solutions are not unique. We give one possible pair for each.

[47] $g(x) = \log_6 x;\ h(x) = 3x + 1.$ [48] $g(x) = \ln x;\ h(x) = e^x + 1.$

[49] $g(x) = \ln|x|;\ h(x) = x + 2.$ [50] $g(x) = |x|;\ h(x) = \ln x - 1.$

[51] $g(x) = \ln x;\ h(x) = x^2.$ [52] $g(x) = x^2;\ h(x) = \ln x.$

[53] $g(x) = e^x;\ h(x) = x^2 + x - 1.$ [54] $g(x) = e^x;\ h(x) = \log_2 (x + 1).$

[55] a) $-\log(1.2 \times 10^{-3}) = 2.92.$ ● 2.92

b) $7.3 = -\log[H^+] \Rightarrow H^+ = e^{-7.3} = 5.01 \times 10^{-8}.$ ● 5.01×10^{-8}

c) $-\log(5.0 \times 10^{-9}) = 8.30.$ ● a base

[56] a) $S = 10 \log\left(\frac{10 I_0}{I_0}\right) = 10 \log 10 = 10.$ b) $S = 10 \log(100) = 20.$

c) $S = 10 \log(10000) = 40.$

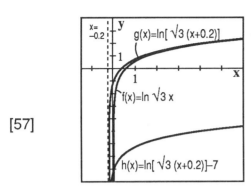

[57]

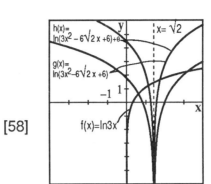

[58]

[59]   a)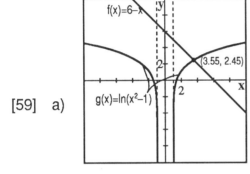

b) $(3.55, 2.45)$

[60]   a)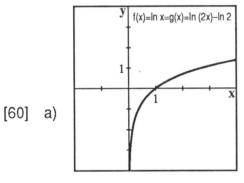

b)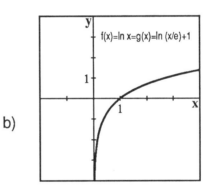

c)

Exercises 4.3

[1] $\log_2 2 = 1$.

[2] $\log_4 \sqrt[4]{4} + 4\log_4 64 = \log_4 4^{\frac{1}{4}} + 4\log_4 4^3 = \frac{1}{4}\log_4 4 + 12\log_4 4 = \frac{1}{4} + 12 = \frac{49}{4}$.

[3] $\log_8 72 - \log_8 9 = \log_8(8 \cdot 9) - \log_8 9 = \log_8 8 + \log_8 9 - \log_8 9 = \log_8 8 = 1$.

[4] $\log_5 30 + \log_5 \frac{5}{6} = \log_5(5 \cdot 6) + \log_5 5 - \log_5 6 = \log_5 5 + \log_5 6 + \log_5 5 - \log_5 6 = 2\log_5 5 = 2$.

[5] $-\frac{1}{4} + \log \sqrt[4]{10} = -\frac{1}{4} + \log 10^{\frac{1}{4}} = -\frac{1}{4} + \frac{1}{4}\log 10 = -\frac{1}{4} + \frac{1}{4} = 0$.

[6] $\log_7 2 - \log_7 14 = \log_7 2 - \log_7(2.7) = \log_7 2 - (\log_7 2 + \log_7 7) = \log_7 2 - \log_7 2 - \log_7 7 =$
     $-\log_7 7 = -1$.

[7] $\ln e^4 - \ln 1 = 4\ln e - 0 = 4$.

[8] $2^{\log_2 16 - \log_2 2^{16}} = 2^{\log_2 2^4 - 16\log_2 2} = 2^{4\log_2 2 - 16} = 2^{4-16} = 2^{-12}$.

[9] $\log_b b^b = b\log_b b = b$. [10] $\log_b b^{1/b} = \frac{1}{b}\log_b b = \frac{1}{b}$.

[11] a) $\log_b 60 = \log_b(5 \cdot 12) = \log_b 5 + \log_b 12 \approx 0.90 + 1.39 = 2.29$. ● 2.29

     b) $\log_b \frac{5}{12} = \log_b 5 - \log_b 12 \approx 0.90 - 1.39 = -0.49$. ● −0.49

     c) $\log_b \sqrt[5]{12} = \log_b 12^{\frac{1}{5}} = \frac{1}{5}\log_b 12 \approx \frac{1}{5}(1.39) = 0.278$. ● 0.278

     d) $\frac{\log_b 5}{\log_b 12} \approx \frac{0.90}{1.39} \approx 0.647$. ● 0.647

[12] $\log_b(x\,y\,z) = \log_b[(x\,y)\,z] = \log_b(x\,y) + \log_b z = \log_b x + \log_b y + \log_b z$.

[13] $4\ln 3x - \ln x = \ln(3x)^4 - \ln x = \ln\frac{(3x)^4}{x} = \ln(81\,x^3)$.

[14] $\log\frac{x^2}{y} + \log\frac{y^2}{2x} = \log\frac{x^2 y^2}{2xy} = \log\frac{xy}{2}$.

[15] $\log_6 \frac{2}{3} - \log_6 4 + \log_6 32 = \log_6 \frac{\frac{2}{3} \cdot 32}{4} = \log_6 \frac{16}{3}$.

[16] $\ln 2x - \ln x^2 - \ln\frac{4}{x} = \ln\frac{2x}{x^2 \cdot \frac{4}{x}} = \ln\frac{1}{2} = \ln 1 - \ln 2 = -\ln 2$.

[17] $\log_b \frac{x+y}{2z} - \log_b \frac{4}{x+y} = \log_b \frac{\frac{x+y}{2z}}{\frac{4}{x+y}} = \log_b \frac{(x+y)^2}{8z}$

[18] $\log_2 \frac{x^2 + 2x - 8}{x^2 - 3x + 2} - \log_2 \frac{x^2 + 6x + 8}{x^2 - 1} = \log_2 \frac{(x+4)(x-2)}{(x-2)(x-1)} - \log_2 \frac{(x+4)(x+2)}{(x+1)(x-1)} =$

    $\log_2 \frac{\frac{(x+4)(x-2)}{(x-2)(x-1)}}{\frac{(x+4)(x+2)}{(x+1)(x-1)}} = \log_2 \left[\frac{(x+4)(x-2)}{(x-2)(x-1)} \cdot \frac{(x+1)(x-1)}{(x+4)(x+2)}\right] = \log_2 \frac{x+1}{x+2}$.

158

[19] $\log_5 x - \frac{1}{2}\left[\log_5 (x-1) + \log_5 (2x+3)\right] = \log_5 x - \frac{1}{2}\log_5\left[(x-1)(2x+3)\right] =$

$\log_5 x - \log_5\left[(x-1)(2x+3)\right]^{1/2} = \log_5 x - \log_5\sqrt{(x-1)(2x+3)} = \log_5 \dfrac{x}{\sqrt{2x^2+x-3}}.$

[20] $\log_b 6 + 5\left[\log_b (2x) - 3\log_b (1-x)\right] = \log_b 6 + 5\left[\log_b (2x) - \log_b (1-x)^3\right] =$

$\log_b 6 + 5\log_b \dfrac{2x}{(1-x)^3} = \log_b 6 + \log_b\left[\dfrac{2x}{(1-x)^3}\right]^5 = \log_b\left[6\cdot\dfrac{32x^5}{(1-x)^{15}}\right] = \log_b \dfrac{192\,x^5}{(1-x)^{15}}.$

[21] $\log_b \dfrac{x^2 y^2}{z^3} = \log_b (x^2 y^2) - \log_b z^3 = \log_b x^2 + \log_b y^2 - \log_b z^3 = 2\log_b x + 2\log_b y - 3\log_b z.$

[22] $\log_b \dfrac{x z^3}{y} = \log_b (x z^3) - \log_b y = \log_b x + \log_b z^3 - \log_b y = \log_b x + 3\log_b z - \log_b y.$

[23] $\log_2 \sqrt[5]{\dfrac{x^2}{yz}} = \log_2\left(\dfrac{x^2}{yz}\right)^{\frac{1}{5}} = \frac{1}{5}\log_2 \dfrac{x^2}{yz} = \frac{1}{5}\left[\log_2 x^2 - \log_2 (yz)\right] =$

$\frac{1}{5}\left[2\log_2 x - (\log_2 y + \log_2 z)\right] = \frac{1}{5}\left(2\log_2 x - \log_2 y - \log_2 z\right).$

[24] $\log_4 x\sqrt[3]{\dfrac{y}{z}} = \log_4\left[x\left(\dfrac{y}{z}\right)^{1/3}\right] = \log_4 x + \log_4\left(\dfrac{y}{z}\right)^{1/3} = \log_4 x + \frac{1}{3}\log_4 \dfrac{y}{z} =$

$\log_4 x + \frac{1}{3}\left(\log_4 y - \log_4 z\right) = \log_4 x + \frac{1}{3}\log_4 y - \frac{1}{3}\log_4 z.$

[25] $\ln \dfrac{2-x}{(x+1)(2x-1)^2} = \ln (2-x) - \ln\left[(x+1)(2x-1)^2\right] = \ln (2-x) - \left[\ln (x+1) + \ln (2x-1)^2\right] =$

$\ln (2-x) - \left[\ln (x+1) + 2\ln (2x-1)\right] = \ln (2-x) - \ln (x+1) - 2\ln (2x-1).$

[26] $\ln \dfrac{1}{x^3-x+2} = \ln 1 - \ln (x^3-x+2) = -\ln (x^3-x+2).$

[27] $\log_b xy\sqrt{z} = \log_b x + \log_b y + \log_b \sqrt{z} = \log_b x + \log_b y + \log_b z^{1/2} = \log_b x + \log_b y + \frac{1}{2}\log_b z.$

[28] $\log_b \sqrt[3]{xy^2\sqrt[3]{z}} = \log_b\left(xy^2 z^{1/3}\right)^{1/3} = \frac{1}{3}\log_b\left(xy^2 z^{1/3}\right) = \frac{1}{3}\left(\log_b x + \log_b y^2 + \log_b z^{1/3}\right) =$

$\frac{1}{3}\left(\log_b x + 2\log_b y + \frac{1}{3}\log_b z\right) = \frac{1}{3}\log_b x + \frac{2}{3}\log_b y + \frac{1}{9}\log_b z.$

[29] $2^x = 10 \Rightarrow \ln 2^x = \ln 10 \Rightarrow x\ln 2 = \ln 10 \Rightarrow x = \dfrac{\ln 10}{\ln 2} \approx 3.32.$ ● $\dfrac{\ln 10}{\ln 2} \approx 3.32$

[30] $e^x = 15 \Rightarrow \ln e^x = \ln 15 \Rightarrow x = \ln 15 \approx 2.71.$ ● $\ln 15 \approx 2.71$

[31] $4^{2-x} = 5^{x-3} \Rightarrow \ln 4^{2-x} = \ln 5^{x-3} \Rightarrow (2-x)\ln 4 = (x-3)\ln 5 \Rightarrow 2\ln 4 - x\ln 4 = x\ln 5 - 3\ln 5 \Rightarrow$

$2\ln 4 + 3\ln 5 = x(\ln 4 + \ln 5) \Rightarrow x = \dfrac{2\ln 4 + 3\ln 5}{\ln 4 + \ln 5} \approx 2.54.$ ● $\dfrac{2\ln 4 + 3\ln 5}{\ln 4 + \ln 5} \approx 2.54$

[32] $3^{-x^2} = \dfrac{1}{8} \Rightarrow \ln 3^{-x^2} = \ln \dfrac{1}{8} \Rightarrow -x^2\ln 3 = \ln 1 - \ln 8 \Rightarrow x^2\ln 3 = \ln 8 \Rightarrow x^2 = \dfrac{\ln 8}{\ln 3} \Rightarrow x = \pm\sqrt{\dfrac{\ln 8}{\ln 3}}.$

● $\pm\sqrt{\dfrac{\ln 8}{\ln 3}} \approx \pm 1.38$

[33] $\log(40x+1) = 4 + \log(x-2) \Rightarrow \log(40x+1) - \log(x-2) = 4 \Rightarrow \log\frac{40x+1}{x-2} = 4 \Rightarrow$

$\frac{40x+1}{x-2} = 10^4 = 10{,}000 \Rightarrow 40x+1 = 10{,}000(x-2) \Rightarrow 9960x = 20001 \Rightarrow x = \frac{20001}{9960} \approx 2.01.$

$\bullet\ \frac{20001}{9960} \approx 2.01$

[34] $\ln x + \ln(x+1) = \ln 6 \Rightarrow \ln[x(x+1)] = \ln 6 \Rightarrow x^2 + x = 6 \Rightarrow x^2 + x - 6 = 0 \Rightarrow x = -3$ or $x = 2$.

$\bullet\ -3, 2$

[35] $\log_2(x^2-7) = 4 \Rightarrow x^2 - 7 = 2^4 = 16 \Rightarrow x^2 = 23 \Rightarrow x = \pm\sqrt{23} \approx \pm 4.80.$ $\bullet\ \pm\sqrt{23} \approx \pm 4.80$

[36] $\log_3(2x^2-1) = 3 \Rightarrow 2x^2 - 1 = 3^3 = 27 \Rightarrow 2x^2 = 28 \Rightarrow x^2 = 14 \Rightarrow x = \pm\sqrt{14} \approx \pm 3.74.$

$\bullet\ \pm\sqrt{14} \approx \pm 3.74$

[37] $\log_x(15-10x) = 2 \Rightarrow 15 - 10x = x^2 \Rightarrow x^2 + 10x - 15 = 0 \Rightarrow x = -5 + 2\sqrt{10} \approx 1.32$

(x must be positive). $\bullet\ -5 + 2\sqrt{10} \approx 1.32$

[38] $\log_x(x^2+3x-3) = 3 \Rightarrow x^2 + 3x - 3 = x^3 \Rightarrow x^3 - x^2 - 3x + 3 = 0 \Rightarrow (x-1)(x^2-3) = 0 \Rightarrow$

$x = 1$, or $x = \pm\sqrt{3}$. Since we do not use 1 as the base of a logarithm and since the base of a

logarithm must be positive, we choose $x = \sqrt{3} \approx 1.73.$ $\bullet\ \sqrt{3} \approx 1.73$

[39] $\log_3|4-x^2| = 2 \Rightarrow |4-x^2| = 9 \Rightarrow 4 - x^2 = -9$ or $4 - x^2 = 9.$ $4 - x^2 = -9 \Rightarrow x^2 = 13 \Rightarrow$

$x = \pm\sqrt{13} \approx \pm 3.61.$ $4 - x^2 = 9 \Rightarrow x^2 = -5 \Rightarrow x = \pm\sqrt{5}i.$ $\bullet\ \pm\sqrt{13} \approx 3.61, \pm\sqrt{5}i$

[40] $\log_5|6-2x^2| = 3 \Rightarrow |6-2x^2| = 5^3 = 125 \Rightarrow 6 - 2x^2 = 125$ or $6 - 2x^2 = -125.$ $6 - 2x^2 = 125 \Rightarrow$

$x^2 = -\frac{119}{2} \Rightarrow x = \pm\sqrt{\frac{119}{2}}\,i.$ $6 - 2x^2 = -125 \Rightarrow x^2 = \frac{131}{2} \Rightarrow x = \pm\sqrt{\frac{131}{2}} \approx \pm 8.09.$

$\bullet\ \pm\sqrt{\frac{131}{2}} \approx \pm 8.09, \pm\sqrt{\frac{119}{2}}\,i$

[41] $\log_5 6 = \frac{\log 6}{\log 5}$

[42] $\ln 4 = \frac{\log 4}{\log e}$

[43] $\log_3 x = \frac{\log x}{\log 3}$

[44] $\log_4(x^2-1) = \frac{\log(x^2-1)}{\log 4}$

[45] $\ln(x-1) = \frac{\log(x-1)}{\log e}$

[46] $\log_b(5-x) = \frac{\log(5-x)}{\log b}$

[47] $\log 15 = \frac{\ln 15}{\ln 10}$

[48] $\log_2 x = \frac{\ln x}{\ln 2}$

[49] $\log_b(x^2-1) = \frac{\ln(x^2-1)}{\ln b}$

[50] $\log_4(x^2-1) = \frac{\ln(x^2-1)}{\ln 4}$

[51] Let $y = \frac{e^x + e^{-x}}{2}$. Multiplying both sides of this equation by $2e^x$, we get $2e^x y = e^{2x} + 1$,

or $e^{2x} - 2e^x y + 1 = 0$. Letting $u = e^x$, we have $u^2 - 2yu + 1 = 0$. Using the quadratic formula,

we find that $u = y \pm \sqrt{y^2-1}$, or $e^x = y \pm \sqrt{y^2-1}$. Hence, $x = \ln\left(y \pm \sqrt{y^2-1}\right)$.

[52] Let $y = \dfrac{e^x - e^{-x}}{2}$. Multiplying both sides of this equation by $2\,e^x$, we get $2\,e^x\,y = e^{2x} - 1$,

or $e^{2x} - 2\,e^x\,y - 1 = 0$. Letting $u = e^x$, we have $u^2 - 2\,yu - 1 = 0$. Solving the latter equation for

u, we find that $u = y \pm \sqrt{y^2 + 1}$. Since $u = e^x$ and since $e^x > 0$ for all x, we have

$u = e^x = y + \sqrt{y^2 + 1}$. Hence, $x = \ln\left(y + \sqrt{y^2 + 1}\right)$.

[53] Let $y = \dfrac{e^x + e^{-x}}{e^x - e^{-x}} \Rightarrow (e^x - e^{-x})\,y = e^x + e^{-x} \Rightarrow e^x y - e^{-x} y = e^x + e^{-x} \Rightarrow e^x y - e^x =$

$e^{-x} y + e^{-x} \Rightarrow e^x (y - 1) = e^{-x}(y + 1) \Rightarrow e^{2x}(y-1) = y + 1 \Rightarrow e^{2x} = \dfrac{y+1}{y-1} \Rightarrow 2x =$

$\ln\left(\dfrac{y+1}{y-1}\right) \Rightarrow x = \dfrac{1}{2}\ln\left(\dfrac{y+1}{y-1}\right).$

[54] $y = \dfrac{e^x - e^{-x}}{e^x + e^{-x}} \Rightarrow (e^x + e^{-x})\,y = e^x - e^{-x} \Rightarrow e^x y + e^{-x} y = e^x - e^{-x} \Rightarrow e^x y - e^x =$

$-\left(e^{-x} + e^{-x} y\right) \Rightarrow e^x (y - 1) = e^{-x}(-1 - y) \Rightarrow e^{2x}(y - 1) = -(y + 1) \Rightarrow e^{2x} = -\dfrac{(y+1)}{y-1} =$

$\dfrac{1 + y}{1 - y} \Rightarrow 2x = \ln\left(\dfrac{1+y}{1-y}\right) \Rightarrow x = \dfrac{1}{2}\ln\left(\dfrac{1+y}{1-y}\right).$

[55] $\dfrac{\log 29}{\log 42} \approx 0.90$; $\log 29 - \log 42 \approx -0.16$.   **[56]** $\sqrt{\ln 6} \approx 1.34$; $\ln \sqrt{6} \approx 0.90$.

[57] $\ln\left(2^{10} + 5\right) \approx 6.94$; $\ln 2^{10} + \ln 5 \approx 8.54$.   **[58]** $\left(\log 47\right)^3 \approx 4.68$; $3 \log 47 \approx 5.02$.

[59]   a)   Let $\log_b x = y$. Then $b^y = x \Rightarrow \log_a b^y = \log_a x \Rightarrow y \log_a b = \log_a x \Rightarrow y = \dfrac{\log_a x}{\log_b x}$.

     b) $\log_b a = \dfrac{\log_a a}{\log_a b} = \dfrac{1}{\log_a b}$.

[60] $\left(\log_2 5\right)\left(\log_5 4\right)\left(\log_6 7\right) = \log_2 5 \left(\dfrac{\log_2 4}{\log_2 5}\right)\left(\log_6 7\right) = 2 \log_6 7$.

[61] Let $M = \log_b p$ and $N = \log_b q$. Then $b^M = p$ and $b^N = q$. Hence, $\dfrac{p}{q} = \dfrac{b^M}{b^N} = b^{M-N}$. Therefore,

$\log_b \dfrac{p}{q} = M - N = \log_b p - \log_b q$.

[62] (See Exercise 52) Let $y = \dfrac{e^x - e^{-x}}{2}$. Interchanging x and y, we get $x = \dfrac{e^y - e^{-y}}{2}$. Multiplying the

latter equation by $2\,e^y$, we have $2\,x\,e^y = e^{2y} - 1$, or $e^{2y} - 2\,x\,e^y - 1 = 0$. Letting $u = e^y$, we see that

$u^2 - 2xu - 1 = 0$. Solving for u, we get $u = \dfrac{2x \pm \sqrt{4x^2 + 4}}{2}$, or $u = x \pm \sqrt{x^2 + 1}$. Since $u = e^y > 0$,

we choose $u = x + \sqrt{x^2 + 1}$. Thus, $e^y = x + \sqrt{x^2 + 1} \Rightarrow y = \ln\left(x + \sqrt{x^2 + 1}\right)$. Hence, if

$f(x) = \ln\left(x + \sqrt{x^2 + 1}\right)$, then $f^{-1}(x) = \dfrac{e^x - e^{-x}}{2}$.

[63]

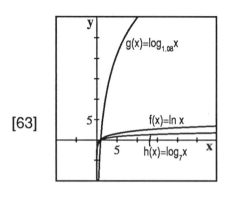

[64] $\ln(2x) - \ln 2 = \ln 2 + \ln x - \ln 2 = \ln x$.

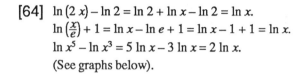

$\ln\left(\frac{x}{e}\right) + 1 = \ln x - \ln e + 1 = \ln x - 1 + 1 = \ln x$.

$\ln x^5 - \ln x^3 = 5 \ln x - 3 \ln x = 2 \ln x$.

(See graphs below).

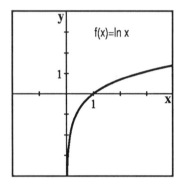

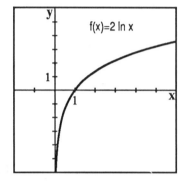

Figure 64, part a   **Figure 64, part b**

Exercises 4.4

[1]   Using the formula $A = P\, e^{rt}$, we have $A = 1000\, e^{(0.07)\, 5} \Rightarrow A = 1419.07$.   ● $1,419.07

[2]   Using the formula $A = P\, e^{rt}$, we have $2500 = P\, e^{(0.08)(10)} \Rightarrow P = \dfrac{2500}{e^{0.8}} \approx 1123.32$.   ● $1,123.32

[3]   The amount of money in Tom's account after 5 years is $A = 5000\left(1 + \dfrac{0.075}{2}\right)^{10} = 5000\,(1.0375)^{10} \approx$

7225.22. After 5 years, the amount of money in Mary's account will be $A = 5000\left(1 + \dfrac{0.075}{365}\right)^{5\,(365)} \approx$

7274.68. Thus, there is approximately $49.46 morc in Mary's account than in Tom's.   ● $49.46

[4]   Solving the formula $A = P\left(1 + \dfrac{r}{n}\right)^{nt}$ for t, we have $\ln A = \ln\left[P\left(1 + \dfrac{r}{n}\right)^{nt}\right] \Rightarrow$

$\ln A = \ln P + \ln\left(1 + \dfrac{r}{n}\right)^{nt} \Rightarrow \ln A - \ln P = nt \ln\left(1 + \dfrac{r}{n}\right) \Rightarrow t = \dfrac{\ln\frac{A}{P}}{n \ln\left(1 + \frac{r}{n}\right)}$.

a) $t = \dfrac{\ln\frac{3}{2}}{4 \ln(1.02)} \approx 5.12$ years.   b) $t = \dfrac{\ln 2}{4 \ln(1.02)} \approx 8.75$ years.

c) $t = \dfrac{\ln 3}{4 \ln(1.02)} \approx 13.87$ years.   ● a) 5.12 years; b) 8.75 years; c) 13.87 years

[5]   Using the formula $y = y_0\, e^{kt}$, we have $\frac{1}{2}\, y_0 = y_0\, e^{1600\,k} \Rightarrow \frac{1}{2} = e^{1600\,k} \Rightarrow \ln \frac{1}{2} = 1600\, k \Rightarrow$

$k = \frac{1}{1600}\, \ln \frac{1}{2} \approx -4.33 \times 10^{-4}$.   ● -4.33×10^{-4}

[6]   **a)** $y = 200 \cdot 3^{1/2} \approx 346$ at 10:00 a.m.; $200 \cdot 3^{3/2} \approx 1039$ at 12:00 noon; $200 \cdot 3^2 = 1800$ at 1:00 p.m.

● 346; 1039; 1800

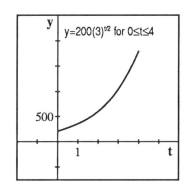

b)

[7]   Since the half–life of carbon–14 is approximately 5600 years, the amount of carbon–14 remaining

in the fossil after t years is given by $y = y_0\, e^{[(-\ln 2)/5600]\,t}$, where y_0 is the amount of carbon–14

present at death (see Example 25). Thus $\frac{1}{5}\, y_0 = y_0\, e^{[(-\ln 2)/5600]\,t} \Rightarrow \frac{1}{5} = e^{[(-\ln 2)/5600]\,t} \Rightarrow$

$\ln \frac{1}{5} = \frac{-\ln 2}{5600}\, t \ \Rightarrow t = \frac{5600\,(-\ln 5)}{-\ln 2} \approx 13002.80$.   ● approximately 13,000 years

[8] $100 = 330\, e^{-1.71\,t} + 70 \Rightarrow 30 = 330\, e^{-1.71\,t} \Rightarrow \frac{1}{11} = e^{-1.71\,t} \Rightarrow -1.71\,t = \ln \frac{1}{11} \Rightarrow t = \frac{\ln 11}{1.71} \approx 1.4$ hrs.

● approximately 1 hr. and 24 min.

[9]   **a)** $V(1) = 0.75\,(15000)\,(0.83)^0 = 11250$.   ● \$11,250

b) $V(3) = 0.75\,(15000)\,(0.83)^2 \approx 7750$.   ● \$7,750

c) $V(5) = 0.75\,(15000)\,(0.83)^4 \approx 5339$.   ● \$5,339

[10] $A = 1500 \left(1 + \frac{0.18}{12}\right)^{12} \approx 1793.43$.   ● \$1,793.43

[11] $1.3 = 5\, e^{-0.12\,t} \Rightarrow \ln \frac{1.3}{5} = -0.12\,t \Rightarrow t = \frac{1}{-0.12}\, \ln \frac{1.3}{5} \approx 11.23$.   ● approximately every 11 hours

[12] $V(7) = 150000\,(1.05)^7 \approx 211065$.   ● approximately \$211,000

[13]   **a)** $0.75 = 1 - e^{-0.05\,t} \Rightarrow e^{-0.05\,t} = 0.25 \Rightarrow -0.05\,t = \ln(0.25) \Rightarrow t = \frac{\ln(0.25)}{-0.05} \approx 27.7$.

● approximately 28 days

b) $0.91 = 1 - e^{-0.05\,t} \Rightarrow e^{-0.05\,t} = 0.10 \Rightarrow -0.05\,t = \ln(0.10) \Rightarrow t = \frac{\ln(0.10)}{-0.05} \approx 46.05$.

● approximately 46 days

[14]

x	0	0.5	1.0	1.5	2.0	3.0
y	1	1.13	1.54	2.35	3.76	10.07

(See graph below).

[15] $P(12) = 3,500,000\, e^{(0.02)(12)} \approx 4449372.03.$ ● 4,449,372

[16]

x	0	0.50	0.75	1.00	1.25	1.50	1.75	2.00
y	0.399	0.352	0.301	0.242	0.183	0.130	0.086	0.054

(See graph below).

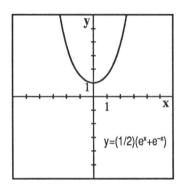

Figure 14

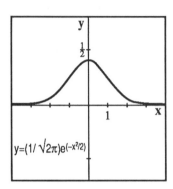

Figure 16

[17] $p(20,000) = 29\, e^{(-0.000034)(20,000)} \approx 14.69.$ ● 14.69 inches of mercury

[18] $I = 2\left(1 - e^{-100\,t}\right)$

t	0	0.005	0.01	0.015	0.02	0.025	0.03	0.035
I	0	0.787	1.264	1.554	1.729	1.836	1.90	1.94

t	0.04	0.045	0.05	0.055	0.06	0.065
I	1.96	1.98	1.99	1.99	1.995	1.99699

(See graph below).

[19]   a) $Q(0) = \dfrac{15}{1 + 14\, e^{(-0.9)(0)}} = \dfrac{15}{15} = 1.$ ● 1,000

b) $Q(2) = \dfrac{15}{1 + 14\, e^{(-0.9)(2)}} \approx 4.526.$ ● approximately 4.526

c)

t	0	1	2	3	4	5
$Q(t)$	1000	2241	4526	7728	10850	12981

(See graph below).

d)   As $t \to \infty,\ Q(t) \to 15,000.$ ● 15,000

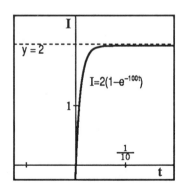

Figure 18

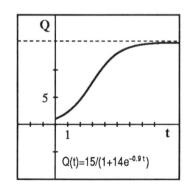

Figure 19, part c

164

[20] $7 = 0.67 \cdot \log(0.37\,E) + 1.46 \Rightarrow \dfrac{5.54}{0.67} = \log(0.37\,E) \Rightarrow 0.37\,E = 10^{5.54/0.67} \Rightarrow$

$E = \dfrac{10^{5.54/0.67}}{0.37} \approx 502 \times 10^6.$ ● 502×10^6 kilowatt – hours

[21] a) $M(8.3) = e^{1.51\,\ln 8.3} - 3.37 = (8.3)^{1.51} - 3.37 \approx 21.05.$ ● 21.05

   b) $M(10) = e^{1.51\,\ln 10} - 3.37 = 10^{1.51} - 3.37 \approx 28.99.$ ● 28.99

[22] a) (1) $p(3000) = 8.63 \ln(3000 - 680) \approx 66.88.$ ● 66.88%

   (2) $p(6000) = 8.63 \ln(6000 - 680) \approx 74.04.$ ● 74.04%

   (3) $p(8000) = 8.63 \ln(8000 - 680) \approx 76.79.$ ● 76.79%

   b)

h	3000	4000	5000	7000	9000	11000	12000
p	66.88	67.97	72.24	75.53	77.90	79.76	80.56

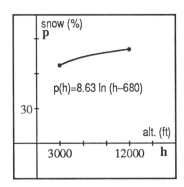

Figure 22, part b

Chapter 4 Review Exercises

[1] $dom(f) = \mathbf{R}$; $ran(f) = (0, \infty)$; horizontal asymptote, $y = 0$. (See graph below).

[2] $dom(f) = \mathbf{R}$; $ran(f) = (0, \infty)$; horizontal asymptote, $y = 0$. (See graph below).

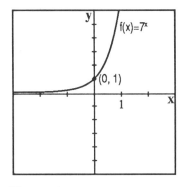

Figure 1

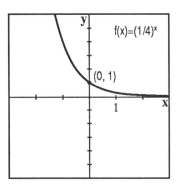

Figure 2

[3] $dom(f) = (0, \infty)$; $ran(f) = \mathbf{R}$; vertical asymptote, $x = 0$. (See graph below).

[4] $dom(f) = (1, \infty)$; $ran(f) = \mathbf{R}$; vertical asymptote, $x = 1$. (See graph below).

[5] $dom(f) = \mathbf{R}$; $ran(f) = [1, \infty)$. (See graph below).

x	0	1	2
$f(x)$	1	2.72	54.6

[6] $dom(f) = \mathbf{R}$; $ran(f) = \mathbf{R}$. (See graph below).

x	0	1	2	–1	–2	–3
$f(x)$	1	21.1	405.4	–0.95	–2.00	–3

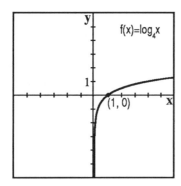

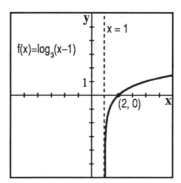

Figure 3   **Figure 4**

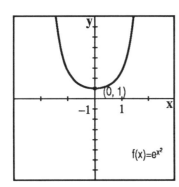

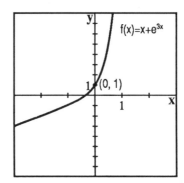

Figure 5   **Figure 6**

[7] $dom(f) = (-4, \infty)$; $ran(f) = \mathbf{R}$; vertical asymptote, x = –4. (See graph below).

[8] $dom(f) = (-\infty, -2) \cup (2, \infty)$; $ran(f) = \mathbf{R}$; vertical asymptotes, x = –2 and $x = 2$. (See graph below).

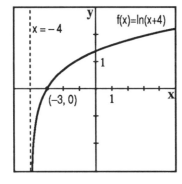

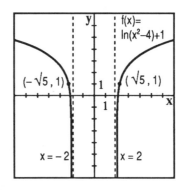

Figure 7   **Figure 8**

[9] $dom(f) = \mathbf{R}$; $ran(f) = (0, \infty)$; horizontal asymptote, y = 0. (See graph below).

[10] $dom(f) = \mathbf{R}$; $ran(f) = (0, 2)$; horizontal asymptotes, y = 0 and y = 2. (See graph below).

x	0	1	2	3	–1	–2	–3
$f(x)$	1	0.58	0.24	0.09	1.46	1.76	1.91

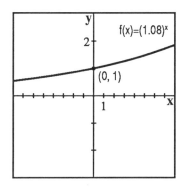

Figure 9

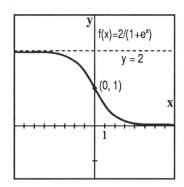

Figure 10

[11] $4^{2x+3} = 7^{x+2} \Rightarrow \ln 4^{2x+3} = \ln 7^{x+2} \Rightarrow (2x+3)\ln 4 = (x+2)\ln 7 \Rightarrow 2x\ln 4 - x\ln 7 =$

$2\ln 7 - 3\ln 4 \Rightarrow x(2\ln 4 - \ln 7) = 2\ln 7 - 3\ln 4 \Rightarrow x = \dfrac{2\ln 7 - 3\ln 4}{2\ln 4 - \ln 7} = \dfrac{\ln 7^2 - \ln 4^3}{\ln 4^2 - \ln 7} = \dfrac{\ln \frac{49}{64}}{\ln \frac{16}{7}} =$

$\left(\ln \frac{49}{64}\right) \cdot \left(\ln \frac{16}{7}\right)^{-1}.$ $\quad\bullet\left\{\left(\ln \frac{49}{64}\right) \cdot \left(\ln \frac{16}{7}\right)^{-1}\right\}$

[12] $2^{5x+3} = 3^{2x+3} \Rightarrow \ln 2^{5x+3} = \ln 3^{2x+3} \Rightarrow (5x+3)\ln 2 = (2x+3)\ln 3 \Rightarrow 5x\ln 2 - 2x\ln 3 =$

$3\ln 3 - 3\ln 2 \Rightarrow x(5\ln 2 - 2\ln 3) = 3\ln 3 - 3\ln 2 \Rightarrow x = \dfrac{3\ln 3 - 3\ln 2}{5\ln 2 - 2\ln 3} = \dfrac{\ln 3^3 - \ln 2^3}{\ln 2^5 - \ln 3^2} = \dfrac{\ln \frac{27}{8}}{\ln \frac{32}{9}} =$

$\left(\ln \frac{27}{8}\right) \cdot \left(\ln \frac{39}{9}\right)^{-1}.$ $\quad\bullet\left\{\left(\ln \frac{27}{8}\right) \cdot \left(\ln \frac{32}{9}\right)^{-1}\right\}$

[13] $\log(5x-1) - \log(x-3) = 2 \Rightarrow \log \frac{5x-1}{x-3} = 2 \Rightarrow \frac{5x-1}{x-3} = 10^2 = 100 \Rightarrow 5x-1 = 100(x-3) \Rightarrow$

$95x = 299 \Rightarrow x = \frac{299}{95}.$ $\quad\bullet\left\{\frac{299}{95}\right\}$

[14] $\log \sqrt[4]{x+1} = \frac{1}{2} \Rightarrow \frac{1}{4}\log(x+1) = \frac{1}{2} \Rightarrow \log(x+1) = 2 \Rightarrow x+1 = 10^2 = 100 \Rightarrow x = 99.$ $\quad\bullet\{99\}$

[15] $e^{\ln|x+2|} = 4 \Rightarrow |x+2| = 4 \Rightarrow x+2 = -4$ or $x+2 = 4.$ $x+2 = -4 \Rightarrow x = -6.$ $x+2 = 4 \Rightarrow x = 2.$

$\quad\bullet\{-6, 2\}$

[16] $e^{1-4x} = e \Rightarrow 1-4x = 1 \Rightarrow x = 0.$ $\quad\bullet\{0\}$

[17] $\ln(x+3) = \ln x + \ln 3 \Rightarrow \ln(x+3) = \ln(3x) \Rightarrow x+3 = 3x \Rightarrow 2x = 3 \Rightarrow x = \frac{3}{2}.$ $\quad\bullet\left\{\frac{3}{2}\right\}$

[18] $\ln 3x = \ln 3 + \ln x$ for $x \in (0, \infty).$ $\quad\bullet(0, \infty)$

[19] $\log_2 \sqrt[3]{2} = \log_2 2^{1/3} = \frac{1}{3}\log_2 2 = \frac{1}{3}.$

[20] $\log_5 75 - \log_5 3 = \log_5 \frac{75}{3} = \log_5 25 = \log_5 5^2 = 2\log_5 5 = 2.$

[21] $\frac{\ln 16}{\ln 4} = \frac{\ln 4^2}{\ln 4} = \frac{2\ln 4}{\ln 4} = 2.$ $\qquad$ [22] $\log_2 2^7 = 7\log_2 2 = 7.$

[23] $3^{\log_3 4} - 6\log_4 \sqrt[3]{4} = 4 - 6\log_4 4^{\frac{1}{3}} = 4 - 2\log_4 4 = 4 - 2 = 2.$

[24] $e^{\ln 2} - e^{\ln 3} + e^{\ln e} = 2 - 3 + e = e - 1.$

[25] $\log x^3 \sqrt[4]{\dfrac{z}{y^3}} = \log x^3 + \log \sqrt[4]{\dfrac{z}{y^3}} = \log x^3 + \log \dfrac{z^{1/4}}{y^{3/4}} = 3 \log x + \log z^{1/4} - \log y^{3/4} =$

       $3 \log x + \dfrac{1}{4} \log z - \dfrac{3}{4} \log y = 3 \log + \dfrac{1}{4} \left(\log z - 3 \log y\right).$

[26] $\ln \dfrac{x^2}{y^3} + 3 \ln y - 4 \ln xy = \ln \dfrac{x^2}{y^3} + \ln y^3 - \ln \left(x y\right)^4 = \ln \left[\dfrac{\dfrac{x^2}{y^3} \cdot y^3}{\left(x y\right)^4}\right] = \ln \dfrac{1}{x^2 y^4} = -\ln \left(x^2 y^4\right).$

[27] a) $I = I_0 b^x \Rightarrow \dfrac{I}{I_0} = b^x \Rightarrow \log\left(\dfrac{I}{I_0}\right) = \log b^x \Rightarrow \log\left(\dfrac{I}{I_0}\right) = x \log b \Rightarrow x = \dfrac{\log\left(\dfrac{I}{I_0}\right)}{\log b}.$

    b) $x = \dfrac{\log\left(\dfrac{0.01 I_0}{I_0}\right)}{\log \frac{1}{3}} = \dfrac{\log 0.01}{\log \frac{1}{3}} \approx 4.19.$ ● 4.19

[28]   Since $pH = -\log\left[H^+\right], \left[H^+\right] = 10^{-pH}.$

    a) $\left[H^+\right] = 10^{-3.1} \approx 7.94 \times 10^{-4}.$ b) $\left[H^+\right] = 10^{-5.6} \approx 2.51 \times 10^{-6}.$

    c) $\left[H^+\right] = 10^{-4.3} \approx 5.01 \times 10^{-5}.$

[29]   Using the formula $S = 10 \log\left(\dfrac{I}{I_0}\right)$, with $I = 0.3$ and $I_0 = 10^{-12}$, we get $S = 10 \log\left(\dfrac{0.3}{10^{-12}}\right) =$

    $114.77.$ ● 114.77 decibels

[30] a) $V\left(10\right) = 75000\, e^{(-0.12)(10)} \approx 22589.57.$ ● \$22589.57

    b) $37500 = 75000\, e^{-0.12 t} \Rightarrow \dfrac{1}{2} = e^{-0.12 t} \Rightarrow \ln \dfrac{1}{2} = -0.12 t \Rightarrow t = \dfrac{\ln 2}{0.12} \approx 5.78.$ ● 5.78 years

[31] a) $\dfrac{1}{3} = 1 - c \ln 2 \Rightarrow c \ln 2 = \dfrac{2}{3} \Rightarrow c = \dfrac{2}{3 \ln 2}.$

    b) $P = 1 - \dfrac{2}{3 \ln 2} \ln 3 \approx -0.05.$ ● 0

[32]   Using the equation $y = y_0\, e^{kt}$ and the fact that the half–life of the substance is m, it follows that

    $k = \dfrac{-\ln 2}{m}.$ Hence, $h = g \cdot e^{(-\ln 2/m)\, t}$ or $h = g \cdot 2^{-t/m}.$ Thus, $\dfrac{h}{g} = 2^{-t/m} \Rightarrow \ln \dfrac{h}{g} = \ln 2^{-t/m} \Rightarrow$

    $\ln \dfrac{h}{g} = -\dfrac{t}{m} \ln 2 \Rightarrow t = \dfrac{-m\left(\ln h - \ln g\right)}{\ln 2}.$ ● $\dfrac{-m\left(\ln h - \ln g\right)}{\ln 2}$

[33]   Using the formula $P = P_0\, e^{rt}$, where $P_0 = 500$, $r = 0.10$, and $t = 10$, we get $P = 500\, e^{1.0} \approx 13359.14.$

    Using the formula $P = P_0\, 2^t$, where $P_0 = 0.01$ and $t = 20$, we have $P = (0.01)\, 2^{20} = 10485.76.$

                     ● Choose the second plan since it's yield is almost \$9,000 greater than the first.

[34] If P_0 dollars are invested at 8.5 percent per year compounded quarterly, the amount of money in the

account at the end of one year is $P = P_0 \left(1 + \frac{0.085}{4}\right)^4 \approx P_0 (1.0877)$. Thus, the effective interest rate

is approximately 8.77 percent. If P_0 dollars are invested at 8.4 percent per year compounded

continuously, the amount of money in the account at the end of one year is

$P = P_0 e^{0.084} \approx P_0 (1.0876)$. Thus, the effective interest rate is approximately 8.76 percent.

● 8.5 percent per year compounded quarterly

[35] a) $25000 = P_0 \left(1 + \frac{0.08}{365}\right)^{3650} \Rightarrow P_0 = 25000 \left(1 + \frac{0.08}{365}\right)^{-3650} \approx 11234.21.$ ● $11,234.21

b) $25000 = P_0 \, e^{(0.08)(10)} \Rightarrow P_0 = 25000 \, e^{(-0.8)} \approx 11234.22.$ ● 11,234.22

[36] $V = V_0 (0.8)^t$, where t is measured in years.

[37] $P = 4 (1 + 0.1)^{25} \approx 43.34.$ ● 43.34 billion

[38] $65 = 75 \left(1 - e^{-0.07x}\right) \Rightarrow \frac{13}{15} = 1 - e^{-0.07x} \Rightarrow e^{-0.07x} = \frac{2}{15} \Rightarrow -0.07\,x = \ln \frac{2}{15} \Rightarrow x = \frac{\ln \frac{2}{15}}{-0.07} \approx$

28.78.   ● approximately 29 weeks.

[39] a) $p(5) = \frac{1000}{1 + 999\, e^{(-0.6)5}} \approx 19.7.$ ● approximately 20 people

b) $500 = \frac{1000}{1 + 999\, e^{-0.6t}} \Rightarrow \frac{1}{2} = \frac{1}{1 + 999\, e^{-0.6t}} \Rightarrow 999\, e^{-0.6t} = 1 \Rightarrow -0.6\,t = \ln \frac{1}{999} \Rightarrow$

$t = \frac{1}{0.6} \ln (999) \approx 11.51.$ ● 11.5 days

[40] a) $75 = 100 \left(1 - e^{50\,c}\right) \Rightarrow \frac{3}{4} = 1 - e^{50\,c} \Rightarrow e^{50\,c} = \frac{1}{4} \Rightarrow 50\,c = \ln \frac{1}{4} \Rightarrow c = \frac{\ln \frac{1}{4}}{50}.$ Thus,

$U = 100 \left(1 - e^{(t/50)(\ln 1/4)}\right) \Rightarrow U = 100 \left[1 - \left(\frac{1}{4}\right)^{t/50}\right].$ ● $U = 100 \left[1 - \left(\frac{1}{4}\right)^{t/50}\right]$

b) $90 = 100 \left[1 - \left(\frac{1}{4}\right)^{t/50}\right] \Rightarrow \left(\frac{1}{4}\right)^{t/50} = 0.1 \Rightarrow \frac{t}{50} \ln \frac{1}{4} = \ln 0.1 \Rightarrow t = 50 \left(\frac{\ln 10}{\ln 4}\right) \approx 83.$ ● 83 days

EXERCISES 5.1

[1] $P(5\pi)$. Since $5\pi = \pi + 2(2\pi)$, $P(5\pi) = P(\pi) = (-1, 0)$.   (See graph below).

[2] $P(-10\pi)$. Since $-10\pi = 5(-2\pi)$, $P(-10\pi) = P(0) = (1, 0)$.   (See graph below).

[3] $P\left(-\frac{2\pi}{3}\right) = \left(-\frac{1}{2}, -\frac{\sqrt{3}}{2}\right)$.   (See graph below).   [4] $P\left(\frac{7\pi}{4}\right) = \left(\frac{\sqrt{2}}{2}, -\frac{\sqrt{2}}{2}\right)$.   (See graph below).

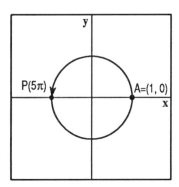

Figure 1

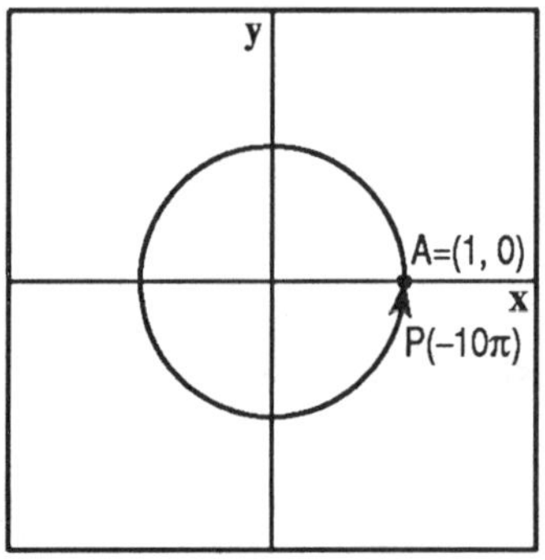

Figure 2

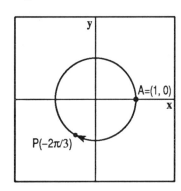

Figure 3

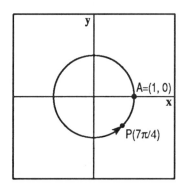

Figure 4

[5] $P\left(-\frac{7\pi}{6}\right) = \left(-\frac{\sqrt{3}}{2}, \frac{1}{2}\right)$.   (See graph below).

[6] $P(9\pi)$. Since $9\pi = \pi + 4(2\pi)$, $P(9\pi) = P(\pi) = (-1, 0)$.   (See graph below).

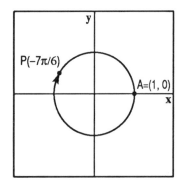

Figure 5

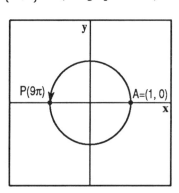

Figure 6

[7] $P\left(-\frac{7\pi}{4}\right) = \left(\frac{\sqrt{2}}{2}, \frac{\sqrt{2}}{2}\right)$.   (See graph below).   [8] $P\left(-\frac{5\pi}{3}\right) = \left(\frac{1}{2}, \frac{\sqrt{3}}{2}\right)$.   (See graph below).

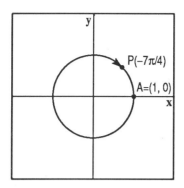

Figure 7

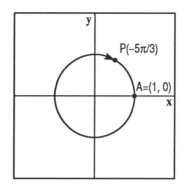

Figure 8

[9] $P\left(\frac{23\,\pi}{3}\right)$. Since $\frac{23\,\pi}{3} = \frac{5\,\pi}{3} + 3\,(2\,\pi)$, $P\left(\frac{23\,\pi}{3}\right) = P\left(\frac{5\,\pi}{3}\right) = \left(\frac{1}{2}, -\frac{\sqrt{3}}{2}\right)$.   (See graph below).

[10] $P\left(-\frac{5\,\pi}{6}\right) = \left(-\frac{\sqrt{3}}{2}, -\frac{1}{2}\right)$.   (See graph below).

[11] $P\left(\frac{50\,\pi}{6}\right)$. Since $\frac{50\,\pi}{6} = \frac{25\,\pi}{3} = \frac{\pi}{3} + 4\,(2\,\pi)$, $P\left(\frac{50\,\pi}{6}\right) = P\left(\frac{\pi}{3}\right) = \left(\frac{1}{2}, \frac{\sqrt{3}}{2}\right)$.   (See graph below).

[12] $P\left(-\frac{2\,\pi}{3}\right) = \left(-\frac{1}{2}, -\frac{\sqrt{3}}{2}\right)$.   (See graph below). [13] $P\left(-\frac{4\,\pi}{3}\right) = \left(-\frac{1}{2}, \frac{\sqrt{3}}{2}\right)$.   (See graph below).

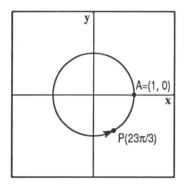

Figure 9

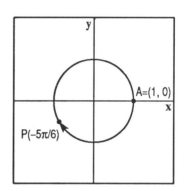

Figure 10

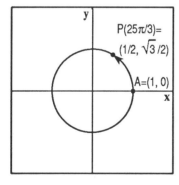

Figure 11

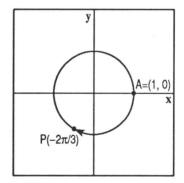

Figure 12

[14] $P\left(\frac{100\,\pi}{4}\right) = P\,(25\,\pi)$. Since $25\,\pi = \pi + 12\,(2\,\pi)$, $P\left(\frac{100\,\pi}{4}\right) = P\,(25\,\pi) = P\,(\pi) = (-1, 0)$.   (See graph below).

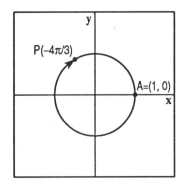

Figure 13

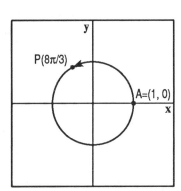

Figure 14

[15] $P\left(\dfrac{22\,\pi}{8}\right) = P\left(\dfrac{11\,\pi}{4}\right)$. Since $\dfrac{11\,\pi}{4} = \dfrac{3\,\pi}{4} + 2\,\pi$, $P\left(\dfrac{22\,\pi}{4}\right) = P\left(\dfrac{11\,\pi}{4}\right) = P\left(\dfrac{3\,\pi}{4}\right) = \left(-\dfrac{\sqrt{2}}{2}, \dfrac{\sqrt{2}}{2}\right)$.   (See graph below).

[16] $P\left(\dfrac{8\,\pi}{3}\right)$. Since $\dfrac{8\,\pi}{3} = \dfrac{2\,\pi}{3} + 2\,\pi$, $P\left(\dfrac{8\,\pi}{3}\right) = P\left(\dfrac{2\,\pi}{3}\right) = \left(-\dfrac{1}{2}, \dfrac{\sqrt{3}}{2}\right)$.   (See graph below).

[17] $P\left(\dfrac{-25\,\pi}{6}\right)$. Since $\dfrac{-25\,\pi}{6} = \dfrac{-\pi}{6} - 2\,(2\,\pi)$, $P\left(\dfrac{-25\,\pi}{6}\right) = P\left(\dfrac{-\pi}{6}\right) = \left(\dfrac{\sqrt{3}}{2}, -\dfrac{1}{2}\right)$.   (See graph below).

[18] $P\left(\dfrac{-50\,\pi}{6}\right) = P\left(\dfrac{-25\,\pi}{3}\right)$. Since $\dfrac{-25\,\pi}{3} = \dfrac{-\pi}{3} - 4\,(2\,\pi)$, $P\left(\dfrac{-50\,\pi}{6}\right) = P\left(\dfrac{-25\,\pi}{3}\right) = P\left(\dfrac{-\pi}{3}\right) = \left(\dfrac{1}{2}, -\dfrac{\sqrt{3}}{2}\right)$.   (See graph below).

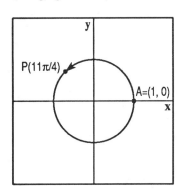

Figure 15

Figure 16

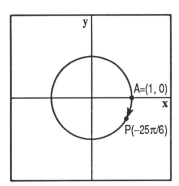

Figure 17

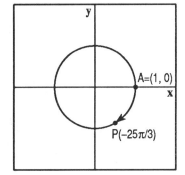

Figure 18

[19] $P(5)$. Since $\dfrac{3\pi}{2} \approx 4.71 < 5 < 6.28 \approx 2\pi$, $P(5)$ is in Q IV.   (See graph below).

[20] $P(6)$. Since $\dfrac{3\pi}{2} \approx 4.71 < 6 < 6.28 \approx 2\pi$, $P(6)$ is in Q IV.   (See graph below).

[21] $P(1)$. Since $0 < 1 < 1.57 \approx \dfrac{\pi}{2}$, $P(1)$ is in Q I.   (See graph below).

[22] $P(-30)$. Since $-30 = 4(-6.28) - 4.88 \approx -4.88 - 4(2\pi)$, it follows that $P(-30) \approx P(-4.88)$.

Since $\dfrac{3\pi}{2} \approx 4.71 < 4.88 < 6.28 \approx 2\pi$, $P(-4.88)$ is in Q I.   (See graph below).

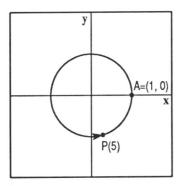

Figure 19

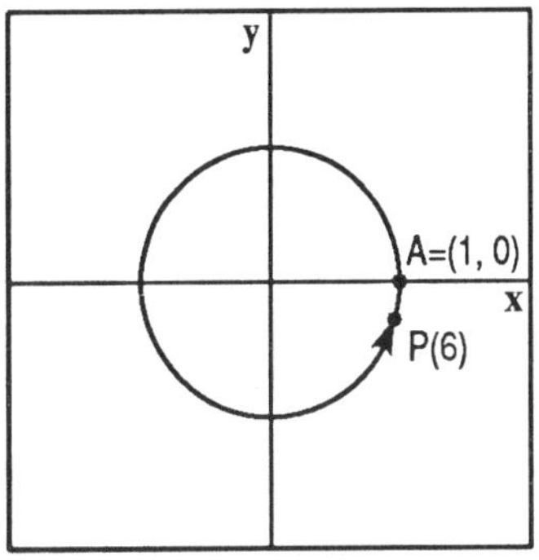

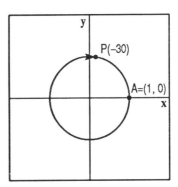

Figure 20

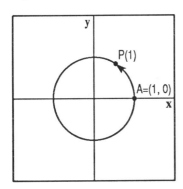

Figure 21

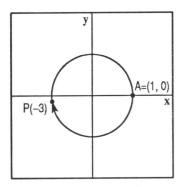

Figure 22

[23] $P(-3)$. Since $\dfrac{\pi}{2} \approx 1.57 < 3 < 3.14 \approx \pi$, $P(-3)$ is in Q II.   (See graph below).

[24] $P(16)$. Since $16 = 2(6.28) + 3.44 \approx 3.44 + 2(2\pi)$, $P(16) \approx P(3.44)$. Since $\pi \approx 3.14 < 3.44 < 4.71 \approx$

$\dfrac{3\pi}{2}$, $P(3.44)$ is in Q III.   (See graph below).

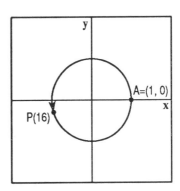

Figure 23

Figure 24

173

[25] $P(100)$. Since $100 = 15(6.28) + 5.8 \approx 5.8 + 15(2\pi)$, $P(100) \approx P(5.8)$. Since $\frac{3\pi}{2} \approx$

      $4.71 < 5.8 < 6.28 \approx 2\pi$, $P(5.8)$ is in Q IV.   (See graph below).

[26] $P(-60)$. Since $-60 = -9(6.28) - 3.48 \approx -3.48 - 9(2\pi)$, it follows that $P(-60) \approx P(-3.48)$. Since

      $\pi \approx 3.14 < 3.48 < 4.71 \approx \frac{3\pi}{2}$, $P(-3.48)$ is in Q II.   (See graph below).

[27] $P(-25)$. Since $-25 = -3(6.28) - 6.16 \approx -3(2\pi) - 6.16$, it follows that $P(-25) \approx P(-6.16)$.

      Since $\frac{3\pi}{2} \approx 4.71 < 6.16 < 6.28 \approx 2\pi$, $P(-6.16)$ is in Q I.   (See graph below).

[28] $P(12)$. Since $12 = 6.28 + 5.72 \approx 2\pi + 5.72$, $P(12) \approx P(5.72)$. Since $\frac{3\pi}{2} \approx 4.71 < 5.72 < 6.28 \approx$

      2π, $P(5.72)$ is in Q IV.   (See graph below).

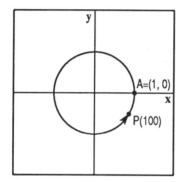

Figure 25

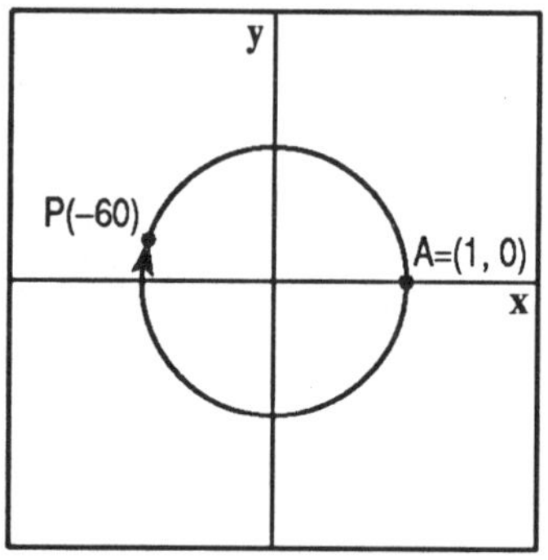

Figure 26

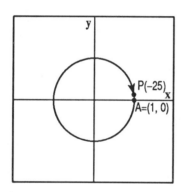

Figure 27

Figure 28

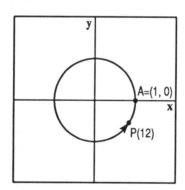

[29] $P\left(-\frac{9\pi}{10}\right)$. Since $\frac{\pi}{2} = \frac{5\pi}{10} < \frac{9\pi}{10} < \frac{10\pi}{10} = \pi$, $P\left(-\frac{9\pi}{10}\right)$ is in Q III.   (See graph below).

[30] $P\left(\frac{13\pi}{7}\right)$. Since $\frac{13\pi}{7} = \frac{26\pi}{14}$ and since $\frac{3\pi}{2} = \frac{21\pi}{14} < \frac{26\pi}{14} < \frac{28\pi}{14} = 2\pi$, $P\left(\frac{13\pi}{7}\right)$ is in Q IV.   (See

      graph below).

[31] $P(15)$. Since $15 = 12.56 + 2.44 \approx 2(2\pi) + 2.44$, $P(15) \approx P(2.44)$.

      Since $\frac{\pi}{2} \approx 1.57 < 2.44 < 3.14 \approx \pi$, $P(2.44)$ is in Q II.   (See graph below).

[32] $P(-8)$. Since $-8 = -6.28 - 1.72 \approx -1.72 - 2\pi$, $P(-8) \approx P(-1.72)$.

      Since $\frac{\pi}{2} \approx 1.57 < 1.72 < 3.14 \approx \pi$, $P(-1.72)$ is in Q III.   (See graph below).

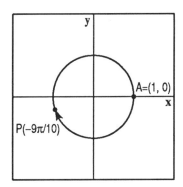

Figure 29

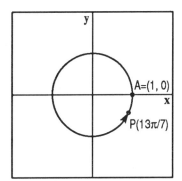

Figure 30

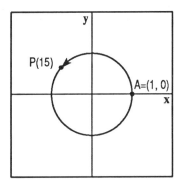

Figure 31

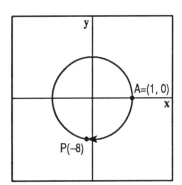

Figure 32

[33] $P\left(-\frac{\pi}{5}\right)$. Since $-\frac{\pi}{5} = -\frac{2\pi}{10}$ and since $0 < \frac{2\pi}{10} < \frac{5\pi}{10} = \frac{\pi}{2}$, $P\left(-\frac{\pi}{5}\right)$ is in Q IV.   (See graph below).

[34] Show that $P\left(\frac{\pi}{6}\right) = \left(\frac{\sqrt{3}}{2}, \frac{1}{2}\right)$.

Let $P\left(\frac{\pi}{6}\right) = (a, b) = B$. Then $P\left(-\frac{\pi}{6}\right) = (a, -b) = C$ as shown in the figure below. Since arc BC has

measure $2\left(\frac{\pi}{6}\right) = \frac{\pi}{3}$ which is one–sixth of the circumference, it follows from plane geometry that

ΔBOC is an equilateral triangle with all sides of length 1 and segment $\overline{OD}$ is the perpendicular

bisector of side $\overline{BC}$. Thus $d(D, B) = \left(\frac{1}{2}\right) d(C, B) = \frac{1}{2}(1) = \frac{1}{2}$. Hence, $b = \frac{1}{2}$. Since (a, b)

lies on the unit circle, $a^2 + b^2 = 1$. Substituting $\frac{1}{2}$ for b, we solve for a as follows: $a^2 + \left(\frac{1}{2}\right)^2 = 1 \Rightarrow$

$a^2 = 1 - \frac{1}{4} = \frac{3}{4} \Rightarrow a = \pm\sqrt{\frac{3}{4}} = \pm\frac{\sqrt{3}}{2}$. Since $P\left(\frac{\pi}{6}\right)$ lies in QI, we know that $a > 0$. So, $a = \frac{\sqrt{3}}{2}$

and it follows that $P\left(\frac{\pi}{6}\right) = \left(\frac{\sqrt{3}}{2}, \frac{1}{2}\right)$.

[35] Find all values of t in $[0, 2\pi]$ for which the y–coordinate of $P(t) = \frac{-\sqrt{3}}{2}$. As we can see from

Figure 5.16 on p. 207, there are precisely two such values of t in $[0, 2\pi]$: $\frac{4\pi}{3}$ and $\frac{5\pi}{3}$.

[36] The x–coordinate of $P(t)$ is $\frac{\sqrt{2}}{2}$. From Figure 5.16 on p. 207, we see that there are precisely two

such values of t in $[0, 2\pi]$: $\frac{\pi}{4}$ and $\frac{7\pi}{4}$.

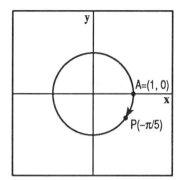

Figure 33   **Figure 34**

[37]   The x–coordinate of $P(t)$ is 1. (See graph below).

As we can see from the figure, there are exactly two such values of t in $[0, 2\pi]$: 0 and 2π.

[38]   The y–coordinate of $P(t)$ is $-\frac{1}{2}$. From Figure 5.16 on p. 207, we see that there are precisely two

such values of t in $[0, 2\pi]$: $\frac{7\pi}{6}$ and $\frac{11\pi}{6}$.

[39]   The y–coordinate of $P(t)$ is ± 1. From the figure below, we see that there are precisely two such

values of t in $[0, 2\pi]$: $\frac{\pi}{2}$ and $\frac{3\pi}{2}$.

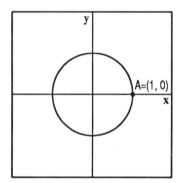   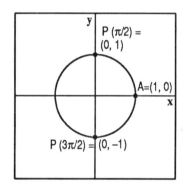

Figure 37   **Figure 39**

[40]   The x–coordinate of $P(t)$ is $-\frac{\sqrt{2}}{2}$. From Figure 5.16 on p. 207, we see that there are exactly two

such values of t in $[0, 2\pi]$: $\frac{3\pi}{4}$ and $\frac{5\pi}{4}$.

[41]   Since $P(t) = \left(\frac{1}{\sqrt{5}}, c\right)$ lies on the unit circle whose equation is $x^2 + y^2 = 1$, $\left(\frac{1}{\sqrt{5}}\right)^2 + c^2 = 1 \Rightarrow$

$\frac{1}{5} + c^2 = 1 \Rightarrow c^2 = \frac{4}{5}$. So, $c = \pm\sqrt{\frac{4}{5}} = \pm\frac{2}{\sqrt{5}}$. Since $P(t)$ lies in QIV, $c < 0$. Hence, $c = \frac{-2}{\sqrt{5}} = -\frac{2\sqrt{5}}{5}$.

[42]   Since $P(t) = \left(c, \frac{\sqrt{2}}{3}\right)$ lies on the unit circle whose equation is $x^2 + y^2 = 1$, $c^2 + \left(\frac{\sqrt{2}}{3}\right)^2 = 1 \Rightarrow$

$c^2 + \frac{2}{9} = 1 \Rightarrow c^2 = \frac{7}{9} \Rightarrow c = \pm\frac{\sqrt{7}}{3}$. Since $P(t)$ lies in QII, $c < 0$. Hence, $c = -\frac{\sqrt{7}}{3}$.

[43]   Since $P(t) = \left(-\frac{1}{3}, c\right)$ lies on the unit circle and since $x^2 + y^2 = 1$ is the equation of the unit circle,

$$\left(-\frac{1}{3}\right)^2 + c^2 = 1 \Rightarrow \frac{1}{9} + c^2 = 1 \Rightarrow c^2 = \frac{8}{9} \Rightarrow c = \pm \frac{2\sqrt{2}}{3}.$$ Since $P(t)$ lies in QII, $c > 0$. Hence, $c = \frac{2\sqrt{2}}{3}$.

[44]   Since $P(t) = \left(-\frac{2}{5}, c\right)$ lies on the unit circle whose equation is $x^2 + y^2 = 1$, $\left(-\frac{2}{5}\right)^2 + c^2 = 1 \Rightarrow$

$\frac{4}{25} + c^2 = 1 \Rightarrow c^2 = \frac{21}{25} \Rightarrow c = \pm \frac{\sqrt{21}}{5}.$ Since $P(t)$ lies in QIII, $c < 0$. Hence, $c = -\frac{\sqrt{21}}{5}$.

[45]   Since $P(t) = \left(\frac{4}{c}, \frac{5}{c}\right)$ lies on the unit circle whose equation is $x^2 + y^2 = 1$, $\left(\frac{4}{c}\right)^2 + \left(\frac{5}{c}\right)^2 = 1 \Rightarrow$

$\frac{16}{c^2} + \frac{25}{c^2} = 1 \Rightarrow \frac{41}{c^2} = 1 \Rightarrow c^2 = 41 \Rightarrow c = \pm\sqrt{41}.$ Since $P(t)$ lies in QI, $c > 0$. Hence, $c = \sqrt{41}$.

[46]   Since $P(t) = \left(\frac{6}{c}, \frac{3}{c}\right)$ lies on the unit circle whose equation is $x^2 + y^2 = 1$, $\left(\frac{6}{c}\right)^2 + \left(\frac{3}{c}\right)^2 = 1 \Rightarrow$

$\frac{36}{c^2} + \frac{9}{c^2} = 1 \Rightarrow \frac{45}{c^2} = 1 \Rightarrow c^2 = 45 \Rightarrow c = \pm 3\sqrt{5}.$ Since $P(t)$ lies in QI, $c > 0$. Hence, $c = 3\sqrt{5}$.

[47]   Since $P(t) = (c, 3c)$ lies on the unit circle whose equation is $x^2 + y^2 = 1$, $(c)^2 + (3c)^2 = 1 \Rightarrow$

$c^2 + 9c^2 = 1 \Rightarrow c^2 = \frac{1}{10} \Rightarrow c = \pm \frac{\sqrt{10}}{10}.$ Since $P(t)$ lies in QIII, $c < 0$. Hence, $c = -\frac{\sqrt{10}}{10}$.

[48]   Since $P(t) = (2c, c)$ lies on the unit circle whose equation is $x^2 + y^2 = 1$, $(2c)^2 + (c)^2 = 1 \Rightarrow$

$4c^2 + c^2 = 1 \Rightarrow 5c^2 = 1 \Rightarrow c^2 = \frac{1}{5} \Rightarrow c = \pm \frac{\sqrt{5}}{5}.$ Since $P(t)$ lies in QIII, $c < 0$. Hence, $c = -\frac{\sqrt{5}}{5}$.

EXERCISES 5.2

[1] $\tan\left(-\frac{7\pi}{2}\right).$ Since $-\frac{7\pi}{2} = -\frac{3\pi}{2} - 2\pi$, $P\left(-\frac{7\pi}{2}\right) = P\left(-\frac{3\pi}{2}\right) = (0,1)$. Hence, $\tan\left(-\frac{7\pi}{2}\right)$ is

undefined (since division by 0 is undefined).

[2] $\cot\left(-\frac{19\pi}{3}\right).$ Since $-\frac{19\pi}{3} = -\frac{\pi}{3} - 3(2\pi)$, $P\left(-\frac{19\pi}{3}\right) = P\left(-\frac{\pi}{3}\right) = \left(\frac{1}{2}, -\frac{\sqrt{3}}{2}\right).$ Therefore,

$$\cot\left(-\frac{19\pi}{3}\right) = \frac{\left(\frac{1}{2}\right)}{\left(-\frac{\sqrt{3}}{2}\right)} = -\frac{1}{\sqrt{3}} = -\frac{\sqrt{3}}{3}.$$

[3] $\sec\left(\frac{11\pi}{4}\right).$ Since $\frac{11\pi}{4} = \frac{3\pi}{4} + 2\pi$, $P\left(\frac{11\pi}{4}\right) = P\left(\frac{3\pi}{4}\right) = \left(-\frac{\sqrt{2}}{2}, \frac{\sqrt{2}}{2}\right).$ Hence, $\sec\left(\frac{11\pi}{4}\right) =$

$\dfrac{1}{\left(-\frac{\sqrt{2}}{2}\right)} = -\sqrt{2}.$

[4] $\csc\left(-\frac{23\pi}{6}\right).$ Since $-\frac{23\pi}{6} = -\frac{11\pi}{6} - 2\pi$, $P\left(-\frac{23\pi}{6}\right) = P\left(-\frac{11\pi}{6}\right) = \left(\frac{\sqrt{3}}{2}, \frac{1}{2}\right).$ Hence,

$\csc\left(-\frac{23\pi}{6}\right) = \dfrac{1}{\left(\frac{1}{2}\right)} = 2.$

[5] $\sin\left(-\frac{7\pi}{3}\right)$. Since $-\frac{7\pi}{3} = -\frac{\pi}{3} - 2\pi$, $P\left(-\frac{7\pi}{3}\right) = P\left(-\frac{\pi}{3}\right) = \left(\frac{1}{2}, -\frac{\sqrt{3}}{2}\right)$. Hence, $\sin\left(-\frac{7\pi}{3}\right) = -\frac{\sqrt{3}}{2}$.

[6] $\cos\left(\frac{21\pi}{4}\right)$. Since $\frac{21\pi}{4} = 2\,(2\pi) + \frac{5\pi}{4}$, $P\left(\frac{21\pi}{4}\right) = P\left(\frac{5\pi}{4}\right) = \left(-\frac{\sqrt{2}}{2}, -\frac{\sqrt{2}}{2}\right)$. Hence,

$\cos\left(\frac{21\pi}{4}\right) = -\frac{\sqrt{2}}{2}$.

[7] $\cos\left(\frac{20\pi}{3}\right)$. Since $\frac{20\pi}{3} = \frac{2\pi}{3} + 3\,(2\pi)$, $P\left(\frac{20\pi}{3}\right) = P\left(\frac{2\pi}{3}\right) = \left(-\frac{1}{2}, \frac{\sqrt{3}}{2}\right)$. Therefore,

$\cos\left(\frac{20\pi}{3}\right) = -\frac{1}{2}$.

[8] $\sin\left(-\frac{13\pi}{3}\right)$. Since $-\frac{13\pi}{3} = -2\,(2\pi) - \frac{\pi}{3}$, $P\left(-\frac{13\pi}{3}\right) = P\left(-\frac{\pi}{3}\right) = \left(\frac{1}{2}, -\frac{\sqrt{3}}{2}\right)$. So,

$\sin\left(-\frac{13\pi}{3}\right) = -\frac{\sqrt{3}}{2}$.

[9] $\cot\left(-11\pi\right)$. Since $-11\pi = -5\,(2\pi) - \pi$, $P\left(-11\pi\right) = P\left(-\pi\right) = \left(-1, 0\right)$. Hence, $\cot\left(-11\pi\right)$ is undefined (since division by zero is undefined).

[10] $\sec\left(\frac{15\pi}{2}\right)$. Since $\frac{15\pi}{2} = 3\,(2\pi) + \frac{3\pi}{2}$, $P\left(\frac{15\pi}{2}\right) = P\left(\frac{3\pi}{2}\right) = \left(0, -1\right)$. Therefore,

$\sec\left(\frac{15\pi}{2}\right)$ is undefined (since division by zero is undefined).

[11] $\tan\left(-\frac{15\pi}{4}\right)$. Since $-\frac{15\pi}{4} = -2\pi - \frac{7\pi}{4}$, $P\left(-\frac{15\pi}{4}\right) = P\left(-\frac{7\pi}{4}\right) = \left(\frac{\sqrt{2}}{2}, \frac{\sqrt{2}}{2}\right)$. Therefore,

$\tan\left(-\frac{15\pi}{4}\right) = \dfrac{\left(\frac{\sqrt{2}}{2}\right)}{\left(\frac{\sqrt{2}}{2}\right)} = 1$.

[12] $\cot\left(\frac{25\pi}{6}\right)$. Since $\frac{25\pi}{6} = 2\,(2\pi) + \frac{\pi}{6}$, $P\left(\frac{25\pi}{6}\right) = P\left(\frac{\pi}{6}\right) = \left(\frac{\sqrt{3}}{2}, \frac{1}{2}\right)$. Hence, $\cot\left(\frac{25\pi}{6}\right) = \dfrac{\left(\frac{\sqrt{3}}{2}\right)}{\left(\frac{1}{2}\right)} = \sqrt{3}$.

[13] $\csc\left(-7\pi\right)$. Since $-7\pi = -3\,(2\pi) - \pi$, $P\left(-7\pi\right) = P\left(-\pi\right) = \left(-1, 0\right)$. Therefore, $\csc\left(-7\pi\right)$ is undefined (since division by zero is undefined).

[14] $\sin\left(-\frac{50\pi}{8}\right)$. Since $-\frac{50\pi}{8} = -\frac{25\pi}{4} = -3\,(2\pi) - \frac{\pi}{4}$, $P\left(-\frac{50\pi}{8}\right) = P\left(-\frac{\pi}{4}\right) = \left(\frac{\sqrt{2}}{2}, -\frac{\sqrt{2}}{2}\right)$. Hence,

$\sin\left(-\frac{50\pi}{8}\right) = -\frac{\sqrt{2}}{2}$.

[15] $\cos\left(-\frac{23\pi}{4}\right)$. Since $-\frac{23\pi}{4} = -2\,(2\pi) - \frac{7\pi}{4}$, $P\left(-\frac{23\pi}{4}\right) = P\left(-\frac{7\pi}{4}\right) = \left(\frac{\sqrt{2}}{2}, \frac{\sqrt{2}}{2}\right)$. So,

$\cos\left(-\frac{23\pi}{4}\right) = \frac{\sqrt{2}}{2}$.

[16] $\sec\left(-\frac{29\pi}{6}\right)$. Since $-\frac{29\pi}{6} = -2(2\pi) - \frac{5\pi}{6}$, $P\left(-\frac{29\pi}{6}\right) = P\left(-\frac{5\pi}{6}\right) = \left(-\frac{\sqrt{3}}{2}, -\frac{1}{2}\right)$. Hence,

$$\sec\left(-\frac{29\pi}{6}\right) = \frac{1}{\left(-\frac{\sqrt{3}}{2}\right)} = -\frac{2}{\sqrt{3}} = -\frac{2\sqrt{3}}{3}.$$

[17] $\sec\left(-\frac{25\pi}{3}\right)$. Since $-\frac{25\pi}{3} = -4(2\pi) - \frac{\pi}{3}$, $P\left(-\frac{25\pi}{3}\right) = P\left(-\frac{\pi}{3}\right) = \left(\frac{1}{2}, -\frac{\sqrt{3}}{2}\right)$. Therefore,

$$\sec\left(-\frac{25\pi}{3}\right) = \frac{1}{\left(\frac{1}{2}\right)} = 2.$$

[18] $\sin(-300\pi)$. Since $-300\pi = -150(2\pi)$, $P(-300\pi) = P(0) = (1, 0)$. Therefore, $\sin(-300\pi) = 0$.

[19] $\csc(250\pi)$. Since $250\pi = 125(2\pi)$, $P(250\pi) = P(0) = (1, 0)$. Therefore, $\csc(250\pi)$ is undefined (since division by zero is undefined).

[20] $\tan\left(-\frac{58\pi}{3}\right)$. Since $-\frac{58\pi}{3} = -9(2\pi) - \frac{4\pi}{3}$, $P\left(-\frac{58\pi}{3}\right) = P\left(-\frac{4\pi}{3}\right) = \left(-\frac{1}{2}, \frac{\sqrt{3}}{2}\right)$. Therefore,

$$\tan\left(-\frac{58\pi}{3}\right) = \frac{\left(\frac{\sqrt{3}}{2}\right)}{\left(-\frac{1}{2}\right)} = -\sqrt{3}.$$

[21] $\cos\left(\frac{51\pi}{6}\right)$. Since $\frac{51\pi}{6} = \frac{17\pi}{2} = \frac{\pi}{2} + 4(2\pi)$, $P\left(\frac{51\pi}{6}\right) = P\left(\frac{\pi}{2}\right) = (0, 1)$. Hence, $\cos\left(\frac{51\pi}{6}\right) = 0$.

[22] $\left(-\frac{1}{\sqrt{5}}, -\frac{2}{\sqrt{5}}\right)$. Since $x^2 + y^2 = \left(-\frac{1}{\sqrt{5}}\right)^2 + \left(-\frac{2}{\sqrt{5}}\right)^2 = \frac{1}{5} + \frac{4}{5} = 1$, the given point lies on the unit

circle and hence is $P(t)$, for some t in **R**. Therefore, $\sin t = -\frac{2}{\sqrt{5}}$, $\cos t = -\frac{1}{\sqrt{5}}$,

$$\tan t = \frac{\left(-\frac{2}{\sqrt{5}}\right)}{\left(-\frac{1}{\sqrt{5}}\right)} = 2, \cot t = \frac{\left(-\frac{1}{\sqrt{5}}\right)}{\left(-\frac{2}{\sqrt{5}}\right)} = \frac{1}{2}, \sec t = \frac{1}{\left(-\frac{1}{\sqrt{5}}\right)} = -\sqrt{5}, \text{ and } \csc t = \frac{1}{\left(-\frac{2}{\sqrt{5}}\right)} = -\frac{\sqrt{5}}{2}.$$

[23] $\left(-\frac{3}{5}, -\frac{4}{5}\right)$. Since $x^2 + y^2 = \left(-\frac{3}{5}\right)^2 + \left(-\frac{4}{5}\right)^2 = \frac{9}{25} + \frac{16}{25} = 1$, the given point lies on the unit circle

and hence is $P(t)$, for some t in **R**. Hence, $\sin t = -\frac{4}{5}$, $\cos t = -\frac{3}{5}$, $\tan t = \frac{\left(-\frac{4}{5}\right)}{\left(-\frac{3}{5}\right)} = \frac{4}{3}$,

$$\cot t = \frac{\left(-\frac{3}{5}\right)}{\left(-\frac{4}{5}\right)} = \frac{3}{4}, \sec t = \frac{1}{\left(-\frac{3}{5}\right)} = -\frac{5}{3}, \text{ and } \csc t = \frac{1}{\left(-\frac{4}{5}\right)} = -\frac{5}{4}.$$

[24] $\left(-\frac{5}{\sqrt{26}},-\frac{1}{\sqrt{26}}\right)$. Since $x^2+y^2=\left(-\frac{5}{\sqrt{26}}\right)^2+\left(-\frac{1}{\sqrt{26}}\right)^2=\frac{25}{26}+\frac{1}{26}=1$, the given point lies on the

unit circle and hence is $P\ (t)$, for some t in **R**. Thus, $\sin t=-\frac{1}{\sqrt{26}}$, $\cos t=-\frac{5}{\sqrt{26}}$, $\tan t=$

$\dfrac{\left(-\frac{1}{\sqrt{26}}\right)}{\left(-\frac{5}{\sqrt{26}}\right)}=\frac{1}{5}$, $\cot t=\dfrac{\left(-\frac{5}{\sqrt{26}}\right)}{\left(-\frac{1}{\sqrt{26}}\right)}=5$, $\sec t=\dfrac{1}{\left(-\frac{5}{\sqrt{26}}\right)}=-\frac{\sqrt{26}}{5}$, and $\csc t=\dfrac{1}{\left(-\frac{1}{\sqrt{26}}\right)}=-\sqrt{26}$.

[25] $\left(-\frac{4}{\sqrt{41}},\frac{5}{\sqrt{41}}\right)$. Since $x^2+y^2=\left(-\frac{4}{\sqrt{41}}\right)^2+\left(\frac{5}{\sqrt{41}}\right)^2=\frac{16}{41}+\frac{25}{41}=1$, the given point lies on the

unit circle and hence is $P\ (t)$, for some t in **R**. Thus, $\sin t=\frac{5}{\sqrt{41}}$, $\cos t=-\frac{4}{\sqrt{41}}$, $\tan t=$

$\dfrac{\left(\frac{5}{\sqrt{41}}\right)}{\left(-\frac{4}{\sqrt{41}}\right)}=-\frac{5}{4}$, $\cot t=\dfrac{\left(-\frac{4}{\sqrt{41}}\right)}{\left(\frac{5}{\sqrt{41}}\right)}=-\frac{4}{5}$, $\sec t=\dfrac{1}{\left(-\frac{4}{\sqrt{41}}\right)}=-\frac{\sqrt{41}}{4}$, and $\csc t=\dfrac{1}{\left(\frac{5}{\sqrt{41}}\right)}=\frac{\sqrt{41}}{5}$.

[26] $\left(-\frac{2}{3},-\frac{\sqrt{5}}{3}\right)$. Since $x^2+y^2=\left(-\frac{2}{3}\right)^2+\left(-\frac{\sqrt{5}}{3}\right)^2=\frac{4}{9}+\frac{5}{9}=1$, the given point lies on the unit circle

and hence is $P\ (t)$, for some t in **R**. Therefore, $\sin t=-\frac{\sqrt{5}}{3}$, $\cos t=-\frac{2}{3}$, $\tan t=\dfrac{\left(-\frac{\sqrt{5}}{3}\right)}{\left(-\frac{2}{3}\right)}=-\frac{\sqrt{5}}{2}$,

$\cot t=\dfrac{\left(-\frac{2}{3}\right)}{\left(-\frac{\sqrt{5}}{3}\right)}=\frac{2}{\sqrt{5}}$, $\sec t=\dfrac{1}{\left(-\frac{2}{3}\right)}=-\frac{3}{2}$, and $\csc t=\dfrac{1}{\left(-\frac{\sqrt{5}}{3}\right)}=-\frac{3}{\sqrt{5}}$.

[27] $\left(\frac{3}{4},\frac{\sqrt{7}}{4}\right)$. Since $x^2+y^2=\left(\frac{3}{4}\right)^2+\left(\frac{\sqrt{7}}{4}\right)^2=\frac{9}{16}+\frac{7}{16}=1$, the given point lies on the unit circle and

hence is $P\ (t)$, for some t in **R**. Hence, $\sin t=\frac{\sqrt{7}}{4}$, $\cos t=\frac{3}{4}$, $\tan t=\dfrac{\left(\frac{\sqrt{7}}{4}\right)}{\left(\frac{3}{4}\right)}=\frac{\sqrt{7}}{3}$,

$\cot t=\dfrac{\left(\frac{3}{4}\right)}{\left(\frac{\sqrt{7}}{4}\right)}=\frac{3}{\sqrt{7}}$, $\sec t=\dfrac{1}{\left(\frac{3}{4}\right)}=\frac{4}{3}$, and $\csc t=\dfrac{1}{\left(\frac{\sqrt{7}}{4}\right)}=\frac{4}{\sqrt{7}}$.

[28] $\left(\frac{5}{6},\frac{\sqrt{11}}{6}\right)$. Since $x^2+y^2=\left(\frac{5}{6}\right)^2+\left(\frac{\sqrt{11}}{6}\right)^2=\frac{25}{36}+\frac{11}{36}=1$, the given point lies on the unit circle and

hence is $P\ (t)$, for some t in **R**. Thus, $\sin t=\frac{\sqrt{11}}{6}$, $\cos t=\frac{5}{6}$, $\tan t=\dfrac{\left(\frac{\sqrt{11}}{6}\right)}{\left(\frac{5}{6}\right)}=\frac{\sqrt{11}}{5}$,

$\cot t=\dfrac{\left(\frac{5}{6}\right)}{\left(\frac{\sqrt{11}}{6}\right)}=\frac{5}{\sqrt{11}}$, $\sec t=\dfrac{1}{\left(\frac{5}{6}\right)}=\frac{6}{5}$, and $\csc t=\dfrac{1}{\left(\frac{\sqrt{11}}{6}\right)}=\frac{6}{\sqrt{11}}$.

[29] $\left(\frac{\sqrt{3}}{5}, -\frac{\sqrt{22}}{5}\right)$. Since $x^2 + y^2 = \left(\frac{\sqrt{3}}{5}\right)^2 + \left(-\frac{\sqrt{22}}{5}\right)^2 = \frac{3}{25} + \frac{22}{25} = 1$, the given point lies on the unit

circle and hence is $P(t)$, for some t in **R**. Therefore, $\sin t = -\frac{\sqrt{22}}{5}$, $\cos t = \frac{\sqrt{3}}{5}$, $\tan t = \dfrac{\left(-\frac{\sqrt{22}}{5}\right)}{\left(\frac{\sqrt{3}}{5}\right)} =$

$-\frac{\sqrt{22}}{\sqrt{3}}$, $\cot t = \dfrac{\left(\frac{\sqrt{3}}{5}\right)}{\left(-\frac{\sqrt{22}}{5}\right)} = -\frac{\sqrt{3}}{\sqrt{22}}$, $\sec t = \dfrac{1}{\left(\frac{\sqrt{3}}{5}\right)} = \frac{5}{\sqrt{3}}$, and $\csc t = \dfrac{1}{\left(-\frac{\sqrt{22}}{5}\right)} = -\frac{5}{\sqrt{22}}$.

[30] $\left(-\frac{\sqrt{2}}{3}, \frac{\sqrt{7}}{3}\right)$. Since $x^2 + y^2 = \left(-\frac{\sqrt{2}}{3}\right)^2 + \left(\frac{\sqrt{7}}{3}\right)^2 = \frac{2}{9} + \frac{7}{9} = 1$, the given point lies on the unit circle

and hence is $P(t)$, for some t in **R**. Thus, $\sin t = \frac{\sqrt{7}}{3}$, $\cos t = -\frac{\sqrt{2}}{3}$, $\tan t = \dfrac{\left(\frac{\sqrt{7}}{3}\right)}{\left(-\frac{\sqrt{2}}{3}\right)} = -\frac{\sqrt{7}}{\sqrt{2}}$,

$\cot t = \dfrac{\left(-\frac{\sqrt{2}}{3}\right)}{\left(\frac{\sqrt{7}}{3}\right)} = -\frac{\sqrt{2}}{\sqrt{7}}$, $\sec t = \dfrac{1}{\left(-\frac{\sqrt{2}}{3}\right)} = -\frac{3}{\sqrt{2}}$, and $\csc t = \dfrac{1}{\left(\frac{\sqrt{7}}{3}\right)} = \frac{3}{\sqrt{7}}$.

[31] $\sin 21.3 \approx 0.6374$

[32] $\cos(-35) \approx -0.9037$

[33] $\sec(-10) = \dfrac{1}{\cos(-10)} \approx -1.192$

[34] $\tan(-100) \approx 0.5872$

[35] $\cos 50 \approx 0.9650$

[36] $\cot 27 = \dfrac{1}{\tan 27} \approx -0.3055$

[37] $\csc(-\sqrt{5}) = \dfrac{1}{\sin(-\sqrt{5})} \approx -1.271$

[38] $\cos[\tan(-3.6)] \approx \cos(-0.493466) \approx 0.8807$

[39] $\tan[\sin(-7)] \approx \tan(-0.656986) \approx -0.7713$

[40] $\sin(\sec 2) \approx \sin(-2.40299) \approx -0.6732$

[41] $\cos(\csc 5) \approx \cos(-1.042835213) \approx 0.5038$

[42] $\tan[\cos(\sqrt{7})] \approx \tan(-0.879568) \approx -1.209$

[43] Show that cotangent is an odd function. If t is any real number in the domain of the cotangent

function, then $\cot(-t) = \dfrac{1}{\tan(-t)} = \dfrac{1}{-\tan t} = -\dfrac{1}{\tan t} = -\cot t$.

[44] Show that secant is an even function. If t is any real number in the domain of the secant function,

then $\sec(-t) = \dfrac{1}{\cos(-t)} = \dfrac{1}{\cos t} = \sec t$.

[45] Show that cosecant is an odd function. If t is any real number in the domain of the cosecant

function, then $\csc(-t) = \dfrac{1}{\sin(-t)} = \dfrac{1}{-\sin t} = -\dfrac{1}{\sin t} = -\csc t$.

[46] Verify that $\cot t = \dfrac{\cos t}{\sin t} = \dfrac{1}{\tan t}$, for $\sin t \neq 0$ and $\cos t \neq 0$. Let $P(t) = (x, y) = (\cos t, \sin t)$

where $x \neq 0$ and $y \neq 0$. Then, by definition 5.1, $\cot t = \dfrac{x}{y} = \dfrac{\cos t}{\sin t}$. Similarly, $\cot t = \dfrac{x}{y} = \dfrac{1}{\frac{y}{x}} = \dfrac{1}{\tan t}$.

[47]   Verify that $\sec t = \frac{1}{\cos t}$, for $\cos t \neq 0$. Let $P(t) = (x, y) = (\cos t, \sin t)$ where $x \neq 0$. Then, by definition 5.1, $\sec t = \frac{1}{x} = \frac{1}{\cos t}$.

[48]   Verify that $\csc t = \frac{1}{\sin t}$, for $\sin t \neq 0$. Let $P(t) = (x, y) = (\cos t, \sin t)$ where $y \neq 0$. Then, using definition 5.1, $\csc t = \frac{1}{y} = \frac{1}{\sin t}$.

[49]   Verify that $\cot^2 t + 1 = \csc^2 t$. We have $\cot^2 t + 1 = \frac{\cos^2 t}{\sin^2 t} + 1 = \frac{\cos^2 t + \sin^2 t}{\sin^2 t} = \frac{1}{\sin^2 t} = \csc^2 t$.

[50]   If $\sin t = -\frac{2}{5}$ and $\tan t > 0$, find (a) $\cos t$, (b) $\cot t$. First, we note that, since $\sin t < 0$ and $\tan t > 0$, $P(t)$ is in QIII. Since $\sin^2 t + \cos^2 t = 1$, and since $\cos t < 0$ for $P(t)$ in QIII, we have $\cos t = -\sqrt{1 - \sin^2 t} = -\sqrt{1 - \left(-\frac{2}{5}\right)^2} = -\frac{\sqrt{21}}{5}$. Hence, (a) $\cos t = -\frac{\sqrt{21}}{5}$, and (b) $\cot t = \frac{\cos t}{\sin t} = \frac{\left(-\frac{\sqrt{21}}{5}\right)}{\left(-\frac{2}{5}\right)} = \frac{\sqrt{21}}{2}$.

[51]   If $\sec t = 10$ and $P(t)$ is in QIV, find (a) $\sin t$, (b) $\tan(-t)$. Since $\sec t = 10$, it follows from basic identity (i) that $\cos t = \frac{1}{10}$. Since $\sin^2 t + \cos^2 t = 1$, and since $\sin t < 0$ for $P(t)$ in QIV, it follows that $\sin t = -\sqrt{1 - \cos^2 t} = -\sqrt{1 - \left(\frac{1}{10}\right)^2} = -\frac{3\sqrt{11}}{10}$. Hence, (a) $\sin t = -\frac{3\sqrt{11}}{10}$. (b) Since tangent is an odd function, $\tan(-t) = -\tan t = -\frac{\sin t}{\cos t} = -\frac{\left(-\frac{3\sqrt{11}}{10}\right)}{\left(\frac{1}{10}\right)} = 3\sqrt{11}$.

[52]   If $\csc t = -3$ and $\cot t > 0$, find (a) $\cos t$, (b) $\sin(-t)$. First, we note that, since $\csc t < 0$ and $\cot t > 0$, $P(t)$ is in QIII. Since $\csc t = -3$, it follows from basic identity (ii) that $\sin t = -\frac{1}{3}$. Since $\sin^2 t + \cos^2 t = 1$, and since $\cos t < 0$ for $P(t)$ in QIII, it follows that $\cos t = -\sqrt{1 - \sin^2 t} = -\sqrt{1 - \left(-\frac{1}{3}\right)^2} = -\frac{2\sqrt{2}}{3}$. Hence, (a) $\cos t = -\frac{2\sqrt{2}}{3}$. (b) Since sine is an odd function, $\sin(-t) = -\sin t = -\left(-\frac{1}{3}\right) = \frac{1}{3}$.

[53]   If $\tan t = -5$ and $\cos t < 0$, find (a) $\sin t$, (b) $\sec(-t)$. First, we note that, since $\tan t < 0$ and $\cos t < 0$, $P(t)$ is in QII. Since $1 + \tan^2 t = \sec^2 t$, and since $\sec t < 0$ for $P(t)$ in QII, $\sec t = -\sqrt{1 + \tan^2 t} = -\sqrt{1 + (-5)^2} = -\sqrt{26}$. Since $\sec t = \frac{1}{\cos t}$, it follows that $\cos t = -\frac{1}{\sqrt{26}}$. Since $\sin^2 t + \cos^2 t = 1$, and since $\sin t > 0$ for $P(t)$ in QII, $\sin t = \sqrt{1 - \cos^2 t} = \sqrt{1 - \left(-\frac{1}{\sqrt{26}}\right)^2} = \frac{5}{\sqrt{26}}$. Hence, (a) $\sin t = \frac{5}{\sqrt{26}}$. (b) Since secant is an even function, $\sec(-t) = \sec t = -\sqrt{26}$.

[54] If $\cot t = -7$ and $P(t)$ is in QII, find (a) $\cos t$, (b) $\csc(-t)$. Since $1 + \cot^2 t = \csc^2 t$, and since $\csc t > 0$ for $P(t)$ in QII, $\csc t = \sqrt{1 + \cot^2 t} = \sqrt{1 + (-7)^2} = 5\sqrt{2}$. Since $\csc t = \dfrac{1}{\sin t}$, it follows that $\sin t = \dfrac{1}{5\sqrt{2}}$. Since $\sin^2 t + \cos^2 t = 1$, and since $\cos t < 0$ for $P(t)$ in QII, $\cos t = -\sqrt{1 - \sin^2 t} = -\sqrt{1 - \left(\dfrac{1}{5\sqrt{2}}\right)^2} = -\dfrac{7}{5\sqrt{2}}$. Hence, (a) $\cos t = -\dfrac{7}{5\sqrt{2}}$. (b) Since cosecant is an odd function, $\csc(-t) = -\csc t = -5\sqrt{2}$.

[55] If $\cos t = -\dfrac{9}{10}$ and $P(t)$ is in QIII, find (a) $\csc(-t)$, (b) $\tan t$. Since $\sin^2 t + \cos^2 t = 1$, and $\sin t < 0$ for $P(t)$ in QIII, $\sin t = -\sqrt{1 - \cos^2 t} = -\sqrt{1 - \left(-\dfrac{9}{10}\right)^2} = -\dfrac{\sqrt{19}}{10}$. Since $\csc t = \dfrac{1}{\sin t}$, $\csc t = -\dfrac{10}{\sqrt{19}}$. Therefore, (a) since cosecant is an odd function, $\csc(-t) = -\csc t = -\left(-\dfrac{10}{\sqrt{19}}\right) = \dfrac{10}{\sqrt{19}}$. (b) We have $\tan t = \dfrac{\sin t}{\cos t} = \dfrac{\left(-\dfrac{\sqrt{19}}{10}\right)}{\left(-\dfrac{9}{10}\right)} = \dfrac{\sqrt{19}}{9}$.

[56] If $\sin t = \dfrac{5}{7}$ and $\cos t < 0$, find (a) $\sec t$, (b) $\cot(-t)$. First we note that since $\sin t > 0$ and $\cos t < 0$, $P(t)$ is in QII. Since $\sin^2 t + \cos^2 t = 1$ and $\cos t < 0$ for $P(t)$ in QII, $\cos t = -\sqrt{1 - \sin^2 t} = -\sqrt{1 - \left(\dfrac{5}{7}\right)^2} = -\dfrac{2\sqrt{6}}{7}$. Since $\sec t = \dfrac{1}{\cos t}$, $\sec t = -\dfrac{7}{2\sqrt{6}}$. Hence, (a) $\sec t = -\dfrac{7}{2\sqrt{6}}$.

(b) Since cotangent is an odd function, $\cot(-t) = -\cot t = -\dfrac{\cos t}{\sin t} = -\dfrac{\left(-\dfrac{2\sqrt{6}}{7}\right)}{\left(\dfrac{5}{7}\right)} = \dfrac{2\sqrt{6}}{5}$.

[57] If $\sec t = 8$ and $\csc t < 0$, find (a) $\sin t$, (b) $\tan(-t)$. First, we note that since $\sec t > 0$ and $\csc t < 0$, $P(t)$ is in QIV. Since $\sec t = \dfrac{1}{\cos t}$, $\cos t = \dfrac{1}{8}$. Since $\sin^2 t + \cos^2 t = 1$ and $\sin t < 0$ for $P(t)$ in QIV, $\sin t = -\sqrt{1 - \cos^2 t} = -\sqrt{1 - \left(\dfrac{1}{8}\right)^2} = -\dfrac{3\sqrt{7}}{8}$. Hence, (a) $\sin t = -\dfrac{3\sqrt{7}}{8}$.

(b) Since tangent is an odd function, $\tan(-t) = -\tan t = -\dfrac{\sin t}{\cos t} = -\dfrac{\left(-\dfrac{3\sqrt{7}}{8}\right)}{\left(\dfrac{1}{8}\right)} = 3\sqrt{7}$.

[58] If $\csc t = -4$ and $P(t)$ is in QIII, find (a) $\sec t$, (b) $\sin(-t)$. Since $\csc t = \dfrac{1}{\sin t}$, $\sin t = -\dfrac{1}{4}$. Since $\sin^2 t + \cos^2 t = 1$ and $\cos t < 0$ for $P(t)$ in QIII, $\cos t = -\sqrt{1 - \left(-\dfrac{1}{4}\right)^2} = -\dfrac{\sqrt{15}}{4}$. So, (a) $\sec t = \dfrac{1}{\cos t} = \dfrac{1}{\left(-\dfrac{\sqrt{15}}{4}\right)} = -\dfrac{4}{\sqrt{15}}$. (b) Since sine is an odd function, $\sin(-t) = -\sin t = -\left(-\dfrac{1}{4}\right) = \dfrac{1}{4}$.

[59] If $\tan t = -\dfrac{5}{6}$ and $P(t)$ is in QII, find (a) $\sin t$, (b) $\cos(-t)$. Since $1 + \tan^2 t = \sec^2 t$ and since $\sec t < 0$ when $P(t) \in$ QII, $\sec t = -\sqrt{1 + \left(-\dfrac{5}{6}\right)^2} = -\dfrac{\sqrt{61}}{6}$. So $\cos t = \dfrac{1}{\sec t} = -\dfrac{6}{\sqrt{61}}$.

(a) Since $\sin^2 t + \cos^2 t = 1$ and since $\sin t > 0$ for $P(t)$ in QII, $\sin t = \sqrt{1 - \left(-\dfrac{6}{\sqrt{61}}\right)^2} = \dfrac{5}{\sqrt{61}}$.

(b) Since cosine is an even function, $\cos(-t) = \cos t = -\dfrac{6}{\sqrt{61}}$.

[60] If $\cot t = -2$ and $\sin t < 0$, find (a) $\cos t$ and (b) $\sin(-t)$. Since $\cot t < 0$ and $\sin t < 0$, $P(t)$ is in QIV. $\cot t = -2 \Rightarrow \tan t = -\frac{1}{2}$. Since $1 + \tan^2 t = \sec^2 t$ and since $\sec t > 0$ when $P(t) \in$ QIV,

$\sec t = \sqrt{1 + \left(-\frac{1}{2}\right)^2} = \frac{\sqrt{5}}{2}$. Hence, (a) $\cos t = \frac{1}{\sec t} = \frac{2}{\sqrt{5}}$. (b) Since $\sin^2 t + \cos^2 t = 1$ and

$\sin t < 0$ when $P(t) \in$ QIV, $\sin t = -\sqrt{1 - \left(\frac{2}{\sqrt{5}}\right)^2} = -\frac{1}{\sqrt{5}}$. So, since sine is an odd function,

$\sin(-t) = -\sin t = -\left(-\frac{1}{\sqrt{5}}\right) = \frac{1}{\sqrt{5}}$.

[61] If $\cos t = 0.2$ and $\tan t < 0$, find (a) $\sin(-t)$ and (b) $\cos(-t)$. Since $\cos t > 0$ and $\tan t < 0$,

$P(t) \in$ QIV. Since $\sin^2 t + \cos^2 t = 1$ and $\sin t < 0$ when $P(t) \in$ QIV, $\sin t = -\sqrt{1 - (0.2)^2} =$

$-\sqrt{1 - \left(\frac{1}{5}\right)^2} = -\frac{2\sqrt{6}}{5}$. So (a) $\sin(-t) = -\sin t = -\left(-\frac{2\sqrt{6}}{5}\right) = \frac{2\sqrt{6}}{5}$. (b) $\cos(-t) =$

$\cos t = 0.2 = \frac{1}{5}$.

[62] $P(t) = \left(\frac{3}{c}, \frac{5}{c}\right) \in$ QIII on the unit circle. Find c and all 6 trigonometric functions of t. Since $P(t)$ is

on the unit circle whose equation is $x^2 + y^2 = 1$, $\left(\frac{3}{c}\right)^2 + \left(\frac{5}{c}\right)^2 = 1 \Rightarrow \frac{34}{c^2} = 1 \Rightarrow c = \pm\sqrt{34}$. Since

$P(t)$ is in QIII, we know that $c < 0$. So, $c = -\sqrt{34}$. Hence, $P(t) = \left(\frac{-3}{\sqrt{34}}, \frac{-5}{\sqrt{34}}\right)$. So, $\sin t = \frac{-5}{\sqrt{34}}$,

$\cos t = \frac{-3}{\sqrt{34}}$, $\tan t = \frac{\left(\frac{-5}{\sqrt{34}}\right)}{\left(\frac{-3}{\sqrt{34}}\right)} = \frac{5}{3}$, $\cot t = \frac{\left(\frac{-3}{\sqrt{34}}\right)}{\left(\frac{-5}{\sqrt{34}}\right)} = \frac{3}{5}$, $\sec t = \frac{1}{\left(\frac{-3}{\sqrt{34}}\right)} = -\frac{\sqrt{34}}{3}$, and

$\csc t = -\frac{\sqrt{34}}{5}$.

[63] $P(t) = \left(-\frac{1}{c}, \frac{\sqrt{2}}{c}\right) \in$ QII. Since $P(t)$ is on the unit circle whose equation is $x^2 + y^2 = 1$,

$\left(-\frac{1}{c}\right)^2 + \left(\frac{\sqrt{2}}{c}\right) = 1 \Rightarrow \frac{3}{c^2} = 1 \Rightarrow c = \pm\sqrt{3}$. Since $P(t) \in QII$, $c > 0$. So, $c = \sqrt{3}$ and hence $P(t) =$

$\left(-\frac{1}{\sqrt{3}}, \frac{\sqrt{2}}{\sqrt{3}}\right)$. So, $\sin t = \frac{\sqrt{2}}{\sqrt{3}} = \frac{\sqrt{6}}{3}$, $\cos t = -\frac{1}{\sqrt{3}} = -\frac{\sqrt{3}}{3}$, $\tan t = \frac{\left(\frac{\sqrt{2}}{\sqrt{3}}\right)}{\left(-\frac{1}{\sqrt{3}}\right)} = -\sqrt{2}$, $\cot t = -\frac{1}{\sqrt{2}}$,

$\sec t = -\sqrt{3}$, and $\csc t = \frac{\sqrt{3}}{\sqrt{2}}$.

[64] Since $\sin^2 t + \cos^2 t = 1$, $\cos^2 t = 1 - \sin^2 t \Rightarrow \cos t = \pm\sqrt{1 - \sin^2 t}$. So, $\tan t = \frac{\sin t}{\cos t} =$

$\pm\frac{\sin t}{\sqrt{1 - \sin^2 t}}$, $\cot t = \pm\frac{\sqrt{1 - \sin^2 t}}{\sin t}$, $\sec t = \pm\frac{1}{\sqrt{1 - \sin^2 t}}$, and $\csc t = \frac{1}{\sin t}$.

[65] Since $\sin^2 t + \cos^2 t = 1$, $\sin t = \pm\sqrt{1 - \cos^2 t}$. So, $\tan t = \pm\frac{\sqrt{1 - \cos^2 t}}{\cos t}$, $\cot t = \pm\frac{\cos t}{\sqrt{1 - \cos^2 t}}$,

$\sec t = \frac{1}{\cos t}$, and $\csc t = \pm\frac{1}{\sqrt{1 - \cos^2 t}}$.

[66] $\sqrt{25 - u^2} = \sqrt{25 - 25\sin^2 t} = \sqrt{25\left(1 - \sin^2 t\right)} = \sqrt{25\cos^2 t} = 5\cos t$.   ($\cos t > 0$ when $P(t) \in$ QIV)

[67] $\sqrt{36 + u^2} = \sqrt{36 + 36\tan^2 t} = \sqrt{36\left(1 + \tan^2 t\right)} = \sqrt{36\sec^2 t} = 6\sec t$.   ($\sec t > 0$ when $P(t) \in$ QI)

[68] $\sqrt{9-u^2} = \sqrt{9-9\sin^2 t} = \sqrt{9\left(1-\sin^2 t\right)} = \sqrt{9\cos^2 t} = 3\cos t.$ $\left(\cos t > 0 \text{ when } P\left(t\right) \in \text{QI}\right)$

[69] $\sqrt{u^2-16} = \sqrt{16\sec^2 t - 16} = \sqrt{16\left(\sec^2 t - 1\right)} = \sqrt{16\tan^2 t} = -4\tan t = 4\tan\left(-t\right),$ since $\tan t < 0$

when $P\left(t\right)$ is in QII, and since tangent is an odd function.

[70] $\sqrt{u^2-100} = \sqrt{100\sec^2 t - 100} = \sqrt{100\left(\sec^2 t - 1\right)} = \sqrt{100\tan^2 t} = -10\tan t = 10\tan\left(-t\right),$ since

$\tan t < 0$ when $P\left(t\right) \in$ QIV and tangent is an odd function.

[71] $\sqrt{81-u^2} = \sqrt{81-81\sin^2 t} = \sqrt{81\left(1-\sin^2 t\right)} = \sqrt{81\cos^2 t} = -9\cos t.$

$\left(\cos t < 0 \text{ when } P\left(t\right) \in \text{QII}\right)$

[72]

t	0.1	0.01	0.001	0.0001	0.00001	0.000001
$\sin t$	0.0998	0.0099998	0.00099999	0.00009999	0.00001	0.000001
$\dfrac{\sin t}{t}$	0.9983	0.9999833	0.99999983	0.999999998	1	1

As t gets closer and closer to 0, $\dfrac{\sin t}{t}$ approaches 1.

[73]

t	0.1	0.01	0.001	0.0001	0.00001	0.000001
$\cos t$	0.9950	0.99995	0.9999995	0.999999995	0.999999999	1
$\dfrac{1-\cos t}{t}$	0.0499583	0.004999958	0.0005	0.00005	0.0000051	0

As t gets closer and closer to 0, $\dfrac{1-\cos t}{t}$ approaches 0.

EXERCISES 5.3

[1]   Since $\cot x = \dfrac{\cos x}{\sin x}$ and *dom* $\left(\cot\right) = \{x \mid x \neq n\,\pi, n \text{ is an integer}\}$, we use the interval $\left(0, \pi\right)$.

x	3.1	3.14	3.141	3.1415	3.14159	3.141592
$\cot x$	-24.03	-627.88	$-1,687.33$	$-10,792.89$	$-376,847.97$	$-1,530,011.17$

As $x \to \pi^-$, $\cot x \to -\infty$. Similarly as $x \to 0^+$, $\cot x \to +\infty$.   (See graph below).

x	$\dfrac{\pi}{6}$	$\dfrac{\pi}{4}$	$\dfrac{\pi}{3}$	$\dfrac{\pi}{2}$	$\dfrac{2\pi}{3}$	$\dfrac{5\pi}{6}$
$\cot x$	$\sqrt{3}$	1	$\dfrac{1}{\sqrt{3}}$	0	$\dfrac{-1}{\sqrt{3}}$	$-\sqrt{3}$

(See graph below).

[2] $\cot x = \dfrac{1}{\tan x}.$

x	0	$\dfrac{\pi}{6}$	$\dfrac{\pi}{4}$	$\dfrac{\pi}{3}$	$\dfrac{\pi}{2}$	$\dfrac{2\pi}{3}$	$\dfrac{5\pi}{6}$	π
$\tan x$	0	$\dfrac{1}{\sqrt{3}}$	1	$\sqrt{3}$	undef.	$-\sqrt{3}$	$\dfrac{-1}{\sqrt{3}}$	0
$\cot x$	undef.	$\sqrt{3}$	1	$\dfrac{1}{\sqrt{3}}$	0	$\dfrac{-1}{\sqrt{3}}$	$-\sqrt{3}$	undef.

(See graph below).

[3] $\csc x = \dfrac{1}{\sin x}.$ (See graph below).

x	0	$\dfrac{\pi}{6}$	$\dfrac{\pi}{3}$	$\dfrac{\pi}{2}$	$\dfrac{2\pi}{3}$	$\dfrac{5\pi}{6}$	π	$\dfrac{7\pi}{6}$	$\dfrac{4\pi}{3}$	$\dfrac{3\pi}{2}$	$\dfrac{5\pi}{3}$	$\dfrac{11\pi}{6}$	2π
$\sin x$	0	$\dfrac{1}{2}$	$\dfrac{\sqrt{3}}{2}$	1	$\dfrac{\sqrt{3}}{2}$	$\dfrac{1}{2}$	0	$-\dfrac{1}{2}$	$-\dfrac{\sqrt{3}}{2}$	-1	$-\dfrac{\sqrt{3}}{2}$	$-\dfrac{1}{2}$	0
$\csc x$	undef.	2	$\dfrac{2}{\sqrt{3}}$	1	$\dfrac{2}{\sqrt{3}}$	2	undef.	-2	$-\dfrac{2}{\sqrt{3}}$	-1	$-\dfrac{2}{\sqrt{3}}$	-2	undef.

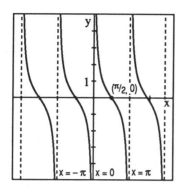

Figure 1

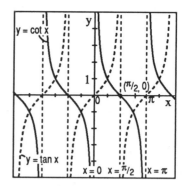

Figure 2

[4]   As we can see from their graphs, there is no largest range value for any of the functions tangent, cotangent, secant, or cosecant. So, the amplitude is undefined for each of these functions.

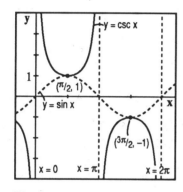

Figure 3

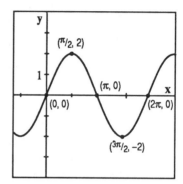

Figure 5

NOTE: On Exercises 5 – 32, the functions can all be expressed in the form $y = a \sin(bx + c)$ or $y = a \cos(bx + c)$, where $b > 0$. Hence, the functions have period $= per = \dfrac{2\pi}{b}$, amplitude $= amp = |a|$, phase shift $= ps = -\dfrac{c}{b}$, and range $= ran = [-|a|, |a|]$. If $a < 0$, the basic sine or cosine wave is reflected across the x – axis, and we indicate this by writing "Reflect." We will also give the interval that corresponds to $[0, 2\pi]$ which we will denote by *Int*.

[5] $y = 2 \sin x$. $Int = [0, 2\pi]$, $per = 2\pi$, $amp = 2$, $ps = 0$, $ran = [-2, 2]$.   (See graph above).

[6] $y = -3 \cos x$. $Int = [0, 2\pi]$, $per = 2\pi$, $amp = 3$, $ps = 0$, $ran = [-3, 3]$, Reflect   (See graph below).

[7] $y = \cos 3x$. $0 \le 3x \le 2\pi \Rightarrow 0 \le x \le \dfrac{2\pi}{3} \Rightarrow Int = \left[0, \dfrac{2\pi}{3}\right]$, $per = \dfrac{2\pi}{3}$, $amp = 1$, $ps = 0$, $ran = [-1, 1]$.

(See graph below).

[8] $y = \sin(-2x) = -\sin 2x$. $0 \le 2x \le 2\pi \Rightarrow 0 \le x \le \pi \Rightarrow Int = [0, \pi]$, $per = \pi$, $amp = 1$, $ps = 0$, $ran = [-1, 1]$, Reflect.   (See graph below).

[9] $y = -4 \sin(-x) = 4 \sin x$. $Int = [0, 2\pi]$, $per = 2\pi$, $amp = 4$, $ps = 0$, $ran = [-4, 4]$.   (See graph below).

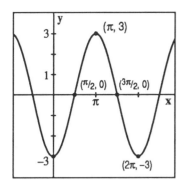

Figure 6

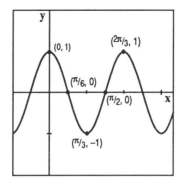

Figure 7

[10] $y = \pi \cos(2x+1).\ 0 \leq 2x+1 \leq 2\pi \Rightarrow -\frac{1}{2} \leq x \leq \pi - \frac{1}{2} \Rightarrow Int = \left[-\frac{1}{2}, \pi - \frac{1}{2}\right], per = \pi, amp = \pi,$

$ps = -\frac{1}{2}, ran = [-\pi, \pi].$ (See graph below).

[11] $y = 2 \sin 3x.\ 0 \leq 3x \leq 2\pi \Rightarrow 0 \leq x \leq \frac{2\pi}{3} \Rightarrow Int = \left[0, \frac{2\pi}{3}\right], per = \frac{2\pi}{3}, amp = 2, ps = 0,$

$ran = [-2, 2].$ (See graph below).

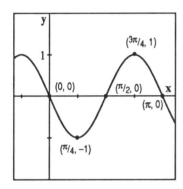

Figure 8

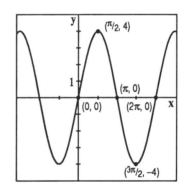

Figure 9

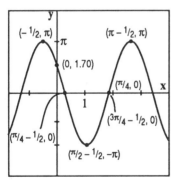

Figure 10

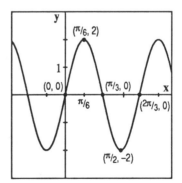

Figure 11

[12] $y = \cos 4x.\ 0 \leq 4x \leq 2\pi \Rightarrow 0 \leq x \leq \frac{\pi}{2} \Rightarrow Int = \left[0, \frac{\pi}{2}\right], per = \frac{\pi}{2}, amp = 1, ps = 0, ran = [-1, 1].$

(See graph below).

[13] $y = -\frac{\pi}{2}\cos(-x) = -\frac{\pi}{2}\cos x.\ Int = [0, 2\pi], per = 2\pi, amp = \frac{\pi}{2}, ps = 0, ran = \left[-\frac{\pi}{2}, \frac{\pi}{2}\right], \text{Reflect.}$

(See graph below).

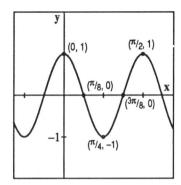

Figure 12

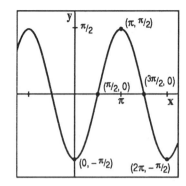

Figure 13

[14] $y = 3 \cos\left(\frac{x}{2} - 1\right)$. $0 \le \frac{x}{2} - 1 \le 2\pi \Rightarrow 2 \le x \le 4\pi + 2 \Rightarrow Int = [2, 4\pi + 2]$, $per = 4\pi$, $amp = 3$, $ps = 2$,

$ran = [-3, 3]$.   (See graph below).

[15] $y = 2 \cos(\pi - x) = 2 \cos(x - \pi)$. $0 \le x - \pi \le 2\pi \Rightarrow \pi \le x \le 3\pi \Rightarrow Int = [\pi, 3\pi]$, $per = 2\pi$, $amp = 2$,

$ps = \pi$, $ran = [-2, 2]$.   (See graph below).

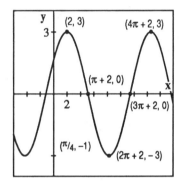

Figure 14

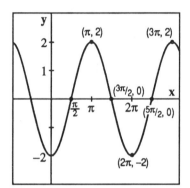

Figure 15

[16] $y = -\sin(-x - 3) = \sin(x + 3)$. $0 \le x + 3 \le 2\pi \Rightarrow -3 \le x \le 2\pi - 3 \Rightarrow Int = [-3, 2\pi - 3]$, $per = 2\pi$,

$amp = 1$, $ps = -3$, $ran = [-1, 1]$.   (See graph below).

[17] $y = \frac{1}{3} \sin\left(\frac{x}{3} + 1\right)$. $0 \le \frac{x}{3} + 1 \le 2\pi \Rightarrow -3 \le x \le 6\pi - 3 \Rightarrow Int = [-3, 6\pi - 3]$, $per = 6\pi$, $amp = \frac{1}{3}$,

$ps = -3$, $ran = \left[-\frac{1}{3}, \frac{1}{3}\right]$.   (See graph below).

[18] $y = \frac{2}{3} \cos\left(\frac{3}{2}x + 1\right)$. $0 \le \frac{3}{2}x + 1 \le 2\pi \Rightarrow -\frac{2}{3} \le x \le \frac{4\pi}{3} - \frac{2}{3} \Rightarrow Int = \left[-\frac{2}{3}, \frac{4\pi}{3} - \frac{2}{3}\right]$, $per = \frac{4\pi}{3}$,

$amp = \frac{2}{3}$, $ps = -\frac{2}{3}$, $ran = \left[-\frac{2}{3}, \frac{2}{3}\right]$.   (See graph below).

[19] $y = -\frac{4}{3} \cos(4 - x) = -\frac{4}{3} \cos(x - 4)$. $0 \le x - 4 \le 2\pi \Rightarrow 4 \le x \le 4 + 2\pi \Rightarrow Int = [4, 4 + 2\pi]$,

$per = 2\pi$, $amp = \frac{4}{3}$, $ps = 4$, $ran = \left[-\frac{4}{3}, \frac{4}{3}\right]$, Reflect.   (See graph below).

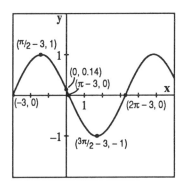

Figure 16

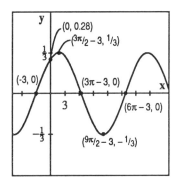

Figure 17

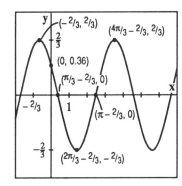

Figure 18

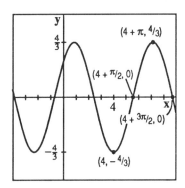

Figure 19

[20] $y = 10 \sin\left(\frac{2x}{3} - 1\right)$. $0 \le \frac{2x}{3} - 1 \le 2\pi \Rightarrow \frac{3}{2} \le x \le \frac{3}{2} + 3\pi \Rightarrow Int = \left[\frac{3}{2}, \frac{3}{2} + 3\pi\right]$, $per = 3\pi$, $amp = 10$,

$ps = \frac{3}{2}$, $ran = [-10, 10]$. (See graph below).

[21] $y = -5 \sin(2x - \pi)$. $0 \le 2x - \pi \le 2\pi \Rightarrow \frac{\pi}{2} \le x \le \frac{3\pi}{2} \Rightarrow Int = \left[\frac{\pi}{2}, \frac{3\pi}{2}\right]$, $per = \pi$, $amp = 5$, $ps = \frac{\pi}{2}$,

$ran = [-5, 5]$, Reflect. (See graph below).

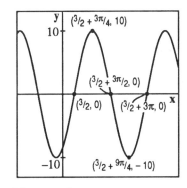

Figure 20

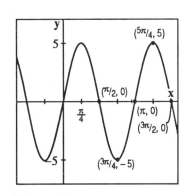

Figure 21

[22] $y = -\pi \cos\left(\frac{x}{\pi} + 1\right)$. $0 \le \frac{x}{\pi} + 1 \le 2\pi \Rightarrow -\pi \le x \le 2\pi^2 - \pi \Rightarrow Int = \left[-\pi, 2\pi^2 - \pi\right]$, $per = 2\pi^2$,

$amp = \pi$, $ps = -\pi$, $ran = [-\pi, \pi]$, Reflect. (See graph below).

189

[23] $y = \frac{1}{2}\cos\left(\frac{x}{2} + 3\right)$. $0 \le \frac{x}{2} + 3 \le 2\pi \Rightarrow -6 \le x \le 4\pi - 6 \Rightarrow Int = [-6, 4\pi - 6]$, $per = 4\pi$, $amp = \frac{1}{2}$,

$ps = -6$, $ran = \left[-\frac{1}{2}, \frac{1}{2}\right]$. (See graph below).

[24] $y = \frac{1}{4}\sin\left(\frac{3x}{4} - \pi\right)$. $0 \le \frac{3x}{4} - \pi \le 2\pi \Rightarrow \frac{4\pi}{3} \le x \le 4\pi - 6 \Rightarrow Int = \left[\frac{4\pi}{3}, 4\pi\right]$, $per = \frac{8\pi}{3}$, $amp = \frac{1}{4}$,

$ps = \frac{4\pi}{3}$, $ran = \left[-\frac{1}{4}, \frac{1}{4}\right]$. (See graph below).

[25] $y = -\sin(\pi - 3x) = \sin(3x - \pi)$. $0 \le 3x - \pi \le 2\pi \Rightarrow \frac{\pi}{3} \le x \le \pi \Rightarrow Int = \left[\frac{\pi}{3}, \pi\right]$, $per = \frac{2\pi}{3}$, $amp = 1$,

$ps = \frac{\pi}{3}$, $ran = [-1, 1]$. (See graph below).

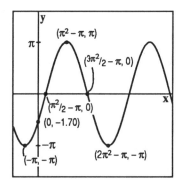

Figure 22

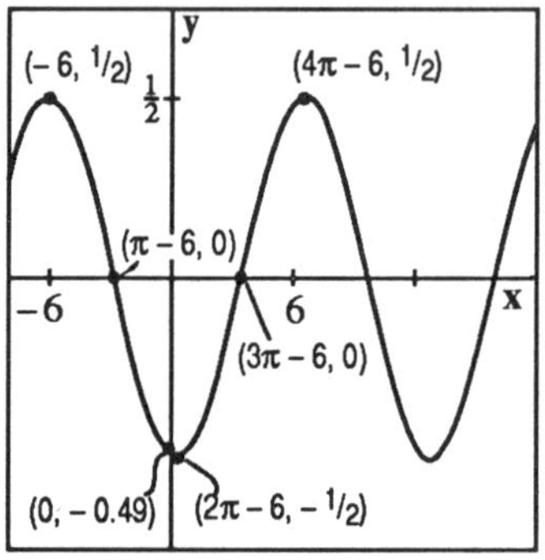

Figure 23

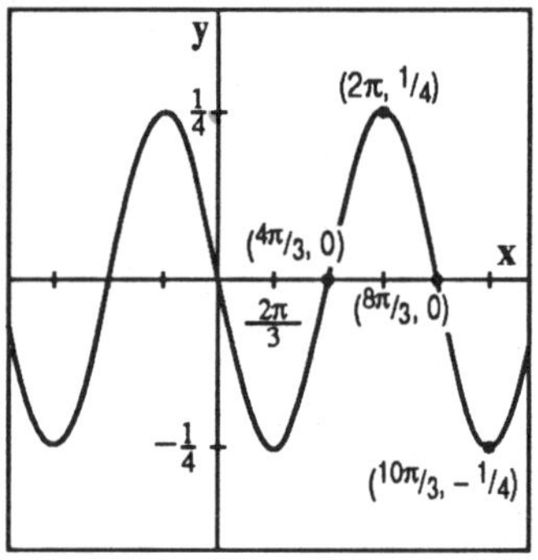

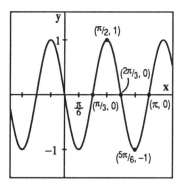

Figure 24

Figure 25

[26] $y = -\cos(\pi - 4x) = -\cos(4x - \pi)$. $0 \le 4x - \pi \le 2\pi \Rightarrow \frac{\pi}{4} \le x \le \frac{3\pi}{4} \Rightarrow Int = \left[\frac{\pi}{4}, \frac{3\pi}{4}\right]$, $per = \frac{\pi}{2}$,

$amp = 1$, $ps = \frac{\pi}{4}$, $ran = [-1, 1]$, Reflect. (See graph below).

[27] $y = 2\sin(2x + 2\pi)$. $0 \le 2x + 2\pi \le 2\pi \Rightarrow -\pi \le x \le 0 \Rightarrow Int = [-\pi, 0]$, $per = \pi$, $amp = 2$, $ps = -\pi$,

$ran = [-2, 2]$. (See graph below).

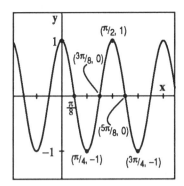

Figure 26

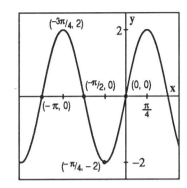

Figure 27

[28] $y = 3\cos(3x + 3).\ 0 \le 3x + 3 \le 2\pi \Rightarrow -1 \le x \le \dfrac{2\pi}{3} - 1 \Rightarrow Int = \left[-1, \dfrac{2\pi}{3} - 1\right],\ per = \dfrac{2\pi}{3},\ amp = 3,$

  $ps = -1,\ ran = [-3, 3].$ (See graph below).

[29] $y = -3\cos(\pi x + 2\pi).\ 0 \le \pi x + 2\pi \le 2\pi \Rightarrow -2 \le x \le 0 \Rightarrow Int = [-2, 0],\ per = 2,\ amp = 3,$

  $ps = -2,\ ran = [-3, 3],$ Reflect. (See graph below).

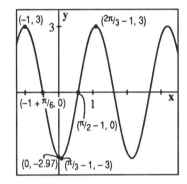

Figure 28

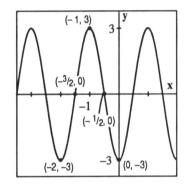

Figure 29

[30] $y = -\pi\cos(\pi x - 2\pi).\ 0 \le \pi x - 2\pi \le 2\pi \Rightarrow 2 \le x \le 4 \Rightarrow Int = [2, 4],\ per = 2,\ amp = \pi,\ ps = 2,$

  $ran = [-\pi, \pi],$ Reflect. (See graph below).

[31] $y = \pi\sin\left(1 - \dfrac{x}{2}\right) = -\pi\sin\left(\dfrac{x}{2} - 1\right).\ 0 \le \dfrac{x}{2} - 1 \le 2\pi \Rightarrow 2 \le x \le 4\pi + 2 \Rightarrow Int = [2, 2 + 4\pi],\ per = 4\pi,$

  $amp = \pi,\ ps = 2,\ ran = [-\pi, \pi],$ Reflect. (See graph below).

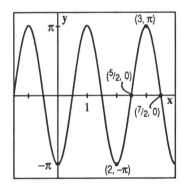

Figure 30

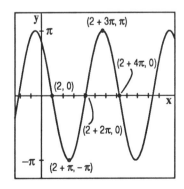

Figure 31

[32] $y = \sqrt{3} \sin\left(2x - \frac{2\pi}{3}\right). \ 0 \le 2x - \frac{2\pi}{3} \le 2\pi \Rightarrow \frac{\pi}{3} \le x \le \frac{4\pi}{3} \Rightarrow Int = \left[\frac{\pi}{3}, \frac{4\pi}{3}\right], per = \pi, amp = \sqrt{3},$

$ps = \frac{\pi}{3}, ran = \left[-\sqrt{3}, \sqrt{3}\right].$ (See graph below).

NOTE: On Exercises 33 – 50, those functions which can be expressed in the form $y = a \sec(bx + c)$ or

$y = a \csc(bx + c)$, where $b > 0$, have period $= per = \frac{2\pi}{b}$, phase shift $= ps = -\frac{c}{b}$, and range $= ran =$

$(-\infty, -|a|] \cup [|a|, \infty)$. If $a < 0$, the basic secant or cosecant graph is reflected across the x – axis, and

we indicate this by writing "Reflect." We will also give the interval that corresponds to $[0, 2\pi]$ which

we will denote by Int. Those functions which can be expressed in the form $y = a \tan(bx + c)$ or

$y = a \cot(bx + c)$, where $b > 0$, have period $= per = \frac{\pi}{b}$, phase shift $= ps = -\frac{c}{b}$, and range $= ran = \mathbf{R}$.

If $a < 0$, the basic tangent or cotangent graph is reflected across the x – axis, and we indicate this by

writing "Reflect." We will also give the interval that corresponds to $\left(-\frac{\pi}{2}, \frac{\pi}{2}\right)$ for the tangent graphs and

$(0, \pi)$ for the cotangent graphs. In either case, we will denote this interval by Int .

[33] $y = -\cot x. \ Int = (0, \pi), per = \pi, ps = 0, ran = \mathbf{R}, \text{Reflect.}$ (See graph below).

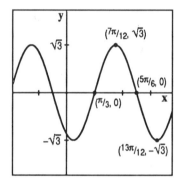

Figure 32

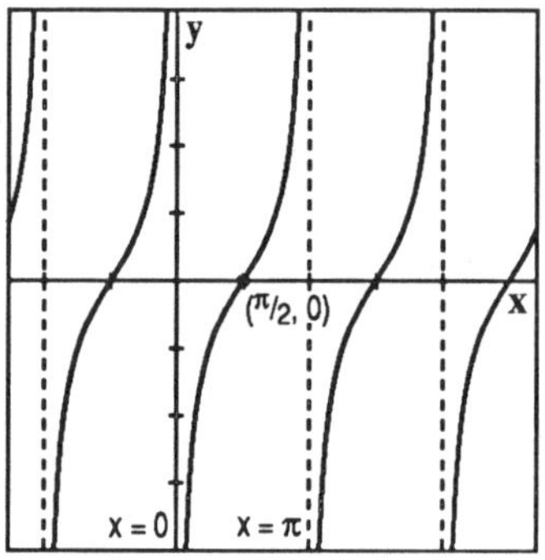

Figure 33

[34] $y = -\tan x. \ Int = \left(-\frac{\pi}{2}, \frac{\pi}{2}\right), per = \pi, ps = 0, ran = \mathbf{R}, \text{Reflect.}$ (See graph below).

[35] $y = \tan 2x. \ -\frac{\pi}{2} < 2x < \frac{\pi}{2} \Rightarrow -\frac{\pi}{4} < x < \frac{\pi}{4} \Rightarrow Int = \left(-\frac{\pi}{4}, \frac{\pi}{4}\right), per = \frac{\pi}{2}, ps = 0, ran = \mathbf{R}.$ (See graph

below).

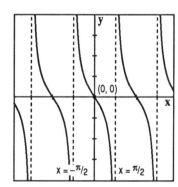

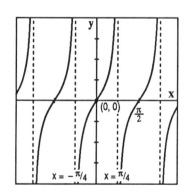

Figure 34

Figure 35

[36] $y = \cot 2x. \ 0 < 2x < \pi \Rightarrow 0 < x < \frac{\pi}{2} \Rightarrow Int = \left(0, \frac{\pi}{2}\right), per = \frac{\pi}{2}, ps = 0, ran = \mathbf{R}.$ (See graph below).

[37] $y = -\tan(2x - 1). -\frac{\pi}{2} < 2x - 1 < \frac{\pi}{2} \Rightarrow \frac{1}{2} - \frac{\pi}{4} < x < \frac{1}{2} + \frac{\pi}{4} \Rightarrow Int = \left(\frac{1}{2} - \frac{\pi}{4}, \frac{1}{2} + \frac{\pi}{4}\right), per = \frac{\pi}{2},$

$ps = \frac{1}{2}, ran = \mathbf{R},$ Reflect (See graph below).

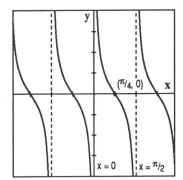

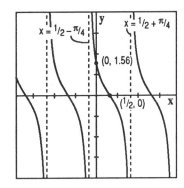

Figure 36   **Figure 37**

[38] $y = -\cot(2x + 1). 0 < 2x + 1 < \pi \Rightarrow -\frac{1}{2} < x < \frac{\pi}{2} - \frac{1}{2} \Rightarrow Int = \left(-\frac{1}{2}, \frac{\pi}{2} - \frac{1}{2}\right), per = \frac{\pi}{2}, ps = -\frac{1}{2},$

$ran = \mathbf{R},$ Reflect (See graph below).

[39] $y = \cot(1 - x) = -\cot(x - 1). 0 < x - 1 < \pi \Rightarrow 1 < x < 1 + \pi \Rightarrow Int = (1, 1 + \pi), per = \pi, ps = 1,$

$ran = \mathbf{R},$ Reflect (See graph below).

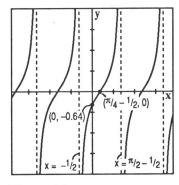

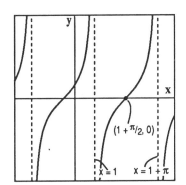

Figure 38   **Figure 39**

[40] $y = \tan(2 - x) = -\tan(x - 2). -\frac{\pi}{2} < x - 2 < \frac{\pi}{2} \Rightarrow 2 - \frac{\pi}{2} < x < 2 + \frac{\pi}{2} \Rightarrow Int = \left(2 - \frac{\pi}{2}, 2 + \frac{\pi}{2}\right), per = \pi,$

$ps = 2, ran = \mathbf{R},$ Reflect (See graph below).

[41] $y = -\tan(1 - 2x) = \tan(2x - 1). -\frac{\pi}{2} < 1 - 2x < \frac{\pi}{2} \Rightarrow \frac{1}{2} - \frac{\pi}{4} < x < \frac{1}{2} + \frac{\pi}{4} \Rightarrow Int = \left(\frac{1}{2} - \frac{\pi}{4}, \frac{1}{2} + \frac{\pi}{4}\right),$

$per = \frac{\pi}{2}, ps = \frac{1}{2}, ran = \mathbf{R}.$ (See graph below).

[42] $y = -\cot(1 - 3x) = \cot(3x - 1). 0 < 3x - 1 < \pi \Rightarrow \frac{1}{3} < x < \frac{1}{3} + \frac{\pi}{3} \Rightarrow Int = \left(\frac{1}{3}, \frac{1}{3} + \frac{\pi}{3}\right), per = \frac{\pi}{3},$

$ps = \frac{1}{3}, ran = \mathbf{R}.$ (See graph below).

[43] $y = \sec 2x. 0 \leq 2x \leq 2\pi \Rightarrow 0 \leq x \leq \pi \Rightarrow Int = [0, \pi], per = \pi, ps = 0, ran = (-\infty, -1] \cup [1, \infty).$ (See graph below).

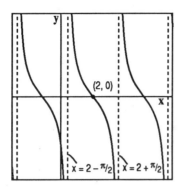

Figure 40

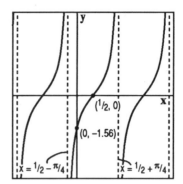

Figure 41

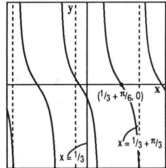

Figure 42

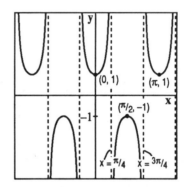

Figure 43

[44] $y = \csc 2x$. $0 \le 2x \le 2\pi \Rightarrow 0 \le x \le \pi \Rightarrow Int = [0, \pi]$, $per = \pi$, $ps = 0$, $ran = (-\infty, -1] \cup [1, \infty)$.   (See graph below).

[45] $y = \sec(x + \pi)$. $0 \le x + \pi \le 2\pi \Rightarrow -\pi \le x \le \pi \Rightarrow Int = [-\pi, \pi]$, $per = 2\pi$, $ps = -\pi$, $ran = (-\infty, -1] \cup [1, \infty)$.   (See graph below).

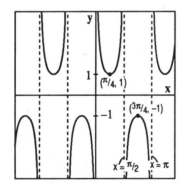

Figure 44

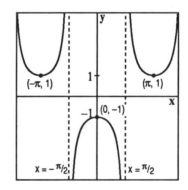

Figure 45

[46] $y = \csc(\pi - x) = -\csc(x - \pi)$. $0 \le x - \pi \le 2\pi \Rightarrow \pi \le x \le 3\pi \Rightarrow Int = [\pi, 3\pi]$, $per = 2\pi$, $ps = \pi$, $ran = (-\infty, -1] \cup [1, \infty)$, Reflect.   (See graph below).

[47] $y = -2\csc\left(x - \dfrac{\pi}{2}\right)$. $0 \le x - \dfrac{\pi}{2} \le 2\pi \Rightarrow \dfrac{\pi}{2} \le x \le \dfrac{5\pi}{2} \Rightarrow Int = \left[\dfrac{\pi}{2}, \dfrac{5\pi}{2}\right]$, $per = 2\pi$, $ps = \dfrac{\pi}{2}$, $ran = (-\infty, -2] \cup [2, \infty)$, Reflect.   (See graph below).

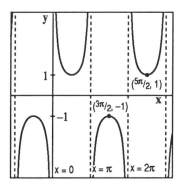

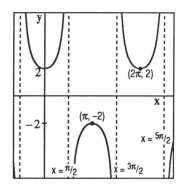

Figure 46   **Figure 47**

[48] $y = -2\sec\left(x + \dfrac{\pi}{2}\right).\ 0 \le x + \dfrac{\pi}{2} \le 2\pi \Rightarrow -\dfrac{\pi}{2} \le x \le \dfrac{3\pi}{2} \Rightarrow Int = \left[-\dfrac{\pi}{2}, \dfrac{3\pi}{2}\right], per = 2\pi, ps = -\dfrac{\pi}{2},$

$ran = (-\infty, -2] \cup [2, \infty),$ Reflect.   (See graph below).

[49] $y = \dfrac{1}{2}\tan\left(\dfrac{x}{2} + 1\right). -\dfrac{\pi}{2} < \dfrac{x}{2} + 1 < \dfrac{\pi}{2} \Rightarrow -2 - \pi < x < -2 + \pi \Rightarrow Int = (-2 - \pi, -2 + \pi), per = 2\pi,$

$ps = -2, ran = \mathbf{R}.$ (See graph below).

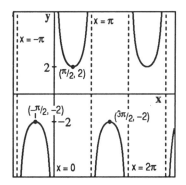

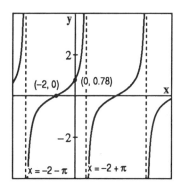

Figure 48   **Figure 49**

[50] $y = -\dfrac{1}{2}\cot\left(\dfrac{x}{2} - 1\right). \ 0 < \dfrac{x}{2} - 1 < \pi \Rightarrow 2 < x < 2\pi + 2 \Rightarrow Int = (2, 2 + 2\pi), per = 2\pi, ps = 2, ran = \mathbf{R},$

Reflect. (See graph below).

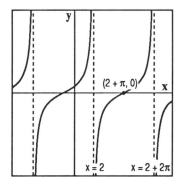

Figure 50

195

[51]　a)　$y = \cos\left(x + \dfrac{\pi}{2}\right)$　　　　　　b)　$y = -\sin x$

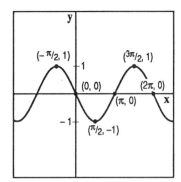

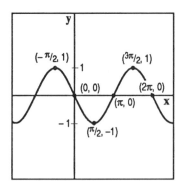

The graphs of $y = \cos\left(x + \dfrac{\pi}{2}\right)$ and $y = -\sin x$ are identical.

[52]　a)　$y = \sec\left(\dfrac{\pi}{2} - x\right)$　　　　　　b)　$y = \csc x$

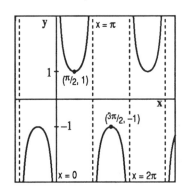

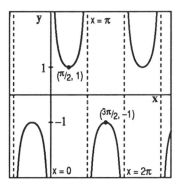

The graphs of $y = \sec\left(\dfrac{\pi}{2} - x\right)$ and $y = \csc x$ are identical.

[53]　a)　$y = \tan\left(x + \dfrac{\pi}{2}\right)$　　　　　　b)　$y = -\cot x$

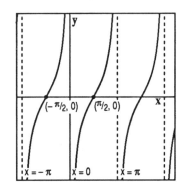

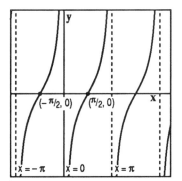

The graphs of $y = \tan\left(x + \dfrac{\pi}{2}\right)$ and $y = -\cot x$ are identical.

[54] Since $f(t) = 20 \cos\left(\sqrt{\frac{5}{7}}\, t\right)$, we see that $a = 20$, $b = \sqrt{\frac{5}{7}}$, and $c = 0$. So,

a) $per = \dfrac{2\pi}{b} = \dfrac{2\pi}{\sqrt{\frac{5}{7}}} = 2\pi \cdot \dfrac{\sqrt{7}}{\sqrt{5}} = \dfrac{2\pi\sqrt{35}}{5}$.

b) $amp = |a| = 20$.

c) $ps = -\dfrac{c}{b} = 0$.

d) $\text{Frequency} = \dfrac{1}{per} = \dfrac{\sqrt{5}}{2\pi\sqrt{7}} = \dfrac{\sqrt{35}}{14\pi}$ cycles per second.

e) $f(t) = 0$ when $20 \cos\left(\sqrt{\frac{5}{7}}\, t\right) = 0 \Rightarrow \cos\left(\sqrt{\frac{5}{7}}\, t\right) = 0$.

From the graph of $y = \cos x$ shown in Figure 5.31 – (b) on page 224, we see that the first value of x (for $x \geq 0$) for which $\cos x = 0$ is $\frac{\pi}{2}$. So, we solve $\sqrt{\frac{5}{7}}\, t = \frac{\pi}{2}$ for t: $\sqrt{\frac{5}{7}}\, t = \frac{\pi}{2} \Rightarrow t = \frac{\pi}{2} \cdot \sqrt{\frac{7}{5}} = \dfrac{\pi\sqrt{7}}{2\sqrt{5}} = \dfrac{\pi\sqrt{35}}{10}$ sec.

[55] Since $f(t) = 3 \sin[\pi(t_0 - 250t)] = 3 \sin(\pi t_0 - 250\pi t) = -3\sin(250\pi t - \pi t_0)$, we see that $a = -3$, $b = 250\pi$, and $c = -\pi t_0$. Thus,

a) $per = \dfrac{2\pi}{b} = \dfrac{2\pi}{250\pi} = \dfrac{1}{125}$.

b) $amp = |a| = |-3| = 3$.

c) $ps = -\dfrac{c}{b} = \dfrac{\pi t_0}{250\pi} = \dfrac{t_0}{250}$.

d) $\text{Frequency} = \dfrac{1}{per} = 125$ cycles per second.

e) $f(t) = 0$ when $-3\sin(250\pi t - \pi t_0) = 0 \Rightarrow \sin(250\pi t - \pi t_0) = 0$.

From the graph of $y = \sin x$ shown in Figure 5.31 – (a) on page 224, we see that the first value of x (for $x \geq 0$) for which $\sin x = 0$ is 0. So, we solve $250\pi t - \pi t_0 = 0$ for t: $250\pi t - \pi t_0 = 0 \Rightarrow$

$250\pi t = \pi t_0 \Rightarrow t = \dfrac{\pi t_0}{250\pi} = \dfrac{t_0}{250}$ sec.

[56] Since $f(t) = 3 \sin\left(\frac{2\pi}{3} t - 1\right)$, we see that $a = 3$, $b = \frac{2\pi}{3}$, and $c = -1$. So,

a) $per = \dfrac{2\pi}{b} = \dfrac{2\pi}{\left(\frac{2\pi}{3}\right)} = 2\pi \cdot \dfrac{3}{2\pi} = 3$.

b) $amp = |a| = 3$.

c) $ps = -\dfrac{c}{b} = \dfrac{1}{\left(\frac{2\pi}{3}\right)} = \dfrac{3}{2\pi}$.

d) $\text{Frequency} = \dfrac{1}{per} = \dfrac{1}{3}$ cycles per second.

e) $f(t) = 0$ when $3\sin\left(\frac{2\pi}{3} t - 1\right) = 0 \Rightarrow \sin\left(\frac{2\pi}{3} t - 1\right) = 0$.

From the graph of $y = \sin x$ shown in Figure 5.31 – (a) on page 224, we see that the second value of x (for $x \geq 0$) for which $\sin x = 0$ is π. So, we solve $\frac{2\pi}{3} t - 1 = \pi$ for t: $\frac{2\pi}{3} t - 1 = \pi \Rightarrow 2\pi t - 3 =$

$3\pi \Rightarrow 2\pi t = 3\pi + 3 \Rightarrow t = \dfrac{3\pi + 3}{2\pi}$ sec.

[57]   Since $g(t) = 100 \cos(10\pi t - 2)$, we see that $a = 100$, $b = 10\pi$, and $c = -2$. So,

    a) $per = \dfrac{2\pi}{b} = \dfrac{2\pi}{10\pi} = \dfrac{1}{5}$.

    b) $amp = |a| = 100$.

    c) $ps = -\dfrac{c}{b} = \dfrac{2}{10\pi} = \dfrac{1}{5\pi}$.

    d)   Frequency $= \dfrac{1}{per} = 5$ cycles per second.

    e) $g(t) = 0$ when $100 \cos(10\pi t - 2) = 0 \Rightarrow \cos(10\pi t - 2) = 0$.

From the graph of $y = \cos x$ shown in Figure 5.31 – (b) on page 224, we see that the second value of

x (for $x \ge 0$) for which $\cos x = 0$ is $\dfrac{3\pi}{2}$. So, we solve $10\pi t - 2 = \dfrac{3\pi}{2}$ for t: $10\pi t - 2 = \dfrac{3\pi}{2} \Rightarrow$

$10\pi t = \dfrac{3\pi}{2} + 2 \Rightarrow 10\pi t = \dfrac{3\pi + 4}{2} \Rightarrow t = \dfrac{3\pi + 4}{20\pi}$ sec.

[58]   Since $f(t) = 60 \sin(120\pi t - \pi)$, we see that $a = 60$, $b = 120\pi$, and $c = -\pi$. So,

    a) $per = \dfrac{2\pi}{b} = \dfrac{2\pi}{120\pi} = \dfrac{1}{60}$.

    b) $amp = |a| = 60$.

    c) $ps = -\dfrac{c}{b} = \dfrac{\pi}{120\pi} = \dfrac{1}{120}$.

    d)   Frequency $= \dfrac{1}{per} = 60$ cycles per second.

    e) $f(t) = 0$ when $60 \sin(120\pi t - \pi) = 0 \Rightarrow \sin(120\pi t - \pi) = 0$.

From the graph of $y = \sin x$ shown in Figure 5.31 – (a) on page 224, we see that the second value of

x (for $x \ge 0$) for which $\sin x = 0$ is π. So, we solve $120\pi t - \pi = \pi$ for t: $120\pi t - \pi = \pi \Rightarrow$

$120\pi t = 2\pi \Rightarrow t = \dfrac{2\pi}{120\pi} = \dfrac{1}{60}$ sec.

[59]   a) $y = \sqrt{2} \sin(\sqrt{3}\, x - 1)$ b) $y = -\sqrt{3} \cos(\sqrt{5}\, x + 1)$

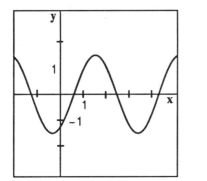
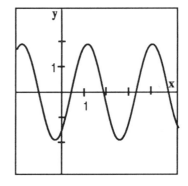

[60]   a) $y = e^x \cos x$ b) $y = \sin(\ln x)$

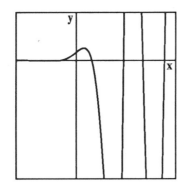
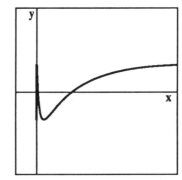

[61] a) $y = \tan(-x)$ b) $y = -\tan x$

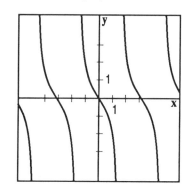

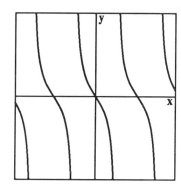

The graphs of $y = \tan(-x)$ and $y = -\tan x$ are identical.

[62] a) $y = \cos(-x)$ b) $y = \cos x$

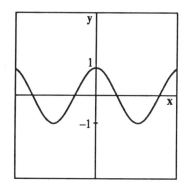

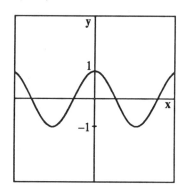

The graphs of $y = \cos(-x)$ and $y = \cos x$ are identical.

EXERCISES 5.4

[1] $\sin^{-1}\left(-\frac{\sqrt{2}}{2}\right) = -\frac{\pi}{4}$.   **[2]** $\cos^{-1}(0) = \frac{\pi}{2}$.

[3] $\cos^{-1}\left(\frac{\sqrt{3}}{2}\right) = \frac{\pi}{6}$.   **[4]** $\arctan(-\sqrt{3}) = -\frac{\pi}{3}$.

[5] $\sin^{-1}\left(-\frac{1}{2}\right) = -\frac{\pi}{6}$.   **[6]** $\arccos\left(\sin\frac{4\pi}{3}\right) = \arccos\left(-\frac{\sqrt{3}}{2}\right) = \frac{5\pi}{6}$.

[7] $\sin\left(\arctan\left(-\frac{\sqrt{2}}{2}\right)\right)$. If $x = \arctan\left(-\frac{\sqrt{2}}{2}\right)$, then $\tan x = -\frac{\sqrt{2}}{2}$ and $-\frac{\pi}{2} < x < \frac{\pi}{2}$. Since $\tan x < 0$,

x is in $\left(-\frac{\pi}{2}, 0\right)$. Since $\tan x = -\frac{\sqrt{2}}{2}$, $\cot x = -\frac{2}{\sqrt{2}} = -\sqrt{2}$. Since x is in $\left(-\frac{\pi}{2}, 0\right)$, $\csc x < 0$. Since

$1 + \cot^2 x = \csc^2 x$ and $\csc x < 0$, it follows that $\csc x = -\sqrt{1 + \cot^2 x} = -\sqrt{1 + (-\sqrt{2})^2} = -\sqrt{3}$.

Since $\csc x = -\sqrt{3}$, $\sin x = -\frac{1}{\sqrt{3}}$. So, $\sin\left[\arctan\left(-\frac{\sqrt{2}}{2}\right)\right] = -\frac{1}{\sqrt{3}}$.

[8] $\arcsin\left(\cos\frac{3\pi}{4}\right) = \arcsin\left(-\frac{\sqrt{2}}{2}\right) = -\frac{\pi}{4}$.   **[9]** $\sin^{-1}\left(\sin\frac{11\pi}{6}\right) = \sin^{-1}\left(-\frac{1}{2}\right) = -\frac{\pi}{6}$.

[10] $\cos^{-1}\left(\cos\frac{5\pi}{4}\right) = \cos^{-1}\left(-\frac{\sqrt{2}}{2}\right) = \frac{3\pi}{4}$.

[11] $\arcsin\left(\sin\left(-\frac{\pi}{6}\right)\right) = -\frac{\pi}{6}$, since $-\frac{\pi}{6}$ is in $\left[-\frac{\pi}{2}, \frac{\pi}{2}\right]$.

[12] $\sin^{-1}\left(\sin\frac{7\pi}{4}\right) = \sin^{-1}\left(-\frac{\sqrt{2}}{2}\right) = -\frac{\pi}{4}$.   [13] $\tan\left(\arctan\left(-25\right)\right) = -25$, since -25 is in **R**.

[14] $\arctan\left(\tan\frac{7\pi}{6}\right) = \arctan\left(\frac{1}{\sqrt{3}}\right) = \frac{\pi}{6}$.

[15] $\sin\left(\arccos\frac{3}{5}\right)$. If $x = \arccos\left(\frac{3}{5}\right)$, then $\cos x = \frac{3}{5}$ and $0 \leq x \leq \pi$. Since x is in $[0, \pi]$, $\sin x \geq 0$ and it

follows from basic identity (v) that $\sin x = \sqrt{1 - \cos^2 x} = \sqrt{1 - \left(\frac{3}{5}\right)^2} = \frac{4}{5}$. So, $\sin\left(\arccos\frac{3}{5}\right) = \frac{4}{5}$.

[16] $\cos\left[\arcsin\left(-\frac{9}{10}\right)\right]$. If $x = \arcsin\left(-\frac{9}{10}\right)$, then $\sin x = -\frac{9}{10}$ and $-\frac{\pi}{2} \leq x \leq \frac{\pi}{2}$. Since x is in $\left[-\frac{\pi}{2}, \frac{\pi}{2}\right]$,

$\cos x \geq 0$ and it follows from basic identity (v) that $\cos x = \sqrt{1 - \sin^2 x} = \sqrt{1 - \left(-\frac{9}{10}\right)^2} = \frac{\sqrt{19}}{10}$.

So, $\cos\left[\arcsin\left(-\frac{9}{10}\right)\right] = \frac{\sqrt{19}}{10}$.

[17] $\cos[\arctan(-2)]$. If $x = \arctan(-2)$, then $\tan x = -2$ and $-\frac{\pi}{2} < x < \frac{\pi}{2}$. Since x is in $\left(-\frac{\pi}{2}, \frac{\pi}{2}\right)$,

$\sec x > 0$ and it follows from the identity $1 + \tan^2 x = \sec^2 x$ that $\sec x = \sqrt{1 + \tan^2 x} =$

$\sqrt{1 + (-2)^2} = \sqrt{5}$. Since $\sec x = \sqrt{5}$, $\cos x = \frac{1}{\sqrt{5}}$. So, $\cos[\arctan(-2)] = \frac{1}{\sqrt{5}}$.

[18] $\sin^{-1}\left(\sin\left(-1.4\right)\right) = -1.4$, since -1.4 is in $\left[-\frac{\pi}{2}, \frac{\pi}{2}\right]$.

[19] $\sin\left[\arccos\left(-\frac{1}{\sqrt{3}}\right)\right]$. If $x = \arccos\left(-\frac{1}{\sqrt{3}}\right)$, then $\cos x = -\frac{1}{\sqrt{3}}$ and $0 \leq x \leq \pi$. Since x is in $[0, \pi]$,

$\sin x \geq 0$ and it follows from basic identity (v) that $\sin x = \sqrt{1 - \cos^2 x} = \sqrt{1 - \left(-\frac{1}{\sqrt{3}}\right)^2} =$

$\sqrt{\frac{2}{3}} = \frac{\sqrt{6}}{3}$. So, $\sin\left[\arccos\left(-\frac{1}{\sqrt{3}}\right)\right] = \frac{\sqrt{6}}{3}$.

[20] $\tan\left[\arccos\left(-\frac{1}{2}\right)\right]$. If $x = \arccos\left(-\frac{1}{2}\right)$, then $\cos x = -\frac{1}{2}$ and $0 \leq x \leq \pi$. So $x \in$ QII, and $\sec x = -2$.

Since $1 + \tan^2 x = \sec^2 x$ and $\tan x < 0$ for $x \in$ QII, $\tan x = -\sqrt{\sec^2 x - 1} = -\sqrt{(-2)^2 - 1} = -\sqrt{3}$.
So, $\tan\left[\arccos\left(-\frac{1}{2}\right)\right] = -\sqrt{3}$.

[21] $\sin\left(\sin^{-1} 4\right)$ is undefined, since 4 is not in $[-1, 1]$.

[22] $\cos\left(\cos^{-1} \pi\right)$ is undefined, since π is not in $[-1, 1]$.

[23] $\tan^{-1}\left(\tan 1.5\right) = 1.5$, since $-\frac{\pi}{2} < 1.5 < \frac{\pi}{2}$.   [24] $\sin^{-1}\left(\sin\sqrt{2}\right) = \sqrt{2}$, since $-\frac{\pi}{2} < \sqrt{2} < \frac{\pi}{2}$.

[25] $\tan\left(\tan^{-1} 100\right) = 100$, since 100 is in **R**.

[26] $\cos\left[\arcsin\left(-\frac{7}{8}\right)\right]$. If $x = \arcsin\left(-\frac{7}{8}\right)$, then $\sin x = -\frac{7}{8}$ and $-\frac{\pi}{2} \leq x \leq \frac{\pi}{2}$. So, $x \in$ QIV. Since

$\cos^2 x + \sin^2 x = 1$ and $\cos x > 0$ for $x \in$ QIV, we see that $\cos x = \sqrt{1 - \sin^2 x} = \sqrt{1 - \left(-\frac{7}{8}\right)^2} =$

$\frac{\sqrt{15}}{8}$. So, $\cos\left[\arcsin\left(-\frac{7}{8}\right)\right] = \frac{\sqrt{15}}{8}$.

[27] $\sec[\arctan 50]$. If $x = \arctan 50$, then $\tan x = 50$ and $-\dfrac{\pi}{2} < x < \dfrac{\pi}{2}$. Since $1 + \tan^2 x = \sec^2 x$ and

$\sec x > 0$ when $-\dfrac{\pi}{2} < x < \dfrac{\pi}{2}$, $\sec x = \sqrt{1 + \tan^2 x} = \sqrt{1 + (50)^2} = \sqrt{2501}$. So, $\sec[\arctan 50] = \sqrt{2501}$.

[28] $\sin^{-1}(0.35) \approx 0.3576$.

[29] $\arccos\left(-\dfrac{\sqrt{11}}{20}\right) \approx 1.737$.

[30] $\arctan 13 \approx 1.494$.

[31] $\arcsin\left(-\dfrac{3}{4}\right) \approx -0.8481$.

[32] $\cos^{-1}\left(-\dfrac{\pi}{10}\right) \approx 1.890$.

[33] $\tan^{-1}(\sqrt{\pi}) \approx 1.057$.

[34] $\sin^{-1}(\sqrt{\pi} - 1) \approx 0.8827$.

[35] $\arccos\left(\dfrac{13}{14}\right) \approx 0.3802$.

[36] $\arctan(-\sqrt{13}) \approx -1.300$.

[37] $f(x) = -\sin^{-1} x$ (See graph below).

[38] $f(x) = 1 + \tan^{-1} x$ (See graph below).

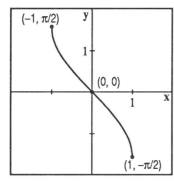

Figure 37

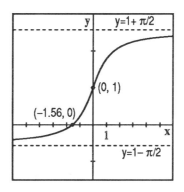

Figure 38

[39] $f(x) = 2 + \cos^{-1} x$ (See graph below).

[40] $f(x) = 2\sin^{-1} x + 1$ (See graph below).

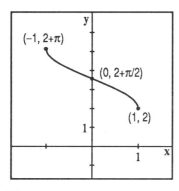

Figure 39

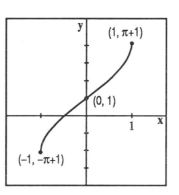

Figure 40

[41] $f(x) = 1 - \tan^{-1} x$ (See graph below).

[42] $f(x) = \sin^{-1} x + 2$ (See graph below).

[43] $f(x) = -\cos^{-1} x$ (See graph below).

[44] $f(x) = 2\tan^{-1} x$ (See graph below).

[45] $f(x) = 1 - \sin^{-1} x$ (See graph below).

[46] $f(x) = \tan^{-1}(x + 1)$ (See graph below).

[47] $f(x) = \cos^{-1}(\cos x) = x$, for all x in $[0, \pi]$.   (See graph below).

[48] $f(x) = \sin\left(\sin^{-1} x\right) = x$, for all x in $[-1, 1]$.   (See graph below).

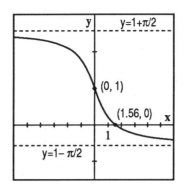

Figure 41

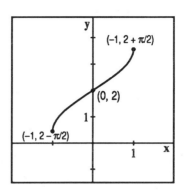

Figure 42

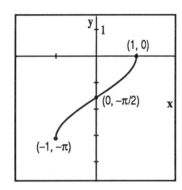

Figure 43

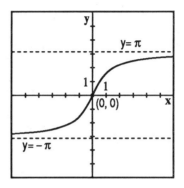

Figure 44

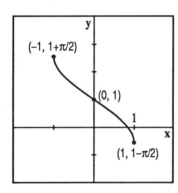

Figure 45

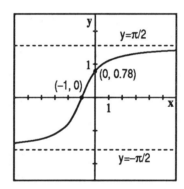

Figure 46

[49] $f(x) = \tan(\arctan x) = x$, for all x in **R**.   (See graph below).

[50] $f(x) = \cos(\cos^{-1} x) = x$, for all x in $[-1, 1]$.   (See graph below).

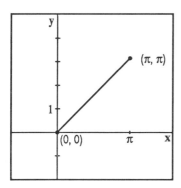

Figure 47

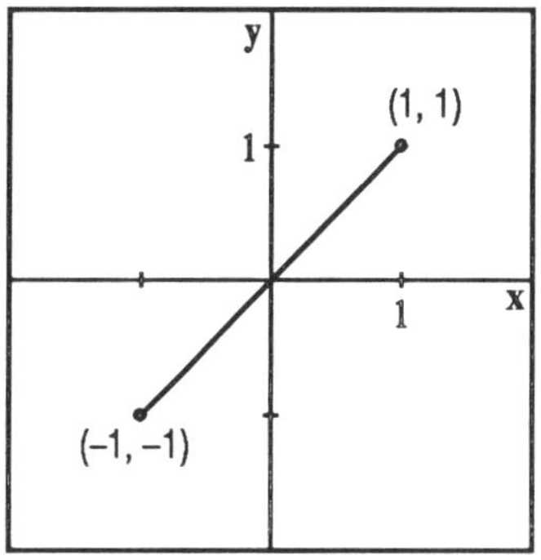

Figure 48

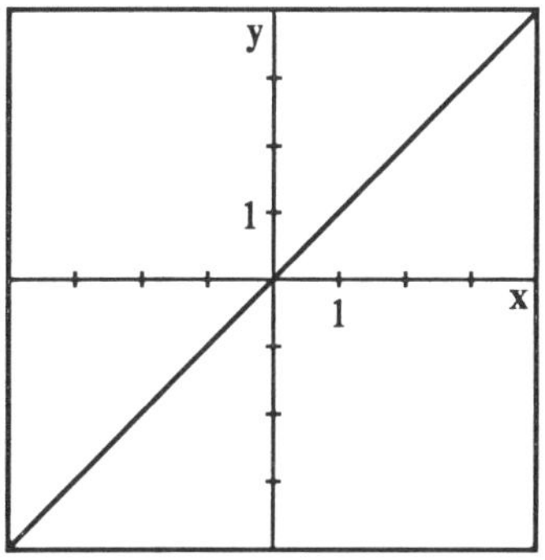

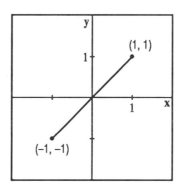

Figure 49

Figure 50

[51] Show that $\tan^{-1}(-x) = -\tan^{-1} x$, for all x in **R**. Let x be in **R** and let $y = \tan^{-1}(-x)$. Then

$\tan y = -x$ and $-\frac{\pi}{2} < y < \frac{\pi}{2}$. Multiplying both sides of the last equation and each part of the last

inequality by -1, we obtain $-\tan y = x$ and $\frac{\pi}{2} > -y > -\frac{\pi}{2}$, or $-\tan y = x$ and $-\frac{\pi}{2} < -y < \frac{\pi}{2}$. Since

$\tan(-y) = -\tan y$, we can substitute $\tan(-y)$ for $-\tan y$ in the last equation to obtain

$\tan(-y) = x$ and $-\frac{\pi}{2} < -y < \frac{\pi}{2}$. It therefore follows from the definition of the inverse tangent

function that $-y = \tan^{-1} x$. Multiplying both sides of the last equation by -1, we obtain

$y = -\tan^{-1} x$. Since $y = \tan^{-1}(-x)$, it thus follows that $\tan^{-1}(-x) = -\tan^{-1} x$. So, the inverse tangent

function is odd.

[52] Show that for all x in $(-1, 1)$, $\tan\left(\sin^{-1} x\right) = \frac{x}{\sqrt{1-x^2}}$. Let x be in $(-1, 1)$ and let $y = \sin^{-1} x$.

Then $\sin y = x$ and $-\frac{\pi}{2} \le y \le \frac{\pi}{2}$. Since $\sin^2 y + \cos^2 y = 1$ and $\cos y \ge 0$ when $-\frac{\pi}{2} \le y \le \frac{\pi}{2}$, $\cos y =$

$\sqrt{1 - \sin^2 y} = \sqrt{1 - x^2}$. Since $x \in (-1, 1)$, we see that $\cos y = \sqrt{1 - x^2} \ne 0$. Therefore, $\tan y =$

$\frac{\sin y}{\cos y} = \frac{x}{\sqrt{1-x^2}}$. So, $\tan\left(\sin^{-1} x\right) = \frac{x}{\sqrt{1-x^2}}$.

[53] Show that $\tan\left(\cos^{-1} x\right) = \frac{\sqrt{1-x^2}}{x}$, for all $x \in [-1, 0) \cup (0, 1]$. Let x be in $[-1, 0) \cup (0, 1]$ and let

$y = \cos^{-1} x$. Then $\cos y = x$ and $0 \le y \le \pi$. Since $\sin^2 y + \cos^2 y = 1$ and $\sin y \ge 0$ when $0 \le y \le \pi$,

$\sin y = \sqrt{1 - \cos^2 y} = \sqrt{1 - x^2}$. Since $\cos y = x \in [-1, 0) \cup (0, 1]$, we see that $\cos y \ne 0$. Therefore,

$\tan y = \frac{\sin y}{\cos y} = \frac{\sqrt{1-x^2}}{x}$. So, $\tan\left(\cos^{-1} x\right) = \frac{\sqrt{1-x^2}}{x}$.

[54] Show that for all x in $\mathbf{R}$, $\cos\left(\tan^{-1}x\right) = \dfrac{1}{\sqrt{1+x^2}}$. Let x be in $\mathbf{R}$ and let $y = \tan^{-1}x$. Then $\tan y = x$

and $-\dfrac{\pi}{2} < y < \dfrac{\pi}{2}$. Since $1 + \tan^2 y = \sec^2 y$ and $\sec y > 0$ when $-\dfrac{\pi}{2} < y < \dfrac{\pi}{2}$, $\sec y = \sqrt{1 + \tan^2 y} =$

$\sqrt{1 + x^2}$. So, $\cos y = \dfrac{1}{\sec y} = \dfrac{1}{\sqrt{1 + x^2}}$. Hence, $\cos\left(\tan^{-1}x\right) = \dfrac{1}{\sqrt{1+x^2}}$.

[55] Since this restricted secant function is one–to–one, it has an inverse function, the inverse secant function $y = \sec^{-1} x$ whose graph is shown below.

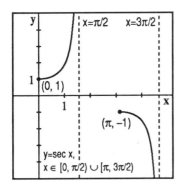 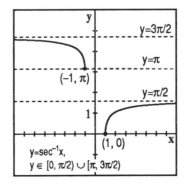

[56] Since this restricted secant function is one–to–one, it has an inverse function, $y = \sec^{-1} x$, whose graph is shown below.

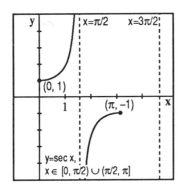 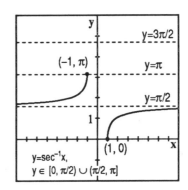

[57] Since this restricted cotangent function is one–to–one, it has an inverse function, $y = \cot^{-1} x$, whose graph is shown below.

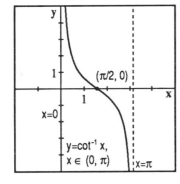 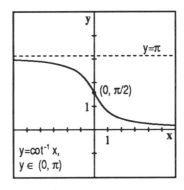

[58]   Since this restricted cosecant function is one–to–one, it has an inverse function, $y = \csc^{-1} x$, whose graph is shown below.

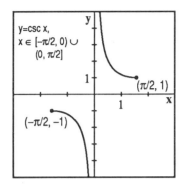 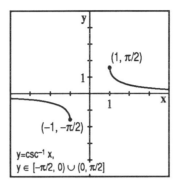

[59]   No, since the cosine function is not one–to–one on this interval.

[60]   Since the inverse sine function is an odd function (see Example 21), it follows that its graph is symmetric with respect to the origin.

[61]   Since the inverse tangent function is an odd function (see Exercise 51), it follows that its graph is symmetric with respect to the origin.

[62] $y = 7 + 8\sin^{-1}(5x+2) \Rightarrow y - 7 = 8\sin^{-1}(5x+2) \Rightarrow \dfrac{y-7}{8} = \sin^{-1}(5x+2) \Rightarrow \sin\left(\dfrac{y-7}{8}\right) =$

$5x + 2 \Rightarrow 5x = \sin\left(\dfrac{y-7}{8}\right) - 2 \Rightarrow x = \dfrac{1}{5}\sin\left(\dfrac{y-7}{8}\right) - \dfrac{2}{5}$.

[63] $y = 4 - 5\cos^{-1}(3x-9) \Rightarrow y - 4 = -5\cos^{-1}(3x-9) \Rightarrow \dfrac{y-4}{-5} = \cos^{-1}(3x-9) \Rightarrow \dfrac{4-y}{5} =$

$\cos^{-1}(3x-9) \Rightarrow \cos\left(\dfrac{4-y}{5}\right) = 3x - 9 \Rightarrow \cos\left(\dfrac{y-4}{5}\right) = 3x - 9 \Rightarrow 3x = \cos\left(\dfrac{y-4}{5}\right) + 9 \Rightarrow$

$x = \dfrac{1}{3}\cos\left(\dfrac{y-4}{5}\right) + 3$.

[64] $y = \pi + 3\tan^{-1}(4x+5) \Rightarrow y - \pi = 3\tan^{-1}(4x+5) \Rightarrow \dfrac{y-\pi}{3} = \tan^{-1}(4x+5) \Rightarrow \tan\left(\dfrac{y-\pi}{3}\right) =$

$4x + 5 \Rightarrow 4x = \tan\left(\dfrac{y-\pi}{3}\right) - 5 \Rightarrow x = \dfrac{1}{4}\tan\left(\dfrac{y-\pi}{3}\right) - \dfrac{5}{4}$.

[65] $y = -3 - 4\sin^{-1}(x-9) \Rightarrow y + 3 = -4\sin^{-1}(x-9) \Rightarrow \dfrac{y+3}{-4} = \sin^{-1}(x-9) \Rightarrow \dfrac{-y-3}{4} =$

$\sin^{-1}(x-9) \Rightarrow \sin\left(\dfrac{-y-3}{4}\right) = x - 9 \Rightarrow -\sin\left(\dfrac{y+3}{4}\right) = x - 9 \Rightarrow x = -\sin\left(\dfrac{y+3}{4}\right) + 9$.

[66] $y = -6 - 7\sin^{-1}(5-3x) \Rightarrow y + 6 = -7\sin^{-1}(5-3x) \Rightarrow -\dfrac{y+6}{7} = \sin^{-1}(5-3x) \Rightarrow$

$\sin\left(-\dfrac{y+6}{7}\right) = 5 - 3x \Rightarrow -\sin\left(\dfrac{y+6}{7}\right) = 5 - 3x \Rightarrow 3x = \sin\left(\dfrac{y+6}{7}\right) + 5 \Rightarrow x = \dfrac{1}{3}\sin\left(\dfrac{y+6}{7}\right) + 5$.

[67] $y = 3 + \tan^{-1}(2-\pi x) \Rightarrow y - 3 = \tan^{-1}(2-\pi x) \Rightarrow \tan(y-3) = 2 - \pi x \Rightarrow \pi x = -\tan(y-3) + 2 \Rightarrow$

$x = \dfrac{2}{\pi} - \dfrac{1}{\pi}\tan(y-3)$.

[68] $y = 2 - \cos^{-1}(\pi - 7x) \Rightarrow y - 2 = -\cos^{-1}(\pi - 7x) \Rightarrow 2 - y = \cos^{-1}(\pi - 7x) \Rightarrow \cos(2 - y) =$

$\pi - 7x \Rightarrow \cos(y - 2) = \pi - 7x \Rightarrow 7x = -\cos(y - 2) + \pi \Rightarrow x = -\frac{1}{7}\cos(y - 2) + \frac{\pi}{7}$.

[69] $h(x) = \sin(\ln x)$. Let $f(x) = \sin x$ and $g(x) = \ln x$. Then $(f \circ g)(x) = f[g(x)] = \sin(\ln x)$.

(other solutions are possible)

[70] $h(x) = e^{\tan x}$. Let $f(x) = e^x$ and $g(x) = \tan x$. Then $(f \circ g)(x) = f[g(x)] = e^{\tan x}$.

(other solutions are possible)

[71] $h(x) = \cos(\tan^{-1} x)$. Let $f(x) = \cos x$ and $g(x) = \tan^{-1} x$. Then $(f \circ g)(x) = f[g(x)] = \cos(\tan^{-1} x)$.

(other solutions are possible)

[72] $h(x) = \ln(\cos x)$. Let $f(x) = \ln x$ and $g(x) = \cos x$. Then $(f \circ g)(x) = f[g(x)] = \ln(\cos x)$.

(other solutions are possible)

[73] $h(x) = e^{\sec x}$. Let $f(x) = e^x$ and $g(x) = \sec x$. Then $(f \circ g)(x) = f[g(x)] = e^{\sec x}$.

(other solutions are possible)

[74] $h(x) = \sin(e^x)$. Let $f(x) = \sin x$ and $g(x) = e^x$. Then $(f \circ g)(x) = f[g(x)] = \sin(e^x)$.

(other solutions are possible)

[75] a) $y = -2\sin^{-1}(x - 1) + 1$.   b) $y = 2\cos^{-1}(2x + 1) - 1$.

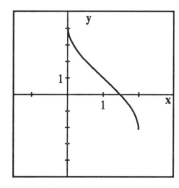
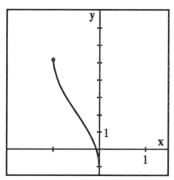

[76] a) $y = 1 - 2\tan^{-1}(2x - 1)$.   b) $y = 3\sin^{-1}(3x - 1) - 2$.

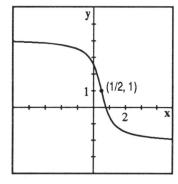
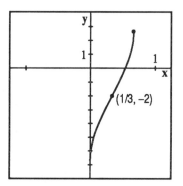

REVIEW EXERCISES

[1] $P(-9\pi)$. Since $-9\pi = -\pi - 4(2\pi)$, $P(-9\pi) = P(-\pi) = (-1, 0)$.   (See graph below).

[2] $P\left(-\dfrac{4\pi}{3}\right)=\left(-\dfrac{1}{2},\dfrac{\sqrt{3}}{2}\right).$ (See graph below).

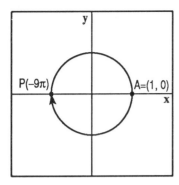

Figure 1

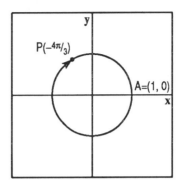

Figure 2

[3] $P\left(\dfrac{17\pi}{3}\right).$ Since $\dfrac{17\pi}{3}=\dfrac{5\pi}{3}+2\,(2\,\pi),$ $P\left(\dfrac{17\pi}{3}\right)=P\left(\dfrac{5\pi}{3}\right)=\left(\dfrac{1}{2},-\dfrac{\sqrt{3}}{2}\right).$ (See graph below).

[4] $P\left(\dfrac{13\pi}{2}\right).$ Since $\dfrac{13\pi}{2}=\dfrac{\pi}{2}+3\,(2\,\pi),$ $P\left(\dfrac{13\pi}{2}\right)=P\left(\dfrac{\pi}{2}\right)=(0,1).$ (See graph below).

[5] $P\left(-\dfrac{25\pi}{4}\right).$ Since $-\dfrac{25\pi}{4}=-\dfrac{\pi}{4}-3\,(2\,\pi),$ $P\left(-\dfrac{25\pi}{4}\right)=P\left(-\dfrac{\pi}{4}\right)=\left(\dfrac{\sqrt{2}}{2},-\dfrac{\sqrt{2}}{2}\right).$ (See graph below).

[6] $P\left(\dfrac{19\pi}{6}\right).$ Since $\dfrac{19\pi}{6}=\dfrac{7\pi}{6}+2\,\pi,$ $P\left(\dfrac{19\pi}{6}\right)=P\left(\dfrac{7\pi}{6}\right)=\left(-\dfrac{\sqrt{3}}{2},-\dfrac{1}{2}\right).$ (See graph below).

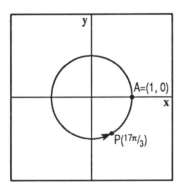

Figure 3

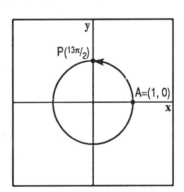

Figure 4

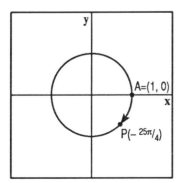

Figure 5

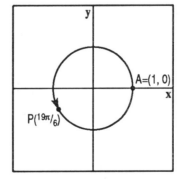

Figure 6

[7] $P\left(-\frac{17\pi}{6}\right)$. Since $-\frac{17\pi}{6} = -\frac{5\pi}{6} - 2\pi$, $P\left(-\frac{17\pi}{6}\right) = P\left(-\frac{5\pi}{6}\right) = \left(-\frac{\sqrt{3}}{2}, -\frac{1}{2}\right)$.   (See graph below).

[8] $P\left(-\frac{19\pi}{4}\right)$. Since $-\frac{19\pi}{4} = -\frac{3\pi}{4} - 2(2\pi)$, $P\left(-\frac{19\pi}{4}\right) = P\left(-\frac{3\pi}{4}\right) = \left(-\frac{\sqrt{2}}{2}, -\frac{\sqrt{2}}{2}\right)$.

(See graph below).

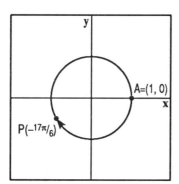

Figure 7

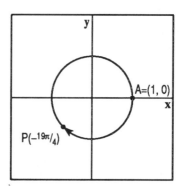

Figure 8

[9] $P\left(\frac{25\pi}{2}\right)$. Since $\frac{25\pi}{2} = \frac{\pi}{2} + 6(2\pi)$, $P\left(\frac{25\pi}{2}\right) = P\left(\frac{\pi}{2}\right) = (0, 1)$.   (See graph below).

[10] $P(28\pi)$. Since $28\pi = 14(2\pi)$, $P(28\pi) = P(0) = (1, 0)$.   (See graph below).

[11] $P(4)$. Since $\pi \approx 3.14 < 4 < 4.71 \approx \frac{3\pi}{2}$, $P(4)$ is in QIII.   (See graph below).

[12] $P(-1)$. Since $0 < 1 < 1.57 \approx \frac{\pi}{2}$, $P(-1)$ is in QIV.   (See graph below).

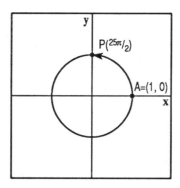

Figure 9

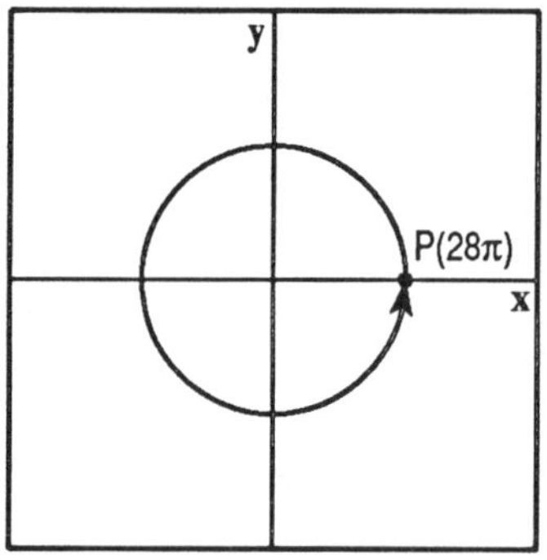

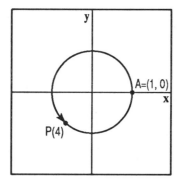

Figure 10

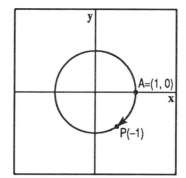

Figure 11

Figure 12

[13] $P(-10)$. Since $-10 = -3.72 - 6.28 \approx -3.72 - 2\pi$, $P(-10) \approx P(-3.72)$. Since

$\pi \approx 3.14 < 3.72 < 4.71 \approx \dfrac{3\pi}{2}$, $P(-3.72)$ is in QII.   (See graph below).

[14] $P(25)$. Since $25 = 6.16 + 3(6.28) \approx 6.16 + 3(2\pi)$, $P(25) \approx P(6.16)$. Since

$\dfrac{3\pi}{2} \approx 4.71 < 6.16 < 6.28 \approx 2\pi$, $P(6.16)$ is in QIV.   (See graph below).

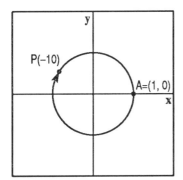

Figure 13

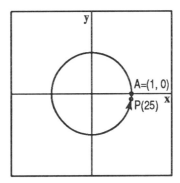

Figure 14

[15] $P(12)$. Since $12 = 5.72 + 6.28 \approx 5.72 + 2\pi$, $P(12) \approx P(5.72)$. Since

$\dfrac{3\pi}{2} \approx 4.71 < 5.72 < 6.28 \approx 2\pi$, $P(5.72)$ is in QIV.   (See graph below).

[16] $P(-8)$. Since $-8 = -1.72 - 6.28 \approx -1.72 - 2\pi$, $P(-8) \approx P(-1.72)$. Since

$\dfrac{\pi}{2} \approx 1.57 < 1.72 < 3.14 \approx \pi$, $P(-1.72)$ is in QIII.   (See graph below).

[17] $P\left(-\dfrac{7\pi}{8}\right)$. Since $\dfrac{\pi}{2} < \dfrac{7\pi}{8} < \pi$, $P\left(-\dfrac{7\pi}{8}\right)$ is in QIII.   (See graph below).

[18] $P\left(-\dfrac{3\pi}{5}\right)$. Since $\dfrac{\pi}{2} < \dfrac{3\pi}{5} < \pi$, $P\left(-\dfrac{3\pi}{5}\right)$ is in QIII.   (See graph below).

[19] $P(-6.1)$. Since $\dfrac{3\pi}{2} \approx 4.71 < 6.1 < 6.28 \approx 2\pi$, $P(-6.1)$ is in QI.   (See graph below).

[20] $P(13)$. Since $13 = 0.44 + 2(6.28) \approx 0.44 + 2(2\pi)$, $P(13) \approx P(0.44)$. Since $0 < 0.44 < 1.57 \approx \dfrac{\pi}{2}$,

$P(0.44)$ is in QI.   (See graph below).

[21] x–coordinate of $P(t)$ is ± 1. From the graph of the unit circle shown in the accompanying figure, we

see that the values of t in $[0, 2\pi)$ are 0 and π.   (See graph below).

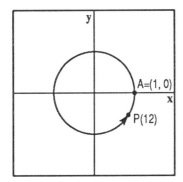

Figure 15

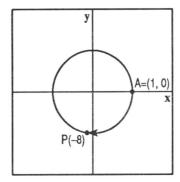

Figure 16

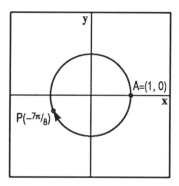

Figure 17

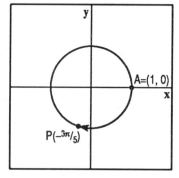

Figure 18

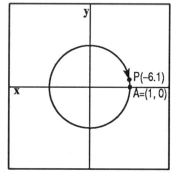

Figure 19

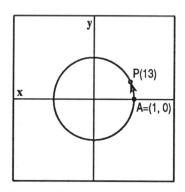

Figure 20

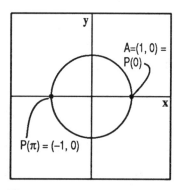

Figure 21

[22]　y–coordinate of $P(t)$ is $\pm \frac{\sqrt{3}}{2}$. We know that the y–coordinate of $P\left(\frac{\pi}{3}\right)$ is $\frac{\sqrt{3}}{2}$. Hence, it follows

that the values of t in $[0, 2\pi)$ are $\frac{\pi}{3}$, $\frac{2\pi}{3}$, $\frac{4\pi}{3}$, and $\frac{5\pi}{3}$.

[23]　y–coordinate of $P(t)$ is -2. If $P(t) = (x, y)$ is on the unit circle, it follows that $-1 \le y \le 1$. Thus
there are no such values of t.

[24]　sum of x and y coordinates is 0. If $P(t) = (x, y)$ is on the unit circle, then $x + y = 0$ when $y = -x$.
Since $P\left(\frac{3\pi}{4}\right) = \left(-\frac{\sqrt{2}}{2}, \frac{\sqrt{2}}{2}\right)$ and $P\left(\frac{7\pi}{4}\right) = \left(\frac{\sqrt{2}}{2}, -\frac{\sqrt{2}}{2}\right)$ and $-\frac{\sqrt{2}}{2} + \frac{\sqrt{2}}{2} = 0$, it follows that

the values of t in $[0, 2\pi)$ are $\frac{3\pi}{4}$ and $\frac{7\pi}{4}$.

[25] x–coordinate of $P(t)$ is $\frac{\sqrt{3}}{2}$. Since we know that the x–coordinate of $P\left(\frac{\pi}{6}\right)$ is $\frac{\sqrt{3}}{2}$, it follows that

the values of t in $[0, 2\pi)$ are $\frac{\pi}{6}$ and $\frac{11\pi}{6}$.

[26] $P(t) = \left(-\frac{2}{\sqrt{5}}, c\right)$; QIII. Since $P(t)$ lies on the unit circle whose equation is $x^2 + y^2 = 1$, it follows

that $\left(-\frac{2}{\sqrt{5}}\right)^2 + c^2 = 1 \Rightarrow \frac{4}{5} + c^2 = 1 \Rightarrow c^2 = \frac{1}{5} \Rightarrow c = \pm\frac{\sqrt{5}}{5}$. Since $P(t) \in$ QIII, $c < 0 \Rightarrow c = -\frac{\sqrt{5}}{5}$.

[27] $P(t) = \left(c, \frac{\sqrt{3}}{5}\right)$; QII. Since the equation of the unit circle is $x^2 + y^2 = 1$, it follows that

$c^2 + \left(\frac{\sqrt{3}}{5}\right)^2 = 1 \Rightarrow c^2 = 1 - \frac{3}{25} \Rightarrow c^2 = \frac{22}{25} \Rightarrow c = \pm\frac{\sqrt{22}}{5}$. Since $P(t) \in$ QII, $c < 0 \Rightarrow c = -\frac{\sqrt{22}}{5}$.

[28] $P(t) = \left(-\frac{2}{3}, c\right)$; QII. Since the equation of the unit circle is $x^2 + y^2 = 1$, it follows that

$\left(-\frac{2}{3}\right)^2 + c^2 = 1 \Rightarrow \frac{4}{9} + c^2 = 1 \Rightarrow c^2 = \frac{5}{9} \Rightarrow c = \pm\frac{\sqrt{5}}{3}$. Since $P(t) \in$ QII, $c > 0 \Rightarrow c = \frac{\sqrt{5}}{3}$.

[29] $P(t) = \left(\frac{15}{c}, \frac{3}{c}\right)$; QIII. Since the equation of the unit circle is $x^2 + y^2 = 1$, it follows that

$\left(\frac{15}{c}\right)^2 + \left(\frac{3}{c}\right)^2 = 1 \Rightarrow \frac{225}{c^2} + \frac{9}{c^2} = 1 \Rightarrow c^2 = 234 \Rightarrow c = \pm 3\sqrt{26}$. Since $P(t) \in$ QIII,

$c < 0 \Rightarrow c = -3\sqrt{26}$.

[30] $P(t) = (4c, c)$; QIII. Since the equation of the unit circle is $x^2 + y^2 = 1$, it follows that

$(4c)^2 + c^2 = 1 \Rightarrow 17c^2 = 1 \Rightarrow c^2 = \frac{1}{17} \Rightarrow c = \pm\frac{\sqrt{17}}{17}$. Since $P(t) \in$ QIII, $c < 0 \Rightarrow c = -\frac{\sqrt{17}}{17}$.

[31] $\tan\left(-\frac{5\pi}{2}\right)$. Since $-\frac{5\pi}{2} = -\frac{\pi}{2} - 2\pi$, $P\left(-\frac{5\pi}{2}\right) = P\left(-\frac{\pi}{2}\right) = (0, -1)$. So, $\tan\left(-\frac{5\pi}{2}\right)$ is undefined.

[32] $\cot\frac{5\pi}{2}$. Since $\frac{5\pi}{2} = \frac{\pi}{2} + 2\pi$, $P\left(\frac{5\pi}{2}\right) = P\left(\frac{\pi}{2}\right) = (0, 1)$. So, $\cot\frac{5\pi}{2} = \frac{0}{1} = 0$.

[33] $\csc\left(-\frac{11\pi}{3}\right)$. Since $-\frac{11\pi}{3} = -\frac{5\pi}{3} - 2\pi$, $P\left(-\frac{11\pi}{3}\right) = P\left(-\frac{5\pi}{3}\right) = \left(\frac{1}{2}, \frac{\sqrt{3}}{2}\right)$. So, $\csc\left(-\frac{11\pi}{3}\right) = \frac{2}{\sqrt{3}}$.

[34] $\sec\frac{10\pi}{3}$. Since $\frac{10\pi}{3} = \frac{4\pi}{3} + 2\pi$, $P\left(\frac{10\pi}{3}\right) = P\left(\frac{4\pi}{3}\right) = \left(-\frac{1}{2}, -\frac{\sqrt{3}}{2}\right)$. So, $\sec\frac{10\pi}{3} = -2$.

[35] $\sin\frac{7\pi}{4} = -\frac{\sqrt{2}}{2}$.   [36] $\cos\left(-\frac{7\pi}{6}\right) = -\frac{\sqrt{3}}{2}$.

[37] $\cos\left(\frac{23\pi}{6}\right)$. Since $\frac{23\pi}{6} = \frac{11\pi}{6} + 2\pi$, $P\left(\frac{23\pi}{6}\right) = P\left(\frac{11\pi}{6}\right) = \left(\frac{\sqrt{3}}{2}, -\frac{1}{2}\right)$. So, $\cos\frac{23\pi}{6} = \frac{\sqrt{3}}{2}$.

[38] $\sin\left(\frac{11\pi}{4}\right)$. Since $\frac{11\pi}{4} = \frac{3\pi}{4} + 2\pi$, $P\left(\frac{11\pi}{4}\right) = P\left(\frac{3\pi}{4}\right) = \left(-\frac{\sqrt{2}}{2}, \frac{\sqrt{2}}{2}\right)$. So, $\sin\frac{11\pi}{4} = \frac{\sqrt{2}}{2}$.

[39] $\sec 13\pi$. Since $13\pi = \pi + 6(2\pi)$, $P(13\pi) = P(\pi) = (-1, 0)$. So, $\sec 13\pi = -1$.

[40] $\csc 20\pi$. Since $20\pi = 10(2\pi)$, $P(20\pi) = P(0) = (1, 0)$. So, $\csc 20\pi$ is undefined.

[41] $\cot\left(-\frac{20\pi}{6}\right)$. Since $-\frac{20\pi}{6} = -\frac{10\pi}{3} = -\frac{4\pi}{3} - 2\pi$, $P\left(-\frac{20\pi}{6}\right) = P\left(-\frac{4\pi}{3}\right) = \left(-\frac{1}{2}, \frac{\sqrt{3}}{2}\right)$.

So, $\cot\left(-\frac{20\pi}{6}\right) = -\frac{1}{\sqrt{3}}$.

[42] $\tan\left(-\frac{30\pi}{4}\right)$. Since $-\frac{30\pi}{4} = -\frac{15\pi}{2} = -\frac{3\pi}{2} - 3(2\pi)$, $P\left(-\frac{30\pi}{4}\right) = P\left(-\frac{3\pi}{2}\right) = (0, 1)$.

So, $\tan\left(-\frac{30\pi}{4}\right)$ is undefined.

[43] $\left(-\frac{\sqrt{6}}{3}, \frac{1}{\sqrt{3}}\right)$. Since $x^2 + y^2 = \left(-\frac{\sqrt{6}}{3}\right)^2 + \left(\frac{1}{\sqrt{3}}\right)^2 = \frac{2}{3} + \frac{1}{3} = 1$, the point is on the unit circle. So,

$\sin t = \frac{1}{\sqrt{3}}$, $\cos t = -\frac{\sqrt{6}}{3}$, $\tan t = \frac{\left(\frac{1}{\sqrt{3}}\right)}{\left(-\frac{\sqrt{6}}{3}\right)} = -\frac{1}{\sqrt{2}}$, $\cot t = -\sqrt{2}$, $\sec t = -\frac{3}{\sqrt{6}}$, and $\csc t = \sqrt{3}$.

[44] $\left(\frac{2}{\sqrt{5}}, -\frac{1}{\sqrt{5}}\right)$. Since $x^2 + y^2 = \left(\frac{2}{\sqrt{5}}\right)^2 + \left(-\frac{1}{\sqrt{5}}\right)^2 = \frac{4}{5} + \frac{1}{5} = 1$, the point is on the unit circle. So,

$\sin t = -\frac{1}{\sqrt{5}}$, $\cos t = \frac{2}{\sqrt{5}}$, $\tan t = \frac{\left(-\frac{1}{\sqrt{5}}\right)}{\left(\frac{2}{\sqrt{5}}\right)} = -\frac{1}{2}$, $\cot t = -2$, $\sec t = \frac{\sqrt{5}}{2}$, and $\csc t = -\sqrt{5}$.

[45] $\left(\frac{\sqrt{22}}{5}, -\frac{\sqrt{3}}{5}\right)$. Since $x^2 + y^2 = \left(\frac{\sqrt{22}}{5}\right)^2 + \left(-\frac{\sqrt{3}}{5}\right)^2 = \frac{22}{25} + \frac{3}{25} = 1$, the point is on the unit circle. So,

$\sin t = -\frac{\sqrt{3}}{5}$, $\cos t = \frac{\sqrt{22}}{5}$, $\tan t = \frac{\left(-\frac{\sqrt{3}}{5}\right)}{\left(\frac{\sqrt{22}}{5}\right)} = -\frac{\sqrt{3}}{\sqrt{22}}$, $\cot t = -\frac{\sqrt{22}}{\sqrt{3}}$, $\sec t = \frac{5}{\sqrt{22}}$, and $\csc t = -\frac{5}{\sqrt{3}}$.

[46] $\left(-\frac{\sqrt{11}}{6}, \frac{5}{6}\right)$. Since $x^2 + y^2 = \left(-\frac{\sqrt{11}}{6}\right)^2 + \left(\frac{5}{6}\right)^2 = \frac{11}{36} + \frac{25}{36} = 1$, the point is on the unit circle. So,

$\sin t = \frac{5}{6}$, $\cos t = -\frac{\sqrt{11}}{6}$, $\tan t = \frac{\left(\frac{5}{6}\right)}{\left(-\frac{\sqrt{11}}{6}\right)} = -\frac{5}{\sqrt{11}}$, $\cot t = -\frac{\sqrt{11}}{5}$, $\sec t = -\frac{6}{\sqrt{11}}$, and $\csc t = \frac{6}{5}$.

[47] $\left(-\frac{\sqrt{5}}{3}, -\frac{2}{3}\right)$. Since $x^2 + y^2 = \left(-\frac{\sqrt{5}}{3}\right)^2 + \left(-\frac{2}{3}\right)^2 = \frac{5}{9} + \frac{4}{9} = 1$, the point is on the unit circle. So,

$\sin t = -\frac{2}{3}$, $\cos t = -\frac{\sqrt{5}}{3}$, $\tan t = \frac{\left(-\frac{2}{3}\right)}{\left(-\frac{\sqrt{5}}{3}\right)} = \frac{2}{\sqrt{5}}$, $\cot t = \frac{\sqrt{5}}{2}$, $\sec t = -\frac{3}{\sqrt{5}}$, and $\csc t = -\frac{3}{2}$.

[48] $\left(\frac{\sqrt{7}}{4}, \frac{3}{4}\right)$. Since $x^2 + y^2 = \left(\frac{\sqrt{7}}{4}\right)^2 + \left(\frac{3}{4}\right)^2 = \frac{7}{16} + \frac{9}{16} = 1$, the point is on the unit circle. So,

$$\sin t = \frac{3}{4}, \cos t = \frac{\sqrt{7}}{4}, \tan t = \frac{\left(\frac{3}{4}\right)}{\left(\frac{\sqrt{7}}{4}\right)} = \frac{3}{\sqrt{7}}, \cot t = \frac{\sqrt{7}}{3}, \sec t = \frac{4}{\sqrt{7}}, \text{ and } \csc t = \frac{4}{3}.$$

[49] $\csc 15 = \frac{1}{\sin 15} \approx 1.538.$

[50] $\sin(-10) \approx 0.5440.$

[51] $\cos(-\sqrt{7}) \approx -0.8796.$

[52] $\cos(\sqrt{13}) \approx -0.8943.$

[53] $\tan 12 \approx -0.6359.$

[54] $\cot(-16) = \frac{1}{\tan(-16)} \approx -3.326.$

[55] $\sec(-21) = \frac{1}{\cos(-21)} \approx -1.826.$

[56] $\csc 30 = \frac{1}{\sin 30} \approx -1.012.$

[57] $\sin(\cot 5) = \sin\left(\frac{1}{\tan 5}\right) \approx -0.2915.$

[58] $\tan(\cos(-\sqrt{5})) \approx -0.7098.$

[59] Since $\sin^2 t + \cos^2 t = 1$ and $\cos t < 0$ when $P(t) \in$ QII, $\cos t = -\sqrt{1 - \sin^2 t} = -\sqrt{1 - \left(\frac{\sqrt{5}}{3}\right)^2} =$

$-\frac{2}{3}$. So, $\tan(-t) = -\tan t = -\frac{\sin t}{\cos t} = -\frac{\left(\frac{\sqrt{5}}{3}\right)}{\left(-\frac{2}{3}\right)} = \frac{\sqrt{5}}{2}.$

[60] Since $\tan t = -5$, $\cot t = -\frac{1}{5}$. Since $1 + \cot^2 t = \csc^2 t$ and since $\csc t < 0$ when $P(t) \in$ QIV, $\csc t =$

$-\sqrt{1 + \left(-\frac{1}{5}\right)^2} = -\frac{\sqrt{26}}{5}$. Since $\csc t = -\frac{\sqrt{26}}{5}$, $\sin t = -\frac{5}{\sqrt{26}}.$

[61] Since $\cos t < 0$ and $\tan t > 0$, $P(t) \in$ QIII. Since $\sin^2 t + \cos^2 t = 1$ and since $\sin t < 0$ when

$P(t) \in$ QIII, $\sin t = -\sqrt{1 - \left(-\frac{1}{7}\right)^2} = -\frac{4\sqrt{3}}{7}$. Since $\sin t = -\frac{4\sqrt{3}}{7}$, $\csc t = -\frac{7}{4\sqrt{3}}.$

[62] Since $\cot t > 0$ and $\cos t < 0$, $P(t) \in$ QIII. Since $1 + \cot^2 t = \csc^2 t$ and $\csc t < 0$ when $P(t) \in$ QIII,

$\csc t = -\sqrt{1 + (10)^2} = -\sqrt{101}$. Since $\csc t = -\sqrt{101}$, $\sin t = -\frac{1}{\sqrt{101}}$. So, $\sin(-t) = -\sin t = \frac{1}{\sqrt{101}}.$

[63] Since $1 + \tan^2 t = \sec^2 t$ and since $\tan t < 0$ when $P(t)$ is in QIV, $\tan t = -\sqrt{\sec^2 t - 1} =$

$-\sqrt{(4)^2 - 1} = -\sqrt{15}$. So, $\cot(-t) = -\cot t = -\left(\frac{1}{-\sqrt{15}}\right) = \frac{1}{\sqrt{15}}.$

[64] Since $\csc t < 0$ and $\cot t > 0$, $P(t) \in$ QIII. Since $\csc t = -\sqrt{3}$, $\sin t = -\frac{1}{\sqrt{3}}$. Since $\sin^2 t + \cos^2 t = 1$

and $\cos t < 0$ when $P(t) \in$ QIII, $\cos t = -\sqrt{1 - \sin^2 t} = -\sqrt{1 - \left(-\frac{1}{\sqrt{3}}\right)^2} = -\frac{\sqrt{2}}{\sqrt{3}}$. Since $\cos t =$

$-\frac{\sqrt{2}}{\sqrt{3}}$, $\sec t = -\frac{\sqrt{3}}{\sqrt{2}}$. Finally, $\sec(-t) = \sec t = -\frac{\sqrt{3}}{\sqrt{2}}.$

[65] Since $P(t)$ is on the unit circle whose equation is $x^2 + y^2 = 1$, $\left(\frac{\sqrt{5}}{c}\right)^2 + \left(\frac{\sqrt{11}}{c}\right)^2 = 1 \Rightarrow \frac{5}{c^2} + \frac{11}{c^2} = 1 \Rightarrow$

$c^2 = 16 \Rightarrow c = \pm 4$. Since $P(t) \in$ QIII, $c = -4$. So, $P(t) = \left(-\frac{\sqrt{5}}{4}, -\frac{\sqrt{11}}{4}\right)$. Hence, $\sin t = -\frac{\sqrt{11}}{4}$,

$\cos t = -\frac{\sqrt{5}}{4}$, $\tan t = \dfrac{\left(-\frac{\sqrt{11}}{4}\right)}{\left(-\frac{\sqrt{5}}{4}\right)} = \frac{\sqrt{11}}{\sqrt{5}}$, $\cot t = \frac{\sqrt{5}}{\sqrt{11}}$, $\sec t = -\frac{4}{\sqrt{5}}$, and $\csc t = -\frac{4}{\sqrt{11}}$.

[66] Since $\sin^2 t + \cos^2 t = 1$ and $\cos t < 0$ when $P(t) \in$ QII, $\sqrt{25 - u^2} = \sqrt{25 - 25\sin^2 t} =$

$\sqrt{25\left(1 - \sin^2 t\right)} = \sqrt{25\cos^2 t} = -5\cos t$.

[67] Since $1 + \tan^2 t = \sec^2 t$, $\tan^2 t = \sec^2 t - 1$. Since $\tan t > 0$ when $P(t) \in$ QIII, $\sqrt{u^2 - 81} =$

$\sqrt{81\sec^2 t - 81} = \sqrt{81\left(\sec^2 t - 1\right)} = \sqrt{81\tan^2 t} = 9\tan t$.

[68] Since $1 + \tan^2 t = \sec^2 t$ and $\sec t > 0$ when $P(t) \in$ QI, $\sqrt{16 + u^2} = \sqrt{16 + 16\tan^2 t} =$

$\sqrt{16\left(1 + \tan^2 t\right)} = \sqrt{16\sec^2 t} = 4\sec t = 4\sec(-t)$.

NOTE: On Exercises 69 – 82, the functions can all be expressed in the form $y = a\sin(bx + c)$ or

$y = a\cos(bx + c)$, where $b > 0$. Hence, the functions have period $= per = \frac{2\pi}{b}$, amplitude $= amp = |a|$,

phase shift $= ps = -\frac{c}{b}$, and range $= ran = [-|a|, |a|]$. If $a < 0$, the basic sine or cosine wave is reflected

across the x – axis, and we indicate this by writing "Reflect." We will also give the interval that

corresponds to $[0, 2\pi]$ which we will denote by *Int.*

[69] $y = 5\cos x$. $Int = [0, 2\pi]$, $per = 2\pi$, $amp = 5$, $ps = 0$, $ran = [-5, 5]$. (See graph below).

[70] $y = -\pi\sin x$. $Int = [0, 2\pi]$, $per = 2\pi$, $amp = \pi$, $ps = 0$, $ran = [-\pi, \pi]$, Reflect. (See graph below).

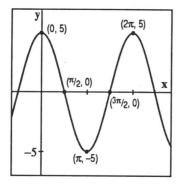

Figure 69

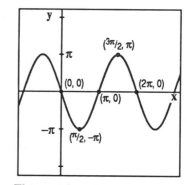

Figure 70

[71] $y = \sin 4x$. $0 \le 4x \le 2\pi \Rightarrow 0 \le x \le \frac{\pi}{2} \Rightarrow Int = \left[0, \frac{\pi}{2}\right]$, $per = \frac{\pi}{2}$, $amp = 1$, $ps = 0$, $ran = [-1, 1]$.

(See graph below).

[72] $y = \cos(-2x) = \cos 2x$. $0 \le 2x \le 2\pi \Rightarrow 0 \le x \le \pi \Rightarrow Int = [0, \pi]$, $per = \pi$, $amp = 1$, $ps = 0$,

$ran = [-1, 1]$. (See graph below).

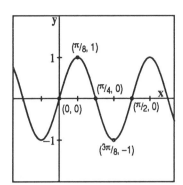

Figure 71

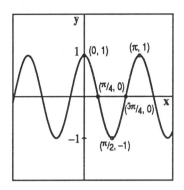

Figure 72

[73] $y = -3\cos\dfrac{x}{3}.\ 0 \le \dfrac{x}{3} \le 2\pi \Rightarrow 0 \le x \le 6\pi \Rightarrow Int = [0, 6\pi],\ per = 6\pi,\ amp = 3,\ ps = 0,\ ran = [-3, 3],$

Reflect. (See graph below).

[74] $y = -2\sin(-2x) = 2\sin 2x.\ 0 \le 2x \le 2\pi \Rightarrow 0 \le x \le \pi \Rightarrow Int = [0, \pi],\ per = \pi,\ amp = 2,\ ps = 0,$

$ran = [-2, 2].$ (See graph below).

[75] $y = 2\sin\dfrac{3x}{2}.\ 0 \le \dfrac{3x}{2} \le 2\pi \Rightarrow 0 \le x \le \dfrac{4\pi}{3} \Rightarrow Int = \left[0, \dfrac{4\pi}{3}\right],\ per = \dfrac{4\pi}{3},\ amp = 2,\ ps = 0,$

$ran = [-2, 2].$ (See graph below).

[76] $y = 3\cos\dfrac{2x}{3}.\ 0 \le \dfrac{2x}{3} \le 2\pi \Rightarrow 0 \le x \le 3\pi \Rightarrow Int = [0, 3\pi],\ per = 3\pi,\ amp = 3,\ ps = 0,$

$ran = [-3, 3].$ (See graph below).

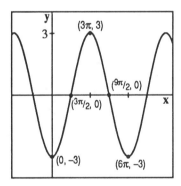

Figure 73

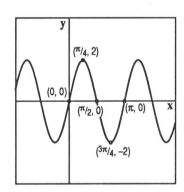

Figure 74

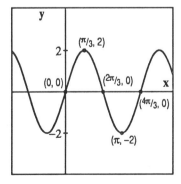

Figure 75

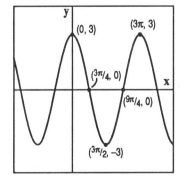

Figure 76

[77] $y = \frac{1}{2} \cos \frac{x}{2}. \ 0 \leq \frac{x}{2} \leq 2\pi \Rightarrow 0 \leq x \leq 4\pi \Rightarrow Int = [0, 4\pi], per = 4\pi, amp = \frac{1}{2}, ps = 0,$

$ran = \left[-\frac{1}{2}, \frac{1}{2} \right].$ (See graph below).

[78] $y = \frac{2}{3} \sin \frac{2x}{3}. \ 0 \leq \frac{2x}{3} \leq 2\pi \Rightarrow 0 \leq x \leq 3\pi \Rightarrow Int = [0, 3\pi], per = 3\pi, amp = \frac{2}{3}, ps = 0,$

$ran = \left[-\frac{2}{3}, \frac{2}{3} \right].$ (See graph below).

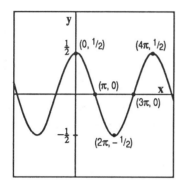 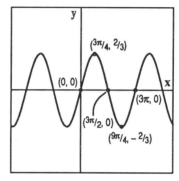

Figure 77   **Figure 78**

[79] $y = -2 \cos(3x + 1). \ 0 \leq 3x + 1 \leq 2\pi \Rightarrow -\frac{1}{3} \leq x \leq \frac{2\pi}{3} - \frac{1}{3} \Rightarrow Int = \left[-\frac{1}{3}, \frac{2\pi}{3} - \frac{1}{3} \right], per = \frac{2\pi}{3},$

$amp = 2, ps = -\frac{1}{3}, ran = [-2, 2], \text{Reflect.}$ (See graph below).

[80] $y = -3 \cos(2x + 1). \ 0 \leq 2x + 1 \leq 2\pi \Rightarrow -\frac{1}{2} \leq x \leq \pi - \frac{1}{2} \Rightarrow Int = \left[-\frac{1}{2}, \pi - \frac{1}{2} \right], per = \pi, amp = 3,$

$ps = -\frac{1}{2}, ran = [-3, 3], \text{Reflect.}$ (See graph below).

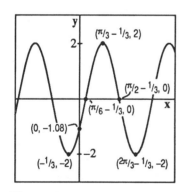 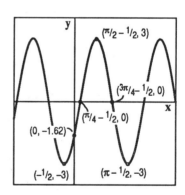

Figure 79   **Figure 80**

[81] $y = 4 \sin(2 - 3x) = -4 \sin(3x - 2). \ 0 \leq 3x - 2 \leq 2\pi \Rightarrow \frac{2}{3} \leq x \leq \frac{2\pi}{3} + \frac{2}{3} \Rightarrow Int = \left[\frac{2}{3}, \frac{2}{3} + \frac{2\pi}{3} \right],$

$per = \frac{2\pi}{3}, amp = 4, ps = \frac{2}{3}, ran = [-4, 4], \text{Reflect.}$ (See graph below).

[82] $y = \frac{1}{3} \cos(4 - 2x) = \frac{1}{3} \cos(2x - 4). \ 0 \leq 2x - 4 \leq 2\pi \Rightarrow 2 \leq x \leq \pi + 2 \Rightarrow Int = [2, 2 + \pi], per = \pi,$

$amp = \frac{1}{3}, ps = 2, ran = \left[-\frac{1}{3}, \frac{1}{3} \right].$ (See graph below).

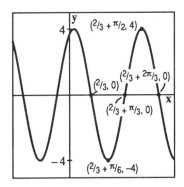

Figure 81

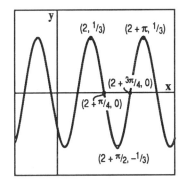

Figure 82

NOTE: On Exercises 83 – 90: Those functions which can be expressed in the form $y = a \sec (b x + c)$ or $y = a \csc (b x + c)$, where $b > 0$, have period $= per = \dfrac{2\pi}{b}$, phase shift $= ps = -\dfrac{c}{b}$, and range $= ran = (-\infty, -|a|] \cup [|a|, \infty)$. If $a < 0$, the basic secant or cosecant graph is reflected across the x – axis, and we indicate this by writing "Reflect." We will also give the interval that corresponds to $[0, 2\pi]$ which we will denote by *Int*. Those functions which can be expressed in the form $y = a \tan (b x + c)$ or $y = a \cot (b x + c)$, where $b > 0$, have period $= per = \dfrac{\pi}{b}$, phase shift $= ps = -\dfrac{c}{b}$, and range $= ran = $ **R**. If $a < 0$, the basic tangent or cotangent graph is reflected across the x – axis, and we indicate this by writing "Reflect." We will also give the interval that corresponds to $\left(-\dfrac{\pi}{2}, \dfrac{\pi}{2}\right)$ for the tangent graphs and $(0. \pi)$ for the cotangent graphs. In either case, we will denote this interval by *Int*.

[83] $y = \cot (-x) = -\cot x$. *Int* $= (0, \pi)$, $per = \pi$, $ps = 0$, $ran = $ **R**, Reflect.   (See graph below).

[84] $y = \tan (-x) = -\tan x$. *Int* $= \left(-\dfrac{\pi}{2}, \dfrac{\pi}{2}\right)$, $per = \pi$, $ps = 0$, $ran = $ **R**, Reflect.   (See graph below).

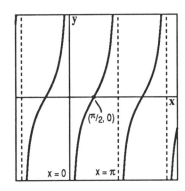

Figure 83

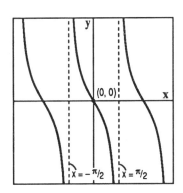

Figure 84

[85] $y = 2 \tan \dfrac{x}{2}$. $-\dfrac{\pi}{2} < \dfrac{x}{2} < \dfrac{\pi}{2} \Rightarrow -\pi < x < \pi \Rightarrow$ *Int* $= (-\pi, \pi)$, $per = 2\pi$, $ps = 0$, $ran = $ **R**.
(See graph below).

[86] $y = 3 \cot \dfrac{x}{3}$. $0 < \dfrac{x}{3} < \pi \Rightarrow 0 < x < 3\pi \Rightarrow$ *Int* $= (0, 3\pi)$, $per = 3\pi$, $ps = 0$, $ran = $ **R**.
(See graph below).

217

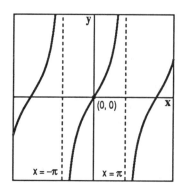

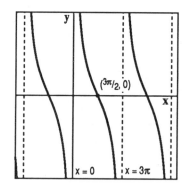

Figure 85   **Figure 86**

[87] $y = \sec(2 - x) = \sec(x - 2).\ 0 \le x - 2 \le 2\pi \Rightarrow 2 \le x \le 2\pi + 2 \Rightarrow Int = [2, 2 + 2\pi],\ per = 2\pi,\ ps = 2,$

$ran = (-\infty, -1] \cup [1, \infty).$ (See graph below).

[88] $y = 2\csc(1 - 2x) = -2\csc(2x - 1).\ 0 \le 2x - 1 \le 2\pi \Rightarrow \frac{1}{2} \le x \le \pi + \frac{1}{2} \Rightarrow Int = \left[\frac{1}{2}, \frac{1}{2} + \pi\right],\ per = \pi,$

$ps = \frac{1}{2},\ ran = (-\infty, -2] \cup [2, \infty),$ Reflect.   (See graph below).

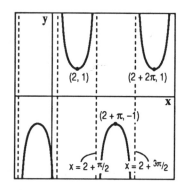

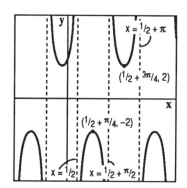

Figure 87   **Figure 88**

[89] $y = -2\csc(2 - 4x) = 2\csc(4x - 2).\ 0 \le 4x - 2 \le 2\pi \Rightarrow \frac{1}{2} \le x \le \frac{1}{2} + \frac{\pi}{2} \Rightarrow Int = \left[\frac{1}{2}, \frac{1}{2} + \frac{\pi}{2}\right],$

$per = \frac{\pi}{2},\ ps = \frac{1}{2},\ ran = (-\infty, -2] \cup [2, \infty).$ (See graph below).

[90] $y = -3\sec(2x + 1).\ 0 \le 2x + 1 \le 2\pi \Rightarrow -\frac{1}{2} \le x \le \pi - \frac{1}{2} \Rightarrow Int = \left[-\frac{1}{2}, \pi - \frac{1}{2}\right],\ per = \pi,\ ps = -\frac{1}{2},$

$ran = (-\infty, -3] \cup [3, \infty),$ Reflect.   (See graph below).

[91]   Since $f(t) = 50\cos(120\pi t - \pi)$, we see that $a = 50,\ b = 120\pi$, and $c = -\pi$. So,

a) $per = \dfrac{2\pi}{b} = \dfrac{2\pi}{120\pi} = \dfrac{1}{60}.$

b) $amp = |a| = 50.$

c) $ps = -\dfrac{c}{b} = \dfrac{\pi}{120\pi} = \dfrac{1}{120}.$

d)   Frequency $= \dfrac{1}{per} = 60$ cycles per second.

218

e) $f(t) = 0$ when $50 \cos(120 \pi t - \pi) = 0 \Rightarrow \cos(120 \pi t - \pi) = 0$.

Since the first value of x (for $x \geq 0$) for which $\cos x = 0$ is $\frac{\pi}{2}$, we solve $120 \pi t - \pi = \frac{\pi}{2}$ for t:

$$120 \pi t - \pi = \frac{\pi}{2} \Rightarrow 120 \pi t = \frac{3\pi}{2} \Rightarrow t = \frac{1}{80} \text{ sec.}$$

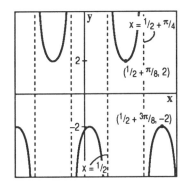

Figure 89

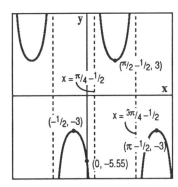

Figure 90

[92]   Since $f(t) = 2 \sin[\pi(t_0 - 360 t)] = 2 \sin(\pi t_0 - 360 \pi t) = -2 \sin(360 \pi t - \pi t_0)$, we see that

$a = -2$, $b = 360 \pi$, and $c = -\pi t_0$. So,

a) $per = \frac{2\pi}{b} = \frac{2\pi}{360\pi} = \frac{1}{180}$.

b) $amp = |a| = 2$.

c) $ps = -\frac{c}{b} = \frac{\pi t_0}{360 \pi} = \frac{t_0}{360}$.

d) $\text{Frequency} = \frac{1}{per} = 180$ cycles per second.

e) $f(t) = 0$ when $-2 \sin(360 \pi t - \pi t_0) = 0 \Rightarrow \sin(360 \pi t - \pi t_0) = 0$.

Since the second value of x (for $x \geq 0$) for which $\sin x = 0$ is π, we solve $360 \pi t - \pi t_0 = \pi$ for t:

$$360 \pi t - \pi t_0 = \pi \Rightarrow 360 \pi t = \pi + \pi t_0 \Rightarrow t = \frac{1 + t_0}{360} \text{ sec.}$$

[93] $\sin^{-1}\left(-\frac{\sqrt{3}}{2}\right) = -\frac{\pi}{3}$.   [94] $\cos^{-1}\left(-\frac{\sqrt{2}}{2}\right) = \frac{3\pi}{4}$.

[95] $\arctan 1 = \frac{\pi}{4}$.   [96] $\arccos\left(-\frac{1}{2}\right) = \frac{2\pi}{3}$.

[97] $\arccos(\sin 2\pi) = \arccos 0 = \frac{\pi}{2}$.   [98] $\arcsin(\cos 2\pi) = \arcsin 1 = \frac{\pi}{2}$.

[99] $\sin^{-1}\left(\cos \frac{3\pi}{4}\right) = \sin^{-1}\left(-\frac{\sqrt{2}}{2}\right) = -\frac{\pi}{4}$.   [100] $\cos^{-1}\left(\sin \frac{5\pi}{6}\right) = \cos^{-1}\left(\frac{1}{2}\right) = \frac{\pi}{3}$.

[101] $\tan^{-1}\left(\cot \frac{5\pi}{6}\right) = \tan^{-1}\left(-\sqrt{3}\right) = -\frac{\pi}{3}$.   [102] $\sin^{-1}\left(\sin \frac{5\pi}{3}\right) = \sin^{-1}\left(-\frac{\sqrt{3}}{2}\right) = -\frac{\pi}{3}$.

[103] $\tan[\arctan(-\pi)] = -\pi$, since $-\pi$ is in **R**.

[104] $\sin\left[\cos^{-1}\left(-\frac{2}{3}\right)\right]$. If $t = \cos^{-1}\left(-\frac{2}{3}\right)$, then $\cos t = -\frac{2}{3}$ and $0 \leq t \leq \pi$. Since $\sin^2 t + \cos^2 t = 1$ and

$\sin t \geq 0$ when $0 \leq t \leq \pi$, $\sin t = \sqrt{1 - \cos^2 t} = \sqrt{1 - \left(-\frac{2}{3}\right)^2} = \sqrt{1 - \frac{4}{9}} = \frac{\sqrt{5}}{3}$. So,

$\sin\left[\cos^{-1}\left(-\frac{2}{3}\right)\right] = \frac{\sqrt{5}}{3}$.

[105] $\sin^{-1}\left(\sin\frac{2\pi}{3}\right) = \sin^{-1}\left(\frac{\sqrt{3}}{2}\right) = \frac{\pi}{3}$.

[106] $\tan^{-1}\left(\tan\frac{7\pi}{6}\right) = \tan^{-1}\left(\frac{1}{\sqrt{3}}\right) = \frac{\pi}{6}$.

[107] $\cos\left[\sin^{-1}\left(\frac{3}{4}\right)\right]$. If $t = \sin^{-1}\left(\frac{3}{4}\right)$, then $\sin t = \frac{3}{4}$ and $-\frac{\pi}{2} \le t \le \frac{\pi}{2}$. Since $\sin^2 t + \cos^2 t = 1$ and

$\cos t \ge 0$ when $-\frac{\pi}{2} \le t \le \frac{\pi}{2}$, $\cos t = \sqrt{1 - \sin^2 t} = \sqrt{1 - \left(\frac{3}{4}\right)^2} = \frac{\sqrt{7}}{4}$. So, $\cos\left[\sin^{-1}\left(\frac{3}{4}\right)\right] = \frac{\sqrt{7}}{4}$.

[108] $\cos\left[\arcsin\left(-\frac{5}{6}\right)\right]$. If $t = \arcsin\left(-\frac{5}{6}\right)$, then $\sin t = -\frac{5}{6}$ and $-\frac{\pi}{2} \le t \le \frac{\pi}{2}$. Since $\sin^2 t + \cos^2 t = 1$ and

$\cos t \ge 0$ when $-\frac{\pi}{2} \le t \le \frac{\pi}{2}$, $\cos t = \sqrt{1 - \sin^2 t} = \sqrt{1 - \left(-\frac{5}{6}\right)^2} = \frac{\sqrt{11}}{6}$.

So, $\cos\left[\arcsin\left(-\frac{5}{6}\right)\right] = \frac{\sqrt{11}}{6}$.

[109] $\sec\left(\tan^{-1} 5\right)$. If $t = \tan^{-1} 5$, then $\tan t = 5$ and $-\frac{\pi}{2} < t < \frac{\pi}{2}$. Since $1 + \tan^2 t = \sec^2 t$ and

$\sec t > 0$ when $-\frac{\pi}{2} < t < \frac{\pi}{2}$, $\sec t = \sqrt{1 + \tan^2 t} = \sqrt{1 + (5)^2} = \sqrt{26}$. So, $\sec\left(\tan^{-1} 5\right) = \sqrt{26}$.

[110] $\sin\left(\arcsin\left(-3\right)\right)$ is undefined, since -3 is not in $[-1, 1]$.

[111] $\arcsin\left(-0.56\right) \approx -0.5944$.

[112] $\cos^{-1}\left(-\frac{\sqrt{5}}{3}\right) \approx 2.412$.

[113] $\tan^{-1} 25 \approx 1.531$.

[114] $\arcsin\left(-\frac{\pi}{12}\right) \approx -0.2649$.

[115] $\cos^{-1}\left(-\frac{15}{16}\right) \approx 2.786$.

[116] $\tan^{-1} \pi \approx 1.263$.

[117] $f(x) = -\cos^{-1} x$ (See graph below).

[118] $f(x) = 1 + \sin^{-1} x$ (See graph below).

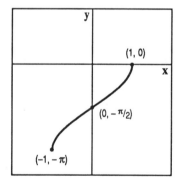

Figure 117

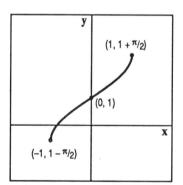

Figure 118

[119] $f(x) = 1 + \tan^{-1} x$ (See graph below).

[120] $f(x) = 1 + \sin^{-1}(-x) = 1 - \sin^{-1} x$ (See graph below).

[121] $f(x) = 2 + \sin^{-1}(-x) = 2 - \sin^{-1} x$ (See graph below).

[122] $f(x) = 3 - \cos^{-1} x$ (See graph below).

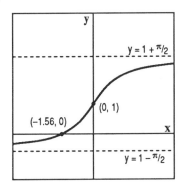

Figure 119

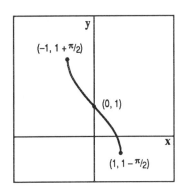

Figure 120

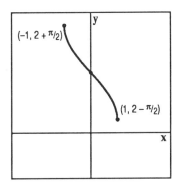

Figure 121

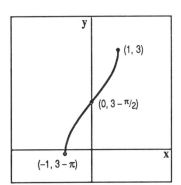

Figure 122

[123] $y = 9 - 10 \sin^{-1}(11x+12) \Rightarrow y - 9 = -10 \sin^{-1}(11x+12) \Rightarrow \dfrac{y-9}{-10} = \sin^{-1}(11x+12) \Rightarrow$

$11x + 12 = \sin\left(\dfrac{y-9}{-10}\right) \Rightarrow 11x + 12 = -\sin\left(\dfrac{y-9}{10}\right) \Rightarrow 11x = 12 - \sin\left(\dfrac{y-9}{10}\right) \Rightarrow$

$x = -\dfrac{12}{11} - \dfrac{1}{11}\sin\left(\dfrac{y-9}{10}\right).$

[124] $y = 2 + 5\cos^{-1}(3x-4) \Rightarrow y - 2 = 5\cos^{-1}(3x-4) \Rightarrow \dfrac{y-2}{5} = \cos^{-1}(3x-4) \Rightarrow \cos\left(\dfrac{y-2}{5}\right) =$

$3x - 4 \Rightarrow 3x = \cos\left(\dfrac{y-2}{5}\right) + 4 \Rightarrow x = \dfrac{1}{3}\cos\left(\dfrac{y-2}{5}\right) + \dfrac{4}{3}.$

[125] $y = -5 + 6\tan^{-1}(\pi x + 2) \Rightarrow y + 5 = 6\tan^{-1}(\pi x + 2) \Rightarrow \dfrac{y+5}{6} = \tan^{-1}(\pi x + 2) \Rightarrow \pi x + 2 =$

$\tan\left(\dfrac{y+5}{6}\right) \Rightarrow \pi x = \tan\left(\dfrac{y+5}{6}\right) - 2 \Rightarrow x = \dfrac{1}{\pi}\tan\left(\dfrac{y+5}{6}\right) - \dfrac{2}{\pi}.$

[126] $y = 1 - 8\sin^{-1}(4x-5) \Rightarrow y - 1 = -8\sin^{-1}(4x-5) \Rightarrow \dfrac{y-1}{-8} = \sin^{-1}(4x-5) \Rightarrow \dfrac{1-y}{8} =$

$\sin^{-1}(4x-5) \Rightarrow \sin\left(\dfrac{1-y}{8}\right) = 4x - 5 \Rightarrow 4x = \sin\left(\dfrac{1-y}{8}\right) + 5 \Rightarrow x = \dfrac{1}{4}\sin\left(\dfrac{1-y}{8}\right) + \dfrac{5}{4}.$

[127] $y = 4 + 7\cos^{-1}(8x-2) \Rightarrow y - 4 = 7\cos^{-1}(8x-2) \Rightarrow \dfrac{y-4}{7} = \cos^{-1}(8x-2) \Rightarrow \cos\left(\dfrac{y-4}{7}\right) =$

$8x - 2 \Rightarrow 8x = \cos\left(\dfrac{y-4}{7}\right) + 2 \Rightarrow x = \dfrac{1}{4} + \dfrac{1}{8}\cos\left(\dfrac{y-4}{7}\right).$

EXERCISES 6.1

[1] $\quad 75°\,32\,'14" = \left(75 + \frac{32}{60} + \frac{14}{3600}\right)^{\circ} \approx 75.537°.$

[2] $\quad 102°\,51\,'19" = \left(102 + \frac{51}{60} + \frac{19}{3600}\right)^{\circ} \approx 102.855°.$

[3] $\quad 187°\,15\,'12" = \left(187 + \frac{15}{60} + \frac{12}{3600}\right)^{\circ} \approx 187.253°.$

[4] $\quad 301°\,23\,'51" = \left(301 + \frac{23}{60} + \frac{51}{3600}\right)^{\circ} \approx 301.398°.$

[5] $\quad 88.123° = 88° + 0.123°.$ Since $0.123° = (0.123)(60') = 7.38' = 7' + 0.38'$ and $0.38' = (0.38)(60") = 22.8" \approx 23",\ 88.123° \approx 88°\,7'23".$

[6] $\quad 99.568° = 99° + 0.568°.$ Since $0.568° = (0.568)(60') = 34.08' = 34' + 0.08'$ and $0.08' = (0.08)(60") = 4.8 \approx 5",\ 99.568° \approx 99°\,34'5".$

[7] $\quad 286.987° = 286° + 0.897°.$ Since $0.897° = (0.897)(60') = 53.82' = 53' + 0.82'$ and $0.82' = (0.82)(60") = 49.2" \approx 49",\ 286.897° \approx 286°\,53'49".$

[8] $\quad 199.543° = 199° + 0.543°.$ Since $0.543° = (0.543)(60') = 32.58' = 32' + 0.58'$ and $0.58' = (0.58)(60") = 34.8" \approx 35",\ 199.543° \approx 199°\,32'35".$

[9] $\quad 95°\,10'13" = \left(95 + \frac{10}{60} + \frac{13}{3600}\right)^{\circ} \approx 95.17° \approx 95.17\left(\frac{\pi}{180}\text{ radians}\right) \approx 1.66\text{ radians.}$ (See graph below).

[10] $\quad 80°\,13'5" = \left(80 + \frac{13}{60} + \frac{5}{3600}\right)^{\circ} \approx 80.218° \approx (80.218)\left(\frac{\pi}{180}\text{ radians}\right) \approx 1.40\text{ radians.}$ (See graph below).

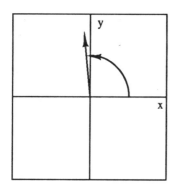

Figure 9   **Figure 10**

[11] $\quad (2\pi)^{\circ} = (2\pi)\left(\frac{\pi}{180}\text{ radians}\right) = \frac{\pi^2}{90}\text{ radians.}$ (See graph below).

[12] $\quad -\pi° = (-\pi)\left(\frac{\pi}{180}\text{ radians}\right) = -\frac{\pi^2}{180}\text{ radians.}$ (See graph below).

[13] $\quad -240° = (-240)\left(\frac{\pi}{180}\text{ radians}\right) = -\frac{4\pi}{3}\text{ radians.}$ (See graph below).

[14] $300° = (300)\left(\frac{\pi}{180} \text{ radians}\right) = \frac{5\pi}{3}$ radians.   (See graph below).

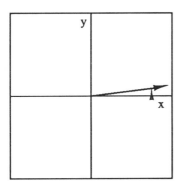

Figure 11

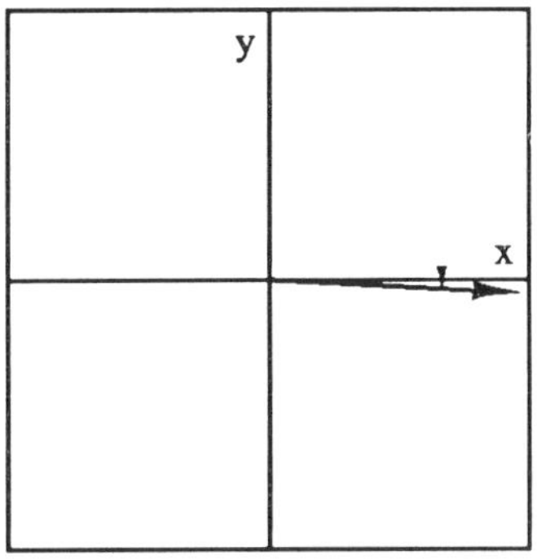

Figure 12

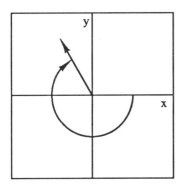

Figure 13

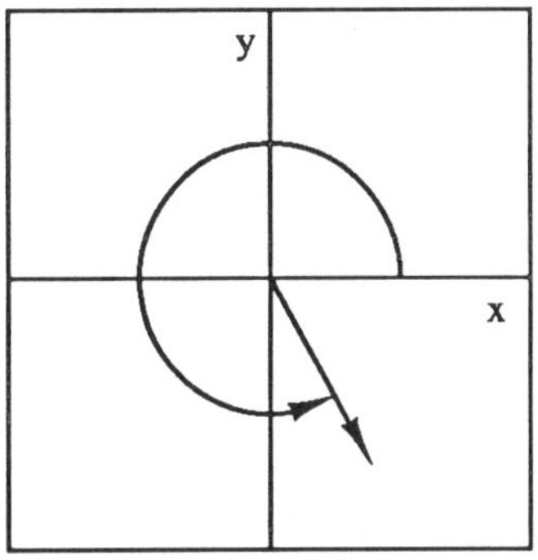

Figure 14

[15] $570° = (570)\left(\frac{\pi}{180} \text{ radians}\right) = \frac{19\pi}{6}$ radians.   (See graph below).

[16] $-135° = (-135)\left(\frac{\pi}{180} \text{ radians}\right) = -\frac{3\pi}{4}$ radians.   (See graph below).

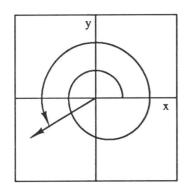

Figure 15

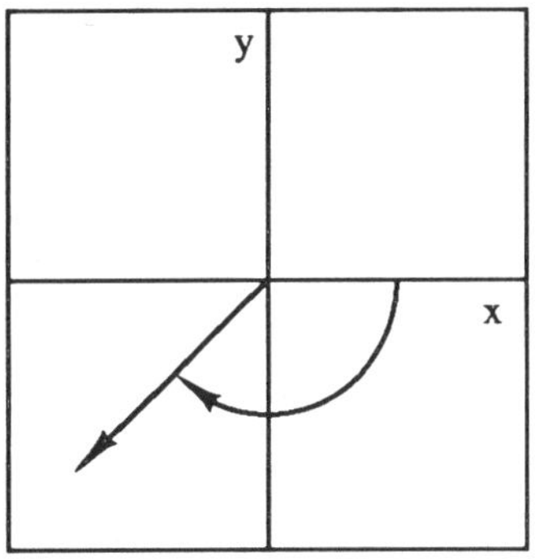

Figure 16

[17] $-315° = (-315)\left(\frac{\pi}{180} \text{ radians}\right) = -\frac{7\pi}{4}$ radians.   (See graph below).

[18] $-495° = (-495)\left(\frac{\pi}{180} \text{ radians}\right) = -\frac{11\pi}{4}$ radians.   (See graph below).

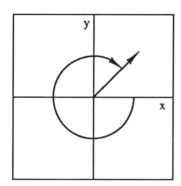

Figure 17

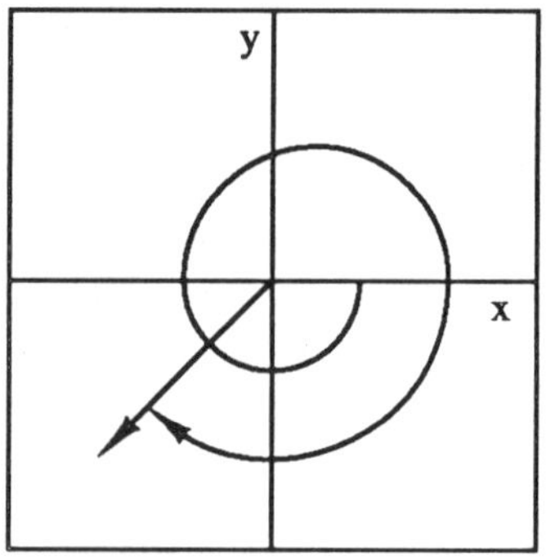

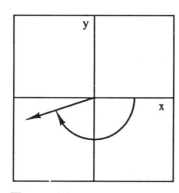

Figure 18

[19] $-\dfrac{9\pi}{10}$ radians $=\left(-\dfrac{9\pi}{10}\right)\left(\dfrac{180}{\pi}\right)^{\circ}=-162^{\circ}$.   (See graph below).

[20] $-\dfrac{7\pi}{8}$ radians $=\left(-\dfrac{7\pi}{8}\right)\left(\dfrac{180}{\pi}\right)^{\circ}=-157.5^{\circ}$.   (See graph below).

[21] 5 radians $=(5)\left(\dfrac{180}{\pi}\right)^{\circ}=\left(\dfrac{900}{\pi}\right)^{\circ}\approx286.479^{\circ}$.   (See graph below).

[22] -3 radians $=(-3)\left(\dfrac{180}{\pi}\right)^{\circ}=\left(\dfrac{-540}{\pi}\right)^{\circ}\approx-171.89^{\circ}$.   (See graph below).

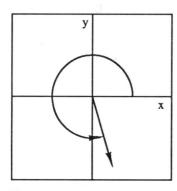

Figure 19

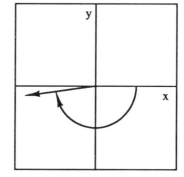

Figure 20

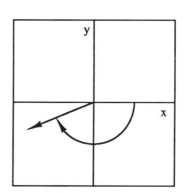

Figure 21

Figure 22

[23] $-\dfrac{11\pi}{6}$ radians $=\left(-\dfrac{11\pi}{6}\right)\left(\dfrac{180}{\pi}\right)^{\circ}=-330^{\circ}$.   (See graph below).

[24] $-\dfrac{7\pi}{4}$ radians $=\left(-\dfrac{7\pi}{4}\right)\left(\dfrac{180}{\pi}\right)^{\circ}=-315^{\circ}$.   (See graph below).

[25] $-\dfrac{9\pi}{4}$ radians $=\left(-\dfrac{9\pi}{4}\right)\left(\dfrac{180}{\pi}\right)^{\circ}=-405^{\circ}$.   (See graph below).

[26] $\dfrac{13\pi}{12}$ radians $=\left(\dfrac{13\pi}{12}\right)\left(\dfrac{180}{\pi}\right)^{\circ}=195^{\circ}$.   (See graph below).

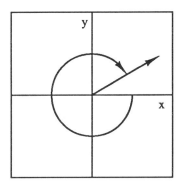

Figure 23

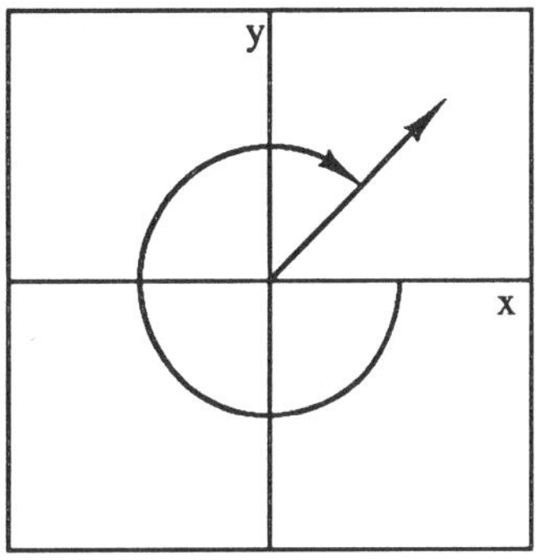

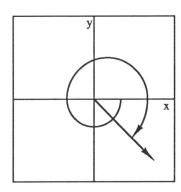

Figure 24

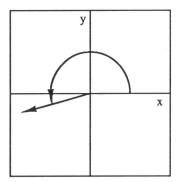

Figure 25

Figure 26

[27] $-\dfrac{11\pi}{24}$ radians $=\left(-\dfrac{11\pi}{24}\right)\left(\dfrac{180}{\pi}\right)^{\circ}=-82.5^{\circ}$.   (See graph below).

[28] $-\dfrac{13\pi}{15}$ radians $=\left(-\dfrac{13\pi}{15}\right)\left(\dfrac{180}{\pi}\right)^{\circ}=-156^{\circ}$.   (See graph below).

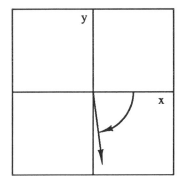

Figure 27

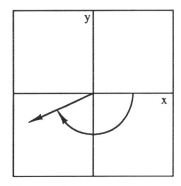

Figure 28

[29] $r = 10$ cm, $\theta = 80°$. $s = r\,\theta = (10)\left[(80)\left(\frac{\pi}{180}\right)\right] = \frac{40\pi}{9}$ cm. $A = \frac{1}{2}r^2\theta = \frac{1}{2}r\,s =$

$\frac{1}{2}(10)\left(\frac{40\pi}{9}\right) = \frac{200\pi}{9}$ cm².

[30] $r = 5$ in, $\theta = 7$ radians. So $s = r\,\theta = (5)(7) = 35$ in, and $A = \frac{1}{2}r^2\theta = \frac{1}{2}r\,s = \frac{1}{2}(5)(35) = 87.5$ in².

[31] $r = 8$ in, $\theta = \pi°$. $s = r\,\theta = (8)\left[(\pi)\left(\frac{\pi}{180}\right)\right] = \frac{2\pi^2}{45}$ in, and $A = \frac{1}{2}r^2\theta = \frac{1}{2}r\,s = \left(\frac{1}{2}\right)(8)\left(\frac{2\pi^2}{45}\right) = \frac{8\pi^2}{45}$ in².

[32] $r = 12$ m, $\theta = 25°$. $s = r\,\theta = (12)\left[(25)\left(\frac{\pi}{180}\right)\right] = \frac{5\pi}{3}$ m, and $A = \frac{1}{2}r^2\theta = \frac{1}{2}r\,s = \frac{1}{2}(12)\left(\frac{5\pi}{3}\right) =$

10π m².

[33] $r = 3$ ft, $\theta = 20$ radians. $s = r\,\theta = (3)(20) = 60$ ft, and $A = \frac{1}{2}r^2\theta = \frac{1}{2}r\,s = \frac{1}{2}(3)(60) = 90$ ft².

[34] $r = \pi$ ft, $\theta = \left(\frac{\pi}{2}\right)°$. $s = r\,\theta = (\pi)\left[\left(\frac{\pi}{2}\right)\left(\frac{\pi}{180}\right)\right] = \frac{\pi^3}{360}$ ft, and $A = \frac{1}{2}r^2\theta = \frac{1}{2}r\,s = \frac{1}{2}(\pi)\left(\frac{\pi^3}{360}\right) = \frac{\pi^4}{720}$ ft².

[35] $\cos 210° = -\frac{\sqrt{3}}{2}$ [36] $\sec(-45°) = \sqrt{2}$

[37] $\sin(-225°) = \frac{\sqrt{2}}{2}$ [38] $\cos 150° = -\frac{\sqrt{3}}{2}$

[39] $\tan 300° = -\sqrt{3}$ [40] $\cot(-240°) = -\frac{1}{\sqrt{3}}$

[41] $\cot(-510°) = \cot(-150°) = \sqrt{3}$ [42] $\tan(-585°) = \tan(-225°) = -1$

[43] $Q = (-1, -4)$. Since $r = d(O, Q) = \sqrt{(-1)^2 + (-4)^2} = \sqrt{17}$, it follows from Theorem 6.2 that

$\sin\theta = \frac{y}{r} = -\frac{4}{\sqrt{17}}$, $\csc\theta = \frac{r}{y} = -\frac{\sqrt{17}}{4}$, $\cos\theta = \frac{x}{r} = -\frac{1}{\sqrt{17}}$, $\sec\theta = \frac{r}{x} = -\sqrt{17}$, $\tan\theta = \frac{y}{x} = 4$, and

$\cot\theta = \frac{x}{y} = \frac{1}{4}$. (See graph below).

[44] $Q = (-3, -2)$. Since $r = \sqrt{(-3)^2 + (-2)^2} = \sqrt{13}$, it follows from Theorem 6.2 that $\sin\theta = \frac{y}{r} = -\frac{2}{\sqrt{13}}$,

$\csc\theta = \frac{r}{y} = -\frac{\sqrt{13}}{2}$, $\cos\theta = \frac{x}{r} = -\frac{3}{\sqrt{13}}$, $\sec\theta = \frac{r}{x} = -\frac{\sqrt{13}}{3}$, $\tan\theta = \frac{y}{x} = \frac{2}{3}$, and $\cot\theta = \frac{x}{y} = \frac{3}{2}$.

(See graph below).

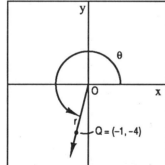

Figure 43   **Figure 44**

[45] $Q = (-1, 5)$. Since $r = \sqrt{(-1)^2 + (-5)^2} = \sqrt{26}$, it follows from Theorem 6.2 that $\sin\theta = \frac{y}{r} = \frac{5}{\sqrt{26}}$,

$\csc\theta = \frac{r}{y} = \frac{\sqrt{26}}{5}$, $\cos\theta = \frac{x}{r} = -\frac{1}{\sqrt{26}}$, $\sec\theta = \frac{r}{x} = -\sqrt{26}$, $\tan\theta = \frac{y}{x} = -5$, and $\cot\theta = \frac{x}{y} = -\frac{1}{5}$.

(See graph below).

[46] $Q = \left(\sqrt{3}, -2\right)$. Since $r = \sqrt{\left(\sqrt{3}\right)^2 + \left(-2\right)^2} = \sqrt{7}$, it follows from Theorem 6.2 that $\sin\theta = \frac{y}{r} = -\frac{2}{\sqrt{7}}$,

$\csc\theta = \frac{r}{y} = -\frac{\sqrt{7}}{2}$, $\cos\theta = \frac{x}{r} = \frac{\sqrt{3}}{\sqrt{7}}$, $\sec\theta = \frac{r}{x} = \frac{\sqrt{7}}{\sqrt{3}}$, $\tan\theta = \frac{y}{x} = -\frac{2}{\sqrt{3}}$, and $\cot\theta = \frac{x}{y} = -\frac{\sqrt{3}}{2}$.

(See graph below).

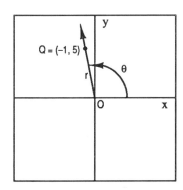

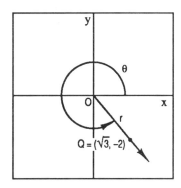

Figure 45   Figure 46

[47] $\cos\theta = \frac{5}{12} = \frac{x}{r} \Rightarrow x = 5$ and $r = 12 \Rightarrow x^2 + y^2 = r^2 \Rightarrow 25 + y^2 = 144 \Rightarrow y^2 = 119 \Rightarrow y = \sqrt{119}$

(in QI, $y > 0$). Therefore, $Q = (5, \sqrt{119})$ is on the terminal side of θ. Hence, $\sin\theta = \frac{y}{r} = \frac{\sqrt{119}}{12}$,

$\tan\theta = \frac{y}{x} = \frac{\sqrt{119}}{5}$, $\cot\theta = \frac{x}{y} = \frac{5}{\sqrt{119}}$, $\sec\theta = \frac{r}{x} = \frac{12}{5}$, $\csc\theta = \frac{r}{y} = \frac{12}{\sqrt{119}}$. (See graph below).

[48] $\theta \in$ QIII and $\cot\theta = \frac{5}{6} = \frac{y}{x} \Rightarrow y = -5$ and $x = -6 \Rightarrow Q = (-6, -5)$ is on the terminal side of $\theta \Rightarrow r =$

$d\,(O, Q) = \sqrt{(-6)^2 + (-5)^2} = \sqrt{61}$. So, $\sin\theta = \frac{y}{r} = -\frac{5}{\sqrt{61}}$, $\csc\theta = \frac{r}{y} = -\frac{\sqrt{61}}{5}$, $\cos\theta = \frac{x}{r} = -\frac{6}{\sqrt{61}}$,

$\sec\theta = \frac{r}{x} = -\frac{\sqrt{61}}{6}$, and $\tan\theta = \frac{x}{y} = \frac{6}{5}$. (See graph below).

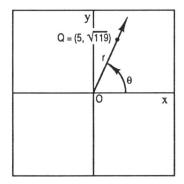

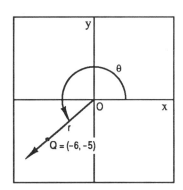

Figure 47   Figure 48

[49] $\sec\theta = 3 = \frac{r}{x} = \frac{3}{1} \Rightarrow x = 1$ and $r = 3 \Rightarrow x^2 + y^2 = r^2 \Rightarrow 1 + y^2 = 9 \Rightarrow y^2 = 8 \Rightarrow y = -2\sqrt{2}$

(in QIV, $y < 0$). Therefore, $Q = (1, -2\sqrt{2})$ is on the terminal side of θ. So, $\sin\theta = \frac{y}{r} = -\frac{2\sqrt{2}}{3}$,

$\cos\theta = \frac{x}{r} = \frac{1}{3}$, $\tan\theta = \frac{y}{x} = -2\sqrt{2}$, $\cot\theta = \frac{x}{y} = -\frac{1}{2\sqrt{2}}$, $\csc\theta = \frac{r}{y} = -\frac{3}{2\sqrt{2}}$. (See graph below).

[50] $\csc\theta = -4 = \frac{r}{y} = \frac{4}{-1} \Rightarrow r = 4$ and $y = -1 \Rightarrow x^2 + y^2 = r^2 \Rightarrow x^2 + 1 = 16 \Rightarrow x^2 = 15 \Rightarrow x = \sqrt{15}$

(in QIV, $x > 0$). Therefore, $Q = (\sqrt{15}, -1)$ is on the terminal side of θ. So, $\sin\theta = \frac{y}{r} = -\frac{1}{4}$,

$\cos\theta = \frac{x}{r} = \frac{\sqrt{15}}{4}$, $\sec\theta = \frac{r}{x} = \frac{4}{\sqrt{15}}$, $\tan\theta = \frac{y}{x} = -\frac{1}{\sqrt{15}}$, $\cot\theta = \frac{x}{y} = -\sqrt{15}$. (See graph below).

[51] $\sin\theta = -\frac{7}{9} = \frac{y}{r} = \frac{-7}{9} \Rightarrow y = -7$ and $r = 9 \Rightarrow x^2 + y^2 = r^2 \Rightarrow x^2 + 49 = 81 \Rightarrow x^2 = 32 \Rightarrow x = -4\sqrt{2}$

(in QIII, $x < 0$). Therefore, $Q = (-4\sqrt{2}, -7)$ is on the terminal side of θ. So, $\cos\theta = \frac{x}{r} = -\frac{4\sqrt{2}}{9}$,

$\tan\theta = \frac{y}{x} = \frac{7}{4\sqrt{2}}$, $\cot\theta = \frac{x}{y} = \frac{4\sqrt{2}}{7}$, $\sec\theta = \frac{r}{x} = -\frac{9}{4\sqrt{2}}$, $\csc\theta = \frac{r}{y} = -\frac{9}{7}$. (See graph below).

[52] $\theta \in$ QII and $\tan\theta = -\frac{9}{10} = \frac{y}{x} = \frac{9}{-10} \Rightarrow x = -10$ and $y = 9 \Rightarrow Q = (-10, 9)$ is on the terminal side of

$\theta \Rightarrow r = d\,(O, Q) = \sqrt{(-10)^2 + (9)^2} = \sqrt{181}$. So, $\sin\theta = \frac{y}{r} = \frac{9}{\sqrt{181}}$, $\csc\theta = \frac{r}{y} = \frac{\sqrt{181}}{9}$, $\cos\theta = \frac{x}{r} =$

$-\frac{10}{\sqrt{181}}$, $\sec\theta = \frac{r}{x} = -\frac{\sqrt{181}}{10}$, and $\cot\theta = \frac{x}{y} = -\frac{10}{9}$. (See graph below).

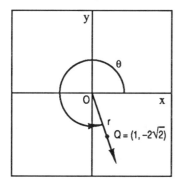

Figure 49   **Figure 50**

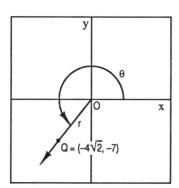

Figure 51   **Figure 52**

[53] $r = 12" = 1$ ft. Since $s = r\theta$ and since one revolution corresponds to a central angle of 2π radians,

1 rev $= r\theta = (1)(2\pi) = 2\pi$ ft. Since 1 mi $= 5280$ ft and 1 rev $= 2\pi$ ft, it follows that the number of

revolutions the wheel makes in 1 mile is $\frac{5280}{2\pi} = \frac{2640}{\pi}$.

[54] $d = 48" \Rightarrow r = 24" = 2$ ft. So, 1 rev $= r\theta = (2)(2\pi) = 4\pi$ ft. Hence, as the wheel travels 4000π feet it

makes $\frac{4000\pi}{4\pi} = 1{,}000$ revolutions.

[55] Let θ (measured in radians) be the central angle joining the cities. Then $s = r\,θ \Rightarrow 500$ mi $=$

$(3960$ mi$)\,θ \Rightarrow θ = \dfrac{500}{3960} = \dfrac{25}{198}.$ (See graph below).

[56] Since 1 revolution of the minute hand takes 60 minutes, 75 min $=$ 1 rev $+$ 15 min $=$ 1 rev $+\dfrac{1}{4}$ rev $=$

$\dfrac{5}{4}$ rev. Since 1 revolution corresponds to a central angle of $2\,π$ radians, the central angle is $θ =$

$\dfrac{5}{4}(2\,π) = \dfrac{5\,π}{2}$ radians. So, $s = r\,θ = (10)\left(\dfrac{5\,π}{2}\right) = 25\,π$ inches. Thus, the tip of the minute hand travels

$25\,π$ inches.

[57] Since $36° = 36\left(\dfrac{π}{180}\right) = \dfrac{π}{5}$ radians and since $s = r\,θ$, it follows that $3\,π = r\left(\dfrac{π}{5}\right) \Rightarrow \dfrac{r}{5} = 3 \Rightarrow r = 15$ ft.

The chain is 15 ft long.   (See graph below).

Figure 55   **Figure 57**

[58] Since a central angle of $2\,π$ radians corresponds to the whole pizza, $\dfrac{2\,π}{9}$ radians is the central angle

of $\dfrac{1}{9}$ of the pizza. Since $A = \dfrac{1}{2}\,r^2\,θ$, $16\,π$ in$^2 = \dfrac{1}{2}\,r^2\left(\dfrac{2\,π}{9}\right) \Rightarrow 16 = \dfrac{r^2}{9} \Rightarrow r^2 = (16)(9) = 144 \Rightarrow$

$r = 12" \Rightarrow d = 2\,r = 24".$ The diameter of the pizza is 24 inches.

[59] Since the hour hand makes one revolution every 12 hours, it follows that in 2 hours and 20 minutes

it makes $\dfrac{2\frac{1}{3}}{12} = \dfrac{\frac{7}{3}}{12} = \dfrac{7}{36}$ of a revolution. Since a central angle of $2\,π$ radians corresponds to one

revolution, the corresponding central angle is $\left(\dfrac{7}{36}\right)(2\,π) = \dfrac{7\,π}{18}$ radians. Since $s = r\,θ = (5)\left(\dfrac{7\,π}{18}\right) =$

$\dfrac{35\,π}{18}$ inches, the tip of the hour hand travles $\dfrac{35\,π}{18}$ inches.

[60] Since $r = \dfrac{d}{2}$, we see that $A = \dfrac{1}{2}\,r^2\,θ = \dfrac{1}{2}\left(\dfrac{d}{2}\right)^2\,θ = \dfrac{1}{2}\left(\dfrac{d^2}{4}\right)\,θ = \dfrac{d^2\,θ}{8}.$

[61] Since the diameter of the pizza is 12" its radius is 6". So, $A = \dfrac{1}{2}\,r^2\,θ \Rightarrow 18\,π = \dfrac{1}{2}(36)\,θ \Rightarrow θ =$

$π$ radians. The central angle the slice determines is $π$ radians.

[62] Since the diameter is 120 inches, the radius is 60 in $=$ 5 ft. Since $s = r\,θ$, $2 = 5\,θ \Rightarrow θ = \dfrac{2}{5}$ radians $\Rightarrow$

$θ = \dfrac{2}{5}\left(\dfrac{180}{π}\right)^{°} = \left(\dfrac{72}{π}\right)^{°}.$

[63] Since $\theta = 12° = 12\left(\dfrac{\pi}{180}\text{ rad}\right) = \dfrac{\pi}{15}$ radians and $s = 21$ cm, we have $s = r\theta \Rightarrow 21 = \left(\dfrac{\pi}{15}\right)r \Rightarrow r =$

$21\left(\dfrac{15}{\pi}\right) = \dfrac{315}{\pi}$ cm.

[64] The central angle generated in one minute has radian measure $\theta = (2000)(2\pi) = 4000\pi$. Hence, the angular velocity of the wheel is $\omega = \dfrac{\theta}{T} = \dfrac{4000\pi}{1} = 4000\pi$ radians per minute. The linear velocity of the wheel is $v = r\,\omega = (6)(4000\pi) = 24{,}000\pi$ feet per minute.

[65] The central angle generated in one minute has radian measure $\theta = (1400)(2\pi) = 2800\pi$. The angular velocity is $\omega = \dfrac{2800\pi}{1} = 2800\pi$ radians per minute. The linear velocity is $v = r\,\omega = (3)(2800\pi) = 8400\pi$ inches per minute.

[66] The central angle generated in one minute has radian measure $\theta = (1200)(2\pi) = 2400\pi$. The angular velocity is $\omega = \dfrac{2400\pi}{1} = 2400\pi$ radians per minute. The linear velocity is $v = r\,\omega = (12)(2400\pi) = 28{,}800\pi$ inches per minute.

[67] The central angle generated in one minute has radian measure $\theta = (100)(2\pi) = 200\pi$. The angular velocity is $\omega = \dfrac{200\pi}{1} = 200\pi$ radians per minute. The linear velocity is $v = r\,\omega = (5)(200\pi) = 1000\pi$ feet per minute.

[68] Since the diameter is 10 in the radius is 5 in $= \dfrac{5}{12}$ ft. Since 60 rpm is 1 revolution per second, the central angle generated in one second has radian measure $\theta = (1)(2\pi) = 2\pi$. So, the angular velocity is $\omega = \dfrac{2\pi}{1} = 2\pi$ radians per second. Hence, the linear velocity of a point on the rim of the wheel is $v = r\,\omega = \left(\dfrac{5}{12}\right)(2\pi) = \dfrac{5\pi}{6}$ feet per second.

[69] Since the diameter is 26 in, the radius is 13 in $= 1\frac{1}{12}$ ft $= \dfrac{13}{12}$ ft. Since the linear velocity is 20 MPH, $v = 20\text{ MPH} = (20)(5280\text{ ft})/\text{hr} = 105{,}600\text{ ft}/\text{hr}$. Since 1 hr $= 3600$ sec, $v = \dfrac{105{,}600}{3600}\text{ ft}/\text{sec} =$

$\dfrac{88}{3}\text{ ft}/\text{sec}$. So, $v = r\,\omega \Rightarrow \dfrac{88}{3}\text{ ft}/\text{sec} = \left(\dfrac{13}{12}\right)\omega \Rightarrow \omega = \dfrac{v}{r} = \dfrac{\frac{88}{3}}{\frac{13}{12}} = \dfrac{88}{3}\cdot\dfrac{12}{13} = \dfrac{352}{13}$ radians per

second. Since 2π radians is the central angle corresponding to one revolution, $\dfrac{352}{13} = (2\pi)x \Rightarrow$

$x = \dfrac{352}{13}\left(\dfrac{1}{2\pi}\right) = \dfrac{176}{13\pi}$ rev per second. So, the wheel makes $\dfrac{176}{13\pi}$ revolutions per second.

[70] Since the radius of the rear wheel is 8 in, it follows from the arc length formula $s = r\theta$ that as the wheel turns through 24 radians the corresponding arc length is $s = r\theta = (8)(24) = 192$ in. Similarly, since the radius of the front wheel is 6 in, we see that $s = r\theta \Rightarrow 192 = 6\theta \Rightarrow \theta = \dfrac{192}{6} = 32$ radians. So, the front wheel turns through 32 radians.

[71] Since the diameter of the pulley is 5 ft, its radius is 2.5 ft. Since 12 ft of cable corresponds to an arc of length 12 ft, it follows that $s = r\theta \Rightarrow 12 = (2.5)\theta \Rightarrow \theta = \frac{24}{5}$ radians. So, the pulley turns through $\frac{24}{5}$ radians.

[72] Since the diameter of the larger pulley is 25 cm, its radius is 12.5 cm. As the larger pulley turns through 12 radians, the corresponding arc length is $s = r\theta = (12.5)(12) = 150$ cm. Similarly, since the diameter of the smaller pulley is 10 cm, its radius is 5 cm. Hence, $s = r\theta \Rightarrow 150 = 5\theta \Rightarrow \theta = 30$ radians. So, the smaller pulley turns through 30 radians.

[73] Proceeding as in the solution of Exercise 72, we see that, as the smaller pulley turns through 8 radians, the corresponding arc length is $s = r\theta = (5)(8) = 40$ cm. Therefore, for the larger pulley, $s = r\theta \Rightarrow 40 = (12.5)\theta \Rightarrow \theta = 3.2$ radians. So, the larger pulley turns through 3.2 radians.

EXERCISES 6.2

[1] $(\tan t + \sec t)(\tan t - \sec t) = \tan^2 t - \sec^2 t = \tan^2 t - (1 + \tan^2 t) = \tan^2 t - 1 - \tan^2 t = -1$.

[2] $\sec t(\cot t - \cos t) = \frac{1}{\cos t}\left(\frac{\cos t}{\sin t} - \cos t\right) = \left(\frac{1}{\cos t}\right)\left(\frac{\cos t}{\sin t}\right) - \left(\frac{1}{\cos t}\right)(\cos t) = \frac{1}{\sin t} - 1 = \csc t - 1$.

[3] $(\sin\theta - \cos\theta)^2 + (\sin\theta + \cos\theta)^2 = (\sin^2\theta - 2\sin\theta\cos\theta + \cos^2\theta) +$

$(\sin^2\theta + 2\cos\theta\sin\theta + \cos^2\theta) = 1 - 2\sin\theta\cos + 1 + 2\sin\theta\cos\theta = 2$.

[4] $\frac{1}{1 + 2\cot^2 t + \cot^4 t} = \frac{1}{(1 + \cot^2 t)^2} = \frac{1}{(\csc^2 t)^2} = \frac{1}{\csc^4 t} = \sin^4 t$.

[5] $\frac{1}{(\csc\theta + \cot\theta)(1 - \cos\theta)} = \frac{1}{\left(\frac{1}{\sin\theta} + \frac{\cos\theta}{\sin\theta}\right)(1 - \cos\theta)} = \frac{1}{\left(\frac{1 + \cos\theta}{\sin\theta}\right)(1 - \cos\theta)} = \frac{1}{\frac{1 - \cos^2\theta}{\sin\theta}} =$

$\frac{1}{\frac{\sin^2\theta}{\sin\theta}} = \frac{1}{\sin\theta} = \csc\theta$.

[6] $\frac{1}{1 + \csc t} + \frac{1}{1 - \csc t} = \frac{(1 - \csc t) + (1 + \csc t)}{1 - \csc^2 t} = \frac{2}{1 - (1 + \cot^2 t)} = \frac{2}{-\cot^2 t} = -2\tan^2 t$.

[7] $(\cot x + \tan x)\cos x = \left(\frac{\cos x}{\sin x} + \frac{\sin x}{\cos x}\right)\cos x = \left(\frac{\cos^2 x + \sin^2 x}{\sin x \cos x}\right)\cos x = \left(\frac{1}{\sin x \cos x}\right)\cos x = \frac{1}{\sin x} =$

$\csc x$.

[8] $\cot^2 t(\sec^2 t - 1) = \cot^2 t(\tan^2 t) = \left(\frac{1}{\tan^2 t}\right)(\tan^2 t) = 1$.

[9] $\frac{\tan^2 v}{\csc^2 v - 1} = \frac{\tan^2 v}{\cot^2 v} = \tan^2 v\left(\frac{\tan^2 v}{1}\right) = \tan^4 v$.

[10] $\frac{\sec^2 t}{1 + \cot^2 t} = \frac{\sec^2 t}{\csc^2 t} = \frac{\frac{1}{\cos^2 t}}{\frac{1}{\sin^2 t}} = \left(\frac{1}{\cos^2 t}\right)\left(\frac{\sin^2 t}{1}\right) = \frac{\sin^2 t}{\cos^2 t} = \tan^2 t$.

[11] $\dfrac{\csc^2\theta - \cot^2\theta}{\cos^2\theta - 1} = \dfrac{(1+\cot^2\theta) - \cot^2\theta}{-\sin^2\theta} = -\dfrac{1}{\sin^2\theta} = -\csc^2\theta.$

[12] $\dfrac{\cos t\,(1+\cot^2 t)}{\cot t} = \dfrac{\cos t\,(\csc^2 t)}{\dfrac{\cos t}{\sin t}} = \dfrac{\cos t\left(\dfrac{1}{\sin^2 t}\right)}{\dfrac{\cos t}{\sin t}} = \dfrac{\cos t}{\sin^2 t}\cdot\dfrac{\sin t}{\cos t} = \dfrac{1}{\sin t} = \csc t.$

[13] $\cos t\,\csc t = \cos t\left(\dfrac{1}{\sin t}\right) = \cot t.$

[14] $\cos^2 x - \sin^2 x = \left(1 - \sin^2 x\right) - \sin^2 x = 1 - 2\sin^2 x.$

[15] $\dfrac{\sin\theta + \cos\theta}{\cos\theta} = \dfrac{\sin\theta}{\cos\theta} + \dfrac{\cos\theta}{\cos\theta} = \tan\theta + 1 = 1 + \tan\theta.$

[16] $\dfrac{\sec v}{\cot v} = \dfrac{\dfrac{1}{\cos v}}{\dfrac{\cos v}{\sin v}} = \dfrac{1}{\cos v}\cdot\dfrac{\sin v}{\cos v} = \dfrac{\sin v}{\cos^2 v} = \dfrac{\sin v}{1 - \sin^2 v}.$

[17] $\dfrac{\sin z}{\csc z} + \dfrac{\cos z}{\sec z} = \dfrac{\sin z}{\dfrac{1}{\sin z}} + \dfrac{\cos z}{\dfrac{1}{\cos z}} = \sin^2 z + \cos^2 z = 1.$

[18] $(\cos\theta - \sin\theta)^2 = \cos^2\theta - 2\sin\theta\cos\theta + \sin^2\theta = 1 - 2\sin\theta\cos\theta.$

[19] $\dfrac{1}{\tan x + \cot x} = \dfrac{1}{\dfrac{\sin x}{\cos x} + \dfrac{\cos x}{\sin x}}\cdot\dfrac{\cos x\sin x}{\cos x\sin x} = \dfrac{\cos x\sin x}{\sin^2 x + \cos^2 x} = \cos x\sin x = \sin x\cos x.$

[20] $\tan x + \cot x = \dfrac{\sin x}{\cos x} + \dfrac{\cos x}{\sin x} = \dfrac{\sin^2 x + \cos^2 x}{\cos x\sin x} = \dfrac{1}{\cos x\sin x} = \sec x\csc x = \csc x\sec x.$

[21] $\sec u - \cos u = \dfrac{1}{\cos u} - \cos u = \dfrac{1 - \cos^2 u}{\cos u} = \dfrac{\sin^2 u}{\cos u} = \dfrac{\sin u}{\cos u}\cdot\sin u = \sin u\tan u.$

[22] $\left(1 - \sin^2 t\right)\sec t = \cos^2 t\cdot\sec t = \cos^2 t\left(\dfrac{1}{\cos t}\right) = \cos t.$

[23] $\dfrac{1 - \cos^2 v}{[1 + \sin(-v)](1 + \sin v)} = \dfrac{1 - \cos^2 v}{(1 - \sin v)(1 + \sin v)} = \dfrac{1 - \cos^2 v}{1 - \sin^2 v} = \dfrac{\sin^2 v}{\cos^2 v} = \tan^2 v.$

[24] $\dfrac{\csc\theta}{\cot\theta + \tan\theta} = \dfrac{\dfrac{1}{\sin\theta}}{\dfrac{\cos\theta}{\sin\theta} + \dfrac{\sin\theta}{\cos\theta}}\cdot\dfrac{\sin\theta\cos\theta}{\sin\theta\cos\theta} = \dfrac{\cos\theta}{\cos^2\theta + \sin^2\theta} = \cos\theta.$

[25] $\dfrac{\sec x - \cos x}{\sec x + \cos x} = \dfrac{\dfrac{1}{\cos x} - \cos x}{\dfrac{1}{\cos x} + \cos x}\cdot\dfrac{\cos x}{\cos x} = \dfrac{1 - \cos^2 x}{1 + \cos^2 x} = \dfrac{\sin^2 x}{1 + \cos^2 x}.$

[26] $\dfrac{1 + \tan^2 x}{\tan^2 x} = \dfrac{1}{\tan^2 x} + \dfrac{\tan^2 x}{\tan^2 x} = \cot^2 x + 1 = \csc^2 x.$

[27] Since $\dfrac{\tan\theta}{\sec\theta + 1} = \dfrac{\dfrac{\sin\theta}{\cos\theta}}{\dfrac{1}{\cos\theta} + 1}\cdot\dfrac{\cos\theta}{\cos\theta} = \dfrac{\sin\theta}{1 + \cos\theta}$, and $\dfrac{1}{\cot\theta + \csc\theta} = \dfrac{1}{\dfrac{\cos\theta}{\sin\theta} + \dfrac{1}{\sin\theta}}\cdot\dfrac{\sin\theta}{\sin\theta} = $

$\dfrac{\sin\theta}{\cos\theta + 1} = \dfrac{\sin\theta}{1 + \cos\theta}$, the identity follows.

[28] $\cos t - \dfrac{\cos t}{1 + \tan(-t)} = \cos t - \dfrac{\cos t}{1 - \tan t} = \cos t - \dfrac{\cos t}{1 - \dfrac{\sin t}{\cos t}} \cdot \dfrac{\cos t}{\cos t} = \cos t - \dfrac{\cos^2 t}{\cos t - \sin t} =$

$\dfrac{\cos^2 t - \cos t \sin t - \cos^2 t}{\cos t - \sin t} = \dfrac{-\cos t \sin t}{\cos t - \sin t} \cdot \dfrac{-1}{-1} = \dfrac{\sin t \cos t}{\sin t - \cos t}.$

[29] $\sec^4 \theta - \tan^4 \theta = \left(\sec^2 \theta - \tan^2 \theta\right)\left(\sec^2 \theta + \tan^2 \theta\right) = (1)\left(\sec^2 \theta + \tan^2 \theta\right) = \left(1 + \tan^2 \theta\right) + \tan^2 \theta =$

$1 + 2\tan^2 \theta.$

[30] $\dfrac{\tan^2 x}{\sec x + 1} = \dfrac{\dfrac{\sin^2 x}{\cos^2 x}}{\dfrac{1}{\cos x} + 1} \cdot \dfrac{\cos^2 x}{\cos^2 x} = \dfrac{\sin^2 x}{\cos x + \cos^2 x} = \dfrac{1 - \cos^2 x}{\cos x(1 + \cos x)} = \dfrac{(1 - \cos x)(1 + \cos x)}{\cos x(1 + \cos x)} =$

$\dfrac{1 - \cos x}{\cos x}.$

[31] $\dfrac{\cos(-w) + \tan w}{\sin w} = \dfrac{\cos w + \tan w}{\sin w} = \dfrac{\cos w}{\sin w} + \dfrac{\tan w}{\sin w} = \cot w + \dfrac{\dfrac{\sin w}{\cos w}}{\sin w} \cdot \dfrac{\cos w}{\cos w} = \cot w + \dfrac{\sin w}{\sin w \cos w} =$

$\cot w + \dfrac{1}{\cos w} = \cot w + \sec w = \cot w + \sec(-w).$

[32] $\dfrac{1}{1 - \cos \theta} + \dfrac{1}{1 + \cos \theta} = \dfrac{(1 + \cos \theta) + (1 - \cos \theta)}{1 - \cos^2 \theta} = \dfrac{2}{\sin^2 \theta} = 2\csc^2 \theta.$

[33] $\left(\tan t + \cot t\right)\left(\sin t + \cos t\right) = \left(\dfrac{\sin t}{\cos t} + \dfrac{\cos t}{\sin t}\right)\left(\sin t + \cos t\right) = \left(\dfrac{\sin^2 t + \cos^2 t}{\cos t \sin t}\right)\left(\sin t + \cos t\right) =$

$\left(\dfrac{1}{\cos t \sin t}\right)\left(\sin t + \cos t\right) = \dfrac{\sin t}{\cos t \sin t} + \dfrac{\cos t}{\cos t \sin t} = \dfrac{1}{\cos t} + \dfrac{1}{\sin t} = \sec t + \csc t.$

[34] Since $\dfrac{1 + \sec x}{\sec x} = \dfrac{1}{\sec x} + \dfrac{\sec x}{\sec x} = \cos x + 1$, and $\dfrac{\sin^2 x}{1 - \cos x} = \dfrac{\sin^2 x}{1 - \cos x} \cdot \dfrac{1 + \cos x}{1 + \cos x} =$

$\dfrac{\sin^2 x(1 + \cos x)}{1 - \cos^2 x} = \dfrac{\sin^2 x(1 + \cos x)}{\sin^2 x} = 1 + \cos x = \cos x + 1$, the identity follows.

[35] $\dfrac{\cos v}{\sin(-v) + 1} = \dfrac{\cos v}{-\sin v + 1} = \dfrac{\cos v}{1 - \sin v} \cdot \dfrac{1 + \sin v}{1 + \sin v} = \dfrac{\cos v(1 + \sin v)}{1 - \sin^2 v} = \dfrac{\cos v(1 + \sin v)}{\cos^2 v} = \dfrac{1 + \sin v}{\cos v}.$

[36] $\left(\tan^2 \theta + 1\right)\left(1 + \cos^2 \theta\right) = \left(\tan^2 \theta + 1\right) + \left(\tan^2 \theta + 1\right)\cos^2 \theta = \tan^2 \theta + 1 + \tan^2 \theta \cos^2 \theta + \cos^2 \theta =$

$\tan^2 \theta + 1 + \left(\dfrac{\sin^2 \theta}{\cos^2 \theta}\right)\cos^2 \theta + \cos^2 \theta = \tan^2 \theta + 1 + \sin^2 \theta + \cos^2 \theta = \tan^2 \theta + 2.$

[37] $\dfrac{\cos \theta \cot \theta}{1 + \sin(-\theta)} - 1 = \dfrac{\cos \theta \left(\dfrac{\cos \theta}{\sin \theta}\right)}{1 - \sin \theta} - 1 = \dfrac{\dfrac{\cos^2 \theta}{\sin \theta} - 1 + \sin \theta}{1 - \sin \theta} \cdot \dfrac{\sin \theta}{\sin \theta} = \dfrac{\cos^2 \theta - \sin \theta + \sin^2 \theta}{\sin \theta(1 - \sin \theta)} =$

$\dfrac{1 - \sin \theta}{\sin \theta(1 - \sin \theta)} = \dfrac{1}{\sin \theta} = \csc \theta.$

[38] $\left(\sec x - \tan x\right)^2 = \left(\dfrac{1}{\cos x} - \dfrac{\sin x}{\cos x}\right)^2 = \dfrac{(1 - \sin x)^2}{\cos^2 x} = \dfrac{(1 - \sin x)^2}{1 - \sin^2 x} = \dfrac{(1 - \sin x)^2}{(1 - \sin x)(1 + \sin x)} = \dfrac{1 - \sin x}{1 + \sin x}.$

[39] $\dfrac{\sin \theta}{1 + \cos \theta} + \dfrac{1 + \cos \theta}{\sin \theta} = \dfrac{(\sin \theta)^2 + (1 + \cos \theta)^2}{\sin \theta(1 + \cos \theta)} = \dfrac{\sin^2 \theta + 1 + 2\cos \theta + \cos^2 \theta}{\sin \theta(1 + \cos \theta)} = \dfrac{2 + 2\cos \theta}{\sin \theta(1 + \cos \theta)} =$

$\dfrac{2(1 + \cos \theta)}{\sin \theta(1 + \cos \theta)} = \dfrac{2}{\sin \theta} = 2\csc \theta.$

[40] $\dfrac{1+\sin t}{1-\sin t} - \dfrac{1-\sin t}{1+\sin t} = \dfrac{(1+\sin t)^2 - (1-\sin t)^2}{(1-\sin t)(1+\sin t)} = \dfrac{(1+2\sin t+\sin^2 t)-1+2\sin t-\sin^2 t}{1-\sin^2 t} =$

$\dfrac{4\sin t}{\cos^2 t} = 4\left(\dfrac{\sin t}{\cos t}\right)\left(\dfrac{1}{\cos t}\right) = 4\tan t\sec t.$

[41] Since $\dfrac{\tan t}{\sec t+1} = \dfrac{\dfrac{\sin t}{\cos t}}{\dfrac{1}{\cos t}+1} \cdot \dfrac{\cos t}{\cos t} = \dfrac{\sin t}{1+\cos t}$, and $\dfrac{1}{\cot t+\csc t} = \dfrac{1}{\dfrac{\cos t}{\sin t}+\dfrac{1}{\sin t}} = \dfrac{1}{\dfrac{\cos t+1}{\sin t}} =$

$\dfrac{\sin t}{\cos t+1} = \dfrac{\sin t}{1+\cos t}$, the identity follows.

[42] Since $(\cot u-\csc u)^2 = \left(\dfrac{\cos u}{\sin u}-\dfrac{1}{\sin u}\right)^2 = \left(\dfrac{\cos u-1}{\sin u}\right)^2 = \dfrac{\cos^2 u-2\cos u+1}{\sin^2 u}$, and $\dfrac{1-\cos u}{1+\cos u} =$

$\dfrac{1-\cos u}{1+\cos u} \cdot \dfrac{1-\cos u}{1-\cos u} = \dfrac{1-2\cos u+\cos^2 u}{1-\cos^2 u} = \dfrac{\cos^2 u-2\cos u+1}{\sin^2 u}$, the identity follows.

[43] $\cos^4\theta + \sin^2\theta = \cos^2\theta(\cos^2\theta) + \sin^2\theta = \cos^2\theta(1-\sin^2\theta) + \sin^2\theta = \cos^2\theta - \cos^2\theta\sin^2\theta +$

$\sin^2\theta = \cos^2\theta - (1-\sin^2\theta)\sin^2\theta + \sin^2\theta = \cos^2\theta - \sin^2\theta + \sin^4\theta + \sin^2\theta = \cos^2\theta + \sin^4\theta.$

[44] By the difference of cubes factoring formula $\dfrac{\sin^3 x-\cos^3 x}{\sin x-\cos x} =$

$\dfrac{(\sin x-\cos x)(\sin^2 x+\sin x\cos x+\cos^2 x)}{\sin x-\cos x} = \sin^2 x+\sin x\cos x+\cos^2 x = 1+\sin x\cos x.$

[45] $\tan^4 m-\sec^4 m = (\tan^2 m-\sec^2 m)(\tan^2 m+\sec^2 m) =$

$[(\sec^2 m-1)-\sec^2 m][(\sec^2 m-1)+\sec^2 m] = (-1)(2\sec^2 m-1) = 1-2\sec^2 m.$

[46] $\dfrac{\sec y}{\sin y}+\cot(-y) = \dfrac{\dfrac{1}{\cos y}}{\sin y}-\cot y = \dfrac{1}{\cos y}\cdot\dfrac{1}{\sin y}-\dfrac{\cos y}{\sin y} = \dfrac{1-\cos^2 y}{\cos y\sin y} = \dfrac{\sin^2 y}{\cos y\sin y} = \dfrac{\sin y}{\cos y} = \tan y.$

[47] $\dfrac{\csc^4 x-1}{\cot^2 x} = \dfrac{(\csc^2 x-1)(\csc^2+1)}{\cot^2 x} = \dfrac{[(1+\cot^2 x)-1][(1+\cot^2 x)+1]}{\cot^2 x} = \dfrac{\cot^2 x(2+\cot^2 x)}{\cot^2 x} =$

$2+\cot^2 x.$

[48] By the sum of the cubes factoring fomula, $\dfrac{\cos^3 t+\sin^3 t}{\cos t+\sin t} =$

$\dfrac{(\cos t+\sin t)(\cos^2 t-\cos t\sin t+\sin^2 t)}{\cos t+\sin t} = \cos^2 t-\cos t\sin t+\sin^2 t = 1-\sin t\cos t.$

[49] $\dfrac{\cos v}{1-\sin v} = \dfrac{\cos v}{1-\sin v}\cdot\dfrac{1+\sin v}{1+\sin v} = \dfrac{\cos v(1+\sin v)}{1-\sin^2 v} = \dfrac{\cos v(1+\sin v)}{\cos^2 v} = \dfrac{1+\sin v}{\cos v}.$

[50] $(\tan x+\cot x)(\sin x\cos x) = \left(\dfrac{\sin x}{\cos x}+\dfrac{\cos x}{\sin x}\right)(\sin x\cos x) =$

$\left(\dfrac{\sin x}{\cos x}\right)(\sin x\cos x) + \left(\dfrac{\cos x}{\sin x}\right)(\sin x\cos x) = \sin^2 x+\cos^2 x = 1.$

[51] $\sin^4 x+2\sin^2 x\cos^2 x+\cos^4 x = (\sin^2 x+\cos^2 x)^2 = (1)^2 = 1.$

[52] $(\sec\theta - \tan\theta)^2 = \left(\dfrac{1}{\cos\theta} - \dfrac{\sin\theta}{\cos\theta}\right)^2 = \left(\dfrac{1 - \sin\theta}{\cos\theta}\right)^2 = \dfrac{(1 - \sin\theta)^2}{\cos^2\theta} = \dfrac{(1 - \sin\theta)^2}{\cos^2\theta} =$

$\dfrac{(1 - \sin\theta)^2}{(1 - \sin\theta)(1 + \sin\theta)} = \dfrac{1 - \sin\theta}{1 + \sin\theta}.$

[53] $\csc^2 t - \csc t \cot t = \csc t \left(\csc t - \cot t\right) = \left(\dfrac{1}{\sin t}\right)\left(\dfrac{1}{\sin t} - \dfrac{\cos t}{\sin t}\right) = \left(\dfrac{1}{\sin t}\right)\left(\dfrac{1 - \cos t}{\sin t}\right) =$

$\dfrac{1 - \cos t}{\sin^2 t} = \dfrac{1 - \cos t}{1 - \cos^2 t} = \dfrac{1 - \cos t}{(1 - \cos t)(1 + \cos t)} = \dfrac{1}{1 + \cos t}.$

[54] $\dfrac{\tan v - \sin v}{\sin^3 v} = \dfrac{\dfrac{\sin v}{\cos v} - \sin v}{\sin^3 v} \cdot \dfrac{\cos v}{\cos v} = \dfrac{\sin v - \sin v \cos v}{\sin^3 v \cos v} = \dfrac{\sin v(1 - \cos v)}{\sin^3 v \cos v} = \dfrac{1 - \cos v}{\sin^2 v \cos v} =$

$\dfrac{1 - \cos v}{(1 - \cos^2 v)\cos v} = \dfrac{1 - \cos v}{(1 - \cos v)(1 + \cos v)\cos v} = \dfrac{1}{(1 + \cos v)\cos v} = \left(\dfrac{1}{\cos v}\right)\left(\dfrac{1}{1 + \cos v}\right) = \dfrac{\sec v}{1 + \cos v}.$

[55] $\dfrac{\cot u - 1}{\cot u + 1} = \dfrac{\dfrac{1}{\tan u} - 1}{\dfrac{1}{\tan u} + 1} \cdot \dfrac{\tan u}{\tan u} = \dfrac{1 - \tan u}{1 + \tan u}.$

[56] $\sec^4 x - \sec^2 x = \sec^2 x\left(\sec^2 x - 1\right) = \left(1 + \tan^2 x\right)\left(\tan^2 x\right) = \tan^2 x + \tan^4 x = \tan^4 x + \tan^2 x.$

[57] $\dfrac{1 + \sec(-t)}{\sin(-t) + \tan(-t)} = \dfrac{1 + \sec t}{-\sin t - \tan t} = -\dfrac{1 + \dfrac{1}{\cos t}}{\sin t + \dfrac{\sin t}{\cos t}} \cdot \dfrac{\cos t}{\cos t} = -\dfrac{\cos t + 1}{\cos t \sin t + \sin t} =$

$-\dfrac{\cos t + 1}{\sin t(\cos t + 1)} = -\dfrac{1}{\sin t} = -\csc t.$

[58] $\dfrac{\cos(-x)}{1 + \sin(-x)} = \dfrac{\cos x}{1 - \sin x} \cdot \dfrac{1 + \sin x}{1 + \sin x} = \dfrac{\cos x(1 + \sin x)}{1 - \sin^2 x} = \dfrac{\cos x(1 + \sin x)}{\cos^2 x} = \dfrac{1 + \sin x}{\cos x} =$

$\dfrac{1}{\cos x} + \dfrac{\sin x}{\cos x} = \sec x + \tan x.$

[59] $\sqrt{\dfrac{1 - \sin\theta}{1 + \sin\theta}} = \sqrt{\dfrac{1 - \sin\theta}{1 + \sin\theta} \cdot \dfrac{1 - \sin\theta}{1 - \sin\theta}} = \sqrt{\dfrac{(1 - \sin\theta)^2}{1 - \sin^2\theta}} = \sqrt{\dfrac{(1 - \sin\theta)^2}{\cos^2\theta}} = \dfrac{|1 - \sin\theta|}{|\cos\theta|}.$

[60] $\sqrt{\dfrac{1 - \cos t}{1 + \cos t}} = \sqrt{\dfrac{1 - \cos t}{1 + \cos t} \cdot \dfrac{1 - \cos t}{1 - \cos t}} = \sqrt{\dfrac{(1 - \cos t)^2}{1 - \cos^2 t}} = \sqrt{\dfrac{(1 - \cos t)^2}{\sin^2 t}} = \dfrac{|1 - \cos t|}{|\sin t|}.$

[61] $\sec^4 y - 4\tan^2 y = \left(\sec^2 y\right)^2 - 4\tan^2 y = \left(1 + \tan^2 y\right)^2 - 4\tan^2 y = \left(1 + 2\tan^2 y + \tan^4 y\right) - 4\tan^2 y =$

$1 - 2\tan^2 y + \tan^4 y = \left(1 - \tan^2 y\right)^2.$

[62] $(\cot u - \csc u)^2 = \left(\dfrac{\cos u}{\sin u} - \dfrac{1}{\sin u}\right)^2 = \left(\dfrac{\cos u - 1}{\sin u}\right)^2 = \dfrac{(\cos u - 1)^2}{\sin^2 u} = \dfrac{(\cos u - 1)(\cos u - 1)}{1 - \cos^2 u} =$

$\dfrac{(\cos u - 1)(\cos u - 1)}{(1 - \cos u)(1 + \cos u)} = \dfrac{(-1)(\cos u - 1)}{1 + \cos u} = \dfrac{1 - \cos u}{1 + \cos u}.$

[63] $\dfrac{\cos w}{1 - \tan w} + \dfrac{\sin w}{1 - \cot w} = \dfrac{\cos w}{1 - \dfrac{\sin w}{\cos w}} + \dfrac{\sin w}{1 - \dfrac{\cos w}{\sin w}} = \dfrac{\cos w}{1 - \dfrac{\sin w}{\cos w}} \cdot \dfrac{\cos w}{\cos w} + \dfrac{\sin w}{1 - \dfrac{\cos w}{\sin w}} \cdot \dfrac{\sin w}{\sin w} =$

$\dfrac{\cos^2 w}{\cos w - \sin w} + \dfrac{\sin^2 w}{\sin w - \cos w} = \dfrac{\cos^2 w - \sin^2 w}{\cos w - \sin w} = \dfrac{(\cos w - \sin w)(\cos w + \sin w)}{\cos w - \sin w} = \cos w + \sin w.$

[64] $\dfrac{\cos\theta - \sin\theta + 1}{\cos\theta + \sin\theta - 1} = \dfrac{(1 - \sin\theta) + \cos\theta}{-(1 - \sin\theta) + \cos\theta} \cdot \dfrac{-(1 - \sin\theta) - \cos\theta}{-(1 - \sin\theta) - \cos\theta} =$

$\dfrac{-(1 - \sin\theta)^2 - \cos\theta(1 - \sin\theta) - \cos\theta(1 - \sin\theta) - \cos^2\theta}{[-(1 - \sin\theta)]^2 - \cos^2\theta} =$

$\dfrac{-(1 - 2\sin\theta + \sin^2\theta) - 2\cos(1 - \sin\theta) - (1 - \sin^2\theta)}{1 - 2\sin\theta + \sin^2\theta - (1 - \sin^2\theta)} =$

$\dfrac{-1 + 2\sin\theta - \sin^2\theta - 2\cos\theta(1 - \sin\theta) - 1 + \sin^2\theta}{1 - 2\sin\theta + \sin^2\theta - 1 + \sin^2\theta} = \dfrac{-2 + 2\sin\theta - 2\cos\theta(1 - \sin\theta)}{-2\sin\theta + 2\sin^2\theta} =$

$\dfrac{-2(1 - \sin\theta) - 2\cos\theta(1 - \sin\theta)}{-2\sin\theta(1 - \sin\theta)} = \dfrac{(1 - \sin\theta)(-2 - 2\cos\theta)}{-2\sin\theta(1 - \sin\theta)} = \dfrac{-2 - 2\cos\theta}{-2\sin\theta} =$

$\dfrac{-2(1 + \cos\theta)}{-2\sin\theta} = \dfrac{\cos\theta + 1}{\sin\theta}.$

[65] $\dfrac{1 + \sin x + \cos x}{1 + \cos x - \sin x} = \dfrac{(1 + \cos x) + \sin x}{(1 + \cos x) - \sin x} \cdot \dfrac{(1 + \cos x) + \sin x}{(1 + \cos x) + \sin x} =$

$\dfrac{(1 + \cos x)^2 + 2\sin x(1 + \cos x) + \sin^2 x}{(1 + \cos x)^2 - \sin^2 x} = \dfrac{1 + 2\cos x + \cos^2 x + 2\sin x(1 + \cos x) + (1 - \cos^2 x)}{(1 + 2\cos x + \cos^2 x) - (1 - \cos^2 x)} =$

$\dfrac{2 + 2\cos x + 2\sin x(1 + \cos x)}{2\cos x + 2\cos^2 x} = \dfrac{2(1 + \cos x) + 2\sin x(1 + \cos x)}{2\cos x(1 + \cos x)} = \dfrac{(1 + \cos x)(2 + 2\sin x)}{2\cos x(1 + \cos x)} =$

$\dfrac{2 + 2\sin x}{2\cos x} = \dfrac{2(1 + \sin x)}{2\cos x} = \dfrac{1 + \sin x}{\cos x} = \dfrac{1}{\cos x} + \dfrac{\sin x}{\cos x} = \sec x + \tan x.$

[66] $\ln|\sec\theta| = \ln\left|\dfrac{1}{\cos\theta}\right| = \ln|\cos\theta|^{-1} = -\ln|\cos\theta|.$

[67] $\ln|\tan\theta| = \ln\left|\dfrac{\sin\theta}{\cos\theta}\right| = \ln|\sin\theta| - \ln|\cos\theta|.$

[68] $\ln|\sec^2\theta - \tan^2\theta| = \ln\left|(1 + \tan^2\theta) - \tan^2\theta\right| = \ln|1| = 0.$

[69] $\ln|\sin x| + \ln|\csc x| = \ln|\sin x \csc x| = \ln\left|\sin x \cdot \dfrac{1}{\sin x}\right| = \ln|1| = 0.$

[70] $\ln|\tan x| + \ln(1 + \cot^2 x) = \ln|\tan x(1 + \cot^2 x)| = \ln|\tan x + \tan x \cot^2 x| =$

$\ln\left|\tan x + \tan x\left(\dfrac{1}{\tan x}\right)\cot x\right| = \ln|\tan x + \cot x|.$

[71] $-\ln|1 + \cos\theta| = \ln|1 + \cos\theta|^{-1} = \ln\left|\dfrac{1}{1 + \cos\theta}\right| = \ln\left|\dfrac{1}{1 + \cos\theta} \cdot \dfrac{1 - \cos\theta}{1 - \cos\theta}\right| = \ln\left|\dfrac{1 - \cos\theta}{1 - \cos^2\theta}\right| =$

$\ln\left|\dfrac{1 - \cos\theta}{\sin^2\theta}\right| = \ln|1 - \cos\theta| - \ln|\sin^2\theta| = \ln|1 - \cos\theta| - 2\ln|\sin\theta|.$

[72] $10^{\log|\cos t|} = |\cos t|.$

[73] $\ln|\cot x| = \ln\left|\dfrac{1}{\tan x}\right| = \ln|\tan x|^{-1} = -\ln|\tan x|.$

[74] $\log|\csc x + \cot x| + \log|\csc x - \cot x| = \log|(\csc x + \cot x)(\csc x - \cot x)| =$

$\log|\csc^2 x - \cot^2 x| = \log\left|(1 + \cot^2 x) - \cot^2 x\right| = \log|1| = 0.$

[75] $\dfrac{\sin x + \cos y}{\cos x + \sin y} = \dfrac{\sin x + \cos y}{\cos x + \sin y} \cdot \dfrac{\cos x - \sin y}{\cos x - \sin y} = \dfrac{(\sin x + \cos y)(\cos x - \sin y)}{\cos^2 x - \sin^2 y} =$

$\dfrac{(\sin x + \cos y)(\cos x - \sin y)}{(1 - \sin^2 x) - (1 - \cos^2 y)} = \dfrac{(\sin x + \cos y)(\cos x - \sin y)}{\cos^2 y - \sin^2 x} =$

$\dfrac{(\sin x + \cos y)(\cos x - \sin y)}{(\cos y - \sin x)(\cos y + \sin x)} = \dfrac{\cos x - \sin y}{\cos y - \sin x}.$

[76] Since $\dfrac{\tan x + \tan y}{\cot x + \cot y} = \dfrac{\tan x + \tan y}{\dfrac{1}{\tan x} + \dfrac{1}{\tan y}} = \dfrac{\tan x + \tan y}{\dfrac{\tan y + \tan x}{\tan x \tan y}} = (\tan x + \tan y)\left(\dfrac{\tan x \tan y}{\tan x + \tan y}\right) = \tan x \tan y,$

and $\dfrac{\tan x \tan y - 1}{1 - \cot x \cot y} = \dfrac{\tan x \tan y - 1}{1 - \left(\dfrac{1}{\tan x}\right)\left(\dfrac{1}{\tan y}\right)} = \dfrac{\tan x \tan y - 1}{\dfrac{\tan x \tan y - 1}{\tan x \tan y}} = (\tan x \tan y - 1)\left(\dfrac{\tan x \tan y}{\tan x \tan y - 1}\right) =$

$\tan x \tan y$, the identity follows.

[77] $(\sin t + \cos t)^2 - (\sin u + \cos u)^2 = (\sin^2 t + 2 \sin t \cos t + \cos^2 t) -$

$(\sin^2 u + 2 \sin u \cos u + \cos^2 u) = (1 + 2 \sin t \cos t) - (1 + 2 \sin u \cos u) = 2 \sin t \cos t -$

$2 \sin u \cos u = 2 (\sin t \cos t - \sin u \cos u).$

[78] $(\sin t - \sin u)^2 + (\cos t - \cos u)^2 = (\sin^2 t - 2 \sin t \sin u + \sin^2 u) +$

$(\cos^2 t - 2 \cos t \cos u + \cos^2 u) = (\sin^2 t + \cos^2 t) + (\sin^2 u + \cos^2 u) - 2 \sin t \sin u -$

$2 \cos t \cos u = 2 - 2 \sin t \sin u - 2 \cos t \cos u = 2 (1 - \sin t \sin u - \cos t \cos u).$

[79] If $x = \dfrac{7\pi}{6}$, then $\sin \dfrac{7\pi}{6} = -\dfrac{1}{2}$ and $\sqrt{1 - \cos^2 \dfrac{7\pi}{6}} = \dfrac{1}{2}$ and $-\dfrac{1}{2} \neq \dfrac{1}{2}.$

[80] If $x = \dfrac{5\pi}{6}$, then $\sec \dfrac{5\pi}{6} = -\dfrac{2}{\sqrt{3}}$ and $\sqrt{1 + \tan^2 \dfrac{5\pi}{6}} = \dfrac{2}{\sqrt{3}}$ and $-\dfrac{2}{\sqrt{3}} \neq \dfrac{2}{\sqrt{3}}.$

[81] If $x = \dfrac{\pi}{6}$, then $\cot \dfrac{\pi}{6} = \sqrt{3}$ and $\tan \dfrac{\pi}{6} = \dfrac{1}{\sqrt{3}}$ and $\sqrt{3} \neq \dfrac{1}{\sqrt{3}}.$

[82] If $x = \dfrac{\pi}{6}$, then $\sin \dfrac{\pi}{6} = \dfrac{1}{2}$ and $\cos \dfrac{\pi}{6} = \dfrac{\sqrt{3}}{2}$ and $\dfrac{1}{2} \neq \dfrac{\sqrt{3}}{2}.$

[83] If $x = \dfrac{\pi}{4}$, then $\cot \dfrac{\pi}{4} = 1$ and $\tan\left(\cot \dfrac{\pi}{4}\right) = \tan(1) \approx 1.56$ and $1.56 \neq 1.$

[84] If $x = \dfrac{\pi}{3}$, then $\cos \dfrac{\pi}{3} = \dfrac{1}{2}$ and $\sec\left(\cos \dfrac{\pi}{3}\right) = \sec\left(\dfrac{1}{2}\right) \approx 1.14$ and $1.14 \neq 1.$

[85] If $x = \dfrac{\pi}{4}$, then $\left(\sin \dfrac{\pi}{4} + \cos \dfrac{\pi}{4}\right)^2 = \left(\dfrac{\sqrt{2}}{2} + \dfrac{\sqrt{2}}{2}\right)^2 = (\sqrt{2})^2 = 2$ and $2 \neq 1.$

[86] If $x = \dfrac{\pi}{4}$, then $\dfrac{\cos \dfrac{\pi}{4}}{1 + \sin \dfrac{\pi}{4}} = \dfrac{\dfrac{\sqrt{2}}{2}}{1 + \dfrac{\sqrt{2}}{2}} = \dfrac{\sqrt{2}}{2 + \sqrt{2}}$ and $\dfrac{1 + \sin \dfrac{\pi}{4}}{\cos \dfrac{\pi}{4}} = \dfrac{1 + \dfrac{\sqrt{2}}{2}}{\dfrac{\sqrt{2}}{2}} = \dfrac{2 + \sqrt{2}}{\sqrt{2}}$ and

$\dfrac{\sqrt{2}}{2 + \sqrt{2}} \neq \dfrac{2 + \sqrt{2}}{\sqrt{2}}.$

[87] If $x = \dfrac{\pi}{3}$, then $\cos \dfrac{\pi}{3} = \dfrac{1}{2}$ and $\ln\left|\dfrac{1}{\dfrac{1}{2}}\right| = \ln 2 \approx 0.69$ and $\dfrac{1}{\ln\left(\dfrac{1}{2}\right)} \approx -1.44$ and $0.69 \neq -1.44.$

[88]   If $x = \frac{\pi}{3}$, then $\sin^2 \frac{\pi}{3} = \left(\frac{\sqrt{3}}{2}\right)^2 = \frac{3}{4}$ and $\cos^2 \frac{\pi}{3} = \left(\frac{1}{2}\right)^2 = \frac{1}{4}$ and $\ln\left(\frac{3}{4}\right) + \ln\left(\frac{1}{4}\right) \approx -1.67$ and $-1.67 \neq 0$.

[89]   Since $-\frac{\pi}{2} \leq \theta \leq \frac{\pi}{2} \Rightarrow \cos\theta \geq 0 \Rightarrow |\cos\theta| = \cos\theta$ and since $a > 0 \Rightarrow |a| = a$, we have $\sqrt{a^2 - u^2} =$

$\sqrt{a^2 - \left(a^2 \sin^2\theta\right)} = \sqrt{a^2 \left(1 - \sin^2\theta\right)} = \sqrt{a^2 \left(\cos^2\theta\right)} = |a \, \cos\theta| = a \, \cos\theta.$

[90]   Since $-\frac{\pi}{2} < \theta < \frac{\pi}{2} \Rightarrow \sec\theta > 0 \Rightarrow |\sec\theta| = \sec\theta$, we have $\sqrt{4 + u^2} = \sqrt{4 + 4\tan^2\theta} =$

$\sqrt{4 \left(1 + \tan^2\theta\right)} = \sqrt{4 \sec^2\theta} = 2|\sec\theta| = 2\sec\theta.$

[91]   Since $-\frac{\pi}{2} < \theta < \frac{\pi}{2} \Rightarrow \sec\theta > 0 \Rightarrow |\sec\theta| = \sec\theta$ and since $a > 0 \Rightarrow |a| = a$, we have $\dfrac{1}{\sqrt{u^2 + a^2}} =$

$\dfrac{1}{\sqrt{a^2 \tan^2\theta + a^2}} = \dfrac{1}{\sqrt{a^2 \left(\tan^2\theta + 1\right)}} = \dfrac{1}{\sqrt{a^2 \sec^2\theta}} = \dfrac{1}{|a \sec\theta|} = \dfrac{1}{a \sec\theta} = \dfrac{1}{a}\cos\theta.$

[92]   Since $0 < \theta < \frac{\pi}{2} \Rightarrow \tan\theta > 0 \Rightarrow |\tan\theta| = \tan\theta$ and since $a > 0 \Rightarrow |a| = a$, we have $\sqrt{u^2 - a^2} =$

$\sqrt{a^2 \sec^2\theta - a^2} = \sqrt{a^2 \left(\sec^2\theta - 1\right)} = \sqrt{a^2 \tan^2\theta} = |a \tan\theta| = a \tan\theta.$

[93]   Since $0 < \theta < \frac{\pi}{2} \Rightarrow \tan\theta > 0 \Rightarrow |\tan\theta| = \tan\theta$, we have $\dfrac{2}{\sqrt{x^2 - 9}} = \dfrac{2}{\sqrt{9\sec^2\theta - 9}} =$

$\dfrac{2}{\sqrt{9 \left(\sec^2\theta - 1\right)}} = \dfrac{2}{\sqrt{9\tan^2\theta}} = \dfrac{2}{3|\tan\theta|} = \dfrac{2}{3\tan\theta} = \dfrac{2}{3}\cot\theta.$

[94]   Since $0 < \theta < \frac{\pi}{2} \Rightarrow \tan\theta > 0 \Rightarrow |\tan\theta| = \tan\theta$, we have $x^2 \sqrt{x^2 - 25} = 25\sec^2\theta \sqrt{25\sec^2\theta - 25} =$

$25\sec^2\theta \sqrt{25 \left(\sec^2\theta - 1\right)} = 25\sec^2\theta \sqrt{25\tan^2\theta} = 25\sec^2\theta \left(5|\tan\theta|\right) = 125\sec^2\theta \tan\theta.$

[95]   Since $-\frac{\pi}{2} < \theta < \frac{\pi}{2} \Rightarrow \sec\theta > 0 \Rightarrow |\sec\theta| = \sec\theta$, we have $\dfrac{1}{\sqrt{9 + x^2}} = \dfrac{1}{\sqrt{9 + 9\tan^2\theta}} =$

$\dfrac{1}{\sqrt{9 \left(1 + \tan^2\theta\right)}} = \dfrac{1}{\sqrt{9\sec^2\theta}} = \dfrac{1}{3|\sec\theta|} = \dfrac{1}{3\sec\theta} = \dfrac{1}{3}\cos\theta.$

[96]   Since $-\frac{\pi}{2} < \theta < \frac{\pi}{2} \Rightarrow \cos\theta > 0 \Rightarrow |\cos\theta| = \cos\theta$, we have $\dfrac{2}{\sqrt{5 - x^2}} = \dfrac{2}{\sqrt{5 - 5\sin^2\theta}} =$

$\dfrac{2}{\sqrt{5 \left(1 - \sin^2\theta\right)}} = \dfrac{2}{\sqrt{5\cos^2\theta}} = \dfrac{2}{\sqrt{5}|\cos\theta|} = \dfrac{2}{\sqrt{5}\cos\theta}.$

[97]   a) $y = \dfrac{\sin^2 2x}{\cos^2 2x}$ b) $y = \tan^2 2x$

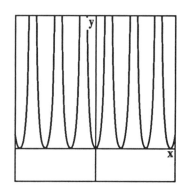

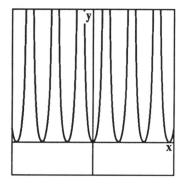

Over the interval $\left(-\dfrac{\pi}{2}, \dfrac{\pi}{2}\right)$ the graphs of $y = \dfrac{\sin^2 2x}{\cos^2 2x}$ and $y = \tan^2 2x$ are identical.

[98] a) $y = \dfrac{\cos^2 2x}{\sin^2 2x}$

b) $y = \dfrac{1}{\tan^2 2x}$

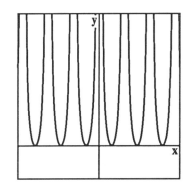

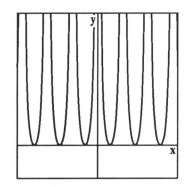

Over the interval $(0, \pi)$ the graphs of $y = \dfrac{\cos^2 2x}{\sin^2 2x}$ and $y = \dfrac{1}{\tan^2 2x}$ are identical.

EXERCISES 6.3

[1] $\sin\left(\dfrac{7\pi}{12}\right) = \sin\left(\dfrac{\pi}{3} + \dfrac{\pi}{4}\right) = \sin\dfrac{\pi}{3}\cos\dfrac{\pi}{4} + \cos\dfrac{\pi}{3}\sin\dfrac{\pi}{4} = \left(\dfrac{\sqrt{3}}{2}\right)\left(\dfrac{\sqrt{2}}{2}\right) + \left(\dfrac{1}{2}\right)\left(\dfrac{\sqrt{2}}{2}\right) = \dfrac{\sqrt{6} + \sqrt{2}}{4}.$

[2] $\cos\left(-15°\right) = \cos\left(30° - 45°\right) = \cos 30° \cos 45° + \sin 30° \sin 45° = \left(\dfrac{\sqrt{3}}{2}\right)\left(\dfrac{\sqrt{2}}{2}\right) + \left(\dfrac{1}{2}\right)\left(\dfrac{\sqrt{2}}{2}\right) = \dfrac{\sqrt{6} + \sqrt{2}}{4}.$

[3] $\tan 15° = \tan\left(45° - 30°\right) = \dfrac{\tan 45° - \tan 30°}{1 + \tan 45° \tan 30°} = \dfrac{1 - \dfrac{1}{\sqrt{3}}}{1 + (1)\left(\dfrac{1}{\sqrt{3}}\right)} = \dfrac{\sqrt{3} - 1}{\sqrt{3} + 1}.$

[4] $\sin 75° = \sin\left(45° + 30°\right) = \sin 45° \cos 30° + \cos 45° \sin 30° = \left(\dfrac{\sqrt{2}}{2}\right)\left(\dfrac{\sqrt{3}}{2}\right) + \left(\dfrac{\sqrt{2}}{2}\right)\left(\dfrac{1}{2}\right) = \dfrac{\sqrt{6} + \sqrt{2}}{4}.$

[5] $\sec\left(\dfrac{5\pi}{12}\right) = \sec\left(\dfrac{\pi}{4} + \dfrac{\pi}{6}\right) = \dfrac{1}{\cos\left(\dfrac{\pi}{4} + \dfrac{\pi}{6}\right)} = \dfrac{1}{\cos\dfrac{\pi}{4}\cos\dfrac{\pi}{6} - \sin\dfrac{\pi}{4}\sin\dfrac{\pi}{6}} = \dfrac{1}{\left(\dfrac{\sqrt{2}}{2}\right)\left(\dfrac{\sqrt{3}}{2}\right) - \left(\dfrac{\sqrt{2}}{2}\right)\left(\dfrac{1}{2}\right)} =$

$\dfrac{1}{\dfrac{\sqrt{6} - \sqrt{2}}{4}} = \dfrac{4}{\sqrt{6} - \sqrt{2}} \cdot \dfrac{\sqrt{6} + \sqrt{2}}{\sqrt{6} + \sqrt{2}} = \dfrac{4\left(\sqrt{6} + \sqrt{2}\right)}{6 - 2} = \dfrac{4\left(\sqrt{6} + \sqrt{2}\right)}{4} = \sqrt{6} + \sqrt{2}.$

[6] $\tan\dfrac{7\pi}{12} = \tan\left(\dfrac{\pi}{3} + \dfrac{\pi}{4}\right) = \dfrac{\tan\dfrac{\pi}{3} + \tan\dfrac{\pi}{4}}{1 - \tan\dfrac{\pi}{3}\tan\dfrac{\pi}{4}} = \dfrac{\sqrt{3} + 1}{1 - (\sqrt{3})(1)} = \dfrac{1 + \sqrt{3}}{1 - \sqrt{3}}.$

[7] $\cos\left(\dfrac{11\pi}{12}\right) = \cos\left(\dfrac{2\pi}{3} + \dfrac{\pi}{4}\right) = \cos\dfrac{2\pi}{3}\cos\dfrac{\pi}{4} - \sin\dfrac{2\pi}{3}\sin\dfrac{\pi}{4} = \left(-\dfrac{1}{2}\right)\left(\dfrac{\sqrt{2}}{2}\right) - \left(\dfrac{\sqrt{3}}{2}\right)\left(\dfrac{\sqrt{2}}{2}\right) = \dfrac{-\sqrt{2} - \sqrt{6}}{2}.$

[8] $\sin 195° = \sin\left(150° + 45°\right) = \sin 150° \cos 45° + \cos 150° \sin 45° = \left(\dfrac{1}{2}\right)\left(\dfrac{\sqrt{2}}{2}\right) + \left(-\dfrac{\sqrt{3}}{2}\right)\left(\dfrac{\sqrt{2}}{2}\right) =$

$\dfrac{\sqrt{2} - \sqrt{6}}{4}.$

[9] $\sin\left(-\dfrac{\pi}{12}\right) = \sin\left(\dfrac{\pi}{6} - \dfrac{\pi}{4}\right) = \sin\dfrac{\pi}{6}\cos\dfrac{\pi}{4} - \cos\dfrac{\pi}{6}\sin\dfrac{\pi}{4} = \left(\dfrac{1}{2}\right)\left(\dfrac{\sqrt{2}}{2}\right) - \left(\dfrac{\sqrt{3}}{2}\right)\left(\dfrac{\sqrt{2}}{2}\right) = \dfrac{\sqrt{2} - \sqrt{6}}{4}.$

[10] $\tan\left(\dfrac{11\,\pi}{12}\right) = \tan\left(\dfrac{2\,\pi}{3} + \dfrac{\pi}{4}\right) = \dfrac{\tan\dfrac{2\,\pi}{3} + \tan\dfrac{\pi}{4}}{1 - \tan\dfrac{2\,\pi}{3}\tan\dfrac{\pi}{4}} = \dfrac{-\sqrt{3} + 1}{1 - \left(-\sqrt{3}\right)(1)} = \dfrac{1 - \sqrt{3}}{1 + \sqrt{3}}.$

[11] $\cos\left(195°\right) = \cos\left(150° + 45°\right) = \cos 150°\cos 45° - \sin 150°\sin 45° = \left(-\dfrac{\sqrt{3}}{2}\right)\left(\dfrac{\sqrt{2}}{2}\right) - \left(\dfrac{1}{2}\right)\left(\dfrac{\sqrt{2}}{2}\right) =$

$\dfrac{-\sqrt{6} - \sqrt{2}}{4}.$

[12] $\cos\left(345°\right) = \cos\left(300° + 45°\right) = \cos 300°\cos 45° - \sin 300°\sin 45° = \left(\dfrac{1}{2}\right)\left(\dfrac{\sqrt{2}}{2}\right) - \left(-\dfrac{\sqrt{3}}{2}\right)\left(\dfrac{\sqrt{2}}{2}\right) =$

$\dfrac{\sqrt{2} + \sqrt{6}}{4}.$

[13] $\cos\dfrac{2\,\pi}{3}\cos\dfrac{\pi}{6} + \sin\dfrac{2\,\pi}{3}\sin\dfrac{\pi}{6} = \cos\left(\dfrac{2\,\pi}{3} - \dfrac{\pi}{6}\right) = \cos\left(\dfrac{4\,\pi - \pi}{6}\right) = \cos\dfrac{\pi}{2} = 0.$

[14] $\sin 40°\cos 10° - \cos 40°\sin 10° = \sin\left(40° - 10°\right) = \sin 30° = \dfrac{1}{2}.$

[15] $\sin 21°\cos 24° + \cos 21°\sin 24° = \sin\left(21° + 24°\right) = \sin 45° = \dfrac{\sqrt{2}}{2}.$

[16] $\cos 220°\cos 80° - \sin 220°\sin 80° = \cos\left(220° + 80°\right) = \cos 300° = \dfrac{1}{2}.$

[17] $\dfrac{\tan 160° - \tan 25°}{1 + \tan 160°\tan 25°} = \tan\left(160° - 25°\right) = \tan\left(135°\right) = -1.$

[18] $\sin 53°\cos 82° + \cos 53°\sin 82° = \sin\left(53° + 82°\right) = \sin 135° = \dfrac{\sqrt{2}}{2}.$

[19] $\sin 173°\cos 53° - \cos 173°\sin 53° = \sin\left(173° - 53°\right) = \sin 120° = \dfrac{\sqrt{3}}{2}.$

[20] $\dfrac{\tan 115° + \tan 110°}{1 - \tan 115°\tan 110°} = \tan\left(115° + 110°\right) = \tan 225° = 1.$

[21] $\cos\dfrac{3\,\pi}{10}\cos\dfrac{\pi}{5} - \sin\dfrac{3\,\pi}{10}\sin\dfrac{\pi}{5} = \cos\left(\dfrac{3\,\pi}{10} + \dfrac{\pi}{5}\right) = \cos\left(\dfrac{3\,\pi + 2\,\pi}{10}\right) = \cos\dfrac{\pi}{2} = 0.$

[22] $\cos 280°\cos 70° + \sin 280°\sin 70° = \cos\left(280° - 70°\right) = \cos 210° = -\dfrac{\sqrt{3}}{2}.$

[23] $\dfrac{\tan 200° + \tan 100°}{1 - \tan 200°\tan 100°} = \tan\left(200° + 100°\right) = \tan 300° = -\sqrt{3}.$

[24] $\dfrac{\tan 260° - \tan 50°}{1 + \tan 260°\tan 50°} = \tan\left(260° - 50°\right) = \tan 210° = \dfrac{1}{\sqrt{3}}.$

[25] $\sin\left(\dfrac{\pi}{2} - v\right)\tan v - \sin v = \cos v\tan v - \sin v = \cos v\left(\dfrac{\sin v}{\cos v}\right) - \sin v = \sin v - \sin v = 0.$

[26] $\tan u\tan\left(\dfrac{\pi}{2} - u\right) - 1 = \tan u\cot u - 1 = \tan u\left(\dfrac{1}{\tan u}\right) - 1 = 1 - 1 = 0.$

[27] $\sin u = \dfrac{1}{4}$ and $u \in \mathrm{QII} \Rightarrow \cos u = -\sqrt{1 - \left(\dfrac{1}{4}\right)^{2}} = -\dfrac{\sqrt{15}}{4}.$

$\cos v = -\dfrac{\sqrt{2}}{3}$ and $v \in \mathrm{QIII} \Rightarrow \sin v = -\sqrt{1 - \left(\dfrac{\sqrt{2}}{3}\right)^{2}} = -\dfrac{\sqrt{7}}{3}.$

a) $\sin\left(u + v\right) = \sin u\cos v + \cos u\sin v = \left(\dfrac{1}{4}\right)\left(-\dfrac{\sqrt{2}}{3}\right) + \left(-\dfrac{\sqrt{15}}{4}\right)\left(-\dfrac{\sqrt{7}}{3}\right) = \dfrac{-\sqrt{2} + \sqrt{105}}{12}.$

b) $\cos\left(u+v\right) = \cos u \cos v - \sin u \sin v = \left(-\frac{\sqrt{15}}{4}\right)\left(-\frac{\sqrt{2}}{3}\right) - \left(\frac{1}{4}\right)\left(-\frac{\sqrt{7}}{3}\right) = \frac{\sqrt{30}+\sqrt{7}}{12}$.

c) $\sin\left(u-v\right) = \sin u \cos v - \cos u \sin v = \left(\frac{1}{4}\right)\left(-\frac{\sqrt{2}}{3}\right) - \left(-\frac{\sqrt{15}}{4}\right)\left(-\frac{\sqrt{7}}{3}\right) = \frac{-\sqrt{2}-\sqrt{105}}{12}$.

d) $\cos\left(u-v\right) = \cos u \cos v + \sin u \sin v = \left(-\frac{\sqrt{15}}{4}\right)\left(-\frac{\sqrt{2}}{3}\right) + \left(\frac{1}{4}\right)\left(-\frac{\sqrt{7}}{3}\right) = \frac{\sqrt{30}-\sqrt{7}}{12}$.

e) $\tan\left(u+v\right) = \dfrac{\sin\left(u+v\right)}{\cos\left(u+v\right)} = \dfrac{\frac{-\sqrt{2}+\sqrt{105}}{12}}{\frac{\sqrt{30}+\sqrt{7}}{12}} = \dfrac{\sqrt{105}-\sqrt{2}}{\sqrt{30}+\sqrt{7}}$.

f) $\tan\left(u-v\right) = \dfrac{\sin\left(u-v\right)}{\cos\left(u-v\right)} = \dfrac{\frac{-\sqrt{2}-\sqrt{105}}{12}}{\frac{\sqrt{30}-\sqrt{7}}{12}} = \dfrac{\sqrt{2}+\sqrt{105}}{\sqrt{7}-\sqrt{30}}$.

g) Since $\sin\left(u+v\right) > 0$ and $\cos\left(u+v\right) > 0$, $u+v \in$ QI.

h) Since $\sin\left(u-v\right) < 0$ and $\cos\left(u-v\right) > 0$, $u-v \in$ QIV.

[28] $\cos u = -\frac{1}{3}$ and $u \in$ QIII $\Rightarrow \sin u = -\sqrt{1-\left(-\frac{1}{3}\right)^2} = -\frac{2\sqrt{2}}{3}$.

$\sin v = \frac{\sqrt{3}}{5}$ and $v \in$ QII $\Rightarrow \cos v = -\sqrt{1-\left(\frac{\sqrt{3}}{5}\right)^2} = -\frac{\sqrt{22}}{5}$.

a) $\sin\left(u+v\right) = \sin u \cos v + \cos u \sin v = \left(-\frac{2\sqrt{2}}{3}\right)\left(-\frac{\sqrt{22}}{5}\right) + \left(-\frac{1}{3}\right)\left(\frac{\sqrt{3}}{5}\right) = \frac{4\sqrt{11}-\sqrt{3}}{15}$.

b) $\cos\left(u+v\right) = \cos u \cos v - \sin u \sin v = \left(-\frac{1}{3}\right)\left(-\frac{\sqrt{22}}{5}\right) - \left(-\frac{2\sqrt{2}}{3}\right)\left(\frac{\sqrt{3}}{5}\right) = \frac{\sqrt{22}+2\sqrt{6}}{15}$.

c) $\sin\left(u-v\right) = \sin u \cos v - \cos u \sin v = \left(-\frac{2\sqrt{2}}{3}\right)\left(-\frac{\sqrt{22}}{5}\right) - \left(-\frac{1}{3}\right)\left(\frac{\sqrt{3}}{5}\right) = \frac{4\sqrt{11}+\sqrt{3}}{15}$.

d) $\cos\left(u-v\right) = \cos u \cos v + \sin u \sin v = \left(-\frac{1}{3}\right)\left(-\frac{\sqrt{22}}{5}\right) + \left(-\frac{2\sqrt{2}}{3}\right)\left(\frac{\sqrt{3}}{5}\right) = \frac{\sqrt{22}-2\sqrt{6}}{15}$.

e) $\tan\left(u+v\right) = \dfrac{\sin\left(u+v\right)}{\cos\left(u+v\right)} = \dfrac{\frac{4\sqrt{11}-\sqrt{3}}{15}}{\frac{\sqrt{22}+2\sqrt{6}}{15}} = \dfrac{4\sqrt{11}-\sqrt{3}}{\sqrt{22}+2\sqrt{6}}$.

f) $\tan\left(u-v\right) = \dfrac{\sin\left(u-v\right)}{\cos\left(u-v\right)} = \dfrac{\frac{4\sqrt{11}+\sqrt{3}}{15}}{\frac{\sqrt{22}-2\sqrt{6}}{15}} = \dfrac{4\sqrt{11}+\sqrt{3}}{\sqrt{22}-2\sqrt{6}}$.

g) Since $\sin\left(u+v\right) > 0$ and $\cos\left(u+v\right) > 0$, $u+v \in$ QI.

h) Since $\sin\left(u-v\right) > 0$ and $\cos\left(u-v\right) < 0$, $u-v \in$ QII.

[29] $\tan u = -\frac{1}{2}$ and $u \in$ QII $\Rightarrow \sec u = -\sqrt{1+\left(-\frac{1}{2}\right)^2} = -\frac{\sqrt{5}}{2} \Rightarrow \cos u = -\frac{2}{\sqrt{5}} \Rightarrow \sin u =$

$\sqrt{1-\left(-\frac{2}{\sqrt{5}}\right)^2} = \frac{1}{\sqrt{5}}$. Similarly, $\cos v = \frac{1}{6}$ and $v \in$ QIV $\Rightarrow \sin v = -\sqrt{1-\left(\frac{1}{6}\right)^2} = -\frac{\sqrt{35}}{6}$.

a) $\sin\left(u+v\right) = \sin u \cos v + \cos u \sin v = \left(\frac{1}{\sqrt{5}}\right)\left(\frac{1}{6}\right) + \left(-\frac{2}{\sqrt{5}}\right)\left(-\frac{\sqrt{35}}{6}\right) = \frac{1+2\sqrt{35}}{6\sqrt{5}}$.

b) $\cos(u+v) = \cos u \cos v - \sin u \sin v = \left(-\frac{2}{\sqrt{5}}\right)\left(\frac{1}{6}\right) - \left(\frac{1}{\sqrt{5}}\right)\left(-\frac{\sqrt{35}}{6}\right) = \frac{-2+\sqrt{35}}{6\sqrt{5}}$.

c) $\sin(u-v) = \sin u \cos v - \cos u \sin v = \left(\frac{1}{\sqrt{5}}\right)\left(\frac{1}{6}\right) - \left(-\frac{2}{\sqrt{5}}\right)\left(-\frac{\sqrt{35}}{6}\right) = \frac{1-2\sqrt{35}}{6\sqrt{5}}$.

d) $\cos(u-v) = \cos u \cos v + \sin u \sin v = \left(-\frac{2}{\sqrt{5}}\right)\left(\frac{1}{6}\right) + \left(\frac{1}{\sqrt{5}}\right)\left(-\frac{\sqrt{35}}{6}\right) = \frac{-2-\sqrt{35}}{6\sqrt{5}}$.

e) $\tan(u+v) = \frac{\sin(u+v)}{\cos(u+v)} = \frac{\frac{1+2\sqrt{35}}{6\sqrt{5}}}{\frac{-2+\sqrt{35}}{6\sqrt{5}}} = \frac{2\sqrt{35}+1}{\sqrt{35}-2}$.

f) $\tan(u-v) = \frac{\sin(u-v)}{\cos(u-v)} = \frac{\frac{1-2\sqrt{35}}{6\sqrt{5}}}{\frac{-2-\sqrt{35}}{6\sqrt{5}}} = \frac{2\sqrt{35}-1}{\sqrt{35}+2}$.

g) Since $\sin(u+v) > 0$ and $\cos(u+v) > 0$, $u+v \in$ QI.

h) Since $\sin(u-v) < 0$ and $\cos(u-v) < 0$, $u-v \in$ QIII.

[30] $\sec u = -\frac{3}{2}$ and $u \in$ QII $\Rightarrow \cos u = -\frac{2}{3} \Rightarrow \sin u = \sqrt{1-\left(-\frac{2}{3}\right)^2} = \frac{\sqrt{5}}{3}$.

$\sin v = -\frac{1}{5}$ and $v \in$ QIII $\Rightarrow \cos v = -\sqrt{1-\left(-\frac{1}{5}\right)^2} = -\frac{2\sqrt{6}}{5}$.

a) $\sin(u+v) = \sin u \cos v + \cos u \sin v = \left(\frac{\sqrt{5}}{3}\right)\left(-\frac{2\sqrt{6}}{5}\right) + \left(-\frac{2}{3}\right)\left(-\frac{1}{5}\right) = \frac{-2\sqrt{30}+2}{15}$.

b) $\cos(u+v) = \cos u \cos v - \sin u \sin v = \left(-\frac{2}{3}\right)\left(-\frac{2\sqrt{6}}{5}\right) - \left(\frac{\sqrt{5}}{3}\right)\left(-\frac{1}{5}\right) = \frac{4\sqrt{6}+\sqrt{5}}{15}$.

c) $\sin(u-v) = \sin u \cos v - \cos u \sin v = \left(\frac{\sqrt{5}}{3}\right)\left(-\frac{2\sqrt{6}}{5}\right) - \left(-\frac{2}{3}\right)\left(-\frac{1}{5}\right) = \frac{-2\sqrt{30}-2}{15}$.

d) $\cos(u-v) = \cos u \cos v + \sin u \sin v = \left(-\frac{2}{3}\right)\left(-\frac{2\sqrt{6}}{5}\right) + \left(\frac{\sqrt{5}}{3}\right)\left(-\frac{1}{5}\right) = \frac{4\sqrt{6}-\sqrt{5}}{15}$.

e) $\tan(u+v) = \frac{\sin(u+v)}{\cos(u+v)} = \frac{\frac{-2\sqrt{30}+2}{15}}{\frac{4\sqrt{6}+\sqrt{5}}{15}} = \frac{2-2\sqrt{30}}{4\sqrt{6}+\sqrt{5}}$.

f) $\tan(u-v) = \frac{\sin(u-v)}{\cos(u-v)} = \frac{\frac{-2\sqrt{30}-2}{15}}{\frac{4\sqrt{6}-\sqrt{5}}{15}} = \frac{2\sqrt{30}+2}{\sqrt{5}-4\sqrt{6}}$.

g) Since $\sin(u+v) < 0$ and $\cos(u+v) > 0$, $u+v \in$ QIV.

h) Since $\sin(u-v) < 0$ and $\cos(u-v) > 0$, $u-v \in$ QIV.

[31] $\csc u = -3$ and $u \in$ QIV $\Rightarrow \sin u = -\frac{1}{3} \Rightarrow \cos u = \sqrt{1-\left(-\frac{1}{3}\right)^2} = \frac{2\sqrt{2}}{3}$. $\cot v = -\frac{2}{3}$ and

$v \in$ QII $\Rightarrow \csc v = \sqrt{1+\left(-\frac{2}{3}\right)^2} = \frac{\sqrt{13}}{3} \Rightarrow \sin v = \frac{3}{\sqrt{13}} \Rightarrow \cos v = -\sqrt{1-\left(\frac{3}{\sqrt{13}}\right)^2} = -\frac{2}{\sqrt{13}}$.

a) $\sin(u+v) = \sin u \cos v + \cos u \sin v = \left(-\dfrac{1}{3}\right)\left(-\dfrac{2}{\sqrt{13}}\right) + \left(\dfrac{2\sqrt{2}}{3}\right)\left(\dfrac{3}{\sqrt{13}}\right) = \dfrac{2+6\sqrt{2}}{3\sqrt{13}}.$

b) $\cos(u+v) = \cos u \cos v - \sin u \sin v = \left(\dfrac{2\sqrt{2}}{3}\right)\left(-\dfrac{2}{\sqrt{13}}\right) - \left(-\dfrac{1}{3}\right)\left(\dfrac{3}{\sqrt{13}}\right) = \dfrac{-4\sqrt{2}+3}{3\sqrt{13}}.$

c) $\sin(u-v) = \sin u \cos v - \cos u \sin v = \left(-\dfrac{1}{3}\right)\left(-\dfrac{2}{\sqrt{13}}\right) - \left(\dfrac{2\sqrt{2}}{3}\right)\left(\dfrac{3}{\sqrt{13}}\right) = \dfrac{2-6\sqrt{2}}{3\sqrt{13}}.$

d) $\cos(u-v) = \cos u \cos v + \sin u \sin v = \left(\dfrac{2\sqrt{2}}{3}\right)\left(-\dfrac{2}{\sqrt{13}}\right) + \left(-\dfrac{1}{3}\right)\left(\dfrac{3}{\sqrt{13}}\right) = \dfrac{-4\sqrt{2}-3}{3\sqrt{13}}.$

e) $\tan(u+v) = \dfrac{\sin(u+v)}{\cos(u+v)} = \dfrac{\dfrac{2+6\sqrt{2}}{3\sqrt{13}}}{\dfrac{-4\sqrt{2}+3}{3\sqrt{13}}} = \dfrac{2+6\sqrt{2}}{3-4\sqrt{2}}.$

f) $\tan(u-v) = \dfrac{\sin(u-v)}{\cos(u-v)} = \dfrac{\dfrac{2-6\sqrt{2}}{3\sqrt{13}}}{\dfrac{-4\sqrt{2}-3}{3\sqrt{13}}} = \dfrac{6\sqrt{2}-2}{4\sqrt{2}+3}.$

g) Since $\sin(u+v) > 0$ and $\cos(u+v) < 0$, $u+v \in$ QII.

h) Since $\sin(u-v) < 0$ and $\cos(u-v) < 0$, $u-v \in$ QIII.

[32] $\cot u = 5$ and $u \in$ QI $\Rightarrow \csc u = \sqrt{1+(5)^2} = \sqrt{26} \Rightarrow \sin u = \dfrac{1}{\sqrt{26}} \Rightarrow \cos u = \sqrt{1-\left(\dfrac{1}{\sqrt{26}}\right)^2} = \dfrac{5}{\sqrt{26}}.$

Similarly, $\sec v = -5$ and $v \in$ QII $\Rightarrow \cos v = -\dfrac{1}{5} \Rightarrow \sin v = \sqrt{1-\left(-\dfrac{1}{5}\right)^2} = \dfrac{2\sqrt{6}}{5}.$

a) $\sin(u+v) = \sin u \cos v + \cos u \sin v = \left(\dfrac{1}{\sqrt{26}}\right)\left(-\dfrac{1}{5}\right) + \left(\dfrac{5}{\sqrt{26}}\right)\left(\dfrac{2\sqrt{6}}{5}\right) = \dfrac{-1+10\sqrt{6}}{5\sqrt{26}}.$

b) $\cos(u+v) = \cos u \cos v - \sin u \sin v = \left(\dfrac{5}{\sqrt{26}}\right)\left(-\dfrac{1}{5}\right) - \left(\dfrac{1}{\sqrt{26}}\right)\left(\dfrac{2\sqrt{6}}{5}\right) = \dfrac{-5-2\sqrt{6}}{5\sqrt{26}}.$

c) $\sin(u-v) = \sin u \cos v - \cos u \sin v = \left(\dfrac{1}{\sqrt{26}}\right)\left(-\dfrac{1}{5}\right) - \left(\dfrac{5}{\sqrt{26}}\right)\left(\dfrac{2\sqrt{6}}{5}\right) = \dfrac{-1-10\sqrt{6}}{5\sqrt{26}}.$

d) $\cos(u-v) = \cos u \cos v + \sin u \sin v = \left(\dfrac{5}{\sqrt{26}}\right)\left(-\dfrac{1}{5}\right) + \left(\dfrac{1}{\sqrt{26}}\right)\left(\dfrac{2\sqrt{6}}{5}\right) = \dfrac{-5+2\sqrt{6}}{5\sqrt{26}}.$

e) $\tan(u+v) = \dfrac{\sin(u+v)}{\cos(u+v)} = \dfrac{\dfrac{-1+10\sqrt{6}}{5\sqrt{26}}}{\dfrac{-5-2\sqrt{6}}{5\sqrt{26}}} = \dfrac{1-10\sqrt{6}}{5+2\sqrt{6}}.$

f) $\tan(u-v) = \dfrac{\sin(u-v)}{\cos(u-v)} = \dfrac{\dfrac{-1-10\sqrt{6}}{5\sqrt{26}}}{\dfrac{-5+2\sqrt{6}}{5\sqrt{26}}} = \dfrac{1+10\sqrt{6}}{5-2\sqrt{6}}.$

g) Since $\sin(u+v) > 0$ and $\cos(u+v) < 0$, $u+v \in$ QII.

h) Since $\sin(u-v) < 0$ and $\cos(u-v) < 0$, $u-v \in$ QIII.

[33] $\cos(270° - \theta) = \cos 270° \cos \theta + \sin 270° \sin \theta = (0)(\cos \theta) + (-1)(\sin \theta) = -\sin \theta.$

[34] $\sec(180° - v) = \dfrac{1}{\cos(180° - v)} = \dfrac{1}{\cos 180° \cos v + \sin 180° \sin v} = \dfrac{1}{(-1)(\cos v) + (0)(\sin v)} = -\sec v.$

[35] $\quad \sin\left(180° + t\right) = \sin 180° \cos t + \cos 180° \sin t = (0)(\cos t) + (-1)(\sin t) = -\sin t.$

[36] $\quad \sin\left(90° - v\right) = \sin 90° \cos v - \cos 90° \sin v = (1)(\cos v) - (0)(\sin v) = \cos v.$

[37] $\quad \sin\left(u - \pi\right) = \sin u \cos \pi - \cos u \sin \pi = (\sin u)(-1) - (\cos u)(0) = -\sin u.$

[38] $\quad \csc\left(u + 270°\right) = \dfrac{1}{\sin\left(u + 270°\right)} = \dfrac{1}{\sin u \cos 270° + \cos u \sin 270°} = \dfrac{1}{(\sin u)(0) + (\cos u)(-1)} =$

$\quad -\sec u.$

[39] $\quad \tan\left(\pi - v\right) = \dfrac{\tan \pi - \tan v}{1 + \tan \pi \tan v} = \dfrac{0 - \tan v}{1 + (0)(\tan v)} = -\tan v.$

[40] $\quad \cos\left(3\pi + v\right) = \cos 3\pi \cos v - \sin 3\pi \sin v = (-1)(\cos v) - (0)(\sin v) = -\cos v.$

[41] $\quad \cos\left(180° + w\right) = \cos 180° \cos w - \sin 180° \sin w = (-1)(\cos w) - (0)(\sin w) = -\cos w.$

[42] $\quad \tan\left(5\pi + v\right) = \dfrac{\tan 5\pi + \tan v}{1 - \tan 5\pi \tan v} = \dfrac{0 + \tan v}{1 - (0)(\tan v)} = \tan v.$

[43] $\quad \tan\left(u + \dfrac{\pi}{6}\right) = \dfrac{\tan u + \tan \dfrac{\pi}{6}}{1 - \tan u \, \tan \dfrac{\pi}{6}} = \dfrac{\tan u + \dfrac{1}{\sqrt{3}}}{1 - (\tan u)\left(\dfrac{1}{\sqrt{3}}\right)} = \dfrac{\sqrt{3} \tan u + 1}{\sqrt{3} - \tan u}.$

[44] $\quad \cot\left(u - \dfrac{\pi}{3}\right) = \dfrac{1}{\tan\left(u - \dfrac{\pi}{6}\right)} = \dfrac{1 + \tan u \tan \dfrac{\pi}{3}}{\tan u - \tan \dfrac{\pi}{3}} = \dfrac{1 + (\tan u)(\sqrt{3})}{\tan u - \sqrt{3}} = \dfrac{1 + \sqrt{3} \tan u}{\tan u - \sqrt{3}}.$

[45] $\quad$ If $u = \sin^{-1}\left(-\dfrac{1}{3}\right)$, then $\sin u = -\dfrac{1}{3}$ and $-\dfrac{\pi}{2} \le u \le \dfrac{\pi}{2} \Rightarrow \cos u = \sqrt{1 - \left(-\dfrac{1}{3}\right)^2} = \dfrac{2\sqrt{2}}{3}.$

$\quad$ If $v = \tan^{-1}\left(\dfrac{1}{4}\right)$, then $\tan v = \dfrac{1}{4}$ and $-\dfrac{\pi}{2} < v < \dfrac{\pi}{2} \Rightarrow v \in \text{QI} \Rightarrow \sec v = \sqrt{1 + \left(\dfrac{1}{4}\right)^2} \Rightarrow$

$\quad \sec v = \dfrac{\sqrt{17}}{4} \Rightarrow \cos v = \dfrac{4}{\sqrt{17}} \Rightarrow \sin v = \sqrt{1 - \left(\dfrac{4}{\sqrt{17}}\right)^2} = \dfrac{1}{\sqrt{17}}.$ So, $\sin\left[\sin^{-1}\left(-\dfrac{1}{3}\right) - \tan^{-1}\left(\dfrac{1}{4}\right)\right] =$

$\quad \sin\left(u - v\right) = \sin u \cos v - \cos u \sin v = \left(-\dfrac{1}{3}\right)\left(\dfrac{4}{\sqrt{17}}\right) - \left(\dfrac{2\sqrt{2}}{3}\right)\left(\dfrac{1}{\sqrt{17}}\right) = \dfrac{-4 - 2\sqrt{2}}{3\sqrt{17}} \approx -0.55.$

[46] $\quad$ If $u = \sin^{-1}\left(\dfrac{2}{3}\right)$, then $\sin u = \dfrac{2}{3}$ and $-\dfrac{\pi}{2} \le u \le \dfrac{\pi}{2} \Rightarrow \cos u = \sqrt{1 - \left(\dfrac{2}{3}\right)^2} = \dfrac{\sqrt{5}}{3}.$ If $v = \cos^{-1}\left(\dfrac{3}{5}\right)$, then

$\quad \cos v = \dfrac{3}{5}$ and $0 \le v \le \pi \Rightarrow \sin v = \sqrt{1 - \left(\dfrac{3}{5}\right)^2} = \dfrac{4}{5} \Rightarrow$ So, $\sin\left[\sin^{-1}\left(\dfrac{2}{3}\right) + \cos^{-1}\left(\dfrac{3}{5}\right)\right] = \sin\left(u + v\right) =$

$\quad \sin u \cos v + \cos u \sin v = \left(\dfrac{2}{3}\right)\left(\dfrac{3}{5}\right) + \left(\dfrac{\sqrt{5}}{3}\right)\left(\dfrac{4}{5}\right) = \dfrac{6 + 4\sqrt{5}}{15} \approx 1.00.$

[47] $\quad$ If $u = \sin^{-1}\left(-\dfrac{2}{3}\right)$, then $\sin u = -\dfrac{2}{3}$ and $-\dfrac{\pi}{2} \le u \le \dfrac{\pi}{2} \Rightarrow \cos u = \sqrt{1 - \left(-\dfrac{2}{3}\right)^2} = \dfrac{\sqrt{5}}{3}.$ If $v = \cos^{-1}\left(\dfrac{3}{4}\right)$,

$\quad$ then $\cos v = \dfrac{3}{4}$ and $0 \le v \le \pi \Rightarrow \sin v = \sqrt{1 - \left(\dfrac{3}{4}\right)^2} = \dfrac{\sqrt{7}}{4}.$ So, $\cos\left[\sin^{-1}\left(-\dfrac{2}{3}\right) + \cos^{-1}\left(\dfrac{3}{4}\right)\right] =$

$\quad \cos\left(u + v\right) = \cos u \, \cos v - \sin u \, \sin v = \left(\dfrac{\sqrt{5}}{3}\right)\left(\dfrac{3}{4}\right) - \left(-\dfrac{2}{3}\right)\left(\dfrac{\sqrt{7}}{4}\right) = \dfrac{3\sqrt{5} + 2\sqrt{7}}{12} \approx 1.00.$

[48] If $u = \sin^{-1}\left(\frac{\sqrt{3}}{4}\right)$, then $\sin u = \frac{\sqrt{3}}{4}$ and $-\frac{\pi}{2} \le u \le \frac{\pi}{2} \Rightarrow \cos u = \sqrt{1 - \left(\frac{\sqrt{3}}{4}\right)^2} = \frac{\sqrt{13}}{4}$. If $v =$

$\cos^{-1}\left(-\frac{\sqrt{3}}{5}\right)$, then $\cos v = -\frac{-\sqrt{3}}{5}$ and $0 \le v \le \pi \Rightarrow \sin v = \sqrt{1 - \left(-\frac{\sqrt{3}}{5}\right)^2} = \frac{\sqrt{22}}{5}$. So,

$\cos\left[\sin^{-1}\left(\frac{\sqrt{3}}{4}\right) - \cos^{-1}\left(-\frac{\sqrt{3}}{5}\right)\right] = \cos(u - v) = \cos u \cos v + \sin u \sin v =$

$\left(\frac{\sqrt{13}}{4}\right)\left(-\frac{\sqrt{3}}{5}\right) + \left(\frac{\sqrt{3}}{4}\right)\left(\frac{\sqrt{22}}{5}\right) = \frac{-\sqrt{39} + \sqrt{66}}{20} \approx 0.09.$

[49] If $u = \sin^{-1}\left(\frac{1}{5}\right)$, then $\sin u = \frac{1}{5}$ and $-\frac{\pi}{2} \le u \le \frac{\pi}{2} \Rightarrow \cos u = \sqrt{1 - \left(\frac{1}{5}\right)^2} = \frac{2\sqrt{6}}{5}$. So, $\tan u = \frac{\sin u}{\cos u} =$

$\frac{1}{2\sqrt{6}}$. If $v = \cos^{-1}\left(\frac{4}{5}\right)$, then $\cos v = \frac{4}{5}$ and $0 \le v \le \pi \Rightarrow \sin v = \sqrt{1 - \left(\frac{4}{5}\right)^2} = \frac{3}{5}$. So, $\tan v =$

$\frac{\sin v}{\cos v} = \frac{3}{4}$. Therefore, $\tan\left[\sin^{-1}\left(\frac{1}{5}\right) + \cos^{-1}\left(\frac{4}{5}\right)\right] = \tan(u + v) = \frac{\tan u + \tan v}{1 - \tan u \tan v} =$

$\dfrac{\frac{1}{2\sqrt{6}} + \frac{3}{4}}{1 - \left(\frac{1}{2\sqrt{6}}\right)\left(\frac{3}{4}\right)} \cdot \left(\frac{8\sqrt{6}}{8\sqrt{6}}\right) = \frac{4 + 6\sqrt{6}}{8\sqrt{6} - 3} \approx 1.13.$

[50] If $u = \cos^{-1}\left(\frac{1}{5}\right)$, then $\cos u = \frac{1}{5}$ and $0 \le u \le \pi \Rightarrow \sin u = \sqrt{1 - \left(\frac{1}{5}\right)^2} = \frac{2\sqrt{6}}{5}$. So, $\tan u = \frac{\sin u}{\cos u} =$

$2\sqrt{6}$. If $v = \tan^{-1}\left(-\frac{2}{3}\right)$, then $\tan v = -\frac{2}{3}$ and $-\frac{\pi}{2} < v < \frac{\pi}{2}$. Thus, $\tan\left[\cos^{-1}\left(\frac{1}{5}\right) - \tan^{-1}\left(-\frac{2}{3}\right)\right] =$

$\tan(u - v) = \frac{\tan u - \tan v}{1 + \tan u \tan v} = \dfrac{2\sqrt{6} - \left(-\frac{2}{3}\right)}{1 + (2\sqrt{6})\left(-\frac{2}{3}\right)} \cdot \left(\frac{3}{3}\right) = \frac{6\sqrt{6} + 2}{3 - 4\sqrt{6}} \approx -2.46.$

[51] $\sin\left(\frac{2\pi}{3} + v\right) = \sin\frac{2\pi}{3}\cos v + \cos\frac{2\pi}{3}\sin v = \left(\frac{\sqrt{3}}{2}\right)(\cos v) + \left(-\frac{1}{2}\right)(\sin v) = \frac{1}{2}\left(\sqrt{3}\cos v - \sin v\right).$

[52] $\sin\left(\frac{5\pi}{6} - v\right) = \sin\frac{5\pi}{6}\cos v - \cos\frac{5\pi}{6}\sin v = \left(\frac{1}{2}\right)(\cos v) - \left(-\frac{\sqrt{3}}{2}\right)(\sin v) = \frac{1}{2}\left(\cos v + \sqrt{3}\sin v\right).$

[53] Since $\cos\left(\frac{\pi}{6} - v\right) = \cos\frac{\pi}{6}\cos v + \sin\frac{\pi}{6}\sin v = \left(\frac{\sqrt{3}}{2}\right)(\cos v) + \left(\frac{1}{2}\right)(\sin v)$ and $\sin\left(\frac{\pi}{3} + v\right) =$

$\sin\frac{\pi}{3}\cos v + \cos\frac{\pi}{3}\sin v = \left(\frac{\sqrt{3}}{2}\right)(\cos v) + \left(\frac{1}{2}\right)(\sin v)$, the identity follows.

[54] $\cos\left(\frac{3\pi}{4} + v\right) = \cos\frac{3\pi}{4}\cos v - \sin\frac{3\pi}{4}\sin v = \left(-\frac{\sqrt{2}}{2}\right)(\cos v) - \left(\frac{\sqrt{2}}{2}\right)(\sin v) = -\frac{\sqrt{2}}{2}\left(\cos v + \sin v\right).$

[55] $\tan\left(u - \frac{3\pi}{2}\right) = \frac{\sin\left(u - \frac{3\pi}{2}\right)}{\cos\left(u - \frac{3\pi}{2}\right)} = \frac{\sin u \cos\frac{3\pi}{2} - \cos u \sin\frac{3\pi}{2}}{\cos u \cos\frac{3\pi}{2} + \sin u \sin\frac{3\pi}{2}} = \frac{(\sin u)(0) - (\cos u)(-1)}{(\cos u)(0) + (\sin u)(-1)} = -\frac{\cos u}{\sin u} =$

$-\cot u.$

[56] $\sin 2v = \sin(v + v) = \sin v \cos v + \cos v \sin v = 2\sin v \cos v.$

245

[57] $\cos 2v = \cos(v+v) = \cos v \cos v - \sin v \sin v = \cos^2 v - \sin^2 v.$

[58] $\cos 2v = \cos(v+v) = \cos v \cos v - \sin v \sin v = \cos^2 v - \sin^2 v = \left(1 - \sin^2 v\right) - \sin^2 v = 1 - 2\sin^2 v.$

[59] $\cos 2v = \cos(v+v) = \cos v \cos v - \sin v \sin v = \cos^2 v - \sin^2 v = \cos^2 v - \left(1 - \cos^2 v\right) = 2\cos^2 v - 1.$

[60]   Using the results of Exercises 56 and 59, we have $\tan 2v = \dfrac{\sin 2v}{\cos 2v} = \dfrac{2\sin v \cos v}{2\cos^2 v - 1} \cdot \dfrac{\dfrac{1}{\cos^2 v}}{\dfrac{1}{\cos^2 v}} =$

$$\frac{\dfrac{2\sin v}{\cos v}}{2 - \dfrac{1}{\cos^2 v}} = \frac{2\tan v}{2 - \sec^2 v}.$$

[61] $\dfrac{\sin(x+h) - \sin x}{h} = \dfrac{\sin x \cos h + \cos x \sin h - \sin x}{h} = \dfrac{\sin x \cos h - \sin x}{h} + \dfrac{\cos x \sin h}{h} =$

$$\sin x \left(\frac{\cos h - 1}{h}\right) + \cos x \left(\frac{\sin h}{h}\right).$$

[62] $\dfrac{\cos(x+h) - \cos x}{h} = \dfrac{\cos x \cos h - \sin x \sin h - \cos x}{h} = \dfrac{\cos x \cos h - \cos x}{h} - \dfrac{\sin x \sin h}{h} =$

$$\cos x \left(\frac{\cos h - 1}{h}\right) - \sin x \left(\frac{\sin h}{h}\right).$$

[63]   Since $\dfrac{\tan u + \tan v}{\tan u + \tan(-v)} = \dfrac{\tan u + \tan v}{\tan u - \tan v}$ and $\dfrac{\sin(u+v)}{\sin(u-v)} = \dfrac{\sin u \cos v + \cos u \sin v}{\sin u \cos v - \cos u \sin v} \cdot \dfrac{\dfrac{1}{\cos u \cos v}}{\dfrac{1}{\cos u \cos v}} =$

$$\frac{\dfrac{\sin u}{\cos u} + \dfrac{\sin v}{\cos v}}{\dfrac{\sin u}{\cos u} - \dfrac{\sin v}{\cos v}} = \frac{\tan u + \tan v}{\tan u - \tan v},$$ the identity follows.

[64] $\dfrac{\cos(u+v)}{\cos(u-v)} = \dfrac{\cos u \cos v - \sin u \sin v}{\cos u \cos v + \sin u \sin v} \cdot \dfrac{\dfrac{1}{\sin u \sin v}}{\dfrac{1}{\sin u \sin v}} = \dfrac{\left(\dfrac{\cos u}{\sin u}\right)\left(\dfrac{\cos v}{\sin v}\right) - 1}{\left(\dfrac{\cos u}{\sin u}\right)\left(\dfrac{\cos v}{\sin v}\right) + 1} = \dfrac{\cot u \cot v - 1}{\cot u \cot v + 1}.$

[65] $\dfrac{\sin(u-v)}{\sin(u+v)} = \dfrac{\sin u \cos v - \cos u \sin v}{\sin u \cos v + \cos u \sin v} \cdot \dfrac{\dfrac{1}{\sin u \cos v}}{\dfrac{1}{\sin u \cos v}} = \dfrac{1 - \left(\dfrac{\cos u}{\sin u}\right)\left(\dfrac{\sin v}{\cos v}\right)}{1 + \left(\dfrac{\cos u}{\sin u}\right)\left(\dfrac{\sin v}{\cos v}\right)} = \dfrac{1 - \cot u \tan v}{1 + \cot u \tan v}.$

[66] $\dfrac{\cos(u+v)}{\sin u \cos v} = \dfrac{\cos u \cos v - \sin u \sin v}{\sin u \cos v} = \dfrac{\cos u \cos v}{\sin u \cos v} - \dfrac{\sin u \sin v}{\sin u \cos v} = \dfrac{\cos u}{\sin u} - \dfrac{\sin v}{\cos v} = \cot u - \tan v.$

[67] $\cos(u+v)\cos(u-v) = (\cos u \cos v - \sin u \sin v)(\cos u \cos v + \sin u \sin v) =$

$\cos^2 u \cos^2 v - \sin^2 u \sin^2 v = \cos^2 u \left(1 - \sin^2 v\right) - \left(1 - \cos^2 u\right)\sin^2 v =$

$\cos^2 u - \cos^2 u \sin^2 v - \sin^2 v + \cos^2 u \sin^2 v = \cos^2 u - \sin^2 v.$

[68] To verify that $\tan(u-v) = \dfrac{\tan u - \tan v}{1 + \tan u \tan v}$, we use the fact that $u - v = u + (-v)$, the sum formula for

tangent, and the fact that $\tan(-v) = -\tan v$: $\tan(u-v) = \tan[u+(-v)] = \dfrac{\tan u + \tan(-v)}{1 - \tan u \tan(-v)} =$

$\dfrac{\tan u - \tan v}{1 + \tan u \tan v}.$

[69] a) $y = \sin(\pi - x)$ b) $y = \sin x$

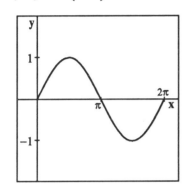 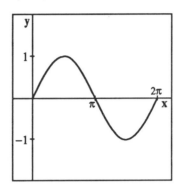

Over the interval $[0, 2\pi)$, the graphs of the functions $y = \sin(\pi - x)$ and $y = \sin x$ are identical.

[70] a) $y = \cos(\pi + x)$ b) $y = -\cos x$

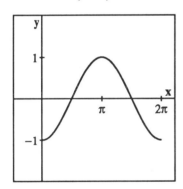 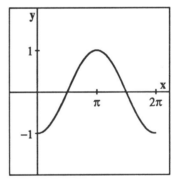

Over the interval $[0, 2\pi)$, the graphs of the functions $y = \cos(\pi + x)$ and $y = -\cos x$ are identical.

[71] a) $y = \cos\left(\dfrac{\pi}{6} - x\right)$ b) $y = \sin\left(\dfrac{\pi}{3} + x\right)$

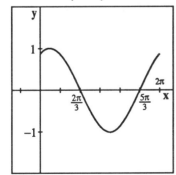 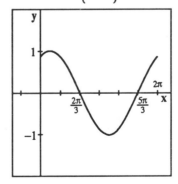

Over the interval $[0, 2\pi)$, the graphs of the functions $y = \cos\left(\dfrac{\pi}{6} - x\right)$ and $y = \sin\left(\dfrac{\pi}{3} + x\right)$ are identical.

[72] a) $y = \sin\left(\frac{\pi}{2} + x\right)$ b) $y = \cos x$

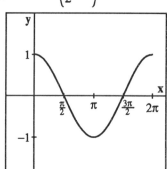   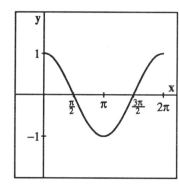

Over the interval $[0, 2\pi)$, the graphs of the functions $y = \sin\left(\frac{\pi}{2} + x\right)$ and $y = \cos x$ are identical.

EXERCISES 6.4

[1] $2 \sin 12 \cos 12 = \sin(2 \cdot 12) = \sin 24$.

[2] $\cos^2 10 - \sin^2 10 = \cos(2 \cdot 10) = \cos 20$.

[3] $2\cos^2 50° - 1 = \cos(2 \cdot 50°) = \cos 100°$.

[4] $\dfrac{2 \tan 80°}{1 - \tan^2 80°} = \tan(2 \cdot 80°) = \tan 160°$.

[5] $\pm\sqrt{\dfrac{1 + \cos 10t}{2}} = \cos\dfrac{10t}{2} = \cos 5t$.

[6] $\pm\sqrt{\dfrac{1 - \cos 6x}{2}} = \sin\dfrac{6x}{2} = \sin 3x$.

[7] $\pm\sqrt{\dfrac{1 - \cos 50x}{1 + \cos 50x}} = \tan\dfrac{50x}{2} = \tan 25x$.

[8] $1 - 2\sin^2 6 = \cos(2 \cdot 6) = \cos 12$.

[9] $\dfrac{1 - \cos 100°}{\sin 100°} = \tan\dfrac{100°}{2} = \tan 50°$.

[10] $\dfrac{\sin 200°}{1 + \cos 200°} = \tan\dfrac{200°}{2} = \tan 100°$.

[11] $\sin v = -\dfrac{\sqrt{3}}{5}$ and $v \in \text{QIV} \Rightarrow \cos v = \sqrt{1 - \left(-\dfrac{\sqrt{3}}{5}\right)^2} = \dfrac{\sqrt{22}}{5}$. So, $\sin 2v = 2 \sin v \cos v =$

$2\left(-\dfrac{\sqrt{3}}{5}\right)\left(\dfrac{\sqrt{22}}{5}\right) = -\dfrac{2\sqrt{66}}{25}$, $\cos 2v = 1 - 2\sin^2 v = 1 - 2\left(-\dfrac{\sqrt{3}}{5}\right)^2 = \dfrac{19}{25}$, and $\tan 2v = \dfrac{\sin 2v}{\cos 2v} =$

$\dfrac{-\dfrac{2\sqrt{66}}{25}}{\dfrac{19}{25}} = -\dfrac{2\sqrt{66}}{19}$. Since $\sin 2v < 0$ and $\cos 2v > 0$, $2v \in \text{QIV}$.

[12] $\cos v = \dfrac{2}{3}$ and $v \in \text{QI} \Rightarrow \sin v = \sqrt{1 - \left(\dfrac{2}{3}\right)^2} = \dfrac{\sqrt{5}}{3}$. So, $\sin 2v = 2 \sin v \cos v = 2\left(\dfrac{\sqrt{5}}{3}\right)\left(\dfrac{2}{3}\right) = \dfrac{4\sqrt{5}}{9}$,

$\cos 2v = 2\cos^2 v - 1 = 2\left(\dfrac{2}{3}\right)^2 - 1 = -\dfrac{1}{9}$, and $\tan 2v = \dfrac{\sin 2v}{\cos 2v} = \dfrac{\dfrac{4\sqrt{5}}{9}}{-\dfrac{1}{9}} = -4\sqrt{5}$. Since $\sin 2v > 0$

and $\cos 2v < 0$, $2v \in \text{QII}$.

[13] $\tan v = -\dfrac{1}{2}$ and $v \in \text{QII} \Rightarrow \sec v = -\sqrt{1 + \left(-\dfrac{1}{2}\right)^2} = -\dfrac{\sqrt{5}}{2} \Rightarrow \cos v = -\dfrac{2}{\sqrt{5}} \Rightarrow \sin v =$

$\sqrt{1 - \left(-\dfrac{2}{\sqrt{5}}\right)^2} = \dfrac{1}{\sqrt{5}}$. So, $\sin 2v = 2 \sin v \cos v = 2\left(\dfrac{1}{\sqrt{5}}\right)\left(-\dfrac{2}{\sqrt{5}}\right) = -\dfrac{4}{5}$, $\cos 2v = \cos^2 v - \sin^2 v =$

$\left(-\dfrac{2}{\sqrt{5}}\right)^2 - \left(\dfrac{1}{\sqrt{5}}\right)^2 = \dfrac{3}{5}$, and $\tan 2v = \dfrac{\sin 2v}{\cos 2v} = \dfrac{-\dfrac{4}{5}}{\dfrac{3}{5}} = -\dfrac{4}{3}$. Since $\sin 2v < 0$ and $\cos 2v > 0$, $2v \in \text{QIV}$.

[14] $\sin v = -\frac{3}{4}$ and $v \in \text{QIII} \Rightarrow \cos v = -\sqrt{1-\left(-\frac{3}{4}\right)^2} = -\frac{\sqrt{7}}{4}$. So, $\sin 2v = 2\sin v \cos v =$

$2\left(-\frac{3}{4}\right)\left(-\frac{\sqrt{7}}{4}\right) = \frac{3\sqrt{7}}{8}$, $\cos 2v = 1 - 2\sin^2 v = 1 - 2\left(-\frac{3}{4}\right)^2 = -\frac{1}{8}$, and $\tan 2v = \frac{\sin 2v}{\cos 2v} =$

$\dfrac{\frac{3\sqrt{7}}{8}}{-\frac{1}{8}} = -3\sqrt{7}$. Since $\sin 2v > 0$ and $\cos 2v < 0$, $2v \in \text{QII}$.

[15] $\sec v = -5$ and $v \in \text{QIII} \Rightarrow \cos v = -\frac{1}{5} \Rightarrow \sin v = -\sqrt{1-\left(-\frac{1}{5}\right)^2} = -\frac{2\sqrt{6}}{5}$. So, $\sin 2v =$

$2\sin v \cos v = 2\left(-\frac{2\sqrt{6}}{5}\right)\left(-\frac{1}{5}\right) = \frac{4\sqrt{6}}{25}$, $\cos 2v = \cos^2 v - \sin^2 v = \left(-\frac{1}{5}\right) - 2\left(-\frac{2\sqrt{6}}{5}\right)^2 = \frac{1}{25} - \frac{24}{25} =$

$-\frac{23}{25}$, and $\tan 2v = \frac{\sin 2v}{\cos 2v} = \dfrac{\frac{4\sqrt{6}}{25}}{-\frac{23}{25}} = -\frac{4\sqrt{6}}{23}$. Since $\sin 2v > 0$ and $\cos 2v < 0$, $2v \in \text{QII}$.

[16] $\csc v = 10$ and $v \in \text{QII} \Rightarrow \sin v = \frac{1}{10} \Rightarrow \cos v = -\sqrt{1-\left(\frac{1}{10}\right)^2} = -\frac{3\sqrt{11}}{10}$. So, $\sin 2v =$

$2\sin v \cos v = 2\left(\frac{1}{10}\right)\left(-\frac{3\sqrt{11}}{10}\right) = -\frac{3\sqrt{11}}{50}$, $\cos 2v = 1 - 2\sin^2 v = 1 - 2\left(\frac{1}{10}\right)^2 = \frac{49}{50}$, and $\tan 2v =$

$\frac{\sin 2v}{\cos 2v} = \dfrac{-\frac{3\sqrt{11}}{50}}{\frac{49}{50}} = -\frac{3\sqrt{11}}{49}$. Since $\sin 2v < 0$ and $\cos 2v > 0$, $2v \in \text{QIV}$.

[17] $\tan v = \frac{4}{3}$ and $\sin v < 0 \Rightarrow v \in \text{QIII} \Rightarrow \sec v = -\sqrt{1+\left(\frac{4}{3}\right)^2} = -\frac{5}{3} \Rightarrow \cos v = -\frac{3}{5} \Rightarrow \sin v =$

$-\sqrt{1-\left(-\frac{3}{5}\right)^2} = -\frac{4}{5}$. So, $\sin 2v = 2\sin v \cos v = 2\left(-\frac{4}{5}\right)\left(-\frac{3}{5}\right) = \frac{24}{25}$, $\cos 2v = \cos^2 v - \sin^2 v =$

$\left(-\frac{3}{5}\right)^2 - \left(-\frac{4}{5}\right)^2 = -\frac{7}{25}$, and $\tan 2v = \frac{\sin 2v}{\cos 2v} = \dfrac{\frac{24}{25}}{-\frac{7}{25}} = -\frac{24}{7}$. Since $\sin 2v > 0$ and $\cos 2v < 0$,

$2v \in \text{QII}$.

[18] $\cot v = -\frac{3}{5}$ and $\cos v > 0 \Rightarrow v \in \text{QIV} \Rightarrow \csc v = -\sqrt{1+\left(-\frac{3}{5}\right)^2} = -\frac{\sqrt{34}}{5} \Rightarrow \sin v = -\frac{5}{\sqrt{34}} \Rightarrow$

$\cos v = \sqrt{1-\left(-\frac{5}{\sqrt{34}}\right)^2} = \frac{3}{\sqrt{34}}$. So, $\sin 2v = 2\sin v \cos v = 2\left(-\frac{5}{\sqrt{34}}\right)\left(\frac{3}{\sqrt{34}}\right) = -\frac{15}{17}$, $\cos 2v =$

$2\cos^2 v - 1 = 2\left(\frac{3}{\sqrt{34}}\right)^2 - 1 = -\frac{8}{17}$, and $\tan 2v = \frac{\sin 2v}{\cos 2v} = \dfrac{-\frac{15}{17}}{-\frac{8}{17}} = \frac{15}{8}$. Since $\sin 2v < 0$ and

$\cos 2v < 0$, $2v \in \text{QIII}$.

[19]　$\csc v = -6$ and $\tan v > 0 \Rightarrow v \in$ QIII. $\csc v = -6 \Rightarrow \sin v = -\frac{1}{6} \Rightarrow \cos v = -\sqrt{1 - \left(-\frac{1}{6}\right)^2} = -\frac{\sqrt{35}}{6}$.

So, $\sin 2v = 2 \sin v \cos v = 2\left(-\frac{1}{6}\right)\left(-\frac{\sqrt{35}}{6}\right) = \frac{\sqrt{35}}{18}$, $\cos 2v = 1 - 2\sin^2 v = 1 - 2\left(-\frac{1}{6}\right)^2 = \frac{17}{18}$,

and $\tan 2v = \frac{\sin 2v}{\cos 2v} = \frac{\frac{\sqrt{35}}{18}}{\frac{17}{18}} = \frac{\sqrt{35}}{17}$. Since $\sin 2v > 0$ and $\cos 2v > 0$, $2v \in$ QI.

[20]　$\sec v = 9$ and $\tan v < 0 \Rightarrow v \in$ QIV. $\sec v = 9 \Rightarrow \cos v = \frac{1}{9} \Rightarrow \sin v = -\sqrt{1 - \left(\frac{1}{9}\right)^2} = -\frac{4\sqrt{5}}{9}$. So,

$\sin 2v = 2 \sin v \cos v = 2\left(-\frac{4\sqrt{5}}{9}\right)\left(\frac{1}{9}\right) = -\frac{8\sqrt{5}}{81}$, $\cos 2v = 2\cos^2 v - 1 = 2\left(\frac{1}{9}\right)^2 - 1 = -\frac{79}{81}$,

and $\tan 2v = \frac{\sin 2v}{\cos 2v} = \frac{-\frac{8\sqrt{5}}{81}}{-\frac{79}{81}} = \frac{8\sqrt{5}}{79}$. Since $\sin 2v < 0$ and $\cos 2v < 0$, $2v \in$ QIII.

[21]　$\cos \frac{5\pi}{8} = \cos\left[\frac{1}{2}\left(\frac{5\pi}{4}\right)\right] = -\sqrt{\frac{1 + \cos \frac{5\pi}{4}}{2}} = -\sqrt{\frac{1 - \frac{\sqrt{2}}{2}}{2}} = -\sqrt{\frac{2 - \sqrt{2}}{4}} = -\frac{\sqrt{2 - \sqrt{2}}}{2}$.

[22]　$\sin\left(\frac{\pi}{8}\right) = \sin\left[\left(\frac{1}{2}\right)\left(\frac{\pi}{4}\right)\right] = \sqrt{\frac{1 - \cos \frac{\pi}{4}}{2}} = \sqrt{\frac{1 - \frac{\sqrt{2}}{2}}{2}} = \sqrt{\frac{2 - \sqrt{2}}{4}} = \frac{\sqrt{2 - \sqrt{2}}}{2}$.

[23]　In Exercise 21, we found that $\cos \frac{5\pi}{8} = -\frac{\sqrt{2 - \sqrt{2}}}{2}$. So, $\cos \frac{5\pi}{16} = \cos\left[\frac{1}{2}\left(\frac{5\pi}{8}\right)\right] =$

$\sqrt{\frac{1 + \cos \frac{5\pi}{8}}{2}} = \sqrt{\frac{1 + \left(\frac{-\sqrt{2 - \sqrt{2}}}{2}\right)}{2}} = \sqrt{\frac{2 - \sqrt{2 - \sqrt{2}}}{4}} = \frac{\sqrt{2 - \sqrt{2 - \sqrt{2}}}}{2}$.

[24]　$\tan(-22.5°) = \tan\left[\left(\frac{1}{2}\right)(-45°)\right] = \frac{\sin(-45°)}{1 + \cos(-45°)} = \frac{-\frac{\sqrt{2}}{2}}{1 + \frac{\sqrt{2}}{2}} = -\frac{\sqrt{2}}{2 + \sqrt{2}}$.

[25]　$\tan\left(\frac{7\pi}{12}\right) = \tan\left[\frac{1}{2}\left(\frac{7\pi}{6}\right)\right] = -\sqrt{\frac{1 - \cos \frac{7\pi}{6}}{1 + \cos \frac{7\pi}{6}}} = -\sqrt{\frac{1 - \left(-\frac{\sqrt{3}}{2}\right)}{1 + \left(-\frac{\sqrt{3}}{2}\right)}} = -\sqrt{\frac{2 + \sqrt{3}}{2 - \sqrt{3}}}$.

[26]　$\cos\left(\frac{7\pi}{12}\right) = \cos\left[\left(\frac{1}{2}\right)\left(\frac{7\pi}{6}\right)\right] = -\sqrt{\frac{1 + \cos \frac{7\pi}{6}}{2}} = -\sqrt{\frac{1 + \left(-\frac{\sqrt{3}}{2}\right)}{2}} = -\frac{\sqrt{2 - \sqrt{3}}}{2}$.

[27]　$\sin 165° = \sin\left(\frac{330°}{2}\right) = \sqrt{\frac{1 - \cos 330°}{2}} = \sqrt{\frac{1 - \frac{\sqrt{3}}{2}}{2}} = \frac{\sqrt{2 - \sqrt{3}}}{2}$.

[28]　$\cos(-75°) = \cos\left(-\frac{150°}{2}\right) = \sqrt{\frac{1 + \cos(-150°)}{2}} = \sqrt{\frac{1 + \left(-\frac{\sqrt{3}}{2}\right)}{2}} = \frac{\sqrt{2 - \sqrt{3}}}{2}$.

[29] $\tan\left(-67.5°\right) = \tan\left(-\dfrac{135°}{2}\right) = -\sqrt{\dfrac{1 - \cos\left(-135°\right)}{1 + \cos\left(-135°\right)}} = -\sqrt{\dfrac{1 - \cos 135°}{1 + \cos 135°}} = -\sqrt{\dfrac{1 + \frac{\sqrt{2}}{2}}{1 - \frac{\sqrt{2}}{2}}} =$

$-\sqrt{\dfrac{2 + \sqrt{2}}{2 - \sqrt{2}}}$.

[30] $\sin\dfrac{5\pi}{8} = \sin\left[\left(\dfrac{1}{2}\right)\left(\dfrac{5\pi}{4}\right)\right] = \sqrt{\dfrac{1 - \cos\frac{5\pi}{4}}{2}} = \sqrt{\dfrac{1 - \left(-\frac{\sqrt{2}}{2}\right)}{2}} = \dfrac{\sqrt{2 + \sqrt{2}}}{2}$.

[31] $\cos\left(\dfrac{7\pi}{8}\right) = \cos\left[\left(\dfrac{1}{2}\right)\left(\dfrac{7\pi}{4}\right)\right] = -\sqrt{\dfrac{1 + \cos\frac{7\pi}{4}}{2}} = -\sqrt{\dfrac{1 + \left(\frac{\sqrt{2}}{2}\right)}{2}} = -\dfrac{\sqrt{2 + \sqrt{2}}}{2}$.

[32] $\tan\left(\dfrac{11\pi}{12}\right) = \tan\left[\left(\dfrac{1}{2}\right)\left(\dfrac{11\pi}{6}\right)\right] = \dfrac{\sin\frac{11\pi}{6}}{1 + \cos\frac{11\pi}{6}} = \dfrac{-\frac{1}{2}}{1 + \frac{\sqrt{3}}{2}} = -\dfrac{1}{2 + \sqrt{3}}$.

[33] $\sin\left(-22.5°\right) = \sin\left(-\dfrac{45°}{2}\right) = -\sqrt{\dfrac{1 - \cos\left(-45°\right)}{2}} = -\sqrt{\dfrac{1 - \left(\frac{\sqrt{2}}{2}\right)}{2}} = -\dfrac{\sqrt{2 - \sqrt{2}}}{2}$.

[34] $\cos\left(112.5°\right) = \cos\left(\dfrac{225°}{2}\right) = -\sqrt{\dfrac{1 + \cos 225°}{2}} = -\sqrt{\dfrac{1 - \frac{\sqrt{2}}{2}}{2}} = -\dfrac{\sqrt{2 - \sqrt{2}}}{2}$.

[35] $\tan 15° = \tan\left(\dfrac{30°}{2}\right) = \sqrt{\dfrac{1 - \cos 30°}{1 + \cos 30°}} = \sqrt{\dfrac{1 - \frac{\sqrt{3}}{2}}{1 + \frac{\sqrt{3}}{2}}} = \sqrt{\dfrac{2 - \sqrt{3}}{2 + \sqrt{3}}}$.

[36] $\sin\left(157.5°\right) = \sin\left(\dfrac{315°}{2}\right) = \sqrt{\dfrac{1 - \cos 315°}{2}} = \sqrt{\dfrac{1 - \frac{\sqrt{2}}{2}}{2}} = \dfrac{\sqrt{2 - \sqrt{2}}}{2}$.

[37] $\sin v = -\dfrac{1}{6}$ and $\dfrac{3\pi}{2} < v < 2\pi \Rightarrow \cos v = \sqrt{1 - \left(-\frac{1}{6}\right)^2} = \dfrac{\sqrt{35}}{6}$. $\dfrac{3\pi}{2} < v < 2\pi \Rightarrow \dfrac{3\pi}{4} < \dfrac{v}{2} < \pi \Rightarrow$

$\dfrac{v}{2} \in$ QII. So, $\sin\dfrac{v}{2} = \sqrt{\dfrac{1 - \cos v}{2}} = \sqrt{\dfrac{1 - \frac{\sqrt{35}}{6}}{2}} = \sqrt{\dfrac{6 - \sqrt{35}}{12}}$, $\cos\dfrac{v}{2} = -\sqrt{\dfrac{1 + \cos v}{2}} =$

$-\sqrt{\dfrac{1 + \frac{\sqrt{35}}{6}}{2}} = -\sqrt{\dfrac{6 + \sqrt{35}}{12}}$, and $\tan\dfrac{v}{2} = \dfrac{\sin\frac{v}{2}}{\cos\frac{v}{2}} = \dfrac{\sqrt{\frac{6 - \sqrt{35}}{12}}}{-\sqrt{\frac{6 + \sqrt{35}}{12}}} = -\dfrac{\sqrt{6 - \sqrt{35}}}{\sqrt{6 + \sqrt{35}}} =$

$-\sqrt{\dfrac{6 - \sqrt{35}}{6 + \sqrt{35}}}$.

[38] $\cos v = \dfrac{5}{6}$ and $0 < v < \dfrac{\pi}{2} \Rightarrow 0 < \dfrac{v}{2} < \dfrac{\pi}{4} \Rightarrow \dfrac{v}{2} \in$ QI. So, $\sin\dfrac{v}{2} = \sqrt{\dfrac{1 - \cos v}{2}} = \sqrt{\dfrac{1 - \frac{5}{6}}{2}} = \sqrt{\dfrac{1}{12}} =$

$\dfrac{1}{2\sqrt{3}}$, $\cos\dfrac{v}{2} = \sqrt{\dfrac{1 + \cos v}{2}} = \sqrt{\dfrac{1 + \frac{5}{6}}{2}} = \sqrt{\dfrac{11}{12}} = \dfrac{\sqrt{11}}{2\sqrt{3}}$, and $\tan\dfrac{v}{2} = \dfrac{\sin\frac{v}{2}}{\cos\frac{v}{2}} = \dfrac{\frac{1}{2\sqrt{3}}}{\frac{\sqrt{11}}{2\sqrt{3}}} = \dfrac{1}{\sqrt{11}}$.

[39] $\cos v = -\frac{3}{8}$ and $\frac{\pi}{2} < v < \pi \Rightarrow \frac{\pi}{4} < \frac{v}{2} < \frac{\pi}{2} \Rightarrow \frac{v}{2} \in$ QI. So, $\sin\frac{v}{2} = \sqrt{\frac{1-\cos v}{2}} = \sqrt{\frac{1-\left(-\frac{3}{8}\right)}{2}} =$

$\sqrt{\frac{11}{16}} = \frac{\sqrt{11}}{4}$, $\cos\frac{v}{2} = \sqrt{\frac{1+\cos v}{2}} = \sqrt{\frac{1+\left(-\frac{3}{8}\right)}{2}} = \sqrt{\frac{5}{16}} = \frac{\sqrt{5}}{4}$, and $\tan\frac{v}{2} = \frac{\sin\frac{v}{2}}{\cos\frac{v}{2}} = \frac{\frac{\sqrt{11}}{4}}{\frac{\sqrt{5}}{4}} =$

$\frac{\sqrt{11}}{\sqrt{5}} = \frac{\sqrt{55}}{5}$.

[40] $\sin v = \frac{5}{7}$ and $\frac{\pi}{2} < v < \pi \Rightarrow \cos v = -\sqrt{1-\left(\frac{5}{7}\right)^2} = -\frac{2\sqrt{6}}{7}$. $\frac{\pi}{2} < v < \pi \Rightarrow \frac{\pi}{4} < \frac{v}{2} < \frac{\pi}{2} \Rightarrow \frac{v}{2} \in$ QI. So,

$\sin\frac{v}{2} = \sqrt{\frac{1-\cos v}{2}} = \sqrt{\frac{1-\left(-\frac{2\sqrt{6}}{7}\right)}{2}} = \sqrt{\frac{7+2\sqrt{6}}{14}}$, $\cos\frac{v}{2} = \sqrt{\frac{1+\cos v}{2}} =$

$\sqrt{\frac{1+\left(-\frac{2\sqrt{6}}{7}\right)}{2}} = \sqrt{\frac{7-2\sqrt{6}}{14}}$, and $\tan\frac{v}{2} = \frac{\sin\frac{v}{2}}{\cos\frac{v}{2}} = \frac{\sqrt{\frac{7+2\sqrt{6}}{14}}}{\sqrt{\frac{7-2\sqrt{6}}{14}}} = \frac{\sqrt{7+2\sqrt{6}}}{\sqrt{7-2\sqrt{6}}} = \sqrt{\frac{7+2\sqrt{6}}{7-2\sqrt{6}}}$.

[41] $\tan v = 3$ and $\pi < v < \frac{3\pi}{2} \Rightarrow \sec v = -\sqrt{1+(3)^2} = -\sqrt{10} \Rightarrow \cos v = -\frac{1}{\sqrt{10}}$. $\pi < v < \frac{3\pi}{2} \Rightarrow$

$\frac{\pi}{2} < \frac{v}{2} < \frac{3\pi}{4} \Rightarrow \frac{v}{2} \in$ QII. So, $\sin\frac{v}{2} = \sqrt{\frac{1-\cos v}{2}} = \sqrt{\frac{1-\left(-\frac{1}{\sqrt{10}}\right)}{2}} = \sqrt{\frac{\sqrt{10}+1}{2\sqrt{10}}}$, $\cos\frac{v}{2} =$

$-\sqrt{\frac{1+\cos v}{2}} = -\sqrt{\frac{1+\left(-\frac{1}{\sqrt{10}}\right)}{2}} = -\sqrt{\frac{\sqrt{10}-\sqrt{10}-1}{2\sqrt{10}}}$, and $\tan\frac{v}{2} = \frac{\sin\frac{v}{2}}{\cos\frac{v}{2}} = \frac{\sqrt{\frac{\sqrt{10}+1}{2\sqrt{10}}}}{-\sqrt{\frac{\sqrt{10}-1}{2\sqrt{10}}}} =$

$-\frac{\sqrt{\sqrt{10}+1}}{\sqrt{\sqrt{10}-1}} = -\sqrt{\frac{\sqrt{10}+1}{\sqrt{10}-1}}$.

[42] $\tan v = 4$ and $\pi < v < \frac{3\pi}{2} \Rightarrow \sec v = -\sqrt{1+(4)^2} = -\sqrt{17} \Rightarrow \cos v = -\frac{1}{\sqrt{17}}$. $\pi < v < \frac{3\pi}{2} \Rightarrow$

$\frac{\pi}{2} < \frac{v}{2} < \frac{3\pi}{4} \Rightarrow \frac{v}{2} \in$ QII. So, $\sin\frac{v}{2} = \sqrt{\frac{1-\cos v}{2}} = \sqrt{\frac{1-\left(-\frac{1}{\sqrt{17}}\right)}{2}} = \sqrt{\frac{\sqrt{17}+1}{2\sqrt{17}}}$, $\cos\frac{v}{2} =$

$-\sqrt{\frac{1+\cos v}{2}} = -\sqrt{\frac{1+\left(-\frac{1}{\sqrt{17}}\right)}{2}} = -\sqrt{\frac{\sqrt{17}-1}{2\sqrt{17}}}$, and $\tan\frac{v}{2} = \frac{\sin\frac{v}{2}}{\cos\frac{v}{2}} = \frac{\sqrt{\frac{\sqrt{17}+1}{2\sqrt{17}}}}{-\sqrt{\frac{\sqrt{17}-1}{2\sqrt{17}}}} =$

$-\frac{\sqrt{\sqrt{17}+1}}{\sqrt{\sqrt{17}-1}} = -\sqrt{\frac{\sqrt{17}+1}{\sqrt{17}-1}}$.

[43] $\sec v = 7$ and $0 < v < \frac{\pi}{2} \Rightarrow \cos v = \frac{1}{7}$. $0 < v < \frac{\pi}{2} \Rightarrow 0 < \frac{v}{2} < \frac{\pi}{4} \Rightarrow \frac{v}{2} \in$ QI. So, $\sin \frac{v}{2} = \sqrt{\frac{1 - \cos v}{2}} =$

$\sqrt{\frac{1 - \left(\frac{1}{7}\right)}{2}} = \sqrt{\frac{6}{14}} = \sqrt{\frac{3}{7}}$, $\cos \frac{v}{2} = \sqrt{\frac{1 + \cos v}{2}} = \sqrt{\frac{1 + \left(\frac{1}{7}\right)}{2}} = \sqrt{\frac{8}{14}} = \frac{2}{\sqrt{7}}$, and $\tan \frac{v}{2} =$

$\dfrac{\sin \frac{v}{2}}{\cos \frac{v}{2}} = \dfrac{\sqrt{\frac{3}{7}}}{\frac{2}{\sqrt{7}}} = \dfrac{\frac{\sqrt{3}}{\sqrt{7}}}{\frac{2}{\sqrt{7}}} = \frac{\sqrt{3}}{2}$.

[44] $\csc v = -3$ and $\frac{3\pi}{2} < v < 2\pi \Rightarrow \sin v = -\frac{1}{3} \Rightarrow \cos v = \sqrt{1 - \left(-\frac{1}{3}\right)^2} = \frac{2\sqrt{2}}{3}$. $\frac{3\pi}{2} < v < 2\pi \Rightarrow$

$\frac{3\pi}{4} < \frac{v}{2} < \pi \Rightarrow \frac{v}{2} \in$ QII. So, $\sin \frac{v}{2} = \sqrt{\frac{1 - \cos v}{2}} = \sqrt{\dfrac{1 - \left(\frac{2\sqrt{2}}{3}\right)}{2}} = \sqrt{\frac{3 - 2\sqrt{2}}{6}}$, $\cos \frac{v}{2} =$

$-\sqrt{\frac{1 + \cos v}{2}} = -\sqrt{\dfrac{1 + \frac{2\sqrt{2}}{3}}{2}} = -\sqrt{\frac{3 + 2\sqrt{2}}{6}}$, and $\tan \frac{v}{2} = \dfrac{\sin \frac{v}{2}}{\cos \frac{v}{2}} = \dfrac{\sqrt{\frac{3 - 2\sqrt{2}}{6}}}{-\sqrt{\frac{3 + 2\sqrt{2}}{6}}} =$

$-\dfrac{\sqrt{3 - 2\sqrt{2}}}{\sqrt{3 + 2\sqrt{2}}} = -\sqrt{\dfrac{3 - 2\sqrt{2}}{3 + 2\sqrt{2}}}$.

[45] If $t = \cos^{-1}(0.2)$, then $\cos t = 0.2 = \frac{1}{5}$ and $0 \le t \le \pi \Rightarrow \sin t = \sqrt{1 - \left(\frac{1}{5}\right)^2} = \frac{2\sqrt{6}}{5}$.

So, $\sin\left[2\cos^{-1}(0.2)\right] = \sin 2t = 2\sin t \cos t = 2\left(\frac{2\sqrt{6}}{5}\right)\left(\frac{1}{5}\right) = \frac{4\sqrt{6}}{25}$.

[46] If $t = \cos^{-1}\left(-\frac{2}{7}\right)$, then $\cos t = -\frac{2}{7}$ and $0 \le t \le \pi \Rightarrow \sin t = \sqrt{1 - \left(-\frac{2}{7}\right)^2} = \frac{3\sqrt{5}}{7} \Rightarrow \tan t = \frac{\sin t}{\cos t} =$

$\dfrac{\frac{3\sqrt{5}}{7}}{-\frac{2}{7}} = -\frac{3\sqrt{5}}{2}$. So, $\tan\left[2\cos^{-1}\left(-\frac{2}{7}\right)\right] = \tan 2t = \dfrac{2\tan t}{1 - \tan^2 t} = \dfrac{2\left(-\frac{3\sqrt{5}}{2}\right)}{1 - \left(-\frac{3\sqrt{5}}{2}\right)^2} = \dfrac{-3\sqrt{5}}{1 - \frac{45}{4}} = \frac{12\sqrt{5}}{41}$.

[47] If $t = \sin^{-1}(-1)$, then $\sin t = -1$ and $-\frac{\pi}{2} \le t \le \frac{\pi}{2} \Rightarrow t = -\frac{\pi}{2}$. So, $\cos\left[2\sin^{-1}(-1)\right] = \cos 2t =$

$\cos\left[2\left(-\frac{\pi}{2}\right)\right] = \cos(-\pi) = -1$.

[48] If $t = \sin^{-1}\left(\frac{1}{4}\right)$, then $\sin t = \frac{1}{4}$ and $-\frac{\pi}{2} \le t \le \frac{\pi}{2} \Rightarrow \cos\left[2\sin^{-1}\left(\frac{1}{4}\right)\right] = \cos 2t = 1 - 2\sin^2 t =$

$1 - 2\left(\frac{1}{4}\right)^2 = \frac{7}{8}$.

[49] If $t = \tan^{-1}\left(-\frac{2}{3}\right)$, then $\tan t = -\frac{2}{3}$ and $-\frac{\pi}{2} < t < \frac{\pi}{2}$. So, $\tan\left[2\tan^{-1}\left(-\frac{2}{3}\right)\right] = \tan 2t = \dfrac{2\tan t}{1 - \tan^2 t} =$

$\dfrac{2\left(-\frac{2}{3}\right)}{1 - \left(-\frac{2}{3}\right)^2} = \dfrac{-\frac{4}{3}}{\frac{5}{9}} = -\frac{12}{5}$.

[50] If $t = \sin^{-1}\left(-\frac{3}{5}\right)$, then $\sin t = -\frac{3}{5}$ and $-\frac{\pi}{2} \le t \le \frac{\pi}{2} \Rightarrow \cos t = \sqrt{1 - \left(-\frac{3}{5}\right)^2} \Rightarrow \cos t = \frac{4}{5}$. So,

$$\sin\left[2\sin^{-1}\left(-\frac{3}{5}\right)\right] = \sin 2t = 2\sin t \cos t = 2\left(-\frac{3}{5}\right)\left(\frac{4}{5}\right) = -\frac{24}{25}.$$

[51] If $t = \sin^{-1}\left(-\frac{6}{7}\right)$, then $\sin t = -\frac{6}{7}$ and $-\frac{\pi}{2} \le t \le \frac{\pi}{2} \Rightarrow \cos t = \sqrt{1 - \left(-\frac{6}{7}\right)^2} \Rightarrow \cos t = \frac{\sqrt{13}}{7}$. So,

$$\sin\left[2\sin^{-1}\left(-\frac{6}{7}\right)\right] = \sin 2t = 2\sin t \cos t = 2\left(-\frac{6}{7}\right)\left(\frac{\sqrt{13}}{7}\right) = -\frac{12\sqrt{13}}{49}.$$

[52] If $t = \sin^{-1}\left(-\frac{5}{6}\right)$, then $\sin t = -\frac{5}{6}$ and $-\frac{\pi}{2} \le t \le \frac{\pi}{2}$. So, $\cos\left[2\sin^{-1}\left(-\frac{5}{6}\right)\right] = \cos 2t = 1 - 2\sin^2 t =$

$$1 - 2\left(-\frac{5}{6}\right)^2 = -\frac{7}{18}.$$

[53] $\tan^2 v = \dfrac{\sin^2 v}{\cos^2 v} = \dfrac{\frac{1-\cos 2v}{2}}{\frac{1+\cos 2v}{2}} = \dfrac{1-\cos 2v}{1+\cos 2v}.$

[54] Since $\cos^2\frac{v}{2} = \dfrac{1 + \cos\left[2\left(\frac{v}{2}\right)\right]}{2} = \dfrac{1+\cos v}{2}$, $\sqrt{\cos^2\frac{v}{2}} = \sqrt{\dfrac{1+\cos v}{2}} \Rightarrow \left|\cos\frac{v}{2}\right| = \sqrt{\dfrac{1+\cos v}{2}} \Rightarrow$

$$\cos\frac{v}{2} = \pm\sqrt{\dfrac{1+\cos v}{2}}.$$

[55] Since $\tan^2\frac{v}{2} = \dfrac{1 - \cos\left[2\left(\frac{v}{2}\right)\right]}{1 + \cos\left[2\left(\frac{v}{2}\right)\right]} = \dfrac{1-\cos v}{1+\cos v}$, $\sqrt{\tan^2\frac{v}{2}} = \sqrt{\dfrac{1-\cos v}{1+\cos v}} \Rightarrow \left|\tan\frac{v}{2}\right| = \sqrt{\dfrac{1-\cos v}{1+\cos v}} \Rightarrow$

$$\tan\frac{v}{2} = \pm\sqrt{\dfrac{1-\cos v}{1+\cos v}}.$$

[56] $\dfrac{1+\tan x}{1-\tan x} = \dfrac{1+\tan x}{1-\tan x} \cdot \dfrac{1-\tan x}{1-\tan x} = \dfrac{1-\tan^2 x}{1 - 2\tan x + \tan^2 x} = \dfrac{1-\tan^2 x}{\left(1+\tan^2 x\right) - 2\tan x} = \dfrac{1-\tan^2 x}{\sec^2 x - 2\tan x} =$

$$\dfrac{1 - \frac{\sin^2 x}{\cos^2 x}}{\frac{1}{\cos^2 x} - 2\frac{\sin x}{\cos x}} \cdot \dfrac{\cos^2 x}{\cos^2 x} = \dfrac{\cos^2 x - \sin^2 x}{1 - 2\sin x \cos x} = \dfrac{\cos 2x}{1 - \sin 2x}.$$

[57] $\cos^4 x = \left(\cos^2 x\right)^2 = \left(\dfrac{1+\cos 2x}{2}\right)^2 = \frac{1}{4}\left(1 + 2\cos 2x + \cos^2 2x\right) = \frac{1}{4}\left[1 + 2\cos 2x + \left(\dfrac{1+\cos 4x}{2}\right)\right] =$

$$\frac{1}{8}\left(2 + 4\cos 2x + 1 + \cos 4x\right) = \frac{1}{8}\left(3 + 4\cos 2x + \cos 4x\right).$$

[58] $\dfrac{\sin 2v}{2\tan v} = \dfrac{2\sin v \cos v}{2\left(\frac{\sin v}{\cos v}\right)} \cdot \dfrac{\cos v}{\cos v} = \dfrac{\sin v \cos^2 v}{\sin v} = \cos^2 v.$

[59] $\cos 3x = \cos\left(2x + x\right) = \cos 2x \cos x - \sin 2x \sin x = \left(2\cos^2 x - 1\right)\cos x - \left(2\sin x \cos x\right)\sin x =$

$2\cos^3 x - \cos x - 2\sin^2 x \cos x = 2\cos^3 x - \cos x - 2\left(1 - \cos^2 x\right)\cos x =$

$2\cos^3 x - \cos x - 2\cos x + 2\cos^3 x = 4\cos^3 x - 3\cos x.$

[60] $\dfrac{1+\tan^2 x}{1-\tan^2 x}=\dfrac{1+\dfrac{\sin^2 x}{\cos^2 x}}{1-\dfrac{\sin^2 x}{\cos^2 x}}\cdot\dfrac{\cos^2 x}{\cos^2 x}=\dfrac{\cos^2 x+\sin^2 x}{\cos^2 x-\sin^2 x}=\dfrac{1}{\cos 2x}=\sec 2x.$

[61] $\left(\sin\dfrac{v}{2}-\cos\dfrac{v}{2}\right)^2-1=\left(\sin^2\dfrac{v}{2}-2\sin\dfrac{v}{2}\cos\dfrac{v}{2}+\cos^2\dfrac{v}{2}\right)-1=\left(1-\sin\left[2\left(\dfrac{v}{2}\right)\right]\right)-1=$

$-\sin v=\sin(-v).$

[62] $\cot 2t=\dfrac{\cos 2t}{\sin 2t}=\dfrac{\cos^2 t-\sin^2 t}{2\sin t\cos t}\cdot\dfrac{\dfrac{1}{\sin^2 t}}{\dfrac{1}{\sin^2 t}}=\dfrac{\dfrac{\cos^2 t}{\sin^2 t}-1}{2\left(\dfrac{\cos t}{\sin t}\right)}=\dfrac{\cot^2 t-1}{2\cot t}.$

[63] $\dfrac{\sin^3 v+\cos^3 v}{\sin v+\cos v}=\dfrac{(\sin v+\cos v)\left(\sin^2 v-\sin v\cos v+\cos^2 v\right)}{\sin v+\cos v}=\sin^2 v+\cos^2 v-\sin v\cos v=$

$(1-\sin v\cos v)\cdot\dfrac{2}{2}=\dfrac{2-2\sin v\cos v}{2}=\dfrac{2-\sin 2v}{2}.$

[64] $\tan\dfrac{v}{2}=\dfrac{1-\cos v}{\sin v}=\dfrac{1}{\sin v}-\dfrac{\cos v}{\sin v}=\csc v-\cot v.$

[65] $\tan v=\tan\left(\dfrac{2v}{2}\right)=\dfrac{\sin 2v}{1+\cos 2v}.$

[66] $\dfrac{2\sin 2v}{\sin 4v}=\dfrac{2\sin 2v}{\sin[2(2v)]}=\dfrac{2\sin 2v}{2\sin 2v\cos 2v}=\dfrac{1}{\cos 2v}=\sec 2v.$

[67] $\cos^4\dfrac{v}{2}-\sin^4\dfrac{v}{2}=\left(\cos^2\dfrac{v}{2}+\sin^2\dfrac{v}{2}\right)\left(\cos^2\dfrac{v}{2}-\sin^2\dfrac{v}{2}\right)=(1)\left(\cos\left[2\left(\dfrac{v}{2}\right)\right]\right)=\cos v.$

[68] $\dfrac{2\tan v}{\sec^2 v}=\dfrac{2\left(\dfrac{\sin v}{\cos v}\right)}{\dfrac{1}{\cos^2 v}}\cdot\dfrac{\cos^2 v}{\cos^2 v}=2\sin v\cos v=\sin 2v.$

[69] $\dfrac{1}{2}\cos v\cot\dfrac{v}{2}\sec^2\dfrac{v}{2}=\dfrac{1}{2}\cos v\left(\dfrac{\cos\dfrac{v}{2}}{\sin\dfrac{v}{2}}\right)\left(\dfrac{1}{\cos^2\dfrac{v}{2}}\right)=\dfrac{\cos v}{2\sin\dfrac{v}{2}\cos\dfrac{v}{2}}=\dfrac{\cos v}{\sin\left[2\left(\dfrac{v}{2}\right)\right]}=\dfrac{\cos v}{\sin v}=\cot v.$

[70] $\dfrac{\cos 3v}{\cos v}-\dfrac{\sin 3v}{\sin v}=\dfrac{\sin v\cos 3v-\cos v\sin 3v}{\sin v\cos v}=\dfrac{\sin(v-3v)}{\sin v\cos v}=\dfrac{\sin(-2v)}{\sin v\cos v}=-\dfrac{\sin 2v}{\sin v\cos v}=$

$-\dfrac{2\sin v\cos v}{\sin v\cos v}=-2.$

[71] $2\sin x\cos x+2\sin x\cos x\tan^2 x=2\sin x\cos x(1+\tan^2 x)=2\sin x\cos x(\sec^2 x)=$

$\sin 2x\sec^2 x=\sec^2 x\sin 2x.$

[72] $\sin^6 x = \left(\sin^2 x\right)^3 = \left(\dfrac{1-\cos 2x}{2}\right)^3 = \dfrac{1}{8}\left(1-\cos 2x\right)\left(1-\cos 2x\right)^2 =$

$\dfrac{1}{8}\left(1-\cos 2x\right)\left(1-2\cos 2x+\cos^2 2x\right) = \dfrac{1}{8}\left(1-\cos 2x\right)\left[1-2\cos 2x+\left(\dfrac{1+\cos 4x}{2}\right)\right] =$

$\dfrac{1}{16}\left(1-\cos 2x\right)\left(2-4\cos 2x+1+\cos 4x\right) = \dfrac{1}{16}\left(1-\cos 2x\right)\left(3-4\cos 2x+\cos 4x\right) =$

$\dfrac{1}{16}\left(3-4\cos 2x+\cos 4x-3\cos 2x+4\cos^2 2x-\cos 2x\cos 4x\right) =$

$\dfrac{1}{16}\left[3-7\cos 2x+\cos 4x+4\left(\dfrac{1+\cos 4x}{2}\right)-\cos 2x\cos 4x\right] =$

$\dfrac{1}{16}\left(3-7\cos 2x+\cos 4x+2+2\cos 4x-\cos 2x\cos 4x\right) =$

$\dfrac{1}{16}\left(5-7\cos 2x+3\cos 4x-\cos 4x\cos 2x\right).$

EXERCISES 6.5

[1] $\cos 2x\cos 3x = \dfrac{1}{2}\left[\cos\left(2x+3x\right)+\cos\left(2x-3x\right)\right] = \dfrac{1}{2}\left[\cos 5x+\cos\left(-x\right)\right] = \dfrac{1}{2}\left(\cos 5x+\cos x\right).$

[2] $\sin 2v\sin 8v = \dfrac{1}{2}\left[\cos\left(2v-8v\right)-\cos\left(2v+8v\right)\right] = \dfrac{1}{2}\left[\cos\left(-6v\right)-\cos 10v\right] =$

$\dfrac{1}{2}\left(\cos 6v-\cos 10v\right).$

[3] $\sin 20°\sin 50° = \dfrac{1}{2}\left[\cos\left(20°-50°\right)-\cos\left(20°+50°\right)\right] = \dfrac{1}{2}\left[\cos\left(-30°\right)-\cos 70°\right] =$

$\dfrac{1}{2}\left(\cos 30°-\cos 70°\right).$

[4] $\cos 15°\cos 10° = \dfrac{1}{2}\left[\cos\left(15°+10°\right)+\cos\left(15°-10°\right)\right] = \dfrac{1}{2}\left(\cos 25°+\cos 5°\right).$

[5] $\sin 3\theta\cos 5\theta = \dfrac{1}{2}\left[\sin\left(3\theta+5\theta\right)+\sin\left(3\theta-5\theta\right)\right] = \dfrac{1}{2}\left[\sin 8\theta+\sin\left(-2\theta\right)\right] = \dfrac{1}{2}\left(\sin 8\theta-\sin 2\theta\right).$

[6] $\cos 6\theta\sin 10\theta = \dfrac{1}{2}\left[\sin\left(6\theta+10\theta\right)-\sin\left(6\theta-10\theta\right)\right] = \dfrac{1}{2}\left[\sin 16\theta-\sin\left(-4\theta\right)\right] =$

$\dfrac{1}{2}\left(\sin 16\theta+\sin 4\theta\right).$

[7] $\cos 10t\sin 8t = \dfrac{1}{2}\left[\sin\left(10t+8t\right)-\sin\left(10t-8t\right)\right] = \dfrac{1}{2}\left(\sin 18t-\sin 2t\right).$

[8] $\sin 12x\cos 6x = \dfrac{1}{2}\left[\sin\left(12x+6x\right)+\sin\left(12x-6x\right)\right] = \dfrac{1}{2}\left(\sin 18x+\sin 6x\right).$

[9] $10\cos 20t\cos 30t = \left(10\right)\left(\dfrac{1}{2}\right)\left[\cos\left(20t+30t\right)+\cos\left(20t-30t\right)\right] = 5\left[\cos 50t+\cos\left(-10t\right)\right] =$

$5\left(\cos 50t+\cos 10t\right).$

[10] $4\sin 5t\sin 9t = \left(4\right)\left(\dfrac{1}{2}\right)\left[\cos\left(5t-9t\right)-\cos\left(5t+9t\right)\right] = 2\left[\cos\left(-4t\right)-\cos 14t\right] =$

$2\left(\cos 4t-\cos 14t\right).$

[11] $\sin 3\theta+\sin 6\theta = 2\sin\dfrac{3\theta+6\theta}{2}\cos\dfrac{3\theta-6\theta}{2} = 2\sin\dfrac{9\theta}{2}\cos\left(-\dfrac{3\theta}{2}\right) = 2\sin\dfrac{9\theta}{2}\cos\dfrac{3\theta}{2}.$

[12] $\sin 5\theta-\sin 10\theta = 2\cos\dfrac{5\theta+10\theta}{2}\sin\dfrac{5\theta-10\theta}{2} = 2\cos\dfrac{15\theta}{2}\sin\left(-\dfrac{5\theta}{2}\right) = -2\cos\dfrac{15\theta}{2}\sin\dfrac{5\theta}{2}.$

[13] $\cos 12° - \cos 62° = -2 \sin \dfrac{12° + 62°}{2} \sin \dfrac{12° - 62°}{2} = -2 \sin 37° \sin(-25°) = 2 \sin 37° \sin 25°.$

[14] $\cos 20° + \cos 38° = 2 \cos \dfrac{20° + 38°}{2} \cos \dfrac{20° - 38°}{2} = 2 \cos 29° \cos(-9°) = 2 \cos 29° \cos 9°.$

[15] $\cos 7x + \cos 8x = 2 \cos \dfrac{7x + 8x}{2} \cos \dfrac{7x - 8x}{2} = 2 \cos \dfrac{15x}{2} \cos\left(-\dfrac{x}{2}\right) = 2 \cos \dfrac{15x}{2} \cos \dfrac{x}{2}.$

[16] $\cos 2t - \cos 5t = -2 \sin \dfrac{2t + 5t}{2} \sin \dfrac{2t - 5t}{2} = -2 \sin \dfrac{7t}{2} \sin\left(-\dfrac{3t}{2}\right) = 2 \sin \dfrac{7t}{2} \sin \dfrac{3t}{2}.$

[17] $\sin 12v - \sin 20v = 2 \cos \dfrac{12v + 20v}{2} \sin \dfrac{12v - 20v}{2} = 2 \cos 16v \sin(-4v) = -2 \cos 16v \sin 4v.$

[18] $\sin 3v + \sin 4v = 2 \sin \dfrac{3v + 4v}{2} \cos \dfrac{3v - 4v}{2} = 2 \sin \dfrac{7v}{2} \cos\left(-\dfrac{v}{2}\right) = 2 \sin \dfrac{7v}{2} \cos \dfrac{v}{2}.$

[19] $\sin 2y + \sin(-3y) = 2 \sin \dfrac{2y - 3y}{2} \cos \dfrac{2y + 3y}{2} = 2 \sin\left(-\dfrac{y}{2}\right) \cos \dfrac{5y}{2} = -2 \sin \dfrac{y}{2} \cos \dfrac{5y}{2}.$

[20] $\cos 6w + \cos(-7w) = 2 \cos \dfrac{6w - 7w}{2} \cos \dfrac{6w + 7w}{2} = 2 \cos\left(-\dfrac{w}{2}\right) \cos \dfrac{13w}{2} = 2 \cos \dfrac{w}{2} \cos \dfrac{13w}{2}.$

[21] $\cos 195° - \cos 105° = -2 \sin \dfrac{195° + 105°}{2} \sin \dfrac{195° - 105°}{2} = -2 \sin 150° \sin 45° =$

$-2\left(\dfrac{1}{2}\right)\left(\dfrac{\sqrt{2}}{2}\right) = -\dfrac{\sqrt{2}}{2}.$

[22] $\cos 195° + \cos 105° = 2 \cos \dfrac{195° + 105°}{2} \cos \dfrac{195° - 105°}{2} = 2 \cos 150° \cos 45° =$

$2\left(-\dfrac{\sqrt{3}}{2}\right)\left(\dfrac{\sqrt{2}}{2}\right) = -\dfrac{\sqrt{6}}{2}.$

[23] $\sin 285° - \sin 195° = 2 \cos \dfrac{285° + 195°}{2} \sin \dfrac{285° - 195°}{2} = 2 \cos 240° \sin 45° =$

$2\left(-\dfrac{1}{2}\right)\left(\dfrac{\sqrt{2}}{2}\right) = -\dfrac{\sqrt{2}}{2}.$

[24] $\sin 285° + \sin 195° = 2 \sin \dfrac{285° + 195°}{2} \cos \dfrac{285° - 195°}{2} = 2 \sin 240° \cos 45° =$

$2\left(-\dfrac{\sqrt{3}}{2}\right)\left(\dfrac{\sqrt{2}}{2}\right) = -\dfrac{\sqrt{6}}{2}.$

[25] $\sin 165° \cos 105° = \dfrac{1}{2}\left[\sin(165° + 105°) + \sin(165° - 105°)\right] = \dfrac{1}{2}\left(\sin 270° + \sin 60°\right) =$

$\dfrac{1}{2}\left(-1 + \dfrac{\sqrt{3}}{2}\right) = \dfrac{1}{2}\left[\dfrac{\sqrt{3} - 2}{2}\right] = \dfrac{\sqrt{3} - 2}{4}.$

[26] $\sin 165° \sin 105° = \dfrac{1}{2}\left[\cos(165° - 105°) - \cos(165° + 105°)\right] = \dfrac{1}{2}\left(\cos 60° - \cos 270°\right) =$

$\dfrac{1}{2}\left(\dfrac{1}{2} - 0\right) = \dfrac{1}{4}.$

[27] $\cos 15° \sin 75° = \dfrac{1}{2}\left[\sin(15° + 75°) - \sin(15° - 75°)\right] = \dfrac{1}{2}\left[\sin 90° - \sin(-60°)\right] =$

$\dfrac{1}{2}\left(\sin 90° + \sin 60°\right) = \dfrac{1}{2}\left(1 + \dfrac{\sqrt{3}}{2}\right) = \dfrac{1}{2}\left(\dfrac{\sqrt{3} + 2}{2}\right) = \dfrac{\sqrt{3} + 2}{4}.$

[28] $\sin 15° \cos 75° = \frac{1}{2}\left[\sin\left(15° + 75°\right) + \sin\left(15° - 75°\right)\right] = \frac{1}{2}\left[\sin 90° + \sin\left(-60°\right)\right] =$

$\frac{1}{2}\left(\sin 90° - \sin 60°\right) = \frac{1}{2}\left(1 - \frac{\sqrt{3}}{2}\right) = \frac{1}{2}\left(\frac{2-\sqrt{3}}{2}\right) = \frac{2-\sqrt{3}}{4}.$

[29] $\cos\frac{5\pi}{12} - \cos\frac{\pi}{12} = -2\sin\left(\frac{\frac{5\pi}{12} + \frac{\pi}{12}}{2}\right)\sin\left(\frac{\frac{5\pi}{12} - \frac{\pi}{12}}{2}\right) = -2\sin\frac{\pi}{4}\sin\frac{\pi}{6} = -2\left(\frac{\sqrt{2}}{2}\right)\left(\frac{1}{2}\right) = -\frac{\sqrt{2}}{2}.$

[30] $\cos 165° - \cos 75° = -2\sin\frac{165° + 75°}{2}\sin\frac{165° - 75°}{2} = -2\sin 120° \sin 45° =$

$-2\left(\frac{\sqrt{3}}{2}\right)\left(\frac{\sqrt{2}}{2}\right) = -\frac{\sqrt{6}}{2}.$

[31] $\cos\frac{\pi}{3}\sin\frac{\pi}{6} = \frac{1}{2}\left[\sin\left(\frac{\pi}{3} + \frac{\pi}{6}\right) - \sin\left(\frac{\pi}{3} - \frac{\pi}{6}\right)\right] = \frac{1}{2}\left(\sin\frac{\pi}{2} - \sin\frac{\pi}{6}\right) = \frac{1}{2}\left(1 - \frac{1}{2}\right) = \frac{1}{4}.$

[32] $\cos\frac{\pi}{3}\cos\frac{\pi}{6} = \frac{1}{2}\left[\cos\left(\frac{\pi}{3} + \frac{\pi}{6}\right) + \cos\left(\frac{\pi}{3} - \frac{\pi}{6}\right)\right] = \frac{1}{2}\left(\cos\frac{\pi}{2} + \cos\frac{\pi}{6}\right) = \frac{1}{2}\left(0 + \frac{\sqrt{3}}{2}\right) = \frac{\sqrt{3}}{4}.$

[33] $\cos\frac{7\pi}{8}\cos\frac{5\pi}{8} = \frac{1}{2}\left[\cos\left(\frac{7\pi}{8} + \frac{5\pi}{8}\right) + \cos\left(\frac{7\pi}{8} - \frac{5\pi}{8}\right)\right] = \frac{1}{2}\left(\cos\frac{3\pi}{2} + \cos\frac{\pi}{4}\right) = \frac{1}{2}\left(0 + \frac{\sqrt{2}}{2}\right) = \frac{\sqrt{2}}{4}.$

[34] $\sin\frac{7\pi}{8}\sin\frac{5\pi}{8} = \frac{1}{2}\left[\cos\left(\frac{7\pi}{8} - \frac{5\pi}{8}\right) - \cos\left(\frac{7\pi}{8} + \frac{5\pi}{8}\right)\right] = \frac{1}{2}\left(\cos\frac{\pi}{4} - \cos\frac{3\pi}{2}\right) = \frac{1}{2}\left(\frac{\sqrt{2}}{2} - 0\right) = \frac{\sqrt{2}}{4}.$

[35] $\cos t - \cos 3t = \cos t - \cos\left(2t + t\right) = \cos t - \left(\cos 2t \cos t - \sin 2t \sin t\right) =$

$\cos t - \cos 2t \cos t + \sin 2t \sin t = \cos t - \cos 2t \cos t + \left(2\sin t \cos t\right)\sin t =$

$\cos t\left(1 - \cos 2t + 2\sin^2 t\right) = \cos t\left[1 - \left(1 - 2\sin^2 t\right) + 2\sin^2 t\right] = \cos t\left(4\sin^2 t\right) = 4\sin^2 t\,\cos t.$

[36] $\dfrac{\sin t + \sin v}{\cos t - \cos v} = \dfrac{2\sin\frac{t+v}{2}\cos\frac{t-v}{2}}{-2\sin\frac{t+v}{2}\sin\frac{t-v}{2}} = -\dfrac{\cos\frac{t-v}{2}}{\sin\frac{t-v}{2}} = -\cot\frac{t-v}{2} = \cot\left[-\left(\frac{t-v}{2}\right)\right] = \cot\frac{v-t}{2}.$

[37] $\dfrac{\sin v - \sin u}{\cos v + \cos u} = \dfrac{2\cos\frac{v+u}{2}\sin\frac{v-u}{2}}{2\cos\frac{v+u}{2}\cos\frac{v-u}{2}} = \dfrac{\sin\frac{v-u}{2}}{\cos\frac{v-u}{2}} = \tan\frac{v-u}{2}.$

[38] $\dfrac{\cos 3x - \cos 5x}{\sin 3x + \sin 5x} = \dfrac{-2\sin\frac{3x+5x}{2}\sin\frac{3x-5x}{2}}{2\sin\frac{3x+5x}{2}\cos\frac{3x-5x}{2}} = -\dfrac{\sin\left(-x\right)}{\cos\left(-x\right)} = -\dfrac{\left(-\sin x\right)}{\cos x} = \dfrac{\sin x}{\cos x} = \tan x.$

[39] $\dfrac{\cos v - \cos u}{\cos v + \cos u} = \dfrac{-2\sin\frac{v+u}{2}\sin\frac{v-u}{2}}{2\cos\frac{v+u}{2}\cos\frac{v-u}{2}} = -\dfrac{\sin\frac{v+u}{2}}{\cos\frac{v+u}{2}}\cdot\dfrac{\sin\frac{v-u}{2}}{\cos\frac{v-u}{2}} = -\tan\frac{v+u}{2}\tan\frac{v-u}{2} =$

$\tan\frac{v+u}{2}\tan\left[-\left(\frac{v-u}{2}\right)\right] = \tan\frac{u+v}{2}\tan\frac{u-v}{2}.$

[40] $\dfrac{\cos u + \cos v}{\sin u + \sin v} = \dfrac{2\cos\frac{u+v}{2}\cos\frac{u-v}{2}}{2\sin\frac{u+v}{2}\cos\frac{u-v}{2}} = \dfrac{\cos\frac{u+v}{2}}{\sin\frac{u+v}{2}} = \cot\frac{u+v}{2} = \cot\frac{v+u}{2}.$

[41] $\dfrac{\cos 50° - \cos 10°}{\sin 10° - \sin 50°} = \dfrac{-2 \sin \frac{50° + 10°}{2} \sin \frac{50° - 10°}{2}}{2 \cos \frac{10° + 50°}{2} \sin \frac{10° - 50°}{2}} = -\dfrac{\sin 30° \sin 20°}{\cos 30° \sin(-20°)} = \dfrac{\sin 30°}{\cos 30°} =$

$\tan 30° = \dfrac{\sqrt{3}}{3}$.

[42] $\dfrac{\cos 6x - \cos 4x}{\sin 6x + \sin 4x} = \dfrac{-2 \sin \frac{6x + 4x}{2} \sin \frac{6x - 4x}{2}}{2 \sin \frac{6x + 4x}{2} \cos \frac{6x - 4x}{2}} = -\dfrac{\sin x}{\cos x} = -\tan x$.

[43] Since $\dfrac{\sin 6x - \sin 10x}{\cos 10x - \cos 6x} = \dfrac{2 \cos \frac{6x + 10x}{2} \sin \frac{6x - 10x}{2}}{-2 \sin \frac{10x + 6x}{2} \sin \frac{10x - 6x}{2}} = -\dfrac{\cos 8x \sin(-2x)}{\sin 8x \sin 2x} =$

$\dfrac{\cos 8x \sin 2x}{\sin 8x \sin 2x} = \dfrac{\cos 8x}{\sin 8x} = \cot 8x$, and $\dfrac{\cos 6x + \cos 10x}{\sin 6x + \sin 10x} = \dfrac{2 \cos \frac{6x + 10x}{2} \cos \frac{6x - 10x}{2}}{2 \sin \frac{6x + 10x}{2} \cos \frac{6x - 10x}{2}} =$

$\dfrac{\cos 8x}{\sin 8x} = \cot 8x$, the identity follows.

[44] $\dfrac{\cos u + \cos 3u}{\sin u + \sin 3u} = \dfrac{2 \cos \frac{u + 3u}{2} \cos \frac{u - 3u}{2}}{2 \sin \frac{u + 3u}{2} \cos \frac{u - 3u}{2}} = \dfrac{\cos 2u}{\sin 2u} = \cot 2u$.

[45] To verify product–to–sum identity (iii), we must show that $\sin u \cos v = \frac{1}{2}[\sin(u + v) + \sin(u - v)]$.

Since $\sin(u + v) = \sin u \cos v + \cos u \sin v$ and $\sin(u - v) = \sin u \cos v - \cos u \sin v$, it follows that

$\sin(u + v) + \sin(u - v) = (\sin u \cos v + \cos u \sin v) + (\sin u \cos v - \cos u \sin v) \Rightarrow$

$\sin(u + v) + \sin(u - v) = 2 \sin u \cos v \Rightarrow \sin u \cos v = \frac{1}{2}[\sin(u + v) + \sin(u - v)]$. To verify

product–to–sum identity (iv), we must show that $\cos u \sin v = \frac{1}{2}[\sin(u + v) - \sin(u - v)]$. We have

$\sin(u + v) - \sin(u - v) = (\sin u \cos v + \cos u \sin v) - (\sin u \cos v - \cos u \sin v) \Rightarrow$

$\sin(u + v) - \sin(u - v) = 2 \cos u \sin v \Rightarrow \cos u \sin v = \frac{1}{2}[\sin(u + v) - \sin(u - v)]$.

[46] $\dfrac{\sin 2\theta + \sin 4\theta}{\cos 2\theta + \cos 4\theta} = \dfrac{2 \sin \frac{2\theta + 4\theta}{2} \cos \frac{2\theta - 4\theta}{2}}{2 \cos \frac{2\theta + 4\theta}{2} \cos \frac{2\theta - 4\theta}{2}} = \dfrac{\sin 3\theta}{\cos 3\theta} = \tan 3\theta$.

[47] $\dfrac{2 \sin 2u}{\sin 3u + \sin u} = \dfrac{2 \sin 2u}{2 \sin \frac{3u + u}{2} \cos \frac{3u - u}{2}} = \dfrac{\sin 2u}{\sin 2u \cos u} = \dfrac{1}{\cos u} = \sec u$.

[48] To verify sum–to–product identity (ii), we must show that $\cos u - \cos v = -2\sin\frac{u+v}{2}\sin\frac{u-v}{2}$.

From product–to–sum identity (ii), it follows that $\sin x \sin y = \frac{1}{2}\left[\cos(x-y)-\cos(x+y)\right]$. (1)

If $u = x+y$ and $v = x-y$, then $u+v = (x+y)+(x-y) = 2x \Rightarrow x = \frac{u+v}{2}$. Similarly, $u - v =$

$(x+y)-(x-y) = 2y \Rightarrow y = \frac{u-v}{2}$. Substituting $\frac{u+v}{2}$ for x, $\frac{u-v}{2}$ for y, u for $x+y$, and v for $x-y$ in

equation (1) above, we have $\sin\frac{u+v}{2}\sin\frac{u-v}{2} = \frac{1}{2}\left(\cos v - \cos u\right) \Rightarrow \cos v - \cos u =$

$2\sin\frac{u+v}{2}\sin\frac{u-v}{2} \Rightarrow \cos u - \cos v = -2\sin\frac{u+v}{2}\sin\frac{u-v}{2}$. To verify sum–to–product identity

(iii), we must show that $\sin u + \sin v = 2\sin\frac{u+v}{2}\cos\frac{u-v}{2}$. From product–to–sum identiry (iii), it

follows that $\sin x \cos y = \frac{1}{2}\left[\sin(x+y)+\sin(x-y)\right]$. (2)

If $u = x+y$ and $v = x-y$, then as before $x = \frac{u+v}{2}$ and $y = \frac{u-v}{2}$. Substituting $\frac{u+v}{2}$ for x, $\frac{u-v}{2}$ for y,

u for $x+y$, and v for $x-y$ in equation (2) above, we obtain $\sin\frac{u+v}{2}\cos\frac{u-v}{2} = \frac{1}{2}\left(\sin u + \sin v\right) \Rightarrow$

$\sin u + \sin v = 2\sin\frac{u+v}{2}\cos\frac{u-v}{2}$. To verify sum–to–product identity (iv), we must show that

$\sin u - \sin v = 2\cos\frac{u+v}{2}\sin\frac{u-v}{2}$. From product–to–sum identity (iv), it follows that

$\cos x \sin y = \frac{1}{2}\left[\sin(x+y)-\sin(x-y)\right]$. (3)

If $u = x+y$ and $v = x-y$, then as before $x = \frac{u+v}{2}$ and $y = \frac{u-v}{2}$. Substituting $\frac{u+v}{2}$ for x, $\frac{u-v}{2}$ for y,

u for $x+y^2$, and v for $x-y^2$ in equation (3) we obtain $\cos\frac{u+v}{2}\sin\frac{u-v}{2} = \frac{1}{2}\left(\sin u - \sin v\right) \Rightarrow$

$\sin u - \sin v = 2\cos\frac{u+v}{2}\sin\frac{u-v}{2}$.

[49] $\dfrac{\cos 6t + \cos 10t}{\sin 6t + \sin 10t} = \dfrac{2\cos\frac{6t+10t}{2}\cos\frac{6t-10t}{2}}{2\sin\frac{6t+10t}{2}\cos\frac{6t-10t}{2}} = \dfrac{\cos 8t}{\sin 8t} = \cot 8t$.

[50] $\dfrac{2\cos 2v}{\cos 3v + \cos v} = \dfrac{2\cos 2v}{2\cos\frac{3v+v}{2}\cos\frac{3v-v}{2}} = \dfrac{\cos 2v}{\cos 2v \cos v} = \dfrac{1}{\cos v} = \sec v$.

[51] $\sin 70° - \sin 110° = 2\cos\dfrac{70°+110°}{2}\sin\dfrac{70°-110°}{2} = 2\cos 90° \sin(-20°) = 2(0)\left[\sin(-20°)\right] = 0$.

[52] $\dfrac{\sin x + \sin 3x}{\cos 3x - \cos x} = \dfrac{2\sin\frac{x+3x}{2}\cos\frac{x-3x}{2}}{-2\sin\frac{3x+x}{2}\sin\frac{3x-x}{2}} = -\dfrac{\sin 2x \cos(-x)}{\sin 2x \sin x} = -\dfrac{\cos x}{\sin x} = -\cot x$.

[53] $\cos(ax+b)\sin(ax+b) = \frac{1}{2}\left[\sin\left[(ax+b)+(ax+b)\right] - \sin\left[(ax+b)-(ax+b)\right]\right] =$

$\frac{1}{2}\left[\sin(2ax+2b) - \sin 0\right] = \frac{1}{2}\left[\sin(2ax+2b) - 0\right] = \frac{1}{2}\sin(2ax+2b)$.

EXERCISES 6.6

[1] $\sin 200° = -\sin 20°$. (See graph below).   [2] $\cos 253° = -\cos 73°$. (See graph below).

[3] $\cos(-256°) = -\cos 76°$. (See graph below).   [4] $\sin(-460°) = -\sin 80°$. (See graph below).

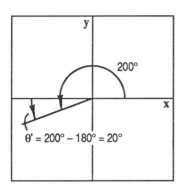

Figure 1

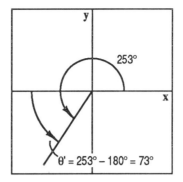

Figure 2

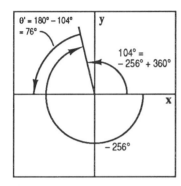

Figure 3

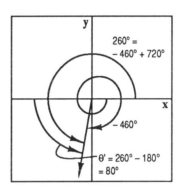

Figure 4

[5] $\tan 11.2 = -\tan 1.36.$ (See graph below).   [6] $\tan 1.68 = -\tan 1.46.$ (See graph below).

[7] $\sin x = \frac{1}{2} \Rightarrow S = \left\{ x \mid x = \frac{\pi}{6} + 2\,k\,\pi \text{ or } x = \frac{5\,\pi}{6} + 2\,k\,\pi, \text{ where } k \text{ is any integer} \right\}.$

[8] $\sin x = -\frac{\sqrt{3}}{2} \Rightarrow S = \left\{ x \mid x = \frac{4\,\pi}{3} + 2\,k\,\pi \text{ or } x = \frac{5\,\pi}{3} + 2\,k\,\pi, \text{ where } k \text{ is any integer} \right\}.$

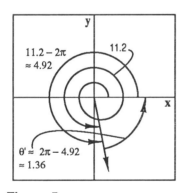

Figure 5

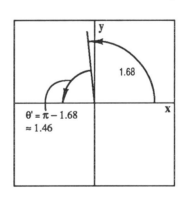

Figure 6

[9] $\tan x = 1 \Rightarrow S = \left\{ x \mid x = \frac{\pi}{4} + k\,\pi, \text{ where } k \text{ is any integer} \right\}.$

[10] $\cot x = -1 \Rightarrow \tan x = -1 \Rightarrow S = \left\{ x \mid x = \frac{3\,\pi}{4} + k\,\pi, \text{ where } k \text{ is any integer} \right\}.$

[11] $\sin x = -\frac{\sqrt{2}}{2} \Rightarrow S = \left\{ x \mid x = \frac{5\,\pi}{4} + 2\,k\,\pi \text{ or } x = \frac{7\,\pi}{4} + 2\,k\,\pi, \text{ where } k \text{ is any integer} \right\}.$

[12] $\csc x = -\sqrt{2} \Rightarrow \sin x = -\dfrac{1}{\sqrt{2}} \Rightarrow S = \left\{ x \mid x = \dfrac{5\pi}{4} + 2k\pi \text{ or } x = \dfrac{7\pi}{4} + 2k\pi, \text{ where } k \text{ is any integer} \right\}.$

[13] $\cos x = -\dfrac{\sqrt{3}}{2} \Rightarrow S = \left\{ x \mid x = \dfrac{5\pi}{6} + 2k\pi \text{ or } x = \dfrac{7\pi}{6} + 2k\pi, \text{ where } k \text{ is any integer} \right\}.$

[14] $\sec x = \dfrac{2}{\sqrt{3}} \Rightarrow \cos x = \dfrac{\sqrt{3}}{2} \Rightarrow S = \left\{ x \mid x = \dfrac{\pi}{6} + 2k\pi \text{ or } x = \dfrac{11\pi}{6} + 2k\pi, \text{ where } k \text{ is any integer} \right\}.$

[15] $\tan x = -\dfrac{1}{\sqrt{3}} \Rightarrow S = \left\{ x \mid x = \dfrac{5\pi}{6} + k\pi, \text{ where } k \text{ is any integer} \right\}.$

[16] $\cos x = -\dfrac{1}{2} \Rightarrow S = \left\{ x \mid x = \dfrac{2\pi}{3} + 2k\pi \text{ or } x = \dfrac{4\pi}{3} + 2k\pi, \text{ where } k \text{ is any integer} \right\}.$

[17] $\sin x = 1 \Rightarrow S = \left\{ x \mid x = \dfrac{\pi}{2} + 2k\pi, \text{ where } k \text{ is any integer} \right\}.$

[18] $\tan x = \sqrt{3} \Rightarrow S = \left\{ x \mid x = \dfrac{\pi}{3} + k\pi, \text{ where } k \text{ is any integer} \right\}.$

[19] $\sec x = -\sqrt{2} \Rightarrow \cos x = -\dfrac{1}{\sqrt{2}} \Rightarrow S = \left\{ x \mid x = \dfrac{3\pi}{4} + 2k\pi \text{ or } x = \dfrac{5\pi}{4} + 2k\pi, \text{ where } k \text{ is any integer} \right\}.$

[20] $\sin x = -1 \Rightarrow S = \left\{ x \mid x = \dfrac{3\pi}{2} + 2k\pi, \text{ where } k \text{ is any integer} \right\}.$

[21] $\cos t = -\dfrac{1}{2} \Rightarrow t = \dfrac{2\pi}{3} \text{ or } t = \dfrac{4\pi}{3}.$ ● $\left\{ \dfrac{2\pi}{3}, \dfrac{4\pi}{3} \right\}$

[22] $\sin t = \dfrac{\sqrt{2}}{2} \Rightarrow t = \dfrac{\pi}{4} \text{ or } t = \dfrac{3\pi}{4}.$ ● $\left\{ \dfrac{\pi}{4}, \dfrac{3\pi}{4} \right\}$

[23] $\tan 2t = -1 \Rightarrow 2t = \dfrac{3\pi}{4} + k\pi \Rightarrow t = \dfrac{3\pi}{8} + k\left(\dfrac{\pi}{2} \right).$ Since t will be in $[0, 2\pi)$ when $k = 0, 1, 2,$ or 3,

$t = \dfrac{3\pi}{8} \text{ or } t = \dfrac{3\pi}{8} + \dfrac{\pi}{2} = \dfrac{7\pi}{8} \text{ or } t = \dfrac{3\pi}{8} + \pi = \dfrac{11\pi}{8} \text{ or } t = \dfrac{3\pi}{8} + \dfrac{3\pi}{2} = \dfrac{15\pi}{8}.$

● $\left\{ \dfrac{3\pi}{8}, \dfrac{7\pi}{8}, \dfrac{11\pi}{8}, \dfrac{15\pi}{8} \right\}$

[24] $\tan 2t = -\dfrac{1}{\sqrt{3}} \Rightarrow 2t = \dfrac{5\pi}{6} + k\pi \Rightarrow t = \dfrac{5\pi}{12} + k\left(\dfrac{\pi}{2} \right).$ Since t will be in $[0, 2\pi)$ when $k = 0, 1, 2,$ or 3,

$t = \dfrac{5\pi}{12} \text{ or } t = \dfrac{5\pi}{12} + \dfrac{\pi}{2} = \dfrac{11\pi}{12} \text{ or } t = \dfrac{5\pi}{12} + \pi = \dfrac{17\pi}{12} \text{ or } t = \dfrac{5\pi}{12} + \dfrac{3\pi}{2} = \dfrac{23\pi}{12}.$

● $\left\{ \dfrac{5\pi}{12}, \dfrac{11\pi}{12}, \dfrac{17\pi}{12}, \dfrac{23\pi}{12} \right\}$

[25] $\sin t = \dfrac{\sqrt{3}}{2} \Rightarrow t = \dfrac{\pi}{3} \text{ or } t = \dfrac{2\pi}{3}.$ ● $\left\{ \dfrac{\pi}{3}, \dfrac{2\pi}{3} \right\}$

[26] $\cos 3t = -1 \Rightarrow 3t = \pi + 2k\pi \Rightarrow t = \frac{\pi}{3} + k\left(\frac{2\pi}{3}\right)$. Since t will be in $[0, 2\pi)$ when $k = 0, 1,$ or $2,$

$t = \frac{\pi}{3}$ or $t = \frac{\pi}{3} + \frac{2\pi}{3} = \pi$ or $t = \frac{\pi}{3} + \frac{4\pi}{3} = \frac{5\pi}{3}$. $\bullet \left\{\frac{\pi}{3}, \pi, \frac{5\pi}{3}\right\}$

[27] $\cos 2t = 0 \Rightarrow 2t = \frac{\pi}{2} + 2k\pi$ or $2t = \frac{3\pi}{2} + 2k\pi \Rightarrow t = \frac{\pi}{4} + k\pi$ or $t = \frac{3\pi}{4} + k\pi$. Since t will be in

$[0, 2\pi)$ when $k = 0$ or 1, $t = \frac{\pi}{4}$ or $t = \frac{3\pi}{4}$ or $t = \frac{\pi}{4} + \pi = \frac{5\pi}{4}$ or $t = \frac{3\pi}{4} + \pi = \frac{7\pi}{4}$.

$\bullet \left\{\frac{\pi}{4}, \frac{3\pi}{4}, \frac{5\pi}{4}, \frac{7\pi}{4}\right\}$

[28] $\sin 2t = 0 \Rightarrow 2t = 0 + 2k\pi$ or $2t = \pi + 2k\pi \Rightarrow t = 0 + k\pi$ or $t = \frac{\pi}{2} + k\pi$. Since t will be in $[0, 2\pi)$

when $k = 0$ or 1, $t = 0$ or $t = \frac{\pi}{2}$ or $t = 0 + \pi = \pi$ or $t = \frac{\pi}{2} + \pi = \frac{3\pi}{2}$. $\bullet \left\{0, \frac{\pi}{2}, \pi, \frac{3\pi}{2}\right\}$

[29] $\sec t = -\frac{2}{\sqrt{3}} \Rightarrow \cos t = -\frac{\sqrt{3}}{2} \Rightarrow t = \frac{5\pi}{6}$ or $t = \frac{7\pi}{6}$. $\bullet \left\{\frac{5\pi}{6}, \frac{7\pi}{6}\right\}$

[30] $\csc t = -2 \Rightarrow \sin t = -\frac{1}{2} \Rightarrow t = \frac{7\pi}{6}$ or $t = \frac{11\pi}{6}$. $\bullet \left\{\frac{7\pi}{6}, \frac{11\pi}{6}\right\}$

[31] $\sin\theta = -\frac{\sqrt{3}}{2} \Rightarrow \theta = 240°$ or $\theta = 300°$. $\bullet \{240°, 300°\}$

[32] $\cos\theta = -\frac{1}{2} \Rightarrow \theta = 120°$ or $\theta = 240°$. $\bullet \{120°, 240°\}$

[33] $\cos\theta = \frac{\sqrt{2}}{2} \Rightarrow \theta = 45°$ or $\theta = 315°$. $\bullet \{45°, 315°\}$

[34] $\sin\theta = -\frac{\sqrt{2}}{2} \Rightarrow \theta = 225°$ or $\theta = 315°$. $\bullet \{225°, 315°\}$

[35] $\tan 2\theta = \sqrt{3} \Rightarrow 2\theta = 60° + k\,(180°) \Rightarrow \theta = 30° + k\,(90°)$. Since $0° \le \theta < 360°$ when $k = 0, 1, 2,$ or 3,

$\theta = 30°$ or $\theta = 30° + 90° = 120°$ or $\theta = 30° + 180° = 210°$ or $\theta = 30° + 270° = 300°$.

$\bullet \{30°, 120°, 210°, 300°\}$

[36] $\cot 2\theta = -\sqrt{3} \Rightarrow \tan 2\theta = -\frac{1}{\sqrt{3}} \Rightarrow 2\theta = 150° + k\,(180°) \Rightarrow \theta = 75° + k\,(90°)$. Since $0° \le \theta < 360°$

when $k = 0, 1, 2,$ or 3, $\theta = 75°$ or $\theta = 75° + 90° = 165°$ or $\theta = 75° + 180° = 255°$ or

$\theta = 75° + 270° = 345°$. $\bullet \{75°, 165°, 255°, 345°\}$

[37] $\csc 2\theta = -2 \Rightarrow \sin 2\theta = -\frac{1}{2} \Rightarrow 2\theta = 210° + k\,(360°)$ or $2\theta = 330° + k\,(360°) \Rightarrow$

$\theta = 105° + k\,(180°)$ or $\theta = 165° + k\,(180°)$. Since $0° \le \theta < 360°$ when $k = 0$ or 1, $\theta = 105°$ or $\theta = 165°$

or $\theta = 105° + 180° = 285°$ or $\theta = 165° + 180° = 345°$. $\bullet \{105°, 165°, 285°, 345°\}$

[38] $\cot^2\theta = \cot\theta \Rightarrow \cot^2\theta - \cot\theta = 0 \Rightarrow \cot\theta\,(\cot\theta - 1) = 0 \Rightarrow \cot\theta = 0$ or $\cot\theta = 1$. If $\cot\theta = 0$, then

$\theta = 90°$ or $\theta = 270°$. If $\cot\theta = 1$, then $\tan\theta = 1 \Rightarrow \theta = 45°$ or $\theta = 225°$. $\bullet \{45°, 90°, 225°, 270°\}$

[39] $\sin 2\theta = \sqrt{2}\cos\theta \Rightarrow 2\sin\theta\cos\theta = \sqrt{2}\cos\theta \Rightarrow 2\sin\theta\cos\theta - \sqrt{2}\cos\theta = 0 \Rightarrow$

$\cos\theta\,(2\sin\theta - \sqrt{2}) = 0 \Rightarrow \cos\theta = 0$ or $\sin\theta = \frac{\sqrt{2}}{2}$. If $\cos\theta = 0$, then $\theta = 90°$ or $\theta = 270°$. If $\sin\theta =$

$\frac{\sqrt{2}}{2}$, then $\theta = 45°$ or $\theta = 135°$. $\bullet \{45°, 90°, 135°, 270°\}$

[40] $\sin 2\theta = -\sqrt{3}\cos\theta \Rightarrow 2\sin\theta\cos\theta = -\sqrt{3}\cos\theta \Rightarrow 2\sin\theta\cos\theta + \sqrt{3}\cos\theta = 0 \Rightarrow$

$\cos\theta(2\sin\theta + \sqrt{3}) = 0 \Rightarrow \cos\theta = 0$ or $\sin\theta = -\frac{\sqrt{3}}{2}$. If $\cos\theta = 0$, then $\theta = 90°$ or $\theta = 270°$. If

$\sin\theta = -\frac{\sqrt{3}}{2}$, then $\theta = 240°$ or $\theta = 300°$. ● $\{90°, 240°, 270°, 300°\}$

[41] $\sin 3t = -1 \Rightarrow 3t = \frac{3\pi}{2} + 2k\pi \Rightarrow t = \frac{\pi}{2} + k\left(\frac{2\pi}{3}\right)$. Since t will be in $[0, 2\pi)$ when $k = 0, 1,$ or $2,$

$t = \frac{\pi}{2}$ or $t = \frac{\pi}{2} + \frac{2\pi}{3} = \frac{7\pi}{6}$ or $t = \frac{\pi}{2} + \frac{4\pi}{3} = \frac{11\pi}{6}$. So, the solutions are $\frac{\pi}{2}, \frac{7\pi}{6},$ and $\frac{11\pi}{6}$.

[42] $\cos 3t = 0 \Rightarrow 3t = \frac{\pi}{2} + 2k\pi$ or $3t = \frac{3\pi}{2} + 2k\pi \Rightarrow t = \frac{\pi}{6} + k\left(\frac{2\pi}{3}\right)$ or $t = \frac{\pi}{2} + k\left(\frac{2\pi}{3}\right)$. Since

t will be in $[0, 2\pi)$ when $k = 0, 1,$ or $2, t = \frac{\pi}{6}$ or $t = \frac{\pi}{2}$ or $t = \frac{\pi}{6} + \frac{2\pi}{3} = \frac{5\pi}{6}$ or $t = \frac{\pi}{2} + \frac{2\pi}{3} = \frac{7\pi}{6}$ or

$t = \frac{\pi}{6} + \frac{4\pi}{3} = \frac{3\pi}{2}$ or $t = \frac{\pi}{2} + \frac{4\pi}{3} = \frac{11\pi}{6}$. Hence, the solutions are $\frac{\pi}{6}, \frac{\pi}{2}, \frac{5\pi}{6}, \frac{7\pi}{6}, \frac{3\pi}{2},$ and $\frac{11\pi}{6}$.

[43] $\sin 2t \sin t = -\cos t \Rightarrow (2\sin t \cos t)\sin t = -\cos t \Rightarrow 2\sin^2 t \cos t + \cos t = 0 \Rightarrow$

$\cos t (2\sin^2 t + 1) = 0 \Rightarrow \cos t = 0$ or $\sin^2 t = -\frac{1}{2}$. If $\cos t = 0$, then $t = \frac{\pi}{2}$ or $t = \frac{3\pi}{2}$. Since $\sin^2 t =$

$-\frac{1}{2}$ has no real solutions, the solutions of the original equation are $\frac{\pi}{2}$ and $\frac{3\pi}{2}$.

[44] $\tan^2 x = \tan x \Rightarrow \tan^2 x - \tan x = 0 \Rightarrow \tan x(\tan x - 1) = 0 \Rightarrow \tan x = 0$ or $\tan x = 1$. If $\tan x = 0$, then

$x = 0$ or $x = \pi$. If $\tan x = 1$, then $x = \frac{\pi}{4}$ or $x = \frac{5\pi}{4}$. So, the solutions are $0, \frac{\pi}{4}, \pi,$ and $\frac{5\pi}{4}$.

[45] $\sin 2t = \sqrt{2}\sin t \Rightarrow 2\sin t \cos t = \sqrt{2}\sin t \Rightarrow 2\sin t \cos t - \sqrt{2}\sin t = 0 \Rightarrow \sin t(2\cos t - \sqrt{2}) =$

$0 \Rightarrow \sin t = 0$ or $\cos t = \frac{\sqrt{2}}{2}$. If $\sin t = 0$, then $t = 0$ or $t = \pi$. If $\cos t = \frac{\sqrt{2}}{2}$, then $t = \frac{\pi}{4}$ or $t = \frac{7\pi}{4}$. So,

the solutions are $0, \frac{\pi}{4}, \pi,$ and $\frac{7\pi}{4}$.

[46] $\cot^2 t = 1 \Rightarrow \tan^2 t = 1 \Rightarrow \tan t = \pm 1$. If $\tan t = 1$, then $t = \frac{\pi}{4}$ or $t = \frac{5\pi}{4}$. If $\tan t = -1$, then $t = \frac{3\pi}{4}$ or

$t = \frac{7\pi}{4}$. So, the solutions are $\frac{\pi}{4}, \frac{3\pi}{4}, \frac{5\pi}{4},$ and $\frac{7\pi}{4}$.

[47] $\sin 2x = \cos x \Rightarrow 2\sin x \cos x = \cos x \Rightarrow 2\sin x \cos x - \cos x = 0 \Rightarrow \cos x(2\sin x - 1) = 0 \Rightarrow$

$\cos x = 0$ or $\sin x = \frac{1}{2}$. If $\cos x = 0$, then $x = \frac{\pi}{2}$ or $x = \frac{3\pi}{2}$. If $\sin x = \frac{1}{2}$, then $x = \frac{\pi}{6}$ or $x = \frac{5\pi}{6}$. So, the

solutions are $\frac{\pi}{6}, \frac{\pi}{2}, \frac{5\pi}{6},$ and $\frac{3\pi}{2}$.

[48] $\tan^2 t = -\tan t \Rightarrow \tan^2 t + \tan t = 0 \Rightarrow \tan t(\tan t + 1) = 0 \Rightarrow \tan t = 0$ or $\tan t = -1$. If $\tan t = 0$, then

$t = 0$ or $t = \pi$. If $\tan t = -1$, then $t = \frac{3\pi}{4}$ or $t = \frac{7\pi}{4}$. So, the solutions are $0, \frac{3\pi}{4}, \pi,$ and $\frac{7\pi}{4}$.

[49] $2\sin^2 t + \sin t = 0 \Rightarrow \sin t(2\sin t + 1) = 0 \Rightarrow \sin t = 0$ or $\sin t = -\frac{1}{2}$. If $\sin t = 0$, then $t = 0$ or $t = \pi$.

If $\sin t = -\frac{1}{2}$, then $t = \frac{7\pi}{6}$ or $t = \frac{11\pi}{6}$. So, the solutions are $0, \pi, \frac{7\pi}{6},$ and $\frac{11\pi}{6}$.

[50] $2\cos^2 x - 3\cos x = 0 \Rightarrow \cos x(2\cos x - 3) = 0 \Rightarrow \cos x = 0$ or $\cos x = \frac{3}{2}$. If $\cos x = 0$, then $x = \frac{\pi}{2}$ or

$x = \frac{3\pi}{2}$. Since $\cos x = \frac{3}{2} > 1$ is impossible, the solutions are $\frac{\pi}{2}$ and $\frac{3\pi}{2}$.

[51] $2\cos^2 t - 9\cos t - 5 = 0 \Rightarrow (2\cos t + 1)(\cos t - 5) = 0 \Rightarrow \cos t = -\frac{1}{2}$ or $\cos t = 5$. If $\cos t = -\frac{1}{2}$,

then $t = \frac{2\pi}{3}$ or $t = \frac{4\pi}{3}$. Since $\cos t = 5 > 1$ is impossible, the solutions are $\frac{2\pi}{3}$ and $\frac{4\pi}{3}$.

[52] $\cot^4 x - 1 = 0 \Rightarrow (2\cot^2 x - 1)(\cot^2 x + 1) = 0 \Rightarrow \cot^2 x = 1$ or $\cot^2 x = -1$. If $\cot^2 x = 1$, then

$\tan^2 x = 1 \Rightarrow \tan x = \pm 1$. If $\tan x = 1$, then $x = \frac{\pi}{4}$ or $x = \frac{5\pi}{4}$. If $\tan x = -1$, then $x = \frac{3\pi}{4}$ or $x = \frac{7\pi}{4}$.

Since $\cot^2 x = -1$ *has no real solutions, the solutions, are* $\frac{\pi}{4}, \frac{3\pi}{4}, \frac{5\pi}{4},$ *and* $\frac{7\pi}{4}$.

[53] $2\sin^2 x - 5\sin x - 3 = 0 \Rightarrow (2\sin x + 1)(\sin x - 3) = 0 \Rightarrow \sin x = -\frac{1}{2}$ or $\sin x = 3$. If $\sin x = -\frac{1}{2}$,

then $x = \frac{7\pi}{6}$ or $x = \frac{11\pi}{6}$. Since $\sin x = 3 > 1$ is impossible, the solutions are $\frac{7\pi}{6}$ and $\frac{11\pi}{6}$.

[54] $6\sin^2 x - \sin x - 1 = 0 \Rightarrow (2\sin x - 1)(3\sin x + 1) = 0 \Rightarrow \sin x = \frac{1}{2}$ or $\sin x = -\frac{1}{3}$. If $\sin x = \frac{1}{2}$, then

$x = \frac{\pi}{6}$ or $x = \frac{5\pi}{6}$. If $\sin x = -\frac{1}{3}$, then since $\sin x < 0$, we know that x lies in either QIII or QIV. In

either case, since $\sin^{-1}\left(-\frac{1}{3}\right) \approx -0.34$, x has $x' \approx 0.34$ as its reference angle. In QIII,

$x = \pi + x' \approx 3.14 + 0.34 = 3.48$. In QIV, $x = 2\pi - x' \approx 6.28 - 0.34 = 5.94$. So, the solutions are

$\frac{\pi}{6}, \frac{5\pi}{6}, 3.5$, and 5.9.

[55] $6\cos^2 t - \sin t - 4 = 0 \Rightarrow 6(1 - \sin^2 t) - \sin t - 4 = 0 \Rightarrow 6\sin^2 t + \sin t - 2 = 0 \Rightarrow (2\sin t - 1).$

$(3\sin t + 2) = 0 \Rightarrow \sin t = \frac{1}{2}$ or $\sin t = -\frac{2}{3}$. If $\sin t = \frac{1}{2}$, then $t = \frac{\pi}{6}$ or $t = \frac{5\pi}{6}$. If $\sin t = -\frac{2}{3}$, then

since $\sin t < 0$, we know that t lies in either QIII or QIV. In either case, since $\sin^{-1}\left(-\frac{2}{3}\right) \approx -0.73$, t

has $t' \approx 0.73$ and its reference angle. In QIII, $t = \pi + t' \approx 3.14 + 0.73 = 3.87$. In QIV, $t = 2\pi - t' \approx$

$6.28 - 0.73 = 5.55$. So, the solutions are $\frac{\pi}{6}, \frac{5\pi}{6}, 3.9$, and 5.6.

[56] $10\cos^2 x + 3\cos x - 1 = 0 \Rightarrow (2\cos x + 1)(5\cos x - 1) = 0 \Rightarrow \cos x = -\frac{1}{2}$ or $\cos x = \frac{1}{5}$.

If $\cos x = -\frac{1}{2}$, then $x = \frac{2\pi}{3}$ or $x = \frac{4\pi}{3}$. If $\cos x = \frac{1}{5}$, then since $\cos x > 0$, we know that x lies in

either QI or QIV. In either case, since $\cos^{-1}\left(\frac{1}{5}\right) \approx 1.37$, x has $x' \approx 1.37$ as its reference angle. In QI,

$x = x' \approx 1.37$. In QIV, $x = 2\pi - x' \approx 6.28 - 1.37 = 4.91$. So, the solutions are $1.4, \frac{2\pi}{3}, \frac{4\pi}{3}$, and 4.9.

[57] $\sin 2t \cos t = \sin t \Rightarrow (2\sin t \cos t)\cos t = \sin t \Rightarrow 2\sin t \cos^2 t - \sin t = 0 \Rightarrow$

$\sin t (2\cos^2 t - 1) = 0 \Rightarrow \sin t = 0$ or $\cos^2 t = \frac{1}{2}$. If $\sin t = 0$, then $t = 0$ or $t = \pi$. If $\cos^2 t = \frac{1}{2}$, then

$\cos t = \pm\frac{\sqrt{2}}{2}$. If $\cos t = \frac{\sqrt{2}}{2}$, then $t = \frac{\pi}{4}$ or $t = \frac{7\pi}{4}$. If $\cos t = -\frac{\sqrt{2}}{2}$, then $t = \frac{3\pi}{4}$ or $t = \frac{5\pi}{4}$. So, the

solutions are $0, \frac{\pi}{4}, \frac{3\pi}{4}, \pi, \frac{5\pi}{4}$, and $\frac{7\pi}{4}$.

[58] $6\sin^2 x + \sin x - 1 = 0 \Rightarrow (2\sin x + 1)(3\sin x - 1) = 0 \Rightarrow \sin x = -\frac{1}{2}$ or $\sin x = \frac{1}{3}$. If $\sin x = -\frac{1}{2}$,

then $x = \frac{7\pi}{6}$ or $x = \frac{11\pi}{6}$. If $\sin x = \frac{1}{3}$, then since $\sin x > 0$, we know that x lies in either QI or QII. In

either case, since $\sin^{-1}\left(\frac{1}{3}\right) \approx 0.34$, x has $x' \approx 0.34$ as its reference angle. In QI, $x = x' \approx 0.34$. In QII,

$x = \pi - x' \approx 3.14 - 0.34 = 2.8$. So, the solutions are $0.3, 2.8, \frac{7\pi}{6}$, and $\frac{11\pi}{6}$.

[59] $\cos 2x = 4 \sin x \Rightarrow 1 - 2 \sin^2 x = 4 \sin x \Rightarrow 2 \sin^2 x + 4 \sin x - 1 = 0 \Rightarrow \sin x = \frac{-2 \pm \sqrt{6}}{2}$. If $\sin x =$

$\frac{-2 + \sqrt{6}}{2} \approx 0.2247$, then since $\sin x > 0$, we know that x lies in either QI or QII. In either case, since

$\sin^{-1}(0.2247) \approx 0.23$, x has $x' \approx 0.23$ as its reference angle. In QI, $x = x' \approx 0.23$. In QII, $x = \pi - x' \approx$

$3.14 - 0.23 = 2.91$. Since $\sin x = \frac{-2 - \sqrt{6}}{2} < -1$ is impossible, the solutions are 0.2 and 2.9.

[60] $15 \tan^2 x + 2 \tan x - 1 = 0 \Rightarrow (5 \tan x - 1)(3 \tan x + 1) = 0 \Rightarrow \tan x = \frac{1}{5}$ or $\tan x = -\frac{1}{3}$.

If $\tan x = \frac{1}{5}$, then since $\tan x > 0$, we know that x lies in either QI or QIII. In either case, since

$\tan^{-1}\left(\frac{1}{5}\right) \approx 0.20$, x has $x' \approx 0.20$ as its reference angle. In QI, $x = x' \approx 0.20$. In QIII, $x = \pi + x' \approx$

$3.14 + 0.20 = 3.34$. If $\tan x = -\frac{1}{3}$, then since $\tan x < 0$, we know that x lies in either QII or QIV. In

either case, since $\tan^{-1}\left(-\frac{1}{3}\right) \approx -0.32$, x has $x' \approx 0.32$ as its reference angle. In QII, $x = \pi - x' \approx$

$3.14 - 0.32 = 2.82$. In QIV, $x = 2\pi - x' \approx 6.28 - 0.32 = 5.96$. So, the solutions are 0.2, 2.8, 3.3, and 6.0.

[61] $3 \tan^2 t - 5 \tan t + 1 = 0 \Rightarrow \tan t = \frac{5 \pm \sqrt{13}}{6}$. If $\tan t = \frac{5 + \sqrt{13}}{6} \approx 1.434$, then since $\tan t > 0$, we

know that t lies in either QI or QIII. In either case, since $\tan^{-1}(1.434) \approx 0.96$, t has $t' \approx 0.96$ as its

reference angle. In QI, $t = t' \approx 0.96$. In QIII, $t = \pi + t' \approx 3.14 + 0.96 = 4.10$. If $\tan t = \frac{5 - \sqrt{13}}{6} \approx$

0.2324, then since $\tan t > 0$, we know that t lies in either QI or QIII. In either case, since

$\tan^{-1}(0.2324) \approx 0.23$, t has $t' \approx 0.23$ as its reference angle. In QI, $t = t' \approx 0.23$. In QIII,

$t = \pi + t' \approx 3.14 + 0.23 = 3.37$. So, the solutions are 0.2, 1.0, 3.4, and 4.1.

[62] $10 \cos^2 x - 3 \cos x - 1 = 0 \Rightarrow (2 \cos x - 1)(5 \cos x + 1) = 0 \Rightarrow \cos x = \frac{1}{2}$ or $\cos x = -\frac{1}{5}$. If $\cos x = \frac{1}{2}$,

then $x = \frac{\pi}{3}$ or $x = \frac{5\pi}{3}$. If $\cos x = -\frac{1}{5}$, then since $\cos x < 0$, we know that x lies in either QII or QIII.

In either case, since $\cos^{-1}\left(-\frac{1}{5}\right) \approx 1.77$, x has $x' = \pi - 1.77 = 1.37$ as its reference angle. In QII,

$x \approx 1.77$. In QIII, $x = \pi + x' \approx 3.14 + 1.37 = 4.51$. So, the solutions are $\frac{\pi}{3}$, 1.8, 4.5, and $\frac{5\pi}{3}$.

[63] $\tan t + \sqrt{3} = \sec t \Rightarrow \tan^2 t + 2\sqrt{3} \tan t + 3 = \sec^2 t \Rightarrow \tan^2 t + 2\sqrt{3} \tan t + 3 = 1 + \tan^2 t \Rightarrow$

$2\sqrt{3} \tan t = -2 \Rightarrow \tan t = -\frac{1}{\sqrt{3}} \Rightarrow t = \frac{5\pi}{6}$ or $t = \frac{11\pi}{6}$. Since $\frac{5\pi}{6}$ is extraneous, the only

solution is $\frac{11\pi}{6}$.

[64] $1 + \sin x = \cos x \Rightarrow 1 + 2 \sin x + \sin^2 x = \cos^2 x \Rightarrow 1 + 2 \sin x + \sin^2 x = 1 - \sin^2 x \Rightarrow$

$2 \sin^2 x + 2 \sin x = 0 \Rightarrow \sin x (\sin x + 1) = 0 \Rightarrow \sin x = 0$ or $\sin x = -1$. If $\sin x = 0$, then $x = 0$ or

$x = \pi$. If $\sin x = -1$, then $x = \frac{3\pi}{2}$. Since π is extraneous, the soultions are 0 and $\frac{3\pi}{2}$.

[65]   Comparing the left side of the equation $\sin x + \sqrt{3} \cos x = 1$ with the left side of the equation

$a \sin v + b \cos v = r \sin (u + v)$ from result (6) on page 296, we see that $a = 1$, $b = \sqrt{3}$, and $v = x$.

From the figure below we see that $r = 2$ and $\cos u = \frac{1}{2}$ and $\sin u = \frac{\sqrt{3}}{2}$. Hence, $u = \frac{\pi}{3}$. So,

$\sin x + \sqrt{3} \cos x = r \sin (u + v) = 2 \sin \left(\frac{\pi}{3} + x \right)$. So, $\sin x + \sqrt{3} \cos x = 1 \Rightarrow 2 \sin \left(\frac{\pi}{3} + x \right) = 1 \Rightarrow$

$\sin \left(\frac{\pi}{3} + x \right) = \frac{1}{2} \Rightarrow \frac{\pi}{3} + x = \frac{\pi}{6}$ or $\frac{\pi}{3} + x = \frac{5\pi}{6} \Rightarrow x = \frac{\pi}{6} - \frac{\pi}{3} = -\frac{\pi}{6}$ or $x = \frac{5\pi}{6} - \frac{\pi}{3} = \frac{\pi}{2}$. Notice that $\frac{\pi}{2}$ is

in $[0, 2\pi)$ but $-\frac{\pi}{6}$ is not in $[0, 2\pi]$. Since the function $f(x) = \left(\frac{\pi}{3} + x \right)$ has period 2π, the value of x

in $[0, 2\pi)$ which corresponds to $-\frac{\pi}{6}$ is $-\frac{\pi}{6} + 2\pi = \frac{11\pi}{6}$. So, the solutions are $\frac{\pi}{2}$ and $\frac{11\pi}{6}$.

(See graph below).

[66]   Comparing the left side of the equation $\sqrt{3} \cos x - \sin x = 1$ with the left side of the equation

$a \sin v + b \cos v = r \sin (u + v)$ from result (6) on page 296, we see that $a = -1$, $b = \sqrt{3}$, and $v = x$.

From the figure below we see that $r = 2$ and $\cos u = -\frac{1}{2}$ and $\sin u = \frac{\sqrt{3}}{2}$. Hence, $u = \frac{2\pi}{3}$. So,

$\sqrt{3} \cos x - \sin x = r \sin (u + v) = 2 \sin \left(\frac{2\pi}{3} + x \right)$. Therefore, $\sqrt{3} \cos x - \sin x = 1 \Rightarrow$

$2 \sin \left(\frac{2\pi}{3} + x \right) = 1 \Rightarrow \sin \left(\frac{2\pi}{3} + x \right) = \frac{1}{2} \Rightarrow \frac{2\pi}{3} + x = \frac{\pi}{6}$ or $\frac{2\pi}{3} + x = \frac{5\pi}{6} \Rightarrow x =$

$\frac{\pi}{6} - \frac{2\pi}{3} = -\frac{\pi}{2}$ or $x = \frac{5\pi}{6} - \frac{2\pi}{3} = \frac{\pi}{6}$. Notice that $\frac{\pi}{6}$ is in $[0, 2\pi)$ but $-\frac{\pi}{2}$ is not in $[0, 2\pi)$. Since the

function $f(x) = \sin \left(\frac{2\pi}{3} + x \right)$ has period 2π, the value of x in $[0, 2\pi)$ which corresponds to $-\frac{\pi}{2}$ is

$-\frac{\pi}{2} + 2\pi = \frac{3\pi}{2}$. So, the solutions are $\frac{\pi}{6}$ and $\frac{3\pi}{2}$.   (See graph below).

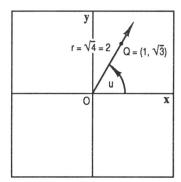

Figure 65

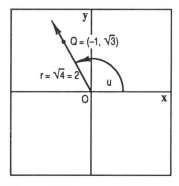

Figure 66

[67] Comparing the left side of the equation $\sin x - \cos x = 1$ with the left side of the equation

$a \sin v + b \cos v = r \sin (u + v)$ from result (6) on page 296, we see that $a = 1$, $b = -1$, and $v = x$.

From the figure below we see that $r = \sqrt{2}$ and $\cos u = \frac{1}{\sqrt{2}}$ and $\sin u = -\frac{1}{\sqrt{2}}$. Hence, $u = \frac{7\pi}{4}$. So,

$\sin x - \cos x = r \sin (u + v) = \sqrt{2} \sin \left(\frac{7\pi}{4} + x\right)$. Hence, $\sin x - \cos x = 1 \Rightarrow \sqrt{2} \sin \left(\frac{7\pi}{4} + x\right) = 1 \Rightarrow$

$\sin \left(\frac{7\pi}{4} + x\right) = \frac{1}{\sqrt{2}} \Rightarrow \frac{7\pi}{4} + x = \frac{\pi}{4}$ or $\frac{7\pi}{4} + x = \frac{3\pi}{4} \Rightarrow x = \frac{\pi}{4} - \frac{7\pi}{4} = -\frac{3\pi}{2}$ or $x = \frac{3\pi}{4} - \frac{7\pi}{4} = -\pi$.

Notice that neither $-\frac{3\pi}{2}$ nor $-\pi$ is in $[0, 2\pi)$. Since the function $f(x) = \sin\left(\frac{7\pi}{4} + x\right)$ has period 2π,

the values of x in $[0, 2\pi)$ which correspond to $-\frac{3\pi}{2}$ and $-\pi$ are $-\frac{3\pi}{2} + 2\pi = \frac{\pi}{2}$ and $-\pi + 2\pi = \pi$.

So, the solutions are $\frac{\pi}{2}$ and π.   (See graph below).

[68] $\sec x - \tan x = -\sqrt{2} \Rightarrow \frac{1}{\cos x} - \frac{\sin x}{\cos x} = -\sqrt{2} \Rightarrow 1 - \sin x = -\sqrt{2} \cos x \Rightarrow \sin x - \sqrt{2} \cos x = 1$.

Comparing the left side of the last equation with the left side of the equation $a \sin v + b \cos v =$

$r \sin (u + v)$ from result (6) on page 296, we see that $a = 1$, $b = -\sqrt{2}$, and $v = x$. From the figure

below we see that $r = \sqrt{3}$ and $\cos u = \frac{1}{\sqrt{3}}$ and $\sin u = -\frac{\sqrt{2}}{\sqrt{3}}$. Since $\sin^{-1}\left(-\frac{\sqrt{2}}{\sqrt{3}}\right) \approx -0.9553$, $u \approx$

$2\pi - 0.9553 \approx 5.33$. So, $\sin x - \sqrt{2} \cos x = r \sin (u + v) \approx \sqrt{3} \sin (5.33 + x)$. Hence,

$\sin x - \sqrt{2} \cos x = 1 \Rightarrow \sqrt{3} \sin (5.33 + x) \approx 1 \Rightarrow \sin (5.33 + x) \approx \frac{1}{\sqrt{3}}$. Since $\sin^{-1}\left(\frac{1}{\sqrt{3}}\right) \approx 0.6155$,

$\sin (5.33 + x) \approx \frac{1}{\sqrt{3}} \Rightarrow 5.33 + x \approx 0.6155$ or $5.33 + x \approx \pi - 0.6155 \approx 2.526 \Rightarrow x \approx 0.6155 - 5.33 \approx$

-4.71 or $x \approx 2.526 - 5.33 \approx -2.80$. Since the function $f(x) = \sin (5.33 + x)$ has period 2π, the values

of x in $[0, 2\pi)$ which correspond to -4.71 and -2.80 are $-4.71 + 2\pi \approx 1.57$ and $-2.80 + 2\pi \approx$

3.48. So, the solutions are 1.6 and 3.5.   (See graph below).

[69] In order to complete the proof of the reference angle theorem, we must consider the cases when

$0 < \theta < \frac{\pi}{2}$, $\pi < \theta < \frac{3\pi}{2}$ and $\frac{3\pi}{2} < \theta < 2\pi$. Clearly, it suffices to show that in all three cases $\sin \theta' =$

$|\sin \theta|$ and $\cos \theta' = |\cos \theta|$. If $0 < \theta < \frac{\pi}{2}$, then $\theta' = \theta \Rightarrow \sin \theta' = \sin \theta = |\sin \theta|$ and $\cos \theta' = \cos \theta =$

$|\cos \theta|$. If $\pi < \theta < \frac{3\pi}{2}$, then in the figure below we see that θ is in standard position and its reference

angle θ' is also in stadard position. Choose any point $P = (x, y)$, other than the origin O, on the

terminal side of θ and let r be the distance from O to P. Now, choose the point $P' = (x', y')$ on the

terminal side of θ' so that the distance from O to P' is also r. Then, as we see from the figure,

triangles OAP and $OA'P'$ are congruent triangles. It follows that $x = -x'$ and $y = -y'$. Hence, by

Theorem 6.2 on page 253, $\sin \theta = \frac{y}{r} = \frac{-y'}{r} = -\sin \theta' \Rightarrow \sin \theta' = -\sin \theta \Rightarrow \sin \theta' = |\sin \theta|$, and

$\cos \theta = \frac{x}{r} = \frac{-x'}{r} = -\cos \theta' \Rightarrow \cos \theta' = -\cos \theta \Rightarrow \cos \theta' = |\cos \theta|$. If $\frac{3\pi}{2} < \theta < 2\pi$, the result follows

in a similar manner.   (See graph below).

[70] We must prove the result: Let a and b be real numbers, not both zero, and let $r = \sqrt{a^2 + b^2}$. Let u be the angle in standard position with the point $Q = (a, b)$ on its terminal side, where $0 \le u < 2\pi$. The for any angle v, $a \sin v + b \cos v = r \sin(u + v)$. In the figure below u is in standard position with the point $Q = (a, b)$ on its terminal side, where $0 \le u < 2\pi$. Since $r = \sqrt{a^2 + b^2} = d(O, Q)$, it follows from Theorem 6.2 on page 253 that $\cos u = \frac{a}{r}$ and $\sin u = \frac{b}{r}$. If v is any angle, then it follows from the sum formula for sine that $\sin(u + v) = \sin u \cos v + \cos u \sin v \Rightarrow \sin(u + v) = \left(\frac{b}{r}\right)\cos v + \left(\frac{a}{r}\right)\sin v \Rightarrow r \sin(u + v) = b \cos v + a \sin v \Rightarrow a \sin v + b \cos v = r \sin(u + v)$. (See graph below).

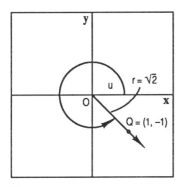

Figure 67

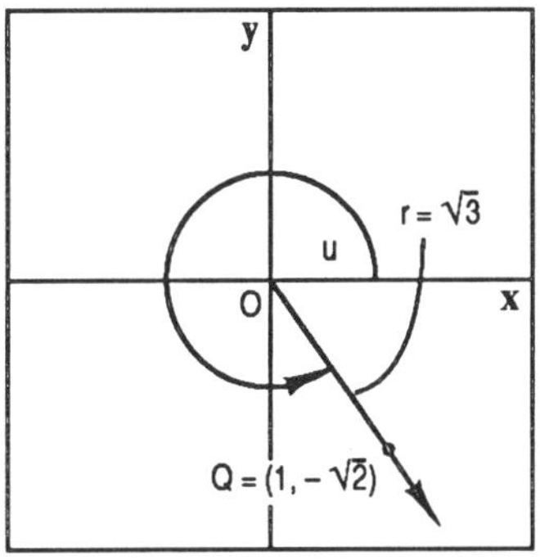

Figure 68

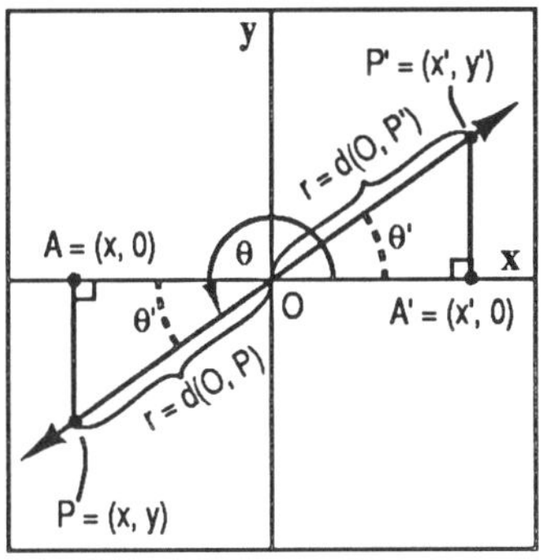

Figure 69

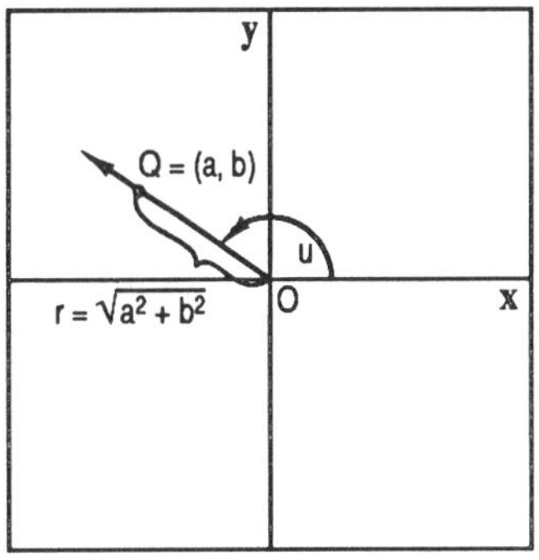

Figure 70

[71] Since $I = f(t) = 60 \cos(120\pi t - \pi)$, $I = 20 \Rightarrow 60 \cos(120\pi t - \pi) = 20 \Rightarrow \cos(120\pi t - \pi) = \frac{1}{3}$. Since $\cos^{-1}\left(\frac{1}{3}\right) \approx 1.23$ and since 1.23 lies in QI, the first time when I = 20 is when $120\pi t - \pi \approx 1.23 \Rightarrow 120\pi t \approx \pi + 1.23 \Rightarrow t \approx \frac{\pi + 1.23}{120\pi} \approx 0.01$ sec.

[72] Since $g(t) = 200 \sin(30\pi t - 3)$, $g(t) = 50 \Rightarrow 200 \sin(30\pi t - 3) = 50 \Rightarrow \sin(30\pi t - 3) = \frac{1}{4}$. If $\theta = 30\pi t - 3$, then we must find the second time when $\sin \theta = \frac{1}{4}$. Since $\sin^{-1}\left(\frac{1}{4}\right) \approx 0.25 \in$ QI, the second time when $\sin \theta = \frac{1}{4}$ is when θ lies in QII and has $\theta' \approx 0.25$ as its reference angle. If θ lies in QII, then $\theta = \pi - \theta' \approx 3.14 - 0.25 \Rightarrow \theta \approx 2.89 \Rightarrow 30\pi t - 3 \approx 2.89 \Rightarrow 30\pi t \approx 5.89 \Rightarrow t \approx \frac{5.89}{30\pi} \approx 0.06$ sec.

[73] a) Since $d = 2 \sin\left[\frac{2\pi}{3}(6-4t)\right]$, $d = 2 \Rightarrow 2 \sin\left[\frac{2\pi}{3}(6-4t)\right] = 2 \Rightarrow \sin\left[\frac{2\pi}{3}(6-4t)\right] = 1 \Rightarrow$

$\sin\left[-\frac{2\pi}{3}(4t-6)\right] = 1 \Rightarrow -\sin\left[\frac{2\pi}{3}(4t-6)\right] = 1 \Rightarrow \sin\left[\frac{2\pi}{3}(4t-6)\right] = -1$. If $\theta =$

$\frac{2\pi}{3}(4t-6)$, then the first time when $\sin\theta = -1$ is when $\theta = \frac{3\pi}{2} \Rightarrow \frac{2\pi}{3}(4t-6) = \frac{3\pi}{2} \Rightarrow$

$4t - 6 = \frac{9}{4} \Rightarrow t = \frac{33}{16} \approx 2.06$ sec.

b) As in part (a), $d = 1 \Rightarrow 2 \sin\left[\frac{2\pi}{3}(6-4t)\right] = 1 \Rightarrow \sin\left[\frac{2\pi}{3}(6-4t)\right] = \frac{1}{2} \Rightarrow \sin\left[\frac{2\pi}{3}(4t-6)\right] =$

$-\frac{1}{2}$. If $\theta = \frac{2\pi}{3}(4t-6)$, then the first time when $\sin\theta = -\frac{1}{2}$ is when $\theta = \frac{7\pi}{6}$ and the second

time is when $\theta = \frac{11\pi}{6} \Rightarrow \frac{2\pi}{3}(4t-6) = \frac{11\pi}{6} \Rightarrow 4t-6 = \frac{11}{4} \Rightarrow t = \frac{35}{16} \approx 2.19$ sec.

[74] a) $d = 2 \Rightarrow 5 \cos\left(\frac{2\pi}{3}t - 3\right) = 2 \Rightarrow \cos\left(\frac{2\pi}{3}t - 3\right) = \frac{2}{5}$. If $\theta = \frac{2\pi}{3}t - 3$, then since $\cos^{-1}\left(\frac{2}{5}\right) \approx$

$1.16 \in$ QI, the second time when $\cos\theta = \frac{2}{5}$ is when θ lies in QIV and has $\theta' \approx 1.16$ as its

reference angle. In QIV, $\theta = 2\pi - \theta' \approx 6.28 - 1.16 \Rightarrow \theta \approx 5.12 \Rightarrow \frac{2\pi}{3}t - 3 \approx 5.12 \Rightarrow \frac{2\pi}{3}t \approx$

$8.12 \Rightarrow t \approx \frac{12.18}{\pi} \approx 3.88$ sec.

b) As in part (a), $d = 3 \Rightarrow 5 \cos\left(\frac{2\pi}{3}t - 3\right) = 2 \Rightarrow \cos\left(\frac{2\pi}{3}t - 3\right) = \frac{2}{5}$. If $\theta = \frac{2\pi}{3}t - 3$, then since

$\cos^{-1}\left(\frac{3}{5}\right) \approx 0.93 \in$ QI, then the first time when $\cos\theta = \frac{2}{5}$ is when θ lies in QI $\Rightarrow \theta \approx 0.93 \Rightarrow$

$\frac{2\pi}{3}t - 3 \approx 0.93 \Rightarrow \frac{2\pi}{3}t \approx 3.93 \Rightarrow t \approx 1.88$ sec.

[75] a) $f(t) = 6 \Rightarrow 8 \sin\left(\sqrt{\frac{3}{10}}\,t\right) = 6 \Rightarrow \sin\left(\sqrt{\frac{3}{10}}\,t\right) = \frac{3}{4}$. If $\theta = \sqrt{\frac{3}{10}}\,t$, then since $\sin^{-1}\left(\frac{3}{4}\right) \approx$

$0.85 \in$ QI, the first time when $f(t) = 6$ is when θ lies in QI $\Rightarrow \theta \approx 0.85 \Rightarrow \sqrt{\frac{3}{10}}\,t \approx 0.85 \Rightarrow$

$t \approx 1.55$ sec.

b) As in part (a), $f(t) = 3 \Rightarrow 8 \sin\left(\sqrt{\frac{3}{10}}\,t\right) = 3 \Rightarrow \sin\left(\sqrt{\frac{3}{10}}\,t\right) = \frac{3}{8}$. If $\theta = \sqrt{\frac{3}{10}}\,t$, then since

$\sin^{-1}\left(\frac{3}{8}\right) \approx 0.38 \in$ QI, the second time when $f(t) = 3$ is when θ lies in QII and has $\theta' \approx 0.38$ as

its reference angle. In QII, $\theta = \pi - \theta' \approx 3.14 - 0.38 \approx 2.76 \Rightarrow \sqrt{\frac{3}{10}}\,t \approx 2.76 \Rightarrow t \approx 5.04$ sec.

[76] $\frac{\sin 25°}{\sin\theta_2} = 1.33 \Rightarrow \sin\theta_2 = \frac{\sin 25°}{1.33} \approx 0.3178 \Rightarrow \theta_2 \approx \sin^{-1}(0.3178) \Rightarrow \theta_2 \approx 18.5°$.

[77] $\frac{\sin\theta_1}{\sin 17.6°} = 1.36 \Rightarrow \sin\theta_1 = 1.36 \sin 17.6° \approx 0.4112 \Rightarrow \theta_1 \approx \sin^{-1}(0.4112) \Rightarrow \theta_1 \approx 24.3°$.

[78] $\frac{\sin 10.2°}{\sin\theta_2} = 1.65 \Rightarrow \sin\theta_2 = \frac{\sin 10.2°}{1.65} \approx 0.1073 \Rightarrow \theta_2 \approx \sin^{-1}(0.1073) \Rightarrow \theta_2 \approx 6.2°$.

[79] $\frac{\sin\theta_1}{\sin 23.7°} = 2.42 \Rightarrow \sin\theta_1 = 2.42 \sin 23.7° \approx 0.9727 \Rightarrow \theta_1 \approx \sin^{-1}(0.9727) \Rightarrow \theta_1 \approx 76.6°$.

[80] $\dfrac{\sin 50.3°}{\sin \theta_2} = 1.63 \Rightarrow \sin \theta_2 = \dfrac{\sin 50.3°}{1.63} \approx 0.4720 \Rightarrow \theta_2 \approx \sin^{-1}(0.4.720) \Rightarrow \theta_2 \approx 28.2°.$

[81] From the graphs, we estimate the x–coordinates of the points of intersection (to the nearest tenth) to be 0.9 and 5.6. Since the second linear equation solved in Example 41 is equivalent to $3 \cos x = 2$, these solutions should be identical to those we found there.   (See graph below).

[82] From the graphs, we estimate the x–coordinates of the points of intersection (to the nearest tenth) to be 2.1 and 4.2. Since the first linear equation solved in Example 41 is equivalent to $2 \cos x = -1$, these solutions should match those we found there.   (See graph below).

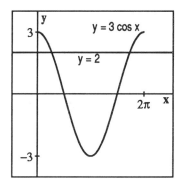

Figure 81

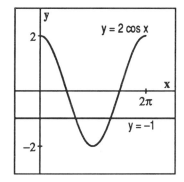

Figure 82

[83] From the graphs, we estimate the x–coordinates of the points of intersection (to the nearest tenth) to be 0.8 and 2.3. These values of x approximate the solutions in $[0, 2\pi)$ to the equation $\sin x = \dfrac{3}{4}$. (See graph below).

[84] From the graphs, we estimate the x–coordinates of the points of intersection (to the nearest tenth) to be 1.7 and 4.6. These values of x approximate the solutions in $[0, 2\pi)$ to the equation $\cos x = -0.15$. (See graph below).

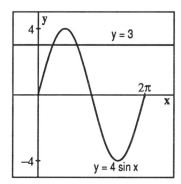

Figure 83

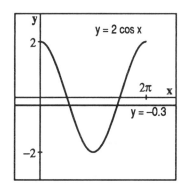

Figure 84

[85] From the graphs, we estimate the x–coordinates of the points of intersection (to the nearest tenth) to be 0.4 and 5.1. These values of x approximate the solutions in $[0, 2\pi)$ to the equation $\cos x = e^{-x/5}$. (See graph below).

[86] From the graphs, we estimate the x–coordinate of the point of intersection (to the nearest tenth) to be 2.2. This value of x approximates the solution in $[0, 2\pi)$ to the equation $\sin x = \ln x$. (See graph below).

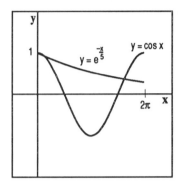

Figure 85   **Figure 86**

REVIEW EXERCISES

[1] $202° \, 43'29'' = \left(202 + \dfrac{43}{60} + \dfrac{29}{3600}\right)° \approx 202.725°.$

[2] $112.453° = 112° + 0.453°.$ Since $0.453° = (0.453)(60') = 27.18' = 27 + 0.18'$ and $0.18' = (0.18)(60'') = 10.8'' \approx 11''$, $112.453° \approx 112° \, 27'11''.$

[3] $100° = (100)\left(\dfrac{\pi}{180} \text{ radians}\right) = \dfrac{5\pi}{9} \text{ radians.}$ (See graph below).

[4] $210° = 210\left(\dfrac{\pi}{180} \text{ radians}\right) = \dfrac{7\pi}{6} \text{ radians.}$ (See graph below).

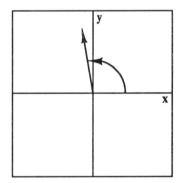

Figure 3   **Figure 4**

[5] $-225° = -225\left(\dfrac{\pi}{180} \text{ radians}\right) = -\dfrac{5\pi}{4} \text{ radians.}$ (See graph below).

[6] $-120° = -120\left(\dfrac{\pi}{180} \text{ radians}\right) = -\dfrac{2\pi}{3} \text{ radians.}$ (See graph below).

[7] $-\dfrac{7\pi}{10} = \left(-\dfrac{7\pi}{10}\right)\left(\dfrac{180}{\pi}\right)° = -126°.$ (See graph below).

[8] $-\dfrac{3\pi}{8} = \left(-\dfrac{3\pi}{8}\right)\left(\dfrac{180}{\pi}\right)^{\circ} = -67.5^{\circ}.$ (See graph below).

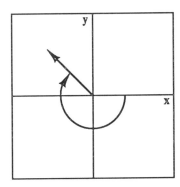

Figure 5

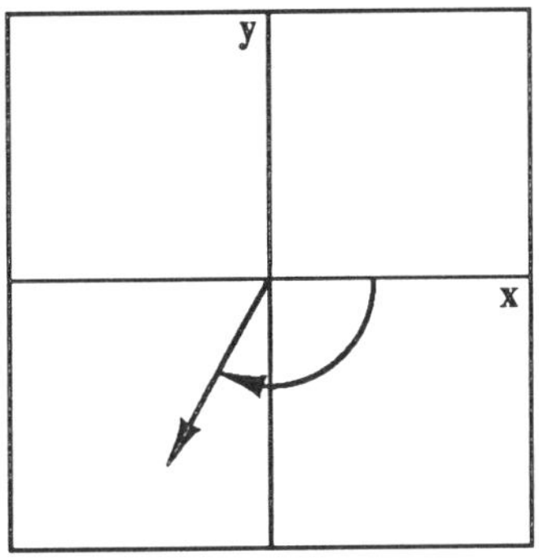

Figure 6

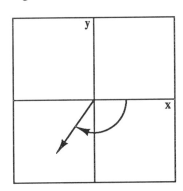

Figure 7

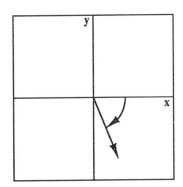

Figure 8

[9] $\dfrac{13\pi}{24} = \left(\dfrac{13\pi}{24}\right)\left(\dfrac{180}{\pi}\right)^{\circ} = 97.5^{\circ}.$ (See graph below).

[10] $\dfrac{13\pi}{15} = \left(\dfrac{13\pi}{15}\right)\left(\dfrac{180}{\pi}\right)^{\circ} = 156^{\circ}.$ (See graph below).

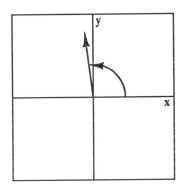

Figure 9

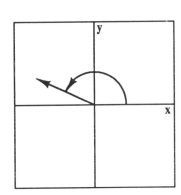

Figure 10

[11] $s = r\,\theta = (12)\left[(35)\left(\dfrac{\pi}{180}\right)\right] = \dfrac{7\pi}{3}$ cm. $A = \dfrac{1}{2}\,r^2\,\theta = \dfrac{1}{2}\,r\,s = \dfrac{1}{2}(12)\left(\dfrac{7\pi}{3}\right) = 14\,\pi$ cm².

[12] $s = r\,\theta = (8)(9) = 72$ in. $A = \dfrac{1}{2}\,r^2\,\theta = \dfrac{1}{2}\,r\,s = \dfrac{1}{2}(8)(72) = 2.88$ in².

273

[13] $\sin(-210°) = \frac{1}{2}$.

[14] $\csc(-60°) = -\frac{2}{\sqrt{3}}$.

[15] $\cos 315° = \frac{\sqrt{2}}{2}$.

[16] $\tan 240° = \sqrt{3}$.

[17] $\theta \in$ QIII and $\tan\theta = \frac{3}{2} = \frac{y}{x} \Rightarrow y = -3$ and $x = -2 \Rightarrow Q = (-2, -3)$ is on the terminal side of $\theta \Rightarrow r =$

$d(O, Q) = \sqrt{(-2)^2 + (-3)^2} = \sqrt{13}$. So, $\sin\theta = \frac{y}{r} = -\frac{3}{\sqrt{13}}$, $\cos\theta = \frac{x}{r} = -\frac{2}{\sqrt{13}}$, $\sec\theta = \frac{r}{x} = -\frac{\sqrt{13}}{2}$,

$\csc\theta = \frac{r}{y} = -\frac{\sqrt{13}}{3}$, and $\cot\theta = \frac{2}{3}$.   (See graph below).

[18] $\theta \in$ QII and $\cos\theta = -\frac{3}{4} = \frac{x}{r} \Rightarrow x = -3$ and $r = 4 \Rightarrow x^2 + y^2 = r^2 \Rightarrow 9 + y^2 = 16 \Rightarrow y^2 = 7 \Rightarrow y = \sqrt{7}$

$(\text{In QII}, y > 0) \Rightarrow Q = (-3, \sqrt{7})$ is on the terminal side of θ. So, $\sin\theta = \frac{y}{r} = \frac{\sqrt{7}}{4}$, $\csc\theta = \frac{r}{y} = \frac{4}{\sqrt{7}}$,

$\sec\theta = \frac{r}{x} = -\frac{4}{3}$, $\tan\theta = \frac{y}{x} = -\frac{\sqrt{7}}{3}$, and $\cot\theta = \frac{x}{y} = -\frac{3}{\sqrt{7}}$.   (See graph below).

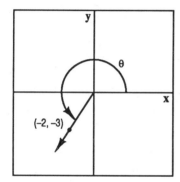

Figure 17

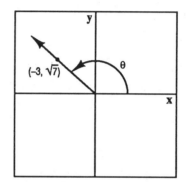

Figure 18

[19] $\theta \in$ QIV and $\sin\theta = -\frac{4}{9} = \frac{y}{r} \Rightarrow y = -4$ and $r = 9 \Rightarrow x^2 + y^2 = r^2 \Rightarrow x^2 + 16 = 81 \Rightarrow x^2 = 65 \Rightarrow x =$

$\sqrt{65}$ $(\text{In QIV}, x > 0) \Rightarrow Q = (\sqrt{65}, -4)$ is on the terminal side of θ. So, $\cos\theta = \frac{x}{r} = \frac{\sqrt{65}}{9}$, $\sec\theta = \frac{r}{x} =$

$\frac{9}{\sqrt{65}}$, $\csc\theta = \frac{r}{y} = -\frac{9}{4}$, $\tan\theta = \frac{y}{x} = -\frac{4}{\sqrt{65}}$, and $\cot\theta = \frac{x}{y} = -\frac{\sqrt{65}}{4}$.   (See graph below).

[20] $\theta \in$ QIII and $\sec\theta = -5 = \frac{r}{x} = \frac{5}{-1} \Rightarrow r = 5$ and $x = -1 \Rightarrow x^2 + y^2 = r^2 \Rightarrow 1 + y^2 = 25 \Rightarrow y^2 = 24 \Rightarrow$

$y = -2\sqrt{6}$ $(\text{In QIII}, y < 0) \Rightarrow Q = (-1, -2\sqrt{6})$ is on the terminal side of θ. So, $\sin\theta = \frac{y}{r} = -\frac{2\sqrt{6}}{5}$,

$\cos\theta = \frac{x}{r} = -\frac{1}{5}$, $\tan\theta = \frac{y}{x} = 2\sqrt{6}$, $\cot\theta = \frac{x}{y} = \frac{1}{2\sqrt{6}}$, and $\csc\theta = \frac{r}{y} = -\frac{5}{2\sqrt{6}}$.   (See graph below).

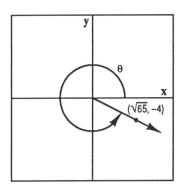

Figure 19 **Figure 20**

[21] $d = 36$ in $\Rightarrow r = 18$ in $= \frac{3}{2}$ ft. Since $s = r\,\theta$ and since one revolution corresponds to a central angle of

$2\,\pi$ radians, it follows that 1 rev $= \left(\frac{3}{2}\right)(2\,\pi) = 3\,\pi$ ft. Since 1 mile $= 5,280$ ft and 1 rev $= 3\,\pi$ ft, the

number of revolutions the wheel makes in 1 mile is $\frac{5280}{3\,\pi} = \frac{1760}{\pi}$.

[22] Let θ (measured in radians) be the central angle joining the two cities. Then $s = r\,\theta \Rightarrow 1000$ mi $=$

$(3960\text{ mi})(\theta) \Rightarrow \theta = \frac{1000}{3960} = \frac{25}{99}.$ (See graph below).

[23] Since 1 revolution of the minute hand takes 60 minutes, 45 min $= \frac{45}{60}$ rev $= \frac{3}{4}$ rev. Since 1 revolution

corresponds to a central angle of $2\,\pi$ radians, the central angle is $\theta = \frac{3}{4}(2\,\pi) = \frac{3\,\pi}{2}$ radians. Since the

minute hand is 12 in $= 1$ ft long, it follows that the distance the tip of the minute hand travels in

45 minutes is $s = r\,\theta = (1)\left(\frac{3\,\pi}{2}\right) = \frac{3\,\pi}{2}$ feet.

[24] Since $7° = 7\left(\frac{\pi}{180}\right) = \frac{7\,\pi}{180}$ radians, $s = r\,\theta \Rightarrow \frac{\pi}{2} = r\left(\frac{7\,\pi}{180}\right) \Rightarrow r = \frac{90}{7}$ ft ≈ 12.86 ft. The chain is about

12.86 ft long.   (See graph below).

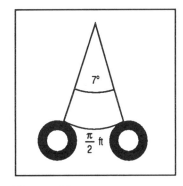

Figure 22 **Figure 24**

[25] Since $18° = 18\left(\frac{\pi}{180}\right) = \frac{\pi}{10}$ radians, it follows that $s = r\,\theta \Rightarrow 30 = r\left(\frac{\pi}{10}\right) \Rightarrow r = \frac{300}{\pi}$ cm.

[26] The central angle generated in one minute has radian measure $\theta = (1000)(2\pi) = 2000\,\pi$. Hence, the angular velocity of the wheel is $\omega = \frac{\theta}{T} = \frac{2000\,\pi}{1} = 2000\,\pi$ radians per minute. The linear velocity is $v = r\,\omega = (5)(2000\,\pi) = 10{,}000\,\pi$ feet per minute.

[27] The central angle generated in one minute has radian measure $\theta = (1500)(2\pi) = 3000\,\pi$. Hence, the angular velocity is $\omega = \frac{\theta}{T} = \frac{3000\,\pi}{1} = 3000\,\pi$ radians per minute. The linear velocity is $v = r\,\omega = (6)(3000\,\pi) = 18{,}000\,\pi$ inches per minute.

[28] $d = 18$ in $\Rightarrow r = 9$ in $= \frac{3}{4}$ ft. Since 50 rpm $= \frac{50}{60} = \frac{5}{6}$ rev per sec, the central angle generated in one second has radian measure $\theta = \left(\frac{5}{6}\right)(2\pi) = \frac{5\pi}{3}$. So, the angular velocity is $\omega = \frac{\theta}{T} = \frac{\frac{5\pi}{3}}{1} = \frac{5\pi}{3}$ radians per sec. Hence, the linear velocity of a point on the rim of the wheel is $v = r\,\omega = \left(\frac{3}{4}\right)\left(\frac{5\pi}{3}\right) = \frac{5\pi}{4}$ feet per sec.

[29] Since $d = 26$ in, $r = 13$ in $= \frac{13}{12}$ ft. Since the linear velocity is 5 MPH, $v = 5$ MPH $= (5)(5280) = 26{,}400$ ft per hr. Since 1 hr $= 3600$ sec, $v = \frac{26{,}400}{3600} = \frac{22}{3}$ ft per sec. So, $v = r\,w \Rightarrow \frac{22}{3} = \left(\frac{13}{12}\right)w \Rightarrow w = \frac{88}{13}$ rad per sec. Since 2π radians is the central angle corresponding to one revolution, $\frac{88}{13} = (2\pi)\,x \Rightarrow x = \frac{44}{13\pi}$ rev per sec.

[30] Since the diameter of the larger pulley is 18 cm, its radius is 9 cm. As the larger pulley turns through 10 radians, the corresponding arc length is $s = r\,\theta = (9)(10) = 90$ cm. Similarly, since the diameter of the smaller pulley is 12 cm, its radius is 6 cm. Hence, $s = r\,\theta \Rightarrow 90 = 6\,\theta \Rightarrow \theta = 15$ radians. So, the smaller pulley turns through 15 radians.

[31] $(\cot t + \csc t)(\cot t - \csc t) = \cot^2 t - \csc^2 t = \cot^2 t - (1 + \cot^2 t) = -1.$

[32] $\csc t\,(\tan t - \sin t) = \frac{1}{\sin t}\left(\frac{\sin t}{\cos t} - \sin t\right) = \frac{1}{\sin t}\,(\sin t)\left(\frac{1}{\cos t} - 1\right) = \frac{1}{\cos t} - 1 = \sec t - 1.$

[33] $\dfrac{1}{1 + 2\tan^2 t + \tan^4 t} = \dfrac{1}{\left(1 + \tan^2 t\right)^2} = \dfrac{1}{\left(\sec^2 t\right)^2} = \dfrac{1}{\sec^4 t} = \cos^4 t.$

[34] $\dfrac{1}{1 + \sec t} + \dfrac{1}{1 - \sec t} = \dfrac{(1 - \sec t) + (1 + \sec t)}{1 - \sec^2 t} = \dfrac{2}{-\tan^2 t} = -2\cot^2 t.$

[35] $x = \frac{2\pi}{3} \Rightarrow \cos\frac{2\pi}{3} = -\frac{1}{2}$ and $\left|\cos\frac{2\pi}{3}\right| = \left|-\frac{1}{2}\right| = \frac{1}{2}$ and $-\frac{1}{2} \neq \frac{1}{2}.$

[36] $x = \frac{\pi}{4} \Rightarrow \sin\frac{\pi}{4} + \cos\frac{\pi}{4} = \frac{\sqrt{2}}{2} + \frac{\sqrt{2}}{2} = \sqrt{2}$ and $\sqrt{2} \neq 1.$

[37] Since $-\frac{\pi}{2} < \theta < \frac{\pi}{2} \Rightarrow \sec\theta > 0 \Rightarrow |\sec\theta| = \sec\theta$, we have $\sqrt{9 + u^2} = \sqrt{9 + 9\tan^2\theta} = \sqrt{9\left(1 + \tan^2\theta\right)} = \sqrt{9\sec^2\theta} = 3\,|\sec\theta| = 3\sec\theta.$

[38] Since $-\frac{\pi}{2} < \theta < \frac{\pi}{2} \Rightarrow \cos\theta > 0 \Rightarrow |\cos\theta| = \cos\theta$, we have $\dfrac{2}{\sqrt{25-x^2}} = \dfrac{2}{\sqrt{25-25\sin^2\theta}} =$

$\dfrac{2}{\sqrt{25\left(1-\sin^2\theta\right)}} = \dfrac{2}{\sqrt{25\cos^2\theta}} = \dfrac{2}{5|\cos\theta|} = \dfrac{2}{5\cos\theta} = \dfrac{2}{5}\sec\theta.$

[39] $\cos\left(\dfrac{7\pi}{12}\right) = \cos\left(\dfrac{\pi}{3} + \dfrac{\pi}{4}\right) = \cos\dfrac{\pi}{3}\cos\dfrac{\pi}{4} - \sin\dfrac{\pi}{3}\sin\dfrac{\pi}{4} = \left(\dfrac{1}{2}\right)\left(\dfrac{\sqrt{2}}{2}\right) - \left(\dfrac{\sqrt{3}}{2}\right)\left(\dfrac{\sqrt{2}}{2}\right) = \dfrac{\sqrt{2}-\sqrt{6}}{4}.$

[40] $\sin(-15°) = \sin(45° - 60°) = \sin 45°\cos 60° - \cos 45°\sin 60° = \left(\dfrac{\sqrt{2}}{2}\right)\left(\dfrac{1}{2}\right) - \left(\dfrac{\sqrt{2}}{2}\right)\left(\dfrac{\sqrt{3}}{2}\right) = \dfrac{\sqrt{2}-\sqrt{6}}{4}.$

[41] $\sin(195°) = \sin(150° + 45°) = \sin 150°\cos 45° + \cos 150°\sin 45° = \left(\dfrac{1}{2}\right)\left(\dfrac{\sqrt{2}}{2}\right) + \left(-\dfrac{\sqrt{3}}{2}\right)\left(\dfrac{\sqrt{2}}{2}\right) =$

$\dfrac{\sqrt{2}-\sqrt{6}}{4}.$

[42] $\tan\left(\dfrac{5\pi}{12}\right) = \tan\left(\dfrac{\pi}{4} + \dfrac{\pi}{6}\right) = \dfrac{\tan\frac{\pi}{4} + \tan\frac{\pi}{6}}{1 - \tan\frac{\pi}{4}\tan\frac{\pi}{6}} = \dfrac{1 + \frac{1}{\sqrt{3}}}{1 - (1)\left(\frac{1}{\sqrt{3}}\right)} = \dfrac{\sqrt{3}+1}{\sqrt{3}-1}.$

[43] $\sin\left(\dfrac{2\pi}{3}\right)\cos\left(\dfrac{\pi}{6}\right) - \cos\left(\dfrac{2\pi}{3}\right)\sin\left(\dfrac{\pi}{6}\right) = \sin\left(\dfrac{2\pi}{3} - \dfrac{\pi}{6}\right) = \sin\left(\dfrac{\pi}{2}\right) = 1.$

[44] $\cos 40°\cos 10° + \sin 40°\sin 10° = \cos(40° - 10°) = \cos 30° = \dfrac{\sqrt{3}}{2}.$

[45] $\cos 201°\cos 24° - \sin 201°\sin 24° = \cos(201° + 24°) = \cos 225° = -\dfrac{\sqrt{2}}{2}.$

[46] $\dfrac{\tan 200° - \tan 80°}{1 + \tan 200°\tan 80°} = \tan(200° - 80°) = \tan 120° = -\sqrt{3}.$

[47] $\sin 32°\cos 58° + \cos 32°\sin 58° = \sin(32° + 58°) = \sin 90° = 1.$

[48] $\dfrac{\tan 110° + \tan 100°}{1 - \tan 110°\tan 100°} = \tan(110° + 100°) = \tan 210° = \dfrac{1}{\sqrt{3}}.$

[49] $\cos u = -\dfrac{1}{4}$ and $u \in \text{QIII} \Rightarrow \sin u = -\sqrt{1 - \left(-\dfrac{1}{4}\right)^2} = -\dfrac{\sqrt{15}}{4}.$ $\sin v = \dfrac{\sqrt{2}}{3}$ and $v \in \text{QII} \Rightarrow \cos v =$

$-\sqrt{1 - \left(\dfrac{\sqrt{2}}{3}\right)^2} = -\dfrac{\sqrt{7}}{3}.$

a) $\sin(u+v) = \sin u\cos v + \cos u\sin v = \left(-\dfrac{\sqrt{15}}{4}\right)\left(-\dfrac{\sqrt{7}}{3}\right) + \left(-\dfrac{1}{4}\right)\left(\dfrac{\sqrt{2}}{3}\right) = \dfrac{\sqrt{105}-\sqrt{2}}{12}.$

b) $\cos(u+v) = \cos u\cos v - \sin u\sin v = \left(-\dfrac{1}{4}\right)\left(-\dfrac{\sqrt{7}}{3}\right) - \left(-\dfrac{\sqrt{15}}{4}\right)\left(\dfrac{\sqrt{2}}{3}\right) = \dfrac{\sqrt{7}+\sqrt{30}}{12}.$

c) $\sin(u-v) = \sin u\cos v - \cos u\sin v = \left(-\dfrac{\sqrt{15}}{4}\right)\left(-\dfrac{\sqrt{7}}{3}\right) - \left(-\dfrac{1}{4}\right)\left(\dfrac{\sqrt{2}}{3}\right) = \dfrac{\sqrt{105}+\sqrt{2}}{12}.$

d) $\cos(u-v) = \cos u\cos v + \sin u\sin v = \left(-\dfrac{1}{4}\right)\left(-\dfrac{\sqrt{7}}{3}\right) + \left(-\dfrac{\sqrt{15}}{4}\right)\left(\dfrac{\sqrt{2}}{3}\right) = \dfrac{\sqrt{7}-\sqrt{30}}{12}.$

e) $\tan(u+v) = \dfrac{\sin(u+v)}{\cos(u+v)} = \dfrac{\frac{\sqrt{105}-\sqrt{2}}{12}}{\frac{\sqrt{7}+\sqrt{30}}{12}} = \dfrac{\sqrt{105}-\sqrt{2}}{\sqrt{7}+\sqrt{30}}.$

f) $\tan(u-v) = \dfrac{\sin(u-v)}{\cos(u-v)} = \dfrac{\frac{\sqrt{105}+\sqrt{2}}{12}}{\frac{\sqrt{7}-\sqrt{30}}{12}} = \dfrac{\sqrt{105}+\sqrt{2}}{\sqrt{7}-\sqrt{30}}.$

g) Since $\sin(u+v) > 0$ and $\cos(u+v) > 0$, $u+v \in$ QI.

h) Since $\sin(u-v) > 0$ and $\cos(u-v) < 0$, $u-v \in$ QII.

[50] $\tan u = \frac{1}{2}$ and $u \in$ QIII $\Rightarrow \sec u = -\sqrt{1+\left(\frac{1}{2}\right)^2} = -\dfrac{\sqrt{5}}{2} \Rightarrow \cos u = -\dfrac{2}{\sqrt{5}} \Rightarrow \sin u =$

$-\sqrt{1-\left(-\dfrac{2}{\sqrt{5}}\right)^2} = -\dfrac{1}{\sqrt{5}}.$ Similarly, $\cos v = \frac{1}{5}$ and $v \in$ QIV $\Rightarrow \sin v = -\sqrt{1-\left(\frac{1}{5}\right)^2} = -\dfrac{2\sqrt{6}}{5}.$

a) $\sin(u+v) = \sin u \cos v + \cos u \sin v = \left(-\dfrac{1}{\sqrt{5}}\right)\left(\dfrac{1}{5}\right) + \left(-\dfrac{2}{\sqrt{5}}\right)\left(-\dfrac{2\sqrt{6}}{5}\right) = \dfrac{-1+4\sqrt{6}}{5\sqrt{5}}.$

b) $\cos(u+v) = \cos u \cos v - \sin u \sin v = \left(-\dfrac{2}{\sqrt{5}}\right)\left(\dfrac{1}{5}\right) - \left(-\dfrac{1}{\sqrt{5}}\right)\left(-\dfrac{2\sqrt{6}}{5}\right) = \dfrac{-2-2\sqrt{6}}{5\sqrt{5}}.$

c) $\sin(u-v) = \sin u \cos v - \cos u \sin v = \left(-\dfrac{1}{\sqrt{5}}\right)\left(\dfrac{1}{5}\right) - \left(-\dfrac{2}{\sqrt{5}}\right)\left(-\dfrac{2\sqrt{6}}{5}\right) = \dfrac{-1-4\sqrt{6}}{5\sqrt{5}}.$

d) $\cos(u-v) = \cos u \cos v + \sin u \sin v = \left(-\dfrac{2}{\sqrt{5}}\right)\left(\dfrac{1}{5}\right) + \left(-\dfrac{1}{\sqrt{5}}\right)\left(-\dfrac{2\sqrt{6}}{5}\right) = \dfrac{-2+2\sqrt{6}}{5\sqrt{5}}.$

e) $\tan(u+v) = \dfrac{\sin(u+v)}{\cos(u+v)} = \dfrac{\frac{-1+4\sqrt{6}}{5\sqrt{5}}}{\frac{-2-2\sqrt{6}}{5\sqrt{5}}} = \dfrac{1-4\sqrt{6}}{2\sqrt{6}+2}.$

f) $\tan(u-v) = \dfrac{\sin(u-v)}{\cos(u-v)} = \dfrac{\frac{-1-4\sqrt{6}}{5\sqrt{5}}}{\frac{-2+2\sqrt{6}}{5\sqrt{5}}} = \dfrac{4\sqrt{6}+1}{2-2\sqrt{6}}.$

g) Since $\sin(u+v) > 0$ and $\cos(u+v) < 0$, $u+v \in$ QII.

h) Since $\sin(u-v) < 0$ and $\cos(u-v) > 0$, $u-v \in$ QIV.

[51] $\sin(270° + \theta) = \sin 270° \cos \theta + \cos 270° \sin \theta = (-1)(\cos \theta) + (0)(\sin \theta) = -\cos \theta.$

[52] $\csc(180° - v) = \dfrac{1}{\sin(180° - v)} = \dfrac{1}{\sin 180° \cos v - \cos 180° \sin v} = \dfrac{1}{(0)(\cos v) - (-1)(\sin v)} =$

$\dfrac{1}{\sin v} = \csc v.$

[53] $\cos\left(u - \dfrac{\pi}{6}\right) = \cos u \cos \dfrac{\pi}{6} + \sin u \sin \dfrac{\pi}{6} = (\cos u)\left(\dfrac{\sqrt{3}}{2}\right) + (\sin u)\left(\dfrac{1}{2}\right) = \dfrac{\sqrt{3}\cos u + \sin u}{2}.$

[54] $\tan\left(u + \dfrac{\pi}{3}\right) = \dfrac{\tan u + \tan \frac{\pi}{3}}{1 - \tan u \tan \frac{\pi}{3}} = \dfrac{\tan u + \sqrt{3}}{1 - (\tan u)(\sqrt{3})} = \dfrac{\tan u + \sqrt{3}}{1 - \sqrt{3}\tan u}.$

[55]   If $u = \sin^{-1}\left(\frac{1}{2}\right)$, then $\sin u = \frac{1}{2}$ and $-\frac{\pi}{2} \le u \le \frac{\pi}{2} \Rightarrow u \in \text{QI} \Rightarrow u = \frac{\pi}{6} \Rightarrow \cos u = \frac{\sqrt{3}}{2}$. If $v =$

   $\tan^{-1}\left(-\frac{1}{4}\right)$, then $\tan v = -\frac{1}{4}$ and $-\frac{\pi}{2} < v < \frac{\pi}{2} \Rightarrow v \in \text{QIV} \Rightarrow \sec v = \sqrt{1 + \left(-\frac{1}{4}\right)^2} = \frac{\sqrt{17}}{4} \Rightarrow$

   $\cos v = \frac{4}{\sqrt{17}} \Rightarrow \sin v = -\sqrt{1 - \left(\frac{4}{\sqrt{17}}\right)^2} = -\frac{1}{\sqrt{17}}$. So, $\cos\left[\sin^{-1}\left(\frac{1}{2}\right) - \tan^{-1}\left(-\frac{1}{4}\right)\right] = \cos(u - v) =$

   $\cos u \cos v + \sin u \sin v = \left(\frac{\sqrt{3}}{2}\right)\left(\frac{4}{\sqrt{17}}\right) + \left(\frac{1}{2}\right)\left(-\frac{1}{\sqrt{17}}\right) = \frac{4\sqrt{3}-1}{2\sqrt{17}} \approx 0.72$.

[56]   If $u = \sin^{-1}\left(-\frac{2}{3}\right)$, then $\sin u = -\frac{2}{3}$ and $-\frac{\pi}{2} \le u \le \frac{\pi}{2} \Rightarrow \cos u = \sqrt{1 - \left(-\frac{2}{3}\right)^2} = \frac{\sqrt{5}}{3}$. If $v =$

   $\cos^{-1}\left(-\frac{1}{2}\right)$, then $\cos v = -\frac{1}{2}$ and $0 \le v \le \pi \Rightarrow v \in \text{QII} \Rightarrow v = \frac{2\pi}{3} \Rightarrow \sin v = \frac{\sqrt{3}}{2}$. So,

   $\sin\left[\sin^{-1}\left(-\frac{2}{3}\right) + \cos^{-1}\left(-\frac{1}{2}\right)\right] = \sin(u + v) = \sin u \cos v + \cos u \sin v = \left(-\frac{2}{3}\right)\left(-\frac{1}{2}\right) + \left(\frac{\sqrt{5}}{3}\right)\left(\frac{\sqrt{3}}{2}\right) =$

   $\frac{2 + \sqrt{15}}{6} \approx 0.98$.

[57] $2\sin 25 \cos 25 = \sin[2(25)] = \sin 50$.   [58] $2\cos^2 83 - 1 = \cos[2(83)] = \cos 166$.

[59] $1 - 2\sin^2 23° = \cos[2(23°)] = \cos 46°$.   [60] $\frac{\sin 100°}{1 + \cos 100°} = \tan\frac{100°}{2} = \tan 50°$.

[61] $\sin v = -\frac{\sqrt{2}}{5}$ and $v \in \text{QIV} \Rightarrow \cos v = \sqrt{1 - \left(-\frac{\sqrt{2}}{5}\right)^2} = \frac{\sqrt{23}}{5}$. So, $\sin 2v = 2\sin v \cos v =$

   $2\left(-\frac{\sqrt{2}}{5}\right)\left(\frac{\sqrt{23}}{5}\right) = -\frac{2\sqrt{46}}{25}$, $\cos 2v = 1 - 2\sin^2 v = 1 - 2\left(-\frac{\sqrt{2}}{5}\right)^2 = \frac{21}{25}$, and $\tan 2v = \frac{\sin 2v}{\cos 2v} =$

   $\frac{-\frac{2\sqrt{46}}{25}}{\frac{21}{25}} = -\frac{2\sqrt{46}}{21}$. Since $\sin 2v < 0$ and $\cos 2v > 0$, $2v \in \text{QIV}$.

[62] $\tan v = \frac{1}{2}$ and $v \in \text{QIII} \Rightarrow \sec v = -\sqrt{1 + \left(\frac{1}{2}\right)^2} = -\frac{\sqrt{5}}{2} \Rightarrow \cos v = -\frac{2}{\sqrt{5}} \Rightarrow \sin v =$

   $-\sqrt{1 - \left(-\frac{2}{\sqrt{5}}\right)^2} = -\frac{1}{\sqrt{5}}$. So, $\sin 2v = 2\sin v \cos v = 2\left(-\frac{1}{\sqrt{5}}\right)\left(-\frac{2}{\sqrt{5}}\right) = \frac{4}{5}$, $\cos 2v =$

   $\cos^2 v - \sin^2 v = \left(-\frac{2}{\sqrt{5}}\right)^2 - \left(-\frac{1}{\sqrt{5}}\right)^2 = \frac{3}{5}$, and $\tan 2v = \frac{\sin 2v}{\cos 2v} = \frac{\frac{4}{5}}{\frac{3}{5}} = \frac{4}{3}$. Since $\sin 2v > 0$ and

   $\cos 2v > 0$, $2v \in \text{QI}$.

[63] $\sin\left(\frac{5\pi}{8}\right) = \sin\left[\left(\frac{1}{2}\right)\left(\frac{5\pi}{4}\right)\right] = \sqrt{\frac{1 - \cos\frac{5\pi}{4}}{2}} = \sqrt{\frac{1 - \left(-\frac{\sqrt{2}}{2}\right)}{2}} = \sqrt{\frac{2 + \sqrt{2}}{4}} = \frac{\sqrt{2 + \sqrt{2}}}{2}$.

[64] $\cos\left(\frac{\pi}{8}\right) = \cos\left[\left(\frac{1}{2}\right)\left(\frac{\pi}{4}\right)\right] = \sqrt{\frac{1 + \cos\frac{\pi}{4}}{2}} = \sqrt{\frac{1 + \frac{\sqrt{2}}{2}}{2}} = \sqrt{\frac{2 + \sqrt{2}}{4}} = \frac{\sqrt{2 + \sqrt{2}}}{2}$.

[65] $\tan(67.5°) = \tan\left(\frac{135°}{2}\right) = \sqrt{\frac{1 - \cos 135°}{1 + \cos 135°}} = \sqrt{\frac{1-\left(-\frac{\sqrt{2}}{2}\right)}{1+\left(-\frac{\sqrt{2}}{2}\right)}} = \sqrt{\frac{2+\sqrt{2}}{2-\sqrt{2}}}.$

[66] $\cos(-22.5°) = \cos\left(-\frac{45°}{2}\right) = \sqrt{\frac{1 + \cos(-45°)}{2}} = \sqrt{\frac{1+\frac{\sqrt{2}}{2}}{2}} = \sqrt{\frac{2+\sqrt{2}}{4}} = \frac{\sqrt{2+\sqrt{2}}}{2}.$

[67] $\cos v = \frac{1}{6}$ and $\frac{3\pi}{2} < v < 2\pi \Rightarrow \frac{3\pi}{4} < \frac{v}{2} < \pi \Rightarrow \frac{v}{2} \in$ QII. So, $\sin \frac{v}{2} = \sqrt{\frac{1 - \cos v}{2}} = \sqrt{\frac{1-\left(\frac{1}{6}\right)}{2}} =$

$\sqrt{\frac{6-1}{12}} = \frac{\sqrt{5}}{2\sqrt{3}}$, $\cos \frac{v}{2} = -\sqrt{\frac{1 + \cos v}{2}} = -\sqrt{\frac{1+\frac{1}{6}}{2}} = -\sqrt{\frac{6+1}{12}} = -\frac{\sqrt{7}}{2\sqrt{3}}$, and $\tan \frac{v}{2} =$

$\frac{\sin \frac{v}{2}}{\cos \frac{v}{2}} = \frac{\frac{\sqrt{5}}{2\sqrt{3}}}{-\frac{\sqrt{7}}{2\sqrt{3}}} = -\frac{\sqrt{5}}{\sqrt{7}}.$

[68] $\tan v = 4$ and $\pi < v < \frac{3\pi}{2} \Rightarrow \sec v = -\sqrt{1 + (4)^2} = -\sqrt{17} \Rightarrow \cos v = -\frac{1}{\sqrt{17}}$. $\pi < v < \frac{3\pi}{2} \Rightarrow$

$\frac{\pi}{2} < \frac{v}{2} < \frac{3\pi}{4} \Rightarrow \frac{v}{2} \in$ QII. So, $\sin \frac{v}{2} = \sqrt{\frac{1 - \cos v}{2}} = \sqrt{\frac{1-\left(-\frac{1}{\sqrt{17}}\right)}{2}} = \sqrt{\frac{\sqrt{17}+1}{2\sqrt{17}}}$, $\cos \frac{v}{2} =$

$-\sqrt{\frac{1 + \cos v}{2}} = -\sqrt{\frac{1+\left(-\frac{1}{\sqrt{17}}\right)}{2}} = -\sqrt{\frac{\sqrt{17}-1}{2\sqrt{17}}}$, and $\tan \frac{v}{2} = \frac{\sin \frac{v}{2}}{\cos \frac{v}{2}} = \frac{\sqrt{\frac{\sqrt{17}+1}{2\sqrt{17}}}}{-\sqrt{\frac{\sqrt{17}-1}{2\sqrt{17}}}} =$

$-\sqrt{\frac{\sqrt{17}+1}{\sqrt{17}-1}}.$

[69] If $t = \cos^{-1}\left(-\frac{1}{2}\right)$, then $\cos t = -\frac{1}{2}$ and $0 \le t \le \pi$. So, $\cos\left[2 \cos^{-1}\left(-\frac{1}{2}\right)\right] = \cos 2t = 2\cos^2 t - 1 =$

$2\left(-\frac{1}{2}\right)^2 - 1 = -\frac{1}{2}.$

[70] If $t = \sin^{-1}\left(\frac{3}{4}\right)$, then $\sin t = \frac{3}{4}$ and $-\frac{\pi}{2} \le t \le \frac{\pi}{2} \Rightarrow \cos t = \sqrt{1-\left(\frac{3}{4}\right)^2} = \frac{\sqrt{7}}{4}$. So, $\sin\left[2\sin^{-1}\left(\frac{3}{4}\right)\right] =$

$\sin 2t = 2 \sin t \cos t = 2\left(\frac{3}{4}\right)\left(\frac{\sqrt{7}}{4}\right) = \frac{3\sqrt{7}}{8}.$

[71] $\sin 3x \sin 2x = \frac{1}{2}\left[\cos(3x - 2x) - \cos(3x + 2x)\right] = \frac{1}{2}(\cos x - \cos 5x).$

[72] $\cos 8v \cos 2v = \frac{1}{2}\left[\cos(8v + 2v) + \cos(8v - 2v)\right] = \frac{1}{2}(\cos 10v + \cos 6v).$

[73] $\cos 50° \sin 20° = \frac{1}{2}\left[\sin(50° + 20°) - \sin(50° - 20°)\right] = \frac{1}{2}(\sin 70° - \sin 30°) = \frac{1}{2}\left(\sin 70° - \frac{1}{2}\right).$

[74] $\sin 10° \cos 15° = \frac{1}{2}\left[\sin(10° + 15°) + \sin(10° - 15°)\right] = \frac{1}{2}\left[\sin 25° + \sin(-5°)\right] = \frac{1}{2}(\sin 25° - \sin 5°).$

[75] $\cos 6\theta + \cos 3\theta = 2 \cos \frac{6\theta + 3\theta}{2} \cos \frac{6\theta - 3\theta}{2} = 2 \cos \frac{9\theta}{2} \cos \frac{3\theta}{2}.$

[76] $\sin 10\theta + \sin 5\theta = 2 \sin \frac{10\theta + 5\theta}{2} \cos \frac{10\theta - 5\theta}{2} = 2 \sin \frac{15\theta}{2} \cos \frac{5\theta}{2}.$

[77] $\sin 62° - \sin 12° = 2\cos\dfrac{62° + 12°}{2}\sin\dfrac{62° - 12}{2} = 2\cos 37°\sin 25°.$

[78] $\cos 38° - \cos 20° = -2\sin\dfrac{38° + 20°}{2}\sin\dfrac{38° - 20°}{2} = -2\sin 29°\sin 9°.$

[79] $\sin 105° - \sin 195° = 2\cos\dfrac{105° + 195°}{2}\sin\dfrac{105° - 195°}{2} = 2\cos 150°\sin(-45°) =$

$-2\cos 150°\sin 45° = -2\left(-\dfrac{\sqrt{3}}{2}\right)\left(\dfrac{\sqrt{2}}{2}\right) = \dfrac{\sqrt{6}}{2}.$

[80] $\sin 195° + \sin 285° = 2\sin\dfrac{195° + 285°}{2}\cos\dfrac{195° - 285°}{2} = 2\sin 240°\cos(-45°) =$

$2\sin 240°\cos 45° = 2\left(-\dfrac{\sqrt{3}}{2}\right)\left(\dfrac{\sqrt{2}}{2}\right) = -\dfrac{\sqrt{6}}{2}.$

[81] $\cos\dfrac{7\pi}{8}\cos\dfrac{\pi}{8} = \dfrac{1}{2}\left[\cos\left(\dfrac{7\pi}{8} + \dfrac{\pi}{8}\right) + \cos\left(\dfrac{7\pi}{8} - \dfrac{\pi}{8}\right)\right] = \dfrac{1}{2}\left(\cos\pi + \cos\dfrac{3\pi}{4}\right) = \dfrac{1}{2}\left[(-1) + \left(-\dfrac{\sqrt{2}}{2}\right)\right] =$

$\dfrac{1}{2}\left(\dfrac{-2-\sqrt{2}}{2}\right) = \left(-\dfrac{1}{4}\right)(2 + \sqrt{2}).$

[82] $\sin\dfrac{3\pi}{8}\cos\dfrac{\pi}{8} = \dfrac{1}{2}\left[\sin\left(\dfrac{3\pi}{8} + \dfrac{\pi}{8}\right) + \sin\left(\dfrac{3\pi}{8} - \dfrac{\pi}{8}\right)\right] = \dfrac{1}{2}\left(\sin\dfrac{\pi}{2} + \sin\dfrac{\pi}{4}\right) = \dfrac{1}{2}\left[(1) + \left(\dfrac{\sqrt{2}}{2}\right)\right] = \dfrac{2 + \sqrt{2}}{4}.$

[83] $\sin x = -\dfrac{1}{2} \Rightarrow S = \left\{x \mid x = \dfrac{7\pi}{6} + 2k\pi \text{ or } x = \dfrac{11\pi}{6} + 2k\pi, \text{ where } k \text{ is any integer}\right\}.$

[84] $\cos x = -\dfrac{\sqrt{3}}{2} \Rightarrow S = \left\{x \mid x = \dfrac{5\pi}{6} + 2k\pi \text{ or } x = \dfrac{7\pi}{6} + 2k\pi, \text{ where } k \text{ is any integer}\right\}.$

[85] $\tan x = -1 \Rightarrow S = \left\{x \mid x = \dfrac{3\pi}{4} + k\pi, \text{ where } k \text{ is any integer}\right\}.$

[86] $\cot x = -\sqrt{3} \Rightarrow \tan x = -\dfrac{1}{\sqrt{3}} \Rightarrow S = \left\{x \mid x = \dfrac{5\pi}{6} + k\pi, \text{ where } k \text{ is any integer}\right\}.$

[87] $\cos 2t = -1 \Rightarrow 2t = \pi + 2k\pi \Rightarrow t = \dfrac{\pi}{2} + k\pi.$ Since t will be in $[0, 2\pi)$ when $k = 0$ or 1, $t = \dfrac{\pi}{2}$ or $t =$

$\dfrac{\pi}{2} + \pi = \dfrac{3\pi}{2}.$ ● $\left\{\dfrac{\pi}{2}, \dfrac{3\pi}{2}\right\}$

[88] $\cos 3t = -\dfrac{1}{2} \Rightarrow 3t = \dfrac{2\pi}{3} + 2k\pi \text{ or } 3t = \dfrac{4\pi}{3} + 2kt \Rightarrow t = \dfrac{2\pi}{9} + k\left(\dfrac{2\pi}{3}\right) \text{ or } t = \dfrac{4\pi}{9} + k\left(\dfrac{2\pi}{3}\right).$

Since t will be in $[0, 2\pi)$ when $k = 0$, 1, or 2, $t = \dfrac{2\pi}{9}$ or $t = \dfrac{4\pi}{9}$ or $t = \dfrac{2\pi}{9} + \dfrac{2\pi}{3} = \dfrac{8\pi}{9}$ or $t =$

$\dfrac{4\pi}{9} + \dfrac{2\pi}{3} = \dfrac{10\pi}{9}$ or $t = \dfrac{2\pi}{9} + \dfrac{4\pi}{3} = \dfrac{14\pi}{9}$ or $t = \dfrac{4\pi}{9} + \dfrac{4\pi}{3} = \dfrac{16\pi}{9}.$

● $\left\{\dfrac{2\pi}{9}, \dfrac{4\pi}{9}, \dfrac{8\pi}{9}, \dfrac{10\pi}{9}, \dfrac{14\pi}{9}, \dfrac{16\pi}{9}\right\}$

[89] $\csc t = -\sqrt{2} \Rightarrow \sin t = -\dfrac{1}{\sqrt{2}} \Rightarrow t = \dfrac{5\pi}{4} \text{ or } t = \dfrac{7\pi}{4}.$ ● $\left\{\dfrac{5\pi}{4}, \dfrac{7\pi}{4}\right\}$

[90] $\sin 2t = -\dfrac{\sqrt{3}}{2} \Rightarrow 2t = \dfrac{4\pi}{3} + 2k\pi \text{ or } 2t = \dfrac{5\pi}{3} + 2k\pi \Rightarrow t = \dfrac{2\pi}{3} + k\pi \text{ or } t = \dfrac{5\pi}{6} + k\pi.$ Since t will

be in $[0, 2\pi)$ when $k = 0$ or 1, $t = \dfrac{2\pi}{3}$ or $t = \dfrac{5\pi}{6}$ or $t = \dfrac{2\pi}{3} + \pi = \dfrac{5\pi}{3}$ or $t = \dfrac{5\pi}{6} + \pi = \dfrac{11\pi}{6}.$

● $\left\{\dfrac{2\pi}{3}, \dfrac{5\pi}{6}, \dfrac{5\pi}{3}, \dfrac{11\pi}{6}\right\}$

[91] $\sin 3\theta = -\frac{\sqrt{2}}{2} \Rightarrow 3\theta = 225° + k(360°)$ or $3\theta = 315° + k(360°) \Rightarrow \theta = 75° + k(120°)$ or $\theta =$

105° + k (120°). Since $0° \le \theta < 360°$ when $k = 0, 1$, or 2, $\theta = 75°$ or $\theta = 105°$ or $\theta = 75° + 120° =$

195° or $\theta = 105° + 120° = 225°$ or $\theta = 75° + 240° = 315°$ or $\theta = 105° + 240° = 345°$.

● $\{75°, 105°, 195°, 225°, 315°, 345°\}$

[92] $\cos\theta = -\frac{1}{\sqrt{2}} \Rightarrow \theta = 135°$ or $\theta = 225°$.

● $\{135°, 225°\}$

[93] $\sin 2\theta = -\sqrt{2}\sin\theta \Rightarrow 2\sin\theta\cos\theta = -\sqrt{2}\sin\theta \Rightarrow 2\sin\theta\cos\theta + \sqrt{2}\sin\theta = 0 \Rightarrow$

$\sin\theta(2\cos\theta + \sqrt{2}) = 0 \Rightarrow \sin\theta = 0$ or $\cos\theta = -\frac{\sqrt{2}}{2}$. If $\sin\theta = 0$, then $\theta = 0°$ or $\theta = 180°$. If $\cos\theta =$

$-\frac{\sqrt{2}}{2}$, then $\theta = 135°$ or $\theta = 225°$.

● $\{0°, 135°, 180°, 225°\}$

[94] $\sin 2\theta = \sqrt{3}\cos\theta \Rightarrow 2\sin\theta\cos\theta = \sqrt{3}\cos\theta \Rightarrow 2\sin\theta\cos\theta - \sqrt{3}\cos\theta = 0 \Rightarrow$

$\cos\theta(2\sin\theta - \sqrt{3}) = 0 \Rightarrow \cos\theta = 0$ or $\sin\theta = \frac{\sqrt{3}}{2}$. If $\cos\theta = 0$, then $\theta = 90°$ or $\theta = 270°$. If $\sin\theta =$

$\frac{\sqrt{3}}{2}$, then $\theta = 60°$ or $\theta = 120°$.

● $\{60°, 90°, 120°, 270°\}$

[95] $\sin 3t = -1 \Rightarrow 3t = \frac{3\pi}{2} + 2k\pi \Rightarrow t = \frac{\pi}{2} + k\left(\frac{2\pi}{3}\right)$. Since t will be in $[0, 2\pi)$ when $k = 0, 1$, or 2,

$t = \frac{\pi}{2}$ or $t = \frac{\pi}{2} + \frac{2\pi}{3} = \frac{7\pi}{6}$ or $t = \frac{\pi}{2} + \frac{4\pi}{3} = \frac{11\pi}{6}$. The solutions are $\frac{\pi}{2}, \frac{7\pi}{6}$, and $\frac{11\pi}{6}$.

[96] $\cos 3t = -\frac{\sqrt{3}}{2} \Rightarrow 3t = \frac{5\pi}{6} + 2k\pi$ or $3t = \frac{7\pi}{6} + 2k\pi \Rightarrow t = \frac{5\pi}{18} + k\left(\frac{2\pi}{3}\right)$ or $t = \frac{7\pi}{18} + k\left(\frac{2\pi}{3}\right)$.

Since t will be in $[0, 2\pi)$ when $k = 0, 1$, or 2, $t = \frac{5\pi}{18}$ or $t = \frac{7\pi}{18}$ or $t = \frac{5\pi}{18} + \frac{2\pi}{3} = \frac{17\pi}{18}$ or $t =$

$\frac{7\pi}{18} + \frac{2\pi}{3} = \frac{19\pi}{18}$ or $t = \frac{5\pi}{18} + \frac{4\pi}{3} = \frac{29\pi}{18}$ or $t = \frac{7\pi}{18} + \frac{4\pi}{3} = \frac{31\pi}{18}$. So, the solutions are $\frac{5\pi}{18}, \frac{7\pi}{18}$,

$\frac{17\pi}{18}, \frac{19\pi}{18}, \frac{29\pi}{18}$, and $\frac{31\pi}{18}$.

[97] $2\sin^2 x\sin x = 0 \Rightarrow \sin x(2\sin x + 1) = 0 \Rightarrow \sin x = 0$ or $\sin x = -\frac{1}{2}$. If $\sin x = 0$, then $x = 0$ or $x =$

π. If $\sin x = -\frac{1}{2}$, then $x = \frac{7\pi}{6}$ or $x = \frac{11\pi}{6}$. So, the solutions are $0, \pi, \frac{7\pi}{6}$, and $\frac{11\pi}{6}$.

[98] $2\cos^2 x - 5\cos x - 3 = 0 \Rightarrow (2\cos x + 1)(\cos x - 3) = 0 \Rightarrow \cos x = -\frac{1}{2}$ or $\cos x = 3$. If $\cos x = -\frac{1}{2}$,

then $x = \frac{2\pi}{3}$ or $x = \frac{4\pi}{3}$. Since $\cos x = 3 > 1$ is impossible, the solutions are $\frac{2\pi}{3}$ and $\frac{4\pi}{3}$.

[99] $3\cot^2 t - 1 = 0 \Rightarrow \cot^2 t = \frac{1}{3} \Rightarrow \cot t = \pm\frac{1}{\sqrt{3}} \Rightarrow \tan t = \pm\sqrt{3}$. If $\tan t = \sqrt{3}$, then $t = \frac{\pi}{3}$ or $t = \frac{4\pi}{3}$. If

$\tan t = -\sqrt{3}$, then $t = \frac{2\pi}{3}$ or $t = \frac{5\pi}{3}$. So, the solutions are $\frac{\pi}{3}, \frac{2\pi}{3}, \frac{4\pi}{3}$, and $\frac{5\pi}{3}$.

[100] $1 + \sin x = \cos x \Rightarrow 1 + 2\sin x + \sin^2 x = \cos^2 x \Rightarrow 1 + 2\sin x + \sin^2 x = 1 - \sin^2 x \Rightarrow$

$2\sin^2 x + 2\sin x = 0 \Rightarrow \sin x(\sin x + 1) = 0 \Rightarrow \sin x = 0$ or $\sin x = -1$. If $\sin x = 0$, then $x = 0$ or $x =$

π. If $\sin x = -1$, then $x = \frac{3\pi}{2}$. Since π extraneous, the solutions are 0 and $\frac{3\pi}{2}$.

[101] $2 \tan^2 x - 3 \tan x - 15 = 0 \Rightarrow \tan x = \frac{3 \pm \sqrt{129}}{4}$. If $\tan x = \frac{3 - \sqrt{129}}{4} \approx -2.089$, then since $\tan x < 0$, x

lies in either QII or QIV. In either case, since $\tan^{-1}(-2.089) \approx -1.12$, x as $x' \approx 1.12$ as its reference

angle. In QII, $x = \pi - x' \approx 3.14 - 1.12 = 2.02$. In QIV, $x = 2\pi - x' \approx 6.28 - 1.12 = 5.16$. If

$\tan x = \frac{3 + \sqrt{129}}{4} \approx 3.589$, then since $\tan x > 0$, x lies in either QI or QIII. In either case, since

$\tan^{-1}(3.589) \approx 1.30$, x has $x' \approx 1.30$ as its reference angle. In QI, $x = x' \approx 1.30$. In QIII, $x = \pi + x' \approx$

$3.14 + 1.30 = 4.44$. So, the solutions are 1.3, 2.0, 4.4, and 5.2.

[102] $15 \cos^2 t - 2 \cos t - 1 = 0 \Rightarrow (5 \cos t + 1)(3 \cos t - 1) = 0 \Rightarrow \cos t = -\frac{1}{5}$ or $\cos t = \frac{1}{3}$. If $\cos t = -\frac{1}{5}$,

then since $\cos t < 0$, t lies in either QII or QIII. In either case, since $\cos^{-1}\left(-\frac{1}{5}\right) \approx 1.77$, t has $t' =$

$\pi - 1.77 \approx 1.37$ as its reference angle. In QII, $t \approx 1.77$. In QIII, $t = \pi + t' \approx 3.14 + 1.37 = 4.51$. If

$\cos t = \frac{1}{3}$, then since $\cos t > 0$, t lies in either QI or QIV. In either case, since $\cos^{-1}\left(\frac{1}{3}\right) \approx 1.23$, t has

$t' \approx 1.23$ as its reference angle. In QI, $t = t' \approx 1.23$. In QIV, $t = 2\pi - t' \approx 6.28 - 1.23 = 5.05$. So, the

solutions are 1.2, 1.8, 4.5, and 5.0.

[103] $2 \sin^2 t - 9 \sin t - 5 = 0 \Rightarrow (2 \sin t + 1)(\sin t - 5) = 0 \Rightarrow \sin t = -\frac{1}{2}$ or $\sin t = 5$. If $\sin t = -\frac{1}{2}$, then

$t = \frac{7\pi}{6}$ or $t = \frac{11\pi}{6}$. Since $\sin t = 5 > 1$ is impossible, the solutions are $\frac{7\pi}{6}$ and $\frac{11\pi}{6}$.

[104] $\cot x - \sqrt{3} = -\csc x \Rightarrow \cot^2 x - 2\sqrt{3} \cot x + 3 = \csc^2 x \Rightarrow \cot^2 x - 2\sqrt{3} \cot x + 3 = 1 + \cot^2 x \Rightarrow$

$-2\sqrt{3} \cot x + 2 = 0 \Rightarrow -2\sqrt{3} \cot x = -2 \Rightarrow \cot x = \frac{1}{\sqrt{3}} \Rightarrow \tan x = \sqrt{3} \Rightarrow x = \frac{\pi}{3}$ or $x = \frac{4\pi}{3}$. Since

$\frac{4\pi}{3}$ is extraneous, the solution is $\frac{\pi}{3}$.

[105] $I = 30 \Rightarrow 50 \sin(100\pi t - 3\pi) = 30 \Rightarrow \sin(100\pi t - 3\pi) = \frac{3}{5}$. Since $\sin^{-1}\left(\frac{3}{5}\right) \approx 0.64$, the first time

when $I = 30$ is when $100\pi t - 3\pi \approx 0.64 \Rightarrow 100\pi t \approx 10.06 \Rightarrow t \approx 0.03$ sec.

[106] $g(t) = 25 \Rightarrow 100 \cos\left(\frac{2}{3}\pi t - 1\right) = 25 \Rightarrow \cos\left(\frac{2}{3}\pi t - 1\right) = \frac{1}{4}$. If $\theta = \frac{2}{3}\pi t - 1$, then since $\cos^{-1}\left(\frac{1}{4}\right) \approx$

$1.32 \in$ QI and $\cos \theta < 0$ when θ lies in either QII or QIII, the second time when $\cos \theta = \frac{1}{4}$ is when θ

lies in QIV and has $\theta' \approx 1.32$ as its reference angle. If θ lies in QIV, $\theta = 2\pi - \theta' \approx 6.28 - 1.32 =$

$4.96 \Rightarrow \frac{2}{3}\pi t - 1 \approx 4.96 \Rightarrow \frac{2}{3}\pi t \approx 5.96 \Rightarrow t \approx 2.85$ sec.

[107] $d = 4 \Rightarrow 6 \sin\left(\frac{3\pi}{4}t - 2\right) = 4 \Rightarrow \sin\left(\frac{3\pi}{4}t - 2\right) = \frac{2}{3}$. If $\theta = \frac{3\pi}{4}t - 2$, then since $\sin^{-1}\left(\frac{2}{3}\right) \approx$

$0.73 \in$ QI, the first time when $\sin \theta = \frac{2}{3}$ is when $\theta \approx 0.73 \Rightarrow \frac{3\pi}{4}t - 2 \approx 0.73 \Rightarrow \frac{3\pi}{4}t \approx 2.73 \Rightarrow$

$t \approx 1.16$ sec.

[108] $\frac{\sin 34°}{\sin \theta_2} = 1.53 \Rightarrow \sin \theta_2 = \frac{\sin 34°}{1.53} \approx 0.3655 \Rightarrow \theta_2 \approx \sin^{-1}(0.3655) \Rightarrow \theta_2 \approx 21.4°$.

[109] $\frac{\sin \theta_1}{\sin 22.7°} = 1.45 \Rightarrow \sin \theta_1 = 1.45 \sin 22.7° \approx 0.5596 \Rightarrow \theta_1 \approx \sin^{-1}(0.5596) \Rightarrow \theta_1 \approx 34.0°$.

[110] $\dfrac{\csc t}{\tan t} = \dfrac{\frac{1}{\sin t}}{\frac{\sin t}{\cos t}} = \dfrac{1}{\sin t} \cdot \dfrac{\cos t}{\sin t} = \dfrac{\cos t}{\sin^2 t} = \dfrac{\cos t}{1 - \cos^2 t}$

[111] $\dfrac{1 + \tan t}{\csc t + \sec t} = \dfrac{1 + \frac{\sin t}{\cos t}}{\frac{1}{\sin t} + \frac{1}{\cos t}} = \dfrac{\frac{\cos t + \sin t}{\cos t}}{\frac{\cos t + \sin t}{\sin t \cos t}} = \dfrac{\cos t + \sin t}{\cos t} \cdot \dfrac{\sin t \cos t}{\cos t + \sin t} = \sin t.$

[112] $\tan x + \cot x = \dfrac{\sin x}{\cos x} + \dfrac{\cos x}{\sin x} = \dfrac{\sin^2 x + \cos^2 x}{\sin x \cos x} = \dfrac{1}{\sin x \cos x}.$

[113] $\left(1 - \cos^2 t\right) \csc t = \sin^2 t \csc t = \sin^2 t \left(\dfrac{1}{\sin t}\right) = \sin t.$

[114] $\dfrac{[1 + \sin(-v)](1 + \sin v)}{1 - \cos^2 v} = \dfrac{(1 - \sin v)(1 + \sin v)}{1 - \cos^2 v} = \dfrac{1 - \sin^2 v}{1 - \cos^2 v} = \dfrac{\cos^2 v}{\sin^2 v} = \cot^2 v.$

[115] $\dfrac{\tan^2 x}{1 + \tan^2 x} = \dfrac{\frac{\sin^2 x}{\cos^2 x}}{1 + \frac{\sin^2 x}{\cos^2 x}} = \dfrac{\frac{\sin^2 x}{\cos^2 x}}{\frac{\cos^2 x + \sin^2 x}{\cos^2 x}} = \dfrac{\frac{\sin^2 x}{\cos^2 x}}{\frac{1}{\cos^2 x}} = \dfrac{\sin^2 x}{\cos^2 x} \cdot \dfrac{\cos^2 x}{1} = \sin^2 x.$

[116] $\dfrac{\sec t + 1}{\tan t} = \dfrac{\sec t}{\tan t} + \dfrac{1}{\tan t} = \dfrac{\frac{1}{\cos t}}{\frac{\sin t}{\cos t}} + \cot t = \dfrac{1}{\cos t} \cdot \dfrac{\cos t}{\sin t} + \cot t = \dfrac{1}{\sin t} + \cot t = \csc t + \cot t =$

$\cot t + \csc t.$

[117] $\dfrac{\sec x + 1}{\tan^2 x} = \dfrac{\sec x + 1}{\sec^2 x - 1} = \dfrac{\sec x + 1}{(\sec x + 1)(\sec x - 1)} = \dfrac{1}{\sec x - 1} = \dfrac{1}{\frac{1}{\cos x} - 1} \cdot \dfrac{\cos x}{\cos x} = \dfrac{\cos x}{1 - \cos x}.$

[118] $\dfrac{1 - \cos x}{\sin^2 x} = \dfrac{1 - \cos x}{1 - \cos^2 x} = \dfrac{1 - \cos x}{(1 - \cos x)(1 + \cos x)} = \dfrac{1}{1 + \cos x} \cdot \dfrac{\frac{1}{\cos x}}{\frac{1}{\cos x}} = \dfrac{\sec x}{\sec x + 1} = \dfrac{\sec x}{1 + \sec x}.$

[119] $\dfrac{\sin(-v) + 1}{\cos v} = \dfrac{-\sin v + 1}{\cos v} = \dfrac{1 - \sin v}{\cos v} \cdot \dfrac{1 + \sin v}{1 + \sin v} = \dfrac{1 - \sin^2 v}{\cos v (1 + \sin v)} = \dfrac{\cos^2 v}{\cos v (1 + \sin v)} = \dfrac{\cos v}{1 + \sin v}.$

[120] $\dfrac{1}{(\sec x - \tan x)^2} = \dfrac{1}{\left(\frac{1}{\cos x} - \frac{\sin x}{\cos x}\right)^2} = \dfrac{1}{\left(\frac{1 - \sin x}{\cos x}\right)^2} = \left(\dfrac{\cos x}{1 - \sin x}\right)^2 = \dfrac{\cos^2 x}{(1 - \sin x)^2} = \dfrac{1 - \sin^2 x}{(1 - \sin x)^2} =$

$\dfrac{(1 + \sin x)(1 - \sin x)}{(1 - \sin x)^2} = \dfrac{1 + \sin x}{1 - \sin x}.$

[121] $\dfrac{\cos^3 t + \sin^3 t}{\cos t + \sin t} = \dfrac{(\cos t + \sin t)(\cos^2 t - \cos t \sin t + \sin^2 t)}{\cos t + \sin t} = \cos^2 t - \cos t \sin t + \sin^2 t =$

$\left(\cos^2 t + \sin^2 t\right) - (\sin t \cos t) \cdot \left(\dfrac{2}{2}\right) = 1 - \dfrac{1}{2}(2 \sin t \cos t) = 1 - \dfrac{1}{2} \sin 2t.$

[122] $\dfrac{\sin^3 v}{\tan v - \sin v} = \dfrac{\sin^3 v}{\frac{\sin v}{\cos v} - \sin v} = \dfrac{\sin^3 v}{\frac{\sin v - \sin v \cos v}{\cos v}} = \dfrac{\sin^3 v}{\sin v \left(\frac{1 - \cos v}{\cos v}\right)} = \dfrac{\sin^2 v}{\frac{1 - \cos v}{\cos v}} = \dfrac{1 - \cos^2 v}{\frac{1 - \cos v}{\cos v}} =$

$\dfrac{(1 - \cos v)(1 + \cos v)}{(1 - \cos v)\left(\frac{1}{\cos v}\right)} = \dfrac{1 + \cos v}{\frac{1}{\cos v}} = \dfrac{1 + \cos v}{\sec v}.$

[123] Since $\dfrac{1+\sin(-x)}{\cos(-x)}=\dfrac{1-\sin x}{\cos x}$ and $\dfrac{1}{\sec x+\tan x}=\dfrac{1}{\dfrac{1}{\cos x}+\dfrac{\sin x}{\cos x}}\cdot\dfrac{\cos x}{\cos x}=\dfrac{\cos x}{1+\sin x}\cdot\dfrac{1-\sin x}{1-\sin x}=$

$\dfrac{\cos(1-\sin x)}{1-\sin^2 x}=\dfrac{\cos x(1-\sin x)}{\cos^2 x}=\dfrac{1-\sin x}{\cos x}$, the identity follows.

[124] $\dfrac{1}{(1+\sec t)(1-\cos t)}=\dfrac{1}{1+\sec t-\cos t-1}=\dfrac{1}{\sec t-\cos t}=\dfrac{1}{\dfrac{1}{\cos t}-\cos t}=\dfrac{1}{\dfrac{1-\cos^2 t}{\cos t}}=$

$\dfrac{\cos t}{1-\cos^2 t}=\dfrac{\cos t}{\sin^2 t}=\dfrac{\cos t}{\sin t}\cdot\dfrac{1}{\sin t}=\cot t\csc t.$

[125] $\dfrac{\cos\theta+\sin\theta-1}{\cos\theta-\sin\theta+1}=\dfrac{\cos\theta+(\sin\theta-1)}{\cos\theta-(\sin\theta-1)}\cdot\dfrac{\cos\theta+(\sin\theta-1)}{\cos\theta+(\sin\theta-1)}=$

$\dfrac{\cos^2\theta+2\cos\theta(\sin\theta-1)+(\sin\theta-1)^2}{\cos^2\theta-(\sin\theta-1)^2}=\dfrac{\cos^2\theta+2\sin\theta\cos\theta-2\cos\theta+\sin^2\theta-2\sin\theta+1}{\cos^2\theta-(\sin^2\theta-2\sin\theta+1)}=$

$\dfrac{(\cos^2\theta+\sin^2\theta)+2\sin\theta\cos\theta-2\cos\theta-2\sin\theta+1}{\cos^2\theta-\sin^2\theta+2\sin\theta-1}=\dfrac{1+2\sin\theta\cos\theta-2\cos\theta-2\sin\theta+1}{(\cos^2\theta-1)-\sin^2\theta+2\sin\theta}=$

$\dfrac{2+2\sin\theta\cos\theta-2\cos\theta-2\sin\theta}{(-\sin^2\theta)-\sin^2\theta+2\sin\theta}=\dfrac{2+2\sin\theta\cos\theta-2\cos\theta-2\sin\theta}{-2\sin^2\theta+2\sin\theta}=$

$\dfrac{2(1+\sin\theta\cos\theta-\cos\theta-\sin\theta)}{2(\sin\theta-\sin^2\theta)}=\dfrac{1+\sin\theta\cos\theta-\cos\theta-\sin\theta}{\sin\theta-\sin^2\theta}=\dfrac{\cos\theta(\sin\theta-1)-(\sin\theta-1)}{\sin\theta(1-\sin\theta)}=$

$\dfrac{(\sin\theta-1)(\cos\theta-1)}{(1-\sin\theta)(\sin\theta)}=\dfrac{(-1)(\cos\theta-1)}{\sin\theta}=\dfrac{1-\cos\theta}{\sin\theta}\cdot\dfrac{1+\cos\theta}{1+\cos\theta}=\dfrac{1-\cos^2\theta}{\sin\theta(1+\cos\theta)}=$

$\dfrac{\sin^2\theta}{\sin\theta(1+\cos\theta)}=\dfrac{\sin\theta}{1+\cos\theta}.$

[126] Since $\dfrac{\cot x+\cot y}{\tan x+\tan y}=\dfrac{\dfrac{\cos x}{\sin x}+\dfrac{\cos y}{\sin y}}{\dfrac{\sin x}{\cos x}+\dfrac{\sin y}{\cos y}}=\dfrac{\dfrac{\cos x\sin y+\cos y\sin x}{\sin x\sin y}}{\dfrac{\sin x\cos y+\cos x\sin y}{\cos x\cos y}}=\dfrac{\cos x\sin y+\cos y\sin x}{\sin x\sin y}\cdot$

$\dfrac{\cos x\cos y}{\cos x\sin y+\cos y\sin x}=\dfrac{\cos x\cos y}{\sin x\sin y}=\dfrac{\cos x}{\sin x}\cdot\dfrac{\cos y}{\sin y}=\cot x\cot y$, and $\dfrac{1-\cot x\cot y}{\tan x\tan y-1}=$

$\dfrac{1-\dfrac{\cos x}{\sin x}\cdot\dfrac{\cos y}{\sin y}}{\dfrac{\sin x}{\cos x}\cdot\dfrac{\sin y}{\cos y}-1}=\dfrac{\dfrac{\sin x\sin y-\cos x\cos y}{\sin x\sin y}}{\dfrac{\sin x\sin y-\cos x\cos y}{\cos x\cos y}}=\dfrac{\sin x\sin y-\cos x\cos y}{\sin x\sin y}\cdot\dfrac{\cos x\cos y}{\sin x\sin y-\cos x\cos y}=$

$\dfrac{\cos x\cos y}{\sin x\sin y}=\dfrac{\cos x}{\sin x}\cdot\dfrac{\cos y}{\sin y}=\cot x\cot y$, the identity follows.

[127] Since $\sin(2\pi-t)=\sin 2\pi\cos t-\cos 2\pi\sin t=(0)(\cos t)-(1)(\sin t)=-\sin t$, and $\cos\left(\dfrac{\pi}{2}+t\right)=$

$\cos\dfrac{\pi}{2}\cos t-\sin\dfrac{\pi}{2}\sin t=(0)(\cos t)-(1)(\sin t)=-\sin t$, the identity follows.

[128] $\dfrac{\sin(v-u)}{\sin u\sin v}=\dfrac{\sin v\cos u-\cos v\sin u}{\sin u\sin v}=\dfrac{\sin v\cos u}{\sin u\sin v}-\dfrac{\cos v\sin u}{\sin u\sin v}=\dfrac{\cos u}{\sin u}-\dfrac{\cos v}{\sin v}=\cot u-\cot v=$

$\cot u+\cot(-v).$

285

[129] $\dfrac{\cos\left(u+v\right)+\cos\left(u-v\right)}{\sin\left(u+v\right)+\sin\left(u-v\right)} = \dfrac{\left(\cos u\cos v - \sin u\sin v\right)+\left(\cos u\cos v + \sin u\sin v\right)}{\left(\sin u\cos v + \cos u\sin v\right)+\left(\sin u\cos v - \cos u\sin v\right)} = \dfrac{2\cos u\cos v}{2\sin u\cos v} =$

$\dfrac{\cos u}{\sin u} = \cot u.$

[130] $\dfrac{\sin\left(u+v\right)}{\sin\left(u-v\right)} = \dfrac{\sin u\cos v + \cos u\sin v}{\sin u\cos v - \cos u\sin v}\cdot\dfrac{\dfrac{1}{\sin u\cos v}}{\dfrac{1}{\sin u\cos v}} = \dfrac{1+\dfrac{\cos u\sin v}{\sin u\cos v}}{1-\dfrac{\cos u\sin v}{\sin u\cos v}} = \dfrac{1+\dfrac{\cos u}{\sin u}\cdot\dfrac{\sin v}{\cos v}}{1-\dfrac{\cos u}{\sin u}\cdot\dfrac{\sin v}{\cos v}} =$

$\dfrac{1+\cot u\tan v}{1-\cot u\tan v}.$

[131] $\dfrac{\sin 3x + \sin 5x}{\cos 3x - \cos 5x} = \dfrac{2\sin\dfrac{3x+5x}{2}\cos\dfrac{3x-5x}{2}}{-2\sin\dfrac{3x+5x}{2}\sin\dfrac{3x-5x}{2}} = \dfrac{2\sin 4x\cos\left(-x\right)}{-2\sin 4x\sin\left(-x\right)} = -\dfrac{\cos x}{-\sin x} = \dfrac{\cos x}{\sin x} = \cot x.$

[132] Since $\dfrac{\cos 6x - \cos 10x}{\sin 10x - \sin 6x} = \dfrac{-2\sin\dfrac{6x+10x}{2}\sin\dfrac{6x-10x}{2}}{2\cos\dfrac{10x+6x}{2}\sin\dfrac{10x-6x}{2}} = \dfrac{-2\sin 8x\sin\left(-2x\right)}{2\cos 8x\sin 2x} =$

$\dfrac{\sin 8x\sin 2x}{\cos 8x\sin 2x} = \dfrac{\sin 8x}{\cos 8x} = \tan 8x,$ and $\dfrac{\sin 6x + \sin 10x}{\cos 6x + \cos 10x} = \dfrac{2\sin\dfrac{6x+10x}{2}\cos\dfrac{6x-10x}{2}}{2\cos\dfrac{6x+10x}{2}\cos\dfrac{6x-10x}{2}} =$

$\dfrac{2\sin 8x\cos\left(-2x\right)}{2\cos 8x\cos\left(-2x\right)} = \dfrac{\sin 8x}{\cos 8x} = \tan 8x,$ the identity follows.

EXERCISES 7.1

[1] $\beta = 90° - \alpha = 18.6°$. $\sin \alpha = \frac{a}{c} \Rightarrow a = c \sin \alpha = 32 \sin 71.4° \approx 30$. $\cos \alpha = \frac{b}{c} \Rightarrow b = c \cos \alpha = 32 \cos 71.4° \approx 10$.

[2] $\alpha = 90° - \beta = 46°40'$. $\tan \alpha = \frac{a}{b} \Rightarrow a = b \tan \alpha = 55.4 \tan 46°40' \approx 58.7$. $\sin \beta = \frac{b}{c} \Rightarrow c = \frac{b}{\sin \beta} = \frac{55.4}{\sin 43°20'} \approx 80.7$.

[3] $\sin \beta = \frac{b}{c} = \frac{115}{824} \approx 0.1396 \Rightarrow \beta \approx 8.0°$. $\alpha = 90° - \beta \approx 82.0°$. $\sin \alpha = \frac{a}{c} \Rightarrow a = c \sin \alpha \approx 824 \sin 82.0° \approx 816$.

[4] $\alpha = 90° - \beta = 70°30'$. $\sin \beta = \frac{b}{c} \Rightarrow b = c \sin \beta = 78.5 \sin 19°30' \approx 26.2$. $\sin \alpha = \frac{a}{c} \Rightarrow a = c \sin \alpha = 78.5 \sin 70°30' \approx 74.0$.

[5] $\tan \alpha = \frac{a}{b} = \frac{2.35}{4.18} \approx 0.5622 \Rightarrow \alpha \approx 29.3°$. $\beta = 90° - \alpha \approx 60.7°$. $\sin \alpha = \frac{a}{c} \Rightarrow c = \frac{a}{\sin \alpha} \approx \frac{2.35}{\sin 29.3°} \approx 4.80$.

[6] $\beta = 90° - \alpha = 29.5°$. $\tan \alpha = \frac{a}{b} \Rightarrow a = b \tan \alpha = 10.2 \tan 60.5° \approx 18.0$. $\sin \beta = \frac{b}{c} \Rightarrow$ $c = \frac{b}{\sin \beta} = \frac{10.2}{\sin 29.5°} \approx 20.7$.

[7] $\alpha = 90° - \beta = 63°10'$. $\tan \beta = \frac{b}{a} \Rightarrow b = a \tan \beta = 3250 \tan 26°50' \approx 1644$. $\sin \alpha = \frac{a}{c} \Rightarrow$ $c = \frac{a}{\sin \alpha} = \frac{3250}{\sin 63°10'} \approx 3642$.

[8] $\beta = 90° - \alpha = 81°50'$. $\sin \alpha = \frac{a}{c} \Rightarrow a = c \sin \alpha = 80.4 \sin 8°10' \approx 11.4$. $\sin \beta = \frac{b}{c} \Rightarrow$ $b = c \sin \beta = 80.4 \sin 81°50' \approx 79.6$.

[9] $\alpha = 90° - \beta = 39°40'$. $\tan \beta = \frac{b}{a} \Rightarrow b = a \tan \beta = 3.14 \tan 50°20' \approx 3.79$. $\sin \alpha = \frac{a}{c} \Rightarrow$ $c = \frac{a}{\sin \alpha} = \frac{3.14}{\sin 39°40'} \approx 4.92$.

[10] $\sin \alpha = \frac{a}{c} = \frac{0.248}{0.565} \approx 0.4389 \Rightarrow \alpha \approx 26.0°$. $\beta = 90° - \alpha \approx 64.0°$. $\sin \beta = \frac{b}{c} \Rightarrow b = c \sin \beta \approx$ $(0.565) \sin 64.0° \approx 0.508$.

[11] $\alpha = 90° - \beta = 44.8°$. $\sin \beta = \frac{b}{c} \Rightarrow b = c \sin \beta = 19,500 \sin 45.2° \approx 13,800$. $\sin \alpha = \frac{a}{c} \Rightarrow$ $a = c \sin \alpha = 19,500 \sin 44.8° \approx 13,700$.

[12] $\tan \alpha = \frac{a}{b} = \frac{17.1}{36.9} \approx 0.4634 \Rightarrow \alpha \approx 24.9°$. $\beta = 90° - \alpha \approx 65.1°$. $\sin \alpha = \frac{a}{c} \Rightarrow c = \frac{a}{\sin \alpha} \approx \frac{17.1}{\sin 24.9°} \approx 40.6$.

[13] $\beta = 90° - \alpha = 62°$. $\tan \alpha = \frac{a}{b} \Rightarrow a = b \tan \alpha = 2.6 \tan 28° \approx 1.4$. $\sin \beta = \frac{b}{c} \Rightarrow c = \frac{b}{\sin \beta} = \frac{2.6}{\sin 62°} \approx 2.9$.

[14] $\sin \beta = \frac{b}{c} = \frac{1250}{2100} \approx 0.5952 \Rightarrow \beta \approx 36.5°$. $\alpha = 90° - \beta \approx 53.5°$. $\sin \alpha = \frac{a}{c} \Rightarrow a = c \sin \alpha \approx 2100 \sin 53.5° \approx 1690$.

[15] $\alpha = 90° - \beta = 79.6°$. $\tan \beta = \frac{b}{a} \Rightarrow a = \frac{b}{\tan \beta} = \frac{72.6}{\tan 10.4°} \approx 396$. $\sin \beta = \frac{b}{c} \Rightarrow c = \frac{b}{\sin \beta} = \frac{72.6}{\sin 10.4°} \approx 402$.

[16] $\tan \alpha = \frac{a}{b} = \frac{78,400}{128,000} \approx 0.6125 \Rightarrow \alpha \approx 31.5°$. $\beta = 90° - \alpha \approx 58.5°$. $\sin \alpha = \frac{a}{c} \Rightarrow c = \frac{a}{\sin \alpha} \approx \frac{78,400}{\sin 31.5°} \approx 150,000$.

[17] $\beta = 90° - \alpha = 30°$. $\tan \alpha = \frac{a}{b} \Rightarrow a = b \tan \alpha = 12 \tan 60° = 12\sqrt{3}$. $\sin \beta = \frac{b}{c} \Rightarrow c = \frac{b}{\sin \beta} = \frac{12}{\sin 30°} = \frac{12}{\frac{1}{2}} = 24$.

[18] $\alpha = 90° - \beta = 45°$. $b = a = 50$. $\sin \alpha = \frac{a}{c} \Rightarrow c = \frac{a}{\sin \alpha} = \frac{50}{\sin 45°} = \frac{50}{\frac{1}{\sqrt{2}}} = 50\sqrt{2}$.

[19] $\alpha = 90° - \beta = 60°$. $\tan \beta = \frac{b}{a} \Rightarrow a = \frac{b}{\tan \beta} = \frac{144}{\tan 30°} = \frac{144}{\frac{1}{\sqrt{3}}} = 144\sqrt{3}$. $\sin \beta = \frac{b}{c} \Rightarrow$

$c = \frac{b}{\sin \beta} = \frac{144}{\sin 30°} = \frac{144}{\frac{1}{2}} = 288$.

[20] $\beta = 90° - \alpha = 45°$. $\sin \alpha = \frac{a}{c} \Rightarrow a = c \sin \alpha = 18 \sin 45° = 18\left(\frac{\sqrt{2}}{2}\right) = 9\sqrt{2}$. $b = a = 9\sqrt{2}$.

[21] $\sin \alpha = \frac{a}{c} = \frac{5\sqrt{3}}{10} = \frac{\sqrt{3}}{2} \Rightarrow \alpha = 60°$. $\beta = 90° - \alpha = 30°$. $\sin \beta = \frac{b}{c} \Rightarrow b = c \sin \beta = 10 \sin 30° = 10\left(\frac{1}{2}\right) = 5$.

[22] $\alpha = 90° - \beta = 30°$. $\sin \alpha = \frac{a}{c} \Rightarrow a = c \sin \alpha = 108 \sin 30° = 108\left(\frac{1}{2}\right) = 54$. $\sin \beta = \frac{b}{c} \Rightarrow b = c \sin \beta = 108 \sin 60° = 108\left(\frac{\sqrt{3}}{2}\right) = 54\sqrt{3}$.

[23] $\beta = 90° - \alpha = 45°$. $b = a = 17$. $\sin \alpha = \frac{a}{c} \Rightarrow c = \frac{a}{\sin \alpha} = \frac{17}{\frac{1}{\sqrt{2}}} = 17\sqrt{2}$.

[24] $\sin \alpha = \frac{a}{c} = \frac{7}{7\sqrt{2}} = \frac{1}{\sqrt{2}} \Rightarrow \alpha = 45°$. $\beta = 90° - \alpha = 45°$. $b = a = 7$.

[25] $\tan 78.5° = \frac{x}{225} \Rightarrow x = 225 \tan 78.5° \approx 1110$. The tower is approximately 1110 feet high. (See figure below).

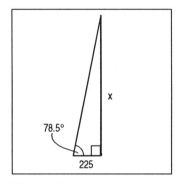

Figure 25

[26] From right triangle ABC, $\sin 45° = \dfrac{12}{x} \Rightarrow x = \dfrac{12}{\sin 45°} = \dfrac{12}{\frac{1}{\sqrt{2}}} = 12\sqrt{2}$. From right triangle ACD,

$d = \sqrt{\left(12\sqrt{2}\right)^2 + \left(12\right)^2} = \sqrt{432} = 12\sqrt{3} \approx 20.8$ The length of each diagonal is approximately 20.8 centimeters (See figure below).

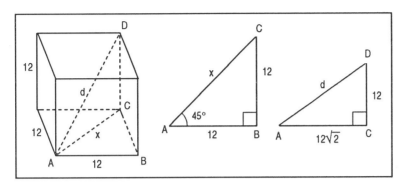

Figure 26

[27] From right triangle ACB, $\tan \theta = \dfrac{9}{30} = 0.3000 \Rightarrow \theta \approx 17°$. (See figure below).

[28] $x = \sqrt{\left(125\right)^2 - \left(20\right)^2} = \sqrt{15{,}225} = 5\sqrt{609} \approx 123$. $x + 6 \approx 129$. The ladder will reach approximately 129 ft. up the front of the building. (See figure below).

[29] $\sin 37.25° = \dfrac{94 - x}{x} \Rightarrow x \sin 37.25° = 94 - x \Rightarrow x + x \sin 37.25° = 94 \Rightarrow$

$x = \dfrac{94}{1 + \sin 37.25°} \approx 58.6$. $188 - 2x = 70.8$. Each of the 2 equal sides measures approximately 58.6 cm and the third side measures approximately 70.8 cm. (See figure below).

[30] From right triangle P_1PP_2, $x = \sqrt{\left(36.5\right)^2 + \left(4\right)^2} = \sqrt{1348.25} \approx 36.7$. The distance between P_1 and P_2 is approximately 36.7 inches. (See figure below).

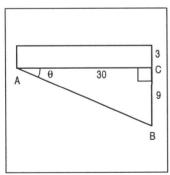

Figure 27

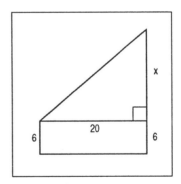

Figure 28

[31] $\cos 18.1° = \dfrac{84 - x}{84} \Rightarrow 84 - x = 84 \cos 18.1° \Rightarrow x = 84 - 84 \cos 18.1° \approx 4.2$. $x + 25.4 = 29.6$. The maximum distance between the pendulum bob and the floor is approximately 29.6 cm. (See figure below).

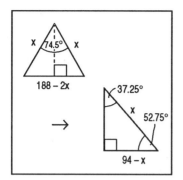

Figure 29

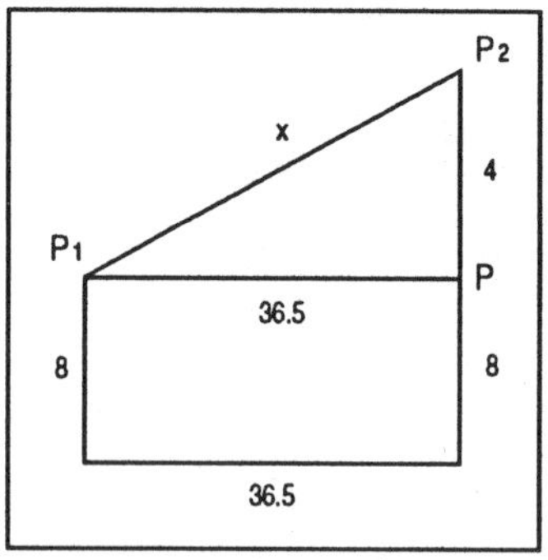

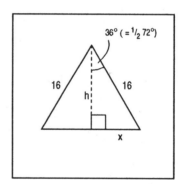

Figure 30

[32] The pentagon consists of 5 isosceles triangles like the one shown in Figure 32. $\sin 36° = \frac{x}{16} \Rightarrow$

$x = 16 \sin 36°$ and $\cos 36° = \frac{h}{16} \Rightarrow h = 16 \cos 36°$. The area of the each triangle is $\frac{1}{2}(2x)(h) =$

$xh = (16 \sin 36°)(16 \cos 36°) = 256 \sin 36° \cos 36°$. The area of the pentagon is

$5(256 \sin 36° \cos 36°)$, which is approximately 608.7 square inches.   (See figure below).

[33] $V = \frac{1}{3}\pi r^2 h \Rightarrow 81 = \frac{\pi}{3}(2.125)^2 h \Rightarrow h = \frac{243}{\pi(2.125)^2} \approx 17.1. \tan\frac{\theta}{2} = \frac{2.125}{h} \approx \frac{2.125}{17.1} \approx 0.1241 \Rightarrow \frac{\theta}{2} \approx$

7.1 $\Rightarrow \theta$ is approximately 14.2°.   (See figure below).

[34] $\tan 23.5'' = \frac{x}{390,000,000} \Rightarrow x = 390,000,000 \tan 23.5'' \approx 44,400 \Rightarrow 2x \approx 88,800$. Thus, the diameter

of Jupiter is approximately 88,800 miles.   (See figure below).

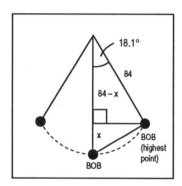

Figure 31

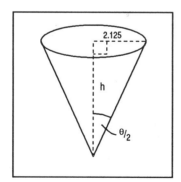

Figure 33

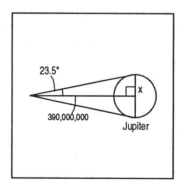

Figure 32

Figure 34

[35] $\cot \frac{\theta}{2} = \frac{h}{\frac{b}{2}} \Rightarrow h = \frac{b}{2} \cot \frac{\theta}{2}$. The area of the triangle is $\left(\frac{b}{2}\right)(h) = \frac{b}{2}\left(\frac{b}{2} \cot \frac{\theta}{2}\right) = \frac{b^2}{4} \cot\left(\frac{1}{2}\theta\right)$.

(See figure below).

[36] $v = gt \sin \theta \Rightarrow 21.2 = 32\,(3.8) \sin \theta \Rightarrow \sin \theta = \frac{21.2}{32\,(3.8)} \approx 0.1743 \Rightarrow \theta \approx 10.0°$. $\sin \theta = \frac{4.5}{x} \Rightarrow$

$x = \frac{4.5}{\sin \theta} \approx \frac{4.5}{\sin 10.0°} \approx 25.8$. The plane's inclined side is approximately 25.8 feet long.

(See figure below).

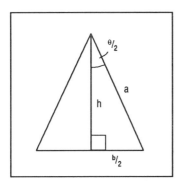

Figure 35

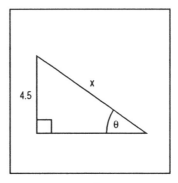

Figure 36

[37] $\tan 36° = \frac{r}{12} \Rightarrow r = 12 \tan 36° \approx 8.7$. The radius of the circle is approximately 8.7 cm.

(See figure below).

[38] When the angle between the pipe (whose length is denoted by $2x$) and the corner is $45°$, we have

the two right triangles shown in Figure 38. $\sin 45° = \frac{6}{x} \Rightarrow x = \frac{6}{\sin 45°} = \frac{6}{\frac{1}{\sqrt{2}}} = 6\sqrt{2} \approx 8.5$.

Since $2x = 12\sqrt{2} \approx 17 < 20$, the 20–foot pipe will not turn the corner.   (See figure below).

[39] $\tan \theta = \frac{48}{30} = 1.6000 \Rightarrow \theta \approx 58°$. The angle of elevation of the sun is approximately $58°$.

(See figure below).

[40] $\tan 47°40' = \frac{x}{135} \Rightarrow x = 135 \tan 47°40' \approx 148 \Rightarrow 200 + x \approx 348$. The taller building is

approximately 348 feet high.   (See figure below).

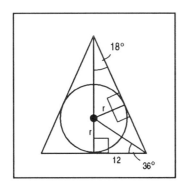

Figure 37

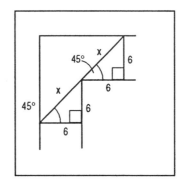

Figure 38

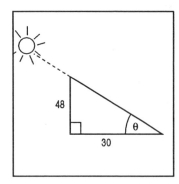

Figure 39

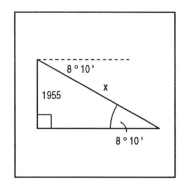

Figure 40

[41] The fire tower is $2470 - 515 = 1955$ meters above the valley. $\sin 8°10' = \dfrac{1955}{x} \Rightarrow$

$x = \dfrac{1955}{\sin 8°10'} \approx 13,760$. The distance from the tower to the fire is approximately 13,760 meters.

(See figure below).

[42] From right $\triangle CDB_2$, $\tan 42°10' = \dfrac{x + 288}{y} \Rightarrow y = \dfrac{x + 288}{\tan 42°10'}$. From right $\triangle ADB_1$, $\tan 33°30' =$

$\dfrac{x}{y + 3600} \Rightarrow x = (y + 3600) \tan 33°30' \Rightarrow \dfrac{x}{\tan 33°30'} = y + 3600 \Rightarrow y = \dfrac{x}{\tan 33°30'} - 3600$.

Since $y = y$, $\dfrac{x + 288}{\tan 42°10'} = \dfrac{x}{\tan 33°30'} - 3600 \Rightarrow x \approx 9640 \Rightarrow x + 288 \approx 9928$. Thus, the height of

the balloon is approximately 9928 feet. (See figure below).

[43] $\tan 8.4° = \dfrac{r}{r + 23,000} \Rightarrow r = (r + 23,000) \tan 8.4° \Rightarrow r = r \tan 8.4° + 23,000 \tan 8.4° \Rightarrow$

$r - r \tan 8.4° = 23,000 \tan 8.4° \Rightarrow r = \dfrac{23,000 \tan 8.4°}{1 - \tan 8.4°} \approx 3990$. Thus, the radius of the earth is

approximately 3990 miles. (See figure below).

[44] In 10 minutes $= 600$ seconds, the balloon travels $(600)(15) = 9000$ feet horizontally and

$(600)(20) = 12,000$ feet vertically. $\tan \theta = \dfrac{12,000}{9000} \approx 1.3333 \Rightarrow \theta \approx 53.1°$. The angle of elevation

of the balloon is approximately 53.1°. (See figure below).

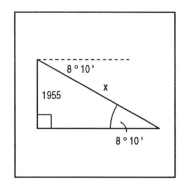

Figure 41

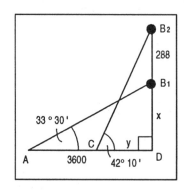

Figure 42

[45] In right $\triangle CAB_1$, $\angle ACB = 90° - 55°20' = 34°40'$. $\tan 34°40' = \dfrac{y}{2400} \Rightarrow y = 2400 \tan 34°40' \approx$

1660. In right triangle CAB_2, $\angle ACB_2 = 90° - 42°30' = 47°30'$. $\tan 47°30' = \dfrac{x + y}{2400} \Rightarrow$

$x + y = 2400 \tan 47°30' \Rightarrow x = 2400 \tan 47°30' - y \approx 2400 \tan 47°30' - 1660 \approx 960$. Thus, the two

boats are approximately 960 feet apart. (See figure below).

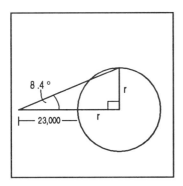

Figure 43

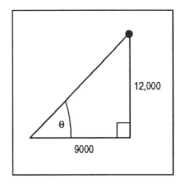

Figure 44

[46] $\tan 72°40' = \dfrac{h}{264} \Rightarrow h = 264 \tan 72°40' \approx 846$. The balloon is approximately 846 feet high.

(See figure below).

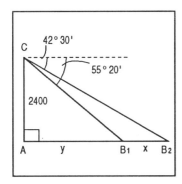

Figure 45

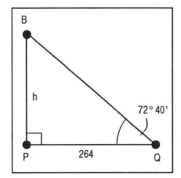

Figure 46

[47] $\dfrac{600}{880} = \dfrac{550}{x} \Rightarrow 600x = 484{,}000 \Rightarrow x = 806\frac{2}{3}$. In 30 seconds, the jet travels $30\left(806\frac{2}{3}\right) =$

24,200 feet. $\sin \theta = \dfrac{5000}{24{,}200} \approx 0.2066 \Rightarrow \theta \approx 11.9°$. Thus, the jet's angle of ascent is

approximately 11.9°.   (See figure below).

[48] From right $\triangle QPW$, $\tan 39.2° = \dfrac{d}{888}$, where d is the distance between point P and the Washington

Monument W. Thus $d = 888 \tan 39.2° \approx 724$ feet. If h denotes the height of the monument, we see

from right triangle PWT that $\tan 37.5° = \dfrac{h}{d} \Rightarrow h = d \tan 37.5° \approx 724 \tan 37.5° \approx 556$. Thus, the

Washington Monument is approximately 556 feet high.   (See figure below).

[49] From right triangle ACP, $\tan 21°40' = \dfrac{x}{6130} \Rightarrow x = 6130 \tan 21°40' \approx 2430$. From right $\triangle BCP$,

$\tan 30°10' = \dfrac{y}{6130} \Rightarrow y = 6130 \tan 30°10' \approx 3560$. Thus $x + y \approx 5990$ and the length of the tunnel

is approximately 5990 feet.   (See figure below).

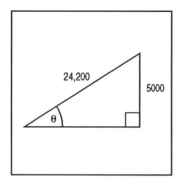

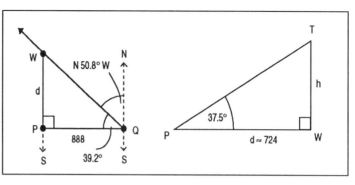

Figure 47   **Figure 48**

[50] Let A and B denote the location of the 2 ships at 1:30 pm. The distance traveled by ship A is $(30)(3.5) = 105$ km and the distance traveled by ship B is $20(3) = 60$ km. Since $35.5° + 54.5° = 90°$, $\angle APB$ is a right angle. From right triangle APB, $d(A, B) = \sqrt{(105)^2 + (60)^2} \approx 121 \Rightarrow$ the two ships are approximately 121 miles apart. Also, from right triangle APB, $\tan \angle A = \frac{60}{105} \approx 0.5714 \Rightarrow \angle A \approx 29.7°$. Since $\angle A + 35.5° \approx 65.2°$, the bearing from ship A to ship B is approximately N 65.2°E.   (See figure below).

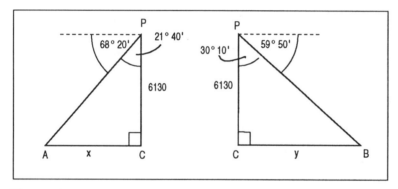

Figure 49

[51] Let S, P and L denote the ship, port and lighthouse, respectively. From right ΔSPQ, $\tan 51°20' = \frac{x}{y} \Rightarrow y = \frac{x}{\tan 51°20'}$. From right ΔSQL, $\tan 10°40' = \frac{14.2 - x}{y} \Rightarrow y = \frac{14.2 - x}{\tan 10°40'}$. $\frac{x}{\tan 51°20'} = \frac{14.2 - x}{\tan 10°40'} \Rightarrow x \tan 10°40' = (14.2 - x)\tan 51°20' \Rightarrow x = \frac{14.2 \tan 51°20'}{\tan 10°40' + \tan 51°20'} \approx 12.3$. From right ΔSQP, $\sin 51°20' = \frac{x}{d} \Rightarrow d = \frac{x}{\sin 51°20'} \approx \frac{12.3}{\sin 51°20'} \approx 15.8$. Thus, the ship is approximately 15.8 miles from its port.   (See figure below).

[52] $\tan \theta = \frac{24.2}{16.0} = 1.5125 \Rightarrow \theta \approx 56.5°$. The rescue ship should take a bearing of approximately S56.5° E.   (See figure below).

[53] $d = \sqrt{(120)^2 + (224)^2} \approx 254$. $\tan \theta = \frac{224}{120} \approx 1.8667 \Rightarrow \theta \approx 61.8°$. $90° - \theta \approx 28.2°$. Thus, the boys should walk approximately 254 paces on a bearing of approximately S28.2°E. (See figure below).

294

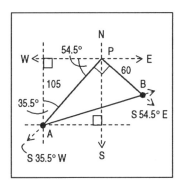

Figure 50

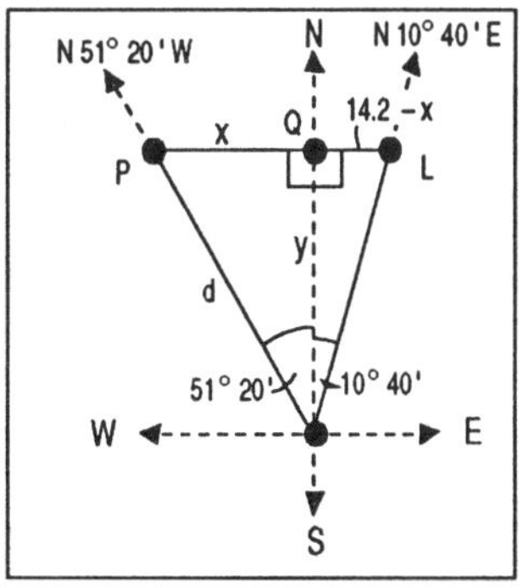

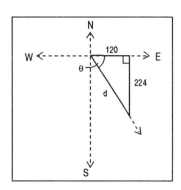

Figure 51

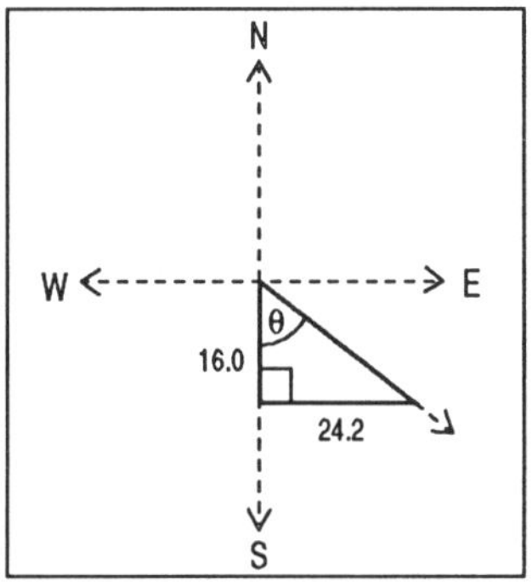

Figure 52

Figure 53

[54] P denotes the private airfield, L denotes Lubbock and A denotes the position of the airplane when it changes direction. From right ΔPQA, $\sin 27°10' = \frac{x}{73.7} \Rightarrow x = 73.7 \sin 27°10' \approx 33.6$. Thus $122.4 - x \approx 88.8$. Also, from right ΔPQA, $\cos 27°10' = \frac{y}{73.7} \Rightarrow y = 73.7 \cos 27°10' \approx 65.6°$. From right ΔPQL, $d = \sqrt{(122.4 - x)^2 + y^2} \approx \sqrt{(88.8)^2 + (65.6)^2} \approx 110$. Thus, the distance between the private airfield and Lubbock is approximately 110 miles.   (See figure below).

[55] $\tan 15°50' = \frac{100}{d} \Rightarrow d = \frac{100}{\tan 15°50'} \approx 353$. Thus, the river is approximately 353 meters wide. (See figure below).

[56] $d = \sqrt{(235)^2 + (340)^2} \approx 413$. $\tan \theta = \frac{340}{235} \approx 1.4468 \Rightarrow \theta \approx 55.3°$. $\theta + 90° \approx 145.3°$. The distance from the plane to the airport is approximately 413 miles, and the azimuth from the airport to the plane is approximately 145.3°.   (See figure below).

[57] A denotes Atlanta, B the position of the plane when it changes course and C the final position of the plane. $d(A, B) = (540)(.75) = 405$ and $d(B, C) = 540(.5) = 270$. $\angle B = 180° - (52°20' + 37°40') = 90°$. From right ΔABC, $d = \sqrt{(405)^2 + (270)^2} \approx 490$.

$\tan \theta = \frac{270}{405} \approx 0.6667 \Rightarrow \theta \approx 33°40'$. $\theta + 127°40' = 161°20'$. Thus, the jet is approximately 490 miles from Atlanta on an azimuth of approximately 161°20'.   (See figure below).

[58] O denotes O'hare Field, H denotes the location of the helicopter and P the location of the plane. $\tan \theta = \frac{650}{188} \approx 3.4574 \Rightarrow \theta \approx 73.9°$. $180° - \theta \approx 106.1°$. Thus, the azimuth from the helicopter to the

plane is approximately 106.1°.   (See figure below).

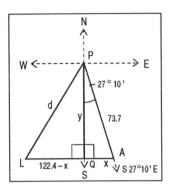

Figure 54

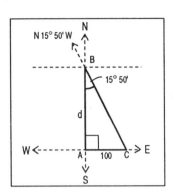

Figure 55

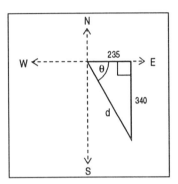

Figure 56

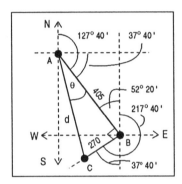

Figure 57

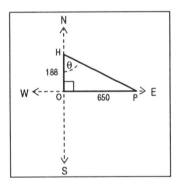

Figure 58

EXERCISES 7.2

[1] $\alpha = 180° - (\beta + \gamma) = 57.5°$. $\dfrac{a}{\sin \alpha} = \dfrac{c}{\sin \gamma} \Rightarrow a = \dfrac{c \sin \alpha}{\sin \gamma} = \dfrac{37.5 \sin 57.5°}{\sin 102.7°} \approx 32.4$. $\dfrac{b}{\sin \beta} = \dfrac{c}{\sin \gamma} \Rightarrow$

$b = \dfrac{c \sin \beta}{\sin \gamma} = \dfrac{37.5 \sin 19.8°}{\sin 102.7°} \approx 13.0$.

[2] $\gamma = 180° - (\alpha + \beta) = 59.2°$. $\dfrac{a}{\sin \alpha} = \dfrac{c}{\sin \gamma} \Rightarrow a = \dfrac{c \sin \alpha}{\sin \gamma} = \dfrac{9.18 \sin 26.3°}{\sin 59.2°} \approx 4.74$. $\dfrac{b}{\sin \beta} = \dfrac{c}{\sin \gamma} \Rightarrow$

$b = \dfrac{c \sin \beta}{\sin \gamma} = \dfrac{9.18 \sin 94.5°}{\sin 59.2°} \approx 10.7$.

[3] $\dfrac{b}{\sin \beta} = \dfrac{c}{\sin \gamma} \Rightarrow \sin \beta = \dfrac{b \sin \gamma}{c} = \dfrac{1250 \sin 37°10'}{875} \approx 0.8631 \Rightarrow \beta_1 \approx 59°40'$ or $\beta_2 =$

$180° - \beta_1 \approx 120°20'$. Triangle 1: $\beta_1 \approx 59°40'$. $\alpha_1 = 180° - (\gamma + \beta_1) \approx 83°10'$. $\dfrac{a_1}{\sin \alpha_1} =$

$\dfrac{c}{\sin \gamma} \Rightarrow a_1 = \dfrac{c \sin \alpha_1}{\sin \gamma} \approx \dfrac{875 \sin 83°10'}{\sin 37°10'} \approx 1440$. Triangle 2 : $\beta_2 \approx 120°20'$. $\alpha_2 =$

$180° - (\gamma + \beta_2) \approx 22°30'$. $\dfrac{a_2}{\sin \alpha_2} = \dfrac{c}{\sin \gamma} \Rightarrow a_2 = \dfrac{c \sin \alpha_2}{\sin \gamma} \approx \dfrac{875 \sin 22°30'}{\sin 37°10'} \approx 554$.

[4] $\beta = 180° - (\alpha + \gamma) = 76.1°$. $\dfrac{a}{\sin \alpha} = \dfrac{b}{\sin \beta} \Rightarrow b = \dfrac{a \sin \beta}{\sin \alpha} = \dfrac{42.6 \sin 76.1°}{\sin 86.4°} \approx 41.4$.

$\dfrac{a}{\sin \alpha} = \dfrac{c}{\sin \gamma} \Rightarrow c = \dfrac{a \sin \gamma}{\sin \alpha} = \dfrac{42.6 \sin 17.5°}{\sin 86.4°} \approx 12.8$.

[5] $\dfrac{a}{\sin \alpha} = \dfrac{b}{\sin \beta} \Rightarrow \sin \beta = \dfrac{b \sin \alpha}{a} = \dfrac{702 \sin 18°40'}{235} \approx 0.9561 \Rightarrow \beta_1 \approx 73°$ and $\beta_2 =$

$180° - \beta_1 \approx 107°$. Triangle 1: $\beta_1 \approx 73°$. $\gamma_1 = 180° - (\alpha + \beta_1) \approx 88°20'$. $\dfrac{a}{\sin \alpha} = \dfrac{c_1}{\sin \gamma_1} \Rightarrow$

$c_1 = \dfrac{a \sin \gamma_1}{\sin \alpha} \approx \dfrac{235 \sin 88°20'}{\sin 18°40'} \approx 734$. Triangle 2: $\beta_2 \approx 107°$. $\gamma_2 = 180° - (\alpha + \beta_2) \approx$

$54°20'$. $\dfrac{a}{\sin \alpha} = \dfrac{c_2}{\sin \gamma_2} \Rightarrow c_2 = \dfrac{a \sin \gamma_2}{\sin \alpha} \approx \dfrac{235 \sin 54°20'}{\sin 18°40'} \approx 597$.

[6] $a = \sqrt{b^2 + c^2 - 2bc \cos \alpha} = \sqrt{(91.2)^2 + (87.1)^2 - 2(91.2)(87.1) \cos 43.6°} \approx 66.3$. $\cos \beta =$

$\dfrac{a^2 + c^2 - b^2}{2ac} \approx \dfrac{(66.3)^2 + (87.1)^2 - (91.2)^2}{2(66.3)(87.1)} \approx 0.3173 \Rightarrow \beta \approx 71.5°$. $\gamma = 180° - (\alpha + \beta) \approx 64.9°$.

[7] $\cos \alpha = \dfrac{b^2 + c^2 - a^2}{2bc} = \dfrac{(81.4)^2 + (68.7)^2 - (25.2)^2}{2(81.4)(68.7)} \approx 0.9576 \Rightarrow \alpha \approx 16.7°$. $\cos \beta = \dfrac{a^2 + c^2 - b^2}{2ac} =$

$\dfrac{(25.2)^2 + (68.7)^2 - (81.4)^2}{2(25.2)(68.7)} \approx -0.3671 \Rightarrow \beta \approx 111.5°$. $\gamma = 180 - (\alpha + \beta) \approx 51.8°$.

[8] $\dfrac{b}{\sin \beta} = \dfrac{c}{\sin \gamma} \Rightarrow \sin \gamma = \dfrac{c \sin \beta}{b} = \dfrac{4.63 \sin 74.4°}{5.15} \approx 0.8659 \Rightarrow \gamma_1 = 60.0°$ or $\gamma_2 = 180 - 60.0° = 120.0°$.

Since $\beta + 120.0° > 180°$, only one triangle is determined and $\gamma = 60.0°$. $\alpha = 180° - (\beta + \gamma) \approx 45.6°$.

$\dfrac{a}{\sin \alpha} = \dfrac{b}{\sin \beta} \Rightarrow a = \dfrac{b \sin \alpha}{\sin \beta} \approx \dfrac{5.15 \sin 45.6°}{\sin 74.4°} \approx 3.82$.

[9] $\gamma = 180° - (\alpha + \beta) = 95°55'$. $\dfrac{a}{\sin \alpha} = \dfrac{b}{\sin \beta} \Rightarrow a = \dfrac{b \sin \alpha}{\sin \beta} = \dfrac{9.75 \sin 5°25'}{\sin 78°40'} \approx 0.94$.

$\dfrac{c}{\sin \gamma} = \dfrac{b}{\sin \beta} \Rightarrow c = \dfrac{b \sin \gamma}{\sin \beta} = \dfrac{9.75 \sin 95°55'}{\sin 78°40'} \approx 9.89$.

[10] $\dfrac{a}{\sin \alpha} = \dfrac{c}{\sin \gamma} \Rightarrow \sin \alpha = \dfrac{a \sin \gamma}{c} = \dfrac{52.3 \sin 27.2°}{36.7} \approx 0.6514 \Rightarrow \alpha_1 = 40.6°$ or $\alpha_2 =$

$180° - \alpha_1 \approx 139.4°$. Triangle 1: $\alpha_1 = 40.6°$. $\beta_1 = 180° - (\gamma + \alpha_1) \approx 112.2°$. $\dfrac{b_1}{\sin \beta_1} =$

$\dfrac{c}{\sin \gamma} \Rightarrow b_1 = \dfrac{c \sin \beta_1}{\sin \gamma} \approx \dfrac{36.7 \sin 112.2°}{\sin 27.2°} \approx 74.3$. Triangle 2: $\alpha_2 = 139.4°$. $\beta_2 =$

$180° - (\gamma + \alpha_2) \approx 13.4°$. $\dfrac{b_2}{\sin \beta_2} = \dfrac{c}{\sin \gamma} \Rightarrow b_2 = \dfrac{c \sin \beta_2}{\sin \gamma} \approx \dfrac{36.7 \sin 13.4°}{\sin 27.2°} \approx 18.6$.

[11] $\frac{a}{\sin \alpha} = \frac{b}{\sin \beta} \Rightarrow \sin \alpha = \frac{a \sin \beta}{b} = \frac{12.0 \sin 31°50'}{7.50} \approx 0.8439 \Rightarrow \alpha_1 \approx 57°30'$ or $\alpha_2 = $

$180° - \alpha_1 \approx 122°30'$. Triangle 1: $\alpha_1 \approx 57°30'$. $\gamma_1 = 180 - (\beta + \alpha_1) \approx 90°40'$. $\frac{b}{\sin \beta} = $

$\frac{c_1}{\sin \gamma_1} \Rightarrow c_1 = \frac{b \sin \gamma_1}{\sin \beta} \approx \frac{7.50 \sin 90°40'}{\sin 31°50'} \approx 14.2$. Triangle 2: $\alpha_2 \approx 122°30'$. $\gamma_2 = $

$180° - (\beta + \alpha_2) \approx 25°40'$. $\frac{b}{\sin \beta} = \frac{c_2}{\sin \gamma_2} \Rightarrow c_2 = \frac{b \sin \gamma_2}{\sin \beta} \approx \frac{7.50 \sin 25°40'}{\sin 31°50'} \approx 6.2$.

[12] $\cos \alpha = \frac{b^2 + c^2 - a^2}{2bc} = \frac{(236)^2 + (357)^2 - (152)^2}{2(236)(357)} \approx 0.9498 \Rightarrow \alpha \approx 18.2°$. $\cos \beta = \frac{a^2 + c^2 - b^2}{2ac} = $

$\frac{(152)^2 + (357)^2 - (236)^2}{2(152)(357)} \approx 0.8740 \Rightarrow \beta \approx 29.1°$. $\gamma = 180° - (\alpha + \beta) \approx 132.7°$.

[13] $\frac{a}{\sin \alpha} = \frac{b}{\sin \beta} \Rightarrow \sin \alpha = \frac{a \sin \beta}{b} = \frac{12.14 \sin 33°42'}{16.35} \approx 0.4120 \Rightarrow \alpha_1 = 24°20'$ or $\alpha_2 = $

$180° - \alpha_1 \approx 155°40'$. Since $\beta + 155°40' > 180°$, only one triangle is determined. Thus, $\alpha = 24°20'$. $\gamma = $

$180° - (\alpha + \beta) \approx 121°58'$. $\frac{b}{\sin \beta} = \frac{c}{\sin \gamma} \Rightarrow c = \frac{b \sin \gamma}{\sin \beta} \approx \frac{16.35 \sin 121°58'}{\sin 33°42'} \approx 25.00$.

[14] $b = \sqrt{a^2 + c^2 - 2ac \cos \beta} = \sqrt{(32.4)^2 + (38.8)^2 - 2(32.4)(38.8) \cos 11.7} \approx 9.7$. $\cos \alpha = $

$\frac{b^2 + c^2 - a^2}{2bc} = \frac{(9.7)^2 + (38.8)^2 - (32.4)^2}{2(9.7)(38.8)} \approx 0.7304 \Rightarrow \alpha \approx 43.1°$. $\gamma = 180° - (\alpha + \beta) \approx 125.2°$.

[15] $c = \sqrt{a^2 + b^2 - 2ab \cos \gamma} = \sqrt{(56.4)^2 + (102.9)^2 - 2(56.4)(102.9) \cos 120°30'} \approx 140.2$. $\cos \alpha = $

$\frac{b^2 + c^2 - a^2}{2bc} \approx \frac{(102.9)^2 + (140.2)^2 - (56.4)^2}{2(102.9)(140.2)} \approx 0.9380 \Rightarrow \alpha \approx 20°20'$. $\beta = 180° - (\alpha + \gamma) \approx 39°10'$.

[16] $\cos \alpha = \frac{b^2 + c^2 - a^2}{2bc} = \frac{(29.8)^2 + (13.4)^2 - (15.2)^2}{2(29.8)(13.4)} \approx 1.0475$. Since $\cos \alpha > 1$, no triangle is possible.

[17] $b = \sqrt{a^2 + c^2 - 2ac \cos \beta} = \sqrt{(2.36)^2 + (1.84)^2 - 2(2.36)(1.84) \cos 115.1°} \approx 3.56$. $\cos \alpha = $

$\frac{b^2 + c^2 - a^2}{2bc} \approx \frac{(3.56)^2 + (1.84)^2 - (2.36)^2}{2(3.56)(1.84)} \approx 0.8007 \Rightarrow \alpha \approx 36.8°$. $\gamma = 180° - (\alpha + \beta) \approx 28.1°$.

[18] $b = \sqrt{a^2 + c^2 - 2ac \cos \beta} = \sqrt{(78.9)^2 + (125.0)^2 - 2(78.9)(125.0) \cos 128°10'} \approx 184.5$. $\cos \alpha = $

$\frac{b^2 + c^2 - a^2}{2bc} \approx \frac{(184.5)^2 + (125.0)^2 - (78.9)^2}{2(184.5)(125.0)} \approx 0.9418 \Rightarrow \alpha \approx 19°40'$. $\gamma = 180° - (\alpha + \beta) \approx 32°10'$.

[19] $\frac{a}{\sin \alpha} = \frac{b}{\sin \beta} \Rightarrow \sin \beta = \frac{b \sin \alpha}{a} = \frac{30 \sin 65°}{24} \approx 1.1329$. Since $\sin \beta > 1$, no triangle is possible.

[20] $\frac{a}{\sin \alpha} = \frac{c}{\sin \gamma} \Rightarrow \sin \alpha = \frac{a \sin \gamma}{c} = \frac{2250 \sin 17.2°}{1372} \approx 0.4849 \Rightarrow \alpha_1 \approx 29.0°$ and $\alpha_2 = 180° - \alpha_1 \approx $

$151.0°$. Triangle 1: $\alpha_1 = 29.0°$. $\beta_1 = 180° - (\alpha_1 + \gamma) \approx 133.8°$. $\frac{b_1}{\sin \beta_1} = \frac{c}{\sin \gamma} \Rightarrow b_1 = \frac{c \sin \beta_1}{\sin \gamma} \approx $

$\frac{1372 \sin 133.8°}{\sin 17.2°} \approx 3350$. Triangle 2: $\alpha_2 = 151.0°$. $\beta_2 = 180° - (\alpha_2 + \gamma) \approx 11.8°$. $\frac{b_2}{\sin \beta_2} = \frac{c}{\sin \gamma} \Rightarrow $

$b_2 = \frac{c \sin \beta_2}{\sin \gamma} \approx \frac{1372 \sin 11.8°}{\sin 17.2°} \approx 950$.

[21] $\dfrac{a}{\sin\alpha}=\dfrac{b}{\sin\beta}\Rightarrow\sin\alpha=\dfrac{a\sin\beta}{b}=\dfrac{4.68\sin 57°50'}{5.03}\approx 0.7876\Rightarrow\alpha_1\approx 52°$ and $\alpha_2=$

$180°-\alpha_1=128°$. Since $\beta+128°>180°$, only 1 triangle is determined. Thus $\alpha\approx 52°$, and

$\gamma=180°-(\alpha+\beta)\approx 70°10'$. $\dfrac{b}{\sin\beta}=\dfrac{c}{\sin\gamma}\Rightarrow c=\dfrac{b\sin\gamma}{\sin\beta}\approx\dfrac{5.03\sin 70°10'}{\sin 57°50'}\approx 5.59$.

[22] $\cos\alpha=\dfrac{b^2+c^2-a^2}{2bc}=\dfrac{(17.3)^2+(95.7)^2-(124.7)^2}{2(17.3)(95.7)}\approx-1.8399$. Since $\cos\alpha<-1$, no triangle is

possible.

[23] $b=\sqrt{a^2+c^2-2ac\cos\beta}=\sqrt{(30.2)^2+(42.8)^2-2(30.2)(42.8)\cos 132.3°}\approx 67.0$. $\cos\alpha=$

$\dfrac{b^2+c^2-a^2}{2bc}\approx\dfrac{(67.0)^2+(42.8)^2-(30.2)^2}{2(67.0)(42.8)}\approx 0.9431\Rightarrow\alpha\approx 19.4°$. $\gamma=180°-(\alpha+\beta)\approx 28.3°$.

[24] $\cos\alpha=\dfrac{b^2+c^2-a^2}{2bc}=\dfrac{(0.3757)^2+(0.4142)^2-(0.2086)^2}{2(0.3757)(0.4142)}\approx 0.8649\Rightarrow\alpha\approx 30.1°$. $\cos\beta=$

$\dfrac{a^2+c^2-b^2}{2ac}=\dfrac{(0.2086)^2+(0.4142)^2-(0.3757)^2}{2(0.2086)(0.4142)}\approx 0.4278\Rightarrow\beta\approx 64.7°$. $\gamma=180°-(\alpha+\beta)\approx 85.2°$.

[25] $A=\dfrac{1}{2}bc\sin\alpha=\dfrac{(46.8)(82.5)\sin 102.4°}{2}\approx 1890$.

[26] $A=\dfrac{1}{2}ab\sin\gamma=\dfrac{(7.18)(10.40)\sin 41°20'}{2}\approx 24.7$

[27] $s=\dfrac{a+b+c}{2}=\dfrac{33+46+65}{2}=72$. $A=\sqrt{s(s-a)(s-b)(s-c)}=\sqrt{72(39)(26)(7)}\approx 715$.

[28] $s=\dfrac{3.48+5.19+7.04}{2}=7.855$. $A=\sqrt{s(s-a)(s-b)(s-c)}=$

$\sqrt{(7.855)(4.375)(2.665)(0.815)}\approx 8.64$.

[29] $A=\dfrac{1}{2}bc\sin\alpha=\dfrac{(2110)(3540)\sin 82.2°}{2}\approx 3{,}700{,}000$.

[30] $A=\dfrac{1}{2}ac\sin\beta=\dfrac{(625)(948)\sin 75°40'}{2}\approx 287{,}000$.

[31] $s=\dfrac{1.48+0.75+1.28}{2}=1.755$. $A=\sqrt{s(s-a)(s-b)(s-c)}=$

$\sqrt{(1.755)(0.275)(1.005)(0.475)}\approx 0.48$.

[32] $A=\dfrac{1}{2}ab\sin\gamma=\dfrac{(32.5)(44.6)\sin 112.9°}{2}\approx 668$.

[33] $A=\dfrac{1}{2}bc\sin\alpha=\dfrac{(256.7)(244.1)\sin 13°35'}{2}\approx 7360$.

[34] $A=\dfrac{1}{2}ac\sin\beta=\dfrac{(14.8)(31.5)\sin 150°}{2}\approx 117$.

[35] $s=\dfrac{1280+1645+1090}{2}=2007.5$. $A=\sqrt{s(s-a)(s-b)(s-c)}=$

$\sqrt{(2007.5)(727.5)(362.5)(917.5)}\approx 697{,}000$.

[36] $s=\dfrac{156{,}400+314{,}000+175{,}500}{2}=322{,}950$.

$A=\sqrt{(322{,}950)(166{,}550)(8950)(147{,}450)}\approx 8{,}425{,}000{,}000$.

[37] $\dfrac{53.5}{\sin\theta}=\dfrac{47.8}{\sin 45°40'}\Rightarrow\sin\theta=\dfrac{53.5\sin 45°40'}{47.8}\approx 0.8006\Rightarrow\theta\approx 53°10'$.   (See figure below).

[38] Let x denote the height of the building. $\dfrac{y}{\sin 68.6^\circ} = \dfrac{22.5}{\sin 6.2^\circ} \Rightarrow y = \dfrac{22.5 \sin 68.6^\circ}{\sin 6.2^\circ} \approx 194$.

$\sin 74.8^\circ = \dfrac{x}{y} \Rightarrow x = y \sin 74.8^\circ \approx 194 \sin 74.8^\circ \approx 187$. Thus, the building is approximately 187 feet high. (See figure below).

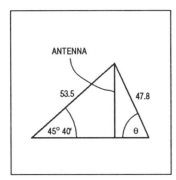

Figure 37

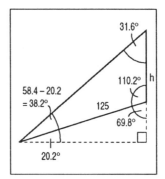

Figure 38

[39] $\dfrac{h}{\sin 18.1^\circ} = \dfrac{135}{\sin 82.5^\circ} \Rightarrow h = \dfrac{135 \sin 18.1^\circ}{\sin 82.5^\circ} \approx 42.3$. Thus, the telephone pole is approximately 42.3 feet high. (See figure below).

[40] Let h denote the height of the flagpole. $\dfrac{h}{\sin 38.2^\circ} = \dfrac{125}{\sin 31.6^\circ} \Rightarrow h = \dfrac{125 \sin 38.2^\circ}{\sin 31.6^\circ} \approx 148$. The flagpole is approximately 148 feet high. (See figure below).

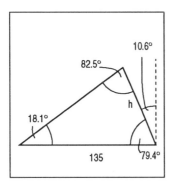

Figure 39

Figure 40

[41] Let P denote the port, A the point at which the ship changes direction and B the final position of the ship. Let x denote the distance between the ship and its port and let θ denote the bearing from the ship to the port. Then $\angle PAB = 48°20' + 12°10' = 60°30'$, and $x =$

$\sqrt{(88.4)^2 + (135)^2 - 2(88.4)(135)\cos 60°30'} \approx 120.$ Also $\dfrac{88.4}{\sin \angle BPA} = \dfrac{x}{\sin \angle PAB} \Rightarrow$

$\sin \angle BPA = \dfrac{88.4 \sin \angle PAB}{x} \approx \dfrac{88.4 \sin 60°30'}{120} \approx 0.6412 \Rightarrow \angle BPA \approx 39°50'.$ Thus,

$\theta = \angle BPA + 48°20' \approx 88°10'.$ The ship is approximately 120 km from its port, and the bearing from the ship to the port is approximately N88°10'E.   (See figure below).

[42] Let F denote the location of the fire, and let x and y denote the fire's distances from A and B, respectively. Then $\angle FAB = 65.3° - 27.5° = 37.8°$, $\angle FBA = 74.2° + 24.7° = 98.9°$, and $\angle BFA =$

$180° - (37.8° + 98.9°) = 43.3°.$ $\dfrac{x}{\sin 98.9°} = \dfrac{16.5}{\sin 43.3°} \Rightarrow x = \dfrac{16.5 \sin 98.9°}{\sin 43.3°} \approx 23.8$ and

$\dfrac{y}{\sin 37.8°} = \dfrac{16.5}{\sin 43.3°} \Rightarrow y = \dfrac{16.5 \sin 37.8°}{\sin 43.3°} \approx 14.7.$ The fire is approximately 23.8 miles from A and approximately 14.7 miles from B.   (See figure below).

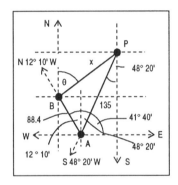

Figure 41

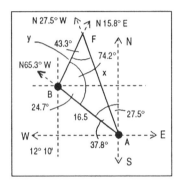

Figure 42

[43] $x = \sqrt{(89.5)^2 + (48.0)^2 - 2(89.5)(48.0)\cos 102.5°} \approx 110.$ $\dfrac{48.0}{\sin \theta} = \dfrac{x}{\sin 102.5°} \Rightarrow$

$\sin \theta = \dfrac{48.0 \sin 102.5°}{x} \approx \dfrac{48.0 \sin 102.5°}{110} \approx 0.4260 \Rightarrow \theta \approx 25.2°.$ $\theta + 12.5° = 37.7°$ is the

approximate angle of elevation of the sun.   (See figure below).

[44] $x = \sqrt{(24.2)^2 + (37.5)^2 - 2(24.2)(37.5)\cos 114°20'} \approx 52.3.$ The support wire is

approximately 52.3 meters long.   (See figure below).

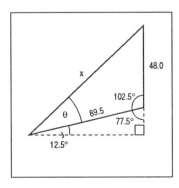

Figure 43

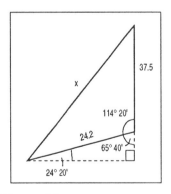

Figure 44

[45] $\dfrac{32.4}{\sin \alpha} = \dfrac{52.5}{\sin 68.4^\circ} \Rightarrow \sin \alpha = \dfrac{32.4 \sin 68.4^\circ}{52.5} \approx 0.5738 \Rightarrow \alpha \approx 35.0^\circ$. $\beta = 180^\circ - (\alpha + 68.4^\circ) \approx$

76.6°, and $\theta = 90^\circ - \beta \approx 13.4^\circ$. Thus, the angle between the pole and the vertical is

approximately 13.4°. (See figure below).

[46] Let P denote the port, A the location of the first ship and B the location of the second ship. Let x

denote the distance between the two ships. Then, $d(P, A) = (32)(2) = 64$ and $d(P, B) =$

$(24)(1.25) = 30$. Thus, $x = \sqrt{(30)^2 + (64)^2 - 2(30)(64) \cos \angle BPA}$ =

$\sqrt{900 + 4096 - 3840 \cos 57.6^\circ} \approx 54.2$. The two ships are approximately 54.2 miles apart.

(See figure below).

[47] $\dfrac{y}{\sin 30^\circ 20'} = \dfrac{100}{\sin 8^\circ 20'} \Rightarrow y = \dfrac{100 \sin 30^\circ 20'}{\sin 8^\circ 20'} \approx 348$. $\sin 38^\circ 40' = \dfrac{x}{y} \Rightarrow x = y \sin 38^\circ 40' \approx$

$348 \sin 38^\circ 40' \approx 217$. Thus, the tower is approximately 217 feet high. (See figure below).

[48] Let C denote the location of the balloon and let h denote its height. Then, $\dfrac{x}{\sin 25.2^\circ} =$

$\dfrac{7.65}{\sin 105.9^\circ} \Rightarrow x = \dfrac{7.65 \sin 25.2^\circ}{\sin 105.9^\circ} \approx 3.39$, and $\sin 48.9^\circ = \dfrac{h}{x} \Rightarrow h = x \sin 48.9^\circ \approx 3.39 \sin 48.9^\circ \approx$

2.55. Thus, the height of the balloon is approximately 2.55 miles. (See figure below).

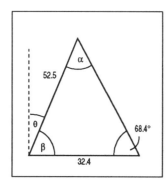

Figure 45

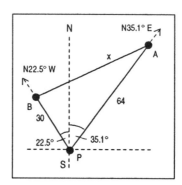

Figure 46

[49] $s = \dfrac{112 + 130 + 164}{2} = 203$. $A = \sqrt{s(s - 112)(s - 130)(s - 164)} =$

$\sqrt{(203)(91)(73)(39)} \approx 7250$. The area of the plot is approximately $7250 \, \text{m}^2$.

[50] $s = \dfrac{625 + 830 + 1220}{2} = 1337.5$. $A = \sqrt{(1337.5)(712.5)(507.5)(117.5)} \approx 238{,}400$.

$\dfrac{238{,}400}{4840} \approx 49.3$. The field contains approximately 49.3 acres.

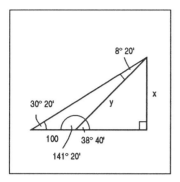

Figure 47

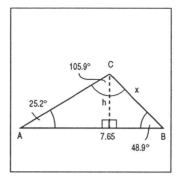

Figure 48

[51] $x = \sqrt{(108.5)^2 + (95.1)^2 - 2(108.5)(95.1)\cos 84.3°} \approx 137.0.$ $\dfrac{95.1}{\sin \angle BAC} =$

$\dfrac{x}{\sin 84.3°} \Rightarrow \sin \angle BAC = \dfrac{95.1 \sin 84.3°}{x} \approx \dfrac{95.1 \sin 84.3°}{137.0} \approx 0.6907 \Rightarrow \angle BAC \approx 43.7°.$ Thus,

$\angle CAD = 72.8° - \angle BAC \approx 29.1°,$ and $\angle ACD = 180° - (99.5° + \angle CAD) \approx 51.4°.$ $\dfrac{y}{\sin \angle ACD} =$

$\dfrac{x}{\sin 99.5°} \Rightarrow y = \dfrac{x \sin \angle ACD}{\sin 99.5°} \approx \dfrac{137.0 \sin 51.4°}{\sin 99.5°} \approx 108.6.$ Area of quadrilateral $ABCD$ = area of

triangle ABC + area of triangle $ACD = \frac{1}{2}(108.5)(95.1)\sin 84.3° + \frac{1}{2}xy \sin \angle CAD \approx$

$\frac{1}{2}(108.5)(95.1)\sin 84.3° + \frac{1}{2}(137.0)(108.6)\sin 29.1° \approx 8750 \text{ in}^2.$ (See figure below).

[52] $A = 2\left[\frac{1}{2}(17.5)(28.8)\sin 48°40'\right] \approx 378.$ The area of the parallelogram is approximately 378 cm². (See figure below).

[53] Let L denote the location of the lighthouse and let x denote the distance between ship B and the

lighthouse. Then, $x = \sqrt{(2.75)^2 + (7.25)^2 - 2(2.75)(7.25)\cos 71°50'} \approx 6.9$ miles. Also, $\dfrac{7.25}{\sin \theta} =$

$\dfrac{x}{\sin 71°50'} \Rightarrow \sin \theta = \dfrac{7.25 \sin 71°50'}{x} \approx \dfrac{7.25 \sin 71°50'}{6.9} \approx 0.9983 \Rightarrow \theta \approx 86.7°.$ $90° - \theta \approx 3.3°,$ and

the bearing from ship B to the lighthouse is approximately $N3.3°W.$ (See figure below).

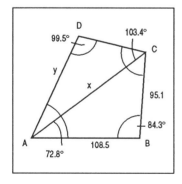

Figure 51

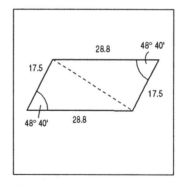

Figure 52

[54] Let x denote the distance between airports A and C. Since $\theta = 65.4°,$ $\angle ABC = 65.4° + 48.7° =$

$114.1°.$ Also, $\angle BAC = 180° - (65.4° + 62.8°) = 51.8°.$ Thus, $\angle ACB = 180° - (114.1° + 51.8°) =$

$14.1°.$ $\dfrac{x}{\sin \angle ABC} = \dfrac{465}{\sin \angle ACB} \Rightarrow x = \dfrac{465 \sin \angle ABC}{\sin \angle ACB} = \dfrac{465 \sin 114.1°}{\sin 14.1°} \approx 1740$ miles.

(See figure below).

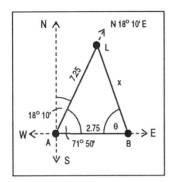

Figure 53

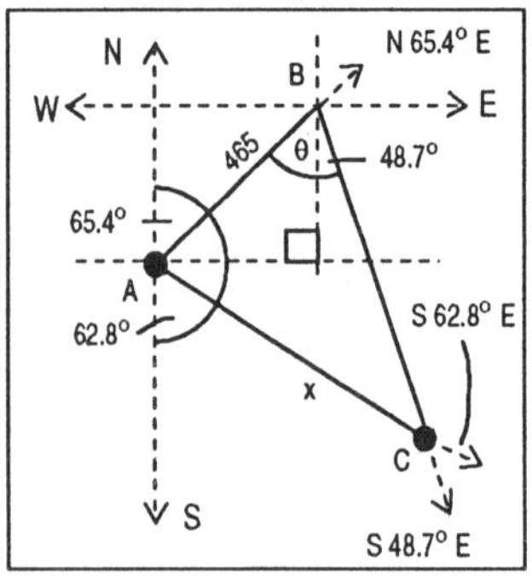

Figure 54

[55] $x = \sqrt{(1250)^2 + (1060)^2 - 2(1250)(1060)\cos 55.2°} \approx 1080$. Thus, the distance across the swamp is approximately 1080 meters.   (See figure below).

[56] $x = \sqrt{(675)^2 + (925)^2 - 2(675)(925)\cos 85.7°} \approx 1100$. Thus, the distance across the lake is approximately 1100 yards.   (See figure below).

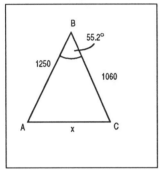

Figure 55

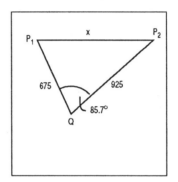

Figure 56

[57] Let L denote the location of the lighthouse, A the first location of the ship and B the second location of the ship. Let x denote the distance to the lighthouse at the second sighting. $\angle B = 81°30' - 27°50' = 53°40'$. Thus, $\angle L = 180° - (53°40' + 27°50' + 36°20') = 62°10'$. $\frac{x}{\sin \angle A} = \frac{8.25}{\sin \angle L} \Rightarrow x = \frac{8.25 \sin 64°10'}{\sin 62°10'} \approx 8.40$ miles.   (See figure below).

[58] Let x denote the height of the tower. $\theta = 180° - (128.70° + 32.45°) = 18.85°$, and $\frac{y}{\sin 128.70°} = \frac{47.2}{\sin 18.85°} \Rightarrow y = \frac{47.2 \sin 128.70°}{\sin 18.85°} \approx 114.0$. Thus, $47.2 + 91.3 = 138.5$, and $x \approx \sqrt{(138.5)^2 + (114.0)^2 - 2(138.5)(114.0)\cos 32.45°} \approx 74.4$ feet.   (See figure below).

[59] Let x denote the length of each side of the triangle. Then $x = \sqrt{(18.4)^2 + (18.4)^2 - 2(18.4)(18.4)\cos 120°} \approx 31.9$ cm. $s = \frac{3x}{2} \approx \frac{3(31.9)}{2} \approx 47.85$. $A = \sqrt{s(s-x)^3} \approx \sqrt{(47.85)(15.95)^3} \approx 441$. The area of the triangle is approximately 441 cm². (See figure below).

[60] Let H denote home plate, P the pitcher's mound, and A, B and C be first, second and third base, respectively. Let x denote the distance from the pitcher's mound to first (or third) base and let y denote the distance from the mound to second base. From $\triangle HPA$, $x =$
$\sqrt{(60.5)^2 + (90)^2 - 2(60.5)(90)\cos 45°} \approx 63.7$ feet. From right $\triangle HAB$, $y + 60.5 =$
$\sqrt{(90)^2 + (90)^2} \approx 127.3 \Rightarrow y \approx 127.3 - 60.5 = 66.8$ feet. (See figure below).

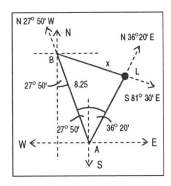

Figure 57

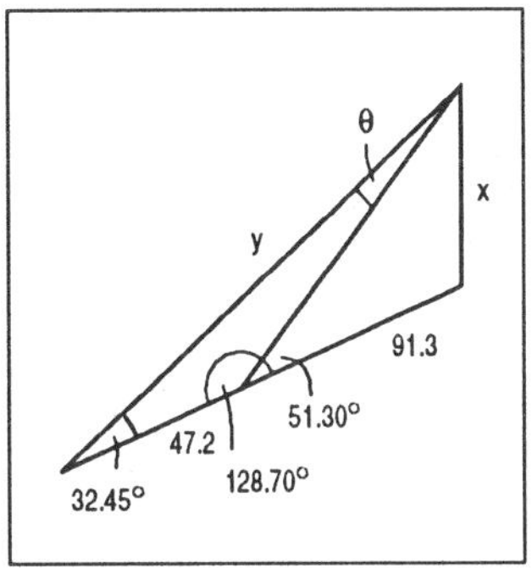

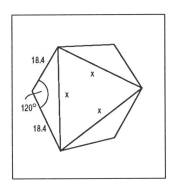

Figure 59

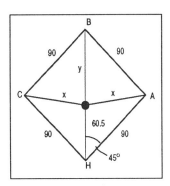

Figure 58

Figure 60

[61] $a = \sqrt{(8)^2 + (20)^2} \approx 21.5$, $b = \sqrt{(8)^2 + (12)^2} \approx 14.4$ and $c = \sqrt{(12)^2 + (20)^2} \approx 23.3$. $\cos\alpha =$
$\dfrac{b^2 + c^2 - a^2}{2bc} \approx \dfrac{(14.4)^2 + (23.3)^2 - (21.5)^2}{2(14.4)(23.3)} \approx 0.4292 \Rightarrow \alpha \approx 65°.$ $A = \frac{1}{2}bc \sin\alpha \approx$
$\frac{1}{2}(14.4)(23.3)\sin 65° \approx 150$. The area of triangle ABC is approximately 150 in². (See figure below).

[62] $\cos\alpha = \dfrac{b^2 + c^2 - a^2}{2bc} \Rightarrow \dfrac{\cos\alpha}{a} = \dfrac{b^2 + c^2 - a^2}{2abc}$. $\cos\beta = \dfrac{a^2 + c^2 - b^2}{2ac} \Rightarrow \dfrac{\cos\beta}{b} = \dfrac{a^2 + c^2 - b^2}{2abc}$, and

$\cos\gamma = \dfrac{a^2 + b^2 - c^2}{2ab} \Rightarrow \dfrac{\cos\gamma}{c} = \dfrac{a^2 + b^2 - c^2}{2abc}$. Thus, $\dfrac{\cos\alpha}{a} + \dfrac{\cos\beta}{b} + \dfrac{\cos\gamma}{c} = \dfrac{b^2 + c^2 - a^2}{2abc} +$

$\dfrac{a^2 + c^2 - b^2}{2abc} + \dfrac{a^2 + b^2 - c^2}{2abc} = \dfrac{a^2 + b^2 + c^2}{2abc}$.

[63] $s = \dfrac{3a}{2} \Rightarrow A = \sqrt{s(s-a)^3} = \sqrt{\left(\dfrac{3a}{2}\right)\left(\dfrac{3a}{2} - a\right)^3} = \sqrt{\dfrac{3a^4}{16}} = \dfrac{a^2\sqrt{3}}{4}$. $A = 36 \Rightarrow$

$\dfrac{a^2\sqrt{3}}{4} = 36 \Rightarrow a^2 = \dfrac{144}{\sqrt{3}} \Rightarrow a = \dfrac{12}{\sqrt[4]{3}}$ inches. (See figure below).

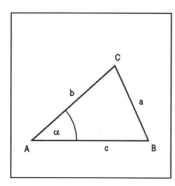

Figure 61

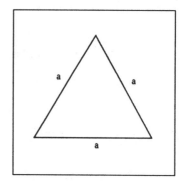

Figure 63

[64] $\theta = 180° - [\alpha + (180° - \beta)] = \beta - \alpha$. $\dfrac{m}{\sin(\beta - \alpha)} = \dfrac{y}{\sin \alpha} \Rightarrow y = \dfrac{m \sin \alpha}{\sin(\beta - \alpha)}$. From right $\triangle BCD$,

$\sin \beta = \dfrac{h}{y} \Rightarrow y = \dfrac{h}{\sin \beta}$. Thus, $\dfrac{h}{\sin \beta} = \dfrac{m \sin \alpha}{\sin(\beta - \alpha)} \Rightarrow h = \dfrac{m \sin \alpha \sin \beta}{\sin(\beta - \alpha)}$. (See figure below).

[65] The area of $\triangle ABC$ is the sum of the areas of triangles ADB, BDC, and ADC, where D is the center of the circle. Since the radius r, drawn from the center D, is perpendicular to side $\overline{AB}$, r is the height of $\triangle ADB$. Thus, the area of $\triangle ADB$ is $\frac{1}{2}cr$. In like manner, the areas of triangles BDC and ADC are given by $\frac{1}{2}ar$, and $\frac{1}{2}br$, respectively. Thus, $A = \frac{1}{2}ar + \frac{1}{2}br + \frac{1}{2}cr = r\left(\dfrac{a+b+c}{2}\right)$.

By Heron's formula, $A = \sqrt{s(s-a)(s-b)(s-c)}$, where $s = \dfrac{a+b+c}{2}$. Hence $rs =$

$\sqrt{s(s-a)(s-b)(s-c)} \Rightarrow r = \dfrac{\sqrt{s(s-a)(s-b)(s-c)}}{s} \Rightarrow r =$

$\sqrt{\dfrac{s(s-a)(s-b)(s-c)}{s^2}} \Rightarrow r = \sqrt{\dfrac{(s-a)(s-b)(s-c)}{s}}$. (See figure below).

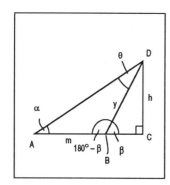

Figure 64

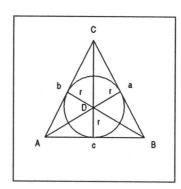

Figure 65

[66] a) We place triangle ABC in a rectangular coordinate system so that α is in standard position and B is on the positive x – axis. We construct segment CD perpendicular to the x – axis and let h denote its length. From right triangle ADC, $\sin \alpha = \dfrac{h}{b}$ and from right triangle BDC, $\sin \beta = \dfrac{h}{a}$. Thus $h = b \sin \alpha$ and $h = a \sin \beta$. It follows that $a \sin \beta = b \sin \alpha$, or $\dfrac{a}{\sin \alpha} = \dfrac{b}{\sin \beta}$. If we place triangle ABC so that β is in standard position and C lies on the positive x – axis, a similar argument will show that $\dfrac{b}{\sin \beta} = \dfrac{c}{\sin \gamma}$. Thus $\dfrac{a}{\sin \alpha} = \dfrac{b}{\sin \beta} = \dfrac{c}{\sin \gamma}$. (See figure below).

b) Applying the Pythagorean Theorem to right triangle BDC of Figure 66, we see that $a^2 = h^2 + [d(D, B)]^2$, or $a^2 = h^2 + [c - d(A, D)]^2$ (*). From right triangle ADC, $\cos \alpha = \dfrac{d(A, D)}{b} \Rightarrow d(A, D) = b \cos \alpha$, and $\sin \alpha = \dfrac{h}{b} \Rightarrow h = b \sin \alpha$. Thus, equation (*) becomes: $a^2 = (b \sin \alpha)^2 + (c - b \cos \alpha)^2$, so that $a^2 = b^2 \sin^2 \alpha + c^2 - 2bc \cos \alpha + b^2 \cos^2 \alpha = b^2 (\sin^2 \alpha + \cos^2 \alpha) + c^2 - 2bc \cos \alpha = b^2 + c^2 - 2bc \cos \alpha$. Hence $a^2 = b^2 + c^2 - 2bc \cos \alpha$.

Similar arguments, with angles β and γ in standard position, will verify that $b^2 = a^2 + c^2 - 2ac \cos \beta$ and $c^2 = a^2 + b^2 - 2ab \cos \gamma$.

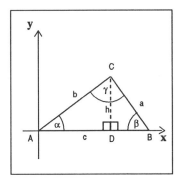

Figure 66

EXERCISES 7.3

[1] $r = \sqrt{(1)^2 + (\sqrt{3})^2} = 2.$ $\cos \theta = \frac{1}{2}$ and $\sin \theta = \frac{\sqrt{3}}{2} \Rightarrow \theta = \frac{\pi}{3}.$ $1 + \sqrt{3} \, i = 2 \operatorname{cis} \frac{\pi}{3}.$

[2] $r = \sqrt{(2)^2 + (-2)^2} = 2\sqrt{2}.$ $\cos \theta = \frac{1}{\sqrt{2}}$ and $\sin \theta = -\frac{1}{\sqrt{2}} \Rightarrow \theta = \frac{7\pi}{4}.$ $2 - 2 \, i = 2\sqrt{2} \operatorname{cis} \frac{7\pi}{4}.$

[3] $r = \sqrt{(-4)^2 + (4)^2} = 4\sqrt{2}.$ $\cos \theta = -\frac{1}{\sqrt{2}}$ and $\sin \theta = \frac{1}{\sqrt{2}} \Rightarrow \theta = \frac{3\pi}{4}.$ $-4 + 4 \, i = 4\sqrt{2} \operatorname{cis} \frac{3\pi}{4}.$

[4] $r = \sqrt{(\sqrt{3})^2 + (-1)^2} = 2.$ $\cos \theta = \frac{\sqrt{3}}{2}$ and $\sin \theta = -\frac{1}{2} \Rightarrow \theta = \frac{11\pi}{6}.$ $\sqrt{3} - i = 2 \operatorname{cis} \frac{11\pi}{6}.$

[5] $r = \sqrt{(-5)^2} = 5.$ $\theta = \frac{3\pi}{2}.$ $-5 \, i = 5 \operatorname{cis} \frac{3\pi}{2}.$

[6] $r = \sqrt{(-1)^2 + (-1)^2} = \sqrt{2}.$ $\cos \theta = -\frac{1}{\sqrt{2}}$ and $\sin \theta = -\frac{1}{\sqrt{2}} \Rightarrow \theta = \frac{5\pi}{4}.$ $-1 - i = \sqrt{2} \operatorname{cis} \frac{5\pi}{4}.$

[7] $r = \sqrt{\left(\frac{\sqrt{3}}{2}\right)^2 + \left(-\frac{1}{2}\right)^2} = 1.$ $\cos \theta = \frac{\sqrt{3}}{2}$ and $\sin \theta = -\frac{1}{2} \Rightarrow \theta = \frac{11\pi}{6}.$ $\frac{\sqrt{3}}{2} - \frac{1}{2} \, i = \operatorname{cis} \frac{11\pi}{6}.$

[8] $r = 9$ and $\theta = 0.$ $9 = 9 \operatorname{cis} 0.$ [9] $r = 0$ and $\theta = 0.$ $0 = 0.$

[10] $r = \frac{1}{2}$ and $\theta = \frac{\pi}{2}.$ $\frac{1}{2} \, i = \frac{1}{2} \operatorname{cis} \frac{\pi}{2}.$

[11] $r = \sqrt{(-3)^2 + (\sqrt{3})^2} = \sqrt{12} = 2\sqrt{3}.$ $\cos \theta = -\frac{3}{2\sqrt{3}} = -\frac{\sqrt{3}}{2}$ and $\sin \theta = \frac{1}{2} \Rightarrow \theta = \frac{5\pi}{6}.$

$-3 + \sqrt{3} \, i = 2\sqrt{3} \operatorname{cis} \frac{5\pi}{6}.$

[12] $r = \sqrt{(\sqrt{2})^2 + (2)^2} = \sqrt{6}.$ $\tan \theta = \frac{2}{\sqrt{2}} = \sqrt{2}$ and θ in QI $\Rightarrow \theta = \tan^{-1} \sqrt{2}.$ $\sqrt{2} + 2 \, i = \sqrt{6} \operatorname{cis} \left(\tan^{-1} \sqrt{2}\right) \approx 2.4 \operatorname{cis} 54.7°.$

[13] $r = \sqrt{(3)^2 + (-2)^2} = \sqrt{13}$. $\tan\theta = -\frac{2}{3}$. If θ' is the reference angle for θ, then $\tan\theta' = |\tan\theta| = \frac{2}{3} \Rightarrow$

$\theta' = \tan^{-1}\frac{2}{3}$. θ in QIV $\Rightarrow \theta = 360° - \theta'$. $3 - 2i = \sqrt{13}\ \text{cis}\left(360° - \tan^{-1}\frac{2}{3}\right) \approx 3.6\ \tan^{-1} 326°$.

[14] $r = \sqrt{(-1)^2 + (1)^2} = \sqrt{2}$. $\cos\theta = -\frac{1}{\sqrt{2}}$ and $\sin\theta = \frac{1}{\sqrt{2}} \Rightarrow \theta = \frac{3\pi}{4}$. $-1 + i = \sqrt{2}\ \text{cis}\ \frac{3\pi}{4}$.

[15] $r = \sqrt{(-\sqrt{3})^2 + (1)^2} = 2$. $\cos\theta = -\frac{\sqrt{3}}{2}$ and $\sin\theta = \frac{1}{2} \Rightarrow \theta = \frac{5\pi}{6}$. $-\sqrt{3} + i = 2\ \text{cis}\ \frac{5\pi}{6}$.

[16] $r = \sqrt{(-1)^2 + (\sqrt{2})^2} = \sqrt{3}$. $\tan\theta = -\sqrt{2}$. If θ' is the reference angle for θ, then $\tan\theta' = |\tan\theta| =$

$\sqrt{2} \Rightarrow \theta' = \tan^{-1}\sqrt{2}$. θ in QII $\Rightarrow \theta = 180° - \theta'$. $-1 + \sqrt{2}\ i = \sqrt{3}\ \text{cis}\left(180° - \tan^{-1}\sqrt{2}\right) \approx 1.7\ \text{cis}\ 125°$.

[17] $r = \sqrt{(-2)^2 + (-2\sqrt{2})^2} = 2\sqrt{3}$. $\tan\theta = \sqrt{2}$. If θ' is the reference angle for θ, then $\tan\theta' =$

$|\tan\theta| = \sqrt{2} \Rightarrow \theta' = \tan^{-1}\sqrt{2}$. θ in QIII $\Rightarrow \theta = 180° + \theta'$. $-2 - 2\sqrt{2}\ i =$

$2\sqrt{3}\ \text{cis}\left(180° + \tan^{-1}\sqrt{2}\right) \approx 3.5\ \text{cis}\ 235°$.

[18] $r = \sqrt{(2\sqrt{3})^2 + (-2)^2} = 4$. $\cos\theta = \frac{\sqrt{3}}{2}$ and $\sin\theta = -\frac{1}{2} \Rightarrow \theta = \frac{11\pi}{6}$. $2\sqrt{3} - 2i = 4\ \text{cis}\ \frac{11\pi}{6}$.

[19] $r = \sqrt{(-5)^2 + (12)^2} = 13$. $\tan\theta = -\frac{12}{5}$. If θ' is the reference angle for θ, then $\tan\theta' = |\tan\theta| = \frac{12}{5} \Rightarrow$

$\theta' = \tan^{-1}\frac{12}{5}$. θ in QII $\Rightarrow \theta = 180° - \theta'$. $-5 + 12i = 13\ \text{cis}\left(180° - \tan^{-1}\frac{12}{5}\right) \approx 13\ \text{cis}\ 113°$.

[20] $r = \sqrt{3^2 + 5^2} = \sqrt{34}$. $\tan\theta = \frac{5}{3}$ and θ in QI $\Rightarrow \theta = \tan^{-1}\frac{5}{3}$. $3 + 5i = \sqrt{34}\ \text{cis}\left(\tan^{-1}\frac{5}{3}\right) \approx 5.8\ \text{cis}\ 59°$.

[21] $r = \frac{\sqrt{3}}{2}$, and $\theta = \pi \Rightarrow -\frac{\sqrt{3}}{2} = \frac{\sqrt{3}}{2}\ \text{cis}\ \pi$.   [22] $r = 9$ and $\theta = \frac{3\pi}{2} \Rightarrow -9i = 9\ \text{cis}\ \frac{3\pi}{2}$.

[23] $2i(\sqrt{3} - i) = 2 + 2\sqrt{3}\ i$. $r = \sqrt{(2)^2 + (2\sqrt{3})^2} = 4$. $\cos\theta = \frac{1}{2}$ and $\sin\theta = \frac{\sqrt{3}}{2} \Rightarrow$

$\theta = \frac{\pi}{3}$. $2 + 2\sqrt{3}\ i = 4\ \text{cis}\ \frac{\pi}{3}$.

[24] $-3i(\sqrt{2} + i) = 3 - 3\sqrt{2}\ i$. $r = \sqrt{(3)^2 + (-3\sqrt{2})^2} = 3\sqrt{3}$. $\tan\theta = -\sqrt{2}$. If θ' is the reference angle for

θ, then $\tan\theta' = |\tan\theta| = \sqrt{2} \Rightarrow \theta' = \tan^{-1}\sqrt{2}$. Since θ is in QIV, $\theta = 360° - \theta'$. Thus, $3 - 3\sqrt{2}\ i =$

$3\sqrt{3}\ \text{cis}\left(360° - \tan^{-1}\sqrt{2}\right) \approx 5.2\ \text{cis}\ 305°$.

[25] $2\ \text{cis}\ \frac{3\pi}{4} = 2\left(-\frac{\sqrt{2}}{2} + \frac{\sqrt{2}}{2}\ i\right) = -\sqrt{2} + \sqrt{2}\ i$.   [26] $3\ \text{cis}\ \frac{\pi}{3} = 3\left(\frac{1}{2} + \frac{\sqrt{3}}{2}\ i\right) = \frac{3}{2} + \frac{3\sqrt{3}}{2}\ i$.

[27] $\sqrt{5}\ \text{cis}\ \frac{\pi}{2} = \sqrt{5}(0 + i) = \sqrt{5}\ i$.   [28] $\sqrt{2}\ \text{cis}\ \frac{3\pi}{2} = \sqrt{2}(0 - i) = -\sqrt{2}\ i$.

[29] $3\sqrt{2}\ \text{cis}\ \frac{2\pi}{3} = 3\sqrt{2}\left(-\frac{1}{2} + \frac{\sqrt{3}}{2}\ i\right) = -\frac{3\sqrt{2}}{2} + \frac{3\sqrt{6}}{2}\ i$.

[30] $8\ \text{cis}\ \frac{5\pi}{4} = 8\left(-\frac{\sqrt{2}}{2} - \frac{\sqrt{2}}{2}\ i\right) = -4\sqrt{2} - 4\sqrt{2}\ i$.

[31] $6\ \text{cis}\ \frac{5\pi}{3} = 6\left(\frac{1}{2} - \frac{\sqrt{3}}{2}\ i\right) = 3 - 3\sqrt{3}\ i$.   [32] $4\ \text{cis}\ \frac{5\pi}{6} = 4\left(-\frac{\sqrt{3}}{2} + \frac{1}{2}\ i\right) = -2\sqrt{3} + 2i$.

[33] $12\ cis\ \frac{7\pi}{6} = 12\left(-\frac{\sqrt{3}}{2} - \frac{1}{2}\ i\right) = -6\sqrt{3} - 6i$.   [34] $\sqrt{17}\ \text{cis}\ 0 = \sqrt{17}(1 + 0i) = \sqrt{17}$.

[35] $\frac{5}{2} \operatorname{cis} \pi = \frac{5}{2}(-1 + 0\,i) = -\frac{5}{2}.$

[36] $\frac{3}{4} \operatorname{cis} \frac{7\pi}{4} = \frac{3}{4}\left(\frac{\sqrt{2}}{2} - \frac{\sqrt{2}}{2}\,i\right) = \frac{3\sqrt{2}}{8} - \frac{3\sqrt{2}}{8}\,i.$

[37] $15 \operatorname{cis} \frac{4\pi}{3} = 15\left(-\frac{1}{2} - \frac{\sqrt{3}}{2}\,i\right) = -\frac{15}{2} - \frac{15\sqrt{3}}{2}\,i.$

[38] $2\sqrt{3} \operatorname{cis} \frac{5\pi}{3} = 2\sqrt{3}\left(\frac{1}{2} - \frac{\sqrt{3}}{2}\,i\right) = \sqrt{3} - 3\,i.$

[39] $\frac{\sqrt{3}}{2} \operatorname{cis} 330^\circ = \frac{\sqrt{3}}{2}\left(\frac{\sqrt{3}}{2} - \frac{1}{2}\,i\right) = \frac{3}{4} - \frac{\sqrt{3}}{4}\,i.$

[40] $24 \operatorname{cis} 225^\circ = 24\left(-\frac{\sqrt{2}}{2} - \frac{\sqrt{2}}{2}\,i\right) = -12\sqrt{2} - 12\sqrt{2}\,i.$

[41] $\cos \frac{5\pi}{12} = \cos\left(\frac{\pi}{6} + \frac{\pi}{4}\right) = \cos \frac{\pi}{6} \cos \frac{\pi}{4} - \sin \frac{\pi}{6} \sin \frac{\pi}{4} = \left(\frac{\sqrt{3}}{2}\right)\left(\frac{\sqrt{2}}{2}\right) - \left(\frac{1}{2}\right)\left(\frac{\sqrt{2}}{2}\right) = \frac{\sqrt{6} - \sqrt{2}}{4}$, and

$\sin \frac{5\pi}{12} = \sin\left(\frac{\pi}{6} + \frac{\pi}{4}\right) = \sin \frac{\pi}{6} \cos \frac{\pi}{4} + \cos \frac{\pi}{6} \sin \frac{\pi}{4} = \left(\frac{1}{2}\right)\left(\frac{\sqrt{2}}{2}\right) + \left(\frac{\sqrt{3}}{2}\right)\left(\frac{\sqrt{2}}{2}\right) = \frac{\sqrt{2} + \sqrt{6}}{4}$. Thus,

$4 \cos \frac{5\pi}{12} = 4\left(\frac{\sqrt{6} - \sqrt{2}}{4} + \frac{\sqrt{2} + \sqrt{6}}{4}\,i\right) = \left(\sqrt{6} - \sqrt{2}\right) + \left(\sqrt{6} + \sqrt{2}\right)i.$

[42] $\cos \frac{7\pi}{12} = \cos\left(\frac{\pi}{3} + \frac{\pi}{4}\right) = \cos \frac{\pi}{3} \cos \frac{\pi}{4} - \sin \frac{\pi}{3} \sin \frac{\pi}{4} = \left(\frac{1}{2}\right)\left(\frac{\sqrt{2}}{2}\right) - \left(\frac{\sqrt{3}}{2}\right)\left(\frac{\sqrt{2}}{2}\right) = \frac{\sqrt{2} - \sqrt{6}}{4}$, and

$\sin \frac{7\pi}{12} = \sin\left(\frac{\pi}{3} + \frac{\pi}{4}\right) = \sin \frac{\pi}{3} \cos \frac{\pi}{4} + \cos \frac{\pi}{3} \sin \frac{\pi}{4} = \left(\frac{\sqrt{3}}{2}\right)\left(\frac{\sqrt{2}}{2}\right) + \left(\frac{1}{2}\right)\left(\frac{\sqrt{2}}{2}\right) = \frac{\sqrt{6} + \sqrt{2}}{4}$. Thus,

$5 \cos \frac{7\pi}{12} = 5\left(\frac{\sqrt{2} - \sqrt{6}}{4} + \frac{\sqrt{6} + \sqrt{2}}{4}\,i\right) = \frac{5\left(\sqrt{2} - \sqrt{6}\right)}{4} + \frac{5\left(\sqrt{2} + \sqrt{6}\right)}{4}\,i.$

[43] $8 \operatorname{cis} 315^\circ = 8\left(\frac{\sqrt{2}}{2} - \frac{\sqrt{2}}{2}\,i\right) = 4\sqrt{2} - 4\sqrt{2}\,i.$

[44] $\cos 105^\circ = \cos\left(60^\circ + 45^\circ\right) = \cos 60^\circ \cos 45^\circ - \sin 60^\circ \sin 45^\circ = \left(\frac{1}{2}\right)\left(\frac{\sqrt{2}}{2}\right) - \left(\frac{\sqrt{3}}{2}\right)\left(\frac{\sqrt{2}}{2}\right) =$

$\frac{\sqrt{2} - \sqrt{6}}{4}$, and $\sin 105^\circ = \sin 60^\circ \cos 45^\circ + \cos 60^\circ \sin 45^\circ = \left(\frac{\sqrt{3}}{2}\right)\left(\frac{\sqrt{2}}{2}\right) + \left(\frac{1}{2}\right)\left(\frac{\sqrt{2}}{2}\right) = \frac{\sqrt{6} + \sqrt{2}}{4}.$

Thus, $10 \operatorname{cis} 105^\circ = 10\left(\frac{\sqrt{2} - \sqrt{6}}{4} + \frac{\sqrt{2} + \sqrt{6}}{4}\,i\right) = \frac{5\left(\sqrt{2} - \sqrt{6}\right)}{2} + \frac{5\left(\sqrt{2} + \sqrt{6}\right)}{2}\,i.$

[45] $\left(2 \operatorname{cis} \frac{\pi}{3}\right)\left(4 \operatorname{cis} \frac{2\pi}{3}\right) = (2)(4) \operatorname{cis}\left(\frac{\pi}{3} + \frac{2\pi}{3}\right) = 8 \operatorname{cis} \pi.$

[46] $\left(6 \operatorname{cis} \frac{3\pi}{2}\right)\left(2 \operatorname{cis} \frac{\pi}{2}\right) = (6)(2) \operatorname{cis}\left(\frac{3\pi}{2} + \frac{\pi}{2}\right) = 12 \operatorname{cis} 0.$

[47] $\left(5 \operatorname{cis} \frac{5\pi}{6}\right)\left(\frac{1}{2} \operatorname{cis} \frac{\pi}{3}\right) = (5)\left(\frac{1}{2}\right) \operatorname{cis}\left(\frac{5\pi}{6} + \frac{\pi}{3}\right) = \frac{5}{2} \operatorname{cis} \frac{7\pi}{6}.$

[48] $\left(12 \operatorname{cis} \frac{2\pi}{3}\right)\left(\frac{\sqrt{3}}{2} \operatorname{cis} \frac{\pi}{6}\right) = (12)\left(\frac{\sqrt{3}}{2}\right) \operatorname{cis}\left(\frac{2\pi}{3} + \frac{\pi}{6}\right) = 6\sqrt{3} \operatorname{cis} \frac{5\pi}{6}.$

[49] $\frac{18 \operatorname{cis} \frac{3\pi}{4}}{3 \operatorname{cis} \frac{\pi}{2}} = \frac{18}{3} \operatorname{cis}\left(\frac{3\pi}{4} - \frac{\pi}{2}\right) = 6 \operatorname{cis} \frac{\pi}{4}.$

[50] $\frac{2 \operatorname{cis} \frac{5\pi}{6}}{12 \operatorname{cis} \frac{2\pi}{3}} = \frac{2}{12} \operatorname{cis}\left(\frac{5\pi}{6} - \frac{2\pi}{3}\right) = \frac{1}{6} \operatorname{cis} \frac{\pi}{6}.$

[51] $\dfrac{3\sqrt{3}\ \text{cis}\ 420°}{6\ \text{cis}\ 210°} = \dfrac{3\sqrt{3}}{6}\ \text{cis}\left(420° - 210°\right) = \dfrac{\sqrt{3}}{2}\ \text{cis}\ 210°.$

[52] $\dfrac{\sqrt{6}\ \text{cis}\ 225°}{\sqrt{2}\ \text{cis}\ 75°} = \dfrac{\sqrt{6}}{\sqrt{2}}\ \text{cis}\left(225° - 75°\right) = \sqrt{3}\ \text{cis}\ 150°.$

[53] $\left(3\sqrt{2}\ \text{cis}\ \dfrac{2\pi}{3}\right)\left(5\ \text{cis}\ \dfrac{\pi}{6}\right) = \left(3\sqrt{2}\right)\left(5\right)\text{cis}\left(\dfrac{2\pi}{3} + \dfrac{\pi}{6}\right) = 15\sqrt{2}\ \text{cis}\ \dfrac{5\pi}{6} = 15\sqrt{2}\left(-\dfrac{\sqrt{3}}{2} + \dfrac{1}{2}i\right) =$

$-\dfrac{15\sqrt{6}}{2} + \dfrac{15\sqrt{2}}{2}i.$

[54] $\left(\dfrac{3}{4}\ \text{cis}\ \dfrac{\pi}{3}\right)\left(8\ \text{cis}\ \dfrac{5\pi}{6}\right) = \left(\dfrac{3}{4}\right)\left(8\right)\text{cis}\left(\dfrac{\pi}{3} + \dfrac{5\pi}{6}\right) = 6\ \text{cis}\ \dfrac{7\pi}{6} = 6\left(-\dfrac{\sqrt{3}}{2} - \dfrac{1}{2}i\right) = -3\sqrt{3} - 3i.$

[55] $\left(16\ \text{cis}\ \dfrac{5\pi}{4}\right)\left(\dfrac{3}{8}\ \text{cis}\ \dfrac{\pi}{2}\right) = \left(16\right)\left(\dfrac{3}{8}\right)\text{cis}\left(\dfrac{5\pi}{4} + \dfrac{\pi}{2}\right) = 6\ \text{cis}\ \dfrac{7\pi}{4} = 6\left(\dfrac{\sqrt{2}}{2} - \dfrac{\sqrt{2}}{2}i\right) = 3\sqrt{2} - 3\sqrt{2}i.$

[56] $\left(2\sqrt{2}\ \text{cis}\ \dfrac{\pi}{6}\right)\left(3\sqrt{2}\ \text{cis}\ \dfrac{5\pi}{3}\right) = \left(2\sqrt{2}\right)\left(3\sqrt{2}\right)\text{cis}\left(\dfrac{\pi}{6} + \dfrac{5\pi}{3}\right) = 12\ \text{cis}\ \dfrac{11\pi}{6} = 12\left(\dfrac{\sqrt{3}}{2} - \dfrac{1}{2}i\right) = 6\sqrt{3} - 6i.$

[57] $\dfrac{\dfrac{\sqrt{3}}{2}\ \text{cis}\ \dfrac{7\pi}{6}}{\dfrac{2}{3}\ \text{cis}\ \dfrac{2\pi}{3}} = \dfrac{\dfrac{\sqrt{3}}{2}}{\dfrac{2}{3}}\ \text{cis}\left(\dfrac{7\pi}{6} - \dfrac{2\pi}{3}\right) = \dfrac{3\sqrt{3}}{4}\ \text{cis}\ \dfrac{\pi}{2} = \dfrac{3\sqrt{3}}{4}\left(0 + i\right) = \dfrac{3\sqrt{3}}{4}i.$

[58] $\dfrac{12\ \text{cis}\ \dfrac{5\pi}{3}}{\dfrac{5}{4}\ \text{cis}\ \dfrac{5\pi}{6}} = \dfrac{12}{\dfrac{5}{4}}\ \text{cis}\left(\dfrac{5\pi}{3} - \dfrac{5\pi}{6}\right) = \dfrac{48}{5}\ \text{cis}\ \dfrac{5\pi}{6} = \dfrac{48}{5}\left(-\dfrac{\sqrt{3}}{2} + \dfrac{1}{2}i\right) = -\dfrac{24\sqrt{3}}{5} + \dfrac{24}{5}i.$

[59] $\dfrac{25\ \text{cis}\ 315°}{2\ \text{cis}\ 75°} = \dfrac{25}{2}\ \text{cis}\left(315° - 75°\right) = \dfrac{25}{2}\ \text{cis}\ 240° = \dfrac{25}{2}\left(-\dfrac{1}{2} - \dfrac{\sqrt{3}}{2}i\right) = -\dfrac{25}{4} - \dfrac{25\sqrt{3}}{4}i.$

[60] $\dfrac{7\sqrt{2}\ \text{cis}\ 220°}{3\ \text{cis}\ 85°} = \dfrac{7\sqrt{2}}{3}\ \text{cis}\left(220° - 85°\right) = \dfrac{7\sqrt{2}}{3}\ \text{cis}\ 135° = \dfrac{7\sqrt{2}}{3}\left(-\dfrac{\sqrt{2}}{2} + \dfrac{\sqrt{2}}{2}i\right) = -\dfrac{7}{3} + \dfrac{7}{3}i.$

[61] $z = r\left(\cos\theta + i\ \sin\theta\right) = r\cos\theta + \left(r\sin\theta\right)i \Rightarrow \bar{z} = r\cos\theta - \left(r\sin\theta\right)i = r\left(\cos\theta - i\ \sin\theta\right).$

[62] $\dfrac{1}{z} = \dfrac{1\ cis\ 0}{r\ cis\ \theta} = \dfrac{1}{r}\ cis\ \left(0 - \theta\right) = \dfrac{1}{r}\left[\cos\left(-\theta\right) + i\ \sin\left(-\theta\right)\right].$ Since $\cos\left(-\theta\right) = \cos\theta$ and $\sin\left(-\theta\right) = -\sin\theta,$

$\dfrac{1}{z} = \dfrac{1}{r}\left(\cos\theta - i\ \sin\theta\right).$

[63] $r\ \text{cis}\left(\theta + \pi\right) = r\left[\cos\left(\theta + \pi\right) + i\ \sin\left(\theta + \pi\right)\right] =$

$r\left[\cos\theta\cos\pi - \sin\theta\sin\pi + i\left(\sin\theta\cos\pi + \cos\theta\sin\pi\right)\right] = r\left(-\cos\theta - i\ \sin\theta\right) =$

$-r\left(\cos\theta + i\ \sin\theta\right) = -r\ \text{cis}\ \theta.$

[64] $z^2 = \left(r\ \text{cis}\ \theta\right)\left(r\ \text{cis}\ \theta\right) = \left(r \cdot r\right)\text{cis}\left(\theta + \theta\right) = r^2\ \text{cis}\ 2\theta$ and $z^3 = z^2 \cdot z = \left(r^2\ \text{cis}\ 2\theta\right)\left(r\ \text{cis}\ \theta\right) =$

$\left(r^2 \cdot r\right)\text{cis}\left(2\theta + \theta\right) = r^3\ \text{cis}\ 3\theta.$ Yes, $z^n = r^n\ \text{cis}\ n\theta.$

[65] Since $|z|$ is the distance between z and the origin, $\left\{z \mid |z| = 1\right\}$ describes the circle of radius 1 with center at the origin. (See figure below).

[66] Since r can be any nonnegative real number, $z = r\ \text{cis}\ \theta$ can be any point on the terminal side of $\theta = \dfrac{\pi}{3}.$ (See figure below).

[67] a) $e^0 = e^{0 + 0i} = e^0\left(\cos 0 + i\ \sin 0\right) = 1\left(1 + 0\right) = 1;\ e^{\pi/2\ i} = e^{0 + \pi/2\ i} = e^0\left(\cos\dfrac{\pi}{2} + i\ \sin\dfrac{\pi}{2}\right) =$

$1\left(0 + i\right) = i;\ e^{\pi i} = e^{0 + \pi i} = e^0\left(\cos\pi + i\ \sin\pi\right) = 1\left(-1 + 0i\right) = -1.$

b) $e^{2\pi i} = e^{0+2\pi i} = e^0 \left(\cos 2\pi + i \sin 2\pi \right) = 1 \left(1 + 0\, i \right) = 1.$ Thus, $e^{z+2\pi i} = e^z \cdot e^{2\pi i} = e^z \cdot 1 = e^z$

c) $\left| e^z \right| = \left| e^{x+yi} \right| = \left| e^x \left(\cos y + i \sin y \right) \right| = \left| e^x \cos y + \left(e^x \sin y \right) i \right| = \sqrt{\left(e^x \cos y \right)^2 + \left(e^x \sin y \right)^2} =$

$\sqrt{e^{2x} \cos^2 y + e^{2x} \sin^2 y} = \sqrt{e^{2x} \left(\cos^2 y + \sin^2 y \right)} = \sqrt{e^{2x}} = e^x.$

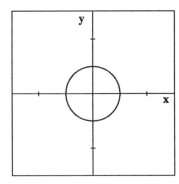

Figure 65

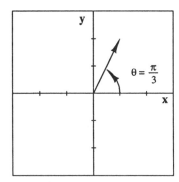

Figure 66

[68] a) $\log i = \log \left(\operatorname{cis} \frac{\pi}{2} \right) = \ln 1 + i \left(\frac{\pi}{2} \right) = 0 + \frac{\pi}{2} i = \frac{\pi}{2} i;\ \log \left(-1 \right) = \log \left(\operatorname{cis} \pi \right) = \ln 1 + i \left(\pi \right) = 0 + \pi i =$

$\pi i;\ \log \left(-i \right) = \log \left(\operatorname{cis} \frac{3\pi}{2} \right) = \ln 1 + i \left(\frac{3\pi}{2} \right) = 0 + \frac{3\pi}{2} i = \frac{3\pi}{2} i.$

b) $e^{\log z}\, e^{\ln r + i\,\theta} = e^{\ln r} \cdot e^{i\,\theta} = r\, e^{\theta i}.$ From the definition of e^z given in Exercise 67, $e^{\,\theta i} = e^0 \left(\cos \theta + i \sin \theta \right) = \operatorname{cis} \theta.$ Thus, $e^{\log z} = r \operatorname{cis} \theta.$

[69] Let $z_1 = r_1 \operatorname{cis} \theta_1,\ z_2 = r_2 \operatorname{cis} \theta_2$ and $z_3 = r_3 \operatorname{cis} \theta_3,$ Then $z_1 \cdot z_2 \cdot z_3 = r_1 \cdot r_2 \cdot r_3 \operatorname{cis} \left(\theta_1 + \theta_2 + \theta_3 \right).$

[70] Let $z_1 = r_1 \operatorname{cis} \theta_1$ and $z_2 = r_2 \operatorname{cis} \theta_2,$ where $z_2 \neq 0.$ Then $\dfrac{z_1}{z_2} = \dfrac{r_1 \operatorname{cis} \theta_1}{r_2 \operatorname{cis} \theta_2} = \dfrac{r_1 \left(\cos \theta_1 + i \sin \theta_1 \right)}{r_2 \left(\cos \theta_2 + i \sin \theta_2 \right)} =$

$\dfrac{r_1 \left(\cos \theta_1 + i \sin \theta_1 \right)}{r_2 \left(\cos \theta_2 + i \sin \theta_2 \right)} \cdot \dfrac{\left(\cos \theta_2 - i \sin \theta_2 \right)}{\left(\cos \theta_2 - i \sin \theta_2 \right)} =$

$\dfrac{r_1}{r_2} \dfrac{\cos \theta_1 \cos \theta_2 - i \cos \theta_1 \sin \theta_2 + i \sin \theta_1 \cos \theta_2 - i^2 \sin \theta_1 \sin \theta_2}{\cos^2 \theta + \sin^2 \theta} =$

$\dfrac{r_1}{r_2} \left[\left(\cos \theta_1 \cos \theta_2 + \sin \theta_1 \sin \theta_2 \right) + i \left(\sin \theta_1 \cos \theta_2 - \cos \theta_1 \sin \theta_2 \right) \right] =$

$\dfrac{r_1}{r_2} \left[\cos \left(\theta_1 - \theta_2 \right) + i \sin \left(\theta_1 - \theta_2 \right) \right] = \dfrac{r_1}{r_2} \operatorname{cis} \left(\theta_1 - \theta_2 \right).$

EXERCISES 7.4

[1] $\left(2 \operatorname{cis} \frac{7\pi}{12} \right)^4 = 2^4 \operatorname{cis} \left(4 \right) \left(\frac{7\pi}{12} \right) = 16 \operatorname{cis} \frac{7\pi}{3} = 16 \left(\frac{1}{2} + \frac{\sqrt{3}}{2} i \right) = 8 + 8\sqrt{3}\, i.$

[2] $\left(3 \operatorname{cis} \frac{5\pi}{12} \right)^6 = 3^6 \operatorname{cis} \left(6 \right) \left(\frac{5\pi}{12} \right) = 729 \operatorname{cis} \frac{5\pi}{2} = 729 \left(0 + i \right) = 729\, i.$

[3] $\left(3 \operatorname{cis} 120° \right)^3 = 3^3 \operatorname{cis} \left(3 \right) \left(120° \right) = 27 \operatorname{cis} 360° = 27 \left(1 + 0\, i \right) = 27.$

[4] $\left(2 \operatorname{cis} 135° \right)^8 = 2^8 \operatorname{cis} \left(8 \right) \left(135° \right) = 256 \operatorname{cis} 1080° = 256 \operatorname{cis} 0° = 256 \left(1 + 0\, i \right) = 256.$

[5] $\left(\text{cis}\,\frac{\pi}{8}\right)^{10} = \text{cis}\,(10)\left(\frac{\pi}{8}\right) = \text{cis}\,\frac{5\pi}{4} = -\frac{\sqrt{2}}{2} - \frac{\sqrt{2}}{2}\,i.$

[6] $\left(\text{cis}\,\frac{3\pi}{8}\right)^{12} = \text{cis}\,(12)\left(\frac{3\pi}{8}\right) = \text{cis}\,\frac{9\pi}{2} = \text{cis}\,\frac{\pi}{2} = 0 + i = i.$

[7] $\left(\sqrt{2}\,\text{cis}\,30°\right)^{9} = \left(\sqrt{2}\right)^{9}\,\text{cis}\,(9)\,(30°) = 16\,\sqrt{2}\,\text{cis}\,270° = 16\,\sqrt{2}\,(0 - i) = -16\,\sqrt{2}\,i.$

[8] $\left(2\,\sqrt{3}\,\text{cis}\,45°\right)^{5} = \left(2\,\sqrt{3}\right)^{5}\,\text{cis}\,(5)\,(45°) = 288\,\sqrt{3}\,\text{cis}\,225° = 288\,\sqrt{3}\left(-\frac{\sqrt{2}}{2} - \frac{\sqrt{2}}{2}\,i\right) =$

$-144\,\sqrt{6} - 144\,\sqrt{6}i.$

[9] $\left(\sqrt[4]{3}\,\text{cis}\,\frac{5\pi}{12}\right)^{16} = \left(\sqrt[4]{3}\right)^{16}\,\text{cis}\,(16)\left(\frac{5\pi}{12}\right) = 81\,\text{cis}\,\frac{20\pi}{3} = 81\,\text{cis}\,\frac{2\pi}{3} = 81\left(-\frac{1}{2} + \frac{\sqrt{3}}{2}\,i\right) = -\frac{81}{2} + \frac{81\sqrt{3}}{2}\,i.$

[10] $\left(\sqrt[5]{2}\,\text{cis}\,\frac{2\pi}{5}\right)^{10} = \left(\sqrt[5]{2}\right)^{10}\,\text{cis}\,(10)\left(\frac{2\pi}{5}\right) = 4\,\text{cis}\,4\pi = 4\,\text{cis}\,0 = 4\,(1 + 0\,i) = 4.$

[11] $\left(\frac{\sqrt{3}}{2}\,\text{cis}\,105°\right)^{6} = \left(\frac{\sqrt{3}}{2}\right)^{6}\,\text{cis}\,(6)\,(105°) = \frac{27}{64}\,\text{cis}\,630° = \frac{27}{64}\,\text{cis}\,270° = \frac{27}{64}\,(0 - i) = -\frac{27}{64}\,i.$

[12] $\left(\frac{\sqrt{2}}{2}\,\text{cis}\,22.5°\right)^{8} = \left(\frac{\sqrt{2}}{2}\right)^{8}\,\text{cis}\,(8)\,(22.5°) = \frac{1}{16}\,\text{cis}\,180° = \frac{1}{16}\,(-1 + 0\,i) = -\frac{1}{16}.$

[13] $\left(1 + \sqrt{3}\,i\right)^{4} = \left(2\,\text{cis}\,\frac{\pi}{3}\right)^{4} = 2^{4}\,\text{cis}\,(4)\left(\frac{\pi}{3}\right) = 16\,\text{cis}\,\frac{4\pi}{3} = 16\left(-\frac{1}{2} - \frac{\sqrt{3}}{2}\,i\right) = -8 - 8\,\sqrt{3}\,i.$

[14] $\left(\sqrt{3} - i\right)^{3} = \left(2\,\text{cis}\,\frac{11\pi}{6}\right)^{3} = 2^{3}\,\text{cis}\,(3)\left(\frac{11\pi}{6}\right) = 8\,\text{cis}\,\frac{11\pi}{2} = 8\,\text{cis}\,\frac{3\pi}{2} = 8\,(0 - i) = -8\,i.$

[15] $\left(\frac{1}{2} - \frac{\sqrt{3}}{2}\,i\right)^{7} = \left(\text{cis}\,\frac{5\pi}{3}\right)^{7} = \text{cis}\,(7)\left(\frac{5\pi}{3}\right) = \text{cis}\,\frac{35\pi}{3} = \text{cis}\,\frac{5\pi}{3} = \frac{1}{2} - \frac{\sqrt{3}}{2}\,i.$

[16] $\left(5 + 5\,i\right)^{4} = \left(5\,\sqrt{2}\,\text{cis}\,\frac{\pi}{4}\right)^{4} = \left(5\,\sqrt{2}\right)^{4}\,\text{cis}\,(4)\left(\frac{\pi}{4}\right) = 2500\,\text{cis}\,\pi = 2500\,(-1 + 0\,i) = -2500.$

[17] $\left(-1 + i\right)^{5} = \left(\sqrt{2}\,\text{cis}\,\frac{3\pi}{4}\right)^{5} = \left(\sqrt{2}\right)^{5}\,\text{cis}\,(5)\left(\frac{3\pi}{4}\right) = 4\,\sqrt{2}\,\text{cis}\,\frac{15\pi}{4} = 4\,\sqrt{2}\,\text{cis}\,\frac{7\pi}{4} = 4\,\sqrt{2}$

$\left(\frac{\sqrt{2}}{2} - \frac{\sqrt{2}}{2}\,i\right) = 4 - 4\,i.$

[18] $\left(1 - i\right)^{8} = \left(\sqrt{2}\,\text{cis}\,\frac{7\pi}{4}\right)^{8} = \left(\sqrt{2}\right)^{8}\,\text{cis}\,(8)\left(\frac{7\pi}{4}\right) = 16\,\text{cis}\,14\pi = 16\,\text{cis}\,0 = 16\,(1 + 0\,i) = 16.$

[19] $\left(-\frac{\sqrt{2}}{2} - \frac{\sqrt{2}}{2}\,i\right)^{6} = \left(\text{cis}\,\frac{5\pi}{4}\right)^{6} = \text{cis}\,(6)\left(\frac{5\pi}{4}\right) = \text{cis}\,\frac{15\pi}{2} = \text{cis}\,\frac{3\pi}{2} = 0 - i = -i.$

[20] $\left(-\frac{\sqrt{2}}{2} + \frac{\sqrt{2}}{2}\,i\right)^{11} = \left(\text{cis}\,\frac{3\pi}{4}\right)^{11} = \text{cis}\,(11)\left(\frac{3\pi}{4}\right) = \text{cis}\,\frac{33\pi}{4} = \text{cis}\,\frac{\pi}{4} = \frac{\sqrt{2}}{2} + \frac{\sqrt{2}}{2}\,i.$

[21] $\left(-3 + 3\,i\right)^{4} = \left(3\,\sqrt{2}\,\text{cis}\,\frac{3\pi}{4}\right)^{4} = \left(3\,\sqrt{2}\right)^{4}\,\text{cis}\,(4)\left(\frac{3\pi}{4}\right) = 324\,\text{cis}\,3\pi = 324\,\text{cis}\,\pi = 324\,(-1 + 0\,i) =$

$-324.$

[22] $(\sqrt{3} - \sqrt{3}\, i)^7 = \left(\sqrt{6} \text{ cis } \frac{7\pi}{4}\right)^7 = (\sqrt{6})^7 \text{ cis } (7)\left(\frac{7\pi}{4}\right) = 216\,\sqrt{6} \text{ cis } \frac{49\pi}{4} = 216\,\sqrt{6} \text{ cis } \frac{\pi}{4} =$

$216\,\sqrt{6}\left(\frac{\sqrt{2}}{2} + \frac{\sqrt{2}}{2}\, i\right) = 216\,\sqrt{3} + 216\,\sqrt{3}\, i.$

[23] $(2\sqrt{3} - 2\, i)^{10} = \left(4 \text{ cis } \frac{11\pi}{6}\right)^{10} = 4^{10} \text{ cis } (10)\left(\frac{11\pi}{6}\right) = 1,048,576 \text{ cis } \frac{55\pi}{3} = 1,048,576 \text{ cis } \frac{\pi}{3} =$

$1,048,576\left(\frac{1}{2} + \frac{\sqrt{3}}{2}\, i\right) = 524,288 + 524,288\,\sqrt{3}\, i.$

[24] $(-5 + 5\sqrt{3}\, i)^{12} = \left(10 \text{ cis } \frac{2\pi}{3}\right)^{12} = 10^{12} \text{ cis } (12)\left(\frac{2\pi}{3}\right) = 10^{12} \text{ cis } 8\pi = 10^{12} \text{ cis } 0 =$

$10^{12}\,(1 + 0\, i) = 10^{12}.$

[25] $(\sqrt{5} \text{ cis } 20°)^7 = (\sqrt{5})^7 \text{ cis } (7)\,(20°) = 125\,\sqrt{5} \text{ cis } 140° \approx 280 \text{ cis } 140°.$

[26] $(3\sqrt{2} \text{ cis } 12°)^6 = (3\sqrt{2})^6 \text{ cis } 6\,(12°) = 5832 \text{ cis } 72°.$

[27] $\left(\frac{7}{4} \text{ cis } \frac{5\pi}{7}\right)^{12} = \left(\frac{7}{4}\right)^{12} \text{ cis } (12)\left(\frac{5\pi}{7}\right) \approx 825 \text{ cis } \frac{60\pi}{7} \approx 825 \text{ cis } 1543° \approx 825 \text{ cis } 103°.$

[28] $\left(\frac{1}{2} \text{ cis } \frac{3\pi}{5}\right)^7 = \left(\frac{1}{2}\right)^7 \text{ cis } (7)\left(\frac{3\pi}{5}\right) = \frac{1}{128} \text{ cis } \frac{21\pi}{5} = \frac{1}{128} \text{ cis } \frac{\pi}{5} = \frac{1}{128} \text{ cis } 36°.$

[29] $(-2\sqrt{2} - 2\, i)^6 = \left[2\sqrt{3} \text{ cis }\left(180° + \tan^{-1} \frac{1}{\sqrt{2}}\right)\right]^6 \approx (2\sqrt{3} \text{ cis } 215.3°)^6 = (2\sqrt{3})^6 \text{ cis } (6)\,(215.3°) \approx$

$1728 \text{ cis } 1292° = 1728 \text{ cis } 212°.$

[30] $(-3 + 4\, i)^{10} = \left[5 \text{ cis }\left(180° - \tan^{-1} \frac{4}{3}\right)\right]^{10} \approx (5 \text{ cis } 126.9°)^{10} = 5^{10} \text{ cis } (10)\,(126.9°) = 5^{10} \text{ cis } 1269° =$

$9,765,625 \text{ cis } 189°.$

[31] $\left(\frac{1}{2} - \frac{1}{3}\, i\right)^4 = \left[\frac{\sqrt{13}}{6} \text{ cis }\left(360° - \tan^{-1} \frac{2}{3}\right)\right]^4 \approx \left(\frac{\sqrt{13}}{6} \text{ cis } 326.3°\right)^4 = \left(\frac{\sqrt{13}}{6}\right)^4 \text{ cis } (4)\,(326.3°) \approx$

$\frac{169}{1296} \text{ cis } 1305° \approx 0.13 \text{ cis } 225°.$

[32] $\left(\frac{1}{4} - \frac{3}{4}\, i\right)^3 = \left[\frac{\sqrt{10}}{4} \text{ cis }\left(360° - \tan^{-1} 3\right)\right]^3 \approx \left(\frac{\sqrt{10}}{4} \text{ cis } 288.4°\right)^3 = \left(\frac{\sqrt{10}}{4}\right)^3 \text{ cis } (3)\,(288.4°) \approx$

$\frac{5\sqrt{10}}{32} \text{ cis } 865° \approx 0.49 \text{ cis } 145°.$

[33] $(3 - 2\, i)^4 = \left[\sqrt{13} \text{ cis }\left(360° - \tan^{-1} \frac{2}{3}\right)\right]^4 \approx (\sqrt{13} \text{ cis } 326.3°)^4 = (\sqrt{13})^4 \text{ cis } (4)\,(326.3°) \approx 169 \text{ cis } 1305° =$

$169 \text{ cis } 225°.$

[34] $(-1 + \sqrt{2}\, i)^6 = \left[\sqrt{3} \text{ cis }\left(180° - \tan^{-1} \sqrt{2}\right)\right]^6 \approx (\sqrt{3} \text{ cis } 125.3°)^6 = (\sqrt{3})^6 \text{ cis } (6)\,(125.3°) \approx 27 \text{ cis } 752° =$
$27 \text{ cis } 32°.$

[35] $(-5 + 12\, i)^3 = \left[13 \text{ cis }\left(180° - \tan^{-1} \frac{12}{5}\right)\right]^3 \approx (13 \text{ cis } 112.6°)^3 = (13)^3 \text{ cis } (3)\,(112.6°) \approx 2197 \text{ cis } 338°.$

[36] $(1 + 3\, i)^5 = \left[\sqrt{10} \text{ cis }\left(\tan^{-1} 3\right)\right]^5 \approx (\sqrt{10} \text{ cis } 71.6°)^5 = (\sqrt{10})^5 \text{ cis } (5)\,(71.6°) = 100\,\sqrt{10} \text{ cis } 358° \approx$
$316.2 \text{ cis } 358°.$

[37] $\sqrt{3} - i = 2 \operatorname{cis} \dfrac{11\pi}{6}$. $u_k = 2^{1/5} \operatorname{cis}\left(\dfrac{\dfrac{11\pi}{6} + 2k\pi}{5}\right) = 2^{1/5} \operatorname{cis}\left[\dfrac{11\pi}{30} + k\left(\dfrac{2\pi}{5}\right)\right] = 2^{1/5} \operatorname{cis}\left(66° + k \cdot 72°\right)$,

for $k = 0, 1, 2, 3$ and 4. Thus, $u_0 = 2^{1/5} \operatorname{cis} 66°$, $u_1 = 2^{1/5} \operatorname{cis} 138°$, $u_2 = 2^{1/5} \operatorname{cis} 210°$,

$u_3 = 2^{1/5} \operatorname{cis} 282°$, and $u_4 = 2^{1/5} \operatorname{cis} 354°$, where $2^{1/5} \approx 1.15$.   (See figure below).

[38] $-1 + \sqrt{3}\,i = 2 \operatorname{cis} \dfrac{2\pi}{3}$. $u_k = 2^{1/3} \operatorname{cis}\left(\dfrac{\dfrac{2\pi}{3} + 2k\pi}{3}\right) = 2^{1/3} \operatorname{cis}\left[\dfrac{2\pi}{9} + k\left(\dfrac{2\pi}{3}\right)\right] = 2^{1/3} \operatorname{cis}\left(40° + k \cdot 120°\right)$,

for $k = 0, 1$ and 2. Thus, $u_0 = 2^{1/3} \operatorname{cis} 40°$, $u_1 = 2^{1/3} \operatorname{cis} 160°$, and $u_2 = 2^{1/3} \operatorname{cis} 280°$,

where $2^{1/3} \approx 1.26$.   (See figure below).

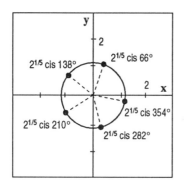

Figure 38

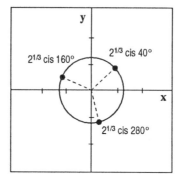

Figure 39

[39] $-\sqrt{2} - \sqrt{2}\,i = 2 \operatorname{cis} \dfrac{5\pi}{4}$. $u_k = 2^{1/3} \operatorname{cis}\left(\dfrac{\dfrac{5\pi}{4} + 2k\pi}{3}\right) = 2^{1/3} \operatorname{cis}\left[\dfrac{5\pi}{12} + k\left(\dfrac{2\pi}{3}\right)\right] = 2^{1/3} \operatorname{cis}\left(75° + k \cdot 120°\right)$,

for $k = 0, 1$ and 2. Thus, $u_0 = 2^{1/3} \operatorname{cis} 75°$, $u_1 = 2^{1/3} \operatorname{cis} 195°$ and $u_2 = 2^{1/3} \operatorname{cis} 315°$, where $2^{1/3} \approx 1.26$.
(See figure below).

[40] $8 - 8\,i = 8\sqrt{2} \operatorname{cis} \dfrac{7\pi}{4}$. $u_k = \left(8\sqrt{2}\right)^{1/4} \operatorname{cis}\left(\dfrac{\dfrac{7\pi}{4} + 2k\pi}{4}\right) = 2^{7/8} \operatorname{cis}\left[\dfrac{7\pi}{16} + k\left(\dfrac{\pi}{2}\right)\right] =$

$2^{7/8} \operatorname{cis}\left(78.75° + k \cdot 90°\right)$, for $k = 0, 1, 2$ and 3. Thus, $u_0 = 2^{7/8} \operatorname{cis} 78.75°$, $u_1 = 2^{7/8} \operatorname{cis} 168.75°$,

$u_2 = 2^{7/8} \operatorname{cis} 258.75°$ and $u_3 = 2^{7/8} \operatorname{cis} 348.75°$, where $2^{7/8} \approx 1.83$.   (See figure below).

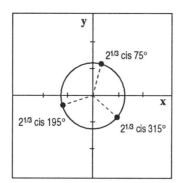

Figure 39

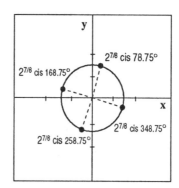

Figure 40

[41] $-2 + 2\sqrt{3}\,i = 4\,\text{cis}\,\dfrac{2\pi}{3}.\ u_k = 4^{1/4}\,\text{cis}\left(\dfrac{\frac{2\pi}{3} + 2k\pi}{4}\right) = 4^{1/4}\,\text{cis}\left[\dfrac{\pi}{6} + k\left(\dfrac{\pi}{2}\right)\right] = 4^{1/4}\,\text{cis}\left(30° + k\cdot 90°\right),$ for

$k = 0,\ 1,\ 2$ and 3. Thus, $u_0 = 4^{1/4}\,\text{cis}\,30°,\ u_1 = 4^{1/4}\,\text{cis}\,120°,\ u_2 = 4^{1/4}\,\text{cis}\,210°,$ and $u_3 = 4^{1/4}\,\text{cis}\,300°,$

where $4^{1/4} \approx 1.41.$ (See figure below).

[42] $-4\sqrt{3} - 4i = 8\,\text{cis}\,\dfrac{7\pi}{6}.\ u_k = 8^{1/5}\,\text{cis}\left(\dfrac{\frac{7\pi}{6} + 2k\pi}{5}\right) = 8^{1/5}\,\text{cis}\left[\dfrac{7\pi}{30} + k\left(\dfrac{2\pi}{5}\right)\right] = 8^{1/5}\,\text{cis}\left(42° + k\cdot 72°\right),$

for $k = 0,\ 1,\ 2,\ 3$ and 4. Thus, $u_0 = 8^{1/5}\,\text{cis}\,42°,\ u_1 = 8^{1/5}\,\text{cis}\,114°,\ u_2 = 8^{1/5}\,\text{cis}\,186°,$

$u_3 = 8^{1/5}\,\text{cis}\,258°$ and $u_4 = 8^{1/5}\,\text{cis}\,330°,$ where $8^{1/5} \approx 1.52.$ (See figure below).

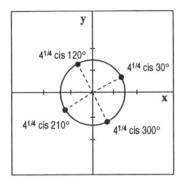

Figure 41

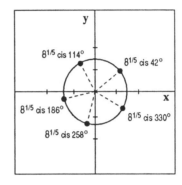

Figure 42

[43] $16i = 16\,\text{cis}\,\dfrac{\pi}{2}.\ u_k = \sqrt{16}\,\text{cis}\left(\dfrac{\frac{\pi}{2} + 2k\pi}{2}\right) = 4\,\text{cis}\left[\dfrac{\pi}{4} + k\pi\right] = 4\,\text{cis}\left(45° + k\cdot 180°\right),$ for $k = 0$ and 1.

Thus, $u_0 = 4\,\text{cis}\,45°$ and $u_1 = 4\,\text{cis}\,225°.$ (See figure below).

[44] $-9i = 9\,\text{cis}\,\dfrac{3\pi}{2}.\ u_k = \sqrt{9}\,\text{cis}\left(\dfrac{\frac{3\pi}{2} + 2k\pi}{2}\right) = 3\,\text{cis}\left[\dfrac{3\pi}{4} + k\pi\right] = 3\,\text{cis}\left(135° + k\cdot 180°\right),$ for $k = 0$ and

1. Thus, $u_0 = 3\,\text{cis}\,135°$ and $u_1 = 3\,\text{cis}\,315°.$ (See figure below).

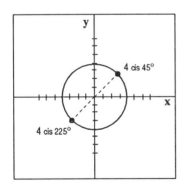

Figure 43

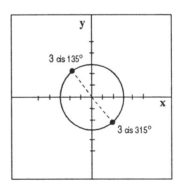

Figure 44

[45] $1 = \text{cis}\,0.\ u_k = \sqrt[6]{1}\,\text{cis}\left(\dfrac{0 + 2k\pi}{6}\right) = 1\,\text{cis}\,k\left(\dfrac{\pi}{3}\right) = \text{cis}\,k\cdot 60°,$ for $k = 0,\ 1,\ 2,\ 3,\ 4$ and 5. Thus,

$u_0 = \text{cis}\,0°,\ u_1 = \text{cis}\,60°,\ u_2 = \text{cis}\,120°,\ u_3 = \text{cis}\,180°,\ u_4 = \text{cis}\,240°,$ and $u_5 = \text{cis}\,300°.$

(See figure below).

[46] $1 = \text{cis } 0.$ $u_k = \sqrt[4]{1}\ \text{cis}\left(\dfrac{0 + 2k\pi}{4}\right) = 1\ \text{cis } k\left(\dfrac{\pi}{2}\right) = \text{cis } k \cdot 90°$, for $k = 0, 1, 2$ and 3. Thus,

$u_0 = \text{cis } 0°$, $u_1 = \text{cis } 90°$, $u_2 = \text{cis } 180°$ and $u_3 = \text{cis } 270°$.   (See figure below).

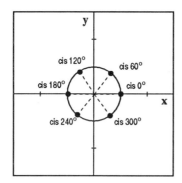
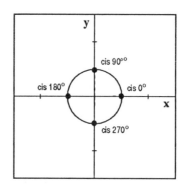

Figure 45   **Figure 46**

[47] $-16 = 16\ \text{cis } \pi.$ $u_k = \sqrt[4]{16}\ \text{cis}\left(\dfrac{\pi + 2k\pi}{4}\right) = 2\ \text{cis } k\left[\dfrac{\pi}{4} + k\left(\dfrac{\pi}{2}\right)\right] = 2\ \text{cis}\left(45° + k \cdot 90°\right)$, for $k = 0, 1,$

2 and 3. Thus, $u_0 = 2\ \text{cis } 45°$, $u_1 = 2\ \text{cis } 135°$, $u_2 = 2\ \text{cis } 225°$ and $u_3 = 2\ \text{cis } 315°$.   (See figure below).

[48] $32 = 32\ \text{cis } 0.$ $u_k = \sqrt[5]{32}\ \text{cis}\left(\dfrac{0 + 2k\pi}{5}\right) = 2\ \text{cis } k\left(\dfrac{2\pi}{5}\right) = 2\ \text{cis } k \cdot 72°$, for $k = 0, 1, 2, 3$ and 4. Thus,

$u_0 = 2\ \text{cis } 0°$, $u_1 = 2\ \text{cis } 72°$, $u_2 = 2\ \text{cis } 144°$, $u_3 = 2\ \text{cis } 216°$, and $u_4 = 2\ \text{cis } 288°$.   (See figure below).

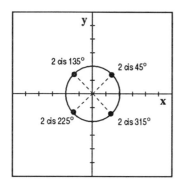
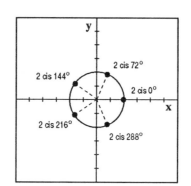

Figure 47   **Figure 48**

[49] $8\ \text{cis }\dfrac{2\pi}{3}.$ $u_k = \sqrt[3]{8}\ \text{cis}\left(\dfrac{\dfrac{2\pi}{3} + 2k\pi}{3}\right) = 2\ \text{cis}\left[\dfrac{2\pi}{9} + k\left(\dfrac{2\pi}{3}\right)\right] = 2\ \text{cis}\left(40° + k \cdot 120°\right)$, for $k = 0, 1, 2$.

Thus, $u_0 = 2\ \text{cis } 40°$, $u_1 = 2\ \text{cis } 160°$ and $u_2 = 2\ \text{cis } 280°$.   (See figure below).

[50] $27\ \text{cis }\dfrac{5\pi}{4}.$ $u_k = \sqrt[3]{27}\ \text{cis}\left(\dfrac{\dfrac{5\pi}{4} + 2k\pi}{3}\right) = 3\ \text{cis}\left[\dfrac{5\pi}{12} + k\left(\dfrac{2\pi}{3}\right)\right] = 3\ \text{cis}\left(75° + k \cdot 120°\right)$, for $k = 0, 1,$

and 2. Thus, $u_0 = 3\ \text{cis } 75°$, $u_1 = 3\ \text{cis } 195°$ and $u_2 = 3\ \text{cis } 315°$.   (See figure below).

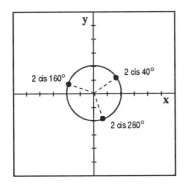

Figure 49

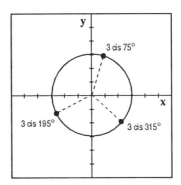

Figure 50

[51] $81 \operatorname{cis} 225°$. $u_k = \sqrt[4]{81} \operatorname{cis} \left[\dfrac{225° + k\left(360°\right)}{4} \right] = 3 \operatorname{cis}\left(56.25° + k \cdot 90°\right)$, for $k = 0, 1, 2$ and 3. Thus, $u_0 =$

$3 \operatorname{cis} 56.25°$, $u_1 = 3 \operatorname{cis} 146.25°$, $u_2 = 3 \operatorname{cis} 236.25°$ and $u_3 = 3 \operatorname{cis} 326.25°$.   (See figure below).

[52] $64 \operatorname{cis} 120°$. $u_k = \sqrt[5]{64} \operatorname{cis} \left[\dfrac{120° + k\left(360°\right)}{5} \right] = 2 \sqrt[5]{2} \operatorname{cis}\left(24° + k \cdot 72°\right)$, for $k = 0, 1, 2, 3$ and 4. Thus,

$u_0 = 2 \sqrt[5]{2} \operatorname{cis} 24°$, $u_1 = 2 \sqrt[5]{2} \operatorname{cis} 96°$, $u_2 = 2 \sqrt[5]{2} \operatorname{cis} 168°$, $u_3 = 2 \sqrt[5]{2} \operatorname{cis} 240°$ and $u_4 = 2 \sqrt[5]{2} \operatorname{cis} 312°$,

where $2 \sqrt[5]{2} \approx 2.30$.   (See figure below).

[53] $32 \operatorname{cis} 230°$. $u_k = \sqrt[5]{32} \operatorname{cis} \left[\dfrac{230° + k\left(360°\right)}{5} \right] = 2 \operatorname{cis}\left(46° + k \cdot 72°\right)$, for $k = 0, 1, 2, 3$ and 4. Thus,

$u_0 = 2 \operatorname{cis} 46°$, $u_1 = 2 \operatorname{cis} 118°$, $u_2 = 2 \operatorname{cis} 190°$, $u_3 = 2 \operatorname{cis} 262°$ and $u_4 = 2 \operatorname{cis} 334°$.

(See figure below).

[54] $125 \operatorname{cis} 87°$. $u_k = \sqrt[3]{125} \operatorname{cis} \left[\dfrac{87° + k\left(360°\right)}{3} \right] = 5 \operatorname{cis}\left(29° + k \cdot 120°\right)$, for $k = 0, 1$ and 2. Thus, $u_0 =$

$5 \operatorname{cis} 29°$, $u_1 = 5 \operatorname{cis} 149°$ and $u_2 = 5 \operatorname{cis} 269°$.   (See figure below).

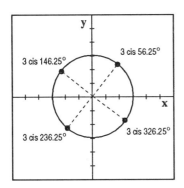

Figure 51

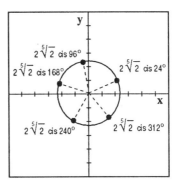

Figure 52

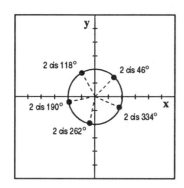

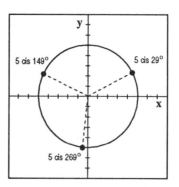

Figure 53   **Figure 54**

[55] $x^3 + 27 = 0 \Rightarrow x^3 = -27$. The solutions are the cube roots of -27. $-27 = 27 \text{ cis } \pi$. $u_k =$

$$\sqrt[3]{27} \text{ cis}\left(\frac{\pi + 2k\pi}{3}\right) = 3 \text{ cis}\left[\frac{\pi}{3} + k\left(\frac{2\pi}{3}\right)\right] = 3 \text{ cis}\left(60° + k \cdot 120°\right), \text{ for } k = 0, 1, \text{ and } 2. \text{ Thus, } u_0 =$$

$3 \text{ cis } 60°$, $u_1 = 3 \text{ cis } 180°$ and $u_2 = 3 \text{ cis } 300°$.

[56] $x^3 - 1 = 0 \Rightarrow x^3 = 1$. The solutions are the cube roots of unity. $1 = 1 \text{ cis } 0°$. $u_k =$

$$\sqrt[3]{1} \text{ cis}\left(\frac{0° + k \, 360°}{3}\right) = \text{cis}\left(k \, 120°\right), \text{ for } k = 0, 1, \text{ and } 2. \text{ Thus, } u_0 = \text{cis } 0°, u_1 = \text{cis } 120° \text{ and}$$

$u_2 = \text{cis } 240°$.

[57] $x^5 - i = 0 \Rightarrow x^5 = i$. The solutions are the fifth roots of i. $i = 1 \text{ cis } \frac{\pi}{2}$. $u_k = \sqrt[5]{1} \text{ cis}\left(\dfrac{\frac{\pi}{2} + 2k\pi}{5}\right) =$

$$\text{cis}\left[\frac{\pi}{10} + k\left(\frac{2\pi}{5}\right)\right] = \text{cis}\left(18° + k \cdot 72°\right), \text{ for } k = 0, 1, 2, 3, \text{ and } 4. \text{ Thus, } u_0 = \text{cis } 18°, u_1 = \text{cis } 90°,$$

$u_2 = \text{cis } 162°$, $u_3 = \text{cis } 234°$ and $u_4 = \text{cis } 306°$.

[58] $x^4 + 2i = 0 \Rightarrow x^4 = -2i$. The solutions are the fourth roots of $-2i$. $-2i = 2 \text{ cis } \frac{3\pi}{2}$. $u_k =$

$$\sqrt[4]{2} \text{ cis}\left(\frac{\frac{3\pi}{2} + 2k\pi}{4}\right) = \sqrt[4]{2} \text{ cis}\left[\frac{3\pi}{8} + k\left(\frac{\pi}{2}\right)\right] = \sqrt[4]{2} \text{ cis}\left(67.5° + k \cdot 90°\right), \text{ for } k = 0, 1, 2, \text{ and } 3. \text{ Thus,}$$

$u_0 = \sqrt[4]{2} \text{ cis } 67.5°$, $u_1 = \sqrt[4]{2} \text{ cis } 157.5°$, $u_2 = \sqrt[4]{2} \text{ cis } 247.5°$ and $u_3 = \sqrt[4]{2} \text{ cis } 337.5°$.

[59] $x^4 - 1 + i = 0 \Rightarrow x^4 = 1 - i$. The solutions are the fourth roots of $1 - i$. $1 - i = \sqrt{2} \text{ cis } \frac{7\pi}{4}$. $u_k =$

$$\sqrt[4]{\sqrt{2}} \text{ cis}\left(\frac{\frac{7\pi}{4} + 2k\pi}{4}\right) = 2^{1/8} \text{ cis}\left[\frac{7\pi}{16} + k\left(\frac{\pi}{2}\right)\right] = 2^{1/8} \text{ cis}\left(78.75° + k \cdot 90°\right), \text{ for } k = 0, 1, 2, \text{ and } 3.$$

Thus, $u_0 = 2^{1/8} \text{ cis } 78.75°$, $u_1 = 2^{1/8} \text{ cis } 168.75°$, $u_2 = 2^{1/8} \text{ cis } 258.75°$ and $u_3 = 2^{1/8} \text{ cis } 348.75°$.

[60] $x^5 - \sqrt{3} + i = 0 \Rightarrow x^5 = \sqrt{3} - i$. The solutions are the fifth roots of $\sqrt{3} - i$. $\sqrt{3} - i = 2 \text{ cis } \frac{11\pi}{6}$. $u_k =$

$$\sqrt[5]{2} \text{ cis}\left(\frac{\frac{11\pi}{6} + 2k\pi}{5}\right) = \sqrt[5]{2} \text{ cis}\left[\frac{11\pi}{30} + k\left(\frac{2\pi}{5}\right)\right] = \sqrt[5]{2} \text{ cis}\left(66° + k \cdot 72°\right), \text{ for } k = 0, 1, 2, 3 \text{ and } 4.$$

Thus, $u_0 = \sqrt[5]{2} \text{ cis } 66°$, $u_1 = \sqrt[5]{2} \text{ cis } 138°$, $u_2 = \sqrt[5]{2} \text{ cis } 210°$, $u_3 = \sqrt[5]{2} \text{ cis } 282°$ and $u_4 = \sqrt[5]{2} \text{ cis } 354°$.

[61] $(1 - \sqrt{3}\, i)^{-4} = \left(2 \operatorname{cis} \dfrac{5\pi}{3}\right)^{-4} = 2^{-4} \operatorname{cis}(-4)\left(\dfrac{5\pi}{3}\right) = \dfrac{1}{16} \operatorname{cis}\left(-\dfrac{20\pi}{3}\right) = \dfrac{1}{16} \operatorname{cis} \dfrac{4\pi}{3} = \dfrac{1}{16}\left(-\dfrac{1}{2} - \dfrac{\sqrt{3}}{2}\, i\right) =$

$-\dfrac{1}{32} - \dfrac{\sqrt{3}}{32}\, i.$

[62] $\dfrac{(1-i)^{-2}}{(-\sqrt{3}+i)^4} = (1-i)^{-2} \cdot (-\sqrt{3}+i)^{-4} \cdot (1-i)^{-2} = \left(\sqrt{2} \operatorname{cis} \dfrac{7\pi}{4}\right)^{-2} = \left(2^{1/2}\right)^{-2} \operatorname{cis}(-2)\left(\dfrac{7\pi}{4}\right) =$

$2^{-1} \operatorname{cis}\left(-\dfrac{7\pi}{2}\right) = \dfrac{1}{2} \operatorname{cis} \dfrac{\pi}{2} = \dfrac{1}{2}(0+i) = \dfrac{1}{2}\, i,$ and $(-\sqrt{3}+i)^{-4} = \left(2 \operatorname{cis} \dfrac{5\pi}{6}\right)^{-4} = 2^{-4} \operatorname{cis}(-4)\left(\dfrac{5\pi}{6}\right) =$

$\dfrac{1}{16} \operatorname{cis}\left(-\dfrac{10\pi}{3}\right) = \dfrac{1}{16} \operatorname{cis} \dfrac{2\pi}{3} = \dfrac{1}{16}\left(-\dfrac{1}{2} + \dfrac{\sqrt{3}}{2}\, i\right) = -\dfrac{1}{32} + \dfrac{\sqrt{3}}{32}\, i.$ Thus $(1-i)^{-2} \cdot (-\sqrt{3}+i)^{-4} =$

$\dfrac{1}{2}\, i\left(-\dfrac{1}{32} + \dfrac{\sqrt{3}}{32}\, i\right) = -\dfrac{\sqrt{3}}{64} - \dfrac{1}{64}\, i.$

[63] The conjugate of $-1+i$, which is $-1-i$, must also be a fourth root of z. Since $-1+i = \sqrt{2} \operatorname{cis} \dfrac{3\pi}{4}$

and $-1-i = \sqrt{2} \operatorname{cis} \dfrac{5\pi}{4}$, and since the four roots are equally spaced around a circle centered at the

origin with radius $\sqrt{2}$, the remaining two fourth roots of z are $\sqrt{2} \operatorname{cis} \dfrac{\pi}{4} = \sqrt{2}\left(\dfrac{1}{\sqrt{2}} + \dfrac{1}{\sqrt{2}}\, i\right) = 1+i$

and $\sqrt{2} \operatorname{cis} \dfrac{7\pi}{4} = \sqrt{2}\left(\dfrac{1}{\sqrt{2}} - \dfrac{1}{\sqrt{2}}\, i\right) = 1-i.$

[64] The conjugate of $\sqrt{3} - i$, namely $\sqrt{3} + i$, must also be a sixth root of z. Since $\sqrt{3} - i = 2 \operatorname{cis} \dfrac{11\pi}{6}$ and

$\sqrt{3} + i = 2 \operatorname{cis} \dfrac{\pi}{6}$, and since the six roots must be equally spaced around a circle of radius 2 centered

at the origin, the remaining four sixth roots must be $2 \operatorname{cis}\left(\dfrac{\pi}{6} + \dfrac{\pi}{3}\right) = 2 \operatorname{cis} \dfrac{\pi}{2} = 2(0+i) = 2\, i,$

$2 \operatorname{cis}\left(\dfrac{\pi}{6} + \dfrac{2\pi}{3}\right) = 2 \operatorname{cis} \dfrac{5\pi}{6} = 2\left(-\dfrac{\sqrt{3}}{2} + \dfrac{1}{2}\, i\right) = -\sqrt{3} + i,\; 2 \operatorname{cis}\left(\dfrac{\pi}{6} + \pi\right) = 2 \operatorname{cis} \dfrac{7\pi}{6} = 2\left(-\dfrac{\sqrt{3}}{2} - \dfrac{1}{2}\, i\right) =$

$-\sqrt{3} - i,$ and $2 \operatorname{cis}\left(\dfrac{\pi}{6} + \dfrac{4\pi}{3}\right) = 2 \operatorname{cis} \dfrac{3\pi}{2} = 2(0 - i) = -2\, i.$

[65] The cube roots of unity are solutions of the equation $x^3 = 1$ or $x^3 - 1 = 0$. If u_k is a cube root of
unity, $u_k^3 - 1 = 0 \Rightarrow (u_k - 1)(u_k^2 + u_k + 1) = 0.$ Thus, if $u_k \neq 1$, u_k must satisfy the equation
$u_k^2 + u_k + 1 = 0.$

[66] If u_k is any nth root of unity, then u_k must satisfy the equation $x^n = 1$ or $x^n - 1 = 0.$ Now, $u_k^n - 1 = 0 \Rightarrow (u_k - 1)(u_k^{n-1} + u_k^{n-2} + \ldots + u_k^2 + u_k + 1) = 0.$ If $u_k \neq 1$, it follows that u_k must satisfy the
equation $u_k^{n-1} + u_k^{n-2} + \ldots + u_k^2 + u_k + 1 = 0.$

[67] $\cos\theta + i \sin\theta = \left(1 - \dfrac{\theta^2}{2!} + \dfrac{\theta^4}{4!} - \dfrac{\theta^6}{6!} + \ldots + \dfrac{(-1)^n \theta^{2n}}{(2n)!} + \ldots\right) +$

$i\left(\theta - \dfrac{\theta^3}{3!} + \dfrac{\theta^5}{5!} - \dfrac{\theta^7}{7!} + \ldots + \dfrac{(-1)^n \theta^{2n+1}}{(2n+1)!} + \ldots\right) = 1 + i\,\theta - \dfrac{\theta^2}{2!} - \dfrac{i\,\theta^3}{3!} + \dfrac{\theta^4}{4!} + \dfrac{i\,\theta^5}{5!} + \ldots + \dfrac{i^n \theta^n}{n!} + \ldots,$

and $e^{i\theta} = 1 + i\,\theta + \dfrac{(i\theta)^2}{2!} + \dfrac{(i\theta)^3}{3!} + \dfrac{(i\theta)^4}{4!} + \dfrac{(i\theta)^5}{5!} + \ldots + \dfrac{(i\theta)^n}{n!} + \ldots =$

$1 + i\,\theta - \dfrac{\theta^2}{2!} - \dfrac{i\,\theta^3}{3!} + \dfrac{\theta^4}{4!} + \dfrac{i\,\theta^5}{5!} + \ldots + \dfrac{i^n \theta^n}{n!} + \ldots.$ Thus $e^{i\theta} = \cos\theta + i \sin\theta$

[68] $e^{i\pi} = \cos\pi + i \sin\pi = -1 + i(0) = -1.$

319

[69] By DeMoivre's Theorem, $(\cos\theta + i\sin\theta)^2 (1 \cdot \operatorname{cis}\theta)^2 = 1^2 \cdot \operatorname{cis} 2 \cdot \theta = \operatorname{cis} 2\theta = \cos 2\theta + (\sin 2\theta) i$.

Also, $(\cos\theta + i\sin\theta)^2 = \cos^2\theta = \cos^2\theta + 2 i\sin\theta\cos\theta - \sin^2\theta =$

$\left(\cos^2\theta - \sin^2\theta\right) + (2\sin\theta\cos\theta) i$. Equating the imaginary parts of the two results, we see that

$\sin 2\theta = 2\sin\theta\cos\theta$.

[70] Consider the equation $x^n - z = 0$, where x and z are complex numbers and $z \neq 0$. Assume that $z = r \operatorname{cis}\theta$ in trigonometric form. Since $x^n - z = 0$ iff $x^n = z$, any solution of the equation $x^n - z = 0$ is also an n th root of z. To show that $u_k = \sqrt[n]{r}\, \operatorname{cis}\left(\frac{\theta + 2k\ \pi}{n}\right)$ is an n th root of z,

for $k = 0, 1, 2, \ldots, n-1$, we must show that $u_k^n = z$, for $k = 0, 1, 2, \ldots, n-1$. If

$u_k = \sqrt[n]{r}\, \operatorname{cis}\left(\frac{\theta + 2\,k\,\pi}{n}\right)$, where k is an integer satisfying $0 \leq k \leq n-1$, then $u_k^n =$

$\left[\sqrt[n]{r}\, \operatorname{cis}\left(\frac{\theta + 2k\ \pi}{n}\right)\right]^n$. By DeMoivre's Theorem, $\left[\sqrt[n]{r}\, \operatorname{cis}\left(\frac{\theta + 2k\ \pi}{n}\right)\right]^n = \left(\sqrt[n]{r}\right)^n \operatorname{cis}\left[n\left(\frac{\theta + 2k\ \pi}{n}\right)\right] =$

$r\operatorname{cis}(\theta + 2k\ \pi)$. Since both the sine and cosine functions have period 2π, it follows that $u_k^n = r\operatorname{cis}\theta = z$. To show that the roots u_k, $k = 0, 1, 2, \ldots n-1$, are distinct, we assume that $u_m = u_p$,

where $0 \leq m \leq n-1$ and $0 \leq p \leq n-1$, and show that $m = p$. $u_m = u_p \Rightarrow \sqrt[n]{r}\, \operatorname{cis}\left(\frac{\theta + 2\,m\,\pi}{n}\right) =$

$\sqrt[n]{r}\, \operatorname{cis}\left(\frac{\theta + 2\,p\,\pi}{n}\right) \Rightarrow \operatorname{cis}\left(\frac{\theta + 2\,m\,\pi}{n}\right) = \operatorname{cis}\left(\frac{\theta + 2\,p\,\pi}{n}\right) \Rightarrow \cos\left(\frac{\theta + 2\,m\,\pi}{n}\right) + i\sin\left(\frac{\theta + 2\,m\,\pi}{n}\right) =$

$\cos\left(\frac{\theta + 2\,p\,\pi}{n}\right) + i\sin\left(\frac{\theta + 2\,p\,\pi}{n}\right)$. Since the sine and cosine functions have period 2π, it follows

that $\frac{\theta + 2\,m\,\pi}{n} = \frac{\theta + 2\,p\,\pi}{n} + q(2\pi)$, for some integer q. Moreover, $\frac{\theta + 2\,m\,\pi}{n} =$

$\frac{\theta + 2\,p\,\pi}{n} + q(2\pi) \Rightarrow \theta + 2\,m\,\pi = \theta + 2\,p\,\pi + n\,q(2\pi) \Rightarrow 2\,m\,\pi = 2\,p\,\pi + n\,q(2\pi) \Rightarrow m = p + n\,q$. To show that $m = p$ we must show that $q = 0$. Since $0 \leq m < n-1$, it follows that $0 \leq m < n$.

Then, $m = p + n\,q \Rightarrow 0 \leq p + n\,q < n$ (∗). From inequality (∗) we see that $p + n\,q < n$. Now,

$p + n\,q < n \Rightarrow n\,q < n - p \Rightarrow n\,q - n < -p$. Also, $p \geq 0 \Rightarrow -p \leq 0$ and the inequalities

$n\,q - n < -p$ and $-p \leq 0$ tell us that $n\,q - n < 0$. From $n\,q - n < 0$ it follows that $n\,q < n$ and

$q < \frac{n}{n} = 1$. Since q is an integer, $q < 1 \Rightarrow q \leq 0$. To show that $q \geq 0$ we proceed in a similar manner.

From inequality (∗), we know that $p + n\,q \geq 0$. Now, $p + n\,q \geq 0 \Rightarrow n\,q \geq -p$. From the inequality

$0 \leq p \leq n-1$, it follows that $p < n$, so that $-p > -n$. Since $n\,q \geq -p$ and $-p > -n$, we see that

$n\,q > -n$. Now, $n\,q > -n \Rightarrow q > -\frac{n}{n} = -1$. Since q is an integer, $q > -1 \Rightarrow q \geq 0$. Thus, $q \leq 0$ and

$q \geq 0$, which means that $q = 0$. Since the equation $x^n - z = 0$ has exactly n roots by Theorem 3.2 on

page 135, the proof is complete.

EXERCISES 7.5

[1] $|\mathbf{v}| = \sqrt{(2)^2 + (-3)^2} = \sqrt{13}$.   (See figure below).

[2] $|\mathbf{v}| = \sqrt{(-1)^2 + (4)^2} = \sqrt{17}$.   (See figure below).

[3] $|\mathbf{v}| = \sqrt{\left(-\frac{3}{2}\right)^2 + \left(-\frac{7}{4}\right)^2} = \frac{\sqrt{85}}{4}$.   (See figure below).

[4] $|\mathbf{v}| = \sqrt{\left(-\frac{5}{2}\right)^2 + \left(-\frac{3}{4}\right)^2} = \frac{\sqrt{109}}{4}$.   (See figure below).

[5] $|\mathbf{v}| = \sqrt{(0)^2 + (-6)^2} = 6$.   (See figure below).

[6] $|\mathbf{v}| = \sqrt{(-5)^2 + (0)^2} = 5$.   (See figure below).

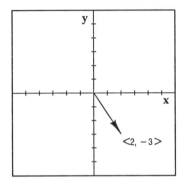

Figure 1

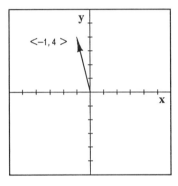

Figure 2

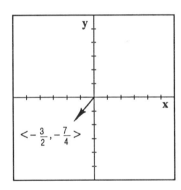

Figure 3

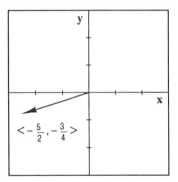

Figure 4

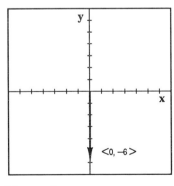

Figure 5

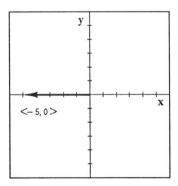

Figure 6

[7] $|\mathbf{v}| = \sqrt{(2\sqrt{3})^2 + (4)^2} = \sqrt{28} = 2\sqrt{7}.$ (See figure below).

[8] $|\mathbf{v}| = \sqrt{\left(-\frac{\sqrt{3}}{2}\right)^2 + \left(\frac{1}{2}\right)^2} = 1.$ (See figure below).

[9] $-2\mathbf{v} = \langle(-2)(-3),(-2)(6)\rangle = \langle 6, -12 \rangle;\ \frac{1}{3}\mathbf{v} = \left\langle \left(\frac{1}{3}\right)(-3), \left(\frac{1}{3}\right)(6) \right\rangle = \langle -1, 2 \rangle.$ (See figure below).

[10] $-2\mathbf{v} = \langle(-2)(6),(-2)(-2)\rangle = \langle -12, 4 \rangle;\ \frac{1}{3}\mathbf{v} = \left\langle \left(\frac{1}{3}\right)(6), \left(\frac{1}{3}\right)(-2) \right\rangle = \left\langle 2, -\frac{2}{3} \right\rangle.$ (See figure below).

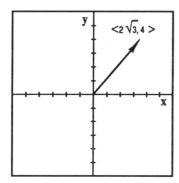

Figure 7

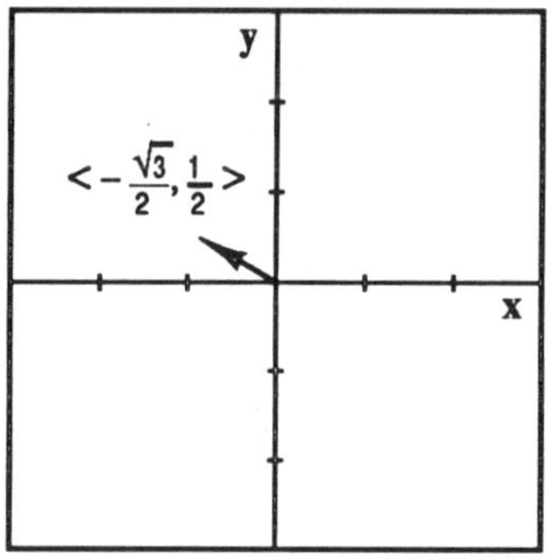

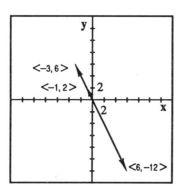

Figure 9

Figure 8

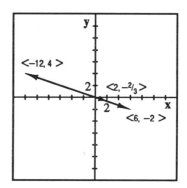

Figure 10

[11] $-2\mathbf{v} = \left\langle (-2)(-2),(-2)\left(-\frac{15}{4}\right) \right\rangle = \left\langle 4, \frac{15}{2} \right\rangle;\ \frac{1}{3}\mathbf{v} = \left\langle \left(\frac{1}{3}\right)(-2), \left(\frac{1}{3}\right)\left(-\frac{15}{4}\right) \right\rangle = \left\langle -\frac{2}{3}, -\frac{5}{4} \right\rangle.$

(See figure below).

[12] $-2\mathbf{v} = \left\langle (-2)\left(-\frac{9}{4}\right),(-2)(-12) \right\rangle = \left\langle \frac{9}{2}, 24 \right\rangle;\ \frac{1}{3}\mathbf{v} = \left\langle \left(\frac{1}{3}\right)\left(-\frac{9}{4}\right), \left(\frac{1}{3}\right)(-12) \right\rangle = \left\langle -\frac{3}{4}, -4 \right\rangle.$

(See figure below).

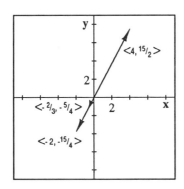

Figure 11

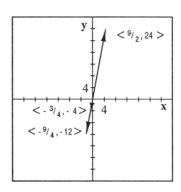

Figure 12

[13] $\mathbf{u} + \mathbf{v} = \langle 1 + (-2), -4 + (-3) \rangle = \langle -1, -7 \rangle$. (See figure below).

[14] $\mathbf{u} + \mathbf{v} = \langle -3 + 2, 1 + (-5) \rangle = \langle -1, -4 \rangle$. (See figure below).

[15] $\mathbf{u} + \mathbf{v} = \left\langle -5 + 6, \frac{7}{2} + (-8) \right\rangle = \left\langle 1, -\frac{9}{2} \right\rangle$. (See figure below).

[16] $\mathbf{u} + \mathbf{v} = \left\langle -\frac{5}{2} + (-6), 4 + (-2) \right\rangle = \left\langle -\frac{17}{2}, 2 \right\rangle$. (See figure below).

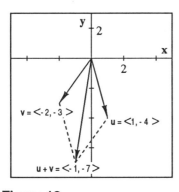

Figure 13

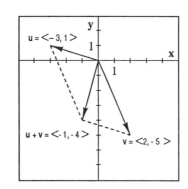

Figure 14

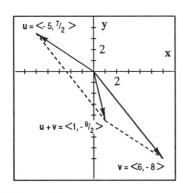

Figure 15

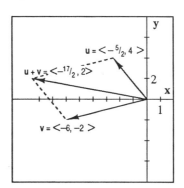

Figure 16

[17] $\mathbf{u} + \mathbf{v} = \left\langle \frac{12}{5} + (-9), 0 + 3 \right\rangle = \left\langle -\frac{33}{5}, 3 \right\rangle$. (See figure below).

[18] $\mathbf{u} + \mathbf{v} = \left\langle 8 + 0, -\frac{3}{4} + (-5) \right\rangle = \left\langle 8, -\frac{23}{4} \right\rangle$. (See figure below).

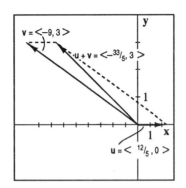

Figure 17

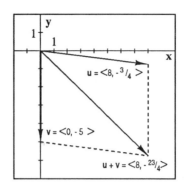

Figure 18

[19] $2\mathbf{u} - \mathbf{v} = 2\langle 3, -4\rangle - \langle -1, 6\rangle = \langle 6, -8\rangle - \langle -1, 6\rangle = \langle 7, -14\rangle.$

[20] $-\mathbf{u} + 3\mathbf{v} = -\langle 3, -4\rangle + 3\langle -1, 6\rangle = \langle -3, 4\rangle + \langle -3, 18\rangle = \langle -6, 22\rangle.$

[21] $-\dfrac{1}{4}\mathbf{u} + \mathbf{v} = -\dfrac{1}{4}\langle 3, -4\rangle + \langle -1, 6\rangle = \left\langle -\dfrac{3}{4}, 1\right\rangle + \langle -1, 6\rangle = \left\langle -\dfrac{7}{4}, 7\right\rangle.$

[22] $3\mathbf{u} - \dfrac{5}{2}\mathbf{v} = 3\langle 3, -4\rangle - \dfrac{5}{2}\langle -1, 6\rangle = \langle 9, -12\rangle - \left\langle -\dfrac{5}{2}, 15\right\rangle = \left\langle \dfrac{23}{2}, -27\right\rangle.$

[23] $\dfrac{\sqrt{2}}{2}\mathbf{u} - \dfrac{\sqrt{2}}{2}\mathbf{v} = \dfrac{\sqrt{2}}{2}\langle 3, -4\rangle - \dfrac{\sqrt{2}}{2}\langle -1, 6\rangle = \left\langle \dfrac{3\sqrt{2}}{2}, -2\sqrt{2}\right\rangle - \left\langle -\dfrac{\sqrt{2}}{2}, 3\sqrt{2}\right\rangle = \langle 2\sqrt{2}, -5\sqrt{2}\rangle.$

[24] $\dfrac{\sqrt{3}}{2}\mathbf{u} + \dfrac{1}{2}\mathbf{v} = \dfrac{\sqrt{3}}{2}\langle 3, -4\rangle + \dfrac{1}{2}\langle -1, 6\rangle = \left\langle \dfrac{3\sqrt{3}}{2}, -2\sqrt{3}\right\rangle + \left\langle -\dfrac{1}{2}, 3\right\rangle = \left\langle \dfrac{3\sqrt{3}-1}{2}, 3 - 2\sqrt{3}\right\rangle.$

[25] $-50\mathbf{u} + 80\mathbf{v} = -50\langle 3, -4\rangle + 80\langle -1, 6\rangle = \langle -150, 200\rangle + \langle -80, 480\rangle = \langle -230, 680\rangle.$

[26] $12\mathbf{u} - 32\mathbf{v} = 12\langle 3, -4\rangle - 32\langle -1, 6\rangle = \langle 36, -48\rangle - \langle -32, 192\rangle = \langle 68, -240\rangle.$

[27] $-3\mathbf{v} - \dfrac{1}{2}\mathbf{u} = -3\langle -1, 6\rangle - \dfrac{1}{2}\langle 3, -4\rangle = \langle 3, -18\rangle - \left\langle \dfrac{3}{2}, -2\right\rangle = \left\langle \dfrac{3}{2}, -16\right\rangle.$

[28] $-\dfrac{7}{2}\mathbf{v} + 2\mathbf{u} = -\dfrac{7}{2}\langle -1, 6\rangle + 2\langle 3, -4\rangle = \left\langle \dfrac{7}{2}, -21\right\rangle + \langle 6, -8\rangle = \left\langle \dfrac{19}{2}, -29\right\rangle.$

[29] $|\mathbf{u}|(\mathbf{u} - \mathbf{v}) = \sqrt{(3)^2 + (-4)^2}\,(\langle 3, -4\rangle - \langle -1, 6\rangle) = 5\langle 4, -10\rangle = \langle 20, -50\rangle.$

[30] $|\mathbf{u}|(\mathbf{u} + \mathbf{v}) = \sqrt{(-1)^2 + (6)^2}\,(\langle 3, -4\rangle + \langle -1, 6\rangle) = \sqrt{37}\langle 2, 2\rangle = \langle 2\sqrt{37}, 2\sqrt{37}\rangle.$

[31] $\tan\theta = \dfrac{-4}{4} = -1$ and θ in QIV $\Rightarrow \theta = \dfrac{7\pi}{4} = 315°.$

[32] $\tan\theta = \dfrac{-3}{-3} = 1$ and θ in QIII $\Rightarrow \theta = \dfrac{5\pi}{4} = 225°.$

[33] $\tan\theta = \dfrac{-1}{-\sqrt{3}} = \dfrac{1}{\sqrt{3}}$ and θ in QIII $\Rightarrow \theta = \dfrac{7\pi}{6} = 210°.$

[34] $\tan\theta = \dfrac{\frac{\sqrt{3}}{2}}{-\frac{1}{2}} = -\sqrt{3}$ and θ in QII $\Rightarrow \theta = \dfrac{2\pi}{3} = 120°.$

[35] $\tan\theta = \dfrac{8}{-6} = -\dfrac{4}{3}.$ If θ' is the reference angle for θ, $\tan\theta' = |\tan\theta| = \dfrac{4}{3} \Rightarrow \theta' = \tan^{-1}\dfrac{4}{3} \approx 53°.$ Since θ is in QII, $\theta = 180° - \theta' = 180° - \tan^{-1}\dfrac{4}{3} \approx 127°.$

[36] $\tan\theta = \dfrac{2}{3}$ and θ in QI $\Rightarrow \theta = \tan^{-1}\dfrac{2}{3} \approx 34°.$

[37] $\tan \theta = \frac{-5}{2} = -\frac{5}{2}$. If θ' is the reference angle for θ, $\tan \theta' = |\tan \theta| = \frac{5}{2} \Rightarrow \theta' = \tan^{-1} \frac{5}{2} \approx 68°$. Since

θ is in QIV, $\theta = 360° - \tan^{-1} \frac{5}{2} \approx 292°$.

[38] $\tan \theta = \dfrac{\frac{3}{4}}{-\frac{1}{2}} = -\frac{3}{2}$. If θ' is the reference angle for θ, $\tan \theta' = |\tan \theta| = \frac{3}{2} \Rightarrow \theta' = \tan^{-1} \frac{3}{2} \approx 56°$. Since

θ is in QII, $\theta = 180° - \theta' = 180° - \tan^{-1} \frac{3}{2} \approx 124°$.

[39] $v_1 = |\mathbf{v}| \cos \theta = 3 \cos \frac{\pi}{3} = 3 \left(\frac{1}{2} \right) = \frac{3}{2}$ and $v_2 = |\mathbf{v}| \sin \theta = 3 \sin \frac{\pi}{3} = 3 \left(\frac{\sqrt{3}}{2} \right) = \frac{3\sqrt{3}}{2} \Rightarrow \mathbf{v} = \left\langle \frac{3}{2}, \frac{3\sqrt{3}}{2} \right\rangle$.

[40] $v_1 = |\mathbf{v}| \cos \theta = 4 \cos \frac{\pi}{4} = 4 \left(\frac{\sqrt{2}}{2} \right) = 2\sqrt{2}$ and $v_2 = |\mathbf{v}| \sin \theta = 4 \sin \frac{\pi}{4} = 4 \left(\frac{\sqrt{2}}{2} \right) = 2\sqrt{2} \Rightarrow$

$\mathbf{v} = \left\langle 2\sqrt{2}, 2\sqrt{2} \right\rangle$.

[41] $v_1 = |\mathbf{v}| \cos \theta = 24 \cos 225° = 24 \left(-\frac{\sqrt{2}}{2} \right) = -12\sqrt{2}$ and $v_2 = |\mathbf{v}| \sin \theta = 24 \sin 225° = 24 \left(-\frac{\sqrt{2}}{2} \right) =$

$-12\sqrt{2} \Rightarrow \mathbf{v} = \left\langle -12\sqrt{2}, -12\sqrt{2} \right\rangle$.

[42] $v_1 = |\mathbf{v}| \cos \theta = 15 \cos 300° = 15 \left(\frac{1}{2} \right) = \frac{15}{2}$ and $v_2 = |\mathbf{v}| \sin \theta = 15 \sin 300° = 15 \left(-\frac{\sqrt{3}}{2} \right) =$

$-\frac{15\sqrt{3}}{2} \Rightarrow \mathbf{v} = \left\langle \frac{15}{2}, -\frac{15\sqrt{3}}{2} \right\rangle$.

[43] $\mathbf{v} = \langle |\mathbf{v}| \cos \theta, |\mathbf{v}| \sin \theta \rangle = \langle 9.52 \cos 137°, 9.52 \sin 137° \rangle \approx \langle -6.96, 6.49 \rangle$.

[44] $\mathbf{v} = \langle |\mathbf{v}| \cos \theta, |\mathbf{v}| \sin \theta \rangle = \langle 37.4 \cos 258°, 37.4 \sin 258° \rangle \approx \langle -7.78, -36.6 \rangle$.

[45] Let $\mathbf{a}$ and $\mathbf{b}$ be vectors representing forces A and B, respectively (see Figure 45). We want to find

the magnitude and direction of the resultant vector $\mathbf{a} + \mathbf{b}$. We see from the figure that the direction

angle for $\mathbf{a}$ is $90° - 15.6° = 74.4°$, while the direction angle for $\mathbf{b}$ is $90° + 78.3° = 168.3°$. Thus, $\mathbf{a}$

and $\mathbf{b}$ have the following algebraic representations $\mathbf{a} = \langle 175 \cos 74.4°, 175 \sin 74.4° \rangle \approx$

$\langle 47.1, 168.6 \rangle$ and $\mathbf{b} = \langle 240 \cos 168.3°, 240 \sin 168.3° \rangle \approx \langle -235.0, 48.7 \rangle$. Thus $\mathbf{a} + \mathbf{b} \approx$

$\langle -187.9, 217.3 \rangle$ and the magnitude of $\mathbf{a} + \mathbf{b}$ is $|\mathbf{a} + \mathbf{b}| \approx \sqrt{(-187.9)^2 + (217.3)^2} \approx 287$. If θ is the

direction angle for $\mathbf{a} + \mathbf{b}$ then $\tan \theta \approx \frac{217.3}{-187.9} \approx -1.1565$. If θ' is the reference angle for θ, then

$\tan \theta' = |\tan \theta| \approx 1.1565 \Rightarrow \theta' \approx \tan^{-1} (1.1565) \approx 49.1°$. Since θ is in QII, $\theta = 180° - \theta' \approx 130.9°$. To

obtain the direction of $\mathbf{a} + \mathbf{b}$, we subtract $90°$ from θ to get $40.9°$. Thus, the resultant of forces A and

B is a force of approximately 287 pounds in the (approximate) direction N $40.9°$ W.

[46]   Let **a** and **b** be vectors representing forces A and B, respectively (see Figure 46). We must find the magnitude and direction of the resultant vector **a** + **b**. We see from figure 46 that the direction angle for **a** is $270° - 32.4° = 237.6°$ and the direction angle for **b** is $90° + 51.5° = 141.5°$. Thus, **a** and **b** have the following algebraic representations **a** = $\langle 44.3 \cos 237.6°, 44.3 \sin 237.6° \rangle \approx$ $\langle -23.7, -37.4 \rangle$ and **b** = $\langle 28.6 \cos 141.5°, 28.6 \sin 141.5° \rangle \approx \langle -22.4, 17.8 \rangle$. Thus **a** + **b** $\approx$ $\langle -46.1, -19.6 \rangle$ and the magnitude of **a** + **b** is $|\mathbf{a}+\mathbf{b}| \approx \sqrt{(-46.1)^2 + (-19.6)^2} \approx 50.1$. If θ is the direction angle for **a** + **b**, then $\tan \theta \approx \dfrac{-19.6}{-46.1} \approx 0.4252$. If θ' is the reference angle for θ, then $\tan \theta' = |\tan \theta| \approx 0.4252 \Rightarrow \theta' \approx \tan^{-1}(0.4252) \approx 23.0°$. Since θ is in QIII, $\theta = 180° + \theta' \approx 203.0°$. To obtain the direction of **a** + **b**, we subtract $203.0°$ from $270°$ to get $67.0°$. Thus, the resultant of forces A and B is a force of approximately 50.1 kilograms in the (approximate) direction S $67.0°$ W.

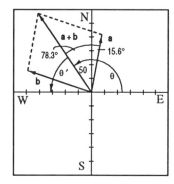

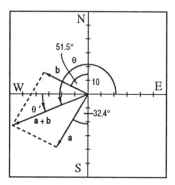

Figure 45   **Figure 46**

[47]   Let **a** and **b** be vectors representing velocities A and B, respectively (see Figure 47). We must find the magnitude and direction of the resultant vector **a** + **b**. We see from figure 47 that the direction angle for **a** is $90° + 41° \, 20' = 131° \, 20'$ and the direction angle for **b** is $270° - 29° \, 30' = 240° \, 30'$. Thus, **a** and **b** have the following algebraic representations **a** = $\langle 815 \cos 131° \, 20', 815 \sin 131° \, 20' \rangle \approx \langle -538.3, 612.0 \rangle$ and **b** = $\langle 520 \cos 240°30', 520 \sin 240°30' \rangle \approx \langle -256.1, -452.6 \rangle$. Thus **a** + **b** $\approx \langle -794.4, 159.4 \rangle$ and the magnitude of **a** + **b** is $|\mathbf{a}+\mathbf{b}| \approx$ $\sqrt{(-794.4)^2 + (159.4)^2} \approx 810$. If θ is the direction angle for **a** + **b**, then $\tan \theta \approx \dfrac{159.4}{-794.4} \approx -0.2007$. If θ' is the reference angle for θ, then $\tan \theta' = |\tan \theta| \approx 0.2007 \Rightarrow \theta' \approx \tan^{-1}(0.2007) \approx 11° \, 20'$. Since θ is in QII, $\theta = 180° - \theta' \approx 168° \, 40'$. To obtain the direction of **a** + **b**, we subtract $90°$ from $168° \, 40'$ to get $78° \, 40'$. Thus, the resultant of velocities A and B is a velocity of approximately 810 kilometers per hour in the (approximate) direction N $78° \, 40'$ W.

[48]   Let **a** and **b** be vectors representing velocities A and B, respectively (see Figure 48). We must find the magnitude and direction of the resultant vector **a** + **b**. We see from the figure that the direction angle for **a** is $270° - 64°\ 45' = 205°\ 15'$ and the direction angle for **b** is $270° + 40°\ 15' = 310°\ 15'$. Thus, **a** and **b** have the following algebraic representations **a** = $\langle 450 \cos 205°\ 15', 450 \sin 205°\ 15' \rangle \approx \langle -407.0, -192.0 \rangle$ and **b** = $\langle 675 \cos 310°15', 675 \sin 310°15' \rangle \approx \langle 436.1, -515.2 \rangle$. Thus **a** + **b** $\approx \langle 29.1, -707.2 \rangle$ and the magnitude of **a** + **b** is $\|\mathbf{a} + \mathbf{b}\| \approx \sqrt{(29.1)^2 + (-707.1)^2} \approx 708$. If θ is the direction angle for **a** + **b**, then $\tan\theta \approx \dfrac{-707.2}{29.1} \approx -24.30$. If θ' is the reference angle for θ, then $\tan\theta' = |\tan\theta| \approx 24.30 \Rightarrow \theta' \approx \tan^{-1}(24.30) \approx 87°\ 45'$. Since θ is in QII, $\theta = 360° - \theta' \approx 272°\ 15'$. To find the direction of **a** + **b**, we subtract $270°$ from $272°\ 15'$ to get $2°\ 15'$. Thus, the resultant of velocities A and B is a velocity of approximately 708 kilometers per hour in the (approximate) direction S $2°\ 15'$ E.

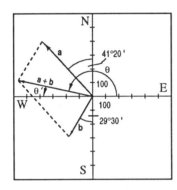

Figure 47

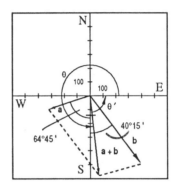

Figure 48

[49] $3\,\mathbf{u} + 4\,\mathbf{v} = 3\left(-\frac{1}{2}\mathbf{i} + 3\,\mathbf{j}\right) + 4\left(5\,\mathbf{i} - \frac{3}{4}\mathbf{j}\right) = \left(-\frac{3}{2}\mathbf{i} + 9\,\mathbf{j}\right) + (20\,\mathbf{i} - 3\,\mathbf{j}) = \frac{37}{2}\mathbf{i} + 6\,\mathbf{j}.$

[50] $2\,\mathbf{u} - 5\,\mathbf{v} = 2\left(-\frac{1}{2}\mathbf{i} + 3\,\mathbf{j}\right) - 5\left(5\,\mathbf{i} - \frac{3}{4}\mathbf{j}\right) = (-\mathbf{i} + 6\,\mathbf{j}) - \left(25\,\mathbf{i} - \frac{15}{4}\mathbf{j}\right) = -26\,\mathbf{i} + \frac{39}{4}\mathbf{j}.$

[51] $-5\,\mathbf{u} + \frac{1}{2}\mathbf{v} = -5\left(-\frac{1}{2}\mathbf{i} + 3\,\mathbf{j}\right) + \frac{1}{2}\left(5\,\mathbf{i} - \frac{3}{4}\mathbf{j}\right) = \left(\frac{5}{2}\mathbf{i} - 15\,\mathbf{j}\right) + \left(\frac{5}{2}\mathbf{i} - \frac{3}{8}\mathbf{j}\right) = 5\,\mathbf{i} - \frac{123}{8}\mathbf{j}.$

[52] $-\frac{4}{3}\mathbf{u} - \frac{1}{3}\mathbf{v} = -\frac{4}{3}\left(-\frac{1}{2}\mathbf{i} + 3\,\mathbf{j}\right) - \frac{1}{3}\left(5\,\mathbf{i} - \frac{3}{4}\mathbf{j}\right) = \left(\frac{2}{3}\mathbf{i} - 4\,\mathbf{j}\right) - \left(\frac{5}{3}\mathbf{i} - \frac{1}{4}\mathbf{j}\right) = -\mathbf{i} - \frac{15}{4}\mathbf{j}.$

[53] $-\sqrt{3}\,\mathbf{u} - \frac{5}{8}\mathbf{v} = -\sqrt{3}\left(-\frac{1}{2}\mathbf{i} + 3\,\mathbf{j}\right) - \frac{5}{8}\left(5\,\mathbf{i} - \frac{3}{4}\mathbf{j}\right) = \left(\frac{\sqrt{3}}{2}\mathbf{i} - 3\sqrt{3}\,\mathbf{j}\right) - \left(\frac{25}{8}\mathbf{i} - \frac{15}{32}\mathbf{j}\right) = $

$\left(\frac{4\sqrt{3} - 25}{8}\right)\mathbf{i} + \left(\frac{15 - 96\sqrt{3}}{32}\right)\mathbf{j}.$

[54] $\frac{\sqrt{2}}{2}\mathbf{v} + \mathbf{u} = \frac{\sqrt{2}}{2}\left(5\,\mathbf{i} - \frac{3}{4}\mathbf{j}\right) + \left(-\frac{1}{2}\mathbf{i} + 3\,\mathbf{j}\right) = \left(\frac{5\sqrt{2}}{2}\mathbf{i} - \frac{3\sqrt{2}}{8}\mathbf{j}\right) + \left(-\frac{1}{2}\mathbf{i} + 3\,\mathbf{j}\right) = $

$\left(\frac{5\sqrt{2} - 1}{2}\right)\mathbf{i} + \left(\frac{24 - 3\sqrt{2}}{8}\right)\mathbf{j}.$

[55]   Let **u**, **v** and **w** represent the instrument velocity, actual velocity, and wind velocity vectors, respectively (see Figure 55). We must find the magnitude and direction of the actual velocity vector. We see from figure 55 that the direction angle for **u** is $270° + 48.6° = 318.6°$ and the direction angle for **w** is $90° - 12.8° = 77.2°$. Thus, **u** and **w** have the following algebraic representations: **u** = $\langle 565 \cos 318.6°, 565 \sin 318.6° \rangle \approx \langle 423.8, -373.6 \rangle$ and **w** = $\langle 48 \cos 77.2°, 48 \sin 77.2° \rangle \approx \langle 10.6, 46.8 \rangle$. Thus **v** = **u** + **w** $\approx \langle 434.4, -326.8 \rangle$ and the magnitude of **v** is $\|\mathbf{v}\| \approx \sqrt{(434.4)^2 + (-326.8)^2} \approx 544$. If θ is the direction angle for **v**, then $\tan \theta \approx \dfrac{-326.8}{434.4} \approx -0.7523$. If θ' is the reference angle for θ, then $\tan \theta' = |\tan \theta| \approx 0.7523 \Rightarrow \theta' \approx \tan^{-1}(0.7523) \approx 37.0°$. Since θ is in QIV, $\theta = 360° - \theta' \approx 323°$. To find the direction of **v**, we subtract $270°$ from θ to get $53°$. Thus, the direction of **v** is approximately S 53° E and its magnitude is approximately 544 miles per hour.

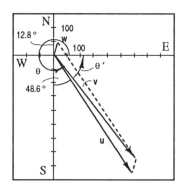   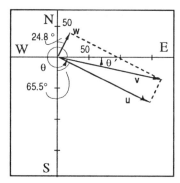

Figure 55   **Figure 56**

[56]   Let **u**, **v** and **w** represent the instrument velocity, actual velocity, and wind velocity vectors, respectively (see Figure 56). We must find the magnitude and direction of the actual velocity vector **v**. Note that wind from the direction s 24.8° W blows in the direction N 24.8° E. We see from figure 56 that the direction angle for **u** is $270° + 65.5° = 335.5°$ while the direction angle for **w** is $90° - 24.8° = 62.2°$. Thus, **u** and **w** have the following algebraic representations: **u** = $\langle 215 \cos 335.5°, 215 \sin 335.5° \rangle \approx \langle 195.6, -89.2 \rangle$ and **w** = $\langle 45 \cos 65.2°, 45 \sin 65.2° \rangle \approx \langle 18.9, 40.8 \rangle$. Thus **v** = **u** + **w** $\approx \langle 214.5, -48.4 \rangle$ and the magnitude of **v** is $\|\mathbf{v}\| \approx \sqrt{(214.5)^2 + (-48.4)^2} \approx 220$. If θ is the direction angle for **v**, then $\tan \theta \approx \dfrac{-48.4}{214.5} \approx -0.2256$. If θ' is the reference angle for θ, then $\tan \theta' = |\tan \theta| \approx 0.2256 \Rightarrow \theta' \approx \tan^{-1}(0.2256) \approx 12.7°$. Since θ is in QIV, $\theta = 360° - \theta' \approx 347.3°$. To find the direction of **v**, we subtract $270°$ from θ to get $77.3°$. Thus, the downed aircraft is approximately 220 miles from the airport on a bearing of S 77.3° E.

[57]   Let **u**, **v** and **w** denote the instrument velocity, actual velocity, and wind velocity vectors, respectively (see Figure 57). We note that the magnitude of the actual velocity vector **v** is $\frac{714}{3} =$ 238 miles per hour. We must find the magnitude and direction of the instrument velocity vector **u**. We see from Figure 57 that the direction angle of **w** is $90° + 24.3° = 114.3°$ and the direction angle for **v** is $270° − 71.2° = 198.8°$. Thus, **w** and **v** have the following algebraic representations $\mathbf{w} = \langle 55 \cos 114.3°, 55 \sin 114.3° \rangle \approx \langle -22.6, 50.1 \rangle$ and $\mathbf{v} = \langle 238 \cos 198.8°, 238 \sin 198.8° \rangle \approx \langle -225.3, -76.7 \rangle$. Since $\mathbf{v} = \mathbf{u} + \mathbf{w}$, it follows that $\mathbf{u} = \mathbf{v} - \mathbf{w} \approx \langle -202.7, -126.8 \rangle$. Thus, the magnitude of **u** is $|\mathbf{u}| \approx \sqrt{(-202.7)^2 + (-126.8)^2} \approx 239$. If θ is the direction angle for **u**, then $\tan \theta \approx \frac{-126.8}{-202.7} \approx 0.6256$. If θ' is the reference angle for θ, then $\tan \theta' = |\tan \theta| \approx 0.6256 \Rightarrow \theta' \approx \tan^{-1}(0.6256) \approx 32.0°$. Since θ is in QIII, $\theta = 180° + \theta' \approx 212°$. To find the direction of **u**, we subtract θ from $270°$ to get $58°$. Thus, the pilot should maintain an airspeed of approximately 239 miles per hour on a bearing of approximately S 58° W.

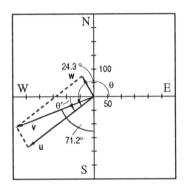

Figure 57   **Figure 58**

[58]   Let **u**, **v** and **w** represent the instrument velocity, actual velocity, and wind velocity vectors, respectively (see Figure 58). We must determine the magnitude and direction of the wind velocity vector **w**. We see from Figure 58 that the direction angle of **u** is $90° + 25.8° = 115.8°$ and the direction angle for **v** is $90° + 17.6° = 107.6°$. Thus, **u** and **v** have the following algebraic representations $\mathbf{u} = \langle 710 \cos 115.8°, 710 \sin 115.8° \rangle \approx \langle -309.0, 639.2 \rangle$ and $\mathbf{v} = \langle 715 \cos 107.6°, 715 \sin 107.6° \rangle \approx \langle -216.2, 681.5 \rangle$. Since $\mathbf{v} = \mathbf{u} + \mathbf{w}$, $\mathbf{w} = \mathbf{v} - \mathbf{u} \approx \langle 92.8, 42.3 \rangle$ and the magnitude of **w** is $|\mathbf{w}| \approx \sqrt{(92.8)^2 + (42.3)^2} \approx 102$. If θ is the direction angle for **w**, then $\tan \theta \approx \frac{42.3}{92.8} \approx 0.4558$. Since θ is in QI, $\theta \approx \tan^{-1}(0.4558) \approx 24.5°$. To find the direction of **w**, we subtract θ from $90°$, obtaining $65.5°$. Thus, the wind is blowing approximately 102 miles per hour in the (approximate) direction N 65.5° E.

[59]   Let **u** and **v** represent the boat's intended and actual velocity vectors and let **w** represent the current's velocity vector (see Figure 59). We see from Figure 59 that $\tan \alpha = \frac{|\mathbf{w}|}{|\mathbf{u}|} = \frac{4.5}{12} = 0.3750$. Since α is acute, $\alpha = \tan^{-1}(0.3750) \approx 21°$.

[60] Let **u** and **v** denote the intended and actual velocity vectors of the boat and let **w** denote the velocity vector of the current (see Figure 60). We see from Figure 60 that $\sin 11.6° = \dfrac{|\mathbf{w}|}{|\mathbf{u}|}$. Thus, $|\mathbf{w}| = |\mathbf{u}| \sin 11.6° = 25 \sin 11.6° \approx 5$, and the speed of the current is approximately 5mph.

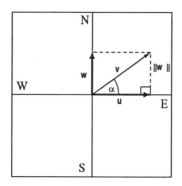

Figure 59 **Figure 60**

[61] Let **u** be the force pulling the truck down the incline, **w** the weight of the truck being pulled downward by the force of gravity and **v** the force pressing the truck into the incline (see Figure 61). We see from right ΔTBC in Figure 61 that $\sin 12.5° = \dfrac{|\mathbf{u}|}{|\mathbf{w}|}$. Thus, $|\mathbf{u}| = |\mathbf{w}| \sin 12.5° = 4850 \sin 12.5° \approx 1050$, so that a force of approximately 1050 pounds is required to keep the truck from rolling down the incline.

[62] Let α denote the angle of the incline. Let **u** denote the force required to keep the tractor from rolling down the incline, **w** the weight of the tractor being pulled downward by the force of gravity and **v** the force pushing the tractor into the incline (see figure 62). We see from the figure that, in right ΔTBC, $\sin \alpha = \dfrac{|u|}{|w|}$. Thus, $\sin \alpha = \dfrac{115}{670} \approx 0.1716 \Rightarrow \alpha \approx \sin^{-1}(0.1716) \approx 9.9°$. This means that the angle of the ramp is approximately 9.9°.

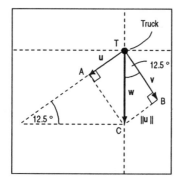
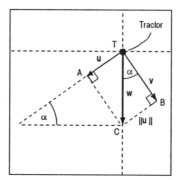

Figure 61 **Figure 62**

[63] Let **u** represent the force required to keep the wheelbarrow from rolling down the incline, **w** the weight of the wheelbarrow being pulled directly downward by the force of gravity and **v** the force pressing the wheelbarrow into the incline (see figure 63). We see from right $\triangle WBC$ in the figure, $\sin 15° = \dfrac{|u|}{|w|}$. Thus, $|w| = \dfrac{|u|}{\sin 15°} = \dfrac{85}{\sin 15°} \approx 330$, and the weight of the wheelbarrow full of dirt is approximately 330 pounds.

[64] Let **u** denote the force required to keep the boat and trailer from rolling down the incline, **w** denote the weight of the boat and trailer being pulled directly downward by the force of gravity and let **v** denote the force pressing the boat and trailer into the incline (see figure 64). We see from right triangle TBC in the figure that $\sin 17.5° = \dfrac{|u|}{|w|}$. Thus, $|w| = \dfrac{|u|}{\sin 17.5°} = \dfrac{825}{\sin 17.5°} \approx 2740$, and the combined weight of the boat and trailer is approximately 2740 pounds.

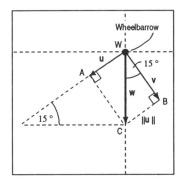

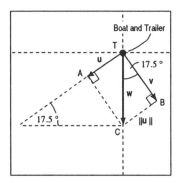

Figure 63　　　　　　　**Figure 64**

[65] If $\mathbf{F_1} + \mathbf{F_2} + \mathbf{F_3} = 0$, then $\mathbf{F_3} = -\mathbf{F_1} - \mathbf{F_2} = -(\mathbf{F_1} + \mathbf{F_2})$. This means that $\mathbf{F_3}$ has the same magnitude, but opposite direction, of the vector sum $\mathbf{F_1} + \mathbf{F_2}$. The vectors $\mathbf{F_1}$, $\mathbf{F_2}$ and $\mathbf{F_1} + \mathbf{F_2}$ are shown in Figure 65. We see from the figure that the direction angle for $\mathbf{F_1}$ is $270° - 42.9° = 227.1°$ and the direction angle for $\mathbf{F_2}$ is $270° + 26.7° = 296.7°$. Thus, $\mathbf{F_1}$ and $\mathbf{F_2}$ have the following algebraic representations $\mathbf{F_1} = \langle 150 \cos 227.1°, 150 \sin 227.1° \rangle \approx \langle -102.1, -109.9 \rangle$ and $\mathbf{F_2} = \langle 230 \cos 296.7°, 230 \sin 296.7° \rangle \approx \langle 103.3, -205.5 \rangle$. It follows that $\mathbf{F_1} + \mathbf{F_2} \approx \langle 1.2, -315.4 \rangle$ and the magnitude of $\mathbf{F_1} + \mathbf{F_2}$ is $|\mathbf{F_1} + \mathbf{F_2}| \approx \sqrt{(1.2)^2 + (-315.4)^2} \approx 315$. If θ is the direction angle for $\mathbf{F_1} + \mathbf{F_2}$, then $\tan \theta \approx \dfrac{-315.4}{1.2} \approx -262.8$. If θ' is the reference angle for θ, $\tan \theta' = |\tan \theta| \approx 262.8 \Rightarrow \theta' \approx \tan^{-1}(262.8) \approx 89.8°$. Since θ is in QIV, $\theta = 360° - \theta' \approx 270.2°$. To find the direction of $\mathbf{F_1} + \mathbf{F_2}$ we subtract $270°$ from θ and get $0.2°$. Thus, the direction of $\mathbf{F_1} + \mathbf{F_2}$ is approximately S $0.2°$ E. Since $\mathbf{F_3}$ has opposite direction, the direction of $\mathbf{F_3}$ is approximately N $0.2°$ W. The magnitude of $\mathbf{F_3}$ is the same as the magnitude of $\mathbf{F_1} + \mathbf{F_2}$, or approximately 315 pounds.

[66] The vectors **u**, **v**, and **w** are shown in Figure 66. We see from the figure that the direction angle for **v** is $180° - 30° = 150°$. The vectors **u**, **v** and **w** have the following algebraic representations $\mathbf{u} = \langle |\mathbf{u}|\cos 50°, |\mathbf{u}|\sin 50° \rangle$, $\mathbf{v} = \langle |\mathbf{v}|\cos 150°, |\mathbf{v}|\sin 150° \rangle$, and $\mathbf{w} = \langle 450\cos 270°, 450\sin 270° \rangle = \langle 450(0), 450(-1) \rangle = \langle 0, -450 \rangle$. Thus, the magnitude of **w** is

$|\mathbf{w}| = \sqrt{(0)^2 + (-450)^2} = 450$. Since the three vectors are in equilibrium, $\mathbf{u} + \mathbf{v} + \mathbf{w} = 0$, or

$\mathbf{u} + \mathbf{v} = -\mathbf{w}$. Thus, $\langle |\mathbf{u}|\cos 50° + |\mathbf{v}|\cos 150°, |\mathbf{u}|\sin 50° + |\mathbf{v}|\sin 150° \rangle = \langle 0, 450 \rangle \Rightarrow$

$|\mathbf{u}|\cos 50° + |\mathbf{v}|\cos 150° = 0 \left(*_1\right)$ and $|\mathbf{u}|\sin 50° + |\mathbf{v}|\sin 150° = 450 \left(*_2\right)$. Since

$\sin 150° = \frac{1}{2}$, equation $\left(*_2\right) \Rightarrow \frac{1}{2}|\mathbf{v}| = 450 - |\mathbf{u}|\sin 50° \Rightarrow |\mathbf{v}| = 900 - 2|\mathbf{u}|\sin 50°$.

Substituting $900 - 2|\mathbf{u}|\sin 50°$ for $|\mathbf{v}|$ in equation $\left(*_1\right) \Rightarrow$

$|\mathbf{u}|\cos 50° + (900 - 2|\mathbf{u}|\sin 50°)\cos 150° = 0 \Rightarrow$

$|\mathbf{u}|\cos 50° + (900 - 2|\mathbf{u}|\sin 50°)\left(-\frac{\sqrt{3}}{2}\right) = 0 \Rightarrow |\mathbf{u}|\cos 50° - 450\sqrt{3} + \sqrt{3}|\mathbf{u}|\sin 50° =$

$0 \Rightarrow |\mathbf{u}|\cos 50° + \sqrt{3}|\mathbf{u}|\sin 50° = 450\sqrt{3} \Rightarrow (\cos 50° + \sqrt{3}\sin 50°)|\mathbf{u}| = 450\sqrt{3} \Rightarrow |\mathbf{u}| =$

$\dfrac{450\sqrt{3}}{\cos 50° + \sqrt{3}\sin 50°} \approx 396$. Thus, as noted earlier, $|\mathbf{v}| = 900 - 2|\mathbf{u}|\sin 50° \approx 293$. The

tension in cable **u** is approximately 396 pounds, and the tension in cable **v** is approximately 293 pounds. Of course, the tension in cable **w** is 450 pounds.

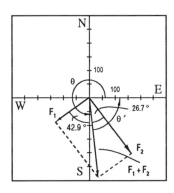

Figure 65

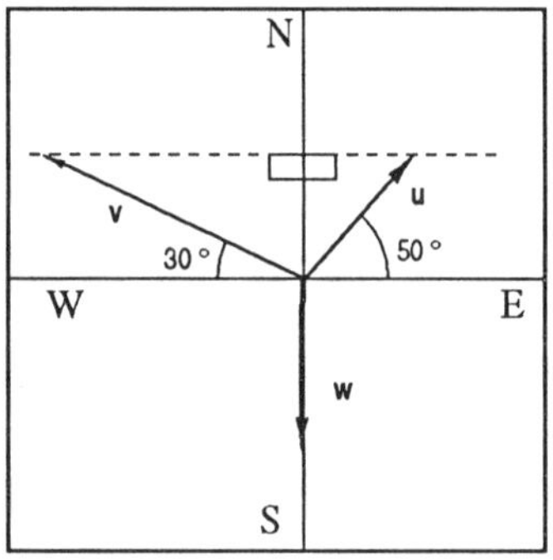

Figure 66

[67] a) $\mathbf{u} \cdot \mathbf{v} = (-4)(2) + (1)(-3) = -11.$ b) $\mathbf{u} \cdot \mathbf{v} = (0)(-4) + (-8)\left(\frac{1}{2}\right) = -4.$

c) $\mathbf{u} \cdot \mathbf{v} = \left(-\frac{5}{2}\right)(6) + \left(-\frac{4}{3}\right)(0) = -15.$

[68] Let $\mathbf{v} = \langle v_1, v_2 \rangle$ be any nonzero vector. Since at least on of v_1 and v_2 is not equal to zero, $v_1^2 + v_2^2 > 0$. Thus $\|\mathbf{v}\| = \sqrt{v_1^2 + v_2^2} > 0$.

[69] a) $|\mathbf{u}| = \sqrt{(-4)^2 + (1)^2} = \sqrt{17}$ and $|\mathbf{v}| = \sqrt{(2)^2 + (-3)^2} = \sqrt{13}$. $\cos \alpha = \dfrac{\mathbf{u} \cdot \mathbf{v}}{|\mathbf{u}||\mathbf{v}|} =$

$\dfrac{-11}{(\sqrt{17})(\sqrt{13})} \approx -0.7399$. Thus, $\alpha \approx \cos^{-1}(-0.7399) \approx 138°$.

b) $|\mathbf{u}| = \sqrt{(0)^2 + (-8)^2} = 8$ and $|\mathbf{v}| = \sqrt{(-4)^2 + \left(\frac{1}{2}\right)^2} = \frac{\sqrt{65}}{2}$. $\cos\alpha = \dfrac{\mathbf{u}\cdot\mathbf{v}}{|\mathbf{u}||\mathbf{v}|} =$

$\dfrac{-4}{(8)\left(\frac{\sqrt{65}}{2}\right)} \approx -0.1240$. Thus, $\alpha \approx \cos^{-1}(-0.1240) \approx 97.1°$.

c) $|\mathbf{u}| = \sqrt{\left(-\frac{5}{2}\right)^2 + \left(-\frac{4}{3}\right)^2} = \frac{17}{6}$ and $|\mathbf{v}| = \sqrt{(6)^2 + (0)^2} = 6$. $\cos\alpha = \dfrac{\mathbf{u}\cdot\mathbf{v}}{|\mathbf{u}||\mathbf{v}|} = \dfrac{-15}{\left(\frac{17}{6}\right)(6)} \approx$

-0.8824. Thus, $\alpha \approx \cos^{-1}(-0.8824) \approx 152°$.

[70]   Let $\mathbf{u} = \langle u_1, u_2 \rangle$, $\mathbf{v} = \langle v_1, v_2 \rangle$, $\mathbf{w} = \langle w_1, w_2 \rangle$ be vectors and let a and b be real numbers.

(ii) $(\mathbf{u} + \mathbf{v}) + \mathbf{w} = \langle u_1 + v_1, u_2 + v_2 \rangle + \langle w_1, w_2 \rangle = \langle u_1 + v_1 + w_1, u_2 + v_2 + w_2 \rangle$, and $\mathbf{u} + (\mathbf{v} + \mathbf{w}) =$

$\langle u_1, u_2 \rangle + \langle v_1 + w_1, v_2 + w_2 \rangle = \langle u_1 + v_1 + w_1, u_2 + v_2 + w_2 \rangle$. *Thus*, $(u + v) + w = u + (v + w)$.

(iii) $(ab)\mathbf{v} = \langle ab\,v_1, ab\,v_2 \rangle$ and $a(b\mathbf{v}) = a\langle b\,v_1, b\,v_2 \rangle = \langle a\,b\,v_1, a\,b\,v_2 \rangle$. Thus $(ab)\mathbf{v} = a(b\mathbf{v})$.

(iv) $a(\mathbf{u} + \mathbf{v}) = a\langle u_1 + v_1, u_2 + v_2 \rangle = \langle a\,u_1 + a\,v_1, a\,u_2 + a\,v_2 \rangle$, and $a\mathbf{u} + a\mathbf{v} =$

$\langle a\,u_1, a\,u_2 \rangle + \langle a\,v_1, a\,v_2 \rangle = \langle a\,u_1 + a\,v_1, a\,u_2 + a\,v_2 \rangle$. Thus $a(\mathbf{u} + \mathbf{v}) = a\mathbf{u} + a\mathbf{v}$.

(v) $(a + b)\mathbf{v} = \langle (a + b)\,v_1, (a + b)\,v_2 \rangle = \langle a\,v_1 + b\,v_1, a\,v_2 + b\,v_2 \rangle$ and $a\mathbf{v} + b\mathbf{v} =$

$\langle a\,v_1, a\,v_2 \rangle + \langle b\,v_1, b\,v_2 \rangle = \langle a\,v_1 + b\,v_1, a\,v_2 + b\,v_2 \rangle$. Thus, $(a + b)\mathbf{v} = a\mathbf{v} + b\mathbf{v}$.

(vi) $\mathbf{v} + 0 = \langle v_1, v_2 \rangle + \langle 0, 0 \rangle = \langle v_1 + 0, v_2 + 0 \rangle = \langle v_1, v_2 \rangle = \mathbf{v}$.

(vii) $1(\mathbf{v}) = 1\langle v_1, v_2 \rangle = \langle 1 \cdot v_1, 1 \cdot v_2 \rangle = \langle v_1, v_2 \rangle = \mathbf{v}$.

(viii) $\mathbf{v} + (-\mathbf{v}) = \langle v_1, v_2 \rangle + \langle -v_1, -v_2 \rangle = \langle v_1 - v_1, v_2 - v_2 \rangle = \langle 0, 0 \rangle = 0$.

(ix) $0(\mathbf{v}) = 0\langle v_1, v_2 \rangle = \langle 0 \cdot v_1, 0 \cdot v_2 \rangle = \langle 0, 0 \rangle = 0$.

(x) $a(0) = a\langle 0, 0 \rangle = \langle a \cdot 0, a \cdot 0 \rangle = \langle 0, 0 \rangle = 0$.

[71]   a) $A = (-3, 4)$, $B = (5, -8)$. If $\mathbf{v} = \overline{AB}$, then $\mathbf{v} = \langle 5 - (-3), -8 - 4 \rangle = \langle 8, -12 \rangle$.

     b) $A = \left(\frac{5}{12}, -2\right)$, $B = \left(7, -\frac{3}{4}\right)$. If $\mathbf{v} = \overline{AB}$, then $\mathbf{v} = \left\langle 7 - \frac{5}{12}, -\frac{3}{4} + 2 \right\rangle = \left\langle \frac{79}{12}, \frac{5}{4} \right\rangle$.

[72]   a) $\mathbf{u} = \dfrac{1}{|\mathbf{v}|} \cdot \mathbf{v} = \dfrac{1}{\sqrt{v_1^2 + v_2^2}} \cdot \langle v_1, v_2 \rangle = \left\langle \dfrac{v_1}{\sqrt{v_1^2 + v_2^2}}, \dfrac{v_2}{\sqrt{v_1^2 + v_2^2}} \right\rangle$. Thus, $|\mathbf{u}| =$

$\sqrt{\left(\dfrac{v_1}{\sqrt{v_1^2 + v_2^2}}\right)^2 + \left(\dfrac{v_2}{\sqrt{v_1^2 + v_2^2}}\right)^2} = \sqrt{\dfrac{v_1^2}{v_1^2 + v_2^2} + \dfrac{v_2^2}{v_1^2 + v_2^2}} = \sqrt{\dfrac{v_1^2 + v_2^2}{v_1^2 + v_2^2}} = \sqrt{1} = 1$.

Thus $\mathbf{u}$ is a unit vector.

b)   1. $\mathbf{v} = \langle 2, -3 \rangle$. $\mathbf{u} = \dfrac{1}{|\mathbf{v}|} \cdot \mathbf{v} = \dfrac{1}{\sqrt{13}} \langle 2, -3 \rangle = \left\langle \dfrac{2}{\sqrt{13}}, -\dfrac{3}{\sqrt{13}} \right\rangle$.

2. $\mathbf{v} = \langle -1, 4 \rangle$. $\mathbf{u} = \dfrac{1}{|\mathbf{v}|} \cdot \mathbf{v} = \dfrac{1}{\sqrt{17}} \langle -1, 4 \rangle = \left\langle -\dfrac{1}{\sqrt{17}}, \dfrac{4}{\sqrt{17}} \right\rangle$.

3. $\mathbf{v} = \left\langle -\dfrac{3}{2}, -\dfrac{7}{4} \right\rangle$. $\mathbf{u} = \dfrac{1}{|\mathbf{v}|} \cdot \mathbf{v} = \dfrac{1}{\frac{\sqrt{85}}{4}} \left\langle -\dfrac{3}{2}, -\dfrac{7}{4} \right\rangle = \left\langle -\dfrac{6}{\sqrt{85}}, -\dfrac{7}{\sqrt{85}} \right\rangle$.

4. $\mathbf{v} = \left\langle -\dfrac{5}{2}, -\dfrac{3}{4} \right\rangle$. $\mathbf{u} = \dfrac{1}{|\mathbf{v}|} \cdot \mathbf{v} = \dfrac{1}{\frac{\sqrt{109}}{4}} \left\langle -\dfrac{5}{2}, -\dfrac{3}{4} \right\rangle = \left\langle -\dfrac{10}{\sqrt{109}}, -\dfrac{3}{\sqrt{109}} \right\rangle$.

5. $\mathbf{v} = \langle 0, -6 \rangle$. $\mathbf{u} = \dfrac{1}{|\mathbf{v}|} \cdot \mathbf{v} = \dfrac{1}{6} \langle 0, -6 \rangle = \langle 0, -1 \rangle$.

6. $\mathbf{v} = \langle -5, 0 \rangle$. $\mathbf{u} = \dfrac{1}{|\mathbf{v}|} \cdot \mathbf{v} = \dfrac{1}{5} \langle -5, 0 \rangle = \langle -1, 0 \rangle$.

7. $\mathbf{v} = \langle 2\sqrt{3}, 4 \rangle$. $\mathbf{u} = \dfrac{1}{|\mathbf{v}|} \cdot \mathbf{v} = \dfrac{1}{2\sqrt{7}} \langle 2\sqrt{3}, 4 \rangle = \left\langle \dfrac{\sqrt{3}}{\sqrt{7}}, \dfrac{2}{\sqrt{7}} \right\rangle$.

8. $\mathbf{v} = \left\langle -\dfrac{\sqrt{3}}{2}, \dfrac{1}{2} \right\rangle$. $\mathbf{u} = \dfrac{1}{|\mathbf{v}|} \cdot \mathbf{v} = \dfrac{1}{1} \left\langle -\dfrac{\sqrt{3}}{2}, \dfrac{1}{2} \right\rangle = \left\langle -\dfrac{\sqrt{3}}{2}, \dfrac{1}{2} \right\rangle$.

EXERCISES 7.6

NOTE : In Exercises 13 – 32, the equation $(r, \theta) = (x, y)$ means that if the polar coordinates of the point P are (r, θ), then the rectangular coordinates of P are (x, y).

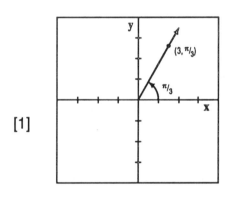

[1]

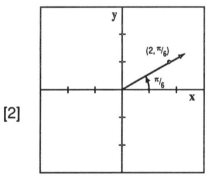

[2]

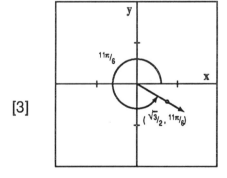

[3]

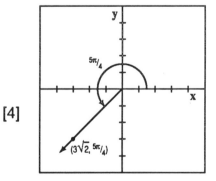

[4]

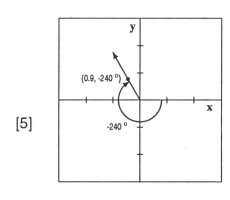

[5]

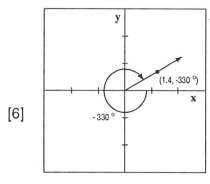

[6]

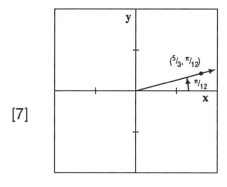

[7]

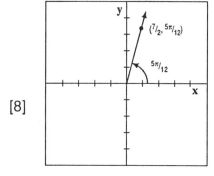

[8]

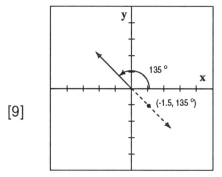

[9]

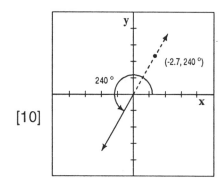

[10]

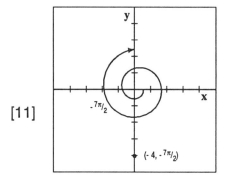

[11]

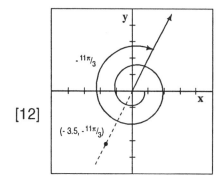

[12]

[13] $\quad x = r \cos \theta = 2 \cos \frac{\pi}{4} = 2 \left(\frac{\sqrt{2}}{2} \right) = \sqrt{2}$ and $y = r \sin \theta = 2 \sin \frac{\pi}{4} = 2 \left(\frac{\sqrt{2}}{2} \right) = \sqrt{2} \Rightarrow \left(2, \frac{\pi}{4} \right) = \left(\sqrt{2}, \sqrt{2} \right).$

[14] $\quad x = 5 \cos \frac{\pi}{3} = 5 \left(\frac{1}{2} \right) = \frac{5}{2}$ and $y = 5 \sin \frac{\pi}{3} = 5 \left(\frac{\sqrt{3}}{2} \right) = \frac{5\sqrt{3}}{2} \Rightarrow \left(5, \frac{\pi}{3} \right) = \left(\frac{5}{2}, \frac{5\sqrt{3}}{2} \right).$

[15] $x = \sqrt{5}\,\cos\dfrac{5\pi}{3} = \sqrt{5}\left(\dfrac{1}{2}\right) = \dfrac{\sqrt{5}}{2}$ and $y = \sqrt{5}\,\sin\dfrac{5\pi}{3} = \sqrt{5}\left(-\dfrac{\sqrt{3}}{2}\right) = -\dfrac{\sqrt{15}}{2} \Rightarrow$

$\left(\sqrt{5},\dfrac{5\pi}{3}\right) = \left(\dfrac{\sqrt{5}}{2},\,-\dfrac{\sqrt{15}}{2}\right).$

[16] $x = \dfrac{\sqrt{3}}{2}\cos\dfrac{7\pi}{6} = \dfrac{\sqrt{3}}{2}\left(-\dfrac{\sqrt{3}}{2}\right) = -\dfrac{3}{4}$ and $y = \dfrac{\sqrt{3}}{2}\sin\dfrac{7\pi}{6} = \dfrac{\sqrt{3}}{2}\left(-\dfrac{1}{2}\right) = -\dfrac{\sqrt{3}}{4} \Rightarrow$

$\left(\dfrac{\sqrt{3}}{2},\dfrac{7\pi}{6}\right) = \left(-\dfrac{3}{4},\,-\dfrac{\sqrt{3}}{4}\right).$

[17] $\cos 75° = \cos(45° + 30°) = \dfrac{\sqrt{6} - \sqrt{2}}{4}$ and $\sin 75° = \sin(45° + 30°) = \dfrac{\sqrt{6} + \sqrt{2}}{4}$. Thus, $x = 4\cos 75° =$

$\sqrt{6} - \sqrt{2}$ and $y = 4\sin 75° = \sqrt{6} + \sqrt{2}$, which means that $(4, 75°) = (\sqrt{6} - \sqrt{2},\, \sqrt{6} + \sqrt{2})$.

[18] $\cos 15° = \cos(45° - 30°) = \dfrac{\sqrt{6} + \sqrt{2}}{4}$ and $\sin 15° = \sin(45° - 30°) = \dfrac{\sqrt{6} - \sqrt{2}}{4}$. Thus, $x = 2\cos 15° =$

$\dfrac{\sqrt{6} + \sqrt{2}}{2}$ and $y = 2\sin 15° = \dfrac{\sqrt{6} - \sqrt{2}}{2}$, which means that $(2, 15°) = \left(\dfrac{\sqrt{6} + \sqrt{2}}{2},\, \dfrac{\sqrt{6} - \sqrt{2}}{+}\right).$

[19] $x = -3\cos 630° = (-3)(0) = 0$ and $y = -3\sin 630° = (-3)(-1) = 3$. Thus, $(-3, 630°) = (0, 3)$.

[20] $x = -4\cos 135° = -4\left(-\dfrac{\sqrt{2}}{2}\right) = 2\sqrt{2}$ and $y = -4\sin 135° = -4\left(\dfrac{\sqrt{2}}{2}\right) = -2\sqrt{2} \Rightarrow$

$(-4, 135°) = (2\sqrt{2},\, -2\sqrt{2}).$

[21] $x = \left(-\dfrac{8}{5}\right)\cos\left(-\dfrac{5\pi}{4}\right) = \left(-\dfrac{8}{5}\right)\left(-\dfrac{\sqrt{2}}{2}\right) = \dfrac{4\sqrt{2}}{5}$ and $y = \left(-\dfrac{8}{5}\right)\sin\left(-\dfrac{5\pi}{4}\right) = \left(-\dfrac{8}{5}\right)\left(\dfrac{\sqrt{2}}{2}\right) = -\dfrac{4\sqrt{2}}{5} \Rightarrow$

$\left(-\dfrac{8}{5}, -\dfrac{5\pi}{4}\right) = \left(\dfrac{4\sqrt{2}}{5},\, -\dfrac{4\sqrt{2}}{5}\right).$

[22] $x = \left(-\dfrac{13}{4}\right)\cos\left(-\dfrac{7\pi}{2}\right) = \left(-\dfrac{13}{4}\right)(0) = 0$ and $y = \left(-\dfrac{13}{4}\right)\sin\left(-\dfrac{7\pi}{2}\right) = \left(-\dfrac{13}{4}\right)(1) = -\dfrac{13}{4} \Rightarrow$

$\left(-\dfrac{13}{4}, -\dfrac{7\pi}{2}\right) = \left(0, -\dfrac{13}{4}\right).$

[23] $r = \sqrt{(2\sqrt{3})^2 + (2)^2} = 4$. $\tan\theta = \dfrac{y}{x} = \dfrac{1}{\sqrt{3}}$ and θ in QI $\Rightarrow \theta = \dfrac{\pi}{6}$. Thus, $(2\sqrt{3}, 2) = \left(4, \dfrac{\pi}{6}\right).$

[24] $r = \sqrt{(3)^2 + (-3)^2} = 3\sqrt{2}$. $\tan\theta = \dfrac{-3}{3} = -1$ and θ in QIV $\Rightarrow \theta = \dfrac{7\pi}{4}$. Thus, $(3, -3) = \left(3\sqrt{2}, \dfrac{7\pi}{4}\right).$

[25] $r = \sqrt{(-4)^2 + (0)^2} = 4$, and $\theta = \pi \Rightarrow (-4, 0) = (4, \pi)$.

[26] $r = \sqrt{(0)^2 + (5)^2} = 5$, and $\theta = \dfrac{\pi}{2} \Rightarrow (0, 5) = \left(5, \dfrac{\pi}{2}\right).$

[27] $r = \sqrt{(3)^2 + (-4)^2} = 5$. $\tan\theta = \dfrac{-4}{3} = -\dfrac{4}{3}$. If θ' is the reference angle for θ, $\tan\theta' = |\tan\theta| = \dfrac{4}{3} \Rightarrow$

$\theta' = \tan^{-1}\dfrac{4}{3} \approx 53°$. Since θ is in QIV, $\theta = 360° - \theta' = 360° - \tan^{-1}\dfrac{4}{3} \approx 307°$. Thus, $(3, -4) =$

$\left(5, 360° - \tan^{-1}\dfrac{4}{3}\right) \approx (5, 307°).$

[28] $r = \sqrt{(-2)^2 + (3)^2} = \sqrt{13}$. $\tan\theta = \dfrac{3}{-2} = -\dfrac{3}{2}$. If θ' is the reference angle for θ, $\tan\theta' = |\tan\theta| = \dfrac{3}{2} \Rightarrow$

$\theta' = \tan^{-1}\dfrac{3}{2} \approx 56°$. Since θ is in QII, $\theta = 180° - \theta' \approx 124°$. Thus, $(-2, 3) =$

$\left(\sqrt{13},\ 180° - \tan^{-1}\dfrac{3}{2}\right) \approx (3.6,\ 124°)$.

[29] $r = \sqrt{(-1)^2 + \left(-\dfrac{\sqrt{2}}{2}\right)^2} = \dfrac{\sqrt{6}}{2}$. $\tan\theta = \dfrac{-\dfrac{\sqrt{2}}{2}}{-1} = \dfrac{\sqrt{2}}{2}$. If θ' is the reference angle for θ, $\tan\theta' =$

$|\tan\theta| = \dfrac{\sqrt{2}}{2} \Rightarrow \theta' = \tan^{-1}\dfrac{\sqrt{2}}{2} \approx 35°$. Since θ is in QIII, $\theta = 180° + \theta' \approx 215°$. Thus, $\left(-1, -\dfrac{\sqrt{2}}{2}\right) =$

$\left(\dfrac{\sqrt{6}}{2},\ 180° + \tan^{-1}\dfrac{\sqrt{2}}{2}\right) \approx (1.2,\ 215°)$.

[30] $r = \sqrt{(-3\sqrt{2})^2 + (-2)^2} = \sqrt{22}$. $\tan\theta = \dfrac{-2}{-3\sqrt{2}} = \dfrac{\sqrt{2}}{3}$. If θ' is the reference angle for θ, $\tan\theta' =$

$|\tan\theta| = \dfrac{\sqrt{2}}{3} \Rightarrow \theta' = \tan^{-1}\dfrac{\sqrt{2}}{3} \approx 25°$. Since θ is in QIII, $\theta = 180° + \theta' \approx 205°$. Thus, $(-3\sqrt{2}, -2) =$

$\left(\sqrt{22},\ 180° + \tan^{-1}\dfrac{\sqrt{2}}{3}\right) \approx (4.7,\ 205°)$.

[31] $r = \sqrt{(2.7)^2 + (-3.6)^2} = 4.5$. $\tan\theta = \dfrac{-3.6}{2.7} = \dfrac{4}{3}$. If θ' is the reference angle for θ, then $\tan\theta' =$

$|\tan\theta| = \dfrac{4}{3} \Rightarrow \theta' = \tan^{-1}\left(\dfrac{4}{3}\right) \approx 53°$. Since θ is in QIV, $\theta = 360° - \theta' \approx 307°$. Thus $(2.7, -3.6) =$

$\left(4.5,\ 360° - \tan^{-1}\dfrac{4}{3}\right) \approx (4.5,\ 307°)$.

[32] $r = \sqrt{\left(-\dfrac{\pi}{3}\right)^2 + \left(\dfrac{\pi}{4}\right)^2} = \sqrt{\dfrac{25\pi^2}{144}} = \dfrac{5\pi}{12}$. $\tan\theta = \dfrac{\dfrac{\pi}{4}}{-\dfrac{\pi}{3}} = -\dfrac{3}{4}$. If θ' is the reference angle for θ, then

$\tan\theta' = |\tan\theta| = \dfrac{3}{4} \Rightarrow \theta' = \tan^{-1}\dfrac{3}{4} \approx 37°$. Since θ is in QII, $\theta = 180° - \theta' \approx 143°$. Thus, $\left(-\dfrac{\pi}{3}, \dfrac{\pi}{4}\right) =$

$\left(\dfrac{5\pi}{12},\ 180° - \tan^{-1}\dfrac{3}{4}\right) \approx (1.3,\ 143°)$.

[33] $\left(5, \dfrac{9\pi}{4}\right), \left(5, \dfrac{17\pi}{4}\right), \left(-5, \dfrac{5\pi}{4}\right), \left(-5, \dfrac{13\pi}{4}\right)$ (See figure below).

[34] $\left(2, \dfrac{7\pi}{3}\right), \left(2, \dfrac{13\pi}{3}\right), \left(-2, \dfrac{4\pi}{3}\right), \left(-2, \dfrac{10\pi}{3}\right)$ (See figure below).

[35] $\left(3, \dfrac{\pi}{6}\right), \left(3, \dfrac{13\pi}{6}\right), \left(-3, \dfrac{19\pi}{6}\right), \left(-3, \dfrac{31\pi}{6}\right)$ (See figure below).

[36] $\left(4, \dfrac{5\pi}{4}\right), \left(4, \dfrac{13\pi}{4}\right), \left(-4, \dfrac{9\pi}{4}\right), \left(-4, \dfrac{17\pi}{4}\right)$ (See figure below).

[37] $\left(2.5, \dfrac{\pi}{3}\right), \left(2.5, \dfrac{7\pi}{3}\right), \left(-2.5, -\dfrac{8\pi}{3}\right), \left(-2.5, -\dfrac{14\pi}{3}\right)$ (See figure below).

[38] $\left(5.2, \dfrac{\pi}{6}\right), \left(5.2, \dfrac{13\pi}{6}\right), \left(-5.2, -\dfrac{17\pi}{6}\right), \left(-5.2, -\dfrac{29\pi}{6}\right)$ (See figure below).

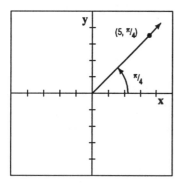

Figure 33

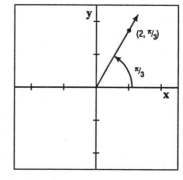

Figure 34

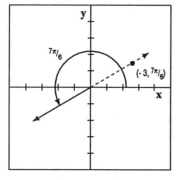

Figure 35

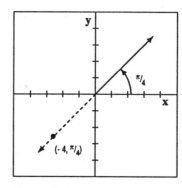

Figure 36

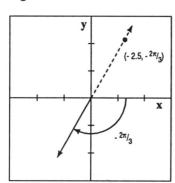

Figure 37

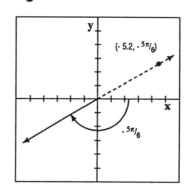

Figure 38

[39] a) $\left(-\sqrt{2}, \dfrac{5\pi}{3}\right)$ satisfies $r^2 - 4\cos\theta = 0$ since $\left(-\sqrt{2}\right)^2 - 4\cos\dfrac{5\pi}{3} = 2 - 4\left(\dfrac{1}{2}\right) = 0$.

b) $\left(2, \dfrac{3\pi}{2}\right)$ does not satisfy $r^2 - 4\cos\theta = 0$ since $(2)^2 - 4\cos\dfrac{3\pi}{2} = 4 - 4\,(0) = 4 \neq 0$.

c) $\left(\sqrt{2}, -\dfrac{7\pi}{3}\right)$ satisfies $r^2 - 4\cos\theta = 0$ since $\left(\sqrt{2}\right)^2 - 4\cos\left(-\dfrac{7\pi}{3}\right) = 2 - 4\left(\dfrac{1}{2}\right) = 0$.

[40] a) $\left(-\dfrac{3}{\sqrt{2}}, \dfrac{\pi}{6}\right)$ does not satisfy $r^2 = 9\sin 2\theta$ since $r^2 = \left(-\dfrac{3}{\sqrt{2}}\right)^2 = \dfrac{9}{2}$ and $9\sin 2\left(\dfrac{\pi}{6}\right) =$

$9\sin\dfrac{\pi}{3} = 9\left(\dfrac{\sqrt{3}}{2}\right) = \dfrac{9\sqrt{3}}{2}$.

338

b) $\left(3, \frac{5\pi}{4}\right)$ satisfies $r^2 = 9 \sin 2\theta$ since $r^2 = 9$ and $9 \sin 2\left(\frac{5\pi}{4}\right) = 9 \sin \frac{5\pi}{2} = 9(1) = 9$.

c) $\left(\frac{3}{\sqrt{2}}, -\frac{7\pi}{12}\right)$ satisfies $r^2 = 9 \sin 2\theta$ since $r^2 = \frac{9}{2}$ and $9 \sin 2\left(-\frac{7\pi}{12}\right) = 9 \sin\left(-\frac{7\pi}{6}\right) = 9\left(\frac{1}{2}\right) = \frac{9}{2}$.

[41] Since $r > 0$, we can square both sides to obtain $r^2 = 0$ or $x^2 + y^2 = 9$.

[42] Since $r > 0$, we can square both sides to obtain $r^2 = 25$ or $x^2 + y^2 = 25$.

[43] Since r can be any real number, the graph of $\theta = \frac{2\pi}{3}$ is the line L formed by the ray $\theta = \frac{2\pi}{3}$ and its

extension through the origin. When $r = 1$, the point $\left(1, \frac{2\pi}{3}\right)$ lies on L. Since $x = r \cos \theta =$

$1 \cdot \cos \frac{2\pi}{3} = -\frac{1}{2}$ and $y = r \sin \theta = 1 \cdot \sin \frac{2\pi}{3} = \frac{\sqrt{3}}{2}$, the point $(x, y) = \left(-\frac{1}{2}, \frac{\sqrt{3}}{2}\right)$ lies on L. Since

$(0, 0)$ is also a point on L, the slope of L is $m = \dfrac{\frac{\sqrt{3}}{2} - 0}{-\frac{1}{2} - 0} = -\sqrt{3}$. Thus, the slope–intercept equation

of L is $y = -\sqrt{3}\, x$.

[44] As in Exercise 43, the graph of $\theta = \frac{7\pi}{6}$ is the line L formed by the ray $\theta = \frac{7\pi}{6}$ and its extension

through the origin. When $r = 1$, the point $\left(1, \frac{7\pi}{6}\right)$ lies on L. Since $x = r \cos \theta = 1 \cos \frac{7\pi}{6} = -\frac{\sqrt{3}}{2}$ and

$y = r \sin \theta = 1 \sin \frac{7\pi}{6} = -\frac{1}{2}$, the point $(x, y) = \left(-\frac{\sqrt{3}}{2}, -\frac{1}{2}\right)$ lies on L. Since $(0, 0)$ also lies on L, the

slope of L is $m = \dfrac{-\frac{1}{2} - 0}{-\frac{\sqrt{3}}{2} - 0} = \frac{1}{\sqrt{3}}$. Thus, the slope–intercept equation of L is $y = \frac{1}{\sqrt{3}}\, x$.

[45] $r = \dfrac{5}{2 \cos \theta + \sin \theta} \Rightarrow 2\, r \cos \theta + r \sin \theta = 5 \Rightarrow 2\, x + y = 5$.

[46] $r = \dfrac{3}{\cos \theta - 5 \sin \theta} \Rightarrow r \cos \theta - 5\, r \sin \theta = 3 \Rightarrow x - 5\, y = 3$.

[47] $r \sin \theta = 4\, r^2 \cos^2 \theta - 1 \Rightarrow y = 4\, x^2 - 1$.

[48] $r^2 \cos^2 \theta = 5 + 2\, r \sin \theta \Rightarrow x^2 = 5 + 2\, y \Rightarrow y = \dfrac{x^2 - 5}{2}$.

[49] $r^2 - 9 = 8\, r \cos \theta \Rightarrow x^2 + y^2 - 9 = 8\, x \Rightarrow x^2 + y^2 - 8\, x - 9 = 0$.

[50] $r^2 = 3 - 2\, r \sin \theta \Rightarrow x^2 + y^2 = 3 - 2\, y \Rightarrow x^2 + y^2 + 2\, y - 3 = 0$.

[51] $2\, x\, y = 1 \Rightarrow 2 \left(r \cos \theta\right)\left(r \sin \theta\right) = 1 \Rightarrow r^2 \left(2 \sin \theta \cos \theta\right) = 1 \Rightarrow r^2 \sin 2\theta = 1 \Rightarrow r^2 = \dfrac{1}{\sin 2\theta}$.

[52] $x^3 = y^2 \Rightarrow \left(r \cos \theta\right)^3 = \left(r \sin \theta\right)^2 \Rightarrow r^3 \cos^3 \theta = r^2 \sin^2 \theta \Rightarrow r \cos^3 \theta = \sin^2 \theta$.

[53] $x^2 + 4\, y^2 = 4 \Rightarrow \left(r \cos \theta\right)^2 + 4 \left(r \sin \theta\right)^2 = 4 \Rightarrow r^2 \cos^2 \theta + 4\, r^2 \sin^2 \theta = 4 \Rightarrow$

$r^2 \left(1 - \sin^2 \theta\right) + 4\, r^2 \sin^2 \theta = 4 \Rightarrow r^2 + 3\, r^2 \sin^2 \theta = 4 \Rightarrow r^2 \left(1 + 3 \sin^2 \theta\right) = 4 \Rightarrow r^2 = \dfrac{4}{1 + 3 \sin^2 \theta}$.

[54] $9\, x^2 - 4\, y^2 = 36 \Rightarrow 9 \left(r \cos \theta\right)^2 - 4 \left(r \sin \theta\right)^2 = 36 \Rightarrow 9\, r^2 \cos^2 \theta - 4\, r^2 \sin^2 \theta = 36 \Rightarrow$

$9\, r^2 \left(1 - \sin^2 \theta\right) - 4\, r^2 \sin^2 \theta = 36 \Rightarrow 9r^2 - 13\, r^2 \sin^2 \theta = 36 \Rightarrow r^2 \left(9 - 13 \sin^2 \theta\right) = 36 \Rightarrow$

$r^2 = \dfrac{36}{9 - 13 \sin^2 \theta}$.

[55] $\left(x^2 + y^2\right)^{3/2} = 8\,x\,y \Rightarrow \left(r^2\right)^{3/2} = 8\left(r\cos\theta\right)\left(r\sin\theta\right) \Rightarrow r^3 = 8\,r^2 \sin\theta\cos\theta \Rightarrow r = 8\sin\theta\cos\theta \Rightarrow$
   $r = 4\left(2\sin\theta\cos\theta\right) \Rightarrow r = 4\sin 2\,\theta.$

[56] $\left(x^2 + y^2\right)^{3/2} - 18\,x\,y = 0 \Rightarrow \left(r^2\right)^{3/2} - 18\left(r\cos\theta\right)\left(r\sin\theta\right) = 0 \Rightarrow r^3 = 18\,r^2 \sin\theta\cos\theta \Rightarrow$
   $r = 18\sin\theta\cos\theta \Rightarrow r = 9\left(2\sin\theta\cos\theta\right) \Rightarrow r = 9\sin 2\,\theta.$

[57] $x^4 + y^4 = 2xy - 2x^2 y^2 \Rightarrow x^4 + 2x^2 y^2 + y^4 = 2xy \Rightarrow \left(x^2 + y^2\right)^2 = 2xy \Rightarrow \left(r^2\right)^2 = 2\left(r\cos\theta\right)\left(r\,\sin\theta\right) \Rightarrow$
   $r^4 = 2\,r^2 \sin\theta\cos\theta \Rightarrow r^2 = 2\sin\theta\cos\theta \Rightarrow r^2 = \sin 2\,\theta.$

[58] $x^4 + y^4 - 8x\,y + 2x^2 y^2 = 0 \Rightarrow x^4 + 2x^2 y^2 + y^4 = 8x\,y \Rightarrow \left(x^2 + y^2\right)^2 = 8x\,y \Rightarrow$
   $\left(r^2\right)^2 = 8\left(r\cos\theta\right)\left(r\sin\theta\right) \Rightarrow r^4 = 8\,r^2 \sin\theta\cos\theta \Rightarrow r^2 = 4\left(2\sin\theta\cos\theta\right) \Rightarrow r^2 = 4\sin 2\,\theta.$

[59]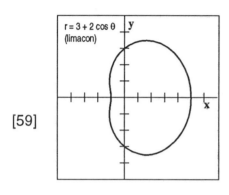

[60]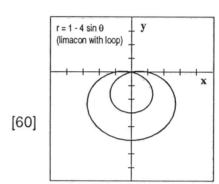

[61]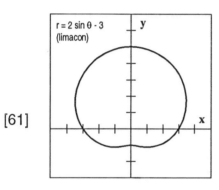

[62]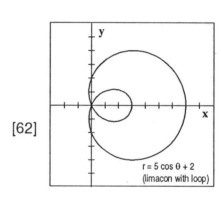

[63]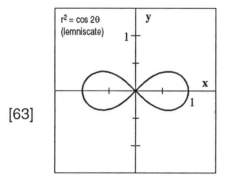

[64]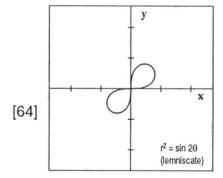

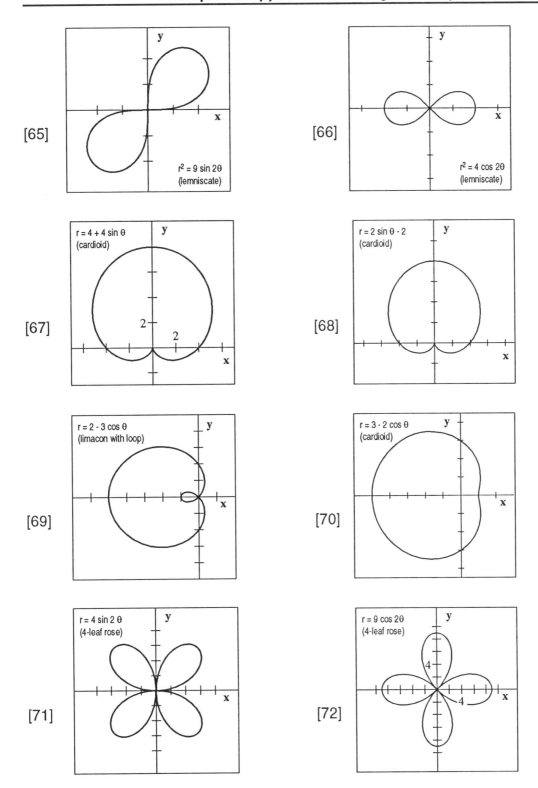

[65] $r^2 = 9 \sin 2\theta$ (lemniscate)

[66] $r^2 = 4 \cos 2\theta$ (lemniscate)

[67] $r = 4 + 4 \sin \theta$ (cardioid)

[68] $r = 2 \sin \theta - 2$ (cardioid)

[69] $r = 2 - 3 \cos \theta$ (limacon with loop)

[70] $r = 3 - 2 \cos \theta$ (cardioid)

[71] $r = 4 \sin 2\theta$ (4-leaf rose)

[72] $r = 9 \cos 2\theta$ (4-leaf rose)

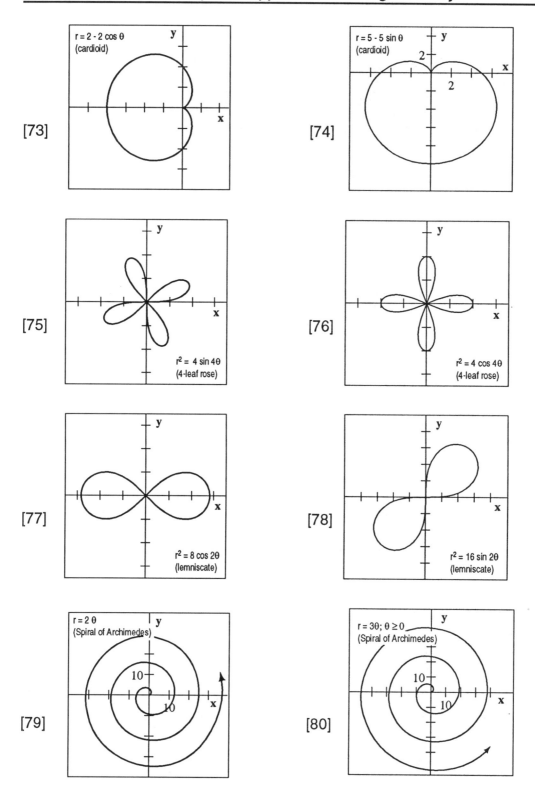

[73] $r = 2 - 2\cos\theta$ (cardioid)

[74] $r = 5 - 5\sin\theta$ (cardioid)

[75] $r^2 = 4\sin 4\theta$ (4-leaf rose)

[76] $r^2 = 4\cos 4\theta$ (4-leaf rose)

[77] $r^2 = 8\cos 2\theta$ (lemniscate)

[78] $r^2 = 16\sin 2\theta$ (lemniscate)

[79] $r = 2\theta$ (Spiral of Archimedes)

[80] $r = 3\theta;\ \theta \geq 0$ (Spiral of Archimedes)

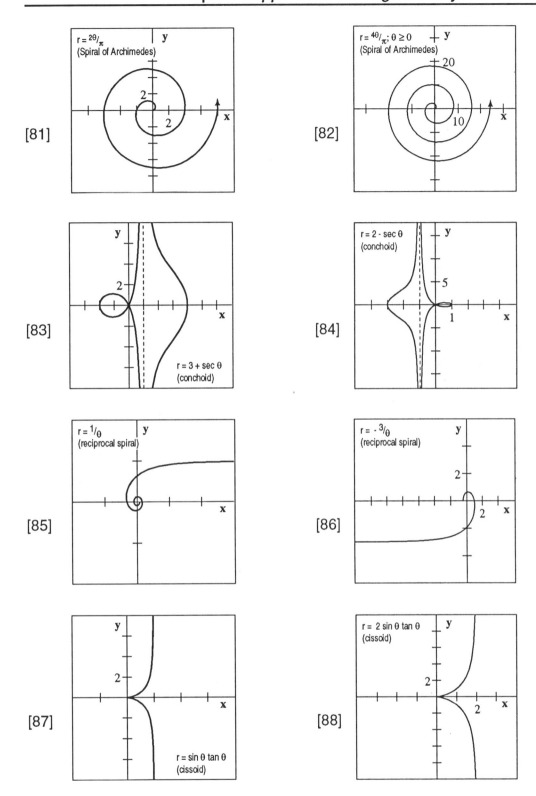

[81] $r = {}^{2\theta}/_{\pi}$ (Spiral of Archimedes)

[82] $r = {}^{4\theta}/_{\pi}; \theta \geq 0$ (Spiral of Archimedes)

[83] $r = 3 + \sec \theta$ (conchoid)

[84] $r = 2 - \sec \theta$ (conchoid)

[85] $r = {}^{1}/_{\theta}$ (reciprocal spiral)

[86] $r = -{}^{3}/_{\theta}$ (reciprocal spiral)

[87] $r = \sin \theta \tan \theta$ (cissoid)

[88] $r = 2 \sin \theta \tan \theta$ (cissoid)

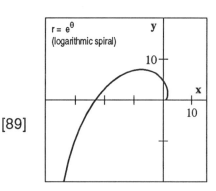

[89]

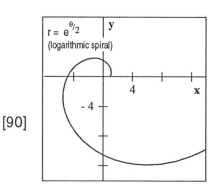

[90]

[91] $-r = 3 \sin 2 \left(-\theta\right) = 3 \sin \left(-2\,\theta\right) = -3 \sin 2\,\theta \Rightarrow r = 3 \sin \theta \Rightarrow$ graph of $r = 3 \sin \theta$ is symmetric with repsect to the y–axis.

[92]   We verify the result for the case illustrated in Figure 92 (the remaining cases can be verified in a similar manner). Applying the Law of Cosines to $\Delta P_1 O P_2$, we see that $\left[d\left(P_1, P_2\right)\right]^2 =$ $\left[d(O, P_1)\right]^2 + \left[d(O, P_2)\right]^2 - 2\left[d(O, P_1)\right]\left[d(O, P_2)\right] \cos\left(\theta_2 - \theta_1\right)$. Thus $d\left(P_1, P_2\right) = \sqrt{r_1^2 + r_2^2 - 2\,r_1\,r_2 \cos\left(\theta_2 - \theta_1\right)}$.

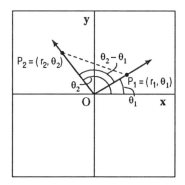

Figure 92

[93] $d\left(P_1, P_2\right) = \sqrt{\left(2\right)^2 + \left(-5\right)^2 - 2\left(2\right)\left(-5\right) \cos\left(\dfrac{4\,\pi}{3} - \dfrac{5\,\pi}{6}\right)} = \sqrt{29 + 20 \cos \dfrac{\pi}{2}} = \sqrt{29 + 20\left(0\right)} = \sqrt{29}$.

For $P_1 = \left(2, \dfrac{5\,\pi}{6}\right)$, $x = 2 \cos \dfrac{5\,\pi}{6} = 2\left(-\dfrac{\sqrt{3}}{2}\right) = -\sqrt{3}$ and $y = 2 \sin \dfrac{5\,\pi}{6} = 2\left(\dfrac{1}{2}\right) = 1$. Thus $P_1 = \left(-\sqrt{3},\, 1\right)$.

For $P_2 = \left(-5, \dfrac{4\,\pi}{3}\right)$, $x = -5 \cos \dfrac{4\,\pi}{3} = \left(-5\right)\left(-\dfrac{1}{2}\right) = \dfrac{5}{2}$ and $y = -5 \sin \dfrac{4\,\pi}{3} = \left(-5\right)\left(-\dfrac{\sqrt{3}}{2}\right) = \dfrac{5\sqrt{3}}{2}$.

Thus $P_2 = \left(\dfrac{5}{2}, \dfrac{5\sqrt{3}}{2}\right)$ and $d\left(P_1, P_2\right) = \sqrt{\left(\dfrac{5}{2} + \sqrt{3}\right)^2 + \left(\dfrac{5\sqrt{3}}{2} - 1\right)^2} =$

$\sqrt{\left(\dfrac{25}{4} + 5\sqrt{3} + 3\right) + \left(\dfrac{75}{4} - 5\sqrt{3} + 1\right)} = \sqrt{25 + 4} = \sqrt{29}$.

[94] $1 - \sin\theta = 1 + \cos\theta \Rightarrow \cos\theta = -\sin\theta \Rightarrow \theta = \frac{3\pi}{4}$ or $\theta = \frac{7\pi}{4}$ $(0 \le \theta < 2\pi)$. If $\theta = \frac{3\pi}{4}$, $r =$

$1 - \sin\frac{3\pi}{4} = 1 - \frac{\sqrt{2}}{2} = \frac{2 - \sqrt{2}}{2} \Rightarrow (r,\theta) = \left(\frac{2-\sqrt{2}}{2}, \frac{3\pi}{4}\right)$. If $\theta = \frac{7\pi}{4}$, $r = 1 - \sin\frac{7\pi}{4} = 1 + \frac{\sqrt{2}}{2} =$

$\frac{2 + \sqrt{2}}{2} \Rightarrow (r,\theta) = \left(\frac{2+\sqrt{2}}{2}, \frac{7\pi}{4}\right)$. We see from the graph shown in Figure 94 that the pole $(0,0)$ is

also a point of intersection of the two graphs. The pole satisfies the equation $r = 1 - \sin\theta$, since

$r = 0$ when $\theta = \frac{\pi}{2}$ and the pole satisfies the equation $r = 1 + \cos\theta$ since $r = 0$ when $\theta = \pi$.

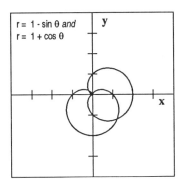

Figure 94

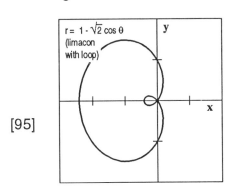

[95]

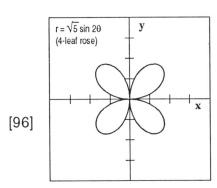

[96]

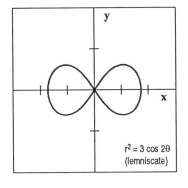

[97]

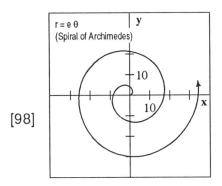

[98]

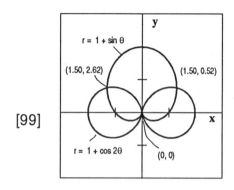

[99]

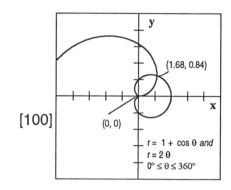

[100]

CHAPTER 7 REVIEW EXERCISES

[1] $\sin \alpha = \frac{a}{c} = \frac{15}{17} \approx 0.8824 \Rightarrow \alpha \approx 62°$. $\beta = 90° - \alpha \approx 28°$. $b = \sqrt{c^2 - a^2} = \sqrt{(17)^2 - (15)^2} = \sqrt{64} = 8$.

[2] $\sin \beta = \frac{b}{c} = \frac{5}{13} \approx 0.3846 \Rightarrow \beta \approx 23°$. $\alpha = 90° - \beta \approx 67°$. $a = \sqrt{(13)^2 - (5)^2} = \sqrt{144} = 12$.

[3] $\beta = 90° - \alpha = 36° \ 10'$. $\tan \alpha = \frac{a}{b} \Rightarrow a = b \tan \alpha = 78.5 \tan 53° \ 50' \approx 107$. $\sin \beta = \frac{b}{c} \Rightarrow$

$c = \frac{b}{\sin \beta} \approx \frac{78.5}{\sin 36° \ 10'} \approx 133$.

[4] $\alpha = 90° - \beta = 65° \ 40'$. $\sin \alpha = \frac{a}{c} \Rightarrow a = c \sin \alpha = 5.14 \sin 65° \ 40' \approx 4.68$. $\sin \beta = \frac{b}{c} \Rightarrow$

$b = c \sin \beta = 5.14 \sin 24° \ 20' \approx 2.12$.

[5] $\tan \alpha = \frac{a}{b} = \frac{1120}{1450} \approx 0.7724 \Rightarrow \alpha \approx 37.7°$. $\beta = 90° - \alpha \approx 52.3°$. $c = \sqrt{(1120)^2 + (1450)^2} \approx 1830$.

[6] $\tan \alpha = \frac{a}{b} = \frac{\frac{3}{4}}{\frac{5}{8}} = \frac{6}{5} = 1.2 \Rightarrow \alpha \approx 50°$. $\beta = 90° - \alpha \approx 40°$. $c = \sqrt{\left(\frac{3}{4}\right)^2 + \left(\frac{5}{8}\right)^2} = \frac{\sqrt{61}}{8}$.

[7] $\alpha = 90° - \beta = 72.9°$. $\tan \alpha = \frac{a}{b} \Rightarrow a = b \tan \alpha = 275 \tan 72.9° \approx 894$. $\sin \beta = \frac{b}{c} \Rightarrow$

$c = \frac{b}{\sin \beta} = \frac{275}{\sin 17.1°} \approx 935$.

[8] $\beta = 90° - \alpha = 16.1°$. $\tan \alpha = \frac{a}{b} \Rightarrow a = b \tan \alpha = 1050 \tan 73.9° \approx 3640$. $\sin \beta = \frac{b}{c} \Rightarrow$

$c = \frac{b}{\sin \beta} = \frac{1050}{\sin 16.1°} \approx 3790$.

[9] $\beta = 180° - (\alpha + \gamma) = 142° \ 10'$. $\frac{a}{\sin \alpha} = \frac{b}{\sin \beta} \Rightarrow a = \frac{b \sin \alpha}{\sin \beta} = \frac{105 \sin 23° \ 20'}{\sin 142° \ 10'} \approx 67.8$. $\frac{b}{\sin \beta} =$

$\frac{c}{\sin \gamma} \Rightarrow c = \frac{b \sin \gamma}{\sin \beta} = \frac{105 \sin 14° \ 30'}{\sin 142° \ 10'} \approx 42.9$.

[10] $\alpha = 180° - (\beta + \gamma) = 33° \ 59'$. $\frac{a}{\sin \alpha} = \frac{b}{\sin \beta} \Rightarrow a = \frac{b \sin \alpha}{\sin \beta} = \frac{428 \sin 33° \ 59'}{\sin 58° \ 42'} \approx 280$. $\frac{b}{\sin \beta} =$

$\frac{c}{\sin \gamma} \Rightarrow c = \frac{b \sin \gamma}{\sin \beta} = \frac{428 \sin 87° \ 19'}{\sin 58° \ 42'} \approx 500$.

[11] $\frac{b}{\sin \beta} = \frac{c}{\sin \gamma} \Rightarrow \sin \gamma = \frac{c \sin \beta}{b} = \frac{10.4 \sin 85.1°}{12.2} \approx 0.8493 \Rightarrow \gamma_1 \approx 58.1°$ or $\gamma_2 \approx 121.9°$. Since

$\beta + \gamma_2 > 180°$, only one triangle is determined and $\gamma \approx 58.1°$. $\alpha = 180° - (\beta + \gamma) \approx 36.8°$. $\frac{a}{\sin \alpha} =$

$\frac{b}{\sin \beta} \Rightarrow a = \frac{b \sin \alpha}{\sin \beta} \approx \frac{12.2 \sin 36.8°}{\sin 85.1°} \approx 7.3$.

[12] $\dfrac{a}{\sin \alpha} = \dfrac{c}{\sin \gamma} \Rightarrow \sin \gamma = \dfrac{c \sin \alpha}{a} = \dfrac{1.84 \sin 26.8°}{2.25} \approx 0.3687 \Rightarrow \gamma_1 \approx 21.6°$ or $\gamma_2 \approx 158.4°$. Since

$\alpha + \gamma_2 > 180°$, only one triangle is determined and $\gamma = 21.6°$. $\beta = 180° - (\alpha + \gamma) \approx 131.6°$.

$\dfrac{a}{\sin \alpha} = \dfrac{b}{\sin \beta} \Rightarrow b = \dfrac{a \sin \beta}{\sin \alpha} \approx \dfrac{2.25 \sin 131.6°}{\sin 26.8°} \approx 3.73$.

[13] $a^2 = b^2 + c^2 - 2bc \cos \alpha \Rightarrow \cos \alpha = \dfrac{b^2 + c^2 - a^2}{2bc} = \dfrac{(46.8)^2 + (59.7)^2 - (73.5)^2}{2(46.8)(59.7)} \approx 0.0630 \Rightarrow$

$\alpha \approx 86.4°$. $\cos \beta = \dfrac{a^2 + c^2 - b^2}{2ac} \approx 0.7721 \Rightarrow \beta \approx 39.5°$. $\gamma = 180° - (\alpha + \beta) \approx 54.1°$.

[14] $\cos \alpha = \dfrac{b^2 + c^2 - a^2}{2bc} = \dfrac{(1075)^2 + (835)^2 - (1140)^2}{2(1075)(835)} \approx 0.3082 \Rightarrow \alpha \approx 72.0°$. $\cos \beta = \dfrac{a^2 + c^2 - b^2}{2ac} \approx$

$0.4419 \Rightarrow \beta \approx 63.8°$. $\gamma = 180° - (\alpha + \beta) \approx 44.3°$.

[15] $a^2 = b^2 + c^2 - 2bc \cos \alpha \Rightarrow a = \sqrt{(17.3)^2 + (14.7)^2 - 2(17.3)(14.7) \text{cis}\, 50°\,10'} \approx 13.8$. $\cos \beta =$

$\dfrac{a^2 + c^2 - b^2}{2ac} \approx \dfrac{(13.8)^2 + (14.7)^2 - (17.3)^2}{2(13.8)(14.7)} \approx 0.2643 \Rightarrow \beta \approx 74°\,40'$. $\gamma = 180° - (\alpha + \beta) \approx 55°\,10'$.

[16] $b^2 = a^2 + c^2 - 2ac \cos \beta \Rightarrow b = \sqrt{(33.2)^2 + (48.4)^2 - 2(33.2)(48.4) \cos 88.5°} \approx 58.0$. $\cos \alpha =$

$\dfrac{b^2 + c^2 - a^2}{2bc} \approx \dfrac{(58.0)^2 + (48.4)^2 - (33.2)^2}{2(58.0)(48.4)} \approx 0.8201 \Rightarrow \alpha \approx 34.9°$. $\gamma = 180° - (\alpha + \beta) \approx 56.6°$.

[17] $\dfrac{a}{\sin \alpha} = \dfrac{b}{\sin \beta} \Rightarrow \sin \beta = \dfrac{b \sin \alpha}{a} = \dfrac{68.2 \sin 75.6°}{52.4} \approx 1.2606$. Since $\sin \beta > 1$, no triangle is possible.

[18] $\dfrac{b}{\sin \beta} = \dfrac{c}{\sin \gamma} \Rightarrow \sin \beta = \dfrac{b \sin \gamma}{c} = \dfrac{817 \sin 52.3°}{670} \approx 0.9648 \Rightarrow \beta_1 \approx 74.8°$ or $\beta_2 = 180° - \beta_1 \approx 105.2°$.

Triangle 1: $\beta_1 \approx 74.8°$, $\alpha_1 = 180° - (\beta_1 + \gamma) \approx 52.9°$. $\dfrac{a_1}{\sin \alpha_1} = \dfrac{c}{\sin \gamma} \Rightarrow a_1 = \dfrac{c \sin \alpha_1}{\sin \gamma} \approx$

$\dfrac{670 \sin 52.9°}{\sin 52.3°} \approx 675$. Triangle 2: $\beta_2 \approx 105.2°$, $\alpha_2 = 180° - (\beta_2 + \gamma) \approx 22.5°$. $\dfrac{a_2}{\sin \alpha_2} = \dfrac{c}{\sin \gamma} \Rightarrow$

$a_2 = \dfrac{c \sin \alpha_2}{\sin \gamma} \approx \dfrac{670 \sin 22.5°}{\sin 52.3°} \approx 324$.

[19] $c^2 = a^2 + b^2 - 2ab \cos \gamma \Rightarrow c = \sqrt{(14.3)^2 + (17.8)^2 - 2(14.3)(17.8) \cos 42°\,40'} \approx 12.1$. $\cos \alpha =$

$\dfrac{b^2 + c^2 - a^2}{2bc} \approx \dfrac{(17.8)^2 + (12.1)^2 - (14.3)^2}{2(17.8)(12.1)} \approx 0.6008 \Rightarrow \alpha \approx 53°$. $\beta = 180° - (\alpha + \gamma) \approx 84°\,20'$.

[20] $b^2 = a^2 + c^2 - 2ac \cos \beta \Rightarrow b = \sqrt{(2.75)^2 + (1.91)^2 - 2(2.75)(1.91) \cos 115°\,50'} \approx 3.97$. $\cos \alpha =$

$\dfrac{b^2 + c^2 - a^2}{2bc} \approx 0.7812 \Rightarrow \alpha \approx 38°\,30'$. $\gamma = 180° - (\alpha + \beta) \approx 25°\,40'$.

[21] $\dfrac{a}{\sin \alpha} = \dfrac{c}{\sin \gamma} \Rightarrow \sin \gamma = \dfrac{c \sin \alpha}{a} = \dfrac{1589 \sin 15.6°}{1016} \approx 0.4206 \Rightarrow \gamma_1 \approx 24.9°$, or $\gamma_2 \approx 180 - 24.9° =$

$155.1°$. Triangle 1: $\gamma_1 = 24.9°$, $\beta_1 = 180° - (\alpha + \gamma_1) \approx 139.5°$. $\dfrac{a}{\sin \alpha} = \dfrac{b_1}{\sin \beta_1} \Rightarrow b_1 = \dfrac{a \sin \beta_1}{\sin \alpha} \approx$

$\dfrac{1016 \sin 139.5°}{\sin 15.6°} \approx 2450$; Triangle 2: $\gamma_2 \approx 155.1°$, $\beta_2 = 180° - (\alpha + \gamma_2) \approx 9.3°$. $\dfrac{a}{\sin \alpha} = \dfrac{b_2}{\sin \beta_2} \Rightarrow b_2 =$

$\dfrac{a \sin \beta_2}{\sin \alpha} \approx \dfrac{1016 \sin 9.3°}{\sin 15.6°} \approx 611$.

[22] $a^2 = b^2 + c^2 - 2\,b\,c\cos\alpha \Rightarrow a = \sqrt{(18.4)^2 + (10.1)^2 - 2(18.4)(10.1)\cos 41.7°} \approx 12.8.\ \cos\beta =$

$\dfrac{a^2 + c^2 - b^2}{2\,a\,c} \approx \dfrac{(12.8)^2 + (10.1)^2 - (18.4)^2}{2(12.8)(10.1)} \approx -0.2812 \Rightarrow \beta \approx 106.3°.\ \gamma = 180° - (\alpha + \beta) \approx 32.0°.$

[23]   600 miles per hour = 10 miles per minute $\Rightarrow$ plane flies $10(5280) = 52,800$ feet in one minute.

$\tan 36.2° = \dfrac{h}{52,800} \Rightarrow h = 52,800\tan 36.2° \approx 38,600.$ The altitude of the plane is approximiately

38,600 feet.   (See figure below).

[24] $\tan\alpha = \dfrac{48.5}{62.4} \approx 0.7772 \Rightarrow \alpha \approx 37.9°.$ The angle of elevation of the sun is approximately $37.9°$.

(See figure below).

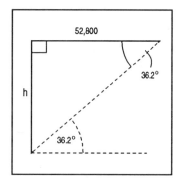

Figure 23   **Figure 24**

[25]   The decagon consists of 10 triangles like the one shown in Figure 25. Thus, $A =$

$10\left[\dfrac{1}{2}(12.2)(12.2)\sin 36°\right] \approx 437.$ The area of the decagon is approximately 437 cm².

(See figure below).

[26] $\sin\alpha = \dfrac{18}{24} = 0.75 \Rightarrow \alpha \approx 48.6°.$ The angle between the ladder and the ground is

approximately $48.6°$.   (See figure below).

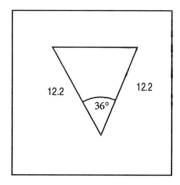
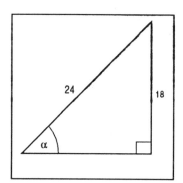

Figure 25   **Figure 26**

[27] $\sin 15° = \dfrac{4.5}{x} \Rightarrow x = \dfrac{4.5}{\sin 15°} \approx 17.$ The length of the ramp is approxmiately 17 feet.

(See figure below).

[28] $\angle ABC = 72.4° + 55.5° = 127.9°$. Thus, by the Law of Cosines, $x =$

$\sqrt{(780)^2 + (540)^2 - 2(780)(540)\cos 127.9°} \approx 1190$. The ship is approximately

1190 miles from port.   (See figure below).

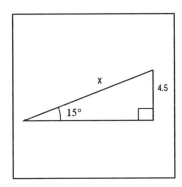

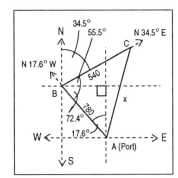

Figure 27   **Figure 28**

[29]   Let L denote the lighthouse and let A and B denote the locations of the ship at 5:00 p.m. and at

7:00 p.m., respectively. Between 5:00 p.m. and 7:00 p.m., the ship travels $2 \cdot 24 = 48$ miles. By the

Law of Sines, $\dfrac{x}{\sin 16.5°} = \dfrac{48}{\sin 92.3°} \Rightarrow x = \dfrac{48 \sin 16.5°}{\sin 92.3°} \approx 13.6$. The distance to the lighthouse is

approximately 13.6 miles.   (See figure below).

[30] $\sin 6.5° = \dfrac{1320}{x} \Rightarrow x = \dfrac{1320}{\sin 6.5°} \approx 11{,}660$. 9 mph $= \dfrac{9(5280)}{60} = 792$ feet per minute. $\dfrac{11{,}660}{792} \approx 14.7$.

It will take the submarine approximately 14.7 seconds to reach a depth of 1320 feet.

(See figure below).

[31]   Let B_1 and B_2 denote the location of the two boats and let x denote the distance between the two

boats. If y denotes $d(A, B_1)$, we see from right $\Delta B_1 CA$ that $\sin 48° 40' = \dfrac{75}{y}$. Thus, $y = \dfrac{75}{\sin 48° 40'} \approx$

100. From the Law of Sines applied to ΔAB_1B_2 we see that $\dfrac{x}{\sin 6° 30'} = \dfrac{y}{\sin 42° 10'}$. Thus $x =$

$\dfrac{y \sin 6° 30'}{\sin 42° 10'} \approx \dfrac{100 \sin 6° 30'}{\sin 42° 10'} \approx 17$. The two boats are approximately 17 feet apart.

(See figure below).

[32]   Let x be the length of diagonal BD. By the Law of Cosines $x =$

$\sqrt{(24.6)^2 + (32.9)^2 - 2(24.6)(32.9)\cos 44° 20'} \approx 23.0$ cm. If y is the length of diagonal AC,

then $y = \sqrt{(32.9)^2 + (24.6)^2 - 2(32.9)(24.6)\cos 135° 40'} \approx 53.3$ cm. The area A of the

parallelogram is $A = 2\left[\dfrac{1}{2}(32.9)(24.6)\sin 44° 20'\right] \approx 566$ cm². (See figure below).

[33] $s = \dfrac{1480 + 1760 + 2050}{2} = 2645$. $A = \sqrt{s(s-a)(s-b)(s-c)} = \sqrt{(2645)(1165)(885)(595)} \approx$

1,270,000. The area of the tract of land is approximately 1,270,000 m².

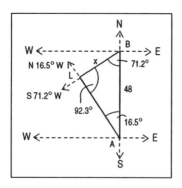

Figure 29

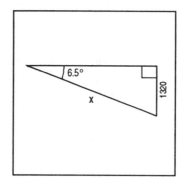

Figure 30

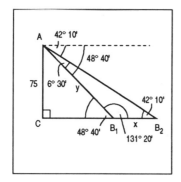

Figure 31

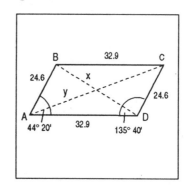

Figure 32

[34] Applying the Law of Sines to $\triangle ABD$, we see that $\dfrac{y}{\sin 34° 20'} = \dfrac{250}{\sin 8° 30'}$. *Thus*, $y = $

$\dfrac{250 \sin 34° 20'}{\sin 8° 30'} \approx 954$. From right $\triangle BCD$, $\sin 42° 50' = \dfrac{x}{y} \Rightarrow x = y \sin 42° 50' \approx 954 \sin 42° 50' \approx$

650. The cliff is approximately 650 feet high. (See figure below).

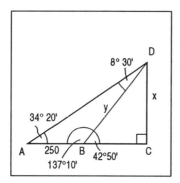

Figure 34

[35] $r = \sqrt{(2)^2 + (-2)^2} = 2\sqrt{2}$. $\cos \theta = \dfrac{1}{\sqrt{2}}$ and $\sin \theta = -\dfrac{1}{\sqrt{2}} \Rightarrow \theta = \dfrac{7\pi}{4}$. $2 - 2i = 2\sqrt{2} \operatorname{cis} \dfrac{7\pi}{4}$.

[36] $r = \sqrt{(-5)^2 + (5)^2} = 5\sqrt{2}$. $\cos \theta = -\dfrac{1}{\sqrt{2}}$ and $\sin \theta = \dfrac{1}{\sqrt{2}} \Rightarrow \theta = \dfrac{3\pi}{4}$. $-5 + 5i = 5\sqrt{2} \operatorname{cis} \dfrac{3\pi}{4}$.

[37] $r = \sqrt{(-\sqrt{3})^2 + (-1)^2} = 2$. $\cos \theta = -\dfrac{\sqrt{3}}{2}$ and $\sin \theta = -\dfrac{1}{2} \Rightarrow \theta = \dfrac{7\pi}{6}$. $-\sqrt{3} - i = 2 \operatorname{cis} \dfrac{7\pi}{6}$.

[38]　$r = \sqrt{\left(-2\sqrt{3}\right)^2 + \left(1\right)^2} = \sqrt{13}$. $\tan\theta = \dfrac{1}{-2\sqrt{3}} = -\dfrac{\sqrt{3}}{6}$. If θ' is the reference angle for θ, then $\tan\theta' =$

$|\tan\theta| = \dfrac{\sqrt{3}}{6} \Rightarrow \theta' = \tan^{-1}\dfrac{\sqrt{3}}{6} \approx 16°$. Since θ is in QII, $\theta = 180° - \theta' \approx 164°$. Thus, $-2\sqrt{3} + i =$

$\sqrt{13}\operatorname{cis}\left(180° - \tan^{-1}\dfrac{\sqrt{3}}{6}\right) \approx 3.6\operatorname{cis}164°$.

[39]　$r = \sqrt{\left(0\right)^2 + \left(-\dfrac{5}{2}\right)^2} = \dfrac{5}{2}$ and $\theta = \dfrac{3\pi}{2}$. Thus, $-\dfrac{5}{2}i = \dfrac{5}{2}\operatorname{cis}\dfrac{3\pi}{2}$.

[40]　$r = 7$ and $\theta = 0° \Rightarrow 7 = 7\operatorname{cis}0°$.

[41]　$r = \sqrt{\left(2\right)^2 + \left(-1\right)^2} = \sqrt{5}$. $\tan\theta = \dfrac{-1}{2} = -\dfrac{1}{2}$. If θ' is the reference angle for θ, then $\tan\theta' =$

$|\tan\theta| = \dfrac{1}{2} \Rightarrow \theta' = \tan^{-1}\dfrac{1}{2} \approx 27°$. Since θ is in QIV, $\theta = 360° - \theta' \approx 333°$. Thus, $2 - i =$

$\sqrt{5}\operatorname{cis}\left(360° - \tan^{-1}\dfrac{1}{2}\right) \approx 2.24\operatorname{cis}333°$.

[42]　$r = \sqrt{\left(-4\right)^2 + \left(-7\right)^2} = \sqrt{65}$. $\tan\theta = \dfrac{-7}{-4} = 1.75$. If θ' is the reference angle for θ, then $\tan\theta' =$

$|\tan\theta| = 1.75 \Rightarrow \theta' = \tan^{-1}\left(1.75\right) \approx 60°$. Since θ is in QIII, $\theta = 180° + \theta' \approx 240°$. Thus, $-4 - 7i =$

$\sqrt{65}\operatorname{cis}\left[180° + \tan^{-1}\left(1.75\right)\right] \approx 8.1\operatorname{cis}240°$.

[43]　$4\operatorname{cis}\dfrac{2\pi}{3} = 4\left(-\dfrac{1}{2} + \dfrac{\sqrt{3}}{2}i\right) = -2 + 2\sqrt{3}\,i$.　　　[44]　$3\operatorname{cis}\dfrac{3\pi}{4} = 3\left(-\dfrac{\sqrt{2}}{2} + \dfrac{\sqrt{2}}{2}i\right) = -\dfrac{3\sqrt{2}}{2} + \dfrac{3\sqrt{2}}{2}\,i$.

[45]　$16\operatorname{cis}\dfrac{5\pi}{4} = 16\left(-\dfrac{\sqrt{2}}{2} - \dfrac{\sqrt{2}}{2}i\right) = -8\sqrt{2} - 8\sqrt{2}\,i$.

[46]　$12\operatorname{cis}\dfrac{7\pi}{6} = 12\left(-\dfrac{\sqrt{3}}{2} - \dfrac{1}{2}i\right) = -6\sqrt{3} - 6i$.　　　[47]　$\dfrac{\sqrt{3}}{2}\operatorname{cis}330° = \dfrac{\sqrt{3}}{2}\left(\dfrac{\sqrt{3}}{2} - \dfrac{1}{2}i\right) = \dfrac{3}{4} - \dfrac{\sqrt{3}}{4}\,i$.

[48]　$7\operatorname{cis}240° = 7\left(-\dfrac{1}{2} - \dfrac{\sqrt{3}}{2}i\right) = -\dfrac{7}{2} - \dfrac{7\sqrt{3}}{2}\,i$.

[49]　$\cos\dfrac{\pi}{12} = \cos\left(\dfrac{\pi}{4} - \dfrac{\pi}{6}\right) = \dfrac{\sqrt{6} + \sqrt{2}}{4}$ and $\sin\dfrac{\pi}{12} = \dfrac{\sqrt{6} - \sqrt{2}}{4}$. Thus, $5\operatorname{cis}\dfrac{\pi}{12} =$

$5\left[\left(\dfrac{\sqrt{6} + \sqrt{2}}{4}\right) + \left(\dfrac{\sqrt{6} - \sqrt{2}}{4}\right)i\right] = \dfrac{5\left(\sqrt{6} + \sqrt{2}\right)}{4} + \dfrac{5\left(\sqrt{6} - \sqrt{2}\right)}{4}\,i$.

[50]　$\cos\dfrac{5\pi}{12} = \cos\left(\dfrac{\pi}{4} + \dfrac{\pi}{6}\right) = \dfrac{\sqrt{6} - \sqrt{2}}{4}$ and $\sin\dfrac{5\pi}{12} = \sin\left(\dfrac{\pi}{4} + \dfrac{\pi}{6}\right) = \dfrac{\sqrt{6} + \sqrt{2}}{4}$. Thus, $\dfrac{13}{2}\operatorname{cis}\dfrac{5\pi}{12} =$

$\dfrac{13}{2}\left[\left(\dfrac{\sqrt{6} - \sqrt{2}}{4}\right) + \left(\dfrac{\sqrt{6} + \sqrt{2}}{4}\right)i\right] = \dfrac{13\left(\sqrt{6} - \sqrt{2}\right)}{8} + \dfrac{13\left(\sqrt{6} + \sqrt{2}\right)}{8}\,i$.

[51]　$\left(3\operatorname{cis}\dfrac{\pi}{5}\right)\left(15\operatorname{cis}\dfrac{\pi}{3}\right) = (3)(15)\operatorname{cis}\left(\dfrac{\pi}{5} + \dfrac{\pi}{3}\right) = 45\operatorname{cis}\dfrac{8\pi}{15}$.

[52]　$\left(4\operatorname{cis}\dfrac{3\pi}{4}\right)\left(12\operatorname{cis}\dfrac{\pi}{2}\right) = (4)(12)\operatorname{cis}\left(\dfrac{3\pi}{4} + \dfrac{\pi}{2}\right) = 48\operatorname{cis}\dfrac{5\pi}{4}$.

[53] $\dfrac{3\sqrt{2}\operatorname{cis}\dfrac{5\pi}{3}}{6\operatorname{cis}\dfrac{\pi}{3}}=\dfrac{3\sqrt{2}}{6}\operatorname{cis}\left(\dfrac{5\pi}{3}-\dfrac{\pi}{3}\right)=\dfrac{\sqrt{2}}{2}\operatorname{cis}\dfrac{4\pi}{3}.$

[54] $\dfrac{2\sqrt{3}\operatorname{cis}225°}{8\operatorname{cis}105°}=\dfrac{2\sqrt{3}}{8}\operatorname{cis}\left(225°-105°\right)=\dfrac{\sqrt{3}}{4}\operatorname{cis}120°.$

[55] $\left(12\operatorname{cis}\dfrac{7\pi}{6}\right)\left(\dfrac{\sqrt{3}}{2}\operatorname{cis}\dfrac{2\pi}{3}\right)=(12)\left(\dfrac{\sqrt{3}}{2}\right)\operatorname{cis}\left(\dfrac{7\pi}{6}+\dfrac{2\pi}{3}\right)6\sqrt{3}\operatorname{cis}\dfrac{11\pi}{6}=6\sqrt{3}\left(\dfrac{\sqrt{3}}{2}-\dfrac{1}{2}i\right)=9-3\sqrt{3}\,i.$

[56] $\left(5\operatorname{cis}\dfrac{2\pi}{3}\right)\left(2\operatorname{cis}\dfrac{5\pi}{6}\right)=(5)(2)\operatorname{cis}\left(\dfrac{2\pi}{3}+\dfrac{5\pi}{6}\right)=10\operatorname{cis}\dfrac{3\pi}{2}=10\left(0-i\right)=-10\,i.$

[57] $\dfrac{14\operatorname{cis}225°}{2\operatorname{cis}75°}=\dfrac{14}{2}\operatorname{cis}\left(225°-75°\right)=7\operatorname{cis}150°=7\left(-\dfrac{\sqrt{3}}{2}+\dfrac{1}{2}i\right)=-\dfrac{7\sqrt{3}}{2}+\dfrac{7}{2}i.$

[58] $\dfrac{3\sqrt{2}\operatorname{cis}315°}{9\operatorname{cis}135°}=\dfrac{3\sqrt{2}}{9}\operatorname{cis}\left(315°-135°\right)=\dfrac{\sqrt{2}}{3}\operatorname{cis}180°=\dfrac{\sqrt{2}}{3}\left(-1+0\,i\right)=-\dfrac{\sqrt{2}}{3}.$

[59] $\left(\sqrt{5}\operatorname{cis}\dfrac{7\pi}{12}\right)^{6}=(\sqrt{5})^{6}\operatorname{cis}(6)\left(\dfrac{7\pi}{12}\right)=125\operatorname{cis}\dfrac{7\pi}{2}=125\left(0-i\right)=-125\,i.$

[60] $\left(\sqrt{7}\operatorname{cis}\dfrac{5\pi}{12}\right)^{8}=(\sqrt{7})^{8}\operatorname{cis}(8)\left(\dfrac{5\pi}{12}\right)=2401\operatorname{cis}\dfrac{10\pi}{3}=2401\operatorname{cis}\dfrac{4\pi}{3}=2401\left(-\dfrac{1}{2}-\dfrac{\sqrt{3}}{2}i\right)=$

$-\dfrac{2401}{2}-\dfrac{2401\sqrt{3}}{2}\,i.$

[61] $\left(\dfrac{\sqrt{2}}{2}\operatorname{cis}20°\right)^{12}=\left(\dfrac{\sqrt{2}}{2}\right)^{12}\operatorname{cis}(12)(20°)=\dfrac{1}{64}\operatorname{cis}240°=\dfrac{1}{64}\left(-\dfrac{1}{2}-\dfrac{\sqrt{3}}{2}i\right)=-\dfrac{1}{128}-\dfrac{\sqrt{3}}{128}\,i.$

[62] $\left(\dfrac{\sqrt{3}}{2}\operatorname{cis}15°\right)^{14}=\left(\dfrac{\sqrt{3}}{2}\right)^{14}\operatorname{cis}(14)(15°)=\dfrac{2187}{16,384}\operatorname{cis}210°=\dfrac{2187}{16,384}\left(-\dfrac{\sqrt{3}}{2}-\dfrac{1}{2}i\right)=$

$-\dfrac{2187\sqrt{3}}{32,768}-\dfrac{2187}{32,768}\,i.$

[63] $\left(\dfrac{\sqrt{2}}{2}-\dfrac{\sqrt{2}}{2}i\right)^{8}=\left(1\operatorname{cis}\dfrac{7\pi}{4}\right)^{8}=1^{8}\operatorname{cis}(8)\left(\dfrac{7\pi}{4}\right)=\operatorname{cis}14\pi=1+0\,i=1.$

[64] $\left(4-4\sqrt{3}\,i\right)^{6}=\left(8\operatorname{cis}\dfrac{5\pi}{3}\right)^{6}=8^{6}\operatorname{cis}(6)\left(\dfrac{5\pi}{3}\right)=262,144\operatorname{cis}10\pi=262,144\left(1+0\cdot i\right)=262,144.$

[65] $\left(\dfrac{\sqrt{3}}{2}-\dfrac{1}{2}i\right)^{5}=\left(1\operatorname{cis}\dfrac{11\pi}{6}\right)^{5}=1^{5}\operatorname{cis}(5)\left(\dfrac{11\pi}{6}\right)=\operatorname{cis}\dfrac{55\pi}{6}=\operatorname{cis}\dfrac{7\pi}{6}=-\dfrac{\sqrt{3}}{2}-\dfrac{1}{2}i.$

[66] $\left(-3+3\,i\right)^{4}=\left(3\sqrt{2}\operatorname{cis}\dfrac{3\pi}{4}\right)^{4}=(3\sqrt{2})^{4}\operatorname{cis}(4)\left(\dfrac{3\pi}{4}\right)=324\operatorname{cis}3\pi=324\left(-1+0\,i\right)=-324.$

[67] $\left(-8-8\,i\right)^{6}=\left(8\sqrt{2}\operatorname{cis}\dfrac{5\pi}{4}\right)^{6}=(8\sqrt{2})^{6}\operatorname{cis}(6)\left(\dfrac{5\pi}{4}\right)=2,097,152\operatorname{cis}\dfrac{15\pi}{2}=2,097,152\operatorname{cis}\dfrac{3\pi}{2}=$

$2,097,152\left(0-i\right)=-2,097,152\,i.$

[68] $\left(\dfrac{5\sqrt{3}}{2}+\dfrac{5}{2}i\right)^{3}=\left(5\operatorname{cis}\dfrac{\pi}{6}\right)^{3}=5^{3}\operatorname{cis}(3)\left(\dfrac{\pi}{6}\right)=125\operatorname{cis}\dfrac{\pi}{2}=125\left(0+i\right)=125\,i.$

[69] $\left(\dfrac{2}{3}\operatorname{cis}32°\right)^{8}=\left(\dfrac{2}{3}\right)^{8}\operatorname{cis}(8)(32°)=\dfrac{256}{6561}\operatorname{cis}256°.$

[70] $\left(\frac{\sqrt{2}}{2} \text{ cis } 24°\right)^6 = \left(\frac{\sqrt{2}}{2}\right)^6 \text{ cis } (6)(24°) = \frac{1}{8} \text{ cis } 144°.$

[71] $\left(-\frac{1}{2} + \frac{3}{4} i\right)^6 = \left[\frac{\sqrt{13}}{4} \text{ cis }\left(180° - \tan^{-1} \frac{3}{2}\right)\right]^6 \approx \left(\frac{\sqrt{13}}{4} \text{ cis } 123.7°\right)^6 = \left(\frac{\sqrt{13}}{4}\right)^6 \text{ cis } (6)(123.7°) =$

$\frac{2197}{4096} \text{ cis } 742.2° \approx 0.536 \text{ cis } 22.2°.$

[72] $(2\sqrt{3} - i)^3 = \left[\sqrt{13} \text{ cis }\left(360° - \tan^{-1} \frac{\sqrt{3}}{6}\right)\right]^3 \approx \left(\sqrt{13} \text{ cis } 343.9°\right)^3 = \left(\sqrt{13}\right)^3 \text{ cis } 3(343.9°) =$

$13\sqrt{13} \text{ cis } 1031.7° \approx 46.9 \text{ cis } 311.7°.$

[73] $(5 - 12i)^4 = \left[13 \text{ cis }\left(360° - \tan^{-1} \frac{12}{5}\right)\right]^4 \approx (13 \text{ cis } 292.6°)^4 = 13^4 \text{ cis } (4)(292.6°) =$

$28{,}561 \text{ cis } 1170.4° = 28{,}561 \text{ cis } 90.4°.$

[74] $(-3 + 4i)^5 = \left[5 \text{ cis }\left(180° - \tan^{-1} \frac{4}{3}\right)\right]^5 \approx (5 \text{ cis } 126.9°)^5 = 5^5 \text{ cis } (5)(126.9°) = 3125 \text{ cis } 634.5° =$

$3125 \text{ cis } 274.5°.$

[75] $2\sqrt{3} - 2i = 4 \text{ cis } \frac{11\pi}{6}.$ $u_k = 4^{1/5} \text{ cis }\left(\frac{\frac{11\pi}{6} + 2k\pi}{5}\right) = 4^{1/5} \text{ cis }\left[\frac{11\pi}{30} + k\left(\frac{2\pi}{5}\right)\right] =$

$4^{1/5} \text{ cis } (66° + k \cdot 72°),$ for $k = 0, 1, 2, 3,$ and 4. Thus $u_0 = 4^{1/5} \text{ cis } 66°,$ $u_1 = 4^{1/5} \text{ cis } 138°,$

$u_2 = 4^{1/5} \text{ cis } 210°,$ $u_3 = 4^{1/5} \text{ cis } 282°$ and $u_4 = 4^{1/5} \text{ cis } 354°,$ where $u_1 = 4^{1/5} \approx 1.3.$

(See figure below).

[76] $-\frac{\sqrt{2}}{2} + \frac{\sqrt{2}}{2} i = 1 \text{ cis } \frac{3\pi}{4}.$ $u_k = \sqrt[4]{1} \text{ cis }\left(\frac{\frac{3\pi}{4} + 2k\pi}{4}\right) = \text{ cis }\left[\frac{3\pi}{16} + k\left(\frac{\pi}{2}\right)\right] = \text{ cis } (33.75° + k \cdot 90°),$

for $k = 0, 1, 2,$ and 3. $u_0 = \text{ cis } 33.75°,$ $u_1 = \text{ cis } 123.75°,$ $u_2 = \text{ cis } 213.75°,$ and $u_3 = \text{ cis } 303.75°.$

(See figure below).

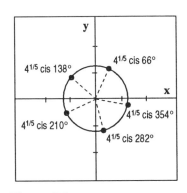

Figure 75

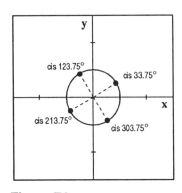

Figure 76

[77] $1 = 1 \text{ cis } 0.$ $u_k = \sqrt[3]{1} \text{ cis }\left(\frac{0 + k \cdot 2\pi}{3}\right) = \text{ cis }\left(k \cdot \frac{2\pi}{3}\right),$ for $k = 0, 1$ and 2. $u_0 = \text{ cis } 0,$ $u_1 = \text{ cis } \frac{2\pi}{3},$

$u_2 = \text{ cis } \frac{4\pi}{3}.$ (See figure below).

[78] $1 = 1 \text{ cis } 0.\ u_k = \sqrt[8]{1} \text{ cis} \left(\dfrac{0 + k \cdot 2\pi}{8} \right) = \text{cis} \left(k \cdot \dfrac{\pi}{4} \right),$ for $k = 0, 1, 2, \ldots, 7.\ u_0 = \text{cis } 0,\ u_1 = \text{cis } \dfrac{\pi}{4},$

$u_2 = \dfrac{\pi}{2},\ u_3 = \text{cis } \dfrac{3\pi}{4},\ u_4 = \text{cis } \pi,\ u_5 = \text{cis } \dfrac{5\pi}{4},\ u_6 = \text{cis } \dfrac{3\pi}{2},$ and $u_7 = \text{cis } \dfrac{7\pi}{4}.$ (See figure below).

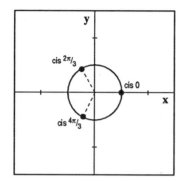

Figure 77

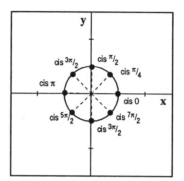

Figure 78

[79] $81 \text{ cis } 72°.\ u_k = \sqrt[4]{81} \text{ cis} \left[\dfrac{72° + k\,(360°)}{4} \right] = 3 \text{ cis} \left(18° + k \cdot 90° \right),$ for $k = 0, 1, 2$ and 3. Thus

$u_0 = 3 \text{ cis } 18°,\ u_1 = 3 \text{ cis } 108°,\ u_2 = 3 \text{ cis } 198°$ and $u_3 = 3 \text{ cis } 288°.$ (See figure below).

[80] $64 \text{ cis } 108°.\ u_k = \sqrt[6]{64} \text{ cis} \left(\dfrac{108° + k \cdot 360°}{6} \right) = 2 \text{ cis} \left(18° + k \cdot 60° \right),$ for $k = 0, 1, 2, 3, 4$ and 5. Thus,

$u_0 = 2 \text{ cis } 18°,\ u_1 = 2 \text{ cis } 78°,\ u_2 = 2 \text{ cis } 138°,\ u_3 = 2 \text{ cis } 198°,\ u_4 = 2 \text{ cis } 258°$ and $u_5 = 2 \text{ cis } 318°.$
(See figure below).

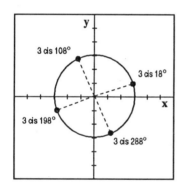

Figure 79

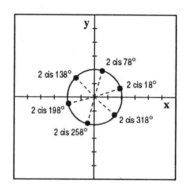

Figure 80

[81] $\|v\| = \sqrt{(-1)^2 + (4)^2} = \sqrt{17}.$ (See figure below).

[82] $\|v\| = \sqrt{(-3)^2 + (-5)^2} = \sqrt{34}.$ (See figure below).

[83] $\|v\| = \sqrt{\left(\dfrac{7}{2} \right)^2 + \left(-\dfrac{13}{4} \right)^2} = \dfrac{\sqrt{365}}{4}.$ (See figure below).

[84] $\|v\| = \sqrt{\left(\dfrac{\sqrt{3}}{2} \right)^2 + \left(-\dfrac{1}{2} \right)^2} = 1.$ (See figure below).

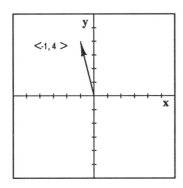

Figure 81

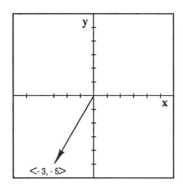

Figure 82

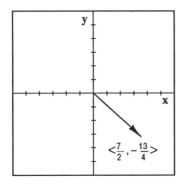

Figure 83

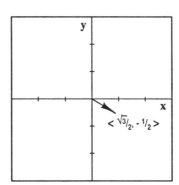

Figure 84

[85]　$u + v = \langle -2 + 4, 3 + 5 \rangle = \langle 2, 8 \rangle.$　(See figure below).

[86]　$u + v = \left\langle -\frac{3}{2} + \frac{5}{2}, -8 + 6 \right\rangle = \langle 1, -2 \rangle.$　(See figure below).

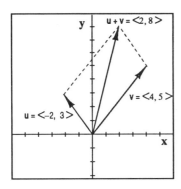

Figure 85

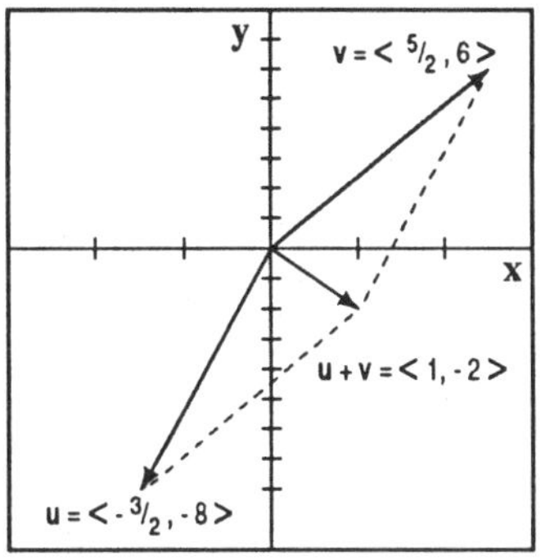

Figure 86

[87]　$-3u + \frac{1}{2}v = \langle 9, -15 \rangle + \langle 1, -2 \rangle = \langle 10, -17 \rangle.$　[88]　$\frac{5}{3}u - 2v = \left\langle -5, \frac{25}{3} \right\rangle - \langle 4, -8 \rangle = \left\langle -9, \frac{49}{3} \right\rangle.$

[89]　$-12u - 4v = \langle 36, -60 \rangle - \langle 8, -16 \rangle = \langle 28, -44 \rangle.$

[90]　$\frac{\sqrt{3}}{2}u - \frac{1}{2}v = \left\langle -\frac{3\sqrt{3}}{2}, \frac{5\sqrt{3}}{2} \right\rangle - \langle 1, -2 \rangle = \left\langle \frac{-2 - 3\sqrt{3}}{2}, \frac{4 + 5\sqrt{3}}{2} \right\rangle.$

[91]　$\tan \theta = \frac{-5}{-5} = 1$ and θ in QIII $\Rightarrow \theta = \frac{5\pi}{4}.$　[92]　$\tan \theta = \frac{-\frac{1}{2}}{\frac{\sqrt{3}}{2}} = -\frac{1}{\sqrt{3}}$ and θ in QIV $\Rightarrow \theta = \frac{11\pi}{6}.$

[93] $\tan \theta = \frac{3}{-4} = -\frac{3}{4}$. If θ' is the reference angle for θ, then $\tan \theta' = |\tan \theta| = \frac{3}{4} \Rightarrow \theta' = \tan^{-1}\frac{3}{4} \approx 37°$.

Since θ is in QII, $\theta = 180° - \tan^{-1}\frac{3}{4} \approx 143°$.

[94] $\tan \theta = \frac{5}{-2} = -\frac{5}{2}$. If θ' is the reference angle for θ, then $\tan \theta' = |\tan \theta| = \frac{5}{2} \Rightarrow \theta' = \tan^{-1}\frac{5}{2} \approx 68°$.

Since θ is in QII, $\theta = 180° - \tan^{-1}\frac{5}{2} \approx 112°$.

[95] $v_1 = |v|\cos\theta = 5\left(-\frac{\sqrt{2}}{2}\right) = -\frac{5\sqrt{2}}{2}$ and $v_2 = |v|\sin\theta = 5\left(\frac{\sqrt{2}}{2}\right) = \frac{5\sqrt{2}}{2} \Rightarrow v = \left\langle -\frac{5\sqrt{2}}{2}, \frac{5\sqrt{2}}{2}\right\rangle$.

[96] $v_1 = |v|\cos\theta = 12\left(-\frac{\sqrt{3}}{2}\right) = -6\sqrt{3}$ and $v_2 = |v|\sin\theta = 12\left(-\frac{1}{2}\right) = -6 \Rightarrow v = \langle -6\sqrt{3}, -6\rangle$.

[97] $2u - 4v = 2\left(\frac{3}{4}i - j\right) - 4\left(-2i + \frac{5}{2}j\right) = \frac{3}{2}i - 2j + 8i - 10j = \frac{19}{2}i - 12j$.

[98] $-\frac{3}{2}u + \frac{13}{4}v = -\frac{3}{2}\left(\frac{3}{4}i - j\right) + \frac{13}{4}\left(-2i + \frac{5}{2}j\right) = -\frac{9}{8}i + \frac{3}{2}j - \frac{13}{2}i + \frac{65}{8}j = -\frac{61}{8}i + \frac{77}{8}j$.

[99] Let **a** and **b** be vectors representing forces **A** and **B**, respectively (see Figure 99). We must find the magnitude and direction of the resultant vector **a** + **b**. We see from the figure that the direction angle for **a** is $270° + 24.2° = 294.2°$ and the direction angle for **b** is $270° - 41.9° = 228.1°$. Thus, **a** and **b** have the following algebraic representations: $\mathbf{a} = \langle 225 \cos 294.2°, 225 \sin 294.2°\rangle \approx$ $\langle 92.2, -205.2\rangle$ and $\mathbf{b} = \langle 280 \cos 228.1°, 280 \sin 228.1\rangle \approx \langle -187.0, -208.4\rangle$. Thus $\mathbf{a} + \mathbf{b} \approx$ $\langle -94.8, -413.6\rangle$ and the magnitude of **a** + **b** is $|\mathbf{a} + \mathbf{b}| \approx \sqrt{(-94.8)^2 + (-413.6)^2} \approx 424$. If θ is the direction angle for **a** + **b**, then $\tan \theta \approx \frac{-413.6}{-94.8} \approx 4.363$. If θ' is the reference angle for θ, then $\tan \theta' = |\tan \theta| = 4.363 \Rightarrow \theta' \approx 77.1°$. Since θ is in QIII, $\theta = 180° + \theta' \approx 257.1°$. To find the direction of **a** + **b**, we subtract $257.1°$ from $270°$ to get $12.9°$. Thus, the resultant of forces **A** and **B** is a force of approximately 424 pounds in the (approximate) direction of S 12.9° W.

[100] Let **a** and **b** be vectors representing velocities **A** and **B**, respectively (see Figure 100). We must find the magnitude and direction of the resultant vector **a** + **b**. We see from the figure that the direction angle for **a** is $90° - 63.2° = 26.8°$ and the direction angle for **b** is $90° - 17.8° = 72.2°$. Thus, **a** and **b** have the following algebraic representations: $\mathbf{a} = \langle 425 \cos 26.8°, 425 \sin 26.8°\rangle \approx$ $\langle 379.3, 191.6\rangle$ and $\mathbf{b} = \langle 310 \cos 72.2°, 310 \sin 72.2\rangle \approx \langle 94.8, 295.2\rangle$. Thus $\mathbf{a} + \mathbf{b} \approx \langle 474.1, 486.8\rangle$ and the magnitude of **a** + **b** is $|\mathbf{a} + \mathbf{b}| \approx \sqrt{(474.1)^2 + (486.8)^2} \approx 680$. If θ is the direction angle for **a** + **b**, then $\tan \theta \approx \frac{486.8}{474.1} \approx 1.027$. Since θ is in QI, $\theta \approx \tan^{-1}(1.027) \approx 45.8°$. To obtain the direction of **a** + **b**, we subtract $45.8°$ from $90°$, getting $44.2°$. Thus, the resultant of velocities **A** and **B** is a velocity of approximately 680 km per hour in the (approximate) direction N 44.2° E.

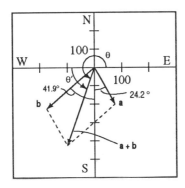

Figure 99

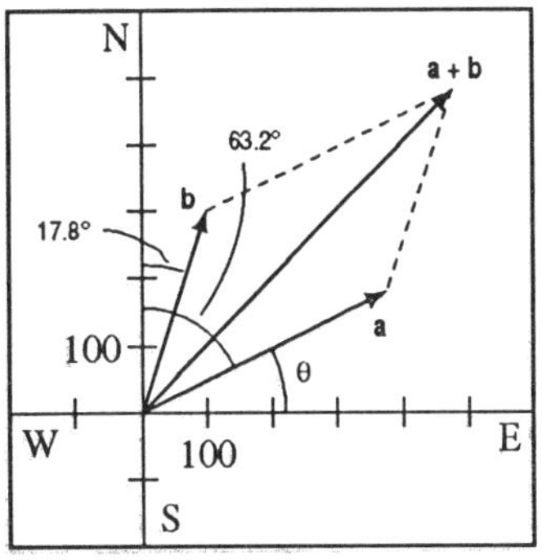

Figure 100

[101] Let **u** be the force pulling the car down the incline, **w** the weight of the car being pulled downward by the force of gravity and **v** the force pressing the car into the incline (see Figure 101).

We see from right $\triangle TBC$ in the figure that $\sin 13.2° = \dfrac{|\mathbf{u}|}{|\mathbf{w}|}$. Thus, $|\mathbf{u}| = |\mathbf{w}| \sin 13.2° =$

2640 sin 13.2° ≈ 603, and a force of approximately 603 pounds is needed to keep the car from rolling down the incline.

[102] *Let **u** and **v** denote to boat's intended and actual velocity vectors and let **w** denote the velocity*

vector of the current (see figure 102). We see from the figure that $\tan \alpha = \dfrac{|\mathbf{w}|}{|\mathbf{u}|} = \dfrac{6.5}{15} \approx 0.4333$.

Since α is acute, $\alpha = \tan^{-1}(0.4333) \approx 23°$.

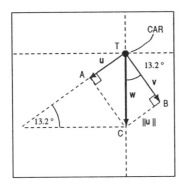

Figure 101

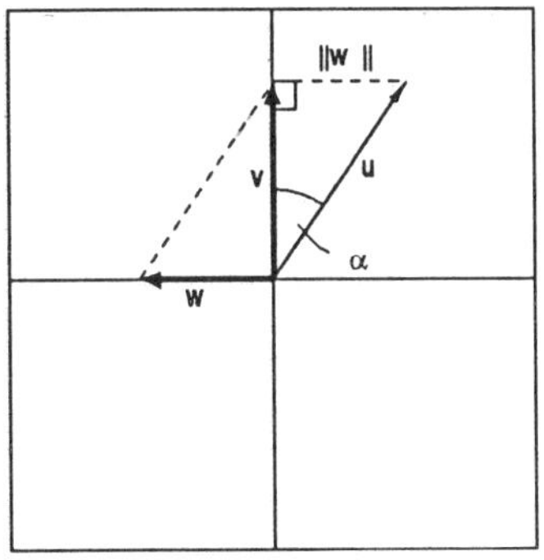

Figure 102

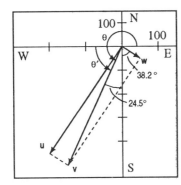

Figure 103

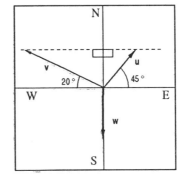

Figure 104

357

[103] Let **u**, **v** and **w** denote the instrument velocity, actual velocity and wind velocity vectors, respectively (see Figure 103). We must find the magnitude and direction of the actual velocity vector **u**. (Note that the wind <u>from</u> the direction N 38.2° W is blowing in the direction S 38.2° E.) We see from the figure that the direction angle for **u** is 270° − 24.5° = 245.5° and the direction angle for **w** is 270° + 38.2° = 308.2°. Thus **u** and **w** have the following algebraic representations **u** = ⟨515 cos 245.5°, 515 sin 245.5°⟩ ≈ ⟨−213.6, −468.6⟩ and **w** = ⟨62.5 cos 308.2°, 62.5 sin 308.2°⟩ ≈ ⟨38.7, −49.1⟩. Thus, **v** = **u** + **w** ≈ ⟨−174.9, −517.7⟩ and the magnitude of **v** is $|\mathbf{v}| \approx \sqrt{(-174.9)^2 + (-517.7)^2} \approx 546$. If θ is the directional angle for **v**, then $\tan\theta \approx \frac{-517.7}{-174.9} \approx 2.960$. If θ′ is the reference angle of θ, than $\tan\theta' = |\tan\theta| \approx 2.960 \Rightarrow \theta' = 71°$. Since θ is in QIII, θ = 180° + θ′ ≈ 251°. To obtain the direction of **v**, we subtract 251° from 270°, getting 19°. Thus, the plane's actual velocity vector has a magnitude of approximately 546 mph in the approximate direction S 19° W.

[104] Let **u**, **v** and **w** represent the 3 cables as shown in Figure 104. We see from the figure that the direction angle for **v** is 180° − 20° = 160°. The vectors **u**, **v** and **w** have the following algebraic representations **u** = ⟨|**u**| cos 45°, |**u**| sin 45°⟩, **v** = ⟨|**v**| cos 160°, |**v**| sin 160°⟩ and **w** = ⟨|**w**| cos 270°, |**w**| sin 270°⟩ = ⟨620 · 0, 620(−1)⟩ = ⟨0, −620⟩. Since the 3 vectors are in equilibrium **u** + **v** + **w** = 0. Thus, **u** + **v** = −**w** ⇒ ⟨|**u**| cos 45° + |**v**| cos 160°, |**u**| sin 45° + |**v**| sin 160°⟩ = ⟨0, 620⟩ ⇒ |**u**| cos 45° + |**v**| cos 160° = 0 $\left(*_1\right)$ and |**u**| sin 45° + |**v**| sin 160° = 620 $\left(*_2\right)$. From equation $\left(*_1\right)$ we see that |**u**| cos 45° = −|**v**| cos 160°, and from equation $\left(*_2\right)$ we see that |**u**| sin 45° = 620 − |**v**| sin 160°. Now, cos 45° = sin 45° ⇒ |**u**| cos 45° = |**u**| sin 45° ⇒ −|**v**| cos 160° = 620 − |**v**| sin 160° ⇒ |**v**| sin 160° − |**v**| cos 160° = 620 ⇒ $|\mathbf{v}| = \frac{620}{\sin 160° - \cos 160°} \approx 484$. Also |**u**| cos 45° = −|**v**| cos 160° ⇒ $|\mathbf{u}| = \frac{-|\mathbf{v}|\cos 160°}{\cos 45°} \approx \frac{-484\cos 160°}{\cos 45°} \approx 643$. Thus, the tension in cable **u** is approximately 643 pounds and the tension in cable **v** is approximately 484 pounds. Of course, the tension in cable **w** is 620 pounds.

[105] $x = r\cos\theta = 3\cos\frac{\pi}{3} = 3\left(\frac{1}{2}\right) = \frac{3}{2}$ and $y = r\sin\theta = 3\sin\frac{\pi}{3} = 3\left(\frac{\sqrt{3}}{2}\right) = \frac{3\sqrt{3}}{2}$. Thus, $\left(3, \frac{\pi}{3}\right) = \left(\frac{3}{2}, \frac{3\sqrt{3}}{2}\right)$.

[106] $x = 3.25\cos 570° = \left(\frac{13}{4}\right)\left(-\frac{\sqrt{3}}{2}\right) = -\frac{13\sqrt{3}}{8}$ and $y = 3.25\sin 570° = \left(\frac{13}{4}\right)\left(-\frac{1}{2}\right) = -\frac{13}{8}$. Thus, $\left(3.25, 570°\right) = \left(-\frac{13\sqrt{3}}{8}, -\frac{13}{8}\right)$.

[107] $x = \left(-\frac{12}{5}\right)\cos 120° = \left(-\frac{12}{5}\right)\left(-\frac{1}{2}\right) = \frac{6}{5}$ and $y = \left(-\frac{12}{5}\right)\sin 120° = \left(-\frac{12}{5}\right)\left(\frac{\sqrt{3}}{2}\right) = -\frac{6\sqrt{3}}{5}$. Thus,

$\left(-\frac{12}{5}, 120°\right) = \left(\frac{6}{5}, -\frac{6\sqrt{3}}{5}\right)$.

[108] $x = (-4)\cos\left(-\frac{3\pi}{4}\right) = (-4)\left(-\frac{\sqrt{2}}{2}\right) = 2\sqrt{2}$ and $y = -4\sin\left(-\frac{3\pi}{4}\right) = (-4)\left(-\frac{\sqrt{2}}{2}\right) = 2\sqrt{2}$. Thus,

$\left(-4, -\frac{3\pi}{4}\right) = \left(2\sqrt{2}, 2\sqrt{2}\right)$.

[109] $r = \sqrt{(-2)^2 + (-2\sqrt{3})^2} = 4$. $\tan\theta = \frac{-2\sqrt{3}}{-2} = \sqrt{3}$ and θ in QIII $\Rightarrow \theta = \frac{4\pi}{3}$. Thus,

$\left(-2, -2\sqrt{3}\right) = \left(4, \frac{4\pi}{3}\right)$.

[110] $r = \sqrt{(-5)^2 + (5)^2} = 5\sqrt{2}$. $\tan\theta = \frac{5}{-5} = -1$ and θ in QII $\Rightarrow \theta = \frac{3\pi}{4}$. Thus, $(-5, 5) = \left(5\sqrt{2}, \frac{3\pi}{4}\right)$.

[111] $r = \sqrt{(-3)^2 + (4)^2} = 5$. $\tan\theta = \frac{4}{-3} = -\frac{4}{3}$. If θ' is the reference angle for θ, then $\tan\theta' = |\tan\theta| = $

$\frac{4}{3} \Rightarrow \theta' \approx 53°$. Since θ is in QII, $\theta = 180° - \theta' \approx 127°$. Thus $(-3, 4) = \left(5, 180° - \tan^{-1}\frac{4}{3}\right) \approx (5, 127°)$.

[112] $r = \sqrt{(0)^2 + \left(-\frac{7}{2}\right)^2} = \frac{7}{2}$, and $\theta = \frac{3\pi}{2} \Rightarrow \left(0, -\frac{7}{2}\right) = \left(\frac{7}{2}, \frac{3\pi}{2}\right)$.

[113] $r = 5 \Rightarrow x^2 + y^2 = 25$

[114] The graph of $\theta = \frac{3\pi}{4}$ is the line L formed by the ray $\theta = \frac{3\pi}{4}$ and its extension through the origin

(pole). When $r = 1$, the point $\left(1, \frac{3\pi}{4}\right)$ belongs to L. Since $x = r\cos\theta = 1\cos\frac{3\pi}{4} = -\frac{\sqrt{2}}{2}$ and $y = $

$r\sin\theta = 1\sin\frac{3\pi}{4} = \frac{\sqrt{2}}{2}$, the point $(x, y) = \left(-\frac{\sqrt{2}}{2}, \frac{\sqrt{2}}{2}\right)$ belongs to L. Since $(0,0)$ also belongs to L,

the slope of L is $m = \dfrac{\frac{\sqrt{2}}{2} - 0}{-\frac{\sqrt{2}}{2} - 0} = -1$. Thus, L is the line $y = -x$.

[115] $r^2\sin 2\theta = 2 \Rightarrow r^2(2\sin\theta\cos\theta) = 2 \Rightarrow (r\sin\theta)(r\cos\theta) = 1 \Rightarrow yx = 1 \Rightarrow y = \frac{1}{x}$.

[116] $r = 12\cos\theta \Rightarrow r^2 = 12r\cos\theta$ (if $r \neq 0$) $\Rightarrow x^2 + y^2 = 12x \Rightarrow x^2 + y^2 - 12x = 0$.

[117] $3xy = 1 \Rightarrow 2(r\cos\theta)(r\sin\theta) = 1 \Rightarrow 3r^2\sin\theta\cos\theta = 1 \Rightarrow r^2 = \dfrac{1}{3\sin\theta\cos\theta}$.

[118] $4x^2 - 9y^2 = 36 \Rightarrow 4(r\cos\theta)^2 - 9(r\sin\theta)^2 = 36 \Rightarrow 4r^2\cos^2\theta - 9r^2\sin^2\theta = 36 \Rightarrow$

$4r^2(1 - \sin^2\theta) - 9r^2\sin^2\theta = 36 \Rightarrow 4r^2 - 13r^2\sin^2\theta = 36 \Rightarrow r^2 = \dfrac{36}{4 - 13\sin^2\theta}$.

[119] $x^2 + 9y^2 = 9 \Rightarrow (r\cos\theta)^2 + 9(r\sin\theta)^2 = 9 \Rightarrow r^2\cos^2\theta + 9r^2\sin^2\theta = 9 \Rightarrow$

$r^2(1 - \sin^2\theta) + 9r^2\sin^2\theta = 9 \Rightarrow r^2 + 8r^2\sin^2\theta = 9 \Rightarrow r^2 = \dfrac{9}{1 + 8\sin^2\theta}$.

[120] $(x^2 + y^2)^{3/2} = 32\,x\,y \Rightarrow (r^2)^{3/2} = 32\,(r\cos\theta)(r\sin\theta) \Rightarrow r^3 = 32\,r^2\sin\theta\cos\theta \Rightarrow$

$r = 16\,(2\sin\theta\cos\theta)$, if $r \neq 0 \Rightarrow r = 16\sin 2\,\theta$.

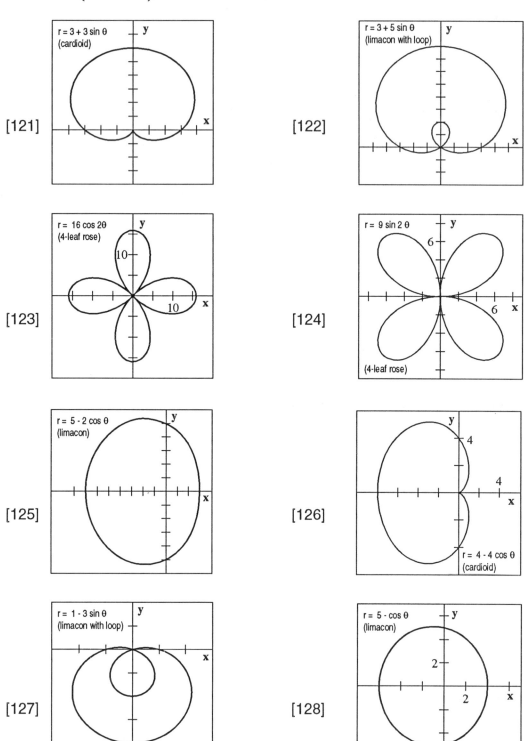

[121]

$r = 3 + 3\sin\theta$
(cardioid)

[122]

$r = 3 + 5\sin\theta$
(limacon with loop)

[123]

$r = 16\cos 2\theta$
(4-leaf rose)

[124]

$r = 9\sin 2\theta$

(4-leaf rose)

[125]

$r = 5 - 2\cos\theta$
(limacon)

[126]

$r = 4 - 4\cos\theta$
(cardioid)

[127]

$r = 1 - 3\sin\theta$
(limacon with loop)

[128]

$r = 5 - \cos\theta$
(limacon)

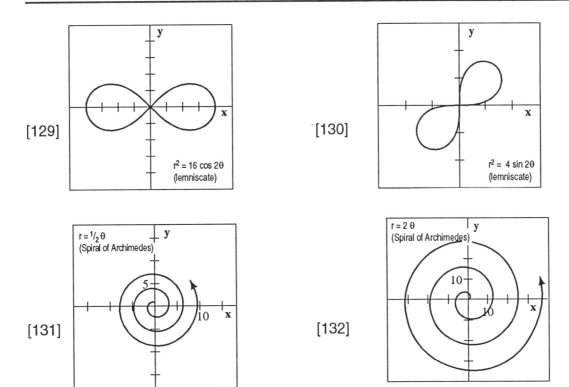

[129]

$r^2 = 16 \cos 2\theta$
(lemniscate)

[130]

$r^2 = 4 \sin 2\theta$
(lemniscate)

[131]

$r = \frac{1}{2}\theta$
(Spiral of Archimedes)

[132]

$r = 2\theta$
(Spiral of Archimedes)

EXERCISES 8.2

[1]   In the equation $x^2 = 16y$, $4p = 16 \Rightarrow p = 4$. vertex, $(0, 0)$; focus, $(0, 4)$; axis, y – axis; directrix,

       $y = -4$. (See figure below).

[2]   In the equation $x^2 = -2y$, $4p = -2 \Rightarrow p = -\frac{1}{2}$. vertex, $(0, 0)$; focus, $\left(0, -\frac{1}{2}\right)$; axis, y – axis;

       directrix, $y = \frac{1}{2}$. (See figure below).

[3]   In the equation $y^2 = 6x$, $4p = 6 \Rightarrow p = \frac{3}{2}$. vertex, $(0, 0)$; focus, $\left(\frac{3}{2}, 0\right)$; axis, x – axis; directrix, $x = -\frac{3}{2}$.

       (See figure below).

[4]   In the equation $y^2 = 10x$, $4p = 10 \Rightarrow p = \frac{5}{2}$. vertex, $(0, 0)$; focus, $\left(\frac{5}{2}, 0\right)$; axis, x – axis; directrix,

       $p = -\frac{5}{2}$. (See figure below).

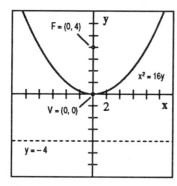

Figure 1   **Figure 2**

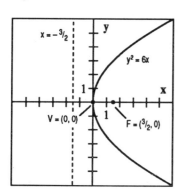   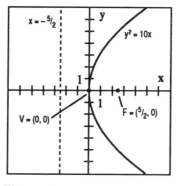

Figure 3   **Figure 4**

[5] $x^2 + 10y = 0 \Rightarrow x^2 = -10y \Rightarrow 4p = -10 \Rightarrow p = -\frac{5}{2}$. vertex, $(0, 0)$; focus, $\left(0, -\frac{5}{2}\right)$; axis, y – axis;

       directrix, $y = \frac{5}{2}$. (See figure below).

[6] $2y^2 - x = 10 \Rightarrow y^2 = \frac{1}{2}(x + 10) \Rightarrow (y - 0)^2 = \frac{1}{2}(x + 10) \Rightarrow 4p = \frac{1}{2} \Rightarrow p = \frac{1}{8}$. vertex, $(-10, 0)$; focus,

       $\left(-\frac{79}{8}, 0\right)$; axis, x – axis; directrix, $x = -\frac{81}{8}$. (See figure below).

[7] $2y^2 - 5x = 0 \Rightarrow y^2 = \frac{5}{2}x \Rightarrow 4p = \frac{5}{2} \Rightarrow p = \frac{5}{8}$. vertex, $(0, 0)$; focus, $\left(\frac{5}{8}, 0\right)$; axis, x – axis; directrix,

       $x = -\frac{5}{8}$. (See figure below).

[8] $3x^2 + 5y = 0 \Rightarrow x^2 = -\frac{5}{3}y \Rightarrow p = -\frac{5}{12}$. vertex, $(0, 0)$; focus, $\left(0, -\frac{5}{12}\right)$; axis, y – axis; directrix,

$y = \frac{5}{12}$. (See figure below).

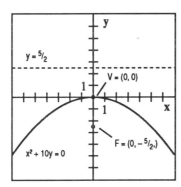

Figure 5

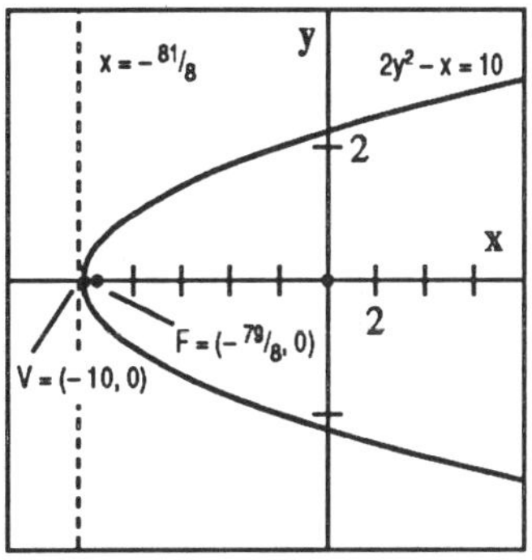

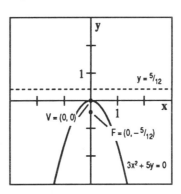

Figure 6

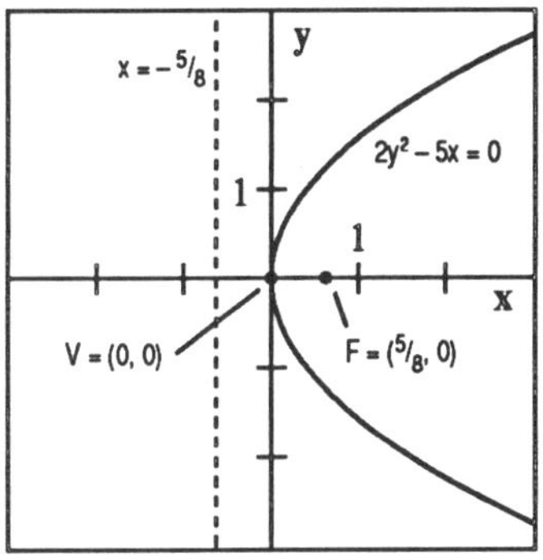

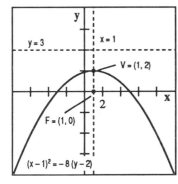

Figure 7

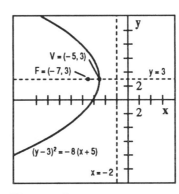

Figure 8

[9] $(x - 1)^2 = -8(y - 2) \Rightarrow 4p = -8 \Rightarrow p = -2$. vertex, $(1, 2)$; focus, $(1, 0)$; axis, $x = 1$; directrix, $y = 4$.

(See figure below).   (See figure below).

[10] $(y - 3)^2 = -8(x + 5) \Rightarrow 4p = -8 \Rightarrow p = -2$. vertex, $(-5, 3)$; focus, $(-7, 3)$; axis, $y = 3$; directrix,

$x = -3$. (See figure below).

Figure 9

Figure 10

[11] $y^2 - 2y - 8x + 25 = 0 \Rightarrow y^2 - 2y = 8x - 25 \Rightarrow y^2 - 2y + 1 = 8x - 24 \Rightarrow (y - 1)^2 = 8(x - 3) \Rightarrow 4p =$

$8 \Rightarrow p = 2$. vertex, $(3, 1)$; focus, $(5, 1)$; axis, $y = 1$; directrix, $x = 1$.   (See figure below).

[12] $x^2 - 2x + 12y + 13 = 0 \Rightarrow x^2 - 2x = -12y - 13 \Rightarrow x^2 - 2x + 1 = -12y - 12 = -12(y + 1) \Rightarrow (x - 1)^2$

       $= -12(y + 1) \Rightarrow 4p = -12 \Rightarrow p = -3.$ vertex, $(1, -1)$; focus, $(1, -4)$; axis, $x = 1$; directrix, $y = 2.$

       (See figure below).

[13] $4x^2 - 10x + y - 1 = 0 \Rightarrow 4x^2 - 10x = -y + 1 \Rightarrow x^2 - \frac{10}{4}x = -\frac{1}{4}y + \frac{1}{4} \Rightarrow x^2 - \frac{10}{4}x + \frac{25}{16} =$

       $-\frac{1}{4}y + \frac{1}{4} + \frac{25}{16} \Rightarrow \left(x - \frac{5}{4}\right)^2 = -\frac{1}{4}y + \frac{24}{16} = -\frac{1}{4}\left(y - \frac{29}{4}\right) \Rightarrow 4p = -\frac{1}{4} \Rightarrow p = -\frac{1}{16}.$ Vertex, $\left(\frac{5}{4}, \frac{29}{4}\right)$;

       focus, $\left(\frac{5}{4}, \frac{115}{16}\right)$; axis, $x = \frac{5}{4}$; directrix, $y = \frac{117}{16}.$

          (See figure below).

[14] $y^2 - 4x - 4y + 8 = 0 \Rightarrow y^2 - 4y = 4x - 8 \Rightarrow y^2 - 4y + 4 = 4x - 4 = 4(x - 1) \Rightarrow (y - 2)^2 = 4(x - 1) \Rightarrow$

       $4p = 4 \Rightarrow p = 1.$ vertex, $(1, 2)$; focus, $(2, 2)$; axis, $y = 2$; directrix, $x = 0$ (the y– axis).

       (See figure below).

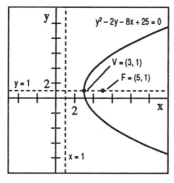

Figure 11

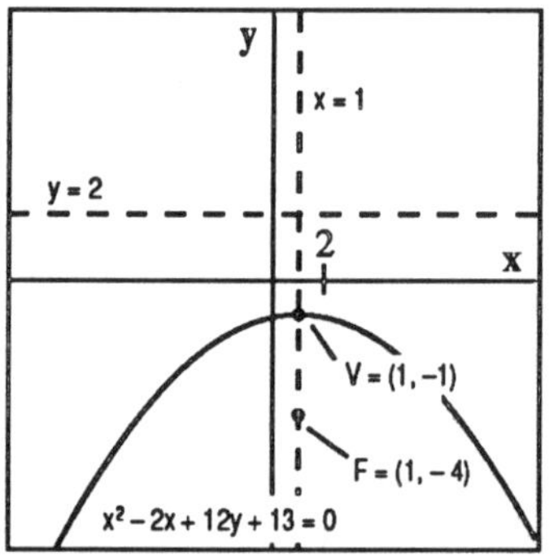

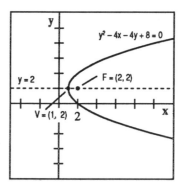

Figure 12

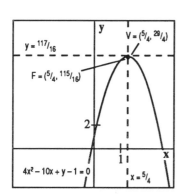

Figure 13

Figure 14

[15] $y - 9 = x^2 + 4 \Rightarrow y - 13 = x^2 \Rightarrow 4p = 1 \Rightarrow p = \frac{1}{4}.$ vertex, $(0, 13)$; focus, $\left(0, \frac{53}{4}\right)$; axis, y– axis;

       directrix, $y = \frac{51}{4}.$ (See figure below).

[16] $y + 5 = x^2 + 4x \Rightarrow y + 9 = x^2 + 4x + 4 \Rightarrow y + 9 = (x + 2)^2 \Rightarrow 4p = 1 \Rightarrow p = \frac{1}{4}.$ vertex, $(-2, -9)$;

       focus, $\left(-2, \frac{-35}{4}\right)$; axis, $x = -2$; directrix, $y = -\frac{37}{4}.$ (See figure below).

[17] $x = y^2 - 6y - 3 \Rightarrow x + 3 = y^2 - 6y \Rightarrow x + 12 = y^2 - 6y + 9 \Rightarrow x + 12 = (y - 3)^2 \Rightarrow 4p = 1 \Rightarrow p = \frac{1}{4}$.

       vertex, $(-12, 3)$; focus, $\left(-\frac{47}{4}, 3\right)$; axis, $y = 3$; directrix, $x = -\frac{49}{4}$.   (See figure below).

[18] $x = y^2 - 2y - 2 \Rightarrow x + 2 = y^2 - 2y \Rightarrow x + 3 = y^2 - 2y + 1 \Rightarrow x + 3 = (y - 1)^2 \Rightarrow 4p = 1 \Rightarrow p = \frac{1}{4}$.

       vertex, $(-3, 1)$; focus, $\left(-\frac{11}{4}, 1\right)$; axis, $y = 1$; directrix, $x = -\frac{13}{4}$.   (See figure below).

[19] $6y = 4x^2 + 4x + 7 \Rightarrow 6y - 7 = 4x^2 + 4x \Rightarrow \frac{3}{2}y - \frac{7}{4} = x^2 + x \Rightarrow \frac{3}{2}y - \frac{3}{2} = x^2 + x + 1 \Rightarrow \frac{3}{2}(y - 1) =$

       $\left(x + \frac{1}{2}\right)^2 \Rightarrow 4p = \frac{3}{2} \Rightarrow p = \frac{3}{8}$. vertex, $\left(-\frac{1}{2}, 1\right)$; focus, $\left(-\frac{1}{2}, \frac{11}{8}\right)$; axis, $x = -\frac{1}{2}$; directrix, $y = \frac{5}{8}$.

       (See figure below).

[20] $2x = y^2 + 2y + 5 \Rightarrow 2x - 5 = y^2 + 2y \Rightarrow 2x - 4 = y^2 + 2y + 1 \Rightarrow 2(x - 2) = (y + 1)^2 \; 4p = 2 \Rightarrow$

       $p = \frac{1}{2}$. vertex, $(2, -1)$; focus, $\left(\frac{5}{2}, -1\right)$; axis, $y = -1$; directrix, $x = \frac{3}{2}$.   (See figure below).

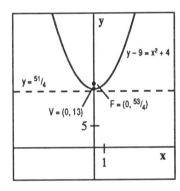

Figure 15

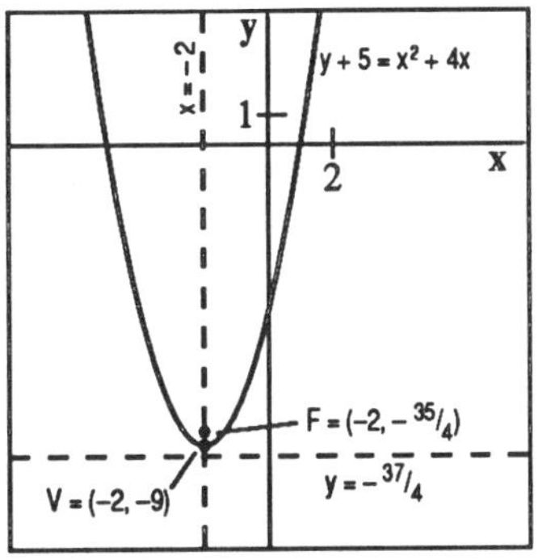

Figure 16

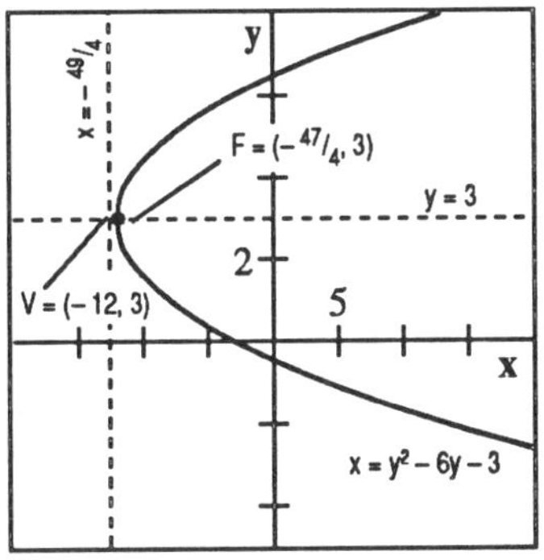

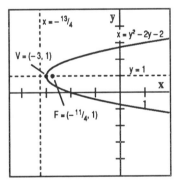

Figure 17

Figure 18

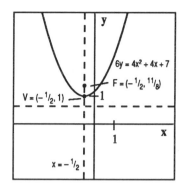

Figure 19

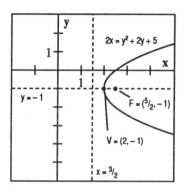

Figure 20

[21] The axis of this parabola is the y – axis. Since the vertex is $(0, 0)$ and since the focus is $\left(0, -\frac{1}{9}\right)$,

$p = -\frac{1}{4}$. Thus, the equation is $x^2 = -y$.   ● $x^2 = -y$.

[22] The axis of this parabola is the y – axis. Since the vertex is $(0, 0)$ and since the focus is $(0, \sqrt{2})$,

$p = \sqrt{2}$. Hence, the equation *is* $x^2 = 4\sqrt{2}\, y$.   ● $x^2 = 4\sqrt{2}\, y$

[23] The axis of this parabola is the line $y = 3$. Since the vertex is $(-2, 3)$ and focus is $(-3, 3)$, $h = -2$

and $h + p = -3 \Rightarrow p = 1$. Thus, the equation is $(y - 3)^2 = -4(x + 2)$.   ● $(y - 3)^2 = -4(x + 2)$.

[24] The axis for this parabola is the line $y = -1$. Since the vertex is $(3, -1)$ and the focus is $(5, -1)$,

$h = 3$ and $h + p = 5 \Rightarrow p = 2$. Thus, the equation is $(y + 1)^2 = 8(x - 3)$. $(y + 1)^2 = 8(x - 3)$.

[25] Since the focus is $(0, 0)$ and the directrix is the line $x = -2$, the vertex is $(-1, 0)$. Hence, $h = -1$ and

$h + p = 0 \Rightarrow p = 1$. Therefore, the equation is $y^2 = 4(x + 1)$.   ● $y^2 = 4(x + 1)$.

[26] Since the focus is $(0, 0)$ and the directrix is the line $y = 6$, the vertex is $(0, 3)$. Thus $k = 3$ and

$k + p = 0 \Rightarrow p = -3$. Therefore, the equation is $x^2 = -12(y - 3)$.   ● $x^2 = -12(y - 3)$.

[27] Since the vertex is $(-4, -2)$ and the directrix is the line $y = -5$, $k = -2$ and $k - p = -5 \Rightarrow p = 3$.

Hence, the equation is $(x + 4)^2 = 12(y + 2)$.   ● $(x + 4)^2 = 12(y + 2)$.

[28] Since the vertex is $(-1, 6)$ and the directrix is $x = 2$, $h = -1$ and $h - p = 2 \Rightarrow p = -3$. Thus, the

equation is $(y - 6)^2 = -12(x + 1)$.   ● $(y - 6)^2 = -12(x + 1)$.

[29] Since the vertex is $(0, 0)$ and the axis is the x – axis, the equation has the form $y^2 = 4px$. Since the

point $P = (-2, 4)$ lies on the parabola, we have $16 = -8p \Rightarrow p = -2$. Hence, the equation is

$y^2 = -8x$.   ● $y^2 = -8x$.

[30] Since the vertex is $(0, 0)$ and the axis is the y – axis, the equation has the form $x^2 = 4py$. Moreover,

since the point $P = (-1, -6)$ lies on the parabola, we have $1 = -24p \Rightarrow p = -\frac{1}{24}$. Hence, the

equation is $x^2 = -\frac{1}{6}y$.   ● $x^2 = -\frac{1}{6}y$.

[31] a)   The directrix l of this parabola is the line $y = -a$. Thus, a point $P = (x, y)$ is on the parabola

with focus $(0, a)$, $a > 0$, and directrix $y = -a$ if an only if $d(P, F) = d(P, l) \Rightarrow$

$\sqrt{(x - 0)^2 + (y - a)^2} = y - (-a) = y + a \Rightarrow x^2 + y^2 - 2ay + a^2 = y^2 + 2ay + a^2 \Rightarrow x^2 - 4ay$.

b) The directrix l of this parabola is the line $y = a$. Hence, a point $P = (x, y)$ is on the parabola with focus $(0, -a)$, $a > 0$, and directrix $y = a$ if and only if $d(P, F) = d(P, l) \Rightarrow$

$$\sqrt{(x-0)^2 + [y - (-a)]^2} = y - a \Rightarrow x^2 + y^2 + 2ay + a^2 = y^2 - 2ay + a^2 \Rightarrow x^2 = -4ay.$$

[32] The directrix l of this parabola is the line $x = h - p$. Therefore, a point $P = (x, y)$ is on this

parabola if and only if $d(P, F) = d(P, l) \Rightarrow \sqrt{[x - (h + p)]^2 + (y - k)^2} = x - (h - p) \Rightarrow$

$[x - h - p]^2 + (y - k)^2 = (x - h + p)^2 \Rightarrow x^2 - 2hx - 2xp - 12hp +$

$h^2 + p^2 + (y - k)^2 = x^2 - 2hx + 2px - 2hp + p^2 \Rightarrow (y - k)^2 = 4px - 4hp \Rightarrow (y - k)^2 = 4p(x - h).$

[33] We introduce a rectangular coordinate system with the origin at the vertex of the parabola as shown

in equation of the form $y^2 = 4px$. Since the point $(4, 6)$ lies on the parabola, $6^2 = 4p(4) \Rightarrow p = \frac{9}{4}$.

Hence, the eyepiece should be placed on the axis $\frac{9}{4}$ feet from the vertex.   (See figure below).

● on the axis $\frac{9}{4}$ feet from the vertex.

[34] We introduce a rectangular coordinate system with the origin at the vertex of the parabola as shown

in the accompanying figure. The equation of this parabola is $y^2 = 4x$. If the point $P = (2, y)$ lies on

this parabola, the $y^2 = 8 \Rightarrow y = 2\sqrt{2}$. Hence, the diameter of the spotlight is $2(2\sqrt{2}) = 4\sqrt{2}$ feet.

(See figure below).   ● $4\sqrt{2}$ *feet.*

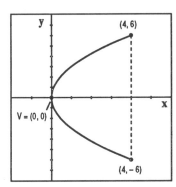

Figure 33   **Figure 34**

[35] The parabolic path is described by the equation $x^2 = -4p(y - 10)$ where $p > 0$. Since the point

$(2, 9)$ lies on this path, $2^2 = -4p(9 - 10) \Rightarrow p = 1 \Rightarrow x^2 = -4(y - 10)$. The water strikes the

ground at the point on the path where $y = 0 \Rightarrow x^2 = 40 \Rightarrow x = \sqrt{40} = 2\sqrt{10} \approx 6.32$ feet.

● approximately 6.32 feet from the vertical line passing through the end of the water pipe.

[36] a) Since $x = 10t$, $t = \frac{x}{10}$. Hence, $y = -16\left(\frac{x}{10}\right)^2 + 64\left(\frac{x}{10}\right) \Rightarrow y = -0.16x^2 + 6.4x$. Therefore, the

path of the projectile is parabolic.

b) (*i*) If $t = 1$, $x = 10$ and $y = 48$ ● $(10, 48)$

(*ii*) If $t = 2$, $x = 20$ and $y = 64$ ● $(20, 64)$

(*iii*) If $t = 3$, $x = 30$ and $y = 48$ ● $(30, 48)$

c) The maximum height of the particle is the y – coordinate of the vertex of the parabola $y =$

$-0.16x^2 + 6.4x$, which is 64.   ● 64 feet.

d) The projectile hits the ground at the time $t > 0$ where $y = 0$. Thus, we solve the equation

$$0 = -16t^2 + 64t \Rightarrow t = 4.$$

● $t = 4$ sec.

[37] a) The axis of the parabola $x^2 = y + 1$ is the y – axis, and the focus is $F = \left(0, -\frac{3}{4}\right)$ since the vertex is $(0, -1)$ and $p = \frac{1}{4}$. Let $Q = (0, b)$ be the point of intersection of the y – axis and the tangent line to this parabola at the point $P = (1, 0)$. Then $d(F, P) = d(F, Q) \Rightarrow$

$$\sqrt{(1-0)^2 + \left[0 - \left(-\tfrac{3}{4}\right)\right]^2} = \left|b - \left(-\tfrac{3}{4}\right)\right| \Rightarrow \sqrt{1 + \tfrac{9}{16}} = \left|b + \tfrac{3}{4}\right| \Rightarrow \tfrac{5}{4} = \left|b + \tfrac{3}{4}\right| \Rightarrow b + \tfrac{3}{4} =$$

$\pm\frac{5}{4} \Rightarrow b = \frac{1}{2}$ or $b = -2$. Since b is less than the y – coordinate of the vertex, $b = -2$. Hence, the tangent line passes through the point $(0, -2)$. Thus, the slope is $\frac{2}{1} = 2$, and the equation is

$y = 2(x - 1) = 2x - 2$. (See figure below).

● $y = 2x - 2$.

b) The axis of the parabola $y^2 = x - 1$ is the x – axis, and the focus is $F = \left(\frac{5}{4}, 0\right)$ since the vertex is $(1, 0)$ and $p = \frac{1}{4}$. The $Q = (b, 0)$ be the point of intersection of the x – axis and the tangent line to this parabola at the point $P = (2, 1)$. Then $d(F, P) = d(F, Q) \Rightarrow \sqrt{\left(2 - \tfrac{5}{4}\right)^2 + (1 - 0)^2} =$

$\left|b - \tfrac{5}{4}\right| \Rightarrow \sqrt{\tfrac{9}{16} + 1} = \left|b - \tfrac{5}{4}\right| \Rightarrow \tfrac{5}{4} = \left|b - \tfrac{5}{4}\right| \Rightarrow b - \tfrac{5}{4} = \pm\tfrac{5}{4} \Rightarrow b = \tfrac{5}{2}$ or $b = 0$. Since b is less than the x – coordinate of the vertex, $b = 0$. Hence, the tangent line passes through the origin. It follows that the slope of the line is $\frac{1}{2}$ and the equation is $y = \frac{1}{2}x$. (See figure below).

● $y = \frac{1}{2}x$.

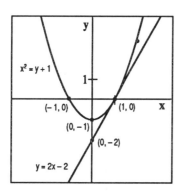

Figure 37a

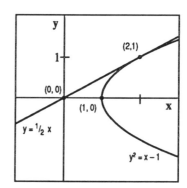

Figure 37b

c) To find the focus of the parabola $4y = 3 + 2x - x^2$, we complete the square in x. $4y = 3 + 2x - x^2 \Rightarrow 4y - 3 = -(x^2 - 2x) \Rightarrow -4y + 3 + 1 = x^2 - 2x + 1 \Rightarrow -4(y-1) = (x-1)^2$. The vertex of this parabola is $(1, 1)$ and the axis is the line $x = 1$. Since $p = -1$, the focus is $F = (1, 0)$. Let $Q = (1, b)$ be the point of intersection of the axis of the parabola, $x = 1$, and the tangent line to this parabola at the point $P = (5, -3)$. Then $d(F, P) = d(F, Q) \Rightarrow$ $\sqrt{(5-1)^2 + (-3-0)^2} = |b-1| \Rightarrow \sqrt{16+9} = |b-0| \Rightarrow 5 = |b| \Rightarrow b = \pm 5$. Since b is greater than the y – coordinate of the vertex, $b = 5$. Hence, the tangent line passes through the point $(1, 5)$. It follows that the slope is $\frac{-3-5}{5-1} = -2$. Hence, the equation of the tangent line is $y + 3 = -2(x - 5) \Rightarrow y = -2x + 7$. (See figure below). $\qquad\bullet\, y = -2x + 7.$

d) The vertex of the parabola $y^2 = -4(x+1)$ is the point $(-1, 0)$ and the axis is the x – axis. Since $p = -1$, the focus $F = (-2, 0)$. Let $Q = (b, 0)$ be the point of intersection of the x – axis and the tangent line to this parabola passing through the point $(-2, 2)$. Then $d(F, P) = d(F, Q) \Rightarrow$ $\sqrt{(-2+2)^2 + (2-0)^2} = |b - (-2)| \Rightarrow \sqrt{4} = |b+2| \Rightarrow 2 = |b+2| \Rightarrow b + 2 = \pm 2 \Rightarrow b = -4$ or $b = 0$. Since b is greater than the x – coordinate of the vertex, $b = 0$. Hence, the tangent line passes through the origin. The slope of the tangent line is $\frac{2}{-2} = -1$ and the equation is $y = -x$. (See figure below). $\qquad\bullet\, y = -x.$

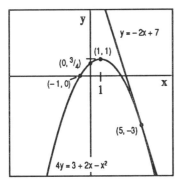

Figure 37c   **Figure 37d**

[38] In the accompanying figure, we have introduced a rectangular coordinate system so that the vertex of the parabolic path is located at the point $(300, 200)$. The equation of this parabola has the form $(x - 300)^2 = -4p(y - 200)$. Since the point $(0, 0)$ lies on this parabola, $(0 - 300)^2 = -4p(0 - 200) \Rightarrow p = -\frac{225}{2}$. Hence, an equation of the parabola is $(x - 300)^2 = -450(y - 200)$. (See figure below). $\qquad\bullet\, (x - 300)^2 = -450(y - 200).$

[39] We introduce a rectangular coordinate system so that the shortest supporting cable lies along the axis of the parabola as shown in the accompanying figure. The equation for the parabola has the form $y = ax^2 + bx + c$. Using the fact that the points $(-150, 100)$, $(0, 50)$ and $(150, 100)$ lie on the parabola, we find that $a = \frac{1}{450}$, $b = 0$ and $c = 50$. Thus, $y = \frac{1}{450}x^2 + 50$. If $x = 50$, then $y \approx 55.6$. (See figure below). $\qquad\bullet\, 55.6\,\text{ft}$

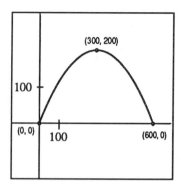

Figure 38

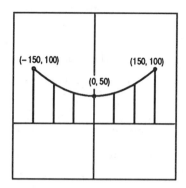

Figure 39

[40]   Referring to the accompanying figure, we know from Definition 8.2 that $d(P_1, F_1) = d(P_1, Q_1)$ and $d(P_2, F_1) = d(P_2, Q_2)$. Moreover, $d(P_2, Q_2) = d(R, Q_1) = d(P_1, Q_1) + d(P_1, R) = d(P_1, F_1) + d(P_1, R)$. It follows that all sound waves will travel the same distance.

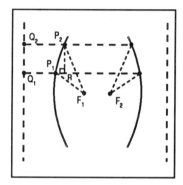

Figure 40

[41]   a)   b)

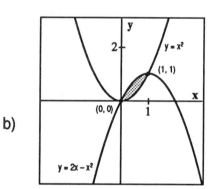

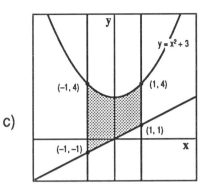

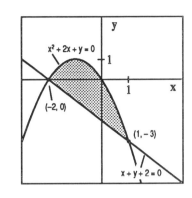

c)

d)

[42] a)   The focus of the parabola $y^2 = 8x$ is the point $(2, 0)$. Hence, the endpoints of the focal chord

have the form $(2, y_0)$, where $y_0^2 = 8\,(2) = 16$. Therefore, the endpoints are $(2, -4)$ and $(2, 4)$,

and thus, the focal width *is* $4 - (-4) = 8$.   ● 8

  b)   The focus of the parabola $x^2 = 4py$ is $(0, p)$. It follows that the endpoints of the focal chord are

$(-2\,|p|, p)$ and $(2\,|p|, p)$. Hence, the length of the focal chord is $2\,|p| - (-2\,|p|) = 4\,|p|$.

The proof is similar for the parabola $y^2 = 4px$.

  c)   From part 42 (b), we know that the endpoints of the focal chord of the parabola are $(-2\,|p|, p)$

and $(2\,|p|, p)$. If $p > 0$, $|p| = p$. Thus, the endpoints are $(\pm 2p, p)$. If $p < 0$, $|p| = -p$. Again,

it follows that the endpoints are $(\pm 2p, p)$. A similar argument can be used for the parabola

$y^2 = 4px$.

[43] Let P be any point of a parabola (shown in the figure below) with focus F and vertex V

such that $P \neq V$. From plane geometry and the definition of a parabola, we know that $d\,(F, Q) =$

$2 \cdot d\,(F, V) < d\,(F, R) < d\,(F, P) + d\,(P, R) = 2d\,(F, P)$. Hence, $d\,(F, V) < d\,(F, P)$.

[44] Let $P = \left(a, \dfrac{a^2}{4p}\right)$ and $Q = \left(b, \dfrac{b^2}{4p}\right)$. Then an equation of the line containing P and Q is $y - \dfrac{a^2}{4p} =$

$\dfrac{b + a}{4p}\,(x - a)$. It follows that the y – coordinate of R is $-\dfrac{ab}{4p} = -\dfrac{4pab}{16p^2}$. Noting that $\triangle PVQ$ is a

right triangle with hypotenuse $\overline{PQ}$, we use the Pythagorean theorem to show that $ab = -\dfrac{a^2 b^2}{16p^2}$, or

$16p^2 = -\dfrac{a^2 b^2}{ab}$. Since the y – coordinate of R is $-\dfrac{4pab}{16p^2}$ and since $16p^2 = -\dfrac{a^2 b^2}{ab}$, it follows that

the y – coordinate of R is $4p$. Hence, $R = (0, 4p)$, and $d\,(V, R) = \sqrt{16p^2} = 4\,|p.|$

[45] a)   Since $A \neq 0$ and $E \neq 0$, the equation $Ax^2 + Dx + Ey + F = 0$ can be written in the form $y =$

$A_0 x^2 + B_0 x + C_0$, where $A_0 = -\dfrac{A}{E}$, $B_0 = -\dfrac{D}{E}$ and $C_0 = -\dfrac{F}{E}$. We know that the graph of this

quadratic equation is a parabola.

  b)   Suppose $E = 0$ and $D^2 - 4AF = 0$. Using the quadratic formula, we find that the solution of

the equation $Ax^2 + Dx + F = 0$ *is* $-\dfrac{D}{2A}$. Hence, $x = -\dfrac{D}{2A}$ and therefore, the graph is a vertical

line.

c)   Suppose $E = 0$ and $D^2 - 4AF > 0$. It follows that $x = \dfrac{-D \pm \sqrt{D^2 - 4AF}}{2A}$; that is, $x =$

$\dfrac{-D + \sqrt{D^2 - 4AF}}{2A}$ or $x = \dfrac{-D - \sqrt{D^2 - 4AF}}{2a}$. The graph of these two equations are vertical

lines.

d)   If $E = 0$ and $D^2 - 4AF < 0$, there are no real number solutions for the equation.

[46]   See the solution to Exercise 45.   [47]   (See figure below).

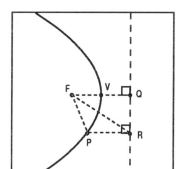

Figure 43

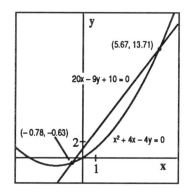

Figure 47

EXERCISES 8.3

[1]   In the equation $\dfrac{x^2}{36} + \dfrac{y^2}{25} = 1$, $a^2 = 36$ and $b^2 = 25 \Rightarrow a = 6$ and $b = 5$. Moreover, $c^2 = a^2 - b^2 =$

11 $\Rightarrow c = \sqrt{11}$. Center, $(0, 0)$; foci, $(\pm \sqrt{11}, 0)$; vertices $(\pm 6, 0)$; endpoints of minor axis, $(0, \pm 5)$.
(See figure below).

[2]   In the equation $\dfrac{x^2}{81} + \dfrac{y^2}{225} = 1$, $a^2 = 225$ and $b^2 = 81 \Rightarrow a = 15$ and $b = 9$. Moreover,

$c^2 = a^2 - b^2 = 144 \Rightarrow c = 12$. Center, $(0, 0)$; foci, $(0, \pm 12)$; vertices, $(0, \pm 15)$; endpoints of minor
axis, $(\pm 9, 0)$.   (See figure below).

[3]   In the equation $\dfrac{x^2}{25} + \dfrac{y^2}{36} = 1$, $a^2 = 36$ and $b^2 = 25 \Rightarrow a = 6$ and $b = 5$. Moreover, $c^2 = a^2 - b^2 =$

11 $\Rightarrow c = \sqrt{11}$. Center, $(0, 0)$; foci, $(0, \pm \sqrt{11})$; vertices, $(0, \pm 6)$; endpoints of minor axis, $(\pm 5, 0)$.
(See figure below).

[4]   In the equation $\dfrac{x^2}{225} + \dfrac{y^2}{81} = 1$, $a^2 = 225$ and $b^2 = 81 \Rightarrow a = 15$ and $b = 9$. Moreover,

$c^2 = a^2 - b^2 = 144 \Rightarrow c = 12$. Center, $(0, 0)$; foci, $(\pm 12, 0)$; vertices, $(\pm 15, 0)$; endpoints of minor
axis, $(0, \pm 9)$.   (See figure below).

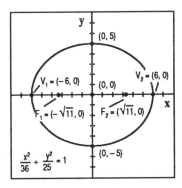

Figure 1

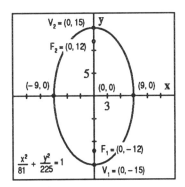

Figure 2

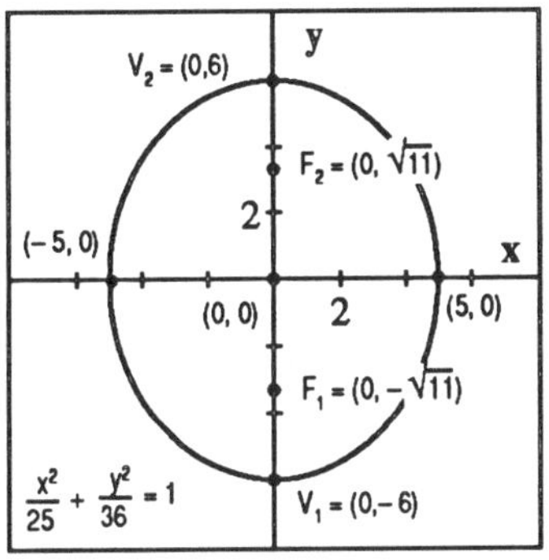

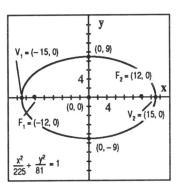

Figure 3

Figure 4

[5]   In the equation $\frac{x^2}{16} + \frac{y^2}{7} = 1$, $a^2 = 16$ and $b^2 = 7 \Rightarrow a = 4$ and $b = \sqrt{7}$. In addition, $c^2 = a^2 - b^2 = 9 \Rightarrow c = 3$. Center, $(0, 0)$; foci, $(\pm 3, 0)$; vertices, $(\pm 4, 0)$; endpoints of the minor axis, $(0, \pm \sqrt{7})$. (See figure below).

[6]   In the equation $\frac{x^2}{3} + \frac{y^2}{15} = 1$, $a^2 = 15$ and $b^2 = 3 \Rightarrow a = \sqrt{15}$ and $b = \sqrt{3}$. Moreover, $c^2 = a^2 - b^2 = 12 \Rightarrow c = 2\sqrt{3}$. Center, $(0, 0)$; foci, $(0, \pm 2\sqrt{3})$; vertices, $(0, \pm 15)$; endpoints of minor axis, $(\pm \sqrt{3}, 0)$. (See figure below).

[7]   In the equation $\frac{(x+2)^2}{9} + \frac{(y-3)^2}{4} = 1$, $a^2 = 9$ and $b^2 = 4 \Rightarrow a = 3$ and $b = 2$. In addition, $c^2 = a^2 - b^2 = 5 \Rightarrow c = \sqrt{5}$. Moreover, $h = -2$ and $k = 3$. Center, $(h, k) = (-2, 3)$; foci, $(h \pm c, k) = (-2 \pm \sqrt{5}, 3)$; vertices, $(h \pm a, k)$, or $(-5, 3)$ and $(1, 3)$; endpoints of minor axis, $(h, k \pm b)$, or $(-2, 1)$ and $(-2, 5)$. (See figure below).

[8]   In the equation $\frac{(x-5)^2}{10} + \frac{(y+1)^2}{12} = 1$, $a^2 = 12$ and $b^2 = 10 \Rightarrow a = 2\sqrt{3}$ and $b = \sqrt{10}$. In addition, $c^2 = a^2 - b^2 = 2 \Rightarrow c = \sqrt{2}$. Moreover, $h = 5$ and $k = -1$. Center, $(h, k) = (5, -1)$; foci, $(h, k \pm c) = (5, -1 \pm \sqrt{2})$; vertices, $(h, k \pm a) = (5, -1 \pm 2\sqrt{3})$; endpoints of minor axis, $(h \pm b, k) = (5 \pm \sqrt{10}, -1)$. (See figure below).

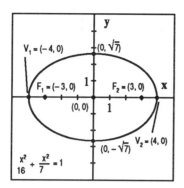

Figure 5

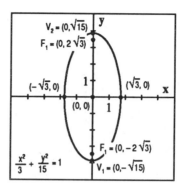

Figure 6

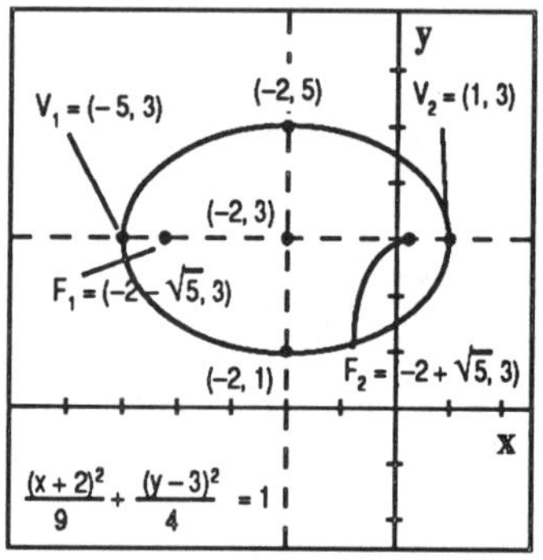

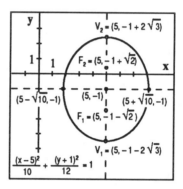

Figure 7

Figure 8

[9] In the equation $x^2 + \dfrac{(y+5)^2}{2} = 1$, $a^2 = 2$ and $b^2 = 1 \Rightarrow a = \sqrt{2}$ and $b = 1$. Moreover,

$c^2 = a^2 - b^2 = 1 \Rightarrow c = 1$. In addition, $h = 0$ and $k = -5$. Center, $(h, k) = (0, -5)$; foci,

$(h, k \pm c)$ or $(0, -6)$ and $(0, -4)$; vertices, $(h, k \pm a) = (0, -5 \pm \sqrt{2})$; endpoints of minor axis,

$(h \pm b, k)$, or $(-1, -5)$ and $(1, -5)$.   (See figure below).

[10] In the equation $\dfrac{(x+\pi)^2}{3} + y^2 = 1$, $a^2 = 3$ and $b^2 = 1 \Rightarrow a = \sqrt{3}$ and $b = 1$. Moreover,

$c^2 = a^2 - b^2 = 2 \Rightarrow c = \sqrt{2}$. Also, $h = -\pi$ and $k = 0$. Center, $(h, k) = (-\pi, 0)$; foci, $(h \pm c, k) =$

$(-\pi \pm \sqrt{2}, 0)$; vertices, $(h \pm a, k) = (-\pi \pm \sqrt{3}, 0)$; endpoints of minor axis, $(h, k \pm b)$ or

$(-\pi, -1)$ and $(-\pi, 1)$.   (See figure below).

[11] $\dfrac{4x^2}{3} + \dfrac{9y^2}{5} = 1 \Rightarrow \dfrac{x^2}{\frac{3}{4}} + \dfrac{y^2}{\frac{5}{9}} = 1 \Rightarrow a^2 = \dfrac{3}{4}$ and $b^2 = \dfrac{5}{9} \Rightarrow a = \dfrac{\sqrt{3}}{2}$ and $b = \dfrac{\sqrt{5}}{3}$. $c^2 = a^2 - b^2 =$

$\dfrac{7}{36} \Rightarrow c = \dfrac{\sqrt{7}}{6}$. Center, $(0, 0)$; foci, $\left(\pm \dfrac{\sqrt{7}}{6}, 0\right)$; vertices, $\left(\pm \dfrac{\sqrt{3}}{2}, 0\right)$; endpoints of minor axis,

$\left(0, \pm \dfrac{\sqrt{5}}{3}\right)$.   (See figure below).

[12] $\dfrac{2x^2}{3} + \dfrac{y^2}{2} = 8 \Rightarrow \dfrac{x^2}{12} + \dfrac{y^2}{16} = 1 \Rightarrow a^2 = 16$ and $b^2 = 12 \Rightarrow a = 4$ and $b = 2\sqrt{3}$. $c^2 = a^2 - b^2 = 4 \Rightarrow$

$c = 2$. Center, $(0, 0)$; foci, $(0, \pm 2)$; vertices, $(0, \pm 4)$; endpoints of minor axis, $(\pm 2\sqrt{3}, 0)$.

(See figure below).

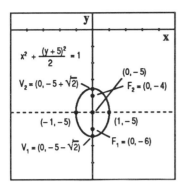

Figure 9

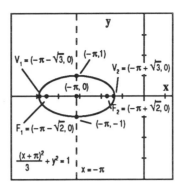

Figure 10

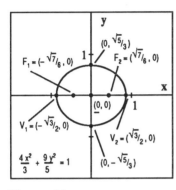

Figure 11

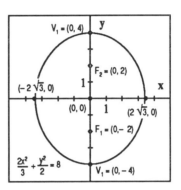

Figure 12

[13] $4x^2 + 3y^2 = 12 \Rightarrow \frac{x^2}{3} + \frac{y^2}{4} = 1 \Rightarrow a^2 = 4$ and $b^2 = 3 \Rightarrow a = 2$ and $b = \sqrt{3}$. $c^2 = a^2 - b^2 = 1 \Rightarrow$

$c = 1$. Center, $(0, 0)$; foci, $(0, \pm 1)$; vertices, $(0, \pm 2)$; endpoints of minor axis, $(\pm \sqrt{3}, 0)$.

(See figure below).

[14] $4x^2 + 5y^2 = 20 \Rightarrow \frac{x^2}{5} + \frac{y^2}{4} = 1 \Rightarrow a^2 = 5$ and $b^2 = 4 \Rightarrow a = \sqrt{5}$ and $b = 2$. $c^2 = a^2 - b^2 = 1 \Rightarrow$

$c = 1$. Center, $(0, 0)$; foci, $(\pm 1, 0)$; vertices, $(\pm \sqrt{5}, 0)$; endpoints of minor axis, $(0, \pm 2)$.

(See figure below).

[15] $9x^2 + 64y^2 = 144 \Rightarrow \frac{x^2}{16} + \frac{y^2}{\frac{9}{4}} = 1 \Rightarrow a^2 = 16$ and $b^2 = \frac{9}{4} \Rightarrow a = 4$ and $b = \frac{3}{2}$. $c^2 = a^2 - b^2 =$

$\frac{55}{4} \Rightarrow c = \frac{\sqrt{55}}{2}$. Center, $(0, 0)$; foci, $\left(\pm \frac{\sqrt{55}}{2}, 0\right)$; vertices, $(\pm 4, 0)$; endpoints of minor axis,

$\left(0, \pm \frac{3}{2}\right)$. (See figure below).

[16] $12x^2 + 25y^2 = 300 \Rightarrow \frac{x}{25} + \frac{y}{12} = 1 \Rightarrow a^2 = 25$ and $b^2 = 12 \Rightarrow a = 5$ and $b = 2\sqrt{3}$. $c^2 = a^2 - b^2 =$

$13 \Rightarrow c = \sqrt{13}$. Center, $(0, 0)$; foci, $(\pm \sqrt{13}, 0)$; vertices, $(\pm 5, 0)$; endpoints of minor axis,

$(0, \pm 2\sqrt{3})$. (See figure below).

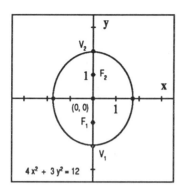

Figure 13

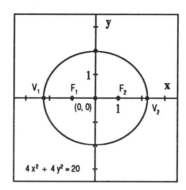

Figure 14

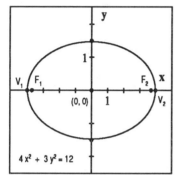

Figure 15

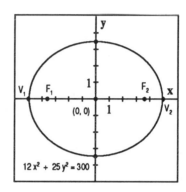

Figure 16

[17] $(x+2)^2 + 5(y-1)^2 = 25 \Rightarrow \dfrac{(x+2)^2}{25} + \dfrac{(y-1)^2}{5} = 1 \Rightarrow a^2 = 25$ and $b^2 = 5 \Rightarrow a = 5$ and $b = \sqrt{5}$

$c^2 = a^2 - b^2 = 20 \Rightarrow c = 2\sqrt{5}$. Also, $h = -2$ and $k = 1$. Center, $(h, k) = (-2, 1)$; foci,

$(h \pm c, k) = (-2 \pm 2\sqrt{5}, 1)$; vertices, $(h \pm a, k)$, or $(-7, 1)$ and $(3, 1)$; endpoints of minor axis,

$(h, k \pm b) = (-2, 1 \pm \sqrt{5})$.   (See figure below).

[18] $9(x+1)^2 + (y-2)^2 = 18 \Rightarrow \dfrac{(x+1)^2}{2} + \dfrac{(y-2)^2}{18} = 1 \Rightarrow a^2 = 18$ and $b^2 = 2 \Rightarrow a = 3\sqrt{2}$ and $b =$

$\sqrt{2}$. $c^2 = a^2 - b^2 = 16 \Rightarrow c = 4$. Moreover, $h = -1$ and $k = 2$. Center, $(h, k) = (-1, 2)$; foci,

$(h, k \pm c)$, or $(-1, -2)$ and $(-1, 6)$; vertices, $(h, k \pm a) = (-1, 2 \pm 3\sqrt{2})$; endpoints of minor

axis, $(h \pm b, k) = (-1 \pm \sqrt{2}, 2)$.   (See figure below).

[19] $5x^2 + 9y^2 - 20x + 36y + 11 = 0 \Rightarrow 5(x^2 - 4x) + 9(y^2 + 4y) = -11 \Rightarrow$

$5(x^2 - 4x + 4) + 9(y^2 + 4y + 4) = 45 \Rightarrow \dfrac{(x-2)^2}{9} + \dfrac{(y+2)^2}{5} = 1 \Rightarrow a^2 = 9$ and $b^2 = 5 \Rightarrow a = 3$

and $b = \sqrt{5}$. $c^2 = a^2 - b^2 = 4 \Rightarrow c = 2$. Also, $h = 2$ and $k = -2$. Center, $(h, k) = (2, -2)$; foci,

$(h \pm c, k)$, or $(0, -2)$ and $(4, -2)$; vertices, $(h \pm a, k)$, or $(-1, -2)$ and $(5, -2)$; endpoints of

minor axis, $(h, k \pm b) = (2, -2 \pm \sqrt{5})$.   (See figure below).

[20] $7x^2 + 16y^2 - 28x - 32y - 68 = 0 \Rightarrow 7(x^2 - 4x) + 16(y^2 - 2y) = 68 \Rightarrow 7(x^2 - 4x + 4) +$

$16(y^2 - 2y + 1) = 112 \Rightarrow \dfrac{(x-2)^2}{16} + \dfrac{(y-1)^2}{7} = 1 \Rightarrow a^2 = 16$ and $b^2 = 7 \Rightarrow a = 4$ and $b = \sqrt{7}$.

$c^2 = a^2 - b^2 = 9 \Rightarrow c = 3$. Also, $h = 2$ and $k = 1$. Center, $(2, 1)$; foci, $(h \pm c, k)$, or $(-1, 1)$ and

$(5, 1)$; vertices, $(h \pm a, k)$, or $(-2, 1)$ and $(6, 1)$; endpoints of the minor axis, $(h, k \pm b) =$

$(2, 1 \pm \sqrt{7})$. (See figure below).

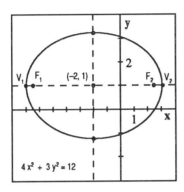

Figure 17

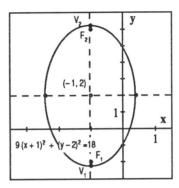

Figure 18

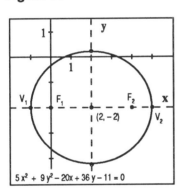

Figure 19

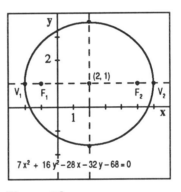

Figure 20

[21] $x^2 + 2y^2 - 4x - 8y - 6 = 0 \Rightarrow x - 4x + 2(y^2 - 4y) = 6 \Rightarrow x - 4x + 4 + 2(y^2 - 4y + 4) = 18 \Rightarrow$

$\dfrac{(x-2)^2}{18} + \dfrac{(y-2)^2}{9} = 1 \Rightarrow a^2 = 18$ and $b^2 = 9 \Rightarrow a = 3\sqrt{2}$ and $b = 3$. $c^2 = a^2 - b^2 = 9 \Rightarrow c = 3$.

Also, $h = 2$ and $k = 2$. Center, $(h, k) = (2, 2)$; foci, $(h \pm c, k)$, or $(-1, 2)$ and $(5, 2)$; vertices,

$(h \pm a, k) = (2 \pm 3\sqrt{2}, 2)$; endpoints of minor axis, $(h, k \pm b)$, or $(2, -1)$ and $(2, 5)$.

(See figure below).

[22] $64x^2 + 48y^2 - 384y - 2304 = 0 \Rightarrow 64x^2 + 48(y^2 - 8y) = 2304 \Rightarrow 64x^2 + 48(y^2 - 8y + 16) =$

$3072 \Rightarrow \dfrac{x^2}{48} + \dfrac{(y-4)^2}{64} = 1 \Rightarrow a^2 = 64$ and $b^2 = 48 \Rightarrow a = 8$ and $b = 4\sqrt{3}$. $c^2 = a^2 - b^2 = 16 \Rightarrow$

$c = 4$. Also, $h = 0$ and $k = 4$. Center, $(h, k) = (0, 4)$; foci, $(h, k \pm c)$, or $(0, 0)$ and $(0, 8)$;

vertices, $(h, k \pm a)$, or $(0, -4)$ and $(0, 12)$; endpoints of the minor axis, $(h \pm b, k) = (\pm 4\sqrt{3}, 4)$.

(See figure below).

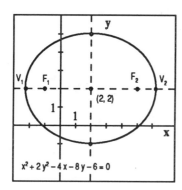

$x^2 + 2y^2 - 4x - 8y - 6 = 0$

Figure 21

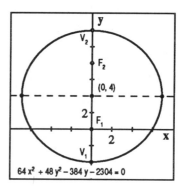

$64x^2 + 48y^2 - 384y - 2304 = 0$

Figure22

[23] $a = 5$ and $c = 3 \Rightarrow a^2 = 25$ and $b^2 = a^2 - c^2 = 16$. The center is $(0, 0)$ and the major axis is the

y – axis. ● $\dfrac{x^2}{16} + \dfrac{y^2}{25} = 1$.

[24] The center is $(0, 0)$ and the major axis is the x – axis. ● $\dfrac{x^2}{25} + \dfrac{y^2}{9} = 1$.

[25] $2a = 6 \Rightarrow a = 3 \Rightarrow a^2 = 9$. The center is $(0, 2)$ and $c = 1$. Hence $b^2 = a^2 - c^2 = 8$. The major axis

is the line $y = 2$. ● $\dfrac{x^2}{9} + \dfrac{(y-2)^2}{8} = 1$.

[26] $a = 3$ and $c = 2 \Rightarrow a^2 = 9$ and $b^2 = a^2 - c^2 = 5$. The major axis is the line $y = 2$.

● $\dfrac{(x-2)^2}{9} + \dfrac{(y-2)^2}{5} = 1$.

[27] $a = 8$ and $b = 3 \Rightarrow a^2 = 64$ and $b^2 = 9$. The major axis is the line $y = 2$. ● $\dfrac{(x-3)^2}{64} + \dfrac{(y-2)^2}{9} = 1$.

[28] The center is $(0, 0)$, $a = 5$ and the major axis is the x – axis. Thus, $\dfrac{x^2}{25} + \dfrac{y^2}{b^2} = 1$. Since the point

$(-3, 2)$ lies on the ellipse, $\dfrac{3^2}{25} + \dfrac{2^2}{b^2} = 1 \Rightarrow \dfrac{4}{b^2} = \dfrac{16}{25}$ or $b^2 = \dfrac{25}{4}$. ● $\dfrac{x^2}{25} + \dfrac{y^2}{\frac{25}{4}} = 1$.

[29] $a = 5$, the center is $(4, 0)$ and the major axis is the line $x = 4$. Hence, $\dfrac{(x-4)^2}{b^2} + \dfrac{y^2}{25} = 1$. Since the

origin lies on the ellipse, $\dfrac{(0-4)^2}{b^2} + \dfrac{0^2}{25} = 1 \Rightarrow b^2 = 16$. ● $\dfrac{(x-4)^2}{16} + \dfrac{y^2}{25} = 1$.

[30] $a = 5$ and $b = \dfrac{3}{2} \Rightarrow a^2 = 25$ and $b^2 = \dfrac{9}{4}$. ● $\dfrac{(x+1)^2}{\frac{9}{4}} + \dfrac{(y+1)^2}{25} = 1$.

[31] Since $2b = 4$, $b = 2 \Rightarrow b^2 = 4$. Moreover, since $(-1, 3)$ is a focus and $(-1, 1)$ is the center, $c = 2$.

Thus, $a^2 = c^2 + b^2 = 8$. ● $\dfrac{(x+1)^2}{4} + \dfrac{(y-1)^2}{8} = 1$.

[32] Since $(4, -3)$ is a vertex and $(4, -5)$ is the center, $a = 2 \Rightarrow a^2 - 4$. Also, $2b = 1 \Rightarrow b = \dfrac{1}{2} \Rightarrow$

$b^2 = \dfrac{1}{4}$. The major axis is $x = 4$. ● $\dfrac{(x-4)^2}{\frac{1}{4}} + \dfrac{(y+3)^2}{4} = 1$.

[33] Let $2a$ denote the sum of the distances from any point $P = (x, y)$ on the ellipse to the foci, $F_1 = (0, -c)$ and $F_2 = (0, c)$. It follows that $c < a$. By Definition 8.3, $d(P, F_1) + d(P, F_2) = 2a \Rightarrow$

$\sqrt{(x-0)^2 + (y+c)^2} + \sqrt{(x-0)^2 + (y-c)^2} = 2a \Rightarrow \sqrt{x^2 + (y+c)^2} = 2a -$

$\sqrt{x^2 + (y-c)^2} \Rightarrow \left(\sqrt{x^2 + (y+c)^2}\right)^2 = \left(2a - \sqrt{x^2 + (y-c)^2}\right)^2 \Rightarrow x^2 + y^2 + 2cy + c^2 =$

$4a^2 - 4a\sqrt{x^2 + (y-c)^2} + x^2 + y^2 - 2cy + c^2 \Rightarrow 4a^2 - 4cy = 4a\sqrt{x^2 + (y-c)^2} \Rightarrow$

$(a^2 - cy)^2 = \left(a\sqrt{x^2 + (y-c)^2}\right)^2 \Rightarrow a^4 - 2a^2cy + c^2y^2 = a^2x^2 + a^2y^2 - 2a^2cy + a^2c^2 \Rightarrow$

$a^2x^2 + (a^2 - c^2)y^2 = a^4 - a^2c^2 = a^2(a^2 - c^2) \Rightarrow \dfrac{x^2}{a^2 - c^2} + \dfrac{y^2}{a^2} = 1$. Since $a > c > 0$, $a^2 - c^2 > 0$.

Letting $b^2 = a^2 - c^2$, we have $\dfrac{x^2}{b^2} + \dfrac{y^2}{a^2} = 1$.

[34] Let (h, k) be the center and let $F_1 = (h - c, k)$ and $F_2 = (h + c, k)$ be the foci. Let $2a$ denote the sum of the distances from any point $P = (x, y)$ on the ellipse to the foci, F_1 and F_2. It follows that $c < a$. From Definition 8.3, $d(P, F_1) + d(P, F_2) = 2a \Rightarrow \sqrt{[x-(h-c)]^2 + (y-k)^2} +$

$\sqrt{[x-(h+c)]^2 + (y-k)^2} = 2a \Rightarrow \sqrt{[x-(h-c)]^2 + (y-k)^2} = 2a -$

$\sqrt{[x-(h+c)]^2 + (y-k)^2} \Rightarrow \left(\sqrt{[x-(h-c)]^2 + (y-k)^2}\right)^2 =$

$\left(2a - \sqrt{[x-(h+c)]^2 + (y-k)^2}\right)^2 \Rightarrow x^2 - 2hx + 2cx + h^2 - 2ch + c^2 + y^2 - 2ky + k^2 =$

$4a^2 - 4a\sqrt{[x-(h+c)]^2 + (y-k)^2} + x^2 - 2hx - 2cx + h^2 + 2ch + c^2 + y^2 - 2ky + k^2 \Rightarrow$

$a^2 + ch - cx = a\sqrt{[x-(h+c)]^2 + (y-k)^2} \Rightarrow (a^2 + ch - cx)^2 =$

$\left(a\sqrt{[x-(h+c)]^2 + (y-k)^2}\right)^2 \Rightarrow a^4 + 2a^2ch - 2a^2cx - 2c^2hx + c^2x^2 + c^2h^2 = a^2x^2 -$

$2a^2xh - 2a^2cx + 2a^2ch + a^2c^2 + a^2h^2 + a^2(y-k)^2 \Rightarrow a^4 - a^2c^2 + c^2(x^2 - 2hx + h^2) =$

$a^2(x^2 - 2xh + h^2) + a^2(y-k)^2 \Rightarrow a^2(a^2 - c^2) = (a^2 - c^2)(x-h)^2 + a^2(y-k)^2 \Rightarrow$

$\dfrac{(x-h)^2}{a^2} + \dfrac{(y-k)^2}{a^2 - c^2}$. Since $a > c > 0$, $a^2 - c^2 > 0$. Letting $b^2 = a^2 - c^2$, the desired result.

[35] We introduce a Cartesian coordinate system as shown in the figure below. It follows that

an equation for this elliptical arch is $\dfrac{x^2}{900} + \dfrac{y^2}{400} = 1$. The height of the arch 15 feet from its center

is $\dfrac{(15)^2}{900} + \dfrac{y^2}{400} = 1 \Rightarrow y = 20\sqrt{\dfrac{675}{900}} \approx 17.32$ feet. ● $\dfrac{x^2}{900} + \dfrac{y^2}{400} = 1$, 17.32 feet.

[36] Choose $a = 2010$ and $b = 1800$. Then, an equation for the elliptical orbit is $\dfrac{x^2}{(2010)^2} + \dfrac{y^2}{(1800)^2} = 1$, where a rectangular system as been positioned so that the origin is located at the center of the moon.

[37] Introducing a Cartesian coordinate system with the origin located at the center of the ellipse and the x-axis along the major axis, one vertex is $(750, 0)$, and $(0, 640)$ is an endpoint of the minor axis. Thus, we choose $a = 750$ and $b = 640$ and hence, $c^2 = a^2 - b^2 = 152900 \Rightarrow c = \sqrt{152900} \approx 391$. Hence $a - c \approx 750 - 391 \approx 359$ feet.
● approximately 359 feet.

[38]   Completing the square in both x and y in the equation $Ax^2 + Cy^2 + Dx + Ey + F = 0$, we get

$$A\left(x^2 + \frac{D}{A}x\right) + C\left(y^2 + \frac{E}{C}x\right) = -F \Rightarrow A\left(x^2 + \frac{D}{A}x + \frac{D^2}{4A^2}\right) + C\left(y^2 + \frac{E}{C}x + \frac{E^2}{4C^2}\right) =$$

$$\frac{D^2}{4A} + \frac{E^2}{4C} - F \Rightarrow A\left(x + \frac{D}{2A}\right)^2 + C\left(y + \frac{E}{2C}\right)^2 = \frac{D^2}{4A} + \frac{E^2}{4C} - F = r.$$

a)   If $r > 0$, the graph is an ellipse since $A \neq C$.

b)   If $r = 0$, the graph is the point $\left(-\frac{D}{2A}, -\frac{E}{2C}\right)$.

c)   If $r < 0$, the graph contains no points.

[39]   Since $c^2 = a^2 - b^2$, $c = \sqrt{a^2 - b^2}$. The point $\left(\sqrt{a^2 - b^2}, y\right)$ is on the ellipse $\frac{x^2}{a^2} + \frac{y^2}{b^2} = 1$ if

and only if $\frac{\left(\sqrt{a^2 - b^2}\right)^2}{a^2} + \frac{y^2}{b^2} = 1$. Thus, $\frac{a^2 - b^2}{a^2} + \frac{y^2}{b^2} = 1 \Rightarrow \frac{y^2}{b^2} = 1 - \frac{a^2 - b^2}{a^2} = \frac{b^2}{a^2} \Rightarrow y^2 =$

$\frac{b^4}{a^2} \Rightarrow y = \frac{b^2}{a}$. Hence, the focal length is $\frac{2b^2}{a}$.   (See figure below).

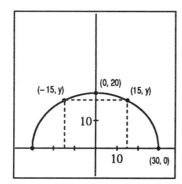

Figure 35   **Figure 39**

[40]   a) $(3)^2 + 3(1)^2 = 9 + 3 = 12$.

b) $x^2 + 3y^2 = 12 \Rightarrow \frac{x^2}{12} + \frac{y^2}{4} = 1$. Therefore, the equation of the tangent line to the ellipse at the

point $(3, 1)$ is $\frac{3x}{12} + \frac{y}{4} = 1$, or $y = -x + 4$.

[41] $Ax^2 + Cy^2 + F = 0 \Rightarrow Ax^2 + Cy^2 = -F$. If $A > 0$, then $C > 0$ and $-F > 0$. If $A < 0$, then

$C < 0$ and $-F < 0$. In either case, $\frac{x^2}{\frac{-F}{A}} + \frac{y^2}{\frac{-F}{C}} = 1$ where $-\frac{F}{A} > 0$ and $-\frac{F}{C} > 0$.

a)   If $A \neq C$, the equation $\frac{x^2}{\frac{-F}{A}} + \frac{y^2}{\frac{-F}{C}} = 1$ is the equation of an ellipse with center $(0, 0)$.

b)   If $A = C$, the equation $\frac{x^2}{\frac{-F}{A}} + \frac{y^2}{\frac{-F}{C}} = 1$ is the equation of a circle with center $(0, 0)$.

[42] $Ax^2 + Cy^2 + Dx + Ey + F = 0 \Rightarrow A\left(x^2 + \dfrac{D}{A}x + \dfrac{D^2}{4A^2}\right) + C\left(y^2 + \dfrac{E}{C}y + \dfrac{E^2}{4C^2}\right) =$

$\dfrac{D^2}{4A} + \dfrac{E^2}{4C} - F \Rightarrow A\left(x + \dfrac{D}{2A}x\right)^2 + C\left(y + \dfrac{E}{C}\right)^2 = \dfrac{D^2}{4A} + \dfrac{E^2}{4C} - F.$ It follows that

a)   The graph is an ellipse if A and $\dfrac{D^2}{4A} + \dfrac{E^2}{4C} - F$ have the same sign.

b)   The graph is a point if $\dfrac{D^2}{4A} + \dfrac{E^2}{4C} - F = 0$.

c)   The graph contains no point if A and $\dfrac{D^2}{4A} + \dfrac{E^2}{4C} - F$ have opposite signs.

[43] a)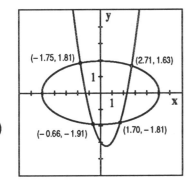

b) $(-0.66, -1.91), (-1.75, 1.81), (1.70, -1.81), (2.71, 1.65)$

[44] a)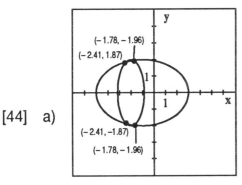

b) $(-2.41, -1.87), (-2.41, 1.87), (-1.78, -1.96), (-1.78, 1.96)$

EXERCISES 8.4

[1]   In the equation $\dfrac{x^2}{16} - \dfrac{y^2}{9} = 1$, $a^2 = 16$ and $b^2 = 9 \Rightarrow a = 4$ *and* $b = 3$. $c^2 = a^2 + b^2 \Rightarrow c^2 = 25 \Rightarrow$
$c = 5$. Center, $(0, 0)$; vertices, $(\pm 4, 0)$; foci, $(\pm 5, 0)$; length of transverse axis, $2a = 8$; length of conjugate axis, $2b = 6$.   (See figure below).

[2]   In the equation $\dfrac{y^2}{16} - \dfrac{x^2}{9} = 1$, $a^2 = 16$ and $b^2 = 9 \Rightarrow a = 4$ *and* $b = 3$. $c^2 = a^2 + b^2 \Rightarrow c^2 = 15 \Rightarrow$
$c = 5$. Center, $(0, 0)$; vertices, $(0, \pm 4)$; foci, $(0, \pm 5)$; length of the transverse axis, $2a = 8$; length of the conjugate axis, $2b = 6$.   (See figure below).

381

[3]   In the equation $\frac{y^2}{25} - \frac{x^2}{144} = 1$, $a^2 = 25$ and $b^2 = 144 \Rightarrow a = 5$ and $b = 12$. $c^2 = a^2 + b^2 = 169 \Rightarrow$

c = 13. Center, $(0, 0)$; vertices, $(0, \pm 5)$; foci, $(0, \pm 13)$; length of transverse axis, $2a = 10$; length

of conjugate axis, $2b = 24$.   (See figure below).

[4]   In the equation $x^2 - \frac{y^2}{49} = 1$, $a^2 = 1$ and $b^2 = 49 \Rightarrow a = 1$ and $b = 7$. $c^2 = a^2 + b^2 = 50 \Rightarrow c =$

$5\sqrt{2}$. Center, $(0, 0)$; vertices, $(\pm 1, 0)$; foci, $(\pm 5\sqrt{2}, 0)$; length of transverse axis, $2a = 2$; length of

conjugate axis, $2b = 14$.   (See figure below).

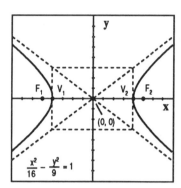

Figure 1

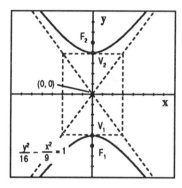

Figure 2

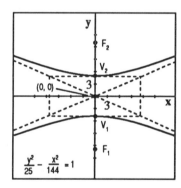

Figure 3

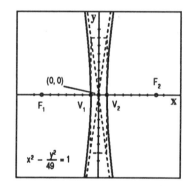

Figure 4

[5] $9x^2 - 4y^2 = 25 \Rightarrow \frac{x^2}{\frac{25}{9}} - \frac{y^2}{\frac{25}{4}} = 1 \Rightarrow a^2 = \frac{25}{9}$ and $b^2 = \frac{25}{4} \Rightarrow a = \frac{5}{3}$ and $b = \frac{5}{2}$. $c^2 = a^2 + b^2 =$

$\frac{325}{36} \Rightarrow c = \frac{\sqrt{13}}{6}$. Center, $(0, 0)$; vertices, $\left(\pm \frac{5}{3}, 0\right)$; foci, $\left(\pm \frac{5\sqrt{13}}{6}, 0\right)$; length of transverse axis,

$2a = \frac{10}{3}$; length of conjugate axis, $2b = 5$.   (See figure below).

[6] $y^2 - 25x^2 = 16 \Rightarrow \frac{y^2}{16} - \frac{x^2}{\frac{16}{25}} = 1 \Rightarrow a^2 = 16$ and $b^2 = \frac{16}{25} \Rightarrow a = 4$ and $b = \frac{4}{5}$. $c^2 = a^2 + b^2 =$

$\frac{416}{25} \Rightarrow c = \frac{4\sqrt{26}}{5}$. Center, $(0, 0)$; vertices, $(0, \pm 4)$; foci, $\left(0, \pm \frac{4\sqrt{26}}{5}\right)$; length of transverse axis,

$2a = 8$; length of conjugate axis, $2b = \frac{8}{5}$.   (See figure below).

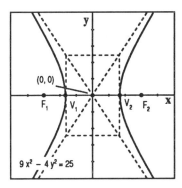

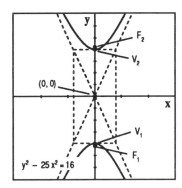

Figure 5   **Figure 6**

[7] $(x+3)^2 - 6y^2 = 6 \Rightarrow \dfrac{(x+3)^2}{6} - y^2 = 1 \Rightarrow a^2 = 6$ and $b^2 = 1 \Rightarrow a = \sqrt{6}$ and $b = 1$. $c^2 = a^2 + b^2 = 7 \Rightarrow c = \sqrt{7}$. Also, $h = -3$ and $k = 0$. Center, $(h, k) = (-3, 0)$; vertices, $(h \pm a, k) = (-3 \pm \sqrt{6}, 0)$; foci, $(h \pm c, k) = (-3 \pm \sqrt{7}, 0)$; length of transverse axis, $2a = 2\sqrt{6}$; length of the conjugate axis, $2b = 2$.   (See figure below).

[8] $2x^2 - 5(y+1) = 25 \Rightarrow \dfrac{x^2}{\frac{25}{2}} - \dfrac{(y+1)^2}{5} = 1 \Rightarrow a^2 = \dfrac{25}{2}$ and $b^2 = 25 \Rightarrow a = \dfrac{5\sqrt{2}}{2}$ and $b = \sqrt{5}$.

$c^2 = a^2 + b^2 = \dfrac{35}{2} \Rightarrow c = \dfrac{\sqrt{70}}{2}$. Moreover, $h = 0$ and $k = -1$. Center, $(h, k) = (0, -1)$; vertices, $(h \pm a, k) = \left(\pm \dfrac{5\sqrt{2}}{2}, -1\right)$; foci, $(h \pm c, k) = \left(\pm \dfrac{\sqrt{70}}{2}, -1\right)$; length of transverse axis, $2a = 5\sqrt{2}$; length of minor axis, $2b = 2\sqrt{5}$.   (See figure below).

[9]   In the equation $(y-3)^2 - \dfrac{(x+2)^2}{4} = 1$, $a^2 = 1$ and $b^2 = 4 \Rightarrow a = 1$ and $b = 2$. $c^2 = a^2 + b^2 = 5 \Rightarrow c = \sqrt{5}$. Also, $h = -2$ and $k = 3$. Center, $(h, k) = (-2, 3)$; vertices, $(h, k \pm a)$, or $(-2, 2)$ and $(-2, 4)$; foci, $(h, k \pm c) = (-2, 3 \pm \sqrt{5})$; length of transverse axis, $2a = 2$; length of conjugate axis, $2b = 4$. (See figure below).

[10]   In the equation $\dfrac{(x-1)^2}{2} - \dfrac{(y+2)^2}{4} = 1$, $a^2 = 2$ and $b^2 = 4 \Rightarrow a = \sqrt{2}$ and $b = 2$. $c^2 = a^2 + b^2 = 6 \Rightarrow c = \sqrt{6}$. Also, $h = 1$ and $k = -2$. Center, $(h, k) = (1, -2)$; vertices, $(h \pm a, k) = (1 \pm \sqrt{2}, -2)$; foci, $(h \pm c, k) = (1 \pm \sqrt{6}, -2)$; length of transverse axis, $2a = 2\sqrt{2}$; length of minor axis, $2b = 4$.   (See figure below).

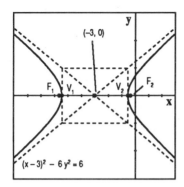

Figure 7

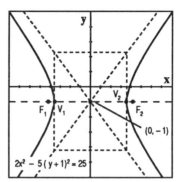

Figure 8

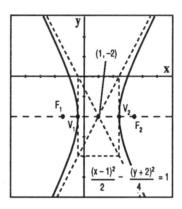

Figure 9   **Figure 10**

[11] In the equation $(x-4)^2 - (y-2)^2 = 1$, $a^2 = b^2 = 1 \Rightarrow a = b = 1$. $c^2 = a^2 + b^2 = 2 \Rightarrow c = \sqrt{2}$. Also, $h = 4$ and $k = 2$. Center, $(h, k) = (4, 2)$; vertices, $(h \pm a, k)$, or $(3, 2)$ and $(5, 2)$; foci, $(h \pm c, k) = (4 \pm \sqrt{2}, 2)$; length of transverse axis, $2a = 2$; length of conjugate axis, $2b = 2$. (See figure below).

[12] $(y+1)^2 - (x+1)^2 = 4 \Rightarrow \dfrac{(y+1)^2}{4} - \dfrac{(x+1)^2}{4} = 1 \Rightarrow a^2 = b^2 = 4 \Rightarrow a = b = 2$. $c^2 = a^2 + b^2 = 8 \Rightarrow c = 2\sqrt{2}$. Also, $h = -1$ and $k = -1$. Center, $(h, k) = (-1, -1)$; vertices, $(h, k \pm a)$, or $(-1, -3)$ and $(-1, 1)$; foci, $(h, k \pm c) = (-1, -1 \pm 2\sqrt{2})$; length of transverse axis, $2a = 4$; length of conjugate axis, $2b = 4$.   (See figure below).

[13] $4(x+2)^2 - 25(y-1)^2 = 100 \Rightarrow \dfrac{(x+2)^2}{25} - \dfrac{(y-1)^2}{4} = 1 \Rightarrow a^2 = 25$ and $b^2 = 4 \Rightarrow a = 5$ and $b = 2$. $c^2 = a^2 + b^2 = 29 \Rightarrow c = \sqrt{29}$. Moreover, $h = -2$ and $k = 1$. Center, $(h, k) = (-2, 1)$; vertices, $(h \pm a, k)$, or $(-7, 1)$ and $(3, 1)$; foci, $(h \pm c, k) = (-2 \pm \sqrt{29}, 1)$; length of transverse axis, $2a = 10$; length of conjugate axis, $2b = 4$.   (See figure below).

[14] $16(y-1)^2 - 9(x-5)^2 = 144 \Rightarrow \dfrac{(y-1)^2}{9} - \dfrac{(x-5)^2}{16} = 1 \Rightarrow a^2 = 9$ and $b^2 = 16 \Rightarrow a = 3$ and $b = 4$. $c^2 = a^2 + b^2 = 25 \Rightarrow c = 5$. Also, $h = 1$ and $k = 5$. Center, $(h, k) = (1, 5)$; vertices, $(h, k \pm a)$, or $(1, 2)$ and $(1, 8)$; foci, $(h, k \pm c)$, or $(1, 0)$ and $(1, 10)$; length of transverse axis; $2a = 6$; length of conjugate axis, $2b = 8$.   (See figure below).

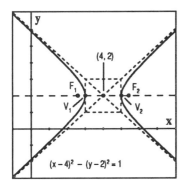

Figure 11

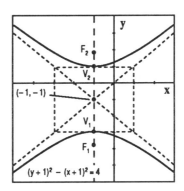

Figure 12

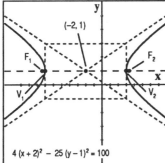

Figure 13

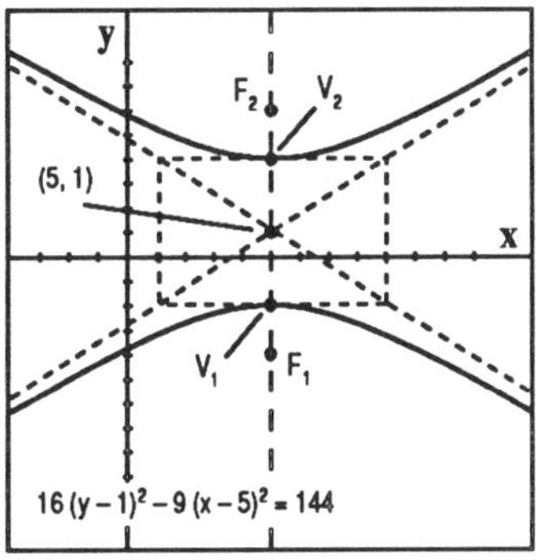

Figure 14

[15] $9x^2 - 16y^2 + 36x + 32y + 164 = 0 \Rightarrow 9\left(x^2 + 4x\right) = 16\left(y^2 - 2y\right) = -164 \Rightarrow 9\left(x^2 + 4x + 4\right) -$

$16\left(y^2 - 2y + 1\right) = -164 + 36 - 16 \Rightarrow 9\left(x + 2\right)^2 - 16\left(y - 1\right)^2 = -144 \Rightarrow \dfrac{\left(y - 1\right)^2}{9} - \dfrac{9\left(x + 2\right)^2}{16} =$

$1 \Rightarrow a^2 = 9$ and $b^2 = 16 \Rightarrow a = 3$ and $b = 4$. $c^2 = a^2 + b^2 = 25 \Rightarrow c = 5$. Also, $h = -2$ and $k = 1$.

Center, $(-2, 1)$; vertices, $(h, k \pm a)$, or $(-2, -2)$ and $(-2, 4)$; foci, $(h, k \pm c)$, or $(-2, -4)$ and

$(-2, 6)$; length of transverse axis, $2a = 6$; length of conjugate axis, $2b = 8$.   (See figure below).

[16] $9x^2 - 4y^2 - 72x + 8y + 176 = 0 \Rightarrow 9\left(x^2 - 8x\right) - 4\left(y^2 - 2y\right) = -176 \Rightarrow 9\left(x^2 - 8x + 16\right) -$

$4\left(y^2 - 2y + 1\right) = -176 + 144 - 4 \Rightarrow 9\left(x - 4\right)^2 - 4\left(y - 1\right)^2 = -36 \Rightarrow \dfrac{\left(y - 1\right)^2}{9} - \dfrac{\left(x - 4\right)^2}{4} = 1 \Rightarrow$

$a^2 = 9$ and $b^2 = 4 \Rightarrow a = 3$ and $b = 2$. $c^2 = a^2 + b^2 = 13 \Rightarrow c = \sqrt{13}$. Also, $h = 4$ and $k = 1$.

Center, $(h, k) = (4, 1)$; vertices, $(h, k \pm a)$, or $(4, -2)$ and $(4, 4)$; foci, $(h, k \pm c) = (4, 1 \pm \sqrt{13})$;

length of transverse axis, $2a = 6$; length of conjugate axis, $2b = 4$.   (See figure below).

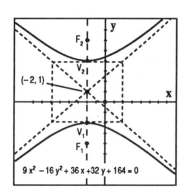

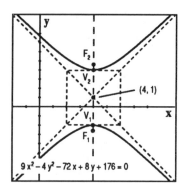

Figure 15 **Figure 16**

[17] $2y^2 - 3x^2 - 4y + 12x + 8 = 0 \Rightarrow 2(y^2 - 2y) - 3(x^2 - 4x) = -8 \Rightarrow 2(y^2 - 2y + 1) -$

$3(x^2 - 4x + 4) = -8 + 2 - 12 \Rightarrow 2(y-1)^2 - 3(x-2)^2 = -18 \Rightarrow \dfrac{(x+2)^2}{6} - \dfrac{(y-1)^2}{9} = 1 \Rightarrow a^2 = 6$

and $b^2 = 9 \Rightarrow a = \sqrt{6}$ and $b = 3$. $c^2 = a^2 + b^2 = 15 \Rightarrow c = \sqrt{15}$. Also, $h = -2$ and $k = 1$. Center,

$(2, 1)$; vertices, $(h \pm a, k) = (2 \pm \sqrt{6}, 1)$; foci, $(h \pm c, k) = (2 \pm \sqrt{15}, 1)$; length of transverse axis,

$2a = 2\sqrt{6}$; length of the conjugate axis, $2b = 6$.   (See figure below).

[18] $y^2 - 4x^2 + 4y - 8x - 9 = 0 \Rightarrow y^2 + 4y - 4(x^2 + 2x) = 9 \Rightarrow (y^2 + 4y + 4) - 4(x^2 + 2x + 1) =$

$9 + 4 - 4 \Rightarrow (y+2)^2 - 4(x+1)^2 = 9 \Rightarrow \dfrac{(y+2)^2}{9} - \dfrac{(x+1)^2}{\frac{9}{4}} = 1 \Rightarrow a^2 = 9$ and $b^2 = \frac{9}{4} \Rightarrow a = 3$

and $b = \frac{3}{2}$. $c^2 = a^2 + b^2 = \frac{45}{4} \Rightarrow c = \frac{3}{2}\sqrt{5}$. Also, $h = -1$ and $k = -2$. Center, $(h, k) =$

$(-1, -2)$; vertices, $(h, k \pm a)$, or $(-1, -5)$ and $(-1, 1)$; foci, $(h, k \pm c) = \left(-1, -2 \pm \frac{3}{2}\sqrt{5}\right)$;

(See figure below).

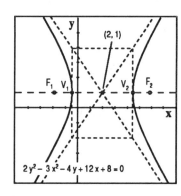

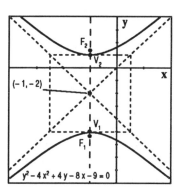

Figure 17 **Figure 18**

[19] $12x^2 - 3y^2 + 24y - 84 = 0 \Rightarrow 12x^2 - 3(y^2 - 8y) = 84 \Rightarrow 12x^2 - 3(y^2 - 8y + 16) = 84 - 48 \Rightarrow$

$12x^2 - 3(y-4)^2 = 36 \Rightarrow \dfrac{x^2}{3} - \dfrac{(y-4)^2}{12} = 1 \Rightarrow a^2 = 3$ and $b^2 = 12 \Rightarrow a = \sqrt{3}$ and $b = 2\sqrt{3}$. $c^2 =$

$a^2 + b^2 = 15 \Rightarrow c = \sqrt{15}$. Also, $h = 0$ and $k = 4$. Center, $(0, 4)$; vertices, $(h \pm a, k) = (\pm \sqrt{3}, 4)$;

foci, $(h \pm c, k) = (\pm \sqrt{15}, 4)$; length of transverse axis, $2a = 2\sqrt{3}$; length of the conjugate axis,

$2b = 4\sqrt{3}$.   (See figure below).

[20] $4x^2 - 16x - 9y^2 - 54y - 101 = 0 \Rightarrow 4(x^2 - 4x) - 9(y^2 + 6y) = 101 \Rightarrow 4(x^2 - 4x + 4) -$

$9(y^2 + 6y + 9) = 101 + 16 - 81 \Rightarrow 4(x-2)^2 - 9(y+3)^2 = 36 \Rightarrow \dfrac{(x-2)^2}{9} - \dfrac{(y+3)^2}{4} = 1 \Rightarrow$

$a^2 = 9$ and $b^2 = 4 \Rightarrow a = 3$ and $b = 2$. $c^2 = a^2 + b^2 = 13 \Rightarrow c = \sqrt{13}$. Also, $h = 2$ and $k = -3$.

Center, $(h, k) = (2, -3)$; vertices, $(h \pm a, k)$, or $(-1, -3)$ and $(5, -3)$; foci, $(h \pm c, k) =$

$(2 \pm \sqrt{13}, -3)$; length of transverse axis, $2a = 6$; length of conjugate axis, $2b = 4$.

(See figure below).

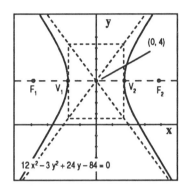
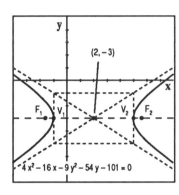

Figure 19   **Figure 20**

[21] For this hyperbola, $a = 4$ and $c = 6 \Rightarrow a^2 = 16$ and $c^2 = 36$. $b^2 = c^2 - a^2 = 20$. The transverse axis

lies along the line y – axis since 0 is the x – coordinate of the vertices and foci.   ● $\dfrac{y^2}{16} - \dfrac{x^2}{20} = 1$.

[22] For this hyperbola $a = 1$ and $c = 3 \Rightarrow a^2 = 1$ and $c^2 = 9$. $b^2 = c^2 - a^2 = 8$. The transverse axis lies

along the x – axis since 0 is the y – coordinate of the vertices and foci.   ● $x^2 - \dfrac{y^2}{8} = 1$.

[23] The transverse axis of this hyperbola lies along the line $x = 2$, and the center is $(2, 3)$ since the

vertices are $(2, 1)$ and $(2, 5)$. It follows that $a = 2 \Rightarrow a^2 = 4$. Moreover, $c = 4 \Rightarrow c^2 = 16$. $b^2 =$

$c^2 - a^2 = 12$.   ● $\dfrac{(y-3)^2}{4} - \dfrac{(x-2)^2}{12} = 1$.

[24] The transverse axis of this hyperbola lies along line $y = -1$, and the center is $(-2, -1)$ since the

vertices are $(-4, -1)$ and $(0, -1)$. It follows that $a = 2 \Rightarrow a^2 = 4$. Moreover, $c = 3 \Rightarrow c^2 = 9$. $b^2 =$

$c^2 - a^2 = 5$.   ● $\dfrac{(x+2)^2}{4} - \dfrac{(y+1)^2}{5} = 1$.

[25] The transverse axis of this hyperbola lies along the line $x = 1$, and the center is $(1, 5)$ since the foci

are $(1, 2)$ and $(1, 8)$. It follows that $c = 3 \Rightarrow c^2 = 9$. Since the length of the conjugate axis is 4,

$b = 2 \Rightarrow b^2 = 4$. Thus $a^2 = c^2 - b^2 = 5$.   ● $\dfrac{(y-5)^2}{5} - \dfrac{(x-1)^2}{4} = 1$.

[26] The transverse axis of this hyperbola lies along the line $y = -4$, and the center is $(1, -4)$ since the

foci are $(-3, -4)$ and $(5, -4)$. It follows that $c = 4 \Rightarrow c^2 = 16$. Since the length of the conjugate axis

is 7, $b = \dfrac{7}{2} \Rightarrow b^2 = \dfrac{49}{4}$. $a^2 = c^2 - b^2 = \dfrac{15}{4}$.   ● $\dfrac{(x-1)^2}{\frac{15}{4}} - \dfrac{(y+4)^2}{\frac{49}{4}} = 1$.

[27] The y – axis contains the transverse axis, and the center is $(0, 0)$ since the vertices are $(0, \pm\sqrt{5})$. It

follows that $a = \sqrt{5} \Rightarrow a^2 = 5$. Thus, the equation has the form $\frac{x^2}{5} - \frac{x^2}{b^2} = 1$. Since the point $(4, 3)$

lies on the hyperbola, $\frac{(3)^2}{5} - \frac{(4)^2}{b^2} = 1 \Rightarrow b^2 = 20$.   ● $\frac{y^2}{5} - \frac{x^2}{20} = 1$.

[28] The transverse axis lies along the x – axis and the center is $(0, 0)$ since the vertices are $(\pm 2, 0)$. It

follows that $a = 2 \Rightarrow a^2 = 4$. Thus, the equation has the form $\frac{x^2}{4} - \frac{y^2}{b^2} = 1$. Since the point $(\sqrt{20}, 8)$

lies on the hyperbola, $\frac{(\sqrt{20})^2}{4} - \frac{(8)^2}{b^2} = 1 \Rightarrow b^2 = 16$.   ● $\frac{x^2}{4} - \frac{y^2}{16} = 1$.

[29] Since the transverse axis has length 2, $a = 1 \Rightarrow a^2 = 1$. Also, since the conjugate axis has length 6,

$b = 3 \Rightarrow b^2 = 9$.   ● $(x+2)^2 - \frac{(y-4)^2}{9} = 1$.

[30] Since the transverse axis has length 10, $a = 5 \Rightarrow a^2 = 25$. Since the conjugate axis has length 5,

$b = \frac{5}{2} \Rightarrow b^2 = \frac{25}{4}$.   ● $\frac{(y+3)^2}{25} - \frac{(x+1)^2}{\frac{25}{4}} = 1$.

[31] $\frac{(0+2)^2}{9} - \frac{(0-1)^2}{4} = c \Rightarrow \frac{4}{9} - \frac{1}{4} = c \Rightarrow c = \frac{7}{36}$.   ● $\frac{7}{36}$.

[32] a) If $c > 9$, $c - 9 > 0$, and the graph is an ellipse with center at the origin and major axis lying

along the x – axis.

     b) If $0 < c < 9$, $c - 9 < 0$, and the graph is a hyperbola with center at the origin and transverse

axis lying along the x – axis.

     c) If $c < 0$, $c - 9 < 0$, and the graph contains no points.

[33] $x^2 - 4x - 16y - 28 = 0 \Rightarrow y = \frac{x^2}{16} - \frac{1}{4}x - \frac{7}{4}$.   ● parabola.

[34] $x^2 + y^2 + 2x - 4y - 15 = 0 \Rightarrow (x^2 + 2x) + (y^2 - 4y) = 15 \Rightarrow (x^2 + 2x + 1) + (y^2 - 4y + 4) = 20 \Rightarrow$

$(x+1)^2 + (y-2)^2 = 20$.   ● circle.

[35] $9x^2 + 4y^2 - 54x + 16y + 29 = 0 \Rightarrow 9(x^2 - 6x) + 4(y^2 + 4y) = -29 \Rightarrow 9(x^2 - 6x + 9) +$

$4(y^2 + 4y + 4) = 68 \Rightarrow 9(x-3)^2 + 4(y+2)^2 = 68 \Rightarrow \frac{(x-3)^2}{\frac{68}{9}} + \frac{(y+2)^2}{17} = 1$.   ● ellipse.

[36] $x^2 + 9y^2 + 6x - 18y + 9 = 0 \Rightarrow (x^2 + 6x) + 9(y^2 - 2y) = -9 \Rightarrow \frac{(x+3)^2}{9} + (y+1)^2$ ● ellipse.

[37] $10x^2 - 8y^2 - 60x - 32y - 22 = 0 \Rightarrow 10(x^2 - 6x) - 8(y^2 + 4y) = 22 \Rightarrow 10(x^2 - 6x + 9) -$

$8(y^2 + 4y + 4) = 80 \Rightarrow 10(x-3)^2 - 8(y+2)^2 = 80 \Rightarrow \frac{(x-3)^2}{8} - \frac{(y+2)^2}{109} = 1$.   ● hyperbola.

[38] $9x^2 + 8y^2 - 54x + 9 = 0 \Rightarrow 9(x^2 - 6x) + 8y^2 = -9 \Rightarrow 9(x^2 - 6x + 9) + 8y^2 = 72 \Rightarrow$

$9(x-3)^2 + 8y^2 = 72 \Rightarrow \frac{(x-3)^2}{8} + \frac{y^2}{9} = 1$.   ● ellipse.

[39] $x^2 + y^2 + 8x - 2y + 6 - 0 \Rightarrow (x^2 + 8x) + (y^2 - 2y) = -6 \Rightarrow (x^2 + 8x + 16) + (y^2 - 2y + 1) = 11 \Rightarrow$

$(x + 4)^2 + (y - 2)^2 = 11.$ ● circle.

[40] $y^2 - 4x - 4y + 8 \Rightarrow x = \frac{1}{4}y^2 - y + 2.$ ● parabola.

[41] From the accompanying figure we see that one of the asymptotes passes through the points $(-a, b)$

and $(0, 0)$. Since the slope of this asymptote is $-\frac{b}{a}$, the equation is $y - 0 = -\frac{b}{a}(x - 0)$, or $y = -\frac{b}{a}x$.

The other asymptote passes through the points $(0, 0)$ and (a, b). Since the slope of this asymptote is

$\frac{b}{a}$, the equation is $y - 0 = \frac{b}{a}(x - 0)$ or $y = \frac{b}{a}x$. (See figure below).

[42] In triangle F_1F_2P (see figure shown in the text), the length of any side is less than the sum of the

lengths of the other two sides. Thus, $d(P, F_1) < d(F_1, F_2) + d(P, F_2)$ and

$d(P, F_1) < d(F_1, F_2) + d(P, F_2)$. Hence, $d(P, F_1) - d(P, F_2) < d(F_1, F_2)$ and

$d(P, F_2) - d(P, F_1) < d(F_1, F_2)$. It follows that $|d(P, F_1) - d(P, F_2)| < d(F_1, F_2)$.

Thus, $2a < 2c \Rightarrow a < c$.

[43] We introduce a rectangular coordinate system as shown in the accompanying figure. Since the radio

signal travels 0.2 miles per microsecond, and since the signal from tower R_1 is received by the ship

250 microseconds after the signal from tower R_2, the ship is 50 miles closer to tower R_2 than to

tower R_1. Since the ship lies on one branch of a hyperbola with foci at R_1 and R_2, we take $2a = 50$

and thus, $a = 25$. Now $c = 150$. It follows that $b^2 = c^2 - a^2 = 21{,}875$. Hence, the equation of the

hyperbola is $\frac{y^2}{625} - \frac{x^2}{21875} = 1$. (See figure below).

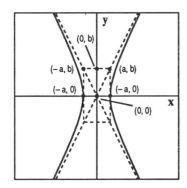

Figure 41

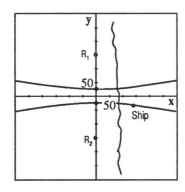

Figure 43

[44] Let (c, y_0) be the endpoint of the focal chord shown in the accompanying figure. Then, $\frac{c^2}{a^2} - \frac{y_0^2}{b^2} =$

1. Thus, $\frac{y_0^2}{b^2} = \frac{c^2}{a^2} - 1 \Rightarrow \frac{y_0^2}{b} = \frac{c^2 - a^2}{a^2} = \frac{b^4}{a^2} = \frac{b^2}{a^2} \Rightarrow y_0 = \frac{b^2}{a}$. Hence, the length of the focal chord is

$\frac{2b^2}{a}$. (See figure below).

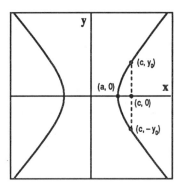

Figure 44

[45] Let $P = (x, y)$ be any point such that $d(P_0, P)$ is twice the distance from P to the line $y = -4$,

which is $|y - (-4)| = |y + 4|$. Then, $d(P_0, P) = 2|y + 4| \Rightarrow \sqrt{(x - 0)^2 + (y - 4)^2} = 2|y + 4| \Rightarrow$

$\left(\sqrt{x^2 + (y-4)^2}\right)^2 = (2|y + 4|)^2 \Rightarrow x^2 + y^2 - 8y + 16 = 4y^2 + 32y + 64 \Rightarrow x^2 - 3y^2 - 40y - 48 =$

$0 \Rightarrow x^2 - 3\left(y^2 + \frac{40}{3}y\right) = 48 \Rightarrow x^2 - 3\left(y^2 + \frac{40}{3}y + \frac{400}{9}\right) = 48 - \frac{400}{3} \Rightarrow x^2 - 3\left(y + \frac{20}{3}\right)^2 =$

$-\frac{256}{3} \Rightarrow \dfrac{\left(y + \frac{20}{3}\right)^2}{\frac{256}{9}} - \dfrac{x^2}{\frac{256}{3}} = 1.$ A hyperbola.   ● $x^2 - 3y^2 - 40y - 48 = 0.$

[46] $Ax^2 + Cy^2 + F = 0 \Rightarrow Ax^2 + Cy^2 = -F \Rightarrow -\frac{A}{F}x^2 - \frac{C}{F}y^2 = 1.$ Since $AC < 0, -\frac{A}{F}$ and $-\frac{C}{F}$ have

opposite signs. It follows that the graph is a hyperbola with center $(0, 0)$.

[47] A point $P = (x, y)$ lies on the hyperbola with center (h, k) and foci $F_1 = (h - c, k)$ and $F_2 =$

$(h + c, k)$ if and only if $d(F_1, P) + d(F_2, P) = \pm 2a$. Thus, $\sqrt{[x - (h - c)]^2 + (y - k)^2} +$

$\sqrt{[x - (h + c)]^2 + (y - k)^2} = \pm 2a \Rightarrow \sqrt{[x - (h - c)]^2 + (y - k)^2} =$

$\pm 2a - \sqrt{[x - (h + c)]^2 + (y - k)^2} \Rightarrow \left(\sqrt{[x - (h - c)]^2 + (y - k)^2}\right)^2 =$

$\left(\pm 2a - \sqrt{[x - (h + c)]^2 + (y - k)^2}\right)^2 \Rightarrow [x - (h - c)]^2 + (y - k)^2 = 4a^2 \pm$

$4a\sqrt{[x - (h + c)]^2 + (y - k)^2} + [x - (h + c)]^2 + (y - k)^2 \Rightarrow x^2 + c^2 + h^2 + 2cx - 2hx - 2ch =$

$4a^2 \pm 4a\sqrt{[x - (h + c)]^2 + (y - k)^2} + x^2 + c^2 + h^2 - 2cx - 2hx + 2ch \Rightarrow a^2 - cx + ch =$

$\pm a\sqrt{[x - (h + c)]^2 + (y - k)^2} \Rightarrow (a^2 - cx + ch)^2 = \left(\pm a\sqrt{[x - (h + c)]^2 + (y - k)^2}\right)^2 \Rightarrow$

$a^4 + c^2x^2 + c^2h^2 - 2a^2cx - 2c^2hx + 2a^2ch = a^2x^2 - 2a^2cx - 2a^2hx + 2a^2ch + a^2c^2 + a^2h^2 +$

$a^2(y - k)^2 \Rightarrow a^2(x^2 - 2hx + h^2) - c^2(x^2 - 2hx + h^2) + a^2(y - k)^2 = a^4 - a^2c^2 \Rightarrow$

$(a^2 - c^2)(x - h)^2 + a^2(y - k)^2 = a^2(a^2 - c^2) \Rightarrow \dfrac{(x - h)^2}{a^2} + \dfrac{(y - k)^2}{a^2 - c^2} = 1 \Rightarrow \dfrac{(x - h)^2}{a^2} -$

$\dfrac{(y - k)^2}{c^2 - a^2} = 1.$ Since $0 < a < c, c^2 - a^2 > 0.$ Letting $b^2 = c^2 - a^2$, we get the desired result.

[48] A point $P = (x, y)$ lies on the hyperbola with center (h, k) and foci $F_1 = (h, k - c)$ and $F_2 =$

$(h, k + c)$ if and only if $d(F_1, P) + d(F_2, P) = \pm 2a$. Thus, $\sqrt{(x - h)^2 + [y - (k - c)]^2} +$

$\sqrt{(x - h)^2 + [y - (k + c)]^2} = \pm 2a.$ Proceeding in a manner similar to that in exercise 47, we

obtain the desired result.

[49] $Ax^2 + Cy^2 + Dx + Ey + F = 0 \Rightarrow A\left(x^2 + \frac{D}{A}x\right) + C\left(y^2 + \frac{E}{C}y\right) = -F \Rightarrow$

$A\left(x^2 + \frac{D}{A}x + \frac{D^2}{4A^2}\right) + C\left(y^2 + \frac{E}{C}y + \frac{E^2}{4C^2}\right) = \frac{D^2}{4A} + \frac{E^2}{4C} - F \Rightarrow A\left(x + \frac{D}{2A}\right)^2 + C\left(y + \frac{E}{2C}\right)^2 =$

$\frac{D^2}{4A} + \frac{E^2}{4C} - F.$

a) If $\frac{D^2}{4A} + \frac{E^2}{4C} - F = 0$, the graph is a hyperbola since $AC < 0$.

b) If $\frac{D^2}{4A} + \frac{E^2}{4C} - F = 0$, the equation $A\left(x + \frac{D}{2A}\right)^2 + C\left(y + \frac{E}{2C}\right)^2 = \frac{D^2}{4A} + \frac{E^2}{4C} - F$ reduces to

$A\left(x + \frac{D}{2A}\right)^2 + C\left(y - \frac{E}{2C}\right)^2 = 0$ or $A\left(x + \frac{D}{2A}\right)^2 = -C\left(y + \frac{E}{2C}\right)^2 = 0.$ Letting $x' =$

$x + \frac{D}{2A}$ and $y' = y + \frac{E}{2C}$, we get $A(x')^2 = -C(y')^2$ or $(y')^2 = -\frac{A}{C}(x')^2.$ Since $AC < 0,$

$-\frac{A}{C} > 0.$ Thus $y' = \pm\sqrt{-\frac{A}{C}(x')^2} = \pm\sqrt{-\frac{A}{C}}\,|x'|.$ It follows that the graph consists of two

intersecting lines.

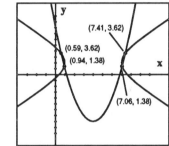

[50] a)

b) $(0.59, 3.62), (0.94, 1.38), (7.06, 1.38), (7.41, 3.62)$

EXERCISES 8.5

[1] In the *equation* $x^2 - xy + y^2 = 2$, $A = 1$, $B = -1$ and $C = 2$. Hence $\cot 2\theta = \frac{A - C}{B} = 0.$ Thus,

$2\theta = 90° \Rightarrow \theta - 45°.$ Since $\cos 45° = \sin 45° = \frac{\sqrt{2}}{2},$ it follows that $x = \frac{\sqrt{2}}{2}x' - \frac{\sqrt{2}}{2}y'$ and

$y = \frac{\sqrt{2}}{2}x' + \frac{\sqrt{2}}{2}y'.$ Substituting these values for x and y into the equation, we get

$\left(\frac{\sqrt{2}}{2}x' - \frac{\sqrt{2}}{2}y'\right)^2 - \left(\frac{\sqrt{2}}{2}x' - \frac{\sqrt{2}}{2}y'\right) \cdot \left(\frac{\sqrt{2}}{2}x' + \frac{\sqrt{2}}{2}y'\right) + \left(\frac{\sqrt{2}}{2}x' + \frac{\sqrt{2}}{2}y'\right)^2 = 2 \Rightarrow$

$\frac{1}{2}(x')^2 - x'y' + \frac{1}{2}(y')^2 - \frac{1}{2}(x')^2 - \frac{1}{2}x'y' + \frac{1}{2}x'y' + \frac{1}{2}(y')^2 + \frac{1}{2}(x')^2 + x'y' + \frac{1}{2}(y') = 2 \Rightarrow$

$\frac{1}{2}(x')^2 + \frac{3}{2}(y')^2 = 2 \Rightarrow \frac{(x')^2}{4} + \frac{(y')^2}{\frac{4}{3}} = 1.$

(See figure below). ● ellipse.

[2] $\theta = 45°$. Since $\cos 45° = \sin 45° = \frac{\sqrt{2}}{2}$, it follows that $x = \frac{\sqrt{2}}{2} x' - \frac{\sqrt{2}}{2} y'$ and $y = \frac{\sqrt{2}}{2} x' + \frac{\sqrt{2}}{2} y'$.

Substituting these values of x and y into the equation $xy + 1 = 0$, we get

$$\left(\frac{\sqrt{2}}{2} x' - \frac{\sqrt{2}}{2} y'\right)\left(\frac{\sqrt{2}}{2} x' + \frac{\sqrt{2}}{2} y'\right) + 1 = 0 \Rightarrow \frac{1}{2}(x')^2 + \frac{1}{2} x'y' - \frac{1}{2}x'y' - \frac{1}{2}(y')^2 = -1 \Rightarrow$$

$$\frac{(y')^2}{2} - \frac{(x')^2}{2} = 1. \quad \text{(See figure below).} \qquad \bullet \text{ hyperbola}$$

[3]   In the equation $x^2 + 2xy + y^2 + x - y = 0$, $A = 1$, $B = 2$ *and* $C = 1$. Hence, $\cot 2\theta = \frac{A-C}{B} = 0$,

and therefore, $2\theta = 90° \Rightarrow \theta = 45°$. Since $\cos 45° = \sin 45° = \frac{\sqrt{2}}{2}$, $x = \frac{\sqrt{2}}{2}x' - \frac{\sqrt{2}}{2}y'$ and $y =$

$\frac{\sqrt{2}}{2}x' + \frac{\sqrt{2}}{2}y'$. Substituting these values of x and y into the equation $x^2 + 2xy + y^2 + x - y = 0$

yield the equation $\left(\frac{\sqrt{2}}{2}x' - \frac{\sqrt{2}}{2}y'\right)^2 + 2\left(\frac{\sqrt{2}}{2}x' - \frac{\sqrt{2}}{2}y'\right)\left(\frac{\sqrt{2}}{2}x' + \frac{\sqrt{2}}{2}y'\right) + \left(\frac{\sqrt{2}}{2}x' + \frac{\sqrt{2}}{2}y'\right)^2 +$

$\frac{\sqrt{2}}{2}x' - \frac{\sqrt{2}}{2}y' - \left(\frac{\sqrt{2}}{2}x' + \frac{\sqrt{2}}{2}y'\right) = 0 \Rightarrow \frac{1}{2}(x')^2 - x'y' + \frac{1}{2}(y')^2 + (x')^2 + x'y' - x'y' - (y')^2 +$

$\frac{1}{2}(x')^2 + x'y' + \frac{1}{2}(y')^2 + \frac{\sqrt{2}}{2}x' - \frac{\sqrt{2}}{2}y' - \frac{\sqrt{2}}{2}x' - \frac{\sqrt{2}}{2}y' = 0 \Rightarrow 2(x')^2 - \sqrt{2}\, y' = 0 \Rightarrow (x')^2 =$

$\frac{\sqrt{2}}{2}y'$. (See figure below). $\qquad \bullet \text{ parabola}$

[4]   In the equation $8x^2 - 4xy + 5y^2 = 36$, $A = 8$, $B = -4$ and $C = 5$. Therefore, $\cot 2\theta = \frac{A-C}{B} =$

$\frac{8-5}{-4} = -\frac{3}{4}$. Since $\cot 2\theta < 0$, we choose $90° < 2\theta < 180°$, and it follows that $\cos 2\theta = -\frac{3}{5}$.

Moreover, we find that $\cos \theta = \frac{\sqrt{5}}{5}$ and $\sin \theta = \frac{2\sqrt{5}}{5}$. Letting $x = \frac{\sqrt{5}}{5}x' - \frac{2\sqrt{5}}{5}y'$ and $y =$

$\frac{2\sqrt{5}}{5}x' + \frac{\sqrt{5}}{5}y'$, and substituting these values of x and y into the equation $8x^2 - 4xy + 5y^2 = 36$,

we get $8\left(\frac{\sqrt{5}}{5}x' - \frac{2\sqrt{5}}{5}y'\right)^2 - 4\left(\frac{\sqrt{5}}{5}x' - \frac{2\sqrt{5}}{5}y'\right)\left(\frac{2\sqrt{5}}{5}x' + \frac{\sqrt{5}}{5}y'\right) + 5\left(\frac{2\sqrt{5}}{5}x' + \frac{\sqrt{5}}{5}y'\right)^2 =$

$36 \Rightarrow \frac{8}{5}(x')^2 - \frac{32}{5}x'y' + \frac{32}{5}(y')^2 - \frac{8}{5}(x')^2 - \frac{4}{5}x'y' + \frac{16}{5}x'y' + \frac{8}{5}(y')^2 + 4(x')^2 + 4x'y' + (y')^2 =$

$36 \Rightarrow 4(x')^2 + 9(y')^2 = 36 \Rightarrow \frac{(x')^2}{9} + \frac{(y')^2}{4} = 1. \quad \text{(See figure below).} \qquad \bullet \text{ ellipse.}$

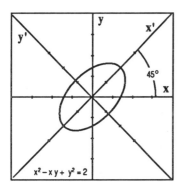

Figure 1

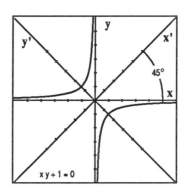

Figure 2

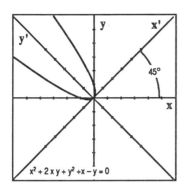

Figure 3

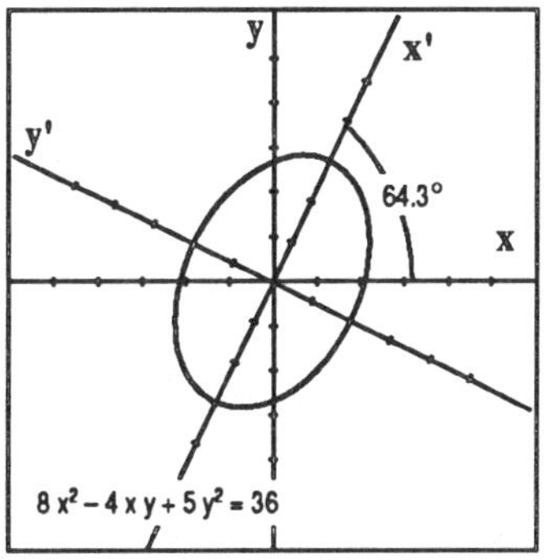

Figure 4

[5] In the equation $2x^2 + 3xy - 2y^2 = 25$, $A = 2$, $B = 3$ and $C = -2$. Therefore, $\cot 2\theta = \dfrac{A-C}{B} =$ $\dfrac{2-(-2)}{-3} = \dfrac{4}{3}$. Since $\cot 2\theta > 0$, $0° < 2\theta < 90°$, and it follows that $\cos 2\theta = \dfrac{4}{5}$. Moreover, we find that $\cos \theta = \sqrt{\dfrac{9}{10}}$ and $\sin \theta = \sqrt{\dfrac{1}{10}}$. Letting $x = \sqrt{\dfrac{9}{10}}x' - \sqrt{\dfrac{1}{10}}y'$ and $y = \sqrt{\dfrac{1}{10}}x' + \sqrt{\dfrac{9}{10}}y'$, and substituting these values for x and y into the equation $2x^2 - 3xy - 2y^2 = 25$ yields the equation $2\left(\sqrt{\dfrac{9}{10}}x' - \sqrt{\dfrac{1}{10}}y'\right)^2 + 3\left(\sqrt{\dfrac{9}{10}}x' - \sqrt{\dfrac{1}{10}}y'\right) \cdot$ $\left(\sqrt{\dfrac{1}{10}}x' + \sqrt{\dfrac{9}{10}}y'\right) - 2\left(\sqrt{\dfrac{1}{10}}x' + \sqrt{\dfrac{9}{10}}y'\right)^2 = 25 \Rightarrow \dfrac{9}{5}(x')^2 - \dfrac{12}{10}x'y' + \dfrac{1}{5}(y')^2 +$ $\dfrac{9}{10}(x')^2 + \dfrac{27}{10}x'y' - \dfrac{3}{10}x'y' - \dfrac{9}{10}(y')^2 - \dfrac{1}{5}(x')^2 - \dfrac{12}{10}x'y' - \dfrac{9}{5}(y')^2 = 25 \Rightarrow \dfrac{25}{10}(x')^2 - \dfrac{25}{10}(y')^2 =$ $25 \Rightarrow \dfrac{(x')^2}{10} - \dfrac{(y')^2}{10} = 1$. (See figure below). ● hyperbola.

[6] In the equation $8x^2 + 12xy + 13y^2 = 885$, $A = 8$, $B = 12$ and $C = 13$. Hence, $\cot 2\theta = \dfrac{A-C}{B} =$ $\dfrac{8-13}{12} = -\dfrac{5}{12}$. Thus, we choose $90° < 2\theta < 180°$, and it follows that $\cos 2\theta = -\dfrac{5}{13}$. Moreover, $\cos \theta = \sqrt{\dfrac{4}{13}}$ and $\sin \theta = \sqrt{\dfrac{9}{13}}$. Substituting $\sqrt{\dfrac{4}{13}}x' - \sqrt{\dfrac{9}{13}}y'$ for x and $\sqrt{\dfrac{9}{13}}x' + \sqrt{\dfrac{4}{13}}y'$ for y in the equation $8x^2 + 12xy + 13y^2 = 885$ yields the equation $8\left(\sqrt{\dfrac{4}{13}}x' - \sqrt{\dfrac{9}{13}}y'\right)^2 +$ $12\left(\sqrt{\dfrac{4}{13}}x' - \sqrt{\dfrac{9}{13}}y'\right)\left(\sqrt{\dfrac{9}{13}}x' + \sqrt{\dfrac{4}{13}}y'\right) + 13\left(\sqrt{\dfrac{9}{13}}x' + \sqrt{\dfrac{4}{13}}y'\right)^2 = 885 \Rightarrow$ $\dfrac{32}{13}(x')^2 - \dfrac{96}{13}x'y' + \dfrac{72}{13}(y')^2 + \dfrac{12}{13}(x')^2 + \dfrac{48}{13}x'y' - \dfrac{108}{13}x'y' - \dfrac{72}{13}(y')^2 + 9(x')^2 + 12x'y' +$ $4(y')^2 = 885 \Rightarrow 17(x')^2 + 4(y')^2 = 885 \Rightarrow \dfrac{(x')^2}{\dfrac{885}{17}} + \dfrac{(y')^2}{\dfrac{885}{4}} = 1$. (See figure below). ● ellipse.

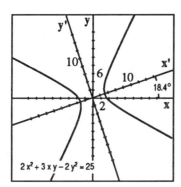

Figure 5   **Figure 6**

[7]   In the equation $x^2 - xy + y^2 = 6$, $A = 1$, $B = -1$ and $C = 1$. Therefore, $\cot 2\theta = 90°$ and $\theta = 45°$.

Since $\cos 45° = \sin 45° = \frac{\sqrt{2}}{2}$, $x = \frac{\sqrt{2}}{2}x' - \frac{\sqrt{2}}{2}y'$ and $y = \frac{\sqrt{2}}{2}x' + \frac{\sqrt{2}}{2}y'$. Substituting these values

of x and y into the equation $x^2 - xy + y^2 = 6$ yields the equation $\left(\frac{\sqrt{2}}{2}x' - \frac{\sqrt{2}}{2}y'\right)^2 -$

$\left(\frac{\sqrt{2}}{2}x' - \frac{\sqrt{2}}{2}y'\right)\left(\frac{\sqrt{2}}{2}x' + \frac{\sqrt{2}}{2}y'\right) + \left(\frac{\sqrt{2}}{2}x' + \frac{\sqrt{2}}{2}y'\right)^2 = 6 \Rightarrow \frac{1}{2}(x')^2 - x'y' + \frac{1}{2}(y')^2 - \frac{1}{2}(x')^2 -$

$\frac{1}{2}x'y' + \frac{1}{2}x'y' + \frac{1}{2}(y')^2 + \frac{1}{2}(x')^2 + x'y' + \frac{1}{2}(y')^2 = 6 \Rightarrow \frac{1}{2}(x')^2 + \frac{3}{2}(y')^2 = 6 \Rightarrow$

$\frac{(x')^2}{12} + \frac{(y')^2}{4} = 1$.   (See figure below).   ● ellipse.

[8]   In the equation $73x^2 - 72xy + 52y^2 - 30x - 40y - 75 = 0$, $A = 73$ $B = -72$ and $C - 52$. Hence,

$\cot 2\theta = \frac{A-C}{B} = \frac{73-52}{-72} = -\frac{7}{24}$. Thus, we choose $90° < 2\theta < 180°$ and it follows that $\cos 2\theta =$

$-\frac{7}{25}$. Moreover, $\cos\theta = \frac{3}{5}$ and $\sin\theta = \frac{4}{5}$. Substituting $\frac{3}{5}x' - \frac{4}{5}y'$ for x and $\frac{4}{5}x' + \frac{3}{5}y'$ for y in the

equation $73x^2 - 72xy + 52y^2 - 30x - 40y - 75 = 0$ yields $73\left(\frac{3}{5}x' - \frac{4}{5}y'\right)^2 - 72\left(\frac{3}{5}x' - \frac{4}{5}y'\right) \cdot$

$\left(\frac{4}{5}x' + \frac{3}{5}y'\right) + 52\left(\frac{4}{5}x' + \frac{3}{5}y'\right)^2 - 30\left(\frac{3}{5}x' - \frac{4}{5}y'\right) - 40\left(\frac{4}{5}x' + \frac{3}{5}y'\right) - 75 = 0 \Rightarrow \frac{657}{25}(x')^2 -$

$\frac{1752}{25}x'y' + \frac{1168}{25}(y')^2 - \frac{864}{25}(x')^2 - \frac{648}{25}x'y' + \frac{1152}{25}x'y' + \frac{864}{25}(y')^2 + \frac{832}{25}(x')^2 + \frac{1248}{25}x'y' +$

$\frac{468}{25}(y')^2 - 18x' + 24y' - 32x' - 24y' - 75 = 0 \Rightarrow 25(x')^2 + 100(y')^2 - 50x' - 75 = 0 \Rightarrow$

$25\left[(x')^2 - 2x'\right] + 100(y')^2 = 75 \Rightarrow 25\left[(x')^2 - 2x' + 1\right] + 100(y')^2 = 100 \Rightarrow \frac{(x-1)^2}{4} + (y')^2 = 1.$

(See figure below).   ● *ellipse.*

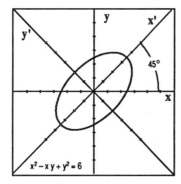   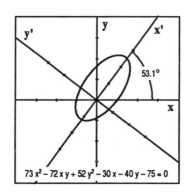

Figure 7   **Figure 8**

[9] In the equation $11x^2 + 4\sqrt{3}xy + 7y^2 - 1 = 0$, $A = 11$, $B = -4\sqrt{3}$ and $C = 7$. Thus, $\cot 2\theta = \dfrac{A-C}{B} = \dfrac{11-7}{4\sqrt{3}} = \dfrac{\sqrt{3}}{3}$. Hence, $2\theta = 60° \Rightarrow \theta = 30°$. Since $\cos 30° = \dfrac{\sqrt{3}}{2}$ and $\sin 30° = \dfrac{1}{2}$, we choose $x = \dfrac{\sqrt{3}}{2}x' - \dfrac{1}{2}y'$ and $y = \dfrac{1}{2}x' + \dfrac{\sqrt{3}}{2}y'$. Substituting these values for x and y into the equation $11x^2 + 4\sqrt{3}xy + 7y^2 - 1 = 0$, we get $11\left(\dfrac{\sqrt{3}}{2}x' - \dfrac{1}{2}y'\right)^2 + 4\sqrt{3}\left(\dfrac{\sqrt{3}}{2}x' - \dfrac{1}{2}y'\right) \cdot$ $\left(\dfrac{1}{2}x' + \dfrac{\sqrt{3}}{2}y'\right) + 7\left(\dfrac{1}{2}x' + \dfrac{\sqrt{3}}{2}y'\right)^2 - 1 = 0 \Rightarrow \dfrac{33}{4}(x')^2 - \dfrac{11\sqrt{3}}{2}x'y' + \dfrac{11}{4}(y')^2 + 3(x')^2 +$ $3\sqrt{3}x'y' - \sqrt{3}x'y' - 3(y')^2 + \dfrac{7}{4}(x')^2 + \dfrac{7\sqrt{3}}{2}x'y' + \dfrac{21}{4}(y')^2 - 1 = 0 \Rightarrow 13(x')^2 + 5(y')^2 = 1 \Rightarrow$ $\dfrac{(x')^2}{\frac{1}{13}} + \dfrac{(y')^2}{\frac{1}{5}} = 1.$ (See figure below). ● ellipse.

[10] In the equation $34x^2 - 24xy + 41y^2 - 40x - 30y - 25 = 0$, $A = 34$, $B = -24$ and $C = 41$. It follows that $\cot 2\theta = \dfrac{A-C}{B} = \dfrac{34-41}{-24} = \dfrac{7}{24}$. Hence, we choose $0° < 2\theta < 90°$ and it can be shown that $\cos 2\theta = \dfrac{7}{25}$. Thus, $\cos\theta = \dfrac{4}{5}$ and $\sin\theta = \dfrac{3}{5}$. Letting $x = \dfrac{4}{5}x' - \dfrac{3}{5}y'$ and $y = \dfrac{3}{5}x' + \dfrac{4}{5}y'$, and substituting these values for x and y into the equation $34x^2 - 24xy + 41y^2 - 40x - 30y - 25 = 0$ yields the equation $34\left(\dfrac{4}{5}x' - \dfrac{3}{5}y'\right)^2 - 24\left(\dfrac{4}{5}x' - \dfrac{3}{5}y'\right)\left(\dfrac{3}{5}x' + \dfrac{4}{5}y'\right) + 41\left(\dfrac{3}{5}x' + \dfrac{4}{5}y'\right)^2 -$ $40\left(\dfrac{4}{5}x' - \dfrac{3}{5}y'\right) - 30\left(\dfrac{3}{5}x' + \dfrac{4}{5}y'\right) - 25 = 0 \Rightarrow \dfrac{544}{25}(x')^2 - \dfrac{816}{25}x'y' + \dfrac{306}{25}(y')^2 - \dfrac{288}{25}(x')^2 -$ $\dfrac{384}{25}x'y' + \dfrac{216}{25}x'y' + \dfrac{288}{25}(y')^2 + \dfrac{369}{25}(x')^2 + \dfrac{984}{25}x'y' + \dfrac{656}{25}(y')^2 - 32x' + 24y' - 18x' - 24y' -$ $25 = 0 \Rightarrow 25(x')^2 + 50(y')^2 - 50x' - 25 = 0 \Rightarrow 25\left[(x')^2 - 2x'\right] + 50(y')^2 = 25 \Rightarrow$ $25\left[(x')^2 - 2x' + 1\right] + 50(y')^2 = 50 \Rightarrow \dfrac{(x'-1)^2}{2} + (y')^2 = 1.$ (See figure below). ● ellipse.

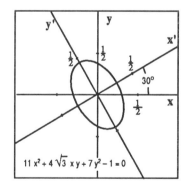

Figure 9   **Figure 10**

[11] In the equation $x^2 + 2\sqrt{3}xy + 3y^2 + 2\sqrt{3}x - 2y = 0$, $A = 1$, $B = 2\sqrt{3}$ and $C = 3$. Therefore, $\cot 2\theta = \dfrac{A-C}{B} = \dfrac{1-3}{2\sqrt{3}} = -\dfrac{\sqrt{3}}{3}$, and it follows that $\theta = 60°$. Since $\cos 60° = \dfrac{1}{2}$ and $\sin 60° = \dfrac{\sqrt{3}}{2}$, we let $x = \dfrac{1}{2}x' - \dfrac{\sqrt{3}}{2}y'$ and $y = \dfrac{\sqrt{3}}{2}x' + \dfrac{1}{2}y'$. Substituting these values of x and y into the equation $x^2 + 2\sqrt{3}xy + 3y^2 + 2\sqrt{3}x - 2y = 0$, we get $\left(\dfrac{1}{2}x' - \dfrac{\sqrt{3}}{2}y'\right)^2 + 2\sqrt{3}\left(\dfrac{1}{2}x' - \dfrac{\sqrt{3}}{2}y'\right) \cdot$ $\left(\dfrac{\sqrt{3}}{2}x' + \dfrac{1}{2}y'\right) + 3\left(\dfrac{\sqrt{3}}{2}x' + \dfrac{1}{2}y'\right)^2 + 2\sqrt{3}\left(\dfrac{1}{2}x' - \dfrac{\sqrt{3}}{2}y'\right) - 2\left(\dfrac{\sqrt{3}}{2}x' + \dfrac{1}{2}y'\right) = 0 \Rightarrow \dfrac{1}{4}(x')^2 -$ $\dfrac{\sqrt{3}}{2}x'y' + \dfrac{3}{4}(y')^2 + \dfrac{3}{2}(x')^2 + \dfrac{\sqrt{3}}{2}x'y' - \dfrac{3\sqrt{3}}{2}x'y' - \dfrac{3}{2}(y')^2 + \dfrac{9}{4}(x')^2 + \dfrac{3\sqrt{3}}{2}x'y' + \dfrac{3}{4}(y')^2 +$ $\sqrt{3}x' - 3y' - 0 \Rightarrow 4(x')^2 - 4y' = 0 \Rightarrow (x')^2 = y'.$ (See figure below). ● parabola.

[12] In the equation $36x^2 + 96xy + 64y^2 + 20x - 15y + 25 = 0$, $A = 36$, $B = 96$ and $C = 64$. Hence, $\cot 2\theta = \frac{A-C}{B} = \frac{36-64}{96} = -\frac{7}{24}$, and we choose $90° < 2\theta < 180°$. It follows that $\cos 2\theta = -\frac{7}{25}$, and hence, $\cos \theta = \frac{3}{5}$ and $\sin \theta = \frac{4}{5}$. Substituting $\frac{3}{5}x' - \frac{4}{5}y'$ for x and $\frac{4}{5}x' + \frac{3}{5}y'$ for y in the equation $36x^2 + 96xy + 64y^2 + 20x - 15y + 25 = 0$ yields the equation $36\left(\frac{3}{5}x' - \frac{4}{5}y'\right)^2 + 96\left(\frac{3}{5}x' - \frac{4}{5}y'\right)\left(\frac{4}{5}x' + \frac{3}{5}y'\right) + 64\left(\frac{4}{5}x' + \frac{3}{5}y'\right)^2 + 20\left(\frac{3}{5}x' - \frac{4}{5}y'\right) - 15\left(\frac{4}{5}x' + \frac{3}{5}y'\right) + 25 = 0 \Rightarrow \frac{324}{25}(x')^2 - \frac{864}{25}x'y' + \frac{576}{25}(y')^2 + \frac{1152}{25}(x')^2 + \frac{864}{25}x'y' - \frac{1536}{25}x'y' - \frac{1152}{25}(y')^2 + \frac{1024}{25}(x')^2 + \frac{1536}{25}x'y' + \frac{576}{25}(y')^2 + 12x' - 16y' - 12x' - 9y' + 25 = 0 \Rightarrow 100(x')^2 - 25y' + 25 = 0 \Rightarrow (x')^2 = \frac{1}{4}(y' - 1)$. (See figure below). ● parabola.

[13] In the equation $2\sqrt{3}xy - 2y^2 - \sqrt{3}x - y = 0$, $A = 0$, $B = 2\sqrt{3}$ and $C = -2$. Thus, $\cot 2\theta = \frac{A-C}{B} = \frac{0-(-2)}{2\sqrt{3}} = \frac{\sqrt{3}}{3}$. It follows that $2\theta = 60° \Rightarrow \theta = 30°$. Since $\cos 30° = \frac{\sqrt{3}}{2}$ and $\sin 30° = \frac{1}{2}$, we substitute $\frac{\sqrt{3}}{2}x' - \frac{1}{2}y'$ for x and $\frac{1}{2}x' + \frac{\sqrt{3}}{2}y'$ for y in the equation $2\sqrt{3}xy - 2y^2 - \sqrt{3}x - y = 0$. This substitution yields the equation $2\sqrt{3}\left(\frac{\sqrt{3}}{2}x' - \frac{1}{2}y'\right)\left(\frac{1}{2}x' + \frac{\sqrt{3}}{2}y'\right) - 2\left(\frac{1}{2}x' + \frac{\sqrt{3}}{2}y'\right)^2 - \sqrt{3}\left(\frac{\sqrt{3}}{2}x' - \frac{1}{2}y'\right) - \left(\frac{1}{2}x' + \frac{\sqrt{3}}{2}y'\right) = 0 \Rightarrow \frac{3}{2}(x')^2 + \frac{3\sqrt{3}}{2}x'y' - \frac{\sqrt{3}}{2}x'y' - \frac{3}{2}(y')^2 - \frac{1}{2}(x')^2 - \sqrt{3}x'y' - \frac{3}{2}(y')^2 - \frac{3}{2}x' - \frac{\sqrt{3}}{2}y' - \frac{1}{2}x' + \frac{\sqrt{3}}{2}y' = 0 \Rightarrow (x')^2 - 3(y')^2 - 2x = 0 \Rightarrow [(x')^2 - 2x + 1] - 3(y')^2 = 1 \Rightarrow (x' - 1)^2 - \frac{(y')^2}{\frac{1}{3}} = 1$. (See figure below). ● hyperbola.

[14] In the equation $3x^2 + 10xy + 3y^2 - 2x - 14y - 5 = 0$, $A = 3$, $B = 10$ and $C = 3$. Hence, $\cot 2\theta = \frac{A-C}{B} = 0 \Rightarrow 2\theta = 90° \; \theta = 45°$. Since $\cos 45° = \sin 45° = \frac{\sqrt{2}}{2}$ we let $x = \frac{\sqrt{2}}{2}x' - \frac{\sqrt{2}}{2}y'$ and $y = \frac{\sqrt{2}}{2}x' + \frac{\sqrt{2}}{2}y'$. Substituting these values of x and y into the equation $3x^2 + 10xy + 3y^2 - 2x - 14y - 5 = 0$, we get $3\left(\frac{\sqrt{2}}{2}x' - \frac{\sqrt{2}}{2}y'\right)^2 + 10\left(\frac{\sqrt{2}}{2}x' - \frac{\sqrt{2}}{2}y'\right)\left(\frac{\sqrt{2}}{2}x' + \frac{\sqrt{2}}{2}y'\right) + 3\left(\frac{\sqrt{2}}{2}x' + \frac{\sqrt{2}}{2}y'\right)^2 - 2\left(\frac{\sqrt{2}}{2}x' - \frac{\sqrt{2}}{2}y'\right) - 14\left(\frac{\sqrt{2}}{2}x' + \frac{\sqrt{2}}{2}y'\right) - 5 = 0 \Rightarrow \frac{3}{2}(x')^2 - 3x'y' + \frac{3}{2}(y')^2 + 5(x')^2 + 5x'y' - 5x'y' - 5(y')^2 + \frac{3}{2}(x')^2 + 3x'y' + \frac{3}{2}(y')^2 - \sqrt{2}x' + \sqrt{2}y' - 7\sqrt{2}x' - 7\sqrt{2}y' - 5 = 0 \Rightarrow 8(x')^2 - 2(y')^2 - 8\sqrt{2}x' - 6\sqrt{2}y' = 5 \Rightarrow 8[(x')^2 - \sqrt{2}x'] - 2[(y')^2 + 3\sqrt{2}y'] = 5 \Rightarrow 8\left[(x')^2 - \sqrt{2}x' + \frac{1}{2}\right] - 2\left[(y')^2 + 3\sqrt{2}y' + \frac{9}{2}\right] = 5 + 4 - 9 = 0 \Rightarrow 4\left(x' - \frac{\sqrt{2}}{2}\right)^2 = \left(y' = \frac{3\sqrt{2}}{2}\right)^2$.

(See figure below). ● two intersecting lines (See Exercise 49 on page 398).

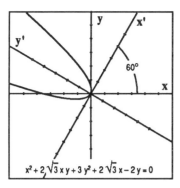

Figure 11

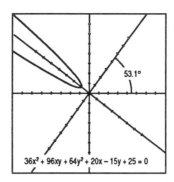

Figure 12

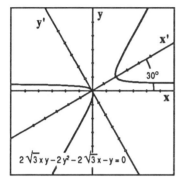

Figure 13

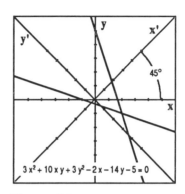

Figure 14

[15] In the equation $3x^2 - 6xy + 3y^2 + 2x = 7$, $A = 3$, $B = -6$ and $C = 3$. Therefore, $\cot 2\theta = \dfrac{A-C}{B} =$

0, and hence, we choose $\theta = 45°$. Since $\cos 45° = \sin 45° = \dfrac{\sqrt{2}}{2}$. we let $x = \dfrac{\sqrt{2}}{2}x' - \dfrac{\sqrt{2}}{2}y'$ and $y =$

$\dfrac{\sqrt{2}}{2}x' + \dfrac{\sqrt{2}}{2}y'$. Substituting the values of x and y into the equation $3x^2 - 6xy + 3y^2 + 2x = 7$, we

get $3\left(\dfrac{\sqrt{2}}{2}x' - \dfrac{\sqrt{2}}{2}y'\right)^2 - 6\left(\dfrac{\sqrt{2}}{2}x' - \dfrac{\sqrt{2}}{2}y'\right)\left(\dfrac{\sqrt{2}}{2}x' + \dfrac{\sqrt{2}}{2}y'\right) + 3\left(\dfrac{\sqrt{2}}{2}x' + \dfrac{\sqrt{2}}{2}y'\right)^2 +$

$2\left(\dfrac{\sqrt{2}}{2}x' - \dfrac{\sqrt{2}}{2}y'\right) = 7 \Rightarrow \dfrac{3}{2}(x')^2 - 3x'y' + \dfrac{3}{2}(y')^2 - 3(x')^2 - 3x'y' + 3x'y' + 3(y')^2 + \dfrac{3}{2}(x')^2 +$

$3x'y' + \dfrac{3}{2}(y')^2 + \sqrt{2}\,x' - \sqrt{2}\,y' = 7 \Rightarrow 6(y')^2 + \sqrt{2}\,x' - \sqrt{2}\,y' = 7 \Rightarrow 6\left[(y')^2 + \dfrac{\sqrt{2}}{6}\,y'\right] = -2\sqrt{x} +$

$7 \Rightarrow 6\left[y' + \dfrac{\sqrt{2}}{6}y + \dfrac{1}{72}\right] = -\sqrt{2}\,x + 7 + \dfrac{1}{12} \Rightarrow \left(y' + \dfrac{\sqrt{2}}{12}\right)^2 = -\dfrac{\sqrt{2}}{6}\left(x' - \dfrac{85\sqrt{2}}{24}\right).$ (See figure below).

● parabola.

[16] In the equation $4x^2 - 12xy + 9y^2 - 52x + 26y + 81 = 0$, $A = 4$, $B = -12$ and $C = 9$. Hence, $\cot 2\theta =$

$\frac{A-C}{B} = \frac{4-9}{-12} = \frac{5}{12}$. Since $\cot 2\theta > 0$, it follows that $\cos 2\theta = \frac{5}{13}$. Moreover, $\cos \theta = \sqrt{\frac{9}{13}}$ and

$\sin \theta = \sqrt{\frac{4}{13}}$. Letting $x = \sqrt{\frac{9}{13}}\,x' - \sqrt{\frac{4}{13}}\,y'$ and $y = \sqrt{\frac{4}{13}}\,x' + \sqrt{\frac{9}{13}}\,y'$, the equation

$4x^2 - 12xy + 9y^2 - 52x + 26y + 81 = 0$ can be written $4\left(\sqrt{\frac{9}{13}}\,x' - \sqrt{\frac{4}{13}}\,y'\right)^2 -$

$12\left(\sqrt{\frac{9}{13}}\,x' - \sqrt{\frac{4}{13}}\,y'\right)\left(\sqrt{\frac{4}{13}}\,x' + \sqrt{\frac{9}{13}}\,y'\right) + 9\left(\sqrt{\frac{4}{13}}\,x' + \sqrt{\frac{9}{13}}\,y'\right)^2 -$

$52\left(\sqrt{\frac{9}{13}}\,x' - \sqrt{\frac{4}{13}}\,y'\right) + 26\left(\sqrt{\frac{4}{13}}\,x' + \sqrt{\frac{9}{13}}\,y'\right) + 81 = 0 \Rightarrow \frac{36}{13}(x')^2 - \frac{48}{13}x'y' + \frac{16}{13}(y')^2 -$

$\frac{72}{13}(x')^2 - \frac{108}{13}x'y' + \frac{48}{13}x'y' + \frac{72}{13}(y')^2 + \frac{36}{13}(x')^2 + \frac{108}{13}x'y' + \frac{81}{13}(y')^2 - 12\sqrt{13}\,x' + 8\sqrt{13}\,y' +$

$6\sqrt{13}\,y' + 81 = 0 \Rightarrow 13\,(y')^2 - 8\sqrt{13}\,x' + 14\sqrt{13}\,y' + 81 = 0 \Rightarrow 13\left[(y')^2 + \frac{14\sqrt{13}}{13}y'\right] =$

$8\sqrt{13}\,x' - 8 \Rightarrow 13\left[(y')^2 + \frac{14\sqrt{13}}{13}y' + \frac{49}{13}\right] = 8\sqrt{13}\,x' - 32 \Rightarrow \left(y' + \frac{7\sqrt{13}}{13}\right)^2 =$

$\frac{8\sqrt{13}}{13}\left(x' - \frac{4\sqrt{13}}{13}\right)$. (See figure below).   ● parabola.

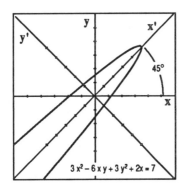

$3x^2 - 6xy + 3y^2 + 2x = 7$

Figure 15   **Figure 16**

$4x^2 - 12xy + 9y^2 - 52x + 26y + 81 = 0$

[17] In the equation $x^2 - 2\sqrt{3}\,xy + 3y^2 - 16\sqrt{3}\,x - 16y = 0$, $A = 1$, $B = -2\sqrt{3}$ and $C = 3$, $\cot 2\theta =$

$\frac{A-C}{B} = \frac{1-3}{-2\sqrt{3}} = \frac{\sqrt{3}}{3}$. Hence, we choose $2\theta = 60° \Rightarrow \theta = 30°$. Since $\cos 30° = \frac{\sqrt{3}}{2}$ and $\sin 30° =$

$\frac{1}{2}$, we let $x = \frac{\sqrt{3}}{2}x' - \frac{1}{2}y'$ and $y = \frac{1}{2}x' + \frac{\sqrt{3}}{2}y'$. Substituting these values of x and y into the

equation $x^2 - 2\sqrt{3}\,xy + 3y^2 - 16\sqrt{3}\,x - 16y = 0$, we get $\left(\frac{\sqrt{3}}{2}x' - \frac{1}{2}y'\right)^2 - 2\sqrt{3}\left(\frac{\sqrt{3}}{2}x' - \frac{1}{2}y'\right) \cdot$

$\left(\frac{1}{2}x' + \frac{\sqrt{3}}{2}y'\right) + 3\left(\frac{1}{2}x' + \frac{\sqrt{3}}{2}y'\right)^2 - 16\sqrt{3}\left(\frac{\sqrt{3}}{2}x' - \frac{1}{2}y'\right) - 16\left(\frac{1}{2}x' + \frac{\sqrt{3}}{2}y'\right) = 0 \Rightarrow \frac{3}{4}(x')^2 -$

$\frac{\sqrt{3}}{2}x'y' + \frac{1}{4}(y')^2 - \frac{3}{2}(x')^2 - \frac{3\sqrt{3}}{2}x'y' + \frac{\sqrt{3}}{2}x'y' + \frac{3}{2}(y')^2 + \frac{3}{4}(x')^2 + \frac{3\sqrt{3}}{2}x'y' + \frac{9}{4}(y')^2 -$

$24x' + 8\sqrt{3}\,y' - 8x' - 8\sqrt{3}\,y' = 0 \Rightarrow 4\,(y')^2 = 32x' \Rightarrow (y')^2 = 8x'$.   (See figure below).   ● parabola.

[18] In the equation $31x^2 + 10\sqrt{3}\,xy + 21y^2 - (124 - 40\sqrt{3})x + (168 - 20\sqrt{3})y + 316 - 80\sqrt{3} = 0$,

$A = 31$, $B = 10\sqrt{3}$ and $C = 21$. Thus, $\cot 2\theta = \dfrac{A-C}{B} = \dfrac{31-21}{10\sqrt{3}} = \dfrac{\sqrt{3}}{3}$. Therefore, we choose $2\theta =$

$60° \Rightarrow \theta = 30°$. Since $\cos 30° = \dfrac{\sqrt{3}}{2}$ and $\sin 30° = \dfrac{1}{2}$, we let $x = \dfrac{\sqrt{3}}{2}x' - \dfrac{1}{2}y'$ and $y = \dfrac{1}{2}x' + \dfrac{\sqrt{3}}{2}y'$.

Substituting these values of x and y into the equation $31x^2 + 10\sqrt{3}\,xy + 21y^2 - (124 - 40\sqrt{3})x +$

$(168 - 20\sqrt{3})y + 316 - 80\sqrt{3} = 0 \Rightarrow 31\left(\dfrac{\sqrt{3}}{2}x' - \dfrac{1}{2}y'\right)^2 + 10\sqrt{3}\left(\dfrac{\sqrt{3}}{2}x' - \dfrac{1}{2}y'\right)\left(\dfrac{1}{2}x' + \dfrac{\sqrt{3}}{2}y'\right) +$

$21\left(\dfrac{1}{2}x' + \dfrac{\sqrt{3}}{2}y'\right)^2 - (124 - 40\sqrt{3})\left(\dfrac{\sqrt{3}}{2}x' - \dfrac{1}{2}y'\right) + (168 - 20\sqrt{3})\left(\dfrac{1}{2}x' + \dfrac{\sqrt{3}}{2}y'\right) + 316 - 80\sqrt{3} =$

$0 \Rightarrow \dfrac{93}{4}(x')^2 - \dfrac{31\sqrt{3}}{2}x'y' + \dfrac{31}{4}(y')^2 + \dfrac{15}{2}(x')^2 + \dfrac{15\sqrt{3}}{2}x'y' - \dfrac{5\sqrt{3}}{2}x'y' - \dfrac{15}{2}(y')^2 +$

$\dfrac{21}{4}(x')^2 + \dfrac{21\sqrt{3}}{2}x'y' + \dfrac{63}{4}(y')^2 - 62\sqrt{3}\,x' + 60x' + 62y' - 20\sqrt{3}\,y' + 84\,x' - 10\sqrt{3}\,x' +$

$84\sqrt{3}\,y' - 30y' + 316 - 80\sqrt{3} = 0 \Rightarrow 36(x')^2 + 16(y')^2 + (144 - 72\sqrt{3})x' + (32 + 64\sqrt{3})y' +$

$316 - 80\sqrt{3} = 0 \Rightarrow 36\left[(x')^2 + (4 - 2\sqrt{3})x'\right] + 16\left[(y')^2 + (2 + 4\sqrt{3})y'\right] = 80\sqrt{3} - 316 \Rightarrow$

$36\left[(x')^2 + (4 - 2\sqrt{3})x' + 7 - 4\sqrt{3}\right] + 16\left[(y')^2 + (2 + 4\sqrt{3})y' + 7 + 4\sqrt{3}\right] = 80\sqrt{3} - 316 + 252 -$

$144\sqrt{3} + 112 + 64\sqrt{3} = 48 \Rightarrow 36(x' + 2 - \sqrt{3})^2 + 16(y' + 2 + 2\sqrt{3})^2 = 48 \Rightarrow$

$\dfrac{(x' + 2 - \sqrt{3})^2}{\frac{4}{3}} + \dfrac{(y' + 2 + 2\sqrt{2})^2}{3} = 1.$ (See figure below).   ● ellipse.

[19] In the equation $5x^2 - 6xy - 3y^2 - 22x + 6y = 25$, $A = 5$, $B = -6$ and $C = -3$. Therefore, $\cot 2\theta =$

$\dfrac{A-C}{B} = \dfrac{5 - (-3)}{-6} = -\dfrac{4}{3}$, and it follows that $\cos 2\theta = -\dfrac{4}{5}$ since $90° < 2\theta < 180°$. Moreover, $\cos\theta =$

$\sqrt{\dfrac{1}{10}}$ and $\sin\theta = \sqrt{\dfrac{9}{10}}$. Thus, we let $x = \sqrt{\dfrac{1}{10}}\,x' - \sqrt{\dfrac{9}{10}}\,y'$ and $y = \sqrt{\dfrac{9}{10}}\,x' + \sqrt{\dfrac{1}{10}}\,y'$.

Substituting these values of x and y into the equation $5x^2 - 6xy - 3y^2 - 22x + 6y = 24$, we get

$5\left(\sqrt{\dfrac{1}{10}}\,x' - \sqrt{\dfrac{9}{10}}\,y'\right)^2 - 6\left(\sqrt{\dfrac{1}{10}}\,x' - \sqrt{\dfrac{9}{10}}\,y'\right)\left(\sqrt{\dfrac{9}{10}}\,x' + \sqrt{\dfrac{1}{10}}\,y'\right) -$

$3\left(\sqrt{\dfrac{9}{10}}\,x' + \sqrt{\dfrac{1}{10}}\,y'\right)^2 - 22\left(\sqrt{\dfrac{1}{10}}\,x' - \sqrt{\dfrac{9}{10}}\,y'\right) + 6\left(\sqrt{\dfrac{9}{10}}\,x' + \sqrt{\dfrac{1}{10}}\,y'\right) = 24 \Rightarrow$

$\dfrac{1}{2}(x')^2 - 3x'y' + \dfrac{9}{2}(y')^2 - \dfrac{9}{5}(x')^2 - \dfrac{3}{5}x'y' + \dfrac{27}{5}x'y' + \dfrac{9}{5}(y')^2 - \dfrac{27}{10}(x')^2 - \dfrac{9}{5}x'y' - \dfrac{3}{10}(y')^2 -$

$\dfrac{11\sqrt{10}}{5}x' + \dfrac{33\sqrt{10}}{5}y' + \dfrac{9\sqrt{10}}{5}x' + \dfrac{3\sqrt{10}}{5}y' = 24 \Rightarrow -4(x')^2 + 6(y')^2 - \dfrac{2\sqrt{10}}{5}x' +$

$\dfrac{36\sqrt{10}}{5}y' = 24 \Rightarrow 6\left[(y')^2 + \dfrac{6\sqrt{10}}{5}y'\right] - 4\left[(x')^2 + \dfrac{\sqrt{10}}{10}x'\right] = 24 \Rightarrow 6\left[(y')^2 + \dfrac{6\sqrt{10}}{5}y' + \dfrac{81}{5}\right] -$

$4\left[(x')^2 + \dfrac{\sqrt{10}}{10}x' + \dfrac{1}{40}\right] = 24 + \dfrac{108}{5} - \dfrac{1}{10} = \dfrac{455}{10} \Rightarrow \dfrac{\left(y' + \dfrac{3\sqrt{10}}{5}\right)^2}{\frac{91}{12}} - \dfrac{\left(x' + \dfrac{\sqrt{10}}{20}\right)^2}{\frac{91}{8}} = 1.$

(See figure below).   ● hyperbola.

[20] In the equation $161x^2 + 480xy - 161y^2 - 510x - 272y = 0$, $A = 161$, $B = 480$ and $C = -161$. Thus

$\cos 2\theta = \dfrac{A-C}{B} = \dfrac{161-(-161)}{480} = \dfrac{161}{240}$, and it follows that $\cos 2\theta = \dfrac{161}{289}$. Hence, $\cos\theta = \dfrac{15}{17}$ and

$\sin\theta = \dfrac{8}{17}$. Letting $x = \dfrac{15}{17}x' - \dfrac{8}{17}y'$ and $y = \dfrac{8}{17}x' + \dfrac{15}{17}y'$ and substituting these values of x and y

into the equation $161x^2 + 480xy - 161y^2 - 510x - 272y = 0$, we get $161\left(\dfrac{15}{17}x' - \dfrac{8}{17}y'\right)^2 +$

$480\left(\dfrac{15}{17}x' - \dfrac{8}{17}y'\right)\left(\dfrac{8}{17}x' + \dfrac{15}{17}y'\right) - 161\left(\dfrac{8}{17}x' + \dfrac{15}{17}y'\right)^2 - 510\left(\dfrac{15}{17}x' - \dfrac{8}{17}y'\right) -$

$272\left(\dfrac{8}{17}x' + \dfrac{15}{17}y'\right) = 0 \Rightarrow \dfrac{36225}{289}(x')^2 - \dfrac{38640}{289}x'y' + \dfrac{10304}{289}(y')^2 + \dfrac{57600}{289}(x')^2 + \dfrac{10800}{289}x'y' -$

$\dfrac{30720}{289}x'y' - \dfrac{57600}{289}(y')^2 - \dfrac{10304}{289}(x')^2 - \dfrac{38640}{289}x'y' - \dfrac{36225}{289}(y')^2 - 450x' + 240\,y' - 128x' -$

$240y' = 0 \Rightarrow 289\,(x')^2 - 289\,(y')^2 - 578x' = 0 \Rightarrow 289\left[(x')^2 - 2x'\right] - 289\,(y')^2 = 0 \Rightarrow$

$289\left[(x')^2 - 2x + 1\right] - 289\,(y')^2 = 289 \Rightarrow (x'-1)^2 - (y')^2 = 1.$ (See figure below).   ● hyperbola.

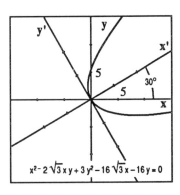

Figure 17

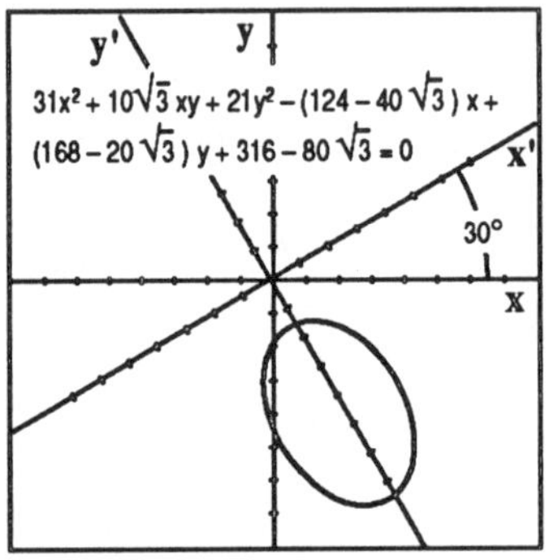

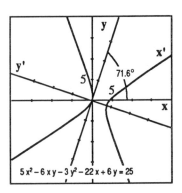

Figure 19

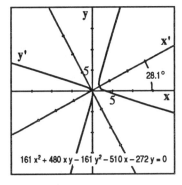

Figure 18

Figure 20

[21] Substituting $x'\cos\theta - y'\sin\theta$ for x and $x'\sin\theta + y'\cos\theta$ for y in the equation $Ax^2 + Bxy + Cy^2 + Dx + Ey + F = 0$ we get

$A\left(x'\cos\theta - y'\sin\theta\right)^2 + B\left(x'\cos\theta - y'\sin\theta\right)\left(x'\sin\theta + y'\cos\theta\right) + C\left(x'\sin\theta + y'\cos\theta\right)^2 + D\left(x'\cos\theta - y'\sin\theta\right) + F\left(x'\sin\theta + y'\cos\theta\right) + F = 0$ (1)

a) To find b, we must determine the coefficient of the $x'y'$ – term in equation (1) above. We

proceed as follows. $b = -2A\sin\theta\cos\theta + B\cos^2\theta - B\sin^2\theta + 2C\sin\theta\cos\theta = 2(C-A)\sin\theta\cos\theta + B\left(\cos^2\theta - \sin^2\theta\right).$

b) Since a is the coefficient of the $(x')^2$ – term in equation (1), $a = A \cos^2 \theta + B \sin\theta \cos\theta + C \sin^2 \theta$. Also, since c is the coefficient of the $(y')^2$ – term in equation (1), $c = A \sin^2 \theta - B \sin\theta \cos\theta + C \cos^2 \theta$. Using the double angle formulas, we can write $b^2 = (C-A)^2 \sin^2 2\theta + 2B(C-A) \sin 2\theta \cos 2\theta + B^2 \cos^2 2\theta = C^2 \sin^2 2\theta - 2AC \sin^2 2\theta + A^2 \sin^2 2\theta - 2AB \sin 2\theta \cos 2\theta + 2BC \sin 2\theta \cos 2\theta + B^2 \cos^2 2\theta$. $-4ac = -4(A \cos^2 \theta + B \sin\theta \cos\theta + C \sin^2 \theta) \cdot (A \sin^2 \theta - B \sin\theta \cos\theta + C \cos^2 \theta) = -4A^2 \sin^2 \theta \cos^2 \theta + 4AB \sin\theta \cos^3 \theta - 4AC \cos^4 \theta - 4AB \sin^3 \theta \cos\theta + 4B^2 \sin^2 \theta \cos^2 \theta - 4BC \sin\theta \cos^3 \theta - 4AC \sin^4 \theta + 4BC \sin^3 \theta \cos\theta - 4C^2 \sin^2 \theta \cos^2 \theta = -A^2 \sin^2 2\theta + 2AB \sin 2\theta \cos 2\theta - 2BC \sin 2\theta \cos 2\theta - 4AC \cos^4 \theta - 4AC \sin^4 \theta + B^2 \sin^2 2\theta - C^2 \sin^2 2\theta$. It follows that $b^2 - 4ac = B^2 \cos^2 2\theta + B^2 \sin^2 2\theta - 4AC \cos^4 \theta - 2AC \sin^2 2\theta - 4AC \sin^4 \theta = B^2 (\cos^2 2\theta + \sin^2 2\theta) - 4AC \cos^4 \theta - 8 AC \cos^2\theta \sin^2 \theta - 4AC \sin^4 \theta = B^2 - 4AC (\cos^4 \theta + 2 \cos^2 \theta \sin^2 \theta + \sin^4 \theta) = B^2 - 4AC [\cos^2 \theta + \sin^2 \theta]^2 = B^2 - 4AC$.

c) By inspecting equation (1), we see that f must be equation to F. From part (b) we know that $a + c = A \cos^2 \theta + B \sin\theta \cos\theta + C \sin^2 \theta + A \sin^2 \theta - B \sin\theta \cos\theta + C \cos^2 \theta = A(\cos^2 \theta + \sin^2 \theta) + C(\sin^2 \theta + \cos^2 \theta) = A + C$.

[22] From Exercise 21 above, we know that the substitution of $x' \cos\theta - y' \sin\theta$ for x and $x'\sin\theta + y'\cos\theta$ for y into the equation $Ax^2 + Bxy + Cy^2 + Dx + Ey + F = 0$ yields an equation of the form $a(x')^2 + bx'y' + c(y')^2 + dx' + ey + f = 0$, where $b = 2(C-A)(\sin\theta \cos\theta) + B(\cos^2 \theta - \sin^2 \theta) = (C-A) \sin 2\theta + B \cos 2\theta$. Choosing θ, such that $0° < \theta < 90°$ and $\cot 2\theta = \dfrac{A-C}{B}$, we have $b = -B \cot 2\theta \sin 2\theta + B \cos 2\theta = 0$. From exercise 21 (a), we know that $B^2 - 4AC = b^2 - 4ac$. Since by our choice of θ, $b = 0$, $B^2 - 4AC = -4ac$. Excluding the degenerate cases, if :

a) $B^2 - 4AC = -4ac = 0$, then at least one of a and $c = 0$, and thus, the graph is a parabola;

b) $B^2 - 4AC = -4ac < 0$, then $ac > 0$, and thus, a and c are both positive or they are both negative. It follows that the graph is an ellipse or a circle;

c) $B^2 - 4AC = -4ac > 0$, $ac < 0$, and hence, $a \neq 0$, $c \neq 0$ and a and c have opposite signs. It follows that the graph is a hyperbola.

[23] $B^2 - 4AC = (-1)^2 - 4(2)(3) = 1 - 24 = -23$.   ● ellipse (or circle).

[24] $B^2 - 4AC = (-1)^2 - 4(2)(3) = -23$.   ● ellipse (or circle).

[25] $B^2 - 4AC = (-5)^2 - 4(7)(-6) = 25 + 168 = 193 > 0$.   ● hyperbola.

[26] $B^2 - 4AC = 4^2 - 4(4)(1) = 16 - 16 = 0$.   ● parabola.

[27] $B^2 - 4AC = (72)^2 - 4(27)(48) = 0$.   ● parabola.

[28] $B^2 - 4AC = 3^2 - 4(2)(6) = 9 - 48 < -39 < 0$.   ● ellipse.

[29] $B^2 - 4AC = 4^2 - 4(1)(4) = 0$.   ● parabola.

[30] $B^2 - 4AC = (12)^2 - 4(12)(3) = 0.$ ● parabola.

[31] $B^2 - 4AC = (-1)^2 - 4(0)(1) = 1.$ ● hyperbola.

[32] $B^2 - 4AC = 4^2 - 4(4)(1) = 0.$ ● parabola.

[33]
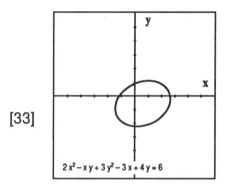

$2x^2 - xy + 3y^2 - 3x + 4y = 6$

[34]
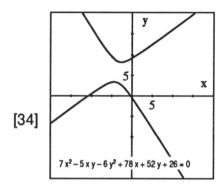

$7x^2 - 5xy - 6y^2 + 78x + 52y + 26 = 0$

[35]
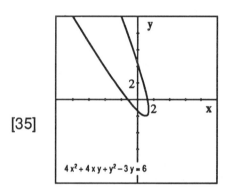

$4x^2 + 4xy + y^2 - 3y = 6$

[36]
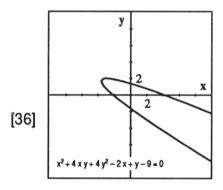

$x^2 + 4xy + 4y^2 - 2x + y - 9 = 0$

[37]
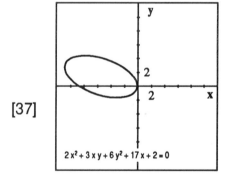

$2x^2 + 3xy + 6y^2 + 17x + 2 = 0$

[38]
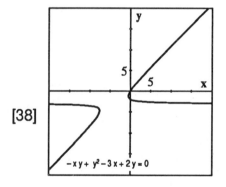

$-xy + y^2 - 3x + 2y = 0$

[39]
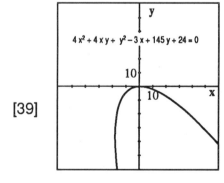

$4x^2 + 4xy + y^2 - 3x + 145y + 24 = 0$

EXERCISES 8.6

[1]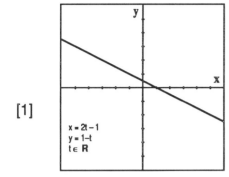

$x = 2t - 1$
$y = 1 - t$
$t \in \mathbf{R}$

[2]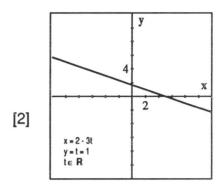

$x = 2 - 3t$
$y = t = 1$
$t \in \mathbf{R}$

[3]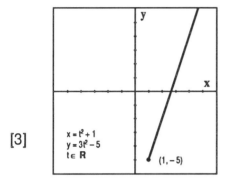

$x = t^2 + 1$
$y = 3t^2 - 5$
$t \in \mathbf{R}$

$(1, -5)$

[4]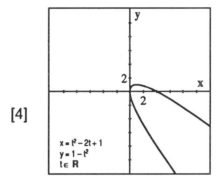

$x = t^2 - 2t + 1$
$y = 1 - t^2$
$t \in \mathbf{R}$

[5]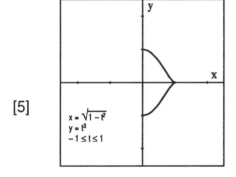

$x = \sqrt{1 - t^2}$
$y = t^3$
$-1 \le t \le 1$

[6]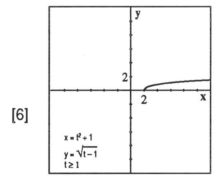

$x = t^2 + 1$
$y = \sqrt{t - 1}$
$t \ge 1$

[7]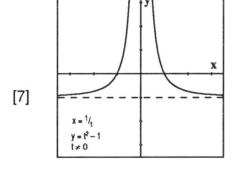

$x = 1/t$
$y = t^2 - 1$
$t \ne 0$

[8]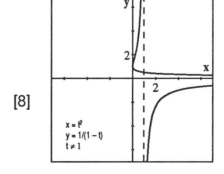

$x = t^2$
$y = 1/(1 - t)$
$t \ne 1$

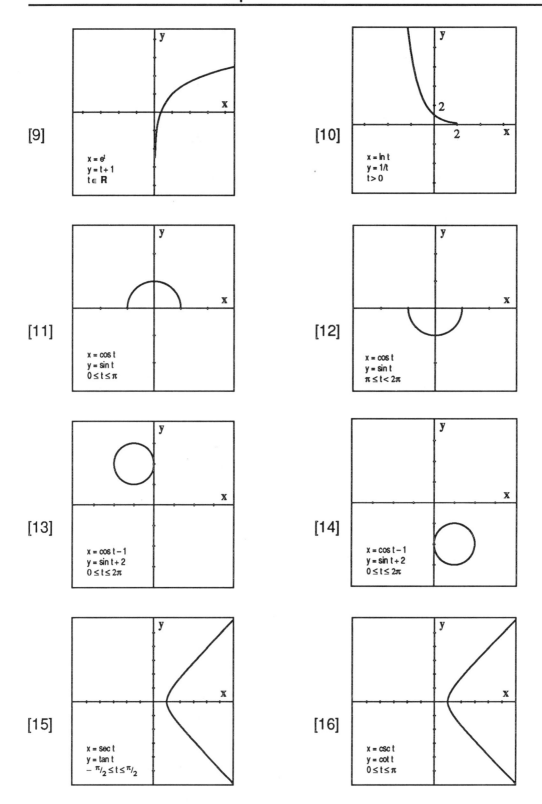

[9]

$x = e^t$
$y = t + 1$
$t \in \mathbf{R}$

[10]

$x = \ln t$
$y = 1/t$
$t > 0$

[11]

$x = \cos t$
$y = \sin t$
$0 \le t \le \pi$

[12]

$x = \cos t$
$y = \sin t$
$\pi \le t < 2\pi$

[13]

$x = \cos t - 1$
$y = \sin t + 2$
$0 \le t \le 2\pi$

[14]

$x = \cos t - 1$
$y = \sin t + 2$
$0 \le t \le 2\pi$

[15]

$x = \sec t$
$y = \tan t$
$-\pi/2 \le t \le \pi/2$

[16]

$x = \csc t$
$y = \cot t$
$0 \le t \le \pi$

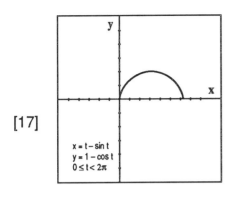

[17]

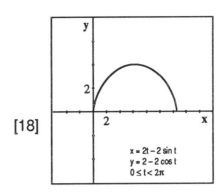

[18]

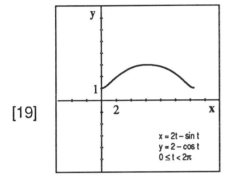

[19]

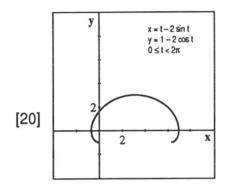

[20]

[21] $x = 2t + 1 \Rightarrow t = \frac{x-1}{2}$. $y = 4 - 3\left(\frac{x-1}{2}\right) \Rightarrow y = -\frac{3}{2}x + \frac{11}{2}$.   (See figure below).

[22] $x = 3 - 2t \Rightarrow t = \frac{3-x}{2}$. $y = \frac{1}{2}\left(\frac{3-x}{2}\right) + 1 \Rightarrow y = -\frac{1}{4}x + \frac{7}{4}$.   (See figure below).

[23] $x = \frac{t+1}{3} \Rightarrow t = 3x - 1$. $y = (3x - 1)^2 + 5(3x - 1) - 1 \Rightarrow y = 9x^2 + 9x - 5$.   (See figure below).

[24] $x = \frac{t-1}{3} \Rightarrow t = 3x + 1$. $y = (3x + 1)^2 - 2(3x + 1) + 3 \Rightarrow y = 9x^2 + 2$.   (See figure below).

[25] $x = t^3 \Rightarrow t = \sqrt[3]{x}$. $y = \left(\sqrt[3]{x}\right)^2 - 1 \Rightarrow y = x^{2/3} - 1$.   (See figure below).

[26] $x = t^2 \Rightarrow t = \pm\sqrt{x}$. $y = \left(\pm\sqrt{x}\right)^3 + 1 \Rightarrow y = \pm x^{3/2} + 1$.   (See figure below).

[27] $x = 2\cos t \Rightarrow \cos t = \frac{x}{2}$. $y = 2\sin t \Rightarrow \sin t = \frac{y}{2}$. $\cos^2 t + \sin^2 t = 1 \Rightarrow \frac{x^2}{4} + \frac{y^2}{4} = 1 \Rightarrow x^2 + y^2 = 4$.

(See figure below).

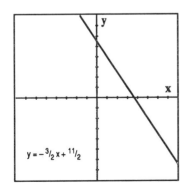

Figure 21

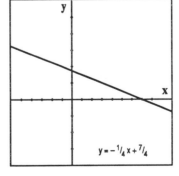

Figure 22

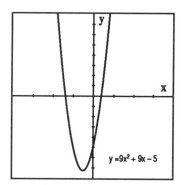

Figure 23

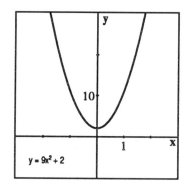

Figure 24

[28] $x = 3 \cos t \Rightarrow \cos t = \frac{x}{3}.$ $y = 3 \sin t \Rightarrow \sin t = \frac{y}{3}.$ $\cos^2 t + \sin^2 t = 1 \Rightarrow \frac{x^2}{9} + \frac{y^2}{9} = 1 \Rightarrow$

$x^2 + y^2 = 9;$ $y \geq 0.$ (See figure below).

[29] $y = 3^{t+1} = 3^t \cdot 3 = 3x,$ $x > 0.$ (See figure below).

[30] $x = 2^{t-1} = 2^t \cdot 2^{-1} = \frac{2^t}{2} \Rightarrow 2^t = 2x.$ $y = 2^t - 2 \Rightarrow y = 2x - 2,$ $x > 0.$ (See figure below).

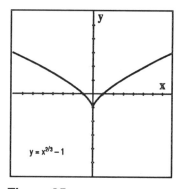

Figure 25

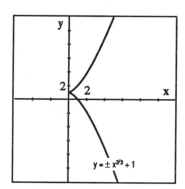

Figure 26

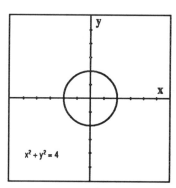

Figure 27

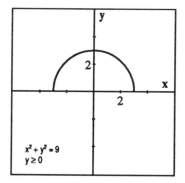

Figure 28

406

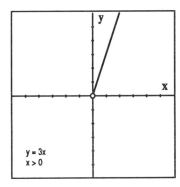

Figure 29

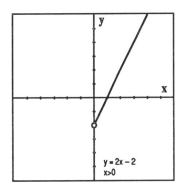

Figure 30

[31] $y = e^{-t} + 1 = \frac{1}{e^t} + 1 \Rightarrow y = \frac{1}{x} + 1, x > 0.$ (See figure below).

[32] $x = e^{-t} = \frac{1}{e^t} \Rightarrow e^t = \frac{1}{x}. y = e^t - 2 \Rightarrow y = \frac{1}{x} - 2, x > 0.$ (See figure below).

[33] $x = 3 \cos t \Rightarrow \cos t = \frac{x}{3}. \cos^2 t + \sin^2 t = 1 \Rightarrow \frac{x^2}{9} + y^2 = 1.$ (See figure below).

[34] $x = 2 \cos t \Rightarrow \cos t = \frac{x}{2}. y = 3 \sin t \Rightarrow \sin t = \frac{y}{3}. \cos^2 t + \sin^2 t = 1 \Rightarrow \frac{x^2}{4} + \frac{y^2}{9} = 1.$

(See figure below).

[35] $x = 3 \cos t \Rightarrow \cos t = \frac{x}{3}. y = 2 \sin t - 1 \Rightarrow \sin t = \frac{y+1}{2}. \cos^2 t + \sin^2 t = 1 \Rightarrow \frac{x^2}{9} + \frac{(y+1)^2}{4} = 1.$

(See figure below).

[36] $x = 2 \cos t - 3 \Rightarrow \cos t = \frac{x+3}{2}. y = 4 \sin t + 1 \Rightarrow \sin t = \frac{y-1}{4}. \cos^2 t + \sin^2 t = 1 \Rightarrow$

$\frac{(x+3)^2}{4} + \frac{(y-1)^2}{16} = 1.$ (See figure below).

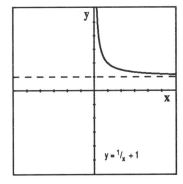

Figure 31

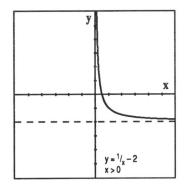

Figure 32

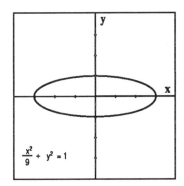

$\frac{x^2}{9} + y^2 = 1$

Figure 33

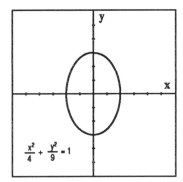

$\frac{x^2}{4} + \frac{y^2}{9} = 1$

Figure 34

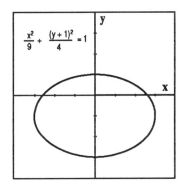

$\frac{x^2}{9} + \frac{(y+1)^2}{4} = 1$

Figure 35

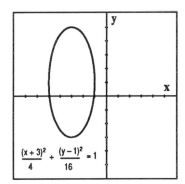

$\frac{(x+3)^2}{4} + \frac{(y-1)^2}{16} = 1$

Figure 36

[37] $y = \sqrt{x^2 - 1}$, $|x| \geq 1$.   (See figure below).

[38] $x = t - 1 \Rightarrow t = x + 1 \Rightarrow y = \sqrt{t} \Rightarrow y = \sqrt{x+1}$, $x \geq -1$.   (See figure below).

[39] $x^2 + y^2 = \left(\frac{2}{t^2+1}\right)^2 + \left(\frac{2t}{t^2+1}\right)^2 = \frac{4 + 4t^2}{(t^2+1)^2} = \frac{4}{(t^2+1)} = 2x \Rightarrow x^2 + y^2 = 2x \Rightarrow x^2 - 2x + y^2 =$

$0 \Rightarrow x^2 - 2x + 1 + y^2 = 1 \Rightarrow (x-1)^2 + y^2 = 1$.   (See figure below).

[40] $x^2 + y^2 = \left(\frac{4t}{t^2+4}\right)^2 + \left(\frac{8}{t^2+4}\right)^2 = \frac{16t^2 + 64}{(t^2+4)^2} = \frac{16}{t^2+4} = 2y \Rightarrow x^2 + y^2 = 2y \Rightarrow x^2 + y^2 - 2y = 0 \Rightarrow$

$x^2 + y^2 - 2y + 1 = 1 \Rightarrow x^2 + (y-1)^2 = 1$.   (See figure below).

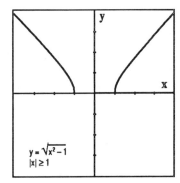

$y = \sqrt{x^2 - 1}$
$|x| \geq 1$

Figure 37

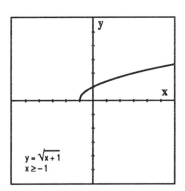

$y = \sqrt{x+1}$
$x \geq -1$

Figure 38

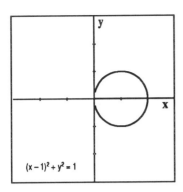

Figure 39

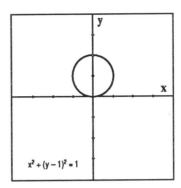

Figure 40

[41] $y = (v_0 \sin \alpha)t - 16t^2$. When $y = 0$, $(1250 \sin 28.6°)t - 16t^2 = 0 \Rightarrow t(1250 \sin 28.6° - 16t) =$

0 $\Rightarrow t = 0$ or $t \approx 37.4$. The cannonball is airborne for approximately

37.4 seconds. $x = (1250 \cos 28.6°)t \approx (1250 \cos 28.6°)(37.4) \approx 41,040$. The cannonball lands

approximately 41,040 feet from the cannon.

[42] $y = (v_0 \sin \alpha)t - 16t^2$. When $y = 0$, $(680 \sin 67.5°)t - 16t^2 = 0 \Rightarrow t(680 \sin 67.5° - 16t) = 0 \Rightarrow$

$t = 0$ or $t \approx 39.3$. The bullet is in the air for approximately 39.3 seconds. $x = (680 \cos 67.5°)t \approx$

$(680 \cos 67.5°)(39.3) \approx 10,200$. The bullet lands approximately 10,200 feet from the rifle.

[43]   The slope of the line through O and P is $t = \dfrac{y-0}{x-0} = \dfrac{y}{x}$, and $y^2 = 4px \Rightarrow x = \dfrac{y^2}{4p}$. Thus, $t = \dfrac{y}{\dfrac{y^2}{4p}} =$

$\dfrac{4p}{y}$, and $y = \dfrac{4p}{t}$. Also, $y^2 = 4px \Rightarrow \left(\dfrac{4p}{t}\right)^2 = 4px \Rightarrow \dfrac{16p^2}{t^2} = 4px \Rightarrow x = \dfrac{4p}{t^2}$. Hence, $x = \dfrac{4p}{t^2}$ and

$y = \dfrac{4p}{t}$.   (See figure below).

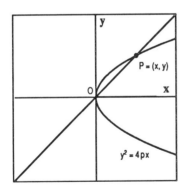

Figure 43

[44] $x = x_1 + (x_2 - x_1)t \Rightarrow t = \dfrac{x - x_1}{x_2 - x_1}$ and $y = y_1 + (y_2 - y_1)t \Rightarrow t = \dfrac{y - y_1}{y_2 - y_1}$. $\dfrac{x - x_1}{x_2 - x_1} = \dfrac{y - y_1}{y_2 - y_1} \Rightarrow$

$(x_2 - x_1)(y_2 - y_1) = (x - x_1)(y_2 - y_1) \Rightarrow y - y_1 = \dfrac{y_2 - y_1}{x_2 - x_1}(x - x_1)$. Since $m = \dfrac{y_2 - y_1}{x_2 - x_1}$ is the slope

of the line containing P_1 and P_2, we have $y - y_1 = m(x - x_1)$, the point − slope equation of the line.

[45]   Let $P' = (x', y')$ be the point at which the line through P and C intersects the circle. We know from the solution of Example 22 in this section that $x' = rt - r \sin t$ and $y' = r - r \cos t$, where t is the radian measure of $\angle P'CD$ and $P'PB$. From right $\triangle P'BP$, we see that $\sin t = \dfrac{d(P',B)}{r-a}$ and $\cos t = \dfrac{d(B,P)}{r-a}$. Thus $d(P',B) = (r-a)\sin t$ and $d(B,P) = (r-a)\cos t$. From the figure we see that $x = x' + d(P',B)\,(*_1)$ and $y = y' + d(B,P)\,(*_2)$. Replacing x' by $rt - r\sin t$ and $d(P',B)$ by $(r-a)\sin t$, equation $(*_1)$ becomes $x = rt - r\sin t + (r-a)\sin t$. Thus, $x = rt - r\sin t + r\sin t - a\sin t \Rightarrow x = rt - a\sin t$. Similarly, replacing y' by $r - r\cos t$ and $d(B,P)$ by $(r-a)\cos t$ in equation $(*_2)$ yields $y = r - r\cos t + (r-a)\cos t$. Thus $y = r - r\cos t + r\cos t - a\cos t \Rightarrow y = r - a\cos t$.   (See figure below).

[46]   Let $P' = (x', y')$ be the point at which the line through P and C intersects the circle. We know from the solution of Example 22 of this section that $x' = rt - r\sin t$ and $y' = r - r\cos t$. From right $\triangle PBP'$ we see that $\sin t = \dfrac{d(P,B)}{a-r}$ and $\cos t = \dfrac{d(B,P')}{a-r}$, where t is the radian measure of $\angle P'CD$ and $\angle PP'B$. Thus, $d(P,B) = (a-r)\sin t$ and $d(B,P') = (a-r)\cos t$. From the figure, we see that $x' = x + d(P,B)\,(*_1)$ and $y' = y + d(B,P')\,(*_2)$. Replacing x' by $rt - r\sin t$ and $d(P,B)$ by $(a-r)\sin t$ in equation $(*_1)$, we get $rt - r\sin t = x + (a-r)\sin t$. Then $rt - r\sin t = x + a\sin t - r\sin t \Rightarrow x = rt - a\sin t$. Similarly, replacing y' by $r - r\cos t$ and $d(B,P)$ by $(a-r)\cos t$ in equation $(*_2)$ yields $r - r\cos t = y + (a-r)\cos t$. Thus, $r - r\cos t = y + a\cos t - r\cos t$, and $y = r - a\cos t$.   (See figure below).

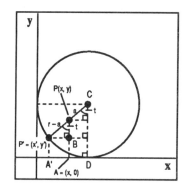

Figure 45   **Figure 46**

[47] $\sin t = x^{1/3}$ and $\cos t = y^{1/3}$. $\sin^2 t + \cos^2 t = 1 \Rightarrow x^{2/3} + y^{2/3} = 1$.   (See figure below).

[48] $\tan t = x + 1$ and $\sec t = y - 2$. $1 + \tan^2 t = \sec^2 t \Rightarrow 1 + (x+1)^2 = (y-2)^2 \Rightarrow -(x+1)^2 + (y-2)^2 = 1$ or $(y-2)^2 - (x+1)^2 = 1$.   (See figure below).

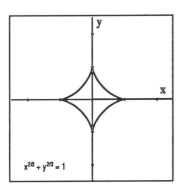

Figure 47

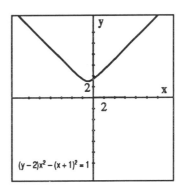

Figure 48

CHAPTER 8 REVIEW EXERCISES

[1]   The equation $x^2 = 2y$ is the equation of a parabola with $4p = 2 \Rightarrow p = \frac{1}{2}$.   (See figure below).

● parabola; vertex, $(0, 0)$; focus, $\left(0, \frac{1}{2}\right)$; axis, $x-$axis; directrix, $y = -\frac{1}{2}$.

[2]   The equation $y^2 = -3x$ is the equation of a parabola with $4p = -3 \Rightarrow p = -\frac{3}{4}$.   (See figure below).

● parabola; vertex, $(0, 0)$; focus, $\left(-\frac{3}{4}, 0\right)$; directrix, $x = \frac{3}{4}$.

[3]   The equation $x^2 - \frac{y^2}{2} = 1$ is the equation of a hyperbola, where $a^2 = 1 \Rightarrow a = 1$, and $b^2 = 2 \Rightarrow$

$b = \sqrt{2}$. $c^2 = a^2 + b^2 = 3 \Rightarrow c = \sqrt{3}$.   (See figure below).

● hyperbola; center, $(0, 0)$; vertices, $(\pm 1, 0)$; foci, $(\pm \sqrt{3}, 0)$; endpoints of conjugate axis, $(0, \pm \sqrt{2})$.

[4]   In the equation $x^2 + \frac{y^2}{2} = 1$, $a^2 = 2 \Rightarrow a = \sqrt{2}$ and $b^2 = 1 \Rightarrow b = 1$. The given equation defines an

ellipse, and therefore, $c^2 = a^2 - b^2 = 1 \Rightarrow c = 1$.   (See figure below).

● ellipse; center, $(0, 0)$; vertices, $(0, \pm \sqrt{2})$; foci, $(0, \pm 1)$; endpoints of minor axis, $(\pm 1, 0)$.

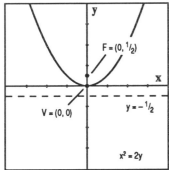

Figure 1

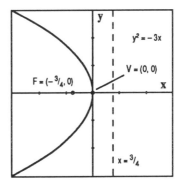

Figure 2

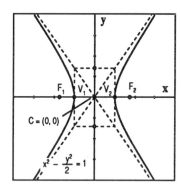

Figure 3

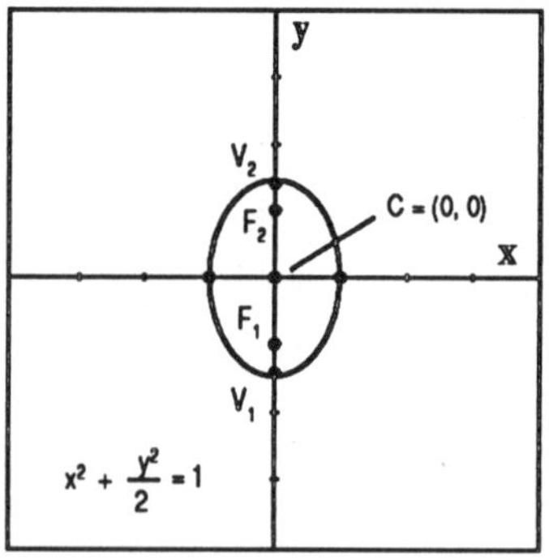

Figure 4

[5]   In the equation $\frac{x^2}{10} + \frac{y^2}{6} = 1$, $a^2 = 10 \Rightarrow a = \sqrt{10}$ and $b^2 = 6 \Rightarrow b = \sqrt{6}$. $c^2 = a^2 - b^2 = 10 - 6 =$

4 $\Rightarrow c = 2$.   (See figure below).

● ellipse; center, $(0, 0)$; vertices, $(\pm\sqrt{10}, a)$; foci, $(\pm 2, 0)$; endpoints of minor axis, $(0, \pm\sqrt{6})$.

[6]   In the equation $y^2 - \frac{x^2}{4} = 1$, $a^2 = 1 \Rightarrow a = 1$ and $b^2 = 4 \Rightarrow b = 2$. The given equation defines a

hyperbola, and thus, $c^2 = a^2 + b^2 = 5 \Rightarrow c = \sqrt{5}$.   (See figure below).

● hyperbola; center, $(0, 0)$; vertices, $(0, \pm 1)$; foci, $(0, \pm\sqrt{5})$; endpoints of conjugate axis, $(\pm 2, 0)$.

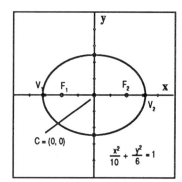

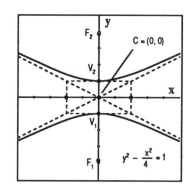

Figure 5

Figure 6

[7] $4x^2 + 4y^2 = 9 \Rightarrow x^2 + y^2 = \frac{9}{4}$.   (See figure below).   ● circle; center, $(0, 0)$; radius, $\frac{3}{2}$.

[8] $2x^2 - 3y^2 = 5 \Rightarrow \dfrac{x^2}{\frac{5}{2}} - \dfrac{y^2}{\frac{5}{3}} = 1$. The graph is a hyperbola, where $a^2 = \frac{5}{2}$ and $b^2 = \frac{5}{3}$. Thus $a =$

$\frac{\sqrt{10}}{2}$ and $b = \frac{\sqrt{15}}{3}$. $c^2 = a^2 + b^2 = \frac{25}{6} \Rightarrow c \frac{5\sqrt{6}}{6}$.   (See figure below).

● hyperbola; center, $(0, 0)$; vertices,

$\left(\pm\frac{\sqrt{10}}{2}, 0\right)$; foci, $\left(\pm\frac{5\sqrt{6}}{6}, 0\right)$; endpoints of the conjugate axis, $\left(0, \pm\frac{\sqrt{15}}{3}\right)$.

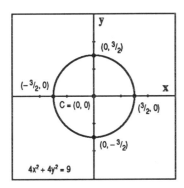

Figure 7

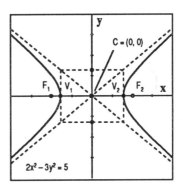

Figure 8

[9] $9x^2 - y^2 = 1 \Rightarrow \dfrac{x^2}{\frac{1}{9}} - y^2 = 1$. The graph is a hyperbola, where $a^2 = \frac{1}{9}$ and $b^2 = 1$. Hence, $a = \frac{1}{3}$ and

$b = 1$, and thus, $c^2 = a^2 + b^2 = \dfrac{10}{9} \Rightarrow c = \dfrac{\sqrt{10}}{3}$.   (See figure below).

● hyperbola; center, $(0, 0)$; vertices, $\left(\pm \frac{1}{3}, 0\right)$;   foci, $\left(\pm \frac{\sqrt{10}}{3}, 0\right)$; endpoints of conjugate axis, $(0, \pm 1)$.

[10] $2x^2 + y - 4x = 0 \Rightarrow x^2 - 2x = -\frac{1}{2}y \Rightarrow x^2 - 2x + 1 = -\frac{1}{2}(y - 2) \Rightarrow (x - 1)^2 = -\frac{1}{2}(y - 2)$. The

graph is a parabola, where $4p = -\frac{1}{2} \Rightarrow p = -\frac{1}{8}$.   (See figure below).

● parabola; vertex, $(1, 2)$; focus, $\left(1, \frac{15}{8}\right)$; axis, $x = 1$; directrix, $y = \dfrac{17}{8}$.

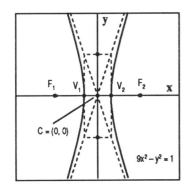

Figure 9

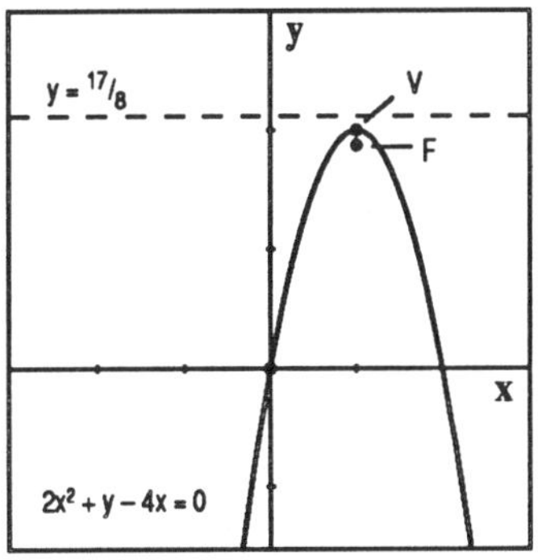

Figure 10

[11]   The graph of the equation $\dfrac{(x + 2)^2}{9} - \dfrac{(y - 5)^2}{16} = 1$ is a hyperbola, where $a^2 = 9$ and $b^2 = 16$. Hence,

$a = 3$ and $b = 4$, and $c^2 = a^2 + b^2 = 25 \Rightarrow c = 5$.   (See figure below).

● hyperbola; center, $(-2, 5)$; vertices, $(-5, 5)$

and $(1, 5)$; foci, $(-7, 5)$ and $(3, 5)$; endpoints of conjugate axis, $(-2, 1)$ and $(-2, 9)$.

[12]   The graph of the equation $(x - 2)^2 + \dfrac{(y + 2)^2}{4} = 1$ is an ellipse, where $a^2 = 4$ and $b^2 = 1$. Hence, $a =$

2 and $b = 1$ and therefore, $c^2 = a^2 - b^2 = 3 \Rightarrow c = \sqrt{3}$.   (See figure below).

$(2, -4)$ and $(2, 0)$; foci, $(2, -2 \pm \sqrt{3})$; endpoints of minor axis, $(1, -2)$ and $(3, -2)$.

[13] $(x+2)^2 = 4y - 8 \Rightarrow (x+2)^2 = 4(y-2)$. The graph is a parabola, where $4p = 4 \Rightarrow p = 1$.

  (See figure below). ● parabola; vertex, $(-2, 2)$; focus, $(-2, 3)$; axis, $x = -2$; directrix, $y = 1$.

[14] $2y = 4x^2 - 1 \Rightarrow \frac{1}{2}\left(y + \frac{1}{2}\right) = x^2$. The graph is a parabola, where $4p = \frac{1}{2} \Rightarrow p = \frac{1}{8}$. (See figure below).

  ● parabola; vertex, $\left(0, -\frac{1}{2}\right)$; focus, $\left(0, -\frac{3}{8}\right)$; axis, x – axis; directrix, $y = -\frac{5}{8}$.

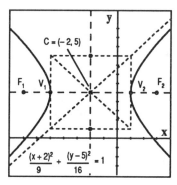

Figure 11

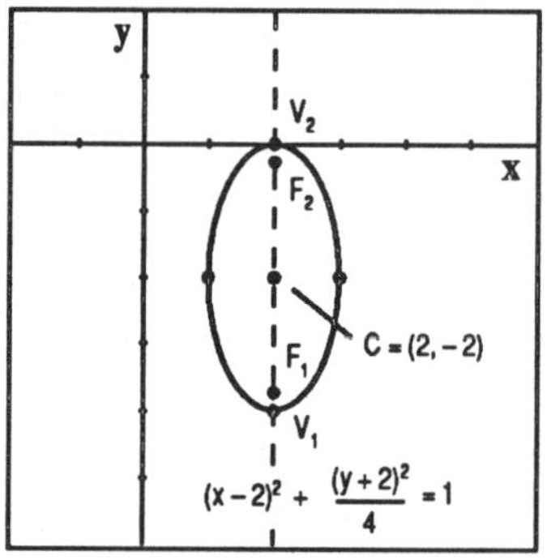

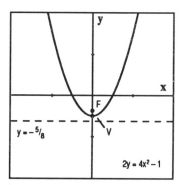

Figure 12

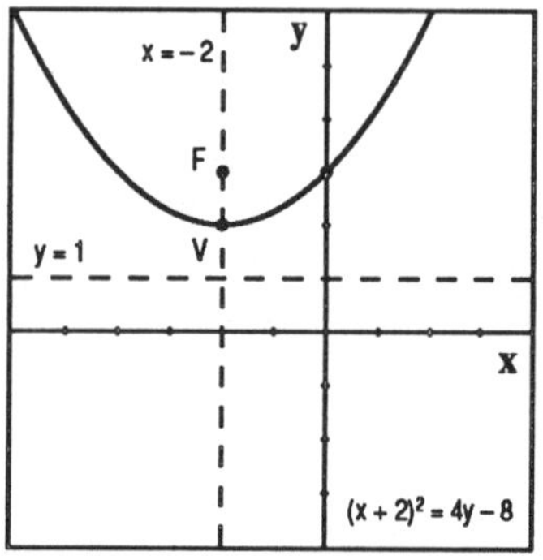

Figure 13

Figure 14

[15] $x^2 + 4y^2 - 6x + 5 = 0 \Rightarrow (x^2 - 6x + 9) + 4y^2 = 4 \Rightarrow \frac{(x-3)^2}{4} + y^2 = 1$. The graph is an ellipse, where

  $a^2 = 4$ and $b^2 = 1$. Therefore, $a = 2$ and $b = 1$, and $c^2 = a^2 - b^2 = 3 \Rightarrow c = \sqrt{3}$. (See figure below).

  ● ellipse; center, $(3, 0)$; vertices, $(1, 0)$ and $(5, 0)$; foci, $(3 \pm \sqrt{3}, 0)$; endpoints of minor axis,

  $(3, -1)$ and $(3, 1)$.

[16] $25(x+3)^2 + 16(y-2)^2 = 400 \Rightarrow \frac{(x+3)^2}{16} + \frac{(y-2)^2}{25} = 1$. The graph is an ellipse, where $a^2 = 25$

  and $b^2 = 16$. Hence, $a = 5$, $b = 4$, and $c^2 = a^2 - b^2 = 9 \Rightarrow c = 3$. (See figure below).

  ● ellipse; center, $(-3, 2)$; vertices, $(-3, -3)$ and $(-3, 7)$; foci, $(-3, -1)$ and $(-3, 5)$; endpoints of

  minor axis, $(-6, 2)$ and $(0, 2)$.

[17] $49x^2 - 9y^2 + 98x - 36y - 428 = 0 \Rightarrow 49(x^2 + 2x) - 9(y^2 + 4y) = 428 \Rightarrow 49(x^2 + 2x + 1) -$

  $9(y^2 + 4y + 4) = 441 \Rightarrow \frac{(x+1)^2}{9} - \frac{(y+2)^2}{49} = 1$. The graph is a hyperbola where $a^2 = 9$ and

  $b^2 = 49$. It follows that $a = 3$, $b = 7$ and $c^2 = a^2 + b^2 = 58 \Rightarrow c = \sqrt{58}$. (See figure below).

  ● hyperbola; center, $(-1, -2)$; vertices, $(-4, -2)$ and $(2, -2)$; foci, $(-1 \pm \sqrt{58}, -2)$; endpoints of

  conjugate axis, $(-1, -9)$ and $(-1, 5)$.

[18] $9x^2 + 4y^2 - 36x + 8y + 4 = 0 \Rightarrow 9(x^2 - 4x) + 4(y^2 + 2y) = -4 \Rightarrow 9(x^2 - 4x + 4) +$

$4(y^2 + 2y + 1) = 36 \Rightarrow \dfrac{(x-2)^2}{4} + \dfrac{(y+1)^2}{9} = 1$. The graph is an ellipse, where $a^2 = 9$ and $b^2 = 4$.

Hence, $a = 3$, $b = 2$ and $c^2 = a^2 - b^2 = 5 \Rightarrow c = \sqrt{5}$.   (See figure below).

● ellipse; center, $(2, -1)$; vertices, $(2, -4)$ and $(2, 2)$; foci, $(2, -1 \pm \sqrt{5})$; endpoints of minor axis,

$(0, -1)$ and $(4, -1)$.

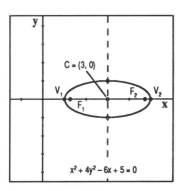

Figure 15

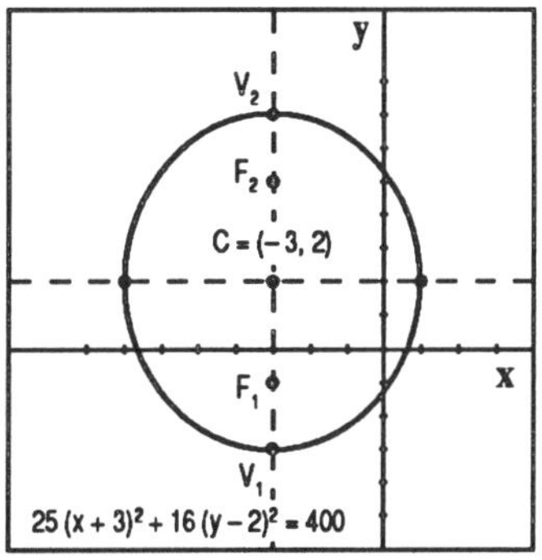

Figure 16

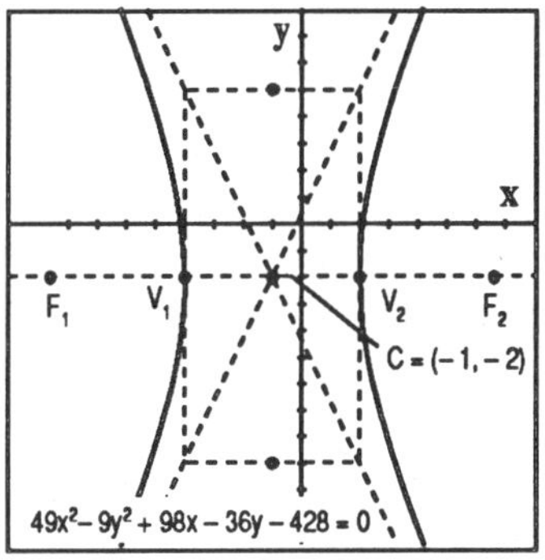

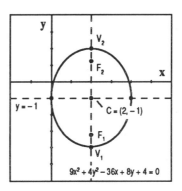

Figure 17

Figure 18

[19] $3y^2 - x^2 + 4x - 12y = 19 \Rightarrow 3(y^2 - 4y) - (x^2 - 4x) = 19 \Rightarrow 3(y^2 - 4y + 4) - (x^2 - 4x + 4) = 27 \Rightarrow$

$\dfrac{(y-2)^2}{9} - \dfrac{(x-2)^2}{27} = 1$. The graph is a hyperbola, where $a^2 = 9$ and $b^2 = 27$. Hence, $a = 3$,

$b = 3\sqrt{3}$ and $c^2 = a^2 + b^2 = 36 \Rightarrow c = 6$.   (See figure below).

● hyperbola; center, $(2, 2)$; vertices, $(2, -1)$

and $(2, 5)$; foci, $(2, -4)$ and $(2, 8)$; endpoints of conjugate axis, $(2 \pm 3\sqrt{3}, 2)$.

[20] $16x^2 + 25y^2 - 32x + 100y = 284 \Rightarrow 16(x^2 - 2x) + 25(y^2 + 4y) = 284 \Rightarrow 16(x^2 - 2x + 1) +$

$25(y^2 + 4y + 4) = 400 \Rightarrow \dfrac{(x-1)^2}{25} + \dfrac{(y+2)^2}{16} = 1$. The graph is an ellipse, where $a^2 = 25$ and

$b^2 = 16$. Therefore, $a = 5$, $b = 4$ and $c^2 = a^2 - b^2 = 9 \Rightarrow c = 3$.   (See figure below).

● ellipse; center, $(1, -2)$; vertices, $(-4, -2)$ and $(6, -2)$; foci, $(-2, -2)$ and $(4, -2)$; endpoints of

minor axis, $(1, -6)$ and $(1, 2)$.

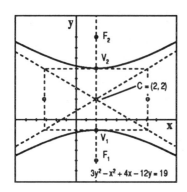

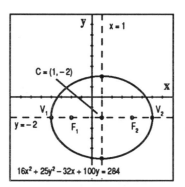

Figure 19   **Figure 20**

[21]   Since the focus is $(-3, -1)$ and the directrix is the line $x = -5$, the vertex is $(-4, -1)$. Hence, $p = 1$

and the equation is $(y + 1)^2 = 4 (x = 4)$.   ● $(y + 1)^2 = 4 (x + 4)$.

[22]   Since the vertices are $(-4, -3)$ and $(-2, 3)$, the center is $(-3, 3)$ and hence, $a = 1 \Rightarrow a^2 = 1$.

Moreover, since the length of the minor axis is $2b = 1$, $b = \frac{1}{2} \Rightarrow b^2 = \frac{1}{4}$.   ● $(x + 3)^2 + \dfrac{(y - 3)^2}{\frac{1}{4}} = 1$.

[23]   Since the center is $(0, 0)$ and a vertex is $(0, 2)$, $a = 2 \Rightarrow a^2 = 4$. Moreover, since a focus is $(0, -4)$,

$c = 4 \Rightarrow c^2 = 16$. Hence $b^2 = c^2 - a^2 = 12$.   ● $\dfrac{y^2}{4} - \dfrac{x^2}{12} = 1$.

[24]   Since the vertex is $(2, 3)$ and the focus is $(2, -1)$, $p = -4$ and the axis is $x = 2$.

                                                    ● $(x - 2)^2 = -16 (y - 3)$.

[25]   Since the major axis has length 10, $a = 5 \Rightarrow a^2 = 25$. Also, since the minor axis has length 5,

$b = \frac{5}{2} \Rightarrow b^2 = \frac{25}{4}$.   ● $\dfrac{(x + 1)^2}{25} + \dfrac{(y + 4)^2}{\frac{25}{4}} = 1$.

[26]   Since the vertices are $(1, 4)$ and $(1, -3)$, the center is $\left(1, \frac{1}{2}\right)$ and hence, $a = \frac{7}{2} \Rightarrow a^2 = \frac{49}{4}$. Since the

intersection of the line $y = 4$ and the asymptote $3x - 2y - 2 = 0$ is $\left(\frac{10}{3}, 4\right)$, it follows that $b = \frac{7}{3} \Rightarrow$

$b^2 = \frac{49}{9}$.   ● $\dfrac{\left(y - \frac{1}{2}\right)^2}{\frac{49}{4}} - \dfrac{(x - 1)^2}{\frac{49}{4}} = 1$.

[27]   Since the vertex is $(-1, -1)$ and since the parabola is symmetric with respect the line $x = -1$, the

equation has the form $(x + 1)^2 = 4p (y + 1)$. Moreover, since the parabola passes through the point

$(-3, 6)$, $(-3 + 1)^2 = 4p (6 + 1) \Rightarrow p = \frac{1}{7}$.   ● $(x + 1)^2 = \frac{4}{7} (y + 1)$.

[28]   Since the length of the major axis is three times the length of the minor axis, $2a = 3\,(2b) \Rightarrow a = 3b$.

Also, since the point $(0, c)$ is a focus, and the center is $(0, 0)$, the equation has the form $\dfrac{x^2}{9b^2} + \dfrac{y^2}{b^2} =$

1. Moreover, since the point $(-3, 3)$ lies on the ellipse, $\dfrac{(-3)^2}{9b^2} + \dfrac{(3)^2}{b^2} = 1 \Rightarrow \dfrac{1}{b^2} + \dfrac{9}{b^2} = 1 \Rightarrow$

$\dfrac{10}{b^2} = 1 \Rightarrow b^2 = 10$. Hence, $\dfrac{x^2}{90} + \dfrac{y^2}{10} = 1$. $\bullet\ \dfrac{x^2}{90} + \dfrac{y^2}{10} = 1.$

[29]   In the equation $5x^2 - 2xy + 5y^2 = 6$, $A = 5$, $B = -2$ and $C = 5 \Rightarrow \cot 2\theta = 0 \Rightarrow 2\theta = 90° \Rightarrow \theta = 45°$.

Letting $x = \dfrac{\sqrt{2}}{2}x' + \dfrac{\sqrt{2}}{2}y'$ and $y = \dfrac{\sqrt{2}}{2}x' + \dfrac{\sqrt{2}}{2}y'$ and substituting these values of x and y into the

given equation yields the equation $4\,(x')^2 + 6\,(y')^2 = 6 \Rightarrow \dfrac{(x')^2}{\frac{3}{2}} + (y')^2 = 1.$ (See figure below).

$\bullet$ ellipse.

[30]   In the equation $5x^2 - 2xy - 5y^2 = 6$, $A = 5$, $B = -2$ and $C = -5 \Rightarrow \cot 2\theta = -5 \Rightarrow \cos 2\theta =$

$-\dfrac{5\sqrt{26}}{26} \Rightarrow \cos \theta = \sqrt{\dfrac{26 - 5\sqrt{26}}{52}}$ and $\sin \theta\ \sqrt{\dfrac{26 + 5\sqrt{26}}{52}}$. Substituting $\sqrt{\dfrac{26 - 5\sqrt{26}}{52}}\ x' -$

$\sqrt{\dfrac{26 + 5\sqrt{26}}{52}}\ y'$ for x and $\sqrt{\dfrac{26 + 5\sqrt{26}}{52}}\ x' + \sqrt{\dfrac{26 - 5\sqrt{26}}{52}}\ y'$ for y in the original equation

and simplifying yields the equation $\sqrt{26}\,(y')^2 - \sqrt{26}\,(x')^2 = 6 \Rightarrow \dfrac{(y')^2}{\frac{6}{\sqrt{26}}} - \dfrac{(x')^2}{\frac{6}{\sqrt{26}}} - 1.$ (See figure below).

$\bullet$ hyperbola.

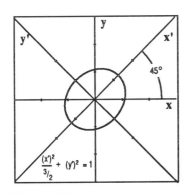

Figure 29

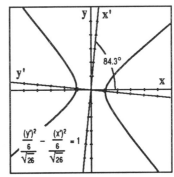

Figure 30

[31]   In the equation $6x^2 + 5xy - 6y^2 - 7 = 0$, $A = 6$, $B = 5$ and $C = -6 \Rightarrow \cot 2\theta = \dfrac{12}{5} \Rightarrow \cos 2\theta = \dfrac{12}{13} \Rightarrow$

$\cos \theta = \dfrac{5\sqrt{26}}{26}$ and $\sin \theta = \dfrac{\sqrt{26}}{26}$. Substituting $\dfrac{5\sqrt{26}}{26}\ x' - \dfrac{\sqrt{26}}{26}\ y'$ for x and $\dfrac{\sqrt{26}}{26}\ x' + \dfrac{5\sqrt{26}}{26}\ y'$ for

y in the given equation and simplifying yields the equation $\dfrac{(x')^2}{\frac{182}{169}} - \dfrac{(y')^2}{\frac{182}{169}} = 1.$ (See figure below).

$\bullet$ hyperbola.

[32]   In the equation $6x^2 + 5xy - 6y^2 + 7 = 0$, $A = 6$, $B = 5$ and $C = -6 \Rightarrow \cot 2\theta = \frac{12}{5} \Rightarrow \cos 2\theta = \frac{12}{13} \Rightarrow$

$\cos \theta = \frac{5\sqrt{26}}{26}$ and $\sin \theta = \frac{\sqrt{26}}{26}$. Substituting $\frac{5\sqrt{26}}{26} x' - \frac{\sqrt{26}}{26} y'$ for x and $\frac{\sqrt{26}}{26} x' + \frac{5\sqrt{26}}{26} y'$ for y

in the given equation and simplifying yields the equation $\dfrac{(x')^2}{\frac{182}{169}} - \dfrac{(y')^2}{\frac{182}{169}} = 1.$ (See figure below).

● hyperbola.

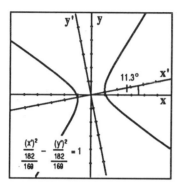

Figure 31   Figure 32

[33]   In the equation $x^2 - 4xy + 4y^2 - 4 = 0$, $A = 1$, $B = -4$ and $C = 4 \Rightarrow \cot 2\theta = \frac{3}{4} \Rightarrow \cos 2\theta = \frac{3}{5} \Rightarrow$

$\cos \theta = \frac{2\sqrt{5}}{5}$ and $\sin \theta = \frac{\sqrt{5}}{5}$. Substituting $\frac{2\sqrt{5}}{5} x' - \frac{\sqrt{5}}{5} y'$ for x and $\frac{\sqrt{5}}{5} x' + \frac{2\sqrt{5}}{5} y'$ for y in the

given equation and simplifying yields $y' = \pm \frac{2}{\sqrt{5}}.$ (See figure below).

● two parallel lines.

[34]   In the equation $9x^2 + 24xy + 16y^2 + 90x - 130y = 0$, $A = 9$, $B = 24$ and $C = 16 \Rightarrow \cot 2\theta = -\frac{7}{24} \Rightarrow$

$\cos 2\theta = -\frac{7}{25} \Rightarrow \cos \theta = \frac{3}{5}$ and $\sin \theta = \frac{4}{5}$. Substituting $\frac{3}{5} x' - \frac{4}{5} y'$ for x and $\frac{4}{5} x' + \frac{3}{5} y'$ for y in the

given equation and simplifying yields the equation $(x' - 1)^2 = 6 \left(y' + \frac{1}{6} \right).$ (See figure below).

● parabola.

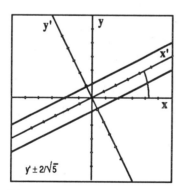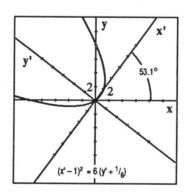

Figure 33   Figure 34

[35] In the equation $xy + x - 2y = 3$, $A = C = 0$ and $B = 1 \Rightarrow \cot 2\theta = 0 \Rightarrow 2\theta = 90° \Rightarrow \theta = 45° \Rightarrow$ $\cos\theta = \sin\theta = \frac{\sqrt{2}}{2}$. Substituting $\frac{\sqrt{2}}{2}x' - \frac{\sqrt{2}}{2}y'$ for x and $\frac{\sqrt{2}}{2}x' + \frac{\sqrt{2}}{2}y'$ for y in the given

equation and simplifying yields the equation $\dfrac{\left(x' - \frac{\sqrt{2}}{2}\right)^2}{2} - \dfrac{\left(y' + \frac{3\sqrt{2}}{2}\right)^2}{2} = 1$. (See figure below).

● hyperbola.

[36] In the equation $xy - 2x + y - 10 = 0$, $A = C = 0$, $B = 1 \Rightarrow \cot 2\theta = 0 \Rightarrow 2\theta = 90° \Rightarrow \theta = 45° \Rightarrow$ $\cos\theta = \sin\theta = \frac{\sqrt{2}}{2}$. Substituting $\frac{\sqrt{2}}{2}x' - \frac{\sqrt{2}}{2}y'$ for x and $\frac{\sqrt{2}}{2}x' + \frac{\sqrt{2}}{2}y'$ for y in the given equation

and simplifying yields the equation $\dfrac{\left(x' - \frac{\sqrt{2}}{2}\right)^2}{16} - \dfrac{\left(y' - \frac{3\sqrt{2}}{2}\right)^2}{16} = 1$. (See figure below). ● hyperbola.

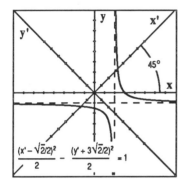

Figure 35   **Figure 36**

[37] In the equation $16x^2 + 24xy + 9y^2 - 130x + 90y = 0$, $A = 16$, $B = 24$ and $C = 9 \Rightarrow \cot 2\theta = \frac{7}{24} \Rightarrow$ $\cos 2\theta = \frac{7}{25} \Rightarrow \cos\theta = \frac{4}{5}$ and $\sin\theta = \frac{3}{5}$. Substituting $\frac{4}{5}x' - \frac{3}{5}y'$ for x and $\frac{3}{5}x' + \frac{4}{5}y'$ for y in the

given equation and simplifying yields $(x' - 1)^2 = -6\left(y' - \frac{1}{6}\right)$. (See figure below). ● parabola.

[38] In the equation $17x^2 - 12xy + 8y^2 - 68x + 24y - 12 = 0$, $A = 17$, $B = -12$ and $C = 8 \Rightarrow \cot 2\theta =$ $-\frac{3}{4} \Rightarrow \cos 2\theta = -\frac{3}{5} \Rightarrow \cos\theta = \frac{\sqrt{5}}{5}$ and $\sin\theta = \frac{2\sqrt{5}}{5}$. Substituting $\frac{\sqrt{5}}{5}x' - \frac{2\sqrt{5}}{5}y'$ for x and

$\frac{2\sqrt{5}}{5}x' + \frac{\sqrt{5}}{5}y'$ for y in the given equation yields the equation $\dfrac{\left(x' - \frac{2\sqrt{5}}{5}\right)^2}{16} + \dfrac{\left(y' + \frac{4\sqrt{5}}{5}\right)^2}{4} = 1$.

(See figure below). ● ellipse.

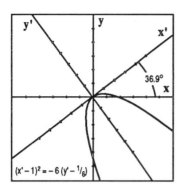

$(x'-1)^2 = -6(y' - ^1/_6)$

Figure 37

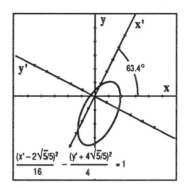

$\dfrac{(x'-2\sqrt{5}/5)^2}{16} - \dfrac{(y'+4\sqrt{5}/5)^2}{4} = 1$

Figure 38

[39]

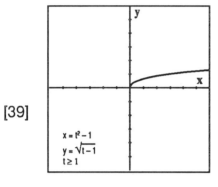

$x = t^2 - 1$
$y = \sqrt{t - 1}$
$t \geq 1$

[40]

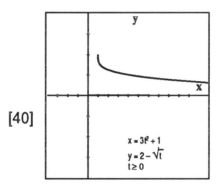

$x = 3t^2 + 1$
$y = 2 - \sqrt{t}$
$t \geq 0$

[41]

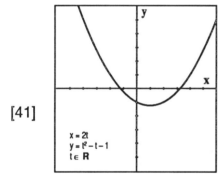

$x = 2t$
$y = t^2 - t - 1$
$t \in \mathbf{R}$

[42]

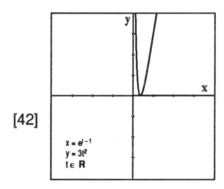

$x = e^{t-1}$
$y = 3t^2$
$t \in \mathbf{R}$

[43]

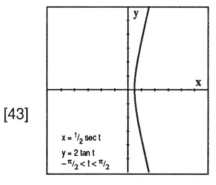

$x = ^1/_2 \sec t$
$y = 2 \tan t$
$-^\pi/_2 < t < ^\pi/_2$

[44]

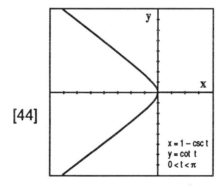

$x = 1 - \csc t$
$y = \cot t$
$0 < t < \pi$

[45] $\quad t = x^2 \Rightarrow y = 3x^2 - 2, \ x \geq 0 \quad$ (See figure below).

[46] $\quad t = \dfrac{x-1}{2} \Rightarrow y = \sqrt{\dfrac{x-1}{2} + 1}$, or $y = \sqrt{\dfrac{x+1}{2}} \quad$ (See figure below).

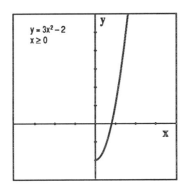

Figure 45

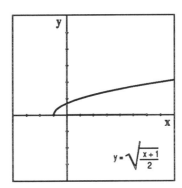

Figure 46

[47] $t = x + 1 \Rightarrow y = \dfrac{2}{x+1}$ (See figure below).

[48] $\dfrac{1}{t} = x + 1 \Rightarrow t = \dfrac{1}{x+1}$. Thus, $y = \left(\dfrac{1}{x+1}\right)^2 - 4$.   (See figure below).

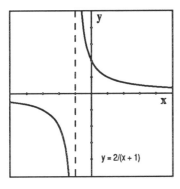

Figure 47

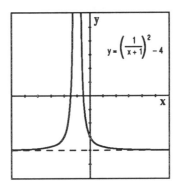

Figure 48

[49] $\sin t = \dfrac{4-x}{4}$ and $\cos t = \dfrac{4-y}{4}$. $\sin^2 t + \cos^2 t = 1 \Rightarrow \left(\dfrac{4-x}{4}\right)^2 + \left(\dfrac{4-y}{4}\right)^2 = 1 \Rightarrow$

$\dfrac{(x-4)^2}{16} + \dfrac{(y-4)^2}{16} = 1 \Rightarrow (x-4)^2 + (y-4)^2 = 16$.   (See figure below).

[50] $t^2 = 1 - x \Rightarrow t = \pm\sqrt{1-x}$. Thus, $y = \pm 3\sqrt{1-x} - 1$.   (See figure below).

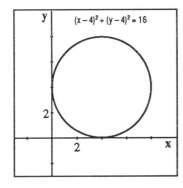

Figure 49

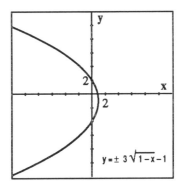

Figure 50

421

[51] $y = \left(v_0 \sin \alpha\right) t - 16t^2 = \left(500 \sin 52.5°\right) t - 16t^2$. When $y = 0$, $\left(500 \sin 52.5°\right) t - 16t^2 = 0 \Rightarrow$

$\left(500 \sin 52.5° - 16t\right) t = 0 \Rightarrow t = 0 \text{ or } t \approx 24.8$. The projectile is airborne for approximately

24.8 seconds. $x = \left(v_0 \cos \alpha\right) t \approx \left(500 \cos 52.5°\right)\left(24.8\right) \approx 7540$. The range of the projectile is

approximately 7540 feet.

[52] $x = a \cos t + h \Rightarrow \cos t = \frac{x-h}{a}$. $y = b \sin t + k \Rightarrow \sin t = \frac{y-k}{b}$. $\cos^2 t + \sin^2 t = 1 \Rightarrow$

$\frac{(x-h)^2}{a^2} + \frac{(y-k)^2}{b^2} = 1$. The latter equation is the equation of an ellipse with center at $C = (h, k)$,

having a major axis of length $2a$ and a minor axis of length $2b$.

[53] Since the vertex is $(2, -3)$ and the directrix is $y = -1$, $p = -2$, and thus the focus is $(2, -5)$. A point

$P = (x, y)$ lies on the parabola if and only if $\sqrt{(x-2)^2 + (y+5)^2} = |y - (-1)| \Rightarrow$

$(x-2)^2 + y^2 + 10y + 25 = y^2 + 2y + 1 \Rightarrow (x-2)^2 = -8y - 24 \Rightarrow (x-2)^2 = -8(y+3)$.

$\bullet \ (x-2)^2 = -8(y+3)$.

[54] If $P = (x, y)$ lies in the set, $\sqrt{x^2 + (y-4)^2} = \frac{2}{3}|y-9| \Rightarrow x^2 + y^2 - 8y + 16 = \frac{4}{9}y^2 - 8y + 36 \Rightarrow$

$x^2 + \frac{5}{9}y^2 = 20 \Rightarrow \frac{x^2}{20} + \frac{y^2}{36} = 1$. $\qquad \bullet \ \frac{x^2}{20} + \frac{y^2}{36} = 1$.

[55] $y^2 - 5x^2 = 25 \Rightarrow \frac{y^2}{25} - \frac{x^2}{5} = 1 \Rightarrow a^2 = 25$ and $b^2 = 5 \Rightarrow a = 5$ and $b = \sqrt{5}$. $c^2 = a^2 + b^2 = 30 \Rightarrow$

$c = \sqrt{30}$. Hence, the vertices are $(0, \pm 5)$ and the foci are $(0, \pm \sqrt{30})$. Therefore, the vertices of the

ellipse are $(0, \pm \sqrt{30})$ and the foci are $(0, \pm 5)$. Thus $a^2 = 30$ and $b^2 = a^2 - c^2 = 30 - 25 = 5$.

(See figure below). $\qquad \bullet \ \frac{x^2}{5} + \frac{y^2}{30} = 1$.

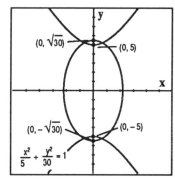

Figure 55

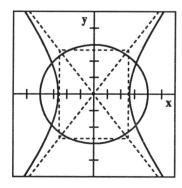

Figure 56

[56] $9x^2 + 16y^2 = 144 \Rightarrow \frac{x^2}{16} + \frac{y^2}{9} = 1 \Rightarrow a^2 = 16$ and $b^2 = 9$. Hence, $c^2 = 16 - 9 = 7$. It follows that the

vertices of the ellipse are $(\pm 4, 0)$ and the foci are $(\pm \sqrt{7}, 0)$. Since the foci of the hyperbola are

$(\pm 4, 0)$, $c = 4 \Rightarrow c^2 = 16$. Also, since the vertices of the hyperbola are $(\pm \sqrt{7}, 0)$, $a = \sqrt{7} \Rightarrow a^2 =$

7. Now $b^2 = c^2 - a^2 = 16 - 7 = 9$. Therefore, the equation of the hyperbola is $\frac{x^2}{7} - \frac{y^2}{9} = 1$.

EXERCISES 9.1

Note: The notation E_1, E_2, E_3 ... refers to the first equation, the second, the third ...

[1] Solving E_1 for x and substituting into E_2 yields $2(-1-2y)-3y=12 \Rightarrow -7y=14 \Rightarrow y=-2$; $x=3$.   ● $\{(3,-2)\}$

[2] Solving E_1 for x and substituting into E_2 yields $5(1-2y)-4y=-23 \Rightarrow -14y=-28 \Rightarrow y=2$; $x=-3$.   ● $\{(-3,2)\}$

[3] Solving E_2 for x and substituting into E_1 yields $2(2+3y)+6y=-3 \Rightarrow 12y=-7 \Rightarrow y=-\frac{7}{12}$; $x=\frac{1}{4}$.   ● $\left\{\left(\frac{1}{4},-\frac{7}{12}\right)\right\}$

[4] Solving E_1 for v and substituting into E_2 yields $2u+5(3u-5)=9 \Rightarrow 17u=34 \Rightarrow u=2$; $v=1$.   ● $\{(2,1)\}$

[5] Solving E_1 for s and substituting into E_2 yields $2r+5\left(-\frac{5}{2}r+\frac{3}{2}\right)=3 \Rightarrow -21r=-9 \Rightarrow r=\frac{3}{7}$; $s=\frac{3}{7}$.   ● $\left\{\left(\frac{3}{7},\frac{3}{7}\right)\right\}$

[6] Solving E_2 for x and substituting into E_1 yields $3(4y)+5y=0 \Rightarrow 17y=0 \Rightarrow y=0$; $x=0$.   ● $\{(0,0)\}$

[7] Solving E_1 for x, we get $x=\frac{1}{3}(-4y+7)$. Substituting into E_2 yields $6\left[\frac{1}{3}(-4y+7)\right]+8y=14 \Rightarrow$ $14=14$. Hence, y is arbitrary. Letting $y=t$, $x=\frac{1}{3}(-4t+7)$.   ● $\left\{\left(\frac{1}{3}(-4t+7),t\right)\middle| t\in\mathbf{R}\right\}$

[8] Solving E_1 for b and substituting into E_2 yields $4a-(2a-3)=4 \Rightarrow a=\frac{1}{2}$; $b=-2$.   ● $\left\{\left(\frac{1}{2},-2\right)\right\}$

[9] Solving E_2 for m and substituting into E_1 yields $\frac{1}{5}(-n)-\frac{1}{3}n=\frac{1}{5} \Rightarrow n=-\frac{3}{8}$; $m=\frac{3}{8}$.   ● $\left\{\left(\frac{3}{8},-\frac{3}{8}\right)\right\}$

[10] Solving E_1 for r and substituting into E_2 yields $16\left[\frac{1}{8}(-16s+20)\right]+50s=55 \Rightarrow s=\frac{5}{6}$; $r=\frac{5}{6}$.   ● $\left\{\left(\frac{5}{6},\frac{5}{6}\right)\right\}$

[11] Solving E_2 for y and substituting into E_1 yields $3x+7(6x-5)=0 \Rightarrow x=\frac{7}{9}$; $y=-\frac{1}{3}$.   ● $\left\{\left(\frac{7}{9},-\frac{1}{3}\right)\right\}$

[12] Solving E_2 for y and substituting into E_1 yields $1.8x+1.2(-1.5y+0.5)=4 \Rightarrow 0.6=4$, which is false.   ● $\varnothing$

[13] Multiplying E_2 by 2 yields $\begin{cases} 3x+4y=3 & (E_3) \\ 2x-4y=-8 & (E_4) \end{cases}$. Adding E_3 to E_4 we get $5x=-5 \Rightarrow$ $x=-1$; $y=\frac{3}{2}$.   ● $\left\{\left(-1,\frac{3}{2}\right)\right\}$

[14] Multiplying E_1 by -1 yields $\begin{cases} -x+2y=-7 & (E_3) \\ 3x-2y= 9 & (E_2) \end{cases}$. Adding E_3 to E_2 we get $2x=2 \Rightarrow$

$x=1; y=-3.$ $\qquad\qquad$ ● $\{(1,-3)\}$

[15] Multiplying E_1 by 6 and E_2 by -12 yields $\begin{cases} 4x-3y= 18 & (E_3) \\ -4x+3y=-18 & (E_4) \end{cases}$. Since E_3 and E_4 are

equivalent, we have $4x-3y=18 \Rightarrow y=\frac{4}{3}x-6$. Letting $x=t$, we have $y=\frac{4}{3}t-6$.

● $\left\{\left(t, \frac{4}{3}t-6\right)\middle| t \in \mathbf{R}\right\}$

[16] Multiplying E_1 by -40 and E_2 by 100 yields $\begin{cases} -16u-120v=-260 & (E_3) \\ 75u-200v= 850 & (E_4) \end{cases}$. Adding E_3 to E_4 we

get $59u=590 \Rightarrow u=10; v=-5.$ $\qquad\qquad$ ● $\{(10,-5)\}$

[17] Multiplying E_1 by -2 yields $\begin{cases} -4x+6y=-14 & (E_3) \\ 4x-6y= 5 & (E_2) \end{cases}$. Adding E_3 to E_2 yields $0=-9$,

which is false. $\qquad\qquad$ ● $\varnothing$

[18] Multiplying E_2 by -2 yields $\begin{cases} 4s+3t= 5 & (E_1) \\ -4s+16t=10 & (E_3) \end{cases}$. Adding E_1 to E_3 we get $19t=15 \Rightarrow$

$t=\frac{15}{19}; s=\frac{25}{38}.$ $\qquad\qquad$ ● $\left\{\left(\frac{25}{38}, \frac{15}{19}\right)\right\}$

[19] Multiplying E_1 by 300 and E_2 by 500 yields $\begin{cases} 36m+15n=210 & (E_3) \\ 55m-15n=125 & (E_4) \end{cases}$. Adding E_3 to E_4 we find

that $91m=335 \Rightarrow m=\frac{335}{91}; n=\frac{470}{91}.$ $\qquad\qquad$ ● $\left\{\left(\frac{335}{91}, \frac{470}{91}\right)\right\}$

[20] Multiplying E_1 by -200 and E_2 by 100 yields $\begin{cases} -2x-6y= 18 & (E_3) \\ 2x-80y=230 & (E_4) \end{cases}$. Adding E_3 to E_4 yields

$-86y=248 \Rightarrow y=-\frac{124}{43}; x=-\frac{15}{43}.$ $\qquad\qquad$ ● $\left\{\left(-\frac{15}{43}, -\frac{124}{43}\right)\right\}$

[21] Multiplying E_2 by -2 yields $\begin{cases} 14x-10y=2 & (E_1) \\ -14x+10y=6 & (E_3) \end{cases}$. Adding E_1 to E_3, we get $0=8$,

which is false. $\qquad\qquad$ ● $\varnothing$

[22] Multiplying E_1 by 3 and E_2 by 4 yields $\begin{cases} 12a+9b= 33 & (E_3) \\ -12a+8b=-16 & (E_4) \end{cases}$. Adding E_3 to E_4, we have

$17b=17 \Rightarrow b=1; a=2.$ $\qquad\qquad$ ● $\{(2,1)\}$

[23] Multiplying E_2 by 2 yields $\begin{cases} 2x-4y= k-1 & (E_1) \\ 6x+4y=6k & (E_3) \end{cases}$. Adding E_1 to E_3 yields $8x=7k-1 \Rightarrow$

$x=\dfrac{7k-1}{8}; \; y=\dfrac{3k+3}{16}.$ ● $\left\{\dfrac{7k-1}{8}, \dfrac{3k+3}{16}\right\}$

[24] Multiplying E_1 by 2 yields $\begin{cases} \frac{4}{3}x-y=10 & (E_3) \\ 4x+y=54 & (E_2) \end{cases}$. Adding E_3 to E_2 yields $\frac{16}{3}x=64 \Rightarrow$

$x=12; \; y=6.$ ● $\{12, 6\}$

Note in Exercises 25 – 44, the symbol $k\,E_i$ denotes the multiplication of equation E_i by the constant k. Also the symbol $k\,E_i + E_j$ is used to indicate that $k\,E_i$ is added to E_j.

[25] $\begin{cases} x+3y+z= 6 \\ 3x+ y-z=-2 \\ 2x+2y-z= 1 \end{cases}$ $\begin{matrix} -3E_1+E_2 \\ -2E_1+E_3 \\ \Rightarrow \end{matrix}$ $\begin{cases} x+ 3y+ z= 6 & (E_1) \\ -8y-4z=-20 & (E_4) \\ -4y-3z=-11 & (E_5) \end{cases}$ $\begin{matrix} -\frac{1}{2}E_4+E_5 \\ \Rightarrow \end{matrix}$

$\begin{cases} x+ 3y+ z= 6 & (E_1) \\ -8y-4z=-20 & (E_4) \\ -z= -1 & (E_6) \end{cases}$ $E_6 \Rightarrow z=1; E_4 \Rightarrow y=2; E_1 \Rightarrow x=-1.$ ● $\{(-1, 2, 1)\}$

[26] $\begin{cases} 3x+y+2z= 1 \\ 4x-y+4z= 4 \\ x+y-2z=-5 \end{cases}$ Interchanging the equations $\Rightarrow$ $\begin{cases} x+y-2z=-5 & (E_1) \\ 3x+y+2z= 1 & (E_2) \\ 4x-y+4z= 4 & (E_3) \end{cases}$ $\begin{matrix} -3E_1+E_2 \\ -4E_1+E_3 \\ \Rightarrow \end{matrix}$

$\begin{cases} x+ y- 2z=-5 & (E_1) \\ -2y+ 8z= 16 & (E_4) \\ -5y+12z= 24 & (E_5) \end{cases}$ $\begin{matrix} -\frac{5}{2}E_4+E_5 \\ \Rightarrow \end{matrix}$ $\begin{cases} x+ y- 2z= -5 & (E_1) \\ -2y+ 8z= 16 & (E_4) \\ -8z=-16 & (E_6) \end{cases}$

$E_6 \Rightarrow z=2; E_4 \Rightarrow y=0; E_1 \Rightarrow x=-1.$ ● $\{(-1, 0, 2)\}$

[27] $\begin{cases} 2a+8b+ 6c= 20 \\ 4a+2b- 2c=-2 \\ 6a-4b+10c= 24 \end{cases}$ $\begin{matrix} -2E_1+E_2 \\ -3E_1+E_3 \\ \Rightarrow \end{matrix}$ $\begin{cases} 2a+ 8b+ 6c= 20 & (E_1) \\ -14b-14c=-42 & (E_4) \\ -28b- 8c=-36 & (E_5) \end{cases}$ $\begin{matrix} -2E_4+E_5 \\ \Rightarrow \end{matrix}$

$\begin{cases} 2a+ 8b+ 6c= 20 & (E_1) \\ -14b-14c=-42 & (E_4) \\ 20c= 48 & (E_6) \end{cases}$ $E_6 \Rightarrow c=\frac{12}{5}; E_4 \Rightarrow b=\frac{3}{5}; E_1 \Rightarrow a=\frac{2}{5}.$ ● $\left\{\left(\frac{2}{5}, \frac{3}{5}, \frac{12}{5}\right)\right\}$

[28] $\begin{cases} 2x-3y+3z=-15 \\ 3x+2y-5z= 19 \\ 5x-4y-2z= -z \end{cases}$ $\begin{matrix} -\frac{3}{2}E_1+E_2 \\ -\frac{5}{2}E_1+E_3 \\ \Rightarrow \end{matrix}$ $\begin{cases} 2x- 3y+ 3z=-15 & (E_1) \\ \frac{13}{2}y-\frac{19}{2}z= \frac{83}{2} & (E_4) \\ \frac{7}{2}y-\frac{19}{2}z= \frac{71}{2} & (E_5) \end{cases}$ $\begin{matrix} -E_4+E_5 \\ \Rightarrow \end{matrix}$

$\begin{cases} 2x- 3y+ 3z=-15 & (E_1) \\ \frac{13}{2}y-\frac{19}{2}z= \frac{83}{2} & (E_4) \\ -3y = -6 & (E_6) \end{cases}$ $E_6 \Rightarrow y=2; E_4 \Rightarrow z=-3; E_1 \Rightarrow x=0.$ ● $\{(0, 2, -3)\}$

425

[29] $\begin{cases} 5x - 3y + 2z = 3 \\ 2x + 4y - z = 7 \\ x - 11y + 4z = 3 \end{cases}$ Interchanging the equations $\Longrightarrow$ $\begin{cases} x - 11y + 4z = 3 & (E_1) & -5E_1 + E_2 \\ 5x - 3y + 2z = 3 & (E_2) & -2E_1 + E_3 \\ 2x + 4y - z = 7 & (E_3) & \Longrightarrow \end{cases}$

$\begin{cases} x - 11y + 4z = 3 & (E_1) \\ \quad 52y - 18z = -12 & (E_4) \\ \quad 26y - 9z = 1 & (E_5) \end{cases}$ $-\frac{1}{2}E_4 + E_5 \Longrightarrow$ $\begin{cases} x - 11y + 4z = 3 & (E_1) \\ \quad 52y - 18z = -12 & (E_4) \\ \quad\quad\quad\quad 0 = 7 & (E_6) \end{cases}$ E_6 is false. $\quad\bullet\,\varnothing$

[30] $\begin{cases} x + y - z = 7 & -4E_1 + E_2 \\ 4x + y - 5z = 4 & -6E_1 + E_3 \\ 6x + y + 3z = 18 & \Longrightarrow \end{cases}$ $\begin{cases} x + y - z = 7 & (E_1) & -\frac{5}{3}E_4 + E_5 \\ \quad -3y - z = -24 & (E_4) \\ \quad -5y + 9z = -24 & (E_5) & \Longrightarrow \end{cases}$

$\begin{cases} x + y - z = 7 & (E_1) \\ \quad -3y - z = -24 & (E_4) \\ \quad\quad \frac{32}{3}z = 16 & (E_6) \end{cases}$ $E_6 \Rightarrow z = \frac{3}{2}; E_4 \Rightarrow y = \frac{15}{2}; E_1 \Rightarrow x = 1.$ $\quad\bullet\,\left\{\left(1, \frac{15}{2}, \frac{3}{2}\right)\right\}$

[31] $\begin{cases} 4u + v - w = 7 \\ 3u - v + 2w = 7 \\ 5u + 3v - 4w = 2 \end{cases}$ $\begin{array}{c} -\frac{3}{4}E_1 + E_2 \\ -\frac{5}{4}E_1 + E_3 \\ \Longrightarrow \end{array}$ $\begin{cases} 4u + v - w = 6 & (E_1) \\ \quad -\frac{7}{4}v + \frac{1}{4}w = \frac{5}{2} & (E_4) \\ \quad \frac{7}{4}v - \frac{11}{4}w = -\frac{11}{2} & (E_5) \end{cases}$ $\begin{array}{c} E_4 + E_5 \\ \Longrightarrow \end{array}$

$\begin{cases} 4u + v - w = 6 & (E_1) \\ \quad -\frac{7}{4}v + \frac{11}{4}w = \frac{5}{2} & (E_4) \\ \quad\quad\quad 0 = -3 & (E_6) \end{cases}$ E_6 is false. $\quad\bullet\,\varnothing$

[32] $\begin{cases} x - 2y + 3z = 4 & -2E_1 + E_2 \\ 2x + y - 4z = 3 & 3E_1 + E_3 \\ -3x + 4y - z = -2 & \Longrightarrow \end{cases}$ $\begin{cases} x - 2y + 3z = 4 & (E_1) & \frac{2}{5}E_4 + E_5 \\ \quad 5y - 10z = -5 & (E_4) \\ \quad -2y + 8z = 10 & (E_5) & \Longrightarrow \end{cases}$

$\begin{cases} x - 2y + 3z = 4 & (E_1) \\ \quad 5y - 10z = -5 & (E_4) \\ \quad\quad\quad 4z = 8 & (E_6) \end{cases}$ $E_6 \Rightarrow z = 2; E_4 \Rightarrow y = 3; E_1 \Rightarrow x = 4.$ $\quad\bullet\,\{(4, 3, 2)\}$

[33] $\begin{cases} 2y + 5z = 6 \\ x \quad - 2z = 4 \\ 2x + 4y = -2 \end{cases}$ $\begin{array}{c} -2E_1 + E_3 \\ \Longrightarrow \end{array}$ $\begin{cases} 2y + 5z = 6 & (E_1) \\ x \quad - 2z = 4 & (E_2) \\ 2x \quad - 10z = -14 & (E_4) \end{cases}$ $\begin{array}{c} -2E_2 + E_4 \\ \Longrightarrow \end{array}$

$\begin{cases} 2y + 5z = 6 & (E_1) \\ x \quad - 2z = 4 & (E_2) \\ \quad -6z = -22 & (E_5) \end{cases}$ $E_5 \Rightarrow z = \frac{11}{3}; E_2 \Rightarrow x = \frac{34}{3}; E_1 \Rightarrow y = -\frac{37}{6}.$ $\quad\bullet\,\left\{\left(\frac{34}{3}, -\frac{37}{6}, \frac{11}{3}\right)\right\}$

[34] $\begin{cases} p + q - 3r = -1 \\ q - r = 0 \\ -p + 2q = 1 \end{cases}$ $\xRightarrow{\substack{E_1 + E_3}}$ $\begin{cases} p + q - 3r = -1 & (E_1) \\ q - r = 0 & (E_2) \\ 3q - 3r = 0 & (E_4) \end{cases}$ $\xRightarrow{\substack{-3E_2 + E_4}}$

$\begin{cases} p + q - 3r = -1 & (E_1) \\ q - r = 0 & (E_2) \\ 0 = 0 & (E_5) \end{cases}$ $E_2 \Rightarrow q = r; E_1 \Rightarrow p = 2r - 1.$ $\qquad \bullet \{(2t - 1, t, t) \mid t \in \mathbf{R}\}$

[35] $\begin{cases} x + 4y - z = 10 \\ 3x + 2y + z = 4 \\ 2x - 3y + 2z = -7 \end{cases}$ $\xRightarrow[\substack{\Longrightarrow}]{\substack{-2E_1 + E_2 \\ -2E_1 + E_3}}$ $\begin{cases} x + 4y - z = 10 & (E_1) \\ -10y + 4z = -26 & (E_4) \\ -11y + 4z = -27 & (E_5) \end{cases}$ $\xRightarrow{\substack{-E_4 + E_5}}$

$\begin{cases} x + 4y - z = 10 & (E_1) \\ -10y + 4z = -26 & (E_4) \\ -y = -1 & (E_6) \end{cases}$ $E_6 \Rightarrow y = 1; E_4 \Rightarrow z = -4; E_1 \Rightarrow x = 2.$ $\qquad \bullet \{(2, 1, -4)\}$

[36] $\begin{cases} x - 2y + 3z = 9 \\ -x + 3y = -4 \\ 2x - 5y + 5z = 17 \end{cases}$ $\xRightarrow[\substack{\Longrightarrow}]{\substack{E_1 + E_2 \\ -2E_1 + E_3}}$ $\begin{cases} x - 2y + 3z = 9 & (E_1) \\ y + 3z = 5 & (E_4) \\ -y - z = -1 & (E_5) \end{cases}$ $\xRightarrow{\substack{E_4 + E_5}}$

$\begin{cases} x - 2y + 3z = 9 & (E_1) \\ y + 3z = 5 & (E_4) \\ 2z = 4 & (E_6) \end{cases}$ $E_6 \Rightarrow z = 2; E_4 \Rightarrow y = -1; E_1 \Rightarrow x = 1.$ $\qquad \bullet \{(1, -1, 2)\}$

[37] $\begin{cases} 2r + 4s + 3t = 6 \\ r - 3s + 2t = -7 \\ -r + 2s - t = 5 \end{cases}$ $\xRightarrow[\substack{\Longrightarrow}]{\text{Interchanging the equations}}$ $\begin{cases} r - 3s + 2t = -7 & (E_1) \\ 2r + 4s + 3t = 6 & (E_2) \\ -r + 2s - t = 5 & (E_3) \end{cases}$ $\xRightarrow[\substack{\Longrightarrow}]{\substack{-2E_1 + E_2 \\ E_1 + E_3}}$

$\begin{cases} r - 3s + 2t = -7 & (E_1) \\ 10s - t = 20 & (E_4) \\ -s + t = -2 & (E_5) \end{cases}$ $\xRightarrow{\substack{E_4 + E_5}}$ $\begin{cases} r - 3s + 2t = -7 & (E_1) \\ 10s - t = 20 & (E_4) \\ 9s = 18 & (E_6) \end{cases}$

$E_6 \Rightarrow s = 2; E_4 \Rightarrow t = 0; E_1 \Rightarrow r = -1.$ $\qquad \bullet \{(-1, 2, 0)\}$

[38] $\begin{cases} x + 10y - 11z = 0 \\ 2x - y + 5z = 5 \\ x + 3y - 2z = 1 \end{cases}$ $\xRightarrow[\substack{\Longrightarrow}]{\substack{-2E_1 + E_2 \\ -E_1 + E_3}}$ $\begin{cases} x + 10y - 11z = 0 & (E_1) \\ -21y + 27z = 5 & (E_4) \\ -7y + 9z = 1 & (E_5) \end{cases}$ $\xRightarrow{\substack{-\frac{1}{3}E_4 + E_5}}$

$\begin{cases} x + 10y - 11z = 0 & (E_1) \\ -21y + 27z = 5 & (E_4) \\ 0 = -\frac{2}{3} & (E_6) \end{cases}$ E_6 is false. $\qquad \bullet \varnothing$

[39] $\begin{cases} x+ y- z=0 \\ 2x-4y+3z=0 \\ x-7y+6z=0 \end{cases}$ $\begin{matrix} -2E_1+E_2 \\ -E_1+E_3 \\ \Rightarrow \end{matrix}$ $\begin{cases} x+ y- z=0 & (E_1) \\ -6y+5z=0 & (E_4) \\ -8y+7z=0 & (E_5) \end{cases}$ $\begin{matrix} -\frac{4}{3}E_4+E_5 \\ \Rightarrow \end{matrix}$

$\begin{cases} x+ y- z=0 & (E_1) \\ -6y+5z=0 & (E_4) \\ \frac{1}{3}z=0 & (E_6) \end{cases}$ $E_6 \Rightarrow z=0; E_4 \Rightarrow y=0; E_1 \Rightarrow x=0.$ $\quad\bullet\ \{(0,0,0)\}$

[40] $\begin{cases} 2x +6z=-9 \\ 3x-2y+11z=-16 \\ 3x- y+7z=-11 \end{cases}$ $\begin{matrix} \frac{1}{2}E_1 \\ \Rightarrow \end{matrix}$ $\begin{cases} x +3z=-\frac{9}{2} & (E_4) \\ 3x-2y+11z=-16 & (E_2) \\ 3x- y+7z=-11 & (E_3) \end{cases}$ $\begin{matrix} -3E_4+E_2 \\ -3E_4+E_3 \\ \Rightarrow \end{matrix}$

$\begin{cases} x +3z=-\frac{9}{2} & (E_4) \\ -2y+2z=-\frac{5}{2} & (E_5) \\ -y-2z=\frac{5}{2} & (E_6) \end{cases}$ $\begin{matrix} E_5+E_6 \\ \Rightarrow \end{matrix}$ $\begin{cases} x +3z=-\frac{9}{2} & (E_4) \\ -2y+2z=-\frac{5}{2} & (E_5) \\ -3y=0 & (E_7) \end{cases}$

$E_7 \Rightarrow y=0; E_5 \Rightarrow z=-\frac{5}{4}; E_4 \Rightarrow x=-\frac{3}{4}.$ $\qquad\bullet\ \left\{\left(-\frac{3}{4}, 0, -\frac{5}{4}\right)\right\}$

[41] Let $u=\frac{1}{x}$ and $v=\frac{1}{y}$. $\begin{cases} u+ v=1 \\ 2u+4v=0 \end{cases}$ $\begin{matrix} -2E_1+E_2 \\ \Rightarrow \end{matrix}$ $\begin{cases} u+ v=1 & (E_1) \\ 2v=-2 & (E_2) \end{cases}$

$E_2 \Rightarrow v=-1 \Rightarrow y=-1; E_1 \Rightarrow u=2 \Rightarrow \frac{1}{x}=\frac{1}{2}.$ $\qquad\bullet\ \left\{\left(\frac{1}{2}, -1\right)\right\}$

[42] Let $x=\frac{1}{u}$ and $y=\frac{1}{v}$. $\begin{cases} x-y=3 \\ 3x-y=-1 \end{cases}$ $\begin{matrix} -3E_1+E_2 \\ \Rightarrow \end{matrix}$ $\begin{cases} x- y=1 & (E_1) \\ 2y=-10 & (E_3) \end{cases}$

$E_3 \Rightarrow y=-5 \Rightarrow v=-\frac{1}{5}; E_1 \Rightarrow x=-2 \Rightarrow u=-\frac{1}{2}.$ $\qquad\bullet\ \left\{\left(-\frac{1}{2}, -\frac{1}{5}\right)\right\}$

[43] Let $p=\frac{1}{x}, q=\frac{1}{y}$ and $r=\frac{1}{z}$. $\begin{cases} p+q-r=-3 \\ p-q-r=1 \\ p-2q+2r=12 \end{cases}$ $\begin{matrix} -E_1+E_2 \\ -E_1+E_2 \\ \Rightarrow \end{matrix}$ $\begin{cases} p+ q- r=-3 & (E_1) \\ -2q=4 & (E_4) \\ -3q+3r=15 & (E_5) \end{cases}$

$E_4 \Rightarrow q=-2 \Rightarrow y=-\frac{1}{2}; E_5 \Rightarrow r=3 \Rightarrow z=\frac{1}{3}; E_1 \Rightarrow p=2 \Rightarrow x=\frac{1}{2}.$ $\quad\bullet\ \left\{\left(\frac{1}{2}, -\frac{1}{2}, \frac{1}{3}\right)\right\}$

[44] Let $p=\frac{1}{u}, q=\frac{1}{v}$ and $r=\frac{1}{z}$. $\begin{cases} p+q-r=1 \\ 2p-q-2r=-8 \\ -p-q+r=-5 \end{cases}$ $\begin{matrix} -2E_1+E_2 \\ E_1+E_3 \\ \Rightarrow \end{matrix}$ $\begin{cases} p+ q-r=1 & (E_1) \\ -3q=-10 & (E_4) \\ 0=-4 & (E_5) \end{cases}$

E_5 is false. $\qquad\qquad\qquad\qquad\qquad\qquad\qquad\qquad\qquad\qquad\bullet\ \varnothing$

[45]   Let w represent the width and l represent the length. Then $\begin{cases} w - l = -6 \\ 2w + 2l = 38 \end{cases}$ $\begin{matrix} -E_1 + E_2 \\ \Longrightarrow \end{matrix}$

$\begin{cases} w - l = -6 \, (E_1) \\ 4 = 50 \, (E_3) \end{cases}$ $E_3 \Rightarrow l = 12.5 \text{ m}; E_1 \Rightarrow w = 6.5 \text{ m}.$ ● 6.5 m, 12.5 m

[46] $\begin{cases} ax + by = e \\ cx + dy = f \end{cases}$ $\begin{matrix} -\frac{c}{a}E_1 + E_2 \\ \Longrightarrow \end{matrix}$ $\begin{cases} ax + \qquad by = \qquad e \; (E_1) \\ \left(\dfrac{ad - bc}{a}\right)y = \dfrac{af - ce}{a} \, (E_3) \end{cases}$ $E_3 \Rightarrow y = \dfrac{af - ce}{ad - bc};$

$E_1 \Rightarrow x = \dfrac{-by + e}{a} \Rightarrow x = \dfrac{-b\left(\dfrac{af - ce}{ad - bc}\right) + e}{a} = \dfrac{\dfrac{-abf + bce + ade - bce}{ad - bc}}{a} = \dfrac{ad - bf}{ad - bc}$

[47] $\begin{cases} 2a - b = 1 & \text{using} \, (2, -1) \\ -3a + 4b = 1 & \text{using} \, (-3, 4) \end{cases}$ $\begin{matrix} 4E_1 + E_2 \\ \Longrightarrow \end{matrix}$ $\begin{cases} 2a - b = 1 \, (E_1) \\ 5a \quad = 5 \, (E_3) \end{cases}$ $E_3 \Rightarrow a = 1; E_1 \Rightarrow b = 1.$

● $(1, 1)$

[48]   Let x represent the amount of the mixture to be drained, and let y represent the amount of the old

mixture that remains. Then $\begin{cases} x + \quad y = 15 \\ x + 0.25y = 7.5 \end{cases}$ $\begin{matrix} \underline{\text{total amount of mixture}} \\ \underline{\text{amount of anitfreeze desired}} \end{matrix}$ $\begin{matrix} -E_1 + E_2 \\ \Longrightarrow \end{matrix}$

$\begin{cases} x + \quad y = \quad 15 \; (E_1) \\ -0.75y = -7.5 \; (E_3) \end{cases}$ $E_3 \Rightarrow y = 10; E_1 \Rightarrow x = 5.$ ● 5 liters

[49]   Let x represent the number of standard models produced and y the number of deluxe models

produced. $\begin{cases} 10x + 12y = 118 & \underline{\text{Cost equation}} \\ 30x + 50y = 480 & \underline{\text{Time equation}} \end{cases}$ $\begin{matrix} -3E_1 + E_2 \\ \Longrightarrow \end{matrix}$ $\begin{cases} 10x + 12y = 118 \; (E_1) \\ 14y = 126 \; (E_3) \end{cases}$

$E_3 \Rightarrow y = 9; E_1 \Rightarrow x = 1.$ The cost of manufacturing 1 standard model and 9 deluxe models is

$118 while the time required is 8 hours.   ● 1 standard model, 9 deluxe models

[50]   Let x represent the amount invested at 8%, y at 10% and z at 14%.

$\begin{cases} x + \quad y + \quad z \equiv 100{,}000 \\ 0.08x + 0.10y + 0.14z = 10{,}000 \\ 0.08x - 0.10y \qquad = -800 \end{cases}$ $\begin{matrix} -0.08E_1 + E_2 \\ -0.08E_1 + E_3 \\ \Longrightarrow \end{matrix}$

$\begin{cases} x + \quad y + \quad z = 100{,}000 \; (E_1) \\ 0.02y + 0.06z = \quad 2000 \; (E_4) \\ 0.18y - 0.08z = -8800 \; (E_5) \end{cases}$ $\begin{matrix} 9E_4 + E_5 \\ \Longrightarrow \end{matrix}$ $\begin{cases} x + \quad y + \quad z = 100{,}000 \; (E_1) \\ 0.02y + 0.06z = \quad 2000 \; (E_4) \\ 0.46z = \quad 9200 \; (E_6) \end{cases}$

$E_6 \Rightarrow z = 20{,}000; E_4 \Rightarrow y = 40{,}000; E_1 \Rightarrow x = 40{,}000.$

● $40,000 at 8%; $40,000 at 10%; $20,000 at 14%

[51] x–chicken; y–rice; z–broccoli.
$$\begin{cases} 2x + 5y + 6z = 47 & \text{iron} \\ 5x + 6y = 38 & \text{fat} \\ 11x + 20y + 4z = 120 & \text{calories} \end{cases} \quad \begin{array}{l} -\frac{5}{2}E_1 + E_2 \\ -\frac{11}{2}E_1 + E_3 \\ \Rightarrow \end{array}$$

$$\begin{cases} 2x + 5y + 6z = 47 & (E_1) \\ -\frac{13}{2}y - 15z = -\frac{159}{2} & (E_4) \\ -\frac{15}{2}y - 29z = -\frac{277}{2} & (E_5) \end{cases} \quad \begin{array}{l} -30E_4 \\ 26E_5 \\ \Rightarrow \end{array} \quad \begin{cases} 2x + 5y + 6z = 47 & (E_1) \\ 195y + 450z = 2385 & (E_6) \\ -195y - 754z = -3601 & (E_7) \end{cases}$$

$$\begin{array}{l} E_6 + E_7 \\ \Rightarrow \end{array} \begin{cases} 2x + 5y + 6z = 47 & (E_1) \\ 195y + 450z = 2385 & (E_6) \\ -304z = -1216 & (E_8) \end{cases} E_8 \Rightarrow z = 4; E_6 \Rightarrow y = 3; E_1 \Rightarrow x = 4.$$

● 4 oz. chicken; 3 oz. rice; 4 oz. broccoli

[52]   Let x represent the proportion of the pool filled by the first pump running for one hour,

y–the second; z–the third.
$$\begin{cases} 4x + 4y + 4z = 1 \\ 5x + 6y = 1 \\ x + 6y + 6z = 1 \end{cases} \quad \begin{array}{l} \text{all pumps running} \\ \text{only first two pumps running} \\ \text{all pumps running for one hour} \\ \text{and then first pump shut off} \end{array} \quad \begin{array}{l} -4E_3 + E_1 \\ -6E_3 + E_2 \\ \Rightarrow \end{array}$$

$$\begin{cases} -20y - 20z = -3 & (E_4) \\ -30y - 36z = -5 & (E_5) \\ x + 6y + 6z = 1 & (E_3) \end{cases} \quad \begin{array}{l} -\frac{2}{3}E_5 + E_4 \\ \Rightarrow \end{array} \quad \begin{cases} 4z = \frac{1}{3} & (E_6) \\ -30y - 36z = -5 & (E_5) \\ x + 6y + 6z = 1 & (E_3) \end{cases}$$

$E_6 \Rightarrow z = \frac{1}{12}; E_5 \Rightarrow y = \frac{1}{15}; E_3 \Rightarrow x = \frac{1}{10}.$ ● 1st pump, 10 hrs; 2nd pump, 15 hrs; 3rd pump, 12 hrs

[53] $$\begin{cases} a - b + c = 8 & \underline{\text{using } (-1, 8)} \\ c = 4 & \underline{\text{using } (0, 4)} \\ a + b + c = 6 & \underline{\text{using } (1, 6)} \end{cases} \quad \text{From } E_2 \text{ we know that } c = 4. \text{ Hence, } \begin{cases} a - b = 4 & (E_4) \\ a + b = 2 & (E_5) \end{cases} \begin{array}{l} E_4 + E_5 \\ \Rightarrow \end{array}$$

$$\begin{cases} a - b = 4 & (E_4) \\ 2a = 6 & (E_6) \end{cases} \quad E_6 \Rightarrow a = 3; E_4 \Rightarrow b = -1. \qquad\qquad ● a = 3, b = 1, c = 4.$$

[54] $$\begin{cases} \frac{1}{A} + \frac{1}{B} = \frac{1}{20} \\ \frac{1}{B} + \frac{1}{C} = \frac{1}{40} \\ \frac{1}{A} + \frac{1}{C} = \frac{1}{30} \end{cases} \quad \text{Let } x = \frac{1}{A}, y = \frac{1}{B} \text{ and } z = \frac{1}{C}. \quad \begin{cases} x + y = \frac{1}{20} \\ y + z = \frac{1}{40} \\ x + z = \frac{1}{30} \end{cases} \quad \begin{array}{l} -E_1 + E_3 \\ \Rightarrow \end{array}$$

$$\begin{cases} x + y = \frac{1}{20} & (E_1) \\ y + z = \frac{1}{40} & (E_2) \\ -y + z = -\frac{1}{60} & (E_4) \end{cases} \quad \begin{array}{l} -E_2 + E_4 \\ \Rightarrow \end{array} \quad \begin{cases} x + y = \frac{1}{20} & (E_1) \\ y + z = \frac{1}{40} & (E_2) \\ 2z = \frac{1}{120} & (E_5) \end{cases}$$

$E_5 \Rightarrow z = \frac{1}{240} \Rightarrow C = 240; E_2 \Rightarrow y = \frac{5}{240} \Rightarrow B = 48; E_1 \Rightarrow x = \frac{7}{240}.$

● $A = \frac{240}{7}$ ohms; $B = 48$ ohms; $C = 240$ ohms

EXERCISES 9.2

[1] $\begin{bmatrix} 1 & -1 & 1 \\ 2 & 0 & -3 \\ 1 & 1 & -5 \end{bmatrix}$; $\left[\begin{array}{ccc|c} 1 & -1 & 1 & 4 \\ 2 & 0 & -3 & -8 \\ 1 & 1 & -5 & 0 \end{array}\right]$ 　　 [2] $\begin{bmatrix} -2 & 0 & -3 \\ 4 & 1 & 1 \\ -1 & 0 & -1 \end{bmatrix}$; $\left[\begin{array}{ccc|c} -2 & 0 & -3 & 0 \\ 4 & 1 & 1 & 5 \\ -1 & 0 & -1 & 6 \end{array}\right]$

[3] $\begin{bmatrix} 6 & 0 & -4 & 5 \\ 0 & 4 & 0 & -1 \\ -2 & 0 & 5 & 0 \\ 1 & -1 & 1 & -3 \end{bmatrix}$; $\left[\begin{array}{cccc|c} 6 & 0 & -4 & 5 & 10 \\ 0 & 4 & 0 & -1 & 0 \\ -2 & 0 & 5 & 0 & 3 \\ 1 & -1 & 1 & -3 & -4 \end{array}\right]$

[4] $\begin{bmatrix} 1 & -2 & 1 & -1 \\ 0 & 1 & 1 & 4 \\ 1 & 0 & -1 & 0 \\ 0 & 1 & 0 & -1 \\ 1 & -1 & 1 & 1 \end{bmatrix}$; $\left[\begin{array}{cccc|c} 1 & -2 & 1 & -1 & 0 \\ 0 & 1 & 1 & 4 & 0 \\ 1 & 0 & -1 & 0 & 0 \\ 0 & 1 & 0 & -1 & 0 \\ 1 & -1 & 1 & 1 & 0 \end{array}\right]$

[5] $\begin{bmatrix} 6 & -8 & 10 & -4 \\ 4 & 0 & -1 & -6 \end{bmatrix}$; $\left[\begin{array}{cccc|c} 6 & -8 & 10 & -4 & 5 \\ 4 & 0 & -1 & -6 & 10 \end{array}\right]$

[6] $\begin{bmatrix} 1 & -2 \\ 2 & 1 \\ -1 & -1 \\ 4 & 1 \end{bmatrix}$; $\left[\begin{array}{cc|c} 1 & -2 & 5 \\ 2 & 1 & 6 \\ -1 & -1 & 4 \\ 4 & 1 & 0 \end{array}\right]$

[7] No. The first nonnzero element in the second row is not zero.

[8] 'No. The first nonzero element in the second row occurs in the third column, while the first nonzero element in the third row occurs in the second column.

[9] Yes.

[10] No. The first nonzero element in the second row occurs in the fifth column, while the first nonzero element in the third row occurs in the second column.

[11] No. The third row contains all zeros, while the fourth row contains a nonzero element.

[12] No. The first nonzero element in the first row now occurs in the fifth column, while the first nonzero element of the second row occurs in the fourth column.

[13] $\left[\begin{array}{cc|c} 2 & -1 & -4 \\ 1 & 2 & 3 \\ 3 & -1 & -1 \end{array}\right] \overset{R_1 \leftrightarrow R_2}{\sim} \left[\begin{array}{cc|c} 1 & 2 & 3 \\ 2 & -1 & -4 \\ 3 & -1 & -1 \end{array}\right] \overset{-2R_1 + R_2 \to R_2}{\underset{-3R_1 + R_3 \to R_3}{\sim}} \left[\begin{array}{cc|c} 1 & 2 & 3 \\ 0 & -5 & -10 \\ 0 & -7 & -10 \end{array}\right] \overset{-\frac{1}{5}R_2 \to R_2}{\sim}$

$\left[\begin{array}{cc|c} 1 & 2 & 3 \\ 0 & 1 & 2 \\ 0 & -7 & -10 \end{array}\right] \overset{7R_2 + R_3 \to R_3}{\sim} \left[\begin{array}{cc|c} 1 & 2 & 3 \\ 0 & 1 & 2 \\ 0 & 0 & 4 \end{array}\right] \overset{\frac{1}{4}R_3 \to R_3}{\sim} \left[\begin{array}{cc|c} 1 & 2 & 3 \\ 0 & 1 & 2 \\ 0 & 0 & 1 \end{array}\right]$

[14] $\left[\begin{array}{ccc|c} 2 & -1 & 3 & 4 \\ 1 & 2 & 1 & 2 \\ 0 & -5 & 1 & 0 \end{array}\right] \overset{R_1 \leftrightarrow R_2}{\sim} \left[\begin{array}{ccc|c} 1 & 2 & 1 & 2 \\ 2 & -1 & 3 & 4 \\ 0 & -5 & 1 & 0 \end{array}\right] \overset{-2R_1 + R_2 \to R_2}{\sim}$

$\left[\begin{array}{ccc|c} 1 & 2 & 1 & 2 \\ 0 & -5 & 1 & 0 \\ 0 & -5 & 1 & 0 \end{array}\right] \overset{-\frac{1}{5}R_2 \to R_2}{\sim} \left[\begin{array}{ccc|c} 1 & 2 & 1 & 2 \\ 0 & 1 & -\frac{1}{5} & 0 \\ 0 & -5 & 1 & 0 \end{array}\right] \overset{5R_2 + R_3 \to R_3}{\sim} \left[\begin{array}{ccc|c} 1 & 2 & 1 & 2 \\ 0 & 1 & -\frac{1}{5} & 0 \\ 0 & 0 & 0 & 0 \end{array}\right]$

[15] $\begin{bmatrix} 2 & 1 & -3 & | & 0 \\ 3 & 2 & -4 & | & 2 \\ 1 & -1 & -3 & | & -6 \end{bmatrix} \begin{array}{l} R_1 \leftrightarrow R_3 \\ \sim \end{array} \begin{bmatrix} 1 & -1 & -3 & | & -6 \\ 3 & 2 & -4 & | & 2 \\ 2 & 1 & -3 & | & 0 \end{bmatrix} \begin{array}{l} -3R_1 + R_2 \to R_2 \\ -2R_1 + R_3 \to R_3 \\ \sim \end{array}$

$\begin{bmatrix} 1 & -1 & -3 & | & -6 \\ 0 & 5 & 5 & | & 20 \\ 0 & 3 & 3 & | & 12 \end{bmatrix} \begin{array}{l} \frac{1}{5}R_2 \to R_2 \\ \sim \end{array} \begin{bmatrix} 1 & -1 & -3 & | & -6 \\ 0 & 1 & 1 & | & 4 \\ 0 & 3 & 3 & | & 12 \end{bmatrix} \begin{array}{l} -3R_2 + R_3 \to R_3 \\ \sim \end{array}$

$\begin{bmatrix} 1 & -1 & -3 & | & -6 \\ 0 & 1 & 1 & | & 4 \\ 0 & 0 & 0 & | & 0 \end{bmatrix}$

[16] $\begin{bmatrix} 1 & 3 & -1 & 2 & | & 1 \\ -2 & -6 & 2 & 0 & | & 2 \\ 0 & 0 & 1 & -1 & | & 7 \\ 2 & 6 & -2 & 2 & | & 0 \end{bmatrix} \begin{array}{l} 2R_1 + R_2 \to R_2 \\ -2R_1 + R_4 \to R_4 \\ \sim \end{array} \begin{bmatrix} 1 & 3 & -1 & 2 & | & 1 \\ 0 & 0 & 0 & 4 & | & 4 \\ 0 & 0 & 1 & -1 & | & 7 \\ 0 & 0 & 0 & -2 & | & -2 \end{bmatrix} \begin{array}{l} R_2 \leftrightarrow R_3 \\ \sim \end{array}$

$\begin{bmatrix} 1 & 3 & -1 & 2 & | & 1 \\ 0 & 0 & 1 & -1 & | & 7 \\ 0 & 0 & 0 & 4 & | & 4 \\ 0 & 0 & 0 & -2 & | & -2 \end{bmatrix} \begin{array}{l} \frac{1}{4}R_3 \to R_3 \\ \sim \end{array} \begin{bmatrix} 1 & 3 & -1 & 2 & | & 1 \\ 0 & 0 & 1 & -1 & | & 7 \\ 0 & 0 & 0 & 1 & | & 1 \\ 0 & 0 & 0 & -2 & | & -2 \end{bmatrix} \begin{array}{l} 2R_3 + R_4 \to R_4 \\ \sim \end{array}$

$\begin{bmatrix} 1 & 3 & -1 & 2 & | & 1 \\ 0 & 0 & 1 & -1 & | & 7 \\ 0 & 0 & 0 & 1 & | & 1 \\ 0 & 0 & 0 & 0 & | & 0 \end{bmatrix}$

[17] $\begin{bmatrix} 1 & 1 & 3 & -4 & | & 1 \\ -5 & 2 & -1 & -3 & | & 0 \\ 7 & -2 & -6 & 1 & | & 5 \end{bmatrix} \begin{array}{l} 5R_1 + R_2 \to R_2 \\ -7R_1 + R_3 \to R_3 \\ \sim \end{array} \begin{bmatrix} 1 & 1 & 3 & -4 & | & 1 \\ 0 & 7 & 14 & -23 & | & 5 \\ 0 & -9 & -27 & 29 & | & -2 \end{bmatrix} \begin{array}{l} -\frac{1}{7}R_2 \to R_2 \\ \sim \end{array}$

$\begin{bmatrix} 1 & 1 & 3 & -4 & | & 1 \\ 0 & 1 & 2 & -\frac{23}{7} & | & \frac{5}{7} \\ 0 & -9 & -27 & 29 & | & -2 \end{bmatrix} \begin{array}{l} 9R_2 + R_3 \to R_3 \\ \sim \end{array} \begin{bmatrix} 1 & 1 & 3 & -4 & | & 1 \\ 0 & 1 & 2 & -\frac{23}{7} & | & \frac{5}{7} \\ 0 & 0 & -9 & -\frac{4}{7} & | & \frac{31}{7} \end{bmatrix} \begin{array}{l} -\frac{1}{10}R_3 \to R_3 \\ \sim \end{array}$

$\begin{bmatrix} 1 & 1 & 3 & -4 & | & 1 \\ 0 & 1 & 2 & -\frac{23}{7} & | & \frac{5}{7} \\ 0 & 0 & 1 & \frac{4}{63} & | & -\frac{31}{63} \end{bmatrix}$

[18]
$$\begin{bmatrix} 1 & 1 & 2 & 1 & 5 & | & 2 \\ 2 & 0 & 1 & 0 & 4 & | & 3 \\ 1 & 1 & 0 & 5 & 2 & | & -1 \\ 6 & -1 & -1 & 2 & 4 & | & -3 \\ -1 & 1 & -1 & 0 & 1 & | & 5 \end{bmatrix} \begin{array}{l} -2R_1 + R_2 \to R_2 \\ -R_1 + R_3 \to R_3 \\ -6R_1 + R_4 \to R_4 \\ R_1 + R_5 \to R_5 \end{array} \begin{bmatrix} 1 & 1 & 2 & 1 & 5 & | & 2 \\ 0 & -2 & -3 & -2 & -6 & | & -1 \\ 0 & 0 & -2 & 4 & -3 & | & -3 \\ 0 & -7 & -13 & -4 & -26 & | & -15 \\ 0 & 2 & 1 & 1 & 6 & | & 7 \end{bmatrix} \begin{array}{l} -\frac{1}{2}R_2 \to R_2 \\ \sim \end{array}$$

$$\begin{bmatrix} 1 & 1 & 2 & 1 & 5 & | & 2 \\ 0 & 1 & \frac{3}{2} & 1 & 3 & | & \frac{1}{2} \\ 0 & 0 & -2 & 4 & -3 & | & -3 \\ 0 & -7 & -13 & -4 & -26 & | & -15 \\ 0 & 2 & 1 & 1 & 6 & | & 7 \end{bmatrix} \begin{array}{l} 7R_2 + R_4 \to R_4 \\ -2R_2 + R_5 \to R_5 \\ \sim \end{array} \begin{bmatrix} 1 & 1 & 2 & 1 & 5 & | & 2 \\ 0 & 1 & \frac{3}{2} & 1 & 3 & | & \frac{1}{2} \\ 0 & 0 & -2 & 4 & -3 & | & -3 \\ 0 & 0 & -\frac{5}{2} & 3 & -5 & | & -\frac{23}{2} \\ 0 & 0 & -2 & -1 & 0 & | & 6 \end{bmatrix} \begin{array}{l} -\frac{1}{2}R_3 \to R_3 \\ \sim \end{array}$$

$$\begin{bmatrix} 1 & 1 & 2 & 1 & 5 & | & 2 \\ 0 & 1 & \frac{3}{2} & 1 & 3 & | & \frac{1}{2} \\ 0 & 0 & 1 & -2 & \frac{3}{2} & | & \frac{3}{2} \\ 0 & 0 & -\frac{5}{2} & 3 & -5 & | & -\frac{23}{2} \\ 0 & 0 & -2 & -1 & 0 & | & 6 \end{bmatrix} \begin{array}{l} \frac{5}{2}R_3 + R_4 \to R_4 \\ 2R_3 + R_5 \to R_5 \\ \sim \end{array} \begin{bmatrix} 1 & 1 & 2 & 1 & 5 & | & 2 \\ 0 & 1 & \frac{3}{2} & 1 & 3 & | & \frac{1}{2} \\ 0 & 0 & 1 & -2 & \frac{3}{2} & | & \frac{3}{2} \\ 0 & 0 & 0 & -2 & -\frac{5}{4} & | & -\frac{31}{4} \\ 0 & 0 & 0 & -5 & 3 & | & 9 \end{bmatrix} \begin{array}{l} -\frac{1}{2}R_4 \to R_4 \\ \sim \end{array}$$

$$\begin{bmatrix} 1 & 1 & 2 & 1 & 5 & | & 2 \\ 0 & 1 & \frac{3}{2} & 1 & 3 & | & \frac{1}{2} \\ 0 & 0 & 1 & -2 & \frac{3}{2} & | & \frac{3}{2} \\ 0 & 0 & 0 & 1 & \frac{5}{8} & | & \frac{31}{8} \\ 0 & 0 & 0 & -5 & 3 & | & 9 \end{bmatrix} \begin{array}{l} 5R_4 + R_5 \to R_5 \\ \sim \end{array} \begin{bmatrix} 1 & 1 & 2 & 1 & 5 & | & 2 \\ 0 & 1 & \frac{3}{2} & 1 & 3 & | & \frac{1}{2} \\ 0 & 0 & 1 & -2 & \frac{3}{2} & | & \frac{3}{2} \\ 0 & 0 & 0 & 1 & \frac{5}{8} & | & \frac{31}{8} \\ 0 & 0 & 0 & 0 & \frac{49}{8} & | & \frac{227}{8} \end{bmatrix} \begin{array}{l} \frac{8}{49}R_5 \to R_5 \\ \sim \end{array}$$

$$\begin{bmatrix} 1 & 1 & 2 & 1 & 5 & | & 2 \\ 0 & 1 & \frac{3}{2} & 1 & 3 & | & \frac{1}{2} \\ 0 & 0 & 1 & -2 & \frac{3}{2} & | & \frac{3}{2} \\ 0 & 0 & 0 & 1 & \frac{5}{8} & | & \frac{31}{8} \\ 0 & 0 & 0 & 0 & 1 & | & \frac{227}{49} \end{bmatrix}$$

[19] The corresponding system of equations is $\begin{cases} x & = 2 \\ y & = 0 \\ & z = 1 \end{cases}$. ● $\{(2, 0, 1)\}$

[20] The corresponding system of equations is $\begin{cases} x + y + 4z = 1 \\ y + z = 1 \\ z = 2 \end{cases}$. $E_3 \Rightarrow z = 2; E_2 \Rightarrow y = -1$;

$E_1 \Rightarrow x = -6$. ● $\{(-6, -1, 2)\}$

[21] The corresponding system of equations is $\begin{cases} x &= 1 \\ y &= 1 \\ 0 &= 2 \end{cases}$. $E_3 \Rightarrow 0 = 2$, which is false. ● $\varnothing$

[22] The corresponding system of equations is $\begin{cases} x + 2y &= 1 \\ y &= -2 \\ 0 &= 2 \end{cases}$. $E_3 \Rightarrow 0 = 2$, which is false. ● $\varnothing$

[23] The corresponding system of equations is $\begin{cases} x + y + z &= 1 \\ y + z + w &= 0 \\ z + 2w &= 1 \end{cases}$. $E_3 \Rightarrow z = 1 - 2w$;

$E_2 \Rightarrow y = -1 + w$, $E_1 \Rightarrow x = 1 + w$. ● $\{(1 + t, -1 + t, 1 - 2t, t) \mid t \in \mathbf{R}\}$

[24] The corresponding system of equations is $\begin{cases} x + z &= 3 \\ y + w &= 2 \\ z + w &= -4 \end{cases}$. $E_3 \Rightarrow z = -4 - w$; $E_2 \Rightarrow y = 2 - w$;

$E_1 \Rightarrow x = 7 + w$. ● $\{(7 + t, 2 - t, -4 - t, t) \mid t \in \mathbf{R}\}$

[25] $\begin{bmatrix} 1 & -1 & -1 & | & 0 \\ 2 & 1 & -3 & | & 1 \\ -1 & 4 & 5 & | & 11 \end{bmatrix} \begin{array}{c} -2R_1 + R_2 \to R_3 \\ R_1 + R_3 \to R_3 \\ \sim \end{array} \begin{bmatrix} 1 & -1 & -1 & | & 0 \\ 0 & 3 & -1 & | & 1 \\ 0 & 3 & 4 & | & 11 \end{bmatrix} \begin{array}{c} \frac{1}{3}R_2 \to R_2 \\ \sim \end{array}$

$\begin{bmatrix} 1 & -1 & -1 & | & 0 \\ 0 & 1 & -\frac{1}{3} & | & \frac{1}{3} \\ 0 & 3 & 4 & | & 11 \end{bmatrix} \begin{array}{c} -3R_2 + R_3 \to R_3 \\ \sim \end{array} \begin{bmatrix} 1 & -1 & -1 & | & 0 \\ 0 & 1 & -\frac{1}{3} & | & \frac{1}{3} \\ 0 & 0 & 5 & | & 10 \end{bmatrix} \begin{array}{c} \frac{1}{5}R_3 \to R_3 \\ \sim \end{array} \begin{bmatrix} 1 & -1 & -1 & | & 0 \\ 0 & 1 & -\frac{1}{3} & | & \frac{1}{3} \\ 0 & 0 & 1 & | & 2 \end{bmatrix}$

R_3: $z = 2$; R_2: $y - \frac{1}{3}z = \frac{1}{3} \Rightarrow y = 1$; R_1: $x - y - z = 0 \Rightarrow x = 3$. ● $\{(3, 1, 2)\}$; consistent, independent

[26] $\begin{bmatrix} 1 & 0 & 1 & | & 3 \\ 0 & 1 & -1 & | & -2 \\ 1 & -1 & 0 & | & 1 \end{bmatrix} \begin{array}{c} -R_1 + R_3 \to R_3 \\ \sim \end{array} \begin{bmatrix} 1 & 0 & 1 & | & 3 \\ 0 & 1 & -1 & | & -2 \\ 0 & -1 & -1 & | & -2 \end{bmatrix} \begin{array}{c} R_2 + R_3 \to R_3 \\ \sim \end{array}$

$\begin{bmatrix} 1 & 0 & 1 & | & 3 \\ 0 & 1 & -1 & | & -2 \\ 0 & 0 & -2 & | & -4 \end{bmatrix} \begin{array}{c} -\frac{1}{2}R_3 \to R_3 \\ \sim \end{array} \begin{bmatrix} 1 & 0 & 1 & | & 3 \\ 0 & 1 & -1 & | & -2 \\ 0 & 0 & 1 & | & 2 \end{bmatrix}$; R_3: $z = 2$; R_2: $y - z = -z \Rightarrow y = 0$;

R_1: $x + z = 3 \Rightarrow x = 1$. ● $\{(1, 0, 2)\}$; consistent, independent

[27] $\begin{bmatrix} 2 & 2 & 1 & | & 1 \\ 4 & 4 & -3 & | & 1 \\ 6 & 2 & -5 & | & 1 \end{bmatrix} \begin{matrix} \frac{1}{2}R_1 \to R_1 \\ \sim \end{matrix} \begin{bmatrix} 1 & 1 & \frac{1}{2} & | & \frac{1}{2} \\ 4 & 4 & -3 & | & 1 \\ 6 & 2 & -5 & | & 1 \end{bmatrix} \begin{matrix} -4R_1 + R_2 \to R_2 \\ -6R_1 + R_3 \to R_3 \\ \sim \end{matrix}$

$\begin{bmatrix} 1 & 1 & \frac{1}{2} & | & \frac{1}{2} \\ 0 & 0 & -5 & | & -1 \\ 0 & -4 & -8 & | & -2 \end{bmatrix} \begin{matrix} R_2 \leftrightarrow R_3 \\ \sim \end{matrix} \begin{bmatrix} 1 & 1 & \frac{1}{2} & | & \frac{1}{2} \\ 0 & -4 & -8 & | & -2 \\ 0 & 0 & -5 & | & -1 \end{bmatrix} \begin{matrix} -\frac{1}{4}R_2 \to R_2 \\ \sim \end{matrix}$

$\begin{bmatrix} 1 & 1 & \frac{1}{2} & | & \frac{1}{2} \\ 0 & 1 & 2 & | & \frac{1}{2} \\ 0 & 0 & -5 & | & -1 \end{bmatrix} \begin{matrix} -\frac{1}{5}R_3 \to R_3 \\ \sim \end{matrix} \begin{bmatrix} 1 & 1 & \frac{1}{2} & | & \frac{1}{2} \\ 0 & 1 & 2 & | & \frac{1}{2} \\ 0 & 0 & 1 & | & \frac{1}{5} \end{bmatrix} \quad R_3: z = \frac{1}{5}; R_2: y + 2z = \frac{1}{2} \Rightarrow y = \frac{1}{10};$

$R_1: x + y - \frac{1}{2}z = \frac{1}{2} \Rightarrow x = \frac{3}{10}.$ $\qquad\qquad \bullet \left\{ \left(\frac{3}{10}, \frac{1}{10}, \frac{1}{5} \right) \right\};$ consistent; independent

[28] $\begin{bmatrix} -3 & -1 & 2 & | & -1 \\ 1 & 0 & -1 & | & 0 \\ -4 & 3 & 1 & | & -3 \end{bmatrix} \begin{matrix} R_1 \leftrightarrow R_2 \\ \sim \end{matrix} \begin{bmatrix} 1 & 0 & -1 & | & 0 \\ -3 & -1 & 2 & | & -1 \\ -4 & 3 & 1 & | & -3 \end{bmatrix} \begin{matrix} 3R_1 + R_2 \to R_2 \\ 4R_1 + R_3 \to R_3 \\ \sim \end{matrix}$

$\begin{bmatrix} 1 & 0 & -1 & | & 0 \\ 0 & -1 & -1 & | & -1 \\ 0 & 3 & -3 & | & -3 \end{bmatrix} \begin{matrix} -R_2 \to R_2 \\ \sim \end{matrix} \begin{bmatrix} 1 & 0 & -1 & | & 0 \\ 0 & 1 & 1 & | & 1 \\ 0 & 3 & -3 & | & -3 \end{bmatrix} \begin{matrix} -3R_2 + R_3 \to R_3 \\ \sim \end{matrix}$

$\begin{bmatrix} 1 & 0 & -1 & | & 0 \\ 0 & 1 & 1 & | & 1 \\ 0 & 0 & -6 & | & -6 \end{bmatrix} \begin{matrix} -\frac{1}{6}R_3 \to R_3 \\ \sim \end{matrix} \begin{bmatrix} 1 & 0 & -1 & | & 0 \\ 0 & 1 & 1 & | & 1 \\ 0 & 0 & 1 & | & 1 \end{bmatrix}; R_3: z = 1; R_2: y + z = 1 \Rightarrow y = 0;$

$R_1: x - z = 0 \Rightarrow x = 1.$ $\qquad\qquad \bullet \{(1, 0, 1)\};$ consistent; independent

[29] $\begin{bmatrix} -1 & 1 & -1 & | & -6 \\ 1 & 3 & 4 & | & 21 \\ 2 & -1 & -2 & | & -8 \end{bmatrix} \begin{matrix} -R_1 \to R_1 \\ \sim \end{matrix} \begin{bmatrix} 1 & -1 & 1 & | & 6 \\ 1 & 3 & 4 & | & 21 \\ 2 & -1 & -2 & | & -8 \end{bmatrix} \begin{matrix} -R_1 + R_2 \to R_2 \\ -2R_1 + R_3 \to R_3 \\ \sim \end{matrix}$

$\begin{bmatrix} 1 & -1 & 1 & | & 6 \\ 0 & 4 & 3 & | & 15 \\ 0 & 1 & -4 & | & -20 \end{bmatrix} \begin{matrix} \frac{1}{4}R_2 \to R_2 \\ \sim \end{matrix} \begin{bmatrix} 1 & -1 & 1 & | & 6 \\ 0 & 1 & \frac{3}{4} & | & \frac{15}{4} \\ 0 & 1 & -4 & | & -20 \end{bmatrix} \begin{matrix} -R_2 + R_3 \to R_3 \\ \sim \end{matrix}$

$\begin{bmatrix} 1 & -1 & 1 & | & 6 \\ 0 & 1 & \frac{3}{4} & | & \frac{15}{4} \\ 0 & 0 & -\frac{19}{4} & | & -\frac{95}{4} \end{bmatrix} \begin{matrix} -\frac{4}{19}R_3 \to R_3 \\ \sim \end{matrix} \begin{bmatrix} 1 & -1 & 1 & | & 6 \\ 0 & 1 & \frac{3}{4} & | & \frac{15}{4} \\ 0 & 0 & 1 & | & 5 \end{bmatrix} \quad R_3: z = 5; R_2: y + \frac{3}{4}z = \frac{15}{4} \Rightarrow y = 0;$

$R_1: x - y + z = 6 \Rightarrow x = 1.$ $\qquad\qquad \bullet \{(1, 0, 5)\};$ consistent; independent

[30]
$$\begin{bmatrix} -1 & -1 & 1 & | & 6 \\ 3 & -2 & 4 & | & 4 \\ 5 & 3 & -2 & | & 21 \end{bmatrix} \begin{matrix} -R_1 \to R_1 \\ \sim \end{matrix} \begin{bmatrix} 1 & 1 & -1 & | & -6 \\ 3 & -2 & 4 & | & 4 \\ 5 & 3 & -2 & | & 21 \end{bmatrix} \begin{matrix} -3R_1 + R_2 \to R_2 \\ -5R_1 + R_3 \to R_3 \\ \sim \end{matrix}$$

$$\begin{bmatrix} 1 & 1 & -1 & | & -6 \\ 0 & -5 & 7 & | & 22 \\ 0 & -2 & 3 & | & 9 \end{bmatrix} \begin{matrix} -\frac{1}{5}R_2 \to R_2 \\ \sim \end{matrix} \begin{bmatrix} 1 & 1 & -1 & | & -6 \\ 0 & 1 & -\frac{7}{5} & | & -\frac{22}{5} \\ 0 & -2 & 3 & | & 9 \end{bmatrix} \begin{matrix} 2R_2 + R_3 \to R_3 \\ \sim \end{matrix}$$

$$\begin{bmatrix} 1 & 1 & -1 & | & -6 \\ 0 & 1 & -\frac{7}{5} & | & -\frac{22}{5} \\ 0 & 0 & \frac{1}{5} & | & \frac{1}{5} \end{bmatrix} \begin{matrix} 5R_3 \to R_3 \\ \sim \end{matrix} \begin{bmatrix} 1 & 1 & -1 & | & -6 \\ 0 & 1 & -\frac{7}{5} & | & -\frac{22}{5} \\ 0 & 0 & 1 & | & 1 \end{bmatrix}$$

$R_3: z = 1;\ R_2: y - \frac{7}{5}z = -\frac{22}{5} \Rightarrow y = -3;$

$R_1: x + y - z = -6 \Rightarrow x = -2.$ ● $\{(-2, -3, 1)\}$; consistent, independent

[31]
$$\begin{bmatrix} 1 & 1 & 1 & | & 2 \\ 1 & 3 & -3 & | & 5 \\ 1 & 2 & -1 & | & 4 \end{bmatrix} \begin{matrix} -R_1 + R_2 \to R_2 \\ -R_1 + R_3 \to R_3 \\ \sim \end{matrix} \begin{bmatrix} 1 & 1 & 1 & | & 2 \\ 0 & 2 & -4 & | & 3 \\ 0 & 1 & -2 & | & 2 \end{bmatrix} \begin{matrix} \frac{1}{2}R_2 \to R_2 \\ \sim \end{matrix}$$

$$\begin{bmatrix} 1 & 1 & 1 & | & 2 \\ 0 & 1 & -2 & | & \frac{3}{2} \\ 0 & 1 & -2 & | & 2 \end{bmatrix} \begin{matrix} -R_2 + R_3 \to R_3 \\ \sim \end{matrix} \begin{bmatrix} 1 & 1 & 1 & | & 2 \\ 0 & 1 & -2 & | & \frac{3}{2} \\ 0 & 0 & 0 & | & \frac{1}{2} \end{bmatrix} \quad R_3: 0 = \frac{1}{2}.\ \text{This statement is false.}$$

● $\varnothing$; inconsistent

[32]
$$\begin{bmatrix} 2 & -5 & 1 & | & 1 \\ 4 & -23 & -3 & | & 5 \\ 1 & 4 & 3 & | & 4 \end{bmatrix} \begin{matrix} R_1 \leftrightarrow R_3 \\ \sim \end{matrix} \begin{bmatrix} 1 & 4 & 3 & | & 4 \\ 4 & -23 & -3 & | & 5 \\ 2 & -5 & 1 & | & 1 \end{bmatrix} \begin{matrix} -4R_1 + R_2 \to R_2 \\ -2R_1 + R_3 \to R_3 \\ \sim \end{matrix}$$

$$\begin{bmatrix} 1 & 4 & 3 & | & 4 \\ 0 & -39 & -15 & | & -11 \\ 0 & -13 & -5 & | & -7 \end{bmatrix} \begin{matrix} -\frac{1}{39}R_2 \to R_2 \\ \sim \end{matrix} \begin{bmatrix} 1 & 4 & 3 & | & 4 \\ 0 & 1 & \frac{15}{39} & | & \frac{11}{39} \\ 0 & -13 & -5 & | & -7 \end{bmatrix} \begin{matrix} -13R_2 + R_3 \to R_3 \\ \sim \end{matrix}$$

$$\begin{bmatrix} 1 & 4 & 3 & | & 4 \\ 0 & 1 & \frac{15}{39} & | & \frac{11}{39} \\ 0 & 0 & 0 & | & -\frac{10}{3} \end{bmatrix} \quad R_3: 0 = \frac{10}{-3}.\ \text{This statement is false.} \quad\quad ● \varnothing; \text{inconsistent}$$

[33] $\begin{bmatrix} 2 & -1 & -1 & | & 1 \\ -3 & 4 & -2 & | & 2 \\ 0 & 11 & -1 & | & 1 \end{bmatrix} \begin{matrix} \frac{1}{2}R_1 \to R_1 \\ \sim \end{matrix} \begin{bmatrix} 1 & -\frac{1}{2} & -\frac{1}{2} & | & \frac{1}{2} \\ -3 & 4 & -2 & | & 2 \\ 0 & 11 & -1 & | & 1 \end{bmatrix} \begin{matrix} 3R_1 + R_2 \to R_2 \\ \sim \end{matrix}$

$\begin{bmatrix} 1 & -\frac{1}{2} & -\frac{1}{2} & | & \frac{1}{2} \\ 0 & \frac{5}{2} & -\frac{7}{2} & | & \frac{7}{2} \\ 0 & 11 & -1 & | & 1 \end{bmatrix} \begin{matrix} \frac{2}{5}R_2 \to R_2 \\ \sim \end{matrix} \begin{bmatrix} 1 & -\frac{1}{2} & -\frac{1}{2} & | & \frac{1}{2} \\ 0 & 1 & -\frac{7}{5} & | & \frac{7}{5} \\ 0 & 11 & -1 & | & 1 \end{bmatrix} \begin{matrix} -11R_2 + R_3 \to R_3 \\ \sim \end{matrix}$

$\begin{bmatrix} 1 & -\frac{1}{2} & -\frac{1}{2} & | & \frac{1}{2} \\ 0 & 1 & -\frac{7}{5} & | & \frac{7}{5} \\ 0 & 0 & \frac{72}{5} & | & -\frac{72}{5} \end{bmatrix} \begin{matrix} \frac{5}{72}R_3 \to R_3 \\ \sim \end{matrix} \begin{bmatrix} 1 & -\frac{1}{2} & -\frac{1}{2} & | & \frac{1}{2} \\ 0 & 1 & -\frac{7}{5} & | & \frac{7}{5} \\ 0 & 0 & 1 & | & -1 \end{bmatrix}$ $R_3: z = -1; R_2: y - \frac{7}{5}z = \frac{7}{5} \Rightarrow y = 0;$

$R_1: x - \frac{1}{2}y - \frac{1}{2}z = \frac{1}{2} \Rightarrow x = 0.$ $\qquad\qquad$ ● $\{(0, 0, -1)\}$; consistent, independent

[34] $\begin{bmatrix} 1 & 2 & 3 & | & 5 \\ 3 & -1 & -2 & | & 0 \\ 2 & 2 & -1 & | & 3 \end{bmatrix} \begin{matrix} -3R_1 + R_2 \to R_2 \\ -2R_1 + R_3 \to R_3 \\ \sim \end{matrix} \begin{bmatrix} 1 & 2 & 3 & | & 5 \\ 0 & -7 & -11 & | & -15 \\ 0 & -2 & -7 & | & -7 \end{bmatrix} \begin{matrix} -\frac{1}{7}R_2 \to R_2 \\ \sim \end{matrix}$

$\begin{bmatrix} 1 & 2 & 3 & | & 5 \\ 0 & 1 & \frac{11}{7} & | & \frac{15}{7} \\ 0 & -2 & -7 & | & -7 \end{bmatrix} \begin{matrix} 2R_2 + R_3 \to R_3 \\ \sim \end{matrix} \begin{bmatrix} 1 & 2 & 3 & | & 5 \\ 0 & 1 & \frac{11}{7} & | & \frac{15}{7} \\ 0 & 0 & -\frac{27}{7} & | & -\frac{19}{7} \end{bmatrix} \begin{matrix} -\frac{7}{27}R_3 \to R_3 \\ \sim \end{matrix} \begin{bmatrix} 1 & 2 & 3 & | & 5 \\ 0 & 1 & \frac{11}{7} & | & \frac{15}{7} \\ 0 & 0 & 1 & | & \frac{19}{27} \end{bmatrix}$

$R_3: z = \frac{19}{27}; R_2: y + \frac{11}{7}z = \frac{15}{7} \Rightarrow y = \frac{28}{27}; R_1: x + 2y + 3z = 5 \Rightarrow x = \frac{22}{27}.$

● $\left\{\left(\frac{22}{27}, \frac{28}{27}, \frac{19}{27}\right)\right\}$; consistent, independent

[35] $\begin{bmatrix} 2 & -1 & -5 & | & -5 \\ 2 & 2 & -1 & | & -3 \\ -2 & 3 & 2 & | & 12 \end{bmatrix} \begin{matrix} \frac{1}{2}R_1 \to R_1 \\ \sim \end{matrix} \begin{bmatrix} 1 & -\frac{1}{2} & -\frac{5}{2} & | & -\frac{5}{2} \\ 2 & 2 & -1 & | & -3 \\ -2 & 3 & 2 & | & 12 \end{bmatrix} \begin{matrix} -2R_1 + R_2 \to R_2 \\ 2R_1 + R_3 \to R_3 \\ \sim \end{matrix}$

$\begin{bmatrix} 1 & -\frac{1}{2} & -\frac{5}{2} & | & -\frac{5}{2} \\ 0 & 3 & 4 & | & 2 \\ 0 & 2 & -3 & | & 7 \end{bmatrix} \begin{matrix} \frac{1}{3}R_2 \to R_2 \\ \sim \end{matrix} \begin{bmatrix} 1 & -\frac{1}{2} & -\frac{5}{2} & | & -\frac{5}{2} \\ 0 & 1 & \frac{4}{3} & | & \frac{2}{3} \\ 0 & 2 & -3 & | & 7 \end{bmatrix} \begin{matrix} -2R_2 + R_3 \to R_3 \\ \sim \end{matrix}$

$\begin{bmatrix} 1 & -\frac{1}{2} & -\frac{5}{2} & | & -\frac{5}{2} \\ 0 & 1 & \frac{4}{3} & | & \frac{2}{3} \\ 0 & 0 & -\frac{17}{3} & | & \frac{17}{3} \end{bmatrix} \begin{matrix} -\frac{3}{17}R_3 \to R_3 \\ \sim \end{matrix} \begin{bmatrix} 1 & -\frac{1}{2} & -\frac{5}{2} & | & -\frac{5}{2} \\ 0 & 1 & \frac{4}{3} & | & \frac{2}{3} \\ 0 & 0 & 1 & | & -1 \end{bmatrix}$ $R_3: z = -1;$

$R_2: y + \frac{4}{3}z = \frac{2}{3} \Rightarrow y = 2; R_1: x - \frac{1}{2}y - \frac{5}{2}z = -\frac{5}{2} \Rightarrow x = -4.$ $\quad$ ● $\{(-4, 2, -1)\}$; consistent, independent

[36] $\begin{bmatrix} 3 & 1 & 1 & | & -7 \\ 1 & -1 & 4 & | & -19 \\ 1 & 3 & -2 & | & 21 \end{bmatrix} \begin{matrix} R_1 \leftrightarrow R_2 \\ \sim \end{matrix} \begin{bmatrix} 1 & -1 & 4 & | & -19 \\ 3 & 1 & 1 & | & -7 \\ 1 & 3 & -2 & | & 21 \end{bmatrix} \begin{matrix} -3R_1 + R_2 \rightarrow R_2 \\ -R_1 + R_3 \rightarrow R_3 \\ \sim \end{matrix}$

$\begin{bmatrix} 1 & -1 & 4 & | & -19 \\ 0 & 4 & -11 & | & 50 \\ 0 & 4 & -6 & | & 40 \end{bmatrix} \begin{matrix} \frac{1}{4}R_2 \rightarrow R_2 \\ \sim \end{matrix} \begin{bmatrix} 1 & -1 & 4 & | & -19 \\ 0 & 1 & -\frac{11}{4} & | & \frac{25}{2} \\ 0 & 4 & -6 & | & 40 \end{bmatrix} \begin{matrix} -4R_2 + R_3 \rightarrow R_3 \\ \sim \end{matrix}$

$\begin{bmatrix} 1 & -1 & 4 & | & -19 \\ 0 & 1 & -\frac{11}{4} & | & \frac{25}{2} \\ 0 & 0 & 5 & | & -10 \end{bmatrix} \begin{matrix} \frac{1}{5}R_3 \rightarrow R_3 \\ \sim \end{matrix} \begin{bmatrix} 1 & -1 & 4 & | & -19 \\ 0 & 1 & -\frac{11}{4} & | & \frac{25}{2} \\ 0 & 0 & 1 & | & -2 \end{bmatrix} \quad R_3: z = -2;$

$R_2: y - \frac{1}{4}z = \frac{25}{2} \Rightarrow y = 7; \ R_1: x - y + 4z = -19 \Rightarrow x = -4.$ ● $\{(-4, 7, -2)\}$; consistent; independent

[37] $\begin{bmatrix} \frac{1}{2} & 1 & 1 & | & -\frac{1}{4} \\ 1 & -2 & 4 & | & -3 \\ -\frac{1}{10} & 1 & -\frac{4}{5} & | & 1 \end{bmatrix} \begin{matrix} R_1 \leftrightarrow R_2 \\ \sim \end{matrix} \begin{bmatrix} 1 & -2 & 4 & | & -3 \\ \frac{1}{2} & 1 & 1 & | & -\frac{1}{4} \\ -\frac{1}{10} & 1 & -\frac{4}{5} & | & 1 \end{bmatrix} \begin{matrix} -\frac{1}{2}R_1 + R_2 \rightarrow R_2 \\ \frac{1}{10}R_1 + R_3 \rightarrow R_3 \\ \sim \end{matrix}$

$\begin{bmatrix} 1 & -2 & 4 & | & -3 \\ 0 & 2 & -1 & | & \frac{5}{4} \\ 0 & \frac{4}{5} & -\frac{2}{5} & | & \frac{7}{10} \end{bmatrix} \begin{matrix} \frac{1}{2}R_2 \rightarrow R_2 \\ \sim \end{matrix} \begin{bmatrix} 1 & -2 & 4 & | & -3 \\ 0 & 1 & -\frac{1}{2} & | & \frac{5}{8} \\ 0 & \frac{4}{5} & -\frac{2}{5} & | & \frac{7}{10} \end{bmatrix} \begin{matrix} -\frac{4}{5}R_2 + R_3 \rightarrow R_3 \\ \sim \end{matrix} \begin{bmatrix} 1 & -2 & 4 & | & -3 \\ 0 & 1 & -\frac{1}{2} & | & \frac{5}{8} \\ 0 & 0 & 0 & | & \frac{1}{5} \end{bmatrix}$

$R_3: 0 = \frac{1}{5}.$ This equation is false. ● $\varnothing$; inconsistent

[38] $\begin{bmatrix} \frac{3}{2} & -\frac{1}{2} & 2 & | & 2 \\ \frac{1}{6} & \frac{2}{3} & -1 & | & \frac{1}{2} \\ \frac{3}{2} & -\frac{8}{3} & \frac{17}{3} & | & \frac{25}{6} \end{bmatrix} \begin{matrix} \frac{2}{3}R_1 \leftrightarrow R_1 \\ \sim \end{matrix} \begin{bmatrix} 1 & -\frac{1}{3} & \frac{4}{3} & | & \frac{1}{3} \\ \frac{1}{6} & \frac{2}{3} & -1 & | & \frac{1}{2} \\ \frac{3}{2} & -\frac{8}{3} & \frac{17}{3} & | & \frac{25}{6} \end{bmatrix} \begin{matrix} -\frac{1}{6}R_1 + R_2 \rightarrow R_2 \\ -\frac{3}{2}R_1 + R_3 \rightarrow R_3 \\ \sim \end{matrix}$

$\begin{bmatrix} 1 & -\frac{1}{3} & \frac{4}{3} & | & \frac{1}{3} \\ 0 & \frac{13}{18} & -\frac{11}{9} & | & \frac{4}{9} \\ 0 & -\frac{13}{6} & \frac{11}{3} & | & \frac{11}{3} \end{bmatrix} \begin{matrix} \frac{18}{13}R_2 \rightarrow R_2 \\ \sim \end{matrix} \begin{bmatrix} 1 & -\frac{1}{3} & \frac{4}{3} & | & \frac{1}{3} \\ 0 & 1 & -\frac{22}{13} & | & \frac{8}{117} \\ 0 & -\frac{13}{6} & \frac{11}{3} & | & \frac{11}{3} \end{bmatrix} \begin{matrix} \frac{13}{6}R_2 + R_3 \rightarrow R_3 \\ \sim \end{matrix}$

$\begin{bmatrix} 1 & -\frac{1}{3} & \frac{4}{3} & | & \frac{1}{3} \\ 0 & 1 & -\frac{22}{13} & | & \frac{8}{117} \\ 0 & 0 & 0 & | & \frac{103}{27} \end{bmatrix} \quad R_3: 0 = \frac{103}{27}.$ This equation is false. ● $\varnothing$; inconsistent

[39] $\begin{bmatrix} 1 & 1 & -3 & | & 0 \\ 1 & -1 & 2 & | & 3 \\ 1 & -1 & -1 & | & 12 \end{bmatrix}$ $\begin{matrix} -R_1 + R_2 \to R_2 \\ -R_1 + R_3 \to R_3 \\ \sim \end{matrix}$ $\begin{bmatrix} 1 & 1 & -3 & | & 0 \\ 0 & -2 & 5 & | & 3 \\ 0 & -2 & 2 & | & 12 \end{bmatrix}$ $\begin{matrix} -\frac{1}{2} R_2 \to R_2 \\ \sim \end{matrix}$

$\begin{bmatrix} 1 & 1 & -3 & | & 0 \\ 0 & 1 & -\frac{5}{2} & | & -\frac{3}{2} \\ 0 & -2 & 2 & | & 12 \end{bmatrix}$ $\begin{matrix} 2R_2 + R_3 \to R_3 \\ \sim \end{matrix}$ $\begin{bmatrix} 1 & 1 & -3 & | & 0 \\ 0 & 1 & -\frac{5}{2} & | & -\frac{3}{2} \\ 0 & 0 & -3 & | & 9 \end{bmatrix}$ $\begin{matrix} -\frac{1}{3} R_3 \to R_3 \\ \sim \end{matrix}$ $\begin{bmatrix} 1 & 1 & -3 & | & 0 \\ 0 & 1 & -\frac{5}{2} & | & -\frac{3}{2} \\ 0 & 0 & 1 & | & -3 \end{bmatrix}$

$R_3: z = -3$; $R_2: y - \frac{5}{2} z = -\frac{3}{2} \Rightarrow y = -9$; $R_1: x + y - 3z = 0 \Rightarrow x = 0$.

● $\{(0, -9, -3)\}$; consistent, independent

[40] $\begin{bmatrix} 1 & -1 & -1 & | & 0 \\ 2 & 0 & 1 & | & 0 \\ 1 & 1 & 2 & | & 0 \end{bmatrix}$ $\begin{matrix} -2R_1 + R_2 \to R_2 \\ -R_1 + R_3 \to R_3 \\ \sim \end{matrix}$ $\begin{bmatrix} 1 & -1 & -1 & | & 0 \\ 0 & 2 & 3 & | & 0 \\ 0 & 2 & 3 & | & 0 \end{bmatrix}$ $\begin{matrix} \frac{1}{2} R_2 \to R_2 \\ \sim \end{matrix}$

$\begin{bmatrix} 1 & -1 & -1 & | & 0 \\ 0 & 1 & \frac{3}{2} & | & 0 \\ 0 & 2 & 3 & | & 0 \end{bmatrix}$ $\begin{matrix} -2R_2 + R_3 \to R_3 \\ \sim \end{matrix}$ $\begin{bmatrix} 1 & -1 & -1 & | & 0 \\ 0 & 1 & \frac{3}{2} & | & 0 \\ 0 & 0 & 0 & | & 0 \end{bmatrix}$ $R_3: 0 = 0$; $R_2: y + \frac{3}{2} z = 0 \Rightarrow y = -\frac{3}{2} z$;

$R_1: x - y - z = 0 \Rightarrow x = -\frac{1}{2} z$. ● $\left\{ \left(-\frac{1}{2} t, -\frac{3}{2} t, t \right) \middle| t \in \mathbf{R} \right\}$; consistent, dependent

[41] $\begin{bmatrix} 1 & -2 & 3 & | & -4 \\ 3 & -6 & 10 & | & 14 \\ 5 & -8 & 19 & | & -21 \\ 2 & -4 & 7 & | & -9 \end{bmatrix}$ $\begin{matrix} -3R_1 + R_2 \to R_2 \\ -5R_1 + R_3 \to R_3 \\ -2R_1 + R_4 \to R_4 \\ \sim \end{matrix}$ $\begin{bmatrix} 1 & -2 & 3 & | & -4 \\ 0 & 0 & 1 & | & 26 \\ 0 & 2 & 4 & | & -1 \\ 0 & 0 & 1 & | & -1 \end{bmatrix}$ $\begin{matrix} R_2 \leftrightarrow R_2 \\ \sim \end{matrix}$

$\begin{bmatrix} 1 & -2 & 3 & | & -4 \\ 0 & 2 & 4 & | & -1 \\ 0 & 0 & 1 & | & 26 \\ 0 & 0 & 1 & | & -1 \end{bmatrix}$ $\begin{matrix} \frac{1}{2} R_2 \to R_2 \\ \sim \end{matrix}$ $\begin{bmatrix} 1 & -2 & 3 & | & -4 \\ 0 & 1 & 2 & | & -\frac{1}{2} \\ 0 & 0 & 1 & | & 26 \\ 0 & 0 & 1 & | & -1 \end{bmatrix}$ $\begin{matrix} -R_3 + R_4 \to R_4 \\ \sim \end{matrix}$ $\begin{bmatrix} 1 & -2 & 3 & | & -4 \\ 0 & 1 & 2 & | & -\frac{1}{2} \\ 0 & 0 & 1 & | & 26 \\ 0 & 0 & 0 & | & -27 \end{bmatrix}$

$R_4: 0 = -27$. This equation is false. ● $\varnothing$; inconsistent

[42] $\begin{bmatrix} 2 & -1 & 3 & | & -2 \\ 1 & -2 & -1 & | & 3 \\ 3 & 5 & -2 & | & 2 \\ 4 & 3 & 1 & | & 6 \end{bmatrix} \begin{array}{l} R_1 \to R_2 \\ \sim \end{array} \begin{bmatrix} 1 & -2 & -1 & | & 3 \\ 2 & -1 & 3 & | & 2 \\ 3 & 5 & -2 & | & 2 \\ 4 & 3 & 1 & | & 6 \end{bmatrix} \begin{array}{l} -2R_1 + R_2 \to R_2 \\ -3R_1 + R_3 \to R_3 \\ -4R_1 + R_4 \to R_4 \\ \sim \end{array}$

$\begin{bmatrix} 1 & -2 & -1 & | & 3 \\ 0 & 3 & 5 & | & -4 \\ 0 & 11 & 1 & | & -7 \\ 0 & 11 & 5 & | & -6 \end{bmatrix} \begin{array}{l} \frac{1}{3}R_2 \to R_2 \\ \sim \end{array} \begin{bmatrix} 1 & -2 & -1 & | & 3 \\ 0 & 1 & \frac{5}{3} & | & -\frac{4}{3} \\ 0 & 11 & 1 & | & -7 \\ 0 & 11 & 5 & | & -6 \end{bmatrix} \begin{array}{l} -11R_2 + R_3 \to R_3 \\ -11R_2 + R_4 \to R_4 \\ \sim \end{array}$

$\begin{bmatrix} 1 & -2 & -1 & | & 3 \\ 0 & 1 & \frac{5}{3} & | & -\frac{4}{3} \\ 0 & 0 & -\frac{52}{3} & | & \frac{23}{3} \\ 0 & 0 & -\frac{40}{3} & | & \frac{26}{3} \end{bmatrix} \begin{array}{l} -\frac{3}{52}R_3 \to R_3 \\ \sim \end{array} \begin{bmatrix} 1 & -2 & -1 & | & 3 \\ 0 & 1 & \frac{5}{3} & | & -\frac{4}{3} \\ 0 & 0 & 1 & | & \frac{23}{52} \\ 0 & 0 & -\frac{40}{3} & | & \frac{26}{3} \end{bmatrix} \begin{array}{l} \frac{40}{3}R_3 + R_4 \to R_4 \\ \sim \end{array}$

$\begin{bmatrix} 1 & -2 & -1 & | & 3 \\ 0 & 1 & \frac{5}{3} & | & -\frac{4}{3} \\ 0 & 0 & 1 & | & \frac{23}{52} \\ 0 & 0 & 0 & | & \frac{568}{39} \end{bmatrix}$ $R_4: 0 = \frac{568}{39}$. This equation is false. ● $\varnothing$; inconsistent

[43] $\begin{bmatrix} 1 & -3 & 4 & 2 & | & 1 \\ 2 & -3 & 5 & -2 & | & -1 \\ -1 & 2 & -3 & 1 & | & 4 \end{bmatrix} \begin{array}{l} -2R_1 + R_2 \to R_2 \\ R_1 + R_3 \to R_3 \\ \sim \end{array} \begin{bmatrix} 1 & -3 & 4 & 2 & | & 1 \\ 0 & 3 & -3 & -6 & | & -3 \\ 0 & -1 & 1 & 3 & | & 5 \end{bmatrix} \begin{array}{l} \frac{1}{3}R_2 \to R_2 \\ \sim \end{array}$

$\begin{bmatrix} 1 & -3 & 4 & 2 & | & 1 \\ 0 & 1 & -1 & -2 & | & -1 \\ 0 & -1 & 1 & 3 & | & 5 \end{bmatrix} \begin{array}{l} R_2 + R_3 \to R_3 \\ \sim \end{array} \begin{bmatrix} 1 & -3 & 4 & 2 & | & 1 \\ 0 & 1 & -1 & -2 & | & -1 \\ 0 & 0 & 0 & 1 & | & 4 \end{bmatrix}$ $R_3: w = 4;$

$R_2: u - v - 2w = -1 \Rightarrow u = v + 7$; $R_1: s - 3u + 4v + 2w = 1 \Rightarrow s = 14 - v$.

● $\{(14 - t, t + 7, t, 4) \mid t \in \mathbf{R}\}$; consistent, dependent

[44] $\begin{bmatrix} 1 & -1 & -1 & 1 & | & 0 \\ 3 & 2 & -1 & 2 & | & -2 \\ -1 & -1 & 4 & 3 & | & 1 \end{bmatrix} \begin{array}{l} -3R_1 + R_2 \to R_2 \\ R_1 + R_3 \to R_3 \\ \sim \end{array} \begin{bmatrix} 1 & -1 & -1 & 1 & | & 0 \\ 0 & 5 & 2 & -1 & | & -2 \\ 0 & -2 & 3 & 4 & | & 1 \end{bmatrix} \begin{array}{l} \frac{1}{5}R_2 \to R_2 \\ \sim \end{array}$

$\begin{bmatrix} 1 & -1 & -1 & 1 & | & 0 \\ 0 & 1 & \frac{2}{5} & -\frac{1}{5} & | & -\frac{2}{5} \\ 0 & -2 & 3 & 4 & | & 1 \end{bmatrix} \begin{array}{l} 2R_2 + R_3 \to R_3 \\ \sim \end{array} \begin{bmatrix} 1 & -1 & -1 & 1 & | & 0 \\ 0 & 1 & \frac{2}{5} & -\frac{1}{5} & | & -\frac{2}{5} \\ 0 & 0 & \frac{19}{5} & \frac{18}{5} & | & \frac{1}{5} \end{bmatrix} \begin{array}{l} \frac{5}{19}R_3 \to R_3 \\ \sim \end{array}$

$\begin{bmatrix} 1 & -1 & -1 & 1 & | & 0 \\ 0 & 1 & \frac{2}{5} & -\frac{1}{5} & | & -\frac{2}{5} \\ 0 & 0 & 1 & \frac{18}{19} & | & \frac{1}{19} \end{bmatrix}$ $R_3: v + \frac{18}{19}w = \frac{1}{19} \Rightarrow v = -\frac{17}{19}w$; $R_2: u + \frac{2}{5}v - \frac{1}{5}w = -\frac{1}{5} \Rightarrow$

$u = \frac{1}{95}(-19 + 53w)$; $R_1: s - u - v + w = 0 \Rightarrow s = \frac{1}{95}(-19 - 127w)$.

● $\left\{\left(\frac{1}{95}(-19 - 127t), \frac{1}{95}(-19 + 53t), -\frac{17}{19}t, t\right) \mid t \in \mathbf{R}\right\}$; consistent, dependent

440

[45]
$$\begin{bmatrix} 1 & 0 & 1 & 0 & | & 0 \\ -2 & -1 & -1 & 1 & | & 0 \\ 1 & -2 & 1 & 0 & | & -2 \\ 0 & 1 & -1 & 1 & | & 4 \end{bmatrix} \begin{array}{l} 2R_1+R_2 \to R_2 \\ -1R_1+R_3 \to R_3 \\ \sim \end{array} \begin{bmatrix} 1 & 0 & 1 & 0 & | & 0 \\ 0 & -1 & 1 & 1 & | & 0 \\ 0 & -2 & 0 & 0 & | & -2 \\ 0 & 1 & -1 & 1 & | & 4 \end{bmatrix} \begin{array}{l} -R_2 \to R_2 \\ \sim \end{array}$$

$$\begin{bmatrix} 1 & 0 & 1 & 0 & | & 0 \\ 0 & 1 & -1 & -1 & | & 0 \\ 0 & -2 & 0 & 0 & | & -2 \\ 0 & 1 & -1 & 1 & | & 4 \end{bmatrix} \begin{array}{l} 2R_2+R_3 \to R_3 \\ -R_2+R_3 \to R_3 \\ \sim \end{array} \begin{bmatrix} 1 & 0 & 1 & 0 & | & 0 \\ 0 & 1 & -1 & -1 & | & 0 \\ 0 & 0 & -2 & -2 & | & -2 \\ 0 & 0 & 0 & 2 & | & 4 \end{bmatrix} \begin{array}{l} -\frac{1}{2}R_3 \to R_3 \\ \sim \end{array}$$

$$\begin{bmatrix} 1 & 0 & 1 & 0 & | & 0 \\ 0 & 1 & -1 & -1 & | & 0 \\ 0 & 0 & 1 & 1 & | & 1 \\ 0 & 0 & 0 & 2 & | & 4 \end{bmatrix} \begin{array}{l} \frac{1}{2}R_4 \to R_4 \\ \sim \end{array} \begin{bmatrix} 1 & 0 & 1 & 0 & | & 0 \\ 0 & 1 & -1 & -1 & | & 0 \\ 0 & 0 & 1 & 1 & | & 1 \\ 0 & 0 & 0 & 1 & | & 2 \end{bmatrix}$$

R_4: $w=2$; R_3: $v+w=1 \Rightarrow v=-1$; R_2: $u-v-w=0 \Rightarrow u=1$; R_1: $s+v=0 \Rightarrow s=1$.

● $\{(1, 1, -1, 2)\}$; consistent, independent

[46]
$$\begin{bmatrix} 1 & -1 & -1 & 1 & | & 0 \\ 1 & 0 & 2 & -3 & | & 6 \\ 0 & 2 & 0 & 4 & | & -6 \\ 0 & 1 & -1 & -1 & | & -1 \end{bmatrix} \begin{array}{l} -R_1+R_2 \to R_2 \\ \sim \end{array} \begin{bmatrix} 1 & -1 & -1 & 1 & | & 0 \\ 0 & 1 & 3 & -4 & | & 6 \\ 0 & 2 & 0 & 4 & | & -6 \\ 0 & 1 & -1 & -1 & | & -1 \end{bmatrix} \begin{array}{l} -2R_2+R_3 \to R_3 \\ -R_2+R_4 \to R_4 \\ \sim \end{array}$$

$$\begin{bmatrix} 1 & -1 & -1 & 1 & | & 0 \\ 0 & 1 & 3 & -4 & | & 6 \\ 0 & 0 & -6 & 12 & | & -18 \\ 0 & 0 & -4 & 3 & | & -7 \end{bmatrix} \begin{array}{l} -\frac{1}{6}R_3 \to R_3 \\ \sim \end{array} \begin{bmatrix} 1 & -1 & -1 & 1 & | & 0 \\ 0 & 1 & 3 & -4 & | & 6 \\ 0 & 0 & 1 & -2 & | & 3 \\ 0 & 0 & -4 & 3 & | & -7 \end{bmatrix} \begin{array}{l} 4R_3+R_4 \to R_4 \\ \sim \end{array}$$

$$\begin{bmatrix} 1 & -1 & -1 & 1 & | & 0 \\ 0 & 1 & 3 & -4 & | & 6 \\ 0 & 0 & 1 & -2 & | & 3 \\ 0 & 0 & 0 & -5 & | & 5 \end{bmatrix} \begin{array}{l} -\frac{1}{5}R_4 \to R_4 \\ \sim \end{array} \begin{bmatrix} 1 & -1 & -1 & 1 & | & 0 \\ 0 & 1 & 3 & -4 & | & 6 \\ 0 & 0 & 1 & -2 & | & 3 \\ 0 & 0 & 0 & 1 & | & -1 \end{bmatrix}$$

R_4: $s=-1$; R_3: $r-2s=3 \Rightarrow r=1$; R_2: $q+3r-4s=6 \Rightarrow q=-1$; R_1: $p-q-r+s=0 \Rightarrow p=1$.

● $\{(1, -1, 1, -1)\}$; consistent, independent

[47]
$$\begin{bmatrix} 1 & -1 & -3 & | & 5 \\ -3 & 2 & 1 & | & -9 \end{bmatrix} \begin{array}{l} 3R_1+R_2 \to R_2 \\ \sim \end{array} \begin{bmatrix} 1 & -1 & -3 & | & 5 \\ 0 & -1 & -8 & | & 6 \end{bmatrix} \begin{array}{l} -R_2 \to R_2 \\ \sim \end{array} \begin{bmatrix} 1 & -1 & -3 & | & 5 \\ 0 & 1 & 8 & | & -6 \end{bmatrix}$$

R_2: $y+8z=-6 \Rightarrow y=-6-8z$; R_1: $x-y-3z=5 \Rightarrow x=-1-5z$.

● $\{(-1-5t, -6-8t, t) \mid t \in \mathbf{R}\}$; consistent, dependent

[48]
$$\begin{bmatrix} 6 & 5 & -2 & | & -2 \\ 5 & -1 & -1 & | & 5 \end{bmatrix} \begin{array}{l} \frac{1}{6}R_1+R_2 \to R_2 \\ \sim \end{array} \begin{bmatrix} 1 & \frac{5}{6} & -\frac{1}{3} & | & -\frac{1}{3} \\ 5 & -1 & -1 & | & 5 \end{bmatrix} \begin{array}{l} -5R_1 \to R_1 \\ \sim \end{array}$$

$$\begin{bmatrix} 1 & \frac{5}{6} & -\frac{1}{3} & | & -\frac{1}{3} \\ 0 & -\frac{31}{6} & \frac{2}{3} & | & \frac{20}{3} \end{bmatrix} \begin{array}{l} -\frac{6}{31}R_2 \to R_2 \\ \sim \end{array} \begin{bmatrix} 1 & \frac{5}{6} & -\frac{1}{3} & | & -\frac{1}{3} \\ 0 & 1 & -\frac{4}{31} & | & -\frac{40}{31} \end{bmatrix}$$

R_2: $y-\frac{4}{31}z=-\frac{40}{31} \Rightarrow y=-\frac{40}{31}+\frac{4}{31}z$; R_1: $x+\frac{5}{6}y-\frac{1}{3}z=-\frac{1}{3} \Rightarrow x=\frac{23}{31}+\frac{7}{31}z$.

● $\left\{\left(\frac{1}{31}(23+7t), \frac{1}{31}(-40+4t), t\right) \mid t \in \mathbf{R}\right\}$; consistent, dependent

441

[49] $\begin{bmatrix} 1 & -1 & | & 1 \\ 4 & -5 & | & 7 \\ 5 & -2 & | & 4 \end{bmatrix} \begin{matrix} -4R_1 + R_2 \to R_2 \\ -5R_1 + R_3 \to R_3 \\ \sim \end{matrix} \begin{bmatrix} 1 & -1 & | & 1 \\ 0 & -1 & | & 3 \\ 0 & 3 & | & -1 \end{bmatrix} \begin{matrix} -R_2 \to R_2 \\ \sim \end{matrix} \begin{bmatrix} 1 & -1 & | & 1 \\ 0 & 1 & | & -3 \\ 0 & 3 & | & -1 \end{bmatrix} \begin{matrix} -3R_2 + R_3 \to R_2 \\ \sim \end{matrix}$

$\begin{bmatrix} 1 & -1 & | & 1 \\ 0 & 1 & | & -3 \\ 0 & 0 & | & 8 \end{bmatrix}$ R_3: $0 = 8$. This equation is false.   ● $\varnothing$; inconsistent

[50] $\begin{bmatrix} 3 & -2 & | & 2 \\ -1 & 4 & | & -3 \\ 5 & -6 & | & -2 \end{bmatrix} \begin{matrix} R_1 \leftrightarrow R_2 \\ \sim \end{matrix} \begin{bmatrix} -1 & 4 & | & -3 \\ 3 & -2 & | & 2 \\ 5 & -6 & | & -2 \end{bmatrix} \begin{matrix} -R_1 \to R_1 \\ \sim \end{matrix} \begin{bmatrix} 1 & -4 & | & 3 \\ 3 & -2 & | & 2 \\ 5 & -6 & | & -2 \end{bmatrix} \begin{matrix} -3R_1 + R_2 \to R_2 \\ -5R_1 + R_3 \to R_3 \\ \sim \end{matrix}$

$\begin{bmatrix} 1 & -4 & | & 3 \\ 0 & 10 & | & -7 \\ 0 & 14 & | & -17 \end{bmatrix} \begin{matrix} \frac{1}{10}R_2 \to R_2 \\ \sim \end{matrix} \begin{bmatrix} 1 & -4 & | & 3 \\ 0 & 1 & | & -\frac{7}{10} \\ 0 & 14 & | & -17 \end{bmatrix} \begin{matrix} -14R_2 + R_3 \to R_3 \\ \sim \end{matrix} \begin{bmatrix} 1 & -4 & | & 3 \\ 0 & 1 & | & -\frac{7}{10} \\ 0 & 0 & | & -\frac{36}{5} \end{bmatrix}$

R_3: $0 = -\frac{36}{5}$. This equation is false.   ● $\varnothing$; inconsistent

[51] $\begin{bmatrix} 1 & 2 & 1 & | & 8 \\ -1 & 3 & -2 & | & 1 \\ 3 & 4 & -7 & | & 10 \end{bmatrix} \begin{matrix} R_1 + R_2 \to R_2 \\ -3R_1 + R_3 \to R_3 \\ \sim \end{matrix} \begin{bmatrix} 1 & 2 & 1 & | & 8 \\ 0 & 5 & -1 & | & 9 \\ 0 & -2 & -10 & | & -14 \end{bmatrix} \begin{matrix} \frac{1}{5}R_2 \to R_2 \\ \sim \end{matrix}$

$\begin{bmatrix} 1 & 2 & 1 & | & 8 \\ 0 & 1 & -\frac{1}{5} & | & \frac{9}{5} \\ 0 & -2 & -10 & | & -14 \end{bmatrix} \begin{matrix} 2R_2 + R_3 \to R_3 \\ \sim \end{matrix} \begin{bmatrix} 1 & 2 & 1 & | & 8 \\ 0 & 1 & -\frac{1}{5} & | & \frac{9}{5} \\ 0 & 0 & -\frac{52}{5} & | & -\frac{52}{5} \end{bmatrix} \begin{matrix} -\frac{52}{5}R_3 \to R_3 \\ \sim \end{matrix}$

$\begin{bmatrix} 1 & 2 & 1 & | & 8 \\ 0 & 1 & -\frac{1}{5} & | & \frac{9}{5} \\ 0 & 0 & 1 & | & 1 \end{bmatrix} \begin{matrix} \frac{1}{5}R_3 + R_2 \to R_2 \\ \sim \end{matrix} \begin{bmatrix} 1 & 2 & 1 & | & 8 \\ 0 & 1 & 0 & | & 2 \\ 0 & 0 & 1 & | & 1 \end{bmatrix} \begin{matrix} -R_3 + R_1 \to R_1 \\ \sim \end{matrix}$

$\begin{bmatrix} 1 & 2 & 0 & | & 7 \\ 0 & 1 & 0 & | & 2 \\ 0 & 0 & 1 & | & 1 \end{bmatrix} \begin{matrix} -2R_2 + R_1 \to R_1 \\ \sim \end{matrix} \begin{bmatrix} 1 & 0 & 0 & | & 3 \\ 0 & 1 & 0 & | & 2 \\ 0 & 0 & 1 & | & 1 \end{bmatrix}$ R_1: $x = 3$; R_2: $y = 2$; R_3: $z = 1$.

● $\{(3, 2, 1)\}$; consistent, independent

[52] $\begin{bmatrix} 2 & -1 & 1 & | & 1 \\ 1 & 1 & 1 & | & 1 \\ -2 & 4 & -1 & | & 5 \end{bmatrix}$ $\begin{matrix} R_1 \leftrightarrow R_2 \\ \sim \end{matrix}$ $\begin{bmatrix} 1 & 1 & 1 & | & 1 \\ 2 & -1 & 1 & | & 1 \\ -2 & 4 & -1 & | & 5 \end{bmatrix}$ $\begin{matrix} -2R_1 + R_2 \rightarrow R_2 \\ 2R_1 + R_3 \rightarrow R_3 \\ \sim \end{matrix}$

$\begin{bmatrix} 1 & 1 & 1 & | & 1 \\ 0 & -3 & -1 & | & -1 \\ 0 & 6 & 1 & | & 7 \end{bmatrix}$ $\begin{matrix} -\frac{1}{3}R_2 \rightarrow R_2 \\ \sim \end{matrix}$ $\begin{bmatrix} 1 & 1 & 1 & | & 1 \\ 0 & 1 & \frac{1}{3} & | & \frac{1}{3} \\ 0 & 6 & 1 & | & 7 \end{bmatrix}$ $\begin{matrix} -6R_2 + R_3 \rightarrow R_3 \\ \sim \end{matrix}$

$\begin{bmatrix} 1 & 1 & 1 & | & 1 \\ 0 & 1 & \frac{1}{3} & | & \frac{1}{3} \\ 0 & 0 & -1 & | & 5 \end{bmatrix}$ $\begin{matrix} -R_3 \rightarrow R_3 \\ \sim \end{matrix}$ $\begin{bmatrix} 1 & 1 & 1 & | & 1 \\ 0 & 1 & \frac{1}{3} & | & \frac{1}{3} \\ 0 & 0 & 1 & | & -5 \end{bmatrix}$ $\begin{matrix} -\frac{1}{3}R_3 + R_2 \rightarrow R_2 \\ \sim \end{matrix}$

$\begin{bmatrix} 1 & 1 & 1 & | & 1 \\ 0 & 1 & 0 & | & 2 \\ 0 & 0 & 1 & | & -5 \end{bmatrix}$ $\begin{matrix} -R_3 + R_1 \rightarrow R_1 \\ \sim \end{matrix}$ $\begin{bmatrix} 1 & 1 & 0 & | & 6 \\ 0 & 1 & 0 & | & 2 \\ 0 & 0 & 1 & | & -5 \end{bmatrix}$ $\begin{matrix} -R_2 + R_1 \rightarrow R_1 \\ \sim \end{matrix}$ $\begin{bmatrix} 1 & 0 & 0 & | & 4 \\ 0 & 1 & 0 & | & 2 \\ 0 & 0 & 1 & | & -5 \end{bmatrix}$

R_1: $x = 4$; R_2: $y = 2$; R_3: $y = -5$.   ● $\{(4, 2, -5)\}$; consistent; independent

[53] $\begin{bmatrix} 1 & 1 & 0 & | & 2 \\ 0 & 1 & 1 & | & 3 \\ 1 & 0 & -1 & | & 1 \end{bmatrix}$ $\begin{matrix} -R_1 + R_3 \rightarrow R_3 \\ \sim \end{matrix}$ $\begin{bmatrix} 1 & 1 & 0 & | & 2 \\ 0 & 1 & 1 & | & 3 \\ 0 & -1 & -1 & | & -1 \end{bmatrix}$ $\begin{matrix} R_2 + R_3 \rightarrow R_3 \\ \sim \end{matrix}$ $\begin{bmatrix} 1 & 1 & 0 & | & 2 \\ 0 & 1 & 1 & | & 3 \\ 0 & 0 & 0 & | & 2 \end{bmatrix}$

R_3: $0 = 2$. This equation is false.   ● $\varnothing$; inconsistent

[54] $\begin{bmatrix} 1 & 1 & 1 & -1 & | & 1 \\ 2 & -3 & 1 & -2 & | & -4 \\ 3 & 1 & -2 & 1 & | & 4 \end{bmatrix}$ $\begin{matrix} -2R_1 + R_2 \rightarrow R_2 \\ -3R_1 + R_3 \rightarrow R_3 \\ \sim \end{matrix}$ $\begin{bmatrix} 1 & 1 & 1 & -1 & | & 1 \\ 0 & -5 & -1 & 0 & | & -6 \\ 0 & -2 & -5 & 4 & | & 1 \end{bmatrix}$ $\begin{matrix} -\frac{1}{5}R_2 \rightarrow R_2 \\ \sim \end{matrix}$

$\begin{bmatrix} 1 & 1 & 1 & -1 & | & 1 \\ 0 & 1 & \frac{1}{5} & 0 & | & \frac{6}{5} \\ 0 & -2 & -5 & 4 & | & 1 \end{bmatrix}$ $\begin{matrix} 2R_2 + R_3 \rightarrow R_3 \\ \sim \end{matrix}$ $\begin{bmatrix} 1 & 1 & 1 & -1 & | & 1 \\ 0 & 1 & \frac{1}{5} & 0 & | & \frac{6}{5} \\ 0 & 0 & -\frac{23}{5} & 4 & | & \frac{17}{5} \end{bmatrix}$ $\begin{matrix} -\frac{5}{23}R_3 \rightarrow R_3 \\ \sim \end{matrix}$

$\begin{bmatrix} 1 & 1 & 1 & -1 & | & 1 \\ 0 & 1 & \frac{1}{5} & 0 & | & \frac{6}{5} \\ 0 & 0 & 1 & -\frac{20}{23} & | & -\frac{17}{23} \end{bmatrix}$ $\begin{matrix} -\frac{1}{5}R_3 + R_2 \rightarrow R_2 \\ -R_3 + R_1 \rightarrow R_1 \\ \sim \end{matrix}$ $\begin{bmatrix} 1 & 1 & 0 & -\frac{3}{23} & | & \frac{40}{23} \\ 0 & 1 & 0 & \frac{4}{23} & | & \frac{155}{115} \\ 0 & 0 & 1 & -\frac{20}{23} & | & -\frac{17}{23} \end{bmatrix}$ $\begin{matrix} -R_2 + R_1 \rightarrow R_1 \\ \sim \end{matrix}$

$\begin{bmatrix} 1 & 0 & 0 & -\frac{7}{23} & | & \frac{45}{115} \\ 0 & 1 & 0 & \frac{4}{23} & | & \frac{155}{115} \\ 0 & 0 & 1 & -\frac{20}{23} & | & -\frac{17}{23} \end{bmatrix}$ R_3: $r - \frac{20}{23}s = \frac{17}{23} \Rightarrow r = -\frac{17}{23} + \frac{20}{23}s$,

R_2: $q + \frac{4}{23}s = \frac{155}{115} \Rightarrow q = \frac{155}{115} - \frac{4}{23}s$; R_1: $p = \frac{7}{23}s = \frac{45}{115} \Rightarrow p = \frac{45}{115} + \frac{7}{23}s$.

● $\left\{ \left(\frac{1}{115}(45 + 35t), \frac{1}{115}(155 - 20t), \frac{1}{23}(-17 + 20t), t \right) \middle| t \in \mathbf{R} \right\}$; consistent; dependent

443

[55] $\begin{bmatrix} 3 & -1 & -1 & 4 & | & 2 \\ 6 & 3 & -1 & 4 & | & 3 \\ 9 & 1 & 0 & -8 & | & 6 \end{bmatrix}$ $\frac{1}{3}R_1 \to R_1$ $\sim$ $\begin{bmatrix} 1 & -\frac{1}{3} & -\frac{1}{3} & \frac{4}{3} & | & \frac{2}{3} \\ 6 & 3 & -1 & 4 & | & 3 \\ 9 & 1 & 0 & -8 & | & 6 \end{bmatrix}$ $\begin{array}{c} -6R_1 + R_2 \to R_2 \\ -9R_1 + R_3 \to R_3 \\ \sim \end{array}$

$\begin{bmatrix} 1 & -\frac{1}{3} & -\frac{1}{3} & \frac{4}{3} & | & \frac{2}{3} \\ 0 & 5 & 1 & -4 & | & -1 \\ 0 & 4 & 3 & -20 & | & 0 \end{bmatrix}$ $\frac{1}{5}R_2 \to R_2$ $\sim$ $\begin{bmatrix} 1 & -\frac{1}{3} & -\frac{1}{3} & \frac{4}{3} & | & \frac{2}{3} \\ 0 & 1 & \frac{1}{5} & -\frac{4}{5} & | & -\frac{1}{5} \\ 0 & 4 & 3 & -20 & | & 0 \end{bmatrix}$ $\begin{array}{c} -4R_2 + R_3 \to R_3 \\ \sim \end{array}$

$\begin{bmatrix} 1 & -\frac{1}{3} & -\frac{1}{3} & \frac{4}{3} & | & \frac{2}{3} \\ 0 & 1 & \frac{1}{5} & -\frac{4}{5} & | & -\frac{1}{5} \\ 0 & 0 & \frac{11}{5} & -\frac{84}{5} & | & \frac{4}{5} \end{bmatrix}$ $\frac{5}{11}R_3 \to R_3$ $\sim$ $\begin{bmatrix} 1 & -\frac{1}{3} & -\frac{1}{3} & \frac{4}{3} & | & \frac{2}{3} \\ 0 & 1 & \frac{1}{5} & -\frac{4}{5} & | & -\frac{1}{5} \\ 0 & 0 & 1 & -\frac{84}{11} & | & \frac{4}{11} \end{bmatrix}$ $\begin{array}{c} -\frac{1}{5}R_3 + R_2 \to R_2 \\ \frac{1}{3}R_3 + R_1 \to R_1 \\ \sim \end{array}$

$\begin{bmatrix} 1 & -\frac{1}{3} & 0 & -\frac{40}{33} & | & \frac{26}{33} \\ 0 & 1 & 0 & \frac{8}{11} & | & -\frac{3}{11} \\ 0 & 0 & 1 & -\frac{84}{11} & | & \frac{4}{11} \end{bmatrix}$ $\frac{1}{3}R_2 + R_1 \to R_1$ $\sim$ $\begin{bmatrix} 1 & 0 & 0 & -\frac{32}{33} & | & \frac{23}{33} \\ 0 & 1 & 0 & \frac{8}{11} & | & -\frac{3}{11} \\ 0 & 0 & 1 & -\frac{84}{11} & | & \frac{4}{11} \end{bmatrix}$

$R_1 : p - \frac{32}{33}s = \frac{23}{33} \Rightarrow p = \frac{23}{33} + \frac{32}{33}s;\ R_2 : q + \frac{8}{11}s = -\frac{3}{11} \Rightarrow q = -\frac{3}{11} - \frac{8}{11}s;$

$R_3 : r - \frac{84}{11}s = \frac{4}{11} \Rightarrow r = \frac{4}{11} + \frac{84}{11}s.$

● $\left\{ \left(\frac{23}{33} + \frac{32}{33}t, -\frac{3}{11} - \frac{8}{11}t, \frac{4}{11} + \frac{84}{11}t, t \right) \middle| t \in \mathbf{R} \right\}$; consistent, dependent

[56] $\begin{bmatrix} 2 & -3 & 1 & -1 & 1 & | & 0 \\ 4 & -6 & 2 & -3 & -1 & | & -5 \\ -2 & 3 & -2 & 2 & -1 & | & 3 \end{bmatrix}$ $\frac{1}{2}R_1 \to R_1$ $\sim$ $\begin{bmatrix} 1 & -\frac{3}{2} & \frac{1}{2} & -\frac{1}{2} & \frac{1}{2} & | & 0 \\ 4 & -6 & 2 & -3 & -1 & | & -5 \\ -2 & 3 & -2 & 2 & -1 & | & 3 \end{bmatrix}$ $\begin{array}{c} -4R_1 + R_2 \to R_2 \\ 2R_1 + R_3 \to R_3 \\ \sim \end{array}$

$\begin{bmatrix} 1 & -\frac{3}{2} & \frac{1}{2} & -\frac{1}{2} & \frac{1}{2} & | & 0 \\ 0 & 0 & 0 & -1 & -3 & | & -5 \\ 0 & 0 & -1 & 1 & 0 & | & 3 \end{bmatrix}$ $R_2 \leftrightarrow R_3$ $\sim$ $\begin{bmatrix} 1 & -\frac{3}{2} & \frac{1}{2} & -\frac{1}{2} & \frac{1}{2} & | & 0 \\ 0 & 0 & -1 & 1 & 0 & | & 3 \\ 0 & 0 & 0 & -1 & -3 & | & -5 \end{bmatrix}$ $\begin{array}{c} -R_2 \leftrightarrow R_2 \\ \sim \end{array}$

$\begin{bmatrix} 1 & -\frac{3}{2} & \frac{1}{2} & -\frac{1}{2} & \frac{1}{2} & | & 0 \\ 0 & 0 & 1 & -1 & 0 & | & -3 \\ 0 & 0 & 0 & -1 & -3 & | & -5 \end{bmatrix}$ $-R_3 \leftrightarrow R_3$ $\sim$ $\begin{bmatrix} 1 & -\frac{3}{2} & \frac{1}{2} & -\frac{1}{2} & \frac{1}{2} & | & 0 \\ 0 & 0 & 1 & -1 & 0 & | & -3 \\ 0 & 0 & 0 & 1 & 3 & | & 5 \end{bmatrix}$ $\begin{array}{c} R_3 + R_2 \to R_2 \\ \frac{1}{2}R_3 + R_1 \to R_1 \\ \sim \end{array}$

$\begin{bmatrix} 1 & -\frac{3}{2} & \frac{1}{2} & 0 & 2 & | & \frac{5}{2} \\ 0 & 0 & 1 & 0 & 3 & | & 2 \\ 0 & 0 & 0 & 1 & 3 & | & 5 \end{bmatrix}$ $-\frac{1}{2}R_2 + R_1 \to R_1$ $\sim$ $\begin{bmatrix} 1 & -\frac{3}{2} & 0 & 0 & \frac{1}{2} & | & \frac{3}{2} \\ 0 & 0 & 1 & 0 & 3 & | & 2 \\ 0 & 0 & 0 & 1 & 3 & | & 5 \end{bmatrix}$ $R_3 : v + 3w = 5 \Rightarrow v = 5 - 3w;$

$R_2 : u + 3w = 2 \Rightarrow u = 2 - 3w;\ R_1 : r - \frac{3}{2}s + \frac{1}{2}w = \frac{3}{2} \Rightarrow r = \frac{3}{2} + \frac{3}{2}s - \frac{1}{2}w.$

● $\left\{ \left(\frac{3}{2} + \frac{3}{2}z - \frac{1}{2}t, z, 2 - 3t, 5 - 3t, t \right) \middle| t, z \in \mathbf{R} \right\}$; consistent, dependent

[57] $\begin{cases} 9\,a^2 - 3\,b + c = -15 \\ a^2 + b + c = -3 \\ 4\,a^2 + 2\,b + c = -5 \end{cases}$ $\begin{array}{l} f(-3) = -15 \\ f(1) = -3 \\ f(2) = -5 \end{array}$. Interchanging E_1 and E_2, the argumented matrix for the

resulting system of equations is $\left[\begin{array}{ccc|c} 1 & 1 & 1 & -3 \\ 9 & -3 & 1 & -15 \\ 4 & 2 & 1 & -5 \end{array}\right]\begin{array}{l} -9\,R_1 + R_2 \to R_2 \\ -4\,R_1 + R_3 \to R_3 \\ \sim \end{array}$

$\left[\begin{array}{ccc|c} 1 & 1 & 1 & -3 \\ 0 & -12 & -8 & 12 \\ 0 & -2 & -3 & 7 \end{array}\right]\begin{array}{l} -\frac{1}{12}R_2 \to R_2 \\ \sim \end{array} \left[\begin{array}{ccc|c} 1 & 1 & 1 & -3 \\ 0 & 1 & \frac{2}{3} & -1 \\ 0 & -2 & -3 & 7 \end{array}\right]\begin{array}{l} 2\,R_2 + R_3 \to R_3 \\ \sim \end{array}$

$\left[\begin{array}{ccc|c} 1 & 1 & 1 & -3 \\ 0 & 1 & \frac{2}{3} & -1 \\ 0 & 0 & -\frac{5}{3} & 5 \end{array}\right]\begin{array}{l} -\frac{3}{5}R_3 \to R_3 \\ \sim \end{array} \left[\begin{array}{ccc|c} 1 & 1 & 1 & -3 \\ 0 & 1 & \frac{2}{3} & -1 \\ 0 & 0 & 1 & -3 \end{array}\right] R_3\colon c = -3;\ R_2\colon b + \frac{2}{3}c = -1 \Rightarrow b = 1;$

$R_1\colon a + b + c = -3 \Rightarrow a = -1.$ ● $f(x) = -x^2 + x - 3.$

[58] $\begin{cases} 4\,a - 2\,b + c = -14 \\ 9\,a + 3\,b + c = 16 \\ 25\,a + 5\,b + c = 28 \end{cases}$ $\begin{array}{l} f(-2) = -14 \\ f(3) = 16 \\ f(5) = 28 \end{array}$ $\left[\begin{array}{ccc|c} 4 & -2 & 1 & -14 \\ 9 & 3 & 1 & 16 \\ 25 & 5 & 1 & 28 \end{array}\right]\begin{array}{l} \frac{1}{4}R_1 \to R_1 \\ \sim \end{array}$

$\left[\begin{array}{ccc|c} 1 & -\frac{1}{2} & \frac{1}{4} & -\frac{7}{2} \\ 9 & 3 & 1 & 16 \\ 25 & 5 & 1 & 28 \end{array}\right]\begin{array}{l} -9\,R_1 + R_2 \to R_2 \\ -25\,R_1 + R_3 \to R_3 \\ \sim \end{array} \left[\begin{array}{ccc|c} 1 & -\frac{1}{2} & \frac{1}{4} & -\frac{7}{2} \\ 0 & \frac{15}{2} & -\frac{5}{4} & \frac{95}{2} \\ 0 & \frac{35}{2} & -\frac{21}{4} & \frac{231}{2} \end{array}\right]\begin{array}{l} \frac{2}{15}R_2 \to R_2 \\ \sim \end{array}$

$\left[\begin{array}{ccc|c} 1 & -\frac{1}{2} & \frac{1}{4} & -\frac{7}{2} \\ 0 & 1 & -\frac{1}{6} & \frac{19}{3} \\ 0 & \frac{35}{2} & -\frac{21}{4} & \frac{231}{2} \end{array}\right]\begin{array}{l} -\frac{35}{2}R_2 + R_3 \to R_3 \\ \sim \end{array} \left[\begin{array}{ccc|c} 1 & -\frac{1}{2} & \frac{1}{4} & -\frac{7}{2} \\ 0 & 1 & -\frac{1}{6} & \frac{19}{3} \\ 0 & 0 & -\frac{7}{3} & \frac{14}{3} \end{array}\right]\begin{array}{l} -\frac{3}{7}R_3 \to R_3 \\ \sim \end{array}$

$\left[\begin{array}{ccc|c} 1 & -\frac{1}{2} & \frac{1}{4} & -\frac{7}{2} \\ 0 & 1 & -\frac{1}{6} & \frac{19}{3} \\ 0 & 0 & 1 & -2 \end{array}\right] R_3\colon c = -2;\ R_2\colon b - \frac{1}{6}c = \frac{19}{3} \Rightarrow b = 6;\ R_1\colon a - \frac{1}{2}b + \frac{1}{4}c = -\frac{7}{2} \Rightarrow a = 0.$

Hence, $f(x) = 6\,x - 2$, which is a linear function.

[59] $\begin{cases} a + b + c = -1 \\ 4\,a + c = -2 \\ 6\,a - 2\,b + c = 0 \end{cases}$ $\begin{array}{l} \underline{\text{Using}\,(1,1,-1)} \\ \underline{\text{Using}\,(4,0,-2)} \\ \underline{\text{Using}\,(6,-2,0)} \end{array}$ $\left[\begin{array}{ccc|c} 1 & 1 & 1 & -1 \\ 4 & 0 & 1 & -2 \\ 6 & -2 & 1 & 0 \end{array}\right]\begin{array}{l} -4\,R_1 + R_2 \to R_2 \\ -6\,R_1 + R_3 \to R_3 \\ \sim \end{array}$

$\left[\begin{array}{ccc|c} 1 & 1 & 1 & -1 \\ 0 & -4 & -3 & 2 \\ 0 & -8 & -5 & 6 \end{array}\right]\begin{array}{l} -\frac{1}{4}R_2 \to R_2 \\ \sim \end{array} \left[\begin{array}{ccc|c} 1 & 1 & 1 & -1 \\ 0 & 1 & \frac{3}{4} & -\frac{1}{2} \\ 0 & -8 & -5 & 6 \end{array}\right]\begin{array}{l} 8\,R_2 + R_3 \to R_3 \\ \sim \end{array} \left[\begin{array}{ccc|c} 1 & 1 & 1 & -1 \\ 0 & 1 & \frac{3}{4} & -\frac{1}{2} \\ 0 & 0 & 1 & 2 \end{array}\right]$

$R_3\colon c = 2;\ R_2\colon b + \frac{3}{4}c = -\frac{1}{2} \Rightarrow b = -2;\ R_1\colon a + b + c = -1 \Rightarrow a = -1.$ ● $z = -x - 2\,y + 2$

[60] $\begin{cases} 5a + b + c = -2 \\ -2a + 3b + c = 0 \\ a + c = 2 \end{cases}$ Using $(5, 1, -2)$
Using $(-2, 3, 0)$. Interchanging E_1 and E_3, we get the following
Using $(1, 0, 2)$

argumented matrix $\begin{bmatrix} 1 & 0 & 1 & | & 2 \\ -2 & 3 & 1 & | & 0 \\ 5 & 1 & 1 & | & -2 \end{bmatrix}$ $\begin{matrix} 2R_1 + R_2 \to R_2 \\ -5R_1 + R_3 \to R_3 \\ \sim \end{matrix}$ $\begin{bmatrix} 1 & 0 & 1 & | & 2 \\ 0 & 3 & 3 & | & 4 \\ 0 & 1 & -4 & | & -12 \end{bmatrix}$ $\frac{1}{3}R_2 \to R_2$
$\sim$

$\begin{bmatrix} 1 & 0 & 1 & | & 2 \\ 0 & 1 & 1 & | & \frac{4}{3} \\ 0 & 1 & -4 & | & -12 \end{bmatrix}$ $\begin{matrix} -R_2 + R_3 \to R_3 \\ \sim \end{matrix}$ $\begin{bmatrix} 1 & 0 & 1 & | & 2 \\ 0 & 1 & 1 & | & \frac{4}{3} \\ 0 & 0 & -5 & | & -\frac{40}{3} \end{bmatrix}$ $\begin{matrix} -\frac{1}{5}R_3 \to R_3 \\ \sim \end{matrix}$ $\begin{bmatrix} 1 & 0 & 1 & | & 2 \\ 0 & 1 & 1 & | & \frac{4}{3} \\ 0 & 0 & 1 & | & \frac{8}{3} \end{bmatrix}$

$R_3: c = \frac{8}{3}$; $R_2: b + c = \frac{4}{3} \Rightarrow b = -\frac{4}{3}$; $R_1: a + c = 2 \Rightarrow a = -\frac{2}{3}$. ● $z = -\frac{2}{3}x - \frac{4}{3}y + \frac{8}{3}$

[61] a) $\begin{bmatrix} 1 & -1 & -3 & | & k \\ 2 & -3 & 4 & | & 0 \\ 3 & -4 & 1 & | & 1 \end{bmatrix}$ $\begin{matrix} -2R_1 + R_2 \to R_2 \\ -3R_1 + R_3 \to R_3 \\ \sim \end{matrix}$ $\begin{bmatrix} 1 & -1 & -3 & | & k \\ 0 & -1 & 10 & | & -2k \\ 0 & -1 & 10 & | & 1-3k \end{bmatrix}$ $\begin{matrix} -1R_2 \to R_2 \\ \sim \end{matrix}$

$\begin{bmatrix} 1 & -1 & -3 & | & k \\ 0 & 1 & -10 & | & 2k \\ 0 & -1 & 10 & | & 1-3k \end{bmatrix}$ $\begin{matrix} R_2 + R_3 \to R_2 \\ \sim \end{matrix}$ $\begin{bmatrix} 1 & -1 & -3 & | & k \\ 0 & 1 & -10 & | & 2k \\ 0 & 0 & 0 & | & 1-k \end{bmatrix}$ $R_3: 0 = 1-k \Rightarrow k = 1$;

$R_2: y - 10z = 2 \Rightarrow y = 2 + 10z$; $R_1: x - y - 3z = 1 \Rightarrow x = 3 + 13z$. If $z = 0$, we get $(3, 2, 0)$.

If $z = 1$, we get $(16, 12, 1)$. ● $k = 1$; $(3, 2, 0)$; $(16, 12, 1)$

b) From part (a) we know that $R_3 \Rightarrow 0 = 1 - k$. If $k \neq 1$, $1 - k \neq 0$ and

the equation $0 = 1 - k$ is false.

[62] $\begin{bmatrix} 1 & 1 & 1 & | & k \\ k & 1 & 2 & | & 2 \\ 1 & -k & 1 & | & 4 \end{bmatrix}$ $\begin{matrix} -kR_1 + R_2 \to R_2 \\ -R_1 + R_3 \to R_3 \\ \sim \end{matrix}$ $\begin{bmatrix} 1 & 1 & 1 & | & k \\ 0 & 1-k & 2-k & | & 2-k^2 \\ 0 & 1-k & 0 & | & 4-k \end{bmatrix}$ $\begin{matrix} -R_2 + R_3 \to R_3 \\ \sim \end{matrix}$

$\begin{bmatrix} 1 & 1 & 1 & | & k \\ 0 & 1-k & 2-k & | & 2-k^2 \\ 0 & 0 & k-2 & | & k^2-k-2 \end{bmatrix}$ $\begin{matrix} \frac{1}{k-2}R_3 \to R_3 \\ \text{(If } k \neq 2) \\ \sim \end{matrix}$ $\begin{bmatrix} 1 & 1 & 1 & | & k \\ 0 & 1-k & 2-k & | & 2-k^2 \\ 0 & 0 & 1 & | & k+1 \end{bmatrix}$ $R_3: z = k + 1$;

$R_2: (1-k)y + (z-k)z = 2-k^2 \Rightarrow y = -\frac{k}{1-k}$; $R_3: x + y + z = k \Rightarrow x = \frac{2k-1}{1-k}$. Hence $k \neq 1$ and $k \neq 2$.

● $\{k \mid k \in \mathbf{R}, k \neq 1, 2\}$

[63] Let x represent the amount borrowed at 8%, which is also the amount borrowed at 10% and

y the amount borrowed at 12%. $\begin{cases} 2x + y = 500,000 \\ 0.18x + 0.12y = 48000 \end{cases}$. Solving E_1 for y and substituting into

E_2 yields $0.18x + 0.12(5000000 - 2x) = 48000 \Rightarrow x = 200,000$; $y = 100,000$.

● $200,000 at 8%; $200,000 at 10%; $100,000 at 12%

[64]
$$\begin{bmatrix} 4 & 5 & 4 & 3 & | & 298 \\ 5 & 6 & 7 & 6 & | & 450 \\ 10 & 8 & 10 & 9 & | & 682 \\ 15 & 12 & 9 & 15 & | & 924 \end{bmatrix} \begin{matrix} R_1 \leftrightarrow R_2 \\ \sim \end{matrix} \begin{bmatrix} 5 & 6 & 7 & 6 & | & 450 \\ 4 & 5 & 4 & 3 & | & 298 \\ 10 & 8 & 10 & 9 & | & 682 \\ 15 & 12 & 9 & 15 & | & 924 \end{bmatrix} \begin{matrix} -R_2 + R_1 \rightarrow R_1 \\ \sim \end{matrix}$$

$$\begin{bmatrix} 1 & 1 & 3 & 3 & | & 152 \\ 4 & 5 & 4 & 3 & | & 298 \\ 10 & 8 & 10 & 9 & | & 682 \\ 15 & 12 & 9 & 15 & | & 924 \end{bmatrix} \begin{matrix} -4R_1 + R_2 \rightarrow R_2 \\ -10R_1 + R_3 \rightarrow R_3 \\ -15R_1 + R_4 \rightarrow R_4 \\ \sim \end{matrix} \begin{bmatrix} 1 & 1 & 3 & 3 & | & 152 \\ 0 & 1 & -8 & -9 & | & -310 \\ 0 & -2 & -20 & -21 & | & -838 \\ 0 & -3 & -36 & -30 & | & -1356 \end{bmatrix} \begin{matrix} 2R_2 + R_3 \rightarrow R_3 \\ 3R_2 + R_4 \rightarrow R_4 \\ \sim \end{matrix}$$

$$\begin{bmatrix} 1 & 1 & 3 & 3 & | & 152 \\ 0 & 1 & -8 & -9 & | & -310 \\ 0 & 0 & -36 & -39 & | & -1458 \\ 0 & 0 & -60 & -57 & | & -2286 \end{bmatrix} \begin{matrix} -\frac{1}{36}R_3 \rightarrow R_3 \\ \sim \end{matrix} \begin{bmatrix} 1 & 1 & 3 & 3 & | & 152 \\ 0 & 1 & -8 & -9 & | & -310 \\ 0 & 0 & 1 & \frac{13}{12} & | & \frac{81}{2} \\ 0 & 0 & -60 & -57 & | & -2286 \end{bmatrix} \begin{matrix} 60R_3 \rightarrow R_4 \\ \sim \end{matrix}$$

$$\begin{bmatrix} 1 & 1 & 3 & 3 & | & 152 \\ 0 & 1 & -8 & -9 & | & -310 \\ 0 & 0 & 1 & \frac{13}{12} & | & \frac{81}{2} \\ 0 & 0 & 0 & 8 & | & 144 \end{bmatrix} \begin{matrix} \frac{1}{4}R_4 \rightarrow R_4 \\ \sim \end{matrix} \begin{bmatrix} 1 & 1 & 3 & 3 & | & 152 \\ 0 & 1 & -8 & -9 & | & -310 \\ 0 & 0 & 1 & \frac{13}{12} & | & \frac{81}{2} \\ 0 & 0 & 0 & 1 & | & 18 \end{bmatrix} R_4: x_4 = 18;$$

$R_3: x_3 + \frac{13}{12}x_4 = \frac{81}{2} \Rightarrow x_3 = 21; R_2: x_2 - 8x_3 - 9x_2 = -310 \Rightarrow x_2 = 20;$

$R_1: x_1 + x_2 + 3x_3 + 3x_4 = 152 \Rightarrow x = 15.$ ● $x_1 = 15; x_2 = 20; x_3 = 21; x_4 = 18$

[65]   Let x represent the longest side; y, the next longest; z, the shortest. $\begin{cases} x + y + z = 74 & \underline{\text{Perimeter}} \\ x - y = 6 \\ x - z = 10 \end{cases}$

$$\begin{bmatrix} 1 & 1 & 1 & | & 74 \\ 1 & -1 & & | & 6 \\ 1 & & -1 & | & 10 \end{bmatrix} \begin{matrix} -R_1 + R_2 \rightarrow R_2 \\ -R_1 + R_3 \rightarrow R_3 \\ \sim \end{matrix} \begin{bmatrix} 1 & 1 & 1 & | & 74 \\ 0 & -2 & -1 & | & -68 \\ 0 & -1 & -2 & | & -64 \end{bmatrix} \begin{matrix} -\frac{1}{2}R_2 \rightarrow R_2 \\ \sim \end{matrix}$$

$$\begin{bmatrix} 1 & 1 & 1 & | & 74 \\ 0 & 1 & \frac{1}{2} & | & 34 \\ 0 & -1 & -2 & | & -64 \end{bmatrix} \begin{matrix} R_2 + R_3 \rightarrow R_3 \\ \sim \end{matrix} \begin{bmatrix} 1 & 1 & 1 & | & 74 \\ 0 & 1 & \frac{1}{2} & | & 34 \\ 0 & 0 & -\frac{3}{2} & | & -30 \end{bmatrix} \begin{matrix} -\frac{2}{3}R_3 \rightarrow R_3 \\ \sim \end{matrix} \begin{bmatrix} 1 & 1 & 1 & | & 74 \\ 0 & 1 & \frac{1}{2} & | & 34 \\ 0 & 0 & 1 & | & 20 \end{bmatrix}$$

$R_3: z = 20; R_2: y + \frac{1}{2}z = 34 \Rightarrow y = 24; R_1: x + y + z = 74 \Rightarrow x = 30.$ ● $30, 24, 20$

[66]   Let x represent the number of grams of food group I; y, the number of grams of food group II;

z, the number of grams of food group III. First diet: $\begin{cases} 0.12x + 0.05y + 0.06z = 47 \\ 0.06x + 0.05y + 0.01z = 20 \\ 0.08x + 0.02y + 0.04z = 30 \end{cases}$. It can be shown

that the solution set of this system of equations is $\{(200, 100, 300)\}$. Hence, the nutritionist chooses

200 grams from group I, 100 grams from group II and 300 grams from group III. Second diet:

$\begin{cases} 0.12x + 0.05y + 0.06z = 53 \\ 0.06x + 0.05y + 0.01z = 25 \\ 0.08x + 0.02y + 0.04z = 34 \end{cases}$. The solution set is $\{(300, 100, 200)\}$. Choose 300 grams from

group I, 100 grams from group II and 200 grams from group III.

[67] **a)** Intersection A: $x_1 + 360 = x_2 + 450 \Rightarrow x_1 - x_2 = 90$. Intersection B: $x_4 + 300 = x_1 + 270 \Rightarrow$

$x_1 - x_4 = 30$. Intersection C: $x_2 + 400 = x_3 + 390 \Rightarrow x_2 - x_3 = -10$. Intersection D: $x_3 + 400 =$

$$x_4 + 350 \Rightarrow x_3 - x_4 = -50. \begin{cases} x_1 - x_2 & = 90 \\ x_1 \quad - x_4 = 30 \\ x_2 - x_3 & = -10 \\ x_3 - x_4 = -50 \end{cases}$$

b) $\begin{bmatrix} 1 & -1 & 0 & 0 & | & 90 \\ 1 & 0 & 0 & -1 & | & 30 \\ 0 & 1 & -1 & 0 & | & -10 \\ 0 & 0 & 1 & -1 & | & -50 \end{bmatrix} \begin{array}{c} -R_1 + R_2 \to R_4 \\ \sim \end{array} \begin{bmatrix} 1 & -1 & 0 & 0 & | & 90 \\ 0 & 1 & 0 & -1 & | & -60 \\ 0 & 1 & -1 & 0 & | & -10 \\ 0 & 0 & 1 & -1 & | & -50 \end{bmatrix} \begin{array}{c} -R_2 + R_3 \to R_3 \\ \sim \end{array}$

$\begin{bmatrix} 1 & -1 & 0 & 0 & | & 90 \\ 0 & 1 & 0 & -1 & | & -60 \\ 0 & 0 & -1 & 1 & | & 50 \\ 0 & 0 & 1 & -1 & | & -50 \end{bmatrix} \begin{array}{c} -R_3 \to R_3 \\ \sim \end{array} \begin{bmatrix} 1 & -1 & 0 & 0 & | & 90 \\ 0 & 1 & 0 & -1 & | & -60 \\ 0 & 0 & 1 & -1 & | & 50 \\ 0 & 0 & 1 & -1 & | & -50 \end{bmatrix} \begin{array}{c} -R_3 + R_4 \\ \sim \end{array}$

$\begin{bmatrix} 1 & -1 & 0 & 0 & | & 90 \\ 0 & 1 & 0 & -1 & | & -60 \\ 0 & 0 & 1 & -1 & | & -50 \\ 0 & 0 & 0 & 0 & | & 0 \end{bmatrix} \begin{array}{l} R_3: x_3 - x_4 = -50 \Rightarrow x_3 = x_4 - 50 \\ R_2: x_2 - x_4 = -60 \Rightarrow x_2 = x_4 - 60 \\ R_1: x_1 - x_2 = 90 \Rightarrow x_1 = x_4 + 30 \end{array}$

● $\{t + 30, t - 60, t - 50, t \mid t \text{ is a positive integer} \geq 60\}$

[68] Let x represent the amount of federal tax paid, let y represent the amount of state tax and let z

represent the amount of local tax. Then, $\begin{cases} 0.30(5{,}000{,}000 - y - z) = x & \underline{\text{Federal tax}} \\ 0.10(5{,}000{,}000 - x - z) = y & \underline{\text{State tax}} \quad \text{Hence,} \\ 0.05(5{,}000{,}000 - x - y) = z & \underline{\text{Local tax}} \end{cases}$

$\begin{cases} x + 0.30y + 0.30z = 1{,}500{,}000 \\ 0.10x + y + 0.10z = 500{,}000 \quad \text{It follows that } x \approx \$1{,}345{,}750, y \approx \$348{,}898 \text{ and } z \approx \$165{,}267. \\ 0.05x + 0.05y + z = 250{,}000 \end{cases}$

EXERCISES 9.3

[1] Equating corresponding components, we have $x = -5$ and $y = -2$ ● $x = -5; y = -2$

[2] Equating corresponding components, we have $x = -4$ and $-y = y + 2 \Rightarrow y = -1$.   ● $x = -4; y = -1$

[3] Equating corresponding components, we get $4x = 2$ and $-16x = -8 \Rightarrow x = \frac{1}{2}$, and $y + 2 = 6 \Rightarrow$

$y = 4$.   ● $x = \frac{1}{2}; y = 4$

[4] Equating corresponding components, we get $2x + 3 = x + 1, -5 = 2x - 1, y - 4 = 2y + 5$ and

$-y + 3 = 12$. It follows that $x = -2$ and $y = -9$.   ● $x = -2; y = -9$

[5] $A + B = \begin{bmatrix} 3 & 5 \\ -13 & 12 \end{bmatrix}; A - B = \begin{bmatrix} 5 & 5 \\ 1 & -8 \end{bmatrix}; 3a = \begin{bmatrix} 12 & 15 \\ -18 & 6 \end{bmatrix};$

$3A - 4B = \begin{bmatrix} 12 & 15 \\ -18 & 6 \end{bmatrix} - \begin{bmatrix} -4 & 0 \\ -28 & 40 \end{bmatrix} = \begin{bmatrix} 16 & 15 \\ 10 & -34 \end{bmatrix}$

[6] $\quad 3A = \begin{bmatrix} -3 & 3 & 3 \\ 6 & 0 & 3 \\ -3 & 0 & 6 \end{bmatrix}$

[7] $\quad 3A = \begin{bmatrix} 6 & 3 & 18 & 5 \\ 3 & 6 & -12 & 9 \end{bmatrix}$

[8] $\quad A+B = \begin{bmatrix} 8 & 2 & 1 & 5 \\ 1 & 1 & 1 & -1 \\ 2 & 5 & 0 & 5 \end{bmatrix}; A-B = \begin{bmatrix} 4 & 2 & 1 & 5 \\ 1 & -1 & 1 & -1 \\ 2 & 5 & 0 & -3 \end{bmatrix}; 3A = \begin{bmatrix} 18 & 6 & 3 & 15 \\ 3 & 0 & 3 & -3 \\ 6 & 15 & 0 & 3 \end{bmatrix};$

$3A-4B = \begin{bmatrix} 18 & 6 & 3 & 15 \\ 3 & 0 & 3 & -3 \\ 6 & 15 & 0 & 3 \end{bmatrix} - \begin{bmatrix} 8 & 0 & 0 & 0 \\ 0 & 4 & 0 & 0 \\ 0 & 0 & 0 & 16 \end{bmatrix} = \begin{bmatrix} 10 & 6 & 3 & 15 \\ 3 & -4 & 3 & -3 \\ 6 & 15 & 0 & -13 \end{bmatrix}$

[9] $\quad A+B = \begin{bmatrix} 3 & 6 & -1 & -7 \\ 0 & 2 & 1 & -2 \\ 0 & 0 & -3 & 8 \\ 0 & 0 & 0 & 3 \end{bmatrix}; A-B = \begin{bmatrix} -1 & 6 & -1 & -7 \\ 0 & 0 & 1 & -2 \\ 0 & 0 & 3 & 8 \\ 0 & 0 & 0 & -5 \end{bmatrix}; 3A = A+B = \begin{bmatrix} 3 & 18 & -3 & -21 \\ 0 & 3 & 3 & -6 \\ 0 & 0 & 0 & 24 \\ 0 & 0 & 0 & -3 \end{bmatrix};$

$3A-4B = \begin{bmatrix} 3 & 18 & -3 & -21 \\ 0 & 3 & 3 & -6 \\ 0 & 0 & 0 & 24 \\ 0 & 0 & 0 & -3 \end{bmatrix} - \begin{bmatrix} 8 & 0 & 0 & 0 \\ 0 & 4 & 0 & 0 \\ 0 & 0 & -12 & 0 \\ 0 & 0 & 0 & 16 \end{bmatrix} = \begin{bmatrix} -5 & 18 & -3 & -21 \\ 0 & -1 & 3 & -6 \\ 0 & 0 & 12 & 24 \\ 0 & 0 & 0 & -19 \end{bmatrix}$

[10] $\quad A+B = \begin{bmatrix} 6 & 4 \\ -5 & 2 \\ 7 & 6 \\ -4 & -1 \\ 16 & -4 \end{bmatrix}; A-B = \begin{bmatrix} 2 & -4 \\ 5 & 2 \\ 5 & 4 \\ 2 & -7 \\ 4 & -14 \end{bmatrix}; 3A = \begin{bmatrix} 12 & 0 \\ 0 & 6 \\ 18 & 15 \\ -3 & -12 \\ 30 & -27 \end{bmatrix};$

$3A-4B = \begin{bmatrix} 12 & 0 \\ 0 & 6 \\ 18 & 15 \\ -3 & -12 \\ 30 & -27 \end{bmatrix} - \begin{bmatrix} 8 & 16 \\ -20 & 0 \\ 4 & 4 \\ -12 & 12 \\ 24 & 20 \end{bmatrix} = \begin{bmatrix} 4 & -16 \\ 20 & 6 \\ 14 & 11 \\ 9 & -24 \\ 6 & -47 \end{bmatrix}$

[11] 0

[12] 0

[13] 4

[14] 37

[15] 0

[16] 1

[17] $\quad AB = \begin{bmatrix} 2 & 0 \\ 4 & 0 \\ -6 & 0 \end{bmatrix}$

[18] Neither product is defined.

[19] $\quad AB = A$

[20] $\quad AB = \begin{bmatrix} 2 & 1 & 6 \\ 0 & 0 & -2 \\ 0 & 0 & 4 \end{bmatrix}; BA = \begin{bmatrix} 2 & -1 & 17 \\ 0 & 0 & 14 \\ 0 & 0 & 4 \end{bmatrix}$

[21] $BA = \begin{bmatrix} -5 & -6 & 0 & -7 \\ 5 & 6 & 0 & 7 \\ -1 & 2 & -3 & 4 \\ 6 & 4 & 3 & 3 \\ -6 & -4 & -3 & -3 \end{bmatrix}$
[22] Neither product is defined.

[23] $AB = \begin{bmatrix} 1 & 0 & 2 \\ -1 & 1 & 0 \\ -1 & 1 & -1 \end{bmatrix}; BA = \begin{bmatrix} 1 & 1 & 0 \\ 1 & 0 & 1 \\ -1 & 0 & 0 \end{bmatrix}$
[24] $AB = \begin{bmatrix} 2 & \frac{5}{3} & 0 \\ 0 & -\frac{2}{3} & 0 \\ 3 & \frac{10}{3} & 1 \end{bmatrix}; BA = \begin{bmatrix} 1 & 0 & 0 \\ 2 & 2 & 1 \\ \frac{5}{3} & 0 & -\frac{2}{3} \end{bmatrix}$

[25] $AB = \begin{bmatrix} -2 & 1 & 7 \\ 1 & 3 & -1 \\ 6 & -1 & 5 \end{bmatrix}; BA = \begin{bmatrix} 8 & 0 & 0 & 4 \\ 1 & 3 & 1 & 6 \\ -5 & 4 & 1 & 7 \\ 4 & -2 & 0 & -6 \end{bmatrix}$

[26] $BA = \begin{bmatrix} 1 & 0 & 2 & 0 & -4 \\ -1 & 2 & 2 & -2 & 1 \\ 0 & 1 & 1 & 1 & 0 \end{bmatrix}$

[27] $\overline{X} = -A - 3B = \begin{bmatrix} -1 & 2 \\ -3 & 0 \\ 4 & -1 \end{bmatrix} - \begin{bmatrix} 9 & 0 \\ -6 & 3 \\ 0 & -15 \end{bmatrix} = \begin{bmatrix} -10 & 2 \\ 3 & -3 \\ 4 & 14 \end{bmatrix}$

[28] $3\overline{X} = 2A - B \Rightarrow \overline{X} = \frac{2}{3}A - \frac{1}{3}B = \begin{bmatrix} \frac{2}{3} & -\frac{4}{3} \\ 2 & 0 \\ -\frac{8}{3} & \frac{2}{3} \end{bmatrix} - \begin{bmatrix} 1 & 0 \\ -\frac{2}{3} & \frac{1}{3} \\ 0 & -\frac{5}{3} \end{bmatrix}$

[29] $2\overline{X} + 4B = 5A \Rightarrow \overline{X} = \frac{5}{2}A - 2B = \begin{bmatrix} \frac{5}{2} & -5 \\ \frac{15}{2} & 0 \\ -10 & \frac{5}{2} \end{bmatrix} - \begin{bmatrix} 6 & 0 \\ -4 & 2 \\ 0 & -10 \end{bmatrix} = \begin{bmatrix} -\frac{7}{2} & -5 \\ \frac{23}{2} & -2 \\ -10 & \frac{25}{2} \end{bmatrix}$

[30] $4\overline{X} - A + 2B = [0]_{3,2} \Rightarrow \overline{X} = \frac{1}{4}A - \frac{1}{2}B = \begin{bmatrix} \frac{1}{4} & -\frac{1}{2} \\ \frac{3}{4} & 0 \\ -1 & \frac{1}{4} \end{bmatrix} - \begin{bmatrix} \frac{3}{2} & 0 \\ -1 & \frac{1}{2} \\ 0 & -\frac{5}{2} \end{bmatrix} = \begin{bmatrix} -\frac{5}{4} & -\frac{1}{2} \\ \frac{7}{4} & -\frac{1}{2} \\ -1 & \frac{11}{4} \end{bmatrix}$

[31] $AB = \begin{bmatrix} 1 & -2 \\ 0 & 0 \end{bmatrix}\begin{bmatrix} 2 & 1 \\ -1 & 0 \end{bmatrix} = \begin{bmatrix} 4 & 1 \\ 0 & 0 \end{bmatrix}; AC = \begin{bmatrix} 1 & -2 \\ 0 & 0 \end{bmatrix}\begin{bmatrix} 2 & 2 \\ -1 & \frac{1}{2} \end{bmatrix} = \begin{bmatrix} 4 & 1 \\ 0 & 0 \end{bmatrix}.$

Thus, $A \neq \begin{bmatrix} 0 & 0 \\ 0 & 0 \end{bmatrix}$ and $AB = AC$, but $B \neq C$. However, if a, b and c are real numbers such that $ab = ac$ and $a \neq 0$, then $b = c$.

[32] $(AB)^2 = \begin{bmatrix} 4 & 1 \\ 0 & 0 \end{bmatrix}\begin{bmatrix} 4 & 1 \\ 0 & 0 \end{bmatrix} = \begin{bmatrix} 16 & 4 \\ 0 & 0 \end{bmatrix}. A^2 = \begin{bmatrix} 1 & -2 \\ 0 & 0 \end{bmatrix}\begin{bmatrix} 1 & -2 \\ 0 & 0 \end{bmatrix} = \begin{bmatrix} 1 & -2 \\ 0 & 0 \end{bmatrix}.$

$B^2 = \begin{bmatrix} 2 & 1 \\ -1 & 0 \end{bmatrix}\begin{bmatrix} 2 & 1 \\ -1 & 0 \end{bmatrix} = \begin{bmatrix} 3 & 2 \\ -2 & -1 \end{bmatrix}. A^2 + B^2 = \begin{bmatrix} 7 & 4 \\ 0 & 0 \end{bmatrix}.$ Thus, $(AB)^2 \neq A^2 B^2$.

[33] (See Exercise 31). $A^2 - B^2 = \begin{bmatrix} -2 & -4 \\ 2 & 1 \end{bmatrix}; A + B = \begin{bmatrix} 3 & -1 \\ -1 & 0 \end{bmatrix}; A - B = \begin{bmatrix} -1 & -3 \\ 1 & 0 \end{bmatrix};$

$(A + B)(A - B) = \begin{bmatrix} -4 & -9 \\ 1 & 3 \end{bmatrix}; \therefore A^2 - B^2 \neq (A + B)(A - B)$

[34] (See Exercises 31, 32 and 33). $A^2 + 2AB + B^2 = \begin{bmatrix} 1 & -2 \\ 0 & 0 \end{bmatrix} + \begin{bmatrix} 8 & 2 \\ 0 & 0 \end{bmatrix} + \begin{bmatrix} 3 & 2 \\ -2 & 1 \end{bmatrix} =$

$\begin{bmatrix} 12 & 2 \\ -2 & 1 \end{bmatrix}. (A + B)(A + B) = \begin{bmatrix} 3 & -1 \\ -1 & 0 \end{bmatrix}\begin{bmatrix} 3 & -1 \\ -1 & 0 \end{bmatrix} = \begin{bmatrix} 10 & -3 \\ -3 & 1 \end{bmatrix}; \therefore A^2 + 2A$

$B + B^2 \neq (A + B)(A + B)$

[39] $A = \begin{bmatrix} 1 & 1 \\ 1 & -2 \end{bmatrix}; \bar{X} = \begin{bmatrix} x \\ y \end{bmatrix}; B = \begin{bmatrix} 6 \\ 5 \end{bmatrix}$ [40] $A = \begin{bmatrix} 2 & -1 \\ -3 & 2 \end{bmatrix}; \bar{X} = \begin{bmatrix} x \\ y \end{bmatrix}; B = \begin{bmatrix} -7 \\ 4 \end{bmatrix}$

[41] $A = \begin{bmatrix} 1 & 0 & -1 \\ 0 & 1 & 4 \\ 3 & 0 & -5 \end{bmatrix}; \bar{X} = \begin{bmatrix} t \\ u \\ v \end{bmatrix}; B = \begin{bmatrix} -1 \\ 2 \\ 6 \end{bmatrix}$

[42] $A = \begin{bmatrix} 1 & -1 & -1 & 1 \\ 0 & 2 & 1 & 0 \end{bmatrix}; \bar{X} = \begin{bmatrix} t \\ u \\ v \\ w \end{bmatrix}; B = \begin{bmatrix} 0 \\ 0 \end{bmatrix}$

[43] $A = \begin{bmatrix} 1 & 1 \\ 2 & 4 \\ 1 & -5 \end{bmatrix}; \bar{X} = \begin{bmatrix} x \\ y \end{bmatrix}; B = \begin{bmatrix} 4 \\ 2 \\ -8 \end{bmatrix}$

[44] $A = \begin{bmatrix} 1 & 1 & -1 & 1 \\ 0 & 1 & 0 & -1 \\ 1 & 1 & 4 & 0 \\ 0 & 1 & 1 & 0 \\ 1 & 0 & -1 & 0 \end{bmatrix}; \bar{X} = \begin{bmatrix} t \\ v \\ w \\ x \end{bmatrix}; B = \begin{bmatrix} 0 \\ 0 \\ 0 \\ 0 \\ 0 \end{bmatrix}$

[45] a) $A + B = \begin{bmatrix} 2 & -9 \\ 10 & 4 \end{bmatrix} + \begin{bmatrix} -7 & -3 \\ 6 & 1 \end{bmatrix} = \begin{bmatrix} -5 & -12 \\ 16 & 5 \end{bmatrix}; (A + B)^T = \begin{bmatrix} -5 & 16 \\ -12 & 5 \end{bmatrix}; A^T =$

$\begin{bmatrix} 2 & 10 \\ -9 & 4 \end{bmatrix}; B^T = \begin{bmatrix} -7 & 6 \\ -3 & 1 \end{bmatrix}; A^T + B^T = \begin{bmatrix} 2 & 10 \\ -9 & 4 \end{bmatrix} + \begin{bmatrix} -7 & 6 \\ -3 & 1 \end{bmatrix} = \begin{bmatrix} -5 & 16 \\ -12 & 5 \end{bmatrix}.$

$\therefore (A + B)^T = A^T + B^T.$

b) $(A^T)^T = \begin{bmatrix} 2 & 10 \\ -9 & 4 \end{bmatrix}^T = \begin{bmatrix} 2 & -9 \\ 10 & 4 \end{bmatrix} = A$.

c) $AB = \begin{bmatrix} 2 & -9 \\ 10 & 4 \end{bmatrix}\begin{bmatrix} -7 & -3 \\ 6 & 1 \end{bmatrix} = \begin{bmatrix} -68 & -15 \\ -46 & -26 \end{bmatrix}$. Thus, $(AB)^T = \begin{bmatrix} -68 & -46 \\ -15 & -26 \end{bmatrix}$.

$B^T A^T = \begin{bmatrix} -7 & 6 \\ -3 & 1 \end{bmatrix}\begin{bmatrix} 2 & 10 \\ -9 & 4 \end{bmatrix} = \begin{bmatrix} -68 & -46 \\ -15 & -26 \end{bmatrix}$. $\therefore (AB)^T = B^T A^T$.

[46] a) $(AB)C = \left(\begin{bmatrix} a & b \\ c & d \end{bmatrix}\begin{bmatrix} e & f \\ g & h \end{bmatrix}\right)\begin{bmatrix} i & j \\ k & l \end{bmatrix} = \begin{bmatrix} ae+bg & af+bh \\ ce+dg & cf+dh \end{bmatrix}\begin{bmatrix} i & j \\ k & l \end{bmatrix} =$

$\begin{bmatrix} (ae+bg)i+(af+bh)k & (ae+bg)j+(af+bh)l \\ (ce+dg)i+(cf+dh)k & (ce+dg)j+(cf+dh)l \end{bmatrix} =$

$\begin{bmatrix} aei+bgi+afk+bhk & aej+bgi+afl+bhl \\ cei+dgi+cfk+dhk & cej+dgi+cfl+dhl \end{bmatrix}$.

$A(BC) = \begin{bmatrix} a & b \\ c & d \end{bmatrix}\left(\begin{bmatrix} e & f \\ g & h \end{bmatrix}\begin{bmatrix} i & j \\ k & l \end{bmatrix}\right) = \begin{bmatrix} a & b \\ c & d \end{bmatrix}\begin{bmatrix} ei+fk & ej+fl \\ gi+hk & gi+hl \end{bmatrix} =$

$\begin{bmatrix} a(ei+fk)+b(gi+hk) & a(ej+fl)+b(gi+hl) \\ c(ei+fk)+d(gi+hk) & c(ej+fl)+d(gi+hl) \end{bmatrix} =$

$\begin{bmatrix} aei+afk+bgi+bhk & aej+afl+bgi+bhl \\ cei+cfk+dgi+dhk & cej+cfl+dgi+dhl \end{bmatrix} =$

Comparing corresponding components, we see that $A(BC) = (AB)C$.

b) $(A+B)C = \left(\begin{bmatrix} a & b \\ c & d \end{bmatrix}+\begin{bmatrix} e & f \\ g & h \end{bmatrix}\right)\begin{bmatrix} i & j \\ k & l \end{bmatrix} = \begin{bmatrix} a+e & b+f \\ c+g & d+h \end{bmatrix}\begin{bmatrix} i & j \\ k & l \end{bmatrix} =$

$\begin{bmatrix} (a+e)i+(b+f)k & (a+e)j+(b+f)l \\ (c+g)i+(d+h)k & (c+g)j+(d+h)l \end{bmatrix} = \begin{bmatrix} ai+ei+bk+fk & aj+ej+bl+fl \\ ci+gi+dk+hk & cj+gj+dl+hl \end{bmatrix}$.

$AC = \begin{bmatrix} a & b \\ c & d \end{bmatrix}\begin{bmatrix} i & j \\ k & l \end{bmatrix} = \begin{bmatrix} ai+bk & aj+bl \\ ci+dk & cj+dl \end{bmatrix}$; $BC = \begin{bmatrix} e & f \\ g & h \end{bmatrix}\begin{bmatrix} i & j \\ k & l \end{bmatrix} = \begin{bmatrix} ei+fk & ej+fl \\ gi+hk & gj+hl \end{bmatrix}$;

$AC+BC = \begin{bmatrix} ai+bk+ei+fk & aj+bl+ej+fl \\ ci+dk+gi+hk & cj+dl+gj+hl \end{bmatrix}$. Comparing corresponding

components, we see that $(A+B)C = AC+BC$.

[47] $\begin{cases} x'\cos\theta - y'\sin\theta = x \\ x'\sin\theta + y'\cos\theta = y \end{cases}$ <u>Equations of rotations</u>. Multiplying E_1 by $\cos\theta$ and E_2 by $\sin\theta$ we have

$\begin{cases} x'\cos^2\theta - y'\sin\theta\cos\theta = x\cos\theta \\ x'\sin^2\theta + y'\cos\theta\sin\theta = y\sin\theta \end{cases} \Rightarrow x'(\cos^2\theta + \sin\theta^2) = x\cos\theta + y\sin\theta \Rightarrow x\cos\theta + y\sin\theta =$

x'. Now multiplying the first equation of rotation by $\sin\theta$ and the second by $\cos\theta$, it can be shown

that $-x\sin\theta + y\cos\theta = y'$.

a) $\begin{bmatrix} -\frac{\sqrt{2}}{2} & \frac{\sqrt{2}}{2} \\ -\frac{\sqrt{2}}{2} & -\frac{\sqrt{2}}{2} \end{bmatrix} \begin{bmatrix} -2 \\ 3 \end{bmatrix} = \begin{bmatrix} \frac{5\sqrt{2}}{2} \\ -\frac{\sqrt{2}}{2} \end{bmatrix}$ 　　　● $\left(\frac{5\sqrt{2}}{2}, \frac{\sqrt{2}}{2} \right)$

b) $\begin{bmatrix} -\frac{\sqrt{3}}{2} & -\frac{1}{2} \\ \frac{1}{2} & -\frac{\sqrt{3}}{2} \end{bmatrix} \begin{bmatrix} -2 \\ 3 \end{bmatrix} = \begin{bmatrix} \sqrt{3} & -\frac{3}{2} \\ -1 & -\frac{3\sqrt{3}}{2} \end{bmatrix}$ 　　● $\left(\sqrt{3} - \frac{3}{2}, -1 - \frac{3\sqrt{3}}{2} \right)$

c) $\begin{bmatrix} \cos 50° & \sin 50° \\ -\sin 50° & \cos 50° \end{bmatrix} \begin{bmatrix} -2 \\ 3 \end{bmatrix} = \begin{bmatrix} -\cos 50° + 3\sin 50° \\ 2\sin 50° + 3\cos 50° \end{bmatrix}$

● $\left(-2\cos 50° + 3\sin 50°, 2\sin 50° + 3\cos 50° \right) \approx \left(1.01, 3.46 \right)$

[48]　a)　6043　　　　　　　　　　b)　350

　　　c)　45　　　　　　　　　　d)　6845

[49]　a)

$$\begin{array}{c} \text{units required} \\ \begin{array}{ccc} \text{wood} & \text{plastic} & \text{glass} \end{array} \\ P = \begin{bmatrix} 60 & 5 & 0 \\ 10 & 2 & 10 \end{bmatrix} \begin{array}{l} \text{Table A} \\ \text{Table B} \end{array} \end{array}$$

b)

$$\begin{array}{c} \text{unit cost} \\ \begin{array}{ccc} \text{factory 1} & \text{factory 2} & \text{factory 3} \end{array} \\ C = \begin{bmatrix} 60 & 5 & 0 \\ 10 & 2 & 10 \\ 12 & 10 & 13 \end{bmatrix} \begin{array}{l} \text{wood} \\ \text{plastic} \\ \text{glass} \end{array} \end{array}$$

c)

$$\begin{array}{c} \begin{array}{ccc} \text{factory 1} & \text{factory 2} & \text{factory 3} \end{array} \\ PC = \begin{bmatrix} 3650 & 310 & 50 \\ 740 & 154 & 150 \end{bmatrix} \begin{array}{l} \text{Table A} \\ \text{Table B} \end{array} \end{array}$$

● Table A : factory 1, $ 3650
　　　　　 factory 2,　$ 310
　　　　　 factory 3,　　$ 50
Table B : factory 1,　$ 740
　　　　　 factory 2,　$ 154
　　　　　 factory 3,　$ 150

EXERICES 9.4

[5] $\begin{bmatrix} 6 & 1 & | & 1 & 0 \\ -4 & 2 & | & 0 & 1 \end{bmatrix}$ $\overset{-\frac{1}{6}R_1 \to R_1}{\sim}$ $\begin{bmatrix} 1 & \frac{1}{6} & | & \frac{1}{6} & 0 \\ -4 & 2 & | & 0 & 1 \end{bmatrix}$ $\overset{4R_1 + R_2 \to R_2}{\sim}$ $\begin{bmatrix} 1 & \frac{1}{6} & | & \frac{1}{6} & 0 \\ 0 & \frac{8}{3} & | & \frac{2}{3} & 1 \end{bmatrix}$

$\overset{\frac{3}{8}R_2 \to R_2}{\sim}$ $\begin{bmatrix} 1 & \frac{1}{6} & | & \frac{1}{6} & 0 \\ 0 & 1 & | & \frac{1}{4} & \frac{3}{8} \end{bmatrix}$ $\overset{-\frac{1}{6}R_2 \to R_1}{\sim}$ $\begin{bmatrix} 1 & 0 & | & \frac{1}{8} & -\frac{1}{16} \\ 0 & 1 & | & \frac{1}{4} & \frac{3}{8} \end{bmatrix}$ $\bullet \begin{bmatrix} \frac{1}{8} & -\frac{1}{16} \\ \frac{1}{4} & \frac{3}{8} \end{bmatrix}$

[6] $\begin{bmatrix} -1 & 3 & | & 1 & 0 \\ 2 & -6 & | & 0 & 1 \end{bmatrix}$ $\overset{-R_1 \to R_1}{\sim}$ $\begin{bmatrix} 1 & -3 & | & -1 & 0 \\ 2 & -6 & | & 0 & 1 \end{bmatrix}$ $\overset{2R_1 + R_2 \to R_2}{\sim}$ $\begin{bmatrix} 1 & -3 & | & -1 & 0 \\ 0 & 0 & | & 2 & 1 \end{bmatrix}$

$\bullet$ matrix is singular

[7] $\begin{bmatrix} 1 & 1 & | & 1 & 0 \\ -1 & 2 & | & 0 & 1 \end{bmatrix}$ $\overset{R_1 + R_2 \to R_2}{\sim}$ $\begin{bmatrix} 1 & 1 & | & 1 & 0 \\ 0 & 3 & | & 1 & 1 \end{bmatrix}$ $\overset{\frac{1}{3}R_2 \to R_2}{\sim}$ $\begin{bmatrix} 1 & 1 & | & 1 & 0 \\ 0 & 1 & | & \frac{1}{3} & \frac{1}{3} \end{bmatrix}$

$\overset{-R_2 + R_1 \to R_1}{\sim}$ $\begin{bmatrix} 1 & 0 & | & \frac{2}{3} & -\frac{1}{3} \\ 0 & 1 & | & \frac{1}{3} & \frac{1}{3} \end{bmatrix}$ $\bullet \begin{bmatrix} \frac{2}{3} & -\frac{1}{3} \\ \frac{1}{3} & \frac{1}{3} \end{bmatrix}$

[8] $\begin{bmatrix} 5 & 5 & | & 1 & 0 \\ 2 & -3 & | & 0 & 1 \end{bmatrix}$ $\overset{\frac{1}{5}R_1 \to R_1}{\sim}$ $\begin{bmatrix} 1 & 1 & | & \frac{1}{5} & 0 \\ 2 & -3 & | & 0 & 1 \end{bmatrix}$ $\overset{-2R_1 + R_2 \to R_2}{\sim}$ $\begin{bmatrix} 1 & 1 & | & \frac{1}{5} & 0 \\ 0 & -5 & | & -\frac{2}{5} & 1 \end{bmatrix}$

$\overset{\frac{1}{5}R_2 \to R_2}{\sim}$ $\begin{bmatrix} 1 & 1 & | & \frac{1}{5} & 0 \\ 0 & 1 & | & \frac{2}{25} & -\frac{1}{5} \end{bmatrix}$ $\overset{-R_2 + R_1 \to R_1}{\sim}$ $\begin{bmatrix} 1 & 0 & | & \frac{3}{25} & \frac{1}{5} \\ 0 & 1 & | & \frac{2}{25} & -\frac{1}{5} \end{bmatrix}$ $\bullet \begin{bmatrix} \frac{3}{25} & \frac{1}{5} \\ \frac{2}{25} & -\frac{1}{5} \end{bmatrix}$

[9]
$$\begin{bmatrix} 1 & 2 & 1 & | & 1 & 0 & 0 \\ -1 & 4 & 10 & | & 0 & 1 & 0 \\ 1 & 1 & 3 & | & 0 & 0 & 1 \end{bmatrix} \begin{matrix} R_1+R_2\to R_2 \\ -R_1+R_3\to R_3 \\ \sim \end{matrix} \begin{bmatrix} 1 & 2 & 1 & | & 1 & 0 & 0 \\ 0 & 6 & 11 & | & 1 & 1 & 0 \\ 0 & -1 & 2 & | & -1 & 0 & 1 \end{bmatrix} \begin{matrix} \frac{1}{6}R_2\to R_2 \\ \sim \end{matrix}$$

$$\begin{bmatrix} 1 & 2 & 1 & | & 1 & 0 & 0 \\ 0 & 1 & \frac{11}{6} & | & \frac{1}{6} & \frac{1}{6} & 0 \\ 0 & -1 & 2 & | & -1 & 0 & 1 \end{bmatrix} \begin{matrix} R_2+R_3\to R_3 \\ \sim \end{matrix} \begin{bmatrix} 1 & 2 & 1 & | & 1 & 0 & 0 \\ 0 & 1 & \frac{11}{6} & | & \frac{1}{6} & \frac{1}{6} & 0 \\ 0 & 0 & \frac{23}{6} & | & -\frac{5}{6} & \frac{1}{6} & 1 \end{bmatrix}$$

$$\begin{matrix} \frac{6}{23}R_3\to R_3 \\ \sim \end{matrix} \begin{bmatrix} 1 & 2 & 1 & | & 1 & 0 & 0 \\ 0 & 1 & \frac{11}{6} & | & \frac{1}{6} & \frac{1}{6} & 0 \\ 0 & 0 & 1 & | & -\frac{5}{23} & \frac{1}{23} & \frac{6}{23} \end{bmatrix} \begin{matrix} -\frac{11}{6}R_3+R_2\to R_2 \\ -R_3+R_1\to R_1 \\ \sim \end{matrix}$$

$$\begin{bmatrix} 1 & 2 & 1 & | & \frac{28}{23} & -\frac{1}{23} & -\frac{6}{23} \\ 0 & 1 & 0 & | & \frac{13}{23} & \frac{2}{23} & -\frac{11}{23} \\ 0 & 0 & 1 & | & -\frac{5}{23} & \frac{1}{23} & \frac{6}{23} \end{bmatrix} \begin{matrix} -2R_2+R_1\to R_1 \\ \sim \end{matrix} \begin{bmatrix} 1 & 2 & 1 & | & \frac{2}{23} & -\frac{5}{23} & \frac{16}{23} \\ 0 & 1 & 0 & | & \frac{13}{23} & \frac{2}{23} & -\frac{11}{23} \\ 0 & 0 & 1 & | & -\frac{5}{23} & \frac{1}{23} & \frac{6}{23} \end{bmatrix}$$

$$\bullet \begin{bmatrix} \frac{2}{23} & -\frac{5}{23} & \frac{16}{23} \\ \frac{13}{23} & \frac{2}{23} & -\frac{11}{23} \\ -\frac{5}{23} & \frac{1}{23} & \frac{6}{23} \end{bmatrix}$$

[10]
$$\begin{bmatrix} 1 & 0 & 1 & | & 1 & 0 & 0 \\ 0 & 1 & 2 & | & 0 & 1 & 0 \\ 3 & 5 & 4 & | & 0 & 0 & 1 \end{bmatrix} \begin{matrix} -3R_1+R_3\to R_3 \\ \sim \end{matrix} \begin{bmatrix} 1 & 0 & 1 & | & 1 & 0 & 0 \\ 0 & 1 & 2 & | & 0 & 1 & 0 \\ 0 & 5 & 1 & | & -3 & 0 & 1 \end{bmatrix}$$

$$\begin{matrix} -5R_2\to R_3 \\ \sim \end{matrix} \begin{bmatrix} 1 & 0 & 1 & | & 1 & 0 & 0 \\ 0 & 1 & 2 & | & 0 & 1 & 0 \\ 0 & 0 & -9 & | & -3 & -5 & 1 \end{bmatrix} \begin{matrix} -\frac{1}{9}R_3\to R_3 \\ \sim \end{matrix} \begin{bmatrix} 1 & 0 & 1 & | & 1 & 0 & 0 \\ 0 & 1 & 2 & | & 0 & 1 & 0 \\ 0 & 0 & 1 & | & \frac{1}{3} & \frac{5}{9} & -\frac{1}{9} \end{bmatrix}$$

$$\begin{matrix} -2R_3+R_2\to R_2 \\ \sim \end{matrix} \begin{bmatrix} 1 & 0 & 1 & | & 1 & 0 & 0 \\ 0 & 1 & 0 & | & -\frac{2}{3} & -\frac{1}{9} & \frac{2}{9} \\ 0 & 0 & 1 & | & \frac{1}{3} & \frac{5}{9} & -\frac{1}{9} \end{bmatrix} \begin{matrix} -R_3+R_1\to R_1 \\ \sim \end{matrix} \begin{bmatrix} 1 & 0 & 1 & | & \frac{2}{3} & -\frac{5}{9} & \frac{1}{9} \\ 0 & 1 & 0 & | & -\frac{2}{3} & -\frac{1}{9} & \frac{2}{9} \\ 0 & 0 & 1 & | & \frac{1}{3} & \frac{5}{9} & -\frac{1}{9} \end{bmatrix}$$

$$\bullet \begin{bmatrix} \frac{2}{3} & -\frac{5}{9} & \frac{1}{9} \\ -\frac{2}{3} & -\frac{1}{9} & \frac{2}{3} \\ \frac{1}{3} & \frac{5}{9} & -\frac{1}{9} \end{bmatrix}$$

[11] $\begin{bmatrix} 1 & -1 & 2 & | & 1 & 0 & 0 \\ 0 & 4 & 8 & | & 0 & 1 & 0 \\ 0 & 0 & 1 & | & 0 & 0 & 1 \end{bmatrix}$ $\begin{matrix} \frac{1}{4}R_2 \to R_2 \\ \sim \end{matrix}$ $\begin{bmatrix} 1 & -1 & 2 & | & 1 & 0 & 0 \\ 0 & 1 & 2 & | & 0 & \frac{1}{4} & 0 \\ 0 & 0 & 1 & | & 0 & 0 & 1 \end{bmatrix}$ $\begin{matrix} -2R_3 + R_2 \to R_2 \\ -2R_3 + R_1 \to R_1 \\ \sim \end{matrix}$

$\begin{bmatrix} 1 & -1 & 0 & | & 1 & 0 & -2 \\ 0 & 1 & 0 & | & 0 & \frac{1}{4} & -2 \\ 0 & 0 & 1 & | & 0 & 0 & 1 \end{bmatrix}$ $\begin{matrix} R_2 + R_1 \to R_1 \\ \sim \end{matrix}$ $\begin{bmatrix} 1 & 0 & 0 & | & 1 & \frac{1}{4} & -4 \\ 0 & 1 & 0 & | & 0 & \frac{1}{4} & -2 \\ 0 & 0 & 1 & | & 0 & 0 & 1 \end{bmatrix}$ $\quad\bullet\ \begin{bmatrix} 1 & \frac{1}{4} & -4 \\ 0 & \frac{1}{4} & -2 \\ 0 & 0 & 1 \end{bmatrix}$

[12] We see that the matrix is singular by inspection. $\quad\bullet$ Matrix is singular

[13] $\begin{bmatrix} 1 & -1 & 1 & | & 1 & 0 & 0 \\ 1 & 1 & -2 & | & 0 & 1 & 0 \\ -1 & 3 & 4 & | & 0 & 0 & 1 \end{bmatrix}$ $\begin{matrix} -R_1 + R_2 \to R_2 \\ R_1 + R_3 \to R_3 \\ \sim \end{matrix}$ $\begin{bmatrix} 1 & -1 & 1 & | & 1 & 0 & 0 \\ 0 & 2 & -3 & | & -1 & 1 & 0 \\ 0 & 2 & 5 & | & 1 & 0 & 1 \end{bmatrix}$ $\begin{matrix} \frac{1}{2}R_2 \to R_2 \\ \sim \end{matrix}$

$\begin{bmatrix} 1 & -1 & 1 & | & 1 & 0 & 0 \\ 0 & 1 & -\frac{3}{2} & | & -\frac{1}{2} & \frac{1}{2} & 0 \\ 0 & 2 & 5 & | & 1 & 0 & 1 \end{bmatrix}$ $\begin{matrix} -2R_2 + R_3 \to R_3 \\ \sim \end{matrix}$ $\begin{bmatrix} 1 & -1 & 1 & | & 1 & 0 & 0 \\ 0 & 1 & -\frac{3}{2} & | & -\frac{1}{2} & \frac{1}{2} & 0 \\ 0 & 0 & 8 & | & 2 & -1 & 1 \end{bmatrix}$ $\begin{matrix} \frac{1}{8}R_3 \to R_3 \\ \sim \end{matrix}$

$\begin{bmatrix} 1 & -1 & 1 & | & 1 & 0 & 0 \\ 0 & 1 & -\frac{3}{2} & | & -\frac{1}{2} & \frac{1}{2} & 0 \\ 0 & 0 & 1 & | & \frac{1}{4} & -\frac{1}{8} & \frac{1}{8} \end{bmatrix}$ $\begin{matrix} \frac{3}{2}R_3 + R_2 \to R_2 \\ -R_3 + R_1 \to R_1 \\ \sim \end{matrix}$ $\begin{bmatrix} 1 & -1 & 1 & | & \frac{3}{4} & \frac{1}{8} & -\frac{1}{8} \\ 0 & 1 & 0 & | & -\frac{1}{8} & \frac{5}{16} & \frac{3}{16} \\ 0 & 0 & 1 & | & \frac{1}{4} & -\frac{1}{8} & \frac{1}{8} \end{bmatrix}$

$\begin{matrix} R_2 + R_1 \to R_1 \\ \sim \end{matrix}$ $\begin{bmatrix} 1 & 0 & 0 & | & \frac{5}{8} & \frac{7}{16} & -\frac{1}{16} \\ 0 & 1 & 0 & | & -\frac{1}{8} & \frac{5}{16} & \frac{3}{16} \\ 0 & 0 & 1 & | & \frac{1}{4} & -\frac{1}{8} & \frac{1}{8} \end{bmatrix}$ $\qquad\bullet\ \begin{bmatrix} \frac{5}{8} & \frac{7}{16} & -\frac{1}{16} \\ -\frac{1}{8} & \frac{5}{16} & \frac{3}{16} \\ \frac{1}{4} & -\frac{1}{8} & \frac{1}{8} \end{bmatrix}$

[14] $\begin{bmatrix} 1 & 0 & 0 & | & 1 & 0 & 0 \\ 2 & 1 & 0 & | & 0 & 1 & 0 \\ 1 & 5 & 2 & | & 0 & 0 & 1 \end{bmatrix}$ $\begin{matrix} -2R_1 + R_2 \to R_2 \\ -R_1 + R_3 \to R_3 \\ \sim \end{matrix}$ $\begin{bmatrix} 1 & 0 & 0 & | & 1 & 0 & 0 \\ 0 & 1 & 0 & | & -2 & 1 & 0 \\ 0 & 5 & 2 & | & -1 & 0 & 1 \end{bmatrix}$ $\begin{matrix} -5R_2 + R_3 \to R_3 \\ \sim \end{matrix}$

$\begin{bmatrix} 1 & 0 & 0 & | & 1 & 0 & 0 \\ 0 & 1 & 0 & | & -2 & 1 & 0 \\ 0 & 0 & 2 & | & 9 & -5 & 1 \end{bmatrix}$ $\begin{matrix} \frac{1}{2}R_3 \to R_3 \\ \sim \end{matrix}$ $\begin{bmatrix} 1 & 0 & 0 & | & 1 & 0 & 0 \\ 0 & 1 & 0 & | & -2 & 1 & 0 \\ 0 & 0 & 1 & | & \frac{9}{2} & -\frac{5}{2} & \frac{1}{2} \end{bmatrix}$

$\qquad\bullet\ \begin{bmatrix} 1 & 0 & 0 \\ -2 & 1 & 0 \\ \frac{9}{2} & -\frac{5}{2} & \frac{1}{2} \end{bmatrix}$

[15] $\begin{bmatrix} 1 & 0 & 1 & | & 1 & 0 & 0 \\ 1 & 1 & 3 & | & 0 & 1 & 0 \\ 0 & 1 & 3 & | & 0 & 0 & 1 \end{bmatrix}$ $\begin{matrix} -R_1+R_2 \to R_2 \\ \sim \end{matrix}$ $\begin{bmatrix} 1 & 0 & 1 & | & 1 & 0 & 0 \\ 0 & 1 & 2 & | & -1 & 1 & 0 \\ 0 & 1 & 3 & | & 0 & 0 & 1 \end{bmatrix}$ $\begin{matrix} -R_2+R_3 \to R_3 \\ \sim \end{matrix}$

$\begin{bmatrix} 1 & 0 & 1 & | & 1 & 0 & 0 \\ 0 & 1 & 2 & | & -1 & 1 & 0 \\ 0 & 0 & 1 & | & 1 & -1 & 1 \end{bmatrix}$ $\begin{matrix} -2R_3+R_2 \to R_2 \\ -R_3+R_1 \to R_1 \\ \sim \end{matrix}$ $\begin{bmatrix} 1 & 0 & 0 & | & 0 & 1 & -1 \\ 0 & 1 & 0 & | & -3 & 3 & -2 \\ 0 & 0 & 1 & | & 1 & -1 & 1 \end{bmatrix}$ $\bullet \begin{bmatrix} 0 & 1 & -1 \\ -3 & 3 & -2 \\ 1 & -1 & 1 \end{bmatrix}$

[16] $\begin{bmatrix} 1 & 2 & 1 & | & 1 & 0 & 0 \\ -1 & 0 & 1 & | & 0 & 1 & 0 \\ 0 & 4 & 3 & | & 0 & 0 & 1 \end{bmatrix}$ $\begin{matrix} R_1+R_2 \to R_2 \\ -R_1+R_3 \to R_3 \\ \sim \end{matrix}$ $\begin{bmatrix} 1 & 2 & 1 & | & 1 & 0 & 0 \\ 0 & 2 & 2 & | & 1 & 1 & 0 \\ 0 & 2 & 2 & | & -1 & 0 & 1 \end{bmatrix}$ $\begin{matrix} \frac{1}{2}R_2 \to R_2 \\ \sim \end{matrix}$

$\begin{bmatrix} 1 & 2 & 1 & | & 1 & 0 & 0 \\ 0 & 1 & 1 & | & \frac{1}{2} & \frac{1}{2} & 0 \\ 0 & 2 & 2 & | & -1 & 0 & 1 \end{bmatrix}$ $\begin{matrix} -2R_2+R_3 \to R_3 \\ \sim \end{matrix}$ $\begin{bmatrix} 1 & 2 & 1 & | & 1 & 0 & 0 \\ 0 & 1 & 1 & | & \frac{1}{2} & \frac{1}{2} & 0 \\ 0 & 0 & 0 & | & -2 & -1 & 1 \end{bmatrix}$

By inspecting the third row, we that the matrix is singular. $\bullet$ matrix is singular

[17] $\begin{bmatrix} 1 & 0 & 1 & 1 & | & 1 & 0 & 0 & 0 \\ -1 & 0 & 0 & 1 & | & 0 & 1 & 0 & 0 \\ 1 & 0 & 2 & -1 & | & 0 & 0 & 1 & 0 \\ 1 & 1 & -1 & 1 & | & 0 & 0 & 0 & 1 \end{bmatrix}$ $\begin{matrix} R_1+R_2 \to R_2 \\ -R_1+R_3 \to R_3 \\ -R_1+R_4 \to R_4 \\ \sim \end{matrix}$ $\begin{bmatrix} 1 & 0 & 1 & 1 & | & 1 & 0 & 0 & 0 \\ 0 & 1 & 1 & 2 & | & 1 & 1 & 0 & 0 \\ 0 & 0 & 1 & -2 & | & -1 & 0 & 1 & 0 \\ 0 & 1 & -2 & 0 & | & -1 & 0 & 0 & 1 \end{bmatrix}$

$\begin{matrix} -R_2+R_4 \to R_4 \\ \sim \end{matrix}$ $\begin{bmatrix} 1 & 0 & 1 & 1 & | & 1 & 0 & 0 & 0 \\ 0 & 1 & 1 & 2 & | & 1 & 1 & 0 & 0 \\ 0 & 0 & 1 & -2 & | & -1 & 0 & 1 & 0 \\ 0 & 0 & -3 & -2 & | & -2 & -1 & 3 & 1 \end{bmatrix}$ $\begin{matrix} 3R_3+R_4 \to R_4 \\ \sim \end{matrix}$

$\begin{bmatrix} 1 & 0 & 1 & 1 & | & 1 & 0 & 0 & 0 \\ 0 & 1 & 1 & 2 & | & 1 & 1 & 0 & 0 \\ 0 & 0 & 1 & -2 & | & -1 & 0 & 1 & 0 \\ 0 & 0 & 0 & -8 & | & -5 & -1 & 3 & 1 \end{bmatrix}$ $\begin{matrix} -\frac{1}{8}R_4 \to R_4 \\ \sim \end{matrix}$ $\begin{bmatrix} 1 & 0 & 1 & 1 & | & 1 & 0 & 0 & 0 \\ 0 & 1 & 1 & 2 & | & 1 & 1 & 0 & 0 \\ 0 & 0 & 1 & -2 & | & -1 & 0 & 1 & 0 \\ 0 & 0 & 0 & 1 & | & \frac{5}{8} & \frac{1}{8} & -\frac{3}{8} & -\frac{1}{8} \end{bmatrix}$

$$
\begin{matrix} 2R_4+R_3 \to R_3 \\ -2R_4+R_2 \to R_2 \\ -R_4+R_1 \to R_1 \\ \sim \end{matrix}
\left[\begin{array}{cccc|cccc}
1 & 0 & 1 & 0 & \frac{3}{8} & -\frac{1}{8} & \frac{3}{8} & \frac{1}{8} \\
0 & 1 & 1 & 0 & -\frac{1}{4} & \frac{3}{4} & \frac{3}{4} & \frac{1}{4} \\
0 & 0 & 1 & 0 & \frac{1}{4} & \frac{1}{4} & \frac{1}{4} & -\frac{1}{4} \\
0 & 0 & 0 & 1 & \frac{5}{8} & \frac{1}{8} & -\frac{3}{8} & -\frac{1}{8}
\end{array}\right]
\begin{matrix} -R_3+R_2 \to R_2 \\ -R_3+R_1 \to R_1 \\ \sim \end{matrix}
$$

$$
\left[\begin{array}{cccc|cccc}
1 & 0 & 0 & 0 & \frac{1}{8} & -\frac{3}{8} & \frac{1}{8} & \frac{3}{8} \\
0 & 1 & 0 & 0 & -\frac{1}{2} & \frac{1}{2} & \frac{1}{2} & \frac{1}{2} \\
0 & 0 & 1 & 0 & \frac{1}{4} & \frac{1}{4} & \frac{1}{4} & -\frac{1}{4} \\
0 & 0 & 0 & 1 & \frac{5}{8} & \frac{1}{8} & -\frac{3}{8} & -\frac{1}{8}
\end{array}\right]
\qquad\bullet
\left[\begin{array}{cccc}
\frac{1}{8} & -\frac{3}{8} & \frac{1}{8} & \frac{3}{8} \\
-\frac{1}{2} & \frac{1}{2} & \frac{1}{2} & \frac{1}{2} \\
\frac{1}{4} & \frac{1}{4} & \frac{1}{4} & -\frac{1}{4} \\
\frac{5}{8} & \frac{1}{8} & -\frac{3}{8} & -\frac{1}{8}
\end{array}\right]
$$

$$
[18]\;
\left[\begin{array}{cccc|cccc}
1 & 0 & 0 & 0 & 1 & 0 & 0 & 0 \\
1 & 3 & 0 & 0 & 0 & 1 & 0 & 0 \\
-1 & 1 & 2 & 0 & 0 & 0 & 1 & 0 \\
1 & 2 & 1 & 1 & 0 & 0 & 0 & 1
\end{array}\right]
\begin{matrix} -R_1+R_2 \to R_2 \\ R_1+R_3 \to R_3 \\ -R_1+R_4 \to R_4 \\ \sim \end{matrix}
\left[\begin{array}{cccc|cccc}
1 & 0 & 0 & 0 & 1 & 0 & 0 & 0 \\
0 & 3 & 0 & 0 & -1 & 1 & 0 & 0 \\
0 & 1 & 2 & 0 & 1 & 0 & 1 & 0 \\
0 & 2 & 1 & 1 & -1 & 0 & 0 & 1
\end{array}\right]
\begin{matrix} \frac{1}{3}R_2 \to R_2 \\ \sim \end{matrix}
$$

$$
\left[\begin{array}{cccc|cccc}
1 & 0 & 0 & 0 & 1 & 0 & 0 & 0 \\
0 & 1 & 0 & 0 & \frac{1}{3} & \frac{1}{3} & 0 & 0 \\
0 & 1 & 2 & 0 & 1 & 0 & 1 & 0 \\
0 & 2 & 1 & 1 & -1 & 0 & 0 & 1
\end{array}\right]
\begin{matrix} -R_2+R_3 \to R_3 \\ -2R_2+R_4 \to R_4 \\ \sim \end{matrix}
\left[\begin{array}{cccc|cccc}
1 & 0 & 0 & 0 & 1 & 0 & 0 & 0 \\
0 & 1 & 0 & 0 & -\frac{1}{3} & \frac{1}{3} & 0 & 0 \\
0 & 0 & 2 & 0 & \frac{4}{3} & -\frac{1}{3} & 1 & 0 \\
0 & 0 & 1 & 1 & -\frac{1}{3} & -\frac{2}{3} & 0 & 1
\end{array}\right]
$$

$$
\begin{matrix} \frac{1}{2}R_3 \to R_3 \\ \sim \end{matrix}
\left[\begin{array}{cccc|cccc}
1 & 0 & 0 & 0 & 1 & 0 & 0 & 0 \\
0 & 1 & 0 & 0 & -\frac{1}{3} & \frac{1}{3} & 0 & 0 \\
0 & 0 & 1 & 0 & \frac{2}{3} & -\frac{1}{6} & \frac{1}{2} & 0 \\
0 & 0 & 1 & 1 & -\frac{1}{3} & -\frac{2}{3} & 0 & 1
\end{array}\right]
\begin{matrix} -R_3+R_4 \to R_4 \\ \sim \end{matrix}
$$

$$
\left[\begin{array}{cccc|cccc}
1 & 0 & 0 & 0 & 1 & 0 & 0 & 0 \\
0 & 1 & 0 & 0 & -\frac{1}{3} & \frac{1}{3} & 0 & 0 \\
0 & 0 & 1 & 0 & \frac{2}{3} & -\frac{1}{6} & \frac{1}{2} & 0 \\
0 & 0 & 0 & 1 & -1 & -\frac{1}{2} & -\frac{1}{2} & 1
\end{array}\right]
\qquad\bullet
\left[\begin{array}{cccc}
1 & 0 & 0 & 0 \\
-\frac{1}{3} & \frac{1}{3} & 0 & 0 \\
\frac{2}{3} & -\frac{1}{6} & \frac{1}{2} & 0 \\
-1 & -\frac{1}{2} & -\frac{1}{2} & 1
\end{array}\right]
$$

[19] $\begin{bmatrix} 0.1 & 0.1 & 0.1 & 1 & 0 & 0 \\ -0.2 & 0.2 & -0.2 & 0 & 1 & 0 \\ 0.1 & 0.2 & 0.3 & 0 & 0 & 1 \end{bmatrix}$ $\begin{matrix} 10\,R_1 \to R_1 \\ 10\,R_2 \to R_2 \\ 10\,R_3 \to R_3 \\ \sim \end{matrix}$ $\begin{bmatrix} 1 & 1 & 1 & 10 & 0 & 0 \\ -2 & 2 & -2 & 0 & 10 & 0 \\ 1 & 2 & 3 & 0 & 0 & 10 \end{bmatrix}$

$\begin{matrix} 2\,R_1 + R_2 \to R_2 \\ -R_1 + R_3 \to R_3 \\ \sim \end{matrix}$ $\begin{bmatrix} 1 & 1 & 1 & 10 & 0 & 0 \\ 0 & 4 & 0 & 20 & 10 & 0 \\ 0 & 1 & 2 & -10 & 0 & 10 \end{bmatrix}$ $\frac{1}{4}R_2 \to R_2$ $\sim$ $\begin{bmatrix} 1 & 1 & 1 & 10 & 0 & 0 \\ 0 & 1 & 0 & 5 & \frac{5}{2} & 0 \\ 0 & 1 & 2 & -10 & 0 & 10 \end{bmatrix}$

$\begin{matrix} -R_2 + R_3 \to R_3 \\ \sim \end{matrix}$ $\begin{bmatrix} 1 & 1 & 1 & 10 & 0 & 0 \\ 0 & 1 & 0 & 5 & \frac{5}{2} & 0 \\ 0 & 0 & 2 & -15 & -\frac{5}{2} & 10 \end{bmatrix}$ $\frac{1}{2}R_3 \to R_3$ $\sim$ $\begin{bmatrix} 1 & 1 & 1 & 10 & 0 & 0 \\ 0 & 1 & 0 & 5 & \frac{5}{2} & 0 \\ 0 & 0 & 1 & -\frac{15}{2} & -\frac{5}{4} & 5 \end{bmatrix}$

$\begin{matrix} -R_3 + R_1 \to R_1 \\ \sim \end{matrix}$ $\begin{bmatrix} 1 & 1 & 1 & \frac{35}{2} & \frac{5}{4} & -5 \\ 0 & 1 & 0 & 5 & \frac{5}{2} & 0 \\ 0 & 0 & 1 & -\frac{15}{2} & -\frac{5}{4} & 5 \end{bmatrix}$ $\begin{matrix} -R_2 + R_1 \to R_1 \\ \sim \end{matrix}$

$\begin{bmatrix} 1 & 0 & 0 & \frac{25}{2} & -\frac{5}{4} & -5 \\ 0 & 1 & 0 & 5 & \frac{5}{2} & 0 \\ 0 & 0 & 1 & -\frac{15}{2} & -\frac{5}{4} & 5 \end{bmatrix}$ $\qquad$ $\bullet \begin{bmatrix} \frac{25}{2} & -\frac{5}{4} & -5 \\ 5 & \frac{5}{2} & 0 \\ -\frac{15}{2} & -\frac{5}{4} & 5 \end{bmatrix}$

[20] $\begin{bmatrix} 0.1 & 0.2 & 0.3 & 1 & 0 & 0 \\ -0.1 & -0.3 & 0.1 & 0 & 1 & 0 \\ 0.2 & 0.1 & 0.4 & 0 & 0 & 1 \end{bmatrix}$ $\begin{matrix} 10\,R_1 \to R_1 \\ 10\,R_2 \to R_2 \\ 10\,R_3 \to R_3 \\ \sim \end{matrix}$ $\begin{bmatrix} 1 & 2 & 3 & 10 & 0 & 0 \\ -1 & -3 & 1 & 0 & 10 & 0 \\ 2 & 1 & 4 & 0 & 0 & 10 \end{bmatrix}$

$\begin{matrix} R_1 + R_2 \to R_2 \\ -2\,R_1 + R_3 \to R_3 \\ \sim \end{matrix}$ $\begin{bmatrix} 1 & 2 & 3 & 10 & 0 & 0 \\ 0 & -1 & 4 & 10 & 10 & 0 \\ 0 & -3 & -2 & -20 & 0 & 10 \end{bmatrix}$ $\begin{matrix} -R_2 \to R_2 \\ \sim \end{matrix}$

$\begin{bmatrix} 1 & 2 & 3 & 10 & 0 & 0 \\ 0 & 1 & -4 & -10 & -10 & 0 \\ 0 & -3 & -2 & -20 & 0 & 10 \end{bmatrix}$ $\begin{matrix} 3\,R_2 + R_3 \to R_3 \\ \sim \end{matrix}$ $\begin{bmatrix} 1 & 2 & 3 & 10 & 0 & 0 \\ 0 & 1 & -4 & -10 & -10 & 0 \\ 0 & 0 & -14 & -50 & -30 & 10 \end{bmatrix}$

$-\frac{1}{14}R_3 \to R_3$ $\sim$ $\begin{bmatrix} 1 & 2 & 3 & 10 & 0 & 0 \\ 0 & 1 & -4 & -10 & -10 & 0 \\ 0 & 0 & 1 & \frac{25}{7} & \frac{15}{7} & -\frac{5}{7} \end{bmatrix}$ $\begin{matrix} 4\,R_3 + R_2 \to R_2 \\ -3\,R_3 + R_1 \to R_1 \\ \sim \end{matrix}$

459

$$
\begin{bmatrix}
1 & 2 & 0 \\
0 & 1 & 0 \\
0 & 0 & 1
\end{bmatrix}
\left|
\begin{matrix}
-\dfrac{5}{7} & -\dfrac{45}{7} & \dfrac{15}{7} \\
\dfrac{30}{7} & -\dfrac{10}{7} & -\dfrac{20}{7} \\
\dfrac{25}{7} & \dfrac{15}{7} & -\dfrac{5}{7}
\end{matrix}
\right]
\begin{array}{c} -2R_2+R_1 \to R_1 \\ \sim \end{array}
\begin{bmatrix}
1 & 0 & 0 \\
0 & 1 & 0 \\
0 & 0 & 1
\end{bmatrix}
\left|
\begin{matrix}
-\dfrac{65}{7} & -\dfrac{25}{7} & \dfrac{55}{7} \\
\dfrac{30}{7} & -\dfrac{10}{7} & -\dfrac{20}{7} \\
\dfrac{25}{7} & \dfrac{15}{7} & -\dfrac{5}{7}
\end{matrix}
\right]
$$

$$
\bullet \left[
\begin{matrix}
-\dfrac{65}{7} & -\dfrac{25}{7} & \dfrac{55}{7} \\
\dfrac{30}{7} & -\dfrac{10}{7} & -\dfrac{20}{7} \\
\dfrac{25}{7} & \dfrac{15}{7} & -\dfrac{5}{7}
\end{matrix}
\right]
$$

[21] $\begin{bmatrix} 6 & 1 \\ -4 & 2 \end{bmatrix}\begin{bmatrix} x \\ y \end{bmatrix}=\begin{bmatrix} 4 \\ -3 \end{bmatrix}\Rightarrow\begin{bmatrix} x \\ y \end{bmatrix}=\begin{bmatrix} \frac{1}{8} & -\frac{1}{16} \\ \frac{1}{4} & \frac{3}{8} \end{bmatrix}\begin{bmatrix} 4 \\ -3 \end{bmatrix}=\begin{bmatrix} \frac{11}{16} \\ -\frac{1}{8} \end{bmatrix}$ $\bullet \left\{\left(\frac{11}{16},-\frac{1}{8}\right)\right\}$

[22] $\begin{bmatrix} 6 & 1 \\ -4 & 2 \end{bmatrix}\begin{bmatrix} x \\ y \end{bmatrix}=\begin{bmatrix} 1 \\ 5 \end{bmatrix}\Rightarrow\begin{bmatrix} x \\ y \end{bmatrix}=\begin{bmatrix} \frac{1}{8} & -\frac{1}{16} \\ \frac{1}{4} & \frac{3}{8} \end{bmatrix}\begin{bmatrix} 1 \\ 5 \end{bmatrix}=\begin{bmatrix} -\frac{3}{16} \\ \frac{17}{8} \end{bmatrix}$ $\bullet \left\{\left(-\frac{3}{16},\frac{17}{8}\right)\right\}$

[23] $\begin{bmatrix} \frac{5}{8} & \frac{7}{16} & \frac{1}{16} \\ -\frac{1}{8} & \frac{5}{16} & \frac{3}{16} \\ \frac{1}{4} & -\frac{1}{8} & \frac{1}{8} \end{bmatrix}\begin{bmatrix} -1 \\ 0 \\ 2 \end{bmatrix}=\begin{bmatrix} -\frac{1}{2} \\ \frac{1}{2} \\ 0 \end{bmatrix}$ $\bullet \left\{\left(-\frac{1}{2},\frac{1}{2},0\right)\right\}$

[24] $\begin{bmatrix} \frac{5}{8} & \frac{7}{16} & \frac{1}{16} \\ -\frac{1}{8} & \frac{5}{16} & \frac{3}{16} \\ \frac{1}{4} & -\frac{1}{8} & \frac{1}{8} \end{bmatrix}\begin{bmatrix} 2 \\ 1 \\ 2 \end{bmatrix}=\begin{bmatrix} \frac{29}{16} \\ \frac{7}{16} \\ \frac{5}{8} \end{bmatrix}$ $\bullet \left\{\left(\frac{29}{16},\frac{7}{16},\frac{5}{8}\right)\right\}$

[25] $\begin{bmatrix} \frac{1}{8} & -\frac{3}{8} & \frac{1}{8} & \frac{3}{8} \\ -\frac{1}{2} & \frac{1}{2} & \frac{1}{2} & \frac{1}{2} \\ \frac{1}{4} & \frac{1}{4} & \frac{1}{4} & -\frac{1}{4} \\ \frac{5}{8} & \frac{1}{8} & -\frac{3}{8} & -\frac{1}{8} \end{bmatrix}\begin{bmatrix} 1 \\ 2 \\ 0 \\ 1 \end{bmatrix}=\begin{bmatrix} -\frac{1}{4} \\ 1 \\ \frac{1}{2} \\ \frac{3}{4} \end{bmatrix}$ $\bullet \left\{\left(-\frac{1}{4},1,\frac{1}{2},\frac{3}{4}\right)\right\}$

[26] $\begin{bmatrix} \frac{1}{8} & -\frac{3}{8} & \frac{1}{8} & \frac{3}{8} \\ -\frac{1}{2} & \frac{1}{2} & \frac{1}{2} & \frac{1}{2} \\ \frac{1}{4} & \frac{1}{4} & \frac{1}{4} & -\frac{1}{4} \\ \frac{5}{8} & \frac{1}{8} & -\frac{3}{8} & -\frac{1}{8} \end{bmatrix}\begin{bmatrix} 4 \\ 1 \\ -1 \\ 0 \end{bmatrix}=\begin{bmatrix} 0 \\ -2 \\ 1 \\ 3 \end{bmatrix}$ $\bullet \left\{\left(0,-2,1,3\right)\right\}$

[27] $\begin{bmatrix} 12.5 & -1.25 & -50 \\ 5.0 & 2.5 & 0 \\ -7.5 & -1.25 & 5.0 \end{bmatrix} \begin{bmatrix} 0.7 \\ 0.3 \\ 0.1 \end{bmatrix} = \begin{bmatrix} 7.875 \\ 4.25 \\ -5.25 \end{bmatrix}$ $\quad\bullet \left\{ \left(7.875, 4.25, -5.25 \right) \right\}$

[28] $\begin{bmatrix} 12.5 & -1.25 & -50 \\ 5.0 & 2.5 & 0 \\ -7.5 & -1.25 & 5.0 \end{bmatrix} \begin{bmatrix} 0.2 \\ 0.1 \\ 0.1 \end{bmatrix} = \begin{bmatrix} 1.875 \\ 1.25 \\ -1.125 \end{bmatrix}$ $\quad\bullet \left\{ \left(1.875, 1.25, -1.125 \right) \right\}$

[29] Suppose $AB = BA = I_n$ and $AC = CA = I_n$. Then $B = B I_n = B (AC) = (BA)C = I_n C = C$.

[30] Let $A = \begin{bmatrix} 1 & 0 \\ 0 & 0 \end{bmatrix}$ and $B = \begin{bmatrix} 0 & 0 \\ 0 & 1 \end{bmatrix}$ then $AB = \begin{bmatrix} 0 & 0 \\ 0 & 0 \end{bmatrix}$.

[31] Suppose $AB = [0]_{n,n}$ and A^{-1} exisit. Then $A^{-1}(AB) = A^{-1}[0]_{n,n} = [0]_{n,n}$.

Now $A^{-1}(AB) = \left(A^{-1}A \right)B = I_n B = B$. Hence, $B = [0]_{n,n}$.

[32] Using the matrices $A = \begin{bmatrix} 6 & 1 \\ -4 & 2 \end{bmatrix}$ and $B = \begin{bmatrix} 1 & 1 \\ -1 & 2 \end{bmatrix}$ (from Exercise 5 and 7, respectively),

$AB = \begin{bmatrix} 5 & 8 \\ -6 & 0 \end{bmatrix}$, and it can be shown that $(AB)^{-1} = \begin{bmatrix} 0 & -\dfrac{1}{6} \\ \dfrac{1}{8} & \dfrac{5}{48} \end{bmatrix}$, while

$A^{-1}B^{-1} = \begin{bmatrix} \dfrac{1}{8} & -\dfrac{1}{16} \\ \dfrac{1}{4} & \dfrac{3}{8} \end{bmatrix} \begin{bmatrix} \dfrac{2}{3} & -\dfrac{1}{3} \\ \dfrac{1}{3} & \dfrac{1}{3} \end{bmatrix} = \begin{bmatrix} -\dfrac{5}{48} & -\dfrac{1}{16} \\ \dfrac{7}{24} & \dfrac{1}{24} \end{bmatrix}$.

[33] $(AB)B^{-1}A^{-1} = A \left(B B^{-1} \right)A^{-1} = (A I_n)A^{-1} = A A^{-1} = I_n$. Also $(B^{-1}A^{-1})(AB) =$

$B^{-1}(A^{-1}A)B = (B^{-1}I_n)B = B^{-1}B = I_n$. Therefore $(AB)^{-1} = B^{-1}A^{-1}$.

[34] $A^{-1}A = I_n = A A^{-1}$. Hence $(A^{-1})^{-1} = A$.

[35] If $A = \begin{bmatrix} -1 & 0 \\ 0 & 1 \end{bmatrix}$, $A^{-1} = \begin{bmatrix} -1 & 0 \\ 0 & 1 \end{bmatrix}$. Also, if $B = \begin{bmatrix} 1 & 0 \\ 0 & -1 \end{bmatrix}$, $B^{-1} = \begin{bmatrix} 1 & 0 \\ 0 & -1 \end{bmatrix}$. Now

$A + B = \begin{bmatrix} 0 & 0 \\ 0 & 0 \end{bmatrix}$, which is singular.

[36] Suppose A is a nonsigular matrix.

a) If $AB = AC$, then $A^{-1}(AB) = A^{-1}(AC)$. Now $A^{-1}(AB) = (A^{-1}A)B = I_n B = B$.

Also, $A^{-1}(AC) = (A^{-1}A)C = I_n C = C$. Hence $B = C$.

b) The proof is parallel to the proof given in part (a).

[37] a) Since the graph of f contains the points $(-1, 1)$, $(0, -1)$ and $(1, 2)$, we know

$\begin{cases} a - b + c = 1 \\ c = -1 \\ a + b + c = 2 \end{cases} \Rightarrow \begin{cases} a - b = 2 \\ a + b = 3 \end{cases} \Rightarrow a = \dfrac{5}{2}, \text{ and } b = \dfrac{1}{2}.$ $\quad\bullet a = \dfrac{5}{2}; b = \dfrac{1}{2}; c = -1$

b) Since the graph of f contains the points $(-1, -3)$, $(0, 4)$ and $(1, -1)$, we have

$$\begin{cases} a - b + c = -3 \\ \quad\quad\; c = \;\;4 \\ a + b + c = -1 \end{cases} \Rightarrow \begin{cases} a - b = -7 \\ a + b = -5 \end{cases} \Rightarrow a = -6, \text{ and } b = 1. \qquad \bullet\; a = -6;\, b = 1;\, c = 4$$

[38] $CA = \dfrac{1}{ad - bc} \begin{bmatrix} d & -b \\ -c & a \end{bmatrix} \begin{bmatrix} a & b \\ c & d \end{bmatrix} = \dfrac{1}{ad - bc} \begin{bmatrix} ad - bc & 0 \\ 0 & ad - bc \end{bmatrix} = \begin{bmatrix} 1 & 0 \\ 0 & 1 \end{bmatrix}.$

$AC = \begin{bmatrix} a & b \\ c & d \end{bmatrix} \left(\dfrac{1}{ad - bc} \begin{bmatrix} d & -b \\ -c & a \end{bmatrix} \right) = \dfrac{1}{ad - bc} \begin{bmatrix} a & b \\ c & d \end{bmatrix} \begin{bmatrix} d & -b \\ -c & a \end{bmatrix} =$

$\dfrac{1}{ad - bc} \begin{bmatrix} ad - bc & 0 \\ 0 & ad - bc \end{bmatrix} = \begin{bmatrix} 1 & 0 \\ 0 & 1 \end{bmatrix}.$

[39] T H E P R E S I D E N T I S . H E R E
20 8 5 27 16 18 5 19 9 4 5 14 20 27 9 19 27 8 5 18 5

Multiply each of the following matrices on the right by $A: \begin{bmatrix} 20 & 8 & 5 \end{bmatrix}; \begin{bmatrix} 27 & 16 & 18 \end{bmatrix};$

$\begin{bmatrix} 5 & 19 & 9 \end{bmatrix}; \begin{bmatrix} 4 & 5 & 14 \end{bmatrix}; \begin{bmatrix} 20 & 27 & 9 \end{bmatrix}; \begin{bmatrix} 19 & 27 & 8 \end{bmatrix}; \begin{bmatrix} 5 & 18 & 5 \end{bmatrix}.$

[40] The inverse of A is $\begin{bmatrix} -1 & 2 & 1 \\ -5 & 8 & 2 \\ 7 & -11 & -3 \end{bmatrix}$. The message is : Mathematics is the queen of the sciences.

EXERCISE 9.5

[1] $M_{11} = \det \begin{bmatrix} -1 & 1 \\ 2 & -1 \end{bmatrix} = 1 - 2 = -1;\; A_{11} = (-1)^{1+1}(-1) = -1$

$M_{12} = \det \begin{bmatrix} -1 & 1 \\ 2 & -1 \end{bmatrix} = 1 - 2 = -1;\; A_{11} = (-1)^{1+1}(-1) = -1$

$M_{13} = \det \begin{bmatrix} 0 & -1 \\ -1 & 2 \end{bmatrix} = -1;\; A_{13} = (-1)^{1+3}(-1) = -1$

$M_{21} = \det \begin{bmatrix} 1 & -1 \\ 2 & -1 \end{bmatrix} = 1;\; A_{21} = (-1)^{2+1}(1) = -1$

$M_{22} = \det \begin{bmatrix} 1 & -1 \\ -1 & -1 \end{bmatrix} = -2;\; A_{22} = (-1)^{2+2}(-2) = -2$

$M_{23} = \det \begin{bmatrix} 1 & 1 \\ -1 & 2 \end{bmatrix} = 3;\; A_{23} = (-1)^{2+3}(3) = -3$

$M_{31} = \det \begin{bmatrix} 1 & -1 \\ -1 & 1 \end{bmatrix} = 0;\; A_{31} = (-1)^{3+1}(0) = 0$

$M_{32} = \det \begin{bmatrix} 1 & -1 \\ 0 & 1 \end{bmatrix} = 1;\; A_{32} = (-1)^{3+2}(1) = -1$

$M_{33} = \det \begin{bmatrix} 1 & 1 \\ 0 & -1 \end{bmatrix} = -1;\; A_{33} = (-1)^{3+3}(-1) = -1$

[2] $M_{11} = \det\begin{bmatrix} 1 & 0 \\ -2 & -1 \end{bmatrix} = -1; A_{11} = (-1)^{1+1}(-1) = -1$

$M_{12} = \det\begin{bmatrix} 2 & 0 \\ 1 & -1 \end{bmatrix} = -2; A_{12} = (-1)^{1+2}(-2) = 2$

$M_{13} = \det\begin{bmatrix} 2 & 1 \\ 1 & -2 \end{bmatrix} = -5; A_{13} = (-1)^{1+3}(-5) = -5$

$M_{21} = \det\begin{bmatrix} -1 & 3 \\ -2 & -1 \end{bmatrix} = 7; A_{21} = (-1)^{2+1}(7) = -7$

$M_{22} = \det\begin{bmatrix} 1 & 3 \\ 1 & -1 \end{bmatrix} = -4; A_{22} = (-1)^{2+2}(-4) = -4$

$M_{23} = \det\begin{bmatrix} 1 & -1 \\ 1 & -2 \end{bmatrix} = -1; A_{23} = (-1)^{2+3}(-1) = 1$

$M_{31} = \det\begin{bmatrix} -1 & 3 \\ 1 & 0 \end{bmatrix} = -3; A_{31} = (-1)^{3+1}(-3) = -3$

$M_{32} = \det\begin{bmatrix} 1 & 3 \\ 2 & 0 \end{bmatrix} = -6; A_{32} = (-1)^{3+2}(-6) = 6$

$M_{33} = \det\begin{bmatrix} 1 & -1 \\ 2 & 1 \end{bmatrix} = 3; A_{33} = (-1)^{3+3}(3) = 3$

[3] $M_{11} = \det\begin{bmatrix} -3 & 0 & 1 \\ 0 & 4 & 0 \\ 0 & 0 & 5 \end{bmatrix} = -3(-1)^{1+1}\det\begin{bmatrix} 4 & 0 \\ 0 & 5 \end{bmatrix} = -60; A_{11} = (-1)^{1+1}(-60) = -60$

$M_{12} = \det\begin{bmatrix} 0 & 0 & 0 \\ 0 & 4 & 0 \\ 0 & 0 & 5 \end{bmatrix} = 0; A_{12} = 0$ $M_{13} = \det\begin{bmatrix} 0 & -3 & 0 \\ 0 & 0 & 4 \\ 0 & 0 & 0 \end{bmatrix} = 0; A_{13} = 0$

$M_{14} = \det\begin{bmatrix} 0 & -3 & 0 \\ 0 & 0 & 4 \\ 0 & 0 & 5 \end{bmatrix} = 0; A_{14} = 0$ $M_{21} = \det\begin{bmatrix} 0 & 0 & 0 \\ 0 & 4 & 0 \\ 0 & 0 & 5 \end{bmatrix} = 0; A_{21} = 0$

$M_{22} = \det\begin{bmatrix} 2 & 0 & 0 \\ 0 & 4 & 0 \\ 0 & 0 & 5 \end{bmatrix} = -2(-1)^{1+1}\det\begin{bmatrix} 4 & 0 \\ 0 & 5 \end{bmatrix} = 40; A_{22} = (-1)^{2+2}(40) = 40$

$M_{23} = \det\begin{bmatrix} 2 & 0 & 0 \\ 0 & 0 & 0 \\ 0 & 0 & 5 \end{bmatrix} = 0; A_{23} = 0$ $M_{24} = \det\begin{bmatrix} 2 & 0 & 0 \\ 0 & 0 & 4 \\ 0 & 0 & 0 \end{bmatrix} = 0; A_{24} = 0$

$M_{31} = \det\begin{bmatrix} 0 & 0 & 0 \\ -3 & 0 & 0 \\ 0 & 0 & 5 \end{bmatrix} = 0; A_{31} = 0$ $M_{32} = \det\begin{bmatrix} 2 & 0 & 0 \\ 0 & 0 & 0 \\ 0 & 0 & 5 \end{bmatrix} = 0; A_{32} = 0$

$M_{33} = \det\begin{bmatrix} 2 & 0 & 0 \\ 0 & -3 & 0 \\ 0 & 0 & 5 \end{bmatrix} = 2(-1)^{1+1}\det\begin{bmatrix} -3 & 0 \\ 0 & 5 \end{bmatrix} = -30; A_{33} = (-1)^{3+3}(-30) = -30$

$M_{34} = A_{34} = M_{41} = A_{41} = M_{42} = A_{42} = M_{43} = A_{43} = 0$

$M_{44} = \det\begin{bmatrix} 2 & 0 & 0 \\ 0 & -3 & 0 \\ 0 & 0 & 4 \end{bmatrix} = 2(-1)^{1+1}\det\begin{bmatrix} -3 & 0 \\ 0 & 4 \end{bmatrix} = -24; A_{44} = (-1)^{4+4}(-24) = -24$

[4] $M_{11} = \det\begin{bmatrix} -7 & 0 & 0 \\ 1 & -5 & 0 \\ 0 & 0 & 1 \end{bmatrix} = 1(-1)^{3+3} \det\begin{bmatrix} -7 & 0 \\ 1 & -5 \end{bmatrix} = 35;\ A_{11} = 35$

$M_{12} = \det\begin{bmatrix} 0 & 0 & 0 \\ 0 & -5 & 0 \\ 0 & 0 & 1 \end{bmatrix} = 0;\ A_{12} = 0$

$M_{13} = \det\begin{bmatrix} 0 & -7 & 0 \\ 1 & 1 & 0 \\ 0 & 0 & 1 \end{bmatrix} = 1(-1)^{1+3} \det\begin{bmatrix} 0 & -7 \\ 1 & 1 \end{bmatrix} = 7;\ A_{13} = (-1)^{1+3}7 = 7$

$M_{14} = \det\begin{bmatrix} 0 & -7 & 0 \\ 0 & 1 & -5 \\ 0 & 0 & 0 \end{bmatrix} = 0;\ A_{14} = 0$

Proceeding as above, we find that : $M_{21} = 0;\ A_{21} = 0;\ M_{22} = 5;\ A_{22} = 5;\ M_{23} = -1;\ A_{23} = 1;$

$M_{24} = 0;\ A_{24} = 0;\ M_{31} = 0;\ A_{31} = 0;\ M_{32} = 0;\ A_{32} = 0;\ M_{33} = 7;\ A_{33} = 7;\ M_{34} = 0;\ A_{34} = 0;$

$M_{41} = 0;\ A_{41} = 0;\ M_{42} = 0;\ A_{42} = 0;\ M_{43} = 0;\ A_{43} = 0;\ M_{44} = -35;\ A_{44} = -35.$

[5] $\det\begin{bmatrix} 6 & -1 \\ 4 & 5 \end{bmatrix} = 30 - (-4) = 34$ ● 34

[6] $\det\begin{bmatrix} 11 & -6 \\ 2 & 7 \end{bmatrix} = 77 - (-12) = 89$ ● 89

[7] $\det\begin{bmatrix} \sin x & \cos x \\ \cos x & -\sin x \end{bmatrix} = -\sin^2 x - \cos^2 x = -(\sin^2 x + \cos^2 x) = -1$ ● -1

[8] $\det\begin{bmatrix} e^x & xe^x \\ e^x & xe^x + e^x \end{bmatrix} = xe^{2x} + e^{2x} - xe^{2x} = e^{2x}$ ● e^{2x}

[9] $\det\begin{bmatrix} 1 & 1 & -1 \\ 0 & -1 & 1 \\ -1 & 2 & -1 \end{bmatrix} \underset{=}{R_2 + R_1 \rightarrow R_1} \det\begin{bmatrix} 1 & 0 & 0 \\ 0 & -1 & 1 \\ -1 & 2 & -1 \end{bmatrix} (\text{expanding by } R_1) = 1$

$\det\begin{bmatrix} -1 & 1 \\ 2 & -1 \end{bmatrix} = -1$ ● -1

[10] $\det\begin{bmatrix} -6 & 1 & 4 \\ 2 & -1 & 5 \\ -1 & -1 & 1 \end{bmatrix} \underset{=}{R_2 + R_1 \rightarrow R_1} \det\begin{bmatrix} -4 & 0 & 9 \\ 2 & -1 & 5 \\ -1 & -1 & 1 \end{bmatrix} (\text{expanding by } R_1) = -4 \cdot$

$\det\begin{bmatrix} -1 & 5 \\ -1 & 1 \end{bmatrix} + 9 \det\begin{bmatrix} 2 & -1 \\ -1 & -1 \end{bmatrix} = -4(4) + 9(-3) = -43$ ● -43

[11] $\det\begin{bmatrix} 3 & 2 & 1 \\ 1 & 1 & -1 \\ 4 & 2 & 5 \end{bmatrix} \underset{=}{R_2 + R_1 \rightarrow R_1} \det\begin{bmatrix} 4 & 3 & 0 \\ 1 & 1 & -1 \\ 4 & 2 & 5 \end{bmatrix} (\text{expanding by } R_1) = 4 \det\begin{bmatrix} 1 & -1 \\ 2 & 5 \end{bmatrix} +$

$3 \det\begin{bmatrix} 1 & -1 \\ 4 & 5 \end{bmatrix} = -4(7) - 3(9) = 1$ ● 1

[12] $\det\begin{bmatrix} -6 & 1 & 4 \\ -2 & 1 & -5 \\ -1 & 1 & 1 \end{bmatrix} \underset{=}{-R_2 + R_1 \rightarrow R_1} \det\begin{bmatrix} -4 & 0 & 9 \\ -2 & 1 & -5 \\ -1 & 1 & 1 \end{bmatrix} (\text{expanding by } R_1) = -4$

$\det\begin{bmatrix} 1 & -5 \\ 1 & 1 \end{bmatrix} + 9 \det\begin{bmatrix} -2 & 1 \\ -1 & 1 \end{bmatrix} = -4(6) + 9(-1) = -33$ ● -33

[13] $\det\begin{bmatrix} 1 & 1 & 2 \\ 4 & 3 & 6 \\ 1 & -1 & -2 \end{bmatrix} \underset{=}{R_3 + R_1 \to R_1} \det\begin{bmatrix} 2 & 0 & 0 \\ 4 & 3 & 6 \\ 1 & -1 & -2 \end{bmatrix}$ (expanding by R_1) $= 2 \cdot$

$\det\begin{bmatrix} 3 & 6 \\ -1 & -2 \end{bmatrix} = 2(0) = 0$ ● 0

[14] $\det\begin{bmatrix} 2 & -1 & 2 \\ 1 & 3 & 2 \\ 5 & 1 & 6 \end{bmatrix} \underset{=}{-R_2 + R_1 \to R_1} \det\begin{bmatrix} 1 & -4 & 0 \\ 1 & 3 & 2 \\ 5 & 1 & 6 \end{bmatrix}$ (expanding by R_1) $= 1 \det\begin{bmatrix} 3 & 2 \\ 1 & 6 \end{bmatrix} +$

$4 \det\begin{bmatrix} 1 & 2 \\ 5 & 6 \end{bmatrix} = 1(16) + 4(-4) = 0$ ● 0

[15] $\det\begin{bmatrix} 2 & 0 & 0 & 0 \\ 0 & -3 & 0 & 0 \\ 0 & 0 & 4 & 0 \\ 0 & 0 & 0 & 5 \end{bmatrix}$ (expanding by R_1) $= 2 \det\begin{bmatrix} -3 & 0 & 0 \\ 0 & 4 & 0 \\ 0 & 0 & 5 \end{bmatrix}$ (expanding by R_1) $=$

$2(-3) \det\begin{bmatrix} 4 & 0 \\ 0 & 5 \end{bmatrix} 2(-3)(20) = -120$ ● -120

[16] $\det\begin{bmatrix} -1 & 0 & 0 & 0 \\ 0 & -7 & 0 & 0 \\ 1 & 1 & -5 & 0 \\ 2 & -2 & 0 & 1 \end{bmatrix}$ (expanding by R_1) $= -1 \det\begin{bmatrix} -7 & 0 & 0 \\ 1 & -5 & 0 \\ -2 & 0 & 1 \end{bmatrix}$ (expanding by R_1) $=$

$(-1)(-7) \det\begin{bmatrix} -5 & 0 \\ 0 & 1 \end{bmatrix} = (-1)(-7)(-5) = -35$ ● -35

[17] $\det\begin{bmatrix} i & j & k \\ 1 & 2 & -1 \\ 1 & 3 & 0 \end{bmatrix}$ (expanding by R_1) $= i \det\begin{bmatrix} 2 & -1 \\ 3 & 0 \end{bmatrix} - j \det\begin{bmatrix} 1 & -1 \\ 1 & 0 \end{bmatrix} + k \det\begin{bmatrix} 1 & 2 \\ 1 & 3 \end{bmatrix} =$

$3i - j + k$

[18] $\det\begin{bmatrix} i & j & k \\ 2 & 1 & 0 \\ 1 & -1 & 1 \end{bmatrix}$ (expanding by R_1) $= i \det\begin{bmatrix} 1 & 0 \\ -1 & 1 \end{bmatrix} - j \det\begin{bmatrix} 2 & 0 \\ 1 & 1 \end{bmatrix} +$

$k \det\begin{bmatrix} 2 & 1 \\ 1 & -1 \end{bmatrix} = i - 2j - 3k$

[19] $\det\begin{bmatrix} 1 & 1 & -1 & 2 \\ 4 & 1 & 1 & 0 \\ 5 & -1 & 0 & 1 \\ 0 & 0 & 1 & -2 \end{bmatrix} \underset{=}{2C_3 + C_4 \to C_4} \det\begin{bmatrix} 1 & 1 & -1 & 0 \\ 4 & 1 & 1 & 2 \\ 5 & -1 & 0 & 1 \\ 0 & 0 & 1 & 0 \end{bmatrix}$ (expanding by R_4) $=$

$1(-1) \det\begin{bmatrix} 1 & 1 & 0 \\ 4 & 1 & 2 \\ 5 & -1 & 1 \end{bmatrix}$ (expanding by R_1) $= (-1) \left(1 \det\begin{bmatrix} 1 & 2 \\ -1 & 1 \end{bmatrix} - 1 \det\begin{bmatrix} 4 & 2 \\ 5 & 1 \end{bmatrix} \right) =$

$-1[3 - 1(-6)] = -9$ ● -9

[20] $\det\begin{bmatrix} 2 & 4 & 0 & 1 \\ 0 & 1 & 2 & -1 \\ 1 & 1 & 0 & 2 \\ 1 & 0 & 1 & 0 \end{bmatrix}$ (expanding by R_4) $=(-1)\det\begin{bmatrix} 4 & 0 & 1 \\ 1 & 2 & -1 \\ 1 & 0 & 2 \end{bmatrix}+(-1)\cdot$

$\det\begin{bmatrix} 2 & 4 & 1 \\ 0 & 1 & -1 \\ 1 & 1 & 2 \end{bmatrix}=(-1)\left(4\det\begin{bmatrix} 2 & -1 \\ 0 & 2 \end{bmatrix}+1\det\begin{bmatrix} 1 & 2 \\ 1 & 0 \end{bmatrix}\right)+$

$(-1)\left(2\det\begin{bmatrix} 1 & -1 \\ 1 & 2 \end{bmatrix}+1\det\begin{bmatrix} 4 & 1 \\ 1 & -1 \end{bmatrix}\right)=$

$(-1)[4(4)+(1)(-2)]+(-1)(6-5)=(-1)(14)-1=-15$ ● -15

[21] $\det\begin{bmatrix} 2-x & 1 \\ -1 & -x \end{bmatrix}=(2-x)(-x)-(-1)(1)=-2x+x^2+1=x^2-2x+1$

[22] $\det\begin{bmatrix} 1-x & -3 \\ 2 & 3-x \end{bmatrix}=(1-x)(3-x)-(2)(-3)=3-x-3x+x^2+6=x^2-4x+9$

● x^2-4x+9

[23] $\det\begin{bmatrix} x-5 & -8 & -16 \\ -4 & x-1 & -8 \\ 4 & 4 & x+11 \end{bmatrix}$ (*expanding by R_1*) $=(x-5)\det\begin{bmatrix} x-1 & -8 \\ 4 & x+11 \end{bmatrix}+$

$8\det\begin{bmatrix} -4 & -8 \\ 4 & x+11 \end{bmatrix}+16\det\begin{bmatrix} -4 & x-1 \\ 4 & 4 \end{bmatrix}=$

$(x-5)[(x-1)(x+11)+32]+8[-4(x+11)+32]-16[-16-4(x-1)]=$

$(x-5)[x^2+10x+21]+8(-4x-12)-16(-12-4x)=x^3+5x^2+3x-9$ ● x^3+5x^2+3x-9

[24] $\det\begin{bmatrix} x-2 & 0 & -1 & -2 \\ 0 & x-2 & 1 & -3 \\ 0 & 0 & x+3 & -1 \\ 0 & 0 & 0 & x-4 \end{bmatrix}$ (expanding by C_1) $=(x-2)\det\begin{bmatrix} x-2 & 1 & -1 \\ 0 & x+3 & -1 \\ 0 & 0 & x-4 \end{bmatrix}$

(expanding by C_1) $=(x-2)^2\det\begin{bmatrix} x+3 & -1 \\ 0 & x-4 \end{bmatrix}=(x-2)^2(x+3)(x-4)$

● $(x-2)^2(x+3)(x-4)$

[25] a) $x^2-2x+1=0\Rightarrow x=1$ ● 1

b) $x^2-4x+9=0\Rightarrow x=\dfrac{4\pm\sqrt{16-36}}{2}$ or $x=2\pm\sqrt{5}i$ ● $2\pm\sqrt{5}i$

c) $x^3+5x^2+3x-9=0\Rightarrow(x-1)(x^2+6x+9)=0\Rightarrow x=-3$ or $x=1$ ● $-3,1$

d) $(x-2)^2(x+3)(x-4)=0\Rightarrow x=-3$, or $x=2$, or $x=4$ ● $-3,2,4$

[26] The formula can be established by using the definition of the determinant and evaluating each matrix

by expanding by the first column of each.

[27] $\det A=4(5)(-1)(-3)(7)=420$ [28] $\det A=1(-1)(-4)(5)(-2)=-40$

[29] $\det A=1$ [30] $\det A=4^4=256$

[31] a)   The value of the determinant is the product of the element on the main diagonal.

b) $\det \begin{bmatrix} 3 & 0 & 0 & 0 \\ 1 & 2 & 0 & 0 \\ 4 & 0 & 2 & 0 \\ 3 & -1 & 1 & 3 \end{bmatrix}$ (expanding by R_1) $= 3 \det \begin{bmatrix} 2 & 0 & 0 \\ 0 & 2 & 0 \\ -1 & 1 & 3 \end{bmatrix}$ (expanding by R_1) $=$

$3 (2) \det \begin{bmatrix} 2 & 0 \\ 1 & 3 \end{bmatrix} = 3 (2) (2) (3) = 36$

c) (See Exercise 26) $a_{11} \cdot a_{22} \cdot \ldots \cdot a_{nn}$.

[32] a) Expanding by the row (or column) that has 0 as each of its entries it is clear that the value of the determinant is zero.

b) If two rows (or columns) are identical, a row (or column) can be introduced by subtracting each entry in one of the rows (or columns) from the corresponding entry in the other row (or column). By Theorem 9.5, we know that the determinant of the new matrix is equal to the determinant of the original. From part (a), we know that the determinant is zero.

c) A sequence of elementary row (or column) operations will introduce a row (or column) that has 0 as each of its entries. It follows that the determinant of the original matrix is zero.

[33] Theorem 9.5 (part *i*)

[34] If a row of a matrix is a multiple of another row, the determinant is zero.

[35] Theorem 9.5 (part *ii*)   [36] Theorem 9.5 (part *ii*)

[37] Theorem 9.5 (part *iii*)   [38] Theorem 9.5 (part *iii*)

[39] $\det \begin{bmatrix} 6 & -1 \\ 4 & 5 \end{bmatrix} = 34$; $x = \dfrac{1}{34} \det \begin{bmatrix} 4 & -1 \\ -7 & 5 \end{bmatrix} = \dfrac{13}{34}$; $y = \dfrac{1}{34} \det \begin{bmatrix} 6 & 4 \\ 4 & -7 \end{bmatrix} = -\dfrac{29}{17}$ $\bullet \left\{ \left(\dfrac{13}{34}, -\dfrac{29}{17} \right) \right\}$

[40] $\det \begin{bmatrix} 11 & -6 \\ 2 & 7 \end{bmatrix} = 89$; $x = \dfrac{1}{89} \det \begin{bmatrix} -3 & -6 \\ 5 & 7 \end{bmatrix} = \dfrac{9}{89}$; $y = \dfrac{1}{89} \det \begin{bmatrix} 11 & -3 \\ 2 & 5 \end{bmatrix} = \dfrac{61}{89}$ $\bullet \left\{ \left(\dfrac{9}{89}, \dfrac{61}{89} \right) \right\}$

[41] $\det \begin{bmatrix} 8 & -9 \\ 15 & 8 \end{bmatrix} = 199$; $x = \dfrac{1}{199} \det \begin{bmatrix} 14 & -9 \\ 5 & 8 \end{bmatrix} = \dfrac{157}{199}$; $y = \dfrac{1}{199} \det \begin{bmatrix} 8 & 14 \\ 15 & 5 \end{bmatrix} = -\dfrac{170}{199}$

$\bullet \left\{ \left(\dfrac{157}{199}, -\dfrac{170}{199} \right) \right\}$

[42] $\det \begin{bmatrix} 19 & -42 \\ 52 & 14 \end{bmatrix} = 2450$; $x = \dfrac{1}{2450} \det \begin{bmatrix} 10 & -42 \\ -7 & 14 \end{bmatrix} = -\dfrac{11}{175}$; $y = \dfrac{1}{2450} \det \begin{bmatrix} 19 & 10 \\ 52 & -7 \end{bmatrix} =$

$-\dfrac{653}{2450}$ $\bullet \left\{ \left(-\dfrac{11}{175}, -\dfrac{653}{2450} \right) \right\}$

[43] $\det\begin{bmatrix} 1 & 1 & -1 \\ 0 & -1 & 1 \\ -1 & 2 & -1 \end{bmatrix} \begin{matrix} R_1 + R_1 \to R_2 \\ = \end{matrix} \det\begin{bmatrix} 1 & 0 & 0 \\ 0 & -1 & 1 \\ -1 & 2 & -1 \end{bmatrix} = 1 \cdot \det\begin{bmatrix} -1 & 1 \\ 2 & -1 \end{bmatrix} = -1;$

$x = -\det\begin{bmatrix} 2 & 1 & -1 \\ 5 & -1 & 1 \\ 0 & 2 & -1 \end{bmatrix} \begin{matrix} R_2 + R_1 \to R_1 \\ = \end{matrix} -\det\begin{bmatrix} 7 & 0 & 0 \\ 5 & -1 & 1 \\ 0 & 2 & -1 \end{bmatrix} = -7\det\begin{bmatrix} -1 & 1 \\ 2 & -1 \end{bmatrix} = 7;$

$y = -\det\begin{bmatrix} 1 & 2 & -1 \\ 0 & 5 & 1 \\ -1 & 0 & -1 \end{bmatrix} \begin{matrix} R_1 + R_3 \to R_3 \\ = \end{matrix} -\det\begin{bmatrix} 1 & 2 & -1 \\ 0 & 5 & 1 \\ 0 & 2 & -2 \end{bmatrix} = -1\det\begin{bmatrix} 5 & 1 \\ 2 & -2 \end{bmatrix} = 12;$

$z = -\det\begin{bmatrix} 1 & 1 & 2 \\ 0 & -1 & 5 \\ -1 & 2 & 0 \end{bmatrix} \begin{matrix} R_1 + R_3 \to R_3 \\ = \end{matrix} -\det\begin{bmatrix} 1 & 1 & 2 \\ 0 & -1 & 5 \\ 0 & 3 & 2 \end{bmatrix} = -1\det\begin{bmatrix} -1 & 5 \\ 3 & 2 \end{bmatrix} = 17.$

$\bullet\{(7, 12, 17)\}$

[44] $\det\begin{bmatrix} 1 & -1 & 3 \\ 2 & 1 & 0 \\ 1 & -2 & -1 \end{bmatrix} \begin{matrix} 3R_3 + R_1 \to R_1 \\ = \end{matrix} -\det\begin{bmatrix} 4 & -7 & 0 \\ 2 & 1 & 0 \\ 1 & -2 & -1 \end{bmatrix} = -1 \cdot \det\begin{bmatrix} 4 & -7 \\ 2 & 1 \end{bmatrix} = -18;$

$x = -\frac{1}{18}\det\begin{bmatrix} 1 & -1 & 3 \\ 1 & 1 & 0 \\ 1 & -2 & -1 \end{bmatrix} \begin{matrix} -R_1 + R_2 \to R_2 \\ -R_1 + R_3 \to R_3 \\ = \end{matrix}$

$-\frac{1}{18}\det\begin{bmatrix} 1 & -1 & 3 \\ 0 & 2 & -3 \\ 0 & -1 & -4 \end{bmatrix} =$

$-\frac{1}{18} \cdot \det\begin{bmatrix} 2 & -3 \\ -1 & -4 \end{bmatrix} = \frac{11}{18}; \quad y = -\frac{1}{18}\det\begin{bmatrix} 1 & 1 & 3 \\ 2 & 1 & 0 \\ 1 & 1 & -1 \end{bmatrix} \begin{matrix} 3R_3 + R_1 \to R_1 \\ = \end{matrix}$

$-\frac{1}{18}\det\begin{bmatrix} 4 & 4 & 0 \\ 2 & 1 & 0 \\ 1 & 1 & -1 \end{bmatrix} = -\frac{1}{18}(-1)\det\begin{bmatrix} 4 & 4 \\ 2 & 1 \end{bmatrix} = -\frac{2}{9}; \quad z =$

$-\frac{1}{18}\det\begin{bmatrix} 1 & -1 & 1 \\ 2 & 1 & 1 \\ 1 & -2 & 1 \end{bmatrix} \begin{matrix} -R_3 + R_1 \to R_2 \\ = \end{matrix}$

$-\frac{1}{18} \cdot \det\begin{bmatrix} 0 & 1 & 0 \\ 2 & 1 & 1 \\ 1 & -2 & -1 \end{bmatrix} = -\frac{1}{18}(-1)\det\begin{bmatrix} 2 & 1 \\ 1 & 1 \end{bmatrix} = \frac{1}{18}$
$\bullet\left\{\left(\frac{11}{18}, -\frac{2}{9}, \frac{1}{18}\right)\right\}$

468

[45] $\det \begin{bmatrix} 5 & 6 & -1 \\ 6 & -8 & -4 \\ -3 & 4 & 2 \end{bmatrix} \begin{matrix} -4R_1 + R_2 \to R_2 \\ 2R_1 + R_3 \to R_3 \\ = \end{matrix}$

$\det \begin{bmatrix} 5 & 6 & -1 \\ -14 & -32 & 0 \\ 7 & 16 & 0 \end{bmatrix} = -1 \det \begin{bmatrix} -14 & -32 \\ 7 & 16 \end{bmatrix} = 0;$

Hence, Cramer's Rule can not be used to find the solution set. $\begin{bmatrix} 5 & 6 & -1 & | & 1 \\ 6 & -8 & -4 & | & -3 \\ -3 & 4 & 2 & | & 0 \end{bmatrix} \begin{matrix} -R_1 \to R_1 \\ \sim \end{matrix}$

$\begin{bmatrix} -5 & -6 & 1 & | & -1 \\ 6 & -8 & -4 & | & -3 \\ -3 & 4 & 2 & | & 0 \end{bmatrix} \begin{matrix} R_2 + R_1 \to R_1 \\ \sim \end{matrix} \begin{bmatrix} 1 & -14 & -3 & | & -4 \\ 6 & -8 & -4 & | & -3 \\ -3 & 4 & 2 & | & 0 \end{bmatrix} \begin{matrix} -6R_1 + R_2 \to R_2 \\ 3R_1 + R_3 \to R_3 \\ \sim \end{matrix}$

$\begin{bmatrix} 1 & -14 & -3 & | & -4 \\ 0 & 76 & 14 & | & 21 \\ 0 & -38 & -7 & | & -12 \end{bmatrix} \begin{matrix} \frac{1}{76}R_2 \to R_2 \\ \sim \end{matrix} \begin{bmatrix} 1 & -14 & -3 & | & -4 \\ 0 & 1 & \frac{7}{38} & | & \frac{21}{76} \\ 0 & -38 & -7 & | & -12 \end{bmatrix} \begin{matrix} 38R_2 + R_3 \to R_3 \\ \sim \end{matrix}$

$\begin{bmatrix} 1 & -14 & -3 & | & -4 \\ 0 & 1 & \frac{7}{38} & | & \frac{21}{76} \\ 0 & 0 & 0 & | & -\frac{3}{2} \end{bmatrix} R_3 : 0 = -\frac{3}{2}.$ This equation is false.   ● $\varnothing$

[46] $\det \begin{bmatrix} 10 & 10 & -10 \\ 15 & -13 & 10 \\ \frac{5}{2} & \frac{5}{2} & \frac{5}{2} \end{bmatrix} \begin{matrix} -R_3 + R_1 \to R_1 \\ = \end{matrix} \det \begin{bmatrix} 0 & 0 & 0 \\ 15 & -13 & 10 \\ \frac{5}{2} & \frac{5}{2} & \frac{5}{2} \end{bmatrix} = 0.$ Hence, Cramer's Rule can not be

used to find the solution set. $\begin{bmatrix} 10 & 10 & -10 & | & 3 \\ 15 & -13 & 10 & | & -1 \\ \frac{5}{2} & \frac{5}{2} & -\frac{5}{2} & | & 4 \end{bmatrix} \begin{matrix} R_1 \leftrightarrow R_3 \\ \sim \end{matrix} \begin{bmatrix} \frac{5}{2} & \frac{5}{2} & -\frac{5}{2} & | & 3 \\ 15 & -13 & 10 & | & -1 \\ 10 & 10 & -10 & | & 3 \end{bmatrix} \begin{matrix} \frac{2}{5}R_1 \to R_1 \\ \sim \end{matrix}$

$\begin{bmatrix} 1 & 1 & -1 & | & \frac{6}{5} \\ 15 & -13 & 10 & | & -1 \\ 10 & 10 & -10 & | & 3 \end{bmatrix} \begin{matrix} -15R_1 + R_2 \to R_2 \\ -10R_1 + R_3 \to R_3 \\ \sim \end{matrix} \begin{bmatrix} 1 & 1 & -1 & | & \frac{6}{5} \\ 0 & -28 & 25 & | & -19 \\ 0 & 0 & 0 & | & -9 \end{bmatrix} R_3 : 0 = -9.$ This equation is

false.   ● $\varnothing$

[47] $\det\begin{bmatrix} -1 & 3 & -4 \\ 1 & 1 & -6 \\ 4 & 6 & -1 \end{bmatrix} \begin{matrix} R_1 + R_2 \to R_2 \\ 4R_1 + R_3 \to R_3 \\ = \end{matrix}$

$\det\begin{bmatrix} -1 & 3 & -4 \\ 0 & 4 & -10 \\ 0 & 18 & -17 \end{bmatrix} = -1 \cdot \det\begin{bmatrix} 4 & -10 \\ 18 & -17 \end{bmatrix} = -112.$

$x = -\frac{1}{112} \cdot \det\begin{bmatrix} 8 & 3 & -4 \\ -1 & 1 & -6 \\ 2 & 6 & -1 \end{bmatrix} \begin{matrix} 8R_1 + R_2 \to R_2 \\ 2R_2 + R_3 \to R_3 \\ = \end{matrix}$

$-\frac{1}{112} \cdot \det\begin{bmatrix} 0 & 11 & -52 \\ -1 & 1 & -6 \\ 0 & 8 & -13 \end{bmatrix} = -\frac{1}{112} \cdot (-1) \cdot (-1)^3 \cdot \det\begin{bmatrix} 11 & -52 \\ 8 & -13 \end{bmatrix} = -\frac{39}{16}.$

$y = -\frac{1}{112} \cdot \det\begin{bmatrix} -1 & 8 & -4 \\ 1 & -1 & -6 \\ 4 & 2 & -1 \end{bmatrix} \begin{matrix} R_1 + R_2 \to R_2 \\ 4R_1 + R_3 \to R_3 \\ = \end{matrix}$

$-\frac{1}{112} \cdot \det\begin{bmatrix} -1 & 8 & -4 \\ 0 & 7 & -10 \\ 0 & 34 & -17 \end{bmatrix} = -\frac{1}{112} \cdot (-1) \cdot \det\begin{bmatrix} 7 & -10 \\ 34 & -17 \end{bmatrix} = \frac{221}{112}.$

$z = -\frac{1}{112} \cdot \det\begin{bmatrix} -1 & 3 & 8 \\ 1 & 1 & -1 \\ 4 & 6 & 2 \end{bmatrix} \begin{matrix} R_1 + R_2 \to R_2 \\ 4R_1 + R_3 \to R_3 \\ = \end{matrix} -\frac{1}{112} \cdot \det\begin{bmatrix} -1 & 3 & 8 \\ 0 & 4 & 7 \\ 0 & 18 & 34 \end{bmatrix} =$

$-\frac{1}{112} \cdot (-1) \cdot \det\begin{bmatrix} 4 & 7 \\ 18 & 34 \end{bmatrix} = \frac{5}{56}.$ $\qquad \bullet \left\{ \left(-\frac{39}{16}, \frac{221}{112}, \frac{5}{56} \right) \right\}$

[48] $\det\begin{bmatrix} 2 & -1 & -1 \\ 1 & 2 & 1 \\ 3 & 1 & -1 \end{bmatrix} \begin{matrix} -2R_2 + R_1 \to R_1 \\ -3R_2 + R_3 \to R_3 \\ = \end{matrix}$

$\det\begin{bmatrix} 0 & -5 & -3 \\ 1 & 2 & 1 \\ 0 & -5 & -4 \end{bmatrix} = 1 \cdot (-1)^3 \cdot \det\begin{bmatrix} -5 & -3 \\ -5 & -4 \end{bmatrix} = -5. \; x = -\frac{1}{5}.$

$\det\begin{bmatrix} 1 & -1 & -1 \\ 0 & 2 & 1 \\ 5 & 1 & -1 \end{bmatrix} \begin{matrix} -5R_1 + R_3 \to R_3 \\ = \end{matrix} -\frac{1}{5} \cdot \det\begin{bmatrix} 1 & -1 & -1 \\ 0 & 2 & 1 \\ 0 & 6 & 4 \end{bmatrix} = -\frac{1}{5} \cdot 1 \cdot \det\begin{bmatrix} 2 & 1 \\ 6 & 4 \end{bmatrix} = -\frac{2}{5}. \; y = -\frac{1}{5}.$

$\det\begin{bmatrix} 2 & 1 & -1 \\ 1 & 0 & 1 \\ 3 & 5 & -1 \end{bmatrix} \begin{matrix} -2R_2 + R_1 \to R_1 \\ -3R_2 + R_3 \to R_3 \\ = \end{matrix}$

$-\frac{1}{5} \cdot \det\begin{bmatrix} 0 & 1 & -3 \\ 1 & 0 & 1 \\ 0 & 5 & -4 \end{bmatrix} = -\frac{1}{5} \cdot 1 \cdot (-1)^3 \cdot \det\begin{bmatrix} 1 & -3 \\ 5 & -4 \end{bmatrix} = \frac{11}{5}.$

$z = -\frac{1}{5} \cdot \det\begin{bmatrix} 2 & -1 & 1 \\ 1 & 2 & 0 \\ 3 & 1 & 5 \end{bmatrix} \begin{matrix} -5R_1 + R_3 \to R_3 \\ = \end{matrix} -\frac{1}{5} \cdot \det\begin{bmatrix} 2 & -1 & 1 \\ 1 & 2 & 0 \\ -7 & 6 & 0 \end{bmatrix} = -\frac{1}{5} \cdot 1 \cdot \det\begin{bmatrix} 1 & 2 \\ -7 & 6 \end{bmatrix} = -4.$

$\qquad \bullet \left\{ \left(-\frac{2}{5}, \frac{11}{5}, -4 \right) \right\}$

[49]   The four fourth roots of unity are $1, i, -1, -i$. Now

$$\det\begin{bmatrix} 1 & i & -1 & -i \\ i & -1 & -i & 1 \\ -1 & -i & 1 & i \\ -i & 1 & i & -1 \end{bmatrix} \underset{=}{R_1+R_3} \det\begin{bmatrix} 1 & i & -1 & -i \\ i & -1 & -i & 1 \\ 0 & 0 & 0 & 0 \\ -i & 1 & i & -1 \end{bmatrix} = 0.$$

[50] $\dfrac{1}{2}\left|\det M\right| = \dfrac{1}{2}\cdot\left|\det\begin{bmatrix} x_1 & y_1 & 1 \\ x_2 & y_2 & 1 \\ x_3 & y_3 & 1 \end{bmatrix}\right| \underset{=}{\substack{-R_1+R_2\to R_2 \\ -R_1+R_3\to R_3}} \dfrac{1}{2}\cdot\left|\det\begin{bmatrix} x_1 & y_1 & 1 \\ x_2-x_1 & y_2-y_1 & 0 \\ x_3-x_1 & y_3-y_1 & 0 \end{bmatrix}\right| = \dfrac{1}{2}\cdot$

$\left|1\cdot\det\begin{bmatrix} x_2-x_1 & y_2-y_1 \\ x_3-x_1 & y_3-y_1 \end{bmatrix}\right| = \dfrac{1}{2}\left|(x_2-x_1)(y_3-y_1)-(x_3-x_1)(y_2-y_1)\right| =$

$\dfrac{1}{2}\left|x_2y_3-x_2y_1-x_1y_3+x_1y_1-x_3y_2+x_3y_1+x_1y_2-x_1y_1\right| =$

$\dfrac{1}{2}\left|x_2y_3-x_2y_1-x_1y_3-x_3y_2+x_3y_1+x_1y_2\right|.$ (1)

The area of $\Delta P_1P_2P_3$ shown in the accompanying figure is the area of the trapezoid $Q_2P_2P_3Q_3 +$ the area of trapezoid $Q_1P_1P_2Q_2$ – the area of trapezoid $Q_1P_1P_3Q_3$. The area of trapezoid

$Q_2P_2P_3Q_3$ is $\dfrac{1}{2}(x_1-x_{x3})(y_2+y_3)$, the area of the trapezoid $Q_1P_1P_2Q_2$ is $\dfrac{1}{2}(x_1-x_2)(y_1+y_2)$.

Lastly, the area of trapezoid $Q_1P_1P_3Q_3$ is $\dfrac{1}{2}(x_1-x_3)(y_1+y_3)$. Hence, the area of $\Delta P_1P_2P_3 =$

$\dfrac{1}{2}\left[(x_2-x_3)(y_2+y_3)+(x_1-x_2)(y_1+y_2)-(x_1-x_3)(y_1+y_3)\right] =$

$\dfrac{1}{2}(x_2y_2+x_2y_3-x_3y_2-x_3y_3+x_1y_1+x_1y_2-x_2y_1-x_2y_2-x_1y_1-x_1y_3+x_3y_1+x_3y_3) =$

$\dfrac{1}{2}\left(x_2y_3-x_3y_2+x_1y_2-x_2y_1-x_1y_3+x_3y_1\right).$ (2)

The result follows from expressions (1) and (2) above.   (See figure below).

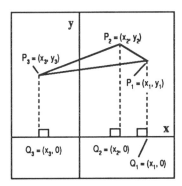

Figure 50

[51] $\quad \dfrac{1}{2}\left|\det\begin{bmatrix} 4 & 2 & 1 \\ -1 & -10 & 1 \\ 1 & 6 & 1 \end{bmatrix}\right|$ $\quad \begin{array}{c} -R_1 + R_2 \rightarrow R_2 \\ -R_1 + R_3 \rightarrow R_3 \\ = \end{array}$ $\quad \dfrac{1}{2}\cdot\left|\det\begin{bmatrix} 4 & 2 & 1 \\ -5 & -12 & 0 \\ -3 & 4 & 0 \end{bmatrix}\right|$

$\dfrac{1}{2}\left|1\cdot\det\begin{bmatrix} -5 & -12 \\ -3 & 4 \end{bmatrix}\right| = 28.$ $\qquad\qquad\qquad\qquad$ ● 28 sq units

[52] $\quad \dfrac{1}{2}\cdot\left|\det\begin{bmatrix} 1 & -1 & 1 \\ -6 & 2 & 1 \\ 4 & 7 & 1 \end{bmatrix}\right|$ $\quad \begin{array}{c} -R_1 + R_2 \rightarrow R_2 \\ -R_1 + R_3 \rightarrow R_3 \\ = \end{array}$

$\dfrac{1}{2}\cdot\left|\det\begin{bmatrix} 1 & -1 & 1 \\ -7 & 3 & 0 \\ 3 & 8 & 0 \end{bmatrix}\right|$ $\quad = \dfrac{1}{2}\cdot\left|1\cdot\det\begin{bmatrix} -7 & 3 \\ 3 & 8 \end{bmatrix}\right| = \dfrac{65}{2}$ $\qquad$ ● $\dfrac{65}{2}$ sq units

[53] $\quad \dfrac{1}{2}\cdot\left|\det\begin{bmatrix} \frac{1}{2} & -\frac{1}{3} & 1 \\ 2 & -\frac{1}{10} & 1 \\ 4 & \frac{2}{3} & 1 \end{bmatrix}\right|$ $\quad \begin{array}{c} -R_1 + R_2 \rightarrow R_2 \\ -R_1 + R_3 \rightarrow R_3 \\ = \end{array}$

$\dfrac{1}{2}\cdot\left|\det\begin{bmatrix} \frac{1}{2} & -\frac{1}{3} & 0 \\ \frac{3}{2} & \frac{7}{30} & 0 \\ \frac{7}{2} & 1 & 0 \end{bmatrix}\right| = \dfrac{1}{2}\left|1\cdot\det\begin{bmatrix} \frac{3}{2} & \frac{7}{30} \\ \frac{7}{2} & 1 \end{bmatrix}\right| = \dfrac{41}{120}$ $\qquad$ ● $\dfrac{41}{120}$ sq units

[54] $\quad \dfrac{1}{2}\cdot\left|\det\begin{bmatrix} 0.4 & -0.6 & 1 \\ 1.6 & 2.3 & 1 \\ 4.1 & -0.9 & 1 \end{bmatrix}\right|$ $\quad \begin{array}{c} -R_1 + R_2 \rightarrow R_2 \\ -R_1 + R_3 \rightarrow R_3 \\ = \end{array}$

$\dfrac{1}{2}\cdot\left|\det\begin{bmatrix} 0.4 & -0.6 & 1 \\ 1.2 & 2.9 & 0 \\ 3.7 & -0.3 & 0 \end{bmatrix}\right| = \dfrac{1}{2}\left|\cdot\det\begin{bmatrix} 1.2 & 2.9 \\ 3.7 & -0.3 \end{bmatrix}\right| = 5.545$ $\qquad$ ● 5.545

EXERCISES 9.6

[1]   Solving E_1 for x and substituting into E_2 yields $(4-y)-y^2+1=0 \Rightarrow y^2+y-5=0 \Rightarrow y =$

$\dfrac{-1 \pm \sqrt{21}}{2}$. If $y = \dfrac{-1-\sqrt{21}}{2}$, then $x = \dfrac{9+\sqrt{21}}{2}$. If $y = \dfrac{-1+\sqrt{21}}{2}$, then $x = \dfrac{9-\sqrt{21}}{2}$.

(See figure below).
$$\bullet \left\{ \left(\frac{9-\sqrt{21}}{2}, \frac{-1+\sqrt{21}}{2} \right), \left(\frac{9+\sqrt{21}}{2}, \frac{-1-\sqrt{21}}{2} \right) \right\}$$

[2]   Solving E_2 for y and substituting into E_1 yields $x^2 + (2x-1) - 3 = 0 \Rightarrow x^2 + 2x - 4 = 0 \Rightarrow x =$

$-1 \pm \sqrt{5}$. If $x = -1 - \sqrt{5}, y = -3 - 2\sqrt{5}$. If $x = -1 + \sqrt{5}, y = -3 + 2\sqrt{5}$.   (See figure below).

$$\bullet \left\{ \left(-1-\sqrt{5}, -3-2\sqrt{5} \right), \left(-1+\sqrt{5}, -3+2\sqrt{5} \right) \right\}$$

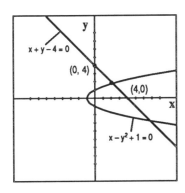

Figure 1   **Figure 2**

[3]   From E_1 and E_2 we know that $x - 3 = \sqrt{x-1}$. Squaring each side of the latter equation yields $x^2 -$

$6x + 9 = x - 1 \Rightarrow x^2 - 7x + 10 = 0 \Rightarrow x = -2 \ or \ x = 5$. Since x can not be equal to 2, we consider

only 5. If $x = 5$, then $y = 2$. You should verify that the solution set is $\{(5,2)\}$.   (See figure below).

$$\bullet \{(5,2)\}$$

[4]   Solving E_2 for x and substituting into E_1 yields $(y+1) + (y-2)^2 = 0 \Rightarrow y^2 - 3y + 5 = 0 \Rightarrow y =$

$\dfrac{3}{2} \pm \dfrac{\sqrt{11}}{2} i$. If $y = \dfrac{3}{2} - \dfrac{\sqrt{11}}{2} i$, then $x = \dfrac{5}{2} - \dfrac{\sqrt{11}}{2} i$. If $y = \dfrac{3}{2} + \dfrac{\sqrt{11}}{2} i$, then $x = \dfrac{5}{2} + \dfrac{\sqrt{11}}{2} i$.

(See figure below).
$$\bullet \left\{ \left(\frac{5}{2} - \frac{\sqrt{11}}{2} i, \frac{3}{2} - \frac{\sqrt{11}}{2} i \right), \left(\frac{5}{2} + \frac{\sqrt{11}}{2} i, \frac{3}{2} + \frac{\sqrt{11}}{2} i \right) \right\}$$

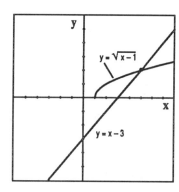

Figure 3

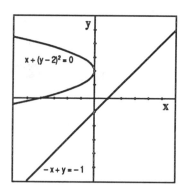

Figure 4

[5] $\begin{cases} x^2 + (y-1)^2 = 1 \\ (x+1)^2 + y^2 = 1 \end{cases} \Rightarrow \begin{matrix} x^2 + y^2 - 2y = 0 \\ x^2 + y^2 + 2x = 0 \end{matrix} \Rightarrow y = -x.$ Substituting $-x$ for y in the equation $(x+1)^2 +$

$(-x)^2 = 1 \Rightarrow 2x^2 + 2x = 0 \Rightarrow x = 0$ or $x = -1$. If $x = 0, y = 0$. If $x = -1, y = 1$.

(See figure below). $\bullet \left\{ (-1, 1), (0, 0) \right\}$

[6]   Solving E_2 for y and substituting into E_1 yields $-y + (y+1)^2 = 16 \Rightarrow y^2 + y - 15 = 0 \Rightarrow y =$

$\dfrac{-1 \pm \sqrt{61}}{2}$. From the equation $(x-2)^2 + y = 0$, we know that $y \le 0$. Hence, we choose $y = \dfrac{-1 - \sqrt{61}}{2}$

and substitute into the equation $(x-2)^2 + y = 0$. We get $(x-2)^2 + \left(\dfrac{-1-\sqrt{61}}{2}\right) = 0 \Rightarrow x^2 - 4x +$

$\dfrac{7 - \sqrt{61}}{2} = 0 \Rightarrow x = \dfrac{4 \pm \sqrt{2 + 2\sqrt{61}}}{2}.$ (See figure below).

$$\bullet \left\{ \left(\frac{4 - \sqrt{2 + 2\sqrt{61}}}{2}, \frac{-1 - \sqrt{61}}{2} \right), \left(\frac{4 + \sqrt{2 + 2\sqrt{61}}}{2}, \frac{-1 - \sqrt{61}}{2} \right) \right\}$$

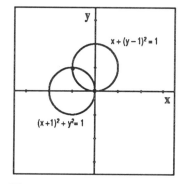

Figure 5

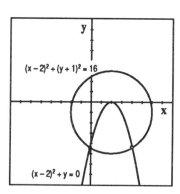

Figure 6

[7]   Solving E_1 for x and substituting into E_2 yields $y^3 - y^2 = 0 \Rightarrow y^2(y-1) = 0 \Rightarrow y = 0$ or $y = 1$. If $y =$

$0, x = 0$ and if $y = 1, x = 1$.   (See figure below). $\bullet \left\{ (0, 0)(1, 1) \right\}$

[8] $\begin{cases} 4x^2 + y = 4 \\ x^4 - y = 1 \end{cases} \Rightarrow x^4 + 4x^2 - 5 = 0.$ Letting $u = x^2$, we have $u^2 + 4u - 5 = 0 \Rightarrow (u + 5)(u - 1) =$

$0 \Rightarrow u = -5$ or $u = 1$. Hence $x^2 = -5$ or $x^2 = 1$. $x^2 = -5 \Rightarrow x = \pm \sqrt{5}\,i$. $x^2 = 1 \Rightarrow x = \pm 1$. From the

equation $x^4 - y = 1$, we know that if $x = \pm 1$, $y = 0$. Moreover, if $x = \pm \sqrt{5}\,i$, it follows that $y = 24$.

(See figure below). ● $\{(-1, 0), (1, 0), (-\sqrt{5}\,i, 24), (\sqrt{5}\,i, 24)\}$

[9] Solving E_2 for y and substituting into E_1 yields $x(-8 - 4x) = 4 \Rightarrow x^2 + 2x + 1 = 0 \Rightarrow (x + 1)^2 =$

$0 \Rightarrow x = -1$. From the equation $xy = 4$, we see that $y = -4$. (See figure below). ● $\{(-1, -4)\}$

[10] Solving E_1 for y and substituting into E_2 yields $\dfrac{(x - 1)^2}{16} - \dfrac{(x - 6)^2}{9} = -1 \Rightarrow 9(x - 1)^2 - 16\cdot$

$(x - 6)^2 = -144 \Rightarrow 7x^2 - 174x + 423 = 0 \Rightarrow x = \dfrac{87 \pm 48\sqrt{2}}{7}; y = \dfrac{87 \pm 48\sqrt{2}}{2}.$ (See figure below).

● $\left\{ \left(\dfrac{87 - 48\sqrt{2}}{7}, \dfrac{87 - 48\sqrt{2}}{7} \right), \left(\dfrac{87 + 48\sqrt{2}}{7}, \dfrac{87 + 48\sqrt{2}}{7} \right) \right\}$

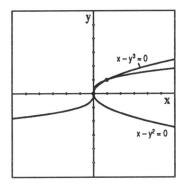

Figure 7

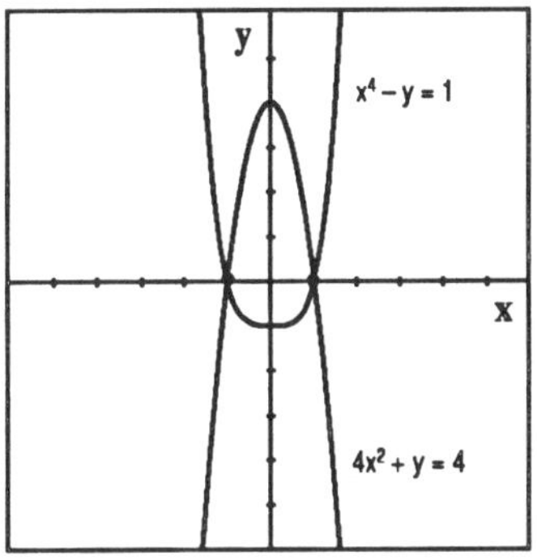

Figure 8

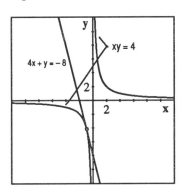

Figure 9

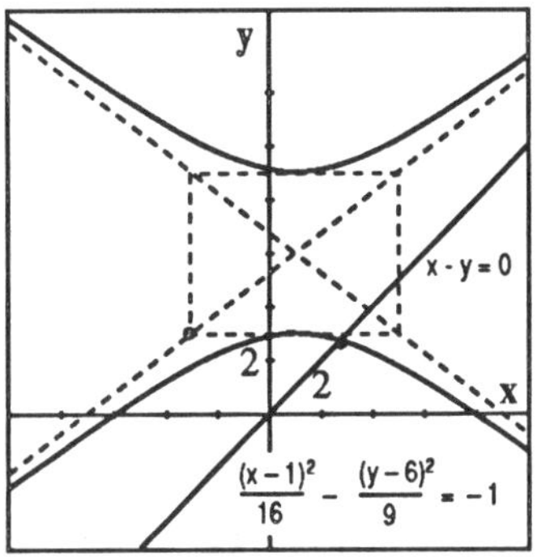

Figure 10

[11] $\begin{cases} x - y^4 = 0 \\ x + y^4 = 0 \end{cases} \Rightarrow 2x = 2 \Rightarrow x = 1.$ If $x = 1$, then $y^4 = 1 \Rightarrow y = -1$ or, $y = 1$, or $y = -i$, or $y = i$.

(See figure below). ● $\{(1, -1), (1, 1), (1, -i), (1, i)\}$

[12] Solving E_1 for y and substituting into E_2 yields $x^2 - 2x(-x) + (-x)^2 = 10 \Rightarrow 4x^2 = 10 \Rightarrow x = \pm$

$\frac{\sqrt{10}}{2}$. If $x = -\frac{\sqrt{10}}{2}$, then $y = \frac{\sqrt{10}}{2}$. If $x = \frac{\sqrt{10}}{2}$, then $y - \frac{\sqrt{10}}{2}$.   (See figure below).

(See figure below).   ● $\left\{ \left(-\frac{\sqrt{10}}{2}, \frac{\sqrt{10}}{2} \right), \left(\frac{\sqrt{10}}{2}, -\frac{\sqrt{10}}{2} \right) \right\}$

[13] $\begin{cases} y = x^3 \\ y = 32\sqrt{x} \end{cases} \Rightarrow x^3 = 32\sqrt{x}$. Letting $u = \sqrt{x}$, it follows that $u^6 - 32u = 0 \Rightarrow u(u^5 - 32) = 0 \Rightarrow u =$

0 or $u = 2 \Rightarrow x = 0$ or $x = 4$. If $x = 0$, then $y = 0$ and if $x = 4$, then $y = 64$.   (See figure below).

(See figure below).   ● $(0, 0), (4, 64)$

[14] $\begin{cases} x - y^2 = 0 \\ 2x - y^2 = 6 \end{cases} \Rightarrow x = 6 \Rightarrow y = \pm\sqrt{6}$.   (See figure below).   ● $(6, -\sqrt{6}), (6, \sqrt{6})$

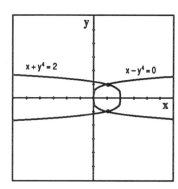

Figure 11

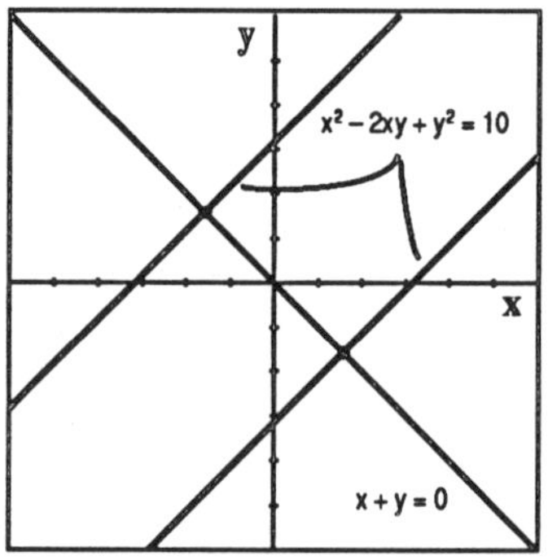

Figure 12

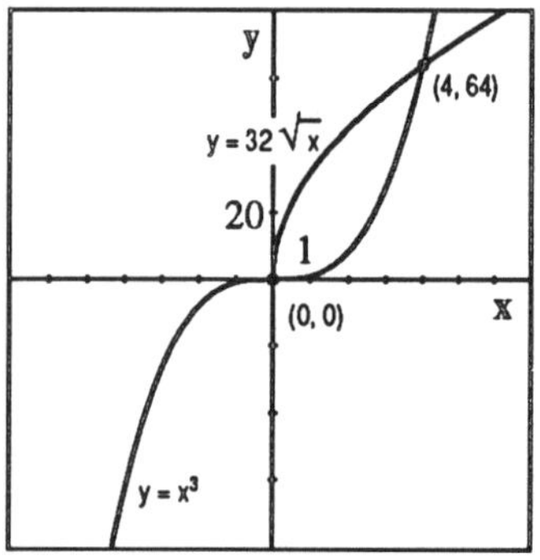

Figure 13

Figure 14

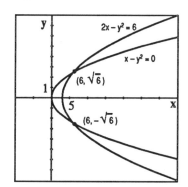

[15] Solving E_1 for x^2 and substituting into E_2 yields $y^{\frac{1}{3}} - y = 0 \Rightarrow y^3 - y = 0 \Rightarrow y(y^2 - 1) = 0 \Rightarrow$

$y = 0$ or $y = \pm 1$. Since $x^2 - y = 0$, $x^2 = 0$ or $x^2 = \pm 1$. The equation $x^2 = -1$ has no solutions which

are real numbers. The points of intersection are $(-1, 1), (0, 0), (1, 1)$.   (See figure below).

● $(-1, 1), (0, 0), (1, 1)$

[16] Substituting x^3 for y in E_3 yields $x^3 = 2x^3 + x^2 - 2x \Rightarrow x^3 + x^2 - 2x = 0 \Rightarrow x(x^2 + x - 2) = 0 \Rightarrow$

$x(x + 2)(x - 1) == 0 \Rightarrow x = -2$ or $x = 0$ or $x = 1$. Since $y = x^3$, it follows that if $x = -2$, $y = -8$; if

$x = 0, y = 0$; if $x = 1, y = 1$.   (See figure below).   ● $(-2, -8), (0, 0), (1, 1)$

[17]   Substituting e^x for y in E_2 yields $e^x = e^{2x} - 1 \Rightarrow e^{2x} - e^x - 1 = 0$. Letting $u = e^x$, we get $u^2 - u - 1 = 0 \Rightarrow u = \dfrac{1 \pm \sqrt{5}}{2}$. Since $u = e^x > 0$ for all x, we choose $u = \dfrac{1 + \sqrt{5}}{2} \Rightarrow x = \ln\left(\dfrac{1 + \sqrt{5}}{2}\right)$. Hence $y = e^x = \dfrac{1 + \sqrt{5}}{2}$.   (See figure below).

$\bullet \left(\ln\left(\dfrac{1 + \sqrt{5}}{2}\right), \dfrac{1 + \sqrt{5}}{2}\right)$

[18]   Substituting 2^x for y in E_2 yields $2^x = 4^x - 6 \Rightarrow \left(2^2\right)^x - 2^x - 6 = 0 \Rightarrow \left(2^x\right)^2 - 2^x - 6 = 0$. Letting $u = 2^x$, we have $u^2 - u - 6 = 0 \Rightarrow (u - 3)(u + 2) = 0 \Rightarrow u = 3$ or $u = -2$. Since $u = 2^x > 0$ for every real number x, we require that $2^x = 3 \Rightarrow x = \dfrac{\ln 3}{\ln 2}$. It follows from the equation $y = 2^x$ that $y = 2^{(\ln 3)/(\ln 2)} = 3$.

(See figure below).

$\bullet \left(\dfrac{\ln 3}{\ln 2}, 3\right)$

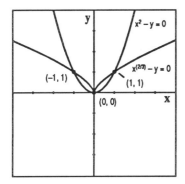

Figure 15

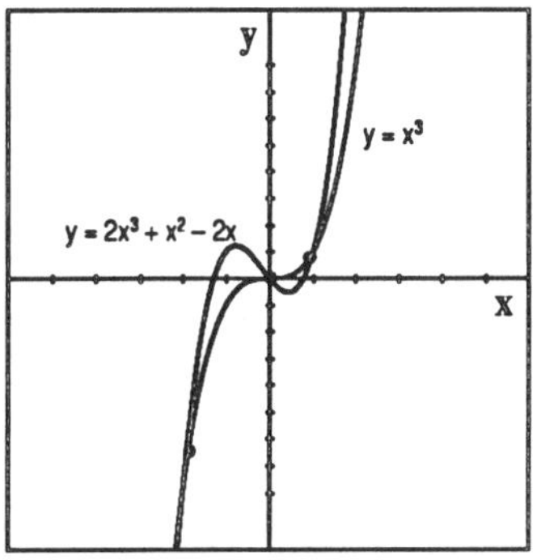

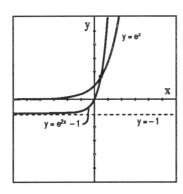

Figure 17

Figure 16

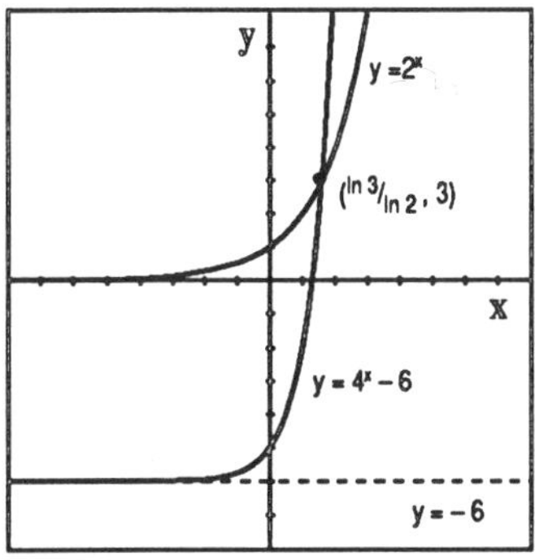

Figure 18

[19]   Substituting $\ln x$ for y in E_2 yields $\ln x = \ln(10 - x) \Rightarrow x = 10 - x \Rightarrow x = 5$. Hence, $y = \ln 5$.
(See figure below).

$\bullet (5, \ln 5)$

[20]   Substituting $\ln|2x|$ for y in E_2 yields $\ln|2x| = \ln|x + 1| \Rightarrow |2x| = |x + 1| \Rightarrow 2x = \pm(x + 1) \Rightarrow$
$x = -\dfrac{1}{3}$ or $x = 1$. If $x = -\dfrac{1}{3}$, $y = \ln\dfrac{2}{3}$, and if $x = 1$, $y = \ln 2 = 0$.   (See figure below).

$\bullet \left\{\left(-\dfrac{1}{3}, \ln\dfrac{2}{3}\right), (1, \ln 2)\right\}$

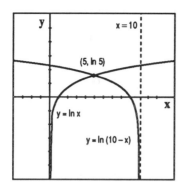

Figure 19

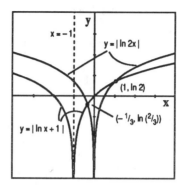

Figure 20

[21]   Solving E_1 for x and substituting into E_2 yields $-3y^2 + y^2 - 12y = 5 \Rightarrow 2y^2 + 12y + 5 = 0 \Rightarrow y =$

$\dfrac{-6 \pm \sqrt{26}}{2}$. If $y = \dfrac{-6 - \sqrt{26}}{2}$, $x = \dfrac{-93 - 18\sqrt{26}}{2}$ and if $y = \dfrac{-6 + \sqrt{26}}{2}$, $x = \dfrac{-93 + 18\sqrt{26}}{2}$.

$$\bullet \left\{ \left(\frac{-93 - 18\sqrt{26}}{2}, \frac{-6 - \sqrt{26}}{2} \right), \left(\frac{-93 + 18\sqrt{26}}{2}, \frac{-6 + \sqrt{26}}{2} \right) \right\}$$

[22] $\begin{cases} x - y^2 = 0 \\ x + y^2 = 32 \end{cases} \Rightarrow 2x = 32 \Rightarrow x = 16$. Since $x = y^2$, $y = \pm 4$. $\quad \bullet \{(16, -4), (16, 4)\}$

[23]   Factoring E_1 we have $y(x - y) = 0 \Rightarrow y = 0$ or $x = y$. Subsituting 0 for y in E_2 yields $x^2 = 10 \Rightarrow x = \pm \sqrt{10}$. Hence, $(-\sqrt{10}, 0)$ and $(\sqrt{10}, 0)$ satisfy the given system of equations. Substituting x for y in

E_2 gives us the equation $x^2 + 2x^2 + x^2 = 10 \Rightarrow 4x^2 = 10 \Rightarrow x = \pm \dfrac{\sqrt{10}}{2}$. It follows that

$\left(-\dfrac{\sqrt{10}}{2}, -\dfrac{\sqrt{10}}{2} \right)$ and $\left(\dfrac{\sqrt{10}}{2}, \dfrac{\sqrt{10}}{2} \right)$ are also solutions.

$$\bullet \left\{ \left(-\frac{\sqrt{10}}{2}, -\frac{\sqrt{10}}{2} \right), \left(\frac{\sqrt{10}}{2}, \frac{\sqrt{10}}{2} \right), (-\sqrt{10}, 0), (\sqrt{10}, 0) \right\}$$

[24]   Factoring E_1 we have $x(x - y) = 0 \Rightarrow x = 0$ or $x = y$. Substituting 0 for x in E_2, we get $2y^2 = 4 \Rightarrow y^2 = 2 \Rightarrow y = \pm \sqrt{2}$. Hence, $(0, -\sqrt{2})$ and $(0, \sqrt{2})$ are solution of the given system. Substituting y for x in E_2 yields $y^2 - y^2 + 2y^2 = 4 \Rightarrow y^2 = 2 \Rightarrow y = \pm \sqrt{2}$. Hence, $(-\sqrt{2}, -\sqrt{2})$ and $(\sqrt{2}, \sqrt{2})$ are also

solutions. $\quad \bullet \{(0, -\sqrt{2}), (0, \sqrt{2}), (-\sqrt{2}, -\sqrt{2}), (\sqrt{2}, \sqrt{2})\}$

[25]   Solving E_1 for x^2 and substituting into E_2 yields $16(10 - y^2) + 9y^2 = 144 \Rightarrow 7y^2 = 16 \Rightarrow y =$

$\pm \sqrt{\dfrac{16}{7}} = \pm \dfrac{4\sqrt{7}}{7}$. Since $x^2 = 10 - y^2$, it follows that $x = \pm \sqrt{\dfrac{54}{7}}$.

$$\bullet \left\{ \left(-\sqrt{\frac{54}{7}}, \frac{-4\sqrt{7}}{7} \right), \left(-\sqrt{\frac{54}{7}}, \frac{4\sqrt{7}}{7} \right), \left(\sqrt{\frac{54}{7}}, \frac{-4\sqrt{7}}{7} \right), \left(\sqrt{\frac{54}{7}}, \frac{4\sqrt{7}}{7} \right) \right\}$$

[26]   Since $xy = 1$, $x \neq 0$ and $y \neq 0$. Solving the equation $xy = 1$ for y, we have $y = \dfrac{1}{x}$. Substituting $\dfrac{1}{x}$ for

y in E_1, we get $x^2 + \dfrac{1}{x^2} = 4 \Rightarrow x^4 + 1 = 4x^2 \Rightarrow x^4 - 4x^2 + 1 = 0$. Letting $u = x^2$ yields $u^2 - 4u + 1 =$

$0 \Rightarrow u = 2 \pm \sqrt{3}$. Hence, $x = \pm \sqrt{2 \pm \sqrt{3}}$. Since $y = \dfrac{1}{x}$, it follows that the solution set is

$$\bullet \left\{ \left(-\sqrt{2 - \sqrt{3}}, \frac{1}{-\sqrt{2 - \sqrt{3}}} \right), \left(-\sqrt{2 + \sqrt{3}}, \frac{1}{-\sqrt{2 + \sqrt{3}}} \right), \left(\sqrt{2 - \sqrt{3}}, \frac{1}{\sqrt{2 - \sqrt{3}}} \right), \left(\sqrt{2 + \sqrt{3}}, \frac{1}{\sqrt{2 + \sqrt{3}}} \right) \right\}$$

[27] Letting $u = \frac{1}{x}$ and $v = \frac{1}{y}$, we get $\begin{cases} 3u - 4v = 2 \\ 5u + 7v = 1 \end{cases} \Rightarrow \begin{cases} -15u + 20v = -10 \\ 15u + 21v = 3 \end{cases} \Rightarrow 41v = -7 \Rightarrow v =$

$-\frac{7}{41} \Rightarrow y = -\frac{41}{7}$. From the equation $3u - 4v = 2$, we see that $u = \frac{54}{123} \Rightarrow x = \frac{41}{18}$.   ● $\left\{ \left(\frac{41}{18}, -\frac{41}{7} \right) \right\}$

[28] Let $u = \frac{1}{x}$ and $v = \frac{1}{y}$, then $\begin{cases} 2u - 6v = 3 \\ -3u + 5v = 2 \end{cases} \Rightarrow \begin{cases} 6u - 18v = 9 \\ -6u + 10v = 4 \end{cases} \Rightarrow -9v = 13 \Rightarrow v = -\frac{13}{9} \Rightarrow y =$

$-\frac{9}{13}$. From the equation $2u - 6v = 3$, it follows that $u = -\frac{17}{6} \Rightarrow x = -\frac{6}{17}$.   ● $\left\{ \left(-\frac{6}{17}, -\frac{9}{13} \right) \right\}$

[29] Let $u = \frac{1}{x^2}$ and $v = \frac{1}{y^2}$, then $\begin{cases} 4u + 3v = 26 \\ 3u - 11v = -7 \end{cases} \Rightarrow \begin{cases} -12u - 9v = -78 \\ 12u - 44v = -28 \end{cases} \Rightarrow -53v = -106 \Rightarrow v =$

$2 \Rightarrow y^2 = \frac{1}{2} \Rightarrow y = \pm\sqrt{\frac{1}{2}}$. From the equation $4u + 3v = 26$, it follows that $u = 5 \Rightarrow x^2 = \frac{1}{5} \Rightarrow x =$

$\pm\sqrt{\frac{1}{5}}$.   ● $\left\{ \left(-\sqrt{\frac{1}{5}}, -\sqrt{\frac{1}{2}} \right), \left(-\sqrt{\frac{1}{5}}, \sqrt{\frac{1}{2}} \right), \left(\sqrt{\frac{1}{5}}, -\sqrt{\frac{1}{2}} \right), \left(\sqrt{\frac{1}{5}}, \sqrt{\frac{1}{2}} \right) \right\}$

[30] Letting $u = \frac{1}{x^2}$ and $v = \frac{1}{y^2}$, we get $\begin{cases} u + v = 2 \\ 2u + 3v = 9 \end{cases} \Rightarrow \begin{cases} -2u - 2v = -4 \\ 2u + 3v = 9 \end{cases} \Rightarrow v = 5 \Rightarrow y^2 = \frac{1}{5} \Rightarrow y =$

$\pm\sqrt{\frac{1}{5}}$. From the equation $u + v = 2$, it follows that $u = -3 \Rightarrow x^2 = -\frac{1}{3} \Rightarrow x = \pm\sqrt{\frac{1}{3}}\, i$.

● $\left\{ \left(-\sqrt{\frac{1}{3}}\, i, -\sqrt{\frac{1}{5}} \right), \left(-\sqrt{\frac{1}{3}}\, i, \sqrt{\frac{1}{5}} \right), \left(\sqrt{\frac{1}{3}}\, i, -\sqrt{\frac{1}{5}} \right), \left(\sqrt{\frac{1}{3}}\, i, \sqrt{\frac{1}{5}} \right) \right\}$

[31] Adding E_1 and E_2, we get $3x^2 = 20 \Rightarrow x^2 = \frac{20}{3} \Rightarrow x = \pm\sqrt{\frac{20}{3}}$. Substituting $\frac{20}{3}$ for x^2 in E_2,

we get $\frac{20}{3} - y^2 = 16 \Rightarrow y^2 = -\frac{28}{3} \Rightarrow y = \pm\sqrt{\frac{28}{3}}\, i$.

● $\left\{ \left(-\sqrt{\frac{20}{3}}, -\sqrt{\frac{28}{3}}\, i \right), \left(-\sqrt{\frac{20}{3}}, \sqrt{\frac{28}{3}}\, i \right), \left(\sqrt{\frac{20}{3}}, -\sqrt{\frac{28}{3}}\, i \right), \left(\sqrt{\frac{20}{3}}, \sqrt{\frac{28}{3}}\, i \right) \right\}$

[32] Subtracting E_1 from E_2, we get $y^2 + y = -5$, or $y^2 + y + 5 = 0 \Rightarrow y = \frac{-1 \pm \sqrt{19}\, i}{2}$. It follows

from E_1 that $x^2 = \frac{11 \pm \sqrt{19}\, i}{2} \Rightarrow x = \pm\sqrt{\frac{11 \pm \sqrt{19}\, i}{2}}$.

● $\left\{ \left(-\sqrt{\frac{11 - \sqrt{19}\, i}{2}}, \frac{-1 - \sqrt{19}\, i}{2} \right), \left(\sqrt{\frac{11 - \sqrt{19}\, i}{2}}, \frac{-1 - \sqrt{19}\, i}{2} \right), \right.$

$\left. \left(-\sqrt{\frac{11 + \sqrt{19}\, i}{2}}, \frac{-1 + \sqrt{19}\, i}{2} \right), \left(\sqrt{\frac{11 + \sqrt{19}\, i}{2}}, \frac{-1 + \sqrt{19}\, i}{2} \right) \right\}$

[33] Adding E_1 and E_2, we get $\frac{5}{4}x^2 = 2 \Rightarrow x^2 = \frac{8}{5} \Rightarrow x = \pm\sqrt{\frac{8}{5}} = \pm 2\sqrt{\frac{2}{5}}$. From E_1, it follows that

$y^2 = x^2 - 1 \Rightarrow y^2 = \frac{3}{5} \Rightarrow y = \pm\sqrt{\frac{3}{5}}$.

● $\left\{ \left(-2\sqrt{\frac{2}{5}}, -\sqrt{\frac{3}{5}} \right), \left(-2\sqrt{\frac{2}{5}}, \sqrt{\frac{3}{5}} \right), \left(2\sqrt{\frac{2}{5}}, -\sqrt{\frac{3}{5}} \right), \left(2\sqrt{\frac{2}{5}}, \sqrt{\frac{3}{5}} \right) \right\}$

[34] Solving E_1 for y^2 and substituting into E_2 yields $x^2 + \dfrac{1 - \dfrac{x^2}{9}}{9} = 1 \Rightarrow 81\,x^2 + 9 - x^2 = 81 \Rightarrow 80\,x^2 =$

$72 \Rightarrow x^2 = \dfrac{9}{10} \Rightarrow x = \pm\dfrac{3\sqrt{10}}{10}$. It follows from E_1 that $y^2 = \dfrac{9}{10} \Rightarrow y = \pm\dfrac{3\sqrt{10}}{10}$.

$\bullet \left\{ \left(-\dfrac{3\sqrt{10}}{10}, -\dfrac{3\sqrt{10}}{10}\right), \left(-\dfrac{3\sqrt{10}}{10}, \dfrac{3\sqrt{10}}{10}\right), \left(\dfrac{3\sqrt{10}}{10}, -\dfrac{3\sqrt{10}}{10}\right), \left(\dfrac{3\sqrt{10}}{10}, \dfrac{3\sqrt{10}}{10}\right) \right\}$

[35] Adding E_1 and E_2 yields $x^2 = 16 \Rightarrow x = \pm 4$. If $x = -4$, we know from E_2 that $-4y - y^2 = 6 \Rightarrow y^2 + 4y + 6 = 0 \Rightarrow y = -2 \pm \sqrt{2}\,i$. It follows that $(-4, -2 - \sqrt{2}\,i)$, and $(-4, 2 + \sqrt{2}\,i)$ are solutions. If $x = 4$, we know from E_2 that $4y - y^2 = 6 \Rightarrow y^2 - 4y + 6 = 0 \Rightarrow y = 2 \pm \sqrt{2}\,i$. Hence, $(4, 2 - \sqrt{2}\,i)$ and $(4, 2 + \sqrt{2}\,i)$ are solutions. $\bullet \{(-4, -2 - \sqrt{2}\,i), (-4, -2 + \sqrt{2}\,i), (4, 2 - \sqrt{2}\,i), (4, 2 + \sqrt{2}\,i)\}$

[36] Substituting 3 for $x^2 + y^2$ E_1 yields $-xy + 3 = 2 \Rightarrow xy = 1 = y = \dfrac{1}{x}$. Substituting $\dfrac{1}{x}$ for y in E_2

yields $x^2 + \left(\dfrac{1}{x}\right)^2 = 3 \Rightarrow x^4 + 1 = 3x^2 \Rightarrow x^4 - 3x^2 + 1 = 0$. Letting $u = x^2$, we have $u^2 - 3u + 1 =$

$0 \Rightarrow u = \dfrac{3 \pm \sqrt{5}}{2} \Rightarrow x^2 = \dfrac{3 \pm \sqrt{5}}{2} \Rightarrow x = \pm\sqrt{\dfrac{3 \pm \sqrt{5}}{2}}$. Using the equation $y = \dfrac{1}{x}$, we determine y as

follows: If $x = -\sqrt{\dfrac{3 + \sqrt{5}}{2}}$, then $y = -\sqrt{\dfrac{2}{3 - \sqrt{5}}}$, if $x = \sqrt{\dfrac{3 - \sqrt{5}}{2}}$, then $y = \sqrt{\dfrac{2}{3 - \sqrt{5}}}$:

Lastly, if $x = \sqrt{\dfrac{3 + \sqrt{5}}{2}}$, then $y = \sqrt{\dfrac{2}{3 + \sqrt{5}}}$.

$\bullet \left\{ \left(-\sqrt{\dfrac{3 - \sqrt{5}}{2}}, -\sqrt{\dfrac{2}{3 - \sqrt{5}}}\right), \left(-\sqrt{\dfrac{3 + \sqrt{5}}{2}}, -\sqrt{\dfrac{2}{3 + \sqrt{5}}}\right)\right.$

$\left. \left(\sqrt{\dfrac{3 - \sqrt{5}}{2}}, \sqrt{\dfrac{2}{3 - \sqrt{5}}}\right), \left(\sqrt{\dfrac{3 + \sqrt{5}}{2}}, \sqrt{\dfrac{2}{3 + \sqrt{5}}}\right) \right\}$

[37] Adding E_1 and E_2 yields the equation $2x = 4 \Rightarrow x = 2$. Substituting 2 for x in E_1 yields $2 = \log_3(y + 1) = 3 \Rightarrow \log_3(y + 1) = 1 \Rightarrow y + 1 = 3 \Rightarrow y = 2$. Hence, the solution set is $\{(2, 2)\}$.

$\bullet \{(2, 2)\}$

[38] Solving E_1 for x and substituting into E_2 yields $-\log_4(y + 6) + 5\log_4(y + 6) = 4 \Rightarrow \log_4(y + 6) = 1 \Rightarrow y + 6 = 4 \Rightarrow y = -2$. It follows from E_1 that $x = -1$. Hence, the solution set $\{(-1, -2)\}$.

$\bullet \{(-1, -2)\}$

[39] Solving E_2 for y and substituting into E_1 yields $e^{2x} - (-2e^x)e^x = 9 \Rightarrow 3e^{2x} = 9 \Rightarrow e^{2x} = 3 \Rightarrow 2x = \ln 3 \Rightarrow x = \dfrac{1}{2}\ln 3$. From E_2, it follows that $y = -2\sqrt{3}$. Thus, the solution set is $\left\{\left(\dfrac{1}{2}\ln 3, -2\sqrt{3}\right)\right\}$.

$\bullet \left\{\left(\dfrac{1}{2}\ln 3, -2\sqrt{3}\right)\right\}$

[40] Adding E_1 and E_2, we get $4y = 8 \Rightarrow y = 2$. Substituting 2 for y in E_1 yields $3^x + 2 = 6 \Rightarrow 3^x = 4 \Rightarrow x = \dfrac{\ln 4}{\ln 3}$. Thus, the solution set is $\left\{\left(\dfrac{\ln 4}{\ln 3}, 2\right)\right\}$. $\bullet \left\{\left(\dfrac{\ln 4}{\ln 3}, 2\right)\right\}$

[41] Let x be the radius of the larger circle, and let y represent the radius of the smaller circle, then

$$\begin{cases} \pi x^2 - \pi y^2 = 40\pi & \text{Difference in Areas} \\ xy = 4 & \text{Difference in Radii} \end{cases}$$ Solving E_2 for x and substituting into E_1 yields π.

$$(y+4)^2 - \pi y^2 = 40\pi \Rightarrow 8y + 16 = 40 \Rightarrow y = 3.$$ ● 3 in.

[42] Let x and y represent the two numbers. Then $\begin{cases} xy = 28 \\ 2x^2 - 4y = 4 \end{cases}$. Solving E_2 for y and substituting into

E_1 yields $x\left(\dfrac{1}{2}x^2 - 1\right) = 28 \Rightarrow x^3 - 2x = 56 \Rightarrow x^3 - 2x - 56 = 0$. Using methods from Chapter 3, *it*

can be shown that $x = 4$ or $x = -2 \pm \sqrt{10}\,i$. Choosing $x = 4$, we see that $y = 7$. If $x = -2 \pm \sqrt{10}\,i$,

$y = \dfrac{28}{-2 \pm \sqrt{10}\,i}.$　　　　 ● $4, 7; -2 - \sqrt{10}\,i, \dfrac{28}{-2-\sqrt{10}\,i}; -2 + \sqrt{10}\,i, \dfrac{28}{-2+\sqrt{10}\,i}$

[43] Let x and y represent the sides of the triangle as shown in the accompanying figure, then

$$\begin{cases} \dfrac{1}{2}xy = 12 & \text{Area of Triangle} \\ x^2 + y^2 = 52 & \text{Pythogarean Theorem} \end{cases}$$ Solving E_1 for y and substituting into E_2 yields

$x^2 + \left(\dfrac{24}{x}\right)^2 = 52 \Rightarrow x^4 + 24 = 52x^2 \Rightarrow x^4 - 52x^2 + 576 = 0 \Rightarrow (u-16)(u-36) = 0 \Rightarrow u = 16$ or

$u = 36$. If $u = 16, x^2 = 16 \Rightarrow x = \pm 4$. If $u = 36, x^2 = 36 \Rightarrow x = \pm 6$. It follows that the lengths of the

legs are 4 meters and 6 meters.　(See figure below).　　　　　　● 4 m, 6 m

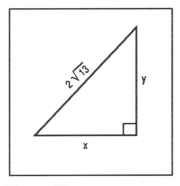

Figure 43

[44] Let x and y represent the radii of the two circles, then $\begin{cases} x + y = 5 & \text{Sum of the Radii} \\ \pi x^2 + \pi y^2 = \dfrac{29}{2}\pi & \text{Sum of the Areas} \end{cases}$.

Solving E_1 for y and substituting into E_2 yields $\pi x^2 + \pi(5-x)^2 = \dfrac{29}{2}\pi \Rightarrow 4x^2 - 20x + 21 = 0 \Rightarrow$

$x = \dfrac{7}{2}$ or $x = \dfrac{3}{2}$. If $x = \dfrac{7}{2}, y = \dfrac{3}{2}$, and if $x = \dfrac{3}{2}, y = \dfrac{7}{2}$.　　　　● $\dfrac{3}{2}$ m, $\dfrac{7}{2}$ m

[45] Let n be the number of people who were originally scheduled for the flight at a cost of c dollars per person. Then $\begin{cases} nc = 31,500 & \underline{\text{Assuming } n \text{ people}} \\ (n+5)(c-15) = 31,500 & \underline{\text{Assuming } n \pm 5 \text{ people}} \end{cases}$ E_2 can be expressed as $nc -$

$15n + 5c - 75 = 31,500$. Substracting E_1 yields $n(15 + 3n) = 31,500 \Rightarrow n^2 + 5n - 10,500 = 0 \Rightarrow$

$(n+105)(n-100) = 0 \Rightarrow n = -105$ or $n = 100$. Since $n > 0$, we choose $n = 100$. From E_1 it

follows that $c = 315$.   ● $100, \$315$

[46] Let h be the height of the cylinder, and let n be the radius of the base, then

$\begin{cases} h - 4r = -2 & \underline{\text{Difference in Height and Radius}} \\ \pi r^2 h = 224 & \underline{\text{Volume of Cylinder}} \end{cases}$. Solving E_1 for h and

substituting into E_2 yields $\pi r^2 (4r-2) = 224 \Rightarrow 4\pi r^3 - 2\pi r^2 = 224 \Rightarrow 2\pi r^3 - \pi r^2 - 112 = 0$.

Using methods from Section 3.5 it can be shown that $r \approx 2.79$. From E_1, it follows that $h \approx 9.16$.

● height ≈ 9.16 inches, radius ≈ 2.79 inches

[47] From the accompanying figure, we see that $x^2 + y^2 = 10$. Also since the area of the rectangle is

12 square centimeters, $4xy = 12$, or $xy = 3$. It follows that we must find the solution set of the

following system of equations $\begin{cases} x^2 + y^2 = 10 \\ xy = 3 \end{cases}$ Solving E_2 for y and substituting into E_1, we get $x^2 +$

$\left(\frac{3}{x}\right)^2 = 10 \Rightarrow x^4 - 10x^2 + 9 = 0$. Letting $u = x^2$, we have $u^2 - 10u + 9 = 0 \Rightarrow (u-9)(u-1) = 0 \Rightarrow$

$u = 9$ or $u = 1$. If $u = 9, x^2 = 9 \Rightarrow x = \pm 3$. If $u = 1, x^2 = 1 \Rightarrow x = \pm 1$. Since $x > 0, x = 1$ or $x = 3$. If

$x = 1, y = 3$ and if $x = 3, y = 1$. Thus, the width of the rectangle is 2 cm and the length is 6 cm.

(See figure below).   ● 2 cm, 6 cm

[48] From the accompanying figure we see that $x^2 + y^2 = 25$ and since the volume of the cylinder is 72π,

$\pi x^2 (2y) = 72\pi$, or $x^2 y = 36$. It follows that we must find the solution set of the following system

of equations $\begin{cases} x^2 + y^2 = 25 \\ x^2 y = 36 \end{cases}$. Solving E_1 for x^2 and substituting into E_2 yields $(25 - y^2)y = 36 \Rightarrow$

$y^3 - 25y + 36 = 0 \Rightarrow y = 4$ or $y = -2 \pm \sqrt{13}$. If $y = 4$, then $x = 3$. If $y = -2 + \sqrt{13}$, then $x =$

$2\sqrt{2 + \sqrt{13}}$. (Note that $y = -2 - \sqrt{13} < 0$). (See figure below).

● height $= 4$ cm and radius $= 3$ cm, or height $= -2 \pm \sqrt{13}$ cm and raduis $= 2\sqrt{2 + \sqrt{13}}$ cm

[49] We are seeking the slope of the tangent line to the parabola at the point $(2, 5)$. Thus, we require that

the following system of equations has a unique solution, namely $(2, 5)$: $\begin{cases} y = x^2 + 1 \\ y - 5 = m(x-2) \end{cases}$. Solving

E_2 for y and substituting into E_1 yields $mx - 2m + 5 = x^2 + 1 \Rightarrow x^2 - mx + 2m - 4 = 0$. The

equation $x^2 - mx + 2m - 4 = 0$ has a unique solution if and only if $m^2 - 8m + 16 = 0 \Rightarrow (m-4)^2 =$

$0 \Rightarrow m = 4$.   ● 4

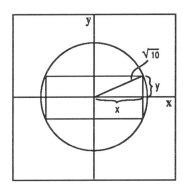

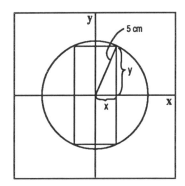

Figure 47   **Figure 48**

[50]   (See Exercise 49 above). We require that the following system of equation has a unique solution:

$$\begin{cases} y = x^2 - 2x \\ y - 3 = m(x+1) \end{cases}$$ Solving E_2 for y and substituting into E_1 yields $mx + m + 3 = x^2 - 2x \Rightarrow x^2 +$

$(-2 - m)x - m - 3 = 0$. The latter equation has a unique solution if and only if $(-2-m)^2 - 4 \cdot$

$(-m - 3) = 0 \Rightarrow m^2 + 8m - 16 = 0 \Rightarrow (m+4)^2 = 0 \Rightarrow m = -4$.   ● -4

EXERCISES 9.7

NOTE: We will use the notation, Test point $(a, b) \to$ True (False), to indicate that the point (a, b) has been substituted into the inequality and makes it true (false). Hence, we have shaded the region containing (not containing) the point (a, b).

[1]   Test point $(0, 0) \to$ True   (See figure below).   [2]   Test point $(0, 0) \to$ True   (See figure below).

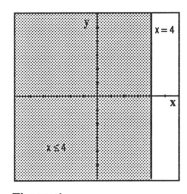

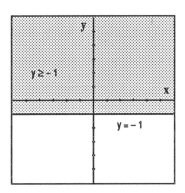

Figure 1   **Figure 2**

[3]   Test point $(0, 1) \to$ False   (See figure below).   [4]   Test point $(0, 1) \to$ True   (See figure below).

[5]   Test point $(0, 0) \to$ True   (See figure below).   [6]   Test point $(0, 0) \to$ False   (See figure below).

[7]   Test point $(0, 0) \to$ False   (See figure below).   [8]   Test point $(0, 0) \to$ True   (See figure below).

[9]   This region is bounded and convex. The vertices are $(0, 4)$, $(0, 6)$, $\left(\frac{3}{2}, 1\right)$, and $\left(\frac{10}{3}, 1\right)$.

(See figure below).

[10]   This region is unbounded and convex. The vertices are $(50, 180)$ and $(100, 130)$.   (See figure below).

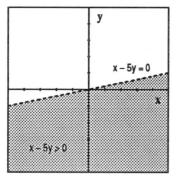

Figure 3

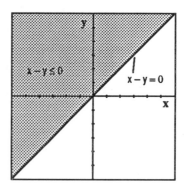

Figure 4

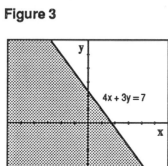

Figure 5

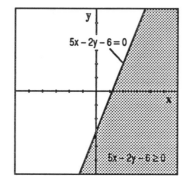

Figure 6

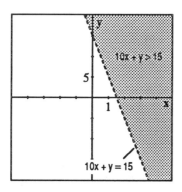

Figure 7

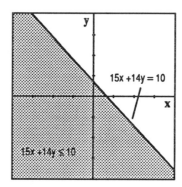

Figure 8

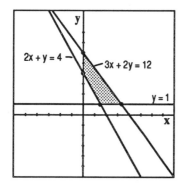

Figure 9

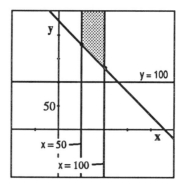

Figure 10

484

[11]   This region is bounded and convex. The vertices are $(0, 3)$, $(0, 6)$, $(4, 0)$, $(5, 2)$ and $(5, 6)$.
(See figure below).

[12]   This region is bounded and convex. The vertices are $(0, 10)$, $(0, 50)$, $(10, 10)$ and $(10, 40)$.
(See figure below).

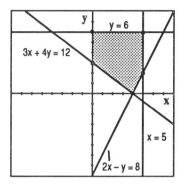

Figure 11   **Figure 12**

[13]   This region is bounded and convex. The vertices are $(0, 0)$, $\left(0, \frac{500}{7}\right)$, $(20, 60)$ and $\left(\frac{400}{11}, 0\right)$.
(See figure below).

[14]   This region is bounded and convex. The vertices are $(0, 0)$, $\left(0, \frac{8}{5}\right)$, $(3, 1)$ and $\left(\frac{9}{2}, 0\right)$.
(See figure below).

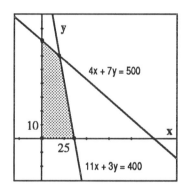

Figure 13   **Figure 14**

[15]   This region is unbounded and convex. The vertices are $(0, 4)$, $\left(\frac{5}{3}, \frac{2}{3}\right)$ and $(3, 0)$.   (See figure below).

[16]   This region is bounded and convex. The vertices are $(0, 0)$, $(0, 4)$, $(3, 2)$ and $\left(\frac{11}{3}, 0\right)$.
(See figure below).

[17]   This region is bounded and convex. The vertices are $(0, 10)$, $(0, 20)$, $(5, 5)$ and $(15, 15)$.
(See figure below).

[18]   This region is unbounded and convex. The vertices are $(0, 70)$, $\left(\frac{44}{7}, \frac{270}{7}\right)$ and $\left(\frac{15}{2}, \frac{65}{2}\right)$.
(See figure below).

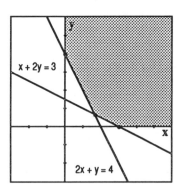

Figure 15

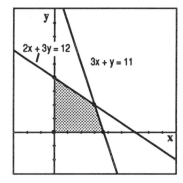

Figure 16

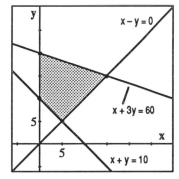

Figure 17

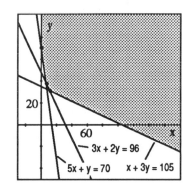

Figure 18

[19]   This region is unbounded and convex. The vertices are $(0, 10)$, $(0, 15)$ and $(4, 15)$.   (See figure below).

[20]   This region is unbounded and convex. The vertices are $\left(\frac{3}{2}, \frac{1}{2}\right)$ and $(2, 0)$.   (See figure below).

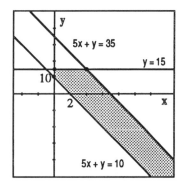

Figure 19

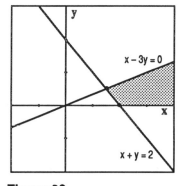

Figure 20

[21]

Vertex	Value of $f(x, y) = 2x + y$
$(0, 4)$	4
$(0, 6)$	6
$\left(\frac{3}{2}, 1\right)$	4
$\left(\frac{10}{3}, 1\right)$	$\frac{23}{3}$

The maximum value of f is $\frac{23}{2}$ and the minimum value is 4.   ● $\frac{23}{2}$, 4

[22]

Vertex	Value of $f(x,y)=x+4y$
$(0,4)$	16
$(0,6)$	24
$\left(\frac{3}{2},1\right)$	$\frac{11}{2}$
$\left(\frac{10}{3},1\right)$	$\frac{22}{3}$

The maximum value of f is 24 and the minimum value is $\frac{11}{2}$. 　●$24,\frac{11}{2}$

[23]

Vertex	Value of $f(x,y)=2x+4y$
$(0,4)$	16
$(0,6)$	24
$\left(\frac{3}{2},1\right)$	7
$\left(\frac{10}{3},1\right)$	14

The maximum value of f is 24 and the minimum value is 7. 　●$24,7$

[24]

Vertex	Value of $f(x,y)=4x+5y$
$(0,4)$	20
$(0,6)$	30
$\left(\frac{3}{2},1\right)$	11
$\left(\frac{10}{3},1\right)$	$\frac{53}{3}$

The maximum value of f is 30 and the minimum value is 11. 　●$30,11$

[25]

Vertex	Value of $f(x,y)=x-y$
$(0,3)$	-3
$(0,6)$	-6
$(4,0)$	4
$(5,2)$	3
$(5,6)$	-1

The maximum value of f is 4 and the minimum value is -6. 　●$4,-6$

[26]

Vertex	Value of $f(x,y)=6x-y$
$(0,3)$	-3
$(0,6)$	-6
$(4,0)$	24
$(5,2)$	28
$(5,6)$	24

The maximum value of f is 28 and the minimum value is -6. 　●$28,-6$

[27]

Vertex	Value of $f(x,y)=6x-y$
$(0,3)$	3
$(0,6)$	6
$(4,0)$	-16
$(5,2)$	-18
$(5,6)$	-14

The maximum value of f is 6 and the minimum value is -18. 　●$6,-18$

[28]

Vertex	Value of $f(x,y)=x-2y$
$(0,3)$	-6
$(0,6)$	-12
$(4,0)$	-4
$(5,2)$	-9
$(5,6)$	-17

The maximum value of f is -4 and the minimum value is -17.   ● $-4,-17$

[29] Let m be a positive real number. Then the point $(m,0)$ is an element of the solution set in Exercise 15. Now $f(m,0)=6m+4(0)=6m>m$. Therefore, f has no maximum value on the solution set in Exercise 15. This result does not contradict Theorem 9.7 since the solution set is not bounded.

[30] Let m be a negative real number. Then the point $(-m,0)$ is on the solution set of Exercise 15. Moreover, $f(-m,0)=-10(-m)=10m<m$. Hence, f has no minimum in the solution set in Exercise 15.

[31]

Pairs of vertices	Equation of line determined by vertices
$(1,6),(4,-6)$	$4x+y=10$
$(1,6),(5,7)$	$4x+y=27$
$(5,7),(8,-5)$	$x-4y=-23$
$(4,-6),(8,-5)$	$x-4y=28$

It follows that a system of linear inequalities that determines the region is $\begin{cases} 4x+y\ge 10 \\ 4x+y\le 27 \\ x-4y\ge -23 \\ x-4y\le 28 \end{cases}$

(See figure below).

[32]

Pairs of vertices	Equation of line determined by vertices
$(1,6),(3,8)$	$x-y=-5$
$(1,6),(4,-6)$	$4x+y=10$
$(4,-6),(6,-4)$	$x-y=10$
$(3,8),(6,-4)$	$4x+y=20$

It follows that a system of linear inequalities that determines the region is $\begin{cases} x-y\ge -5 \\ 4x+y\ge 10 \\ x-y\le 10 \\ 4x+y\le 20 \end{cases}$

(See figure below).

[33]

Pairs of vertices	Equation of line determined by vertices
$(-1,-3),(-1,4)$	$x=-1$
$(-1,-3),(3,-7)$	$x+y=-4$
$(-1,4),(3,10)$	$3x-2y=-11$
$(3,-7),(3,10)$	$x=3$

It follows that a system of linear inequalities that determines the region is $\begin{cases} x\ge -1 \\ x+y\ge -4 \\ 3x-2y\ge -11 \\ x\le 3 \end{cases}$

(See figure below).

[34]

Pairs of vertices	Equation of line determined by vertices
$(-3, 4), (1, 10)$	$3x - 2y = -17$
$(-3, 4), (-1, -2)$	$3x + y = -5$
$(-1, -2), (4, -6)$	$4x + 5y = -14$
$(4, -6), (7, 3)$	$3x - y = 18$
$(1, 10), (7, 3)$	$7x + 6y = 67$

It follows that a system of linear inequalities that determines the region is $\begin{cases} 3x - 2y \geq -17 \\ 3x + y \geq -5 \\ 4x + 5y \geq -14 \\ 3x - y \leq 18 \\ 7x + 6y \leq 67 \end{cases}$

(See figure below).

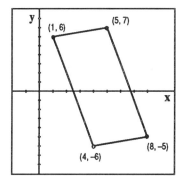

Figure 31

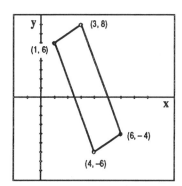

Figure 32

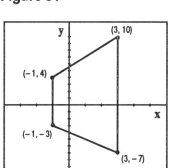

Figure 33

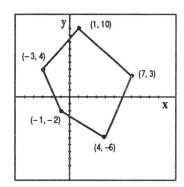

Figure 34

[35] Let x represent the number of minutes of advertisement on television and y be number of minutes of advertisement on radio. Then the cost function is defined by $C(x, y) = 3000x + 400y$. We want to minimize the cost function on the region determined by the following system of inequalities:

$\begin{cases} x \geq 0 \\ y \geq 0 \\ 0.08x + 0.01y \geq 3.10 \quad \text{\underline{number of people reached}} \\ x + y \leq 100 \quad \text{\underline{total amount (30, 70) of advertisement}} \end{cases}$

Vertex	$C(x, y) = 3000x + 400y$
$(0, 310)$	$124,000$
$(30, 70)$	$118,000$
$(100, 0)$	$300,000$

(See figure below). ● 30 min. of television; 70 min. of radio

489

[36] Let x be the amount of money invested in type A bonds and y the amount invested in type B bonds. We want to maximize the interest function, which is defined by $I(x,y) = 0.096x + 0.12y$, on the region defined by the following system of linear inequalities.

$$\begin{cases} x \geq 0 \\ y \geq 0 \\ x + y \leq 10{,}000 \quad \underline{\text{total investment}} \\ 0.40x - y \geq -200 \end{cases}$$

Vertex	$I(x,y) = 0.096x + 0.12y$
(0,0)	0
(0, 200)	24
(7000, 3000)	1032
(10000, 0)	960

(See figure below).

● Invest $ 7000 in type A bonds; $ 3000 in type B bonds

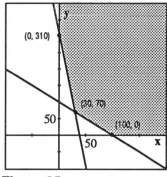

(0, 310)

(30, 70)

50

(100, 0)

50

x

Figure 35

y

(7000, 3000)

2000

(0, 200)

x

2000   (10000, 0)

(0, 0)

Figure 36

[37] Let x denote the number of acres of pine to be harvested and y the number of acres of fir to be harvested. We wish to minimize the cost function, which is defined by $C(x,y) = 800x + 1500y$,

on the region determined by the following system of linear inequalities.

$$\begin{cases} x \geq 0 \\ y \geq 0 \\ 6x + 2y \geq 12 \quad \underline{\text{pulpwood}} \\ 4x + 2y \geq 10 \quad \underline{\text{resin}} \\ 2x + 4y \geq 4 \quad \underline{\text{lumber}} \end{cases}$$

Vertex	$C(x,y) = 800x + 1500y$
(0, 6)	9000
(1, 3)	5300
$\left(\frac{5}{2}, 0\right)$	2000

(See figure below).

● 2.5 acres of pine; 0 acres of fir

[38] Let x be the number of ounces of type A food and y the number of ounces of type B food. We wish to minimize the cost function, which is defined by $C(x,y) = 0.40x + 0.35y$, on the region defined by

the following system of linear inequalities.

$$\begin{cases} x \geq 0 \\ y \geq 0 \\ 150x + 60y \geq 3000 \quad \underline{\text{calories}} \\ 21x + 42y \geq 1260 \quad \underline{\text{vitamins}} \\ x + y \leq 60 \quad \underline{\text{total ounces per day}} \end{cases}$$

Vertex	$C(x,y) = 0.40x + 0.35y$
(0, 50)	17.50
(0, 60)	21.00
(10, 25)	12.75
(60, 0)	24.00

● 10 oz. of food type A and 25 oz. of food type B

(See figure below).

490

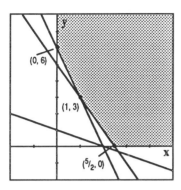

Figure 37

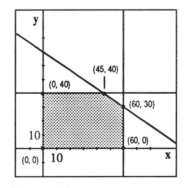

Figure 38

[39]   Let x be the number of A – frame birdhouses manufactured and y be the number of bungalow birdhouses. We want to maximize the profit function, defined by $P(x, y) = 4.00x + 5.00y$, on the region defined by the following system of linear inequalities   (See figure below).

$$\begin{cases} x \geq 0 \\ y \geq 0 \\ 2x + 3y \leq 180 \quad \underline{\text{number of square feet of cedar}} \\ 3x + 4y \leq 250 \quad \underline{\text{number of square feet of plywood}} \\ x + y \leq 75 \quad \underline{\text{number of birdhouses built}} \end{cases}$$

Vertex	$P(x, y) = 4.00x + 5.00y$
$(0, 0)$	0.00
$(0, 60)$	300.00
$(30, 40)$	320.00
$(50, 25)$	325.00
$(75, 0)$	300.00

● 50 A – frames and 25 bugalows, $ 325.00

[40]   Let x be the number of video recorders manufactured and y the number of televisions manufactured. We wish to maximize the profit function, defined by $P(x, y) = 35x + 50y$, on the region defined by the following system of linear inequalities. $\begin{cases} 0 \leq x \leq 60 \quad \underline{\text{number of recorders manufactured}} \\ 0 \leq y \leq 40 \quad \underline{\text{number of televisions manufactured}} \\ 4x + 6y \leq 420 \quad \underline{\text{number of worker–hours needed}} \end{cases}$

Vertex	$P(x, y) = 35x + 50y$
$(0, 0)$	0
$(0, 40)$	2000
$(45, 40)$	3575
$(60, 0)$	2100
$(60, 30)$	3600

(See figure below).   ● 60 video recorders and 30 televisions, $ 3600

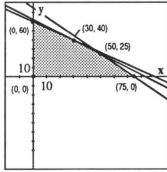

Figure 39

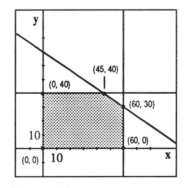

Figure 40

[41] Let x represent the number of professors employed and y represent the number of research assistants.

We wish to minimize the cost function, defined by $C(x, y) = 500x + 400y$, on the region defined

by the following system of linear inequalities $\begin{cases} x \geq 0 \\ y \geq 0 \\ 3x + 4y \geq 25 \quad \underline{\text{number of hours spent in field}} \\ 5x + 3y \geq 27 \quad \underline{\text{number of hours spent in the lab}} \\ x + y \leq 8 \quad \underline{\text{total number of people employed}} \end{cases}$

Vertex	$C(x, y) = 500x + 400y$
$\left(\frac{3}{2}, \frac{13}{2}\right)$	3350
$(3, 4)$	3100
$(7, 1)$	3900

(See figure below). ● 3 professors and 4 assistants; $ 3100

[42] Let x represent the number, measured in millions, of barrels of gasoline produced per day and y the

number of, in millions, of barrels of diesel fuel produced. We wish to maximize the profit function,

defined by $P(x, y) = 2.20x + 2.00y$, on the region defined by the following system of linear

inequalities $\begin{cases} 0 \leq x \leq 2 \\ 0 \leq y \leq 2.5 \\ x + y \leq 3 \ \underline{\text{total number of barrels produced}} \end{cases}$

Vertex	$P(x, y) = 2.20x + 2.00y$ (in millions)
$(0, 0)$	0
$(0, 2.5)$	5
$(0.5, 2.5)$	6.1
$(2, 0)$	4.4
$(2, 1)$	6.4

(See figure below).

● 2,000,000 barrels of gasoline and 1,000,000 barrels of diesel fuel; $ 6,400,000

[43] Let x represent the number of type -1 shifts and y represent the number of type -2 shifts. We wants

to minimize the cost function, defined by $C(x, y) = 90x + 50y$, on the region defined by the

following system of linear inequalities $\begin{cases} x \geq 0 \\ y \geq 0 \\ 3x + 6y \leq 36 \quad \underline{\text{total number of employees}} \\ 3x + 2y \geq 24 \quad \underline{\text{number of additional hours}} \end{cases}$

Vertex	$C(x, y) = 90x + 50y$
$(6, 3)$	690
$(8, 0)$	720
$(12, 0)$	1080

(See figure below). ● 6 type -1 shifts and 3 types -2 shifts; $ 690

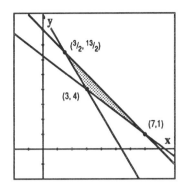

Figure 41

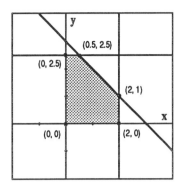

Figure 42

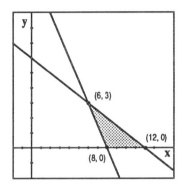

Figure 43

CHAPTER 9 REVIEW EXERCISES

[1] Solving E_1 for x and substituting into E_2 yields $4(y+4)+y=-6 \Rightarrow 5y+16=-6 \Rightarrow 5y=$
$-22 \Rightarrow y=-22 \Rightarrow y=-\frac{22}{5} \Rightarrow x=-\frac{2}{5}$. ● $\left(-\frac{2}{5},-\frac{22}{5}\right)$

[2] Solving E_1 for x and substituting into E_2 yields $2x+3(4x-4)=5 \Rightarrow 14x-12=5 \Rightarrow 14x=$
$17 \Rightarrow x=\frac{17}{14}; y=\frac{6}{7}$. ● $\left(\frac{17}{14},\frac{6}{7}\right)$

[3] Solving E_1 for x and substituting into E_2 yields $3x+8(5x+6)=0 \Rightarrow 43x+48=0 \Rightarrow x=$
$-\frac{48}{43} \Rightarrow y=\frac{18}{43}$. ● $\left(-\frac{48}{43},\frac{18}{43}\right)$

[4] Solving E_1 for x and substituting into E_2 yields $3(b-6)+3b-5=0 \Rightarrow 6b-23=0 \Rightarrow b=$
$\frac{23}{6}; a=-\frac{13}{6}$. ● $\left(\frac{23}{6},-\frac{13}{6}\right)$

[5] Solving E_1 for x and substituting into E_2 yields $3(y+1)^2+y=13 \Rightarrow 3(y^2+2y+1)+y=13 \Rightarrow$
$3y^2+7y-10=0 \Rightarrow (3y+10)(y-1)=0 \Rightarrow 3y+10=0 \text{ or } y-1=0 \Rightarrow y=-\frac{10}{3} \text{ or } y=1.$ If $y=$
$-\frac{10}{3}, x=-\frac{7}{3}.$ If $y=1, x=2.$ ● $\left\{\left(-\frac{7}{3},-\frac{10}{3}\right),(2,1)\right\}$

[6] Solving E_1 for x and substituting into E_2 yields $(y^2)^2-y^2=2 \Rightarrow y^4-y^2=2 \Rightarrow y^4-y^2-2=0.$
Letting $u=y^2$, we get $u^2-u-2=0 \Rightarrow (u-2)(u+1)=0 \Rightarrow u=2 \text{ or } u=-1.$ Hence $y^2=2$
or $y^2=-1. y^2=2 \Rightarrow y=\pm\sqrt{2}. y^2=-1 \Rightarrow y=\pm i. y^2=2 \Rightarrow x=2. y^2=-1 \Rightarrow x=-1.$
● $\left\{(2,-\sqrt{2}),(2,\sqrt{2}),(-1,-i),(-1,i)\right\}$

493

[7] Solving E_1 for x and substituting into E_2 yields $(2y)^2 - 3y^2 = 9 \Rightarrow y^2 = 9 \Rightarrow y = \pm 3; x = \pm 6.$

● $\{(-6, -3), (6, 3)\}$

[8] Solving E_1 for c and substituting into E_2 yields $3\left(\frac{3}{4}d\right)d = 1 \Rightarrow \frac{9}{4}d^2 = 1 \Rightarrow d^2 = \frac{4}{9} \Rightarrow d = \pm\frac{2}{3};$

$c = \pm\frac{1}{2}.$

● $\left\{\left(-\frac{1}{2}, -\frac{2}{3}\right), \left(\frac{1}{2}, \frac{2}{3}\right)\right\}$

[9] Solving E_2 for x_2 and substituting into E_1 yields $(6y - 2) + y^2 - 25 = 0 \Rightarrow y^2 + 6y - 27 = 0 \Rightarrow$

$(y + 9)(y - 3) = 0 \Rightarrow y = -9$ or $y = 3.$ If $y = 9, x = \pm 2\sqrt{14}\,i$. If $y = 3, x = \pm 4.$

● $\{(-2\sqrt{14}\,i, -9), (2\sqrt{14}\,i, -9), (-4, 3), (4, 3)\}$

[10] Solving E_2 for y^2 and substituting into E_1 yields $x^2 + (x^2 + 7) - 25 = 0 \Rightarrow 2x^2 - 18 = 0 \Rightarrow x^2 = 9 \Rightarrow$

$x = \pm 3 \Rightarrow y = \pm 4.$

● $\{(-3, -4), (-3, 4), (3 - 4), (3, 4)\}$

[11] $E_1 + (-1)E_2$ yields $x \Rightarrow -1 \Rightarrow y = 5.$

● $\{(-1, 5)\}$

[12] $(-3)E_1 + 2E_2$ yields the equation $5y = -5 \Rightarrow y = -1 \Rightarrow x = -2.$

● $\{(-2, -1)\}$

[13] $-60E_1 + 160E_2$ yields the equation $-109y = 670 \Rightarrow y = -\frac{670}{109} \Rightarrow x = \frac{1326}{109}.$

● $\left\{\left(\frac{1326}{109}, -\frac{670}{109}\right)\right\}$

[14] $10E_1 - 10E_2$ yields the equation $2a = 2 \Rightarrow a = 1 \Rightarrow b = 4.$

● $(1, 4)$

[15] $3E_1 + (-2)E_2$ yields the equation $-2x^2 + 3x = -18 \Rightarrow 2x^2 - 3x - 18 = 0 \Rightarrow x = \frac{3 \pm \sqrt{153}}{4} \Rightarrow$

$y = \pm\sqrt{\frac{3 \pm \sqrt{153}}{8}}.$

● $\left\{\left(\frac{3 - \sqrt{153}}{4}, -\sqrt{\frac{\sqrt{153} - 3}{8}}\,i\right), \left(\frac{3 + \sqrt{153}}{4}, \sqrt{\frac{\sqrt{153} - 3}{8}}\,i\right),\right.$

$\left.\left(\frac{3 + \sqrt{153}}{4}, -\sqrt{\frac{3 + \sqrt{153}}{8}}\right), \left(\frac{3 + \sqrt{153}}{4}, \sqrt{\frac{3 + \sqrt{153}}{8}}\right)\right\}$

[16] $E_1 + (-1)E_2$ yields the equation $y^2 - y = 26 \Rightarrow y^2 - y - 26 = 0 \Rightarrow y = \frac{1 \pm \sqrt{105}}{2} \Rightarrow x =$

$\pm\sqrt{\frac{15 \pm \sqrt{105}}{2}}.$

● $\left\{\left(-\sqrt{\frac{15 - \sqrt{105}}{2}}, \frac{1 - \sqrt{105}}{2}\right), \left(\sqrt{\frac{15 - \sqrt{105}}{2}}, \frac{1 - \sqrt{105}}{2}\right),\right.$

$\left.\left(-\sqrt{\frac{15 + \sqrt{105}}{2}}, \frac{1 + \sqrt{105}}{2}\right), \left(\sqrt{\frac{15 + \sqrt{105}}{2}}, \frac{1 + \sqrt{105}}{2}\right)\right\}$

[17] $E_1 + (-1)E_2$ yields the equation $\frac{4}{y} = \frac{2}{5} \Rightarrow y = 10 \Rightarrow x = 6.$

● $\{(6, 10)\}$

[18] $E_1 + (-6)E_2$ yields the equation $\frac{10}{y} = \frac{5}{8} \Rightarrow y = 16 \Rightarrow x = 8.$

● $\{(8, 16)\}$

[19] $\begin{cases} x + y + 3z = 2 & (E_1) \\ x \quad\;\; z = 3 & (E_2) \\ 3x + y \quad = -2 & (E_3) \end{cases} \overset{(-3)E_2 + E_1}{\Longrightarrow} \begin{cases} -2x + y \quad = -7 & (E_4) \\ x \quad\;\; +z = 3 & (E_2) \\ 3x + y \quad = -2 & (E_3) \end{cases} \overset{(-1)E_1 + E_3}{\Longrightarrow}$

$\begin{cases} -2x + y \quad = -7 & (E_4) \\ x \quad\;\; +z = 3 & (E_2) \\ 5x \quad\quad = 5 & (E_5) \end{cases}$ $\begin{array}{l} E_5 : 5x = 5 \Rightarrow x = 1 \\ E_2 : x + z = 3 \Rightarrow z = 3 - x \Rightarrow z = 2 \\ E_4 : -2x + y = -7 \Rightarrow y = 2x - 7 \Rightarrow y = -5 \end{array}$

● $\{(1, -5, 2)\}$

[20] $\begin{cases} r - s + 3t = 2 & (E_1) \ (-4)E_1 + E_2 \\ 4r + 7s - t = 7 & (E_2) \ (-1)E_1 + E_3 \\ r + 4s - 2t = 3 & (E_3) \quad \Rightarrow \end{cases}$

$\begin{cases} r - s + 3t = 2 & (E_1) \ (-5)E_4 \\ 11s - 13t = -1 & (E_4) \ (11)E_5 \\ 5s - 5t = 1 & (E_5) \quad \Rightarrow \end{cases}$ $\begin{cases} r - s + 3t = 2 & (E_1) \\ -55s + 65t = 5 & (E_6) \\ 55s - 55t = 11 & (E_7) \end{cases}$

$\begin{matrix} E_6 + E_7 \\ \Rightarrow \end{matrix}$ $\begin{cases} r - s + 3t = 2 & (E_1) \\ -55s + 65t = 5 & (E_6) \\ 10t = 16 & (E_8) \end{cases}$

$E_8 : 10t = 16 \Rightarrow t = \dfrac{8}{5}$

$E_6 : -55s + 65t = 5 \Rightarrow s = -\dfrac{1}{55}(5 - 65t) \Rightarrow s = \dfrac{9}{5}$

$E_1 : r - s + 3t = 2 \Rightarrow r = 2 + s - 3t \Rightarrow r = -1$

$\bullet \left\{ \left(-1, \dfrac{9}{5}, \dfrac{8}{5} \right) \right\}$

[21] $\begin{bmatrix} 1 & 2 & -1 & 6 \\ -2 & 1 & -3 & 13 \\ -3 & 2 & -3 & 16 \end{bmatrix}$ $\begin{matrix} 2R_1 + R_2 \to R_2 \\ 3R_1 + R_3 \to R_3 \\ \sim \end{matrix}$ $\begin{bmatrix} 1 & 2 & -1 & 6 \\ 0 & 5 & -5 & 25 \\ 0 & 8 & -6 & 34 \end{bmatrix}$

$\begin{matrix} \frac{1}{5}R_2 \to R_2 \\ \sim \end{matrix}$ $\begin{bmatrix} 1 & 2 & -1 & 6 \\ 0 & 1 & -1 & 5 \\ 0 & 8 & -6 & 34 \end{bmatrix}$ $\begin{matrix} -8R_2 + R_3 \to R_3 \\ \sim \end{matrix}$

$\begin{bmatrix} 1 & 2 & -1 & 6 \\ 0 & 1 & -1 & 5 \\ 0 & 0 & -2 & 6 \end{bmatrix}$ $\begin{matrix} -\frac{1}{2}R_3 \to R_3 \\ \sim \end{matrix}$ $\begin{bmatrix} 1 & 2 & -1 & 6 \\ 0 & 1 & -1 & 5 \\ 0 & 0 & 1 & -3 \end{bmatrix}$

$R_3 : z = -3$

$R_2 : y - z = 5 \Rightarrow y = 2$

$R_1 : x + 2y - z = 6 \Rightarrow x = -1$

$\bullet \{(-1, 2, -3)\}$

[22] $\begin{bmatrix} 3 & -2 & 4 & 6 \\ 2 & 3 & -5 & -8 \\ -5 & 4 & -3 & -7 \end{bmatrix}$ $\begin{matrix} \frac{1}{3}R_1 \to R_1 \\ \sim \end{matrix}$ $\begin{bmatrix} 1 & -\frac{2}{3} & \frac{4}{3} & 2 \\ 2 & 3 & -5 & -8 \\ -5 & 4 & -3 & -7 \end{bmatrix}$

$\begin{matrix} -2R_1 + R_2 \to R_2 \\ 5R_1 + R_3 \to R_3 \\ \sim \end{matrix}$ $\begin{bmatrix} 1 & -\frac{2}{3} & \frac{4}{3} & 2 \\ 0 & \frac{3}{13} & -\frac{23}{3} & -12 \\ 0 & \frac{2}{3} & \frac{11}{3} & 3 \end{bmatrix}$ $\begin{matrix} \frac{3}{13}R_2 \to R_2 \\ \sim \end{matrix}$

$\begin{bmatrix} 1 & -\frac{2}{3} & \frac{4}{3} & 2 \\ 0 & 1 & -\frac{23}{3} & -\frac{36}{13} \\ 0 & \frac{2}{3} & \frac{11}{3} & 3 \end{bmatrix}$ $\begin{matrix} -\frac{2}{3}R_2 + R_3 \to R_3 \\ \sim \end{matrix}$ $\begin{bmatrix} 1 & -\frac{2}{3} & \frac{4}{3} & 2 \\ 0 & 1 & -\frac{23}{13} & -\frac{36}{13} \\ 0 & 0 & \frac{63}{13} & \frac{63}{13} \end{bmatrix}$ $\begin{matrix} \frac{13}{63}R_3 \to R_3 \\ \sim \end{matrix}$

$\begin{bmatrix} 1 & -\frac{2}{3} & \frac{4}{3} & 2 \\ 0 & 1 & -\frac{23}{3} & -\frac{36}{13} \\ 0 & 1 & 1 & 1 \end{bmatrix}$

$R_3 : z = 1$

$R_2 : y - \dfrac{23}{13}z = -\dfrac{36}{13} \Rightarrow y = -1$

$R_1 : x - \dfrac{2}{3}y + \dfrac{4}{3}z = 2 \Rightarrow x = 0$

$\bullet \{(0, -1, 1)\}$

495

[23] $\begin{bmatrix} 1 & 1 & -1 & | & 100 \\ 10 & 4 & 5 & | & 680 \\ 1 & -2 & 0 & | & 0 \end{bmatrix} \begin{matrix} -10R_1+R_2 \to R_2 \\ -R_1+R_3 \to R_3 \\ \sim \end{matrix} \begin{bmatrix} 1 & 1 & -1 & | & 100 \\ 0 & -6 & 15 & | & -320 \\ 0 & -3 & 1 & | & -100 \end{bmatrix} \begin{matrix} \frac{1}{6}R_2 \to R_2 \\ \sim \end{matrix}$

$\begin{bmatrix} 1 & 1 & -1 & | & 100 \\ 0 & 1 & -\frac{5}{2} & | & \frac{160}{3} \\ 0 & -3 & 1 & | & -100 \end{bmatrix} \begin{matrix} 3R_2+R_3 \to R_3 \\ \sim \end{matrix} \begin{bmatrix} 1 & 1 & -1 & | & 100 \\ 0 & 1 & -\frac{5}{2} & | & \frac{160}{3} \\ 0 & 0 & \frac{13}{2} & | & 60 \end{bmatrix} \begin{matrix} -\frac{2}{13}R_3 \to R_3 \\ \sim \end{matrix}$

$\begin{bmatrix} 1 & 1 & -1 & | & 100 \\ 0 & 1 & -\frac{5}{2} & | & \frac{160}{3} \\ 0 & 0 & 1 & | & -\frac{120}{13} \end{bmatrix}$ $\begin{matrix} R_3: z = -\dfrac{120}{13} \\ R_2: y - \dfrac{5}{2}z = \dfrac{160}{3} \Rightarrow y = \dfrac{1180}{39} \\ R_1: x+y-z = 100 \Rightarrow x = \dfrac{2360}{39} \end{matrix}$ $\bullet \left\{ \left(\dfrac{2360}{39}, \dfrac{1180}{39}, -\dfrac{120}{13} \right) \right\}$

[24] $\begin{bmatrix} 1 & 1 & 1 & | & 140 \\ 3 & 4 & 8 & | & 840 \\ 1 & -2 & 0 & | & 0 \end{bmatrix} \begin{matrix} -3R_1+R_2 \to R_2 \\ -R_1+R_3 \to R_3 \\ \sim \end{matrix} \begin{bmatrix} 1 & 1 & 1 & | & 140 \\ 0 & 1 & 5 & | & 420 \\ 1 & -3 & -1 & | & -140 \end{bmatrix} \begin{matrix} 3R_2+R_3 \to R_3 \\ \sim \end{matrix}$

$\begin{bmatrix} 1 & 1 & 1 & | & 140 \\ 0 & 1 & 5 & | & 420 \\ 0 & 0 & 14 & | & 1120 \end{bmatrix} \begin{matrix} \frac{1}{14}R_3 \to R_3 \\ \sim \end{matrix} \begin{bmatrix} 1 & 1 & 1 & | & 140 \\ 0 & 1 & 5 & | & 420 \\ 0 & 0 & 1 & | & 80 \end{bmatrix} \begin{matrix} R_3: c = 80 \\ R_2: b+5c = 420 \Rightarrow b = 20 \\ R_1: a+b+c = 140 \Rightarrow a = 40 \end{matrix}$

$\bullet \{(40, 20, 80)\}$

[25] $\begin{bmatrix} 1 & 2 & -3 & 2 & | & 0 \\ 2 & 5 & -8 & 6 & | & 5 \\ 3 & 4 & -5 & 2 & | & 4 \end{bmatrix} \begin{matrix} -2R_1+R_2 \to R_2 \\ -3R_1+R_3 \to R_3 \\ \sim \end{matrix} \begin{bmatrix} 1 & 2 & -3 & 2 & | & 0 \\ 0 & 1 & -2 & 2 & | & 5 \\ 0 & -2 & 4 & -4 & | & 4 \end{bmatrix} \begin{matrix} 2R_1+R_3 \to R_3 \\ \sim \end{matrix}$

$\begin{bmatrix} 1 & 2 & -3 & 2 & | & 0 \\ 0 & 1 & -2 & 2 & | & 5 \\ 0 & 0 & 0 & 0 & | & 14 \end{bmatrix} R_3: 0 = 14.$ This equation is false. Hence the solution set is $\varnothing$. $\bullet \varnothing$

[26] $\begin{bmatrix} 1 & 2 & 2 & | & 2 \\ 3 & -1 & -1 & | & 5 \\ 2 & -5 & 3 & | & -4 \\ 1 & 4 & 6 & | & 0 \end{bmatrix} \begin{matrix} -3R_1+R_2 \to R_2 \\ -2R_1+R_3 \to R_3 \\ -R_1+R_4 \to R_4 \\ \sim \end{matrix} \begin{bmatrix} 1 & 2 & 2 & | & 2 \\ 0 & -7 & -7 & | & -1 \\ 0 & -9 & -1 & | & -8 \\ 0 & 2 & 4 & | & -2 \end{bmatrix} \begin{matrix} -\frac{1}{7}R_2 \to R_2 \\ \sim \end{matrix} \begin{bmatrix} 1 & 2 & 2 & | & 2 \\ 0 & 1 & 1 & | & \frac{1}{7} \\ 0 & -9 & -1 & | & -8 \\ 0 & 2 & 4 & | & -2 \end{bmatrix}$

$\begin{matrix} 9R_2+R_3 \to R_3 \\ -2R_2+R_4 \to R_4 \\ \sim \end{matrix} \begin{bmatrix} 1 & 2 & 2 & | & 2 \\ 0 & 1 & 1 & | & \frac{1}{7} \\ 0 & 0 & 8 & | & -\frac{47}{7} \\ 0 & 0 & 2 & | & -\frac{16}{7} \end{bmatrix} \begin{matrix} \frac{1}{8}R_3 \to R_3 \\ \sim \end{matrix} \begin{bmatrix} 1 & 2 & 2 & | & 2 \\ 0 & 1 & 1 & | & \frac{1}{7} \\ 0 & 0 & 1 & | & -\frac{47}{56} \\ 0 & 0 & 2 & | & -\frac{16}{7} \end{bmatrix} \begin{matrix} -2R_3+R_4 \to R_4 \\ \sim \end{matrix}$

$\begin{bmatrix} 1 & 2 & 2 & | & 2 \\ 0 & 1 & 1 & | & \frac{1}{7} \\ 0 & 0 & 1 & | & -\frac{47}{56} \\ 0 & 0 & 0 & | & -\frac{17}{28} \end{bmatrix} R_4: 0 = -\dfrac{17}{28}.$ This equation is false. Hence, the solution set is $\varnothing$. $\bullet \varnothing$

496

[27] $BA = \begin{bmatrix} 6 & -9 & 4 & 5 \\ 5 & -1 & -1 & 2 \\ 0 & -9 & 6 & 3 \end{bmatrix}$; AB is undefined

[28] $AB = \begin{bmatrix} 4 & 0 & 2 & -1 \\ -3 & 4 & -1 & 3 \\ 13 & 4 & 5 & 6 \end{bmatrix}$; BA is undefined

[29] $AB = \begin{bmatrix} 6 & 5 \\ 0 & -4 \\ 13 & 21 \\ 13 & 23 \end{bmatrix}$; BA is undefined

[30] $AB = \begin{bmatrix} 1 & 4 & 2 & 0 \\ 0 & 2 & 2 & 0 \\ 1 & -2 & 2 & 4 \\ 0 & -1 & 0 & 2 \end{bmatrix}$; $BA = \begin{bmatrix} 2 & 4 & 2 & 0 \\ -1 & 0 & 1 & 0 \\ 1 & 2 & 3 & 4 \\ 0 & -1 & -1 & 2 \end{bmatrix}$

[31] $\left[\begin{array}{cc|cc} 1 & 2 & 1 & 0 \\ -1 & 10 & 0 & 1 \end{array}\right] \overset{R_1 + R_2 \to R_2}{\sim} \left[\begin{array}{cc|cc} 1 & 2 & 1 & 0 \\ 0 & 12 & 1 & 1 \end{array}\right] \overset{\frac{1}{12}R_2 \to R_2}{\sim} \left[\begin{array}{cc|cc} 1 & 2 & 1 & 0 \\ 0 & 1 & \frac{1}{12} & \frac{1}{12} \end{array}\right]$

$\overset{-2R_2 + R_1 \to R_1}{\sim} \left[\begin{array}{cc|cc} 1 & 0 & \frac{5}{6} & -\frac{1}{6} \\ 0 & 1 & \frac{1}{12} & \frac{1}{12} \end{array}\right]$
$\bullet \left[\begin{array}{cc} \frac{5}{6} & -\frac{1}{6} \\ \frac{1}{12} & \frac{1}{12} \end{array}\right]$

[32] $\left[\begin{array}{cc|cc} 4 & 6 & 1 & 0 \\ -3 & 8 & 0 & 1 \end{array}\right] \overset{\frac{1}{4}R_1 \to R_1}{\sim} \left[\begin{array}{cc|cc} 1 & \frac{3}{2} & \frac{1}{4} & 0 \\ -3 & 8 & 0 & 1 \end{array}\right] \overset{3R_1 + R_2 \to R_2}{\sim} \left[\begin{array}{cc|cc} 1 & \frac{3}{2} & \frac{1}{4} & 0 \\ 0 & \frac{25}{2} & \frac{3}{4} & 1 \end{array}\right]$

$\overset{\frac{2}{25} + R_2 \to R_2}{\sim} \left[\begin{array}{cc|cc} 1 & \frac{3}{2} & \frac{1}{4} & 0 \\ 0 & 1 & \frac{3}{50} & \frac{2}{25} \end{array}\right] \overset{-\frac{3}{2}R_2 + R_1 \to R_1}{\sim} \left[\begin{array}{cc|cc} 1 & 0 & \frac{4}{25} & -\frac{3}{25} \\ 0 & 1 & \frac{3}{50} & \frac{2}{25} \end{array}\right]$

$\bullet \left[\begin{array}{cc} \frac{4}{25} & -\frac{3}{25} \\ \frac{3}{50} & \frac{2}{25} \end{array}\right]$

[33] $\left[\begin{array}{ccc|ccc} 1 & -1 & 3 & 1 & 0 & 0 \\ -1 & 0 & -1 & 0 & 1 & 0 \\ 2 & -1 & 4 & 0 & 0 & 1 \end{array}\right] \overset{R_1 + R_2 \to R_2}{\underset{-2R_1 + R_3 \to R_3}{\sim}} \left[\begin{array}{ccc|ccc} 1 & -1 & 3 & 1 & 0 & 0 \\ 0 & -1 & 2 & 1 & 1 & 0 \\ 0 & 1 & -2 & -2 & 0 & 1 \end{array}\right] \overset{-R_2 \to R_2}{\sim}$

$\left[\begin{array}{ccc|ccc} 1 & -1 & 3 & 1 & 0 & 0 \\ 0 & 1 & -2 & -1 & -1 & 0 \\ 0 & 1 & -2 & -2 & 0 & 1 \end{array}\right] \overset{-R_2 + R_3 \to R_3}{\sim} \left[\begin{array}{ccc|ccc} 1 & -1 & 3 & 1 & 0 & 0 \\ 0 & 1 & -2 & -1 & -1 & 0 \\ 0 & 0 & 0 & -1 & 1 & 1 \end{array}\right]$ This matrix is

singular since each entry to the left of the vertical bar in the third row is zero. $\quad \bullet$ Singular

[34]

$$\begin{bmatrix} 3 & 2 & 5 & 1 & 0 & 0 \\ 1 & -3 & 2 & 0 & 1 & 0 \\ 2 & 2 & -1 & 0 & 0 & 1 \end{bmatrix} \overset{R_1 \leftrightarrow R_2}{\sim} \begin{bmatrix} 1 & -3 & 2 & 0 & 1 & 0 \\ 3 & 2 & 5 & 1 & 0 & 0 \\ 2 & 2 & -1 & 0 & 0 & 1 \end{bmatrix} \begin{matrix} -3R_1 + R_2 \to R_2 \\ -3R_1 + R_3 \to R_3 \\ \sim \end{matrix}$$

$$\begin{bmatrix} 1 & -3 & 2 & 0 & 1 & 0 \\ 0 & 11 & -1 & 1 & -3 & 0 \\ 0 & 8 & -5 & 0 & -2 & 1 \end{bmatrix} \overset{\frac{1}{11}R_2 \to R_2}{\sim} \begin{bmatrix} 1 & -3 & 2 & 0 & 1 & 0 \\ 0 & 1 & -\frac{1}{11} & \frac{1}{11} & -\frac{3}{11} & 0 \\ 0 & 8 & -5 & 0 & -2 & 1 \end{bmatrix} \begin{matrix} -8R_2 + R_3 \to R_3 \\ \sim \end{matrix}$$

$$\begin{bmatrix} 1 & -3 & 2 & 0 & 1 & 0 \\ 0 & 1 & -\frac{1}{11} & \frac{1}{11} & -\frac{3}{11} & 0 \\ 0 & 0 & -\frac{47}{11} & -\frac{8}{11} & -\frac{2}{11} & 1 \end{bmatrix} \overset{-\frac{11}{47}R_3 \to R_3}{\sim} \begin{bmatrix} 1 & -3 & 2 & 0 & 1 & 0 \\ 0 & 1 & -\frac{1}{11} & \frac{1}{11} & -\frac{3}{11} & 0 \\ 0 & 0 & 1 & \frac{8}{47} & -\frac{2}{11} & -\frac{11}{47} \end{bmatrix}$$

$$\overset{\frac{1}{11}R_3 + R_2 \to R_2}{\sim} \begin{bmatrix} 1 & -3 & 2 & 0 & 1 & 0 \\ 0 & 1 & 0 & \frac{5}{47} & -\frac{13}{47} & -\frac{1}{47} \\ 0 & 0 & 1 & \frac{8}{47} & -\frac{2}{47} & -\frac{11}{47} \end{bmatrix} \begin{matrix} -2R_3 + R_1 \to R_1 \\ \sim \end{matrix}$$

$$\begin{bmatrix} 1 & -3 & 0 & -\frac{16}{47} & \frac{51}{47} & \frac{22}{47} \\ 0 & 1 & 0 & \frac{5}{47} & -\frac{13}{47} & -\frac{1}{47} \\ 0 & 0 & 1 & \frac{8}{47} & -\frac{2}{47} & -\frac{11}{47} \end{bmatrix} \overset{3R_2 + R_1 \to R_1}{\sim} \begin{bmatrix} 1 & 0 & 0 & -\frac{1}{47} & \frac{12}{47} & \frac{19}{47} \\ 0 & 1 & 0 & \frac{5}{47} & -\frac{13}{47} & -\frac{1}{47} \\ 0 & 0 & 1 & \frac{8}{47} & -\frac{2}{47} & -\frac{11}{47} \end{bmatrix}$$

$$\bullet \begin{bmatrix} -\frac{1}{47} & \frac{12}{47} & \frac{19}{47} \\ \frac{5}{47} & -\frac{13}{47} & -\frac{1}{47} \\ \frac{8}{47} & -\frac{2}{47} & -\frac{11}{47} \end{bmatrix}$$

[35]

$$\begin{bmatrix} 1 & 2 & 0 & 1 & 0 & 0 \\ 0 & 1 & 1 & 0 & 1 & 0 \\ 0 & 0 & 4 & 0 & 0 & 1 \end{bmatrix} \overset{\frac{1}{4}R_3 \to R_3}{\sim} \begin{bmatrix} 1 & 2 & 0 & 1 & 0 & 0 \\ 0 & 1 & 1 & 0 & 1 & 0 \\ 0 & 0 & 1 & 0 & 0 & \frac{1}{4} \end{bmatrix} \begin{matrix} -R_3 + R_2 \to R_2 \\ \sim \end{matrix}$$

$$\begin{bmatrix} 1 & 2 & 0 & 1 & 0 & 0 \\ 0 & 1 & 0 & 0 & 1 & -\frac{1}{4} \\ 0 & 0 & 1 & 0 & 0 & \frac{1}{4} \end{bmatrix} \overset{-2R_2 + R_1 \to R_1}{\sim} \begin{bmatrix} 1 & 0 & 0 & 1 & -2 & \frac{1}{2} \\ 0 & 1 & 0 & 0 & 1 & -\frac{1}{4} \\ 0 & 0 & 1 & 0 & 0 & \frac{1}{4} \end{bmatrix} \quad \bullet \begin{bmatrix} 1 & -2 & \frac{1}{2} \\ 0 & 1 & -\frac{1}{4} \\ 0 & 0 & \frac{1}{4} \end{bmatrix}$$

[36] $\begin{bmatrix} -3 & 2 & 1 \\ 0 & 1 & 3 \\ 4 & -6 & 0 \end{bmatrix}\begin{matrix} 1 & 0 & 0 \\ 0 & 1 & 0 \\ 0 & 0 & 1 \end{matrix}$ $-\frac{1}{3}R_1 \to R_1$ $\begin{bmatrix} 1 & -\frac{2}{3} & -\frac{1}{3} \\ 0 & 1 & 3 \\ 4 & -6 & 0 \end{bmatrix}\begin{matrix} -\frac{1}{3} & 0 & 0 \\ 0 & 1 & 0 \\ 0 & 0 & 1 \end{matrix}$ $-4R_1 + R_3 \to R_3$

$\begin{bmatrix} 1 & -\frac{2}{3} & -\frac{1}{3} \\ 0 & 1 & 3 \\ 0 & -\frac{10}{3} & \frac{4}{3} \end{bmatrix}\begin{matrix} -\frac{1}{3} & 0 & 0 \\ 0 & 1 & 0 \\ \frac{4}{3} & 0 & 1 \end{matrix}$ $\frac{10}{3}R_2 + R_3 \to R_3$ $\begin{bmatrix} 1 & -\frac{2}{3} & -\frac{1}{3} \\ 0 & 1 & 3 \\ 0 & 0 & \frac{34}{3} \end{bmatrix}\begin{matrix} -\frac{1}{3} & 0 & 0 \\ 0 & 1 & 0 \\ \frac{4}{3} & \frac{10}{3} & 1 \end{matrix}$ $\frac{3}{34}R_3 \to R_3$

$\begin{bmatrix} 1 & -\frac{2}{3} & -\frac{1}{3} \\ 0 & 1 & 3 \\ 0 & 0 & 1 \end{bmatrix}\begin{matrix} -\frac{1}{3} & 0 & 0 \\ 0 & 1 & 0 \\ \frac{2}{17} & \frac{5}{17} & \frac{3}{34} \end{matrix}$ $\begin{matrix} -3R_3 + R_2 \to R_2 \\ \frac{1}{3}R_3 + R_1 \to R_1 \end{matrix}$ $\begin{bmatrix} 1 & -\frac{2}{3} & 0 \\ 0 & 1 & 0 \\ 0 & 0 & 1 \end{bmatrix}\begin{matrix} -\frac{15}{51} & \frac{5}{51} & \frac{1}{34} \\ -\frac{6}{17} & \frac{2}{17} & -\frac{9}{34} \\ \frac{2}{17} & \frac{5}{17} & \frac{3}{34} \end{matrix}$

$\frac{2}{3}R_2 + R_1 \to R_1$ $\begin{bmatrix} 1 & 0 & 0 \\ 0 & 1 & 0 \\ 0 & 0 & 1 \end{bmatrix}\begin{matrix} -\frac{9}{17} & \frac{3}{17} & -\frac{5}{34} \\ -\frac{6}{17} & \frac{2}{17} & -\frac{9}{34} \\ \frac{2}{17} & \frac{5}{17} & \frac{3}{34} \end{matrix}$ $\bullet \begin{bmatrix} -\frac{9}{17} & \frac{3}{17} & -\frac{5}{34} \\ -\frac{6}{17} & \frac{2}{17} & -\frac{9}{34} \\ \frac{2}{17} & \frac{5}{17} & \frac{3}{34} \end{bmatrix}$

[37] $\begin{bmatrix} 1 & 1 & -1 & 4 \\ 1 & 0 & 4 & -1 \\ 6 & 0 & 0 & 5 \\ 2 & 1 & -1 & 3 \end{bmatrix}\begin{matrix} 1 & 0 & 0 & 0 \\ 0 & 1 & 0 & 0 \\ 0 & 0 & 1 & 0 \\ 0 & 0 & 0 & 1 \end{matrix}$ $\begin{matrix} -R_1 + R_2 \to R_2 \\ -6R_1 + R_3 \to R_3 \\ -2R_1 + R_4 \to R_4 \end{matrix}$ $\begin{bmatrix} 1 & 1 & -1 & 4 \\ 0 & -1 & 5 & -5 \\ 0 & -6 & 6 & -19 \\ 0 & -1 & 1 & -5 \end{bmatrix}\begin{matrix} 1 & 0 & 0 & 0 \\ -1 & 1 & 0 & 0 \\ -6 & 0 & 1 & 0 \\ -2 & 0 & 0 & 1 \end{matrix}$

$-R_2 \to R_2$ $\begin{bmatrix} 1 & 1 & -1 & 4 \\ 0 & 1 & -5 & 5 \\ 0 & -6 & 6 & -19 \\ 0 & -1 & 1 & -5 \end{bmatrix}\begin{matrix} 1 & 0 & 0 & 0 \\ 1 & -1 & 0 & 0 \\ -6 & 0 & 1 & 0 \\ -2 & 0 & 0 & 1 \end{matrix}$ $\begin{matrix} 6R_2 + R_3 \to R_3 \\ R_2 + R_4 \to R_4 \end{matrix}$

$\begin{bmatrix} 1 & 1 & -1 & 4 \\ 0 & 1 & -5 & 5 \\ 0 & 0 & -24 & 11 \\ 0 & 0 & -4 & 0 \end{bmatrix}\begin{matrix} 1 & 0 & 0 & 0 \\ 1 & -1 & 0 & 0 \\ 0 & -6 & 1 & 0 \\ -1 & -1 & 0 & 1 \end{matrix}$ $-\frac{1}{24}R_3 \to R_3$ $\begin{bmatrix} 1 & 1 & -1 & 4 \\ 0 & 1 & -5 & 5 \\ 0 & 0 & 1 & -\frac{11}{24} \\ 0 & 0 & -4 & 0 \end{bmatrix}\begin{matrix} 1 & 0 & 0 & 0 \\ 1 & -1 & 0 & 0 \\ 0 & \frac{1}{4} & -\frac{1}{24} & 0 \\ -1 & -1 & 0 & 1 \end{matrix}$

$4R_3 + R_4 \to R_4$ $\begin{bmatrix} 1 & 1 & -1 & 4 \\ 0 & 1 & -5 & 5 \\ 0 & 0 & 1 & -\frac{11}{24} \\ 0 & 0 & 0 & -\frac{11}{6} \end{bmatrix}\begin{matrix} 1 & 0 & 0 & 0 \\ 1 & -1 & 0 & 0 \\ 0 & \frac{1}{4} & -\frac{1}{24} & 0 \\ -1 & 0 & -\frac{1}{6} & 1 \end{matrix}$ $-\frac{6}{11}R_4 \to R_4$

$\begin{bmatrix} 1 & 1 & -1 & 4 \\ 0 & 1 & -5 & 5 \\ 0 & 0 & 1 & -\frac{11}{24} \\ 0 & 0 & 0 & 1 \end{bmatrix}\begin{matrix} 1 & 0 & 0 & 0 \\ 1 & -1 & 0 & 0 \\ 0 & \frac{1}{4} & -\frac{1}{24} & 0 \\ \frac{6}{11} & 0 & \frac{1}{11} & -\frac{6}{11} \end{matrix}$ $\begin{matrix} \frac{11}{24}R_4 + R_3 \to R_3 \\ -5R_4 + R_2 \to R_2 \\ -4R_4 + R_1 \to R_1 \end{matrix}$

$$\left[\begin{array}{cccc|cccc} 1 & 1 & -1 & 0 & -\dfrac{13}{11} & 0 & -\dfrac{4}{11} & \dfrac{24}{11} \\ 0 & 1 & -5 & 0 & -\dfrac{19}{11} & -1 & -\dfrac{5}{11} & \dfrac{30}{11} \\ 0 & 0 & 1 & 0 & \dfrac{1}{4} & \dfrac{1}{4} & 0 & -\dfrac{1}{4} \\ 0 & 0 & 0 & 1 & \dfrac{6}{11} & 0 & \dfrac{1}{11} & -\dfrac{6}{11} \end{array}\right] \begin{array}{l} 5R_3+R_2\to R_2 \\ R_3+R_3\to R_3 \\ \sim \end{array}$$

$$\left[\begin{array}{cccc|cccc} 1 & 1 & 0 & 0 & -\dfrac{41}{44} & \dfrac{1}{4} & -\dfrac{4}{11} & \dfrac{85}{44} \\ 0 & 1 & 0 & 0 & -\dfrac{21}{44} & \dfrac{1}{4} & -\dfrac{5}{11} & \dfrac{65}{44} \\ 0 & 0 & 1 & 0 & \dfrac{1}{4} & \dfrac{1}{4} & 0 & -\dfrac{1}{4} \\ 0 & 0 & 0 & 1 & \dfrac{6}{11} & 0 & \dfrac{1}{11} & -\dfrac{6}{11} \end{array}\right] \begin{array}{l} -R_2+R_1\to R_1 \\ \sim \end{array}$$

$$\left[\begin{array}{cccc|cccc} 1 & 1 & 0 & 0 & -\dfrac{5}{11} & 0 & \dfrac{1}{11} & \dfrac{5}{11} \\ 0 & 1 & 0 & 0 & -\dfrac{21}{44} & \dfrac{1}{4} & -\dfrac{5}{11} & \dfrac{65}{44} \\ 0 & 0 & 1 & 0 & \dfrac{1}{4} & \dfrac{1}{4} & 0 & -\dfrac{1}{4} \\ 0 & 0 & 0 & 1 & \dfrac{6}{11} & 0 & \dfrac{1}{11} & -\dfrac{6}{11} \end{array}\right] \qquad \bullet \left[\begin{array}{cccc} -\dfrac{5}{11} & 0 & \dfrac{1}{11} & \dfrac{5}{11} \\ -\dfrac{21}{44} & \dfrac{1}{4} & -\dfrac{5}{11} & \dfrac{65}{44} \\ \dfrac{1}{4} & \dfrac{1}{4} & 0 & -\dfrac{1}{4} \\ \dfrac{6}{11} & 0 & \dfrac{1}{11} & -\dfrac{6}{11} \end{array}\right]$$

[38] $$\left[\begin{array}{cccc|cccc} 5 & 4 & 2 & 1 & 1 & 0 & 0 & 0 \\ 2 & 3 & 1 & -2 & 0 & 1 & 0 & 0 \\ -5 & -7 & -3 & 9 & 0 & 0 & 1 & 0 \\ 1 & -2 & -1 & 4 & 0 & 0 & 0 & 1 \end{array}\right] \begin{array}{l} R_1\leftrightarrow R_4 \\ \sim \end{array} \left[\begin{array}{cccc|cccc} 1 & -2 & -1 & 4 & 0 & 0 & 0 & 1 \\ 2 & 3 & 1 & -2 & 0 & 1 & 0 & 0 \\ -5 & -7 & -3 & 9 & 0 & 0 & 1 & 0 \\ 5 & 4 & 2 & 1 & 1 & 0 & 0 & 0 \end{array}\right]$$

$$\begin{array}{l} -2R_1+R_2\to R_2 \\ 5R_1+R_3\to R_3 \\ -5R_1+R_4\to R_4 \\ \sim \end{array} \left[\begin{array}{cccc|cccc} 1 & -2 & -1 & 4 & 0 & 0 & 0 & 1 \\ 0 & 7 & 3 & -10 & 0 & 1 & 0 & -2 \\ 0 & -17 & -8 & 29 & 0 & 0 & 1 & 5 \\ 0 & 14 & 7 & -19 & 1 & 0 & 0 & -5 \end{array}\right] \begin{array}{l} \dfrac{1}{7}R_2\to R_2 \\ \sim \end{array}$$

$$\left[\begin{array}{cccc|cccc} 1 & -2 & -1 & 4 & 0 & 0 & 0 & 1 \\ 0 & 7 & \dfrac{3}{7} & -\dfrac{10}{7} & 0 & \dfrac{1}{7} & 0 & -\dfrac{2}{7} \\ 0 & -17 & -8 & 29 & 0 & 0 & 1 & 5 \\ 0 & 14 & 7 & -19 & 1 & 0 & 0 & -5 \end{array}\right] \begin{array}{l} 17R_2+R_3\to R_3 \\ -14R_2+R_4\to R_4 \\ \sim \end{array}$$

$$\left[\begin{array}{cccc|cccc} 1 & -2 & -1 & 4 & 0 & 0 & 0 & 1 \\ 0 & 1 & \dfrac{3}{7} & -\dfrac{10}{7} & 0 & \dfrac{1}{7} & 0 & -\dfrac{2}{7} \\ 0 & 0 & -\dfrac{5}{7} & \dfrac{33}{7} & 0 & \dfrac{17}{5} & 1 & \dfrac{1}{7} \\ 0 & 0 & 1 & 1 & 1 & -2 & 0 & -1 \end{array}\right] \begin{array}{l} -\dfrac{7}{5}R_3\to R_3 \\ \sim \end{array}$$

$$\left[\begin{array}{cccc|cccc} 1 & -2 & -1 & 4 & 0 & 0 & 0 & 1 \\ 0 & 1 & \dfrac{3}{7} & -\dfrac{10}{7} & 0 & \dfrac{1}{7} & 0 & -\dfrac{2}{7} \\ 0 & 0 & 1 & -\dfrac{33}{5} & 0 & -\dfrac{17}{5} & -\dfrac{7}{5} & -\dfrac{1}{5} \\ 0 & 0 & 1 & 1 & 1 & -2 & 0 & -1 \end{array}\right] \begin{array}{l} -R_3+R_4\to R_4 \\ \sim \end{array}$$

500

$$\begin{bmatrix} 1 & -2 & -1 & 4 & 0 & 0 & 0 & 1 \\ 0 & 1 & \frac{3}{7} & -\frac{10}{7} & 0 & \frac{1}{7} & 0 & -\frac{2}{7} \\ 0 & 0 & 1 & -\frac{33}{7} & 0 & -\frac{17}{5} & -\frac{7}{5} & -\frac{1}{5} \\ 0 & 0 & 0 & \frac{38}{5} & 1 & \frac{7}{5} & \frac{7}{5} & -\frac{4}{5} \end{bmatrix} \begin{matrix} \frac{5}{38}R_4 \to R_4 \\ \\ \\ \sim \end{matrix}$$

$$\begin{bmatrix} 1 & -2 & -1 & 4 & 0 & 0 & 0 & 1 \\ 0 & 1 & \frac{3}{7} & -\frac{10}{7} & 0 & \frac{1}{7} & 0 & -\frac{2}{7} \\ 0 & 0 & 1 & -\frac{33}{5} & 0 & -\frac{17}{5} & -\frac{7}{5} & -\frac{1}{5} \\ 0 & 0 & 0 & 1 & \frac{5}{38} & \frac{7}{38} & \frac{7}{38} & -\frac{2}{19} \end{bmatrix} \begin{matrix} \frac{33}{5}R_4+R_3 \to R_3 \\ \frac{10}{7}R_4+R_2 \to R_2 \\ -4R_4+R_1 \to R_1 \\ \sim \end{matrix}$$

$$\begin{bmatrix} 1 & -2 & -1 & 0 & -\frac{10}{19} & -\frac{14}{19} & -\frac{14}{19} & \frac{27}{19} \\ 0 & 1 & \frac{3}{7} & 0 & \frac{25}{133} & \frac{54}{133} & \frac{5}{19} & -\frac{58}{133} \\ 0 & 0 & 1 & 0 & \frac{33}{38} & -\frac{83}{38} & -\frac{7}{38} & -\frac{17}{19} \\ 0 & 0 & 0 & 1 & \frac{5}{38} & \frac{7}{38} & \frac{7}{38} & -\frac{2}{19} \end{bmatrix} \begin{matrix} -\frac{3}{7}R_3+R_2 \to R_2 \\ R_3+R_1 \to R_1 \\ \\ \sim \end{matrix}$$

$$\begin{bmatrix} 1 & -2 & 0 & 0 & \frac{13}{38} & -\frac{111}{38} & -\frac{35}{38} & \frac{10}{19} \\ 0 & 1 & 0 & 0 & -\frac{7}{38} & \frac{51}{38} & \frac{13}{38} & -\frac{1}{19} \\ 0 & 0 & 1 & 0 & \frac{33}{38} & -\frac{83}{38} & -\frac{7}{38} & -\frac{17}{19} \\ 0 & 0 & 0 & 1 & \frac{5}{38} & \frac{7}{38} & \frac{7}{38} & -\frac{2}{19} \end{bmatrix} \begin{matrix} 2R_2+R_1 \to R_1 \\ \\ \sim \end{matrix}$$

$$\begin{bmatrix} 1 & 0 & 0 & 0 & -\frac{1}{38} & -\frac{9}{38} & -\frac{9}{38} & \frac{8}{19} \\ 0 & 1 & 0 & 0 & -\frac{7}{38} & \frac{51}{38} & \frac{13}{38} & -\frac{1}{19} \\ 0 & 0 & 1 & 0 & \frac{33}{38} & -\frac{83}{38} & -\frac{7}{38} & -\frac{17}{19} \\ 0 & 0 & 0 & 1 & \frac{5}{38} & \frac{7}{38} & \frac{7}{38} & -\frac{2}{19} \end{bmatrix} \qquad \bullet \begin{bmatrix} -\frac{1}{38} & -\frac{9}{38} & -\frac{9}{38} & \frac{8}{19} \\ -\frac{7}{38} & \frac{51}{38} & \frac{13}{38} & -\frac{1}{19} \\ \frac{33}{38} & -\frac{83}{38} & -\frac{7}{38} & -\frac{17}{19} \\ \frac{5}{38} & \frac{7}{38} & \frac{7}{38} & -\frac{2}{19} \end{bmatrix}$$

[39] $\det(A) = (10)(-13) - (14)(15) = -130 - 210 = -340$ $\qquad\qquad \bullet -340$

[40] $\det(A) = (-4)(3) - (0)(1) = -12 - 0 = -12$ $\qquad\qquad \bullet -12$

[41] Expanding by the third row, we get $\det A = a_{31}A_{31} + a_{32}A_{32} + a_{33}A_{33} = 8(-27) + 7(3) + 0 = -195$
$\qquad\qquad\qquad\qquad \bullet -195$

[42] Expanding by the third row, we get $\det A = a_{31}A_{31} + a_{32}A_{32} + a_{33}A_{33} = 0 + (1)(38) + (-9)(2) = 20$
$\qquad\qquad\qquad\qquad \bullet 20$

[43] Expanding by the first column, we get $\det A = a_{11}A_{11} + a_{21}A_{21} + a_{31}A_{31} = 1(-12) + 0 + 0 = -12$
$\qquad\qquad\qquad\qquad \bullet -12$

[44] Expanding by the first row, we get $\det A = a_{11}A_{11} + a_{12}A_{12} + a_{13}A_{13} = 9(-10) + (-2)(8) + 6(2)$
$= -94$
$\qquad\qquad\qquad\qquad \bullet -94$

[45] $\det\begin{bmatrix} 3 & -4 \\ 5 & 7 \end{bmatrix} = 3(7)-(-4)(5)=41.\, x=\frac{1}{41}\cdot\det\begin{bmatrix} 8 & -4 \\ 15 & 7 \end{bmatrix}=\frac{1}{41}\cdot[8(7)-(-4)(15)]=\frac{116}{41};\, y=$

$\frac{1}{41}\cdot\det\begin{bmatrix} 3 & 8 \\ 5 & 15 \end{bmatrix}=\frac{1}{41}\cdot[3(15)-(5)(8)]=\frac{5}{41}$ ● $\left\{\left(\frac{116}{41},\frac{5}{41}\right)\right\}$

[46] $\det\begin{bmatrix} 7 & -9 \\ -3 & 10 \end{bmatrix}=7(10)-(-9)(-3)=43.\, x=\frac{1}{43}\cdot\det\begin{bmatrix} 14 & -9 \\ 5 & 10 \end{bmatrix}=\frac{1}{43}\cdot[14(10)-(-9)(5)]=$

$\frac{185}{43};\, y=\frac{1}{43}\cdot\det\begin{bmatrix} 7 & 14 \\ -3 & 5 \end{bmatrix}=\frac{1}{43}\cdot[(7)(5)-(14)(-3)]=\frac{77}{43}$ ● $\left\{\left(\frac{185}{43},\frac{77}{43}\right)\right\}$

[47] $\det\begin{bmatrix} 1 & 1 & 1 \\ 2 & 0 & -1 \\ 3 & 2 & 0 \end{bmatrix}\underset{=}{R_1+R_2\to R_2}\det\begin{bmatrix} 1 & 1 & 1 \\ 3 & 1 & 0 \\ 3 & 2 & 0 \end{bmatrix}=(\text{expanding by column 3})\,1\cdot[(3)(2)-(1)(3)]$

$=3.\, x=\frac{1}{3}\cdot\det\begin{bmatrix} 14 & 1 & 1 \\ 5 & 0 & -1 \\ -6 & 2 & 0 \end{bmatrix}\underset{=}{R_1+R_2\to R_2}\frac{1}{3}\cdot\det\begin{bmatrix} 14 & 1 & 1 \\ 19 & 1 & 0 \\ -6 & 2 & 0 \end{bmatrix}=(\text{expanding by column 3})$

$\frac{1}{3}\cdot1\cdot[(19)(2)-(1)(-6)]=\frac{44}{3};\, y=\frac{1}{3}\cdot\det\begin{bmatrix} 1 & 14 & 1 \\ 2 & 5 & -1 \\ 3 & -6 & 0 \end{bmatrix}\underset{=}{R_1+R_2\to R_2}\frac{1}{3}\cdot\det\begin{bmatrix} 1 & 14 & 1 \\ 3 & 19 & 0 \\ 3 & -6 & 0 \end{bmatrix}=$

$(\text{expanding by column 3})\frac{1}{3}\cdot1\cdot[(3)(-6)-(19)(3)]=-25;\, z=\frac{1}{3}\cdot\det\begin{bmatrix} 1 & 1 & 14 \\ 2 & 0 & 5 \\ 3 & 2 & -6 \end{bmatrix}=(\text{expanding}$

by row 2) $\frac{1}{3}\cdot\{(-2)[(1)(-6)-(14)(2)]-(5)[(1)(2)-(1)(3)]\}=\frac{73}{3}$ ● $\left\{\left(\frac{44}{3},-25,\frac{73}{3}\right)\right\}$

[48] $\det\begin{bmatrix} 6 & 2 & -1 \\ 5 & -3 & 2 \\ 4 & -8 & 5 \end{bmatrix}\underset{\sim}{\begin{array}{l}2R_1+R_2\to R_2\\ 5R_1+R_3\to R_3\end{array}}\det\begin{bmatrix} 6 & 2 & -1 \\ 17 & 1 & 0 \\ 34 & 2 & 0 \end{bmatrix}=0\,(\text{row 3 is a multiple of row 2}).\text{ Hence,}$

Cramer's Rule can not be used. $\left[\begin{array}{ccc|c} 6 & 2 & -1 & 7 \\ 5 & -3 & 2 & -4 \\ 4 & -8 & 5 & 3 \end{array}\right]\underset{\sim}{\frac{1}{6}R_1\to R_1}\left[\begin{array}{ccc|c} 1 & \frac{1}{3} & -\frac{1}{6} & \frac{7}{6} \\ 5 & -3 & 2 & -4 \\ 4 & -8 & 5 & 3 \end{array}\right]$

$\begin{array}{l}-5R_1+R_2\to R_2\\ -4R_1+R_3\to R_3\\ \sim\end{array}\left[\begin{array}{ccc|c} 1 & \frac{1}{3} & -\frac{1}{6} & \frac{7}{6} \\ 0 & -\frac{14}{3} & \frac{17}{6} & -\frac{59}{6} \\ 0 & \frac{28}{3} & \frac{17}{3} & -\frac{5}{3} \end{array}\right]\underset{\sim}{-\frac{3}{14}R_2\to R_2}\left[\begin{array}{ccc|c} 1 & \frac{1}{3} & -\frac{1}{6} & \frac{7}{6} \\ 0 & 1 & \frac{17}{28} & -\frac{59}{28} \\ 0 & \frac{28}{3} & \frac{17}{3} & -\frac{5}{3} \end{array}\right]$

$\underset{\sim}{-\frac{28}{3}R_2+R_3\to R_3}\left[\begin{array}{ccc|c} 1 & \frac{1}{3} & -\frac{1}{6} & \frac{7}{6} \\ 0 & 1 & \frac{17}{28} & -\frac{59}{28} \\ 0 & 0 & 0 & 18 \end{array}\right]R_3:0=18.\text{ The equation is false. Thus, the solution set is }\varnothing.$

● $\varnothing$

[49] $\det \begin{bmatrix} 1 & 0 & -1 \\ 0 & 1 & 2 \\ 3 & 0 & 5 \end{bmatrix} = (\text{expanding by column 2}) = 1\,[(1)\,(5)-(-1)\,(3)] = 8.\ x = \dfrac{1}{8}.$

$\det \begin{bmatrix} 2 & 0 & -1 \\ 4 & 1 & 2 \\ -1 & 0 & 5 \end{bmatrix} = (\text{expanding by column 2})\ \dfrac{1}{8}\cdot 1\cdot[(2)\,(5)-(-1)\,(-1)] = \dfrac{9}{8};\ y = \dfrac{1}{8}.$

$\det \begin{bmatrix} 1 & 2 & -1 \\ 0 & 4 & 2 \\ 3 & -1 & 5 \end{bmatrix} \begin{array}{c} -3R_1+R_3\to R_3 \\ = \end{array} \dfrac{1}{8}\cdot\det \begin{bmatrix} 1 & 2 & -1 \\ 0 & 4 & 2 \\ 0 & -7 & 8 \end{bmatrix} = (\text{expanding by column 1})\ \dfrac{1}{8}\cdot 1\cdot$

$[(4)\,(8)-(2)\,(-7)] = \dfrac{23}{4};\ z = \dfrac{1}{8}\cdot\det \begin{bmatrix} 1 & 0 & 2 \\ 0 & 1 & 4 \\ 3 & 0 & -1 \end{bmatrix} = (\text{expanding by column 2})\ \dfrac{1}{8}\cdot 1\cdot$

$[(1)\,(-1)-(2)\,(3)] = -\dfrac{7}{8}.$ $\bullet\ \left\{\left(\dfrac{9}{8},\dfrac{23}{4},-\dfrac{7}{8}\right)\right\}$

[50] $\det \begin{bmatrix} 1 & -3 & 1 \\ -2 & -1 & 3 \\ 6 & 3 & -1 \end{bmatrix} \begin{array}{c} 2R_1+R_2\to R_2 \\ -\frac{1}{6}R_1+R_3\to R_3 \\ = \end{array}$

$\det \begin{bmatrix} 1 & -3 & 1 \\ 0 & -7 & 5 \\ 0 & 21 & -7 \end{bmatrix} = (\text{expanding by column 1})\ 1\cdot$

$[(-7)\,(-7)-(5)\,(21)] = -56.\ x = -\dfrac{1}{56}\cdot\det \begin{bmatrix} 4 & -3 & 1 \\ -5 & -1 & 3 \\ 2 & 3 & -1 \end{bmatrix} \begin{array}{c} -3R_1+R_2\to R_2 \\ R_1+R_3\to R_3 \\ = \end{array} -\dfrac{1}{56}\cdot$

$\det \begin{bmatrix} 4 & -3 & 1 \\ -17 & 8 & 0 \\ 6 & 0 & 0 \end{bmatrix} = (\text{expanding by row 3})\ -\dfrac{1}{56}\cdot 6\cdot[(-3)\,(0)-(1)\,(8)] = \dfrac{6}{7};\ y = -\dfrac{1}{56}\cdot$

$\det \begin{bmatrix} 1 & 4 & 1 \\ -2 & -5 & 3 \\ 6 & 2 & -1 \end{bmatrix} \begin{array}{c} -3R_1+R_2\to R_2 \\ R_1+R_3\to R_3 \\ = \end{array}$

$-\dfrac{1}{56}\cdot\det \begin{bmatrix} 1 & 4 & 1 \\ -5 & -17 & 0 \\ 7 & 6 & 0 \end{bmatrix} = (\text{expanding by column 3})$

$-\dfrac{1}{56}\cdot 1\cdot[(-5)\,(6)-(-17)\,(7)] = -\dfrac{89}{56};\ z = -\dfrac{1}{56}\cdot\det \begin{bmatrix} 1 & -3 & 4 \\ -2 & -1 & -5 \\ 6 & 3 & 2 \end{bmatrix} \begin{array}{c} 2R_1+R_2\to R_2 \\ -6R_1+R_3\to R_3 \\ = \end{array}$

$-\dfrac{1}{56}\cdot 1\cdot[(-7)(-22)-(3)\,(21)] = \dfrac{13}{8}.$ $\bullet\ \left\{\left(\dfrac{6}{7},-\dfrac{89}{56},\dfrac{13}{8}\right)\right\}$

[51]   Unbounded; vertex: $\left(\dfrac{24}{7},\dfrac{6}{7}\right)$ (See figure below).

[52]   Unbounded; vertex: $(-2, 14)$ (See figure below).

[53]   Unbounded; vertices: $(0, 1), (1, 0), (4, 0)$ (See figure below).

[54]   Unbounded; vertex : $(0, 6)$ (See figure below).

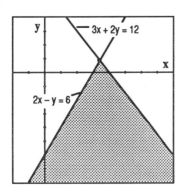

Figure 51

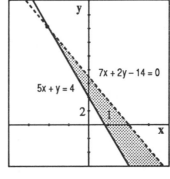

Figure 52

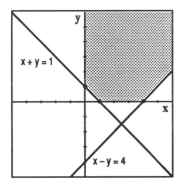

Figure 53

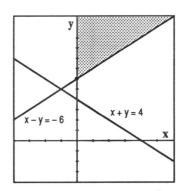

Figure 54

[55]   Bounded; vertices : $(0, 0), (0, 2), (1, 4), (4, 0), (6, 2)$ (See figure below).

[56]   Bounded; vertices : $(0, 2), \left(\frac{7}{5}, \frac{24}{5}\right), (2, 3), \left(\frac{5}{2}, 0\right)$ (See figure below).

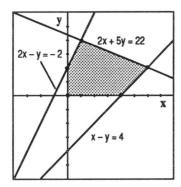

Figure 55

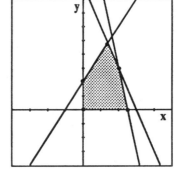

Figure 56

[57]

Vertex	$f(x, y) = 5x + 2y$
$(0, 0)$	0
$(0, 2)$	4
$(1, 4)$	13
$(4, 0)$	20
$(6, 2)$	34

● maximum 34

[58]

Vertex	$f(x, y) = 2x - y$
$(0, 0)$	0
$(0, 2)$	-2
$\left(\frac{7}{5}, \frac{24}{5}\right)$	-2
$\left(\frac{5}{2}, 0\right)$	5
$(2, 3)$	1

● minimum -2

[59] Let x denote the numerator and y denote the denominator of the original fraction. Then $x + y = 12$.

If the numerator of the fraction $\frac{x}{y}$ is decreased by 2 and the denominator increased by 1, then the new

fraction is $\frac{x - 2}{y + 1}$. From the information given, $\frac{x}{y} \cdot \frac{x - 2}{y + 1} = \frac{7}{6} \Rightarrow \frac{x^2 - 2x}{y^2 + y} = \frac{7}{6} \Rightarrow 6x^2 - 12x = 7y^2 +$

$7y \Rightarrow 6x^2 - 7y^2 - 12x - 7y = 0$. We must solve the following system of equations.

$\begin{cases} x + y = 12 \\ 6x^2 - 7y^2 - 12x - 7y = 0 \end{cases}$ Using the method of substitution it can be shown that the solution set of

the above system is $\{(7, 5), (156, -144)\}$. Since x and y must positive, we choose $x = 7$ and $y = 5$.

● 7, 5

[60] From the accompanying figure, we see that the area of the garden is given by $(2x + 3)(2y + 3) =$

450. Also, the area of paths is $3(3 + 2y) + 2(3x) = 126 \Rightarrow 6x + 6y = 117$. Hence, we must find

the solution set for the system $\begin{cases} 6x + 6y = 117 \\ (2x + 3)(2y + 3) = 450 \end{cases}$ Solving E_1 for y yields the equation $y =$

$\frac{39}{2} - x$. Substituting $\frac{39}{2} - x$ for y in E_2 yields the equation $(2x + 3)(42 - 2x) = 450$, or $2x^2 - 39x +$

$162 = 0$. Using the quadratic formula, we find that $x = 6$ or $x = 13.5$. If $x = 6, y = 13.5$ and if $x =$

$13.5, y = 6$. It follows that the length of the garden is 30 feet and the width is 15 feet.

● maximum 34   ● 30 ft, 15 ft

[61] From the accompanying figure, we see that the area of the lot is $x y = 6000$ and the area of the

sidewalk is $5(x + 5) + 5y = 975$ or $x + y = 190$. Thus, we must find the solution set for

$\begin{cases} x + y = 190 \,(E_1) \\ x y = 6000 \,(E_2) \end{cases}$. Solving E_1 for x and substituting into E_2 yields $x(190 - x) = 6000 \Rightarrow 190x -$

$x^2 = 6000 \Rightarrow x^2 - 190 + 6000 = 0 \Rightarrow (x - 40)(x - 150) = 0 \Rightarrow x = 40$ or $x = 150$. If $x = 40, y =$

150 and if $x = 150, y = 40$. Hence, the length is 150 ft and the width is 40 ft.

● maximum 34   ● 150 ft, 40 ft

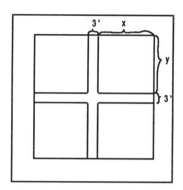

Figure 60

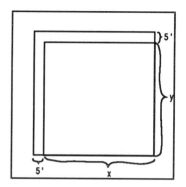

Figure 61

[62]   We must determine the values of a, b and c. We have $f(1) = a + b + c = 544$, $f(2) = 4a + 2b + c = 544$ and $f(3) = 9a + 3b + c = 222$. To find a, b and c, we find the solution set of the system

$$\begin{cases} a + b + c = 544 \\ 4a + 2b + c = 544 \\ 9a + 3b + c = 222 \end{cases} \text{ using Gauss–Jordan elimination as follows } \begin{bmatrix} 1 & 1 & 1 & 544 \\ 4 & 2 & 1 & 544 \\ 9 & 3 & 1 & 222 \end{bmatrix}$$

$$\begin{matrix} -4R_1 + R_2 \to R_2 \\ -9R_1 + R_2 \to R_2 \\ \sim \end{matrix} \begin{bmatrix} 1 & 1 & 1 & 544 \\ 0 & -2 & -3 & -1632 \\ 0 & -6 & -8 & -4674 \end{bmatrix} \begin{matrix} -\frac{1}{2}R_2 \to R_2 \\ \sim \end{matrix} \begin{bmatrix} 1 & 1 & 1 & 544 \\ 0 & 1 & \frac{3}{2} & 816 \\ 0 & -6 & -8 & -4674 \end{bmatrix} \begin{matrix} 6R_2 + R_3 \to R_3 \\ \sim \end{matrix}$$

$$\begin{bmatrix} 1 & 1 & 1 & 544 \\ 0 & 1 & \frac{3}{2} & 816 \\ 0 & 0 & 1 & 222 \end{bmatrix} \begin{matrix} -\frac{3}{2}R_3 + R_2 \to R_2 \\ -R_3 + R_1 \to R_1 \\ \sim \end{matrix} \begin{bmatrix} 1 & 1 & 0 & 322 \\ 0 & 1 & 0 & 483 \\ 0 & 0 & 1 & 222 \end{bmatrix} \begin{matrix} -R_2 + R_1 \to R_1 \\ \sim \end{matrix} \begin{bmatrix} 1 & 0 & 0 & -161 \\ 0 & 1 & 0 & 483 \\ 0 & 0 & 1 & 222 \end{bmatrix}$$

It follows that $a = -161$, $b = 483$ and $c = 222$, and therefore, $f(t) = -161t^2 + 483t + 222$.

Solving the equation $161t^2 - 483t - 222 = 0$ using the quadratic formula, we find that $t \approx 3.40$ or $t \approx -0.40$. Discarding the negative value of t, we have $t \approx 3.40$.   ● approximately 3.40 sec.

[63]   Substituting the coordinates of each of the three points into the equation $Ax^2 + Ay^2 + Bx + Cy + D = 0$ and simplifying yields the system $\begin{cases} 2A - B + C + D = 0 \\ 4A + 2C + D = 0 \\ 18A + 3B + 3C + D = 0. \end{cases}$ Using Gaussian–Jordan

elimination to solve this system, we get $\begin{bmatrix} 2 & -1 & 1 & 1 & 0 \\ 4 & 0 & 2 & 1 & 0 \\ 18 & 3 & 3 & 1 & 0 \end{bmatrix} \begin{matrix} \frac{1}{2}R_2 \to R_2 \\ \sim \end{matrix} \begin{bmatrix} 1 & -\frac{1}{2} & \frac{1}{2} & \frac{1}{2} & 0 \\ 4 & 0 & 2 & 1 & 0 \\ 18 & 3 & 3 & 1 & 0 \end{bmatrix}$

$$\begin{matrix} -4R_1 + R_2 \to R_2 \\ -18R_1 + R_3 \to R_3 \\ \sim \end{matrix} \begin{bmatrix} 1 & -\frac{1}{2} & \frac{1}{2} & \frac{1}{2} & 0 \\ 0 & 2 & 0 & -1 & 0 \\ 0 & 12 & -6 & -8 & 0 \end{bmatrix} \begin{matrix} \frac{1}{2}R_2 \to R_2 \\ \sim \end{matrix} \begin{bmatrix} 1 & -\frac{1}{2} & \frac{1}{2} & \frac{1}{2} & 0 \\ 0 & 1 & 0 & -\frac{1}{2} & 0 \\ 0 & 12 & -6 & -8 & 0 \end{bmatrix}$$

$$-12\,R_2 + R_3 \to R_3 \atop \sim \begin{bmatrix} 1 & -\frac{1}{2} & \frac{1}{2} & \frac{1}{2} & 0 \\ 0 & 1 & 0 & -\frac{1}{2} & 0 \\ 0 & 0 & -6 & -2 & 0 \end{bmatrix} -\frac{1}{6}R_3 \to R_3 \atop \sim \begin{bmatrix} 1 & -\frac{1}{2} & \frac{1}{2} & \frac{1}{2} & 0 \\ 0 & 1 & 0 & -\frac{1}{2} & 0 \\ 0 & 0 & 1 & \frac{1}{3} & 0 \end{bmatrix}$$

$$-\frac{1}{2}R_3 + R_1 \to R_1 \atop \sim \begin{bmatrix} 1 & -\frac{1}{2} & 0 & \frac{1}{3} & 0 \\ 0 & 1 & 0 & -\frac{1}{2} & 0 \\ 0 & 0 & 1 & \frac{1}{3} & 0 \end{bmatrix} \frac{1}{2}R_2 + R_1 \to R_1 \atop \sim \begin{bmatrix} 1 & 0 & 0 & \frac{1}{12} & 0 \\ 0 & 1 & 0 & -\frac{1}{2} & 0 \\ 0 & 0 & 1 & \frac{1}{3} & 0 \end{bmatrix}$$

$R_3 : C + \frac{1}{3}D = 0 \Rightarrow C = -\frac{1}{3}D.$

$R_2 : B - \frac{1}{2}D - 0 \Rightarrow B = \frac{1}{2}D.$ Letting $D = -12$, we get $C = 4, B = -6$ and $A = 1.$

$R_1 : A + \frac{1}{12}D = 0 \Rightarrow A = -\frac{1}{12}D.$

● $x^2 + y^2 - 6x + 4y - 12 = 0$

[64] Let x represent the amount of 20 % solution used, y represent the amount of 10 % solution and

z the amount of 40 % solution. We must solve the following system of equations:

$$\begin{cases} x & +y & +z = & 20 \\ 0.20x & +0.15y + 0.40z = 0.25(x + y + z) \\ 2x & -y & = & 0 \end{cases} \Rightarrow \begin{cases} x & +y & +z = 20 \\ 5x & +10y - 15z = & 0 \\ 2x & -y & = 0. \end{cases}$$ We proceed as follows:

$$\begin{bmatrix} 1 & 1 & 1 & 20 \\ 5 & 10 & -15 & 0 \\ 2 & -1 & 0 & 0 \end{bmatrix} \begin{matrix} -5R_1 + R_2 \to R_2 \\ -2R_1 + R_3 \to R_3 \\ \sim \end{matrix} \begin{bmatrix} 1 & 1 & 1 & 20 \\ 0 & 5 & -20 & -100 \\ 0 & -3 & -2 & -40 \end{bmatrix} \begin{matrix} \frac{1}{5}R_2 \to R_2 \\ \sim \end{matrix}$$

$$\begin{bmatrix} 1 & 1 & 1 & 20 \\ 0 & 1 & -4 & -20 \\ 0 & -3 & -2 & -40 \end{bmatrix} 3R_2 + R_3 \to R_3 \atop \sim \begin{bmatrix} 1 & 1 & 1 & 20 \\ 0 & 1 & -4 & -20 \\ 0 & 0 & -14 & -100 \end{bmatrix} -\frac{1}{14}R_3 \to R_3 \atop \sim \begin{bmatrix} 1 & 1 & 1 & 20 \\ 0 & 1 & -4 & -20 \\ 0 & 0 & 1 & \frac{50}{7} \end{bmatrix}$$

$$\begin{matrix} -4R_3 + R_2 \to R_2 \\ -R_3 + R_1 \to R_1 \\ \sim \end{matrix} \begin{bmatrix} 1 & 1 & 0 & \frac{90}{7} \\ 0 & 1 & 0 & \frac{60}{7} \\ 0 & 0 & 1 & \frac{50}{7} \end{bmatrix} -R_2 + R_1 \to R_1 \atop \sim \begin{bmatrix} 1 & 1 & 0 & \frac{30}{7} \\ 0 & 1 & 0 & \frac{60}{7} \\ 0 & 0 & 1 & \frac{50}{7} \end{bmatrix} R_3 : z = \frac{50}{7}; R_2 : y = \frac{60}{7};$$

$R_1 : x = \frac{30}{7}.$

$\frac{30}{7} \approx 4.29$ liters of 20 % solution,

● $\frac{60}{7} \approx 8.57$ liters of 15 % solution,

$\frac{50}{7} \approx 7.14$ liters of 40 % solution

[65] Let x represent the number of type – one frames manufactured and y represent the number of type –

two frames manufactured. We must minimize the cost function defined by $C(x) = 3.60x +$

10.80y subject to the following constraints $\begin{cases} x \geq 0 \\ y \geq 0 \\ 2x + y \geq 48 \\ x + 2y \geq 60 \\ x + 6y \geq 108 \end{cases}$. The set of feasible solutions is shown

in the accompanying figure.

Vertex	Value of $C(x) = 3.60x + 10.80y$
(0, 48)	518.40
(12, 24)	302.40
(36, 12)	259.20
(108, 0)	388.80

She should manufacture

36 type – one wooden frames and 12 type – two frames at a cost of $ 259.20.   ● maximum 34

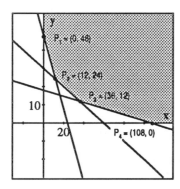

Figure 65

EXERCISES10.1

[1] $\{3n-5\}; -2, 1, 4, 25$

[2] $\{2n^2+1\}; 3, 9, 19, 201$

[3] $\{\pi\}; \pi, \pi, \pi, \pi$

[4] $\left\{\frac{5}{4}\right\}; \frac{5}{4}, \frac{5}{4}, \frac{5}{4}, \frac{5}{4}$

[5] $\left\{1-\frac{1}{2^n}\right\}; \frac{1}{2}, \frac{3}{4}, \frac{7}{8}, \frac{1023}{1024}$

[6] $\left\{\frac{1}{3^n}-2\right\}; -\frac{5}{3}, -\frac{17}{9}, -\frac{53}{27}, \frac{118{,}097}{59{,}049}$

[7] $\left\{\frac{n^2-2}{n+1}\right\}; -\frac{1}{2}, \frac{2}{3}, \frac{7}{4}, \frac{98}{11}$

[8] $\left\{\frac{3n^2}{2n-1}\right\}; 3, 4, \frac{27}{5}, \frac{300}{19}$

[9] $\{(n-1)(n+1)\}; 0, 3, 8, 99$

[10] $\{n(2n+1)\}; 3, 10, 21, 210$

[11] $\left\{\left(1+\frac{1}{n}\right)^n\right\}; 2, \frac{9}{4}, \frac{64}{27}, (1.1)^{10}$

[12] $\left\{\left(2-\frac{2}{n}\right)^n\right\}; 0, 1, \frac{64}{27}, (1.8)^{10}$

[13] $\left\{\frac{(-1)^n}{2n-3}\right\}; 1, 1, -\frac{1}{3}, \frac{1}{17}$

[14] $\left\{\frac{(-1)^{n-1}}{\sqrt{n}}\right\}; 1, -\frac{1}{\sqrt{2}}, \frac{1}{\sqrt{3}}, -\frac{1}{\sqrt{10}}$

[15] $\left\{\frac{\sin\left(\frac{n\pi}{2}\right)}{n^2}\right\}; 1, 0, -\frac{1}{9}, 0$

[16] $\left\{\frac{\cos(n\pi)}{2^n}\right\}; -\frac{1}{2}, \frac{1}{4}, -\frac{1}{8}, \frac{1}{1024}$

[17] $\left\{\frac{(n+2)!}{(n-1)!}\right\}; 6, 24, 60, 1320$

[18] $\left\{\frac{(2n+2)!}{(2n)!}\right\}; 12, 30, 56, 462$

[19] $\left\{\frac{\log n}{\sqrt{n}}\right\}; 0, \frac{\log 2}{\sqrt{2}}, \frac{\log 3}{\sqrt{3}}, \frac{\log 10}{\sqrt{10}}$

[20] $\left\{\frac{\ln n}{e^n}\right\}; 0, \frac{\ln 2}{e^2}, \frac{\ln 3}{e^3}, \frac{\ln 10}{e^{10}}$

[21] $\frac{20!}{(20-3)! \cdot 3!} = \frac{20 \cdot 19 \cdot 18 \cdot 17!}{17! \; 3 \cdot 2 \cdot 1} = 1140$

[22] $\frac{15!}{5! \; 10!} = \frac{15 \cdot 14 \cdot 13 \cdot 12 \cdot 11 \cdot 10!}{5 \cdot 4 \cdot 3 \cdot 2 \cdot 1 \cdot 10!} = 3003$

[23] $\{(n+1)!\}; \dfrac{a_{k+1}}{a_k} = \dfrac{(k+2)!}{(k+1)!} = \dfrac{(k+2) \cdot (k+1)!}{(k+1)!} = k+2$

[24] $\{(2n-1)!\}; \dfrac{a_{k+1}}{a_k} = \dfrac{(2k+1)!}{(2k-1)!} = \dfrac{(2k+1)(2k)(2k-1)!}{(2k-1)!} = 4k^2+2k$

NOTE: In Exercises 25 – 34, the answers are listed in the order a_1, a_2, a_3, a_4 and a_5.

[25] $a_1 = 2, a_n = 2a_{n-1}-1; 2, 3, 5, 9, 17$

[26] $a_1 = 5, a_n = 3a_{n-1}-1; 5, 14, 41, 122, 365$

[27] $a_1 = \frac{1}{2}, a_n = \frac{a_{n-1}}{2}; \frac{1}{2}, \frac{1}{4}, \frac{1}{8}, \frac{1}{16}, \frac{1}{32}$

[28] $a_1 = \frac{2}{3}, a_n = \frac{a_{n-1}}{3}; \frac{2}{3}, \frac{2}{9}, \frac{2}{27}, \frac{2}{81}, \frac{2}{243}$

[29] $a_1 = 4, a_n = a_{n-1}^{1/2}; 4, 2, 2^{1/2}, 2^{1/4}, 2^{1/8}$

[30] $a_1 = -1, a_n = (a_{n-1})^n; -1, 1, 1, 1, 1$

[31] $a_1 = -1, a_n = (-1)^{n-1}a_{n-1}-5; -1, -4, -9, 4, -1$

[32] $a_1 = 3, a_n = (-1)^n a_{n-1}-5; 3, -2, -3, -8, 3$

[33] $a_1 = 1, a_2 = 3, a_n = \frac{a_{n-1}+a_{n-2}}{2}; 1, 3, 2, \frac{5}{2}, \frac{9}{4}$

[34] $a_1 = 1, a_2 = 2, a_n = (-1)^{n+1}\frac{a_{n-2}}{a_{n-1}}; 1, 2, \frac{1}{2}, -4, -\frac{1}{8}$

[35] $\displaystyle\sum_{k=1}^{50} 2k$

[36] $\displaystyle\sum_{k=1}^{10} \frac{1}{k^2}$

[37] $\displaystyle\sum_{k=1}^{5} kx^{k-1}$

[38] $\displaystyle\sum_{k=0}^{n} \frac{1}{x^k}$

[39] $\displaystyle\sum_{k=1}^{n} k^k$

[40] $\displaystyle\sum_{k=0}^{4} (-1)^k \frac{2^k}{3^k}$

[41] $\displaystyle\sum_{k=1}^{n}(-1)^{k+1}\frac{e^{k}}{k+1}$

[42] $\displaystyle\sum_{k=1}^{10}\frac{\log k}{2^{k-1}}$

[43] $\displaystyle\sum_{k=1}^{10}\frac{k}{k-\dfrac{1}{k+1}}$

[44] $\displaystyle\sum_{k=1}^{10}\frac{1}{(2k)(2k+1)}$

[45] $\displaystyle\sum_{i=1}^{6}(i+4)=(1+4)+(2+4)+(3+4)+(4+4)+(5+4)+(6+4)=45$

[46] $\displaystyle\sum_{i=1}^{5}(i-3)=(1-3)+(2-3)+(3-3)+(4-3)+(5-3)=0$

[47] $\displaystyle\sum_{i=1}^{4}(2i-1)^{2}=1^{2}+3^{2}+5^{2}+7^{2}=84$

[48] $\displaystyle\sum_{i=1}^{3}\left(3i^{2}-7\right)=-4+5+20=21$

[49] $\displaystyle\sum_{k=1}^{5}\frac{1}{2^{k-1}}=1+\frac{1}{2}+\frac{1}{4}+\frac{1}{8}+\frac{1}{16}=\frac{31}{16}$

[50] $\displaystyle\sum_{k=0}^{4}\left(\frac{2}{5}\right)^{k}=1+\frac{2}{5}+\frac{4}{25}+\frac{8}{125}+\frac{16}{625}=\frac{1031}{625}$

[51] $\displaystyle\sum_{k=0}^{6}\sin\left(\frac{k\pi}{3}\right)=\sin 0+\sin\frac{\pi}{3}+\sin\frac{2\pi}{3}+\sin\pi+\sin\frac{4\pi}{3}+\sin\frac{5\pi}{3}+\sin 2\pi=0$

[52] $\displaystyle\sum_{k=0}^{8}\cos\left(\frac{k\pi}{2}\right)=\cos 0+\cos\frac{\pi}{2}+\cos\pi+\cos\frac{3\pi}{2}+\cos 2\pi+\cos\frac{5\pi}{2}+\cos 3\pi+\cos\frac{7\pi}{2}+\cos 4\pi=1$

[53] $\displaystyle\sum_{k=1}^{5}(-1)^{k-1}\ln e^{k-1}=(-1)^{0}\ln e^{0}+(-1)^{1}\ln e^{1}+(-1)^{2}\ln e^{2}+(-1)^{3}\ln e^{3}+(-1)^{4}\ln e^{4}=$

$0-1+2-3+4=2$

[54] $\displaystyle\sum_{k=1}^{4}(-1)^{k}e^{\ln k}=(-1)^{1}e^{\ln 1}+(-1)^{2}e^{\ln 2}+(-1)^{3}e^{\ln 3}+(-1)^{4}e^{\ln 4}=-1+2-3+4=2$

[55] $\displaystyle\sum_{k=1}^{300}\pi=300\pi$

[56] $\displaystyle\sum_{k=1}^{100}e^{2}=100e^{2}$

[57] $\displaystyle\sum_{i=1}^{5}50i^{2}=50\sum_{i=1}^{5}i^{2}=50\left(1^{2}+2^{2}+3^{2}+4^{2}+5^{2}\right)=2750$

[58] $\displaystyle\sum_{i=1}^{6}\frac{i}{100}=\frac{1}{100}\sum_{i=1}^{6}i=\frac{1}{100}(1+2+3+4+5+6)=\frac{21}{100}$

[59] $\displaystyle\sum_{j=1}^{8}(5j+32{,}000)=\sum_{j=1}^{8}5j+\sum_{j=1}^{8}32{,}000=5\sum_{j=1}^{8}j+8(32{,}000)=5(1+2+3+4+5+6+7+8)+$

$256{,}000=256{,}180$

[60] $\displaystyle\sum_{j=1}^{7}(3j-1700)=\sum_{j=1}^{7}3j-\sum_{j=1}^{7}1700=3\sum_{j=1}^{7}j-\sum_{j=1}^{7}1700=3(1+2+3+4+5+6+7)-7(1700)=$

$-11{,}816$

[61] $a_{1}=1,a_{2}=1,a_{n}=a_{n-1}+a_{n-2},$ *for* $n\geq 3;$ $a_{3}=2,a_{4}=3,a_{5}=5,a_{6}=8,a_{7}=13,a_{8}=21,a_{9}=34$
and $a_{10}=55.$

[62] $a_1 = \dfrac{N}{2}, a_n = \dfrac{1}{2}\left(a_{n-1} + \dfrac{N}{a_{n-1}}\right); a_1 = \dfrac{5}{2} = 2.5, a_2 = \dfrac{1}{2}\left(2.5 + \dfrac{5}{2.5}\right) = 2.25, a_3 = \dfrac{1}{2}\left(2.25 + \dfrac{5}{2.25}\right) \approx$

$2.236, a_4 \approx 2.236, a_5 \approx 2.236$

[63] $a_1 = 1, a_n = n\,a_{n-1}; a_1 = 1, a_2 = 2, a_3 = 6, a_4 = 24, a_5 = 120, a_6 = 720.$ $1 + \displaystyle\sum_{i=1}^{6} \dfrac{1}{a_i} \approx 2.718$

[64] $\displaystyle\sum_{k=2}^{20}\left[\dfrac{1}{(k-1)^2} - \dfrac{1}{k^2}\right] = \left(1 - \dfrac{1}{4}\right) + \left(\dfrac{1}{4} - \dfrac{1}{9}\right) + \left(\dfrac{1}{9} - \dfrac{1}{16}\right) + \cdots +$

$\left[\dfrac{1}{(17)^2} - \dfrac{1}{(18)^2}\right] + \left[\dfrac{1}{(18)^2} - \dfrac{1}{(19)^2}\right] + \left[\dfrac{1}{(19)^2} - \dfrac{1}{(20)^2}\right] = 1 - \dfrac{1}{400} = \dfrac{399}{400}$

[65] a) $a_1 = 10{,}000\left(1 + \dfrac{0.075}{4}\right)^4 \approx 10{,}771.36, a_2 = 10{,}000\left(1 + \dfrac{0.075}{4}\right)^8 \approx 11{,}602.22, a_3 =$

$10{,}000\left(1 + \dfrac{0.075}{4}\right)^{12} \approx 12{,}497.16, a_4 = 10{,}000\left(1 + \dfrac{0.075}{4}\right)^{16} \approx 13{,}461.14.$

  b) $a_{10} = 10{,}000\left(1 + \dfrac{0.075}{4}\right)^{40} \approx 21{,}023.49.$ The amount after 10 years is

approximately $21,023.49.

[66] $a_1 = 20{,}000, a_2 = 20{,}000 - (0.25)(20{,}000) = 15{,}000, a_3 = 15{,}000 - (0.25)(15{,}000) = 11{,}250, a_4 =$
$11{,}250 - (0.25)(11{,}250) = 8437.5, a_5 = 8437.5 - (0.25)(8437.5) = 6328.125.$ The value of the car
after 4 years is approximately $6328.13.

[67] $a_1 = 25{,}000, a_2 = 25{,}000 + (0.06)(25{,}000) = 26{,}500, a_3 = 26{,}500 + (0.06)(26{,}500) = 28{,}090, a_4 =$
$28{,}090 + (0.06)(28{,}090) = 29{,}775.4, a_5 = 29{,}775.4 + (0.06)(29{,}775.4) = 31{,}561.924, a_6 =$
$31{,}561.924 + (0.06)(31{,}561.924) = 33{,}455.63944.$ After the fifth year, the salary is approximately
$33,455.64.

[68] $a_1 = 70{,}000, a_2 = 70{,}000 + (0.10)(70{,}000) = 77{,}000, a_3 = 77{,}000 + (0.10)(77{,}000) = 84{,}700, a_4 =$
$84{,}700 + (0.10)(84{,}700) = 93{,}170, a_5 = 93{,}170 + (0.10)(93{,}170) = 102{,}487.$ After the fourth year,
the salary is $102,487.

[69] For example, $\displaystyle\sum_{i=1}^{2} 3 = 6$ and $\displaystyle\sum_{i=1}^{2} i = 3$, and $\displaystyle\sum_{i=1}^{2} 3i = 9$. In this case, $\displaystyle\sum_{i=1}^{2} 3i = 9 \neq 18 = \left(\displaystyle\sum_{i=1}^{2} 3\right)\left(\displaystyle\sum_{i=1}^{2} i\right).$

[70] (ii) $\displaystyle\sum_{i=1}^{n}(a_i - b_i) = (a_1 - b_1) + (a_2 - b_2) + \cdots + (a_n - b_n) = a_1 - b_1 + a_2 - b_2 + \cdots + a_n - b_n =$

$(a_1 + a_2 + \cdots + a_n) - (b_1 + b_2 + \cdots + b_n) = \displaystyle\sum_{i=1}^{n} a_i - \displaystyle\sum_{i=1}^{n} b_i.$

(iii) $\displaystyle\sum_{i=1}^{n} c\,a_i = c\,a_1 + c\,a_2 + \cdots + c\,a_n = c\,(a_1 + a_2 + \cdots + a_n) = c\displaystyle\sum_{i=1}^{n} a_i.$

EXERCISES 10.2

[1] $\{4n - 1\}; a_n - a_{n-1} = (4n - 1) - (4n - 5) = 4.$ arithmetic sequence with $d = 4$.

[2] $\{2n + 1\}; a_n - a_{n-1} = (2n + 1) - (2n - 1) = 2.$ arithmetic sequence with $d = 2$.

[3] $\{n(n + 1)\}; a_n - a_{n-1} = [n(n + 1)] - [(n - 1)n] = n^2 + n - n^2 + n = 2n.$ Not an arithmetic
sequence.

[4] $\{3n^2 - 2\}$; $a_n - a_{n-1} = (3n^2 - 2) - [3(n-1)^2 - 2] = (3n^2 - 2) - (3n^2 - 6n + 3 - 2) = 6n - 3$. not an arithmetic sequence.

[5] $\{\frac{3}{2}n + 7\}$; $a_n - a_{n-1} = (\frac{3}{2}n + 7) - [\frac{3}{2}(n-1) + 7] = \frac{3}{2}$. arithmetic sequence with $d = \frac{3}{2}$.

[6] $\{7 - \frac{2}{5}n\}$; $a_n - a_{n-1} = (7 - \frac{2}{5}n) - [7 - \frac{2}{5}(n-1)] = -\frac{2}{5}$. arithmetic sequence with $d = -\frac{2}{5}$

[7] $\{e^{n-1}\}$; $a_n - a_{n-1} = e^{n-1} - e^{n-2}$. not an arithmetic sequence.

[8] $\{1 + \ln n\}$; $a_n - a_{n-1} = (1 + \ln n) - [1 + \ln(n-1)] = \ln n - \ln(n-1) = \ln \frac{n}{n-1}$. not an arithmetic sequence.

[9] $\{\log 10^n\}$; $\log 10^n = n \Rightarrow a_n - a_{n-1} = n - (n-1) = 1$. arithmetic sequence with $d = 1$.

[10] $\{\ln 2^n\}$; $a_n - a_{n-1} = \ln 2^n - \ln 2^{n-1} = \ln \frac{2^n}{2^{n-1}} = \ln 2$. arithmetic sequence with $d = \ln 2$.

[11] $4, 7, 10, \cdots$; $a_n = a_1 + (n-1)d = 4 + (n-1)(3) = 3n + 1$

[12] $2, 7, 12, \cdots$; $a_n = 2 + (n-1)(5) = 5n - 3$ [13] $5, 1, -3, \cdots$; $a_n = 5 + (n-1)(-4) = 9 - 4n$

[14] $1, -4, -9, \cdots$; $a_n = 1 + (n-1)(-5) = 6 - 5n$

[15] $2, \frac{5}{2}, 3, \cdots$; $a_n = 2 + (n-1)(\frac{1}{2}) = \frac{1}{2}n + \frac{3}{2} = \frac{n+3}{2}$

[16] $0, \frac{1}{2}, 1, \cdots$; $a_n = 0 + (n-1)(\frac{1}{2}) = \frac{1}{2}n - \frac{1}{2} = \frac{n-1}{2}$

[17] $-\frac{1}{3}, \frac{1}{3}, 1, \cdots$; $a_n = -\frac{1}{3} + (n-1)(\frac{2}{3}) = \frac{2}{3}n - 1$ [18] $-\frac{7}{4}, \frac{1}{2}, \frac{11}{4}, \cdots$; $a_n = -\frac{7}{4} + (n-1)(\frac{9}{4}) = \frac{9}{4}n - 4$

[19] $\ln 2, \ln 4, \ln 8, \cdots$; $a_n = \ln 2 + (n-1)\ln 2 = n \ln 2$

[20] $\log 5, \log 25, \log 125, \cdots$; $a_n = \log 5 + (n-1)\log 5 = n \log 5$

[21] $\{7n - 1\}$; $a_{40} = a_1 + 39d = 6 + (39)(7) = 279$ [22] $\{\frac{2}{3}n + 4\}$; $a_{40} = \frac{14}{3} + (39)(\frac{2}{3}) = \frac{92}{3}$

[23] $\{5 + \frac{1}{2}n\}$; $a_{40} = \frac{11}{2} + (39)(\frac{1}{2}) = 25$ [24] $\{12 - 1.5n\}$; $a_{40} = 10.5 + (39)(-1.5) = -48$

[25] $11, 16, 21, \cdots$; $a_{40} = 11 + (39)(5) = 206$ [26] $-4, 3, 10, \cdots$; $a_{40} = -4 + (39)(7) = 269$

[27] $-2, \frac{3}{2}, 5, \cdots$; $a_{40} = -2 + (39)(\frac{7}{2}) = \frac{269}{2}$ [28] $-\frac{5}{3}, 2, \frac{17}{3}, \cdots$; $a_{40} = -\frac{5}{3} + (39)(\frac{11}{3}) = \frac{424}{3}$

[29] $a_1 = -\frac{12}{5}$ and $d = \frac{3}{5}$; $a_{40} = -\frac{12}{5} + (39)(\frac{3}{5}) = 21$

[30] $a_1 = \frac{3}{4}$ and $d = -\frac{9}{4}$; $a_{40} = \frac{3}{4} + (39)(-\frac{9}{4}) = -87$

[31] $a_3 = -1$ and $a_5 = -9$; $a_5 = a_3 + 2d \Rightarrow -9 = -1 + 2d \Rightarrow 2d = -8 \Rightarrow d = -4$. $a_1 = a_3 - 2d = -1 - (2)(-4) = 7$. Thus, $a_{40} = 7 + (39)(-4) = -149$.

[32] $a_2 = 2$ and $a_6 = -10$; $a_6 = a_2 + 4d \Rightarrow -10 = 2 + 4d \Rightarrow d = -3$. $a_1 = a_2 - d = 2 - (-3) = 5$. Thus, $a_{40} = 5 + (39)(-3) = -112$.

[33] $2, 8, 14, \cdots$; $S_{10} = 10a_1 + \frac{(10)(9)}{2}d = (10)(2) + (45)(6) = 290$

[34] $-12, -8, -4, \cdots$; $S_{10} = (10)(-12) + (45)(4) = 60$

[35] $1.3, 0.6, -0.1, \cdots$; $S_{10} = (10)(1.3) + (45)(-0.7) = -18.5$

[36] $4, 5.2, 6.4, \cdots$; $S_{10} = (10)(4) + (45)(1.2) = 94$ [37] $\{5n - 2\}$; $S_{10} = (10)(3) + (45)(5) = 255$

[38] $\left\{\frac{3}{2}n - 4\right\}; S_{10} = (10)\left(-\frac{5}{2}\right) + (45)\left(\frac{3}{2}\right) = \frac{85}{2} = 42.5$

[39] $a_1 = -4$ and $d = \frac{3}{2}; S_{10} = (10)(-4) + (45)\left(\frac{3}{2}\right) = \frac{55}{2} = 27.5$

[40] $a_1 = 7$ and $d = -\frac{8}{5}; S_{10} = (10)(7) + (45)\left(-\frac{8}{5}\right) = -2$

[41] $a_1 = \log 2$ and $d = \log 2; S_{10} = (10)(\log 2) + (45)(\log 2) = 55 \log 2$

[42] $a_1 = \ln 100$ and $d = \ln 10; S_{10} = (10)(\ln 100) + (45)(\ln 10) = (10)\left(\ln 10^2\right) + 45 \ln 10 =$

$10(2 \ln 10) + 45 \ln 10 = 65 \ln 10$

[43] $a_3 = \frac{5\pi}{2}$ and $a_{10} = 13\pi; a_{10} = a_3 + 7d \Rightarrow 13\pi = \frac{5\pi}{2} + 7d \Rightarrow 7d = \frac{21\pi}{2} \Rightarrow d = \frac{3\pi}{2}$. Also, $a_1 =$

$a_3 - 2d = \frac{5\pi}{2} - (2)\left(\frac{3\pi}{2}\right) = -\frac{\pi}{2}$. Thus, $S_{10} = (10)\left(-\frac{\pi}{2}\right) + (45)\left(\frac{3\pi}{2}\right) = \frac{125\pi}{2}$

[44] $a_2 = 2\sqrt{3}$ and $a_7 = \frac{19\sqrt{3}}{2}; a_7 = a_2 + 5d \Rightarrow \frac{19\sqrt{3}}{2} = 2\sqrt{3} + 5d \Rightarrow 5d = \frac{15\sqrt{3}}{2} \Rightarrow d = \frac{3\sqrt{3}}{2}$. Also,

$a_1 = a_2 - d = 2\sqrt{3} - \frac{3\sqrt{3}}{2} = \frac{\sqrt{3}}{2}$. Thus, $S_{10} = (10)\left(\frac{\sqrt{3}}{2}\right) + (45)\left(\frac{3\sqrt{3}}{2}\right) = \frac{145\sqrt{3}}{2}$

[45] $a_1 = 4$ and $a_5 = 28$. $a_5 = a_1 + 4d \Rightarrow 28 = 4 + 4d \Rightarrow d = 6$. Thus, $m_1 = 10, m_2 = 16$ and $m_3 = 22$.

[46] $a_1 = 3$ and $a_6 = 16$. $16 = 3 + 5d \Rightarrow d = \frac{13}{5}$. Thus, $m_1 = \frac{28}{5}, m_2 = \frac{41}{5}, m_3 = \frac{54}{5}$ and $m_4 = \frac{67}{5}$.

[47] $a_1 = 2$ and $a_{10} = 7$. $7 = 2 + 9d \Rightarrow d = \frac{5}{9}$. Thus, $m_1 = \frac{23}{9}, m_2 = \frac{28}{9}, m_3 = \frac{11}{3}, m_4 = \frac{38}{9}, m_5 = \frac{43}{9}, m_6 =$

$\frac{16}{3}, m_7 = \frac{53}{9}$ and $m_8 = \frac{58}{9}$.

[48] $a_1 = 1$ and $a_8 = 15$. $15 = 1 + 7d \Rightarrow d = 2$. Thus, $m_1 = 3, m_2 = 5, m_3 = 7, m_4 = 9, m_5 = 11$ and

$m_6 = 13$.

[49] $a_1 = \frac{1}{3}$ and $a_7 = 4$. $4 = \frac{1}{3} + 6d \Rightarrow d = \frac{11}{18}$. Thus, $m_1 = \frac{17}{18}, m_2 = \frac{14}{9}, m_3 = \frac{13}{6}, m_4 = \frac{25}{9}$ and

$m_5 = \frac{61}{18}$.

[50] $a_1 = \frac{5}{2}$ and $a_9 = 12$. $12 = \frac{5}{2} + 8d \Rightarrow d = \frac{19}{16}$. Thus, $m_1 = \frac{59}{16}, m_2 = \frac{39}{8}, m_3 = \frac{97}{16}, m_4 = \frac{29}{4}$,

$m_5 = \frac{135}{16}, m_6 = \frac{77}{8}$ and $m_7 = \frac{173}{16}$.

[51] $a_1 = -11$ and $a_6 = 2$. $2 = -11 + 5d \Rightarrow d = \frac{13}{5}$. Thus, $m_1 = -\frac{42}{5}, m_2 = -\frac{29}{5}, m_3 = -\frac{16}{5}$ and

$m_4 = -\frac{3}{5}$.

[52] $a_1 = -8$ and $a_7 = 3$. $3 = -8 + 6d \Rightarrow d = \frac{11}{6}$. Thus, $m_1 = -\frac{37}{6}, m_2 = -\frac{13}{3}, m_3 = -\frac{5}{2}, m_4 = -\frac{2}{3}$ and

$m_5 = \frac{7}{6}$.

[53] $2 + k \cdot 3 = 1000 \Rightarrow 3k = 998 \Rightarrow k = \frac{998}{3}$. Since k is not an integer, 1000 is not a term of the

sequence.

[54] $-3 + k \cdot 7 = 270 \Rightarrow 7k = 273 \Rightarrow k = 39$. Thus, $270 = -3 + 39(7)$ is a term of the sequence.

[55] $S_6 = \frac{6}{2}(a_1 + a_6) \Rightarrow 27 = 3(-8 + a_6) \Rightarrow 9 = a_6 - 8 \Rightarrow a_6 = 17$. Also, $a_6 = a_1 + 5d \Rightarrow 17 =$

$-8 + 5d \Rightarrow d = 5$. Thus, $a_n = a_1 + (n - 1)d = -8 + (n - 1)5 = 5n - 13$.

[56] $S_{10} = \frac{10}{2}(a_1 + a_{10}) \Rightarrow -30 = 5(6 + a_{10}) \Rightarrow a_{10} = -12$. Also $a_{10} = a_1 + 9d \Rightarrow -12 = 6 + 9d \Rightarrow$
$d = -2$. Thus, $a_n = a_1 + (n-1)d = 6 + (n-1)(-2) = 8 - 2n$.

[57] $(x-d) + x + (x+d) = 30 \Rightarrow 3x = 30 \Rightarrow x = 10$. Thus, $(10-d)(10)(10+d) = 510 \Rightarrow$
$100 - d^2 = 51 \Rightarrow d^2 = 49 \Rightarrow d = 7$. The three terms are 3, 10 and 17.

[58] $(x-d) + x + (x+d) = 6 \Rightarrow 3x = 6 \Rightarrow x = 2$. Thus, $(2-d)^2 + 2^2 + (2+d)^2 = 44 \Rightarrow 12 + 2d^2 =$
$44 \Rightarrow d^2 = 16 \Rightarrow d = 4$. The three terms are -2, 2 and 6.

[59] $a_n = a_1 + (n-1)d \Rightarrow 2 = \frac{1}{4} + (n-1)\left(\frac{1}{8}\right) \Rightarrow \frac{1}{8}n = \frac{15}{8} \Rightarrow n = 15$. On the 15th day the jogging
distance will be 2 miles.

[60] $a_n = a_1 + (n-1)d \Rightarrow 90 = 10 + (n-1)5 \Rightarrow 5n = 85 \Rightarrow n = 17$. In the 17th week, participants will
be cycling 90 minutes each day.

[61] $S_n = na_1 + \frac{n(n-1)}{2}d \Rightarrow S_n = (24)(18) + \frac{(24)(23)}{2}(3) = 1260$. Thus, there are 1260 seats in
the theater.

[62] $a_n = a_1 + (n-1)d \Rightarrow 16 = 48 + (n-1)(-1) \Rightarrow n = 33$. Thus, $S_n = S_{33} = \frac{33}{2}(48 + 16) = 1056$.
There are 1056 logs in the pile.

[63] $a_1 = 1$ and $d = 2 \Rightarrow S_n = n(1) + \frac{n(n-1)}{2}(2) = n^2$. Thus $\sum_{i=1}^{n}(2i-1) = n^2$.

[64] $a_m - a_k = [a_1 + (m-1)d] - [a_1 + (k-1)d] = (m-1)d - (k-1)d = md - kd =$
$(m-k)d \Rightarrow \frac{a_m - a_k}{m-k} = d$.

EXERCISES 10.3

[1] $\left\{\left(\frac{1}{4}\right)^n\right\}; \dfrac{a_{k+1}}{a_k} = \dfrac{\left(\frac{1}{4}\right)^{n+1}}{\left(\frac{1}{4}\right)^n} = \dfrac{1}{4}$. geometric sequence with $r = \frac{1}{4}$

[2] $\left\{\left(-\frac{1}{3}\right)^n\right\}; \dfrac{\left(-\frac{1}{3}\right)^{k+1}}{\left(-\frac{1}{3}\right)^k} = -\dfrac{1}{3}$. geometric sequence with $r = -\frac{1}{3}$

[3] $\left\{\dfrac{1}{2n}\right\}; \dfrac{\frac{1}{2k+2}}{\frac{1}{2k}} = \dfrac{2k}{2k+2}$. not a geometric sequence

[4] $\{n^2\}; \dfrac{(k+1)^2}{k^2} = \dfrac{k^2 + 2k + 1}{k^2}$. not a geometric sequence

[5] $\left\{\left(-\frac{2}{5}\right)^{n-1}\right\}; \dfrac{\left(-\frac{2}{5}\right)^{k}}{\left(-\frac{2}{5}\right)^{k-1}} = -\frac{2}{5}.$ geometric sequence with $r = -\frac{2}{5}$

[6] $\left\{\left(\frac{3}{4}\right)^{n}\right\}; \dfrac{\left(\frac{3}{4}\right)^{k+1}}{\left(\frac{3}{4}\right)^{k}} = \frac{3}{4}.$ geometric sequence with $r = \frac{3}{4}$

[7] $\left\{e^{nx}\right\}; \dfrac{e^{(k+1)x}}{e^{kx}} = \dfrac{e^{kx+x}}{e^{kx}} = e^{x}.$ geometric sequence with $r = e^{x}$

[8] $\left\{\pi^{(n-1)x}\right\}; \dfrac{\pi^{kx}}{\pi^{(k-1)x}} = \dfrac{\pi^{kx}}{\pi^{kx-x}} = \pi^{x}.$ geometric sequence with $r = \pi^{x}$

[9] $\left\{\log 2^{2^{n}}\right\}; \log 2^{2^{n}} = 2^{n} \log 2.$ Thus, $\dfrac{a_{k+1}}{a_{k}} = \dfrac{2^{k+1} \log 2}{2^{k} \log 2} = 2.$ geometric sequence with $r = 2$

[10] $\left\{\ln 10^{n^{2}}\right\}; \ln 10^{n^{2}} = n^{2} \ln 10.$ Thus, $\dfrac{a_{k+1}}{a_{k}} = \dfrac{(k+1)^{2} \ln 10}{k^{2} \ln 10} = \dfrac{k^{2}+2k+1}{k^{2}}.$

not a geometric sequence

[11] $3, 6, 12, \cdots; a_{n} = a_{1}r^{n-1} = 3\left(2^{n-1}\right)$ [12] $2, -1, \frac{1}{2}, \cdots; a_{n} = 2\left(-\frac{1}{2}\right)^{n-1}$

[13] $-4, 3, -\frac{9}{4}, \cdots; a_{n} = (-4)\left(-\frac{3}{4}\right)^{n-1}$ [14] $5, 3, \frac{9}{5}, \cdots; a_{n} = 5\left(\frac{3}{5}\right)^{n-1}$

[15] $1.2, 0.24, 0.048, \cdots; a_{n} = (1.2)(0.2)^{n-1}$ [16] $-5, 1, -0.2, \cdots; a_{n} = (-5)(-0.2)^{n-1}$

[17] $1, -x^{2}, x^{4}, \cdots; a_{n} = (1)\left(-x^{2}\right)^{n-1} = \left(-x^{2}\right)^{n-1}$ [18] $1, \frac{x}{3}, \frac{x^{2}}{9}, \cdots; a_{n} = (1)\left(\frac{x}{3}\right)^{n-1} = \left(\frac{x}{3}\right)^{n-1}$

[19] $\ln 10, \ln 100, \ln 10{,}000, \cdots; r = \dfrac{\ln 100}{\ln 10} = \dfrac{\ln 10^{2}}{\ln 10} = \dfrac{2 \ln 10}{\ln 10} = 2.$ Thus, $a_{n} = (\ln 10) 2^{n-1}.$

[20] $\log 5, \log 25, \log 625, \cdots; r = \dfrac{\log 25}{\log 5} = \dfrac{\log 5^{2}}{\log 5} = \dfrac{2 \log 5}{\log 5} = 2.$ Thus, $a_{n} = (\log 5) 2^{n}.$

[21] $a_{1} = 3$ and $a_{5} = 48.$ $a_{5} = a_{1}r^{4} \Rightarrow 48 = 3r^{4} \Rightarrow r^{4} = 16 \Rightarrow r = \pm 2.$ Thus, $a_{3} = 3 \cdot (\pm 2)^{2} = 12.$

[22] $a_{1} = 7$ and $a_{5} = 112.$ $112 = 7r^{4} \Rightarrow r^{4} = 16 \Rightarrow r = \pm 2.$ Thus, $a_{4} = 7 \cdot 2^{3} = 56,$ or $a_{4} = 7 \cdot (-2)^{3} = -56.$

[23] $a_{2} = -6$ and $a_{5} = \frac{81}{4}.$ $-6 = a_{1}r \Rightarrow r = -\dfrac{6}{a_{1}}.$ Thus, $a_{5} = a_{1}r^{4} \Rightarrow \frac{81}{4} = a_{1}\left(-\dfrac{6}{a_{1}}\right)^{4} \Rightarrow \frac{81}{4} = \dfrac{1296}{a_{1}^{3}} \Rightarrow$

$a_{1}^{3} = 64 \Rightarrow a_{1} = 4$ and $r = -\dfrac{6}{a_{1}} = -\frac{3}{2}.$ It follows that $a_{4} = 4\left(-\frac{3}{2}\right)^{3} = -\frac{27}{2}.$

[24] $a_2 = 4$ and $a_6 = \frac{4}{81}$. $4 = a_1 r \Rightarrow r = \frac{4}{a_1}$. Thus, $a_6 = a_1 r^5 \Rightarrow \frac{4}{81} = a_1 \left(\frac{4}{a_1}\right)^5 \Rightarrow \frac{4}{81} = \frac{1024}{a_1^4} \Rightarrow a_1^4 =$

20,736 $\Rightarrow a_1 = \pm 12$, and $r = \frac{4}{\pm 12} = \pm \frac{1}{3}$. When $a_1 = 12$ and $r = \pm \frac{1}{3}$, $a_3 = 12\left(\pm \frac{1}{3}\right)^2 = \frac{4}{3}$. When $a_1 =$

-12 and $r = \pm \frac{1}{3}$, $a_3 = (-12)\left(\pm \frac{1}{3}\right)^2 = -\frac{4}{3}$.

[25] $a_3 = \frac{4}{3}$ and $a_6 = -\frac{32}{81}$. $\frac{4}{3} = a_1 r^2 \Rightarrow a_1 = \frac{4}{3r^2}$. Also, $-\frac{32}{81} = a_1 r^5 \Rightarrow -\frac{32}{81} = \left(\frac{4}{3r^2}\right)(r^5) \Rightarrow -\frac{32}{81} =$

$\frac{4r^3}{3} \Rightarrow r^3 = -\frac{8}{27} \Rightarrow r = -\frac{2}{3}$. Thus, $a_1 = \frac{4}{3\left(-\frac{2}{3}\right)^2} = 3$ and $a_2 = 3\left(-\frac{2}{3}\right) = -2$.

[26] $a_2 = -4$ and $a_6 = -\frac{64}{81}$. $-4 = a_1 r \Rightarrow a_1 = -\frac{4}{r}$. Thus, $-\frac{64}{81} = a_1 r^5 \Rightarrow -\frac{64}{81} = \left(-\frac{4}{r}\right) r^5 \Rightarrow r^4 =$

$\frac{16}{81} \Rightarrow r = \pm \frac{2}{3}$. If $r = \frac{2}{3}$, $a_1 = -\frac{4}{r} = -6$, and $a_4 = (-6)\left(\frac{2}{3}\right)^3 = -\frac{16}{9}$. If $r = -\frac{2}{3}$, $a_1 = -\frac{4}{r} = 6$

and $a_4 = 6\left(-\frac{2}{3}\right)^3 = -\frac{16}{9}$.

[27] $8, 4, 2, \cdots$; $S_{10} = \frac{a_1\left(1 - r^{10}\right)}{1 - r} = \frac{8\left[1 - \left(\frac{1}{2}\right)^{10}\right]}{1 - \frac{1}{2}} = \frac{1023}{64}$.

[28] $7, 14, 28, \cdots$; $S_8 = \frac{a_1\left(1 - r^8\right)}{1 - r} = \frac{7\left(1 - 2^8\right)}{1 - 2} = 1785$.

[29] $300, -30, 3, \cdots$; $S_{12} = \frac{300\left[1 - \left(-\frac{1}{10}\right)^{12}\right]}{1 - \left(-\frac{1}{10}\right)} = \frac{3000\left(1 - \frac{1}{10^{12}}\right)}{11} \approx \frac{3000}{11}$.

[30] $162, -54, 18, \cdots$; $S_{10} = \frac{162\left[1 - \left(-\frac{1}{3}\right)^{10}\right]}{1 - \left(-\frac{1}{3}\right)} = \frac{243\left(1 - \frac{1}{3^{10}}\right)}{2} = \frac{29,524}{243}$.

[31] $\frac{1}{2}, \frac{3}{8}, \frac{9}{32}, \cdots$; $S_6 = \frac{\frac{1}{2}\left[1 - \left(\frac{3}{4}\right)^6\right]}{1 - \frac{3}{4}} = 2\left(1 - \frac{729}{4096}\right) = \frac{3367}{2048}$.

[32] $4, 3, \frac{9}{4}, \cdots$; $S_{12} = \frac{4\left[1 - \left(\frac{3}{4}\right)^{12}\right]}{1 - \frac{3}{4}} = 16\left(1 - \frac{531,441}{16,777,216}\right) = \frac{16,245,775}{1,048,576}$.

[33] $\ln 4, \ln 2, \ln \sqrt{2}, \cdots$; $S_{10} = \frac{\ln 4\left[1 - \left(\frac{1}{2}\right)^{10}\right]}{1 - \frac{1}{2}} = (2 \ln 4)\left(1 - \frac{1}{1024}\right) = (2 \ln 2^2)\left(\frac{1023}{1024}\right) =$

$(4 \ln 2)\left(\frac{1023}{1024}\right) = \frac{1023 \ln 2}{256}$.

[34] $\log 27, \log 3, \log \sqrt[3]{3}, \cdots; S_5 = \dfrac{\log 27 \left[1 - \left(\frac{1}{3}\right)^5\right]}{1 - \frac{1}{3}} = \dfrac{\log 3^3 \left(1 - \frac{1}{243}\right)}{\frac{2}{3}} = \left(\frac{3}{2}\right)(3 \log 3)\left(\frac{242}{243}\right) =$

$\dfrac{121 \log 3}{27}$.

[35] $10, 10^{x+1}, 10^{2x+1}, \cdots; a_1 = 10$ and $r = 10^x$. Thus, $S_8 = \dfrac{10\left[1 - \left(10^x\right)^8\right]}{1 - 10^x} = \dfrac{10\left(1 - 10^{8x}\right)}{1 - 10^x}$.

[36] $1, -\dfrac{x}{3}, \dfrac{x^2}{9}, \cdots; a_1 = 1$ and $r = -\dfrac{x}{3}$. Thus, $S_7 = \dfrac{1\left[1 - \left(-\frac{x}{3}\right)^7\right]}{1 - \left(-\frac{x}{3}\right)} = \dfrac{1 + \frac{x^7}{2187}}{1 + \frac{x}{3}} = \dfrac{2187 + x^7}{2187} \cdot \dfrac{3}{3 + x} =$

$\dfrac{x^7 + 2187}{729x + 2187}$.

[37] $a_1 = 2$ and $a_5 = 162$. $a_5 = a_1 r^4 \Rightarrow 162 = 2r^4 \Rightarrow r^4 = 81 \Rightarrow r = \pm 3$. If $r = 3$, $m_1 = 6$, $m_2 = 18$ and

$m_3 = 54$. If $r = -3$, $m_1 = -6$, $m_2 = 18$ and $m_3 = -54$.

[38] $a_1 = 10$ and $a_5 = 20$. $20 = 10r^4 \Rightarrow r = \pm \sqrt[4]{2}$. If $r = \sqrt[4]{2}$, $m_1 = 10\sqrt[4]{2}$, $m_2 = 10\sqrt[4]{4}$ and $m_3 = 10\sqrt[4]{8}$.

If $r = -\sqrt[4]{2}$, $m_1 = -10\sqrt[4]{2}$, $m_2 = 10\sqrt[4]{4}$ and $m_3 = -10\sqrt[4]{8}$.

[39] $a_1 = -27$ and $a_6 = \dfrac{32}{9}$. $\dfrac{32}{9} = -27r^5 \Rightarrow r^5 = -\dfrac{32}{243} \Rightarrow r = -\dfrac{2}{3}$. Thus, $m_1 = 18$, $m_2 = -12$, $m_3 = 8$

and $m_4 = -\dfrac{16}{3}$.

[40] $a_1 = 2$ and $a_6 = 0.00064$. $0.00064 = 2r^5 \Rightarrow r^5 = 0.00032 \Rightarrow r = 0.2$. Thus, $m_1 = 0.4$, $m_2 = 0.08$,

$m_3 = 0.016$, and $m_4 = 0.0032$.

[41] $a_1 = 4$ and $a_5 = 0.0324$. $0.0324 = 4r^4 \Rightarrow r^4 = .0081 \Rightarrow r = \pm 0.3$. If $r = 0.3$, $m_1 = 1.2$, $m_2 = 0.36$

and $m_3 = 0.108$. If $r = -0.3$, $m_1 = -1.2$ $m_2 = 0.36$ and $m_3 = -0.108$.

[42] $a_1 = -24$ and $a_5 = -\dfrac{243}{2}$. $-\dfrac{243}{2} = -24r^4 \Rightarrow r^4 = \dfrac{243}{48} = \dfrac{81}{16} \Rightarrow r = \pm \dfrac{3}{2}$. If $r = \dfrac{3}{2}$, $m_1 = -36$, $m_2 =$

-54 and $m_3 = -81$. If $r = -\dfrac{3}{2}$, $m_1 = 36$, $m_2 = -54$ and $m_3 = 81$.

[43] Since $|r| = \dfrac{2}{3} < 1$, the series has sum $\dfrac{a_1}{1 - r} = \dfrac{6}{1 - \frac{2}{3}} = 18$.

[44] Since $|r| = \dfrac{3}{4} < 1$, the series has sum $\dfrac{12}{1 - \frac{3}{4}} = 48$.

[45] Since $|r| = \dfrac{2}{3} < 1$, the series has sum $\dfrac{1}{1 + \frac{2}{3}} = \dfrac{3}{5}$.

[46] Since $|r| = \dfrac{5}{6} < 1$, the series has sum $\dfrac{1}{1 + \frac{5}{6}} = \dfrac{6}{11}$.

[47] Since $|r| = 0.3 < 1$, the series has sum $\dfrac{4}{1 - 0.3} = \dfrac{40}{7}$.

[48] Since $|r| = 0.2 < 1$, the series has sum $\dfrac{0.3}{1 - 0.2} = \dfrac{3}{8}$.

[49] Since $|r| = \sqrt{3} > 1$, the series has no sum.   **[50]** Since $|r| = \sqrt{2} > 1$, the series has no sum.

[51] Since $|r| = \frac{1}{e} < 1$, the series has sum $\dfrac{\frac{1}{e}}{1 - \frac{1}{e}} = \dfrac{1}{e - 1}$.

[52] Since $|r| = \frac{\pi}{4} < 1$, the series has sum $\dfrac{2}{1 - \frac{\pi}{4}} = \dfrac{8}{4 - \pi}$.

[53] $0.141414\cdots = 0.14 + 0.0014 + 0.000014 + \cdots$. Since $|r| = 0.01 < 1$, the sum of the series is

$\dfrac{0.14}{1 - 0.01} = \dfrac{0.14}{0.99} = \dfrac{14}{99}$.

[54] $0.272727\cdots = 0.27 + 0.0027 + 0.000027 + \cdots$. Since $|r| = 0.01 < 1$, the sum of the series is

$\dfrac{0.27}{1 - 0.01} = \dfrac{27}{99}$.

[55] $2.632632\cdots = 2 + 0.632632\cdots$. Now, $0.632632\cdots = 0.632 + 0.000632 + \cdots$. Since $|r| = 0.001 < 1$,

the sum of the series is $\dfrac{0.632}{1 - 0.001} = \dfrac{632}{999}$. Thus, $2.632632\cdots = 2 + \dfrac{632}{999} = \dfrac{2630}{999}$.

[56] $3.158158\cdots$; $0.158158\cdots = 0.158 + 0.000158 + \cdots$. Since $|r| = 0.001 < 1$, the series has sum

$\dfrac{0.158}{1 - 0.001} = \dfrac{158}{999}$. Thus, $3.158158\cdots = 3 + \dfrac{158}{999} = \dfrac{3155}{999}$.

[57] $10.50219219\cdots$; $0.00219219\cdots = 0.00219 + 0.00000219 + \cdots$. Since $|r| = 0.001 < 1$, the sum of

this series is $\dfrac{0.00219}{1 - 0.001} = \dfrac{0.00219}{0.999} = \dfrac{219}{99,900}$. Thus, $10.50219219\cdots = 10.50 + \dfrac{219}{99,900} = \dfrac{21}{2} +$

$\dfrac{73}{33,300} = \dfrac{349,723}{33,300}$.

[58] $1.36445445\cdots$; $0.00445445\cdots = 0.00445 + 0.00000445 + \cdots$. Since $|r| = 0.001 < 1$, the series has

sum $\dfrac{0.00445}{1 - 0.001} = \dfrac{445}{99,900}$. Thus, $1.36445445\cdots = 1.36 + 0.00445445\cdots = \dfrac{34}{25} + \dfrac{445}{99,900} = \dfrac{136,309}{99,900}$.

[59] $3.20482048\cdots$; $0.20482048\cdots = 0.2048 + 0.00002048 + \cdots$. Since $|r| = 0.0001 < 1$, the sum of this

series is $\dfrac{0.2048}{1 - 0.0001} = \dfrac{2048}{9999}$. Thus, $3.20482048\cdots \ 3 + \dfrac{2048}{9999} = \dfrac{32,045}{9999}$.

[60] $12.06930693\cdots$; $0.06930693\cdots = 0.0693 + 0.00000693 + \cdots$. Since $|r| = 0.0001 < 1$, the sum of

this series is $\dfrac{0.0693}{1 - 0.0001} = \dfrac{693}{9999} = \dfrac{7}{101}$. Thus, $12.06930693\cdots = 12 + \dfrac{7}{101} = \dfrac{1219}{101}$.

[61] $\left(\frac{a}{r}\right)(a)(ar) = -216 \Rightarrow a^3 = -216 \Rightarrow a = -6$. $\frac{a}{r} + a + ar = 7 \Rightarrow \dfrac{a + ar + ar^2}{r} = 7 \Rightarrow$

$\dfrac{-6 - 6r - 6r^2}{r} = 7 \Rightarrow 6r^2 + 13r + 6 = 0 \Rightarrow (3r + 2)(2r + 3) = 0 \Rightarrow r = -\frac{3}{2}$ or $r = -\frac{2}{3}$.

[62] $\left(\frac{a}{r}\right)(a)(ar) = 8 \Rightarrow a^3 = 8 \Rightarrow a = 2$. $\frac{a}{r} + a + ar = \dfrac{26}{3} \Rightarrow \dfrac{2}{r} + 2 + 2r = \dfrac{26}{3} \Rightarrow 3r^2 - 10r + 3 =$

$0 \Rightarrow r = \frac{1}{3}$ or $r = 3$.

[63] $a_1 = 500(1.08)$. $S_n = \dfrac{a_1(1 - r^n)}{1 - r} \Rightarrow S_{12} = \dfrac{500(1.08)\left[1 - (1.08)^{12}\right]}{1 - 1.08} \approx \$10,247.65$.

(See Example 24)

[64] $S_{72} = \dfrac{50(1.0075)\left[1 - (1.0075)^{72}\right]}{1 - 1.0075} \approx \$4,785.98$.

[65] $a_1 = (24,000)(.75) = 18,000$, $a_2 = (24,000)(.75)^2$, $\cdots \Rightarrow a_8 = (24,000)(.75)^8 \approx \$2402.71$.

[66] $a_1 = (36,000)(.80) = 28,800$, $a_2 = (36,000)(.80)^2$, $\cdots \Rightarrow a_{10} = (36,000)(.80)^{10} \approx \$3865.47$.

[67] Downward bounces: $20 + 10 + 5 + \cdots \Rightarrow S_d = \dfrac{20}{1 - \frac{1}{2}} = 40$. Upward bounces: $10 + 5 + \frac{5}{2} + \Rightarrow S_u =$

$\dfrac{10}{1 - \frac{1}{2}} = 20$. The total distance the ball travels is 60 feet.

[68] Downward bounces: $60 + \left(12 + 8 + \frac{16}{3} + \cdots\right) \Rightarrow S_d = 60 + \dfrac{12}{1 - \frac{2}{3}} = 60 + 36 = 96$. Upward bounces:

$12 + 8 + \frac{16}{3} + \cdots \Rightarrow S_u = \dfrac{12}{1 - \frac{2}{3}} = 36$. The total distance the ball travels is $96 + 36 = 132$ feet.

[69] Let $\{a_n\}$ be a geometric sequence of positive terms with common ratio r. Let $\{b_n\}$ be defined by

$b_n = \ln a_n$. Then $b_{n+1} - b_n = \ln a_{n+1} - \ln a_n = \ln a_1 r^n - \ln a_1 r^{n-1} = \ln \dfrac{a_1 r^n}{a_1 r^{n-1}} = \ln r$. Thus

$\{b_n\}$ is an arithmetic sequence with common difference $d = \ln r$.

[70] $P_n = (a_1) \cdot (a_1 r) \cdot (a_1 r^2) \cdot \cdots \cdot (a_1 r^{n-1}) = a_1^n r^{(1+2+\cdots+n-1)}$. By example 16 of section 10.2,

$1 + 2 + \cdots + n - 1 = \dfrac{(n-1)n}{2} = \dfrac{n^2 - n}{2}$. Thus $P_n = a_1^n r^{n^2 - n/2}$.

[71] $|r| < 1 \Rightarrow |2x| < 1 \Rightarrow -1 < 2x < 1 \Rightarrow -\frac{1}{2} < x < \frac{1}{2} \Rightarrow x \in \left(-\frac{1}{2}, \frac{1}{2}\right)$.

[72] $|r| < 1 \Rightarrow \left|\frac{3x}{5}\right| < 1 \Rightarrow -1 < \frac{3x}{5} < 1 \Rightarrow -5 < 3x < 5 \Rightarrow -\frac{5}{3} < x < \frac{5}{3} \Rightarrow x \in \left(-\frac{5}{3}, \frac{5}{3}\right)$.

[73] $A = \frac{1}{4} + \left(\frac{1}{4}\right)\left(\frac{3}{4}\right) + \left(\frac{1}{4}\right)\left(\frac{9}{16}\right) + \cdots = \frac{1}{4} + \frac{3}{16} + \frac{9}{64} + \cdots$ is a geometric series with $a_1 = \frac{1}{4}$ and $r = \frac{3}{4}$.

Thus, $A = \dfrac{\frac{1}{4}}{1 - \frac{3}{4}} = 1$. The total area will be shaded.

[74] $A = \frac{1}{8} + \left(\frac{1}{8}\right)\left(\frac{1}{2}\right) + \left(\frac{1}{8}\right)\left(\frac{1}{4}\right) + \cdots = \frac{1}{8} + \frac{1}{16} + \frac{1}{32} + \cdots$ is a geometric series with $a_1 = \frac{1}{8}$ and $r = \frac{1}{2}$.

Thus, $A = \dfrac{\frac{1}{8}}{1 - \frac{1}{2}} = \frac{1}{4}$.

EXERCISES 10.4

NOTE: In Exercises 1 – 32, S will denote the set of all positive integers for which the given statement is true. We will use P.M.I. to abbreivate the principle of mathematical induction.

[1] First, since $\frac{1}{1 \cdot 2} = \frac{1}{2}$ and $\frac{1}{1+1} = \frac{1}{2}$, we see that $1 \in S$. If we assume that $k \in S$, then $\frac{1}{1 \cdot 2} +$

$\frac{1}{2 \cdot 3} + \cdots + \frac{1}{k(k+1)} = \frac{k}{k+1}$. To show that $k + 1 \in S$, we must show that $\frac{1}{1 \cdot 2} + \frac{1}{2 \cdot 3} + \cdots +$

$\frac{1}{(k+1)(k+2)} = \frac{k+1}{k+2}$. Starting on the left side we obtain the right side as follows: $\frac{1}{1 \cdot 2} +$

$\frac{1}{2 \cdot 3} + \cdots + \frac{1}{(k+1)(k+2)} = \left[\frac{1}{1 \cdot 2} + \frac{1}{2 \cdot 3} + \cdots + \frac{1}{k(k+1)}\right] + \frac{1}{(k+1)(k+2)} =$

$\frac{k}{k+1} + \frac{1}{(k+1)(k+2)} = \frac{k(k+2)+1}{(k+1)(k+2)} = \frac{k^2 + 2k + 1}{(k+1)(k+2)} = \frac{(k+1)(k+1)}{(k+1)(k+2)} = \frac{k+1}{k+2}$.

By P.M.I., $S = \mathbf{N}$.

[2] First, since $1(1+1) = 2$ and $\dfrac{1(2)(3)}{3} = 2$, we see that $1 \in S$. If $k \in S$, then $1 \cdot 2 + 2 \cdot 3 + \cdots +$

$k(k+1) = \dfrac{k(k+1)(k+2)}{3}$. To show that $k+1 \in S$, we must show that $1 \cdot 2 + 2 \cdot 3 + \cdots +$

$(k+1)(k+2) = \dfrac{(k+1)(k+2)(k+3)}{3}$. We have $1 \cdot 2 + 2 \cdot 3 + \cdots + (k+1)(k+2) =$

$[1 \cdot 2 + 2 \cdot 3 + \cdots + k(k+1)] + (k+1)(k+2) = \dfrac{k(k+1)(k+2)}{3} + (k+1)(k+2) =$

$\dfrac{k(k+1)(k+2) + 3(k+1)(k+2)}{3} = \dfrac{(k+1)(k+2)(k+3)}{3}$. So, by P.M.I., $S = \mathbf{N}$.

[3] First, since $3(1) - 2 = 1$ and $\dfrac{1(3-1)}{2} = 1$, we see that $1 \in S$. If $k \in S$, then $1 + 4 + \cdots + (3k-2) =$

$\dfrac{k(3k-1)}{2}$. To show that $k+1 \in S$, we must show that $1 + 4 + \cdots + [3(k+1) - 2] =$

$\dfrac{(k+1)[3(k+1)-1]}{2}$, or that $1 + 4 + \cdots + (3k+1) = \dfrac{(k+1)(3k+2)}{2}$. We have $1 + 4 + \cdots +$

$(3k+1) = [1 + 4 + \cdots + (3k-2)] + (3k+1) = \dfrac{k(3k-1)}{2} + (3k+1) = \dfrac{k(3k-1) + 2(3k+1)}{2} =$

$\dfrac{3k^2 + 5k + 2}{2} = \dfrac{(k+1)(3k+2)}{2}$. By P.M.I., $S = \mathbf{N}$.

[4] First, since $2^1 = 2$ and $2^{1+1} - 2 = 4 - 2 = 2$, we see that $1 \in S$. If $k \in S$, then $2^1 + 2^2 + \cdots + 2^k =$

$2^{k+1} - 2$. To show that $k+1 \in S$, we must show that $2^1 + 2^2 + \cdots + 2^{k+1} = 2^{k+2} - 2$. We have

$2^1 + 2^2 + \cdots + 2^{k+1} = (2^1 + 2^2 + \cdots + 2^k) + 2^{k+1} = (2^{k+1} - 2) + 2^{k+1} = 2 \cdot 2^{k+1} - 2 = 2^{k+2} - 2$. By

P.M.I., $S = \mathbf{N}$.

[5] First, since $[2(1) - 1]^2 = 1^2 = 1$ and $\dfrac{1[2(1)+1][2(1)-1]}{3} = \dfrac{3}{3} = 1$, $1 \in S$. If $k \in S$, then $1^2 +$

$3^2 + \cdots + (2k-1)^2 = \dfrac{k(2k+1)(2k-1)}{3}$. To show that $k+1 \in S$, we must show that $1^2 + 3^2$

$+ \cdots + [2(k+1) - 1]^2 = \dfrac{(k+1)[2(k+1)+1][2(k+1)-1]}{3}$, or that $1^2 + 3^2 + \cdots + (2k+1)^2 =$

$\dfrac{(k+1)(2k+3)(2k+1)}{3}$. We have $1^2 + 3^2 + \cdots + (2k+1)^2 = [1^2 + 3^2 + \cdots + (2k-1)^2] +$

$(2k+1)^2 = \dfrac{k(2k+1)(2k-1)}{3} + (2k+1)^2 = (2k+1)\left[\dfrac{k(2k-1)}{3} + (2k+1)\right] = (2k+1) \cdot$

$\left(\dfrac{2k^2 + 5k + 3}{3}\right) = (2k+1)\left[\dfrac{(2k+3)(k+1)}{3}\right] = \dfrac{(k+1)(2k+3)(2k+1)}{3}$. By P.M.I., $S = \mathbf{N}$.

[6] First, since $1^3 = 1$ and $\dfrac{(1)^2(1+1)^2}{4} = \dfrac{4}{4} = 1$, we see that $1 \in S$. *If* $k \in S$, the $1^3 + 2^3 + \cdots + k^3 =$

$\dfrac{k^2(k+1)^2}{4}$. To show that $k+1 \in S$, we must show that $1^3 + 2^3 + \cdots + (k+1)^3 =$

$\dfrac{(k+1)^2(k+2)^2}{4}$. We have $1^3 + 2^3 + \cdots + (k+1)^3 = (1^3 + 2^3 + \cdots + k^3) + (k+1)^3 =$

$\dfrac{k^2(k+1)^2}{4} + (k+1)^3 = (k+1)^2\left[\dfrac{k^2}{4} + (k+1)\right] = (k+1)^2\left(\dfrac{k^2 + 4k + 4}{4}\right) = \dfrac{(k+1)^2(k+2)^2}{4}$.

By P.M.I., $S = \mathbf{N}$.

[7] First, since $1 \cdot 1! = 1 \cdot 1 = 1$ and $(1+1)! - 1 = 2! - 1 = 2 - 1 = 1$, we see that $1 \in S$. If $k \in S$, then $1 \cdot 1! + 2 \cdot 2! + \cdots + k \cdot k! = (k+1)! - 1$. To show that $k + 1 \in S$, we must show that $1 \cdot 1! + 2 \cdot 2! + \cdots + (k+1) \cdot (k+1)! = (k+2)! - 1$. We have $1 \cdot 1! + 2 \cdot 2! + \cdots + (k+1) \cdot (k+1)! = (1 \cdot 1! + 2 \cdot 2! + \cdots + k \cdot k!) + (k+1) \cdot (k+1)! = [(k+1)! - 1] + (k+1) \cdot (k+1)! = (k+1)! + (k+1) \cdot (k+1)! - 1 = (k+1)![1 + (k+1)] - 1 = (k+1)!(k+2) - 1 = (k+2)! - 1$. By P.M.I., $S = \mathbf{N}$

[8] First, since $5(1) = 5$ and $\dfrac{5(1)(1+1)}{2} = 5$, we that $1 \in S$. If $k \in S$, then $5 + 10 + \cdots + 5k = \dfrac{5k(k+1)}{2}$. To show that $k + 1 \in S$, we must show that $5 + 10 + \cdots + 5(k+1) = \dfrac{5(k+1)(k+2)}{2}$. We have $5 + 10 + \cdots + 5(k+1) = (5 + 10 + \cdots + 5k) + 5(k+1) = \dfrac{5k(k+1)}{2} + 5(k+1) = 5(k+1)\left(\dfrac{k}{2} + 1\right) = 5(k+1)\left(\dfrac{k+2}{2}\right) = \dfrac{5(k+1)(k+2)}{2}$. By P.M.I., $S = \mathbf{N}$.

[9] First, since $3^1 = 3$ and $2^1 = 2$ and $3 > 2$, we see that $1 \in S$. If $k \in S$, then $3^k > 2^k$. To show that $k + 1 \in S$, we must show that $3^{k+1} > 2^{k+1}$. We have $3^k > 2^k \Rightarrow 3 \cdot 3^k > 3 \cdot 2^k \Rightarrow 3^{k+1} > 3 \cdot 2^k$. Since $3 \cdot 2^k = (2+1) \cdot 2^k = 2 \cdot 2^k + 1 \cdot 2^k = 2^{k+1} + 2^k > 2^{k+1}$, it follows that $3 \cdot 2^k > 2^{k+1}$. Thus, $3^{k+1} > 3 \cdot 2^k$ and $3 \cdot 2^k > 2^{k+1} \Rightarrow 3^{k+1} > 2^{k+1}$. By P.M.I., $S = \mathbf{N}$.

[10] First, since $\left(\dfrac{a}{b}\right)^1 = \dfrac{a}{b}$ and $\dfrac{a^1}{b^1} = \dfrac{a}{b}$, we see that $1 \in S$. If $k \in S$, then $\left(\dfrac{a}{b}\right)^k = \dfrac{a^k}{b^k}$. To show that $k + 1 \in S$, we must show that $\left(\dfrac{a}{b}\right)^{k+1} = \dfrac{a^{k+1}}{b^{k+1}}$. We have $\left(\dfrac{a}{b}\right)^{k+1} = \left(\dfrac{a}{b}\right)^k \cdot \left(\dfrac{a}{b}\right) = \left(\dfrac{a^k}{b^k}\right) \cdot \left(\dfrac{a}{b}\right) = \dfrac{a^k \cdot a}{b^k \cdot b} = \dfrac{a^{k+1}}{b^{k+1}}$. By P.M.I., $S = \mathbf{N}$.

[11] First, since $1 + 4 = 5$ and $6(1)^2 = 6$ and $5 < 6$, we see that $1 \in S$. If $k \in S$, then $k + 4 < 6k^2$. To show that $k + 1 \in S$, we must show that $(k+1) + 4 < 6(k+1)^2$, or that $k + 5 < 6(k+1)^2$. We have $k + 4 < 6k^2 \Rightarrow k + 5 < 6k^2 + 1$. Since $6(k+1)^2 = 6(k^2 + 2k + 1) = 6k^2 + 12k + 6 = (6k^2 + 1) + (12k + 5) > 6k^2 + 1$, it follows that $6(k+1)^2 > 6k^2 + 1$. Thus, $k + 5 < 6k^2 + 1$ and $6k^2 + 1 < 6(k+1)^2 \Rightarrow k + 5 < 6(k+1)^2$. By P.M.I., $S = \mathbf{N}$.

[12] First, since $[2(1) - 1]^3 = 1^3 = 1$ and $(1)^2[2(1)^2 - 1] = (1)(1) = 1$, we see that $1 \in S$. If $k \in S$, then $1 + 27 + \cdots + (2k-1)^3 = k^2(2k^2 - 1)$. To show that $k + 1 \in S$, we must show that $1 + 27 + \cdots + [2(k+1) - 1]^3 = (k+1)^2[2(k+1)^2 - 1]$, or that $1 + 27 + \cdots + (2k+1)^3 = (k+1)^2 \cdot [2(k+1)^2 - 1]$. We have $1 + 27 + \cdots + (2k+1)^3 = [1 + 27 + \cdots + (2k-1)^3] + (2k+1)^3 = k^2(2k^2 - 1) + (2k+1)^3 = 2k^4 - k^2 + (8k^3 + 12k^2 + 6k + 1) = 2k^4 + 8k^3 + 11k^2 + 6k + 1$. Using synthetic division, we find that $2k^4 + 8k^3 + 11k^2 + 6k + 1 = (k+1)^2(2k^2 + 4k + 1)$. Since $2k^2 + 4k + 1 = 2k^2 + 4k + 1 + (1 - 1) = (2k^2 + 4k + 2) - 1 = 2(k^2 + 2k + 1) - 1 = 2(k+1)^2 - 1$, it follows that $2k^4 + 8k^3 + 11k^2 + 6k + 1 = (k+1)^2[2(k+1)^2 - 1]$. Therefore, $1 + 27 + \cdots + (2k+1)^3 = (k+1)^2[2(k+1)^2 - 1]$. By P.M.I., $S = \mathbf{N}$.

[13] First, since $\dfrac{1}{[2(1)-1][2(1)+1]} = \dfrac{1}{(1)(3)} = \dfrac{1}{3}$ and $\dfrac{1}{2(1)+1} = \dfrac{1}{3}$, we see that $1 \in S$. If $k \in S$,

then $\dfrac{1}{1\cdot 3} + \dfrac{1}{3\cdot 5} + \cdots + \dfrac{1}{(2k-1)(2k+1)} = \dfrac{k}{2k+1}$. To show that $k+1 \in S$, we must show that

$\dfrac{1}{1\cdot 3} + \dfrac{1}{3\cdot 5} + \cdots + \dfrac{1}{[2(k+1)-1][2(k+1)+1]} = \dfrac{k+1}{2(k+1)+1}$, or that $\dfrac{1}{1\cdot 3} + \dfrac{1}{3\cdot 5} + \cdots +$

$\dfrac{1}{(2k+1)(2k+3)} = \dfrac{k+1}{2k+3}$. We have $\dfrac{1}{1\cdot 3} + \dfrac{1}{3\cdot 5} + \cdots + \dfrac{1}{(2k+1)(2k+3)} =$

$\left[\dfrac{1}{1\cdot 3} + \dfrac{1}{3\cdot 5} + \cdots + \dfrac{1}{(2k-1)(2k+1)}\right] + \dfrac{1}{(2k+1)(2k+3)} = \dfrac{k}{2k+1} + \dfrac{1}{(2k+1)(2k+3)} =$

$\dfrac{k(2k+3)+1}{(2k+1)(2k+3)} = \dfrac{2k^2+3k+1}{(2k+1)(2k+3)} = \dfrac{(2k+1)(k+1)}{(2k+1)(2k+3)} = \dfrac{k+1}{2k+3}$. By P.M.I., $S = \mathbf{N}$.

[14] First, since $[2(1)]^3 = 8$ and $2(1)^2(1+1)^2 = 8$, we see that $1 \in S$. If $k \in S$, then $2^3 + 4^3 + \cdots +$

$(2k)^3 = 2k^2(k+1)^2$. To show that $k+1 \in S$, we must show that $2^3 + 4^3 + \cdots + [2(k+1)]^3 =$

$2(k+1)^2(k+2)^2$. We have $2^3 + 4^3 + \cdots + [2(k+1)]^3 = [2^3 + 4^3 + \cdots + (2k)^3] + [2(k+1)]^3 =$

$2k^2(k+1)^2 + [2(k+1)]^3 = 2k^2(k+1)^2 + 8(k+1)^3 = 2(k+1)^2[k^2 + 4(k+1)] =$

$2(k+1)^2(k^2+4k+4) = 2(k+1)^2(k+2)^2$. By P.M.I., $S = \mathbf{N}$.

[15] First, since $1^2 + 1 = 2$ and $2 > 1$, we see that $1 \in S$. If $k \in S$, then $k^2 + 1 > k$. To show that $k+1 \in S$,

we must show that $(k+1)^2 + 1 > k+1$. Since $k^2 + 1 > k \Rightarrow (k^2+1) + 2k + 1 > (k) + 2k + 1$ and

since $k + 2k + 1 = (k+1) + 2k > k+1$, it follows that $(k^2+1) + 2k + 1 > k+1$. Therefore,

$(k+1)^2 + 1 = (k^2 + 2k + 1) + 1 = (k^2+1) + 2k + 1 > k+1 \Rightarrow (k+1)^2 + 1 > k+1$.

By P.M.I., $S = \mathbf{N}$.

[16] First, since $(1)(1+1)^2 = 4$ and $\dfrac{(1)(1+1)(1+2)[3(1)+5]}{12} = \dfrac{48}{12} = 4$, we see that $1 \in S$. If $k \in S$,

then $1\cdot 2^2 + 2\cdot 3^2 + \cdots + k(k+1)^2 = \dfrac{k(k+1)(k+2)(3k+5)}{12}$. To show that $k+1 \in S$, we

must show that $1\cdot 2^2 + 2\cdot 3^2 + \cdots + (k+1)(k+2)^2 = \dfrac{(k+1)(k+2)(k+3)[3(k+1)+5]}{12}$, or

that $1\cdot 2^2 + 2\cdot 3^2 + \cdots + (k+1)(k+2)^2 = \dfrac{(k+1)(k+2)(k+3)(3k+8)}{12}$. We have $1\cdot 2^2 +$

$2\cdot 3^2 + \cdots + (k+1)(k+2)^2 = [1\cdot 2^2 + 2\cdot 3^2 + \cdots + k(k+1)^2] + (k+1)(k+2)^2 =$

$\dfrac{k(k+1)(k+2)(3k+5)}{12} + (k+1)(k+2)^2 = (k+1)(k+2)\left[\dfrac{k(3k+5)}{12} + (k+2)\right] =$

$(k+1)(k+2)\left[\dfrac{3k^2 + 5k + 12(k+2)}{12}\right] = (k+1)(k+2)\left(\dfrac{3k^2 + 17k + 24}{12}\right) = (k+1)(k+2)\cdot$

$\left[\dfrac{(k+3)(3k+8)}{12}\right] = \dfrac{(k+1)(k+2)(k+3)(3k+8)}{12}$. By P.M.I., $S = \mathbf{N}$.

[17] First, since $4(1) - 3 = 1$ and $(1)[2(1)-1] = 1$, we see that $1 \in S$. If $k \in S$, then $1 + 5 + \cdots +$

$(4k-3) = k(2k-1)$. To show that $k+1 \in S$, we must show that $1 + 5 + \cdots + [4(k+1)-3] =$

$(k+1)[2(k+1)-1]$, or that $1 + 5 + \cdots + (4k+1) = (k+1)(2k+1)$. We have $1 + 5 + \cdots +$

$(4k+1) = [1 + 5 + \cdots + (4k-3)] + (4k+1) = k(2k-1) + (4k+1) = 2k^2 - k + 4k + 1 =$

$2k^2 + 3k + 1 = (k+1)(2k+1)$. By P.M.I., $S = \mathbf{N}$.

[18] First, since $[3(1)-2]^2 = 1$ and $\dfrac{(1)\left[6(1)^2 - 3(1) - 1\right]}{2} = \dfrac{2}{2} = 1$, we see that $1 \in S$. If $k \in S$, then

$1^2 + 4^2 + \cdots + (3k-2)^2 = \dfrac{k\left(6k^2 - 3k - 1\right)}{2}$. To show that $k+1 \in S$, we must show that $1^2 +$

$4^2 + \cdots + [3(k+1) - 2]^2 = \dfrac{(k+1)\left[6(k+1)^2 - 3(k+1) - 1\right]}{2}$, or that $1^2 + 4^2 + \cdots + (3k+1)^2 =$

$\dfrac{(k+1)\left(6k^2 + 9k + 2\right)}{2}$. We have $1^2 + 4^2 + \cdots + (3k+1)^2 = \left[1^2 + 4^2 + \cdots + (3k-2)^2\right] +$

$(3k+1)^2 = \dfrac{k\left(6k^2 - 3k - 1\right)}{2} + (3k+1)^2 = \dfrac{6k^3 - 3k^2 - k + 2\left(9k^2 + 6k + 1\right)}{2} =$

$\dfrac{6k^3 + 15k^2 + 11k + 2}{2} = \dfrac{(k+1)\left(6k^2 + 9k + 2\right)}{2}$. (We used synthetic division to factor the

numerator.) By P.M.I., $S = \mathbf{N}$.

[19] First, since $(1)(1+1)(1+2) = 6$ and $\dfrac{(1)(1+1)(1+2)(1+3)}{4} = \dfrac{24}{4} = 6$, we see that $1 \in S$. If

$k \in S$, then $1 \cdot 2 \cdot 3 + 2 \cdot 3 \cdot 4 + \cdots + k(k+1)(k+2) = \dfrac{k(k+1)(k+2)(k+3)}{4}$. To show that

$k+1 \in S$, we must show that $1 \cdot 2 \cdot 3 + 2 \cdot 3 \cdot 4 + \cdots + (k+1)(k+2)(k+3) =$

$\dfrac{(k+1)(k+2)(k+3)(k+4)}{4}$. We have $1 \cdot 2 \cdot 3 + 2 \cdot 3 \cdot 4 + \cdots + (k+1)(k+2)(k+3) =$

$[1 \cdot 2 \cdot 3 + 2 \cdot 3 \cdot 4 + \cdots + k(k+1)(k+2)] + (k+1)(k+2)(k+3) =$

$\dfrac{k(k+1)(k+2)(k+3)}{4} + (k+1)(k+2)(k+3) = (k+1)(k+2)(k+3)\left(\dfrac{k}{4} + 1\right) =$

$(k+1)(k+2)(k+3)\left(\dfrac{k+4}{4}\right) = \dfrac{(k+1)(k+2)(k+3)(k+4)}{4}$. By P.M.I., $S = \mathbf{N}$.

[20] First, since $3^1 = 3$ and $\dfrac{3^{1+1} - 3}{2} = \dfrac{6}{2} = 3$, we see that $1 \in S$. If $k \in S$, then $3 + 3^2 + \cdots + 3^k =$

$\dfrac{3^{k+1} - 3}{2}$. To show that $k+1 \in S$, we must show that $3 + 3^2 + \cdots + 3^{k+1} = \dfrac{3^{k+2} - 3}{2}$. We have

$3 + 3^2 + \cdots + 3^{k+1} = (3 + 3^2 + \cdots + 3^k) + 3^{k+1} = \dfrac{3^{k+1} - 3}{2} + 3^{k+1} = \dfrac{3^{k+1} - 3 + 2 \cdot 3^{k+1}}{2} =$

$\dfrac{3 \cdot 3^{k+1} - 3}{2} = \dfrac{3^{k+2} - 3}{2}$. By P.M.I., $S = \mathbf{N}$.

[21] First, since $[2(1) - 1][2(1)] = 2$ and $\dfrac{(1)(1+1)[4(1) - 1]}{3} = \dfrac{6}{3} = 2$, we see that $1 \in S$. If $k \in S$,

then $1 \cdot 2 + 3 \cdot 4 + \cdots + (2k-1)(2k) = \dfrac{k(k+1)(4k-1)}{3}$. To show that $k+1 \in S$, we must

show that $1 \cdot 2 + 3 \cdot 4 + \cdots + [2(k+1) - 1][2(k+1)] = \dfrac{(k+1)(k+2)[4(k+1) - 1]}{3}$, or that

$1 \cdot 2 + 3 \cdot 4 + \cdots + (2k+1)[2(k+1)] = \dfrac{(k+1)(k+2)(4k+3)}{3}$. We have $1 \cdot 2 + 3 \cdot 4 + \cdots +$

$(2k+1)[2(k+1)] = [1 \cdot 2 + 3 \cdot 4 + \cdots + (2k-1)(2k)] + (2k+1)[2(k+1)] =$

$\dfrac{k(k+1)(4k-1)}{3} + (2k+1)[2(k+1)] = (k+1)\left[\dfrac{k(4k-1)}{3} + 2(2k+1)\right] = (k+1) \cdot$

$\left(\dfrac{4k^2 - k + 12k + 6}{3}\right) = (k+1)\left(\dfrac{4k^2 + 11k + 6}{3}\right) = (k+1)\left[\dfrac{(k+2)(4k+3)}{3}\right] =$

$\dfrac{(k+1)(k+2)(4k+3)}{3}$. By P.M.I., $S = \mathbf{N}$.

[22] First, since $(a \cdot b)^1 = a \cdot b = ab$ and $a^1 \cdot b^1 = a \cdot b = ab$, we see that $1 \in S$. If $k \in S$, then $(a \cdot b)^k = a^k \cdot b^k$. To show that $k + 1 \in S$, we must show that $(a \cdot b)^{k+1} = a^{k+1} \cdot b^{k+1}$. We have $(a \cdot b)^{k+1} = (a \cdot b)^k \cdot (a \cdot b) = (a^k \cdot b^k) \cdot (a \cdot b) = (a^k \cdot a)(b^k \cdot b) = a^{k+1} \cdot b^{k+1}$. By P.M.I., $S = \mathbf{N}$.

[23] First, since $8(1) = 8$ and $[2(1) + 1]^2 = 9$ and $8 < 9$, we see that $1 \in S$. If $k \in S$, then $8 + 16 + \cdots + 8k < (2k + 1)^2$. To show that $k + 1 \in S$, we must show that $8 + 16 + \cdots + 8(k + 1) < [2(k + 1) + 1]^2$ or that $8 + 16 + \cdots + 8(k + 1) < (2k + 3)^2$. Since $8 + 16 + \cdots + 8(k + 1) = (8 + 16 + \cdots + 8k) + 8(k + 1)$, and since $8 + 16 + \cdots + 8k < (2k + 1)^2 \Rightarrow (8 + 16 + \cdots + 8k) + 8(k + 1) < (2k + 1)^2 + 8(k + 1)$, it follows that $8 + 16 + \cdots + 8(k + 1) < (2k + 1)^2 + 8(k + 1)$. Since $(2k + 1)^2 + 8(k + 1) = (4k^2 + 4k + 1) + 8k + 8 = 4k^2 + 12k + 9 = (2k + 3)^2$, it follows that $8 + 16 + \cdots + 8(k + 1) < (2k + 3)^2$. By P.M.I., $S = \mathbf{N}$.

[24] First, since $[r \operatorname{cis} \theta]^1 = r \operatorname{cis} \theta$ and $r^1 \operatorname{cis} [(1)\theta] = r \operatorname{cis} \theta$, we see that $1 \in S$. If $k \in S$, then $[r \operatorname{cis} \theta]^k = r^k \operatorname{cis}(k\theta)$. To show that $k + 1 \in S$, we must show that $[r \operatorname{cis} \theta]^{k+1} = r^{k+1} \operatorname{cis}[(k + 1)\theta]$. We have $[r \operatorname{cis} \theta]^{k+1} = [r \operatorname{cis} \theta]^k (r \operatorname{cis} \theta) = [r^k \operatorname{cis}(k\theta)](r \operatorname{cis} \theta) = (r^k \cdot r) \operatorname{cis}(k\theta + \theta) = r^{k+1} \operatorname{cis}[(k + 1)\theta]$. By P.M.I., $S = \mathbf{N}$.

[25] First, since $\sin[x + (1)\pi] = \sin(x + \pi) = \sin x \cos \pi + \cos x \sin \pi = (\sin x)(-1) + (\cos x)(0) = -\sin x$ and since $(-1)^1 \sin x = -\sin x$, we see that $1 \in S$. If $k \in S$, then $\sin(x + k\pi) = (-1)^k \sin x$. To show that $k + 1 \in S$, we must show that $\sin[x + (k + 1)\pi] = (-1)^{k+1} \sin x$. We have $\sin[x + (k + 1)\pi] = \sin[(x + k\pi) + \pi] = \sin(x + k\pi) \cos \pi + \cos(x + k\pi) \sin \pi = [\sin(x + k\pi)](-1) + [\cos(x + k\pi)](0) = [\sin(x + k\pi)](-1) = [(-1)^k \sin x](-1) = (-1)^{k+1} \sin x$. By P.M.I., $S = \mathbf{N}$.

[26] First, since $\cos[x + (1)\pi] = \cos(x + \pi) = \cos x \cos \pi - \sin x \sin \pi = (\cos x)(-1) - (\sin x)(0) = -\cos x$ and since $(-1)^1 \cos x = -\cos x$, we see that $1 \in S$. If $k \in S$, then $\cos(x + k\pi) = (-1)^k \cdot \cos x$. To show that $k + 1 \in S$, we must show that $\cos[x + (k + 1)\pi] = (-1)^{k+1} \cos x$. We have $\cos[x + (k + 1)\pi] = \cos[(x + k\pi) + \pi] = \cos(x + k\pi) \cos \pi - \sin(x + k\pi) \sin \pi = [\cos(x + k\pi)](-1) - [\sin(x + k\pi)](0) = [\cos(x + k\pi)](-1) = [(-1)^k \cos x](-1) = (-1)^{k+1} \cos x$. By P.M.I., $S = \mathbf{N}$.

[27] First, since $(1)(1 + 1) = 2$ and 2 is a factor of 2, we see that $1 \in S$. If $k \in S$, then 2 is a factor of $k(k + 1)$. To show that $k + 1 \in S$, we must show that 2 is a factor of $(k + 1)(k + 2)$. Since $(k + 1)(k + 2) = k(k + 1) + 2(k + 1)$ and since 2 is a factor of $k(k + 1)$ and since 2 is a factor of $2(k + 1)$, it follows that 2 is a factor of $(k + 1)(k + 2)$. By P.M.I., $S = \mathbf{N}$.

[28] First, since $(1)^2 - (1) + 2 = 2$ and since 2 is a factor of 2, we see that $1 \in S$. If $k \in S$, then 2 is a factor of $k^2 - k + 2$. To show that $k + 1 \in S$, we must show that 2 is a factor of $(k + 1)^2 - (k + 1) + 2$. Since $(k + 1)^2 - (k + 1) + 2 = (k^2 + 2k + 1) - k - 1 + 2 = (k^2 - k + 2) + 2k$ and since 2 is a factor of $k^2 - k + 2$ and 2 is a factor of $2k$, it follows that 2 is a factor of $(k + 1)^2 - (k + 1) + 2$. By P.M.I., $S = \mathbf{N}$.

[29] First, since $(1)(1+1)(1+2)=6$ and 3 is a factor of 6, we see that $1 \in S$. If $k \in S$, then 3 is a factor of $k(k+1)(k+2)$. To show that $k+1 \in S$, we must show that 3 is a factor of $(k+1)(k+2) \cdot (k+3)$. Since $(k+1)(k+2)(k+3) = k(k+1)(k+2) + 3(k+1)(k+2)$ and since 3 is a factor of $k(k+1)(k+2)$ and since 3 is a factor of $3(k+1)(k+2)$, it follows that 3 is a factor of $(k+1)(k+2)(k+3)$. By P.M.I., $S = \mathbf{N}$.

[30] First, since $5^1 - 3^1 = 2$ and since 2 is a factor of 2, we see that $1 \in S$. If $k \in S$, then 2 is a factor of $5^k - 3^k$. To show that $k+1 \in S$, we must show that 2 is a factor of $5^{k+1} - 3^{k+1}$. We have $5^{k+1} - 3^{k+1} = 5^{k+1} - 3^{k+1} + (5 \cdot 3^k - 5 \cdot 3^k) = (5^{k+1} - 5 \cdot 3^k) + (5 \cdot 3^k - 3^{k+1}) = 5(5^k - 3^k) + 3^k(5-3) = 5(5^k - 3^k) + 3^k(2)$. Since 2 is a factor of $5^k - 3^k$ and hence a factor of $5(5^k - 3^k)$ and since 2 is a factor of $3^k(2)$, it follows that 2 is a factor of $5^{k+1} - 3^{k+1}$. By P.M.I., $S = \mathbf{N}$.

[31] First, since $4^1 - 1 = 3$ and since 3 is a factor of 3, we see that $1 \in S$. If $k \in S$, then 3 is a factor of $4^k - 1$. To show that $k+1 \in S$, we must show that 3 is a factor of $4^{k+1} - 1$. We have $4^{k+1} - 1 = 4 \cdot 4^k - 1 = 4 \cdot 4^k - 4 + 3 = 4(4^k - 1) + 3$. Since 3 is a factor of $4^k - 1$ and hence of $4(4^k - 1)$ and since 3 is a factor of 3, it follows that 3 is a factor of $4^{k+1} - 1$. By P.M.I., $S = \mathbf{N}$.

[32] First, since $3^{2(1)} - 1 = 8$ and since 8 is a factor of 8, we see that $1 \in S$. If $k \in S$, then 8 is a factor of $3^{2k} - 1$. To show that $k+1 \in S$, we must show that 8 is a factor of $3^{2(k+1)} - 1$. We have $3^{2(k+1)} - 1 = 3^{2k+2} - 1 = 3^{2k} \cdot 3^2 - 9 + 8 = 9(3^{2k} - 1) + 8$. Since 8 is a factor of $3^{2k} - 1$ and hence a factor of $9(3^{2k} - 1)$ and since 8 is a factor of 8, it follows that 8 is a factor of $3^{2(k+1)} - 1$. By P.M.I., $S = \mathbf{N}$.

[33] Let $S = \{n \in \mathbf{N} \mid x_n = n(n+1)\}$. Since $x_1 = 2$ and since $(1)(1+1) = 2$, we see that $1 \in S$. If $k \in S$, then $x_k = k(k+1)$. To show that $k+1 \in S$, we must show that $x_{k+1} = (k+1)(k+2)$. Since $k + 1 \geq 2$, it follows that $x_{k+1} = x_{(k+1)-1} + 2(k+1) = x_k + 2(k+1) = k(k+1) + 2(k+1) = (k+1)(k+2)$. By the principle of mathematical induction, $S = \mathbf{N}$.

[34] Let $S = \left\{ n \in \mathbf{N} \mid 2 + 4 + 6 + \cdots + 2n = \dfrac{(2n+1)^2}{4} \right\}$.

a) If $k \in S$, then $2 + 4 + \cdots + 2k = \dfrac{(2k+1)^2}{4}$. To show that $k+1 \in S$, we must show that
$$2 + 4 + \cdots + 2(k+1) = \frac{[2(k+1)+1]^2}{4}, \text{ or that } 2 + 4 + \cdots + 2(k+1) = \frac{(2k+3)^2}{4}. \text{ We have}$$
$$2 + 4 + \cdots + 2(k+1) = (2 + 4 = \cdots + 2k) + 2(k+1) = \frac{(2k+1)^2}{4} + 2(k+1) =$$
$$\frac{(4k^2 + 4k + 1) + (8k + 8)}{4} = \frac{4k^2 + 12k + 9}{4} = \frac{(2k+3)^2}{4}. \text{ So, if } k \in S, \text{ then } k+1 \in S.$$

b) No, it does not follow that $S = \mathbf{N}$. For example, $1 \notin S$ since $2 \neq \frac{9}{4}$. In fact, $S = \varnothing$.

[35] Let $S = \{n \in \mathbf{N} \mid 4^n > n^4\}$. Since $4^5 = 1024$ and $5^4 = 625$ and $1024 > 625$, we see that $5 \in S$. Let $k \geq 5$. If $k \in S$, then $4^k > k^4$. To show that $k + 1 \in S$, we must show that $4^{k+1} > (k+1)^4$. Since $4^k > k^4, 4 \cdot 4^k > 4 \cdot k^4 \Rightarrow 4^{k+1} > 4 \cdot k^4$. Since $k \geq 5, k^3 \cdot k \geq 5k^3 > 4k^3 \Rightarrow k^4 > 4k^3$. Similarly, $k \geq 5 \Rightarrow k^2 \geq 25 \Rightarrow k^2 \cdot k^2 \geq 25k^2 > 6k^2 \Rightarrow k^4 > 6k^2$, and $k \geq 5 \Rightarrow k^3 \geq 125 \Rightarrow k \cdot k^3 \geq 125k > 4k + 1 \Rightarrow k^4 > 4k + 1$. Therefore, $4 \cdot k^4 = k^4 + (k^4 + k^4 + k^4) > k^4 + [4k^3 + 6k^2 + (4k + 1)] \Rightarrow 4 \cdot k^4 > k^4 + 4k^3 + 6k^2 + 4k + 1 = (k+1)^4 \Rightarrow 4 \cdot k^4 > (k+1)^4$. Since $4^{k+1} > 4 \cdot k^4$ and $4 \cdot k^4 > (k+1)^4$, it follows that $4^{k+1} > (k+1)^4$. By the principle of mathematical induction, S contains every positive integer $n \geq 5$.

[36] Let $S = \{n \in \mathbf{N} \mid n! > 2^n\}$. Since $4! = 24$ and $2^4 = 16$ and $24 > 16$, we see that $4 \in S$. Let $k \geq 4$. If $k \in S$, then $k! > 2^k$. To show that $k + 1 \in S$, we must show that $(k+1)! > 2^{k+1}$. Since $k! > 2^k$, $(k+1)k! > (k+1)2^k \Rightarrow (k+1)! > (k+1)2^k$. Since $k \geq 4, k+1 \geq 5 > 2 \Rightarrow k+1 > 2 \Rightarrow (k+1)2^k > 2 \cdot 2^k \Rightarrow (k+1)2^k > 2^{k+1}$. Since $(k+1)! > (k+1)2^k$ and since $(k+1)2^k > 2^{k+1}$, it follows that $(k+1)! > 2^{k+1}$. By the principle of mathematical induction, S contains every positive integer $n \geq 4$.

[37] Let $S = \{n \in \mathbf{N} \mid 3^n > 2^n + 1\}$. Since $3^2 = 9$ and $2^2 + 1 = 5$ and $9 > 5$, we see that $2 \in S$. Let $k \geq 2$. If $k \in S$, then $3^k > 2^k + 1$. To show that $k + 1 \in S$, we must show that $3^{k+1} > 2^{k+1} + 1$. Since $3^k > 2^k + 1, 3 \cdot 3^k > 3(2^k + 1) \Rightarrow 3^{k+1} > 3 \cdot 2^k + 3$. Since $3 > 2, 3 \cdot 2^k > 2 \cdot 2^k \Rightarrow 3 \cdot 2^k > 2^{k+1}$. Since $3 \cdot 2^k > 2^{k+1}$ and $3 > 1$, it follows that $3 \cdot 2^k + 3 > 2^{k+1} + 1$. Since $3^{k+1} > 3 \cdot 2^k + 3$ and $3 \cdot 2^k + 3 > 2^{k+1} + 1$, it follows that $3^{k+1} > 2^{k+1} + 1$. By the principle of mathematical induction, S contains every positive integer $n \geq 2$.

[38] Let $S = \{n \in \mathbf{N} \mid 2^n > n^2\}$. Since $2^5 = 32$ and $5^2 = 25$ and $32 > 25$, we see that $5 \in S$. Let $k \geq 5$. If $k \in S$, then $2^k > k^2$. To show that $k + 1 \in S$, we must show that $2^{k+1} > (k+1)^2$. Since $2^k > k^2$, $2 \cdot 2^k > 2k^2 \Rightarrow 2^{k+1} > 2k^2$. Now, $k \geq 5 \Rightarrow k \cdot k \geq 5k = 2k + 3k \Rightarrow k^2 \geq 2k + 3k$. Similarly, $k \geq 5 \Rightarrow 3k \geq 15 > 1 \Rightarrow 3k > 1 \Rightarrow 2k + 3k > 2k + 1$. Thus, $k^2 \geq 2k + 3k$ and $2k + 3k > 2k + 1 \Rightarrow k^2 > 2k + 1 \Rightarrow k^2 + k^2 > k^2 + 2k + 1 \Rightarrow 2k^2 > (k+1)^2$. Since $2^{k+1} > 2k^2$ and $2k^2 > (k+1)^2$, it follows that $2^{k+1} > (k+1)^2$. By the principle of mathematical induction, S contains every positive integer $n \geq 5$.

[39] Let $S = \{n \in \mathbf{N} \mid 2^{n+3} < (n+1)!\}$. Since $2^{5+3} = 2^8 = 256$ and $(5+1)! = 6! = 720$ and $256 < 720$, we see that $5 \in S$. Let $k \geq 5$. If $k \in S$, then $2^{k+3} < (k+1)!$. To show that $k + 1 \in S$, we must show that $2^{k+4} < (k+2)!$. Since $2^{k+3} < (k+1)!, 2 \cdot 2^{k+3} < 2(k+1)! \Rightarrow 2^{k+4} < 2(k+1)!$. Now, $(k+2)! = (k+2)(k+1)! = k(k+1)! + 2(k+1)! > 2(k+1)! \Rightarrow (k+2)! > 2(k+1)! \Rightarrow 2(k+1)! < (k+2)!$. Since $2^{k+4} < 2(k+1)!$ and $2(k+1)! < (k+2)!$, it follows that $2^{k+4} < (k+2)!$. By the principle of mathematical induction, S contains every positive integer $n \geq 5$.

[40]   Let $S = \{n \in \mathbf{N} \mid 4^n < n!\}$. Since $4^9 = 262,144$ and $9! = 362,880$ and $262,144 < 362,880$, we see that $9 \in S$. Let $k \geq 9$. If $k \in S$, then $4^k < k!$. To show that $k + 1 \in S$, we must show that $4^{k+1} < (k+1)!$. Since $4^k < k!$, $4 \cdot 4^k < 4k! \Rightarrow 4^{k+1} < 4k!$. Now, since $k \geq 9$, $k + 1 \geq 10 > 4 \Rightarrow k + 1 > 4 \Rightarrow (k+1)k! > 4k! \Rightarrow 4k! < (k+1)!$. Since $4^{k+1} < 4k!$ and $4k! < (k+1)!$, it follows that $4^{k+1} < (k+1)!$. By the principle of mathematical induction, S contains every positive integer $n \geq 9$.

[41]   a)   Let $S = \{n \in \mathbf{N} \mid f_1 + f_3 + f_5 + \cdots + f_{2n-1} = f_{2n}\}$. Since $f_{2(1)-1} = f_1 = 1$ and $f_{2(1)} = f_2 = 1$ and $f_1 = f_2$, we see that $1 \in S$. If $k \in S$, then $f_1 + f_3 + \cdots + f_{2k-1} = f_{2k}$. To show that $k + 1 \in S$, we must show that $f_1 + f_3 + \cdots + f_{2(k+1)-1} = f_{2(k+1)}$, or that $f_1 + f_3 + \cdots + f_{2k+1} = f_{2k+2}$. We have $f_1 + f_3 + \cdots + f_{2k+1} = (f_1 + f_3 + \cdots + f_{2k-1}) + f_{2k+1} = f_{2k} + f_{2k+1} = f_{2k+2}$. By the principle of mathematical induction, $S = \mathbf{N}$.

   b)   Let $S = \{n \in \mathbf{N} \mid f_2 + f_4 + f_6 + \cdots + f_{2n} = f_{2n+1} - 1\}$. Since $f_{2(1)} = f_2 = 1$ and $f_{2(1)+1} - 1 = f_3 - 1 = 2 - 1 = 1$, we see that $1 \in S$. If $k \in S$, then $f_2 + f_4 + \cdots + f_{2k} = f_{2k+1} - 1$. To show that $k + 1 \in S$, we must show that $f_2 + f_4 + \cdots + f_{2(k+1)} = f_{2(k+1)+1} - 1$, or that $f_2 + f_4 + \cdots + f_{2k+2} = f_{2k+3} - 1$. We have $f_2 + f_4 + \cdots + f_{2k+2} = (f_2 + f_4 + \cdots + f_{2k}) + f_{2k+2} = (f_{2k+1} - 1) + f_{2k+2} = (f_{2k+1} + f_{2k+2}) - 1 = f_{2k+3} - 1$. By the principle of mathematical induction, $S = \mathbf{N}$.

   c)   Let $S = \{n \in \mathbf{N} \mid f_1 + f_2 + f_3 + \cdots + f_n = f_{n+2} - 1\}$. Since $f_1 = 1$ and $f_{1+2} - 1 = f_3 - 1 = 2 - 1 = 1$, we see that $1 \in S$. If $k \in S$, then $f_1 + f_2 + \cdots + f_k = f_{k+2} - 1$. To show that $k + 1 \in S$, we must show that $f_1 + f_2 + \cdots + f_{k+1} = f_{k+3} - 1$. We have $f_1 + f_2 + \cdots + f_{k+1} = (f_1 + f_2 + \cdots + f_k) + f_{k+1} = (f_{k+2} - 1) + f_{k+1} = (f_{k+2} + f_{k+1}) - 1 = f_{k+3} - 1$. By the principle of mathematical induction, $S = \mathbf{N}$.

   d)   Let $S = \{n \in \mathbf{N} \mid f_n \cdot f_{n+3} - f_{n+1} \cdot f_{n+2} = (-1)^{n+1}\}$. Since $f_1 \cdot f_{1+3} - f_{1+1} \cdot f_{1+2} = f_1 \cdot f_4 - f_2 \cdot f_3 = (1)(3) - (1)(2) = 1$ and $(-1)^{1+1} = (-1)^2 = 1$, we see that $1 \in S$. If $k \in S$, then $f_k \cdot f_{k+3} - f_{k+1} \cdot f_{k+2} = (-1)^{k+1}$. To show that $k + 1 \in S$, we must show that $f_{k+1} \cdot f_{k+4} - f_{k+2} \cdot f_{k+3} = (-1)^{k+2}$. We have $f_{k+1} \cdot f_{k+4} - f_{k+2} \cdot f_{k+3} = f_{k+1} \cdot (f_{k+2} + f_{k+3}) - (f_k + f_{k+1}) \cdot f_{k+3} = f_{k+1} \cdot f_{k+2} + f_{k+1} \cdot f_{k+3} - f_k \cdot f_{k+3} - f_{k+1} \cdot f_{k+3} = f_{k+1} \cdot f_{k+2} - f_k \cdot f_{k+3} = -(f_k \cdot f_{k+3} - f_{k+1} \cdot f_{k+2}) = -[(-1)^{k+1}] = (-1)^{k+2}$. By the principle of mathematical induction, $S = \mathbf{N}$.

[42]   Let $S = \{n \in \mathbf{N} \mid x_n = 2n - 1\}$. Since $x_1 = 1$ and $2(1) - 1 = 1$, we see that $1 \in S$. If $k \in S$, then $x_k = 2k - 1$. To show that $k + 1 \in S$, we must show that $x_{k+1} = 2(k+1) - 1$, or that $x_{k+1} = 2k + 1$. Since $k + 1 \geq 2$, it follows that $x_{k+1} = x_{(k+1)-1} + 2 = x_k + 2 = (2k - 1) + 2 = 2k + 1$. By the principle of mathematical induction, $S = \mathbf{N}$.

[43]   Let $S = \{n \in \mathbf{N} \mid x_n = 2^{2n-1}\}$. Since $x_1 = 2$ and $2^{2(1)-1} = 2^1 = 2$, we see that $1 \in S$. If $k \in S$, then $x_k = 2^{2k-1}$. To show that $k + 1 \in S$, we must show that $x_{k+1} = 2^{2(k+1)-1}$, or that $x_{k+1} = 2^{2k+1}$. Since $k + 1 \geq 2$, it follows that $x_{k+1} = 4x_{(k+1)-1} = 4x_k = 4(2^{2k-1}) = 2^2 \cdot 2^{2k-1} = 2^{2k+1}$. By the principle of mathematical induction, $S = \mathbf{N}$.

[44]   Let $S = \{n \in \mathbf{N} \mid x - 1 \text{ is a factor of } x^n - 1\}$. Since $x^1 - 1 = x - 1$ and $x - 1$ is a factor of $x - 1$, we

see that $1 \in S$. If $k \in S$, then $x - 1$ is a factor $x^k - 1$. To show that $k + 1 \in S$, we must show that $x - 1$ is a factor of $x^{k+1} - 1$. We have $x^{k+1} - 1 = x^{k+1} - 1 + (x - x) = \left(x^{k+1} - x\right) + (x - 1) = x\left(x^k - 1\right) + (x - 1)$. Since $x - 1$ is a factor of $x^k - 1$ and hence of $x\left(x^k - 1\right)$ and since $x - 1$ is a factor of $x - 1$,

it follows that $x - 1$ is a factor of $x^{k+1} - 1$. By the principle of mathematical induction, $S = \mathbf{N}$.

[45]   In Exercise 6 earlier in this section we used the principle of mathematical induction to prove that for

every positive integer n, $1^3 + 2^3 + 3^3 + \cdots + n^3 = \dfrac{n^2(n+1)^2}{4}$. In Example 28 on page 521, we used

the principle of mathematical induction to prove that for every positive integer n, $1 + 2 + 3 + \cdots + n$

$= \dfrac{n(n+1)}{2}$. Squaring both sides of the last equation, we get $(1 + 2 + 3 + \cdots + n)^2 = \dfrac{n^2(n+1)^2}{4}$.

Since $1^3 + 2^3 + 3^3 + \cdots + n^3 = \dfrac{n^2(n+1)^2}{4} = (1 + 2 + 3 + \cdots n)^2$, it follows that for every positive

integer n, $1^3 + 2^3 + 3^3 + \cdots + n^3 = (1 + 2 + 3 + \cdots + n)^2$.

EXERCISES 10.5

[1] $\binom{8}{3} = \frac{8!}{3!\,5!} = 56$

[2] $\binom{7}{4} = \frac{7!}{4!\,3!} = 35$

[3] $\binom{12}{8} = \frac{12!}{8!\,4!} = 495$

[4] $\binom{13}{10} = \frac{13!}{10!\,3!} = 286$

[5] $\binom{n}{n-1} = \frac{n!}{(n-1)!\,1!} = n$

[6] $\binom{10}{2} = \frac{10!}{2!\,8!} = 45$

[7] $(a+b)^8 = \binom{8}{0}a^8 + \binom{8}{1}a^7b + \binom{8}{2}a^6b^2 + \binom{8}{3}a^5b^3 + \binom{8}{4}a^4b^4 + \binom{8}{5}a^3b^5 + \binom{8}{6}a^2b^6 +$

$\binom{8}{7}ab^7 + \binom{8}{8}b^8 = a^8 + 8a^7b + 28a^6b^2 + 56a^5b^3 + 70a^4b^4 + 56a^3b^5 + 28a^2b^6 +$

$8ab^7 + b^8$.

[8] $(a-b)^8 = a^8 - 8a^7b + 28a^6b^2 - 56a^5b^3 + 70a^4b^4 - 56a^3b^5 + 28a^2b^6 - 8ab^7 + b^8$.

[9] $(x - 2y)^5 = x^5 - 10x^4y + 40x^3y^2 - 80x^2y^3 + 80xy^4 - 32y^5$.

[10] $(2x - y)^6 = 64x^6 - 192x^5y + 240x^4y^2 - 160x^3y^3 + 60x^2y^4 - 12xy^5 + y^6$.

[11] $\left(\sqrt{x} - \frac{2}{\sqrt{x}}\right)^4 = x^2 - 8x + 24 - \frac{32}{x} + \frac{16}{x^2}$.

[12] $\left(3\sqrt{x} - \frac{4}{\sqrt{x}}\right)^5 = 243\sqrt{x^5} - 1{,}620\sqrt{x^3} + 4{,}320\sqrt{x} - \frac{5{,}760}{\sqrt{x}} + \frac{3{,}840}{\sqrt{x^3}} - \frac{1{,}024}{\sqrt{x^5}}$.

[13] $\left(\frac{x}{2} - 4y\right)^3 = \frac{x^3}{8} - 3x^2y + 24xy^2 - 64y^3$

[14] $(5x + 4y)^4 = 625x^4 + 2{,}000x^3y + 2{,}400x^2y^2 + 1{,}280xy^3 + 256y^4$

[15] $\left(6x + \frac{y}{3}\right)^6 = 46{,}656x^6 + 15{,}552x^5y + 2{,}160x^4y^2 + 160x^3y^3 + \frac{20}{3}x^2y^4 + \frac{4}{27}xy^5 + \frac{y^6}{729}$

[16] $\left(3y - \frac{x}{6}\right)^3 = 27y^3 - \frac{9}{2}xy^2 + \frac{1}{4}x^2y - \frac{1}{216}x^3$ [17] $\left(x^{-1} - 2x\right)^5 = \frac{1}{x^5} - \frac{10}{x^3} + \frac{40}{x} - 80x + 80x^3 - 32x^5$

[18] $\left(4x - 2x^{-2}\right)^5 = 1{,}024x^5 - 2{,}560x^2 + \dfrac{2{,}560}{x} - \dfrac{1{,}280}{x^4} + \dfrac{320}{x^7} - \dfrac{32}{x^{10}}$

NOTE: In Exercises 19 – 22, we use summation notation to represent the first three terms of each expansion.

[19] $\sum_{k=0}^{2}\binom{10}{k}(x)^{10-k}(2y)^k = \binom{10}{0}x^{10} + \binom{10}{1}x^9(2y)^1 + \binom{10}{2}x^8(2y)^2 = x^{10} + 20x^9y + 180x^8y^2.$

[20] $\sum_{k=0}^{2}\binom{12}{k}(x)^{12-k}(-4y)^k = \binom{12}{0}x^{12} + \binom{12}{1}x^{11}(-4y)^1 + \binom{12}{2}x^{10}(-4y)^2 = x^{12} - 48x^{11}y +$

1,056$x^{10}y^2$.

[21] $\sum_{k=0}^{2}\binom{15}{k}(2x)^{15-k}(-3y)^k = \binom{15}{0}(2x)^{15} + \binom{15}{1}(2x)^{14}(-3y)^1 + \binom{15}{2}(2x)^{13}(-3y)^2 =$

32,768$x^{15} - 737{,}280x^{14}y + 7{,}741{,}440x^{13}y^2.$

[22] $\sum_{k=0}^{2}\binom{13}{k}(2x)^{13-k}(3y)^k = \binom{13}{0}(2x)^{13} + \binom{13}{1}(2x)^{12}(3y)^1 + \binom{13}{2}(2x)^{11}(3y)^2 = 8{,}192x^{13} +$

159,744$x^{12}y + 1{,}437{,}696x^{11}y^2.$

NOTE: In Exercises 23 – 36, we use the formula for the k th term in the expansion of $(a + b)^n$:

$\binom{n}{k-1}a^{n-(k-1)}b^{k-1}.$

[23] $(2x - y)^{12}$; third term $= \binom{12}{2}(2x)^{10}(-y)^2 = 67{,}584x^{10}y^2$

[24] $(3x - 2y)^{10}$; eighth term $= \binom{10}{7}(3x)^3(-2y)^7 = -414{,}720x^3y^7$

[25] $\left(\frac{x}{3} - 6y\right)^{11}$; fourth term $= \binom{11}{3}\left(\frac{x}{3}\right)^8(-6y)^3 = -\frac{440}{81}x^8y^3$

[26] $\left(2x + \frac{y}{4}\right)^9$; fourth term $= \binom{9}{3}(2x)^6\left(\frac{y}{4}\right)^3 = 84x^6y^3$

[27] $\left(\sqrt{x} + \frac{10}{\sqrt{x}}\right)^9$; sixth term $= \binom{9}{5}(\sqrt{x})^4\left(\frac{10}{\sqrt{x}}\right)^5 = \frac{12{,}600{,}000}{\sqrt{x}}$

[28] $\left(\sqrt{x} - \frac{3}{\sqrt{x}}\right)^7$; third term $= \binom{7}{2}(\sqrt{x})^5\left(-\frac{3}{\sqrt{x}}\right)^2 = 189\sqrt{x^3}$

[29] $(5x + y)^{13}$; fifth term $= \binom{13}{4}(5x)^9(y)^4 = 1{,}396{,}484{,}375x^9y^4$

[30] $(6x + 2y)^8$; fifth term $= \binom{8}{4}(6x)^4(2y)^4 = 1{,}451{,}520x^4y^4$

[31] $(x^3 - 3y^2)^{11}$; Consider only the variable y in the kth term of the expansion $\binom{11}{k-1}(x^3)^{11-(k-1)}.$

$(-3y^2)^{k-1}$: $(y^2)^{k-1} = y^6 \Rightarrow 2k - 2 = 6 \Rightarrow k = 4$; 4th term $= \binom{11}{3}(x^3)^8(-3y^2)^3 = -4{,}455x^{24}y^6.$

[32] $(x^2 - 3y^3)^9$; Consider only the variable x in the kth term of the expansion $\binom{9}{k-1}(x^2)^{9-(k-1)}.$

$(-3y^3)^{k-1}$: $(x^2)^{9-(k-1)} = x^8 \Rightarrow 20 - 2k = 8 \Rightarrow k = 6$; 6th term $= \binom{9}{5}(x^2)^4(-3y^3)^5 =$

$-30{,}618x^8y^{15}.$

[33] $(2x^2 + 4y^3)^8$; Consider only the variable x in the kth term of the expansion $\binom{8}{k-1}(2x^2)^{8-(k-1)}$.

$(4y^3)^{k-1} : (x^2)^{8-(k-1)} = x^8 \Rightarrow 18 - 2k = 8 \Rightarrow k = 5$; 5th term $= \binom{8}{4}(2x^2)^4(4y^3)^4 =$

$286,720x^8y^{12}$.

[34] $(2x^3 - 3y^2)^{15}$; Consider only the variable y in the kth term of the expansion $\binom{15}{k-1}$.

$(2x^3)^{15-(k-1)}(-3y^2)^{k-1} : (y^2)^{k-1} = y^6 \Rightarrow 2k - 2 = 6 \Rightarrow k = 4$; 4th term $= \binom{15}{3}(2x^3)^{12} \cdot$

$(-3y^2)^3 = -50,319,360x^{36}y^6$.

[35] $(\sqrt{y} - \sqrt{x})^{10}$; Consider only the variable y in the kth term of the expansion $\binom{10}{k-1}(\sqrt{y})^{10-(k-1)}$.

$(-\sqrt{x})^{k-1} : (\sqrt{y})^{10-(k-1)} = y^3 \Rightarrow \frac{11-k}{2} = 3 \Rightarrow k = 5$; 5th term $= \binom{10}{4}(\sqrt{y})^6(-\sqrt{x})^4 = 210x^2y^3$.

[36] $(\sqrt{x} + \sqrt{y})^8$; Consider only the variable y in the kth term of the expansion $\binom{8}{k-1}(\sqrt{x})^{8-(k-1)}$.

$(\sqrt{y})^{k-1} : (\sqrt{y})^{k-1} = y^2 \Rightarrow \frac{k-1}{2} = 2 \Rightarrow k = 5$; 5th term $= \binom{8}{4}(\sqrt{x})^4(\sqrt{y})^4 = 70x^2y^2$.

[37] Let n and r be integers where $0 \le r \le n$.

a) Show that $\binom{n}{r} = \binom{n}{n-r}$: Since $\binom{n}{r} = \frac{n!}{r!(n-r)!}$ and $\binom{n}{n-r} = \frac{n!}{(n-r)![n-(n-r)]!} =$

$\frac{n!}{(n-r)!(r)!} = \frac{n!}{r!(n-r)!}$, the result follows.

b) Show that $\binom{n}{0} = 1$: We have $\binom{n}{0} = \frac{n!}{0!(n-0)!} = \frac{n!}{(1)(n!)} = 1$.

c) Show that $\binom{n}{n} = 1$: We have $\binom{n}{n} = \frac{n!}{n!(n-n)!} = \frac{1}{0!} = \frac{1}{1} = 1$.

d) Show that $\binom{n}{r} + \binom{n}{r-1} = \binom{n+1}{r}$: We have $\binom{n}{r} + \binom{n}{r-1} = \frac{n!}{r!(n-r)!} +$

$\frac{n!}{(r-1)![n-(r-1)]!} = \frac{n!}{r!(n-r)!} + \frac{n!}{(r-1)!(n-r+1)!} = \frac{n!(n-r+1)+n!(r)}{r!(n-r+1)!} =$

$\frac{n![(n-r+1)+r]}{r![(n+1)-r]!} = \frac{n!(n+1)}{r![(n+1)-r]!} = \frac{(n+1)!}{r![(n+1)-r]!} = \binom{n+1}{r}$.

[38] a) Since $\binom{1}{0}a^1 + \binom{1}{1}b^1 = 1 \cdot a + 1 \cdot b = a + b = (a+b)^1$, we see that $1 \in S$.

b) Since $k \in S$, $(a+b)^k = \binom{k}{0}a^k + \binom{k}{1}a^{k-1}b + \cdots + \binom{k}{k-1}ab^{k-1} + \binom{k}{k}b^k \Rightarrow$

$(a+b)^{k+1} = \left[\binom{k}{0}a^{k+1} + \binom{k}{1}a^k b + \cdots + \binom{k}{k-1}a^2b^{k-1} + \binom{k}{k}ab^k\right] +$

$\left[\binom{k}{0}a^k b + \binom{k}{1}a^{k-1}b^2 + \cdots + \binom{k}{k-1}ab^k + \binom{k}{k}b^{k+1}\right] = \binom{k}{0}a^{k+1} + \left[\binom{k}{1}+\binom{k}{0}\right]a^k b +$

$\left[\binom{k}{2}+\binom{k}{1}\right]a^{k-1}b^2 + \cdots + \left[\binom{k}{k-1}+\binom{k}{k-2}\right]a^2b^{k-1} + \left[\binom{k}{k}+\binom{k}{k-1}\right]ab^k + \binom{k}{k}b^{k+1} =$

$\binom{k+1}{0}a^{k+1} + \binom{k+1}{1}a^k b + \binom{k+1}{2}a^{k-1}b^2 + \cdots + \binom{k+1}{k-1}a^2b^{k-1} + \binom{k+1}{k}ab^k +$

$\binom{k+1}{k+1}b^{k+1}$. Therefore, $k + 1 \in S$. By the principle of mathematical induction, $S = \mathbf{N}$.

[39] $2^n = (1+1)^n = \binom{n}{0}(1)^n + \binom{n}{1}(1)^{n-1}(1) + \binom{n}{2}(1)^{n-2}(1)^2 + \cdots + \binom{n}{n}(1)^n = \binom{n}{0} + \binom{n}{1} +$

       $\binom{n}{2} + \cdots + \binom{n}{n}.$

[40] $\binom{n-1}{r-1} + \binom{n-1}{r} = \dfrac{(n-1)!}{(r-1)![(n-1)-(r-1)]!} + \dfrac{(n-1)!}{r![(n-1)-r]!} = \dfrac{(n-1)!}{(r-1)!(n-r)!} +$

       $\dfrac{(n-1)!}{r![(n-r)-1]!} = \dfrac{(n-1)![r+(n-r)]}{r!(n-r)!} = \dfrac{(n-1)!(n)}{r!(n-r)!} = \dfrac{n!}{r!(n-r)!} = \binom{n}{r}.$

[41] $(2+3i)^5 = (2)^5 + (5)(2)^4(3i) + (10)(2)^3(3i)^2 + (10)(2)^2(3i)^3 + (5)(2)(3i)^4 + (3i)^5 =$

       $122 - 597i$

[42] $(i-4i)^6 = (1)^6 + (6)(1)^5(-4i) + (15)(1)^4(-4i)^2 + (20)(1)^3(-4i)^3 + (15)(1)^2(-4i)^4 +$

       $(6)(1)(-4i)^5 + (-4i)^6 = -495 - 4{,}888i$

[43] $(2-3i)^4 = (2)^4 + (4)(2)^3(-3i) + (6)(2)^2(-3i)^2 + (4)(2)(-3i)^3 + (-3i)^4 = -119 + 120i$

[44] $(5+2i)^5 = (5)^5 + (5)(5)^4(2i) + (10)(5)^3(2i)^2 + (10)(5)^2(2i)^3 + (5)(5)(2i)^4 + (2i)^5 =$

       $-1{,}475 + 4{,}282i$

[45] $\left(\sqrt{2} + \sqrt{2}\,i\right)^8 = \left(\sqrt{2}\right)^8 + (8)\left(\sqrt{2}\right)^7\left(\sqrt{2}\,i\right) + (28)\left(\sqrt{2}\right)^6\left(\sqrt{2}\,i\right)^2 + (56)\left(\sqrt{2}\right)^5\left(\sqrt{2}\,i\right)^3 +$

       $(70)\cdot\left(\sqrt{2}\right)^4\left(\sqrt{2}\,i\right)^4 + (56)\left(\sqrt{2}\right)^3\left(\sqrt{2}\,i\right)^5 + (28)\left(\sqrt{2}\right)^2\left(\sqrt{2}\,i\right)^6 + (8)\left(\sqrt{2}\right)\left(\sqrt{2}\,i\right)^7 + \left(\sqrt{2}\,i\right)^8 = 256$

[46] $\left(\sqrt{2} - \sqrt{2}\,i\right)^8 = \left(\sqrt{2}\right)^8 + (8)\left(\sqrt{2}\right)^7\left(-\sqrt{2}\,i\right) + (28)\left(\sqrt{2}\right)^6\left(-\sqrt{2}\,i\right)^2 + (56)\left(\sqrt{2}\right)^5\left(-\sqrt{2}\,i\right)^3 +$

       $(70)\left(\sqrt{2}\right)^4\left(-\sqrt{2}\,i\right)^4 + (56)\left(\sqrt{2}\right)^3\left(-\sqrt{2}\,i\right)^5 + (28)\left(\sqrt{2}\right)^2\left(-\sqrt{2}\,i\right)^6 + (8)\left(\sqrt{2}\right)\left(-\sqrt{2}\,i\right)^7 +$

       $\left(-\sqrt{2}\,i\right)^8 = 256$

[47]   Since the eighth row of Pascal's triangle *is* 1, 7, 21, 35, 35, 21, 7, 1, it follows that $(2x-y)^7 =$

       $(1)(2x)^7 + (7)(2x)^6(-y) + (21)(2x)^5(-y)^2 + (35)(2x)^4(-y)^3 + (35)(2x)^3(-y)^4 +$

       $(21)(2x)^2(-y)^5 + (7)(2x)(-y)^6 + (1)(-y)^7 = 128x^7 - 448x^6y + 672x^5y^2 - 560x^4y^3 +$

       $280x^3y^4 - 84x^2y^5 + 14xy^6 - y^7$

[48]   Since the ninth row of Pascal's triangle is 1, 8, 28, 56, 70, 56, 28, 8, 1, it follows that $(x-2y)^8 =$

       $(1)(x)^8 + (8)(x)^7(-2y) + (28)(x)^6(-2y)^2 + (56)(x)^5(-2y)^3 + (70)(x)^4(-2y)^4 + (56)(x)^3 \cdot$

       $(-2y)^5 + (28)(x)^2(-2y)^6 + (8)(x)(-2y)^7 + (1)(-2y)^8 = x^8 - 16x^7y + 112x^6y^2 - 448x^5y^3 +$

       $1{,}120x^4y^4 - 1{,}792x^3y^5 + 1{,}792x^2y^6 - 1{,}024xy^7 + 256y^8$

EXERCISES 10.6

NOTE: In Exercises 1 – 34, the solutions given have the following format: In the first line, the given rational expression is shown first as it appeared in the exercise and then with its denominator factored into linear or irreducible (over **R**) quadratic factors. On the extreme right side of the equation in the first line is its partial fraction decomposition. In the second line, we give the equation obtained when both sides of the equation in the first line are multiplied by the least common denominator (L.C.D.). In the remainder of the solution, we determine the values of $A, B, C, \cdots$ etc.

[1] $\dfrac{4}{x^2 - 4} = \dfrac{4}{(x-2)(x+2)} = \dfrac{A}{x-2} + \dfrac{B}{x+2}$

    $4 = A(x+2) + B(x-2)$

    If $x = 2$, then $4 = 4A \Rightarrow A = 1$. *If* $x = -2$, then $4 = -4B \Rightarrow B = -1$.

    ● $\dfrac{1}{x-2} - \dfrac{1}{x+2}.$

[2] $\dfrac{3x-3}{x^2-9} = \dfrac{3x-3}{(x-3)(x+3)} = \dfrac{A}{x-3} + \dfrac{B}{x+3}$

$3x - 3 = A(x+3) + B(x-3)$

If $x = -3$, then $-12 = -6B \Rightarrow B = 2$. If $x = 3$, then $6 = 6A \Rightarrow A = 1$.

● $\dfrac{1}{x-3} + \dfrac{2}{x+3}$

[3] $\dfrac{-x-7}{x^2-x-6} = \dfrac{-x-7}{(x+2)(x-3)} = \dfrac{A}{x+2} + \dfrac{B}{x-3}$

$-x - 7 = A(x-3) + B(x+2)$

If $x = 3$, then $-10 = 5B \Rightarrow b = -2$. If $x = -2$, then $-5 = -5A \Rightarrow A = 1$.

● $\dfrac{1}{x+2} - \dfrac{2}{x-3}$

[4] $\dfrac{5x}{x^2+x-6} = \dfrac{5x}{(x+3)(x-2)} = \dfrac{A}{x+3} + \dfrac{B}{x-2}$

$5x = A(x-2) + B(x+3)$

If $x = 2$, then $10 = 5B \Rightarrow B = 2$. If $x = -3$, then $-15 = -5A \Rightarrow A = 3$.

● $\dfrac{3}{x+3} + \dfrac{2}{x-2}$

[5] $\dfrac{x+9}{x^2-3x-10} = \dfrac{x+9}{(x+2)(x-5)} = \dfrac{A}{x+2} + \dfrac{B}{x-5}$

$x + 9 = A(x-5) + B(x+2)$

If $x = 5$, then $14 = 7B \Rightarrow B = 2$. If $x = -2$, then $7 = -7A \Rightarrow A = -1$.

● $\dfrac{-1}{x+2} + \dfrac{2}{x-5}$

[6] $\dfrac{3x-13}{x^2-2x-3} = \dfrac{3x-13}{(x+1)(x-3)} = \dfrac{A}{x+1} + \dfrac{B}{x-3}$

$3x - 13 = A(x-3) + B(x+1)$

If $x = 3$, then $-4 = 4B \Rightarrow B = -1$. If $x = -1$, then $-16 = -4A \Rightarrow A = 4$.

● $\dfrac{4}{x+1} - \dfrac{1}{x-3}$

[7] $\dfrac{x^2+7x+1}{(x-1)(x^2+x-2)} = \dfrac{x^2+7x+1}{(x-1)^2(x+2)} = \dfrac{A}{x-1} + \dfrac{B}{(x-1)^2} + \dfrac{C}{x+2}$

$x^2 + 7x + 1 = A(x-1)(x+2) + B(x+2) + C(x-1)^2 \qquad (E_1)$

$ = (A+C)x^2 + (A+B-2C)x + (-2A+2B+C)$

In E_1, *if* $x = 1$, then $9 = 3B \Rightarrow B = 3$. If $x = -2$, then $-9 = 9C \Rightarrow C = -1$. Equating the coefficients

of x^2, we obtain $1 = A + C \Rightarrow 1 = A + (-1) \Rightarrow A = 2$.

● $\dfrac{2}{x-1} + \dfrac{3}{(x-1)^2} - \dfrac{1}{x+2}$

[8] $\dfrac{2x^2+14x+15}{(x+2)(x^2-x-6)} = \dfrac{2x^2+14x+15}{(x+2)^2(x-3)} = \dfrac{A}{x+2} + \dfrac{B}{(x+2)^2} + \dfrac{C}{x-3}$

$2x^2 + 14x + 15 = A(x+2)(x-3) + B(x-3) + C(x+2)^2 \qquad (E_1)$

$ = (A+C)x^2 + (-A+B+4C)x + (-6A-3B+4C)$

In E_1, if $x = 3$, then $75 = 25C \Rightarrow C = 3$. If $x = -2$, then $-5 = -5B \Rightarrow B = 1$. Equating the

coefficients of x^2, we obtain $2 = A + C \Rightarrow 2 = A + (3) \Rightarrow A = -1$.

● $\dfrac{-1}{x+2} + \dfrac{1}{(x+2)^2} + \dfrac{3}{x-3}$

[9] $\dfrac{7x^2-17x+3}{(x-2)(x^2-x-2)} = \dfrac{7x^2-17x+3}{(x-2)^2(x+1)} = \dfrac{A}{x-2} + \dfrac{B}{(x-2)^2} + \dfrac{C}{x+1}$

$7x^2-17x+3 = A(x-2)(x+1) + B(x+1) + C(x-2)^2$ $\qquad (E_1)$
$\qquad\qquad = (A+C)x^2 + (-A+B-4C)x + (-2A+B+4C)$

In E_1, if $x=2$, then $-3 = 3B \Rightarrow B = -1$. If $x=-1$, then $27 = 9C \Rightarrow C = 3$. Equating the

coefficients of x^2, we obtain $7 = A + C \Rightarrow 7 = A + (3) \Rightarrow A = 4$. $\qquad$ ● $\dfrac{4}{x-2} - \dfrac{1}{(x-2)^2} + \dfrac{3}{x+1}$

[10] $\dfrac{4x^2+42x+98}{(x+5)(x^2+4x-5)} = \dfrac{4x^2+42x+98}{(x+5)^2(x-1)} = \dfrac{A}{x+5} + \dfrac{B}{(x+5)^2} + \dfrac{C}{x-1}$

$42x^2+42x+98 = A(x+5)(x-1) + B(x-1) + C(x+5)^2$ $\qquad (E_1)$
$\qquad\qquad = (A+C)x^2 + (4A+B+10C)x + (-5A-B+25C)$

In E_1, if $x=1$, then $144 = 36C \Rightarrow C = 4$. If $x=-5$, then $-12 = -6B \Rightarrow B = 2$. Equating the

coefficients of x^2, we obtain $4 = A + C \Rightarrow 4 = A + (4) \Rightarrow A = 0$. $\qquad$ ● $\dfrac{2}{(x+5)^2} + \dfrac{4}{x-1}$

[11] $\dfrac{9x^2-43x+30}{x^3-10x^2} = \dfrac{9x^2-43x+30}{x^2(x-10)} = \dfrac{A}{x} + \dfrac{B}{x^2} + \dfrac{C}{x-10}$

$9x^2-43x+30 = A(x)(x-10) + B(x-10) + C(x^2)$ $\qquad (E_1)$
$\qquad\qquad = (A+C)x^2 + (-10A+B)x + (-10B)$

In E_1, if $x=0$, then $30 = -10B \Rightarrow B = -3$. If $x=10$, then $500 = 100C \Rightarrow C = 5$. Equating the

coefficients of x^2, we obtain $9 = A + C \Rightarrow 9 = A + (5) \Rightarrow A = 4$. $\qquad$ ● $\dfrac{4}{x} - \dfrac{3}{x^2} + \dfrac{5}{x-10}$

[12] $\dfrac{x^2+x-3}{x^4-x^3} = \dfrac{x^2+x-3}{x^3(x-1)} = \dfrac{A}{x} + \dfrac{B}{x^2} + \dfrac{C}{x^3} + \dfrac{D}{x-1}$

$x^2+x-3 = A(x^2)(x-1) + B(x)(x-1) + C(x-1) + D(x^3)$ $\qquad (E_1)$
$\qquad\qquad = (A+D)x^3 + (-A+B)x^2 + (-B+C)x + (-C)$

In E_1, if $x=0$, then $-3 = -C \Rightarrow C = 3$. If $x=1$, then $-1 = D \Rightarrow D = -1$. Equating the coefficients

of x^3, we obtain $0 = A + D \Rightarrow 0 = A + (-1) \Rightarrow A = 1$. Equating the coefficients of x^2, we obtain

$1 = -A + B \Rightarrow 1 = -(1) + B \Rightarrow B = 2$. $\qquad$ ● $\dfrac{1}{x} + \dfrac{2}{x^2} + \dfrac{3}{x^3} - \dfrac{1}{x-1}$

[13] $\dfrac{5x^2-6x+15}{x^3+3x} = \dfrac{5x^2-6x+15}{x(x^2+3)} = \dfrac{A}{x} + \dfrac{Bx+C}{x^2+3}$

$5x^2-6x+15 = A(x^2+3) + (Bx+C)(x)$ $\qquad (E_1)$
$\qquad\qquad = (A+B)x^2 + (C)x + (3A)$

In E_1, if $x=0$, then $15 = 3A \Rightarrow A = 5$. Equating the coefficients of x^2, we obtain $5 = A + B \Rightarrow$

$5 = (5) + B \Rightarrow B = 0$. Equating the coefficients of x, we obtain $-6 = C \Rightarrow C = -6$. $\qquad$ ● $\dfrac{5}{x} - \dfrac{6}{x^2+3}$

[14] $\dfrac{-2x^2+5x-2}{x^3+x}=\dfrac{-2x^2+5x-2}{x(x^2+1)}=\dfrac{A}{x}+\dfrac{Bx+C}{x^2+1}$

$-2x^2+5x-2=A\left(x^2+1\right)+(Bx+C)(x)$ (E_1)

$=(A+B)x^2+(C)x+(A)$

In E_1, if $x=0$, then $-2=A\Rightarrow A=-2$. Equating the coefficients of x^2, we obtain $-2=A+B\Rightarrow$

$-2=(-2)+B\Rightarrow B=0$. Equating the coefficients of x, we obtain $5=C\Rightarrow C=5$. $\bullet\;\dfrac{-2}{x}+\dfrac{5}{x^2+1}$

[15] $\dfrac{-2x^2-5}{x^4+x^2}=\dfrac{-2x^2-5}{x^2(x^2+1)}=\dfrac{A}{x}+\dfrac{B}{x^2}+\dfrac{Cx+D}{x^2+1}$

$-2x^2-5=A(x)\left(x^2+1\right)+B\left(x^2+1\right)+(Cx+D)\left(x^2\right)$ (E_1)

$=(A+C)x^3+(B+D)x^2+(A)x+(B)$

In E_1, if $x=0$, then $-5=B\Rightarrow B=-5$. Equating the coefficients of like powers of x, we have

$x^3:\; 0=A+C\Rightarrow 0=(0)+C\Rightarrow C=0$ $x^2:\; -2=B+D\Rightarrow -2=(-5)+D\Rightarrow D=3$

$x:\; 0=A\Rightarrow A=0$ $x^0:\; -5=B$

$\bullet\;\dfrac{-5}{x^2}+\dfrac{3}{x^2+1}$

[16] $\dfrac{-x^2+2x+6}{x^4+2x^2}=\dfrac{-x^2+2x+6}{x^2(x^2+2)}=\dfrac{A}{x}+\dfrac{B}{x^2}+\dfrac{Cx+D}{x^2+2}$

$-x^2+2x+6=A(x)\left(x^2+2\right)+B\left(x^2+2\right)+(Cx+D)\left(x^2\right)$

$=(A+C)x^3+(B+D)x^2+(2A)x+(2B)$

Equating the coefficients of like powers of x, we have

$x^3:\; 0=A+C\Rightarrow 0=(1)+C\Rightarrow C=-1$ $x^2:\; -1=B+D\Rightarrow -1=(3)+D\Rightarrow D=-4$

$x:\; 2=2A\Rightarrow A=1$ $x^0:\; 6=2B\Rightarrow B=3$

$\bullet\;\dfrac{1}{x}+\dfrac{3}{x^2}-\dfrac{x+4}{x^2+2}$

[17] $\dfrac{-5x^2+4}{x^4+2x^2}=\dfrac{-5x^2+4}{x^2(x^2+2)}=\dfrac{A}{x}+\dfrac{B}{x^2}+\dfrac{Cx+D}{x^2+2}$

$-5x^2+4=A(x)\left(x^2+2\right)+B\left(x^2+2\right)+(Cx+D)\left(x^2\right)$

$=(A+C)x^3+(B+D)x^2+(2A)x+(2B)$

Equating the coefficients of like powers of x, we obtain

$x^3:\; 0=A+C\Rightarrow 0=(0)+C\Rightarrow C=0$ $x^2:\; -5=B+D\Rightarrow -5=(2)+D\Rightarrow D=-7$

$x:\; 0=2A\Rightarrow A=0$ $x^0:\; 4=2B\Rightarrow B=2$

$\bullet\;\dfrac{2}{x^2}-\dfrac{7}{x^2+2}$

534

[18] $\dfrac{3x^3 - 8x^2 + 5x - 6}{(x-1)^2(x^2+1)} = \dfrac{A}{x-1} + \dfrac{B}{(x-1)^2} + \dfrac{Cx+D}{x^2+1}$

$3x^3 - 8x^2 + 5x - 6 = A(x-1)(x^2+1) + B(x^2+1) + (Cx+D)(x-1)^2$ $\qquad (E_1)$

$= (A+C)x^3 + (-A+B-2C+D)x^2 + (A+C-2D)x + (-A+B+D)$

In E_1, if $x = 1$, then $-6 = 2B \Rightarrow B = -3$. Equating the coefficients of like powers of x, we have

x^3: $3 = A + C$ $\qquad (E_2)$

x^2: $-8 = -A + B - 2C + D$ $\qquad (E_3)$

x^1: $5 = A + C - 2D$ $\qquad (E_4)$

x^0: $-6 = -A + B + D$ $\qquad (E_5)$

Using E_2, we can substitute 3 for $A+C$ in E_4: $5 = 3 - 2D \Rightarrow D = -1$. Substituting -1 for D and

-3 for B in E_5, we have $-6 = -A - 3 - 1 \Rightarrow A = 2$. Substituting 2 for A in E_2, we get $3 = 2 + C \Rightarrow$

$C = 1$. $\qquad\qquad\qquad\qquad\qquad$ ● $\dfrac{2}{x-1} - \dfrac{3}{(x-1)^2} + \dfrac{x-1}{x^2+1}$

$\dfrac{x^2 + 5x - 5}{(x^2+6)^2} = \dfrac{Ax+B}{x^2+6} + \dfrac{Cx+D}{(x^2+6)^2}$

[19] $\quad x^2 + 5x - 5 = (Ax+B)(x^2+6) + Cx + D$

$= (A)x^3 + (B)x^2 + (6A+C)x + (6B+D)$

Equating the coefficients of like powers of x, we have

$x^3: 0 = A \Rightarrow A = 0$

$x^2: 1 = B \Rightarrow B = 1$

$x: 5 = 6A + C \Rightarrow 5 = 6(0) + C \Rightarrow C = 5$ $\qquad$ ● $\dfrac{1}{x^2+6} + \dfrac{5x-11}{(x^2+6)^2}$

$x^0: -5 = 6B + D \Rightarrow -5 = 6(1) + D \Rightarrow D = -11$

[20] $\quad \dfrac{2x^3 + 3x^2 + 9x + 11}{(x^2+4)^2} = \dfrac{Ax+B}{x^2+4} + \dfrac{Cx+D}{(x^2+4)^2}$

$2x^3 + 3x^2 + 9x + 11 = (Ax+B)(x^2+4) + Cx + D$

$= (A)x^3 + (B)x^2 + (4A+C)x + (4B+D)$

Equating the coefficients of like powers of x, we have

x^3: $2 = A \Rightarrow A = 2$ $\qquad\qquad\qquad$ x^2: $3 = B \Rightarrow B = 3$

x: $9 = 4A + C \Rightarrow 9 = 4(2) + C \Rightarrow C = 1$ $\qquad$ x^0: $11 = 4B + D \Rightarrow 11 = 4(3) + D \Rightarrow D = -1$

● $\dfrac{2x+3}{x^2+4} + \dfrac{x-1}{(x^2+4)^2}$

[21] $\dfrac{2x^3 - x^2 + 2x - 1}{(2x^2 + x + 1)^2} = \dfrac{Ax + B}{2x^2 + x + 1} + \dfrac{Cx + D}{(2x^2 + x + 1)^2}$

$2x^3 - x^2 + 2x - 1 = (Ax + B)(2x^2 + x + 1) + (Cx + D)$
$= (2A)x^3 + (A + 2B)x^2 + (A + B + C)x + (B + D)$

Equating the coefficients of like powers of x, we have

x^3: $2 = 2A \Rightarrow A = 1$ $\qquad\qquad$ x^2: $-1 = A + 2B \Rightarrow -1 = (1) + 2B \Rightarrow B = -1$

x: $2 = A + B + C \Rightarrow 2 = (1) + (-1) + C \Rightarrow C = 2$ $\qquad$ x^0: $-1 = B + D \Rightarrow -1 = (-1) + D \Rightarrow D = 0$

$\bullet\ \dfrac{x - 1}{2x^2 + x + 1} + \dfrac{2x}{(2x^2 + x + 1)^2}$

[22] $\dfrac{x^3 + 3x^2 + 10x + 1}{(x^2 + 3x + 10)^2} = \dfrac{Ax + B}{x^2 + 3x + 10} + \dfrac{Cx + D}{(x^2 + 3x + 10)^2}$

$x^3 + 3x^2 + 10x + 1 = (Ax + B)(x^2 + 3x + 10) + Cx + D$
$= (A)x^3 + (3A + B)x^2 + (10A + 3B + C)x + (10B + D)$

Equating the coefficients of like powers of x, we have

x^3: $1 = A \Rightarrow A = 1$ $\qquad\qquad$ x^2: $3 = 3A + B \Rightarrow 3 = 3(1) + B \Rightarrow B = 0$

x: $10 = 10A + 3B + C \Rightarrow 10 = 10(1) + 3(0) + C \Rightarrow C = 0$

x^0: $1 = 10B + D \Rightarrow 1 = 10(0) + D \Rightarrow D = 1$ $\qquad$ $\bullet\ \dfrac{x}{x^2 + 3x + 10} + \dfrac{1}{(x^2 + 3x + 10)^2}$

[23] $\dfrac{3x^3 - x^2 + 13x - 2}{x^4 + 9x^2 + 20} = \dfrac{3x^3 - x^2 + 13x - 2}{(x^2 + 4)(x^2 + 5)} = \dfrac{Ax + B}{x^2 + 4} + \dfrac{Cx + D}{x^2 + 5}$

$3x^3 - x^2 + 13x - 2 = (Ax + B)(x^2 + 5) + (Cx + D)(x^2 + 4)$
$= (A + C)x^3 + (B + D)x^2 + (5A + 4C)x + (5B + 4D)$

Equating the coefficients of like powers of x, we have

x^3: $3 = A + C$ $\qquad\qquad\qquad\qquad$ (E_1)

x^2: $-1 = B + D$ $\qquad\qquad\qquad\qquad$ (E_2)

x: $13 = 5A + 4C$ $\qquad\qquad\qquad\qquad$ (E_3)

x^0: $-2 = 5B + 4D$ $\qquad\qquad\qquad\qquad$ (E_4)

Using E_1 and E_3, we get $\begin{cases} A + C = 3 & (E_1) \\ 5A + 4C = 13 & (E_3) \end{cases}$

$E_3 + (-4)E_1 \Rightarrow A = 1 \Rightarrow (1) + C = 3 \Rightarrow C = 2$.

Similarly, using E_2 and E_4, we get $\begin{cases} B + D = -1 & (E_2) \\ 5B + 4D = -2 & (E_4) \end{cases}$

$E_4 + (-4)E_2 \Rightarrow B = 2 \Rightarrow (2) + D = -1 \Rightarrow D = -3$.

$\bullet\ \dfrac{x + 2}{x^2 + 4} + \dfrac{2x - 3}{x^2 + 5}$

[24] $\dfrac{2x^3+6x+4}{x^4+6x^2+8} = \dfrac{2x^3+6x+4}{(x^2+4)(x^2+2)} = \dfrac{Ax+B}{x^2+4} + \dfrac{Cx+D}{x^2+2}$

$2x^3+6x+4 = (Ax+B)(x^2+2)+(Cx+D)(x^2+4)$

$\qquad = (A+C)x^3+(B+D)x^2+(2A+4C)x+(2B+4D)$

Equating the coefficients of like powers of x, we have

x^3: $2 = A+C$ $\hspace{4cm}$ (E_1)

x^2: $0 = B+D$ $\hspace{4cm}$ (E_2)

x: $6 = 2A+4C$ $\hspace{3.7cm}$ (E_3)

x^0: $4 = 2B+4D$ $\hspace{3.5cm}$ (E_4)

From E_1 and E_3: $\begin{cases} A+\;\;C = 2 \;(E_1) \\ 2A+4C = 6 \;(E_3) \end{cases}$

$E_3+(-2)E_1 \Rightarrow 2C = 2 \Rightarrow C = 1 \Rightarrow A+(1) = 2 \Rightarrow A = 1.$

From E_2 and E_4: $\begin{cases} B\;\;+\;\;D \;=\; 0 \;(E_2) \\ 2B\;+4D \;=\; 4 \;(E_4) \end{cases}$

$E_4+(-2)E_2 \Rightarrow 2D = 4 \Rightarrow D = 2 \Rightarrow B+(2) = 0 \Rightarrow B = -2.$ $\hspace{1cm}$ ● $\dfrac{x-2}{x^2+4} + \dfrac{x+2}{x^2+2}$

[25] By first dividing and then factoring, we have

$(3x+4)+\dfrac{-3x^2+6x+16}{x(x^2+8)} = (3x+4)+\dfrac{A}{x}+\dfrac{Bx+C}{x^2+8}$

$-3x^2+6x+16 = A(x^2+8)+(Bx+C)(x) = (A+B)x^2+(C)x+(8A)$

Equating the coefficients of like powers of x, we have

x^2: $-3 = A+B \Rightarrow -3 = (2)+B \Rightarrow B = -5$

x: $6 = C \Rightarrow C = 6$ $\hspace{2.5cm}$ x^0: $16 = 8A \Rightarrow A = 2$ $\hspace{1cm}$ ● $(3x+4)+\dfrac{2}{x}+\dfrac{-5x+6}{x^2+8}$

[26] By first dividing and then factoring, we have

$(2x-1)+\dfrac{-x^2-x+18}{x(x^2+9)} = (2x-1)+\dfrac{A}{x}+\dfrac{Bx+C}{x^2+9}$

$-x^2-x+18 = A(x^2+9)+(Bx+C)(x)-x^2-x+18 = A(x^2+9)+(Bx+C)(x)$

Equating the coefficients of like powers of x, we have

x^2: $-1 = A+B \Rightarrow -1 = (2)+B \Rightarrow B = -3$

x: $-1 = C \Rightarrow C = -1$ $\hspace{2cm}$ x^0: $18 = 9A \Rightarrow A = 2$ $\hspace{1cm}$ ● $(2x-1)+\dfrac{2}{x}-\dfrac{3x+1}{x^2+9}$

[27] By first dividing and then factoring, we have

$(2x+1)+\dfrac{-x-7}{(x-3)(x+2)} = (2x+1)+\dfrac{A}{x-3}+\dfrac{B}{x+2}$ $\hspace{1.5cm}$ $-x-7 = A(x+2)+B(x-3)$

If $x = -2$, then $-5 = -5B \Rightarrow B = 1.$ If $x = 3$, then $-10 = 5A \Rightarrow A = -2.$ $\hspace{0.5cm}$ ● $(2x+1)-\dfrac{2}{x-3}+\dfrac{1}{x+2}$

[28] By first dividing and then factoring, we have

$$(3x-1)+\frac{5x}{(x+3)(x-2)}=(3x-1)+\frac{A}{x+3}+\frac{B}{x-2}$$

$$5x=A(x-2)+B(x+3) \qquad\qquad \text{If } x=2, \text{ then } 10=5B \Rightarrow B=2.$$

If $x=-3$, then $-15=-5A \Rightarrow A=3$.
$$\bullet\ (3x-1)+\frac{3}{x+3}+\frac{2}{x-2}$$

[29] $\dfrac{x^2+2x+3}{(x+2)^3}=\dfrac{A}{x+2}+\dfrac{B}{(x+2)^2}+\dfrac{C}{(x+2)^3}$

$$x^2+2x+3=A(x+2)^2+B(x+2)+C$$

$$=(A)x^2+(4A+B)x+(4A+2B+C)$$

Equating the coefficients of like powers of x, we have

x^2: $1=A \Rightarrow A=1$ $\qquad\qquad$ x: $2=4A+B \Rightarrow 2=4(1)+B \Rightarrow B=-2$

x^0: $3=4A+2B+C \Rightarrow 3=4(1)+2(-2)+C \Rightarrow C=3$ $\qquad \bullet\ \dfrac{1}{x+2}-\dfrac{2}{(x+2)^2}+\dfrac{3}{(x+2)^3}$

[30] $\dfrac{4x+11}{(x+3)^3}=\dfrac{A}{x+3}+\dfrac{B}{(x+3)^2}+\dfrac{C}{(x+3)^3}$

$$4x+11=A(x+3)^2+B(x+3)+C$$

$$=(A)x^2+(6A+B)x+(9A+3B+C)$$

Equating the coefficients of like powers of x, we have

x^2: $0=A \Rightarrow A=0$ $\qquad\qquad$ x: $4=6A+B \Rightarrow 4=6(0)+B \Rightarrow B=4$

x^0: $11=9A+3B+C \Rightarrow 11=9(0)+3(4)+C \Rightarrow C=-1$ $\qquad \bullet\ \dfrac{4}{(x+3)^2}-\dfrac{1}{(x+3)^3}$

[31] Using factoring by grouping (or synthetic division) to factor the denominator, we have

$$\frac{x^2-2x+7}{x^3-x^2+5x-5}=\frac{x^2-2x+7}{(x-1)(x^2+5)}=\frac{A}{x-1}+\frac{Bx+C}{x^2+5}$$

$$x^2-2x+7=A(x^2+5)+(Bx+C)(x-1) \qquad\qquad (E_1)$$
$$=(A+B)x^2+(-B+C)x+(5A-C)$$

In E_1, if $x=1$, then $6=6A \Rightarrow A=1$. Equating the coefficients of like powers of x and substituting 1 for A, we have

x^2: $1=A+B \Rightarrow 1=(1)+B \Rightarrow B=0$

x: $-2=-B+C \Rightarrow -2=-(0)+C \Rightarrow C=-2$ $\qquad\qquad \bullet\ \dfrac{1}{x-1}-\dfrac{2}{x^2+5}$

[32] Using the factoring by grouping (or synthetic division) to factor the denominator, we have

$$\frac{2x^2-5x+24}{x^3-2x^2+7x-14}=\frac{2x^2-5x+24}{(x-2)(x^2+7)}=\frac{A}{x-2}+\frac{Bx+C}{x^2+7}$$

$$2x^2-5x+24=A(x^2+7)+(Bx+C)(x-2) \qquad\qquad (E_1)$$
$$=(A+B)x^2+(-2B+C)x+(7A-2C)$$

In E_1, if $x=2$, then $22=11A \Rightarrow A=2$. Equating the coefficients of like powers of x and

substituting 2 for A, we have $\qquad\qquad x^2$: $2=A+B \Rightarrow 2=(2)+B \Rightarrow B=0$

x: $-5=-2B+C \Rightarrow -5=-2(0)+C \Rightarrow C=-5$ $\qquad \bullet\ \dfrac{2}{x-2}-\dfrac{5}{x^2+7}$

[33] $\dfrac{5x^2+3x+7}{x^3+3x^2+6x+4} = \dfrac{5x^2+3x+7}{(x+1)(x^2+2x+4)} = \dfrac{A}{x+1} + \dfrac{Bx+C}{x^2+2x+4}$

$5x^2 + 3x + 7 = A\left(x^2 + 2x + 4\right) + (Bx + C)(x + 1)$ (E_1)

$\qquad\qquad\quad = (A + B)x^2 + (2A + B + C)x + (4A + C)$

In E_1, if $x = -1$, then $9 = 3A \Rightarrow A = 3$. Equating the coefficients of like powers of x and substituting

3 for A, we have $\qquad\qquad\qquad\qquad$ x^2: $5 = A + B \Rightarrow 5 = (3) + B \Rightarrow B = 2$

x: $3 = 2A + B + C \Rightarrow 3 = 2(3) + (2) + C \Rightarrow C = -5$ $\qquad$ $\bullet\ \dfrac{3}{x+1} + \dfrac{2x-5}{x^2+2x+4}$

[34]   Using synthetic division to factor the denominator, we have

$\dfrac{9x^2-7x-32}{2x^3-9x^2+7x+6} = \dfrac{9x^2-7x-32}{(x-2)(x-3)(2x+1)} = \dfrac{A}{x-2} + \dfrac{B}{x-3} + \dfrac{C}{2x+1}$

$9x^2 - 7x - 32 = A(x-3)(2x+1) + B(x-2)(2x+1) + C(x-2)(x-3)$

If $x = 2$, then $-10 = -5A \Rightarrow A = 2$. If $x = 3$, then $28 = 7B \Rightarrow B = 4$.

If $x = -\frac{1}{2}$, then $-\frac{105}{4} = \left(\frac{35}{4}\right)C \Rightarrow C = -3$. $\qquad$ $\bullet\ \dfrac{2}{x-2} + \dfrac{4}{x-3} - \dfrac{3}{2x+1}$

CHAPTER 10: REVIEW EXERCISES

NOTE : In Exercises 1 – 6, the answers are given in the order a_1, a_2, a_3, a_4 and a_8.

[1] $\left\{\dfrac{n+1}{2n-1}\right\}$; $2, 1, \dfrac{4}{5}, \dfrac{5}{7}$ and $\dfrac{3}{5}$ $\qquad\qquad$ [2] $\left\{\dfrac{(-1)^{n-1}}{2^n}\right\}$; $\dfrac{1}{2}, -\dfrac{1}{4}, \dfrac{1}{8}, -\dfrac{1}{16}$ and $-\dfrac{1}{256}$

[3] $\left\{\dfrac{\cos \frac{n\pi}{2}}{n^2}\right\}$; $0, -\dfrac{1}{4}, 0, \dfrac{1}{16}$ and $\dfrac{1}{64}$ $\qquad$ [4] $\left\{\left(-\dfrac{3}{2}\right)^n\left(1-\dfrac{1}{n}\right)\right\}$; $0, \dfrac{9}{8}, -\dfrac{9}{4}, \dfrac{243}{64}$ and $\dfrac{45{,}927}{2048}$

[5] $\left\{\dfrac{(-1)^n x^n}{n!}\right\}$; $-x, \dfrac{x^2}{2}, -\dfrac{x^3}{6}, \dfrac{x^4}{24}$ and $\dfrac{x^8}{40{,}320}$ $\qquad$ [6] $\left\{\dfrac{2n!\ln n}{(n-1)!}\right\}$; $0, 4\ln 2, 6\ln 3, 6\ln 4$ and $16\ln 8$

[7] $a_1 = 3, a_n = \dfrac{a_{n-1}}{2} + 1$; $a_2 = \dfrac{5}{2}, a_3 = \dfrac{9}{4}, a_4 = \dfrac{17}{8}$

[8] $a_1 = \dfrac{5}{2}, a_n = \dfrac{(-1)^n a_{n-1}}{n+1}$; $a_2 = \dfrac{5}{6}, a_3 = -\dfrac{5}{24}, a_4 = -\dfrac{1}{24}$

[9] $a_1 = x, a_n = \dfrac{(-1)^n (a_{n-1})^2}{n^2}$; $a_2 = \dfrac{x^2}{4}, a_3 = -\dfrac{x^4}{144}, a_4 = \dfrac{x^8}{331{,}776}$

[10] $a_1 = 1, a_2 = 3, a_n = 5a_{n-2} - \dfrac{a_{n-1}}{2}$; $a_2 = 3, a_3 = \dfrac{7}{2}, a_4 = \dfrac{53}{4}$

[11] $\displaystyle\sum_{k=1}^{9}\left(-\dfrac{1}{2}\right)^{k-1}$ $\qquad\qquad\qquad\qquad$ [12] $\displaystyle\sum_{k=1}^{10}\left(1-\dfrac{1}{2k+1}\right)^{k+1}$

[13] $\displaystyle\sum_{k=1}^{99}\dfrac{\ln(k+1)}{k(k+1)}$ $\qquad\qquad\qquad\qquad$ [14] $\displaystyle\sum_{k=1}^{n}\dfrac{(-1)^{k-1}(x-2)^k}{k!}$

[15] $\displaystyle\sum_{i=1}^{5}\left(\dfrac{2}{3}\right)^{i-1} = \left(\dfrac{2}{3}\right)^0 + \left(\dfrac{2}{3}\right)^1 + \left(\dfrac{2}{3}\right)^2 + \left(\dfrac{2}{3}\right)^3 + \left(\dfrac{2}{3}\right)^4 = \dfrac{211}{81}$

[16] $\displaystyle\sum_{k=1}^{10}\dfrac{(2k-1)^2}{k} = 1 + \dfrac{9}{2} + \dfrac{25}{3} + \dfrac{49}{4} + \dfrac{81}{5} + \dfrac{121}{6} + \dfrac{169}{7} + \dfrac{225}{8} + \dfrac{289}{9} + \dfrac{361}{10} = \dfrac{460{,}981}{2520}$

[17] $\displaystyle\sum_{k=0}^{8}\dfrac{\sin k\pi}{k!} = \dfrac{\sin 0}{0!} + \dfrac{\sin \pi}{1!} + \dfrac{\sin 2\pi}{2!} + \cdots + \dfrac{\sin 8\pi}{8!} = 0$

[18] $\sum_{i=1}^{8} (3i - 4) = -1 + 2 + 5 + 8 + 11 + 14 + 17 + 20 = 76$

[19] $\sum_{k=1}^{10} \left[\frac{1}{(k+1)^2} - \frac{1}{k^2} \right] = \left(\frac{1}{4} - 1 \right) + \left(\frac{1}{9} - \frac{1}{4} \right) + \left(\frac{1}{16} - \frac{1}{9} \right) + \cdots + \left(\frac{1}{81} - \frac{1}{64} \right) + \left(\frac{1}{100} - \frac{1}{81} \right) +$

$\left(\frac{1}{121} - \frac{1}{100} \right) = \frac{1}{121} - 1 = -\frac{120}{121}$

[20] $(30,000)(1.08)^6 \approx \$ 47,606.23$

[21] $5, 11, 17, \ldots; a_n = a_1 + (n-1)d = 5 + (n-1)6 = 6n - 1$

[22] $-\frac{4}{5}, -\frac{7}{5}, -2, \ldots; a_n = -\frac{4}{5} + (n-1)\left(-\frac{3}{5} \right) = -\frac{3}{5}n - \frac{1}{5}$

[23] $8, 5.3, 2.6, \ldots; a_n = 8 + (n-1)(-2.7) = -2.7n + 10.7$

[24] $\frac{7}{3}, \frac{11}{4}, \frac{19}{6}, \ldots; a_n = \frac{7}{3} + (n-1)\frac{5}{12} = \frac{5}{12}n + \frac{23}{12}$

[25] $\frac{3\pi}{2}, \frac{5\pi}{3}, \frac{11\pi}{6}, \ldots; a_n = \frac{3\pi}{2} + (n-1)\frac{\pi}{6} = \frac{\pi}{6}n + \frac{4\pi}{3}$

[26] $\ln 4, \ln 16, \ln 64, \ldots; a_n = \ln 4 + (n-1)\ln 4 = n \ln 4$

[27] $-\frac{3}{2}, 2, \frac{11}{2}, \ldots; a_{20} = -\frac{3}{2} + 19\left(\frac{7}{2} \right) = 65$

[28] $3.2, 0.8, -1.6, \ldots; a_{10} = 3.2 + 9(-2.4) = -18.4$

[29] $\log 2, \log 4, \log 8, \ldots; a_{12} = \log 2 + 11 \log 2 = 12 \log 2$

[30] $x + 1, x + 3, x + 5, \ldots; a_{50} = (x+1) + (49)(2) = x + 99$

[31] $30, 24, 18, \ldots; S_{10} = 10a_1 + \frac{(10)(9)}{2}d = (10)(30) + (45)(-6) = 30$

[32] $-8.1, -6.3, -4.5, \ldots; S_{10} = (10)(-8.1) + (45)(1.8) = 0$

[33] $\frac{5}{2}, \frac{13}{4}, 4, \ldots; S_{10} = (10)\left(\frac{5}{2} \right) + (45)\left(\frac{3}{4} \right) = \frac{235}{4}$

[34] $\ln 2, \ln 4, \ln 8, \ldots; S_{10} = (10)(\ln 2) + (45)(\ln 2) = 55 \ln 2$

[35] $a_1 = -5$ and $a_5 = 13$. $13 = -5 + 4d \Rightarrow d = \frac{9}{2}$. Thus, $m_1 = -\frac{1}{2}, m_2 = 4$ and $m_3 = \frac{17}{2}$.

[36] $a_1 = \frac{3}{4}$ and $a_6 = \frac{17}{2}$. $\frac{17}{2} = \frac{3}{4} + 5d \Rightarrow d = \frac{31}{20}$. Thus, $m_1 = \frac{23}{10}, m_2 = \frac{77}{20}, m_3 = \frac{27}{5}$ and $m_4 = \frac{139}{20}$.

[37] $S_n = \frac{n}{2}(a_1 + a_n) \Rightarrow 55 = \frac{5}{2}(a_1 + 15) \Rightarrow a_1 = 7. a_5 = a_1 + 4d \Rightarrow 15 = 7 + 4d \Rightarrow d = 2$. Thus,

$a_n = a_1 + (n-1)d = 7 + (n-1)2 = 2n + 5$.

[38] $24 = 8 + (n-1)\frac{1}{4} \Rightarrow 96 = 32 + n - 1 \Rightarrow n = 65$. On the 65th day the student will be sleeping 24

hours per day.

[39] $\frac{3}{5}, \frac{9}{25}, \frac{27}{125}, \ldots; a_n = a_1 r^{n-1} \Rightarrow a_n = \left(\frac{3}{5} \right) \left(\frac{3}{5} \right)^{n-1} = \left(\frac{3}{5} \right)^n$

[40] $-8, 14, -\frac{49}{2}, \ldots; r = \frac{14}{-8} = -\frac{7}{4} \Rightarrow a_n = (-8)\left(-\frac{7}{4} \right)^{n-1}$

[41] $3.2, 0.8, 0.2, \ldots; a_n = 3.2 \left(\frac{1}{4} \right)^{n-1}$

[42] $\frac{\sqrt{3}}{2}, -\frac{3}{4}, \frac{3\sqrt{3}}{8}, \ldots; a_n = \frac{\sqrt{3}}{2} \left(-\frac{\sqrt{3}}{2} \right)^{n-1} = (-1)^{n-1} \left(\frac{\sqrt{3}}{2} \right)^n$

[43] $\ln 3, \ln 9, \ln 81, \ldots; a_n = (\ln 3) 2^{n-1} = 2^{n-1} \ln 3$

[44] $\frac{1}{x^2}, \frac{1}{x^5}, \frac{1}{x^8}, \ldots; a_n = \left(\frac{1}{x^2}\right)\left(\frac{1}{x^3}\right)^{n-1} = \frac{1}{x^{3n-1}}$

[45] $1, \frac{1}{10}, \frac{1}{100}, \ldots; S_6 = \frac{1\left[1-(0.1)^6\right]}{1-0.1} = 1.11111$

[46] $1.2, 1.8, 2.7, \ldots; S_8 = \frac{1.2\left[1-(1.5)^8\right]}{1-1.5} \approx 59.109375$

[47] $15, -20, \frac{80}{3}, \ldots; S_8 = \frac{15\left[1-\left(-\frac{4}{3}\right)^8\right]}{1+\frac{4}{3}} = -\frac{42,125}{729}$

[48] $\ln 4, \ln 2, \ln \sqrt{2}, \ldots; S_6 = \frac{\ln 4\left[1-\left(\frac{1}{2}\right)^6\right]}{1-\frac{1}{2}} = 2\ln 4\left(1-\frac{1}{64}\right) = \frac{63\ln 2}{16}$

[49] $a_1 = 4$ and $a_5 = 12$. $a_5 = a_1 r^4 \Rightarrow 12 = 4r^4 \Rightarrow r = \pm 3^{1/4}$. If $r = 3^{1/4}$, $m_1 = 4(3)^{1/4}$, $m_2 = 4(9)^{1/4}$ and $m_3 = 4(27)^{1/4}$. If $r = -3^{1/4}$, $m_1 = -4(3)^{1/4}$, $m_2 = 4(9)^{1/4}$ and $m_3 = -4(27)^{1/4}$.

[50] $a_1 = -8$ and $a_6 = \frac{243}{128}$. $\frac{243}{128} = (-8)r^5 \Rightarrow r^5 = -\frac{243}{1024} \Rightarrow r = -\frac{3}{4}$. Thus, $m_1 = 6$, $m_2 = -\frac{9}{2}$, $m_3 = \frac{27}{8}$ and $m_4 = -\frac{81}{32}$.

[51] $-1 + \frac{1}{3} - \frac{1}{9} + \ldots$; Since $|r| = \frac{1}{3} < 1$, the sum of the series is $\frac{a_1}{1-r} = \frac{-1}{1+\frac{1}{3}} = -\frac{3}{4}$.

[52] $2 + 2\sqrt{2} + 4 + \ldots$; Since $|r| = \sqrt{2} > 1$, the series has no sum.

[53] $\frac{5}{6} - \frac{5}{4} + \frac{15}{8} - \frac{45}{16} + \ldots$; Since $|r| = \frac{3}{2} > 1$, the series has no sum.

[54] $\frac{e}{2} + \frac{e}{4} + \frac{e}{8} + \ldots$; Since $|r| = \frac{1}{2} < 1$, the series has sum $\frac{\frac{e}{2}}{\frac{1}{2}} = e$.

[55] $2.3161616\ldots$; $0.0161616\ldots = 0.016 + 0.00016 + 0.0000016 + \ldots$ is a geometric series with sum $\frac{0.016}{1-0.01} = \frac{16}{990}$. Thus, $2.3161616\ldots = 2.3 + 0.0161616\ldots = \frac{23}{10} + \frac{16}{990} = \frac{2293}{990}$.

[56] $a_2 = \frac{3}{2}$ and $a_5 = \frac{32}{243}$. $\frac{3}{2} = a_1 r \Rightarrow r = \frac{3}{2a_1}$. $\frac{32}{243} = a_1 r^4 \Rightarrow \frac{32}{243} = a_1\left(\frac{3}{2a_1}\right)^4 \Rightarrow \frac{32}{243} = \frac{81}{16a_1^3} \Rightarrow$

$a_1 = \frac{27}{8}$.

[57] $S_{60} = \frac{a_1(1-r^n)}{1-r} = \frac{100(1.0075)\left[1-(1.0075)^{60}\right]}{1-1.0075} \approx \$7598.98$.

[58] Upward bounces: $10 + 6 + \frac{18}{5} + \ldots = \frac{10}{1-\frac{3}{5}} = 25$. Downward bounces: $50 + \left(10 + 6 + \frac{18}{5} + \ldots\right) =$

$50 + 25 = 75$. Thus, the total distance the ball travels is 100 feet.

[59] Let $S = \left\{n \in \mathbf{N} \mid 3 \cdot 1^2 + 3 \cdot 2^2 + \ldots + 3 \cdot n^2 = \dfrac{n(n+1)(2n+1)}{2}\right\}$. Since $3 \cdot 1^2 = 3$ and

$\dfrac{(1)(1+1)[2(1)+1]}{2} = \dfrac{6}{2} = 3$, we see that $1 \in S$. If $k \in S$, then $3 \cdot 1^2 + 3 \cdot 2^2 + \ldots + 3 \cdot k^2 =$

$\dfrac{k(k+1)(2k+1)}{2}$. To show that $k+1 \in S$, we must show that $3 \cdot 1^2 + 3 \cdot 2^2 + \ldots + 3 \cdot (k+1)^2 =$

$\dfrac{(k+1)(k+2)[2(k+1)+1]}{2}$ or that $3 \cdot 1^2 + 3 \cdot 2^2 + \ldots + 3(k+1)^2 = \dfrac{(k+1)(k+2)(2k+3)}{2}$.

We have $3 \cdot 1^2 + 3 \cdot 2^2 + \ldots + 3(k+1)^2 = \left(3 \cdot 1^2 + 3 \cdot 2^2 + \ldots + 3 \cdot k^2\right) + 3(k+1)^2 =$

$\dfrac{k(k+1)(2k+1)}{2} + 3(k+1)^2 = (k+1)\left[\dfrac{k(2k+1)}{2} + 3(k+1)\right] = (k+1)\left(\dfrac{2k^2+7k+6}{2}\right) =$

$(k+1)\left[\dfrac{(k+2)(2k+3)}{2}\right] = \dfrac{(k+1)(k+2)(2k+3)}{2}$. By the principle of mathematical induction,

$S = \mathbf{N}$.

[60] Let $S = \left\{n \in \mathbf{N} \mid 2(-1) + 2(-1)^2 + \ldots + 2(-1)^n = (-1)^n - 1\right\}$. Since $2(-1)^1 = -2$ and $(-1)^1 -$

$1 = -2$, we see that $1 \in S$. If $k \in S$, then $2(-1) + 2(-1)^2 + \ldots + 2(-1)^k = (-1)^k - 1$. To show that

$k+1 \in S$, we must show that $2(-1) + 2(-1)^2 + \ldots + 2(-1)^{k+1} = (-1)^{k+1} - 1$. We have $2(-1) +$

$2(-1)^2 + \ldots + 2(-1)^{k+1} = \left[2(-1) + 2(-1)^2 + \ldots + 2(-1)^k\right] + 2(-1)^{k+1} = \left[(-1)^k - 1\right] +$

$2(-1)^{k+1} = \left[(-1)^k + 2(-1)^{k+1}\right] - 1 = (-1)^k[1 + 2(-1)] - 1 = (-1)^k(-1) - 1 = (-1)^{k+1} - 1$.

By the principle of mathematical induction, $S = \mathbf{N}$.

[61] Let $S = \left\{n \in \mathbf{N} \mid 2^n + 10n < 3^n\right\}$. Since $2^4 + 10(4) = 56$ and $3^4 = 81$ and $56 < 81$, we see that $4 \in S$.

Let $k \geq 4$. If $k \in S$, then $2^k + 10k < 3^k$. To show that $k+1 \in S$, we must show that $2^{k+1} +$

$10(k+1) < 3^{k+1}$. Since $2^k + 10k < 3^k$, $2\left(2^k + 10k\right) < 2\left(3^k\right) \Rightarrow 2^{k+1} + 20k < 2 \cdot 3^k$. Since $2 < 3$,

$2 \cdot 3^k < 3 \cdot 3^k \Rightarrow 2 \cdot 3^k < 3^{k+1}$. Thus, $2^{k+1} + 20k < 2 \cdot 3^k$ and $2 \cdot 3^k < 3^{k+1} \Rightarrow 2^{k+1} + 20k <$

3^{k+1}. Since $k \geq 4$, $10k \geq 40 \Rightarrow 10k > 10 \Rightarrow 10k + 10k > 10k + 10 \Rightarrow 20k > 10(k+1) \Rightarrow$

$10(k+1) < 20k \Rightarrow 2^{k+1} + 10(k+1) < 2^{k+1} + 20k$. Since $2^{k+1} + 10(k+1) < 2^{k+1} + 20k$ and

$2^{k+1} + 20k < 3^{k+1}$, it follows that $2^{k+1} + 10(k+1) < 3^{k+1}$. By the principle of mathematical

induction, S contains every positive integer $n \geq 4$.

[62] Let $S = \left\{n \in \mathbf{N} \mid 2 \text{ is a factor of } 5^n - 1\right\}$. Since $5^1 - 1 = 4$ and 2 is a factor of 4, we see that $1 \in S$. If

$k \in S$, then 2 is a factor of $5^k - 1$. To show that $k+1 \in S$, we must show that 2 is a factor of $5^{k+1} -$

1. We have $5^{k+1} - 1 = 5^{k+1} + 4 - 5 = \left(5^k \cdot 5 - 5\right) + 4 = 5\left(5^k - 1\right) + 4$. Since 2 is a factor of $5^k - 1$

and hence of $5\left(5^k - 1\right)$ and 2 is a factor of 4, it follows that 2 is a factor of $5^{k+1} - 1$. By the principle

of mathematical induction, $S = \mathbf{N}$.

[63] Let $S = \left\{n \in \mathbf{N} \mid 2 \text{ is a factor of } 7^n - 3^n\right\}$. Since $7^1 - 3^1 = 4$ and 2 is a factor of 4, we see that $1 \in S$.

If $k \in S$, then 2 is a factor of $7^k - 3^k$. To show that $k+1 \in S$, we must show that 2 is a factor of

$7^{k+1} - 3^{k+1}$. We have $7^{k+1} - 3^{k+1} = 7^{k+1} - 3^{k+1} + \left(7 \cdot 3^k - 7 \cdot 3^k\right) = \left(7^{k+1} - 7 \cdot 3^k\right) +$

$\left(7 \cdot 3^k - 3^{k+1}\right) = 7\left(7^k - 3^k\right) + 3^k(7 - 3) = 7\left(7^k - 3^k\right) + 3^k(4)$. Since 2 is a factor of $7^k - 3^k$ and

hence a factor of $7\left(7^k - 3^k\right)$ and since 2 is a factor of $3^k(4)$, it follows that 2 is a factor of $7^{k+1} -$

3^{k+1}. By the principle of mathematical induction, $S = \mathbf{N}$.

[64]　Let $S = \left\{n \in \mathbf{N} \mid \dfrac{1}{(2)(5)} + \dfrac{1}{(5)(8)} + \ldots + \dfrac{1}{(3n-1)(3n+2)} = \dfrac{n}{2(3n+2)}\right\}$. Since

$\dfrac{1}{[3(1)-1][3(1)+2]} = \dfrac{1}{10}$ and $\dfrac{1}{2[3(1)+2]} = \dfrac{1}{10}$, we see that $1 \in S$. If $k \in S$, then $\dfrac{1}{(2)(5)} +$

$\dfrac{1}{(5)(8)} + \ldots + \dfrac{1}{(3k-1)(3k+2)} = \dfrac{k}{2(3k+2)}$. To show that $k+1 \in S$, we must show that

$\dfrac{1}{(2)(5)} + \dfrac{1}{(5)(8)} + \ldots + \dfrac{1}{[3(k+1)-1][3(k+1)+2]} = \dfrac{k+1}{2[3(k+1)+2]}$, or that $\dfrac{1}{(2)(5)} +$

$\dfrac{1}{(5)(8)} + \ldots + \dfrac{1}{(3k+2)(3k+5)} = \dfrac{k+1}{2(3k+5)}$. We have $\dfrac{1}{(2)(5)} + \dfrac{1}{(5)(8)} + \ldots +$

$\dfrac{1}{(3k+2)(3k+5)} = \left[\dfrac{1}{(2)(5)} + \dfrac{1}{(5)(8)} + \ldots + \dfrac{1}{(3k-1)(3k+2)}\right] + \dfrac{1}{(3k+2)(3k+5)} =$

$\dfrac{k}{2(3k+2)} + \dfrac{1}{(3k+2)(3k+5)} = \dfrac{k(3k+5)+2}{2(3k+2)(3k+5)} = \dfrac{3k^2+5k+2}{2(3k+2)(3k+5)} =$

$\dfrac{(3k+2)(k+1)}{2(3k+2)(3k+5)} = \dfrac{k+1}{2(3k+5)}$. By the principle of mathematical induction, $S = \mathbf{N}$.

[65]　Let $S = \{n \in \mathbf{N} \mid \ln 1 + \ln 2 + \ldots + \ln n = \ln(n\,!)\}$. Since $\ln 1 = 0$ and $\ln(1!) = \ln 1 = 0$, we see that
$1 \in S$. If $k \in S$, then $\ln 1 + \ln 2 + \ldots + \ln k = \ln(k!)$. To show that $k+1 \in S$, we must show that
$\ln 1 + \ln 2 + \ldots \ln(k+1) = \ln[(k+1)!]$. We have $\ln 1 + \ln 2 + \ldots + \ln(k+1) =$
$(\ln 1 + \ln 2 + \ldots + \ln k) + \ln(k+1) = \ln(k!) + \ln(k+1) = \ln[(k!) \cdot (k+1)] = \ln[(k+1)!]$. By
the principle of mathematical induction, $S = \mathbf{N}$.

[66]　Let $S = \{n \in \mathbf{N} \mid 2^n > 3n\}$. Since $2^4 = 16$ and $3(4) = 12$ and $16 > 12$, we see that $4 \in S$. Let $k \geq 4$.
If $k \in S$, then $2^k > 3k$. To show that $k+1 \in S$, we must show that $2^{k+1} > 3(k+1)$. Since $2^k > 3k$,
$2 \cdot 2^k > 2 \cdot (3k) \Rightarrow 2^{k+1} > 3k + 3k$. Since $k \geq 4$, $3k \geq 12 \Rightarrow 3k > 3 \Rightarrow 3k + 3k > 3k + 3 \Rightarrow$
$3k + 3k > 3(k+1)$. Since $2^{k+1} > 3k + 3k$ and $3k + 3k > 3(k+1)$, it follows that $2^{k+1} >$
$3(k+1)$. By the principle of mathematical induction, S contains every positive integer $n \geq 4$.

[67]　$(2x-y)^5 = 32x^5 - 80x^4y + 80x^3y^2 - 40x^2y^3 + 10xy^4 - y^5$

[68]　$(3x+1)^6 = 729x^6 + 1{,}458x^5 + 1{,}215x^4 + 540x^3 + 135x^2 + 18x + 1$

[69]　$\left(\sqrt{x} + \dfrac{2}{\sqrt{x}}\right)^4 = x^2 + 8x + 24 + \dfrac{32}{x} + \dfrac{16}{x^2}$

[70]　$(x^{-1} + 3y)^4 = \dfrac{1}{x^4} + \dfrac{12y}{x^3} + \dfrac{54y^2}{x^2} + \dfrac{108y^3}{x} + 81y^4$

[71]　$(3x+y)^{10}$; third term $= \dbinom{10}{2}(3x)^8(y)^2 = 295{,}245x^8y^2$

[72]　$(2x-3y)^{11}$; fifth term $= \dbinom{11}{4}(2x)^7(-3y)^4 = 3{,}421{,}440\,x^7y^4$

[73]　$(x^2 - 2y^3)^9$; Consider only the variable y in the kth term of the expansion $\dbinom{9}{k-1}(x^2)^{9-(k-1)}$.

$(-2y^3)^{k-1}$: $(y^3)^{k-1} = y^{12} \Rightarrow 3k - 3 = 12 \Rightarrow k = 5$; 5th term $= \dbinom{9}{4}(x^2)^5(-2y^3)^4 = 2{,}016x^{10}y^{12}$

[74]　$(x^3 + 4y^2)^6$; Consider only the variable x in the kth term of the expansion $\dbinom{6}{k-1}(x^3)^{6-(k-1)}$.

$(4y^2)^{k-1}$: $(x^3)^{6-(k-1)} = x^9 \Rightarrow 21 - 3k = 9 \Rightarrow k = 4$; 4th term $= \dbinom{6}{3}(x^3)^3(4y^2)^3 = 1{,}280x^9y^6$

NOTE: In Exercises 75 – 84, the solutions given have the following format: In the first line, the given rational expression is shown first as it appeared in the exercise and then with its denominator factored into linear or irreducible(over R) quadratic factors. On the extreme right side of the equation in the first line is its partial fraction decomposition. In the second line, we give the equation obtained when both sides of the equation in the first line are multiplied by the least common denominator(L. C. D.). In the remainder of the solution, we determine the values of A, B, C, ... etc.

[75] $\dfrac{11x-8}{x^2-x-6} = \dfrac{11x-8}{(x-3)(x+2)} = \dfrac{A}{x-3}+\dfrac{B}{x+2}$

$11x-8 = A(x+2)+B(x-3)$

If $x=-2$, then $-30=-5B \Rightarrow B=6$. If $x=3$, then $25=5A \Rightarrow A=5$.   ● $\dfrac{5}{x-3}+\dfrac{6}{x+2}$

[76] $\dfrac{-x-37}{x^2+2x-15} = \dfrac{-x-37}{(x-3)(x+5)} = \dfrac{A}{x-3}+\dfrac{B}{x+5}$

$-x-37 = A(x+5)+B(x-3)$

If $x=3$, then $-40=8A \Rightarrow A=-5$. If $x=-5$, then $-32=-8B \Rightarrow B=4$.   ● $\dfrac{-5}{x-3}+\dfrac{4}{x+5}$

[77] $\dfrac{5x^2-41x+98}{(x-5)(x^2-4x-5)} = \dfrac{5x^2-41x+98}{(x-5)^2(x+1)} = \dfrac{A}{x-5}+\dfrac{B}{(x-5)^2}+\dfrac{C}{x+1}$

$5x^2-41x+98 = A(x-5)(x+1)+B(x+1)+C(x-5)^2$ (E_1)

$\qquad = (A+C)x^2+(-4A+B-10C)x+(-5A+B+25C)$

In E_1, if $x=5$, then $18=6B \Rightarrow B=3$. If $x=-1$, then $144=36C \Rightarrow C=4$. Equating the

coefficients of x^2, we have $5=A+C \Rightarrow 5=A+(4) \Rightarrow A=1$.   ● $\dfrac{1}{x-5}+\dfrac{3}{(x-5)^2}+\dfrac{4}{x+1}$

[78] $\dfrac{5x^2+25x+31}{(x+3)(x^2+5x+6)} = \dfrac{5x^2+25x+31}{(x+3)^2(x+2)} = \dfrac{A}{x+3}+\dfrac{B}{(x+3)^2}+\dfrac{C}{x+2}$

$5x^2+25x+31 = A(x+3)(x+2)+B(x+2)+C(x+3)^2$ (E_1)

$\qquad = (A+C)x^2+(5A+B+6C)x+(6A+2B+9C)$

In E_1, if $x=-3$, then $1=-B \Rightarrow B=-1$. If $x=-2$, then $1=C \Rightarrow C=1$. Equating the coefficients

of x^2, we get $5=A+C \Rightarrow 5=A+(1) \Rightarrow A=4$.   ● $\dfrac{4}{x+3}-\dfrac{1}{(x+3)^2}+\dfrac{1}{x+2}$

[79] $\dfrac{2x^3+x^2+33x+15}{(x^2+16)^2} = \dfrac{Ax+B}{x^2+16}+\dfrac{Cx+D}{(x^2+16)^2}$

$2x^3+x^2+33x+15 = (Ax+B)(x^2+16)+Cx+D$

$\qquad = (A)x^3+(B)x^2+(16A+C)x+(16B+D)$

Equating the coefficients of like powers of x, we have

x^3: $2=A \Rightarrow A=2$ x^2: $1=B \Rightarrow B=1$

x : $33=16A+C \Rightarrow 33=16(2)+C \Rightarrow C=1$ x^0: $15=16B+D \Rightarrow 15=16(1)+D \Rightarrow D=-1$

                                         ● $\dfrac{2x+1}{x^2+16}+\dfrac{x-1}{(x^2+16)^2}$

[80] $\dfrac{x^3 + 2x^2 + 6x + 1}{\left(x^2 + 2x + 7\right)^2} = \dfrac{Ax + B}{x^2 + 2x + 7} + \dfrac{Cx + D}{\left(x^2 + 2x + 7\right)^2}$

$$x^3 + 2x^2 + 6x + 1 = (Ax + B)\left(x^2 + 2x + 7\right) + Cx + D$$
$$= (A)x^3 + (2A + B)x^2 + (7A + 2B + C)x + (7B + D)$$

Equating the coefficients of like powers of x, we have

x^3: $1 = A \Rightarrow A = 1$ $\qquad\qquad$ x^2: $2 = 2A + B \Rightarrow 2 = 2(1) + B \Rightarrow B = 0$

x: $6 = 7A + 2B + C \Rightarrow 6 = 7(1) + 2(0) + C \Rightarrow C = -1$

x^0: $1 = 7B + D \Rightarrow 1 = 7(0) + D \Rightarrow D = 1$ $\qquad$ ● $\dfrac{x}{x^2 + 2x + 7} + \dfrac{-x + 1}{\left(x^2 + 2x + 7\right)^2}$

[81] $\dfrac{2x^2 + 5x + 7}{\left(x + 2\right)^3} = \dfrac{A}{x + 2} + \dfrac{B}{\left(x + 2\right)^2} + \dfrac{C}{\left(x + 2\right)^3}$

$$2x^2 + 5x + 7 = A(x + 2)^2 + B(x + 2) + C$$
$$= (A)x^2 + (4A + B)x + (4A + 2B + C)$$

Equating the coefficients of like powers of x, we have

x^2: $2 = A \Rightarrow A = 2$ $\qquad\qquad$ x: $5 = 4A + B \Rightarrow 5 = 4(2) + B \Rightarrow B = -3$

x^0: $7 = 4A + 2B + C \Rightarrow 7 = 4(2) + 2(-3) + C \Rightarrow C = 5$ $\qquad$ ● $\dfrac{2}{x + 2} - \dfrac{3}{\left(x + 2\right)^2} + \dfrac{5}{\left(x + 2\right)^3}$

[82] $\dfrac{5x + 1}{\left(x - 1\right)^3} = \dfrac{A}{x - 1} + \dfrac{B}{\left(x - 1\right)^2} + \dfrac{C}{\left(x - 1\right)^3}$

$$5x + 1 = A(x - 1)^2 + B(x - 1) + C$$
$$= (A)x^2 + (-2A + B)x + (A - B + C)$$

Equating the coefficients of like powers of x, we have

x^2: $0 = A \Rightarrow A = 0$ $\qquad\qquad$ x: $5 = -2A + B \Rightarrow 5 = -2(0) + B \Rightarrow B = 5$

x^0: $1 = A - B + C \Rightarrow 1 = (0) - (5) + C \Rightarrow C = 6$ $\qquad$ ● $\dfrac{5}{\left(x - 1\right)^2} + \dfrac{6}{\left(x - 1\right)^3}$

[83] By first dividing and then factoring, we have

$$\left(2x^2 + 1\right) + \dfrac{4x^2 + 5x + 4}{x\left(x^2 + 1\right)} = \left(2x^2 + 1\right) + \dfrac{A}{x} + \dfrac{Bx + C}{x^2 + 1}$$

$$4x^2 + 5x + 4 = A\left(x^2 + 1\right) + (Bx + C)(x) \qquad (E_1)$$
$$= (A + B)x^2 + (C)x + (A)$$

In E_1, if $x = 0$, then $4 = A \Rightarrow A = 4$. Equating the coefficients of like powers of x and substituting 4

for A, we have $\qquad\qquad\qquad$ x^2: $4 = A + B \Rightarrow 4 = (4) + B \Rightarrow B = 0$

x: $5 = C \Rightarrow C = 5$ $\qquad\qquad$ ● $\left(2x^2 + 1\right) + \dfrac{4}{x} + \dfrac{5}{x^2 + 1}$

[84]   Using factoring by grouping (or synthetic division) to factor the denominator, we have

$$\frac{2x^2 + 3x + 1}{x^3 + 2x - x^2 - 2} = \frac{2x^2 + 3x + 1}{(x-1)(x^2+2)} = \frac{A}{x-1} + \frac{Bx+C}{x^2+2}$$

$$2x^2 + 3x + 1 = A\left(x^2+2\right) + (Bx+C)(x-1) \qquad (E_1)$$
$$= (A+B)x^2 + (-B+C)x + (2A-C)$$

In E_1, if $x = 1$, then $6 = 3A \Rightarrow A = 2$. Equating the coefficients of like powers of x and substituting 2 for A, we have

x^2: $2 = A + B \Rightarrow 2 = (2) + B \Rightarrow B = 0$ $\qquad\qquad$ x: $3 = -B + C \Rightarrow 3 = -(0) + C \Rightarrow C = 3$

$$\bullet \; \frac{2}{x-1} + \frac{3}{x^2+2}$$

[1] $(xy^2)(-2x^3y)^{-2} = (xy^2)\left[\frac{1}{(-2)^2}x^{-6}y^{-2}\right] = \frac{1}{4}x^{-5} = \frac{1}{4x^5}$

[2] $\frac{8x^{-8}y^{-12}}{2x^{-2}y^{-6}} = 4x^{-6}y^{-6} = \frac{4}{x^6y^6}$

[3] $\frac{3a^{-3}b^0c^2}{-6a^3bc^{-2}} = -\frac{1}{2}a^{-6}b^{-1}c^4 = -\frac{c^4}{2a^6b}$

[4] $\left(\frac{2x^{-2}}{y^3}\right)^{-1} = \frac{2^{-1}x^2}{y^{-3}} = \frac{x^2y^3}{2}$

[5] $(4a^2b)^4\left(\frac{-a^3}{2b}\right)^3 = (256a^8b^4)\left(\frac{-a^9}{8b^3}\right) = -32a^{17}b$

[6] $\frac{-12x^{-9}y^{10}}{4x^{-12}y^5} = -3x^3y^5$

[7] $\left(\frac{-5x^{-2}y}{2y^3}\right)^{-2} = \left(\frac{-5}{2x^2y^2}\right)^{-2} = \frac{4x^4y^4}{25}$

[8] $\frac{(rs^5)(r^2s^{-2})^3}{(r^2s^{-4})^3} = \frac{(rs^5)(r^6s^{-6})}{r^6s^{-12}} = \frac{r^7s^{-1}}{r^6s^{-12}} = rs^{11}$

[9] $\left(\frac{1}{2}x^{-4}y^3\right)^{-3}(-3x^{-5}y^0)^2 = (8x^{12}y^{-9})(9x^{-10}) = \frac{72x^2}{y^9}$

[10] $\left(\frac{2m^{-4}n^2p^0}{18m^3n^{-5}}\right)^{-2} = \left(\frac{n^7}{9m^7}\right)^{-2} = \frac{81m^{14}}{n^{14}}$

[11] $\frac{(x+y)^{-3}(x+y)^5}{(x+y)^{-1}} = (x+y)^2 \cdot (x+y) = (x+y)^3$

[12] $\frac{b^{-1}+a^{-1}}{(ab)^{-1}} = \left(\frac{1}{b}+\frac{1}{a}\right)(ab) = \left(\frac{a+b}{ab}\right)(ab) = a+b$

[13] $\sqrt[3]{16t^4u^3v^9} = \sqrt[3]{(8t^3u^3v^9)(2t)} = 2tuv^3\sqrt[3]{2t}$

[14] $\sqrt{\frac{8x^2y}{2x^4y^{-1}}} = \sqrt{\frac{4y^2}{x^2}} = \frac{2y}{x}$

[15] $\sqrt{\left(\frac{2a^{-2}}{b^{-3}}\right)^{-2}} = \left(\frac{2a^{-2}}{b^{-3}}\right)^{-1} = \frac{a^2}{2b^3}$

[16] $\sqrt[5]{-2m^4n^{-1}}\,\sqrt[5]{16m^{-9}n^4} = \sqrt[5]{-32m^{-5}n^3} = -2m^{-1}\sqrt[5]{n^3} = -\frac{2\sqrt[5]{n^3}}{m}$

[17] $\sqrt[3]{\frac{54p^2qr^{-1}}{p^{-1}r^5}} = \sqrt[3]{\frac{54p^3q}{r^6}} = \sqrt[3]{\left(\frac{27p^3}{r^6}\right)(2q)} = \frac{3p\sqrt[3]{2q}}{r^2}$

[18] $\sqrt[4]{(2a-3b)^4} = |2a-3b|$

[19] $\sqrt{\frac{t^3}{2tv^4}}\,\sqrt{\frac{t^{-2}v^3}{(3tv)^2}} = \sqrt{\frac{t^2}{2v^4}}\,\sqrt{\frac{t^{-2}v^3}{9t^2v^2}} = \sqrt{\frac{t^2}{2v^4}}\,\sqrt{\frac{v}{9t^4}} = \sqrt{\frac{1}{18t^2v^3}} = $

$\frac{1}{\sqrt{(9t^2v^2)(2v)}} = \frac{1}{3tv\,\sqrt{2v}}$

[20] $\sqrt[3]{(5-2y)^2}\,\sqrt[3]{(5-2y)^4} = \sqrt[3]{(5-2y)^6} = (5-2y)^2$

[21] $\frac{\sqrt[6]{2p^{10}q^{12}r^{-5}}}{\sqrt[6]{128p^{-2}q^7r}} = \sqrt[6]{\frac{2p^{10}q^{12}r^{-5}}{128p^{-2}q^7r}} = \sqrt[6]{\frac{p^{12}q^5}{64r^6}} = \frac{p^2\sqrt[6]{q^5}}{2r}$

[22] $\sqrt{a^3b}\left(\frac{1}{\sqrt{ab}}-\sqrt{ab}\right) = \sqrt{\frac{a^3b}{ab}}-\sqrt{a^4b^2} = \sqrt{a^2}-\sqrt{a^4b^2} = a-a^2b$

[23] $\sqrt{\sqrt[3]{x^7y^6z^{-12}}} = \sqrt[6]{(x^6y^6z^{-12})(x)} = \frac{xy\sqrt[6]{x}}{z^2}$

[24] $\sqrt{\sqrt{\frac{32ab^{-1}}{2a^5b^{-8}}}} = \sqrt[4]{\frac{16b^7}{a^4}} = \frac{2b\sqrt[4]{b^3}}{a}$

[25] $\frac{5}{\sqrt{3}}\cdot\frac{\sqrt{3}}{\sqrt{3}} = \frac{5\sqrt{3}}{3}$

[26] $\frac{\sqrt[3]{3x}}{\sqrt[3]{4x^2}}\cdot\frac{\sqrt[3]{2x}}{\sqrt[3]{2x}} = \frac{\sqrt[3]{6x^2}}{2x}$

[27] $\frac{2}{3-\sqrt{2}}\cdot\frac{3+\sqrt{2}}{3+\sqrt{2}} = \frac{6+2\sqrt{2}}{7}$

[28] $\frac{4-\sqrt{3}}{4+\sqrt{3}}\cdot\frac{4-\sqrt{3}}{4-\sqrt{3}} = \frac{19-8\sqrt{3}}{13}$

[29] $\frac{2x-1}{\sqrt{x-2}}\cdot\frac{\sqrt{x-2}}{\sqrt{x-2}} = \frac{(2x-1)\sqrt{x-2}}{x-2}$

[30] $\frac{\sqrt{a}+\sqrt{b}}{\sqrt{a}-\sqrt{b}}\cdot\frac{\sqrt{a}+\sqrt{b}}{\sqrt{a}+\sqrt{b}} = \frac{a+2\sqrt{ab}+b}{a-b}$

[31] $\dfrac{\sqrt{a}-2}{3\sqrt{a}}\cdot\dfrac{\sqrt{a}+2}{\sqrt{a}+2}=\dfrac{a-4}{3a+6\sqrt{a}}$

[32] $\dfrac{1-\sqrt{2x}}{\sqrt{5x}}\cdot\dfrac{1+\sqrt{2x}}{1+\sqrt{2x}}=\dfrac{1-2x}{\sqrt{5x}+\sqrt{10x^2}}=\dfrac{1-2x}{\sqrt{5x}+x\sqrt{10x}}$

[33] $\dfrac{\sqrt{m}+\sqrt{n}}{2mn}\cdot\dfrac{\sqrt{m}-\sqrt{n}}{\sqrt{m}-\sqrt{n}}=\dfrac{m-n}{2mn\left(\sqrt{m}-\sqrt{n}\right)}$

[34] $\dfrac{\sqrt{2x+2h+1}-\sqrt{2x+1}}{h}\cdot\dfrac{\sqrt{2x+2h+1}+\sqrt{2x+1}}{\sqrt{2x+2h+1}+\sqrt{2x+1}}=\dfrac{(2x+2h+1)-(2x+1)}{h\left(\sqrt{2x+2h+1}+\sqrt{2x+1}\right)}=$

$\dfrac{2}{\sqrt{2x+2h+1}+\sqrt{2x+1}}$

[35] $\left(\dfrac{9}{16}\right)^{-3/2}=\left(\dfrac{16}{9}\right)^{3/2}=\left(\sqrt{\dfrac{16}{9}}\right)^3=\left(\dfrac{4}{3}\right)^3=\dfrac{64}{27}$

[36] $a^{4/3}a^{-3/2}a^{1/6}=a^{8/6-9/6+1/6}=a^0=1$ [37] $(-27x^{-3}y^{12})^{2/3}=(\sqrt[3]{-27})^2x^{-2}y^8=\dfrac{9y^8}{x^2}$

[38] $\dfrac{r^2s^{-1/2}t^{1/3}}{r^{1/3}s^{-3/4}t^3}=r^{2-1/3}s^{-1/2+3/4}t^{1/3-3}=\dfrac{r^{5/3}s^{1/4}}{t^{8/3}}$

[39] $(p^2q^{-3})^{1/3}(p^{-2}q)^{-1/2}=(p^{2/3}q^{-1})(pq^{-1/2})=\dfrac{p^{5/3}}{q^{3/2}}$

[40] $(-8m^{5/2}n^{-6}p^0)^{4/3}=(\sqrt[3]{-8})^4m^{10/3}n^{-8}=\dfrac{16m^{10/3}}{n^8}$ [41] $\left(\dfrac{x^3y^7z^{-3}}{x^2y^{-5}z^{-10}}\right)^{2/5}=(xy^{12}z^7)^{2/5}=x^{2/5}y^{24/5}z^{14/5}$

[42] $\left(\dfrac{1}{4}\right)^{3/2}\left(-\dfrac{1}{8}\right)^{2/3}=\left(\sqrt{\dfrac{1}{4}}\right)^3\left(\sqrt[3]{-\dfrac{1}{8}}\right)^2=\left(\dfrac{1}{8}\right)\left(\dfrac{1}{4}\right)=\dfrac{1}{32}$ [43] $\left(\dfrac{a^{-2/3}b^{1/6}}{a^{1/2}b^{-5/6}}\right)^6=\left(\dfrac{b}{a^{7/6}}\right)^6=\dfrac{b^6}{a^7}$

[44] $\dfrac{(-27u^3v^0w^6)^{4/3}}{(125u^{3/2}v^6w^{-3})^{2/3}}=\dfrac{(\sqrt[3]{-27})^4u^4w^8}{(\sqrt[3]{125})^2uv^4w^{-2}}=\dfrac{81u^3w^{10}}{25v^4}$

[45] $\dfrac{(2x-1)^{2/5}(2x-1)^{-1/3}}{(2x-1)^{-2/3}}=(2x-1)^{2/5}(2x-1)^{1/3}=(2x-1)^{11/15}$

[46] $\left(\dfrac{x^{-2}y^3}{x^4y^{-3}}\right)^{-1/2}\left(\dfrac{x^4y^{-5}}{xy^4}\right)^{-1/3}=\left(\dfrac{y^6}{x^6}\right)^{-1/2}\left(\dfrac{x^3}{y^9}\right)^{-1/3}=\left(\dfrac{x^6}{y^6}\right)^{1/2}\left(\dfrac{y^9}{x^3}\right)^{1/3}=\left(\dfrac{x^3}{y^3}\right)\left(\dfrac{y^3}{x}\right)=x^2$

[47] $\left(2x^3+\dfrac{1}{2}x^2+4x\right)-\left(3x^3-\dfrac{1}{2}x^2+2x-4\right)=2x^3+\dfrac{1}{2}x^2+4x-3x^3+\dfrac{1}{2}x^2-2x+4=$

$-x^3+x^2+2x+4$

[48] $(2x^3y^2-5xy+x^2y^3)+(3xy-x^2y^3)-(x^3y^2+2xy)=2x^3y^2-5xy+x^2y^3+3xy-x^2y^3-$

$x^3y^2-2xy=x^3y^2-4xy$

[49] $3x^2(2x-1+15x^3)=45x^5+6x^3-3x^2$

[50] $(4m-3)(3m+7)=12m^2+28m-9m-21=12m^2+19m-21$

[51] $(7y+1)(-2y+9)=-14y^2+63y-2y+9=-14y^2+61y+9$

[52] $(ax-y)(ax+2y)=a^2x^2+2axy-axy-2y^2=a^2x^2+axy-2y^2$

[53] $(r+2)(3r^2-12r+4)=3r^3-12r^2+4r+6r^2-24r+8=3r^3-6r^2-20r+8$

[54] $(3u-2v)^2=9u^2-6uv-6uv+4v^2=9u^2-12uv+4v^2$

[55] $(2x-1)(4x^2+1)=8x^3-4x^2+2x-1$

[56] $(x-y)(x^2+xy+y^2)=x^3+x^2y+xy^2-x^2y-xy^2-y^3=x^3-y^3$

[57] $(2x^2 - 3)(4x^2 - 6x + 1) = 8x^4 - 12x^3 + 2x^2 - 12x^2 + 18x - 3 = 8x^4 - 12x^3 - 10x^2 + 18x - 3$

[58] $(2m - 3)^3 = (2m - 3)(4m^2 - 12m + 9) = 8m^3 - 24m^2 + 18m - 12m^2 + 36m - 27 =$

$8m^3 - 36m^2 + 54m - 27$

[59] $(x - 3y + 1)^2 = (x - 3y + 1)(x - 3y + 1) = x^2 - 3xy + x - 3xy + 9y^2 - 3y + x - 3y + 1 =$

$x^2 + 9y^2 + 2x - 6y - 6xy + 1$

[60] $(2x + 1)(3x - 1)(4x + 3) = (2x + 1)(12x^2 + 5x - 3) = 24x^3 + 22x^2 - x - 3$

[61] $a^2 - 2a - 3 = (a + 1)(a - 3)$

[62] $6x^2 - 4x - 16 = 2(3x^2 - 2x - 8) = 2(3x + 4)(x - 2)$

[63] $16 - 9z^2 = (4 + 3z)(4 - 3z)$ [64] $a^2 - 3ab - 4b^2 = (a + b)(a - 4b)$

[65] $6t^2 + 10t + 4 = 2(3t^2 + 5t + 2) = 2(3t + 2)(t + 1)$

[66] $3x^5 - 12x^3 = 3x^3(x^2 - 4) = 3x^3(x + 2)(x - 2)$ [67] $y^3 - 125 = (y - 5)(y^2 + 5y + 25)$

[68] $2m^3 + 16 = 2(m^3 + 8) = 2(m + 2)(m^2 - 2m + 4)$

[69] $x^3 - 2x^2y - x + 2y = (x^3 - 2x^2y) - (x - 2y) = x^2(x - 2y) - (x - 2y) = (x - 2y)(x^2 - 1) =$

$(x - 2y)(x + 1)(x - 1)$

[70] $5ru + 10rv + 2ut + 4vt = (5ru + 10rv) + (2ut + 4vt) = 5r(u + 2v) + 2t(u + 2v) =$

$(u + 2v)(5r + 2t)$

[71] $4u^2 + 12uv + 9v^2 - 9 = (4u^2 + 12uv + 9v^2) - 9 = (2u + 3v)^2 - 9 = (2u + 3v + 3)(2u + 3v - 3)$

[72] $x^2 + 2xy + y^2 - 16 = (x^2 + 2xy + y^2) - 16 = (x + y)^2 - 16 = (x + y + 4)(x + y - 4)$

[73] $y^4 - 16 = (y^2 + 4)(y^2 - 4) = (y^2 + 4)(y + 2)(y - 2)$

[74] $4x^{10} - 5x^5 - 6 = (4x^5 + 3)(x^5 - 2)$ [75] $36a^2 + 60ab^2 + 25b^4 = (6a + 5b^2)^2$

[76] $6r^5s - 3r^3s^2 - 30rs^3 = 3rs(2r^4 - r^2s - 10s^2) = 3rs(2r^2 - 5s)(r^2 + 2s)$

[77] $\dfrac{4a^2 + 12a + 9}{2a^2 + 3a} = \dfrac{(2a + 3)(2a + 3)}{a(2a + 3)} = \dfrac{2a + 3}{a}$

[78] $\dfrac{5v^2 + 10v + 5}{v^3 + v^2 - v - 1} = \dfrac{5(v^2 + 2v + 1)}{(v^3 + v^2) - (v + 1)} = \dfrac{5(v + 1)^2}{v^2(v + 1) - (v + 1)} = \dfrac{5(v + 1)^2}{(v + 1)(v^2 - 1)} = \dfrac{5}{v - 1}$

[79] $\dfrac{x^2 - 1}{3x^3} \cdot \dfrac{x^2 - 5x}{x^3 - 1} = \dfrac{(x + 1)(x - 1) \cdot x(x - 5)}{3x^3(x - 1)(x^2 + x + 1)} = \dfrac{(x + 1)(x - 5)}{3x^2(x^2 + x + 1)}$

[80] $\dfrac{3a + 12}{a^2 + 8a + 16} \cdot \dfrac{a^2 - 16}{3a^2 - 11a - 4} = \dfrac{3(a + 4)(a + 4)(a - 4)}{(a + 4)^2(3a + 1)(a - 4)} = \dfrac{3}{3a + 1}$

[81] $\dfrac{9m^2 - 18m - 16}{6m + 4} \div \dfrac{9m^2 - 64}{6m + 16} = \dfrac{(3m + 2)(3m - 8)}{2(3m + 2)} \cdot \dfrac{2(3m + 8)}{(3m + 8)(3m - 8)} = 1$

[82] $\dfrac{x^2 + x - 6}{5x^2 - 3x - 2} \div \dfrac{x^2 - 9}{5x^2 + 2x} = \dfrac{(x + 3)(x - 2)}{(5x + 2)(x - 1)} \cdot \dfrac{x(5x + 2)}{(x + 3)(x - 3)} = \dfrac{x(x - 2)}{(x - 1)(x - 3)}$

[83] $\dfrac{1}{a^2 - 5a + 6} - \dfrac{1}{a^2 - 4} = \dfrac{1}{(a - 3)(a - 2)} - \dfrac{1}{(a + 2)(a - 2)} = \dfrac{(a + 2) - (a - 3)}{(a - 3)(a + 2)(a - 2)} =$

$\dfrac{5}{(a - 3)(a^2 - 4)}$

[84] $\dfrac{x+1}{x^2+x}+\dfrac{3x}{x^2-1}=\dfrac{x+1}{x(x+1)}+\dfrac{3x}{(x+1)(x-1)}=\dfrac{1}{x}+\dfrac{3x}{x^2-1}=\dfrac{x^2-1+3x^2}{x(x^2-1)}=\dfrac{4x^2-1}{x(x^2-1)}$

[85] $\dfrac{\frac{1}{a^2}-\frac{1}{b^2}}{\frac{a}{b}-\frac{b}{a}}\cdot\dfrac{a^2b^2}{a^2b^2}=\dfrac{b^2-a^2}{a^3b-ab^3}=\dfrac{b^2-a^2}{ab(a^2-b^2)}=-\dfrac{1}{ab}$

[86] $\dfrac{\frac{1}{x}-\frac{2}{x^2}+1}{\frac{1}{x}-\frac{3}{x^2}+2}\cdot\dfrac{x^2}{x^2}=\dfrac{x-2+x^2}{x-3+2x^2}=\dfrac{x^2+x-2}{2x^2+x-3}=\dfrac{(x+2)(x-1)}{(2x+3)(x-1)}=\dfrac{x+2}{2x+3}$

[87] $\dfrac{\frac{1}{(x+h)^2}-\frac{1}{x^2}}{h}\cdot\dfrac{x^2(x+h)^2}{x^2(x+h)^2}=\dfrac{x^2-(x+h)^2}{hx^2(x+h)^2}=\dfrac{x^2-(x^2+2hx+h^2)}{hx^2(x+h)^2}=-\dfrac{2x+h}{x^2(x+h)^2}$

[88] $\dfrac{(x+h)^{-3}-x^{-3}}{h}=\dfrac{\frac{1}{(x+h)^3}-\frac{1}{x^3}}{h}\cdot\dfrac{x^3(x+h)^3}{x^3(x+h)^3}=\dfrac{x^3-(x+h)^3}{hx^3(x+h)^3}=$

$\dfrac{x^3-(x^3+3hx^2+3h^2x+h^3)}{hx^3(x+h)^3}=-\dfrac{3x^2+3hx+h^2}{x^3(x+h)^3}$

[89] $(3x^2+1)^3(7)(5x^2+2)^6(10x)+(5x^2+2)^7(3)(3x^2+1)^2(6x)=2x(3x^2+1)^2(5x^2+2)^6\cdot$

$[35(3x^2+1)+9(5x^2+2)]=2x(3x^2+1)^2(5x^2+2)^6(150x^2+53)$

[90] $(6x+2)^{1/2}(4)(3x^2-2x+1)^3(6x-2)+(3x^2-2x+1)^4\left(\frac{1}{2}\right)(5x+2)^{-1/2}(6)=(6x+2)^{-1/2}\cdot$

$(3x^2-2x+1)^3[4(6x+2)(6x-2)+3(3x^2-2x+1)]=(6x+2)^{-1/2}(3x^2-2x+1)^3\cdot$

$(144x^2-16+9x^2-6x+3)=\dfrac{(3x^2-2x+1)^3(153x^2-6x-13)}{(6x+2)^{1/2}}$

[91] $\dfrac{(3x-1)^{1/3}(5)(3x^2-1)^4(6x)-(3x^2-1)^5\left(\frac{1}{3}\right)(3x-1)^{-2/3}(3)}{\left[(3x-1)^{1/3}\right]^2}=$

$\dfrac{(3x-1)^{-2/3}(3x^2-1)^4[30x(3x-1)-(3x^2-1)]}{(3x-1)^{2/3}}=\dfrac{(3x^2-1)^4(87x^2-30x+1)}{(3x-1)^{4/3}}$

[92] $(3)\left(\dfrac{5-2x^2}{4x+1}\right)^2\cdot\dfrac{(4x+1)(-4x)-(5-2x^2)(4)}{(4x+1)^2}=\dfrac{3(5-2x^2)^2}{(4x+1)^2}\cdot\dfrac{-16x^2-4x-20+8x^2}{(4x+1)^2}=$

$-\dfrac{12(5-2x^2)^2(2x^2+x+5)}{(4x+1)^4}$